Lepton-Photon Symposium
1989

Proceedings of the 1989 International Symposium on Lepton and Photon Interactions at High Energies

August 7–12, 1989
Stanford University

Editor
Michael Riordan

Organized by Stanford Linear Accelerator Center

Sponsored by International Union of Pure and Applied Physics, National Science Foundation, Stanford University, and U. S. Department of Energy

Published by

World Scientific Publishing Co. Pte. Ltd.
P O Box 128, Farrer Road, Singapore 9128
USA office: 687 Hartwell Street, Teaneck, NJ 07666
UK office: 73 Lynton Mead, Totteridge, London N20 8DH

PROCEEDINGS OF THE 1989 INTERNATIONAL SYMPOSIUM ON LEPTON AND PHOTON INTERACTIONS AT HIGH ENERGIES

ISBN 981-02-0104-4

Printed by Singapore National Printers Ltd

FOREWORD

The 1989 International Symposium on Lepton and Photon Interactions at High Energies was held at Stanford from August 7 through August 12. This was the fourteenth meeting in this series, which dates back to 1963, and the third time the Symposium has been hosted by the Stanford Linear Accelerator Center. The earlier meetings coincided with major physics accomplishments at SLAC; the first results from the two-mile accelerator were presented at the 1967 meeting, and the 1975 meeting followed closely the discoveries of the ψ and τ particles at SPEAR. In keeping with this tradition, the latest meeting included the first results from $e^+e^- \rightarrow Z^\circ$ events observed at the SLC. A report to the Symposium indicated that LEP results on the same reaction were close at hand, and subsequent developments have shown that claim to be amply justified.

Other advances of great interest reported to the Symposium included new limits on the top mass from the $p\bar{p}$ colliders, the ϵ'/ϵ results from experiments at CERN and Fermilab, and evidence for $b \rightarrow u$ quark transitions at CESR and DESY. The Standard Model continues to accommodate virtually all measurements in the field, including the wealth of new data presented to this conference. Let us hope that future experiments will soon modify that situation.

SLAC expects to welcome the Lepton-Photon Symposium again in the not-too-distant future, and hopes that results from (one or more) large linear colliders will be presented at that time.

Richard E. Taylor
Symposium Chairman

ACKNOWLEDGMENTS

The success of each Lepton-Photon Symposium depends critically on the invited presentations of the speakers, and we have been particularly fortunate in this regard. The organizers wish to express their deep appreciation to the speakers who worked long and hard to assemble information on the various subjects and transmitted that information to the conferees with clarity and precision. Special thanks are due Professor Frank Scuilli who accepted the difficult task of summarizing the Symposium on only one week's notice.

We also wish to recognize here the contributions of the physicists who acted as scientific secretaries to the speakers and continued to be of assistance through the preparation of their manuscripts. The large number of contributed papers were organized by Dieter Cords and Lance Dixon.

The Directorate of SLAC and the Organizing Committee want to thank all those who worked very hard to provide the framework for the gathering along with food, shelter, transportation and entertainment for the delegates. Much of that effort fell to the Conference Secretariat, which included Maura Chatwell, Effie Clewis, Jeanette Hayden and Karen Rogers. Clive Field and his assistants were responsible for auditorium arrangements and closed-circuit television coverage of the sessions. Dennis Wisinski coordinated the efforts of the SLAC computing services to supply computer facilities to the delegates; we thank the Apollo Division of Hewlett Packard, Digital Equipment Corporation and QMS for their loan of equipment for this purpose. We are indebted to Nina Adelman-Stolar for her help and assistance, and to Bernard Lighthouse who was responsible for the transportation of delegates. Our thanks also must go to the many other members of the SLAC staff who helped in myriad ways with various facets of the Symposium.

I should like to end with some personal thanks to two members of the Organizing Committee. In the months before the Symposium I had some significant health problems, often at crucial points in the organizing activities. Vera Luth and Fred Gilman took over my responsibilities and did much of the work that would otherwise have fallen to me. Vera took on much of the responsibility for all the arrangements, in addition to her already heavy responsibilities as Secretary. Fred was of great help in many ways, and was left to deal with many of the program decisions alone. I am grateful beyond measure to both of them.

Richard E. Taylor, Symposium Chairman

The publication of these Proceedings benefited from the work of many minds and hands. My sincere thanks are due the speakers, who delivered finished manuscripts of their talks amidst the press of many other responsibilities. The scientific secretaries read these papers carefully to ferret out the remaining errors. I am grateful to all of them for this important contribution — especially Alan Breakstone, Raymond Frey, David Kastor and Stephen Wagner, who helped transcribe the discussion sessions after each talk. I am also indebted to Alyssa Higashi and Sharon Jensen, who typed the abstracts of contributed papers using a computer system devised by Louise Addis.

The photographs that grace these pages came from several sources. Harvey Lynch of SLAC snapped many fine photos of the Symposium and its participants. The front and back endpapers are based on exquisite color photographs of Stanford taken by Hugh Steimle and Charles Painter, respectively. Additional Stanford photos were cheerfully supplied by Cindy Romain of the Stanford News Service.

Led by Kevin Johnston and Dianne Land, the SLAC Publications Department put the finishing touches on this volume just before it went to press. Terry Anderson contributed the cover artwork, and Sylvia MacBride did the layout of the photo pages; both helped me with the myriad details involved in the final stages of production. I am also grateful to Vani Bustamante, who carefully retyped several manuscripts.

Finally, my task as Editor would have been far more difficult and much less enjoyable had it not been for the unstinting efforts of Maura Chatwell and Rene Donaldson. Maura coordinated the efforts of the scientific secretaries and kept diligent track of the many individual contributions that went into these Proceedings. Rene took eager charge of the overall design, tracked down the beautiful photographs found in the book, and helped me pull everything together into a volume that participants and libraries will be proud to place on their bookshelves. I thank both of them very much for all their help.

Michael Riordan, Editor

SYMPOSIUM ORGANIZATION

R. Taylor, Chairman

W. Panofksy, Honorary Chairman

F. Gilman, Program Chairman

V. Lüth, Secretary

Local Organizing Committee

D. Burke	SLAC
J. Dorfan	SLAC
G. Feldman	SLAC
F. Gilman	SLAC
T. Himel	SLAC
M. Levi	LBL
V. Lüth	SLAC
B. Lynn	Stanford
M. Peskin	SLAC
C. Prescott	SLAC
M. Riordan	SLAC
T. Schalk	UC Santa Cruz
A. Seiden	UC Santa Cruz
M. Swartz	SLAC
R. E. Taylor	SLAC
G. Trilling	LBL
R. Wagoner	Stanford

International Advisory Committee

M. Davier	LAL, Orsay
M. K. Gaillard	LBL, Berkeley
M. B. Green	Queen Mary College, London
L. V. Gribov	Landau Institute, Moscow
W. Hoogland	NIKEF, Amsterdam
M. Klein	DAW, Zeuthen
L. Maiani	Rome University
A. Michelini	CERN
Ye Minghan	IHEP, Beijing
A. Minten	CERN
A. P. Onuchin	Soviet Academy of Science, Novosibirsk
S. Orito	University of Tokyo
K. K. Phua	University of Singapore
A. Santoro	CPF, Rio de Janiero
D. Schramm	EFI, University of Chicago
F. Scuilli	Columbia University
P. Söding	DESY
K. C. Stanfield	Fermilab
H. Sugawara	KEK, Japan

TABLE OF CONTENTS

Foreword v

Acknowledgments vi

Symposium Organization vii

The Status of Traditional Lepton-Photon Physics

A Perspective on Lepton-Photon Physics 1
W. K. H. Panofsky

Status and Future of Structure Function Measurements 13
J. Feltesse

Polarised Nucleon Structure Functions 41
G. G. Ross

Two-Photon Physics 60
R. N. Cahn

Double Beta-Decay Experiments 83
I. V. Kirpichnikov

Properties of Leptons 89
M. L. Perl

Decays of Strangeness, Charm and Beauty

Weak Decays of Charmed Particles 105
P. E. Karchin

Status and Physics Program of BEPC 122
M. Ye and Z. Zheng

B Physics from CLEO 129
D. L. Kreinick

Recent ARGUS Results on B Meson Decays 139
M. V. Danilov

A Measurement of ϵ'/ϵ by E731 155
B. Winstein

NA31 Results on CP Violation in K Decays, and a Test of CPT 168
D. Fournier

Rare Kaon Decays 184
L. S. Littenberg

Experiments on Electroweak Interactions

Results from TRISTAN 203
A. Maki

Measurement of the Z Boson Resonance Parameters 225
G. J. Feldman

Results from the Mark II at SLC on Decays of the Z° 236
A. J. Weinstein

LEP Report 249
F. Dydak

Neutrino-Electron Scattering 258
J. Panman

W and Z Production and Decay Results from UA2 266
A. R. Weidberg

Electroweak Results from CDF 274
M. Campbell

Theory of Precision Electroweak Experiments 286
G. Altarelli

The Standard Model at the High Energy Frontier

Status of the UA1 Top Search 305
K. Eggert

Search for New Particles with the UA2 Detector 318
L. DiLella

New Particle Searches at CDF 328
P. K. Sinervo

The Production of High P_T Photons and Lepton Pairs 348
J. Huston

QCD and Jets 368
J. Huth

The Physics of QCD Jets 387
V. A. Khoze

The Sky Is the Limit

High Energy Astrophysics 403
J. W. Cronin

Accelerators for Very Energetic Gamma Rays from Neutron Star Systems 420
M. Ruderman

Gauge Fields, Geometry, Strings, and Gravity 433
J. A. Harvey

The Future e^+e^- Colliders 442
G. A. Voss

Summary of the Conference 452
F. Sciulli

Contributed Papers 461

List of Participants 523

Author Index 542

The Status of Traditional
Lepton-Photon Physics

The Status of Traditional Lepton-Photon Physics

Monday, 7 August 1989

A Perspective on Lepton-Photon Physics 1

Speaker: W. K. H. Panofsky

Chairperson: R. Taylor

Scientific Secretary: C. K. Jung

Status and Future of Structure Function Measurements 13

Speaker: J. Feltesse

Chairperson: H. Kendall

Scientific Secretary: F. Le Diberder

Polarised Nucleon Structure Functions 41

Speaker: G. G. Ross

Chairperson: V. Hughes

Scientific Secretary: R. Frey

Two-Photon Physics 60

Speaker: R. N. Cahn

Chairperson: D. Caldwell

Scientific Secretary: N. Roe

Double Beta-Decay Experiments 83

Speaker: I. V. Kirpichnikov

Chairperson: D. Caldwell

Scientific Secretary: W. Koska

Properties of Leptons 89

Speaker, M. L. Perl

Chairperson: D. Stairs

Scientific Secretary: C. Hearty

A PERSPECTIVE ON LEPTON–PHOTON PHYSICS*

WOLFGANG K. H. PANOFSKY
Stanford Linear Accelerator Center
Stanford University, Stanford, California 94309

ABSTRACT

This paper reviews some key experiments of the past in which the same basic physical processes are attacked both through lepton–photon interactions and by using hadron machines as primary tools. Not surprisingly, it is concluded that the basic distinction between lepton–photon physics and elementary particle physics in general is unreal but that the tools and methodology can be very different indeed. A look is then taken into the expected future evolution of particle accelerators. Existing accelerator technologies both for proton and electron colliders are approaching basic limits as the collision energy in the constituent frame is raised. At this time no clear path exists for electron–positron colliders to compete with the SSC as far as energy reach is concerned, but the superior clarity and coverage of phenomena not accessible to hadron colliders makes it absolutely essential that the development of both electron–positron and hadron colliders be pursued vigorously. It is concluded that accelerator R&D effort underway is insufficient if a large hiatus in productivity in particle physics is to be avoided. Electron–positron linear colliders are the most promising approach for the extension of knowledge beyond LEP and beyond the SSC, but the difficulties to reach an electron–positron energy of 15 TeV or beyond in the constituent frame look formidable. Both electron–positron and proton colliders appear to face severe future detector limitations, the former due to electron–positron pair creation during the collision and the latter due to the enormous hadronic background event rates.

INTRODUCTION

The organizers of this conference have assigned me the title of "A Perspective on Lepton–Photon Physics." The advantage of the term "perspective" is that it applies both looking backward and forward; let me start by practicing some hindsight.

The program of this and the preceding conferences makes it abundantly clear that the subject of lepton and photon physics as an isolated topic does not really exist; the more we learn the less valid is that distinction. Traditionally, the separation originated principally through the tools used, rather than the physical interest expressed. Let me illustrate this pattern by reflecting on some past experiments where the same physics has been attacked, starting with hadrons and with photon and lepton beams.

THE 3–3 RESONANCE

The Δ states of the proton were first seen in photon beams from electron synchrotrons. R. R. Wilson and collaborators[1] at Cornell charted the approach to the resonance, and the Cal Tech group took the data over the peak. Copious production of the Δ became evident at the Chicago Synchrotron and the unambiguous identification of the spin parity of the Δ then became possible.

Figures 1(a) and 1(b) juxtapose some graphs from the photoproduction and scattering experiments. I leave it to the audience to judge whether this is lepton–photon physics or hadron physics.

DISCOVERY OF THE π^0

Extensive theoretical conjectures that there should be what is now recognized to be a neutral pion were developed before the war from cosmic ray evidence. At the 184-inch hadron synchrocyclotron at Berkeley, B. J. Moyer and collaborators[2] observed the gamma-ray spectra originating from hadron–hadron collisions in internal targets, and these spectra were clearly consistent with decay of a neutral pion into two photons. However, the real identification of the neutral pion came from the experiments of J. Steinberger and collaborators[3] in the photon beam of the Electron Synchrotron at Berkeley by observing gamma–gamma coincidences from neutral pion decay. This experiment constituted a dramatic demonstration of the decay kinematics unique to the neutral pion. Figures 2(a) and 2(b) show results of these two experiments in juxtaposition.

⋆ Work supported by Department of Energy contract DE–AC03–76SF00515.

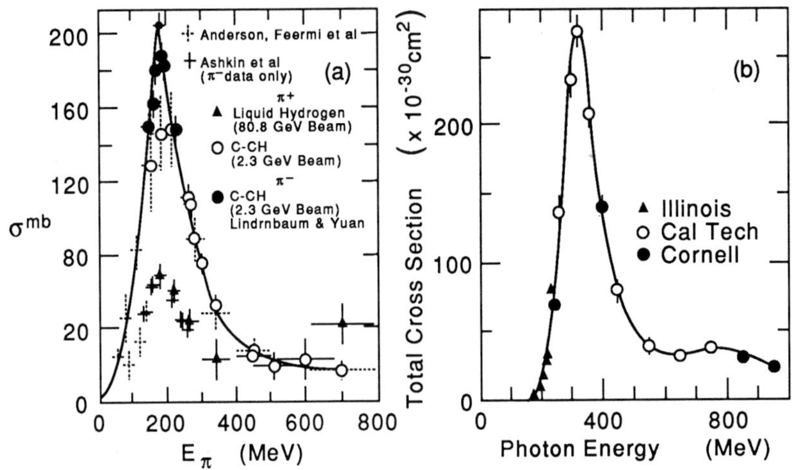

Fig. 1. The Δ resonance.

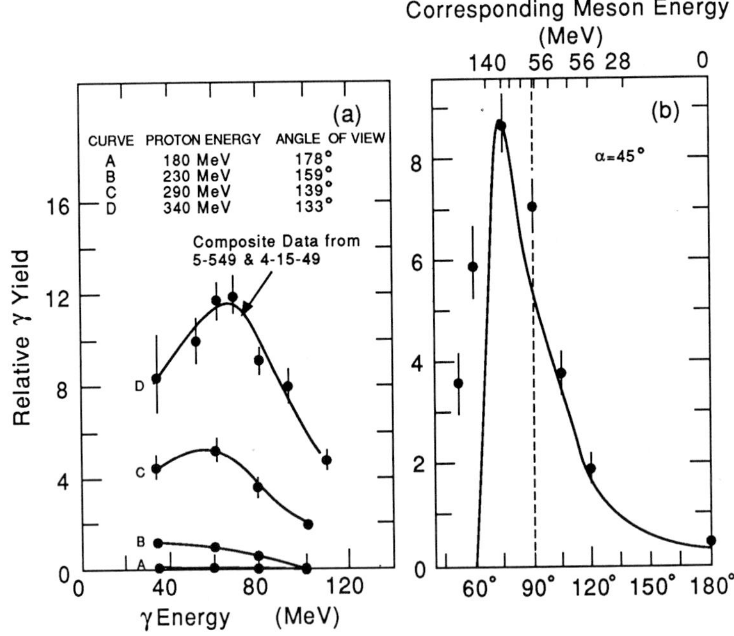

Fig. 2. Experimental history of the π^0.

THE PION PROTON INTERACTION AND THE SPIN PARITY OF THE PION

In 1948, the absorption of negative pions on the proton at rest resulted in a gamma-ray spectrum which proved directly measurable. The pions were produced in an internal target struck by the proton beam of the 184-inch cyclotron at Berkeley.[4] At the same time, the gamma-ray spectrum also revealed the charge exchange process leading to neutral pions. You can judge the progress of instrumentation days by considering a "biomechanical" coincidence circuit that was used to register electron–positron pairs produced by the gamma-rays observed from the chamber in which negative pions were captured. This coincidence circuit consists of a square array of nails arranged in a 15 × 15 matrix. When flashing lights indicated a coincidence between arrival of an electron and a positron, a washer was thrown by the experimenter over the relevant nail in the matrix and the accumulation of the piles of washers in the matrix in-

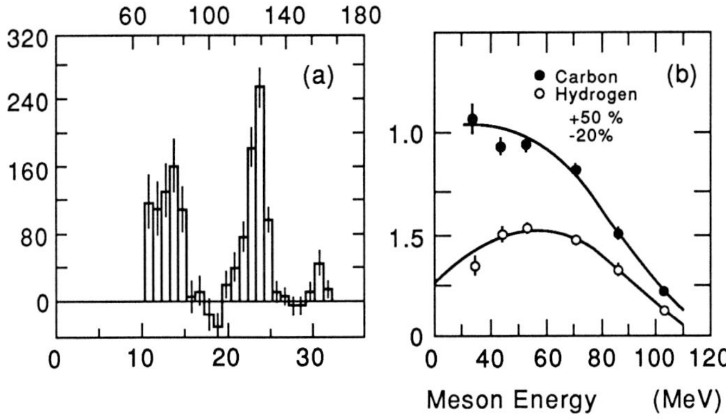

Fig. 3. Photoproduction/absorption of pions.

dicated the spectrum generated. There has been a bit of progress in instrumentation since then! While this might be considered a hadron experiment, the inverse reaction which gives similar information by detailed balancing is the photoproduction of charged pions on the nucleon. Such photoproduction of π^+ and π^- mesons was brought under investigation at the 300-MeV electron synchrotron at Berkeley[5] at the same time at which the π^- absorption experiments were done at the proton synchrocyclotron.

Figures 3(a) and 3(b) give the two results in juxtaposition. These experiments deal with essentially the same basic matrix elements observed in hadron and photon machines measuring mutually inverse processes, albeit at different energies.

PRODUCTION OF VECTOR PARTICLES BY VIRTUAL PHOTONS

This topic has a long history too complex to cover here. I will here only compare production of vector particles by electron–positron annihilation in storage rings with production of the same objects through the Drell–Yan process from hadron–hadron collisions. These two processes have been pursued in parallel throughout. The best known example is, of course, the discovery of the J/Ψ simultaneously in Brookhaven by the observation of lepton pairs from hadron collisions and at SLAC from e^+e^- annihilation. These results are shown in Figs. 4(a) and 4(b). This was followed by the detailed exploration of psion spectroscopy at SLAC. Then there is the discovery of the Υ in the muon pair spectrum generated from a hadron beam, followed by elaboration of the Υ spectroscopy in electron–positron annihilation, first at DESY and then at Cornell.

EXAMINATION OF HADRON STRUCTURE BY INELASTIC LEPTON SCATTERING

The quark hypothesis derived from the interpretation of the rapidly evolving data on resonant states of the nucleons as induced primarily at the Bevatron at Berkeley, followed by work at other hadron machines. A more direct revelation of the quark substructure of the hadrons came from the deep inelastic electron scattering experiments at SLAC carried out by the SLAC/MIT collaboration. This was followed with work at other electron laboratories and then by results from neutrino and muon beams from hadron machines reaching much higher momentum transfers but generally lower statistics. Figure 5 shows a comparative graphical summary.

THE Z^0

Charged and neutral intermediate vector bosons were predicted prior to their experimental discovery as part of the electroweak unification of Weinberg and Salam. Strong experimental indications on the existence of these particles originated from many

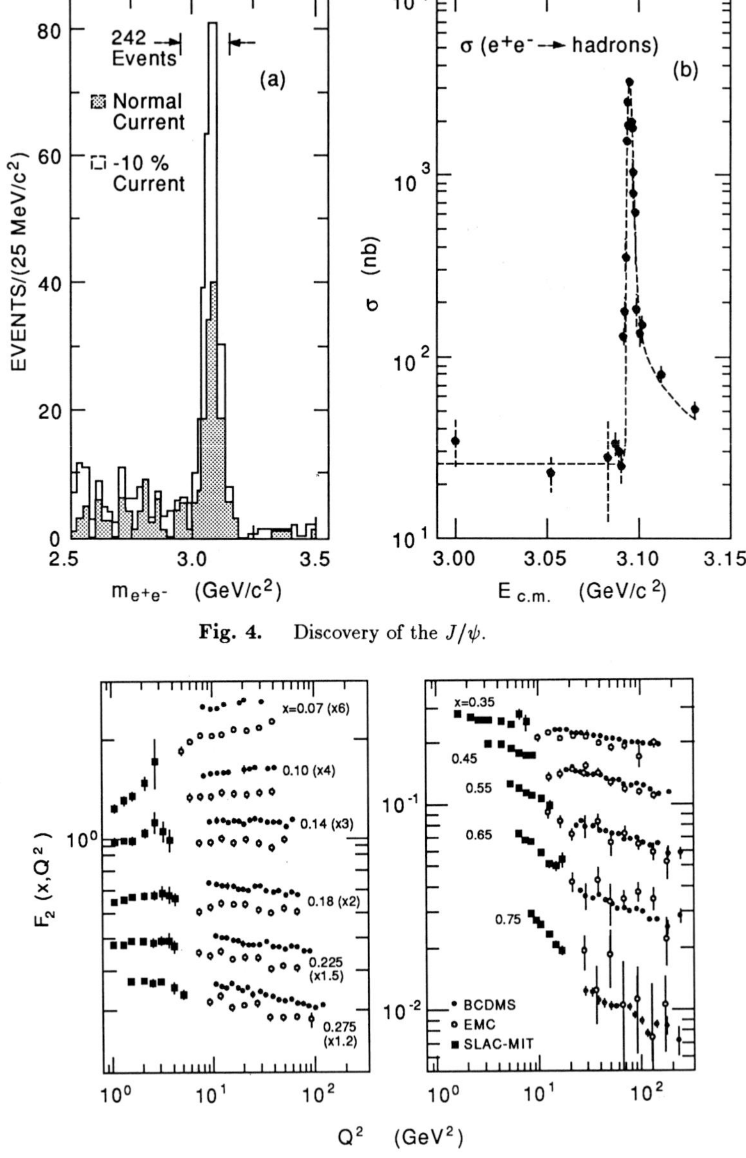

Fig. 4. Discovery of the J/ψ.

Fig. 5 The proton structure function $F_2^P(x, Q^2)$ from deep inelastic scattering of charged leptons on hydrogen targets. Where necessary, the SLAC–MIT and EMC data are interpolated to the x bins of the BCDMS data.

sources including the angular asymmetry of lepton pairs produced in electron–positron annihilations and other evidence of interference between electromagnetic and weak interaction channels. The bosons themselves were discovered at CERN in the S$\bar{\text{p}}$pS and have recently been more copiously produced in the proton–antiproton collider—the Tevatron at Fermilab. Recently, as will be reported later in this conference, well above 100 Z^0's have been observed in e^+e^- annihilations. Figures 6(a) and 6(b) present a comprehensive data summary. Again, the comparison is illuminating: while the electron–positron annihilation data still are very sparse, they give superior mass and width measurements of the Z^0, and both lepton and hadronic decay channels can be detected with high efficiency, while in hadron colliders only the lepton channels can be clearly isolated. On the other hand, lots of information which is not accessible to e^+e^- colliders has been generated, and continues to flow from the higher energy $p\bar{p}$ colliders.

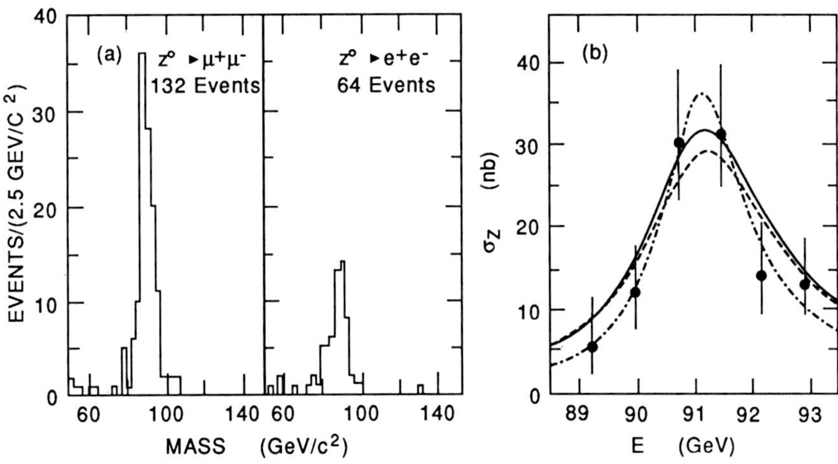

Fig. 6. (a) FNAL/CDF; (b) SLAC/Mark II.

All the previous juxtapositions are, of course, only a sketchy overview of a complex situation, most of which is now ancient history well known to all of you. However, one can reflect on some general observations quite apart from the fact that these dual graphs dramatically show the overall unity of lepton and hadron physics.

One observation is that *discoveries* of the fundamental lepton and quark family members are divided among electron machines, cosmic rays and hadron machines with the electron machines having made a fair share of the discoveries and elaborating on the full spectroscopy of quark mesonic states. Not surprisingly, the electron was discovered by J. J. Thompson in an electron accelerator! The Mu meson was discovered in cosmic rays and the Tau lepton in an electron–positron collider. Direct observation of neutrinos requires extraterrestrial sources, nuclear reactors, or high energy hadron beams.

Following the quark hypothesis devised to explain early phenomena in hadron spectroscopy, the confirmation of the existence of up and down quarks can experimentally be attributed to deep inelastic electron scattering experiments; strangeness was discovered in cosmic rays but the "strange" spectroscopy was elaborated in hadron colliders. The charmed quark became credible from electron–positron annihilation and through the detection of lepton pair spectra from targets bombarded by protons. The basic charmed quark spectroscopy unfolded from electron–positron storage rings. The *b*–quark was discovered in a hadron collider but its spectroscopy elaborated in electron–positron colliders. Overall, it is clear that hadron colliders have generally reached considerably larger momentum transfers and larger collision energies in the constituent frame, while the clarity of data tends to be considerably greater in the lepton–photon domain.

The reason for greater clarity of data and, more specifically, better signal-to-background ratio is, of course, well known. The total hadron–hadron cross sections as a function of energy are nearly constant and are in fact increasing logarithmically with energy, while cross sections for producing new objects of a given mass or leading to momentum transfers of a given magnitude decrease as the square of those masses or momentum transfers. Thus, as interest focuses on these higher mass or momentum transfer events, the signal-to-background ratio for hadron colliders degenerates as the square of the energy. In contrast, in lepton collisions both signals and background decrease quadratically together. Moreover, in electron–positron collisions leading to particles having the same quantum number as the virtual photon produced in the collisions the signal of interest can be larger than any other event. Therefore in high energy electron–positron colliders, the main problem is that of reaching adequate absolute rates rather than data analysis isolating signal from background.

What I call "background" here can, of course, be in itself frequently of scientific interest. After all, yesterday's signal tends to be today's background. The joke: "In an electron–positron collider either you find something new or you find nothing; while in a hadron collider, when you find nothing new you can always study the background" overstates the case. There is, of course, major scientific interest in accumulating systematic data on hadron collisions and on understanding QCD phenomena at an increasing level of detail and precision.

Another point of comparison derives, of course, from our clear quantitative understanding of quantum electrodynamics and almost as clear an understanding of the electroweak interaction. A specific consequence of that understanding is the power of lepton–photon physics to establish "positive denial" of the existence of conjectured objects or processes. If the dynamics of generation of such objects of processes is understood, then nonobservation has specific evidential value. Thus particle searches originating from lepton and photon collisions permit sharper interpretations. In addition experiments uniquely isolating quantum electrodynamics or electroweak processes, which are independent of or at least insensitive to hadronic processes, can be used to examine the limits of validity of quantum electrodynamics and electroweak theory.

The fact that QED and electroweak theory is understood and validated down to distances of at least 10^{-17}cm means that high precision measurements in lepton and photon physics and the examination of small branching ratios can be sensitive to conjectured higher mass states. Therefore, such searches can constitute large mass reach experiments if colliders to reach such masses directly are not available.

The previous brief retrospective view selects some corresponding results from lepton–photon physics and hadron physics, or more precisely, lepton–photon collisions and hadron collisions. From these I should now like to turn to some "perspective" into the future.

Most conferences in high energy physics, including those dedicated to lepton and photon physics, encompass the "standard speech" on the "Standard Model." This speech summarizes the conference saying that no deviations from the Standard Model have as yet been seen including those results reported at the conference. However, the Standard Model cannot be the whole story for many well-known reasons — too many arbitrary constants, no explanation for the number of generations of flavors, no experimental evidence for the existence of specific agents which establish the mass scale among particles of the same basic quantum numbers, and finally, no experimental data which relate gravity to particle physics phenomena. In other words, the "standard speech" persuasively argues that there must be physics beyond the Standard Model. Others at this conference will no doubt address these issues, so I would like to confine any futuristic remarks to the instrumental expectations.

Last year, Carlo Rubbia concluded his summary talk of the previous lepton–photon conference with the phrase "to choose between a machine we know how to build but for which so far no satisfactory detector has been proposed, and a machine for which the present detector technology is adequate but for which no clear machine design exists so far ... ." I would not take that sharp a position as to the alternatives we face. Rather, the question is which parameters are in fact attainable during the next one or two decades. I would, however, agree that the rate of progress of lepton–photon physics, and of elementary particle physics in general, is paced by instrumental developments in the collider arts, particle detection, and data analysis.

I have illustrated from past history the critical role played by electron machines. One can even strengthen the case by pointing out that in the 1970's many of the profound contributions by hadron machines occurred through the use of external lepton beams, that is, neutrinos and muons. In fact, the '70's can well be designated as the decade of the leptons.

The question is how to extrapolate from this to future expectations. There is no question that formally, measured in terms of the energy in the "constituent" frame, *i.e.*, the lepton or quark frame, proton machines will be able to reach much further in the coming decades and can do so more cheaply. There is no expectation that electron–positron collider technology can match the reach of the SSC, as measured by that single parameter. The question is how accessible the resulting information is. Here an enormous amount of work has been done in workshops, at Snowmass summer studies, and through specific contributions by individuals and groups. I will not present even samples of the results of these efforts. In general, such studies generate Monte Carlo data—making assumptions about projected phenomena, be they Higgs particles of various mass, supersymmetric particles, second-generation of W's and Z's, or recurrences of other classes of particles. Background is projected based on known phenomena from the Standard Model and on QCD calculations. In general, such studies project the "reach" measured in terms of the maximum mass of the particles conjectured (Fig. 7). At the same time, such studies specify what type of segmentation of detector is required, in what radiation environment it has to live, and how vast are the imposed data processing requirements.

For the SSC, the numbers are indeed impressive. In rough numbers one starts out with 10^8 interactions *per second,* each generating perhaps 10^6 bytes of information. Trigger systems have to reject all but a few Hertz' worth of event rate. Offline cuts then have to isolate the interesting events which in most cases number in the 100 to 1000 *per year* range

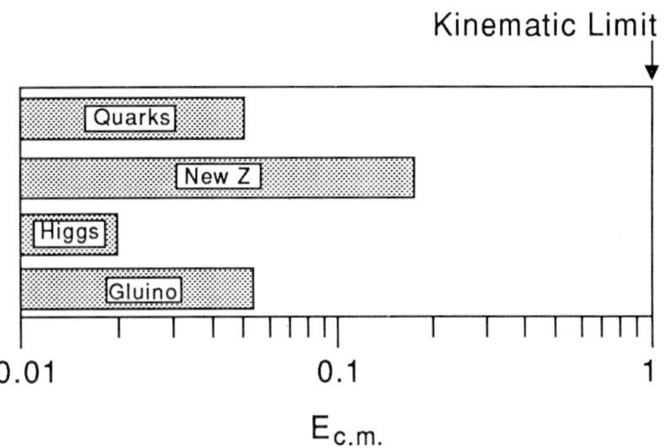

Fig. 7 Discovery "reach" versus kinematic center-of-mass energy for the SSC (Ref. 6).

for the more interesting phenomena. In other words, we are talking about a rejection ratio of about 10^{12}. Of course, this numerology overstates the case somewhat, since many interactions generate events which are totally swallowed by the beam pipe.

In general, analyses indicate that although a giant and expensive effort is required and the challenges to detector design are huge, the problems, addressed as pre-identified issues, are soluble. Yet the lingering doubt remains that although specific analyses aimed at examining the discoverability of conjectured phenomena and yielding detector and data analysis requirements give positive results, this extreme filtering process required for proton accelerators may throw away evidence of the "truly unexpected." Moreover, even ignoring this possibility, there are bars to discoveries in specific regions. For instance, if the Higgs mass lies in the band between twice the Z^0 mass and the Z^0 mass, it will be swamped by the general QCD background, and similar discovery bars can arise at higher masses.

It is interesting to note that one of the specific prominent designs for hadron detectors is to design and build an instrument which essentially throws away most information except that identifying penetrating leptons. Indeed, as we have seen before, this has been a fruitful avenue of discovery in the past, but it also eliminates an enormous amount of data at the source which could be obtained only with hadron machines. Based both on the physics and on history, the question can be legitimately asked whether that information which manifests itself through detecting lepton pairs from hadron collisions is not accessible with greater clarity from electron–positron colliders — *if* such colliders can be built to reach a competitive energy range.

Thus, the net which a high energy proton machine will cast will indeed yield a vast product but some interesting fish will surely get away. But in the foreseeable future proton machines can cast that net further into the ocean. All this does in no way speak against the need for a major assault upon the next generation of hadron machines. It simply means that the lepton frontier *must* be covered.

And this is the crux of the matter. How credible are the projections and studies which extrapolate beyond LEP and the SLC towards the attainment of electron–positron linear colliders of higher energy? There seems to be a consensus that LEP is the last of the highest energy electron–positron colliders based on electron–positron storage rings. The reason is the well-known argument which leads to the cost and radius of such machines increasing roughly quadratically with energy. This conclusion stems from a balance of those costs growing linearly with orbit radius against costs related to radiation loss which vary with the fourth power of the energy divided by the orbit radius. This argument is matched against the conventional wisdom that the cost of linear colliders is linear, that is, it goes up proportionally to the beam energy. If the SSC tunnel were a suitable housing for an electron–positron collider (which it is not!) it would extend the energy frontier by less than a factor of 2 beyond LEP. But does the cost of a "linear" collider scale linearly in practice? And what are the realistic coefficients of the scaling laws?

Scaling law arguments are often used to justify a new type of accelerator or collider. At the end of World War II, Luis Alvarez argued persuasively that a proton linear accelerator would be the machine of the future since at that time cyclotron costs varied roughly as the cube of the energy, and proton linear accelerators would exhibit a linear cost-scaling relationship. Unfortunately for linear proton accelerators but fortunately for physics, the rules were changed; the invention of phase stability by McMillan and Vexler together with the invention of strong focusing by Christophilos and Livingston, and Courant and Snyder, also changed the scaling laws for a circular proton machine to an approximately linear relationship. The higher cost per unit energy made the proton linear accelerator noncompetitive at higher energy, although of course it has remained the injector of choice for all high energy proton machines. Similarly, in comparing linear and circular electron machines, one has to be mindful of future changes both in the coefficients of the scaling relationships as well as in more fundamental respects.

Recently there have been numerous and extensive reviews and workshops dedicated to examining the status and promise of electron–positron linear colliders. Major studies and experimental activities are being pursued in Novosibirsk, in Japan, at CERN, and at SLAC. There is now a general consensus that for some decades the basic accelerator for linear colliders has to be based on "conventional" RF structures, albeit operating at shorter wave length than the customary 10 cm or longer now in use. A fundamental basis for this conclusion is that the average beam powers have to be very large for electron–positron linear colliders going well beyond SLC and LEP energies employing *any* means of particle acceleration. If adequate luminosities are to be attained, this need for high average beam powers in the megawatt range is derived through very general considerations. This, however, demands that the efficiency of power transfer from wall plug to beam be reasonably high. Thus, although very large gradients are in principle attainable by novel methods of acceleration, such as those based on using electromagnetic fields in lasers, plasma wakefields, and other "collective" methods, such approaches look wildly improbable today when overall power efficiency is considered, and also when the demands for highly precise accelerating conditions are to be met.

You will hear later in this conference about the initial results from the SLC, which again demonstrate that even a moderate amount of data from an electron–positron collider gives new physical insight in an energy region which has been accessible to hadron colliders for some considerable length of time. Yet none of these studies on electron–positron linear colliders gives absolute clarity as to what the realistic scaling laws of cost vs. energy of electron–positron colliders will be. I tend to be significantly more pessimistic than many participants in this work as to the energy to which the electron–positron linear collider art can practically be pushed during the next decades.

Many of the parameters required for electron–positron colliders vary with energy in a predictable way. The conventional line of reasoning is to specify that the required luminosity of such devices must increase with the square of the energy, due to the expected variation of the relevant cross sections. At the same time, the beam–beam interaction results in phenomena which impose limits on the number of particles in each individual bunch and on the structure of the bunches which are brought into collisions. First there is beamstrahlung, which degenerates the energy spectrum of the particles. This broadens the particle energy and thereby widens the resonance peaks for producing particles having the same quantum numbers as the single virtual photon. At the same time, the radiative tail resulting from the radiative electron–positron collisions provides an overview over a wide spectrum of electron–positron energies. In other words, the radiative broadening resulting from collisions of electron–positron bunches provides a "self energy scanning" feature of such colliders.

But then there are other consequences of the beam–beam interaction. The photons produced from beamstrahlung can result in electron–positron pair formation both in individual collisions of these photons with the opposing electrons or positrons or by coherent interaction of the photons with the electromagnetic field of the opposing bunch. These electron–positron pairs can lead to an intolerable background in the detectors. The disruption of the particles in one bunch by the electromagnetic field of the opposing bunch will cause the electrons and positrons to spray on the face of the final focusing lenses which produce the high density of interaction required for an adequate luminosity. Although ingenious tricks have been devised to reduce this problem it cannot be totally avoided.

Considerable improvement results from using flat rather than circular beams in the collisions. In that case the relationship between the mean density of colliding particles to the electromagnetic field which each particle sees can be improved. Such a flat beam is not as unnatural an object as it may appear at first glance. The damping rings which are used to reduce the radial momenta of electrons and positrons in linear colliders have the natural characteristic of reducing the emittance perpendicular to the plane of the orbit by a much larger factor than the emittance in the plane of the orbit. If the mixing between the vertical and horizontal phase space of the particles emitted from such damping rings can be held to a low value as these particles are being accelerated and brought into final collisions, then flat beams are the natural product. Yet notwithstanding all these ingenious inventions, there are strong limitations on the number of electrons per pulse which can be usefully employed in the final collisions. Therefore, adequate data rates require either a high pulse repetition rate or a large number of electron–positron bunches within each radiofrequency pulse, or both. The first results in high average beam power and the second results in requirements to minimize the regenerative beam breakup which occurs when many successive intense electron–positron bunches are accelerated in a single radiofrequency pulse. Again, that latter problem has been attacked by ingenious methods. One is to design new accelerating structures which radiate away or otherwise damp the modes which cause the transverse beam breakup. The other is to program the phases of acceleration and the frequencies of higher modes in clever ways to reduce the instabilities.

To obtain the requisite high density in the final electron–positron collisions a focus system has to reduce the total cross section of the beam by a factor much below that currently attained in the SLAC SLC, which in itself has already achieved the spectacularly small beam size corresponding to a radius of roughly 3 μm rms. Is such a further drastic reduction attainable or not in practice and how do the means of attaining such a reduction relate to the scaling laws of cost for a linear collider of the future?

The requirement for the final focus spot to be small puts stringent limits on the radial emittance of the colliding beams, its energy width, as well as the design of the Final Focus System itself. This, in turn, not only puts demands on the design of the damping rings but also puts severe conditions on the emittance growth, both during acceleration and the beam transport after the damping ring has "cooled" the beams radially. Damping rings meeting these requirements have been designed in principle, although specific demands on kicker design, wall impedances, and tolerances are difficult to meet.

The control of emittance growth in acceleration generates a contest between competing design considerations; as the wave length of the linear accelerator becomes larger, then alignment tolerances are relaxed because wake field effects become more serious with a very high inverse power of the aperture through which the beam has to pass. However, if the wave length is shorter, then the RF power wall losses go down and the maximum possible accelerating gradients are higher at shorter operating wavelengths. How important these two factors are is open to question. Most of the RF power requirement is simply the product of the energy storage in the accelerating guide times the pulse repetition frequency; indeed, that energy storage increases as the square of the wavelength. However, if one succeeds in extracting a fair fraction of the stored energy into the beam by the use of multibunch operation during each pulse, then the overall power efficiency is not severely dependent on choice of wavelength. Also, the matter of attainable gradient need not be controlling, since setting the aesthetics of an overly long accelerator aside, and ignoring pre-established site constraints, purely economic considerations would generally not lead to the highest gradient attainable technically.

Under all circumstances the tolerances which specify the level of congruence between the centroid of the beam, the electromagnetic axis of the accelerator, the beam position indicators, and the external focusing elements are extremely serious—much more so than they are in the case of the SLC. To express this in the form of a scaling law one can show that for constant average beam power, beam–beam disruption, and radiative beam–beam energy broadening, the radial invariant emittance (that is the actual radial emittance multiplied by the relativistic γ factor) has to decrease with something like the inverse eighth power of the energy. While some of the assumptions in this extremely steep scaling relationship can be modified, the severity of the emittance requirements rises sharply with energy. This problem reflects, in turn, on the precision of manufacture and alignment of components and on the demands for quality of beam position indicators, correcting elements and feedback loops; these requirements have not as yet been factored into cost estimates; this is difficult to do without detailed design.

There is one further crucial matter. What counts is the total luminosity integrated over long running periods. Therefore, as has been painfully learned during the past years, the matter of reliability is becoming of increasing importance as the complexity and number of components in accelerators and colliders increase, as they must as we go to higher and higher energies. This, in turn, implies that quality standards must be increased. This means as a minimum larger investments in R&D; but it may also imply higher unit costs in construction, counterbalancing the hoped-for cost reductions due to economies of scale.

The existing rough cost scaling considerations pertaining to electron–positron linear colliders are largely based on such tangible data as unit costs of modulators, RF power tubes, past experience with accelerating structures, and digging tunnels. Faced with the extremely steep scaling laws relating to tolerances and the increasing emphasis which has to be placed on reliability, I would not be sure how overall costs grow with energy for a nominally "linear" collider.

All these considerations indicate that there is a relatively clear predictable path, albeit at an R&D effort much larger than is now being invested by the four major centers dedicated to linear collider development, to an energy of perhaps 400–500 GeV in the electron–positron collision frame. Above that, predictions become speculative, both in regard to costs and time scale. However, at these lower energies such a machine would still be an enormously powerful tool for particle physics. How powerful depends, of course, on the masses of the hitherto elusive objects which are predicted "beyond the Standard Model." The detectability and ease of measurement of such objects, to the extent they exist, is, however, excellent all the way up to the kinematic limit. For instance, heavier quarks and leptons produced in pairs will decay into W–bosons in combination with the existing lighter quarks or leptons, and the signature of such processes remains clean.

So, notwithstanding the desire so frequently expressed to establish clear priorities among future colliders, the fact remains that both the proton and electron collider fronts *must* be covered. It is my view that electron–positron colliders cannot hope for a decade or two to match the energy "reach" of the SSC. However, that reach will be beset by limitations set both by the capability of detectors and fundamental gaps in coverage where general QCD background will prevent discoverability of new processes. As history has amply demonstrated, the clarity and usually also the discovery potential of electron machines is expected to remain superior to hadron colliders within the kinematic range accessible to such colliders, but extending that range by a large factor beyond that now expected to be reached by LEP–II is going to be a real battle.

The SSC is rightly billed as a conventional extension of the technology successfully demonstrated at the Fermilab Tevatron. Yet even at the SSC, synchrotron radiation of protons is already becoming a dominant design consideration, since the nine or so kilowatts of photons radiated deposit their energy at liquid helium temperature. Thus, as proton machines "beyond the SSC" are contemplated, many of the design limitations for electron–positron colliders which we have just discussed will also apply to proton machines. Thus the distinction between "hadron physics" and "lepton–photon physics" which already hardly exists in basic particle physics will also tend to disappear for machine design as we contemplate yet another leap in energy. Thus, Rubbia's pronouncement about choosing between a machine we don't know how to build and one we don't know how to use becomes a choice between hadron and electron machines, neither of which we know how to build; and a choice between electron machines we might know how to use if we can live with the blast of electron and positron pairs, and a proton machine we don't know how to use at all. A great deal of accelerator research and development has to be done before particle physics can (either with hadrons or electrons and leptons) penetrate deeply beyond the TeV region. Let me close on this happy note!

REFERENCES

1. T. L. Jenkins, D. Luckey, T. R. Palfrey, and R. R. Wilson, *Phys. Rev.* **95** (1954) 179.
2. Bjorklund, Crandall, Moyer, and York, *Phys. Rev.* **77** (1950) 213.
3. J. Steinberger, W. K. H. Panofsky, and P. R. Steller, *Phys. Rev.* **78** (1950) 802.
4. W. K. H. Panofsky, R. L. Aamodt, and J. Hadley, *Phys. Rev.* **81** (1951) 565.
5. J. Steinberger and A. S. Bishop, *Phys. Rev.* **78** (1950) 494.
6. C. Rubbia, *Proc. of the 13th Int'l Symp. on Lepton and Photon Interactions* (Hamburg, 1987) p. 813.

STATUS AND FUTURE OF STRUCTURE FUNCTION MEASUREMENTS

J. Feltesse

DPhPE, CEN-Saclay
91191 Gif-sur-Yvette Cedex, France.

Abstract

After a brief reminder on the theoretical guides on structure functions in deep inelastic lepton-nucleon scattering, the consistency of the high statistics data sets is discussed and the various structure functions are compared with perturbative QCD predictions. On these many aspects we focus on the results of this year and the future prospects opened by the impending commissioning of HERA.

1. INTRODUCTION

In the field of deep inelastic scattering of leptons on nucleons, we are now in a transition period where almost all analyses of high statistics data sets on fixed-target are completed and where HERA, the electron proton collider at DESY, is under construction to be commissioned at the end of 1990. It is thus a good time to review where we stand, and where we are heading in the measurement of structure functions. The Q^2 evolution is so far the only prediction of QCD, the theory of strong interactions, on structure functions. From this evolution it is however also possible to deduce several sum rules and to predict the x behaviour at very low and very high x. The data accumulated up to last year have been compared and their consistency questioned. [1,2,3,4].

In this paper, after a short reminder of the theoretical predictions, the present experimental status and the prospects are discussed as well as the interpretation in the QCD improved parton model. First, in order to illustrate the experimental difficulties in the measurement of structure functions with neutrinos or charged leptons beams, a few examples of systematic effects are given. The most recent results at CERN, Fermilab and SLAC on light and heavy targets are then presented and critically compared with previous results. The anticipated renewal of the field with the new kinematical domain opened by HERA is also sketched. Finally, before commenting on the QCD tests and the parton distributions determination, a brief status of the nuclear effects in the structure function is presented.

2. THEORETICAL PREDICTIONS

It is important to separate the definitions of structure functions which are free of any model assumptions, and the relations between structure functions and parton distributions which have ambiguities. In this section, we concentrate on the definitions and the predictions of perturbative QCD. The parton distributions are discussed in the last section.

2.1 *Structure functions in lepton-nucleon scattering*

The kinematics of inclusive lepton nucleon scattering is characterized by the centre-of-mass squared energy s and by the well known variables x, y and Q^2 whose Lorentz-invariant definitions are :

$$x = Q^2/2P.q \qquad (1)$$

$$y = P.q/P.p \qquad (2)$$

$$Q^2 = -q^2 = -(p-p')^2 . \qquad (3)$$

In these definitions p, p' and P denote the four-momenta of the incoming and scattered lepton and the incoming nucleon, respectively.

Neglecting the masses of leptons and nucleons and in the approximation of single gauge boson exchange, the most general invariant cross section for the scattering of a lepton off an unpolarised nucleon (see e.g. [5]) can be written as :

$$\frac{d^2\sigma^{L}_{R}}{dxdy} = \frac{1}{4\pi}\sum_{ij}\frac{s}{(Q^2+M_i^2)(Q^2+M_j^2)}\begin{bmatrix} a_i\,a_j \\ b_i\,b_j\end{bmatrix}$$

$$Q^{ij}\,[xy^2\,F_1^{ij} + (1-y)\,F_2^{ij} \mp (y-\frac{y^2}{2})x\,F_3^{ij}\,] \quad (4)$$

where

- L,R indicates the helicity of the incoming lepton.
- The indices i and j label the exchanged gauge bosons.
- M_i and M_j are the gauge boson masses.
- a_i and b_j are the couplings of the gauge bosons to the incident leptons.
- Q^{ij} stand for the couplings of the gauge bosons to the nucleon.
- F^{ij}_{123} are the unknown structure functions.

For the scattering of neutrinos, in charged or neutral current interactions there is only one exchanged boson W or Z^0 and the cross section can be simplified as :

$$\frac{d^2\sigma^{\bar{\nu},\nu}}{dxdy} = \frac{G^2 s}{2\pi}[xy^2F_1 + (1-y)F_2 \pm (y - \frac{y^2}{2})xF_3]\,. \quad (5)$$

For the scattering of electrons or muons, in charged current an identical relation can be written but in neutral current interaction the photon and the Z^0 can be exchanged and there exist two simplified relations. At fixed-target energies the weak interaction is negligible and the structure functions F^{ij}_3 which violates parity is therefore zero. The cross section is simply :

$$\frac{d^2\sigma}{dxdy} = \frac{4\pi\alpha^2 s}{Q^4}\,[xy^2\,F_1 + (1-y)\,F_2\,]\,. \quad (6)$$

At HERA energies both exchanged bosons contribute and the cross section can be written as :

$$\frac{d^2\sigma^{\mp}}{dxdy} = \frac{4\pi\alpha^2 s}{Q^4}[xy^2F_1 + (1-y)F_2$$

$$\pm (y - \frac{y^2}{2})xF_3]\,. \quad (7)$$

In this formulation, the coupling to the virtual photon is factorized, implying that the coupling to the Z_0 is absorbed in the definitions of the structure functions. This contribution has to be unfolded for the extraction of parton distributions or for QCD tests [6].

2.2 QCD predictions

It is well known that from the renormalisation group equations it is possible to deduce differential equations which govern the Q^2 evolution of the structure functions [7]. The comparison of the data with these predictions will be commented on in chapter 5. Hereafter we consider the behaviour at the endpoints (x = 0 and 1) and the sum rules.

2.2.1 Behaviour near x = 1

Assuming that the structure functions behave as

$$F(x,Q^2) \sim A(Q^2)(1-x)^{\nu(Q^2)} \quad (8)$$

for x close to $x = 1$, the following counting rules can be obtained at leading order in α_s [8] :

$$\nu_V = \nu_0 - \frac{16}{33-2n_f}\,Log(\alpha_s(Q^2)) \quad (9)$$

$$\nu_G = \nu_V + 1 \quad (10)$$

$$\nu_{sea} = \nu_V + 2 \quad (11)$$

where the suffixes V,G and sea are refering to the valence, gluon and sea distributions and n_f is the number of flavors. ν_0 is not given by the theory, although one expects $\nu_0 \approx 2$ *to* 3 [9]. These counting rules have not been checked in detail in the data due to the small rates and to the unavoidable systematic errors on structure functions for x close to 1. However no contradiction with the theory has been observed so far. The experimental situation is not expected to improve in the near future.

2.2.2 Behaviour near x = 0

All structure functions can be expressed as the sum of a flavor Singlet (S) and a flavor Non-Singlet (NS) term. Very different behaviour is predicted for these two types of components. It can be rigorously demonstrated that Non-Singlet distributions go to zero as

$$F^{NS}(x,Q^2) \sim x^{\lambda} \quad (12)$$

with $0 < \lambda < 1$ and λ independent of Q^2 [10]. From a Regge analysis it has been inferred that $\lambda = 1/2$ [11].

In contrast to Non-Singlet distributions, it has been demonstrated a long time ago, in the framework of asymptotically free field theories [12], that Singlet distributions rise at low x , faster than $Log(1/x)$ but slower than $(1/x)$. Some years later, by considering the Q^2 evolution of the moments in QCD, the following shape for the Singlet structure functions has been obtained [10] :

$$F^S(x,Q^2) = C\,(Log\,(\frac{Q^2}{\Lambda^2}))^d x^{\lambda} \quad (13)$$

where λ and d are strictly positive. The suggested value of λ is 1/3. More recently, similar strong variations of the gluon structure functions at low x have been deduced directly from the Altarelli-Parisi evolution equations [13] or from the dynamical model, which predicts an x behaviour steeper than $1/\sqrt{x}$ [14]. We should note however that such behaviours cannot continue all the way down to $x = 0$, for this would violate at high enough Q^2 the Froissart bound. Some saturation mechanism must exist which restores unitarity [15]. A possible mechanism is the rescattering process which predicts saturation at $x < 10^{-4}$ [16]. This model is controversial however*.

The experimental status in the low x domain is quite simple : there are no data so far at $x = 10^{-2}$ and $Q^2 > 10\ \text{GeV}^2$. HERA will be the first machine to reach the interesting $x \approx 10^{-4}$ domain at moderate Q^2.

2.2.3 Sum rules

The moments of structure functions are defined as :

$$M_n(Q^2) = \int_0^1 x^{n-2}\, F(x,Q^2)\, dx\ . \tag{14}$$

It can be demonstrated in QCD that, in leading order, moments of order one (resp. two) are Q^2 independent for Non-Singlet (resp. Singlet) structure functions [17]:

$$\int_0^1 \frac{dx}{x}\, F^{NS}(x,Q^2) = J_0^{NS} + J_1^{NS} \frac{\alpha_s(Q^2)}{4\pi} \tag{15}$$

$$\int_0^1 dx\, F^S(x,Q^2) = J_0^S + J_1^S \frac{\alpha_s(Q^2)}{4\pi} \tag{16}$$

where the J_0's are calculated in the free parton model. The next-to-leading order corrections J_1 have been calculated for Non-Singlet distributions [18]. A sample of usual Non-Singlet sum rules is given below :

- Gross and Llewellyn-Smith [19]

$$\int_0^1 \frac{dx}{x}\, [\, xF_3^{\nu} + xF_3^{\bar{\nu}}\,] = 6(1 - \frac{\alpha_s(Q^2)}{\pi}) \tag{17}$$

- Adler [20]

$$\int_0^1 \frac{dx}{x}\, [F_2^{\bar{\nu}p} + F_2^{\nu p}] = 2 \tag{18}$$

- Gottfried [21]

$$\int_0^1 \frac{dx}{x} [F_2^{ep} - F_2^{en}] = \frac{1}{3} [1 + k_0 \frac{\alpha_s(Q^2)}{\pi}] \tag{19}$$

where k_0 is a known minute constant [18]. We can remark that these sum rules were established in the free parton model and need corrections at finite Q^2, with one exception, the Adler sum rule, which is related to an equal time Current-Algebra commutator [20].

The Singlet sum rules for F_2^S are usually called the momentum sum rules. The asymptotic values are not given by the free parton model but by QCD which gives also the Q^2 evolution :

$$M_2(Q^2) = M_2^{\infty} + (M_2(Q_0^2) - M_2^{\infty}) \left(\frac{\alpha_s(Q^2)}{\alpha_s(Q_0^2)} \right)^{d_2} \tag{20}$$

where d_2 are positive constants of the theory [22]. Assuming four flavors, the limit in neutrino scattering is :

$$M_{2,\nu}^{\infty} = \frac{3n_f}{2n_g + 3n_f} = \frac{3}{7} \tag{21}$$

with n_g the number of gluons. In electromagnetic interactions of charged leptons the asymptotic limit is multiplied by the average charge of the quarks in the nucleon :

$$M_{2,e}^{\infty} = \frac{3n_f}{2n_g + 3n_f} < Q_q^2 > = \frac{5}{42}\ . \tag{22}$$

The next-to-leading order corrections to the Q^2 evolution have been calculated and are minute [18]. There is however no absolute prediction at finite Q^2.

The experimental tests of the Non-Singlet sum rules are not very accurate because the structure functions are obtained by subtraction of cross sections. Within the errors, the sum rules are satisfied in the data, as can be seen in a previous compilation [23] and in the most recent results from the CCFR collaboration in neutrino scattering [24] :

$$\int_0^1 \frac{dx}{x}\, F_3(x) = 2.79 \pm 0.17 \tag{23}$$

and from the BCDMS collaboration in muon scattering [28] :

$$\int_0^1 \frac{dx}{x}\, [F_2^p(x) - F_2^n(x)] = 0.34 \pm 0.09\ . \tag{24}$$

---

* See the comment of E. Reya appended to this review.

The experimental determinations of the momentum sum rules are more precise [26,27] and are found to be above the asymptotic limits for four flavors and decreasing with Q^2. This is one of the best pieces of evidence for the existence of gluons [22].

3. DATA SETS

In this section we present the origin of the main data sets and comment on a few experimental problems.

3.1 Data sources

Since two decades, data on structure functions of nucleons have been collected in experiments exposed to 5 different beams : electrons at Stanford, neutrinos and muons at FNAL and at CERN. The experiments with the highest statistics on heavy targets (more than 100k events) are listed in Table 1 where the target material, the number of events and the (Q^2,x) range are given. New data, this year, come from the CCFR collaboration which presents the combined analysis of FNAL experiments E616 and E701 using the Narrow Band Beam (NBB) [24] and from the final publication by the CDHSW collaboration of Wide Band Beam (WBB) data at CERN [29].*

On light targets (H_2 or D_2), there are two distinct types of experiments : the high statistics electron or muon experiments (more than 200k events) and the low statistics neutrino experiments (about 10k events). The neutrino experiments are not superseded by the charged lepton experiments due to their unique sensitivity to the flavor of quarks. The two types of experiments are listed in Table 2. The new data come from the CERN muon experiments. The BCDMS collaboration presents for the first time their deuterium data [28] and the final results of hydrogen data [30]. The EMC has extracted the F_2 structure function on deuterium at very low x from the experiment NA28 primarily designed to study the shadowing in nuclear targets [31]. In addition, a complete reanalysis of all the data taken at SLAC between 1970 and 1985 has been submitted to this conference [32]. All the SLAC experiments have been renormalised to the latest (E140) and corrected for radiative effects using improved versions of external and internal radiative correction procedures. These modifications of the old data turn out to be small and the main merit of this new analysis is to extend significantly the kinematical range of previous publications of SLAC data [33] (Table 2).

---

* The CDHSW data appeared two days after the end of this conference and are included in this review for completeness.

Table 1 : High Statistics Data on Heavy Targets

Collaborations	Beam	Target	Q^2 $(GeV/c)^2$	x	# evts
BCDMS [34]	μ	C	25 - 280	0.25 - 0.80	700 k
BFP [35]	μ	Fe	5 - 220	0.06 - 0.70	690 k
EMC [36]	μ	Fe	3 - 220	0.04 - 0.70	800 k
	$\nu+\bar{\nu}$				
CCFR (R) [37]	NBB	Fe	1 - 200	0.01 - 0.70	150 + 23 k
CCFR [24]	NBB	Fe	"	"	~ 300 k
CCFR	WBB	Fe	Not yet	Not yet	(1500 + 250) k
CDHS [38]	NBB	Fe	1 - 200	0.01 - 0.70	(90 + 25) k
CDHSW [29]	WBB	Fe	0.2 - 200	0.01 - 0.70	(650 + 550) k
CHARM [39]	WBB	Fe	0.4 - 100	0.01 - 0.80	(50 + 110) k

Table 2 : High Statistics Data on n and p Targets

Collaborations	Beam	Target	Q^2 $(GeV/c)^2$	x	# evts
MIT - SLAC [33]	e	H_2, D_2	1 - 15	0.10 - 0.80	~ 2000 k
New Analysis of SLAC exp. [32]	e	H_2, D_2	0.6 - 30	0.06 - 0.90	~ 5000 k
BCDMS [30]	μ	H_2	7 - 260	0.06 - 0.80	1800 k
BCDMS [28]	μ	D_2	8 - 260	0.06 - 0.80	800 k
EMC (NA2) [26]	μ	H_2	3 - 190	0.02 - 0.80	400 k
EMC (NA2) [40]	μ	D_2	6 - 190	0.02 - 0.80	200 k
EMC (NA28) [31]	μ	D_2	0.2 - 8	0.002 - 0.17	25 k
	$\nu+\bar{\nu}$				
BEBC (WA21) [41]	WBB	H_2	2 - 11	0.06 - 0.65	12 k + 7 k
BEBC (WA25) [42]	WBB	D_2	0.2 - 44	0.018 - 0.17	12 k + 11 k
FNAL [43]	WBB	D_2	6 - 30	0.05 - 0.80	6 k + 4 k
CDHS [44]	WBB	H_2	3.3 - 43	0.05 - 0.65	4.5 k + 7.4 k

A comprehensive compilation of all these data which treats properly the systematic errors has never been done, the necessary correlation from bins to bins induced by systematic effects is usually ignored*. This task is feasible but is hard and clearly beyond the scope of this review. Hereafter we just underline the main trends of the data, starting with a few examples of systematic effects one should have in mind for a relevant comparison.

3.2 Experimental problems

To illustrate a few common experimental problems, the systematics in the overall normalisation and in the energy calibration of charged leptons and neutrino scattering experiments are discussed.

3.2.1 Normalisation

In fixed-target experiments, what is usually referred to as normalisation uncertainties are the ones related to the counting of the incoming flux and to the number of nucleons in the fiducial volume. The latter source of error is usually negligible. The former one depends strongly on the type of beams. A priori, the hardest beam to count is the neutrino one. This was the subject of hot debates in the early eighties [45] but is settled since 1985 after one of the opponents has improved its counting technique [46]. The remaining uncertainty is about 3% and similar to what is achieved in the counting of incoming electrons (2%) or muons (3%). The EMC overall uncertainty factor is nevertheless 5% because to the 3% of the muon counting is added a global 4% uncertainty for reconstruction losses which is compatible with being independent of Q^2 and x [26].

3.2.2 Energy scale in e(μ) experiments

It is well known that small absolute calibration errors of the energy of either the incident or the scattered lepton may produce large distortions on the F_2 structure function at low y (i.e. low Q^2 at a given x), especially so at large x due to the $(1-x)^3$ behaviour [26,34]. In a given (Q^2, x) bin the shift is therefore the largest at the highest beam energy (see Figure 1). Hence, for a given experiment, to compare data sets of various beam energies in the same apparatus is a good handle to control the energy scale provided statistics is sufficient at large x [47]. This systematic error can be strongly reduced if the beam and the spectrometer are cross-calibrated with high precision [48]. This was done in SLAC experiments and is actually done in the new generation of muon experiments at CERN (NA28 [31] and NA37 [49]) where the energy scale uncertainties are below one per mil.

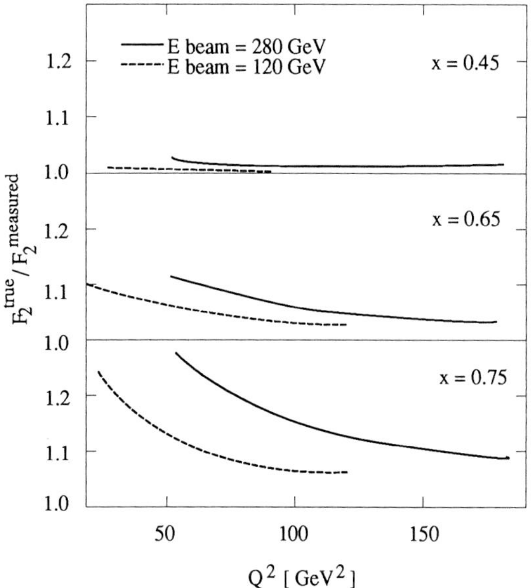

Fig.1 - The effect of a systematic change of 0.5% in the scattered muon energy, for two beam energies E_b = 120 GeV and E_b = 280 GeV, on the measurement of F_2 in muon scattering [47].

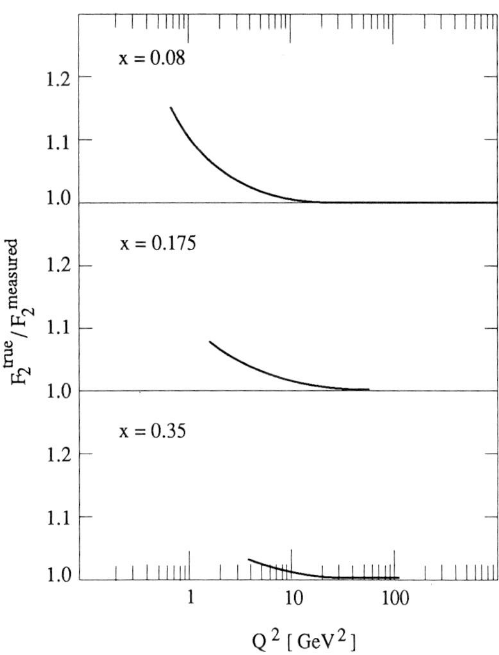

Fig.2 - The effect of a systematic change of 500 MeV in the total hadronic energy on the measurement of F_2 in a neutrino scattering experiment in the Wide Band Beam at CERN [50].

---

* An effort is presently under way in this direction by the Snowmass Structure Function working group [3].

3.2.3 Hadronic energy in ν experiments

It is necessary in a neutrino scattering experiment to measure the total energy flow to reconstruct the kinematic variables. The hadronic energy is measured in a calorimeter which is also crossed by the scattered muon whose energy deposition has therefore to be subtracted. This subtraction may lead to a systematic shift of about 500 MeV on the hadronic energy [29]. The corresponding distortions have been calculated [50] and are displayed in Figure 2. One can observe that in this case the critical domain is the very low x part.

3.3 F_2 on heavy targets

Let us concentrate in this section on the structure function F_2, where the statistics are large enough to discuss systematic effects. The status of the other structure functions will be mentioned in the sections on quark distributions and on QCD tests. The high statistics data on heavy target come from neutrino and muon experiments on iron or carbon. To do a meaningful comparison, which includes neutrino and charged lepton data sets, we have to assume the QCD improved quark parton model and reduce the experimental F_2 to a universal F_2 :

$$F_2 = \sum_{f=1}^{4} q_f + \bar{q}_f \tag{25}$$

and hence correct the neutrino data for charm threshold effect and multiply the muon data by 18/5 (the inverse of the average square charge of the quarks). It is also important to use the same assumption on the longitudinal structure function. Hereafter we have assumed that R (= $(F_2 - 2xF_1)\ /\ 2xF_1$) behaves as predicted by QCD (see section 5). The Q^2 variation of F_2 for the highest statistics experiments are displayed on Figure 3 where the muon data have been interpolated to the x bins of the neutrino experiments. Error bars represent statistical and systematic errors summed in quadrature. At first look, the result is not glorious. The typical dispersion is about $\pm 10\%$. However a detailed examination of these plots shows that in most of the x bins, the Q^2 variation of all the experiments are parallel and that the difference in the data of two experiments does not vary with x . It seems therefore that, if we take the liberty to vary the normalisation of the four data sets, the dispersion should be strongly reduced. Actually this is what we have done in Figure 4 where, for a rough estimate, we have used arbitrarily the data of CDHSW as a reference and then corrected the EMC data by +8% and the CCFR data by −4%, without correcting those of BCDMS. The result is much better, the dispersion of the four data sets is now below $\pm 5\%$ for all x bins with $0.05 \leq x \leq 0.60$. The discrepancies are still quite large in the extreme x bins : $x = 0.015$ and $x = 0.65$. The difficulties in these extreme bins are easy to understand. The lowest x bins are very sensitive to systematic effects in neutrino experiments. At large x, on the other hand, a small error on x is enhanced in F_2 due to the law in $(1-x)^3$. The size of the correction factors is within the errors for flux counting in BCDMS, CCFR and CDHSW and may appear large for EMC, but, as discussed above for this experiment, these overall factors can be related to reconstruction losses and hence be larger. In conclusion, after applying global correction factors, the experimental situation of the structure function F_2 on heavy targets is globally satisfying.

3.4 F_2 on deuteron

On deuterons and protons the high statistics data come from the muon experiments at CERN and from the electron experiments at SLAC. The new data of BCDMS [28] and the result of the reanalysis of SLAC experiments [32] are displayed in Figure 5 together with the previous data of EMC (NA2) [40]. Only statistical errors are shown because the systematic errors of BCDMS are not yet published. For clarity let us first compare the new data of BCDMS and SLAC. The global matching of both sets of data is quite good. A detailed examination shows however two local discrepancies : at $x = 0.275$ where the agreement is marginal within the statistical errors and at $x > 0.5$ where the low Q^2 points of BCDMS are about 10% lower than those of SLAC experiments in the overlap region. This discrepancy is likely to be attributed to systematic effects in the data of BCDMS which are known to be large in this domain but decrease quite fast when Q^2 rises (see Figure 1).

Let us now compare the new data of SLAC experiments with those of the EMC experiment NA2 [40]. It is clear in Figure 5 that the data points of EMC (NA2) are below any reasonable extrapolation of those of SLAC for $x < 0.15$ but could match those of SLAC at larger x although the comparison is not very significant in this domain due to the large statistical errors of EMC. This discrepancy at low x may be related to what has been observed on the ratio of structure functions $F_2^{Fe}\ /\ F_2^{D_2}$ where it turned out that the data of the original EMC effect [40], which rely on the data of NA2 shown in Figure 5, were too high by about 10% at $x < 0.1$ with regard to all subsequent measurements.

As in the case of heavy targets data a much better consistency at low x can be obtained by a simple variation of the normalisation of the three data sets. This is shown in Figure 6, where we have shifted the data of EMC by +8% and those of BCDMS by −2%, using arbitrarily SLAC data as a reference for normalisation.

The new data of NA28 on Deuterium have larger errors and the comparison with NA2 or SLAC data cannot resolve the discrepancy. The main interest of these new data is in the nuclear shadowing study (see section 4) and also in the first measurement at x as low as 0.002 and Q^2 around 1 GeV^2. The Q^2 variation of the structure function F_2 at $x = 0.005$, shown in Figure 7, agrees with extrapolations of conventional quark distributions [52] and is a factor two below the predictions of the dynamical model [14].

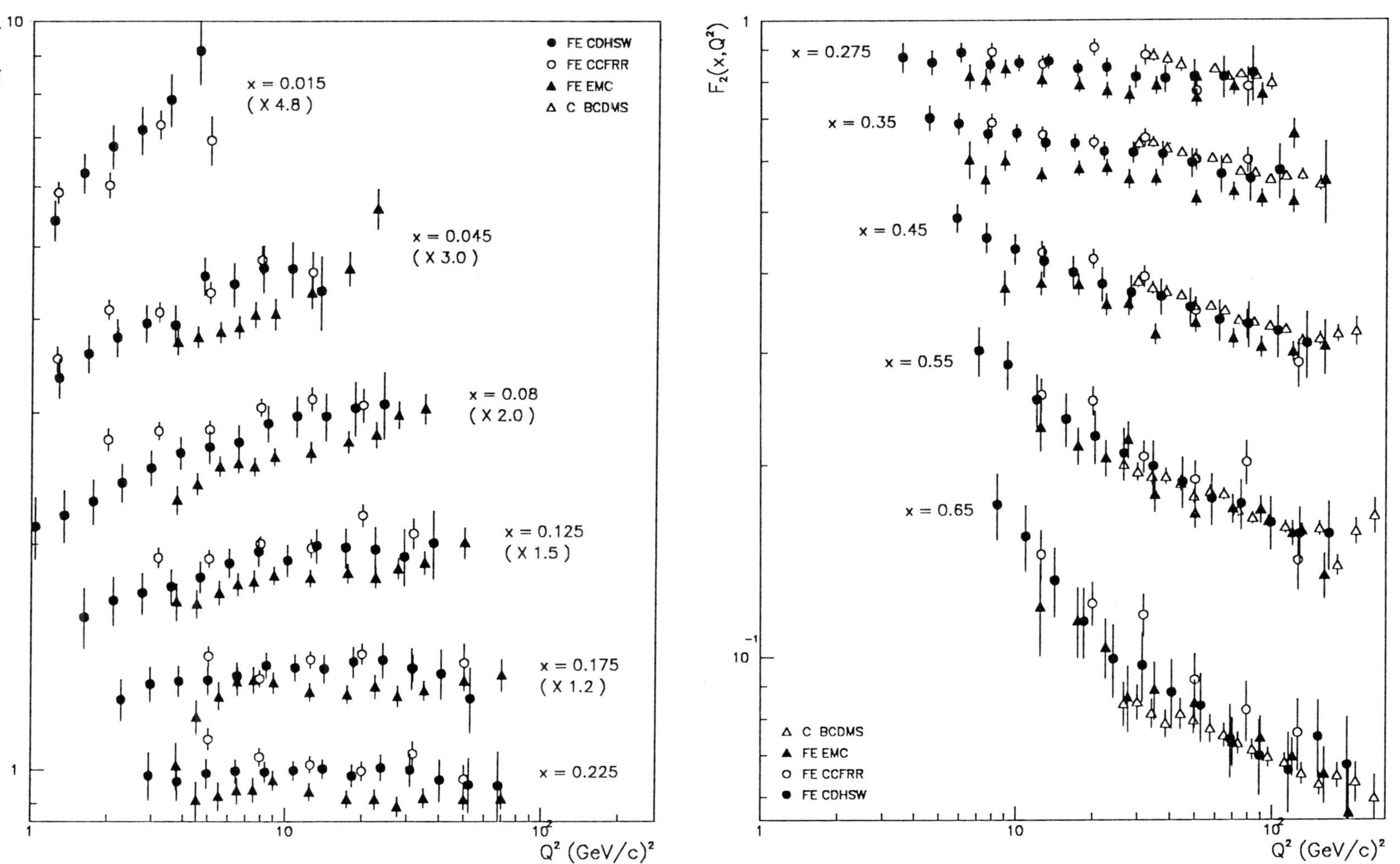

Fig.3 - $F_2(x,Q^2)$ in bins of x as a function of Q^2 on heavy targets. All F_2 are extracted for R = R_{QCD}. The data points of EMC on iron [36] and of BCDMS on carbon [34] are multiplied by 18/5. The data points of CCFR [24] and CDHSW [29] are corrected for charm threshold. Error bars represent statistical and systematic errors summed in quadrature.

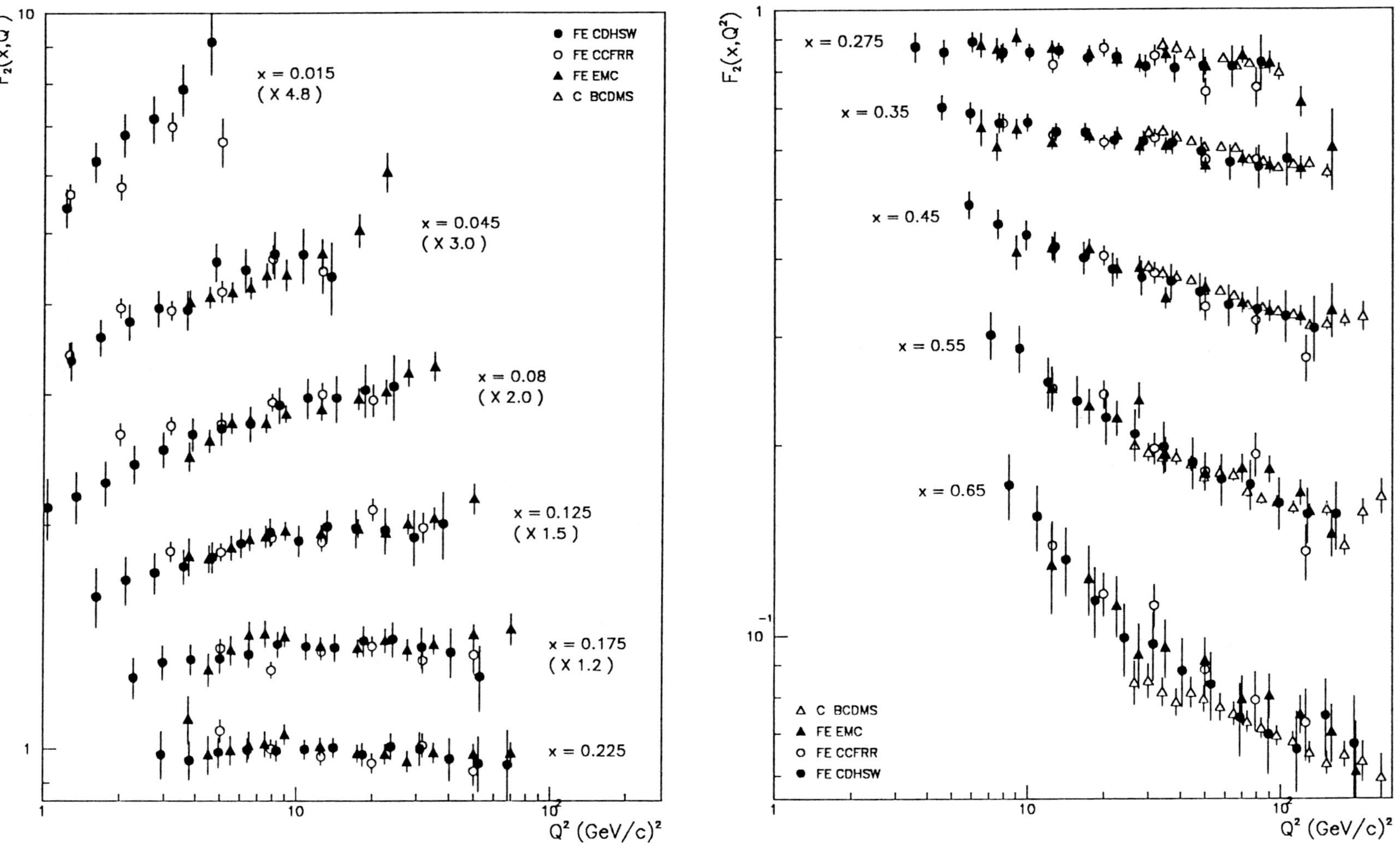

Fig.4 - $F_2(x,Q^2)$ in bins of x as a function of Q^2 on heavy targets. As for Figure 3 but after shifting the data of EMC and of CCFR by +8% and −4%, respectively.

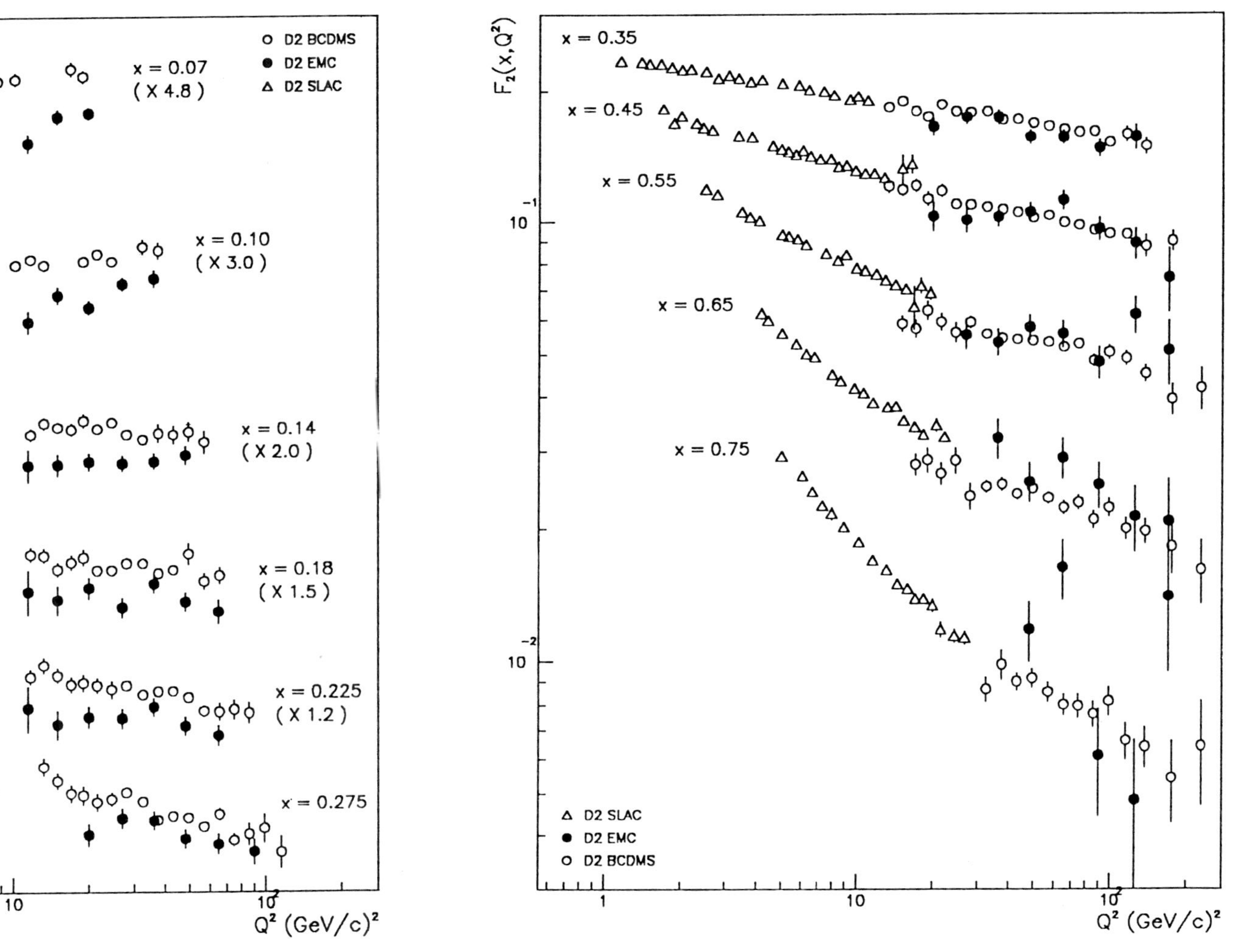

Fig.5 - $F_2(x,Q^2)$ in bins of x as a function of Q^2 on deuterium target. The data of BCDMS [28] and of EMC [40] are extracted for R = R_{QCD}. The new data of SLAC experiments [32] are extracted with the value of R fitted in this new analysis (see Figure 16). Error bars are only statistical.

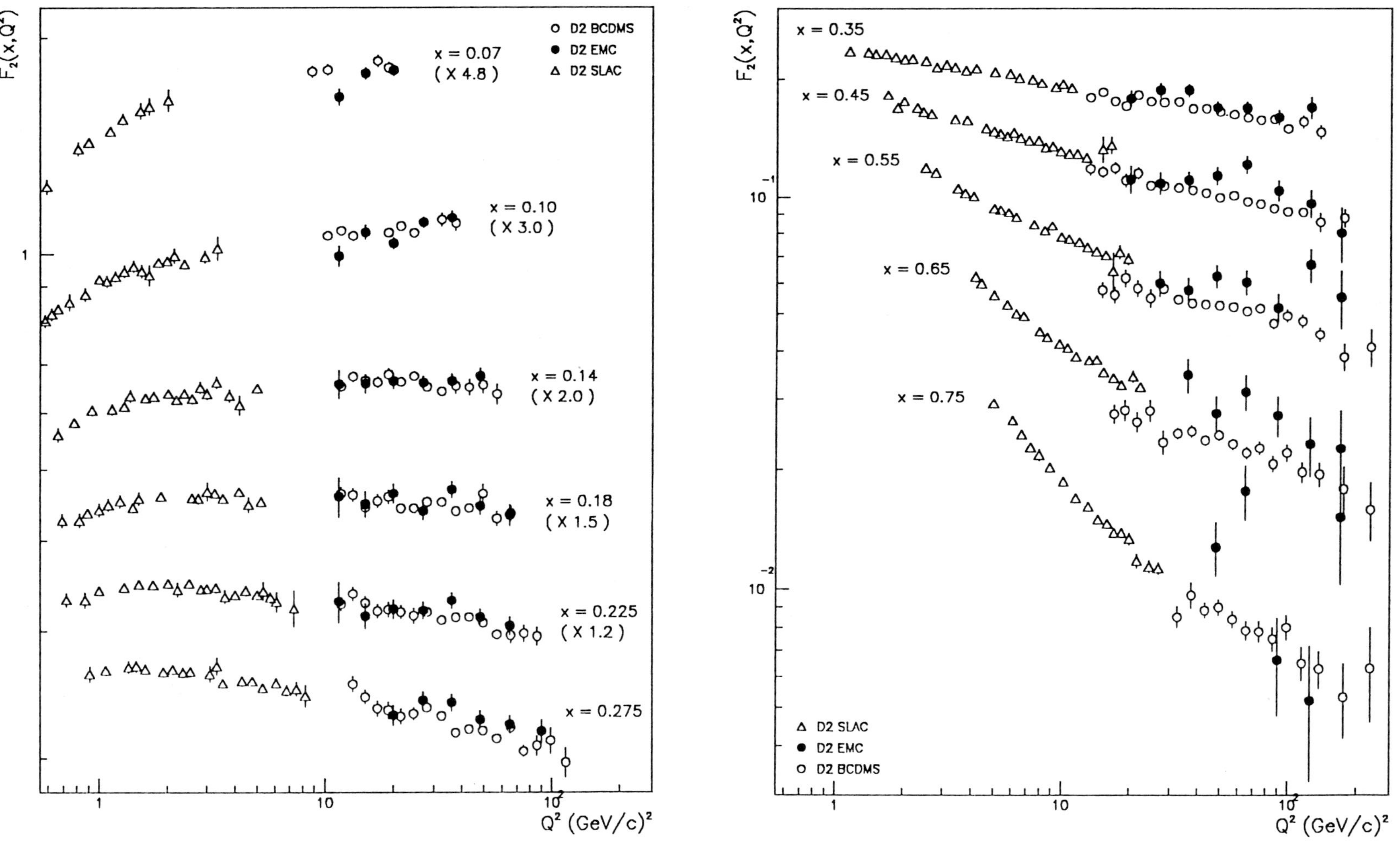

Fig.6 - $F_2(x,Q^2)$ in bins of x as a function of Q^2 on deuterium target. As for Figure 5 but after shifting the data of EMC and BCDMS by +8% and −2%, respectively.

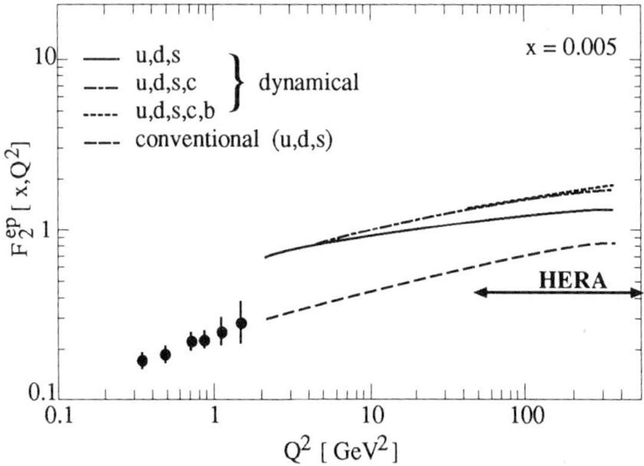

Fig.7 - Comparison of F_2^{ep} at $x = 0.005$ as predicted for the dynamical [14] and conventional distributions [52] to the data points of EMC (NA28) [31]. Error bars represent statistical and systematic errors summed in quadrature.

We have seen that the latter, inspired by perturbative QCD, predicts steep variations with decreasing x at large Q^2 and should be better tested in the kinematical domain of HERA.

3.5 F_2 on proton

For the structure function F_2 on free protons, conclusions similar to those for F_2 on deuterons can be reached from the comparison between the 1985 data of EMC [26], the final results of BCDMS [30] and the new analysis of SLAC experiments [32] (Figure 8), namely :

- there is a clear need to move the points of EMC upwards at low x by about 8 %,
- there is global agreement between SLAC and BCDMS, although
- local discrepancies around $Q^2 = 10$ GeV^2, $x = 0.25$ and around $Q^2 = 15$ GeV^2 and $x = 0.6$ are observed between BCDMS and SLAC.

For the data on free protons the three collaborations have published their systematic errors. It is possible to be more quantitative on the level of agreement. The ideal solution should be to do a fit which includes the correlation between systematics. Such a correlation is not yet available in the new analysis of SLAC experiments. Moreover for a precise comparison, all data sets should be corrected for the same value of R in a given (Q^2,x) bin. In the low x domain where the data of the muon experiments are sensitive to R, we have used in Figure 8 the R value predicted by QCD *. The SLAC data have been extracted with a value of R which at large Q^2 is still above QCD expectations (see section 5.2 and Figure 16), this may generate few percent effects which need further studies. A rough quantitative estimate can however be obtained by adding in quadrature the systematic and statistical errors and doing separate fits to EMC−SLAC and BCDMS−SLAC. From what has been learnt so far, it seems reasonable to fit the Q^2 variation by $(Q^2)^b$ terms (QCD like) multiplied by a simple power-suppressed term (see section 5.1 on QCD tests) :

$$F_2(x_i) = a_i(Q^2)^{b_i}(1 + \frac{c_i}{Q^2}) \qquad (26)$$

where the coefficients a_i, b_i and c_i are parameters to adjust in each bin of x . The resulting χ^2 per degree of freedom are quite good in both cases (around 0.8) because there is little overlap between CERN and SLAC data points. More instructive informations can be drawn from a visual inspection. The curves resulting from the fits of equations (26) are drawn on Figure 9, superimposed to the three sets of data points where global corrections of +8% on the points of EMC and −2% on the points of BCDMS have been applied.* It is striking that in a given x bin, in the whole Q^2 range, both sets of curves do not differ by more than 6% (maximum deviation) except at very large x ($x > 0.6$). This is an average behaviour and it gives a feeling of the overall consistency after application of overall factors. At very large x the discrepancies are clearly larger. However the errors on the EMC points are large, such that the fit (equ.(26)) is not constrained enough and is therefore not conclusive.

3.6 Future prospects of structure function measurements

New analyses of structure functions in fixed-target experiments are still expected in the near future but the new measurements should come mainly from the HERA collider. On heavy targets no new data taking are foreseen, what is to come is the analysis of the large statistics Wide Band Beam data of the CCFR collaboration (see Table 1). On light targets (hydrogen and deuterium) a large amount of data has been taken by NMC, a new CERN muon collaboration, from which only preliminary data on F_2^n / F_2^p have been published so far (see section 6). The data taking is still going on and will improve the present situation. The FNAL Tevatron experiment E665 is also being analyzed and will contribute in the low x part and to the overall normalisation on light targets [54]. We have seen that the present limitations of structure function measurements are mainly systematics in experiments designed in the mid-seventies without being fully aware of the extreme sensitivity to systematics when it comes to QCD tests. An

* It has not been possible to correct the data of EMC in the whole x range because the method to merge the data of different beam energies at large x is not explicitely given. The effect of R is however negligible in the CERN data at large x.

* The points of SLAC have been taken as reference for normalisation correction, some intermediate value between the 3 sets of data could also have been chosen.

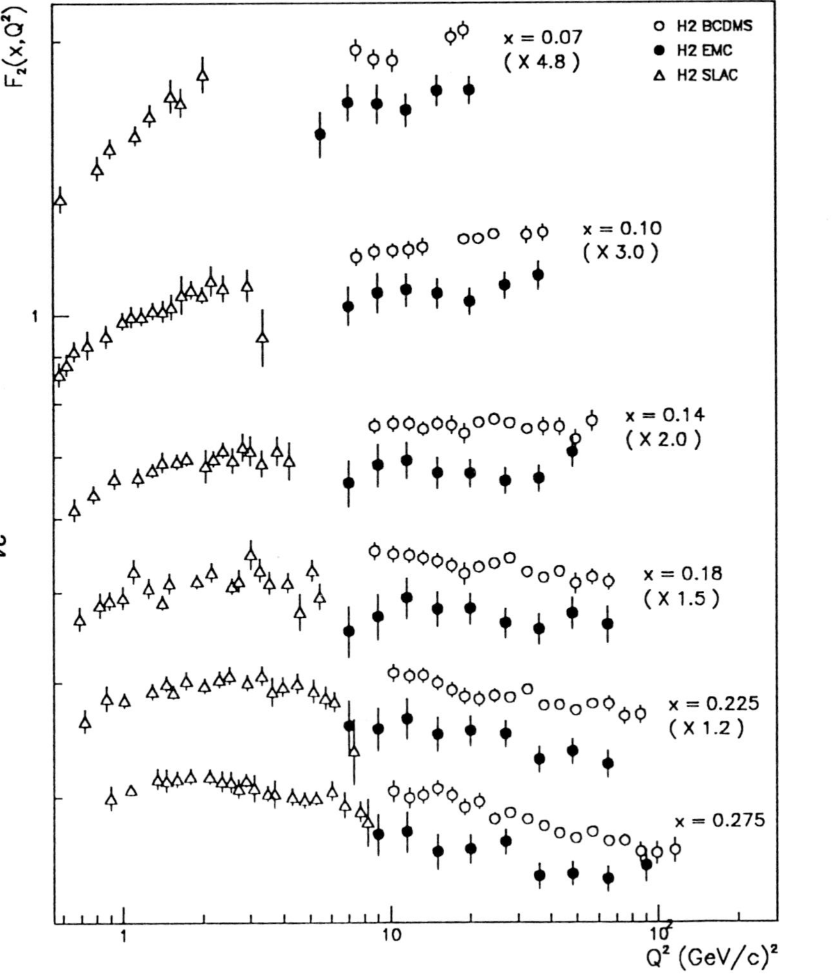

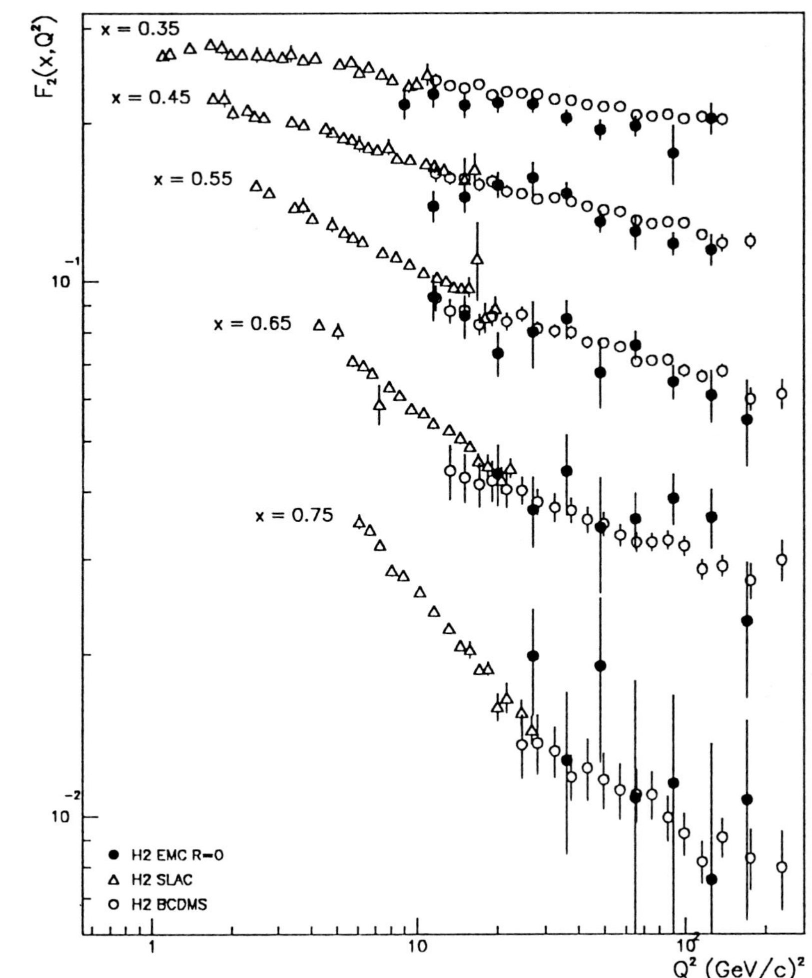

Fig.8 - $F_2(x,Q^2)$ in bins of x as a function of Q^2 on hydrogen target. The data of BCDMS [30] and of EMC [26] are extracted for R = R_{QCD} except the large $x(x > 0.35)$ data of EMC for which R = 0, as explained in the text. The new data of SLAC experiments [32] are extracted with the value of R fitted in this new analysis (see Figure 16). Error bars represent statistical and systematic errors summed in quadrature.

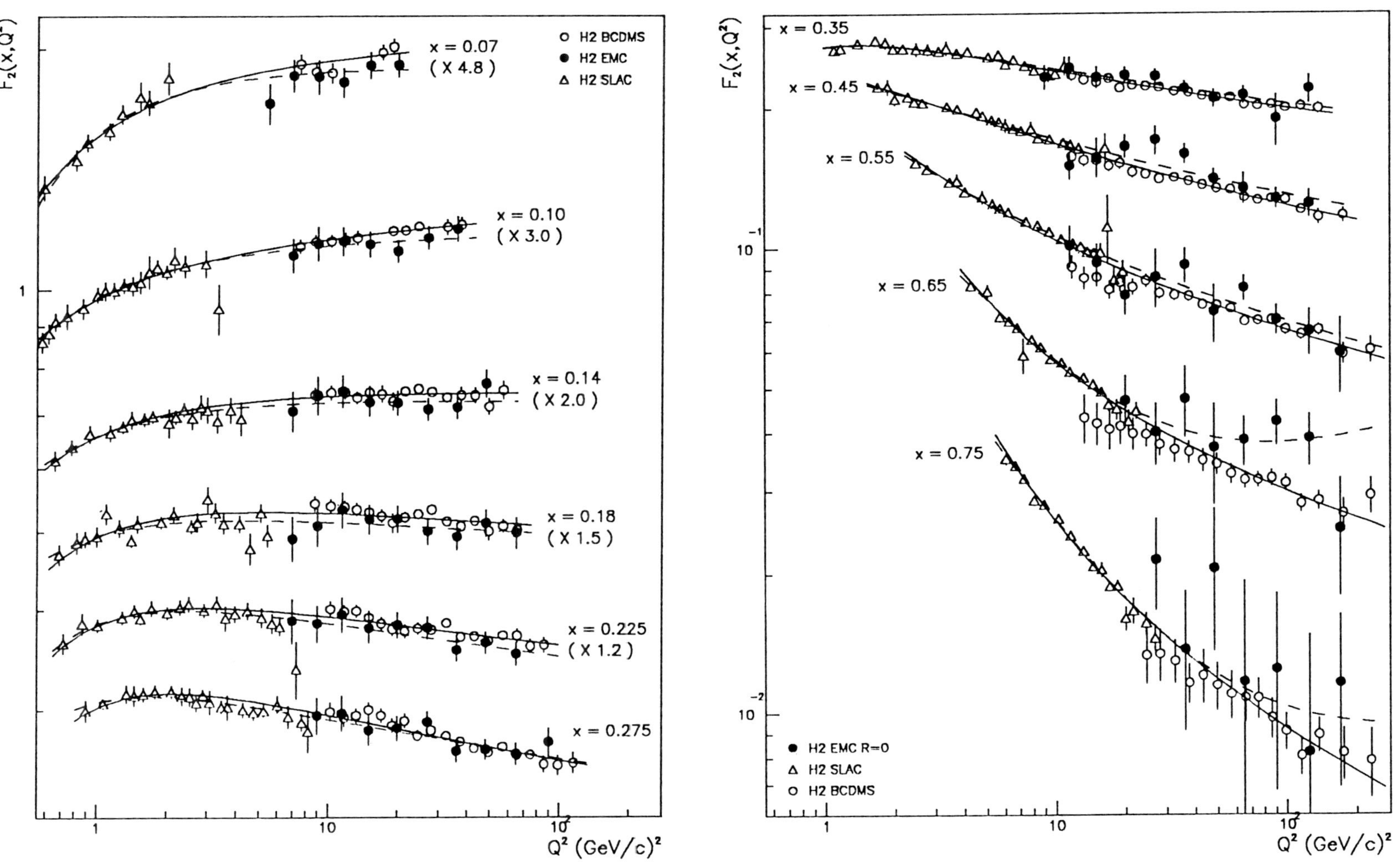

Fig.9 - $F_2(x,Q^2)$ in bins of x as a function of Q^2 on hydrogen target. The data points are as for Figure 8, but after shifting those of EMC and BCDMS by +8% and −2%, respectively. The full and and dashed lines represent the fits with the equations (26) of the data of (SLAC-MIT + BCDMS) and of (SLAC-MIT + EMC), respectively.

ultra precise experiment with muon beams is presently being intended with the goal of improving by an order of magnitude both the statistical and the systematic errors [55].

However, in the immediate future, the field will be rejuvenated by the commissioning of HERA, the first electron-proton collider. At nominal energies of 30 GeV for the electron and 820 GeV for the proton the centre-of-mass energy is 314 GeV. The highest Q^2 reached in the data should be of a few 10^4 GeV^2 at large x and the lowest x is 10^{-4} at $Q^2 = 10\ GeV^2$. We have seen that at these energies it will be possible to measure two structure functions in charged current interactions and three in neutral current interactions. Some quantitative estimates on the statistical accuracy are given in the next sections on QCD tests and parton distributions. As in fixed-target experiments, the overall precision on structure functions is also determined by the accuracy of the kinematic reconstruction of the variables Q^2, x or y. These variables can only be reconstructed from the total hadron flow in charged currents interactions and from both the total hadron flow and the scattered electron in neutral current interactions (Figure 10). The sensitivity to systematic effects is not the same for the two types of reconstruction [56]. A few examples are given in Figure 11 where one can observe that the domain where the systematics can be kept below 10% will not overlap with the present domain of fixed-target experiments. There is at least an order of magnitude in Q^2 between endpoints of both domains. Some overlap can only be established by running HERA at proton and electron beam energies lowered by about a factor three.

A problem which is not related to the measurement but to the extraction of the structure functions from the differential cross sections is the size of the electroweak radiative corrections in the HERA kinematic domain. Several groups have tackled the calculations actively and, hopefully, the results will be ready at the time of the first data [57].

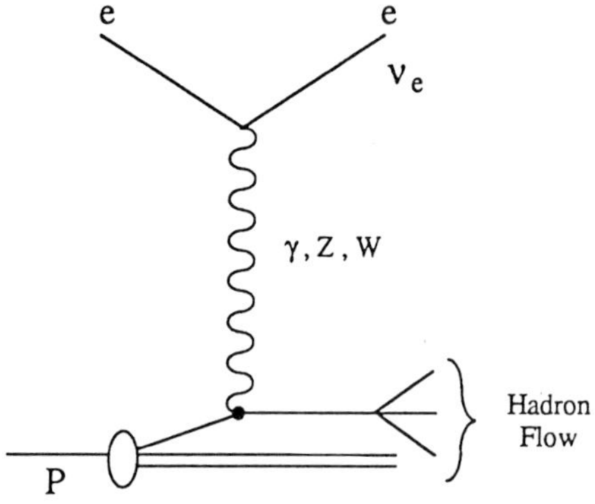

Fig.10 - The Feynman diagram for the first order deep inelastic electron-proton process.

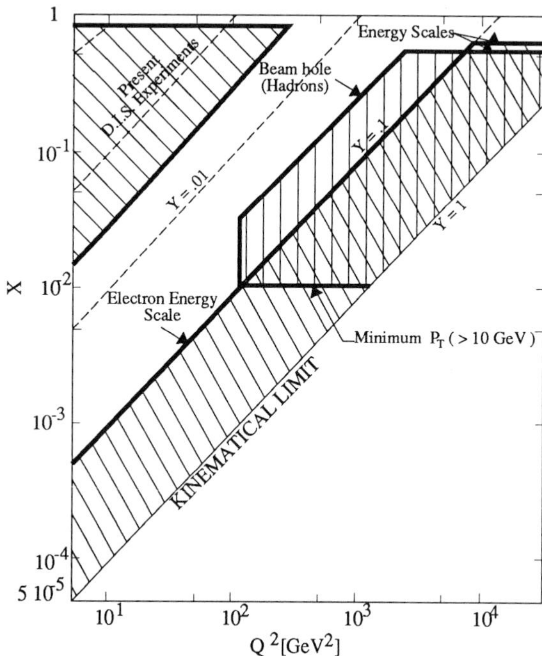

Fig.11 - Experimentally accessible regions for structure functions measurements at HERA nominal beam energies of 30 GeV electrons on 820 GeV protons. The domains covered by vertical and 60 degrees hatched lines specify the regions where differential cross-sections are expected to be measurable with less than 10% systematic errors based on the kinematics reconstruction from the visible hadron flow and the scattered electron, respectively.

4. NUCLEAR EFFECTS

The discovery in 1983 by the EMC [51] that the structure functions of a nucleon embedded in a heavy nucleus are different, after subtraction of simple Fermi motion effects, from those of nucleons loosely bound in a deuteron has casted doubts on the validity of QCD tests and on the parton distributions extracted on heavy targets. The experimental pattern of these nuclear effects is now well established, the salient features are the following :

- A compilation, borrowed from reference [31], of the world data* on the ratio of F_2^A / F_2^D, is presented in Figure 12 for the $\langle A \rangle = 55$ nuclei. One can observe three distinct x domains :

* except the original EMC publication [51] which has large systematics at low x.

1. At low x (x < 0.05), the shadowing region, there is a clear depletion in heavy nuclei which increases with decreasing x and which points towards the value found in photoproduction experiments [58].

2. For 0.05 < x < 0.6, after a small overshoot, sometimes called the anti-shadowing [62], there is a depletion which increases with rising x. This is the original "EMC effect".

3. At high x (x > 0.6), the ratio rises again as predicted by all models on Fermi motion.

- There is no visible Q^2 variation in the shadowing region (Figure 13) down to $Q^2 = 0.3$ GeV^2 and in the "EMC effect" region (Figure 14). The present data are described by a flat variation. A few percent effect is not excluded however.

- There is no indication that the nuclear effect can be related to an increase of the glue [5] or to an increase of the sea within 10 % [59].

- The variation of the longitudinal structure function with A is consistent with zero for 0.2 < x < 0.5 and $5 < Q^2 < 12$ GeV^2 [60] :

$$R^{Fe} - R^{D} = 0.001 \pm 0.018 \pm 0.016 \qquad (27)$$

- The variation of the ratio of structure functions is smooth in $Log(A)$ for x > 0.3 [61].

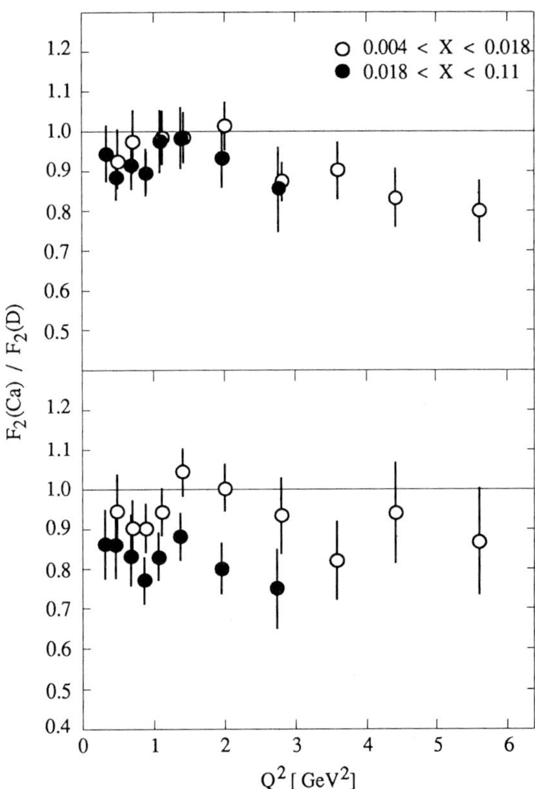

Fig.13 - Structure function ratios versus Q^2 obtained for two intervals of x : 0.004 < x < 0.018 (closed points) and 0.018 < x < 0.110 (open points) measured by the EMC (NA28) [31]. Error bars give statistical and systematic errors summed in quadrature.

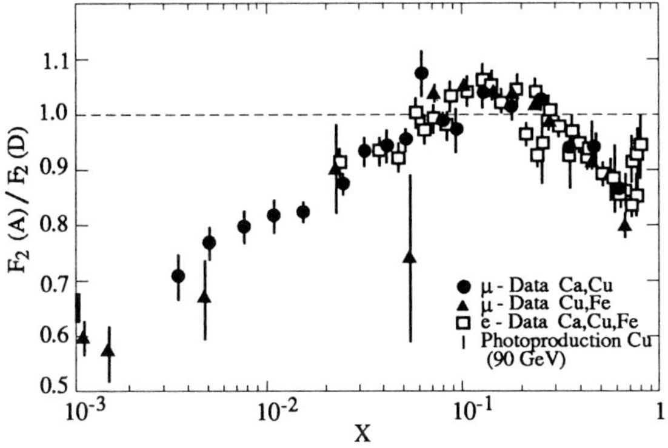

Fig.12 - Compilation of the data on the nuclear to nucleon structure function ratio for the < A > = 55 nuclei, borrowed from reference [31]. The data come from muon [31,63,66,67] and electron [61,64,65] experiments together with the average measurement in real photoproduction [58]. The errors are statistical.

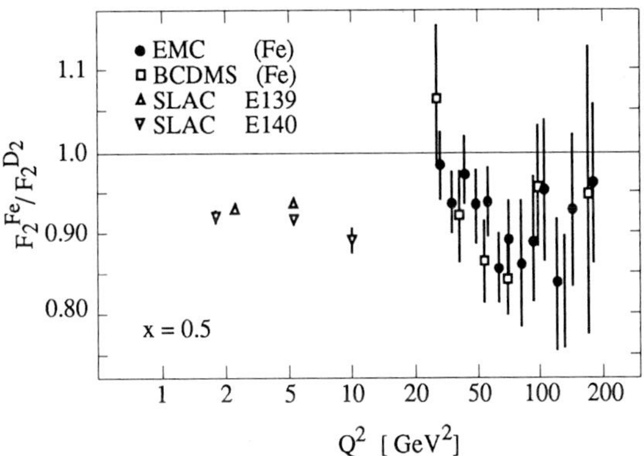

Fig.14 - Structure function ratios at x = 0.5 from electron experiments at low Q^2 [61,60] and muon experiments at high Q^2 [66,67].

With respect to theoretical interpretations, the conventional nuclear process as the Fermi motion is still poorly known. A detailed discussion of all the models is beyond the scope of this review. Many compilations exist in the litterature and, for example, a comprehensive review of the theory of the "EMC effect" can be found in reference [5] and references therein. In the shadowing region, quantitative models have been developed in connection with or prior to the results of EMC (NA28), they are presented in reference [31].

In summary, the nuclear effects which affect the structure functions are sufficiently understood to allow for a simple correction on the x behaviour before extracting parton distributions and should not affect the QCD tests in the present domain of deep inelastic physics.

5. QCD TESTS

The structure functions of the nucleon are one of the masterpieces of QCD, the theory of strong interactions. It is well known that the behaviour in x of the structure functions is not predicted by perturbative QCD. So far only the asymptotic limits can be inferred but are not exactly predicted by QCD (section 2). Given the x dependence of the structure function at $Q^2 = Q_0^2$, then the evolution equations allow to predict it at all Q^2 [7] :

$$\frac{\partial}{\partial LogQ^2} F^{NS}(x,Q^2) = \frac{\alpha_s(Q^2)}{2\pi} \int_x^1 \frac{dy}{y} \times F^{NS}(y,Q^2) P_{qq}^{NS}(\frac{x}{y}) \qquad (28)$$

$$\frac{\partial}{\partial LogQ^2} F^{S}(x,Q^2) = \frac{\alpha_s(Q^2)}{2\pi} \int_x^1 \frac{dy}{y} \times [F^{S}(y,Q^2) P_{qq}^{S}(\frac{x}{y}) + 2n_f G(y,Q^2) P_{qg}^{S}(\frac{x}{y})] \qquad (29)$$

$$\frac{\partial}{\partial LogQ^2} G(x,Q^2) = \frac{\alpha_s(Q^2)}{2\pi} \int_x^1 \frac{dy}{y} \times [F^{S}(y,Q^2) P_{gq}^{S}(\frac{x}{y}) + G(y,Q^2) P_{gg}^{S}(\frac{x}{y})] \qquad (30)$$

where the P_{qq}, P_{gq} and P_{gg} are splitting functions given by the theory and $G(x,Q^2)$ is the gluon density. The related scaling violations proportional to α_s are the most solid and powerful method for testing perturbative QCD [68]. However the apparent experimental situation two years ago led to serious doubts on many analyses of scaling violations [1,2]. An outstanding question in the interpretation of structure functions is : what is the domain of validity of perturbative QCD ? Or, in other words, in which domain in x and Q^2 are the scaling violations described by the leading twist terms of the Altarelli-Parisi equations ? Another important test of QCD is the longitudinal structure function, which has a special interest because it is only at high order of perturbative QCD that it is predicted to vanish as $Log(Q^2)$ and not in $1/Q^2$. This structure function is closely related to the gluon distribution. Let us review successively the experimental status of the higher twist terms, the longitudinal structure function, the scaling violations and the gluon density. We end with an outlook on the future prospects to improve the tests of perturbative QCD by structure function measurements in the coming years.

5.1 Higher twist terms

There is no prescription neither on the x dependence nor on the size and even on the sign of contributions of high order twist terms in the Q^2 evolution of the structure functions. What has been attempted is the isolation in the data of power suppressed terms (twist 4) which contribute to the structure function as :

$$F_2(x,Q^2) = F_2^{LT}(x,Q^2)\,(1 + \frac{\mu^2(x)}{Q^2}). \qquad (31)$$

On free protons, evaluations have been done in neutrino bubble chamber experiments alone, and in analyses which involve low Q^2 electron data and high Q^2 muon data. The resulting $\mu^2(x)$ are displayed in Figure 15. The charged

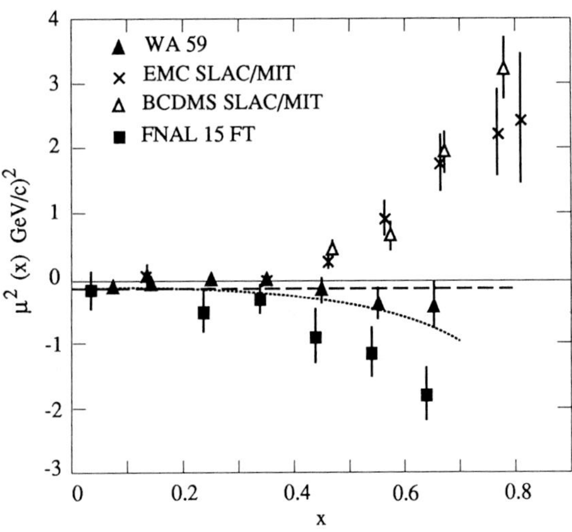

Fig.15 - x dependence of a general higher twist term (equ.31). The data come from WA59 [70], (EMC + SLAC-MIT (1979)) [26], (BCDMS + SLAC-MIT (1979)) [72] and FNAL (15 FT) [69] experiments. Error bars are only statistical.

lepton points have been calculated before the new analysis of the SLAC data. They are obtained by combining the old SLAC-MIT data [33] and the CERN data from EMC [26] or from BCDMS [30] after renormalisations of +7.5% and −2.5% of the two muon experiments respectively. It is clear that in charged lepton scattering the higher twist terms are positive at large x and rise with x. In contrast, in neutrino experiments the situation is more confused : data from FNAL [69] are negative and decrease strongly with x but data from BEBC [70] are just slightly negative. There exist already models to justify why higher twist terms have opposite sign in weak and in electromagnetic interactions [71]. However we should be aware that these terms are extremely sensitive to systematic errors, especially when combining two different data sets. As an example, a change of 1.5 per mil in the energy scale of muon experiments can modify the higher twist term by a factor two at high x [72]. The determination of the higher twist term is likely to be improved using the new SLAC points. In any case, the data demonstrate so far that the higher order twist terms can be neglected (i.e. are below 5%) for $Q^2 > 10\ \mathrm{GeV}^2$ at $x < 0.5$.

5.2 $F_L(x,Q^2)\ (= F_2 - 2xF_1)$

In the naive parton model and in leading order of QCD the longitudinal structure function is zero. In next-to-leading order of QCD, F_L gets contribution in α_s from both quarks and gluons :

$$F_L(x,Q^2) = \frac{\alpha_s(Q^2)}{2\pi} x^2 \int_x^1 \frac{dy}{y^3} \left[\frac{8}{3} F_2(y,Q^2) + 2\,\Sigma_{i=1}^{2n_f} e_i^2 (y-x) G(y,Q^2)\right] \qquad (32)$$

where $(2\,\Sigma_{i=1}^{2n_f} e_i^2)$ is the sum of all coefficients of q and $\overline{q}$ in the naive parton model expression of F_2 / x (for $n_f = 4$ it is 20/9 for neutral current electron scattering and 8 for charged current neutrino scattering).

F_L should vanish smoothly as $\mathrm{Log}(Q^2)$ and gets sizeable contributions at low x ($x < 0.1$) due to the increase of the sea and of the gluon distributions.

The prediction on the Q^2 variation is well satisfied by the data at $Q^2 > 10\ \mathrm{GeV}^2$ (Figure 16). At lower Q^2 the recent data of the SLAC experiment E140 seem to require non perturbative effects which can barely be described by the so-called target mass correction [73].

The prediction on the x behaviour is compatible with the present data which indicate a rise above zero at x around 0.05 by two to three sigmas (Figure 17).

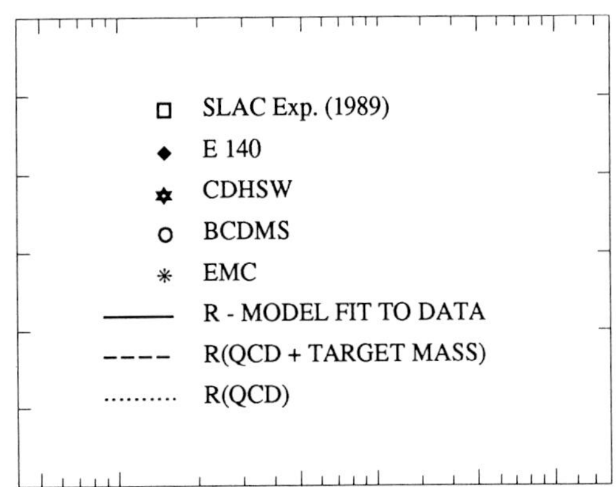

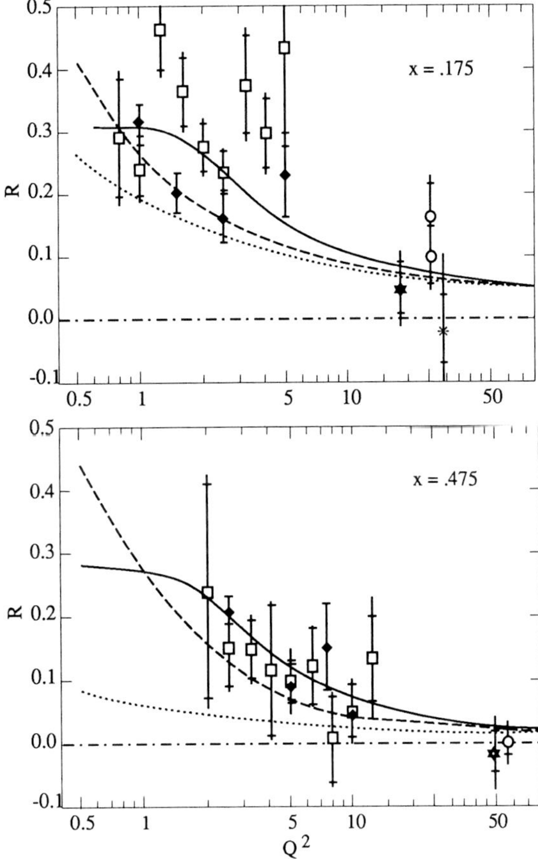

Fig.16 - $R(x,Q^2)\ (= F_L / 2xF_1)$ in two bins of x as a function of Q^2 from the new analysis of SLAC experiments [32] together with the data of the SLAC E140 [60], CDHSW [29], BCDMS [34] and EMC [36]. The dotted and the small dashed lines [32] represent R_{QCD} and R_{QCD} plus target-mass corrections, respectively. The solid line is the fit of the SLAC points used to extract F_2 in the new analysis of SLAC experiments.

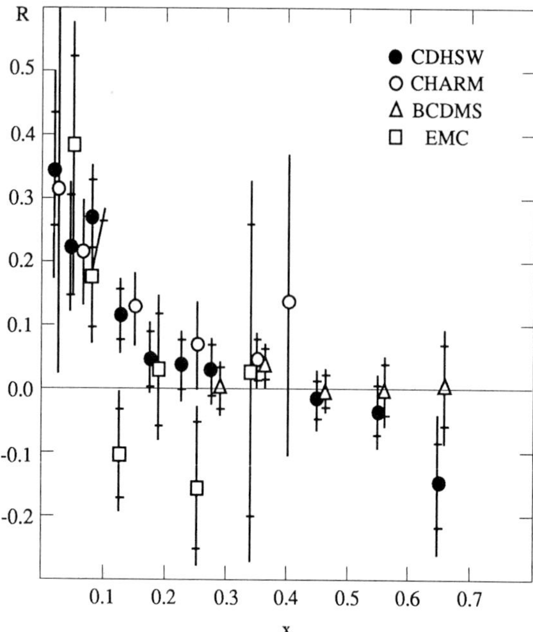

Fig.17 - $R(x,Q^2)$ as a function of x from the BCDMS [34], CDHSW [29], CHARM [74] and EMC [36] experiments. Inner error bars are statistical only. Outer error bars are statistical and systematic added in quadrature.

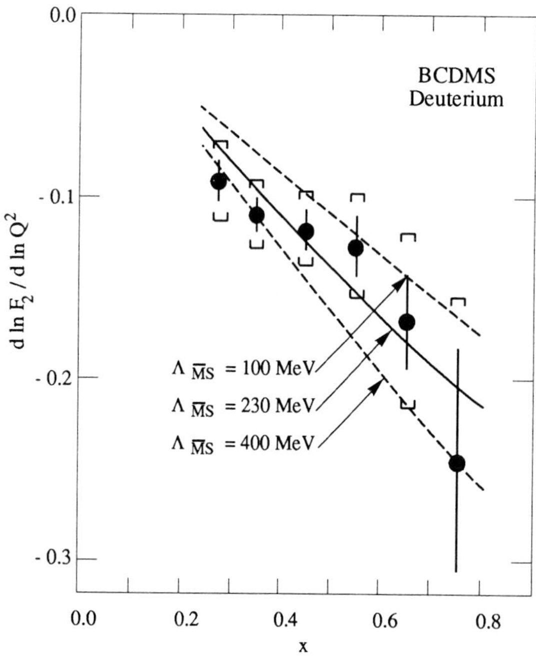

Fig.18

The logarithmic derivatives $dLog\ F_2(x,Q^2)/dLog\ Q^2$ measured in the BCDMS experiment on deuterium [28] at $Q^2 > 20$ GeV^2 and $x > 0.275$. The inner error bars are statistical, the outer error bars show statistical and systematic errors added linearly. The lines show non singlet predictions for 3 values of Λ.

5.3 Scaling violations

The QCD analysis of the new data on the deuteron structure function has been submitted at this symposium by the BCDMS collaboration [28]. A non singlet fit of the F_2^D structure function has been performed in the kinematic range $x \geq 0.275$ and $Q^2 \geq 20$ GeV^2, where the gluon contribution is known to be negligible. The logarithmic derivatives dLog(F_2) / dLog(Q^2), shown on Figure 18, are in agreement with a non singlet QCD fit which yields :

$$\Lambda_{\overline{MS}} = 230 \pm 40 \pm 70 \text{ MeV} . \tag{33}$$

This result is in good agreement with earlier measurements by this collaboration on carbon and hydrogen targets [34,75] :

$$\Lambda_{\overline{MS}} = 220 \pm 15 \pm 50 \text{ MeV} , \tag{34}$$

corresponding to a strong coupling constant of

$$\alpha_s(100\ GeV^2) = 0.1585 \pm 0.0025 \pm 0.0090 . \tag{35}$$

In the full x range the logarithmic derivatives of F_2^D are also in agreement with QCD predictions for $\Lambda_{\overline{MS}} = 250$ MeV (Figure 19). The agreement of the Q^2 evolution of the structure functions on free nucleons with QCD predictions was already observed in previous experiments although with larger errors [26,42]. In addition, as shown in Figure 20, the BCDMS collaboration has compared for the first time [25] the non singlet difference of structure functions $(F_2{}^p - F_2{}^n)$ with QCD predictions, giving

$$\Lambda_{\overline{MS}} = 235^{+140}_{-120} \text{ MeV (stat.)} . \tag{36}$$

On heavy targets the situation is more confused. It was noticed two years ago [1] that, by considering only statistical errors, the χ^2 of the non singlet QCD fits of the structure function F_2 on iron were poor in the two experiments with the smallest statistical errors : EMC and CDHSW. It was shown last year [2] that taking into account the systematic errors in the EMC iron data the discrepancy with the QCD predictions almost disappears. This year, the final results of CDHSW [29] and the preliminary results of CCFR [24] have been presented. The x variation of the the logarithmic derivatives of xF_3 are shown on Figure 21 and Figure 22 together with the QCD predictions. In both figures the Q^2 cut is small (5 or 6 GeV^2). In the CCFR data the agreement with QCD predictions is satisfying giving

$$\Lambda_{\overline{MS}} = 235^{+136}_{-105} \text{ MeV (stat.)} . \tag{37}$$

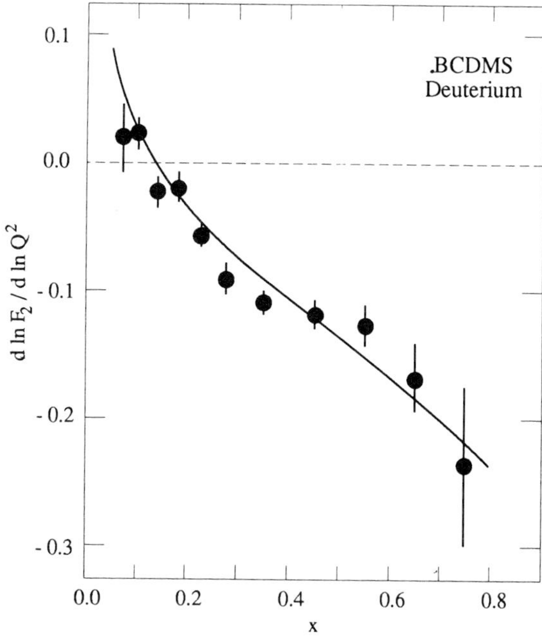

Fig.19 - The same as Figure 18 but for the full x range. The line shows a singlet QCD prediction for $\Lambda_{\overline{MS}}$ = 250 MeV. Only statistical errors are shown.

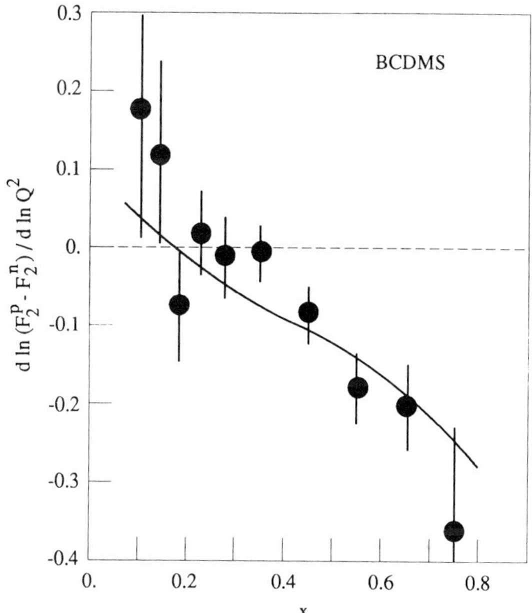

Fig.20

The logarithmic derivatives $dLog\,(F_2{}^p - F_2{}^n)/dLog\,Q^2$ as a function of x measured by the BCDMS collaboration [25]. Only statistical errors are shown. The solid line is a next-to-leading order prediction for $\Lambda_{\overline{MS}}$ = 235 MeV.

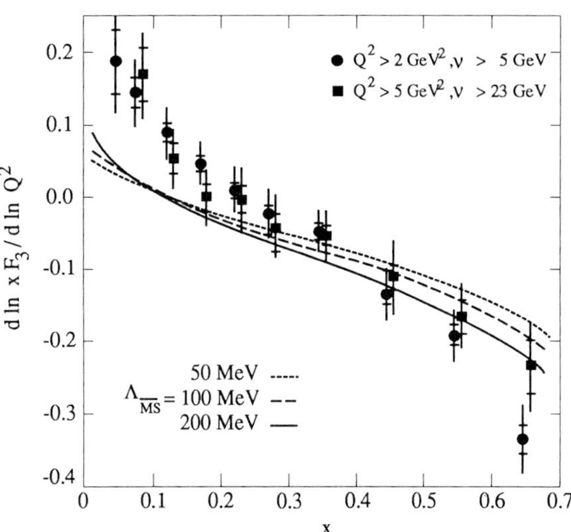

Fig.21

Logarithmic derivatives $dLog\,xF_3(x,Q^2)/dLog\,Q^2$ as a function of x. The data come from the CDHSW experiment and are calculated for $Q^2 > 2GeV^2, \nu > 5$ GeV (circles), and for $Q^2 > 5GeV^2, \nu > 23GeV$ (squares) [29]. The lines are next-to-leading order QCD predictions for 3 values of Λ.

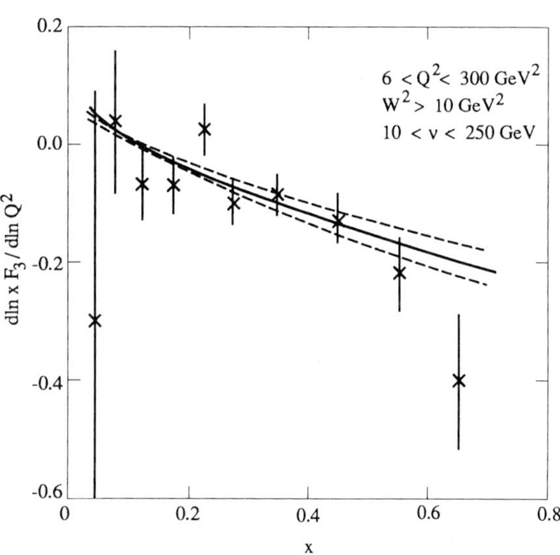

Fig.22

The logarithmic derivatives $dLog\,xF_3(x,Q^2)/dLog\,Q^2$ as a function of x measured by the CCFR collaboration [24]. The curves correspond to $\Lambda_{\overline{MS}}$ = 235 MeV (solid) and $\pm 1\sigma$ (dashed) fits.

In the CDHSW data, where the statistical errors are much smaller, the agreement is quite poor even after adding the correlated systematic errors. The fitted value of the QCD scale parameter is

$$\Lambda_{\overline{MS}} = 100^{+90}_{-60} \text{ MeV (stat.)} . \qquad (38)$$

We can remark that the three recent determinations of $\Lambda_{\overline{MS}}$ in pure Non Singlet combinations of structure functions in neutrino and muon experiments (equ. (36), (37), (38)) correspond to similar errors on the measured value of $\Lambda_{\overline{MS}}$ and are consistent. They are also in agreement with the more precise determination obtained with the F_2 structure function by the BCDMS collaboration (equ.(34))

The CCFR collaboration has also presented a singlet QCD analysis of the F_2 structure function which has also a poorer agreement with QCD predictions. In both cases, xF_3 in CDHSW and F_2 in CCFR, the quality of the fits improves by raising the cuts on Q^2 or W^2. However, no detailed study of the efficiency of the cuts has been published.

In summary, the situation is not yet precisely clarified but there is no clear physics case for a violation of perturbative QCD in the evolution of structure functions on heavy nuclei.

5.4 *Gluon distribution*

We have seen that the gluon distribution contributes directly to the longitudinal structure function, but the data are not precise enough, in the kinematic domain of fixed-target experiments, to do a quantitative measurement. Most of the information on the gluon distribution comes from the evolution equations of the singlet structure function. The usual method consists in assuming a functional form in x for the gluon distribution with one or two free parameters and then to solve simultaneously the two coupled singlet evolution equations (29) and (30). The extraction of the gluon distribution is especially delicate because it depends on many factors :

- The assumed value of $F_L(x,Q^2)$. The contribution of the longitudinal structure function must be subtracted before solving the evolution equations [29].
- The functional form which is fitted [29].
- The Q_0^2 value at which the gluon distribution is parametrized [76].
- The input value of Λ. There is a strong correlation at low x between Λ and the gluon distribution. It is only when the statistics is large enough at large x to constrain Λ that the gluon distribution can be unambiguously determined.
- The order in α_s of the differential equation. It is more correct to work at next-to-leading order [75].

So, a meaningful comparison between various glue distributions should be done after analyses with the same assumptions on the longitudinal structure function, performed at the next-to-leading order, with the same Q_0^2, the same type of parametrisation and the same input value of Λ. All the analyses so far differ in these many aspects. Only a crude comparison can be done as in Figure 23 where various parametrisations are drawn together with the points of BCDMS which, thanks to the high statistics of this experiment, were obtained without forcing a functional form in x [75]. Looking at this figure, one can easily conclude that the gluon distribution is poorly known for x > 0.1 and almost completely unknown at smaller x. As a last remark let us mention that some complementary information can be obtained at larger x from other processes as :

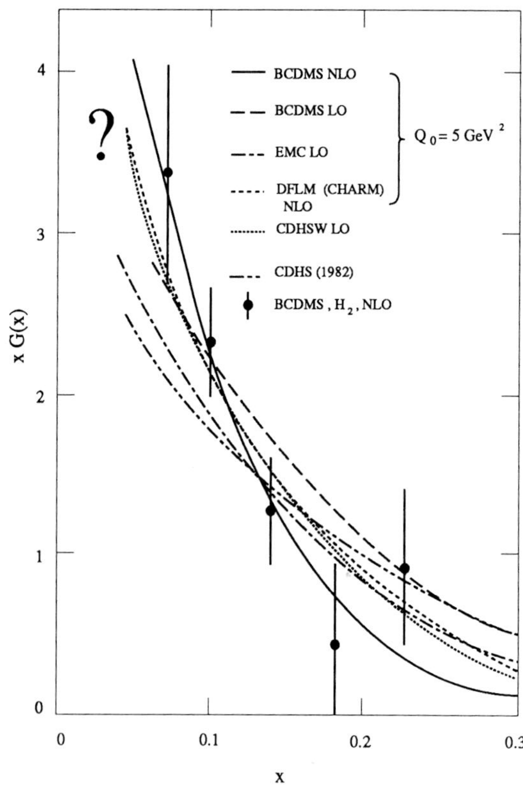

Fig.23 - Compilation of the Gluon distribution $xG(x,Q_0^2)$ as a function of x at fixed Q_0^2. The data points come from the BCDMS experiment on hydrogen [75]. The lines show functional forms fitted at Leading Order (LO) and Next-To-Leading Order (NLO) in the muon experiments on hydrogen BCDMS [75] and EMC [26] together with the neutrino experiments on marble CHARM [52] and on iron CDHS (NBB) [38] and CDHSW (WBB) [29].

$$pp \rightarrow \gamma + x \tag{39}$$

which have their own ambiguities and are discussed in an other review talk [77].

5.5 Future prospects on QCD tests

In the near future some new developments on the comparison between data and perturbative QCD are expected. On fixed-target experiments, the completely new set of data of CCFR in the neutrino wide band beam at FNAL will be analyzed and should contribute to the test of QCD on heavy targets and the new data from NMC should contribute to the glue determination at low x, low Q^2. More spectacular results are likely to come from the commissioning of HERA. At very low x, in a region free of higher twist effects, the limit of perturbative QCD and its interplay with the description in terms of Regge pole will be studied for the first time for the scaling violations and the x behaviour. The important prediction of QCD on the rise of F_L at low x, which is only indicative in the present data, should be firmly checked provided HERA is run at two different centre-of-mass energies (Figure 24). From the scaling violations at $x > 0.25$, the anticipated statistical precision of a nonsinglet fit on Λ (175 $\pm$ 175 MeV) is very modest, but it corresponds to an error of $\pm$ 18% on α_s at an average Q^2 of 3000 GeV^2 [79] which has to be compared with the expected decrease of 40% on α_s from fixed-target to HERA energies (Figure 25). If non perturbative physics has no impact down to $x = 10^{-4}$, the complete x range can be used for $Q^2 > 10$ GeV^2 and, in the restricted domain with small systematics, the anticipated precision on Λ is $\pm$ 25 MeV, a value which is competitive with the best experiments on fixed targets (see subsection 5.3). These estimates have been made by using only the differential evolution equations without any assumptions on the gluon distribution. We have seen that F_L can provide an almost direct measurement of the glue. By unfolding the gluon distribution from equation (32), quantitative estimates have been done [80] for two types of glue compatible with the present data but giving very different predictions in the HERA domain [81]. As an example, the resulting distribution presented in Figure 26 at $Q^2 = 50\ GeV^2$ shows the power of resolution on the gluon distribution at HERA.

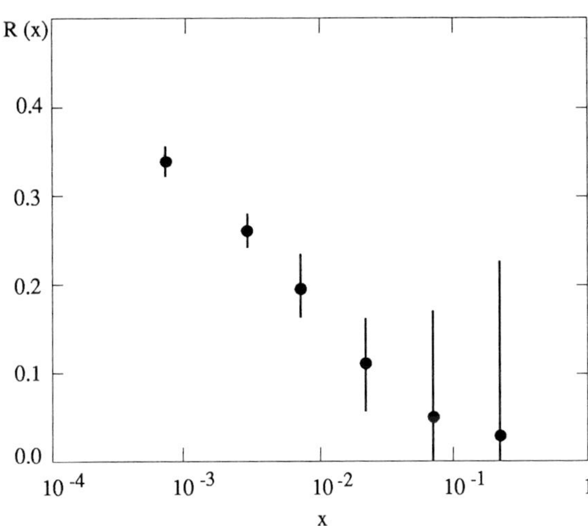

Fig.24 - Monte-Carlo estimations on a measurement of R(x) at HERA assuming proton beam energies of 300 and 820 GeV and the nominal electron beam energy of 30 GeV [78]. The assumed integrated luminosity is L = 100pb^{-1} for each beam setting.

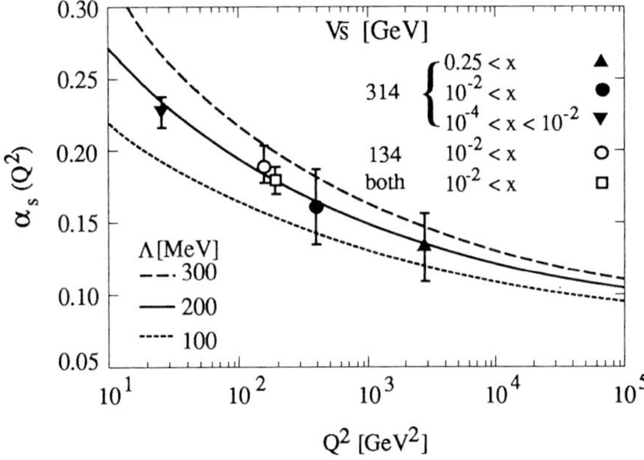

Fig.25 - Monte-Carlo estimates on the running coupling constant $\alpha_s(Q^2)$ from scaling violations in F_2 measured at HERA. The curves are QCD leading order predictions. For the two energies, $\sqrt{s}$ = 314 and 134 GeV, the integrated luminosities are 200 and 100 pb^{-1}, respectively. The errors shown are statistical only.

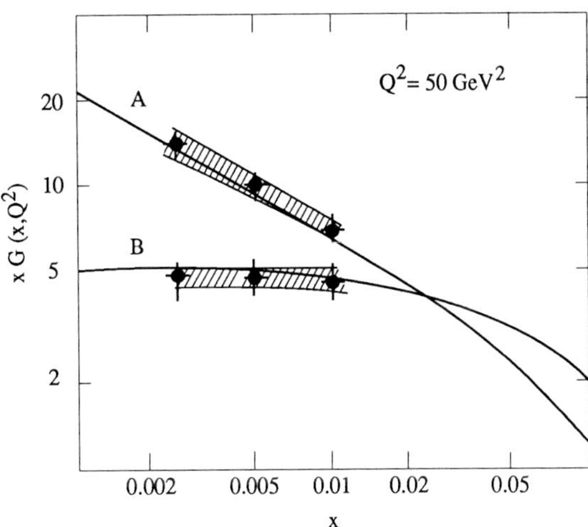

Fig.26 - The gluon distribution (A) $xG(x,Q^2) = 0.676\ x^{-1/2}(1-x)^5$ and (B) $xG(x,Q^2) = 5(1-x)^5$ superimposed on the measurements expected at HERA for Q^2 = 50 GeV^2 [80]. The error bars shown indicate the size of the statistical error expected for 100 pb^{-1} luminosity at each centre-of-mass energy and the shaded band indicates the size of the total systematic error due to finite resolution and overall normalisation.

6. QUARK DISTRIBUTIONS

6.1 Valence quarks

The valence quark distributions can only be isolated by using a combination of cross sections of ν and $\overline{\nu}$ on hydrogen or deuterium targets. The measured distributions have not changed since several years. The d_v is softer than the u_v quarks and the ratio d_v / u_v is well parametrised by the simple law [52] :

$$d_v / u_v = 0.57\,(1-x)\,. \tag{40}$$

In charged lepton scattering experiments, as far as the only exchanged virtual boson is the photon, there is no direct discrimination of quark flavors. By combining cross sections on hydrogen and deuterium, only some combinations of valence quark distributions can be isolated :

$$F_2^{\,p} - F_2^{\,n} = x\,(u_v - d_v) + 2x\,(\overline{u} - \overline{d}) \tag{41}$$

$$\frac{F_2^n}{F_2^p} = \frac{1 + 4\,\dfrac{D}{U}}{4 + \dfrac{D}{U}} \tag{42}$$

$$\text{with } D = d + \overline{d} + \frac{2}{5}\,\overline{s} + \frac{8}{5}\,\overline{c}$$

$$\text{and } U = u + \overline{u} + \frac{2}{5}\,\overline{s} + \frac{8}{5}\,\overline{c}$$

Fitting functional forms for u_v and d_v, constrained to two u and one d valence quark in the proton, and a functional form for antiquarks distribution, the EMC collaboration has also extracted the valence quark distributions [82]. The resulting u_v distribution is in agreement with neutrino experiments but the d_v distribution is a factor two higher for a reason which is not yet understood [23,5]. In a paper submitted at this conference [25], the BCDMS collaboration has presented new results on the ratio $F_2^{\,n}/F_2^{\,p}$ which are shown in Figure 27 together with the prelimimary results of NMC [53] and in Figure 28 together with the previous results of EMC [82] and the older results from the SLAC-MIT collaboration [33]. The agreement between NMC and BCDMS is very good. Both experiments extrapolate well to $F_2^{\,n} / F_2^{\,p} = 1$ at $x = 0$ and to $F_2^{\,n} / F_2^{\,p} = 0.25$ at $x = 1$, corresponding to $d_v / u_v = 0$ at $x = 1$. Within systematics, the BCDMS points are also consistent with those of EMC measured at the same Q^2 but only marginally with those of SLAC-MIT measured at a lower Q^2 (see table 2). It will be interesting to see how this pattern evolves with the new analysis of SLAC experiments at a larger Q^2.

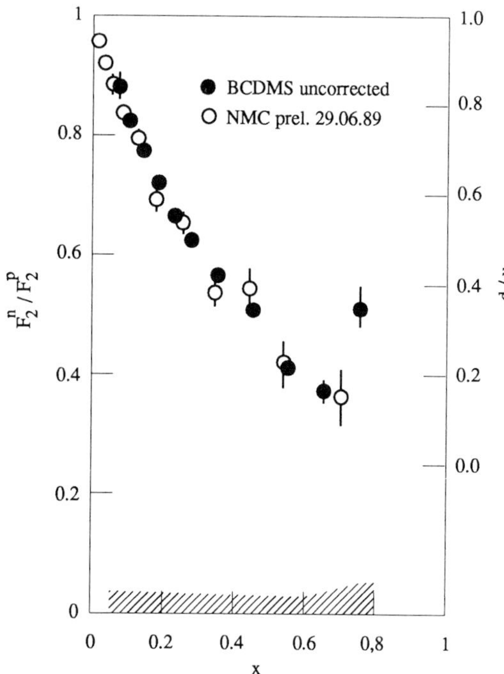

Fig.27 - $F_2^{\,n} / F_2^{\,p}$ as a function of x without correction for Fermi motion. The data come from the BCDMS [25] and NMC [53] experiments. The systematic errors of BCDMS are indicated by the hatched area.

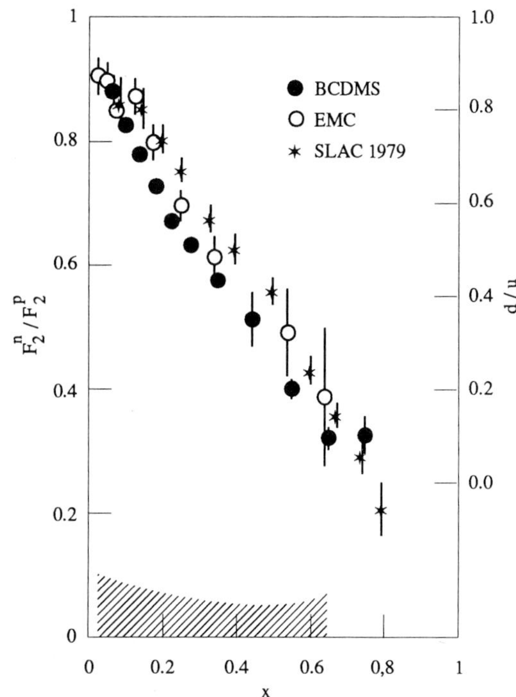

Fig.28 - $F_2^{\,n} / F_2^{\,p}$ as a function of x after correction for Fermi motion. The data come from BCDMS [25], EMC [26] and SLAC-MIT experiments [33]. The hatched area shows the systematic errors of the EMC data.

6.2 Sea quarks

The sea quark distributions have only been studied in neutrino scattering experiments. The flavor composition is still poorly known. The WA25 deuterium collaboration has found no differences between $\bar{u}$ and $\bar{d}$ [83]. The strange quark, studied in multimuon final states in neutrino scattering experiments on iron target, is a factor two smaller than $\bar{u}$ or $\bar{d}$ [84]. There are no recent results on the antiquarks in free nucleons. We have seen in table 2 that the neutrino experiments on light targets have meager statistics and antiquark distributions can only be obtained after assumptions on the size of the longitudinal cross section. However the trends are well established : rise at low x , vanishing at x around 0.4 and no Q^2 variations observed in the range $0.03 < x < 0.2$ and $2 < Q^2 < 20$ GeV^2. Much more precise data have been extracted from neutrino experiments on iron. In the data of CDHSW [29] the statistics were sufficient to extract $\bar{q}$ from the y distribution without further assumptions on F_L :

$$\frac{d^2\sigma^{\bar{\nu}N}}{dxdy} - (1-y)^2\frac{d^2\sigma^{\nu N}}{dxdy} = \sigma_0\{[1-(1-y)^4]\,\bar{q}^{\bar{\nu}} + [(1-y) - (1-y)^3]F_L\} \qquad (43)$$

where $\bar{q}^{\bar{\nu}}(x,Q^2) = x(\bar{u} + \bar{d} + 2\bar{s})$.

The Q^2 evolutions of $\bar{q}$ are shown in Figure 29 for various bins of x. The data of CDHSW on iron indicate clearly a rise of the sea with Q^2 and are barely compatible with those of WA25 on deuterium which are consistent with no Q^2 variation.

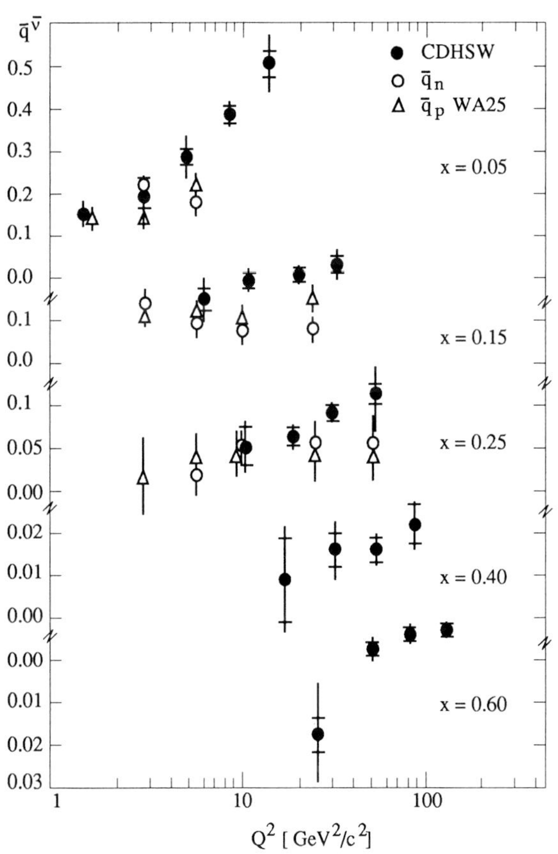

Fig.29 - The antiquark distribution in bins of x as a function of Q^2. The data of CDHSW on iron are determined from the simultaneous fit of F_L and $\bar{q}$ [29]. The data of WA25 on deuterium are determined assuming R = 0 [83].

6.3 Future prospects on quark distributions

The NMC collaboration is still taking data at CERN in a muon scattering experiment on hydrogen and deuterium targets and should improve the knowledge on $(u_v - d_v)$ at low x, but not on the sea quarks. In contrast, at HERA the sensitivity to the quark flavors is naturally obtained in charged current interactions as in neutrino scattering experiments, and in neutral current interactions at Q^2 large enough for a significant contribution of the exchange of the virtual Z^0. The flavor isolation will be done either by a complete combination of the four cross sections, neutral and charged currents with beams of positrons and electrons, or by using a single cross section in a kinematic domain where the contribution of the other flavors is negligible. Monte-Carlo estimates taking into account systematic uncertainties have been performed for a great variety of combinations [85]. As an illustration we present on Figure 30 and Figure 31 the predicted x variation, after QCD evolution, of the up valence quark u_v and of the sea with charge 2/3. These distributions are obtained at an average Q^2 of 3000 GeV^2 and should clearly depart from the present distributions at Q^2 which are two to three orders of magnitude lower.

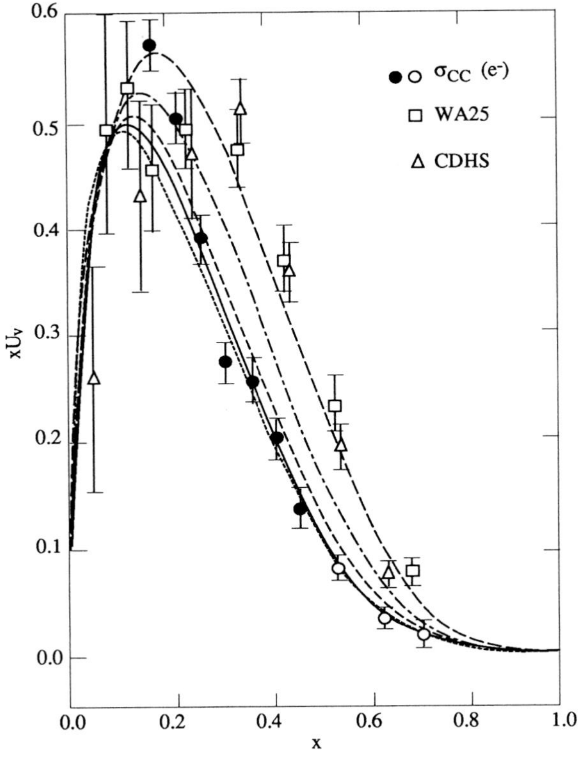

Fig.30 - The valence up-quark distribution $xu_v(x)$ determined from the Charged Current $e\ p$ cross-section averaged over $0.03 < y < 0.3$ as simulated at HERA nominal energies for an integrated liminosity of 200 pb^{-1}. The curves represent $xu_v(x,Q^2)$ at $Q^2 = 10$ (long-dashed), 10^2 (dash-dotted), 10^3 (dashed), 10^4 (dotted) GeV^2 and $<Q^2>$ (full). $<Q^2>$ being the average Q^2 associated with the MC data. Also plotted are the data of WA25 [83] and CDHS [44].

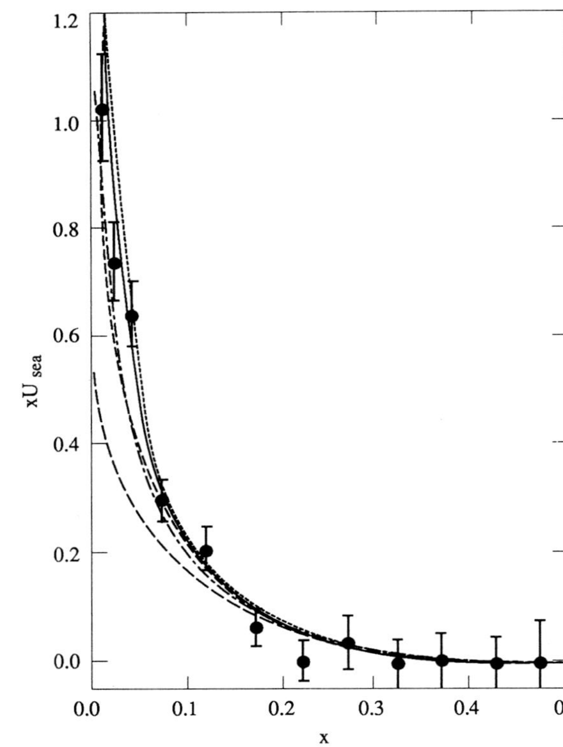

Fig.31 - The up-type sea quark distribution at HERA. Further explanations on the Monte-Carlo simulations and the curves are given in Figure 30.

7. CONCLUSION

In conclusion, we have shown that from all the high statistics data accumulated using electron, muon and neutrino beams, it is possible to obtain a coherent set of data in the kinematic domain $0.05 < x < 0.6$ and $1 < Q^2 < 200$ GeV^2, after application of global correction factors consistent with known uncertainties. In each bin of x, the distance between average lines fitting each set of data is then at most of ± 5 %. This gives hope that a comprehensive compilation which takes into account the correlations in the known systematic effects should yield an average value with an even better precision. The data on deep inelastic scattering have proven to be a very powerful tool to test the theory of strong interactions, QCD, yielding the most precise value on the running coupling constant. The parton distributions extracted from structure function measurements will play an essential role in the future hadron colliders. The new high Q^2 domain, shown in Figure 32, can be predicted through the QCD evolution equations resting on the present data for $x > 0.05$, but the lower x region, where spectacular effects are anticipated at large Q^2, must wait for the measurements at HERA, the first electron-proton collider.

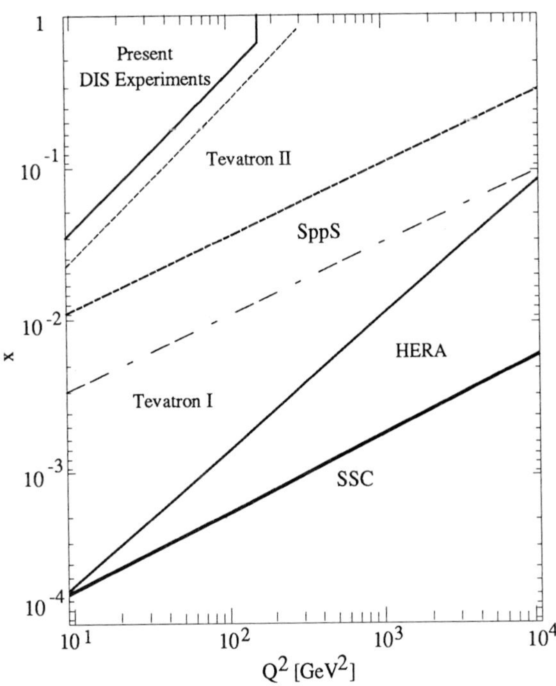

Fig.32 - Road Map of High Energy Physics in the 1990's [3].

8. ACKNOWLEDGMENTS

It is a pleasure to thank B.Vallage and M.Virchaux for discussions and active help in preparing this review, and A.Milsztajn for a critical reading of the manuscript. I have also benefited from important remarks from Michael Riordan. Finally, I would like to thank F. Le Diberder for his kind assistance during the conference and for a very careful reading of the manuscript.

BIBLIOGRAPHY

1. R. Voss, Nucl. Phys. B (Proc.Suppl.) 3 (1988) 581.
2. R.P. Mount, Proceedings of the XXIV International Conference on High Energy Physics (1988) 1007.
3. Wu-Ki Tung et al., FERMILAB-Conf-89/26.
4. J.G. Morfin, FERMILAB-Conf-89/26.
5. T. Sloan, G. Smadja and R. Voss, Phys. Rep. 162 (1988) 45.
6. E. Longo, Proceedings of the workshop experimentation at HERA (1983), DESY HERA-83/20, p.285.
7. G. Altarelli and G. Parisi, Nucl. Phys. B126 (1977) 298.
8. F. Martin, Phys .Rev. D19 (1979) 1382.
9. R.P. Feynman and R.D. Field, Phys. Rev. D15 (1977) 2590.
10. C. Lopez and F.J. Yndurain, Nucl. Phys. B171 (1980) 231.
11. H.D. Abarbanel, M.L. Goldberger and S.B. Treiman, Phys. Rev. Lett.22 (1969) 500.
12. A. de Rujula et al. Phys Rev. D10 (1974) 1649.
13. Wu-ki Tung, Nucl. Phys. B315 (1989) 378, and references therein (see also references cited in ref.[14]).
14. M. Glück, R.M. Godbole and E. Reya, DO-TH 88/18.
15. L.V. Gribov, E.M. Levin and M.G. Ryskin, Phys. Rep. 100 (1983) 1.
16. A.H. Mueller and J. Quiu, Nucl. Phys. B268 (1985) 427; J. Kwiecinski, Z. Phys. C29 (1985) 147.
17. H. Georgi and H.D. Politzer, Phys. Rev. D9 (1974) 416. D.J. Gross and F. Wilczek, Phys. Rev. D9(1974) 980.

18. C. Lopez and F.J. Yndurain, Nucl. Phys. B183 (1981) 157.
M. Gluck and E. Reya, Phys. Rev. D25 (1982) 1211.

19. D.J. Gross and C.H. Llewellyn Smith, Nucl. Phys. B14 (1969) 337.

20. S.L. Adler, Phys. Rev. 143 (1966) 143.

21. K. Gottfried, Phys. Rev. Lett. 18 (1967) 1154.

22. G. Altarelli, Phys. Rep. 81 (1982) 1.

23. J. Feltesse, Proc. Int. Europhysics Conf. on High-Energy Physics, Bari, ed. by L. Nizzi and G. Preparata (1985) 977.

24. CCFR coll., E. Oltman, Nevis preprint # 1417, July 1989.

25. BCDMS coll., A.C. Benvenuti et al., paper No 110 submitted to this conference.

26. EMC coll., J.J. Aubert et al., Nucl. Phys. B259 (1985) 189.

27. F. Eisele, Rep. Prog. Phys. 49 (1986) 233.

28. BCDMS coll., A.C. Benvenuti et al., paper No 109 submitted to this conference.

29. CDHSW coll., P. Berge et al., CERN-EP/89-103, submitted to Z. Phys.

30. BCDMS coll., A.C. Benvenuti et al., Phys. Lett. B223 (1989) 485, and CERN-EP/89-06.

31. EMC coll., M.Arneodo et al., paper No 183 submitted to this conference and to Nucl. Phys. B.

32. L.W. Whitlow et al., paper No 270 submitted to this conference.

33. SLAC-MIT coll., A. Bodek et al., Phys. Rev. D20 (1979) 1471.

34. BCDMS coll., A.C. Benvenuti et al., Phys. Lett. B195 (1987) 91.

35. BFP coll., P. Meyers et al., Phys. Rev. D34, (1986) 1265.

36. EMC coll., J.J. Aubert et al., Nucl. Phys. B272 (1986) 158.

37. CCFRR coll., D.B. Mac Farlane et al., Z. Phys. C26 (1984) 1.

38. CDHS coll., H. Abramowicz et al., Z. Phys. C17 (1983) 283.

39. CHARM coll., F. Bergsma et al., Phys. Lett. B123 (1983) 269.
J.V. Allaby et al., Phys. Lett. B197 (1987) 281.

40. EMC coll., J.J. Aubert et al., Nucl. Phys. B293 (1987) 740.

41. WA21 coll., G.T. Jones et al., paper No 121 submitted to this conference and reference therein.

42. WA25 coll., D. Allasia et al., Z. Phys. C28 (1985) 321

43. T. Kitagaki et al., Phys. Rev. D28 (1983) 436.

44. CDHS coll., H. Abramowicz et al. Z. Phys. C25 (1984) 29

45. F.Eisele, Proc. 21st Int. Conf. on High Energy Physics, Paris, J. Physique 43 suppl. C12 (1982) 3.

46. CDHSW coll., P. Berge et al., Z. Phys. C35 (1987) 443.

47. M. Virchaux, Thèse, Université Paris VII, 1988.

48. F.L. Navarria, C. Zupancic and J. Feltesse, NIM 212 (1983) 125.

49. NMC coll., CERN/SPSC 85-18, SPSC P210 (2/85).

50. B. Vallage, Thèse Université Paris Sud (1987), Note CEA-N-2513.

51. EMC coll., J.J. Aubert et al., Phys. Lett. B123 (1983) 275, See also [40].

52. Diemoz et al., Z. Phys. C39 (1988) 21.

53. NMC coll., M.A.J. Botje, Proc. 19th Int. Symp. on Multiparticle Production, Arles, 1988.

54. E665 coll., see the remark of H. Montgomery appended to this review.

55. C.Guyot et al., unnumbered DPHPE-Saclay report, march 14 th, 1989.

56. J.Feltesse, Proc. of the HERA Workshop, Hamburg, 1987, ed. R.D. Peccei, (DESY Hamburg, 1988) vol 1 p.33.

57. J. Kripfganz, H.J. Mohring, Z. Phys. C38 (1988) 653.
H. Spiesberger, Proc. of the HERA Workshop, Hamburg, 1987, ed. R.D. Peccei, (DESY Hamburg, 1988) vol 2 p.605.
J.Blümlein, Zeuthen preprint PHE 89-08.
D.Yu. Bardin et al., Z. Phys. C42 (1989) 679 and Dubna preprint E2-89-145.

58. D.O. Caldwell et al., Phys. Rev. Lett. 42 (1979) 553 and references therein.

59. A.M. Sarkar-Cooper, Proc. of the 11-th Conference on Neutrino Physics and Astrophysics, Dortmund (1984) 394.

60. SLAC E140 coll., S. Dasu et al., Phys. Rev. Lett. 61 (1988) 1061.

61. SLAC E139 coll., R.G. Arnold et al., Phys. Rev. Lett. 52 (1984) 727.

62. R. Windmolders, Proc. of the XXIV International Conference on High Energy Physics (1988) 267.

63. M.S. Goodman et al., Phys. Rev. Lett. 47 (1981) 293.

64. SLAC E61 coll., S. Stein et al., Phys. Rev. D12 (1975) 1884.

65. SLAC-MIT E87 coll., A. Bodek et al., Phys. Rev. Lett. 50 (1983) 1431.

66. BCDMS coll., A.C. Benvenuti et al., Phys. Lett. B189 (1987) 483.

67. EMC coll. NA2′, J. Ashman et al., Phys. Lett. B202 (1988) 603.

68. G. Altarelli, CERN-TH.5290/89, to be published in Ann. Rev. Nucl. Part. Sci. 39 (1989).

69. FNAL, V.V. Ammosov et al., JETP Lett. 46 (1987) 59.

70. WA59 coll., K. Varvell et al., Z. Phys. C36 (1987) 1.

71. S.P. Cuttrell, S. Wada and B.R. Webber, Nucl. Phys. B188 (1981) 219., E.V. Shuryak and A.I. Vainshtein, Nucl. Phys. B201 (1982) 141.

72. A. Ouraou, Proc. of the Int. Workshop on High P_T Physics and Higher Twists, Nucl. Phys. B (Proc. Suppl.) 7 (1989) 146.

73. H. Georgi and H.D. Politzer, Phys. Rev. D14 (1976) 1829.

74. CHARM coll., F. Bergsma et al., Phys. Lett. B141 (1984) 129.

75. BCDMS coll., A.C. Benvenuti et al., Phys. Lett. B223 (1989) 490.

76. P. Aurenche et al., Phys. Rev. D39 (1989) 3275.

77. J. Huston, Production of lepton pairs and high P_T photons, these proceedings.

78. J. Blümlein, Proc. of the HERA Workshop, Hamburg, 1987, ed. R.D. Peccei, (DESY Hamburg, 1988) vol 1 p.67.

79. J. Blümlein et al., Zeuthen PHE 89-01 and DESY 89-101.

80. A.M. Cooper-Sarkar et al., Z. Phys. C39 (1988) 281.

81. A.D. Martin, R.G. Roberts and W.J. Stirling, Phys. Rev. D37 (1988) 1161.

82. EMC coll., J.J. Aubert et al., Nucl.Phys. B293 (1987) 740.

83. WA25 coll., D. Allasia et al., Phys. Lett. B135 (1984) 231.

84. S. Mishra and F.J. Sciulli, submitted for publication and references therein.

85. G. Ingelman and R. Rückl, DESY 89-025.

DISCUSSION

H. Montgomery, FNAL: In your talk you mentioned the NMC and HERA experiments as being the future sources of structure function data. I would like to mention that E665 at Fermilab also has data using hydrogen, deuterium and xenon targets and will take more next year, on a variety of targets. The group expects to produce structure functions and ratios that extend to rather low x.

J. Feltesse: It was my impression that E665 emphasized the analysis of final-state hadrons, but if it could give some information on F_2 for hydrogen or deuterium, this would be very nice.

H. Montgomery: It is true that final-state hadrons are important to E665, and the luminosity is much less than that of BCDMS. However, some effort has been made to normalize the data, and the luminosity should be sufficient for interesting measurements — especially with the current emphasis on low to moderate x. Results are expected next year.

E. Reya, Dortmund University: I have two remarks. The Mueller–Kwiecinski rescattering (shadowing) effects are small down to $x \simeq 10^{-4}$ provided "flat" standard gluon distributions are used. The effects for steeper gluon inputs, (e.g., $xG \simeq x^{-1/2}$) have not been calculated yet and could very well be much larger.

Secondly, you mentioned higher order corrections to the second moment of F_2. These have been analyzed by several groups; they are exceedingly small — typically less than 2% depending on the scheme chosen.

Polarised Nucleon Structure Functions

Graham G. Ross,
Dept. of Theoretical Physics,
University of Oxford,
1 Keble Road,
OXFORD.
OX1 3NP

1. Introduction

There has been a lot of interest in polarised structure functions since the publication by the European Muon collaboration[1] (EMC) of their polarised measurements extending to higher Q^2 the earlier results obtained at SLAC[2]. It is commonly agreed that these measurements shed light on the distribution of the spin of the nucleon amongst its constituents, but opinions vary considerably about their implications in detail. Indeed it has variously been argued that the data shows the proton is a skyrmion, that gluons are/are not important, even that perturbative QCD is wrong. In this talk I will try to resolve some of the theoretical controversy surrounding these analyses and to establish just what can be learnt from polarised scattering measurements.

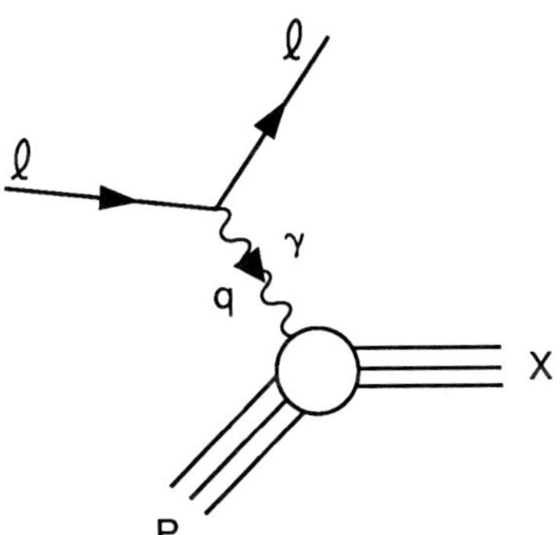

Fig. 1. Virtual photon graph contributing to deep-inelastic lepton-proton scattering.

The starting point is the general structure of deep inelastic scattering from polarised targets. It is usual to split the inclusive charged lepton-nucleon cross section arising from Fig. 1 into two parts, one involving the leptonic tensor $L_{\mu\nu}$ and one involving the hadronic tensor $W^{\mu\nu}$.

$$d\sigma = \frac{\alpha_{em}^2}{\pi} \frac{1}{q^4} L_{\mu\nu} W^{\mu\nu} \frac{d^3k'}{(P.k)E'} \quad (1)$$

The leptonic tensor is known for pointlike interaction with the virtual photon, so measurement of the cross section may be interpreted as measurement of the hadronic tensor. The most general structure for $W^{\mu\nu}$, consistent with gauge invariance, is given by(3)

$$\begin{aligned} \frac{1}{\pi} W_{\mu\nu}(q,p,S) &= \left[-g_{\mu\nu} - \frac{q_\mu q_\nu}{q^2} \right] W_1(\nu,q^2) \\ &+ \frac{1}{M^2} \left[P_\mu - \frac{\nu}{q^2} q^\mu \right] \left[\frac{\nu q_\nu}{p_{q^2}} \right] W_2(\nu,q^2) \\ &+ \frac{i}{m} \epsilon_{\mu\nu\rho\sigma} q_\rho \left[S_\sigma \left[G^1(\nu,q^2) + \frac{1}{m^2} G_2(\nu,q^2) \right] \right. \\ &\left. - (S.q)\, p_\sigma \frac{1}{m^2} G_2(\nu,q^2) \right] \quad (2) \end{aligned}$$

where M is the nucleon mass and S is the nucleon spin. The structure functions W_1 and W_2 are the usual unpolarised structure functions, while G_1 and G_2 are the new polarised structure functions and contain the new information provided by polarised measurements. In the Bjorken limit these structure functions will have the following scaling properties in QCD up to logarithmic corrections in Q^2.

$$\lim_{\nu, Q^2\to\infty} \frac{\nu}{M^2} G_1(\nu,q^2) \equiv g_1(x)$$

$$: \text{for} = \frac{Q^2}{2\nu} \quad \text{constant}$$

$$\lim_{\nu, Q^2\to\infty} \left[\frac{\nu}{M^2}\right]^2 G_2(\nu,q^2) \equiv g_2(x)$$

$$\nu = p.q \tag{3}$$

The EMC and SLAC measurements were made with longitudinally polarised leptons and longitudinally polarised photons. In terms of the measured asymmetry

$$A(x,Q^2) = \frac{d\sigma^{\uparrow\downarrow} - d\sigma^{\uparrow\uparrow}}{d\sigma^{\uparrow\downarrow} + d\sigma^{\uparrow\uparrow}} \tag{4}$$

where the arrows refer to the lepton and photon relative polarisations, the structure function $g_1(x,Q^2)$ is given by

$$g_1(x,Q^2) \underset{Q^2,\nu \text{ large}}{\simeq} \frac{F_2(x,Q^2)\ A(x,Q^2)}{2x(1+R(x,Q^2))} \tag{5}$$

where $F_2(x,Q^2)$ is the unpolarised structure function $(=\nu W_2(x,Q^2))$ and $R(x,Q^2)$ is the usual ratio of longitudinal to transverse scattering cross sections, again measured in unpolarised scattering. With longitudinally polarised protons, $g_2(x,Q^2)$ is not measurable as its contribution to the cross-section is suppressed by M^2/Q^2; in order to determine g_2 it is necessary to measure the scattering of transversely polarised nucleons when it contributes at the same order as g_1. The results of the SLAC and EMC measurements are given in Fig. 2. The recent EMC measurements extend down to x=0.01 with a mean value of $<Q^2>$ = 10.7 GeV^2. The earlier SLAC measurements have a lower $<Q^2>$ but, as may be seen in Fig 1, there is no evidence for Q^2 variation.

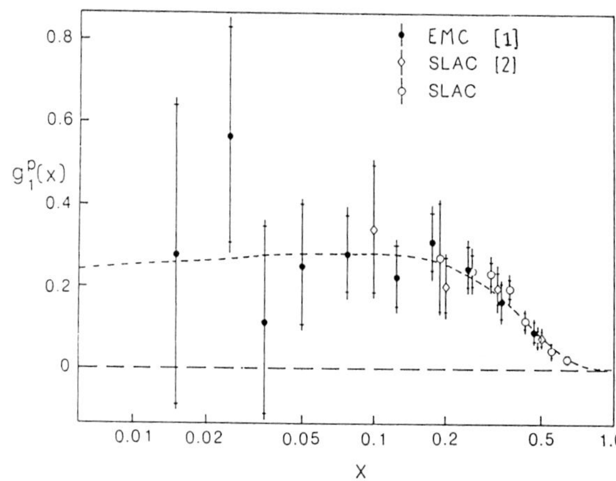

Fig. 2. Experimental results for $g_1{}^P(x)$.

2. The parton model

The interpretation of the measurements in the quark-parton model follow immediately from a calculation of the underlying quark-photon scattering

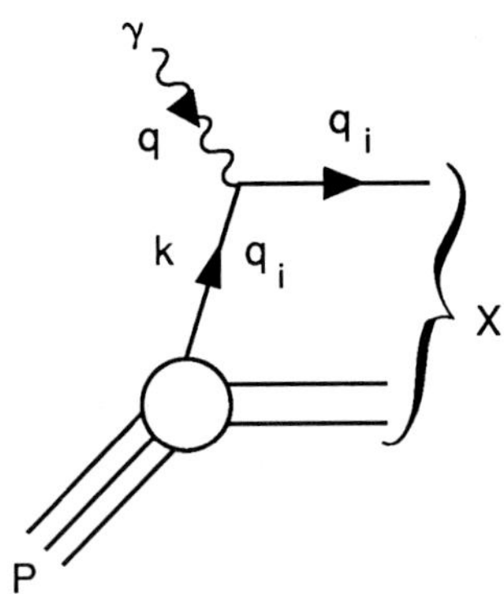

Fig. 3. Quark contribution to virtual-photon-proton deep-inelastic scattering.

of Fig. 3. Evaluating the trace and keeping the polarisation dependent term gives, for massless quarks, and helicity $\frac{1}{2}S_q$

$$W_{\mu\nu}^{\gamma\text{-quark}}(A) = Q_i^2\ \delta((k+q)^2)\ T\gamma[(1-\gamma_5 S_q)\not{k}\ \gamma_\mu(\not{k}+\not{q})\gamma_\nu]$$

$$= i\ Q_i^2\ 2\nu\ \delta(x - \frac{Q^2}{2\nu})\ S_q \epsilon_{\mu\nu\rho\sigma}\ q_\rho\ k\sigma \tag{6}$$

Comparison with eq (2) immediately gives

$$g_1(x) = \frac{1}{2} \sum_i Q_i^2 \, (q_+^i(x) - q_-^i(x)) \equiv \frac{1}{2} \sum_i Q_i^2 \Delta q^i(x) \qquad (7)$$

where $q^i_\pm(x)$ give the probability of funding a quark with helicity $\frac{1}{2}S_q = \pm\frac{1}{2}$ in a proton of helicity $+\frac{1}{2}$, and carrying a fraction x of the proton's momentum. For comparison the unpolarised structure function $F_1(x)$ is given by

$$F_1(x) = \frac{1}{2} \sum_i Q_i^2 \, (q_+^i(x) + q_-^i(x)) \qquad (8)$$

showing that $g_1(x)$ measures a different combination of quark distributions from the unpolarised case. The result of eq(7) may be interpreted immediately; the optical theorem shows g_i is proportional to $(\sigma_{\frac{1}{2}} - \sigma_{3/2})$ where σ_{JZ} is the virtual photo-absorption cross section with angular momentum projection J_Z. Since a spin $\frac{1}{2}$ parton can only absorb a photon when their the helicities are opposite the result, eq (7), immediately follows.

For nucleon polarisations S not entirely longitudinal it is necessary to go to a covariant form of the parton model to preserve the Lorentz covariant form for $W^{\mu\nu}$ of eq (2)(4). Then again for massless quarks, evaluation of Fig. 3 gives

$$\begin{aligned} W_{\mu\nu}^{(\text{Antisymmetric})}(q,p,S) &= i\sum_s \int d^4k \, f(p,k,S,s) \, s \, \epsilon_{\mu\nu\rho\sigma} \, q_\rho \, k_\sigma \, \delta((k+q)^2) \\ &= i\epsilon_{\mu\nu\rho\sigma} \, q_\rho \int d^4k \, \Delta q \, (P.k, k.S) \, K_\sigma \, \delta((k+q)^2) \end{aligned}$$

where the quark distribution Δq can only depend on the only Lorentz invariant variables P.k and k.S. Since the polarisation appears at most linearly the only consistent form for Δq is proportional to $(k.S)\tilde{f}(P.k)$, ie dependent on a single unknown function $\tilde{f}$. As a result, after performing the k integration and comparing with eq (2), it is found that $g_1(x)$ and $g_2(x)$ are related with, for arbitrary nucleon polarisation, $g_1(x)$ given by eq (7) and $g_2(x)$ given by(4)

$$g_2(x) = \int_x^1 \frac{dy}{y} \, g_1(y) - g_1(x) \qquad (10)$$

or, equivalently,

$$\int_x^1 dx \, x^{J-1} \left\{ \frac{J-1}{J} g_1(x) + g_2(x) \right\} = 0 \qquad (11)$$

It had previously been observed(5) that these relations follow if the twist-3 operators, O_2, contributing to the operator-product-expansion for polarised scattering vanish. For massless quarks and gluons there are two classes of operators contributing(6), the twist-two operators

$$\begin{aligned} O_1 : \; & -\bar{\psi}\gamma_5\gamma^\sigma D^{\mu_1} \ldots D^{\mu_n} \psi \\ & F^{\sigma\alpha} D^{\mu_1} \ldots D^{\mu_{n-1}} F_\sigma{}^{\mu_n} \end{aligned} \qquad (12)$$

and the twist-three operators

$$\begin{aligned} O_2 : \; & \bar{\psi}\gamma_5\gamma^\lambda D^\sigma D^{\mu_1} \ldots D^{\mu_{n-1}} \psi \\ & F^{\lambda\alpha} D^{\mu_1} \ldots D^{\mu_{n-1}} F_\alpha{}^\sigma \end{aligned} \qquad (13)$$

where the indices $\sigma, \mu_1 \ldots \mu_n$ are symmetrised and the indices λ, σ are antisymmetrised. Applying the on-shell equations of motion to the operators O_2 shows they vanish up to multi-parton components involving more than two quark or gluon fields(7). Thus measurements of deviations from the sum rules given by eq (11) would imply the existence of these multi-parton correlations characteristic of higher twist operators. It is a novel feature of polarised scattering that higher twist operators contribute at the same order as leading twist two operators in the Bjorken limit. A more physical interpretation of these higher twist operator contributions to g_2 comes from the parton model discussed above where, for massless partons, the validity of the sum, eqs (10) and (11), rules follows if the struck partons are on-shell, deviations from the sum-rules corresponding to off-shell partons the off-shell-ness resulting from the quark binding potential which in QCD, arises due to the gluon interactions (hence the appearance of gluon fields in the form of O_2 after using the equations of motion).

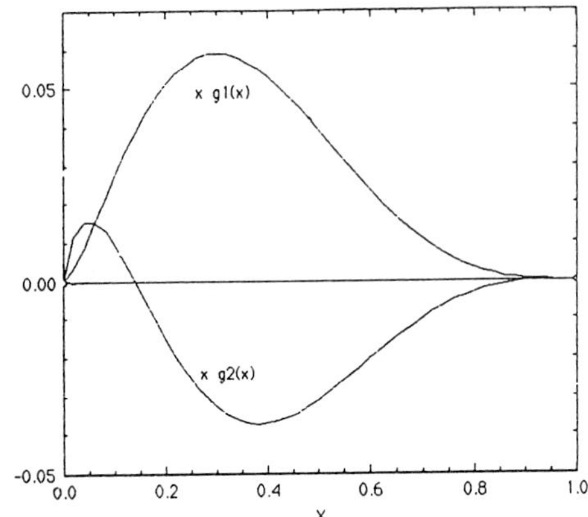

Fig. 4. Prediction for $g_2^P(x)$ given $g_1^P(x)$ which follows if struck partons are on-shell.

The sum rule, eq (10), which applies for free on-shell partons, determines $g_2(x)$ from $g_1(x')$ for $x' \gtrsim x$ and hence the implications of eq (10) follow from the experimental measurement of $g_1(x)$ down to x=0.01. This is shown in Fig. 4. Using the parton model it is straightforward to determine the effects of a non-zero quark mass for in this case (still assuming on-shell quarks) we find[4]

$$g_1(x) = \frac{\pi x}{8} \int dk_\perp^2 \left[1 - \frac{m^2+k_\perp^2}{x^2 M^2} \right] \left[1 - \frac{2m^2}{x^2M^2+m^2+k_\perp^2} \right] . \tilde{f}$$
$$g_1(x) + g_2(x) = \frac{\pi x}{8} \int dk_\perp^2 \left[\frac{k_\perp^2}{x^2M^2} \right] \tilde{f} \tag{14}$$

We note that the form of (g_1+g_2) does not change with quark masses, but requires non-vanishing $k_\perp^2$, ie transverse momentum distribution. The appearance of $k_\perp^2$ is a reflection of the appearance of twist three operators at leading order in polarised structure functions and shows that polarised scattering probes a different aspect of the parton model than unpolarised scattering. The mass terms in eq (14) generate deviations from the original sum rules, eq (10) or eq (11), and must be allowed for before probing for the "off-shell" nature of the stuck partons (as must also be done for target mass effects).

3. <u>Comparision of theory with data</u>

In this section I will consider g_1 only, as g_2 has not yet been measured. Most recent analyses of the data[9] have concentrated on a sum rule which averages over x for this has a direct physical interpretation in terms of the constituent spin structure. From eq (7) we have

$$\int_0^1 dx\, g_1^P(x,Q^2) = \tfrac{1}{2} \sum_i Q_i^2 \, \Delta q_i = \tfrac{1}{2} \left[\frac{4}{9} \Delta u + \frac{1}{9} \Delta d + \frac{1}{9} \Delta s \right] \tag{15}$$

where

$$\Delta q_i \equiv \int_0^1 dx \left[q_{i+}(x) + \bar{q}_{i+}(x) - q_{i-}(x) - \bar{q}_{i-}(x) \right] \tag{16}$$

includes both quark and antiquark contributions.

Thus in the quark parton model the sum rule probes the spin Δq_i carried by quarks of flavour i (plus the antiquark) in the proton. Taking the difference between proton and neutron structure functions gives the Bjorken sum rule[8]

$$\int_0^1 dx \left\{ g_1^P (x,Q^2) - g_1^n (x,Q^2) \right\} = {}^1/_6 \, (\Delta u - \Delta d) = \frac{1}{6} \frac{g_A}{g_V} \tag{17}$$

where an isospin rotation has been used to relate the current J_μ^3 ($\propto \Delta u - \Delta d$) to J_μ^+ for which

$$\langle P \mid \bar{u}\, \gamma_\mu \, (1-\gamma_5) \, d \mid N \rangle = \bar{u}_p \, \gamma_\mu \left[1 - \frac{g_A}{g_V} \gamma_5 \right] u_n \tag{18}$$

On a proton target alone more information is needed. From eq (15)

$$\int_0^1 dx\ g_1{}^P(x,Q^2) = \frac{1}{12}(\Delta u-\Delta d) + \frac{1}{36}(\Delta u+\Delta d-2\Delta s) + \frac{1}{9}(\Delta u+\Delta d+\Delta s) = \frac{1}{12}a_3 + \frac{1}{36}a_8 + \frac{1}{9}a_1 \tag{19}$$

where we have defined a_i via the matrix element of the nonet of axial currents

$$\langle P, S \mid A^{\mu}{}_j \mid P, S\rangle = 2M\, a_j\, S^{\mu} \tag{20}$$

The quantity a_3 is just g_A/g_V as above and a_8 is related via SU(3) to hyperon decays and is given by 3F−D where F and D are the two idependent SU(3) couplings. Some time ago Gourdin[10] and Ellis and Jaffe[10] derived a sum rule from eq (19) making the reasonable (?) assumption that the proton contained no polarised strange quarks ($\Delta S=0$). This meant that eq (19) could be rewritten as

$$\int_0^1 dx\ g_1{}^P(x,Q^2) - \frac{1}{12}\left|\frac{g_A}{g_V}\right| \left[1 + \frac{5}{3}\,\frac{3^F/D-1}{^F/D+1}\right] = +\ 0.189\pm0.005 \tag{21}$$

where a fit to β decays giving $^F/D = 0.631\pm0.018$ has been used. Fig. 5 shows the experimental determination of this moment using the measured $g_1{}^P$ using the continuation to small x (<0.01) is shown in Fig. 2. This is clearly in conflict with the expectation of eq (21). In the quark parton model the failure of this sum rule is an indication that $\Delta s\neq0$, and using the values of g_A/g_V, $^F/D$ and the experimental determination of the LHS of eq (19) gives

$$\tfrac{1}{2}\,\Delta u = 0.391\pm0.016\pm0.023$$

$$\tfrac{1}{2}\,\Delta d = -0.236\pm0.016\mp0.023$$

$$\tfrac{1}{2}\,\Delta s = -0.095\pm0.016\pm0.023 \tag{22}$$

with

$$\tfrac{1}{2}\,\Delta\Sigma = \tfrac{1}{2}(\Delta u+\Delta d+\Delta s) = 0.060\pm0.047\pm0.069$$

Note that in the free quark model with zero angular momentum the latter quantity, $\frac{1}{2}\,\Delta\Sigma$, 'the total contribution to the helicity of the proton carried by quark helicities, is expected to be $\frac{1}{2}$. This is another way of demonstrating the deviation from the naive quark model predictions found in the polarised scattering experiments.

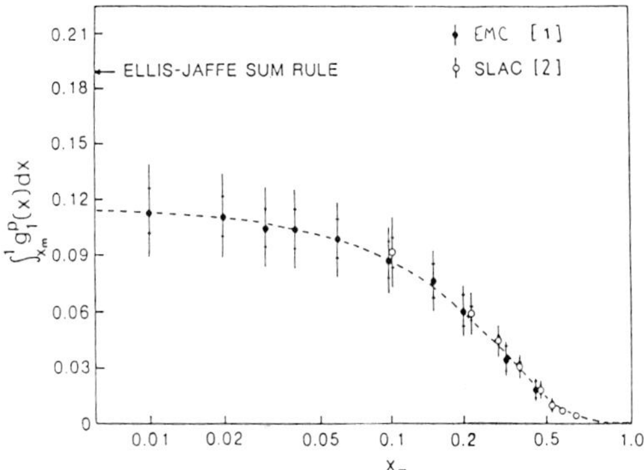

Fig. 5. Experimental determination of the first moment of $g_1{}^P(x)$.

In view of this apparently surprising result one may ask whether there is any supporting evidence for such a deviation. One other process sensitive to the axial vector currents Δq_i is elastic neutrino–proton scattering[12]. This is mediated by Z^0 exchange coupling to the quarks via $(\bar{u}\gamma_\mu\gamma_5 u - \bar{d}\gamma_\mu\gamma_5 d - \bar{s}\gamma_\mu\gamma_5 s)$ and hence proportional to $\Delta u-\Delta d-\Delta s$ in a proton target at momentum transfer $Q^2=0$. In order to make this connection it is necessary to extrapolate the experimental measurements for $Q^2\neq0$ to $Q^2=0$ and this was done[13] using the form factor $(1+Q^2/M_A{}^2)^{-1}$. For M_A given by the world average 1.032±0.036 GeV the result for Δs is −0.15±0.09 apparently lending some support to a non-zero value for Δs. However the errors are strongly correlated[14] and the choice $\Delta s=0$, $M_A=1.06$ GeV gives an equally acceptable fit to the neutrino data and is within one standard deviation of the mean value for M_A. There are other critisims one may make concerning this analysis; for example it assumed that the singlet and octet form factors were the same. We conclude it is not possible to use the present elastic neutrino scattering data to provide additional evidence for $\Delta s \neq 0$ but clearly improved data for this process could provide an important independent measurement of Δs.

Let me return to the results summarised in Fig. 5 and eq (22) to discuss the various uncertainties that may not be represented by the errors quoted. One obvious problem(15) is the sensitivity of the sum rule to small x, below the measured value x_0. Consider the small x behaviour for $g_1P(x)$ of the Regge form

$$g_1P(x) \sim \left[\frac{x}{x_0} \right]^{-\alpha} \tag{23}$$

With this form the contribution from the unmeasured region to the sum rule is given by

$$\int_0^{x_0} g_1P(x)\,dx = \frac{x_0\, g_1P(x_0)}{(1-\alpha)} \tag{24}$$

Since x_0 is very small in practice (x_0=0.01 for the EMC measurements) there is a significant correction only if α is close to one. (The EMC extrapolation assumes $\alpha \simeq 0$). Although there is no evidence in the data for $\alpha \simeq 1$(16) and theoretically $\alpha<1$ is expected(17) it has not been definitely proven that Pomeron cuts can not contribute an α=1 singularity. However the problem can be finessed by writing an "ϵ-sum-rule" in which the average is made over a reduced x-range.

$$\int_\epsilon^1 dx\, g_1P(x,Q^2) = \frac{1}{12}(\Delta u - \Delta d) + \frac{1}{36} (\Delta u + \Delta d - 2\Delta s) + 0\left[\frac{\epsilon}{1-\alpha'}\right] + \frac{1}{9}\int_\epsilon^1 dx\, g_1^{SINGLET}(x,Q^2) \tag{25}$$

In deriving eq (25) the range of the integration has been extended down to 0 for the first two terms on the right-hand-side involving the octet currents. However for these pieces there is no Pomeron singularity so the corrections to this procedure are of $O(\epsilon/(1-\alpha'))$ where $\alpha' \lesssim 0$ and hence are small. The significance of the ϵ sum rule is that it involves only measured quantities on the left-hand side. Using the EMC data shows that $\Delta S|_\epsilon$ for ϵ=0.01 is essentially the same as that given in eq (22) because the continuation assumed by the EMC collaboration below x=ϵ contributes very little to the sum rule. Thus irrespective of the x behaviour of $g_1(x)$ below ϵ, the experimental measurements imply a sizeable strange quark contribution to the proton for x values above 0.01. Any theoretical model generating such a distribution must also satisfy the ϵ sum rule, meaning that the strange quark (or gluon) contribution must not be too soft[18]. I will discuss this constraint in more detail in section 6.

There are several other potential sources of error in the analysis of the Ellis-Jaffe-Gourdin sum rule. Higher twist effects have been proposed as a way to evade the apparent discrepancy with the quark model results[19] but there is no significant Q^2 dependence in the data and it has been argued that higher twist effects are unlikely to be substantial[20]. Another possible source of error in the theoretical derivation of the sum rule is SU(3) breaking but this proves to be a small effect in the sum rule(1) and Lipkin[21] has shown that it is possible to show a discrepancy with the naive quark model predictions without using SU(3). Perhaps the most important source of uncertainty is the choice of F/D ratio. This is determined via a fit to hyperon decays in which there are two conflicting pieces of data determining $g_A/g_V \equiv F+D$. These are the neutron lifetime measurement and the decay asymmetry and as a result of the conflict it turns out the χ^2 distribution determining F/D is rather flat. Recent analyses[22,23] favour a smaller value with $F/D \simeq$ 0.58 and use of this reduces Δs from the value given in eq (22) to $\frac{1}{2}\Delta s = -0.08 \pm .04$, somewhat smaller but still non-zero.

4. <u>Quark-model predictions for $\Delta\Sigma$ and Δs</u>

In the quark model the proton is made up of two up and a down quark. Assuming zero orbital angular momentum, the wave function must be overall symmetric in its flavour and spin components (it is colour antisymmetric) and this is sufficient to determine the wave function completely giving*

$$\psi_p\ (J_Z{=}\tfrac{1}{2}) = \frac{1}{\sqrt{18}}$$
$$\{\ 2u\uparrow u\uparrow d\downarrow + 2d\downarrow\ u\uparrow u\uparrow + 2u\uparrow d\downarrow\ u\uparrow$$
$$-u\downarrow\ d\uparrow u\uparrow - u\uparrow u\downarrow\ d\uparrow - u\downarrow\ u\uparrow d\uparrow$$
$$- d\uparrow u\downarrow\ u\uparrow - u\uparrow d\uparrow u\downarrow - d\uparrow u\uparrow u\downarrow\ \} \qquad (26)$$

and hence the result.

$$\Delta u={}^4/_3,\ \Delta d=-{}^1/_3,\ \Delta s=0 \qquad (27)$$

This is the SU(6) result, obtained here on more general grounds. However we know that SU(6) makes some poor predictions which may be corrected by including strong interaction effects so it is reasonable to ask how these corrections affect the predictions of eq (27) under these corrections. In

	SU(6)	EXPERIMENT	RENORMALISED RESULT
g_A/g_V	1.67	1.25	$(.\tilde{g}^8{}_A)$ 1.23
F	0.67	0.49	$(.\tilde{g}^8{}_A)$ 0.49
D	1	0.77	$(.g^8{}_A)$ 0.74
F/D	0.67	0.63	(.1) 0.67

Table 1. SU(6) predictions before and after renormalisation of the axial vector current.

table 1 we show the original SU(6) predictions for g_A/g_V, F and D and also corrected the prediction for the octet axial vector current which allows for strong interaction effects when changing from current quarks, q, to contituent quarks, Q by a renormalisation constant $g_A{}^8$. (The vector current, being conserved, is not renormalised)

$$\bar{q}\ \gamma_\mu\ \gamma_5\ T^a\ q \rightarrow \tilde{g}_A{}^8\ \bar{Q}\gamma_\mu\gamma_5 T^a\ Q \qquad (28)$$

Clearly a choice of $\tilde{g}_A{}^8 \simeq 0.75$ dramatically improves the comparison between "theory" and experiment. In order to determine the effect of such a renormalisation on eq (27) it is necessary to consider the singlet current too

$$\sum_i q_i\gamma_\mu\gamma_5 q_i \rightarrow \tilde{g}_A{}^0\ \Sigma\ \bar{Q}_i\ \gamma_\mu\ \gamma_5\ Q_i \qquad (29)$$

Using eq (26) which applies to constituent quarks we find for the current quarks probed by the virtual photon[12]

$$\Delta s = {}^1/_3\ (\tilde{g}_A{}^8 - \tilde{g}_A{}^0) \qquad (30)$$

Thus our expectation from the "naive" quark model that Δs so will be modified if $\tilde{g}_A{}^8 \neq \tilde{g}_A{}^0$. One obvious reason why this may happen is due to the fact that the axial anomaly couples only to singlet channels so it is not surprising that many of the explanations of the EMC result rely on significant effects coming from the anomaly[25].

5. QCD - the inclusion of a gluon contribution

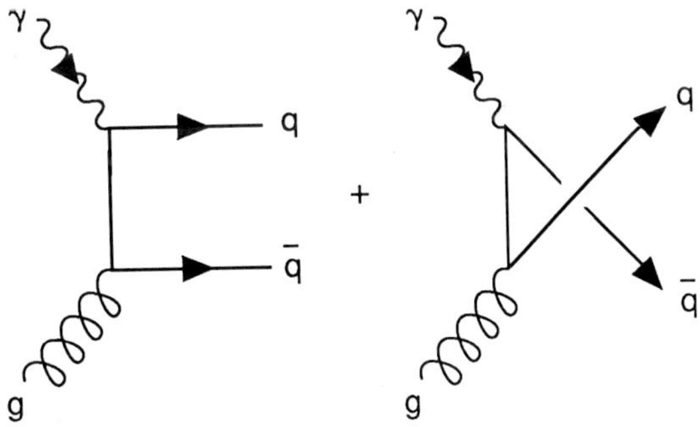

Fig. 6. Photon-gluon scattering graphs.

So far I have discussed polarised structure functions only in the framework of the quark parton model but in unpolarised deep-inelastic scattering it is known that the gluon contributes significantly (it carries ~ 50% of the proton's momentum) so it is natural to ask whether it contributes significantly to

*Corrections involving non-zero orbital momentum have been investigated[24] but lead to similar conclusions to those discussed here.

the proton spin too[26]. To answer this we consider a gluon-parton contribution to g_i of the usual convolution form[27]

$$g_1{}^{P}{}_{,g}(x, Q^2) = \int_x^1 \frac{dy}{y} \Delta g(y, Q^2) . \Delta\sigma^{\gamma g} \left[\frac{x}{y}, Q^2 \right] \quad (31)$$

where

$$\Delta g(y, Q^2) = g_+(y, Q^2) - g_-(y, Q^2) \quad (32)$$

and $g_\pm(y, Q^2)$ is the probability of finding a gluon with momentum y and helicity ± 1 in a proton of helicity $+\frac{1}{2}$. Calculation of Fig. 6 gives

$$\frac{d}{d\cos\theta}(\Delta\sigma^{\gamma g}(x)) = \frac{\alpha_s}{2\pi}\frac{e_i{}^2}{2}(x^2-(1-x)^2) \left[\frac{1}{1-\beta\cos\theta} + \frac{1}{1+\beta\cos\theta} - 1 \right] \quad (33)$$

where the factor $(x^2-(1-x)^2)$ is the polarised gluon splitting function. In eq (33) it is necessary to regulate the infra-red singularity which may conveniently be done by taking the final state quarks off mass shell giving

$$\beta \equiv \frac{|\vec{p}|}{E} \approx 1 - \frac{2p^2x}{Q^2(1-x)} \quad (34)$$

On doing the theta integration the term in square brackets in eq (33) becomes $[\ln p^2/Q^2 + \ln x/(1-x)]$. The moments in x of the convolution, eq (31), split into a product of moments of the two factors and shows that for all except the first moment there is a log p^2/Q^2 correction corresponding to the usual QCD evolution with anomalous dimension given by the moment of the splitting function. For the first moment, however

$$\int_0^1 dx\, (x^2-(1-x)^2) \left\{ \ln p^2/Q^2 + \ln\frac{x}{(1-x)} \right\} = -1 \quad (35)$$

meaning that for it there is a infra-red finite gluon contribution independent of $\ln(P^2/Q^2)$. Many different infra-red regulation prescriptions are possible but the gluon configuration has been shown to be independent of the infrared regularisation prescription[20,39]. Thus for f flavours, including the gluon correction, we have

$$\int_0^1 dx\, g_1^{\text{SINGLET}}(x, Q^2) = \frac{\langle e^2\rangle}{2}\left[\Delta\Sigma - f\frac{\alpha_s}{2\pi}\Delta g \right] \quad (36)$$

where $\langle e^2\rangle = \frac{1}{f}\sum_i e_i{}^2$ and $\Delta g = \int_0^1 dx\, \Delta g(x, Q^2)$,

At first sight the additional gluon contribution might be thought to be very small, for if the strong coupling α_s is to be evaluated at Q^2, $\alpha_s = \alpha_s(Q^2)$, the additional term will be suppressed by $\alpha_s(Q^2)/2\pi$ small at large Q^2. To determine the correct argument of α_s it is necessary to compute QCD radiative corrections at higher order. These are most conveniently summed using the renormalisation group and the net result[27] is that the gluon correction is

$$\Delta\Gamma(Q^2) \equiv \frac{\alpha_s(Q^2)}{2\pi}\Delta g(Q^2) \equiv \frac{\alpha_s(\mu^2)}{2\pi}\Delta g(\mu^2) \quad (37)$$

What has happened is that the reduction of $\alpha_s(Q^2) \sim 1/\ln Q^2$ is exactly compensated by the growth of $\Delta g(Q^2)$ via its anomalous dimension. Thus the gluon contribution persists even at aysmptotic Q^2 and there is no a priori reason to say that it is small.

The absence of a Q^2 dependence is immediately obvious using the operator product expansion for <u>only</u> the operator corresponding to the axial vector current $j^{5\alpha} = \Sigma_i \bar{q}_i \gamma^\alpha \gamma_5 q_i$ contributes to the first moment and, because it is conserved for massless quarks, its anomalous dimension vanishes in leading order. However this result immediately raises the question of the role of the gluon for there is no gauge invariant operator with a single Lorentz index that can be formed using the gluon field strength $F^{\mu\nu,a}$ alone. The candidate operator

$$k_\alpha = \frac{\alpha_s}{2\pi} \epsilon_{\alpha\nu\lambda\sigma} \mathrm{Tr} \left\{ A^\nu \left[F^{\lambda\sigma} - \frac{2}{3} A^\lambda A^\sigma \right] \right\} \tag{38}$$

is gauge variant due to the appearance of the vector field $A^{\nu,a}$. The connection between the operator product approach and the quark and gluon parton model approach follows because the operator $j^{5,\alpha}$ is anomalous and due to the anomaly has gluon as well as quark matrix elements. Thus writing

$$\langle p, s \mid j^{\alpha,5} \mid p,s \rangle = \Delta q^{\text{CURRENT}} s^\alpha \tag{39}$$

we have

$$\Delta g \sum_i \Delta q_i^{\text{CURRENT}} = \sum_i \Delta q_i^{\text{PARTON}} - \frac{f\alpha_s}{2\pi} \text{PARTON} \tag{40}$$

where the gluon piece on the right hand side corresponds to the proton matrix element of the current quark operator proceeding via a gluon intermediate state.

This split has been critised by Manohar and Jaffe[23] who argue that it is sensitive to infra-red effects, that there is no predictive power in the decomposition, and that it is a gauge variant decomposition. To answer these criticisms note that the explicit calculation of the parton graphs[(26)] shows that the transverse momentum of the quarks produced is large, of order $\sqrt{Q^2}$ and so should not be considered as a quark distribution within the proton (which should have a small transverse momentum with a scale less than or of order the nucleon mass) but rather, as the parton interpretation prefers, should be identified as a gluon contribution which scatters off the virtual photon to give two quark jets. The large transverse momentum means the contribution is not sensitive to infra-red effects. Of course, this split would be meaningless it it did not imply some measurable effect elsewhere but that it clearly does for the gluon distribution identified here may also be looked for in other processes. For example, there should be two-jet events in heavy quark production of the same magnitude once the energy is far above the heavy quark threshold and these can be looked for in charm production processes as I will discuss. The parton model analysis implies the gluon component will also contribute to other hard scattering processes such as pp collisions; this is just a natural extension to polarised scattering of the partonic analysis so successful in describing hard processes in unpolarised scattering.

The calculation for heavy quark production ($Q^2 >> m^2$) does clarify the worry of Manohar and Jaffe that the anomalous contribution is infra-red sensitive for it is found in this case[28] there are two distinct regions of transverse momenta contributing, one hard, of order $\sqrt{Q^2}$, and one soft, of order m. The contributions conspire to cancel in the first moment (the anomaly decouples for massive quarks[(27)]) but in the partonic analysis only the hard component is to be identified with a gluon piece producing two quark jets $\gamma+g\to q+\bar{q}$ both with large transverse momentum; the other component corresponds to a single quark jet event. For light, quarks, e.g. the strange quark, this should be considered as modifying the intrinsic quark component of the proton which also gives rise to single quark jets via γ+"q"$\to$q. The final question raised by Manohar and Jaffe concerned invariance of the split under large gauge transformations. Recently Forte[(32)] has shown that the result of integrating out the fermions, including non-perturbative effects, is to give for the axial charge measured in the polarised scattering experiments

$$Q_5^{\text{CURRENT}} = Q_5{}^{0P} - \frac{g^2}{8\pi^2} \int d^3x \, \mathrm{Tr} \left\{ \epsilon^{ijk} \left[A_i \partial_j A_k + \frac{2}{3} g A_i A_j A_k \right] \right\} + 2 \mathrm{ind}\,(i\not{D}_4) \tag{41}$$

where the first term is the usual valence quark contribution to the spin, the second term is the Chern-Simons term and the third (non-local) term, which computes the index of the Dirac operator, arises in such a way as to preserve invariance of the RHS under large gauge transformations. Taking proton matrix elements of this Forte derives the form

$$\Delta\Sigma^{CURRENT} = \Delta\Sigma^{op} - f \langle\Omega_0^{CLASSICAL}\rangle - f \frac{\alpha s}{2\pi} \Delta g \ QUANTUM \tag{42}$$

where the last term on the RHS corresponds to the perturbative "two jet" piece, insensitive to infra-red effects formed by taking the two gluon matrix element of the Chern-Simons form. The second term on the RHS also comes from the Chern-Simons term but is sensitive to long-distance physics. It is clearly to be identified with the "one-jet" gluon piece and, according to the discussion given above, should be included with $\Delta\Sigma^{op}$ in the quark component, $\Delta\Sigma^{PARTON}$. The advantage of this formalism is that it clearly demonstrates how $\Delta\Sigma+\Delta\Gamma$ is invariant under large gauge transformations.

The connection with helicity components may be made for, although the Chern-Simons term is not the angular momentum operator, its two-gluon matrix element evaluated to leading order in perturbation theory is proportional to the gluon helicity as it should be from the definition of eq (32). Beyond leading order however this component will not be purely gluon helicity but involve angular momentum, too, as is obvious since creating two parallel gluons of helicity ±1 out of one cannot be done without orbital angular momentum. Thus as the gluon distribution changes with Q^2 due to radiative corrections the angular momentum will change, too, so that the total angular momentum is left invariant[33].

6. Phenomenological implications

Keeping the second order in the QCD corrections the sum rules have the form[6,34].

$$\int_0^1 dx \left(g_1{}^p(x,Q^2) - g_1{}^n(x,Q^2)\right) = \frac{1}{6}\left[1-\frac{\alpha s(Q^2)}{\pi}\right]\frac{gA}{gV}$$

$$\int_0^1 dx\ g_1{}^p(x,Q^2) - \frac{1}{12} \left\{ \left[(\Delta u-\Delta d) + \frac{1}{3}(\Delta u+\Delta d-2\Delta s) \right] \left[1-\frac{\alpha_s(Q^2)}{\pi} \right] + \frac{4}{3}\left[\Delta u+\Delta d+\Delta s-3\Delta\Gamma \right] \left[1 - \frac{\alpha_s(Q^2)}{\pi} \cdot \frac{33-8f}{33-2f} \right] \right\} \tag{43}$$

From this we see that the dominant effect of gluons is to change Δq_i to $\Delta q_i-\Delta\Gamma$ so eq (22) is modified to read

$$\begin{aligned} \tfrac{1}{2}(\Delta u-\Delta\Gamma) &= 0.391 \pm 0.016 \pm 0.023 \\ \tfrac{1}{2}(\Delta d-\Delta\Gamma) &= -0.236 \pm 0.016 \pm 0.023 \\ \tfrac{1}{2}(\Delta s-\Delta\Gamma) &= -0.095 \pm 0.016 \pm 0.023 \end{aligned} \tag{44}$$

Clearly these three equations are insufficient to determine the quark and gluon distributions unambiguously but we may discuss the viability of the opposite extreme to that of eq (22) in which $\Delta s=0$, as in the original Ellis-Jaffe analysis. Then

$$\begin{aligned} \Delta\Gamma &= 0.19 \pm 0.08 \\ \left.\begin{aligned}\tfrac{1}{2}\Delta u &= 0.49 \pm 0.04 \\ \tfrac{1}{2}\Delta d &= -0.14 \pm 0.04\end{aligned}\right\} & \ \tfrac{1}{2}(\Delta u+\Delta d) = 0.35 \end{aligned} \tag{45}$$

We have from eq (37) and (32)

$$\Delta\Gamma \equiv \frac{\alpha s(Q^2)}{2\pi}\Delta g(Q^2) = \frac{\alpha s(Q^2)}{2\pi}\int_0^1 dx \left[g_+(x,Q^2) - g_-(x,Q^2) \right] \tag{46}$$

and one may ask whether reasonable distributions $g_\pm(x,Q^2)$ may give rise to a value of $\Delta\Gamma$ of the magnitude given in eq (45)[20,35]. Since, in the parton model, $g_\pm(x,Q^2)$ are positive, corresponding to probability densities, we have the constraint

$$\left[g_+(x,Q^2) - g_-(x,Q^2) \right] \leq \left[g_+(x,Q^2) + g_-(x,Q^2) \right] \tag{47}$$

where the right hand side is the unpolarised gluon distribution. In Fig. 7 we show an analysis of Altarelli and Stirling[20] in which they construct gluon distributions satisfying this bound and consistent with the Regge behaviour expected at low x. Their gluon distribution generates $\Delta\Gamma$ = 0.23 (the earlier central value quoted by the EMC) assuming a convolution in which $\Delta\sigma^{\gamma g}(x) = \delta(x-1)$ in eq (31). This ammounts to absorbing the constant piece coming from eq (33) in the polarised gluon distribution. We will comment on the ambiguity in defining $\Delta\sigma^{\gamma g}$ presently. The gluon distributions corresponding to the fit of Fig. 7 are given in Fig. 8 showing that the bound of eq (47) is satisfied. However the distribution is peaked at small x and so that the "ϵ sum rule" gives $\int_{\epsilon=0.01}^{1} dx\,(g^{+}-^{-}g)$ = 0.16, considerably below the initial EMC value and below the revised[1] central EMC value leading to eq (45). This is not a serious problem for the $\Delta s=0$ case can be maintained by slightly changing the gluon distribution but it does illustrate the need to check that any explanation proposed to fit the Ellis–Jaffe–Gourdin sum rule is consistent with the observed x distribution.

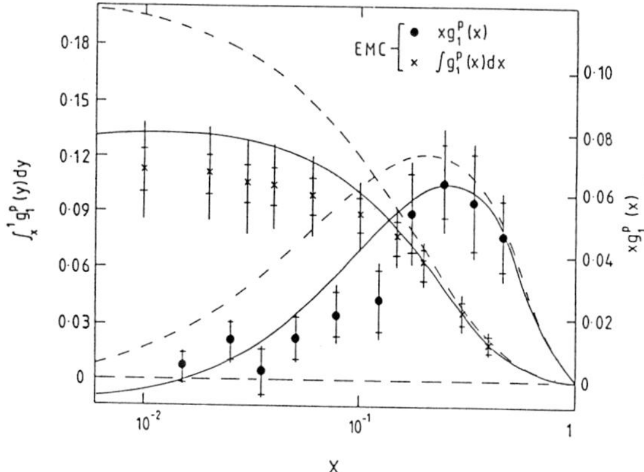

Fig. 7. Possible quark and gluon contributions to $xg_1P(x)$ and $\int_0^1 g_1P(x)dx$ corresponding to $\Delta s=0$. The solid curve is the total contribution and the dashed curve is the contribution from quarks only.

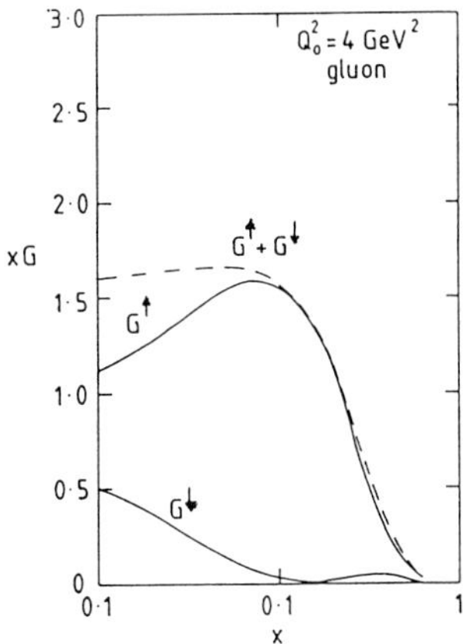

Fig. 8. Polarised gluon distributions coming from the fit of Fig. 7.

Let us return to the question of the correct definition of $\Delta\sigma^{\gamma g}(x)$. The gluon distribution may be re-defined absorbing a constant from $\Delta\sigma^{\gamma g}(x)$ as was done by Alteralli and Stirling. However the constraint of eq (47) is sensitive to such a redefinition and care should be taken to use the same definition of the gluon for both the polarised and unpolarised distributions. The latter is largely determined by deep inelastic photon–proton scattering using a convention in which only the logarithmic piece is absorbed. However as these logarithmic corrections differ between the polarised and unpolarised scattering there is an inherent ambiguity in relating these distributions (the polarised distribution is also sensitive to the renormalisation prescription chosen to regulate the γ_5[36]). The problem is that the parton model result, eq (47), is inconsistent with QCD radiative corrections and, at best, eq (47) should be applied at a unique value of Q^2, presumably of order of the nucleon mass squared. We will not discuss the implications of this ambiguity further, save to mention that typically other choices for $\Delta\sigma^{\gamma g}(x)$ will make the gluon contribution softer and make it more difficult to satisfy the ϵ sum rule[37].

The example just discussed shows that it is possible to choose gluon distributions to fit the sum rule results without a strange quark contribution at all. A less arbitrary choice might be to try to

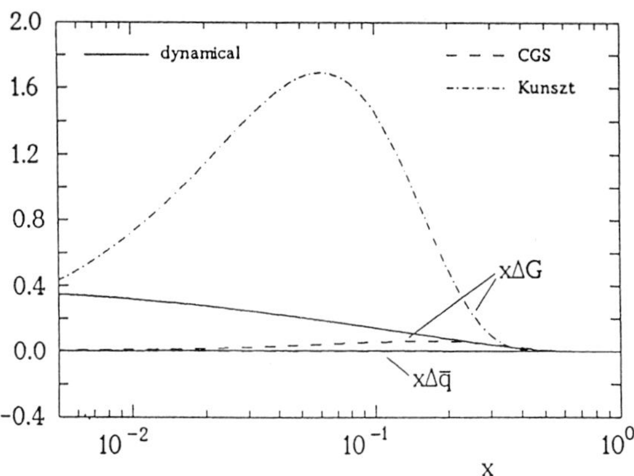

Fig. 9. Dynamical antiquark and gluon contributions determined in ref [38] (Gluck et al.).

generate the gluon and sea quark content of the proton starting with a pure valence quark proton at a (low) scale μ^2 [38]. Gluck and Reya[38] have attempted to do this using the renormalisation group evolution equations which require

$$\Delta\Sigma(Q^2) = \Delta\Sigma(\mu^2)$$
$$(\Delta\Sigma-3\Delta\Gamma)(Q^2) = (\Delta\Sigma-3\Delta\Gamma)(\mu^2) \left[1 - \frac{2}{3\pi}(\alpha_s(\mu^2) - \alpha_s(Q^2))\right] \quad (48)$$

The result of their analysis is shown in Fig. 9. They assume $\Delta g(x,\mu^2) = \Delta\bar{q}(x,\mu^2) = 0$ at μ^2 is given by $\alpha_s(\mu^2)/\alpha_s(Q_0{}^2 \simeq 10\text{GeV}^2) = 14\pm1$, a value they found analysing the unpolarised data in the same way. At $Q^2=10\text{GeV}^2$ they find $\Delta s \simeq 0$, $\Delta\Gamma=0.1$, somewhat smaller than the experimental requirement of of eq (45) but within the experimental error. However, using $\Delta\sigma^{\gamma g}(x)=\delta(x-1)$, one finds that $\Delta\Gamma_{\epsilon=0.01}$ is unacceptably small, reflecting the fact the radiatively generated distribution of gluons is peaked at small x. It may be that the discrepancy can be resolved for a different choice of $\Delta\sigma^{\gamma g}$.

To summarise, let us return to the fit of Fig. 7 and 8. At $Q^2=10\text{GeV}^2$ this corresponds to $\Delta g=6.3$ and so the proton helicity is given by

$$\tfrac{1}{2} = \tfrac{1}{2}\Delta\Sigma + \Delta g + L_z$$
$$= 0.35 + 6.3 - 6.15 \quad (49)$$

The large cancellation between the orbital piece and gluon helicity contribution is predicted[33] by QCD for, as argued above Δg and L_z are correlated through $d/dt\,(\Delta g+L_z) = 0$. The net contribution of gluons to the spin of the proton is just 0.15.

The large values of Δg and L_z separately are simply due to the ln Q^2 growth of the gluon distribution predicted in QCD. At low Q^2 the net contribution is still 0.15 but will arise from values of Δg and L_z which are much smaller, more in line with the quark model expectations.

7. Problems with perturbative QCD?

It has been suggested[39] that the large value of $\Delta s^{\text{CURRENT}}$, needed to explain the experimental measurements, is inconsistent with the unpolarised measurement of the strange quark content and hence the whole framework of perturbative QCD applied to the analysis of deep inelastic scattering fails. In brief the argument starts with the inequality

$$\Delta s \equiv |s_+ + \bar{s}_+ - s_- - \bar{s}_-| \leqslant |s_{+c} + s_{-c} + \bar{s}_{+c} + \bar{s}_{-c}| \quad (50)$$

where, as usual, $s_\pm(\bar{s}_\pm)$ refers to the polarised strange curent quark (antiquark) distribution, including possible gluon effects; and $s_{\pm c}(\bar{s}_{\pm c})$ refer to the unpolarised strange quark distributions defined by

$$\tfrac{1}{2}(s_{+c}+s_{-c}) = s^{\text{UNPOLARISED}}\big|_{\text{NON-POMERON}} \quad (51)$$

and similarly for the anitquark distributions. Some comments concerning eq (50) are in order: since s_+-s_- decouples from the pomeron it is obvious $|s_+-s_-|\leqslant|s_{+c}s_{-c}|$ provided $s_{\pm c}$ are both positive. That this should be so is plausibly argued in ref [34] on the grounds that these distributions are, by duality, each equivalent to positive definite resonant contributions. The same argument applies to the antiquark distributions.

The trick in applying eq (50) is to make the non-Pomeron separation, and in ref [39] this is done writing the unpolarised sea-quark distribution as

$$x\bar{q} = (A+Bx)(1-x)\rho \qquad (52)$$

where the term proportional to A is assumed to be the Pomeron piece and the term proportional to B the non-Pomeron piece. A fit to the unpolarised data is used to determine A and B and application eq (50) yields $|\Delta s| \leq 0.1$ in contradiction with eq (22).

An obvious criticism of this analysis concerns the validity of the non-Pomeron separation. The form of eq (52) assumes that the Regge behaviour characteristic of the various contributions applies at all x up to a common x dependent factor. However there is no firm evidence for such Regge behaviour at large x – indeed this assumption is in contradiction with the QCD evolution equations as is clear from Fig. 9 in which the polarized gluon distribution generated entirely by QCD effects shows <u>no</u> evidence of the fall-off proportional to x which in eq (52) characterises such a non-Pomeron contribution. Of course, QCD should be consistent with the Regge behaviour, but only at a small enough x value at which the higher order QCD corrections, which build up the Pomeron, are important. Even if we accept the Regge behaviour for relatively large values of $x>x_0$, say, the separatioin is ambiguous enough to allow for a sizeable Δs contribution. An example of this was given in ref [40], which chose

$$x\bar{q} = \text{POMERON} + (x\,\bar{q})_{\text{NON-POMERON}}$$

$$\text{where} \quad (x\bar{q})_{\text{NON-POMERON}} = \begin{cases} c\; x^{-\alpha} & x<x_0 \\ x\,\bar{q} & x>x_0 \end{cases} \qquad (53)$$

A fit to unpolarised data gave Δs consistent with eq (22) although the "Pomeron" contribution is constrained to small x, $x<x_0$. This is not obviously unacceptable in comparision with hadronic processes when the effect of other Regge contributions such as the f_1 A_1 ... are included, but even this constraint may be relaxed using a different Pomeron parameterisation and/or the recent values of Δs taking account of the uncertanties in determining the F/D value[14].

In summary, the problems claimed for perturbative QCD in fitting the polarised data seem to me more likely to be problems associated with the separation of the Pomeron and non-Pomeron contribution. Certainly I do not think a strong case has been made to cause us to doubt the applicability of perturbative QCD to hard processes.

8. <u>Predictions for polarised quark and gluon contributions</u>

We have seen that the experimental information is presently insufficient to determine the relative polarised quark and gluon contributions to the singlet channel. However several theoretical attempts have been made to estimate these quantities and here I will briefly discuss two such approaches – for lack of time I will not discuss bag model or other estimates[41,42].

a) <u>Skyrme model estimates</u>

The first estimate[42] of polarised scattering which was able to explain the smallness of the singlet channel moment was based on the Skyrme model. This model describes low energy physics in terms of an effective Lagrangian involving only the light pseudoscaler meson π, K, η, η' as elementary fields. Baryons are identified with soliton solutions made from bosonic degrees of freedom and associated with non-trival topological configurations. These configurations exist only for non-abelian groups and so the baryon, to leading order in $1/N_C$ (N_C is the number of colours) and for vanishing quark masses, has no coupling to the abelian SU(3) singlet sector. ie In this order it is not composed of the η_0 the SU(3) singlet pseudoscalar meson. This is equivalent to taking $\tilde{g}_A{}^0 = 0$ in eq (30) and immediately gives $\Delta\Sigma=0$ and $\Delta s \neq 0$. Inclusion of non-zero quark masses which mix η_0 and η_8 affects this result and gives rise to a non-zero $\Delta\Sigma$. An estimate of these[40] effects gives $\Delta\Sigma = -0.18$, still quite small. The model used for this estimate has a stable η', but recent work has[44] modified the Skyrme Lagrangian to allow for η' decay. This leads to a different estimate for Σ of 0.2±0.1. The general picture that emerges is that Σ should be small in the Skyrme model although its magnitude is not precisely determined. However it should be noted(23) that the SU(3) Skyrme model makes rather poor predictions for the individual values of Δu, Δd and Δs and bad predictions for other singlet matrix elements such as $<p|\bar{s}\ (\gamma^\mu D^\nu+\gamma^\nu D^\mu)\ s|p>$.

One may also ask about the expected value of the gluon contribution. This had no place in the original formulation of the Skyrme model but an attempt was made to parameterise the gluon contribution via a singlet scalar component, the dilaton, and this led to the estimate of $\Delta g=0$. In this model all the spin of the proton resides in orbital angular momentum, somewhat more than half associated with the quarks and the remainder associated with the gluons. However the reliability of the method in estimating the gluon contribution is open to question particularly since, as noted above, the separation into orbital and spin components depends on the scale and $\Delta g=0$, $L^g\neq 0$ at one scale may corresond to non-zero Δg at another scale.

b) Current algebra estimates

The starting point for the current algebra estimates is the anomaly equation

$$\partial^\mu J^0{}_{\mu 5} = \sum_j 2m_j\ \bar{q}_j(i\gamma_5)q_j + N_f \frac{\alpha s}{2\pi}\ \mathrm{Tr}\ G\tilde{G} \qquad (54)$$

Taking matrix elements between proton states gives

$$\Delta\Sigma_C = <p\ |\ \delta^\mu J_{\mu 5}{}^0\ |\ p> =$$
$$2M(\Delta\Sigma_{CA} - f\ \frac{\alpha s}{2\pi}\ \Delta g_{CA})\ \bar{u}\ (i\gamma_5)u \qquad (55)$$

where

$$<p\ |\ \mathrm{Tr}\ G\tilde{G}\ |\ p> = -2M\ \Delta g_{CA}\ \bar{u}\ (i\gamma_5)\ u.$$

It is normally assumed that $\Delta\Sigma_{CA} = \Delta\Sigma_p$ and $\Delta g_{CA} = \Delta g_p$ where the latter are the parton model polarised moments discussed above, but one should beware that this may not be the case as I will discuss shortly.

Cheng and Li(45) tried to estimate Δg by relating the matrix elements of $J_\mu{}^{5\ 0}$ to deviations from the Goldberger–Treimain relation using the relation

$$<p(k)|(m_u-m_d)(\bar{u}i\gamma_5 u+\bar{d}_{i\gamma_5}d)|P(k)> =$$

$$<p(k)|\partial^\mu[\bar{u}\gamma_\mu\gamma_5 u-\bar{d}\gamma_\mu\gamma_5 d]|p(k)>$$

$$-\ (m_u+m_d)<p(k)(\bar{u}i\gamma_5 u-\bar{d}_i\gamma_5 d)|p(k)>$$

$$= (2Mg_A-2f_\pi g_{\pi NN})\ \bar{\psi}(k)\ \gamma_\mu\ \gamma_5\ \psi(k) \qquad (56)$$

where the second term on the right hand side has been evaluated using π^0 dominance. They find, using the EMC measurements of $\Delta\Sigma c$,

$$f\Delta\Gamma^p = (-0.87 \sim -\ 0.48)$$
$$f\Delta\Gamma^n = (-0.06\ -\ +0.06) \qquad (57)$$

ie the gluon contribution to the proton is large and negative, quite the opposite to what was needed to reduce the strange quark content of the proton. However, the derivation of eq (56) is open to question since the LHS vanishes in the absence of isospin breaking while there is no reason to think deviations from the Goldberger–Tremain relation, the RHS, do so. This sensitivity to isospin breaking results in the large fluctuation found in going to $\Delta\Gamma^n$ in eq (57) and presumably reflects the approximation involved in using pion pole dominance.

Hatsuda[46] has recently applied pole dominance in the large N_c approximation to the singlet channel to try to estimate Δg_{CA} directly. He finds

$$-2M\Delta g_{CA} = \sqrt{\frac{3}{2}}\ E_\pi\ (m_{\eta'}{}^2 + m_\eta{}^2\ -\ 2m_K{}^2)$$
$$\left[\frac{g_{\eta' N}}{m_{\eta'}{}^2}\ \cos\theta_3 + \frac{g_{\mu N}}{m_\eta{}^2}\ \sin\theta \pm \frac{g_{\pi N}}{m_{\pi^0}}\ {}_2\theta_2\right] \qquad (58)$$

where the singlet current couples to the field S given by

$$S = \eta' \cos\theta_3 + \eta \sin\theta_3 + \pi^0 \sin\theta_2 \tag{59}$$

and allowance has been made for isospin violating effects. Using known values of the mixing angles he predicts Δg_{CA} (and $\Delta\Sigma_{CA}$) in terms of the (unmeasured) couplings $g_{\eta N}$ and $g_{\eta' N}$. In the limit of $m_{\eta'}{}^2 \to \infty$ and no mixing, Δg_c and $\Delta\Sigma_c$ are zero[46,47]. For the case of the physical η' mass the results are given in Fig. 10 where it may be seen that $(\Delta\Sigma_{CA} - f\ \alpha_s/2\pi\ \Delta g_{CA}) = 0$ occurs for $g_{\eta' N} = 1.5$, $g_{\eta N} = 6.8$ but this results from a cancellation of $\Delta\Sigma_{CA} = -0.2$, $3\Delta\Gamma_{CA} = -0.2$, both small and with $\Delta\Gamma_{CA}$ still the wrong sign to reduce Δs. The significance of this result is not yet clear for, as stressed above, $\Delta\Sigma_{CA}$ and Δg_{CA} may not

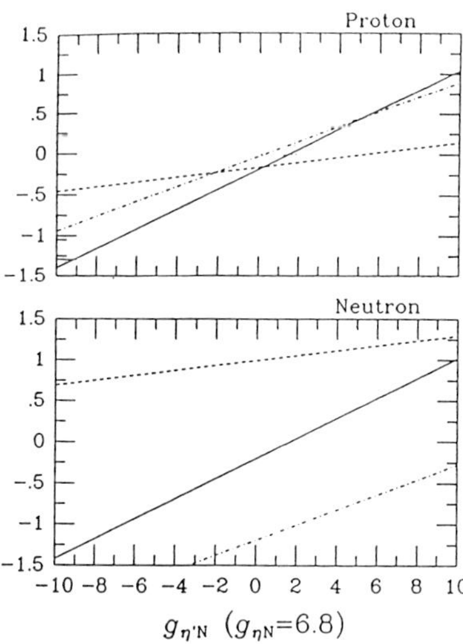

Fig. 10. Current algebra + large N_c prediction for the first moment of $g_1{}^P$ and $g_1{}^N$. The dashed lines denote the quark contributions and the solid lines the total contribution plotted as a function of $g_{\eta' N}$.

correspond to the parton model definition of these quantities. This is made clear from the fact that the estimate of $\Delta\Sigma_{CA}$ relies on it coupling to the η, η' and π^0. However the η' obtains its mass largely due to its coupling to the anomaly and hence $\Delta\Sigma_{CA}$ also couples to the anomaly suggesting it contains a sizeable piece which should be identified with Δg_p. If the anomalous contribution to the η' mass is "switched off" the current algebra estimate for $\Delta\Sigma_{CA}$ will change, giving the quark model result due to the light η' pole. The difference between this value and the result given in Fig. 10 might more reasonably be interpreted as a Δg_p piece, bringing the estimate into line with the form of equation (49).

However it should be stressed that the smallness of the measured combination $(\Delta\Sigma_{CA} - f\alpha_s/\tfrac{1}{2}\pi\ \Delta g_{CA})$ which _is_ independent of the ambiguities in the definition of $\Delta\Sigma$ and Δg, results only for a special choic of $g_{\eta' N}$ which is not explained on the basis of current algebra alone and reflects dynamical properties of the proton (for example we have seen it follows from the skyrmion picture of the proton).

9. Summary and conclusions

We have discussed how polarised structure functions give new information about the distribution of nucleon spin amongst its constituents. The data is consistent with a distribution amongst quark and gluons – in particular I see no conflict with perturbative QCD. Although other schemes have been able to fit the polarised structure functions[48], the perturbative QCD analysis complements the extensive perturbative analysis of deep inelastic unpolarised data. The constituent quark model is inconsisent with the SLAC and EMC polarised strucutre function results unless there is a relative renormalisation of the octet and singlet currents (Cf eq (30)). However, QCD does distinguish between these currents via the anomaly and so can accommodate the experimental result. We have argued that there is a well-defined separation of the contributions to the first moment of the polarised structure functions into quark and gluon contributions but that the data cannot at present distinguish between a gluon and a strange quark contribution.

The situation is illustrated by two extreme examples one in which there is no strange quark contribution to the spin of the proton and one in which there is no gluon contribution. In the former case the spin content of the proton is given (at Q^2=10 GeV^2) by

$$\left.\begin{array}{l}\left.\begin{array}{l}\tfrac{1}{2}\,\Delta u = 0.49\\ \tfrac{1}{2}\,\Delta d = -0.14\end{array}\right\}\ \tfrac{1}{2}\,\Delta\Sigma = 0.35\\ \tfrac{1}{2}\,\Delta s = 0\\ \left.\begin{array}{l}\Delta g = 6.3\\ \Delta L^{g} = -6.16\end{array}\right\}\ \Delta J^{g} = 0.15\end{array}\right\}\ J_p = \tfrac{1}{2} \qquad (60)$$

In the latter case we have

$$\left.\begin{array}{l}\left.\begin{array}{l}\tfrac{1}{2}\,\Delta u = 0.38\\ \tfrac{1}{2}\,\Delta d = -0.25\\ \tfrac{1}{2}\,\Delta s = -0.13\end{array}\right\}\ \tfrac{1}{2}\,\Delta\Sigma = 0\\ \left.\begin{array}{l}\Delta g = 0\\ \Delta L^{g}<\tfrac{1}{4}\\ \Delta L^{q}>\tfrac{1}{4}\end{array}\right\}\ \Delta L = \tfrac{1}{2}\end{array}\right\}\ J_p = \tfrac{1}{2} \qquad (61)$$

where the split into orbital momentum components follows for a specific model. It is obviously important to try to distinguish between these possibilities both theoretically and experimentally. Unfortunately the theoretical expectations are unclear due to the difficulty in extracting predictions for the parton-like quantities of eqs (60) and (61). Thus although the observed smallness of the first moment of the singlet polarised structure function may be explained in current algebra or in models of the proton such as the Skyrme model, it is difficult to use these to determine the quark and gluon components separately. However it should be possible to test for a large gluon contribution by future experiment(20,27,28,37,49,50) and thus distinguish between the two extremes of eq (60) and (61). For, example, heavy quark production proceeds in part via the gluon distribution $\gamma^{*}g\to q\bar{q}$ and a large polarised gluon distribution Δg such as given in eq (50) will give rise to a large contribution to the cross section(20,37,47). This is shown in Fig. (12) in which it may be seen that the gluon

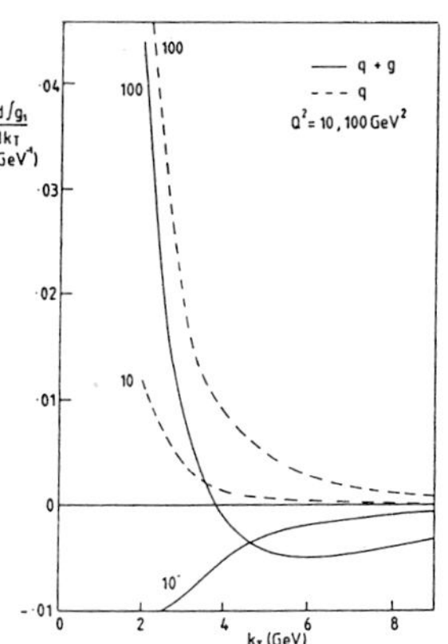

Fig. 11. The transverse momentum distribution $d\int g_1/dk_T$ obtained by convoluting the spin-dependent quark and gluon distributions with the distributions in Fig. 7. The distribution is shown at two Q^2 values, 10 GeV^2 and 100 GeV^2. The dashed lines correspond to the $\gamma^{*}q\to qg$ contribution only.

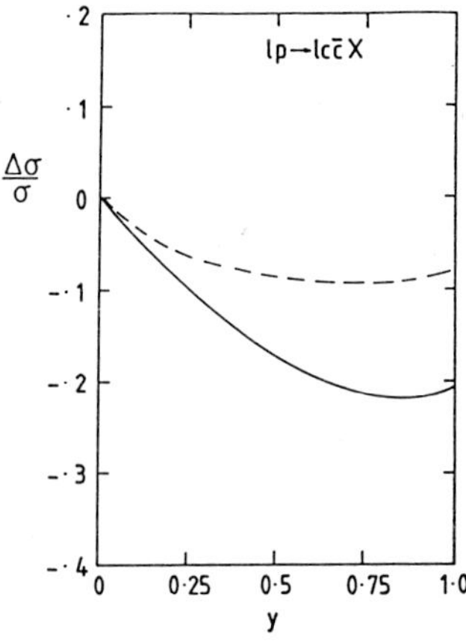

Fig. 12. The charmed quark proton asymmetry in polarized lepton-proton scattering as a function of y. The lepton beam energy is 200 GeV and the charmed quark mass is taken to be 1.5 GeV/c^2. The solid line corresponds to "open" charmed quark production, while the dashed line corresponds to integrating the $c\bar{c}$ invariant mass over the charmonium resonance region only.

contribution $\gamma^{*}g\to q\bar{q}$ gives a sizeable increase to the contribution from the quark contribution $\gamma^{*}q\to qg$ alone. The two contributions may be distinguished, only the gluon contribution gives rise to two-jet events with both quarks at large transverse momentum. This observation gives rise to another method for measuring the polarised gluon distribution via the transverse momentum distribution in inclusive jet production(28). The signal to be expected with and without the gluon contribution of eq (60) is given in Fig. 11(20).

In conclusion we have seen that the polarised structure function measurements represents the first important step in determining the spin structure of the proton, but they need to be complemented by other measurements in order to complete our picture of the underlying nature of the nucleon and to determine the distribution of the nucleon's helicity amongst its constituents.

References

1. EMC, J. Ashman et al., Phys. Let. 206B (1988) 364; CERN Preprint CERN-EP/89-73. V.W. Hughes et al., Phys. Lett. 212B (1988) 511.

2. M.J. Alguard et al., Phys. Rev. Lett. 37 (1976) 1261 and 41 (1987) 70. G. Baum et al., Phys. Rev. Lett. 51 (1983) 1135. G. Baum et al., Phys. Rev. Lett. 45 (1980) 2000.

3. R.L. Heimann, Nucl. Phys. B64 (1973) 429.

4. J.D. Jackson, R.G. Roberts, G.G. Ross, Phys. Lett. 226 (1989) 159.

5. S. Wandzura and F. Wilczek, Phys. Lett. B72 (1977) 195.

6. M.A. Ahmed and G.G. Ross, Nucl. Phys. B111 (19) 441; K. Sasaki, Progr. Theor. Phys. 54 (1975) 1816. A.J.G. Hey and J.E. Mandula, Phys. Rev. D5 (1971) 2670.

7. A.P. Bukhrostov, E.A. Kuraev and L.M. Lipatov, Sov. Phys. JETP 60 (1) (1984) 22. E.V. Shuryak and A.I. Vainshtein, Nucl. Phys. B201 (1982) 141.

8. J.D. Bjorken, Phys. Rev. 148 (1966) 1467; J.D. Bjorken, Phys. Rev. D1 (1970) 1376.

9. H. Cheng and E. Fischbach, Phys. Rev. D19 (1979) 860. D.J.E. Callaway and S.D. Ellis, Phys. Rev. D29 (1984) 567. R. Carlitz and J. Kaur, Phys. Rev. Lett. 38 (1977) 673 and (1977) 1102. J. Kaur, Nucl. Phys. B128 (1977) 219. J. Schwinger, Nucl. Phys. B123 (1977) 223. A. Schäfer, Phys. Lett. 208B (1988) 175.

10. J. Ellis and R.L. Jaffe, Phys. Rev . D9 (1974) 1444; Erratum D10 (1974) 1669. M. Gourdin, Nucl. Phys. B38 (1972) 418.

11. M. Bourquin et al., Z. Phys. C21 (1983) 27.

12. D. Kaplan and A. Manohar, Nucl. Phys. B310 (1988) 527.

13. L.A. Ahrens et al., Phys. Rev. D35 (1987) 785. D.B. Kaplan, Nucl. Phys. B260 (1985) 215; H. Goergi, D.B. Kaplan and L. Randall, Phys. Lett. 169 B (1986) 73.

14. F.E. Close, Oak Ridge National Lab/University of Tennessee preprint 1989. To be published in Nucl. Phys. A and Phys. Rev. D.

15. F.E. Close and R.G. Roberts, Phs. Rev. Lett. 60 (1988) 1471.

16. G.P. Ramsey, J.W. Qui, D. Richards, D. Sivers, Phys. Rev. D39 (1989) 1471.

17. R.L. Heimann, Nucl. Phys. B64 (1973) 429; E.C. Berger, A.C. Irving and C. Sorensen, Phys. Rev. D17 (1978) 2971.

18. C. Sachrajda, private communication. See also ref. [37].

19. M.Anselmino, B.L. Ioffe and E. Leader, NSF-ITP-88-94 (Submitted to the Sov. J. Nucl. Phys.).

20. G. Altarelli and J. Stirling, CERN preprint (1988) CERN.TH.5249.

21. H.J. Lipkin, Argonne National Lab. preprint ANL-HIP-PR-88-48 and Weizmann preprint WIS-89/10/MAY-PH.

22. S.Y. Hsueh et al., Phys. Rev. D38 (1988) 2056.

23. R.L. Jaffe and A. Manohar, MIT preprint (1989) CTPI706.

24. A. Abbas, J. Phys. G15 (1989) 73 and University of Manchester preprint (1989).

25. R.L. Jaffe, Phys. Lett. B193 (1987) 101.

26. L.M. Sehgal, Phys. Rev. D10 (1974) 1663.

27. G. Altarelli and G.G. Ross, Phys. Lett. 212B (1988) 391.

28. R.D. Carlitz, J.C. Collins, A.H. Mueller, Phys. Lett. 214B (1988) 229.

29. A.V. Efremov and O.V. Teryaev, Dubna preprint E2-88-287.

30. M. Anselmino and E. Leader, NSF-ITP-88-142 .

31. S.L. Adler, Phys. Rev. 177 (1969) 2426.
J.S. Bell and R. Jackiw, Nuovo Cim. 60A (1969) 47.

32. S. Forte, Saclay preprints SPLT/89-018 and SPLT/89-106.

33. P. Ratcliffe, Phys. Lett. B192 (1987) 180.
M.B. Einhorn and J. Soffer, Nucl. Phys. B274 (1986) 714.

34. J. Kodaira et al., Phys. Rev. D20 (1979) 627;
J. Kodaira et al., Nucl. Phys. B159 (1979) 99;
J. Kodaira , Nucl. Phys. B165 (1980) 129.

35. Z. Kunst, ETH preprint (1988) ETH-PT.88-15.

36. P. Ratcliffe, Nucl. Phys. B223 (1983) 45.

37. J. Ellis, M. Karliner and C.T. Sachrajda, CERN preprint CERN-TH-5471/89.
A.D. Watson, Z. Phys. C12 (1982) 123.

38. M. Gluck and E. Reya, Dortmund preprint DO-TH88/22 (Z Phys. C - to appear).
J. Kumb, P.J. Mulders and S. Pollock, NIKHEF preprint NIKHELF-P-1 (1989).
M. Glück, E. Reya and W. Vogelsang, Dortmund preprint DO-TH88/3 (1989).

39. G. Preparata and J. Soffer, Phys. Rev. Lett. 61 (1988) 1167.

40. J. Ellis and M. Karliner, Phys. Lett. 213B (1988) 73.

41. C.J. Bencsh and G.A. Miller, University of Washington preprint (1988) 40427-36-N8.
T. Hatsmuda and I. Zahed, Stony Brook preprint (1989).
A. Abbas, University of Manchester preprint (1989).

42. N. Tornqvist, University of Helsinki preprint, (1989), HU-TFT-89-6.
G. Clément and J. Stern, CERN preprint TH.5216 (1988).
H. Dreiner, J. Ellis and R.A. Flores, CERN preprint TH.5251 (1988).

43. S.J. Brodsky, J. Ellis, M. Karliner, Phys. Lett. B206 (1988) 309.

44. Z. Ryzak, Phys. Lett. 217B (1989) 325.

45. T.P. Cheng and Ling-Fong Li, Phys. Rev. Lett. 62 (1989) 1441.

46. T. Hatsuda, Stong Brook preprint 1989 (Nucl. Phys. B. to be published).

47. H. Fritzsch preprint MPI-PAE/PTH-18/89.

48. A. Gianelli, L. Nitti, G. Preparata and P. Sforta, Phys. Lett. 150B (1985) 214.

49. M. Gluck and E. Reya, University of Dortmund preprint DO-TH 88/22 (1988).

50. J.Ph. Guillet, Z. Phys. C39 (1988) 75;
Z. Kunszt, ETH preprint ETH preprint ETH-PT/88-15 (1988) (to appear in Phys. Lett.);
J.L. Cortes and B. Pire, CNRS Preprint A.838.068 (1988).
M. Gluck and E. Reya, Z. Phys. C39 (1988) 569.

DISCUSSION

G. Martinelli, University of Rome: I would like to stress a point that was not so explicit in your talk. Within one measurement you can change the definition of quark or gluon, with the same final result. It is only once you have fixed all your quantities ($\Delta u, \Delta d, \Delta s$ and Δg) that you can make a prediction for a new process and check QCD.

G. Ross: Yes, I think I mentioned that in my talk. The last transparency emphasized the point that if you have a gluon distribution, you consider a process where gluons are dominant relative to quarks and you get a prediction. This was done, for example taking from the paper by Altarelli and Stirling, to show that charm production will have a significant difference depending upon whether you have quarks and gluons, or just quarks. There will be a testable effect.

H. Montgomery, FNAL: What can we expect in the near future for structure functions in polarization asymmetries?

G. Ross: Perhaps our chairman can tell us more about that?

V. Hughes, Yale University: There are several experiments planned, and one at CERN is approved. It uses polarized muons on polarized proton and deuteron targets. It will measure G_1 for both neutrons and protons. This will allow a test of the Bjorken sum rule. The transverse asymmetry, related to G_2, can also be measured, and exploratory measurements are planned. The time scale for this experiment is said to be 1991.

There is a HERA experiment where, of course, they need polarized electrons in their 30 GeV machine. This is in the development stage and awaits tests of electron polarization. At LEP there is a letter of intent, and also some interest at SLAC. At the moment, I don't think any of these experiments propose to look at asymmetries in the final state.

L.V. Gribov, Landau Institute: I don't quite agree with your modest statement that QCD can't contradict the spin problem. For me, it is the opposite — and connected to the question of what is a quark and what is a gluon. What is a parton? A particle with a pointlike interaction with a photon. Before, we knew that only quarks have this local interaction. Now during the last year's discussion we understand that there is a local second-order interaction between the photon and the gluon. It is completely necessary to include this interaction in the parton picture. It is difficult to formulate because the photon-gluon interaction is partly local, partly not. It is not easy to write the guage-invariant form. For example, divergences of this amplitude are completely local and can be well defined. In a sense there is no problem physically. We are happy that there exists a pointlike photon-gluon interaction, and can measure the gluon well, even for large Q^2.

G. Ross: That is a theorist's answer, and I want to emphasize that the values of $\Delta\Sigma$ and Δg are yet to be established.

E. Reya, Dortmund University: I agree that the purely dynamically predicted small-x behavior of $\Delta G(x, Q^2)$ is something to worry about. However, your argument using your "ϵ sum rule" is not very convincing, because an integral from ϵ to 1 with $\epsilon \neq 0$ does *not* represent the appropriate contribution to $g_1(x, Q^2)$. The correct expression is the well-known convolution of ΔG with its Wilson coefficient, where the latter is known to be entirely arbitrary.

By the way, the dynamical $\Delta G(x, Q^2)$ turns over for $x < 10^{-3}$ and tends to 0 for $x \to 0$ as it must, because the integral from 0 to 1 of ΔG always exists.

G. Ross: The ϵ sum applies to a physical quantity and must be true, but I agree that your gluon distribution must be convoluted with a coefficient function in order to calculate the ϵ sum rule result.

S. Ellis, University of Washington: At least one set of authors of the parton-model analysis suggest that the gluon contribution should be associated with different final states than the quark piece, i.e., with larger characteristic transverse transverse momentum. Would you please comment on that?

G. Ross: It has been suggested by Carlitz *et al.* that when the gluon box diagram is worked out, the dominant piece involves $p_T^2 \approx Q^2$. They suggested looking for an effect associated with this large p_T. In the Altarelli and Stirling paper, there are two illustrations of how this effect might occur. This is another example of where one can emphasize the sensitivity to the gluon piece.

Two-Photon Physics

Robert N. Cahn
Lawrence Berkeley Laboratory,
University of California,
Berkeley, CA 94720

ABSTRACT

Recent experimental results in two-photon physics are reviewed. Possibilities for future experimentation at high $\gamma\gamma$ collision energies are discussed.

1 Introduction

Two-photon physics has traditionally explored the low-energy regime of even-spin, even charge-conjugation states, although the range of two-photon physics has expanded in the last several years with the results on spin-one mesons. Despite its limited range, two-photon physics has commanded significant interest because the initial state is particularly simple and well understood. The major results have been in meson spectroscopy, studies of perturbative QCD, and low-energy hadronic phenomenology. This review covers results obtained since the 1987 Lepton-Photon Conference.

Will the future of two-photon physics be confined to the low-energy domain? Straightforward extensions like that available at LEP will not significantly increase the accessible domain. Only entirely new approaches can open up the high-energy domain. Some proposals to do so are discussed in the second part of the review.

Two-photon physics has been reviewed extensively. The report of Olsson [1] at the 1987 Lepton-Photon Conference gives comprehensive coverage up to that time. Two excellent sources are the reviews by Kolanoski and Zerwas [2] and by Cooper [3]. The proceedings of the 1988 Photon-Photon Workshop [4] are the source for many of the results obtained since the last Lepton-Photon meeting.

2 Pseudoscalar Mesons

Modern two-photon physics began with the observation of Francis Low [5] that the rate for the process $e^+e^- \to e^+e^-\pi^0$ was determined by the $\gamma\gamma$ width of the π^0. Measurements of the $\gamma\gamma$ widths of the π^0, η, and η', including some recent results, are shown in Table 2.1.

New results for the $\gamma\gamma$ widths have been reported for the η and η' by the ASP Collaboration at PEP using the $\gamma\gamma$ final state [6], and for the η' by the Mark II Collaboration in several decay modes [7]. The results of the Crystal Ball at DORIS II are particularly stunning for their excellent resolution, as seen in Fig. 2.1.

The $\gamma\gamma$ widths are determined by a matrix element of the electromagnetic current taken twice, between the pseudoscalar in question and the vacuum. The pure neutral states are

$$\begin{aligned}\pi^0 &= \frac{1}{\sqrt{2}}|u\bar{u} - d\bar{d}\rangle \\ \eta_8 &= \frac{1}{\sqrt{6}}|u\bar{u} + d\bar{d} - 2s\bar{s}\rangle \\ \eta_0 &= \frac{1}{\sqrt{3}}|u\bar{u} + d\bar{d} + s\bar{s}\rangle. \end{aligned} \qquad (2.1)$$

The isoscalars can mix

$$\begin{aligned}|\eta\rangle &= \cos\theta|\eta_8\rangle - \sin\theta|\eta_0\rangle \\ |\eta\rangle' &= \sin\theta|\eta_8\rangle + \cos\theta|\eta_0\rangle, \end{aligned} \qquad (2.2)$$

Collaboration	$\gamma\gamma$ width	Technique	Ref.
	π		
Crystal Ball	$7.7 \pm 0.5 \pm 0.5$ eV	$\gamma\gamma$	PR D38,1365(1988)
H. Atherton et al.	$7.25 \pm 0.18 \pm 0.11$ eV	lifetime	PL 158B,81,(1985)
G. Bellettini et al	11.8 ± 1.3 eV	Primakoff	NC 66A, 243, (1970)
A. Browman et al	8.0 ± 0.4 eV	Primakoff	PRL 37,1400(1974)
V. Kryshkin et al.	7.3 ± 0.6 eV	Primakoff	JETP 30, 1037 (1970)
	η		
Crystal Ball	$0.514 \pm 0.017 \pm 0.035$ keV	$\eta \to \gamma\gamma$	PR D38,1365 (1988)
ASP	$0.490 \pm 0.010 \pm 0.048$ keV	$\eta \to \gamma\gamma$	SLAC-PUB 4931
	η'		
JADE	$3.8 \pm 0.26 \pm 0.43$ keV	$\eta' \to \pi^+\pi^-\gamma$	PL 142B,125 (1984)
TPC/2γ	$4.5 \pm 0.3 \pm 0.7$ keV	$\eta' \to \pi^+\pi^-\gamma$	PR D35,2650(1987)
Mark II	$4.7 \pm 0.6 \pm 0.9$ keV	$\eta' \to \eta\pi^+\pi^-$	PRL 59, 2012 (1987)
Crystal Ball	$4.7 \pm 0.5 \pm 0.5$ keV	$\eta' \to \gamma\gamma$	PR D38, 1365 (1988)
JADE	$3.80 \pm 0.13 \pm 0.50$ keV	$\eta' \to \eta\pi^+\pi^-$	Shoresh p. 77.
CELLO	$4.7 \pm 0.2 \pm 1.0$ keV	$\eta' \to \pi^+\pi^-\gamma$	Shoresh p. 85.
ASP	$4.96 \pm 0.23 \pm 0.72$ keV	$\eta' \to \gamma\gamma$	SLAC-PUB 4931
Mark II	$4.61 \pm 0.32 \pm 0.60$ keV	$\eta' \to \rho\gamma$	LBL-26465(rev.)
Mark II	$4.37 \pm 0.62^{+0.98}_{-0.95}$ keV	$\eta' \to \eta\pi^+\pi^-$	LBL-26465(rev.)
Mark II	$4.60 \pm 0.49^{+0.68}_{-0.96}$ keV	$\eta' \to 4\pi$	LBL-26465(rev.)
TPC/2γ	$3.8 \pm 0.7 \pm 0.6$ keV	$\eta' \to \eta\pi^+\pi^-$	PR D38,1 (1988)

Table 2.1: Results on the $\gamma\gamma$ widths of the pseudoscalar mesons.

where we have assumed no other states (e.g., glueballs) are involved. The $\gamma\gamma$ widths of the physical states depend on the mixing angle and on the pseudoscalar decay constants F_π, F_0, and F_8. Perfect SU(3) symmetry requires $F_\pi = F_8$ while nonet symmetry would give, in addition, $F_0 = F_8$. The ratios of the $\gamma\gamma$ widths are given by [9]

$$\frac{\Gamma(\eta \to \gamma\gamma)}{\Gamma(\pi^0 \to \gamma\gamma)} = \left(\frac{m_\eta}{m_\pi}\right)^3 \left(\frac{1}{\sqrt{3}}\frac{F_\pi}{F_8}\cos\theta - 2\sqrt{\frac{2}{3}}\frac{F_\pi}{F_0}\sin\theta\right)^2 \quad (2.3)$$

$$\frac{\Gamma(\eta' \to \gamma\gamma)}{\Gamma(\pi^0 \to \gamma\gamma)} = \left(\frac{m_{\eta'}}{m_\pi}\right)^3 \left(\frac{1}{\sqrt{3}}\frac{F_\pi}{F_8}\sin\theta + 2\sqrt{\frac{2}{3}}\frac{F_\pi}{F_0}\cos\theta\right)^2 . \quad (2.4)$$

Chiral symmetry calculations indicate that $F_\pi/F_8 \approx 0.8$ [10,11]. If we use the measured $\gamma\gamma$ widths we can extract both the mixing angle and the ratio F_π/F_0. With values derived from Table 2.1

$$\Gamma(\pi^0 \to \gamma\gamma) = 7.29 \text{ eV}$$
$$\Gamma(\eta \to \gamma\gamma) = 0.51 \text{ keV}$$
$$\Gamma(\eta' \to \gamma\gamma) = 4.4 \text{ keV}$$

we find that if $F_\pi/F_8 = 1$, then $\theta = -17.5°$ and $F_\pi/F_0 = 0.945$. If instead we take $F_\pi/F_8 = 0.8$, then $\theta = -21.7°$ and $F_\pi/F_0 = 0.968$.

The $\gamma\gamma$ production of the η and η' has been studied as a function of the Q^2, the negative of the mass-squared, of one of the photons by the TPC/2γ Collaboration by single tagging. The data extend as far as $Q^2 = 4$ GeV^2, but most of the data are below $Q^2 = 2$ GeV^2. The data agree with expectations from the vector dominance model and also with a QCD-inspired result of Brodsky and Lepage [12], as

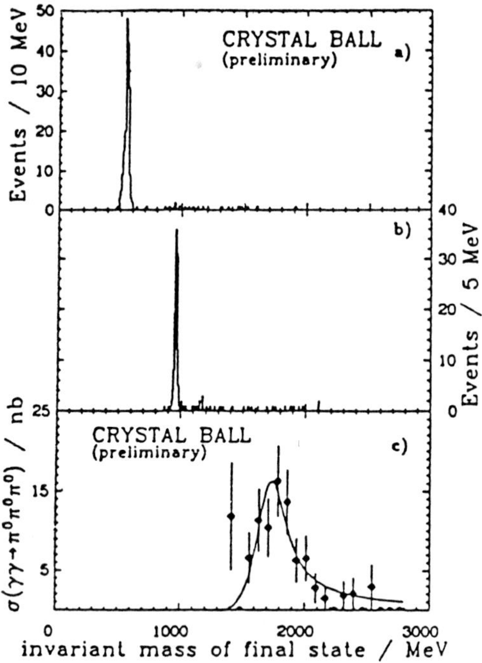

Figure 2.1: Results of the Crystal Ball Collaboration in the 6 γ final state. The top figure shows $\eta \to 3\pi^0 \to 6\gamma$. The middle figure show $\eta' \to \eta\pi^0\pi^0 \to 6\gamma$. The bottom figure shows $\pi_2(1670) \to 3\pi^0 \to 6\gamma$. Ref. [8]

seen in Fig. 2.2.

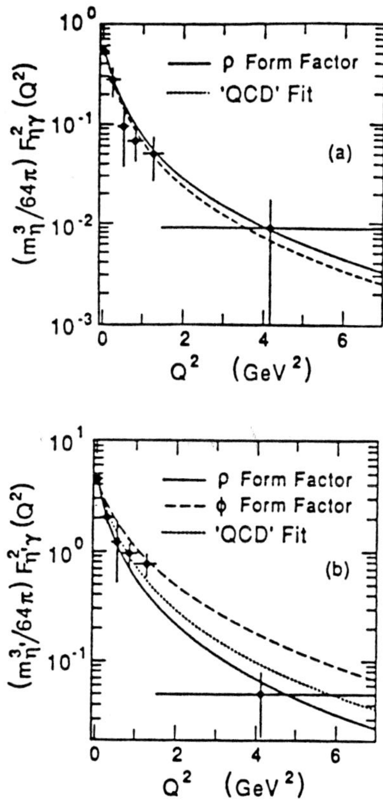

Figure 2.2: Results from the TPC/2γ Collaboration for the form factor squared as a function of Q^2 in the couplings $\eta\gamma\gamma^*$ and $\eta'\gamma\gamma^*$ [13].

3 Tensor Mesons

Numerous measurements of the $\gamma\gamma$ widths of tensor mesons have been made since the last Lepton-Photon Conference, many but not all of which were reported at the Shoresh meeting. A summary appears in Table 3.1. Crystal Ball [14], JADE [15], CELLO [16], and Mark II [17] all reported on the $f_2(1270)$ at Shoresh, with widths near 3.1 keV. A preliminary result from TOPAZ [18] at TRISTAN is substantially lower. The measurements in the $\pi^+\pi^-$ channel are plagued with backgrounds from $e^+e^- \to e^+e^-\mu^+\mu^-$ and $e^+e^- \to e^+e^-e^+e^-$. It is reassuring that the bulk of the measurements in the charged $\pi^+\pi^-$ channel agree with those in the neutral $\pi^0\pi^0$ channel. (It must be borne in mind however, that assigning all the observed events in the appropriate mass range to the f_2 may not be correct as discussed further in Sections 5 and 6 below [17].) The results of the Crystal Ball Collaboration on the all-photon final states are especially impressive, as shown in Fig. 3.1.

The isovector $a_2(1320)$ has been seen in $\gamma\gamma$ collisions in both its 3π and $\eta\pi^0$ decay channels. The data from the Crystal Ball are shown in Fig. 3.1.

If the 2^{++} tensors are described as nonrelativistic quark-antiquark bound states, their production in $\gamma\gamma$ collisions should produce only the helicity ± 2 states and no helicity-0 states. Generally this is assumed in extrapolating the observed decays to the full angular region. Angular distributions from JADE [15] on $f_2 \to \pi^0\pi^0$, from PLUTO [19] on $f_2'(1535) \to K_S^0K_S^0$, and from the Crystal Ball [20] on $a_2(1320)$ support this assumption.

The 2^{++} tensor meson mixing is nearly ideal, so

Collaboration	$\gamma\gamma$ width	Technique	Ref.
	$a_2(1320)$		
CELLO	$1.00 \pm 0.07 \pm 0.19$ keV	$\pi^+\pi^-\pi^0$	This conference
JADE	$0.84 \pm 0.07 \pm 0.15$ keV	$\pi^+\pi^-\pi^0$	Aachen $\gamma\gamma$ (1983)
JADE	$1.09 \pm 0.14 \pm 0.25$ keV	$\eta\pi^0$	Olsson/Shoresh
PLUTO	$1.06 \pm 0.18 \pm 0.19$ keV	$\pi^+\pi^-\pi^0$	PL 149B,427 (1984)
Crystal Ball (DORIS)	$1.14 \pm 0.20 \pm 0.26$ keV	$\eta\pi^0$	PR D33,1847 (1987)
TASSO	$0.90 \pm 0.27 \pm 0.16$ keV	$\pi^+\pi^-\pi^0$	ZfP C31,537 (1986)
TPC/2γ	$0.90 \pm 0.09 \pm 0.22$ keV	$\pi^+\pi^-\pi^0$	1987 EPS Meeting
Mark II	$1.03 \pm 0.13 \pm 0.22$ keV	$\pi^+\pi^-\pi^0$	LBL-26465
	$f_2(1270)$		
TPC/2γ	$3.2 \pm 0.1 \pm 0.4$ keV	$\pi^+\pi^-$	PRL 57, 404 (1986)
Crystal Ball	$3.26 \pm 0.16 \pm 0.28$ keV	$\pi^0\pi^0$	Marsiske/Shoresh
JADE	$3.09 \pm 0.10 \pm 0.38$ keV	$\pi^0\pi^0$	Olsson/Shoresh
CELLO	$3.0 \pm 0.1 \pm 0.5$ keV (prel.)	$\pi^+\pi^-$	Harjes/Shoresh
Mark II	$3.21 \pm 0.09 \pm 0.40$ keV	$\pi^+\pi^-$	Boyer/Shoresh
TOPAZ	$2.25 \pm 0.18 \pm 0.25$ keV (prel.)	$\pi^+\pi^-$	This conference
	$f_2'(1535)$		
TPC/2γ	$0.12 \pm 0.07 \pm 0.04$ keV	K^+K^-	PRL 57, 404 (1986)
ARGUS	0.054 ± 0.013 keV (prel.)	$f_2' \to K^+K^-$	Nilsson/Shoresh
PLUTO	$0.10^{+0.04}_{-0.03}{}^{+0.03}_{-0.02}$ keV	$f_2' \to K_S^0K_S^0$	Feindt/Shoresh
CELLO	$0.11^{+0.03}_{-0.02} \pm 0.02$ keV	$f_2' \to K_S^0K_S^0$	Feindt/Shoresh
	$\pi_2(1670)$		
Crystal Ball	1.4 ± 0.3 keV	$\pi^0\pi^0\pi^0$	Muryn/Shoresh
CELLO	$1.3 \pm 0.3 \pm 0.2$ keV	$\pi^+\pi^-\pi^0$	This conference

Table 3.1: Recent measurements of $\gamma\gamma$ widths of tensor mesons.

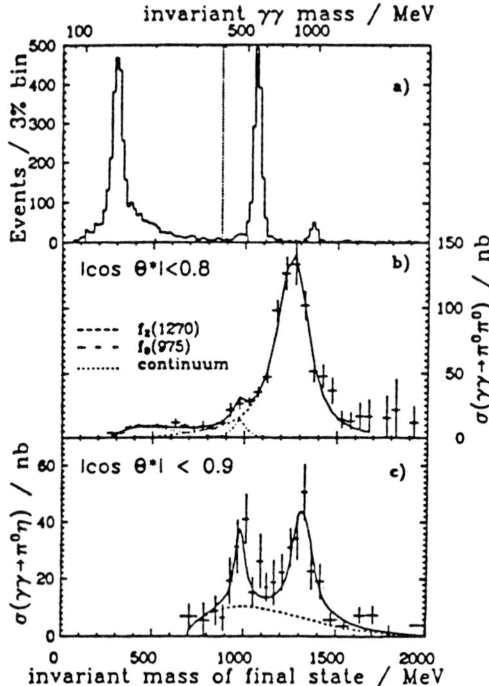

Figure 3.1: Results of the Crystal Ball Collaboration in the 4 γ final state. In a) and b) data for $\gamma\gamma \to \pi^0\pi^0 \to 4\gamma$ are displayed. In c) data for $\gamma\gamma \to \pi^0\eta \to 4\gamma$ are displayed (Ref. [8]). In a) and b) the $f_2(1270)$ is apparent together with a small signal for the $f_0(975)$. In c) the $a_0(980)$ and the $a_2(1320)$ are readily seen.

it is appropriate to describe the physical states as

$$\begin{aligned} f_2 &= \cos\lambda\frac{1}{\sqrt{2}}|u\bar{u}+d\bar{d}\rangle + \sin\lambda|s\bar{s}\rangle \\ f_2' &= -\sin\lambda\frac{1}{\sqrt{2}}|u\bar{u}+d\bar{d}\rangle + \cos\lambda|s\bar{s}\rangle \\ a_2 &= \frac{1}{\sqrt{2}}|u\bar{u}-d\bar{d}\rangle. \end{aligned} \tag{3.1}$$

Ignoring the mass differences and assuming that there is no breaking of the full symmetry (U(3)) of the quarks, we obtain the relations

$$\Gamma(f_2 \to \gamma\gamma)/\Gamma(a_2 \to \gamma\gamma) = 3\sin^2(\lambda+\beta)$$

$$\Gamma(f_2' \to \gamma\gamma)/\Gamma(a_2 \to \gamma\gamma) = 3\cos^2(\lambda+\beta) \tag{3.2}$$

$$3\Gamma(a_2 \to \gamma\gamma) = \Gamma(f_2 \to \gamma\gamma) + \Gamma(f_2' \to \gamma\gamma), \tag{3.3}$$

where

$$\tan\beta = 5/\sqrt{2}. \tag{3.4}$$

From Table 3.1 we obtain the nominal values

$$\begin{aligned} \Gamma(a_2 \to \gamma\gamma) &= 1.0 \text{ keV} \\ \Gamma(f_2 \to \gamma\gamma) &= 3.2 \text{ keV} \\ \Gamma(f_2' \to \gamma\gamma) &= 0.10 \text{ keV}, \end{aligned}$$

which yield $\lambda = 6°$ and are in rather good agreement with Eq. 3.3.

The 2^{-+} axial tensors can be made from a d-wave $q\bar{q}$ state. The $\pi_2(1670)$ (formerly the A_3) decays into $f_2\pi$ and $\rho\pi$ with a width of 250 MeV. It has been observed by the Crystal Ball [21] in $3\pi^0$ (see Fig. 2.1) and by CELLO [22] in $\pi^+\pi^-\pi^0$. The $\gamma\gamma$ width is more than 1 keV, larger than that of the a_2. This is perhaps surprising since in a nonrelativistic quark model the d-wave decay involves the second derivative of the wave function at the origin, rather than just the first derivative, which enters for p-wave states like the a_2.

4 η_c

Measuring the $\gamma\gamma$ width of the η_c has been a prominent challenge because there is a clear prediction for the value based on a nonrelativistic $c\bar{c}$ model:

$$\Gamma(^1S_0 \to \gamma\gamma) = \frac{12\alpha^2}{M^2}\left(\frac{2}{3}\right)^4 |R(0)|^2. \tag{4.1}$$

This can be compared to the prediction for the width of $\psi \to e^+e^-$:

$$\Gamma(^3S_1 \to e^+e^-) = \frac{4\alpha^2}{M^2}\left(\frac{2}{3}\right)^2 |R(0)|^2 = 4.7 \text{ keV}. \tag{4.2}$$

where the value given is the experimental one. A calculation [23] of the effect of the hyperfine interaction, responsible for the η_c - ψ splitting, increases the prediction for the $\gamma\gamma$ width by about 25%, to a little less than 8 keV.

The new TASSO result [24] is larger than the predicted value, but not inconsistent with it (See Table 4.1). The TPC/2γ Collaboration is unique in measuring the $4K$ final state [25]. Its results are similar to those of the Mark II Collaboration [26]. The CLEO Collaboration has reported [27] on both the $K_SK\pi$ and $K^{*0}K\pi$ channels, again finding general agreement with the predicted value.

While the results from $\gamma\gamma$ collisions and from the complementary $p\bar{p}$ experiment, R704, at CERN [28] are consistent with the theoretical prediction, the uncertainties remain disappointingly large. This is clearly one measurement that could benefit from greatly increased statistics.

5 Scalar Mesons

Of the extensively investigated multiplets, the scalar is the most enigmatic. The study of the $\gamma\gamma$ widths of the scalars can provide important insights into the problems raised by the nonstrange 0^{++} mesons, as discussed by Chanowitz at the 1988 Two-Photon Conference [29]. In the nonrelativistic quark model, the $\gamma\gamma$ widths of the scalars and tensors are simply related since both are ^{3}P states:

$$\frac{\Gamma(0^{++} \to \gamma\gamma)}{\Gamma(2^{++} \to \gamma\gamma)} = \frac{15}{4} \times \text{phase space.} \qquad (5.1)$$

However, a glance at Tables 3.1 and 5.1 reveals

$$\frac{\Gamma(a_0(980))}{\Gamma(a_2(1320))} \approx 0.2 - 0.3 \qquad (5.2)$$

$$\frac{\Gamma(f_0(975))}{\Gamma(f_2(1270))} \approx 0.1 \qquad (5.3)$$

in gross violation of Eq. (5.1), even if phase space effects are included.

Not only are the widths of the scalars much too small, their masses seem to be as well. The strange member of the multiplet is apparently the $K^*(1430)$. Indeed there is a candidate scalar, $f_0(1400)$, in the correct mass range. This leaves two problems [29]: where are the higher mass scalars with their few keV $\gamma\gamma$ widths and what are the $f_0(975)$ and the $a_0(980)$? It has been suggested that the two light scalars are actually $qq\bar{q}\bar{q}$ states [30] or $K\bar{K}$ molecules[31]. As for the yet-to-be-observed more massive scalars, there is the provocative suggestion that they are lying underneath the tensors [29]. To determine whether this is so requires a careful partial-wave analysis.

6 $\pi\pi$ Final State

The $\pi\pi$ final state is of interest not only because of the problems in the scalar channel near 1 GeV, but also because of its special simplicity in the low-energy range. There the constraints of analyticity and unitarity provide a useful handle on the amplitude [32]. At the same time, it is an excellent subject for an analysis based on chiral symmetry [33,34].

Morgan and Pennington [35] have emphasized the importance of a proper treatment of the $\gamma\gamma \to \pi\pi$ data that incorporates the effects of unitary and is consistent with what is known about $\pi\pi$ scattering. Preliminary results from such analysis using Mark II and Crystal Ball data [14] are shown in Fig. 6.1. There is a suggestion here of some scalar resonance hiding under the f_2 as hypothesized by Chanowitz, though a complete analysis remains to be done.

7 Glueballs

Glueballs are gluonic bound states that should couple feebly to $\gamma\gamma$, while coupling strongly to *gluon-gluon*. Limits on the $\gamma\gamma$ widths of glueball candidates are given in Table 7.1. Chanowitz [36] has proposed a quantitative measure, stickiness, S, that should help identify glueballs:

$$S_X = \frac{\Gamma(\psi \to \gamma X)}{\Gamma(X \to \gamma\gamma)} \cdot \frac{LIPS(X \to \gamma)}{LIPS(\psi \to \gamma X)}, \qquad (7.1)$$

where $LIPS$ is the Lorentz-invariant phase space factor. The status of the glueball candidates $\eta(1430)$, $f_2(1720)$, and $X(2230)$ has been reviewed by Feindt [37]. The pertinent data are from TPC/2γ [38], PLUTO [39,40], CELLO [41,42], and Mark II [43].

From these numbers and the ψ radiative decays, Feindt derives the stickiness ratios [37]:

$$S_{\pi^0} : S_\eta : S_{\eta'} : S_\iota = 0.02 : 1 : 4 : 80 (95\% C.L.) \qquad (7.2)$$

Collaboration	$\gamma\gamma$ width	Technique	Ref.
		$\eta_c(2980)$	
PLUTO	28 ± 15 keV	$K_S K \pi$	PL 167B,120(1986)
TASSO	$19.9 \pm 6.1 \pm 8.6$ keV	$K_S K\pi$,$K^+K^-\pi^+\pi^-$,4π	ZfP C41,533(1989)
TPC/2γ	$6.4^{+5.0}_{-3.4}$ keV	$4K$	PRL 60,2533(1988)
Mark II	8 ± 6 keV	$K_S K \pi$	Gidal, Berkeley 1986
CLEO	$9.4^{+3.7}_{-3.0} \pm 2.7$ keV	$K_S K \pi$	Cornell June 1989
CLEO	$8.5^{+4.6}_{-4.2} \pm 3.9$ keV	$K^* K \pi$	Cornell June 1989
R704	$4.3^{+3.4}_{-3.7} \pm 2.4$ keV	$p\bar{p}$ annihilation	PL 187B,191(1987)

Table 4.1: Results for the $\gamma\gamma$ width of the η_c.

Collaboration	$\Gamma(\to \gamma\gamma)B(a_0 \to \eta\pi)$	Technique	Ref.
	$a_0(980)$ (δ)		
JADE	$0.29 \pm 0.05 \pm 0.14$ keV	$\eta\pi^0$	Olsson/Shoresh
Crystal Ball	$0.19 \pm 0.07^{+0.10}_{-0.07}$ keV	$\eta\pi^0$	PR D36,2633 (1987)
	$f_0(975)$ (S^*)		
Mark II	$0.24 \pm 0.06 \pm 0.15$ keV	$\pi^+\pi^-$	Boyer/Shoresh
Crystal Ball	$0.31 \pm 0.14 \pm 0.11$ keV	$\pi^0\pi^0$	Marsiske/Shoresh

Table 5.1: Measurements of $\gamma\gamma$ widths of scalar mesons.

and

$$S_{f_2} : S_{f_2'} : S_{f_2(1720)} : S_{X(2230)}; \\ = 1 : 13 :> 28(95\% C.L.) :> 9(95\% C.L.). \tag{7.3}$$

The large apparent stickiness of the f_2' is just a reflection of its small $\gamma\gamma$ width caused by the small charge of its quarks. The large stickiness of the $\iota = \eta(1430)$ is not only impressive, it understates the case! Combining the data from the various experiments for the limit on $\Gamma(\eta(1430) \to \gamma\gamma)B(K\overline{K}\pi)$, Feindt [37] derives an upper limit of 0.75 keV and a corresponding stickiness greater than 128.

An alternative explanation for nonexotic "extra" states is that they are radial excitations. Radial excitations should have $\gamma\gamma$ widths that are smaller, but not very much smaller, than ordinary mesons. The possibility that the radially excited pseudoscalar multiplet consists of $\pi(1300), \kappa(1460), \eta(1280)$, and $\eta(1390)$ has been examined by Chanowitz [29]. A primary difficulty is the Crystal Ball limit

$$\Gamma(\eta(1280, 1390) \to \gamma\gamma)B(\eta\pi\pi) < 0.3 \text{ keV} \tag{7.4}$$

at 90% C.L., which applies to states below 1500 MeV with widths less than 50 MeV. The anticipated $\gamma\gamma$ widths for radially-excited states would be significantly greater [29].

8 Spin-One Mesons

The discovery in 1986 by the TPC/2γ Collaboration of a spin-1 meson produced in $\gamma^*\gamma$ collisions demonstrated the versatility of two-photon experimentation. There are now results from several collaborations, summarized in Table 8.1, on both the $f_1(1425)$ and the previously known $f_1(1285)$ (the old D(1285)).

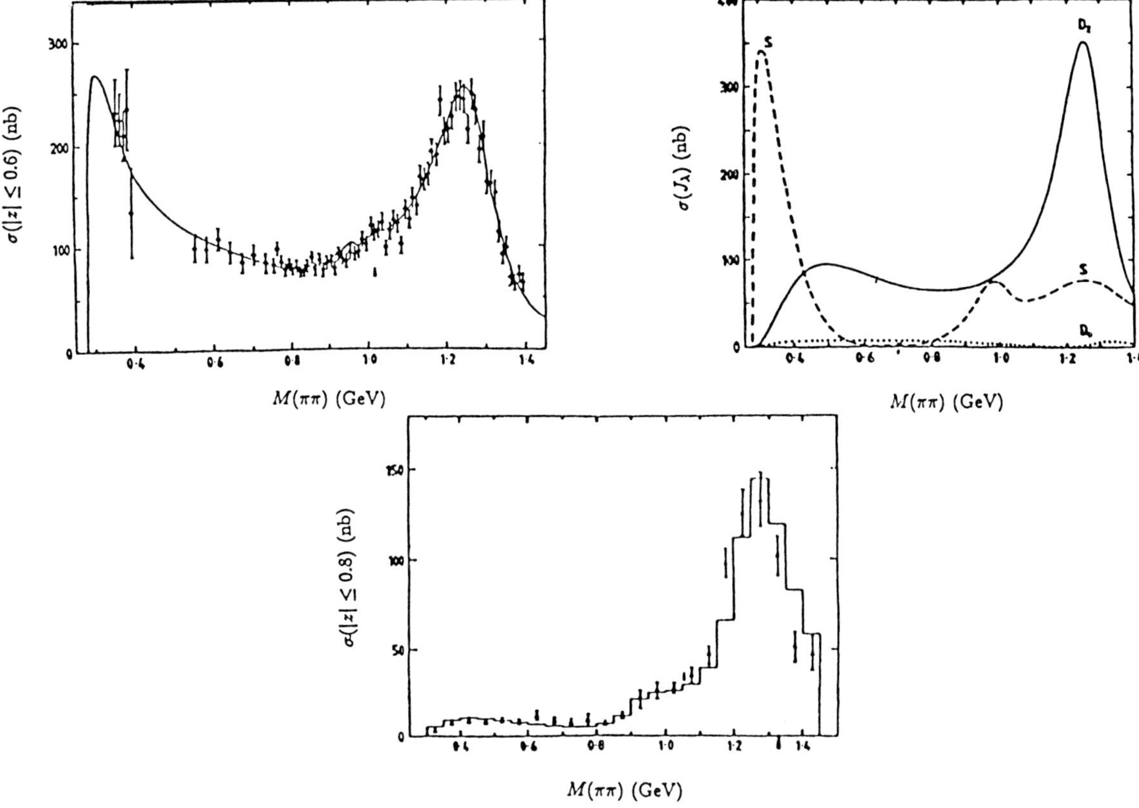

Figure 6.1: Preliminary results of fits by D. Morgan and M. Pennington to Mark II data [17] on $\gamma\gamma \to \pi^+\pi^-$ and Crystal Ball data [14] on $\gamma\gamma \to \pi^0\pi^0$. The figure on the left shows the $\pi^+\pi^-$ data. The middle figure shows the $\pi^0\pi^0$ data. The figure on the right shows the various partial wave contributions to $\pi\pi$ scattering. Beneath the D_2 (f_2, helicity 2) is a contribution from S (f_0).

The pioneering theoretical investigation was done by Renard [44]. While Yang's Theorem forbids a spin-1 particle from decaying into two real photons, decays into a real photon and a virtual one are allowed. Of course the coupling vanishes as the virtual photon becomes real. Thus the experimental signature for a spin-1 meson in $\gamma\gamma$ collisions is the appearance of a resonance when one electron is tagged, and so one photon is measured to be quite virtual. The resonance is essentially absent in the untagged data (or better, antitagged data).

Because the $\gamma^*\gamma$ width vanishes as the virtual photon becomes real, it is necessary to adopt a new measure of the coupling. The conventional one is

$$\tilde{\Gamma} = \lim_{Q^2 \to 0} \frac{M^2}{Q^2} \Gamma(Q^2) \tag{8.1}$$

where M is the mass of the resonance. Unfortunately, there is another convention to establish, that for $\Gamma(Q^2)$. For the same data Mark II and CELLO would report values twice as large as TPC/2γ and JADE. The Mark II - CELLO convention seems preferable [45], and we use it throughout.

For lack of a good alternative, the $\gamma^*\gamma$ width of the f_1 states can be estimated using a nonrelativistic quark model. The same model gives a prediction for the $\gamma\gamma$ widths of the $J = 0$ and $J = 2$ states in the same term, ^3P. If the $q\bar{q}$ wave function is written as $R(r)$ times an angular and spin wave function, with $\int dr\, r^2 R^2(r) = 1$, then all the widths are proportional to $[R'(0)]^2$. In particular,

$$\Gamma(f_2 \to \gamma\gamma) = \frac{576}{5} \alpha^2 \langle e_q^2 \rangle^2 \frac{|R'(0)|^2}{M^4} \tag{8.2}$$

$$\Gamma(f_1 \to \gamma\gamma^*) = 192 \alpha^2 \langle e_q^2 \rangle^2 \frac{|R'(0)|^2}{M^4} \frac{Q^2}{M^2}. \tag{8.3}$$

If we use $\Gamma(f_2 \to \gamma\gamma) = 3.2$ keV, we predict for an f_1 with the same quark content $\tilde{\Gamma} = (5/3) \times 3.2$

Collaboration	$\Gamma B(\to KK\pi)$	Technique	Ref.
		$\eta(1430)$	
TPC/2γ	1.6 keV	$K_S K\pi$	PRL 57, 51 (1986)
Mark II	1.5 keV	$K_S K\pi$	PRL 59, 2016 (1987)
CELLO	1.2 keV	$K_S K\pi$	ZfP 42, 367 (1989)
PLUTO	2.7 keV	$K_S K\pi$	Feindt Thesis
		$f_2(1720)$	
PLUTO	0.07 keV	$K_S K_S$	ZfP 37, 329 (1988)
CELLO	0.11 keV	$K_S K_S$	ZfP 43, 91 (1988)
		$X(2230)$	
PLUTO	0.07 keV	$K_S K_S$	ZfP 37, 329 (1988)
CELLO	0.12 keV	$K_S K_S$	ZfP 43, 91 (1988)

Table 7.1: Limits on the $\gamma\gamma$ widths of glueball candidates. See references for details of assumptions made in determining the upper limits on the $\gamma\gamma$ widths. For the $\eta(1430)$, the Mark II number is obtained by assuming $\Gamma(\eta(1430)) = 60$ MeV.

keV = 5.3 keV. The factor for the quark charges is 1/81 for a pure $s\bar{s}$ state, 25/162 for a pure isoscalar $(u\bar{u}+d\bar{d})/\sqrt{2}$ state, and 1/18 for the isovector $(u\bar{u}-d\bar{d})/\sqrt{2}$. Thus if the state observed at 1425 MeV were the old E that decays predominantly to strange states, the $\gamma^*\gamma$ width would be expected to be much smaller than that of the entirely nonstrange $f_1(1285)$.

Perhaps then the state at 1425 MeV is not the old E. Chanowitz has proposed that it might not even have $J^{PC} = 1^{++}$, but be an exotic 1^{-+} instead [46]. He shows that if the state is a meikton $(u\bar{u}+d\bar{d})g/\sqrt{2}$ it would have a large coupling to $\gamma\gamma$ while still producing the KK^* final state. The 1^{++} and 1^{-+} alternatives can be distinguished by measuring the distribution of the angle between the normal to the decay plane containing the $K\overline{K}\pi$ and the beam direction. If the production is entirely through one transverse and one longitudinal photon, the angular distributions are unique: $1+\cos^2\theta$ for 1^{++} and $1-\cos^2\theta$ for 1^{-+} [47]. Unfortunately, there are contributions from pairs of transverse photons as well, though these are suppressed at low Q^2. Within the nonrelativistic quark model there is a definite connection between the longitudinal–transverse and transverse–transverse production. However, for the exotic state that cannot be a $q\bar{q}$ there is no simple model. Thus to exclude reliably the exotic interpretation, only data at Q^2 small should be used. Data from JADE, CELLO, Mark II, and TPC/2γ prefer the 1^{++} assignment to 1^{-+}, but cannot provide a definitive answer [48]. The data for the $f_1(1285)$ quite convincingly choose $J^P = 1^+$ [48].

The CELLO Collaboration has done an extensive investigation of the $f_1(1425)$ within the limitations imposed by the data sample of 17 candidate events. They find that the final state is dominantly KK^* and use this to provide additional tests of the parity through angular distributions. Their results for the squares of the transverse-transverse and transverse-longitudinal form factors are shown in Fig.8.1. The total cross section constrains the results to lie in a band for each of four Q^2 intervals. The angular distributions reflect the ratio of transverse-transverse to transverse-longitudinal. As $Q^2 \to 0$, the latter must dominate the former by a power of Q^2. For the 1^{++} hypothesis, the transverse-longitudinal component is found to dominate and, except in the highest Q^2 bin, no transverse-transverse component is required and only a 95% C.L. can be plotted.

Collaboration	$\Gamma B(\to KK\pi)$	Technique	Ref.
$f_1(1425)$			
TPC/2γ	$2.6 \pm 1.0 \pm 0.6$ keV	ρ form factor	PR D38,1 (1988)
TPC/2γ	$1.26 \pm 0.48 \pm 0.3$ keV	ϕ form factor	PR D38,1 (1988)
Mark II	$3.2 \pm 1.4 \pm 0.6$ keV	ρ form factor	PRL 59,2016(1987)
Mark II	$2.1 \pm 1.0 \pm 0.4$ keV	ϕ form factor	PRL 59,2016(1987)
JADE	$4.6^{+2.0}_{-1.8} \pm 1.6$ keV	ρ form factor	ZfP C42,355(1989)
JADE	$3.0^{+1.2}_{-1.0} \pm 1.0$ keV	ϕ form factor	ZfP C42,355(1989)
CELLO	$3.0 \pm 0.9 \pm 0.7$ keV	ρ form factor	ZfP C42,367(1989)
CELLO	$1.4 \pm 0.4 \pm 0.3$ keV	ϕ form factor	ZfP C42,367(1989)
$f_1(1285)$ (D)			
TPC/2γ	$4.8 \pm 1.0 \pm 1.0$ keV	$\eta\pi\pi$	PR D38,1 (1988)
Mark II	$9.4 \pm 2.5 \pm 1.7$ keV	$\eta\pi\pi$	PRL 59,2016(1987)
JADE	$3.6 \pm 0.6 \pm 0.8$ keV	$\eta\pi\pi$	Olsson/Shoresh
CELLO	$7.2 \pm 2.2 \pm 2.4$ keV	$\eta\pi\pi$	Ahme/Shoresh

Table 8.1: Measurements of $\gamma\gamma$ widths of spin-one mesons. The Mark II - CELLO convention is used.

In contrast, for the 1^{-+} hypothesis, the transverse-transverse piece predominates, even at the lowest Q^2 where no transverse-longitudinal component is found in the fit. Thus the exotic spin-parity assignment might solve the puzzle of why the $f_1(1425)$ has too large a $\gamma\gamma$ width to be an $s\bar{s}$ state, but it raises the dynamical puzzle of the dominance of transverse-transverse production at low Q^2.

9 Vector-Vector Final States

The vector dominance model suggests that the vector-vector final state in $\gamma\gamma$ interactions ought to be especially interesting. The ARGUS Collaboration has made extensive contributions to the data on these channels. While there are a variety of theoretical models, none is successful in dealing with the entire data sample.

Some of the recent experimental results are summarized in Table 9.1, which shows that roughly speaking there are three distinct categories: those with a large cross section ($\rho^0\rho^0$, $K^{*+}\overline{K}^{*-}$), those with a medium cross section ($\rho^+\rho^-$, $\rho\omega$, $\omega\omega$), and those with a small cross section ($K^{*0}\overline{K}^{*0}$, $\phi\phi$, $\rho\phi$, $\omega\phi$). The energy at which the peak cross section occurs also varies from one final state to another. The striking difference between the charged and neutral $\rho\rho$ channels precludes a simple s-channel resonance explanation.

Among the theoretical models are the t-channel Factorization Model (TCFM) [52], $qq\overline{qq}$ models [53,54], and a QCD-motivated model [55]. The first seeks to identify specific t-channel exchanges and extract them from photoproduction data according to

$$\sigma(\gamma\gamma \to V_1V_2) = \sum_i \frac{\sigma^i(\gamma p \to V_1 p)\sigma^i(\gamma p \to V_2 p)}{\sigma^i(pp \to pp)} \frac{F^2_{\gamma p}}{F_{pp}F_{\gamma\gamma}}, \tag{9.1}$$

where the F_{ij}s are flux factors. The second relies on the predictions for $qq\overline{qq}$ states [30]. Interference between nearly degenerate $qq\overline{qq}$ states can account for some of the intricate behavior in the data. The third considers three perturbative diagrams for $\gamma\gamma \to (q\overline{q}')(q'\overline{q})$ that include one exchange of a gluon. The successes and failures of these mod-

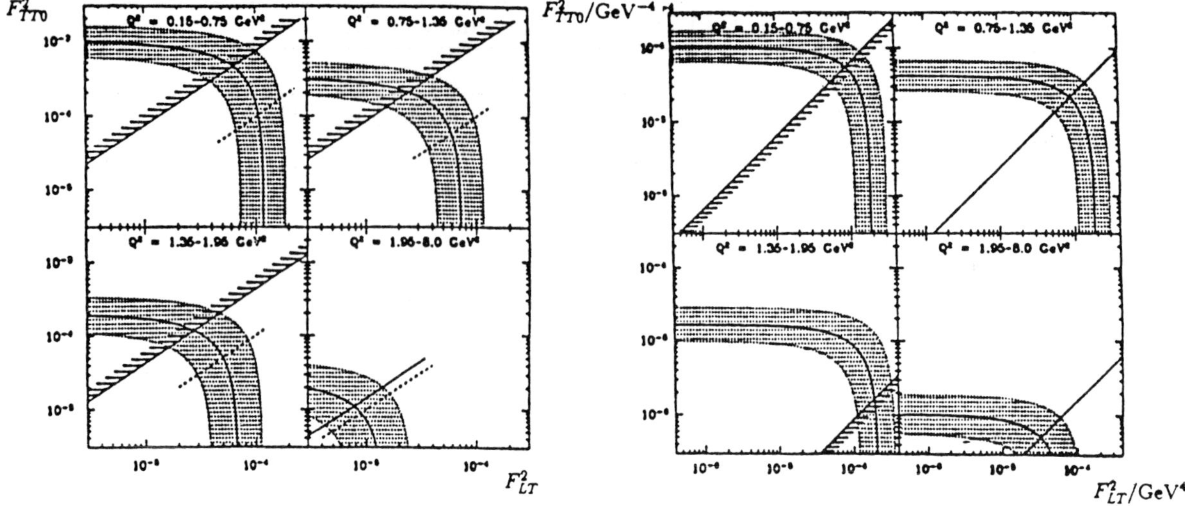

Figure 8.1: Analysis from the CELLO Collaboration for the squares of the form factors F^2_{TT} and F^2_{TL} for the $\gamma^*\gamma$ couplings to the $f_1(1425)$. The left set assumes the state has $J^{PC} = 1^{++}$; the right assumes 1^{-+}. The bands show allowed (1σ) regions. The angular distribution analysis prefers results along the diagonal lines in three of the figures. For the remaining five the horizontal lines indicate regions excluded at 95% C. L. The figures show that the 1^{++} assignment requires dominance by the transverse-longitudinal production while the 1^{-+} requires dominance by the transverse-transverse production, even at low Q^2, in contradiction with expectations.

els have been reviewed previously [2,56,57].

The TCFM does well for the $\rho^0\rho^0$ data, acceptably for $\rho\omega$, and poorly for $\omega\omega$ [57], while pleading ignorance in the instance of $K^*\overline{K}^*$. The $qq\overline{q}\overline{q}$ model has trouble with the $K^*\overline{K}^*$ states. Achasov and Shestakov [58] proposed a K-exchange model to explain the $K^{*0}\overline{K}^{*0}$ data, but it predicts that the cross section for the charged $K^*\overline{K}^*$ final state will be smaller than that for the neutral, which contradicts the data. Li and Liu find that permitting mixing among the $qq\overline{q}\overline{q}$ states enables them to fit the data [59]. The QCD motivated model also is undone by the $K^*\overline{K}^*$ data since it underestimates the absolute cross sections by about a factor of 8. It seems that understanding hadronic dynamics at low energy is difficult indeed, even when the initial state is as simple as $\gamma\gamma$.

One instance of inclusive vector production is notable: D^*. Production of charm in photon-photon collisions, like that of u quarks, is enhanced by a factor of 16 relative to that of d and s. Of course, the threshold is rather high for $\gamma\gamma$ collisions, but evidence has been presented previously for charm production by the JADE Collaboration [49] and the TPC/2γ Collaboration [50]. Now the TASSO Collaboration has reported the observation of inclusive $D^{*\pm}$ production and $D^0\overline{D}^0$ production [51]. The detection of D^* is facilitated by the small mass difference between the D and D^*. Models must be used to compare the observed cross sections to the predictions of the quark model. The TASSO observations indicate a cross section larger than expected, a result that agrees with the JADE results but not those of the TPC/2γ Collaboration.

10 Baryons

New data from ARGUS [60] and the TPC/2γ Collaboration [61] have become available for exclusive states containing baryons. The ARGUS data cover the final states $p\overline{p}$, $p\overline{p}\pi^0$, $p\overline{p}\pi^+\pi^-$, and $p\overline{p}\pi^+\pi^-\pi^0$. The TPC 2$\gamma$ data are for $p\overline{p}\pi^+\pi^-$. Neither group finds an established signal for $\Delta^0\overline{\Delta}^0$, and both find only small signals for $\Delta^{++}\overline{\Delta}^{--}$ compared with the $p\overline{p}$ signal.

This result is contrary to the very naive expectation that the Δ^{++}, having twice the charge of the proton (which is in the same SU(6) multiplet),

Process	Collaboration	Ref.	Features
$\rho^0\rho^0$	PLUTO	ZfP C38,521 (1988)	100 nb @ 1.5 GeV
	TPC/2γ	PR D37, 28 (1988)	120 nb @ 1.3 GeV
$\omega\omega$	ARGUS	PL 198B,577 (1987)	16 nb @ 1.9 GeV
$\phi\phi$	ARGUS	PL 210B,273 (1988)	$<$ few nb
	TPC/2γ	PR D37,28 (1988)	$<$ few nb
$\rho\omega$	ARGUS	PL 196B,101 (1987)	30 nb @ 1.9 GeV
	TPC/2γ	Ronen/Shoresh	36 nb @ 1.8 GeV
$\rho\phi$	TPC/2γ	PR D37,28 (1988)	$<$ few nb
	ARGUS	PL 198B,255 (1988)	$<$ 1.0 nb
$\omega\phi$	ARGUS	PL 210B,273 (1988)	$<$ few nb
$\rho^+\rho^-$	CELLO	PL 218B,493 (1989)	22 nb @ 1.9 GeV
	ARGUS	PL 217B,205 (1989)	30 nb @ 1.6 GeV
$K^{*+}K^{*-}$	ARGUS	PL 212B,528 (1988)	50 nb @ 1.4 GeV
$K^{*0}\overline{K}^{*0}$	ARGUS	PL 198B,255 (1988)	7 nb @ 2 GeV

Table 9.1: Results for the $\gamma\gamma$ production of vector-vector final states.

should be produced 16 times as frequently. It is even further from early predictions of models based on perturbative QCD that the ratio should be 50 [62]. However, a newer model using QCD sum-rule wavefunctions [63], finds a ratio between 0.5 and 2.

11 Structure Functions and Total Cross Section

The most elegant goal of two-photon physics has been to investigate the structure of the photon itself by scattering a virtual photon from a nearly real one. This has attracted enormous theoretical attention since the demonstration by Witten that the structure function of the photon in the high Q^2 limit is completely calculable in QCD [64]. Unfortunately, at subasymptotic energies there are important contributions from nonperturbative hadronic interactions. While these can be modeled, there is inevitably a problem of double counting interactions. Moreover, the perturbative calculation develops a singularity at $x = 0$, where $x = Q^2/2M\nu$ represents the fraction of the target photon's momentum carried by the struck quark. That singularity is of course canceled by one in the nonpertubative calculation, so that the sum, which is physical, is free of singularities.

A controversy continues over whether it is then possible to extract effectively the scale parameter, Λ_{QCD}, from the data available. An optimistic view is taken by Berger and Wagner [65] in their extensive review of two-photon physics. From the data available to them, they concluded

$$\Lambda_{\overline{MS}} = 195 {}^{+60}_{-40} \text{ MeV}. \tag{11.1}$$

A pessimistic view is taken by Field, Kapusta, and Poggioli [66]. They nicely organize the calculation of the structure function so that the cancellation of the singularity is manifest. Their perturbative calculation is cutoff at some particular value of the transverse momentum of the struck quark. For lower values of the transverse momentum the photon is regarded as a hadronic object with its own structure function. The resulting expression is less sensitive to the value of Λ_{QCD} than the originally derived expression of Witten and thus less effective for the purpose of deducing Λ_{QCD} from the data.

Frazer has provided a convenient analysis [67] of

this situation using expressions [68] in the variable x rather than the more opaque ones for the moments. His conclusion is that there is some truth in both the pessimistic and optimistic positions: The sensitivity to Λ_{QCD} is indeed reduced, but at large values of Q^2 and x, say $Q^2 = 100$ GeV^2 and $x = 0.9$, the reduction is not great.

Photon structure function results were submitted to this conference by the TPC/2γ Collaboration [69] and by the AMY Collaboration [70]. The TPC/2γ data are presented for

$$\sigma_{\gamma\gamma}(W,Q^2) = \frac{4\pi^2\alpha}{Q^2} F_2^{\gamma}(x,Q^2), \tag{11.2}$$

with $x = Q^2/(Q^2+W^2)$. The data are compared to the vector dominance model (VDM) and the quark parton model (QPM). The VDM contribution has a Q^2 variation $(1+Q^2/m_V^2)^{-2}$. A generalized vector dominance model, GVDM [71], which has an additional piece varying as $(1+Q^2/m_0^2)^{-1}$, was also examined. A very naive model adds the VDM and QPM contributions to account for low- and high-Q^2 regions. The GVDM already contains a point-like piece to simulate the QPM portion, but it is still possible to imagine adding the GVDM and QPM contributions.

The TPC/2γ data are shown in Figure 11.1 for four regions of the $\gamma\gamma$ c.m. energy, W. The combination GVDM + QPM does not describe the data since it does not fall fast enough with increasing Q^2. The pure VDM describes the data at high W well, but falls too rapidly with increasing Q^2 for low W. Both the GVDM and VDM+QPM describe the data reasonably well except in the lowest W region.

The cross section for $\gamma\gamma$ at $Q^2 = 0$ has also been extracted from the TPC/2γ data (using GVDM). The results are compared with earlier results from PLUTO [72] and the 2γ Collaboration [73] in Figure 11.2. The rise in the cross section in the PLUTO data at low W is not confirmed. In fact, subsequent data from the PLUTO Collaboration itself failed to confirm their earliest results [74].

The AMY Collaboration at TRISTAN has reported on the structure function, F_2, measured at an average value of Q^2 of 67 GeV^2, based on about 40 events [70]. In Fig. 11.2 the data are shown, together with fits based on QPM + VDM, for three values of the cutoff introduced by Field, Kapusta, and Poggioli. It is clear that the data that exist are adequately represented by this form. More data are needed to distinguish definitively between the various models considered in this Figure.

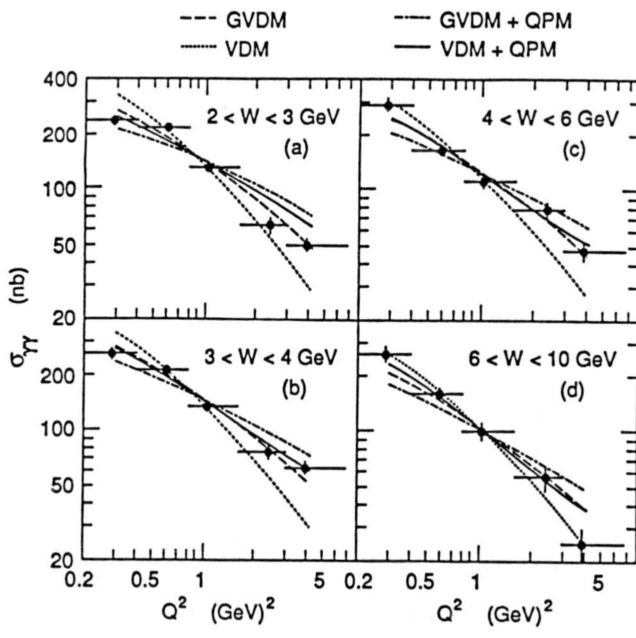

Figure 11.1: Data from the TPC/2γ Collaboration for the $\gamma\gamma$ total cross section compared to simple theoretical models.

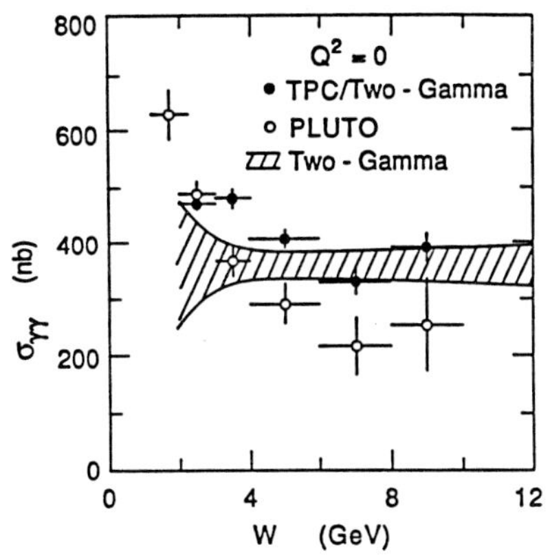

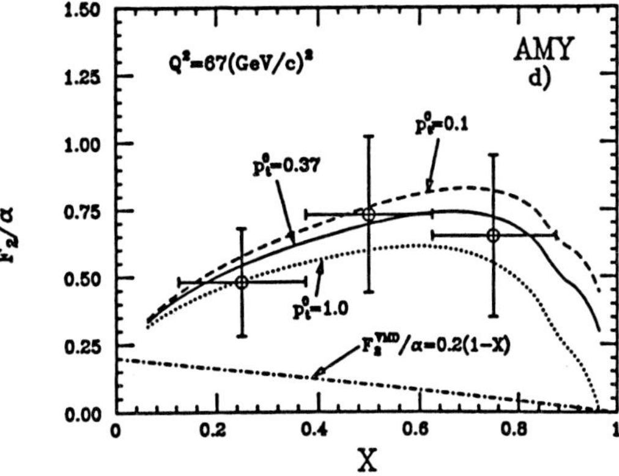

Figure 11.2: Results on the $\gamma\gamma$ cross section at high and low Q^2. On the left, data from the TPC/2γ Collaboration [69] at $Q^2 = 0$ compared with results from PLUTO [72] and the 2γ Collaboration [73]. On the right, data from the AMY Collaboration [70] at an average value of $Q^2 = 67$ GeV^2 compared to fits with QPM + VDM for various values of the p_t cutoff.

12 Prospects for High-Energy $\gamma\gamma$ Collisions

The beginning of LEP and the prospect of LEP II might suggest that we are entering a era of high-energy $\gamma\gamma$ collisions. Unfortunately this is not so. The Low formula for production of a resonance of spin J may be written

$$\sigma(e^+e^- \to e^+e^-R) = \eta^2 \frac{(2J+1)8\pi^2\Gamma(R\to\gamma\gamma)}{m_R^3} \times \left[\left(2+\frac{m_R^2}{s}\right)\ln\frac{s}{m_R^2} - 2\left(1-\frac{m_R^2}{s}\right)\left(3+\frac{m_R^2}{s}\right)\right] \tag{12.1}$$

where nominally

$$\eta = (\alpha/2\pi)(\ln s/m_e^2). \tag{12.2}$$

Actually η reflects the available phase space and is cutoff by form factors at a scale nearer m_R^2 than s. Altogether, at fixed m_R, $\sigma(e^+e^- \to e^+e^-R)$ grows approximately as $\ln s/m_e^2$ as s increases. Of course the event rates depend on the luminosity, $\mathcal{L}$, as well as a σ. The luminosities and energies of some e^+e^- machines are shown in Table 12.1. In Fig. 12.1 we show the number of events that would be produced at four machines with an integrated luminosity of 100 pb^{-1} for a $J = 0$ resonance with $\Gamma(R \to \gamma\gamma) =$ 1 keV. Since resonances like η_c have small branching ratios into reconstructable channels, at least 100 or more events are needed. Increasing the energy from that of CESR to that of LEP expands the range of resonances accessible from about 3 GeV to about 6 GeV. To explore a higher mass range a completely different technique is required.

Machine	Energy	L (cm^{-2}s^{-1})
PEP	15 + 15	$7 \cdot 10^{31} - 3 \cdot 10^{32}$
PETRA	22 + 22	†
DORIS	5 + 5	$4 \cdot 10^{31}$
CESR	5 + 5	$10^{32} - 5 \cdot 10^{32}$
TRISTAN	30 + 30	$2 \cdot 10^{31}$
LEP	50 + 50	$6 \cdot 10^{30} - 1 \cdot 10^{32}$

Table 12.1: Luminosities and energies of selected e^+e^- machines. PETRA is no longer operating.

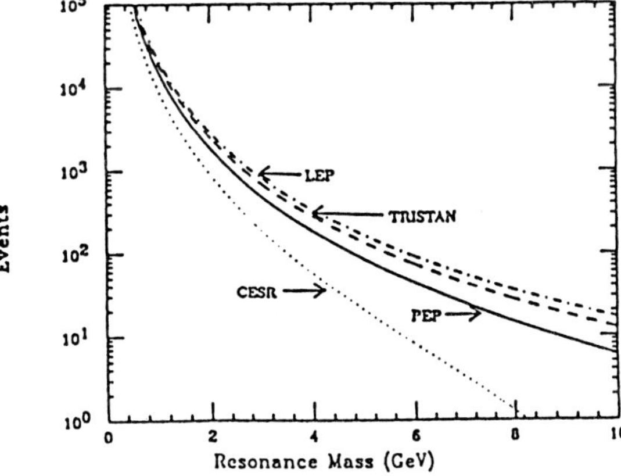

Figure 12.1: The number of events that would be produced with an integrated luminosity of 10^{38} cm^{-2} for a $J = 0$ resonance with $\Gamma(R \to \gamma\gamma) = 1$ keV at four machines.

13 Beamstrahlung

In linear e^+e^- colliders electrons (or positrons) in one bunch are deflected by the field of the other bunch. As a result, the electrons emit synchrotron radiation. Himel and Siegrist [75] estimated the effect by relating it to deflection by a magnetic field whose strength would give the same radius of curvature of the path as in the bunch – bunch collision. They used known results, including quantum corrections, for the deflection in a magnetic field. The problem was taken up by Blankenbecler and Drell [76,77] who did a thorough quantum mechanical calculation, finding results in substantial agreement with Himel and Siegrist. The problem has also been addressed by Jacob and Wu [78,79] and by Bell and Bell [81].

In fact, the general problem of quantum corrections to synchrotron radiation was solved much earlier by Baier and Katkov [82,83]. They showed that the quantum mechanically correct results may be obtained from the classical results by a simple modification of the energy variable, as described below.

To understand the classical result we consider the frame in which one bunch is at rest. The incident particles have momentum $P = m\Gamma = 2m\gamma^2$ in this frame (and momentum $m\gamma$ in the beam-beam cms). The stationary cylindrical bunch has length L, radius B, and uniform charge density. Inside the bunch the electrostatic field is radial with strength

$$|eE| = \frac{2\alpha N}{BL}\frac{b}{B}, \tag{13.1}$$

where N is the number of electrons or positrons in the bunch. The radius of curvature, ρ, of the path of the incident particle is determined by

$$a = \frac{v^2}{\rho} = \frac{eE}{m\Gamma}, \tag{13.2}$$

and the arc subtends an angle

$$\frac{L}{\rho} = \frac{2\alpha N}{mB\Gamma} \equiv \frac{2y}{\Gamma}\frac{b}{B}, \tag{13.3}$$

from the center of curvature, where y is a parameter introduced by Blankenbecler and Drell [76]. A

	SLC	TLC	Super
N	$5 \cdot 10^{10}$	10^{10}	$3 \cdot 10^8$
E	50 GeV	325 GeV	5 TeV
B	10^{-4} cm	$7 \cdot 10^{-6}$ cm	$5 \cdot 10^{-8}$ cm
l_0	10^{-1} cm	$6 \cdot 10^{-}2$ cm	$3 \cdot 10^{-5}$ cm
y	140	400	2000
C	50	1.5	10^{-5}
δ_0	0.014	1.3	$2 \cdot 10^8$

Table 13.1: Parameters for the SLC and two hypothetical e^+e^- colliders [76,77].

second parameter, C, is defined by

$$C = \frac{m^2}{\Gamma(eE)_{max}} = \frac{m^2BL}{2N\alpha\Gamma}. \tag{13.4}$$

Values of these and other parameters for three machines are shown in Table 13.1.

The critical frequency for synchrotron radiation is [84]

$$\omega_c = \frac{3\Gamma^3}{\rho}, \tag{13.5}$$

in terms of which the classical synchrotron radiation spectrum is [84]

$$\frac{dI}{d\omega} = 2\sqrt{3}\alpha\Gamma\frac{\omega}{\omega_c}\int_{2\omega/\omega_c}^{\infty} dx\, K_{5/3}(x). \tag{13.6}$$

To apply this to our problem, two changes are required. First, the result must be multiplied by $(\pi y/2\Gamma)(b/B)$, the fraction the arc makes of a full circle. Second, we must use Eq. (13.6) only up to the maximum energy, ω_q, that the electron could possibly emit:

$$\omega_q = m\Gamma = \frac{CB}{3b}\omega_c. \tag{13.7}$$

If we introduce the dimensionless variable $x = \omega/\omega_q$ and calculate the photon number spectrum, averaging over impact parameter, we find

$$\frac{dN}{dx} = \frac{2\alpha yC}{\sqrt{3}\pi}\xi^2\int_{\xi}^{\infty}\left(\frac{1}{\xi^2} - \frac{1}{z^2}\right)K_{5/3}(z)dz, \tag{13.8}$$

where

$$\xi = 2Cx/3. \tag{13.9}$$

Now the full result of Blankenbecler and Drell for spinless particles [76] is

$$\frac{dN}{dx} = \frac{2\alpha y(1-x)}{x} u^{1/2} \int_u^\infty \left(\frac{3}{2}v - u - \frac{u^4}{2v}\right) Ai(v)dv, \tag{13.10}$$

where Ai is the Airy function and

$$u^3 = [Cx/(1-x)]^2. \tag{13.11}$$

Judicious integrations by parts and use of recursion relations show that this may be written

$$\frac{dN}{dx} = \frac{2\alpha y C}{\sqrt{3}\pi} \xi'^2 \int_{\xi'}^\infty \left(\frac{1}{\xi'^2} - \frac{1}{z^2}\right) K_{5/3}(z)dz. \tag{13.12}$$

where

$$\xi' = \frac{2Cx}{3(1-x)} \tag{13.13}$$

This is in perfect accord with the result of Baier and Katkov, who show that for dN/dx the quantum mechanical result is obtained by replacing x with $x/(1-x)$ [82,83].

For a Dirac particle the result is simply Eq. 13.12 multiplied by $1 + \frac{1}{2}x^2/(1-x)$.

14 Heavy Ions

Heavy ions have been considered recently as a potential source for high-energy $\gamma\gamma$ collisions, possibly using the LHC or SSC. The advantage of heavy ions is that the cross section varies as $Z^4\alpha^4$ rather than just as α^4. Of course there is a cost: the process must be coherent, that is, the nucleus must survive intact.

A straightforward calculation of the flux, including the nuclear form factor, gives [85]

$$\frac{dN}{dx} = \frac{Z^2\alpha}{\pi x} \int_{x^2M^2}^\infty \frac{dQ^2}{Q^2} F(Q^2)^2 \left(1 - \frac{x^2M^2}{Q^2}\right). \tag{14.1}$$

If the source were instead an electron, the form factor would be absent as would the x^2M^2/Q^2 term. Setting the upper limit of integration to s would give $dN/dx \approx (\alpha/\pi x)\ln(s/m_e^2)$, in agreement with the usual Weizsäcker-Williams form

$$\left[\frac{dN}{dx}\right]_{WW} = \left(\frac{\alpha}{\pi x}\ln\frac{s}{4m_e^2}\right)\frac{1}{2}[1+(1-x)^2]. \tag{14.2}$$

Drees et al. approximated

$$F(Q^2)^2 = \exp(-Q^2/Q_0^2), \tag{14.3}$$

with $Q_0 = 55 - 60$ MeV for Pb. For $x << 1$ this gives

$$\left[\frac{dN}{dx}\right]_{DEZ} = \frac{Z^2\alpha}{\pi x}\left(\ln\frac{Q_0^2}{(xM)^2} - 1.577\right). \tag{14.4}$$

In an earlier work Papageorgiu used [86]

$$\left[\frac{dN}{dx}\right]_P = \frac{Z^2\alpha}{\pi x}\ln\frac{1}{(xMR)^2}, \tag{14.5}$$

where R is the nuclear radius, $R \approx 1.2A^{1/3}$f, $1/R \approx 165A^{1/3}$ MeV. To compare these approximations we write

$$\left[\frac{dN}{dx}\right]_{DEZ} = \frac{Z^2\alpha}{\pi x}\left(\ln\frac{1}{(xMR)^2} - 1.577 + \ln Q_0^2R^2\right). \tag{14.6}$$

For Pb, $\ln Q_0^2R^2 = 1.40$, so the DEZ result is just slightly smaller than that of Papageorgiu.

If the nucleus were never disrupted by the collision it would still be necessary to include the electromagnetic form factor. However, here we want to study a rather delicate process, $\gamma\gamma \to X$, in an environment dominated by Pb Pb$\to$ *horrible mess.* This requires that we consider only events in which the nuclei do not physically collide.

Indeed, as regards the nuclei, the process is classical. The equivalent photon approximation should be calculated in impact parameter space, restricting the events to those with impact parameter $b > 2R$. The result is [87]

$$\left[\frac{dN}{dx}\right]_{cl} = \frac{2Z^2\alpha}{\pi x}\left\{\left(\frac{x}{x_0}\right)K_0\left(\frac{x}{x_0}\right)K_1\left(\frac{x}{x_0}\right) - \frac{1}{2}\left(\frac{x}{x_0}\right)^2\left[K_1^2\left(\frac{x}{x_0}\right) - K_0^2\left(\frac{x}{x_0}\right)\right]\right\}. \tag{14.7}$$

where

$$x_0 = \frac{1}{b_{min}M}. \tag{14.8}$$

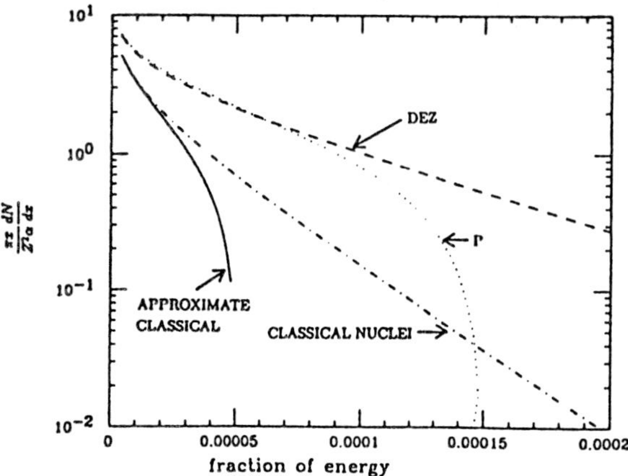

Figure 14.1: Various approximations to the equivalent photon flux from a heavy ion beam. The quantity $(\pi x/Z^2\alpha)(dN/dx)$ is shown as a function of x, the fraction of the ion's momentum given to the photon. The result of Drees et al. [85], Eq. (14.1), is shown as the dashed curve. The result of Papageorgiu, [86], Eq. (14.5), is shown as a dotted curve. The full classical result, Eq. (14.7), is shown as a dot-dashed curve. The low-momentum approximation to the full classical result, Eq. (14.9), is shown as a solid curve.

For very small values of x this becomes

$$\left[\frac{dN}{dx}\right] = \frac{Z^2\alpha}{\pi x}\left[\ln\frac{1}{(xMR)^2} - 2.15\right], \qquad (14.9)$$

a result that is substantially below those of Drees et al. and Papageorgiu. Moreover, as seen in Fig. 14.1, this difference becomes greater as x increases.

15 Linacs and Backscattering

High-energy photon beams have regularly been obtained by scattering laser light from linac beams. Recently H. Sens has promoted this as a technique for $\gamma\gamma$ scattering [88]. The potential is impressive.

For a 3 eV photon colliding with a 50 GeV electron beam, the c.m. energy of the Compton scattering is $\sqrt{s} = \sqrt{m_e^2 + 4E_eE_\gamma} = 0.86$ GeV. With a sufficiently intense laser beam, every electron gets scattered by a photon, so the photon flux is essentially equal to that of the initial electron flux. The

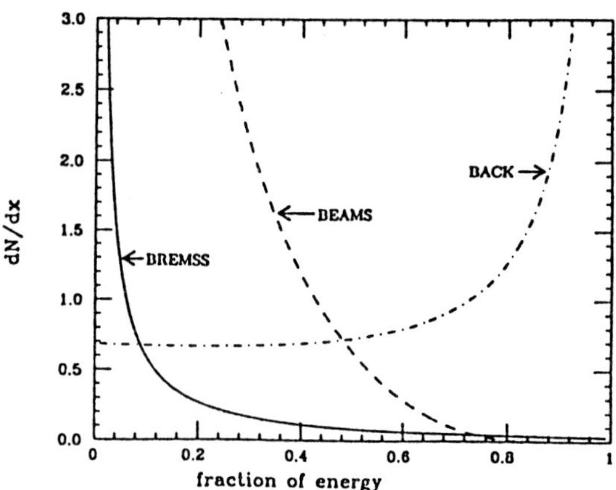

Figure 15.1: Equivalent photon spectra that could be obtained from an electron beam with energy 325 GeV. The solid curve shows the Weizsäcker-Williams spectrum, Eq.(14.2). The beamstrahlung spectrum, the dashed curve, is shown for $C = 1.5, y = 400$ [76,77]. The spectrum from the backscattered laser beam, Eq. (15.1), is shown as the dot-dashed curve for the same electron energy and photon energy $k = 2$ eV.

energy distribution of the photons is determined by the differential cross section for Compton scattering boosted to the appropriate frame. It is easy to show that

$$\frac{dN}{dx} = \frac{1 - x + \frac{1}{1-x} - \frac{4x}{y(1-x)} + \frac{4x^2}{y^2(1-x)^2}}{(1-4/y+8/y^2)\ln(1+y)+1/2+8/y-(1+y)^{-2}/2} \qquad (15.1)$$

where

$$x = k'/E, \quad y = 4Ek/m_e^2, \qquad (15.2)$$

and $E\,(E')$ is the incident (final) electron energy and $k\,(k')$ is the incident (final) photon energy. The collision is assumed to be head-on. This spectrum is compared to that from beamstrahlung and the usual Weizsäcker-Williams spectrum in Fig. 15.1.

16 *WW* Collisions

The Lepton-Photon Conference is by now the Electroweak Conference, and so it makes sense to consider the electroweak variants of $\gamma\gamma$ scattering: WW and ZZ scattering. In 1983 it was realized that this

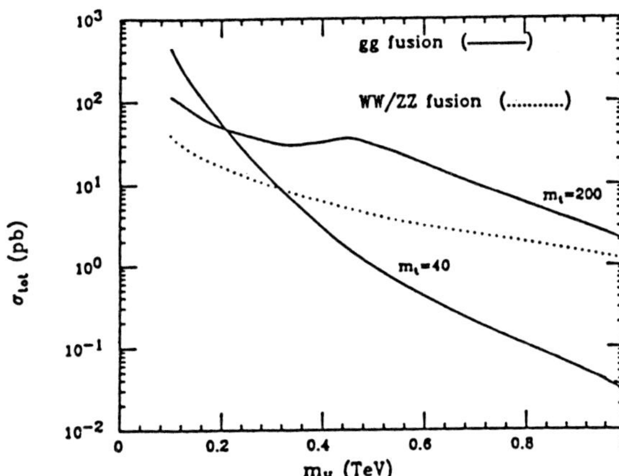

Figure 16.1: The Higgs boson production cross section in pp collisions at $\sqrt{s} = 40$ TeV as a function of the Higgs boson mass. The dotted curve shows the contribution from WW and ZZ fusion. The solid curves are for gluon-gluon fusion for two values of the t quark mass. Ref. [93]

process would dominate the production of an orthodox Higgs boson at the SSC if the mass of the Higgs boson were about 300 GeV or more [89]. Shortly thereafter the equivalent flux of Ws and Zs from a high-momentum quark was calculated [90,91,92], with the result for the longitudinal and transverse bosons

$$\begin{aligned}\frac{dN_L}{dx} &= \frac{g_V^2 + g_A^2}{8\pi^2 x}\ln\frac{s}{m_W^2}[1+(1-x)^2]\\ \frac{dN_T}{dx} &= \frac{g_V^2 + g_A^2}{8\pi^2 x}2(1-x).\end{aligned} \tag{16.1}$$

Here, for WW collisions, $g_V = -g_A = e/(\sin\theta_W 2\sqrt{2})$. It is the longitudinal Ws and Zs that dominate [89]. In the equivalent W approximation $\sigma(ud \to duH)$ via the WW intermediate state is given by [90],

$$\sigma = \frac{1}{16m_W^2}\left(\frac{\alpha}{\sin^2\theta_W}\right)^3 [(1+\tau)\ln(1/\tau) - 2 + 2\tau], \tag{16.2}$$

where $\tau = m_H^2/s$. This competes with another $\gamma\gamma$ analogue, gluon-gluon fusion. The cross section for Higgs boson production in pp scattering at $\sqrt{s} = 40$ TeV is shown in Fig. 16.1.

17 Higgs Boson Production

The best hope for the discovery of a very massive Higgs boson, with $m_H > 200$ GeV, is a very high energy hadron collider like the SSC. The decay $H \to ZZ$ followed by decays of the Z into electrons or muons provides the best signature. However, if the Higgs boson has a mass less than twice the mass of the Z, the task is complicated [94]. If the mass is greater than about 125 GeV it is possible to look for the decay into one real Z and one virtual Z in the charged leptonic channels. Finding a 100 GeV Higgs boson would certainly be difficult at the SSC, since its primary decay would be to $b\bar{b}$, for which the QCD-generated background would be stupendous. This provides motivation for considering the production of a 100 GeV Higgs boson at an electron-positron collider.

The same WW fusion mechanism, $e^+e^- \to \nu\bar{\nu}W^+W^- \to \nu\bar{\nu}H$ would be available at an e^+e^- collider [95]. Using Eq. (16.2), for $\sqrt{s} = 2E_{beam} = 1$ TeV, the cross section is about $3 \cdot 10^{-37}\text{cm}^2$. Despite the absence of a purely hadronic background, there remain some serious problems, especially if the mass of the Higgs boson is near that of the W. A very substantial integrated luminosity is required, perhaps 30 fb^{-1}.

An alternative is $\gamma\gamma$ collisions. The observed cross section is related to the $\gamma\gamma$ luminosity relative to the e^+e^- luminosity,

$$\mathcal{L}_{rel} = \mathcal{L}_{\gamma\gamma}/\mathcal{L}_{e^+e^-}, \tag{17.1}$$

by

$$\sigma_{Higgs\ via\ \gamma\gamma} = \left[\frac{8\pi^2}{m_H^3}\Gamma(H\to\gamma\gamma)\right]\tau\frac{d\mathcal{L}_{rel}}{d\tau} \tag{17.2}$$

Since the $\gamma\gamma$ width of a 100 GeV Higgs boson is a few keV, the factor in square brackets is a few times 30 fb. Since $\tau\frac{d\mathcal{L}_{rel}}{d\tau}$ is of order 1, the resulting cross section is roughly equivalent to that of $e^+e^- \to \nu\bar{\nu}W^+W^- \to \nu\bar{\nu}H$. Fig. 15.1 shows that both beamstrahlung and backscattering produce photon fluxes that are substantial for momenta that are a good fraction of the beam energy. With electron-positron luminosities in the range $10^{34}\,\mathrm{cm}^{-2}\,\mathrm{s}^{-1}$ these techniques could be capable of producing a large number of 100 GeV Higgs bosons.

Drees et al. estimated that lead-lead collisions in the SSC with a luminosity of a few times $10^{26}\,\mathrm{cm}^{-2}\,\mathrm{s}^{-1}$ would be adequate for the purpose. This value may need to be revised upward in view of the reductions displayed in Fig. 14.1.

18 Summary and Prospects

Two-photon physics continues to challenge us both theoretically and experimentally. While the pseudoscalar and tensor multiplets are quite well understood, there are significant puzzles among the scalars and axial vectors. New, high statistics experiments could provide results that would resolve important ambiguities. With the increasing reliability of lattice calculations of the QCD spectrum, there will be more and more interest in understanding both the $q\bar{q}$ and non-$q\bar{q}$ mesons, and two-photon physics can provide unique insights into these particles.

Understanding the dynamics of nonresonant hadronic final states is especially difficult. A profusion of data on the vector-vector final state has demonstrated once again how hard this is. The brave application of perturbative QCD to exclusive final states has achieved mixed results, but these may improve in time. Although the data for the structure function of the photon continue to grow, the time of precision measurements has not yet arrived.

In the short term, the frontier in two-photon physics is the frontier of integrated luminosity. The differences between the energies of the various operating e^+e^- machines have only a small significance. Only accumulated events count. The outstanding work done by many of the collaborations on the $f_1(1425)$ testifies to the potential for two-photon physics. While there may be more immediate interest in B or Z physics, there is work of lasting value to be done with $\gamma\gamma$ collisions.

In the long term, the high-energy frontier can be extended in $\gamma\gamma$ collisions only by the introduction of novel accelerator techniques: beamstrahlung, backscattered laser beams, and heavy ions. It is too soon to know how practical these will prove to be, but it is not too soon to start thinking about them.

References

[1] J. Olsson, *Proc. XIII International Symposium on Lepton and Photon Interactions, Nucl. Phys.* B (Proc. Suppl.) 3, 613 (1988).

[2] H. Kolanoski and P. Zerwas, "Two-Photon Physics" in *High Energy Electron-Positron Physics*, Eds. A. Ali and P. Söding, World Scientific, 1988.

[3] S. Cooper, *Ann. Rev. Nucl. Part. Sci.* **38**, 705 (1988).

[4] *Proceedings of the VIIIth International Workshop on Photon-Photon Collision*, Shoresh, Israel, April 24 - 28, 1988. Ed. U. Karshon World Scientific, 1988.

[5] F. Low, *Phys. Rev.* **120**, 582 (1960).

[6] N. A. Roe et al, "Measurement of the Two-photon Width of the η and η'," SLAC-PUB 4931, accepted for publication in *Phys. Rev.* D.

[7] F. Butler, "Resonant Production in Two Photon Collisions," LBL-26465. The results here must be corrected by a factor ≈ 1.15, see J. H. Boyer in [17].

[8] J. K. Bienlein, "Observed and Unobserved States - Crystal Ball Results on Two-Photon Physics," in *Glueballs, Hybrids, and Exotic Hadrons*, A. I. P. Conference Proceedings No. 185, Ed. S.-U. Chung, p. 486.

[9] See, for example, F. J. Gilman and R. Kauffman, *Phys. Rev.* **D36**, 2761 (1987).

[10] J. F. Donoghue, B. R. Holstein, and Y-C. R. Lin, *Phys. Rev. Lett.* **55**, 2766 (1985).

[11] J. Gasser and H. Leutwyler, *Nucl. Phys.* **B250**, 465 (1985).

[12] S. J. Brodsky and G. P. Lepage, *Phys. Rev.* **D24**, 1808 (1981).

[13] TPC/2γ Collaboration., "Investigation of the Electromagnetic Structure of η and η' Mesons by Two-Photon Interactions, submitted to this conference (paper # 290).

[14] Crystal Ball Collaboration, H. Marsiske, *Proceedings of the VIII th International Workshop on Photon-Photon Collisions*, Shoresh, Israel, April 24 -28, 1988, Ed. U. Karshon, p. 15.

[15] JADE Collaboration, J. Olsson, *Proceedings of the VIII th International Workshop on Photon-Photon Collisions*, Shoresh, Israel, April 24 -28, 1988, Ed. U. Karshon, p. 77.

[16] CELLO Collaboration, J. Harjes, *Proceedings of the VIII th International Workshop on Photon-Photon Collisions*, Shoresh, Israel, April 24 -28, 1988, Ed. U. Karshon, p. 85.

[17] Mark II Collaboration, J. Boyer, *Proceedings of the VIII th International Workshop on Photon-Photon Collisions*, Shoresh, Israel, April 24 - 28, 1988, Ed. U. Karshon, p. 94; J. H. Boyer, "Two Photon Production of Pion Pairs in e^+e^- Collisions at 29 GeV," LBL-27180.

[18] TOPAZ Collaboration, I. Adachi et al., submitted to this conference (paper # 146).

[19] PLUTO Collaboration, M. Feindt, *Proceedings of the VIII th International Workshop on Photon-Photon Collisions*, Shoresh, Israel, April 24 -28, 1988, Ed. U. Karshon, p. 3.

[20] Crystal Ball Collaboration, D. Antreasyan et al., *Phys. Rev.* **33**, 1847 (1986).

[21] Crystal Ball Collaboration, B. Muryn, *Proceedings of the VIII th International Workshop on Photon-Photon Collisions*, Shoresh, Israel, April 24 -28, 1988, Ed. U. Karshon, p. 102.

[22] CELLO Collaboration, H. -J. Behrend et al., submitted to this conference (paper # 229).

[23] H. J. Lipkin, *Proceedings of the VIII th International Workshop on Photon-Photon Collisions*, Shoresh, Israel, April 24 -28, 1988, Ed. U. Karshon, p. 107.

[24] TASSO Collaboration, W. Braunschweig et al., *Z. Phys.* **41**, 533 (1989).

[25] TPC/2γ Collaboration, H. Aihara et al., *Phys. Rev. Lett.* **60**, 2355 (1988).

[26] G. Gidal et al., Proceedings of the XXII International conference on High Energy Physics, Berkeley, CA (1986), World Scientific, Vol. II, p. 1220

[27] CLEO Collaboration, T. Jensen et al., reported at Cornell Meeting on Heavy Quark Physics, June, 1989.

[28] C. Baglin et al., *Phys. Lett.* **187B**, 191 (1987).

[29] M. S. Chanowitz, "Resonances in Photon-Photon Scattering," *Proceedings of the VIII th International Workshop on Photon-Photon Collisions*, Shoresh, Israel, April 24 -28, 1988, Ed. U. Karshon, p. 205.

[30] R. L. Jaffe, *Phys. Rev.* **D15**, 267 (1977); R. L. Jaffe and K. Johnson, *Phys. Lett.* **60B**, 201 (1976).

[31] J. Weinstein and N. Isgur, *Phys. Rev. Lett.* **48**, 659 (1982); *Phys. Rev.* **D27**, 588 (1983).

[32] R. L. Goble, R. Rosenfeld, and J. L. Rosner, *Phys. Rev.* **D39**, 3264 (1989) and references therein.

[33] J. F. Donoghue, B. R. Holstein, and Y. C. Lin, *Phys. Rev.* **D37**, 2423 (1987).

[34] J. Bijnens and F. Cornet, *Nucl. Phys.* **B296**, 557 (1988).

[35] M. R. Pennington, *Proceedings of the VIII th International Workshop on Photon-Photon Collisions*, Shoresh, Israel, April 24 -28, 1988, Ed. U. Karshon, p. 297.

[36] M. S. Chanowitz, Proc. VI International Workshop of Photon-Photon Interactions, Ed. R. Lander, World Scientific, 1984, p.95.

[37] M. Feindt,"1988 CELLO, JADE, and PLUTO - Contributions to 'Exotic' Meson Spectroscopy," in *Glueballs, Hybrids, and Exotic Hadrons*, A. I. P. Conference Proceedings No. 185, Ed. S.-U. Chung, p. 501.

[38] TPC/2γ Collaboration, H. Aihara et al., *Phys. Rev. Lett.* **57**, 51 (1986).

[39] M. Feindt, "Experimentelle Untersuchung der Reaktionen $\gamma\gamma \to K_S^0 K^\pm \pi^\mp$ und $\gamma\gamma \to K_S^0 K_S^0$ mit dem Detektor PLUTO," DESY F14-88-02.

[40] PLUTO Collaboration, Ch. Berger et al., *Z. Phys.* **37**, 329 (1988).

[41] CELLO Collaboration, H.-J. Behrend et al., *Z. Phys.* **42**, 367 (1989).

[42] CELLO Collaboration, H.-J. Behrend et al., *Z. Phys.* **43**, 91 (1989).

[43] Mark II Collaboration, G. Gidal et al., *Phys. Rev. Lett.* **59**, 2016 (1987).

[44] F. M. Renard, *Nuovo Cimento* **80A**, 1 (1984).

[45] R. N. Cahn, "Twos in Two-Photon Physics," *Proceedings of the VIII th International Workshop on Photon-Photon Collisions*, Shoresh, Israel, April 24 -28, 1988, Ed. U. Karshon, p.110.

[46] M. S. Chanowitz, *Phys. Lett.* **B187**, 409 (1987).

[47] R. N. Cahn, *Phys. Rev.* **D35**, 3342 (1987).

[48] See the reviews by G. Gidal "Radiative Widths of Resonances (Experiments)," in *Proceedings of the VIII th International Workshop on Photon-Photon Collisions*, Shoresh, Israel, April 24 -28, 1988, Ed. U. Karshon, p. 182, and "Resonance Formation in Photon-Photon Collisions," in *Glueballs, Hybrids, and Exotic Hadrons*, A. I. P. Conference Proceedings No. 185, Ed. S.-U. Chung, p. 171.

[49] W. Bartel et al., *Phys. Lett.* **184B**, 288 (1987); A. J. Finch, "Inclusive D^* Production in Single Tagged $\gamma\gamma$ Events," in *Proceedings of the VIII th International Workshop on Photon-Photon Collisions*, Shoresh, Israel, April 24 -28, 1988, Ed. U. Karshon, p. 75.

[50] TPC/2γ Collaboration, F. C. Erné, "Exclusive and Inclusive Charm Production in Photon-Photon Collisions," in *Proceedings of the VIII th International Workshop on Photon-Photon Collisions*, Shoresh, Israel, April 24 -28, 1988, Ed. U. Karshon, p. 71.

[51] TASSO Collaboration, W. Braunschweig et al., "Study of Charmed Meson Production in $\gamma\gamma$ Interactions," submitted to this conference (paper # 232).

[52] G. Alexander, U. Maor, and P. G. Williams, *Phys. Rev.* **D26**, 1198 (1982); G. Alexander, A. Levy, and U. Maor, *Z. Phys.* **C30**, 65 (1986).

[53] N. N. Achasov, S. A Deryanin, and G. N. Shestakov, *Phys. Lett.* **B108**, 134 (1982); *Z. Phys.* **C16**, 55 (1982); ibid. **C27**, 99 (1985).

[54] B. A. Li and K. F. Liu, *Phys. Lett.* **B118**, 435 (1982); ibid. **B124**, 55 (1982); *Phys. Rev. Lett.* **51**, 1510 (1983); *Phys. Rev.* **D30**, 613 (1984).

[55] S. J. Brodsky, G. Köpp, and P. M. Zerwas, *Phys. Rev. Lett.* **58**, 443 (1987).

[56] A. W. Nilsson, "Exclusive Final States - Continuum (Experimental)" in *Proceedings of the VIII th International Workshop on Photon-Photon Collisions*, Shoresh, Israel, April 24 -28, 1988, Ed. U. Karshon, p. 261.

[57] U. Maor, "Vector Meson Production in $\gamma\gamma$ Reactions," in *Proceedings of the VIII th International Workshop on Photon-Photon Collisions*, Shoresh, Israel, April 24 -28, 1988, Ed. U. Karshon, p. 282.

[58] N. N. Achason and G. N. Shestakov, *Phys. Lett.* **203B**, 309 (1988).

[59] B. A. Li and K.-F. Liu, "$K^*\overline{K}^*$ Mesonium Production in $\gamma\gamma$ Reactions and Hadronic Collisions," University of Kentucky preprint, UK/89-02, submitted to this conference (paper # 9).

[60] ARGUS Collaboration, "Two-Photon Production of Final States with a $p\overline{p}$ Pair," DESY 88-194.

[61] TPC/2γ Collaboration, H. Aihara et al., "Exclusive Production of $p\overline{p}\pi^+\pi^-$ in Photon Photon Collisions," NIKHEF-H/89-9.

[62] G. R. Farrar, E. Maina, and F. Neri, *Nucl. Phys.* **B259**, 702 (1985).

[63] G. R. Farrar, et al., "Perturbative QCD Predictions for $\gamma\gamma \rightarrow B\overline{B}$, *Proceedings of the VIII th International Workshop on Photon-Photon Collisions*, Shoresh, Israel, April 24 -28, 1988, Ed. U. Karshon, p. 43.

[64] E. Witten, *Nucl. Phys.* **B120**, 189 (1977).

[65] Ch. Berger and W. Wagner, *Phys. Rep. C* **146**, 1 (1987).

[66] J. H. Field, F. Kapusta, and L. Poggioli, *Phys. Lett.* **181B**, 362 (1986); J. H. Field "Structure Functions and Total Cross Sections," in *Proceedings of the VIII th International Workshop on Photon-Photon Collisions*, Shoresh, Israel, April 24 -28, 1988, Ed. U. Karshon, p. 349.

[67] W. Frazer, *Phys. Lett.* **B194**, 287 (1987).

[68] W. Frazer and G. Rossi, *Phys. Rev.* **D25**, 843 (1982).

[69] TPC/2γ Collaboration, H. Aihara et al., "A Measurement of the Total Hadronic Cross Section in Tagged $\gamma\gamma$ Reactions," submitted to this conference (paper # 262).

[70] AMY Collaboration, "A Measurement of the Photon Structure Function F_2 at an Average Q^2 of 67 $(\mathrm{GeV}/c)^2$," submitted to this conference (paper # 107).

[71] J. J. Sakurai and D. Schildknecht, *Phys. Lett.* **40B**, 121 (1972).

[72] Ch. Berger et al., *Phys. Lett.* **149B**, 421 (1984).

[73] D. Bintinger et al., *Phys. Rev. Lett.* **54**, 763 (1985).

[74] M. Feindt, "Recent PLUTO Results on Photon Photon Reactions," in *VIIth International Workshop on Photon-Photon Collisions,* Ed. A. Courau and P. Kessler, World Scientific, 1986, p. 388.

[75] T. Himel and J. Siegrist, "Quantum Effects in Linear Collider Scaling Laws," SLAC-PUB-3572, 2nd International Workshop on Laser Acceleration of Particles, Los Angeles, Jan. 7-18, 1985.

[76] R. Blankenbecler and S. D. Drell, *Phys. Rev.* **D36**, 277 (1987).

[77] R. Blankenbecler and S. D. Drell, *Phys. Rev. Lett.* **61**, 2324 (1988), Erratum *ibid.* **62**, 116 (1989).

[78] M. Jacob and T. T. Wu, *Phys. Lett.* **197B**, 253 (1987).

[79] M. Jacob and T. T. Wu, *Nucl. Phys.* **B303**, 373 (1988).

[80] M. Jacob and T. T. Wu, *Nucl. Phys.* **B303**, 389 (1988).

[81] M. Bell and J. S. Bell, CERN-TH 5174/88.

[82] V. N. Baier and V. M. Katkov, *Phys. Lett.* **25A**, 492 (1967).

[83] V. N. Baier and V. M. Katkov, *Sov. Phys. JETP* **26**, 854 (1968).

[84] J. D. Jackson, *Classical Electrodynamics*, 2nd Edition, Wiley, 1975, p.677.

[85] M. Dress, J. Ellis, and D. Zeppenfeld, *Phys. Lett.* **B223**, 454 (1989).

[86] E. Papageorgiu, *Phys. Rev.* **D40**, 92 (1989).

[87] J. D. Jackson, *Classical Electrodynamics*, 1st Edition, Wiley, 1962, p. 523.

[88] H. Sens, "Critical Issues and Prospects for Future Work," in *Proceedings of the VIII th International Workshop on Photon-Photon Collisions*, Shoresh, Israel, April 24 -28, 1988, Ed. U. Karshon, p.143.

[89] R. N. Cahn and S. Dawson, *Phys. Lett.* **136B**, 196 (1984).

[90] M. S. Chanowitz and M. K. Gaillard, *Phys. Lett.* **142B**, 85 (1984); *Nucl. Phys.* **B261**, 379 (1985).

[91] S. Dawson, *Nucl. Phys.* **B249**, 42 (1985).

[92] G. L. Kane, W. W. Repko, and W. B. Rollnick, *Phys. Lett.* **148B**, 367 (1984).

[93] J. F. Gunion, G. L. Kane, and J. Wudka, *Nucl. Phys.* **299**, 231 (1987).

[94] See, for example, R. N. Cahn, *Rep. Prog. Phys.* **52**, 389 (1989).

[95] See, for example, C. Ahn et al., "Opportunities and Requirements for experimentation at High Energy e^+e^- Collider," SLAC-329, 1988.

DISCUSSION

S. B. Gerasimov, JINR, Dubna: I would like to make a short comment and then ask a question. First, one cannot agree that the spectroscopy of light mesons (including the 0^{-+} states) as seen in the $\gamma\gamma$ reaction is in good shape until at least the first radially excited state is found. There are reasons to expect, for example, the presence of the first radial excitations of the η - η' states around the $\iota(1430)$. These states may have a small $\gamma\gamma$ width and mix with a possible glueball state, forming the $\iota(1430)$ and other 0^{-+} states.

Now the question. What is the status of and perspective for finding the missing states and, in particular, a possibly more complex configuration structure of the $\iota(1430)$.

R. Cahn: Certainly it is important to pursue meson states outside the standard multiplets. We ought to find radially excited states, and without understanding them it will be all the more difficult to identify glueballs convincingly. The Crystal Ball has set a limit

$$\Gamma(\eta(1280,1390) \to \gamma\gamma)B(\eta\pi\pi) < 0.3 \text{ keV}$$

at 90% C.L., which applies to states below 1500 MeV with widths less than 50 MeV. The anticipated $\gamma\gamma$ widths for radially-excited states would be significantly greater, so this is a bit of a puzzle.

DOUBLE BETA-DECAY EXPERIMENTS

I. V. KIRPICHNIKOV
Institute of Theoretical and Experimental Physics,
117259, Moscow, USSR

ABSTRACT

A brief review of the new results in double beta-decay experiments is given. The observation of double neutrino double beta-decay of ^{76}Ge was claimed by the ITEP/YePI collaboration. The UCI group published an improved value for double neutrino-decay lifetime of ^{82}Se. More stringent constraints on theory were provided by the UCSB/LBL and ITEP/YePI measurements of lifetime limits for ^{76}Ge neutrinoless modes of double beta-decay.

INTRODUCTION

There are three principally different modes of double beta-decay, searches for which are being intensively performed by experimentalists:

$$(A,Z) \to (A,Z+2) + 2e^- \quad , \quad (0\nu)\text{-decay}$$
$$(A,Z) \to (A,Z+2) + 2e^- + B \quad , \quad (0\nu,B)\text{-decay}$$
$$(A,Z) \to (A,Z+2) + 2e^- + 2\nu \quad , \quad (2\nu)\text{-decay}$$

The (0ν)-decay violates the law of lepton number conservation and requires the existence of massive Majorana neutrinos (maybe together with right-handed currents). The $(0\nu,B)$-decay is possible if a hypothetical scalar boson, the Majoron, exists.

The (2ν)-decay occurs in second order in the Standard Model of weak interactions. It is not very informative because even its usefulness as a object for test calculations is questionable. Nevertheless, the search for this process has one essential advantage compared with the neutrinoless modes, which makes it very attractive for experimentalists. It seems likely that the (2ν)-decay is the only kind of double beta-decay that really exists.

Table 1 illustrates the general situation. It includes the most interesting results for all three double beta-decay modes.[1–10]

There are several essential improvements of the data compared to the previous conference. Two groups have claimed the laboratory observation of (2ν)-decay. The UCI group has improved background conditions and published a new value for the ^{82}Se lifetime[6]:

$$T_{1/2} \;=\; (1.1^{+0.8}_{-0.3}) \cdot 10^{20} \text{ years} \quad .$$

Now the result is in good agreement with the geochemical data[11,12]:

Table 1: Double beta-decay experiments.

		Lifetime limits, years (68% C.L.)		
		(0ν)	$(0\nu,B)$	(2ν)
^{76}Ge	UCSB/LBL$^{(a)}$	$1.1 \cdot 10^{24}$	$1.4 \cdot 10^{21}$	$8.0 \cdot 10^{19}$
	ITEP/YePI	$1.3 \cdot 10^{24}$	$7.5 \cdot 10^{21}$	$(9^{+3}_{-2}) \cdot 10^{20}$
	Cal/Neu/ SIN	$2.7 \cdot 10^{23}$	$1.7 \cdot 10^{21}$	$1.2 \cdot 10^{20}$
^{48}Ca	Columbia	$2.0 \cdot 10^{21}$	$8.6 \cdot 10^{20}$	$3.6 \cdot 10^{20}$
	Beijing	$3.6 \cdot 10^{21}$		
^{82}Se	UCI	$1.1 \cdot 10^{22}$	$7.3 \cdot 10^{20}$	$(11^{+8}_{-3}) \cdot 10^{19}$
^{100}Mo	LBL/Hol/ UNM	$1.3 \cdot 10^{22}$	$3.3 \cdot 10^{20}$	$3.8 \cdot 10^{19}$
	Baksan	$4.4 \cdot 10^{20}$	$2.8 \cdot 10^{19}$	$6.2 \cdot 10^{18}$
^{136}Xe	Milano	$9 \cdot 10^{21}$	$6 \cdot 10^{20}$	
	Baksan	$3.3 \cdot 10^{21}$	$1.9 \cdot 10^{20}$	$7.8 \cdot 10^{19}$

$^{(a)}$ Limit for (0ν)-decay from the background fluctuations is given. The group also used Bayesian statistics and found $T_{1/2} > 2.3 \cdot 10^{24} y$.

$$T_{1/2} \;=\; (1.45 \pm 0.05) \cdot 10^{20} \text{ years} \quad ,$$
$$T_{1/2} \;=\; (1.2 \pm 0.1) \cdot 10^{20} \text{ years} \quad .$$

The ITEP/YePI collaboration performed the experiment with a germanium detector fabricated of enriched material (85% abundance of ^{76}Ge compared to 7.8% in natural germanium[2]). The same volume detector fabricated of natural Ge was simultaneously used for background measurements. The group estimated the lifetime value for (2ν)-decay of ^{76}Ge as:

$$T_{1/2} \;=\; (0.9^{+0.3}_{-0.2}) \cdot 10^{21} \text{ years} \quad .$$

New results were published by the Milano group for ^{136}Xe and by the Baksan laboratory[10] for ^{100}Mo and ^{136}Xe.[9] The Beijing group improved the lifetime limit[5] for ^{48}Ca, which had not changed for almost 20 years. Progress in experimental methods is evident, and it may be illustrated by presenting the latter results, which were submitted to the Conference.

The Beijing group used two large CaF_2 crystals, 10 kg each, in a scintillation spectrometer. A rather large quantity of ^{48}Ca (a total of 20 g) and low background provided more than a factor of 2 improvement in the lifetime limit compared to the previous best data.[4] These measurements are still in progress; the group intends to add two more identical crystals and get even better results.

All of the above data illustrate the possibilities of the experiments, but no single one of them can compete with the germanium results. The germanium data for the (0ν)-decay lifetime limit provide the most stringent constraints on theories. Semiconductor germanium detectors have unique features that make them almost ideal devices for the search for (0ν)-decay. Excellent energy resolution, better than 0.15%, is combined with almost 100% detection efficiency. Since the source is dispersed throughout the detector, the quantity of material is limited only by the volume of available crystals. There is no problem with detector intrinsic radioactivity because the detectors are fabricated from highly purified materials (zone-refined, admixtures of heavy metals being less than $10^{-14} - 10^{-15}$ g/g). Now two groups claim lifetime limits close to $T_{1/2} = 10^{24}$ years. It means the upper limit for Majorana neutrino mass $< m_\nu >$ is less than several electron volts.

The best value for ^{76}Ge $(0\nu, B)$-decay lifetime value $(T_{1/2} > 7.5 \cdot 10^{21} y)$ was given by the ITEP/YePI group. Together with the ^{82}Se data, $T_{1/2} > 7.3 \cdot 10^{20} y$, it provides the best limit for the Majoron-to-neutrino coupling constant.

Below, I discuss more carefully some of the most recent informative results.

NEUTRINOLESS DOUBLE BETA-DECAY OF ^{76}Ge

Two groups claimed lifetime limits for (0ν)-decay of ^{76}Ge higher than 10^{24} years. These were the UCSB/LBL and the ITEP/YePI collaborations. Since all the other results are much lower, we limit the discussion to the presentation of the above groups' data.

The UCSB/LBL collaboration started in 1985. They use an 8-crystal detector inside a NaI(T1) active shield. The total quantity of Ge is 7.1 kg. Now

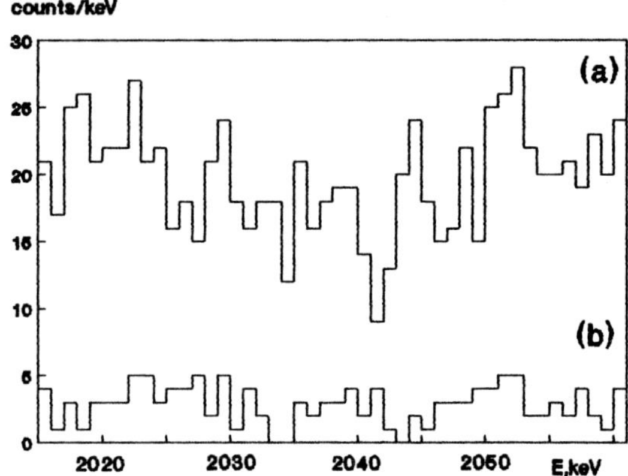

Figure 1: (a) The UCSB/LBL results in the vicinity of the expected (0ν)-decay "spike" (2041 keV) for live time 16.6 y-kg (17.1 $y \cdot$ mole ^{76}Ge). (b) The ITEP/YePI results for 9.5 $y \cdot$ mole.

they have collected the largest live time, and the experiment is still in progress.

Results submitted to this Symposium include the data collected over a 21 kg-year period or 21.5 y-mole (^{76}Ge), and the reported lifetime limit is the $T_{1/2} > 2.3 \cdot 10^{24} y$ (68% C.L.). The data (Fig. 1a) illustrate the common problem for all such experiments: the small expected effect compared to a large background. The number of expected events due to (0ν)-decay is about 4 counts if $T_{1/2} = 2.3 \cdot 10^{24} y$.

Since the mean background is about 20 counts per keV, and the real energy resolution close to 4 keV (FWHM), the effect should be observed over a background of 80 counts. The UCSB/LBL group used special (Bayesian) statistics to achieve better limits. A more conservative estimate from background fluctuations yields

$$T_{1/2} > 1.1 \cdot 10^{24} \text{ years (68\% C.L.)} \quad .$$

The ITEP/YePI group made the next step in experimental techniques. This work has been in progress since 1987. A three-detector system was used, two of the crystals, 106 cm^3 and 109 cm^3, being fabricated of enriched germanium (85% of ^{76}Ge) and one, 115 cm^3, of natural (7.8% of ^{76}Ge). All three crystals were mounted in a common cryostat inside a sodium iodide active shield and the usual (lead, borated polyethylene) passive shield. The apparatus was designed and constructed in ITEP and then moved to the Low Background Laboratory of YePI, Avan salt mine, Yerevan, 245 m underground.

The summed data collected with the enriched crystals for 9.5 $y \cdot$ mole (^{76}Ge) are presented in

Table 2: Comparison of UCSB/LBL and ITEP/YePI detectors.

	Total Weight (g)	Quant. of ^{76}Ge (g)	Bckgrd. cts/mole of ^{76}Ge	Resoln. keV
UCSB	7100	554	1.15	3.2
ITEP/YePI	996	846	0.24	4.4

Fig. 1b. Since the expected effect is proportional to the quantity of ^{76}Ge, and the background to the total volume, this detector has a much better signal-to-noise ratio compared to other devices.

The lifetime value estimated from the mean background fluctuations is equal to

$$T_{1/2} > 1.3 \cdot 10^{24} \text{ years (68\% C.L.)} \quad .$$

The main parameters of the UCSB/LBL and ITEP/YePI detector are given in Table 2.

MAJORON-INDUCED DOUBLE BETA-DECAY

Interest in this mode of double beta-decay drastically increased after the PNL/USC claim in 1987 for the observation of $(0\nu, B)$-decay of ^{76}Ge with $T_{1/2} = (6 \pm 1) \cdot 10^{20}\ y$. Now several groups have given lifetime limits for this decay of Ge that are more than several times $10^{21}\ y$. The possibility of comparing the spectra from enriched and natural germanium crystals has proved to be very useful, and due to this factor, the ITEP/YePI collaboration had the best value of the lifetime limit.

The principal factor limiting the accuracy of the method used by the ITEP/YePI group was the identity of background conditions for all three detectors. Careful analysis of the possible sources of background was performed, including beta-activity due to U and Th mixtures and beta-decays of ^{77}Ge and ^{68}Ga (both the latter activities would be due to cosmogenic processes).

An example of the spectrum measured with an enriched detector is shown in Fig. 2. The level of the beta-activity background was estimated directly from the collected spectra by analysis of the data at high energies (above 2 MeV). A special run with an expanded (up to 8 MeV) energy scale was performed to check on admixtures of U and Th in the crystals (by counting 4–5 MeV alpha-particles).

The ITEP/YePI group concluded that most of the background could be attributed to gamma radiation of nuclei, members of the U and Th series, which were contained in the surrounding constructive materials. Such background events could be

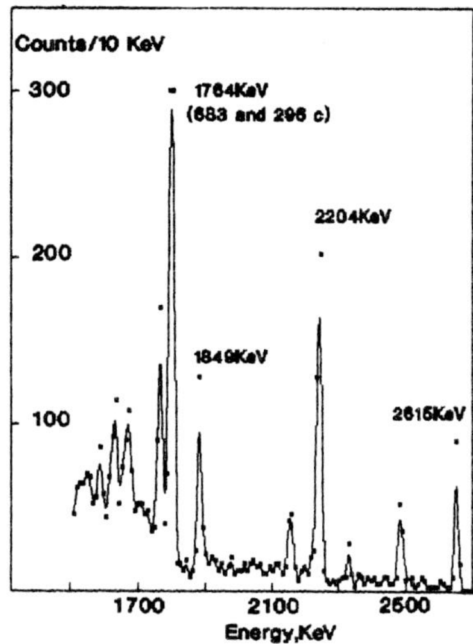

Figure 2: Part of the enriched-detector spectrum for 7612 h leave time in the ITEP/YePI experiment.

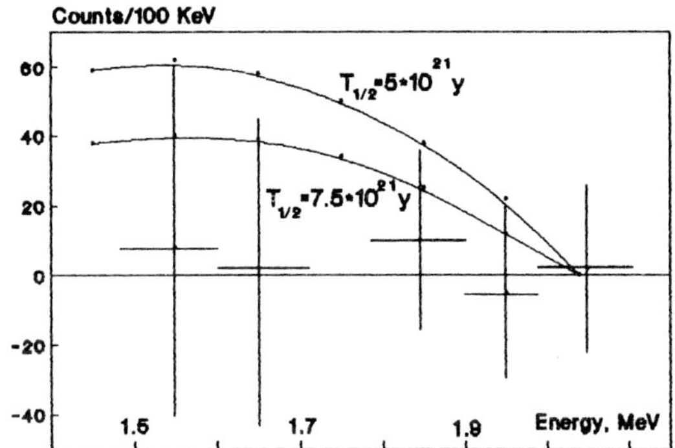

Figure 3: Difference between enriched and control germanium detector spectra in the ITEP/YePI search for Majoron-induced double beta-decay.

Table 3: Limits on Majoron-induced double beta-decay.

		lim $T_{1/2}\ y$	lim $g_B(\times 10^4)$
^{48}Ca	Columbia	$8.6 \cdot 10^{20}$	4.8
^{76}Ge	ITEP/YePI	$7.5 \cdot 10^{21}$	2.1
^{82}Se	UCI	$1.6 \cdot 10^{21}$	1.9
^{100}Mo	LBL/Hol/NM	$1.0 \cdot 10^{20}$	4.0

reconstructed with sufficient accuracy, and the existing uncertainties did not disturb the results.

The final results (Fig. 3) show the difference between the spectra collected by the two enriched detector in 11961 h (summed) and by the control detector in 7501 h. There is no visible effect. The group claimed the lifetime limit

$$T_{1/2} > 7.5 \cdot 10^{21} \text{ years (68\% C.L.)} \quad .$$

The results for Majoron-to-neutrino coupling constant calculations are given in Table 3. The data are taken from D. Caldwell's review,[13] except for the coupling constant for Ge. For simplicity, only one set of matrix elements[16] were used for the estimates given in the table. The estimates cover an order of magnitude when one uses the other existing matrix elements calculations.[14–19]

DOUBLE NEUTRINO DOUBLE BETA-DECAY

Beyond the old geochemical data we have now two claims for laboratory observation of double neutrino double beta-decay.

The UCI group has been performing a search for the (2ν)-decay of ^{82}Se since the end of the 1970s. Their first results were published in 1980.[20] The modern version of their experimental setup included two time projection chambers (TPC) between which a thin aluminized Mylar foil was placed. It contained 14 mg of 97% enriched ^{82}Se with a mean total source thickness of 7 mg/cm^2. Helmholtz coils provided a uniform magnetic field of 715 G perpendicular to the source plane.

The TPC reconstructed the ionization tracks left by electrons. The double beta-decay candidate events were those for which two electron tracks were reconstructed emerging from the same point. Several additional conditions also were imposed to improve the signal-to-background ratio.

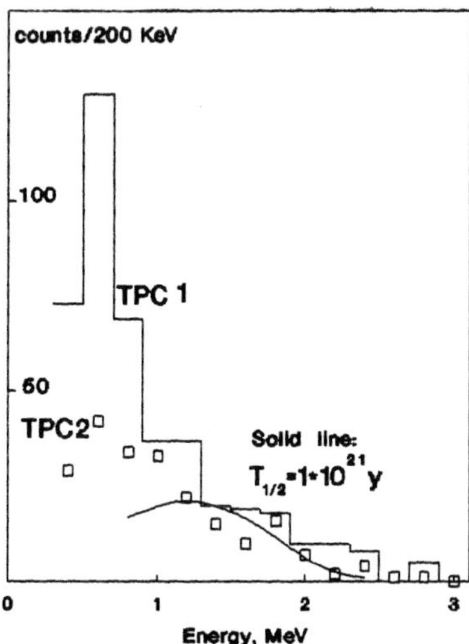

Figure 4: The UCI experimental results for double neutrino double beta-decay.

Two sets of measurements, with 7960 h (TPC 1) and 7763 h (TPC 2) of leave time, were performed. The summed $2e^-$ energy spectra are shown in Fig. 4. A total of 46 events between 1.3 and 2 MeV collected with the TPC 1 were analyzed. The spectra of single electrons for members of the candidate pairs, and distributions of the angles between the tracks were also analyzed.

The group concluded that most of the events above 1.3 MeV were due to double neutrino-decay of ^{82}Se with

$$T_{1/2} = (1.1^{+0.8}_{-0.3}) \cdot 10^{20} \text{years} \quad .$$

Possible backgrounds that could imitate the double beta events were internal conversion of beta-decay electrons, Compton scattering of gamma rays with consequent Möller scattering of the secondary electrons, and double scattering of gamma rays. Direct measurements of the background were not performed. Numerical estimates were made for the first two processes. The background due to the last process was neglected.

It is very difficult to judge the validity of these estimates. Double Compton scattering was the main component of the background in their previous measurements. Comparison of the TPC 1 and TPC 2 data shows the pronounced decrease of the double electron events corresponding to a decrease in the single electron background. It means that the number of imitated double beta events is not negligible, and that the lifetime value was underestimated.

The next result concerns the observation of double neutrino-decay of ^{76}Ge. The ITEP/YePI group compared the spectra measured with the enriched detectors (106 cm^3, 4664 h and 109 cm^3, 4324 h) and with the control natural germanium detector (115 cm^3, 4664 h). Because of significant differences in the abundance of ^{76}Ge (85% and 7.8%), the spectrum of the control detector could be assumed practically to represent the background. The possible signal due to (2ν)-decay was estimated by calculating the difference

$$N(E) \; = \; F(E) - k(E)f(E) \quad ,$$

where $F(E)$ and $f(E)$ are the spectra of the enriched and control detectors, and $k(E)$ is the two-detector relative efficiency for the background.

The values of $k(E)$ for all combinations of the detectors were measured with a set of gamma ray radioactivity sources imitating the background spectra with good accuracy. This was possible because at the energies of interest, 0.7–1.5 MeV, the background is dominated by electrons due to Compton scattering of gamma rays. The most intense was the 1.46 MeV gamma line of ^{40}K. Gamma radiation from nuclei that are members of the U and Th series, ^{60}Co, and ^{137}Cs (below 0.7 MeV) is the main source of the remaining background. The level of intrinsic background due to beta-decay of ^{68}Ga,^{77}Ge, and daughter nuclei of the U and Th chains was negligible.

Preliminary estimates have shown that the signal, if it exists, did not exceed 10–15% of background. Therefore careful calibrations of relative efficiencies had to be performed throughout the entire time of data collection. The final accuracy of the $k(E)$ measurements was about 2%.

The adopted analysis procedure eliminated uncertainties due mainly to the poor knowledge of the detector active volumes. The absolute value of the limit is dependent on the enriched detector volumes, but not the existence of the effect and its confidence level. Additional uncertainty in the result did not exceed the uncertainty in the detector volume measurements, typically 5–10%.

The difference between the summed spectrum of enriched detectors and that of the control detector is shown in Fig. 5. The solid line corresponds to the expected $2e$ spectrum due to (2ν)-decay with $T_{1/2} = 1 \cdot 10^{21}$ y. The data in Fig. 6 show the difference between the spectra of the two identical (enriched) detectors. Any visible effect is absent in this case.

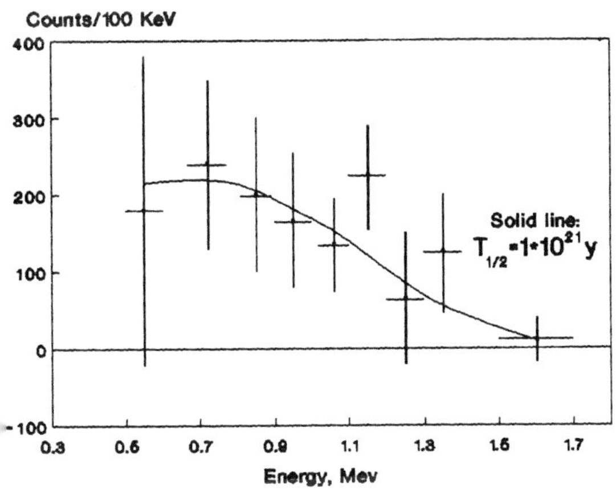

Figure 5: Difference between the summed spectrum of the enriched detectors (8988 h) and the control detector spectrum (4664 h) in the ITEP/YePI search for (2ν) double beta-decay.

Theoretical predictions for (2ν)-decay matrix elements are rather indefinite. Calculations in the frame of QPRPA cover the ranges of lifetime values

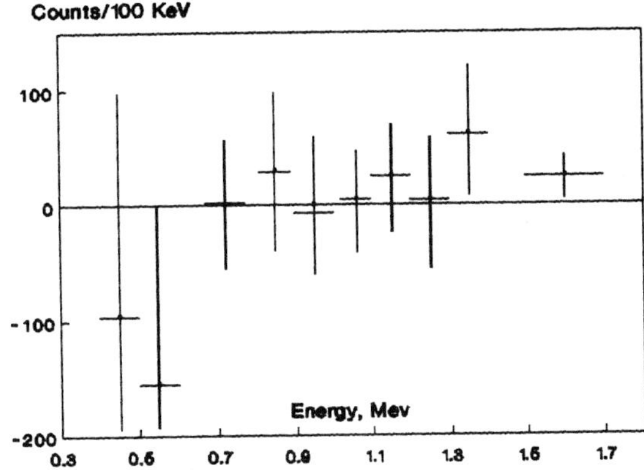

Figure 6: Difference between the spectra of identical (enriched) detectors in the ITEP/YePI experiment. The live times are 4664 h and 4324 h.

$4.4 \cdot 10^{18}$–$6.5 \cdot 10^{20}$ y for ^{82}Se and $1 \cdot 10^{20}$–$4 \cdot 10^{21}$ y for ^{76}Ge .[14–19] The theoretical situation is not understood completely, and new ideas and calculations are necessary.

CONCLUSIONS

More than 50 years have passed between the Maria Goppert-Mayer prediction[21] and the laboratory observation of the new nuclear decay. The existence of double beta-decay with double neutrino emission has never been in doubt, but the success of the experimenters is evident. It should not be long before several projected experiments[7,9,22–24] with ^{82}Se, ^{100}Mo, ^{136}Xe give positive results with much better accuracy. Now the experimentalists hope that that theoretical discrepancies will also be understood and resolved.

The most impressive result of these double beta-decay experiments seems to be the upper limit to a Majorana neutrino mass. Depending on how one calculates the matrix element,[14–19] the estimates yield limits from 0.5 to 5 eV (for $T_{1/2} = 10^{24}$ y). If the QPRPA mechanism does not suppress (0ν)-decay significantly, more stringent limits for the Majorana neutrino mass would be $< m_\nu >< (1-2)$ eV.

Further marked improvement of this limit is probable in the near future due to operation with enriched germanium detectors. Large quantities of highly enriched (up to 90% abundance of ^{76}Ge) material are now available. Yet the problem of background reduction remains crucial, as the sensitivities of these experiments are strongly dependent on the background level. It is difficult to expect results for the Ge half-life higher than 5×10^{24} y, corresponding to $< m_\nu >$ less than (0.5–1) eV.

The known projected experiments with other candidate isotopes[7,9,22–24] are not able to improve this situation significantly. More advantageous would be cryogenic detectors. The main problem here is to make these detectors large enough. What is evident is that by one way or another, the experimenters should go ahead because searching for neutrinoless double beta-decay is a powerful tool for the investigation of many problems.

REFERENCES

1. D. O. Caldwell, R. M. Eisberg, F. S. Goulding, F. R. Magnusson, B. Sadoulet, A. R. Smith, and M. S. Witherell, UCSB–HEP–89–02 and paper No. 219 contributed to this Symposium.
2. A. A. Vasenko, I. V. Kirpichnikov, V. Kuznetsov, A. S. Starostin, G. E. Marcosyan, V. M. Oganesyan, V. S. Pogosov, A. G. Tamanyan, and S. P. Shachysisyan, *Proc. Sec. Int. Symp. on Underground Physics,* Baksan Valley, USSR, August 17–19, 1987; Moscow, "Nauka," 1988. Preprint ITEP–75–88 and report submitted to Int. Europhysics Conf. on High Energy Physics, Madrid, Spain, September 6–13, 1989.
3. P. Fisher, F. Boehm, E. Bovet, J.-P. Egger, H. Henricson, K. Gabathuller, D. Reusser, and J.-L. Vuilliumier, *Phys. Lett.* **B218** (1989) n2.
4. R. K. Bardin, P. J. Gollon, J. D. Ullman, and C. S. Wu, *Nucl. Phys.* **A158** (1970) 337.
5. Ke You *et al.,* preprint BIHEP–EP–89–1, Beijing, China.
6. S. R. Elliot, A. A. Hann, and M. K. Moe, *Phys. Rev. Lett.* **59** (1987) 2020; *Phys. Rev.* **C36** (1987) 2129; UCI preprint Neutrino-88.
7. M. Alson-Garnjost, B. Dougherty, R. Kenny, J. Krivicich, R. Tripp, H. Nickolson, S. Sutton, B. Diterle, J. Cang, and C. Levitt, *Phys. Rev. Lett.* **60** (1988) 1928.
8. A. A. Klimenko, V. V. Kuzminov, and V. M. Lobashev, *Proc. XI Int. Symp. on Neutrino Mass and Related Topics,* Tokyo, March 16–18, 1988 (World Scientific, Singapore, 1988), p. 180.
9. E. Belotti *et al., Phys. Lett.* **B221** (1989) 209.
10. A. S. Barabash, V. V. Kuzminov, V. M. Lobashev, V. M. Novikov, B. M. Ovchinnikov, and A. A. Pomansky, *Phys. Lett.* **B223** (1989) n2, 273.
11. T. Kirsten, E. Heusser, D. Kaether, J. Oehm, E. Pernica, and H. Richter, *Nuclear Beta Decays and Neutrino* (World Scientific, Singapore, 1986), p. 81.
12. W. J. Lin, O. K. Manuel, G. L. Cumming, and D. Krstic, *Nucl. Phys.* **A481** (1988) 477.
13. D. O. Caldwell, *Int. Journ. Mod. Phys.* **A4** (1989) n8, 1851.
14. J. Engel, P. Vogel, and M. R. Zirnbauer, *Phys. Rev.* **C37** (1988) 731.
15. T. Tomoda and A. Fassler, *Phys. Lett.* **B199** (1987) 475.
16. K. Grotz and H. V. Klapdor, *Phys. Lett.* **B153** (1985) 1, and **B157** (1985) 242.
17. W. C. Haxton and G. J. Stephenson, *Prog. Par. Nucl. Sci.* **12** (1984) 409.
18. M. Doi, T. Kotani, and E. Takasugi, *Prog. Theor. Phys. Suppl.* **83** (1985) 1.
19. T. Tomoda, A. Faessler, K. W. Schmidt, and F. Grummer, *Nucl. Phys.* **A452** (1986) 591.
20. M. K. Moe and D. D. Lowenthal, *Phys. Rev.* **C22** (1980) 2186.
21. M. Goeppert-Mayer, *Phys. Rev.* **48** (1935).
22. T. Watanabe, Ph.D. dissertation, University of Osaka, Toyonaka Osaka 560, Japan (Jan. 1989).
23. H. T. Wong *et al.,* Caltech–Neuchatel–PSI preprint (Feb. 1989).
24. M. S. Ainutdinov, V. A. Artemjev, O. Ya. Zeldovich *et al.,* preprint ITEP–111–88.

DISCUSSION

Q. K. Winter, CERN: If you deduce the spectrum and the rate of the (2ν)-decay of Ge from the difference of the spectra of enriched and natural germanium, you assume that the background is not affected by the enrichment. How can you justify this assumption?

A. The detector germanium itself is a very pure material. Nevertheless, we estimated the possible contamination of U and Th experimentally by counting 4–5 MeV alpha-particles. It appears to be less than 10^{-14}–10^{-15} g/g. The background due to this admixture was negligible below 2.5 MeV.

PROPERTIES OF LEPTONS*

MARTIN L. PERL
Stanford Linear Accelerator Center
Stanford University, Stanford, California 94309

ABSTRACT

The properties of the electron, muon, tau, and their neutrinos are reviewed. Three discrepancies in our understanding of those properties are discussed: the lifetime of orthopositronium, the mass spectra of e^+e^- pairs produced in heavy ion collisions, and the 1-charged particle modes problem in tau decays. The review concludes with a discussion of what we need to learn about the tau and the consequent need for a tau-charm factory.

TABLE OF CONTENTS

I. Introduction and History of Leptons
II. Limits on the Existence of New Leptons
III. Future Searches for New Leptons
IV. Electron
V. Muon
VI. Neutrinos: ν_e, ν_μ, ν_τ
VII. Tau
VIII. Tau Physics at a Tau-Charm Factory

I. INTRODUCTION AND HISTORY OF LEPTONS

In a talk given at the XIV International Symposium on Lepton and Photon Interactions I had two purposes: to review our present knowledge of the properties of leptons and to point out mysteries or loose ends in that knowledge. In this shorter written version of the talk I emphasize the second purpose in a context in which one physicist's mystery is another physicist's loose end.

The history of leptons goes back almost 90 years.

A. *Electron, 1895*

The final elucidation of the existence and nature of the electron is due to J. J. Thomson in 1895 using a cathode ray tube. His description of the discovery is in *Phil. Magazine* **44**, 293 (1897), reprinted in part in *Great Experiments in Physics* (Dover Pub., New York, 1987), edited by M. H. Shamos.

Other physicists had worked on the particle nature of the cathode ray—using, for example, magnetic deflection. Thomson was able to consistently use in addition electrostatic deflection by improving the vacuum in the cathode ray tube; a crucial technical advance.

B. *Electron Neutrino, 1930*

W. Pauli suggested the existence of the electron neutrino in 1930. A fascinating description of Pauli's thought was given by L. M. Brown in *Phys. Today*, September, 1978, p. 23.

C. *Muon, 1937*

The classic Letter to the Editor of S. H. Neddemeyer and C. D. Anderson [*Phys. Rev.* **51**, 884 (1937)] describes their discovery of the muon in cosmic rays using a cloud chamber. The Letter is fascinating with its discussion of an "intermediate mass" between the electron and the proton. The understanding of the difference between the muon and the pion required another ten years.

D. *Electron Neutrino, 1956*

In another classic Letter to the Editor, F. Reines and C. L. Cowan, Jr., *Phys. Rev.* **92**, 830 (1956) describe their detection of the electron neutrino. They used a reactor for the source of neutrinos from neutron decay and the most modern counting electronics of the time to detect the e^+ and n from $\overline{\nu}_e + p \to e^+ + n$.

*Work supported by Department of Energy contract DE–AC03–76SF00515.

E. *Muon Neutrino, 1962*

Yet another technique was used by the 1988 Nobel Prize winners, L. Lederman, M. Schwartz, and J. Steinberger to observe the muon neutrino and separate it from the electron neutrino. A neutrino beam from the AGS and thick plate spark chambers were used [G. Danby *et al.*, *Phys. Rev. Lett.* **9**, 36 (1962)].

F. *Tau, 1975*

The electron–positron storage ring and the reaction $e^+ + e^- \rightarrow \tau^+ + \tau^-$ was used to find the τ [M. L. Perl *et al.*, *Phys. Rev. Lett.* **35**, 1489 (1975)]. Important in the history of the discovery of the τ was the development of the theory of heavy lepton decays by Y. S. Tsai [*Phys. Rev.* **D4**, 2821 (1971)] and the heavy lepton searches at ADONE [M. Bernardini *et al.*, *Nuovo Cimento* **17**, 383 (1973) and S. Orioto *et al.*, *Phys. Lett.* **48B**, 165 (1974)].

G. *Tau Neutrino*

The existence and properties of the tau neutrino are deduced from studies of tau decays. The ν_τ has not been detected directly.

In the ninety years since the discovery of the electron, a different experimental technique was used for each discovery. Will we be able to discover new leptons with present experimental methods or will a new method be required? Or are these additional leptons to be discovered?

II. LIMITS ON THE EXISTENCE OF NEW LEPTONS

In connection with the last question, I give in Table 1 some limits[1] as of August 1989 on the existence of additional leptons. This paper is being written before there are complete results from the searches for new leptons in Z^0 decays at LEP and the SLC, hence limits based on Z^0 decay studies are not included. Perhaps the Z^0 decay will be the new experimental technique which allows a new lepton to be discovered.

The search for new lepton pairs (L^-, L^0) requires special care if the L^0 mass m_0 is less than but close to the L^- mass m_-. When $\delta = m_- - m_0$ is less than 4 or 5 GeV/c^2 there is not enough visible energy for the usual search method in $e^+e^- \rightarrow L^+L^-$; and there is not enough missing transverse momentum for the usual search method in $W^+ \rightarrow L^+ + L^0$. Figure 1 gives the limits[2,3] as of August 1989. Once again searches using Z^0 decays will simplify these studies.

Table 1. Experimental lower limits on masses of charged leptons as of August 1989. Searches using Z^0 are not included because this paper was written before definitive results were published. The C.L. is 95% unless otherwise noted.

Lepton Type	Mass Lower Limit, GeV/c^2	Method
Stable	30.1	e^+e^-,TRISTAN
L^-, L^0: $m_0 << m_-$	41.$^{(a)}$	$\bar{p}p$, UA1
L^-, L^0: $m_0 << m_-$	29.9	e^+e^-,TRISTAN
$e^* \rightarrow e + \gamma$	pair: 30.2 single: 60.5	e^+e^-,TRISTAN
$\mu^* \rightarrow \mu + \gamma$	pair: 30.1 single: 53.0	e^+e^-,TRISTAN
$\tau^* \rightarrow \tau + \gamma$	pair: 29.0 single: 50.0	e^+e^-,TRISTAN
$\tilde{e} \rightarrow e + \tilde{\gamma}, m_{\tilde{\gamma}} = 0$	29.5	e^+e^-,TRISTAN
$\tilde{\mu} \rightarrow \mu + \tilde{\gamma}, m_{\tilde{\gamma}} = 0$	25.8	e^+e^-,TRISTAN
$\tilde{\tau} \rightarrow \tau + \tilde{\gamma}, m_{\tilde{\gamma}} = 0$	24.7	e^+e^-,TRISTAN

$^{(a)}$ 90% C.L.

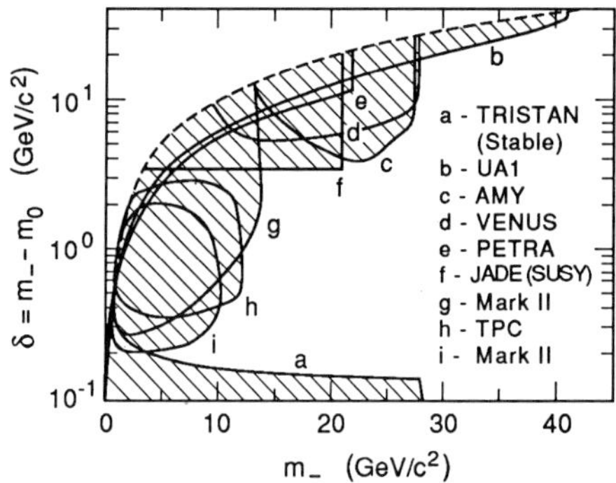

Figure 1: The crosshatched area shows the excluded region in the δ–m_- plane. The figure does not include limits from Z^0 decay studies.

III. FUTURE SEARCHES FOR NEW LEPTONS

Future searches fall into the five classes described next. Here L means L^- or L^0 and m means mass.

A. $m_L \lesssim 45$ GeV/c^2

This region will be completely explored at the SLC and LEP provided $Z^0 \rightarrow L + \overline{L}$.

B. $45 \lesssim m_L \lesssim 80$ GeV/c^2

This region will be completely explored at LEP II provided $Z^0_{virtual} \rightarrow L + \overline{L}$ or $\gamma_{virtual} \rightarrow L + \overline{L}$. In addition, $W^+ \rightarrow L^+ + L^0$ can be used at $\overline{p}p$ colliders to explore part of this range.

C. $m_L \gtrsim 80$ *GeV/c*$^2 \approx m_W$ *at* e^+e^- *Colliders*

When $m_L > m_W$ there are two changes in the e^+e^- search method.

First, the decay is now

$$\begin{aligned} L^- &\to W^-_{real} + L^0 \ , \ W^-_{real} \to \text{decay modes} \ , \\ L^0 &\to W^+_{real} + L^- \ , \ W^+_{real} \to \text{decay modes} \ . \end{aligned} \quad (1)$$

Second, LEP II can only reach 10 or 20 GeV/c^2 into this mass range. The straightforward future search method[4] is to use a TeV-range e^+e^- linear collider for

$$e^+ + e^- \to L^+ + L^- \ . \quad (2)$$

But the cross section is small

$$\sigma = \frac{0.087\,R}{E^2_{cm}\ (\text{TeV})} \ \text{pb} \ , \quad (3a)$$

where, for example,

$$\begin{aligned} E_{cm} = 0.7 \text{ TeV}: \ R_{L^+L^-} = 1.18 \, , \ R_{L^0\overline{L}^0} = 0.32 \ , \\ E_{cm} = 2.0 \text{ TeV}: \ R_{L^+L^-} = 1.17 \, , \ R_{L^0\overline{L}^0} = 0.31 \ . \end{aligned} \quad (3b)$$

D. $m_L \gtrsim 80$ *GeV/c*$^2 \approx m_W$ *at pp Colliders*

The power of search methods at the SSC or an LHC using

$$p + p \to L + \overline{L} + \ldots \quad (4)$$

is still obscure. Hinchliff[5] has given a recent discussion.

E. $e^- + p \ \to \ L_e + \ldots$

Provided there is a lepton L_e with the lepton number of the e^-, searches at HERA[5] can reach $m_{L_e} \sim 200$ GeV/c^2. L_e represents L^-_e, L^0_e, L^{*-} and so forth.

IV. ELECTRON

Our present picture of the electron is that it is simply a stable, Dirac point particle which obeys the conventional theory of the electroweak interaction and has a mass of 0.511 MeV/c^2. There are no confirmed deviations from this picture, but there are three loose ends or mysteries, depending on one's point of view. These are: the comparison of measurement and theory for $g_e - 2$; the lifetime of orthopositronium; and the possible existence of e^+e^- resonant states produced in heavy ion collisions. I conclude this section with comments on the evidence for the electron being a point particle obeying standard electroweak theory.

A. $g_e - 2$

H. Dehmelt and his colleagues[6,7] have used the Penning trap technique to find

$$\begin{aligned} \text{Electron:} \ & \frac{1}{2}(g_{e^-} - 2) \times 10^{12} = \\ & 1\ 159\ 652\ 188.4\,(4.3) \ , \\ \text{Positron:} \ & \frac{1}{2}(g_{e^+} - 2) \times 10^{12} = \\ & 1\ 159\ 652\ 187.9\,(4.3) \ . \end{aligned} \quad (5)$$

The major cause of the experimental error is the uncertainty in calculating the frequency shift[8] due to the $e^\pm$ being in a cavity of finite size.

In a recent paper[9] T. Kinoshita discusses the calculations[9,10] of g_e by his colleagues and himself. The calculations and the errors are sensitive to the value of α, the fine structure constant. Kinoshita[9] gives a most recent calculation:

$$\text{Theory:} \ \frac{1}{2}(g_e - 2) \times 10^{12} = 1\ 159\ 652\ 133\,(29) \ , \quad (6a)$$

and one based on an earlier value of α:

$$\text{Theory:} \ \frac{1}{2}(g_e - 2) \times 10^{12} = 1\ 159\ 652\ 133\,(108) \ . \quad (6b)$$

The agreement between the calculated values in Eqs. (6) and the measurement in Eqs. (5) is marvelous. Still there is a loose end in the dependence of the calculated value of g_e on α and the uncertainty in the precise value of α.

B. *Lifetime of Orthopositronium*

At present there is a mystery in the comparison of theory with the measurement of the lifetime of orthopositronium. The orthopositronium state is 1^3S_1 and the decay is

$$Ps \to \gamma + \gamma + \gamma \ . \quad (7)$$

Westbrook *et al.*[11] of the University of Michigan measure

$$\Gamma_{exp} = 7.0516 \pm 0.0013 \ \mu\text{s}^{-1} \quad ; \quad (8)$$

while present calculations[12,13] give

$$\Gamma_{theor} = 7.03830 \pm 0.00007 \ . \qquad (9)$$

If Eqs. (8) and (9) are accepted than Γ_{exp} is larger than Γ_{theor} by many standard deviations and the measured lifetime would be shorter than the calculated lifetime. There are three possibilities, not mutually exclusive. The measurement may have an unknown systematic error; the one usually mentioned is the correction for the decay occurring in a gas. The University of Michigan experimenters[11] believe their gas density correction is right; in addition they are beginning[11,14] an experiment in which the decay occurs in vacuum.

A second explanation of $\Gamma_{exp} > \Gamma_{theor}$ is that the calculation is wrong. Γ is given by

$$\Gamma = \frac{\alpha^6 mc^2}{\hbar} \frac{2(\pi^2 - 9)}{9\pi} \left[1 + A(\alpha/\pi) + \frac{1}{3}\alpha^2 \ell n\alpha + B(\alpha/\pi)^2 + \ldots\right] \qquad (10)$$

Here $A = -10.28$ but B has not been calculated. To increase Γ_{theor} by 0.013 μs^{-1} making it equal to Γ_{exp}, B would have to be about 300. No one has calculated if B can be that large.

A third explanation of the discrepancy is that the e^+e^- 1^3S_1 system decays about 2×10^{-3} of the time through a process not contained in quantum electrodynamics. At present there is no evidence for such a process.[15]

C. e^+e^- Pairs from Heavy Ion Collisions

In the past half decade two groups of experimenters have studied the reaction

$$A + A' \to e^+ + e^- + (A + A')_{excited} \ . \qquad (11)$$

Here A and A' are heavy ions such as Ta, Th, and U. The collision is carried out at energies close to the Coulomb barrier energy. The experimenters find "resonances" or "lines" in

$$E_{kin,sum} = E_{kin,e^+} + E_{kin,e^-} \ . \qquad (12)$$

Here E_{kin} is kinetic energy. The measurements are difficult because the signal is small compared to the total e^+ and e^- production and because the signal is sensitive to the energy of the incident ion beam and other parameters. In addition the lines are of different strengths.

In a recent paper[16] the EPOS collaboration reports lines at

$$E_{kin,sum} \approx 610, 750, \text{ and } 810 \text{ keV} \ . \qquad (13a)$$

The ORANGE collaboration reports[17] lines at

$$E_{kin,sum} \approx 540, 640, 716, 809, \text{ and } 895 \text{ keV} \ , \qquad (13b)$$

with the most pronounced signal at 809 keV.

When these lines were first observed there were many searches for an elementary particle with the decay $\phi \to e^+ + e^-$ and a mass of $2m_e + E_{kin,sum}$ where m_e is the e mass. As reviewed by Davier[18] these searches have excluded an explanation of the lines by such an elementary particle.

Interest then turned to explanations which assume resonant structures in the e^+e^- system, the structures due to physics outside conventional quantum electrodynamics. There have been many searches for such resonances in Bhabha scattering,

$$e^+ + e^- \to e^+ + e^- \ , \qquad (14)$$

in the barycentric energy range $1.5 \lesssim E_{cms} \lesssim 1.9$ MeV. At present there are no confirmed resonances.[19] Searches have also been made in

$$e^+ + e^- \to \gamma + \gamma \ . \qquad (15)$$

Again there are no confirmed resonances.[20]

These e^+e^- lines observed in heavy ion collisions are the longest standing experimental mystery in the fields of nuclear and particle physics.

D. Electron Structure and Quantum Electrodynamic Tests

At present the most stringent tests of the point particle nature of the electron come from high energy measurements of Bhabha scattering at e^+e^- colliders

$$e^+ + e^- \to e^+ + e^- \ . \qquad (16)$$

There are two popular models for testing for structure. One model uses the form factor concept

$$\sigma(e^+e^- \to e^+e^-) = \sigma_{point}(e^+e^- \to e^+e^-)\, F^2(x) \ , \qquad (17a)$$

$$F(x) = 1 \mp \frac{x}{x - \Lambda^2_\pm} \ . \qquad (17b)$$

Here x means s, the square of the total energy, or t, the square of the four-momentum transfer.

The other model uses the contact interaction composite particle concept with

$$L_{eff} = \pm \frac{g^2}{2\Lambda_{\pm}^{c\,2}} \overline{\psi}_2 \gamma^\mu \psi_2 \overline{\psi}_1 \gamma_\mu \psi_1 \; , \qquad (18a)$$

and

$$\frac{g^2}{4\pi} = 1 \; . \qquad (18b)$$

This assumes a vector-vector interaction, other interactions have analogous forms.

Measurements[21] by the TASSO collaboration at PETRA give the following typical results for 95% C.L. lower limits:

Form Factor	Contact Interaction (VV)	
$\Lambda_+ > 0.4$ TeV	$\Lambda_+^c > 3.6$ TeV	(19)
$\Lambda_- > 0.6$ TeV	$\Lambda_-^c > 7.1$ TeV	

The experimenters[21] point out that the form factor results can be interpreted that "electrons are point-like objects down to distances of 5×10^{-17} cm."

Although the Λ and Λ^c measurements limit the complexity of the electron at high energy and the $g_e - 2$ measurements (Sec. IV.A) limit the complexity of the electron at low energy, it is still possible that anomalies in electron behavior can be detected at moderate energy. An example is the speculation of Hawkins and myself[22,23] about a neutral particle, λ, which might couple only to charged leptons.

Figure 2(a) shows the $e - \lambda - e$ coupling with $\alpha_{\lambda e} = g_{\lambda e}^2/4\pi$, in analogy to $\alpha = e^2/4\pi$ for a photon. Experimentally excluded areas[22,23] on the $\alpha_{\lambda e} - m_\lambda$ are given in Fig. 3 for λ a vector particle of mass m_λ. In the mass range of $20 \lesssim m_\lambda \lesssim 2000$ MeV/c^2, a very sensitive search method[25] uses the reactions

$$e^- + p \to e^- + p + \lambda \; , \qquad \lambda \to e^+ + e^- \qquad (20)$$

of Fig. 2(b). This is one of the goals of the PEGASYS experiment,[25] an internal gas jet target proposed for PEP. Tsai[26] has made the basic λ production calculations. Experiments at HERA can search for the λ in a high range of m_λ.

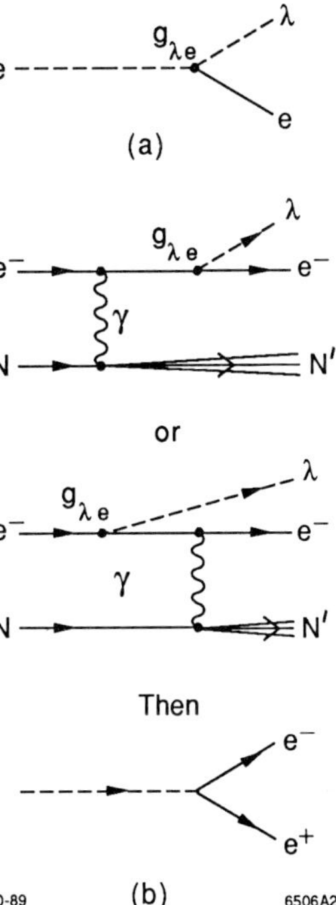

Figure 2: (a) The e-λ-e vertex. (b) Method for searching for the λ using $e^- + N \to e^- + N' + \lambda$ where N and N' are protons or nuclei.

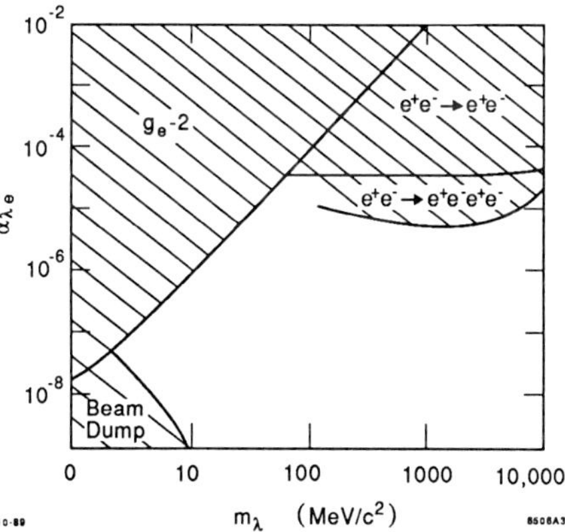

Figure 3: The crosshatched area shows the excluded region in the $\alpha_{\lambda e}$-m_λ plane for vector λ.

V. MUON

Unlike the electron, at this time there are no mysteries or loose ends in the particle physics of the muon. But the ingenuity of the experimenters in muon physics is leading to more precise and more probing studies of the properties of the muon.

A. $g_\mu - 2$

The values of $g_\mu - 2$ from measurement[27] and theory[28] are

$$\text{Experiment}: \quad \frac{1}{2}(g_\mu - 2) \times 10^{10} = 11\,659\,230\,(84) \; . \qquad (20a)$$

$$\text{Theory}: \quad \frac{1}{2}(g_\mu - 2) \times 10^{10} = 11\,659\,194.7\,(14.3) \; . \qquad (20b)$$

Thus experiment and theory agree within the errors.

There are plans to reduce both errors. A future experiment at the Brookhaven AGS would reduce the error from 84×10^{-10} to about 4×10^{-10}. Reduction in the theoretical error requires a better calculation of the hadronic contribution to g_μ; and this in turn requires a more precise measurement of $e^+ + e^- \rightarrow$ hadrons below 1 GeV.[29]

B. *Dynamics of* $\mu^- \rightarrow e^- \overline{\nu}_e \nu_\mu$

The basic question is how well does V-A coupling describe the dynamics of the decay

$$\mu^- \rightarrow e^- + \overline{\nu}_e + \nu_\mu \ . \tag{21}$$

If a general four-fermion interaction is assumed and constrained only by the requirement that it be local, nonderivative and lepton-number conserving, then this dynamics is described by 19 real parameters.[30,31] However, only 10 combinations of these parameters can be measured using present experimental techniques, since these techniques cannot detect neutrinos. Within this restriction, all measurements of μ decay dynamics agree with V-A coupling.

C. *Radiative Muon Decays*

Although the radiative decays

$$\mu^- \rightarrow e^- + \overline{\nu}_e + \nu_\mu + \gamma \ , \tag{22a}$$
$$\mu^- \rightarrow e^- + \overline{\nu}_e + \nu_\mu + e^+ + e^- \ , \tag{22b}$$

have long been known and their theory is well known,[32] the measurements have large errors[33] since they are either old or are adjuncts to studies of forbidden muon decay limits. Is a dedicated study of the decays in Eqs. (22) worthwhile?

D. *Limits on Forbidden Muon Decays*

The 90% C.L. upper limits on the branching ratio for unconventional muon decays are

$$B(\mu^- \rightarrow e^- \gamma) \quad < 4.9 \times 10^{-11} \quad \text{Ref. [34]} \tag{23a}$$

$$B(\mu^- \rightarrow e^- \gamma\gamma) \quad < 7.2 \times 10^{-11} \quad \text{Ref. [34]} \tag{23b}$$

$$B(\mu^- \rightarrow e^- e^+ e^-) \quad < 1.0 \times 10^{-12} \quad \text{Ref. [35]} \tag{23c}$$

Certainly experimenters will look for these decays at yet smaller branching ratios, for example $B(\mu^- \rightarrow e^- \gamma)$ will be studied[36] into the 10^{-12} to 10^{-13} range.

E. *Muon Structure*

All measurements agree with the muon being a Dirac point particle. For example, high energy studies[37] of

$$e^+ + e^- \rightarrow \mu^+ + \mu^- \tag{24}$$

give the 95% C.L. lower limits

Form Factor	Contact Interaction (VV)	
$\Lambda_+ > 0.23$ TeV	$\Lambda^c_+ > 5.8$ TeV	(25)
$\Lambda_- > 0.25$ TeV	$\Lambda^c_- > 4.8$ TeV	

This is from the JADE collaboration using data from PETRA.

F. *Other Probes of the Muon*

Other searches for unconventional properties of the muon are based on the hypothetical reactions

$$\mu^- + \text{nucleus} \rightarrow e^- + \text{nucleus} \ , \tag{26a}$$

$$\mu^- + \text{nucleus} \rightarrow e^+ + \text{nucleus} \ , \tag{26b}$$

$$\mu^+ + e^- \rightarrow \mu^- + e^+ \ . \tag{26c}$$

No evidence for these reactions has been found.[30,31] K. K. Gan used the reaction $e^+e^- \rightarrow \mu^+\mu^-$ to place additional limits on unconventional properties of the muon.[38]

VI. NEUTRINOS

In this section I discuss four subjects: the number of small mass neutrinos; the masses of ν_e, ν_μ, and ν_τ; neutrino oscillations; and our present experimental knowledge of solar neutrinos. In this talk I do not have the time to describe the astrophysical and cosmological constraints on neutrino properties, I refer you to the review by Kolb, Schramm, and Turner.[39]

A. *Number of Small Mass Neutrinos*

We know of only three kinds of neutrinos: ν_e, ν_μ, ν_τ. Before August 1989 astrophysical and cosmological considerations as well as various experiments had shown that there are no more than four or five kinds of light neutrinos. Now measurements of the Z^0 width, Table 2, show that there are only three neutrinos with mass much less than $m_{Z^0}/2$ and conventional weak interaction coupling. An amazing simplicity in nature! All we have in this category are ν_e, ν_μ, and ν_τ.

Table 2. Number of types of small mass neutrinos, N_ν determined from Z^0 width. Experiments are listed in alphabetical order.

Experiment	N_ν
ALEPH	3.3 ± 0.3
DELPHI	2.9 ± 0.7
L3	3.4 ± 0.5
MARK II	2.8 ± 0.6
OPAL	3.1 ± 0.4

B. *Neutrino Masses*

Two general references on neutrino masses are the book by Boehm and Vogel[40] and the review by Langacker.[41]

1. *Electron Neutrino Mass*

There are two new upper limits on m_{ν_e}. At the Los Alamos National Laboratory, T. J. Bowles *et al.*[42] find

$$m_{\nu_e}^2 = (-198 \pm 143) \quad eV^2 , \tag{27a}$$

$$m_{\nu_e} < 13.4 eV \ , \ 95\% \ \text{C.L.} \tag{27b}$$

The 95% C.L. is based on statistical and systematic errors. The other new limit is from the Institute for Nuclear Study of the University of Tokyo; H. Kawakami *et al.*[43] find

$$m_{\nu_e}^2 = (-82 \pm 87)\ \text{eV}^2 \ . \tag{28a}$$

The authors state that "very preliminary analysis" gives

$$m_{\nu_e} < 11\ \text{eV} \quad . \tag{28b}$$

This limit and its C.L. of 95% are based on statistical error only.

Of course, the incentive for these new experiments was the finding of Boris *et al.*[44]

$$m_{\nu_e} = (26 \pm 5)\ \text{eV} \ , \tag{29}$$

with a "model-independent" range of 17 to 40 eV. The upper limits in Eqs. (27) and (28) disagree with the nonzero value in Eq. (29).

For completeness I list three other m_{ν_e} measurements published in 1987 and 1986.

$$m_{\nu_e} < 18\ \text{eV} \ , \quad 95\%\ \text{C.L.} \quad \text{Ref. [45]}$$

$$m_{\nu_e} < 32\ \text{eV} \ , \quad 95\%\ \text{C.L.} \quad \text{Ref. [46]}$$

$$m_{\nu_e} < 27\ \text{eV} \ , \quad 95\%\ \text{C.L.} \quad \text{Ref. [47]}$$

Thus there is no confirmed evidence for a nonzero mass for ν_e.

2. *Muon and Tau Neutrino Masses*

In the past year there have been no changes in the upper limits on m_{ν_μ} and m_{ν_τ}:

$$m_{\nu_\mu} < 0.25\ \text{MeV/c}^2 \ , \quad 90\%\ \text{C.L.} \tag{30}$$

$$m_{\nu_\tau} < \ 35.\ \text{MeV/c}^2 \ , \quad 95\%\ \text{C.L.} \tag{31}$$

If one believes in the hypothesis

$$\frac{m_{\nu_1}}{m_{\nu_2}} = \frac{m_1^2}{m_2^2} \ , \tag{32}$$

then

$$m_{\nu_\mu} \left(\frac{m_e}{m_\mu}\right)^2 < 5.9\ \text{eV} \ , \tag{33a}$$

$$m_{\nu_\tau} \left(\frac{m_e}{m_\tau}\right)^2 < 2.9\ \text{eV} \ . \tag{33b}$$

The upper limits in Eqs. (33) are smaller than the upper limits in Eqs. (27b) and (28b).

C. *Neutrino Oscillations*

There is no *confirmed* evidence for neutrino oscillations.[48] There is a loose end in the findings of Astier *et al.*[49] in a search for $\nu_\mu \to \nu_e$ oscillations.

D. *Solar Neutrinos*

The mystery continues of the difference between measured and calculated solar neutrino rates. In the well known experiment of Davis and his colleagues[50] the capture rate of solar neutrinos in ^{37}Cl is measured in SNU units, where 1 SNU = 10^{36} captures/atom sec. The neutrino energy must be above 0.8 MeV for capture. In a 1988 review Totsuka[51] summarized the ^{37}Cl results:

$$\text{1970–1985 capture rate } = 2.1 \pm 0.3\ \text{SNU} \ , \tag{34a}$$

$$\text{1986–1987 capture rate } = 5.0 \pm 0.8\ \text{SNU} \ . \tag{34b}$$

Conventional solar theory gives[51]

$$\text{Calculated capture rate} = 7.9 \pm 2.6\ \ \text{SNU} \ . \tag{35}$$

Thus to the old mystery of the difference between the measured capture rate in Eqs. (34) and the calculated rate in Eq. (35), has been added the question of the significance of the difference between the two measured rates in Eqs. (34).

The recent results[52] of the Kamiokande II experiment also show a difference

$$\frac{\text{measured flux}}{\text{predicted flux}} = 0.46 \pm 0.13 \pm 0.08 \quad , \qquad (36a)$$

and Koshiba[52] reported to this meeting very recent data:

$$\frac{\text{measured flux}}{\text{predicted flux}} = 0.39 \pm 0.09 \pm 0.06 \quad . \qquad (36b)$$

Kamiokande II is a water Čerenkov detector with much higher energy thresholds than the ^{37}Cl experiment: 9.3 MeV for the data in Eq. (36a) and 7.5 MeV for the data in Eq. (36b).

We don't know if the calculations of neutrino production by the sun are wrong or if the experiments are wrong or if the difference is due to an unconventional property of the ν_e.

VII. TAU

A. *Allowed Decay Modes*

The τ mass of 1784 MeV/c^2 allows a great number of decay modes. Figure 4, reproduced from the *1988 Review of Particle Properties,*[53] shows 40 different decay modes and their branching ratios. Note that even these 40 modes are limited to those with measured values or upper limits for their branching fractions. For example, there are no listings for $\tau^- \to \pi^- n\pi^0 \nu_\tau$, $n = 4, 5, 6 \ldots$ Thus the decay structure is rich and many decay quantities are unmeasured.

Table 3 summarizes the world averages of the major branching fractions from Ref. [53] and from the recent review of Kiesling.[54] The averages agree because they are mostly based on the same set of experiments.

B. *Forbidden Decay Modes*

Figure 5, also from Ref. [53], lists 26 hypothetical decay modes in which lepton number conservation would be violated. There is no evidence for such a violation.

C. *The 1-Charged Particle Decay Modes Problem*

There is a mystery in the 1-charged particle decay modes of the τ, a mystery so puzzling that it is described in the *1988 Review of Particle Properties,*[53] Table 4. Reference [55] discusses the puzzle in detail. The sum of measured 1-charged particle branching ratios is $78.3 \pm 1.9\%$. Theoretical limits[55–57] have been set on unmeasured, 1-charged particle modes such as $\pi^- 3\pi^0 \nu_\tau$, $\pi^- 4\pi^0 \nu_\tau$, $\pi^- 5\pi^0 \nu_\tau$, $\pi^- \eta \nu_\tau$, and $\pi^- \eta 2\pi^0 \nu_\tau$ to give the total 2.2% upper limit in Table 4. This give the sum of individual 1-charged particle branching fractions as $\leq 80.5 \pm 1.9\%$ yet the topological 1-charged particle branching fraction is $86.6 \pm 0.3\%$. Thus 6% out of the 86.6% is unexplained.

There have been several studies[54,55,58–61] of this puzzle, searching for systematic errors or biases in the many experiments which contribute to the world

τ DECAY MODES

τ^+ modes are charge conjugates of the modes below.

	Mode	Fraction (Γ_i/Γ)	Scale/ Conf.Lev.
Γ_1	$\tau^- \to \mu^- \bar{\nu}_\mu \nu_\tau$	$(17.8 \pm 0.4) \times 10^{-2}$	
Γ_2	$\tau^- \to e^- \bar{\nu}_e \nu_\tau$	$(17.5 \pm 0.4) \times 10^{-2}$	
Γ_3	$\tau^- \to 2\pi^+ 3\pi^- \nu_\tau$	$(5.6 \pm 1.6) \times 10^{-4}$	
Γ_4	$\tau^- \to 2\pi^+ 3\pi^- \pi^0 \nu_\tau$	$(5.1 \pm 2.2) \times 10^{-4}$	
Γ_5	$\tau^- \to K^+ K^- \pi^- \nu_\tau$	$(2.2\ ^{+1.7}_{-1.1}) \times 10^{-3}$	
Γ_6	$\tau^- \to K^- \pi^+ \pi^-\ (\geq 0\ \pi^0)\ \nu_\tau$	$(2.2\ ^{+1.6}_{-1.3}) \times 10^{-3}$	
Γ_7	$\tau^- \to K^- \nu_\tau$	$(6.6 \pm 1.9) \times 10^{-3}$	S=1.1
Γ_8	$\tau^- \to \pi^- \omega \nu_\tau$	$(1.6 \pm 0.5) \times 10^{-2}$	
Γ_9	$\tau^- \to \pi^- \rho^0 \nu_\tau$	$(5.4 \pm 1.7) \times 10^{-2}$	
Γ_{10}	$\tau^- \to a_1(1260)^- \nu_\tau$		
Γ_{11}	$\tau^- \to K^-\ (\geq 1$ neutral$)\ \nu_\tau$	$(1.05\ ^{+0.27}_{-0.28}) \times 10^{-2}$	
Γ_{12}	$\tau^- \to \pi^- \nu_\tau$	$(10.8 \pm 0.6) \times 10^{-2}$	
Γ_{13}	$\tau^- \to \pi^+ 2\pi^- \nu_\tau$ (non-resonant)	$< 1.4 \times 10^{-2}$	CL=95%
Γ_{14}	$\tau^- \to \pi^- \eta \nu_\tau$	$< 1.0 \times 10^{-2}$	CL=95%
Γ_{15}	$\tau^- \to \rho^- \nu_\tau$	$(22.3 \pm 1.1) \times 10^{-2}$	
Γ_{16}	$\tau^- \to \pi^+ 2\pi^- \nu_\tau$	$(6.8 \pm 0.6) \times 10^{-2}$	S=1.5
Γ_{17}	$\tau^- \to \pi^+ 2\pi^- \gamma(s)\ \nu_\tau$	$(6.4 \pm 0.6) \times 10^{-2}$	S=1.4
Γ_{18}	$\tau^- \to$ hadron$^-$ $(\geq 2$ hadron$^0)\ \nu_\tau$	$(16.3 \pm 1.3) \times 10^{-2}$	
Γ_{19}	$\tau^- \to 2\pi^+ 3\pi^-\ (\geq 0$ neutrals$)\ \nu_\tau$	$(1.15 \pm 0.27) \times 10^{-3}$	
Γ_{20}	$\tau^- \to K^*(892)^- \nu_\tau$	$(1.43 \pm 0.31) \times 10^{-2}$	
Γ_{21}	$\tau^- \to K_2^*(1430)^- \nu_\tau$	$< 9 \times 10^{-3}$	CL=95%
Γ_{22}	$\tau^- \to K^*(892)^-\ (\geq 0$ neutrals$)\ \nu_\tau$	$(1.4 \pm 0.9) \times 10^{-2}$	
Γ_{23}	$\tau^- \to \pi^- \pi^0$ (non-resonant) ν_τ	$(3.0\ ^{+3.0}_{-1.9}) \times 10^{-3}$	
Γ_{24}	$\tau^- \to \pi^+ 2\pi^- \pi^0 \nu_\tau$	$(4.4 \pm 1.6) \times 10^{-2}$	
Γ_{25}	$\tau^- \to \pi^- 3\pi^0 \nu_\tau$	$(3.0 \pm 2.7) \times 10^{-2}$	
Γ_{26}	$\tau^- \to \pi^- 2\pi^0 \nu_\tau$	$(7.5 \pm 0.9) \times 10^{-2}$	
Γ_{27}	$\tau^- \to K^-$ (2 charged) $(\geq 0$ neutrals$)\ \nu_\tau$	$< 6 \times 10^{-3}$	CL=90%
Γ_{28}	$\tau^- \to \pi^+ 2\pi^- K^0\ (\geq 0\ \gamma)\ \nu_\tau$	$< 2.7 \times 10^{-3}$	CL=90%
Γ_{29}	$\tau^- \to \pi^- \pi^0 \eta \nu_\tau$	$< 2.1 \times 10^{-2}$	CL=95%
Γ_{30}	$\tau^- \to 7$hadrons$^\pm$ $(\geq 0 \pi^0)\ \nu_\tau$	$< 1.9 \times 10^{-4}$	CL=90%
Γ_{31}	$\tau^- \to K^- K^0 \nu_\tau$	$< 2.6 \times 10^{-3}$	CL=95%
Γ_{32}	$\tau^- \to K^- K^0 \pi^0 \nu_\tau$	$< 2.6 \times 10^{-3}$	CL=95%
Γ_{33}	$\tau^- \to \pi^- \eta\ (\geq 0$ neutrals$)\ \nu_\tau$	$< 2.1 \times 10^{-2}$	CL=95%
Γ_{34}	$\tau^- \to \pi^- \eta \eta\ (\geq 0$ neutrals$)\ \nu_\tau$	$< 5 \times 10^{-3}$	CL=90%
Γ_{35}	$\tau^- \to \pi^+ \pi^- \pi^- \eta\ (\geq 0$ neutrals$)\ \nu_\tau$	$< 3.0 \times 10^{-3}$	CL=90%
Γ_{36}	$\tau^- \to$ (1 charged) $(\geq 0$ neutrals$)\ \nu_\tau$		
Γ_{37}	$\tau^- \to$ hadron$^-$ $(\geq 1\ \pi^0)\ \nu_\tau$		
Γ_{38}	$\tau^- \to$ hadron$^-$ $(\geq 0$ neutrals$)\ \nu_\tau$		
Γ_{39}	$\tau^- \to \pi^+ 2\pi^-\ (\geq 0\ \pi^0)\ \nu_\tau$		
Γ_{40}	$\tau^- \to K^-\ (\geq 0$ neutrals$)\ \nu_\tau$		

Figure 4: Allowed decay modes of the τ from the *1988 Review of Particle Properties.*[53]

Table 3. World averages of major τ branching fractions from *Review of Particle Properties*[53] and Kiesling.[54]

Mode τ^-	Branching Fraction in %	
	Review of Particle Properties	Kiesling Review
1-charged particle	86.6 ± 0.3	86.5 ± 0.3
3-charged particle	13.3 ± 0.3	13.4 ± 0.3
5-charged particle	0.12 ± 0.3	0.14 ± 0.04
7-charged particle	< 0.019	
$e^-\overline{\nu}_e\nu_\tau$	17.5 ± 0.4	17.7 ± 0.4
$\mu^-\overline{\nu}_\mu\nu_\tau$	17.8 ± 0.4	17.8 ± 0.4
$\pi^-\nu_\tau$	10.8 ± 0.6	10.8 ± 0.6
$\rho^-\nu_\tau$	22.3 ± 1.1	22.3
$\pi^-2\pi^0\nu_\tau$	7.5 ± 0.9	

LEPTON NUMBER OR LEPTON FAMILY NUMBER VIOLATING MODES

Γ_{41}	$\tau^- \to \mu^-\gamma$	< 5	$\times 10^{-4}$	CL=90%
Γ_{42}	$\tau^- \to e^-\gamma$	< 6	$\times 10^{-4}$	CL=90%
Γ_{43}	$\tau^- \to \mu^-$ charged particles			
Γ_{44}	$\tau^- \to e^-$ charged particles			
Γ_{45}	$\tau^- \to \mu^-\mu^+\mu^-$	< 2.9	$\times 10^{-5}$	CL=90%
Γ_{46}	$\tau^- \to e^-\mu^+\mu^-$	< 3.3	$\times 10^{-5}$	CL=90%
Γ_{47}	$\tau^- \to \mu^-e^+e^-$	< 3.3	$\times 10^{-5}$	CL=90%
Γ_{48}	$\tau^- \to e^-e^+e^-$	< 4	$\times 10^{-5}$	CL=90%
Γ_{49}	$\tau^- \to \mu^-\pi^0$	< 8	$\times 10^{-4}$	CL=90%
Γ_{50}	$\tau^- \to e^-\pi^0$	< 2.1	$\times 10^{-3}$	CL=90%
Γ_{51}	$\tau^- \to \mu^-K^0$	< 1.0	$\times 10^{-3}$	CL=90%
Γ_{52}	$\tau^- \to e^-K^0$	< 1.3	$\times 10^{-3}$	CL=90%
Γ_{53}	$\tau^- \to \mu^-\rho^0$	< 4	$\times 10^{-5}$	CL=90%
Γ_{54}	$\tau^- \to e^-\rho^0$	< 4	$\times 10^{-5}$	CL=90%
Γ_{55}	$\tau^- \to e^-\pi^+\pi^-$	< 4	$\times 10^{-5}$	CL=90%
Γ_{56}	$\tau^- \to e^+\pi^-\pi^-$	< 6	$\times 10^{-5}$	CL=90%
Γ_{57}	$\tau^- \to \mu^-\pi^+\pi^-$	< 4	$\times 10^{-5}$	CL=90%
Γ_{58}	$\tau^- \to \mu^+\pi^-\pi^-$	< 6	$\times 10^{-5}$	CL=90%
Γ_{59}	$\tau^- \to e^-\pi^+K^-$	< 4	$\times 10^{-5}$	CL=90%
Γ_{60}	$\tau^- \to e^+\pi^-K^-$	< 1.2	$\times 10^{-4}$	CL=90%
Γ_{61}	$\tau^- \to \mu^-\pi^+K^-$	< 1.2	$\times 10^{-4}$	CL=90%
Γ_{62}	$\tau^- \to \mu^+\pi^-K^-$	< 1.2	$\times 10^{-4}$	CL=90%
Γ_{63}	$\tau^- \to e^-K^*(892)^0$	< 5	$\times 10^{-5}$	CL=90%
Γ_{64}	$\tau^- \to \mu^-K^*(892)^0$	< 6	$\times 10^{-5}$	CL=90%
Γ_{65}	$\tau^- \to e^+\mu^-\mu^-$	< 4	$\times 10^{-5}$	CL=90%
Γ_{66}	$\tau^- \to \mu^+e^-e^-$	< 4	$\times 10^{-5}$	CL=90%

Figure 5: Decay modes of the τ which violate lepton number conservation from the *1988 Review of Particle Properties*.[53]

average values in Table 4. But no explanation has been found. Nor has any explanation using unconventional physics been confirmed.

The CELLO collaboration has published[62] and submitted to this meeting[63] their recent measurement of τ branching ratios relevant to the 1-charged particle mode problem. Their branching ratios are

Table 4. The 1-charged particle decay mode problem from *Review of Particle Properties*.[53] All branching fractions are averages of experiments except the theoretical limits on unmeasured modes.

Decay Mode	Branching Fraction in %
$e^-\nu\nu$	17.7 ± 0.4
$\mu^-\nu\nu$	17.7 ± 0.4
$\rho^-\nu$	22.5 ± 0.9
$\pi^-\nu$	10.8 ± 0.6
$K^-(\geq 0 \text{ neutrals})\ \nu$	1.71 ± 0.29
$K^{*-}\nu, K^{*-} \to \pi^-(2\pi^0 \text{ or } K_L)$	0.5 ± 0.1
$\pi^-(2\pi^0)\nu$	7.4 ± 1.4
Sum of measured modes	78.3 ± 1.9
Theoretical limits on unmeasured modes	< 2.2
Sum of exclusive modes	80.5 ± 1.9
Measured 1-prong branching fraction	86.6 ± 0.3
Difference	$> 6.1 \pm 1.9$

Table 5. Comparison of τ branching fractions in percent relevant to 1-charged particle mode problem from Behrend *et al.*,[63] the CELLO collaboration, and the *Review of Particle Properties*.[53]

Decay Channel	CELLO	World Avg
$\tau \to e\nu\nu$	$18.4 \pm 0.8 \pm 0.4$	17.5 ± 0.4
$\tau \to \mu\nu\nu$	$17.7 \pm 0.8 \pm 0.4$	17.8 ± 0.4
$\tau \to \pi\nu$	$11.1 \pm 0.9 \pm 0.5$	10.8 ± 0.6
$\tau \to K\nu$		0.7 ± 0.2
$\tau \to \rho\nu$	$22.2 \pm 1.5 \pm 0.7$	22.3 ± 1.1
$\tau \to \pi 2\pi^0\nu$	$10.0 \pm 1.5 \pm 1.1$	7.5 ± 0.9
$\tau \to \pi \geq 3\pi^0\nu$	$3.2 \pm 1.0 \pm 1.0$	3.0 ± 2.7
B_1	$85.0 \pm 2.4 \pm 1.2$	86.6 ± 0.3

compared with the world average branching ratios in Table 5. Compared to the world average values, their $B(\pi^-2\pi^0\nu_\tau)$ is larger by 2.5%, their $B(e^-\overline{\nu}_e\nu_\tau)$ is larger by 0.9%, and their B_1 is smaller by 1.7%. Therefore the discrepancy of about 6% found using world averages, Table 4, disappears in the CELLO data. There is no objective and nonstatistical way to decide between the world averages picture of the 1-charged particle branching ratios and the CELLO picture. If the CELLO collaboration's results are averaged with world-average values, the CELLO results are overwhelmed by the world average values on the basis of statistics.

Thus the 1-charged particle decay modes problem remains unsolved.

D. *Tau Lifetime*

The world average tau lifetime[53] is

$$\tau_\tau(\text{measured}) = (3.04 \pm 0.09) \times 10^{-13}\ \text{sec}\ . \quad (37a)$$

The lifetime calculated[60] from world average values of $B(e^-\overline{\nu}_e\nu_\tau)$ and $B(\mu^-\overline{\nu}_\mu\nu_\tau)$ is

$$\tau_\tau(\text{calculated}) = (2.87 \pm 0.04) \times 10^{-13}\text{sec}\ . \quad (37b)$$

The difference is

$$\tau_\tau(\text{measured}) - \tau_\tau(\text{calculated}) = [0.17 \pm 0.10] \times 10^{-13}\ \text{sec}\ .$$

This is not a significant difference in my opinion.

E. *Tau Structure*

There is no evidence for structure in the τ. Using the notation of Eqs. (17)–(19), Ref. [37] uses measurements of

$$e^+ + e^- \to \tau^+ + \tau^-\ , \quad (38)$$

to give the 95% C.L. lower limits:

Form Factor	Contact Interaction (VV)	
$\Lambda_+ > 0.29$ TeV	$\Lambda_+^c > 4.1$ TeV	(39)
$\Lambda_- > 0.21$ TeV	$\Lambda_-^c > 5.7$ TeV	

F. *What We Want to Know About the Tau*

The new results on the Z^0 width, Sec. VI.A, mean that the τ is the last of the sequential leptons: a charged lepton with a small mass neutrino partner. Eventually other kinds of leptons may be found, but the only leptons upon which we can do research in the foreseeable future are the e, μ, τ and their neutrinos. Research on the τ is crucial because it has the most complex behavior in the lepton family, because of the existing puzzle in the 1-charged particle decay mode, and because the third generation of fermions may have special properties. There are many things we want to know about the τ and ν_τ. As shown in Table 6, the things we want to know will provide powerful probes for new phenomena in particle physics and basic tests of the standard model of particle physics.

Some progress in tau physics will come from the new Beijing e^+e^- storage ring and from CESR, DORIS, PEP, SLC, LEP. But a new kind of facility, a tau-charm factory, is required to carry out powerful and precise tau physics research.

VIII. TAU PHYSICS AT A TAU-CHARM FACTORY

A. *Tau-Charm Factory*

Figure 6 shows the particle physics range and energy range of a tau-charm factory as first proposed by Kirkby.[64] The tau-charm factory is dedicated to

(i) τ physics,

(ii) D physics,

(iii) D_s physics,

(iv) ψ/J, ψ', and other charmonium physics.

Reference [65] and the May 1989 Tau-Charm Factory Workshop proceedings[66] provide a large amount of information on this physics. In addition, Λ_c and other charm baryon physics is accessible.

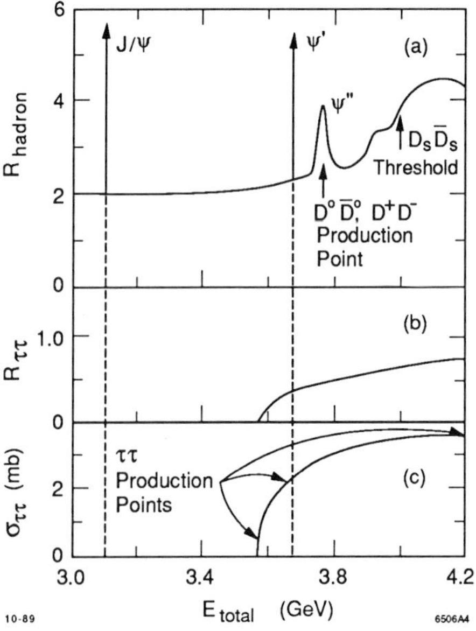

Figure 6: Main particle physics range of a tau-charm factory.

The first collider design was carried out by Jowett.[67] Figure 7 is a schematic and Table 7 gives the parameters of the present design of the machine. The basic design is:

Table 6. What we want to learn about the τ and ν_τ.

Subject	Search for New Physics	Test Standard Model	Tau-Charm Factory
Understand 1-charged particle modes puzzle	√	√	√
Untangle multiple π^0 and η modes in 1-charged particle modes		√	√
Precise measurement of $B(e\nu\nu)$, $B(\mu\nu\nu)$, $B(\pi\nu)$, $B(\rho\nu)$, and their ratio to 0.5%	√	√	√
Precise measurement of Cabibbo-suppressed modes	√	√	√
Full study of dynamics of $\tau \to e\nu\nu$, $\tau \to \mu\nu\nu$ analogous to $\mu \to e\nu\nu$ in detail	√	√	√
Detailed study of 3-, 5-, 7-charged particle modes		√	√
Find and study rare allowed modes such as radiative decays and second-class currents	√	√	√
Explore forbidden decay modes	√	√	√
Precise measurement of τ lifetime	√	√	
Explore ν_τ mass to a few MeV/c^2	√	√	√
Detect ν_τ	√	√	
Study interactions of ν_τ	√	√	
Precise low energy study of $e^+e^- \to \tau^+\tau^-$, $\tau^+\tau^-\gamma$	√	√	√
Precise high energy study of $e^+e^- \to \tau^+\tau^-$, $\tau^+\tau^-\gamma$	√	√	
Study of $Z^0 \to \tau^+\tau^-$	√	√	
Study of $W^- \to \tau^-\overline{\nu}_\tau$	√	√	
Measure $B(D^- \to \tau^-\overline{\nu}_\tau)$	?	√	√
Measure $B(D_s^- \to \tau^-\overline{\nu}\tau)$	?	√	√
Measure $B(B^- \to \tau^-\overline{\nu}_\tau)$	?	√	
Make and study $\tau^+\tau^-$ atom	√	√	?

(i) design luminosity = 10^{33} cm^{-2} s^{-1},

(ii) two-ring, e^+e^-, circular collider,

(iii) equal energies and 0^0 crossing angle,

(iv) dedicated e^+ and e^- injector.

Table 8 gives the tau-charm factory particle yields at $L = 10^{33}$ cm^{-2} s^{-1}.

B. *Tau Physics at a Tau-Charm Factory*

At existing e^+e^- colliders progress in tau physics has been, and continues to be, severely restricted by

(i) insufficient statistics,

(ii) inefficient tagging of τ pairs (3–30%),

(iii) double tagging of τ pairs mostly required,

(iv) impure data sets (5–50% backgrounds),

(v) no direct measurement of backgrounds,

(vi) no control of production energy.

Tau physics research at proposed B-factories will continue to be restricted by problems (ii) through (vi).

Tau physics at a tau-charm factory does not have these restrictions and problems. There are three energies (Fig. 6) at which

$$e^+ + e^- \to \tau^+ + \tau^-$$

data will be acquired: (a) several MeV above the threshold energy; (b) at 3.67 GeV, just below the ψ'; (c) at 4.2 GeV, where $\sigma(e^+e^- \to \tau^+\tau^-)$ is a maximum.

(a) $E_{total} = 2m_\tau + 2$ or 3 MeV/c^2:

(i) Due to Coulomb interaction $\sigma(e^+e^- \to \tau^+\tau^-)$ is not zero at threshold but is about

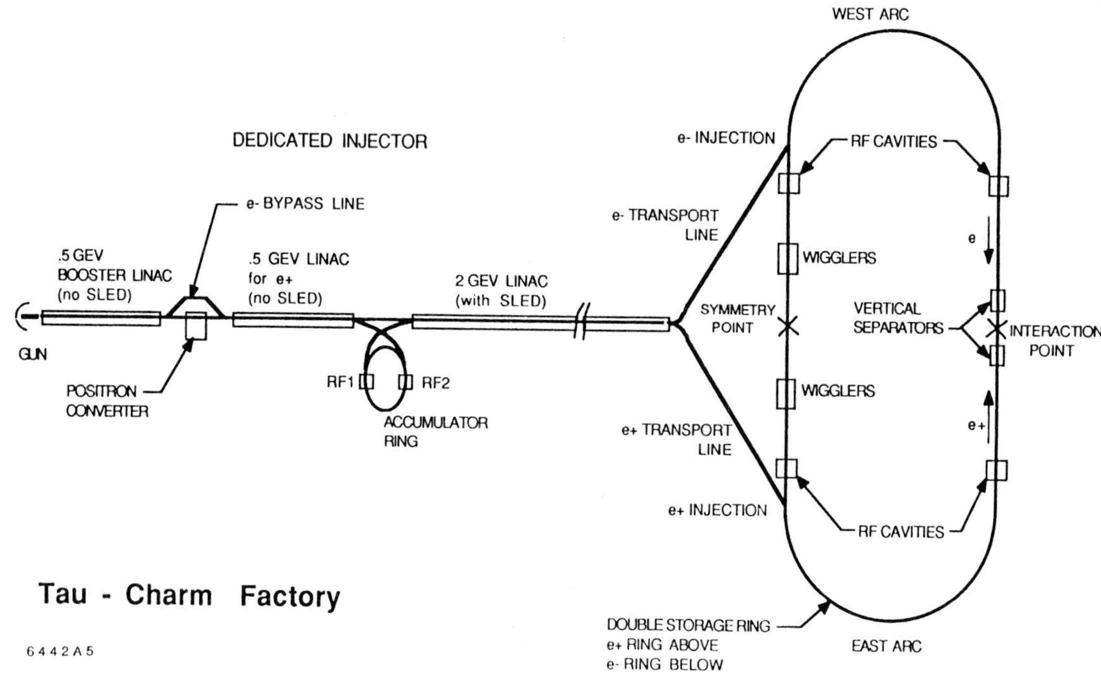

Figure 7: Schematic of a tau-charm factory two-ring collider and injector.

0.2 nb. Just 2 MeV above threshold it is about 0.4 nb. Here $\sigma_{E_{beam}} \approx 1$ MeV.

(ii) Single tagging can be carried out using $\tau^- \rightarrow \pi^- \nu_\tau$ because at threshold this π is kinematically separated[68] from the e^-, μ^-, and K^- of the other non-π^0 single-charged particle modes. Particle identification will also be used to isolate the π.

(iii) Single tagging will also be carried out using

$$\begin{aligned} \tau^- &\rightarrow e^- \\ &\quad + \text{missing energy} , \\ \tau^- &\rightarrow \mu^- \\ &\quad + \text{missing energy} . \end{aligned} \tag{40}$$

(iv) There is no background from D or B decays.

(v) Hadronic backgrounds are directly measured by going below threshold.

(b) $E_{total} = 3.67$ GeV, just below ψ':

(i) $\sigma(e^+e^- \rightarrow \tau^+\tau^-)$ is substantial, 2.0 nb.

(ii) Single tagging will be carried out using the decays in Eq. (40).

(iii) There is no background from D or B decays.

(iv) The light quark background and other backgrounds can be directly measured by reducing E_{total} by just 100 MeV to get below threshold.

(c) $E_{total} \approx 4.2$ GeV:

(i) $\sigma(e^+e^- \rightarrow \tau^+\tau^-)$ is a maximum, 2.8 nb.

(ii) Here $v_\tau/c \approx 0.5$, hence spin-dependant measurements can be carried out, such as the study of the dynamics of the $e\nu\nu$ and $\mu\nu\nu$ decay modes.

(iii) Single tagging will be carried out using the decays in Eq. (40).

(iv) There is no background from B decays.

(v) Other backgrounds can be studied by comparing with measurements at 3.67 GeV and at threshold.

Thus the tau-charm factory will produce very large samples of τ data, about 10^8 τ pairs in a few years; the data will be very pure. Therefore the tau-charm factory provides the best way, and in many cases the only way to answer the many experimental questions

Table 7. General parameters of the tau-charm factory at energy per beam of 2.2 GeV.

Energy	E	2.2 GeV
Circumference	C	329.9 m
Revolution frequency	f_o	0.909 MHz
Bending radius	ρ	12 m
β-function at IP	β_x^*	0.2 m
	β_y^*	0.01 m
Betatron coupling	κ^2	0.05
Betatron tunes	Q_x	8.87
	Q_y	7.76
Momentum compaction	α	0.0396
Natural emittance	ϵ_x	424 nm
Fractional energy spread	σ_ε	5.44 x 10^{-4}
Energy loss per turn	U_0	0.173 MeV
Damping times	τ_x	28 msec
	τ_y	28 msec
	τ_ε	14 msec
RF frequency	f_{RF}	350 MHz
RF voltage	V_{RF}	32 MV
Radiated power per beam	P_{rad}	86 kW
Number of bunches	k_b	21
Bunch separation	S_b	15.71 m
Bunch spacing	τ_b	52.4 ns
Bunch crossing frequency	f_b	19.1 MHz
Total beam current	I	498 mA
Particles per bunch	N_b	1.63 x 10^{11}
r.m.s. bunch length	σ_z	6 mm
Beam sizes at IP	σ_x^*	284 μm
	σ_y^*	14 μm
Beam-beam parameter	ξ_y	0.04
Luminosity	L	1.0 x 10^{33} $cm^{-2}s^{-1}$

Table 8. Tau-charm factory particle yields at 10^{33} cm^{-2} s^{-1}.

e^+e^- Energy	Particle	Produced Events
$J/\psi\,(3.095)$	$J\psi$	2×10^9/month
$\psi'\,(3.680)$	ψ'	8×10^8/month
$2m\,(\tau) + 2$ MeV	$\tau^+\tau^-$	7×10^6 pairs/year
3.67 GeV	$\tau^+\tau^-$	4×10^7 pairs/year
$\psi''\,(3.77)$	D^+D^-	3×10^7 pairs/year
$\psi''\,(3.77)$	$D^0\overline{D}^0$	4×10^7 pairs/year
4.14 GeV	$D_s^\pm D_s^{*\pm}$	10^7 pairs/year
4.2 GeV	$\tau^+\tau^-$	5×10^7 pairs/year

in tau physics, Table 6. Tau-charm factory designs are being considered in

France
Japan
Spain
USSR
USA

and perhaps other countries.

ACKNOWLEDGMENTS

I greatly appreciate conversations with, and review papers from, F. Boehm, L. Brown, K. Hayes, P. Herczeg, V. Khoze, P. Langacker, and A. Rich. For material related to the tau-charm factory, I am greatly indebted to the Tau-Charm Factory Machine Design Group (Appendix A), and the Tau-Charm Collaboration (Appendix B).

APPENDIX A

Tau-Charm Factory Machine Design Group

M. Allen	P. Morton
K. Bane	R. Miller
T. Barklow	R. Mozley
K. Brown	C. Munger
W. Brunk	A. Odian
Z. Chen	M. Perl
D. Coward	R. Ruth
E. Daly	W. Savage
D. Faust	R. Servranckx
S. Haynes	C. Spencer
T. Jenkins	S. St. Lorant
J. Kirkby	K. Thompson
A. Lisin	H. Weidner
G. Loew	D. Wright

APPENDIX B

The Tau-Charm Collaboration

B. Barish, R. Stroynowski, A. Weinstein
California Institute of Technology, Pasadena, CA 91125

P. Burchat, J. J. Gomez-Cadenas, C. A. Heusch, W. Lockman, A. Seiden
University of California at Santa Cruz, Santa Cruz, CA 95064

R. Johnson, B. Meadows, M. Nussbaum, M. Sokoloff, I. Stockdale
University of Cincinnati, Cincinnati, OH 45221

J. M. Jowett, J. Kirkby
CERN, CH-1211, Geneva 23, Switzerland

R. Eisenstein, G. Gladding, J. Izen, J. Thaler
University of Illinois at Urbana-Champaign, Urbana, IL 61801

R. Cowan, M. Fero, R. Yamamoto
Massachusettes Institute of Technology, Cambridge, MA 02139

J. Brau, R. Frey
University of Oregon, Eugene, OR 97403

T. Barklow, R. Bell, K. Brown, K. Bunnell, D. Coward, D. Faust, D. Fryberger, K. K. Gan, R. Gearhart, A. Lisin, G. Loew, R. Miller, C. Munger, A. Odian, J.M. Paterson, M. Perl, C. Rago, P. Saez, S. St. Lorant, W. Savage, R. Schindler, W. Toki, F. Villa, H. Weidner
Stanford Linear Accelerator Center, Stanford University, Stanford, CA 94309

T. Burnett, V. Cook, R. Mir, P. Mockett
University of Washington, Seattle, WA 98195

REFERENCES

1. A. Maki, *Proc. XIV Int. Symp. on Lepton and Photon Interactions,* Stanford, 1989, ed. M. Riordan; S. Uno, *Proc. 17th SLAC Summer Inst. on Particle Physics,* 1989, ed. E. C. Brennan; F. Boehm and P. Vogel, *Physics of Massive Neutrinos* (Cambridge Univ. Press, Cambridge, 1987); K. K. Gan and M. L. Perl, *Int. J. Mod. Phys.* **A3**, 531 (1988).
2. M. L. Perl, *Proc. of the XXIII Int. Conf. on High Energy Physics,* p. 596, Berkeley, 1986; D. Stoker *et al.*, *Phys. Rev.* **D39**, 1811 (1989); R. M. Barnett and H. E. Haber, *Phys. Rev.* **D36**, 2042 (1987); L. G. Mathis, Ph.D. thesis, Univ. of Calif., LBL-25261 (1985).
3. K. Riles, Ph.D. thesis, Stanford Univ., SLAC-342 (1989); K. Riles *et al.*, SLAC-PUB-5043 (1989).
4. C. Ahn *et al.*, SLAC-Report 329 (1988); P. Igo-Kemenes, *Proc. Workshop on Phys. at Future Accelerators,* CERN 87-07, Vol. II, p. 69; R. Kleiss and P. M. Zerwas, *ibid.*, p. 277.
5. I. Hinchliffe, LBL-25963 (1988).
6. R. S. Van Dyck, Jr., P. B. Schwinberg, and H. G. Dehmelt, *Phys. Rev. Lett.* **59**, 26 (1987).
7. H. Dehmelt, *Phys. Scripta* **T22**, 102 (1988).
8. L. S. Brown, DOE-ER-40423-07PB (1988).
9. T. Kinoshita, CERN-TH 5097/88 (1988).
10. T. Kinoshita and W. B. Lindquist, *Phys. Rev.* **D27**, 886 (1983); P. Cvitanovic and T. Kinoshita, *Phys. Rev.* **D10**, 4007 (1974).
11. C. I. Westbrook *et al.*, *Phys. Rev. Lett.* **58**, 1328 (1987).
12. G. S. Adkins, *Ann. Phys.* (NY) **146**, 78 (1983).
13. W. G. Caswell and G. P. Lepage, *Phys. Rev.* **A20**, 36 (1979).
14. A. Rich, private communication.
15. S. Orito *et al.*, *Phys. Rev. Lett.* **63**, 597 (1989).
16. H. Bokemeyer *et al.*, GSI-89-49 (1989).
17. W. Koenig *et al.*, *Phys. Lett.* **B218**, 12 (1989).
18. M. Davier, *Proc. XXIII Int. Conf. on High Energy Physics* (World Scientific, 1987), ed. S. C. Loken.
19. C. Kozhuharov, *Proc. XXIV Int. Conf. on High Energy Physics,* (Springer, Berlin, 1989), eds. R. Kotthaus and J. Kühn, p. 1447.
20. See for example, S. H. Connell *et al.*, *Phys. Rev. Lett.* **601**, 2242 (1988).
21. W. Braunschweig *et al.*, *Z. Phys.* **C37**, 171 (1988).
22. C. A. Hawkins and M. L. Perl, *Phys. Rev.* **D40**, 823 (1989).
23. C. A. Hawkins and M. L. Perl, SLAC-PUB-4949 (1989).
24. M. L. Perl, SLAC-PUB-4881 (1989).
25. PEGASYS *Proposal to Stanford Linear Accelerator Center,* 1988.
26. Y. S. Tsai, *Phys. Rev.* **D40**, 760 (1989).
27. J. Bailey *et al.*, *Nucl. Phys.* **B150**, 1 (1979).
28. V. W. Hughes, *Phys. Scripta* **T22**, 111 (1988).
29. See for example: M. Greco, *Nuovo Cimento* **100A**, 597 (1988).
30. P. Herezeg, *Proc. Rare Decay Symposium,* Vancouver, 1988 (World Scientific, Singapore, 1989); also issued as LA-UR-89-1576 (1989).
31. R. Engfer, *Ann. Rev. Nucl. Part. Sci.* **36**, 327 (1986).
32. T. Kinoshita and A. Sirlin, *Phys. Rev. Lett.* **2**, 177 (1959).

33. See μ branching ratio section of M. Aguilar-Benitez *et al.*, *Review of Particle Properties, Phys. Lett.* **B204**, 1 (1988).
34. R. D. Bolton *et al.*, *Phys. Rev.* **D38**, 2077 (1988).
35. V. Bellgardt *et al.*, *Nucl. Phys.* **B299**, 1 (1988).
36. D. H. White, LA–UR–87–356 (1986).
37. W. Bartel *et al.*, *Z. Phys.* **C30**, 371 (1986).
38. K. K. Gan, *Phys. Lett.* **209**, 95 (1988).
39. E. W. Kolb, D. N. Schramm, and M. S. Turner, *Neutrino Physics* (Cambridge Univ. Press, Cambridge, 1989), ed. K. Winter.
40. F. Boehm and P. Vogel, *Physics of Massive Neutrinos* (Cambridge Univ. Press, Cambridge, 1987).
41. P. Langacker, *New Directions in Neutrino Physics* (Fermilab, 1988); also issued as UPR–0386T (1989).
42. T. J. Bowles *et al.*, LA–UR–89–2010 (1989).
43. H. Kawakami *et al.*, INS–Rep–758, KEK preprint 89–60 (1989).
44. S. Boris *et al.*, *Phys. Rev. Lett.* **58**, 2019 (1987); S. Boris *et al.*, *Pis'ma Zh. Eksp. Teor. Fiz.* **45**, 267 (1987) [*Sov. Phys. JETP Lett.* **45**, 333 (1987)].
45. M. Fritschi *et al.*, *Phys. Lett.* **173B**, 485 (1980).
46. H. Kawakami *et al.*, *Phys. Rev. Lett.* **187B**, 198 (1987).
47. J. F. Wilkerson *et al.*, *Phys. Rev. Lett.* **58**, 2023 (1987).
48. F. Boehm, CALT–63–546 (1989).
49. P. Astier *et al.*, *Phys. Lett.* **B220**, 646 (1989).
50. J. K. Rowley, B. T. Cleveland, and R. Davis, Jr., *Solar Neutrinos and Neutrino Astronomy* (Am. Inst. Phys., New York, 1985), eds. M. L. Cherry, K. Lande, and W. A. Fowler.
51. Y. Totsuka, *Proc. XXIV Int. Conf. on High Energy Physics* (Springer, Berlin, 1989), eds. R. Kotthaus and J. Kühn.
52. K. S. Hirata *et al.*, *Phys. Rev. Lett.* **63**, 16 (1989); M. Koshiba, private communication.
53. M. Aguilar-Benitez *et al.*, *Review of Particle Properties, Phys. Lett.* **B204**, 1 (1988).
54. C. Kiesling, *High Energy Electron–Positron Physics* (World Scientific, Singapore, 1988), eds. A. Ali and P. Soding, p. 177.
55. M. L. Perl, *Les Rencontres de Physique de la Vallée d'Aoste,* 1988 (Editions Frontières, Gif-sur-Yvette Cedex, 1988), ed. M. Greco, p. 257; also issued as SLAC–PUB–4632 (1988).
56. F. J. Gilman and S. H. Rhie, *Phys. Rev.* **D31**, 1066 (1985).
57. F. J. Gilman, *Phys. Rev.* **D35**, 3541 (1987).
58. B. C. Barish and R. Stroynowski, *Phys. Rep.* **157**, 1 (1988).
59. K. K. Gan and M. L. Perl, *Int. J. Mod. Phys.* **A3**, 531 (1988).
60. K. G. Hayes and M. L. Perl, *Phys. Rev.* **D38**, 3351 (1988).
61. K. G. Hayes, M. L. Perl, and B. Efron, *Phys. Rev.* **D39**, 274 (1989).
62. H. J. Behrend *et al.*, *Phys. Lett.* **B222**, 163 (1989).
63. H. J. Behrend *et al.*, Paper No. 228 submitted to *XIV Int. Symp. on Lepton and Photon Interactions,* Stanford, 1989.
64. J. Kirkby, *Int. School of Phys. with Low-Energy Antiprotons,* (Erice, 1987) CERN-EP/87-210-91987; *Particle World* **1**, 27 (1989).
65. B. Barish *et al.*, SLAC–PUB–5053 (1989).
66. *Proc. Tau-Charm Factory Workshop,* (Stanford, 1989) SLAC–PUB–343, ed. L. Beers.
67. J. M. Jowett, CERN LEP–TH/87–56 (1987); CERN LEP–TH/88–22 (1988).
68. J. J. Gomez-Cadenas, C. A. Heusch, and A. Seiden, *Particle World* **1**, 10 (1989).

DISCUSSION

M. Samuels, Oklahoma State: To account for the discrepancy in the ortho-positronium decay-rate, the coefficient B of the next-order correction must be +350. Do you think this is possible, and do you believe the experiment is correct?

M. Perl: I don't know of any conclusion based on calculations as to the maximum size of the B coefficient. The University of Michigan experimenters who have measured the decay rate in gas are planning a new experiment studying the decay rate in vacuum.

M. Davier, Orsay: You may have given the impression that the CELLO analysis was aimed at making the few-percent corrections necessary in some decay channel to solve the τ 1-prong problem. This was not the case, in fact; the results came from a comprehensive study of all decay channels simultaneously analyzed in the sense that *all* $\tau^+\tau^-$ events were assigned to given final states. In this analysis, we could make sure that all τ decays were properly accounted for, and we believe this to be the first consistent treatment of τ decay modes. This is why it may not be correct to compare CELLO results with the "world average", where many pieces of data are put together from independent analyses on at most a few channels, with different and sometimes understated systematic effects.

As a final comment, I think you can also add Orsay to your list of laboratories presently studying a τ-charm factory.

M. Perl: The CELLO results may be the right answer to the problem in the 1-prong decay modes of the τ. But at this time there is no way to evaluate the correctness of the CELLO results relative to the many experiments that comprise the world average. We must wait for measurements with much larger statistics and smaller systematic errors. In the written version of my talk I have included the Orsay laboratory as one of the institutions studying τ-charm factory design and potential.

Decays of Strangeness, Charm and Beauty

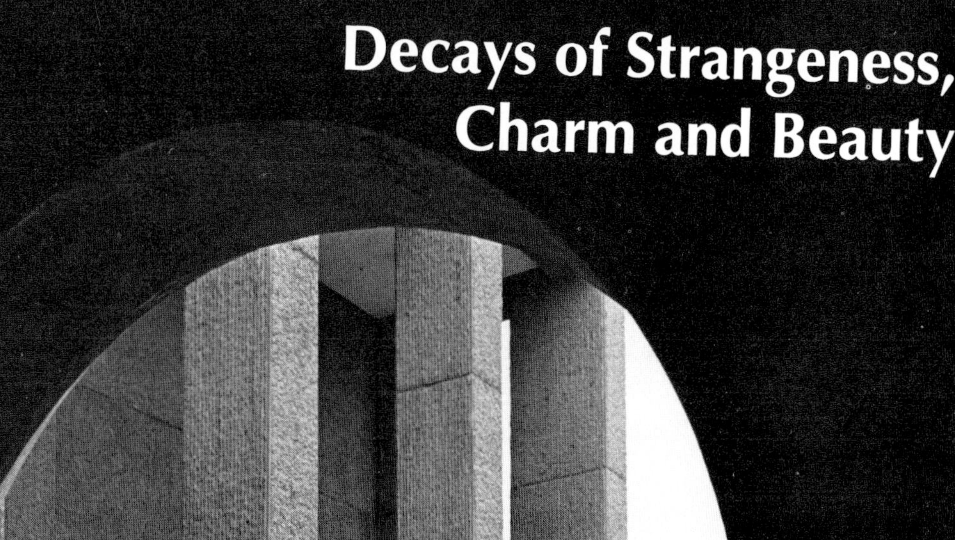

Decays of Strangeness, Charm and Beauty

Tuesday, 8 August 1989

Weak Decays of Charmed Particles 105

Speaker: P. E. Karchin

Chairperson: M. Davier

Scientific Secretary: P. Kim

Status and Physics Program of BEPC 122

Speaker: M. Ye

Chairperson: M. Davier

Scientific Secretary: R. De Sangro

B Physics from CLEO 129

Speaker: D. L. Kreinick

Chairperson: S. Yamada

Scientific Secretary: S. Wagner

Recent ARGUS Results on *B* Meson Decays 139

Speaker: M. V. Danilov

Chairperson: S. Yamada

Scientific Secretary: S. Wagner

A Measurement of ϵ'/ϵ by E731 155

Speaker: B. Winstein

Chairperson: M. Schwartz

Scientific Secretary: S. Dasu

NA31 Results on CP Violation in *K* Decays, and a Test of CPT 168

Speaker: D. Fournier

Chairperson: M. Schwartz

Scientific Secretary: S. Dasu

Rare Kaon Decays 184

Speaker: L. S. Littenberg

Chairperson: N. Cabibbo

Scientific Secretary: D. Pitman

Theory of Weak Decays*

Speaker: G. Martinelli

Chairperson: N. Cabibbo

Scientific Secretary: D. Lewellen

*This paper is not included in these Proceedings.

WEAK DECAYS OF CHARMED PARTICLES

Paul E. Karchin

Yale University
J. W. Gibbs Laboratory
P.O. Box 6666
New Haven, CT 06511

Abstract

The decay rates of charmed particles are generally understood from weak interaction theory. However, the detailed predictions for exclusive channels depend crucially on the quantitative treatment of effects due to strong interactions. Recent experimental measurements strengthen the basis for current models of charm decay but also pose new challenges for the prediction of form factors in semi-leptonic decays, for branching fractions of charmed mesons to many bodies and for the lifetimes and decay modes of the charmed baryons.

OVERVIEW

This review will highlight measurements submitted to this conference as well as recent work in this field, especially on semi-leptonic decays. There are new results presented on $D^0 \to \pi^-\pi^+$ and $D^0 \to K^-K^+$, which strengthen the theoretical interpretation of hadronic charm decays. Other results address the two, three, and four-body decays of charmed mesons, including an example of the analysis of resonance structure for a vector-vector mode. For the charm-strange meson, there are a number of updates, one on the $\phi\pi$ absolute branching fraction from Mark III, and others on $\eta\pi$ and $\eta'\pi$ decays. Brief comments are made concerning D^+ decays and charmed baryons.

A convention used in this paper is that the charge-conjugate state is implicitly included in discussion of a particle and its decay.

SEMILEPTONIC DECAYS

$D^0 \to K^-e^+\nu_e$ and $\pi^-e^+\nu_e$

The simplest semi-leptonic D^0 decays have a clear theoretical interpretation. They proceed via spectator processes in which the charm quark decays into a lower mass quark and a W, which subsequently decays into a positron and a neutrino as shown in Figure 1. The final state quark can result from either the Cabibbo-favored transition, which depends on the CKM element V_{cs}, or the Cabibbo-suppressed transition which depends on V_{cd}. The

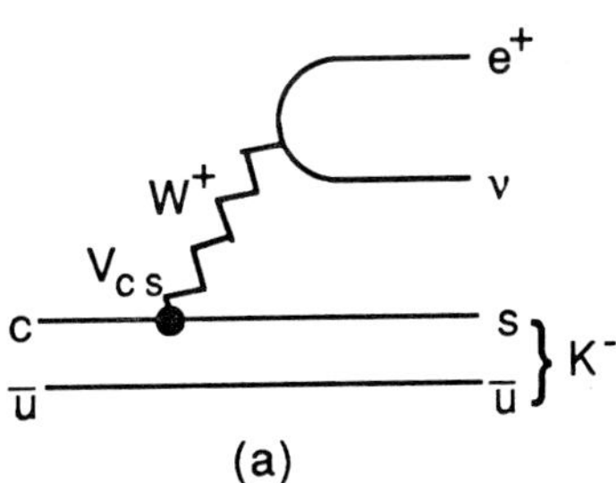

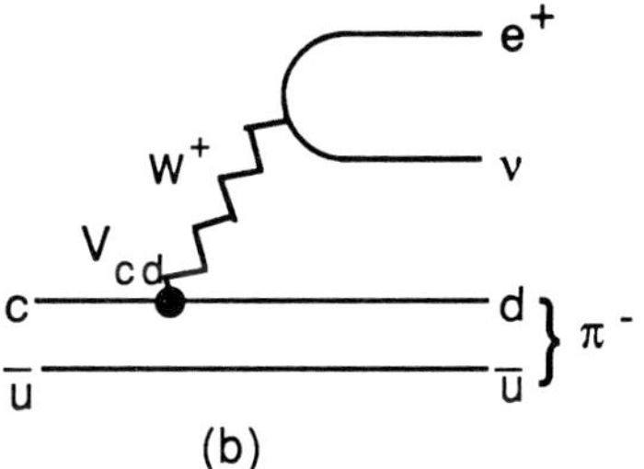

Figure 1: Semi- leptonic decay diagrams for D^0.

semi-leptonic decay rates are given by,

$$\Gamma(D^0 \to K^- e^+ \nu) \sim \mid V_{cs} \mid^2 \mid f_+^K(t=0) \mid^2 \qquad (1)$$

and,

$$\Gamma(D^0 \to \pi^- e^+ \nu) \sim \mid V_{cd} \mid^2 \mid f_+^\pi(t=0) \mid^2 \qquad (2)$$

where f_+^K and f_+^π are the vector form factors for the Kaon and pion, respectively. The form factors are functions of momentum transfer, e.g.,

$$t = q^2 = (\overline{P}_D - \overline{P}_K)^2, \qquad (3)$$

where $\overline{P}_D$ and $\overline{P}_K$ are the 4-momenta of the D and K, respectively. The form factors can be predicted by strong interaction models, and ultimately by lattice QCD calculations.

In principle, the fundamental parameters of the CKM matrix can be measured from these decays. However, in the CKM model, $\mid V_{cs} \mid$ and $\mid V_{cd} \mid$ are essentially the cos and sin of the Cabibbo angle, θ_c, which is well measured from strange particle decays. One tactic for now is to assume that the CKM model is correct and use the measured semi-leptonic rates to determine the form factors. At a future time, when the decay rates can be measured with much greater statistical precision and the form factors are well predicted from QCD, the CKM elements can be extracted from the decay rates to provide a consistency test of the CKM model.

There are some recent publications on these semi-leptonic decays (in April of this year) by Mark III [3] at SPEAR and the photo-production experiment E691 [10] at Fermilab. Both experiments measured the decay $D^0 \to K^- e^+ \nu_e$. The branching fraction, B, averaged from both experiments, is 3.5% ± 0.5%. B can be converted into a decay rate using the relation $\Gamma = B/\tau$ since the lifetime is accurately measured. This leads to the the relation,

$$\mid V_{cs} \mid^2 \mid f_+^K(t=0) \mid^2 = .54 \pm .08. \qquad (4)$$

Assuming the validity of the CKM model relation, $\mid V_{cs} \mid = cos(\theta_c)$, and the measured value, $cos(\theta_c) = .975$, Equation (4) gives $f_+^K(t=0) = .75 \pm .05$, which is in good agreement with a number of theoretical models [29,33,37]. Thus, the $D \to Ke\nu$ data are understood in the sense that if the CKM model coupling is assumed to be true, then the extracted form factor is consistent with theoretical expectations.

Now consider the Cabibbo-suppressed decay. In the Mark III measurement there are seven events for $D^0 \to \pi^- e^+ \nu_e$. The rate for this decay can be compared to that for $Ke\nu$ with the assumption that the form factors for the Kaon and the pion are the same at $q^2 = 0$. This gives,

$$\mid \frac{V_{cd}}{V_{cs}} \mid^2 = .057^{+.038}_{-.015} \pm .005, \qquad (5)$$

which is consistent with $sin^2(\theta_c)$.

$D^+ \to \overline{K^{*0}} e^+ \nu_e$

The decay $D^+ \to K^- \pi^+ e^+ \nu_e$ is more complicated than the decays $D^0 \to Ke\nu$ and $\pi e\nu$. It is possible that the K^- and π^+ come from $\overline{K^{*0}}$ decay. In this case, the simplest semi-leptonic decay of D^+ is to $\overline{K^{*0}} e^+ \nu_e$. Because the K^* has spin 1, there are 3 form factors involved in the decay $D^+ \to \overline{K^{*0}} e^+ \nu_e$ [38]: a vector form factor, V(t), and two axial vector form factors, $A_1(t)$ and $A_2(t)$, where

$$t = q^2 = (\overline{P}_D - \overline{P}_{K^*})^2. \qquad (6)$$

The Kinematics of the decay is described by t and three angles, defined in Figure 2. θ_v is the angle in the K^* rest frame between the π from the K^* decay and the boost direction of the D. θ_e is the angle in the W (e ν) rest frame between the electron and the boost direction of the D. χ is the angle between the plane containing θ_v and the plane containing θ_e. In principle, if there were experimental data with high enough statistics and low enough

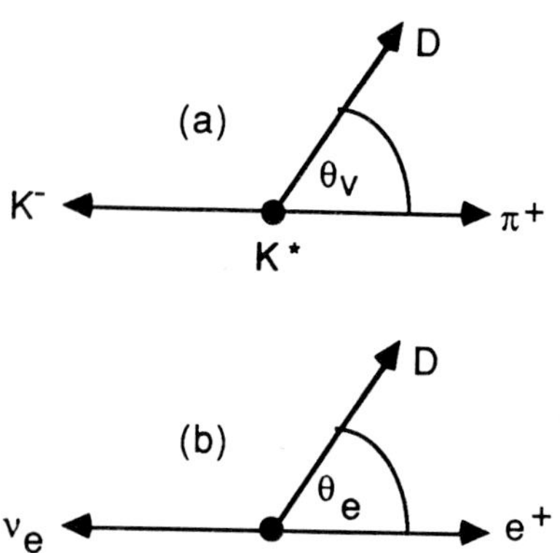

Figure 2: Angles describing the decay $D^+ \to \overline{K^{*0}} e^+ \nu_e$.

background, the three form factors could be measured from the distribution of events in the four Kinematic variables t, θ_v, θ_e, and χ. The status of this experimental program is discussed below.

A measurement of $D^+ \to K^-\pi^+e^+\nu_e$ from E691 was recently published [9]. In this experiment, electrons with energy greater than 12 GeV are well identified and measured with a sampling calorimeter. The K and π are identified with Čerenkov counters and their momenta are measured with a magnetic spectrometer. Since this is a fixed target experiment equipped with a silicon strip microvertex detector, the secondary vertex formed by the K, π, and e can be measured. This vertex is required to be significantly separated from the primary vertex, and the rest frame decay time is required to be more than 0.2 picoseconds. The $K\pi e$ invariant mass distribution will have a characteristic shape due to the missing 4-momentum of the neutrino. The experimental distribution is shown in Figure 3. There are 318 right sign $(K^-\pi^+e^+)$ events. The background can be estimated from the 66 events with the wrong sign signature $K^-\pi^+e^-$. As an additional check on the validity of the D^+ signal, the number of signal events surviving cuts of varying decay time are consistent with the Known lifetime.

The $K^-\pi^+$ mass spectrum, shown in Figure 4a, shows a large K^* signal. A maximum likelihood fit yields 227 ± 20 signal events with K^* and 25 ± 18 events with non-resonant $K\pi$. The fit includes the effect of the small background estimated

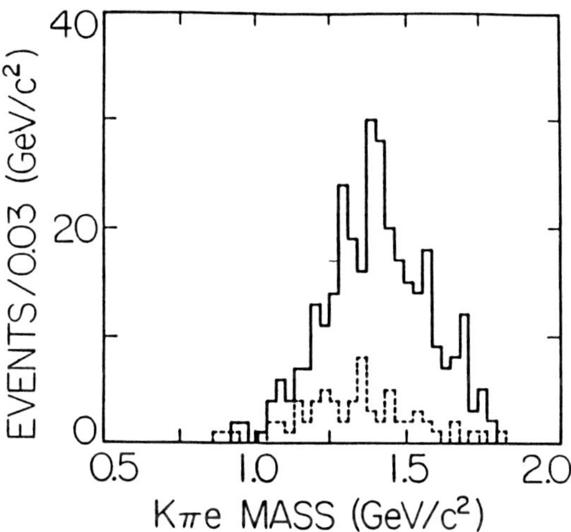

Figure 3: E691 signal for the decay $D^+ \to K^-\pi^+e^+\nu_e$ (solid line) and wrong-sign background (dotted line).

from the wrong sign combinations. As a check, the effect on the $K^-\pi^+$ mass spectrum of tighter cuts on the electron requirement, the Čerenkov requirement for the Kaon, and on the secondary vertex χ^2 is shown in Figure 4b.

Using the E691 result for $B(D^+ \to \overline{K^{*0}}e^+\nu_e)$ / $B(D^+ \to K^-\pi^+\pi^+)$ and the absolute branching fraction $B(D^+ \to K^-\pi^+\pi^+)$ from Mark III gives

$$B(D^+ \to \overline{K^{*0}}e^+\nu_e) = 4.5 \pm 0.7 \pm 0.5\%.$$

Similarly,

$$\begin{aligned} B(D^+ \to (K^-\pi^+)_{NR}e^+\nu_e) &= 0.3 \pm 0.2 \pm 0.2\%, or \\ &< 0.7\% \text{ @ } 90\% \; C.L. \end{aligned}$$

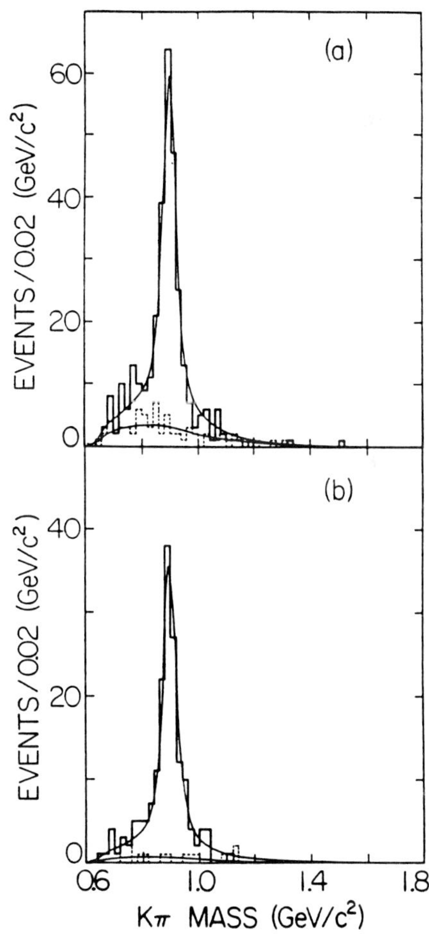

Figure 4: $K\pi$ mass distribution in $D^+ \to K^-\pi^+e^+\nu_e$ from E691 data. Standard cuts are used in (a); tight cuts are used in (b). Solid and dotted histograms are for right and wrong sign combinations, respectively. The solid curves are from maximum likelihood fits.

It is clear that nearly all of the decays $D^+ \to K^-\pi^+e^+\nu_e$ are due to the three body decay $D^+ \to \overline{K^{*0}}e^+\nu_e$.

The distribution of events, W, with angle θ_v, defined previously, depends on the polarization parameter, α, of the K^* according to the relation,

$$W(\theta_v) = 1 + \alpha cos^2(\theta_v), \qquad (7)$$

where the ratio of rates for longitudinal polarization to that for transverse polarization is, $\Gamma_L/\Gamma_T = (1+\alpha)/2$. An experimental difficulty arises in measuring θ_v because the neutrino momentum isn't measured. As a result, there are two solutions for the momentum of the D^+. From a Monte Carlo study it was found that the lower momentum solution is correct about 70% of the time, so this was the one taken. Other than this ambiguity, the experimental resolution in θ_v is dominated by the accuracy of measuring the direction of the D decay from the vector connecting the primary and secondary vertices.

The experimental distribution for $cos(\theta_v)$ is shown in Figure 5a. Only 9% of the events are background and only 3% are due to decays with non-resonant $K\pi$. A fit to the data gives, $\Gamma_L/\Gamma_T = 2.4^{+1.7}_{-0.9} \pm 0.2$, suggesting strong longitudinal polarization, although higher statistics are clearly needed. The distribution of Monte Carlo events generated with 100% longitudinal polarization of the K^* is shown in Figure 5b and looks similar to the experimental data.

This polarization of the K^* was not expected from earlier theoretical models. This discrepancy may be related to the failure of another theoretical expectation that the decays rates for $D^+ \to K^-\pi^+e^+\nu_e$ and $D^0 \to K^-e^+\nu_e$ should be equal. The E691 measurements of these decays, combined with the lifetime measurements from the same experiment, give

$$\frac{\Gamma(D^+ \to K^-\pi^+e^+\nu_e)}{\Gamma(D^0 \to K^-e^+\nu_e)} = .45 \pm .09 \pm .07.$$

More detailed theoretical and experimental work is required to understand the dependence of the $D^+ \to \overline{K^{*0}}e^+\nu_e$ on the four Kinematic variables and the relation to the decay rate.

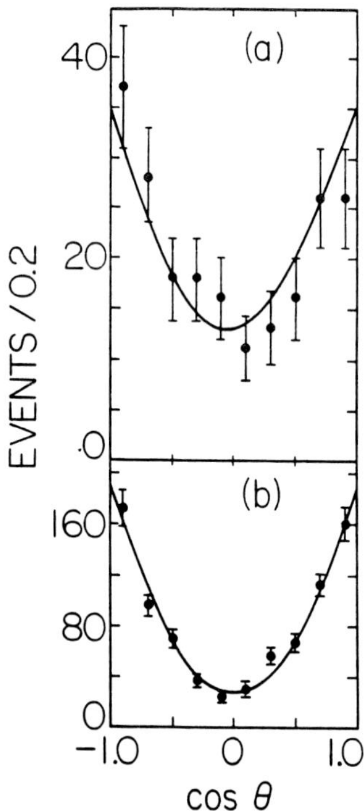

Figure 5: Distribution of $cos(\theta_v)$ from E691 data (a) and Monte Carlo (b) for the decay $D^+ \to \overline{K^{*0}}e^+\nu_e$.

HADRONIC D^0 DECAYS

$D^0 \to \pi^-\pi^+$ and $D^0 \to K^-K^+$

Consider two Cabibbo-suppressed 2-body decay modes: $D^0 \to \pi^+\pi^-$ and K^+K^-. These decays proceed by simple spectator diagrams, shown in Figure 6, The diagrams are of the external type

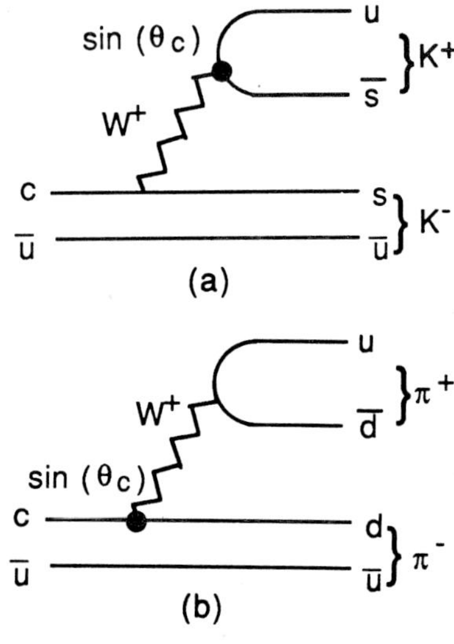

Figure 6: Diagrams for the decays $D^0 \to K^-K^+$ (a) and $D^0 \to \pi^-\pi^+$ (b).

since the decay products of the W don't combine with the spectator quarks. There is a Cabibbo-suppressed vertex either for the W-decay to $u\bar{s}$ (instead of the Cabibbo-favored $u\bar{d}$) or a $c \to d$ quark transition (instead of $c \to s$). If there was perfect SU(3) flavor symmetry in the final state, then the amplitudes of the diagrams would be the same. The relative rate between the two decays would be determined by the difference in phase space. In that case, the ratio of the rate to KK compared to $\pi\pi$ would be .86. Early measurements for this ratio were 3.4 ± 1.8 from Mark II (1979) [1] and 3.7 ± 1.4 from Mark III (1985) [13]. Although the errors are large, these results suggest a rather big SU(3) flavor symmetry breakdown. There are models for hadronic charm decays [16,17,19,20,21,22] which take into account the effect of the final state interactions causing SU(3) breaking. Predictions [18,36] for this ratio are in the range of 1.4 to 2.7. Since it is difficult for models to give a ratio of 3.7, this has been a puzzle for a few years. It is nice that there are new data from ARGUS [6] and CLEO [25] which address this problem.

The CLEO experiment at the CESR e^+e^- storage ring at Cornell detects D pair production in the region of the Υ resonances. Since the higher momentum D's can be reconstructed with lower background, a cut is made requiring that the D carry a momentum fraction x_D of at least half the maximum possible. Particle identification is based on the energy deposit (dE/dx) in the drift chamber. The well Known technique of tagging the D^0 decays coming from D^* is used to reduce the combinatoric background. There is also a requirement that the KK or $\pi\pi$ mass be within two detector resolutions of the Known D mass. The $D^* - D$ mass plots from CLEO are shown in Figure 7. There are clear D^* signals for both KK and $\pi\pi$. To estimate the KK and $\pi\pi$ signals, it is necessary to take into account the backgrounds due to $D^0 \to K^-\pi^+$ when the K is mistaken as a π or the π is mistaken as a K. These backgrounds are estimated to be 20 ± 13 events for KK and 32 ± 11 for $\pi\pi$. The signals are 372 ± 20 for KK and 197 ± 22 for $\pi\pi$. This gives the ratio of branching fractions,

$$\frac{B(D^0 \to K^-K^+)}{B(D^0 \to \pi^-\pi^+)} = 2.22 \pm .036(stat.) \pm 0.30(syst.).$$

This is a more precise determination than before. The new ratio is lower than before which makes it easier to interpret in the hadronic decay models.

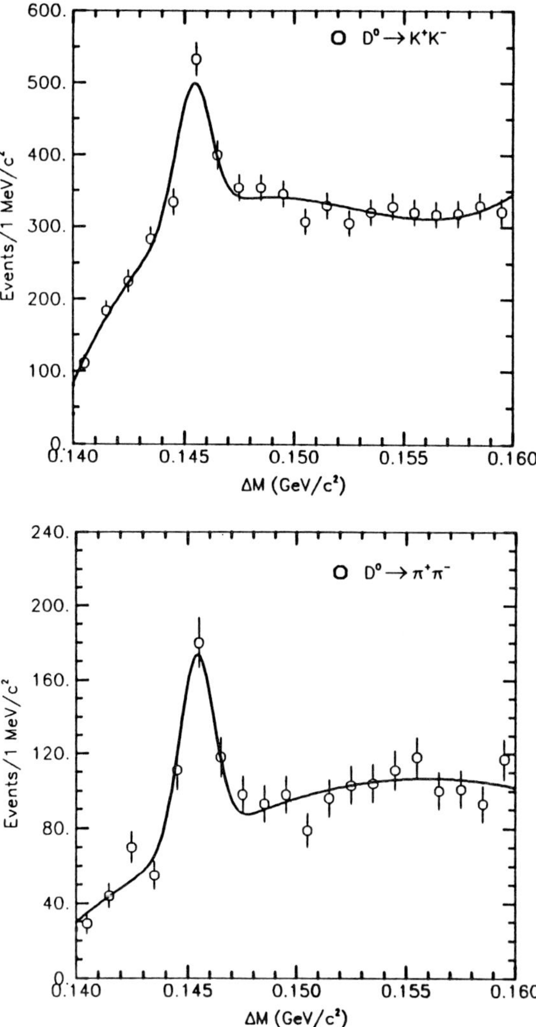

Figure 7: CLEO signals for $D^0 \to K^-K^+$ and $D^0 \to \pi^-\pi^+$.

There are also data on these modes from the ARGUS experiment at the DORIS II storage ring at DESY operating at the Υ resonances. The ARGUS analysis employs cuts on x_D, D mass, and D^* tagging similar to those previously described for the CLEO analysis. For ARGUS, time of flight as well as dE/dx information is used for particle identification. To further reduce the background, cuts are made on the decay angle distribution. In the D^0 rest frame, the angular distribution of one of the two decay products with respect to the D^0 boost is isotropic because the D^0 and its decay products all have spin 0. On the other hand, the combinatoric background tends to peak at large values of $| cos(\theta) |$. A cut on $| cos(\theta) | < 0.85$ thus reduces the background.

The decay angle distribution can be further exploited to reduce the previously discussed backgrounds due to misidentified pions and Kaons from $D^0 \to K^-\pi^+$. These backgrounds are studied using data for $D^0 \to K^-\pi^+$. A scatter plot of the difference between the invariant mass of the misidentified pair of particles and the D^0 mass versus the decay angle in the D^0 rest frame is shown in Figure 8a for false $D^0 \to \pi^-\pi^+$ events and in Figure 8b for false $D^0 \to K^-K^+$ events. The background populates a well defined region at a corner of each plot. In contrast, the signal has a Gaussian distribution in the center of the plot (with σ = 22 MeV/c^2 for $\pi\pi$) and a uniform distribution in $cos(\theta)$. A cut can be made to completely exclude the background while retaining most of the signal. After application of these cuts, the D^* mass plots are shown in Figures 9a and 9b. The number of signal events are 57 ± 9 for $\pi\pi$ and 131 ± 20 for KK. These signals have less background than those from CLEO, but the number of events is smaller.

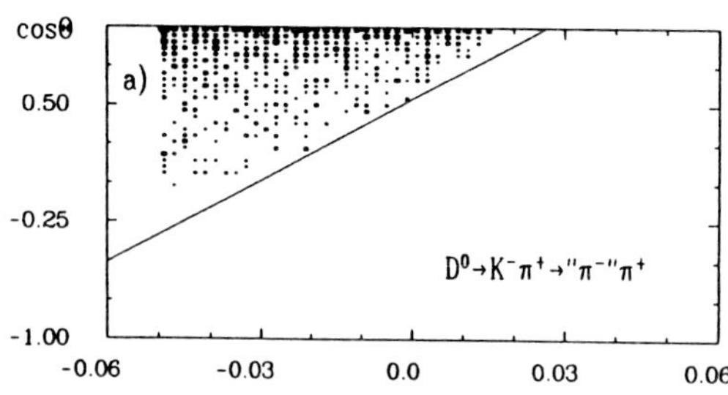

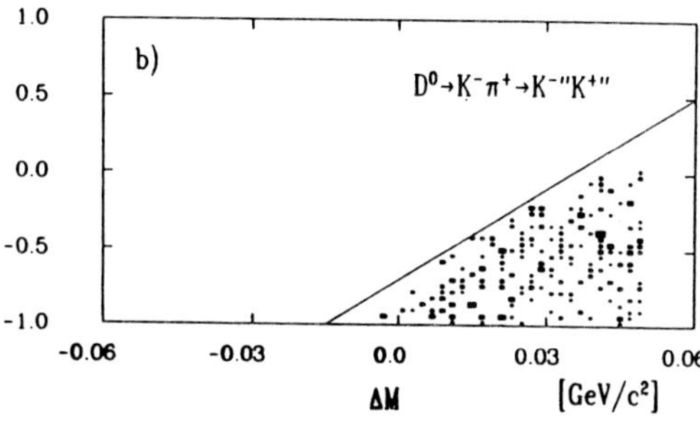

Figure 8: Study of backgrounds to $D^0 \to \pi^-\pi^+$ (a) and $D^0 \to K^-K^+$ (b) from ARGUS.

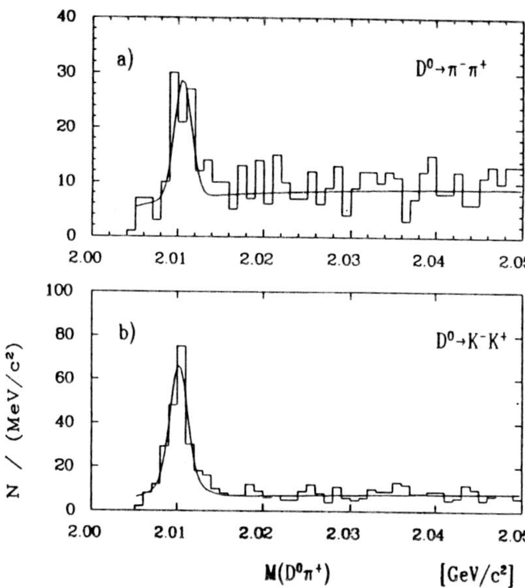

Figure 9: ARGUS signals for $D^0 \to \pi^-\pi^+$ (a) and $D^0 \to K^-K^+$ (b).

The Mark II, Mark III, ARGUS, and CLEO results for the relative branching fractions are summarized in Table 1. The new measurements of $2.5 \pm .7$ and $2.2 \pm .5$ for the KK to $\pi\pi$ ratio are both more accurate and smaller than before, thus removing the worst obstacle to the models of hadronic charm decay. The breaking of SU(3) flavor symmetry is substantial, but is now measured to be in a range that can be accommodated in these models.

Table 1: Relative branching fractions for the decays $D^0 \to \pi^-\pi^+$, K^-K^+, and $K^-\pi^+$.

expt.	$B(K^-K^+)/B(K^-\pi^+)$	$B(\pi^-\pi^+)/B(K^-\pi^+)$	$B(K^-K^+)/B(\pi^-\pi^+)$
Mark II	$.113 \pm .030$	$.033 \pm .015$	3.4 ± 1.8
Mark III	$.122 \pm .018$	$.033 \pm .010$	3.7 ± 1.4
ARGUS	$.10 \pm .02 \pm .01$	$.04 \pm .007 \pm .006$	2.5 ± 0.7
CLEO	$.11 \pm .008 \pm .007$	$.046 \pm .006 \pm .006$	2.2 ± 0.5

$D^0 \to K^0\overline{K^0}$

For the decay $D^0 \to K^0\overline{K^0}$ there are two Cabibbo-suppressed decay diagrams shown in Figure 10. Because the amplitudes have opposite signs the decay rate for this channel is zero in the limit of

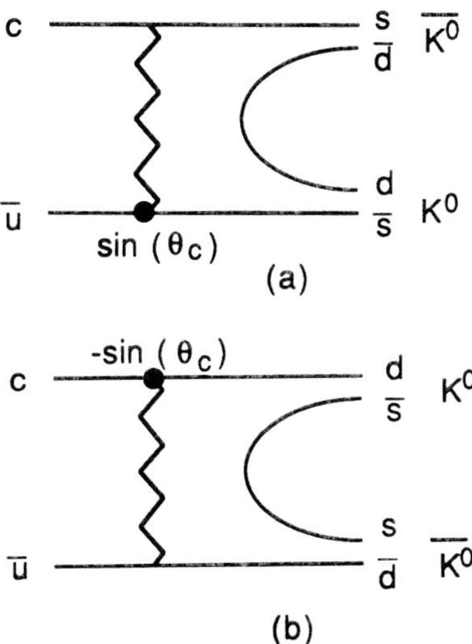

Figure 10: Diagrams for $D^0 \to K^0\overline{K^0}$.

SU(3) flavor symmetry. Any observed rate for this decay is due to the effect of final state interactions. Pham [34] predicted in 1987 that the decay rate is about half that for $D^0 \to K^-K^+$, which is also Cabibbo-suppressed. There is a recent measurement of $D^0 \to K^0\overline{K^0}$ from CLEO [26,30] and a new result from ARGUS [6]. The $K^0_sK^0_s$ mass plot from CLEO is shown in Figure 11a, with no D^* cut and in 11b with a D^* cut. The D^* cut removes most of the background but with loss of signal as well. Still, there is a very clean signal for this unusual decay mode. Other cuts, such as the one exploiting the hard momentum spectrum, have been included in the analysis. Some of the background present without the D^* cut comes from D^+_s and $D^+ \to K^{*+}K^0$. The fit shown does not include the parameterization of those backgrounds since there would be little effect on the signal estimate of 12 ± 5 events. The branching fraction computed from the five clean events of Figure 11b is used for comparison with other experiments in Table 2. As

Table 2: Measurements of $B(D^0 \to K^0\overline{K^0})$.

expt.	normalizing mode	$B(D^0 \to K^0\overline{K^0})$ (in percent)
E400	$D^0 \to K^+K^-$	$.10 \pm .08$
CLEO	$D^0 \to K^0_s\pi^+\pi^-$	$.11^{+.06}_{-.04} \pm .02$
ARGUS	$D^0 \to K^0_s\pi^+\pi^-$	$< .11$ @ 90% C.L.

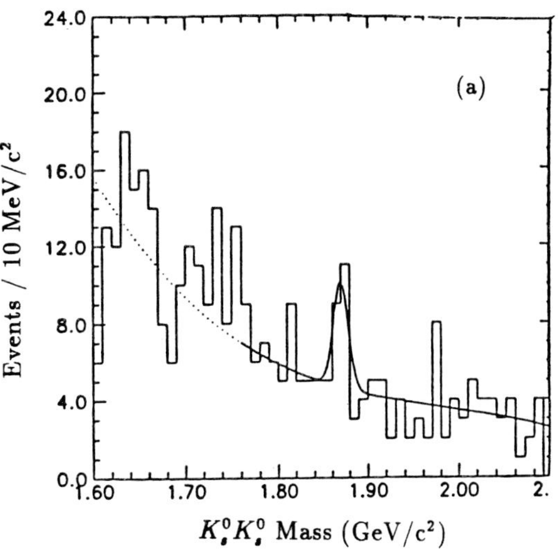

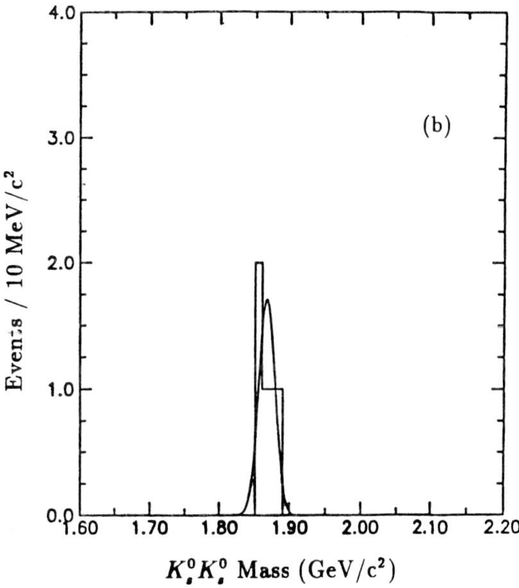

Figure 11: Signals for $D^0 \to K^0_sK^0_s$ from CLEO, (a) without D^* cut, (b) with D^* cut.

shown in the table, there is a previous measurement [27] of .1% for this decay rate from Fermilab experiment E400, using a neutron beam on a fixed target. The E400 result, the recent CLEO measurement of .11% and the new ARGUS upper limit are in good agreement. Measurement of $D^0 \to K^0K^0$ provides a valuable gauge for estimating the strength of the final state rescattering.

$D^0 \to K_s^0 K_s^0 K_s^0$

The decay $D^0 \to K_s^0 K_s^0 K_s^0$ is reminiscent of $D^0 \to \phi K_s^0$, which some may remember as the "smoking gun" for the exchange diagram. Although these decays can proceed by W exchange, as shown in Figure 12, they can also proceed through spectator diagrams with final state interactions. The

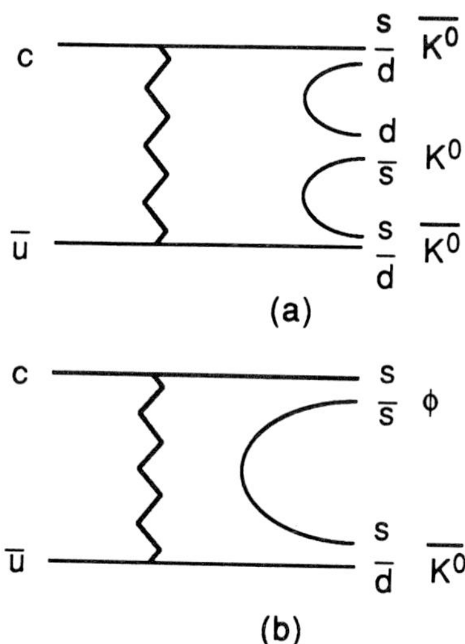

Figure 12: W-exchange diagrams for $D^0 \to \overline{K^0} K^0 \overline{K^0}$ and $D^0 \to \phi \overline{K^0}$.

ARGUS collaboration has reported to this Symposium the first observation of the decay $D^0 \to K_s^0 K_s^0 K_s^0$ [6]. The $3K_s^0$ mass plot is shown in Figure 13. There is a nearly background-free signal of 5.5 ± 2.5 events. The relative branching fraction measured is,

$$\frac{B(D^0 \to K_s^0 K_s^0 K_s^0)}{B(D^0 \to K_s^0 \pi^+ \pi^-)} = .034 \pm .014 \pm .010$$

Using the absolute branching fraction for $D^0 \to K_s^0 \pi^+ \pi^-$ from Mark III, gives $B(D^0 \to K_s^0 K_s^0 K_s^0) = (.11 \pm .05)\%$. To convert this to the branching fraction for $D^0 \to \overline{K^0} K^0 \overline{K^0}$ the quantum statistics of the final state must be taken into account. Then a comparison can be made with B$(D^0 \to K_s^0 \phi) \sim 1\%$.

Resonant Structure of $D^0 \to K^- \pi^+ \pi^- \pi^+$

Most theoretical work on hadronic charm meson decays has been done for two-body decays. While the pseudoscalar-pseudoscalar (PP) and pseudoscalar-vector (PV) decays are well explored experimentally, there is not much information on vector-vector (VV) decays. These decays are measured through study of multi-body channels, e.g., $D^0 \to K^- \pi^+ \pi^- \pi^+$. Because of the complexity of the resonance analysis, a large sample of events is required. Mark III has reconstructed over 1000 $D^0 \to K^- \pi^+ \pi^- \pi^+$ decays [4,28] produced at the $\psi(3770)$ resonance which decays exclusively to $D\overline{D}$. By constraining the invariant mass of the K3π, the individual particle momentum measurements are improved. This gives a better measurement of the invariant mass recoiling against the charm candidate state. The recoil mass is given by,

$$M = \sqrt{E_{beam}^2 - P^2}$$

where P is the total measured momentum of the K3π. The recoil mass distribution is shown in Figure 14. The strong interaction resonances considered in the analysis are listed in Table 3, including their widths and decay modes. In the resonance analysis, a complex amplitude is assigned to each of the resonant channels. Interference can result

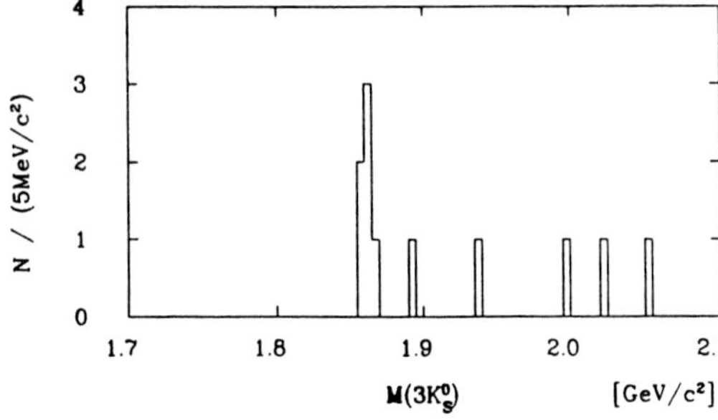

Figure 13: Signal for $D^0 \to K_s^0 K_s^0 K_s^0$ from ARGUS.

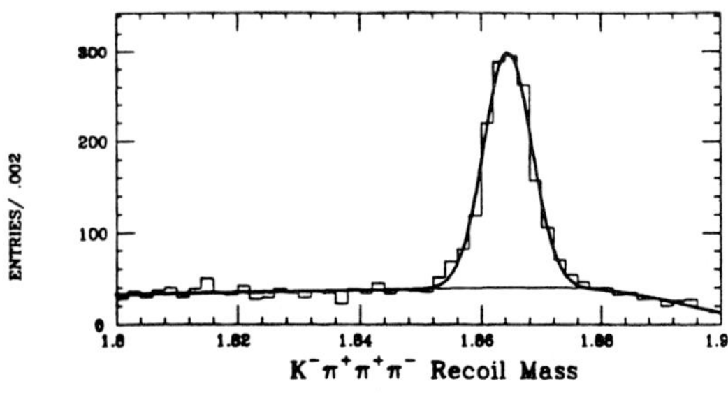

Figure 14: Mark III signal for $D^0 \to K^- \pi^+ \pi^- \pi^+$.

Table 3: Properties of resonances accessible in the decay $D^0 \to K^-\pi^+\pi^-\pi^+$.

resonance	decay mode(s)	width Γ (MeV)
$\overline{K^{*0}}$	$K^-\pi^+$	52
ρ^0	$\pi^+\pi^-$	153
$a_1(1260)^+$	$\rho^0\pi^+$	300-600
$K_1(1270)^-$	$K^-\rho^0, \overline{K^{*0}}\pi^-$	90
$K_1(1400)^-$	$\overline{K^{*0}}\pi^-$	184

from the square of the summed amplitudes. The polarization states must also be considered. The $\overline{K^{*0}}\rho$ final state is assumed to be in an $l = 0$ state where either the $\overline{K^{*0}}$ and ρ are both transversely polarized or the $\overline{K^{*0}}$ and the ρ are both longitudinally polarized.

A global fit is made with all the amplitudes as a function of the kinematic variables. A projection of the global fit can be made onto a particular sub-component. For example, in Figure 15(a), the 3 pion mass plot is shown. Because the a_1 mass at 1.260 GeV is near the edge of phase space, the characteristic Breit-Wigner shape is not visible, and instead a modified structure is seen. Also shown are peaks due to ρ in Figure 15(b) and $\overline{K^{*0}}$ in Figure 16(b). These various projections illustrate the complexity of the analysis.

The results of the fit for the branching fractions to various sub-channels are given in Table 4. The largest component is the $K^-a_1^+(1260)$. The sum of the $\overline{K^{*0}}\rho$ branching fractions from both polarization states is only about 2%. This contribution is in agreement with the model of Bauer, Stech, and Wirbel [16], where a 2.5% branching fraction to $\overline{K^{*0}}\rho$ was predicted. The experimental study of these VV decay modes is just beginning.

Table 4: Mark III results for $D^0 \to K^-\pi^+\pi^-\pi^+$ sub-channels.

mode	branching fraction
$(K^-\pi^+\pi^-\pi^+)_{NR}$	$.026 \pm .004 \pm .010$
$(\overline{K^{*0}}\rho^0)_{long.}$	$.005 \pm .002 \pm .003$
$(\overline{K^{*0}}\rho^0)_{tran.}$	$.016 \pm .003 \pm .003$
$K^-a_1(1260)^+$	$.087 \pm .009 \pm .020$
$K_1(1270)^-\pi^+$	$.019 \pm .004 \pm .005$
$K_1(1400)^-\pi^+$	$.006 \pm .002 \pm .004$

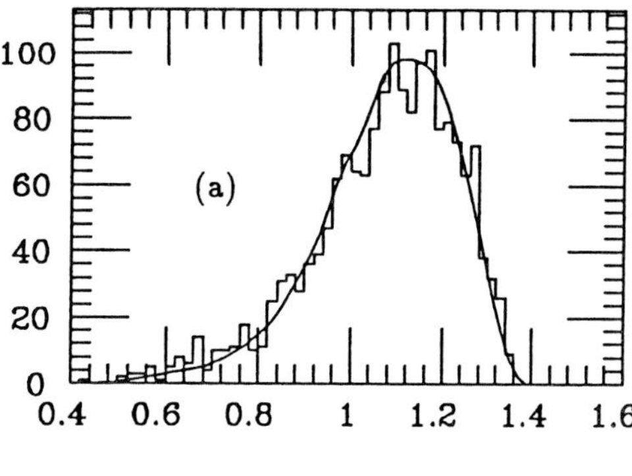

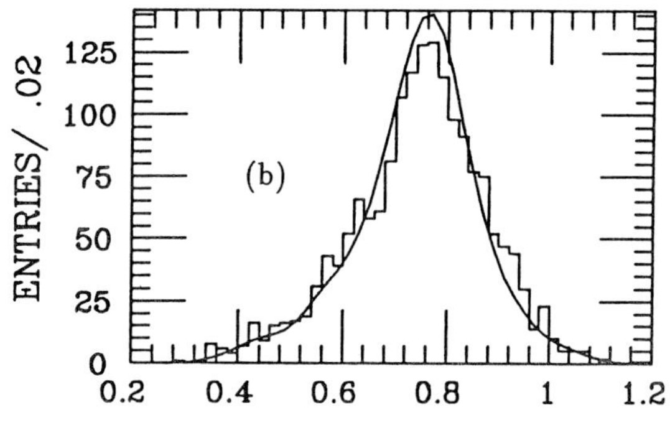

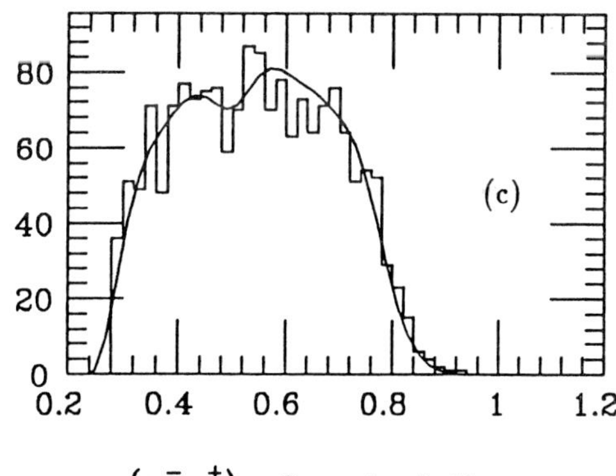

Figure 15: Resonance analysis (Part I) of $D^0 \to K^-\pi^+\pi^-\pi^+$ from Mark III.

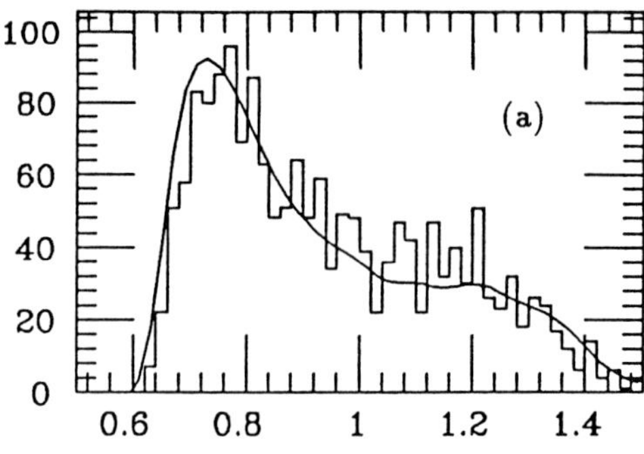

$K^-\pi^-$ Invariant Mass

ENTRIES/ .02

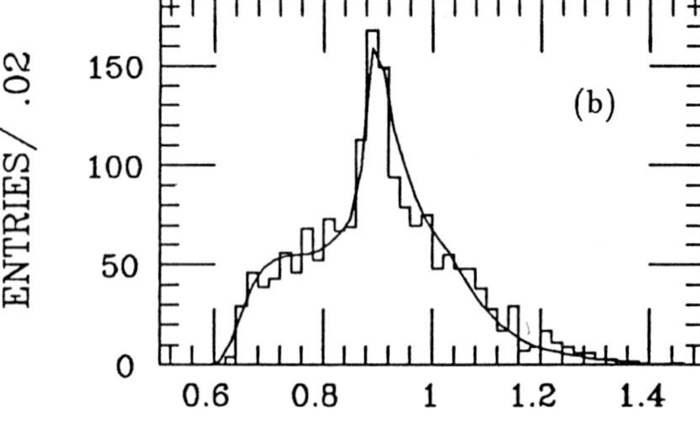

$(K^-\pi^+)_1$ Invariant Mass

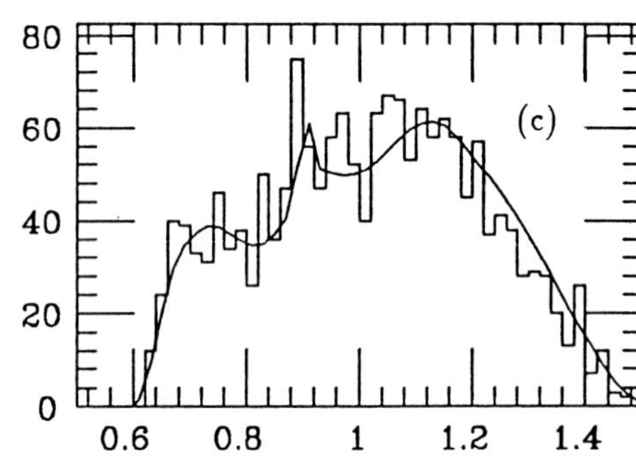

$(K^-\pi^+)_2$ Invariant Mass

Figure 16: Resonance analysis (Part II) of $D^0 \to K^-\pi^+\pi^-\pi^+$ from Mark III.

Miscellaneous New Measurements

A variety of new measurements of Cabibbo-favored D^0 decays by ARGUS [6] and CLEO [25] are listed in the first two columns of Table 5. Some of these

Table 5: Measurements of Cabibbo- favored D^0 branching fractions (in percent).

mode	ARGUS	CLEO	Mark III
$\overline{K^0}\pi^0$	$1.7 \pm 0.4 \pm 0.3$	$2.3 \pm 0.4 \pm 0.5$	$1.8 \pm 0.2 \pm 0.2$
$\overline{K^0}\eta$	$1.4 \pm 0.5 \pm 0.3$		$1.6 \pm 0.6 \pm 0.4$
$\overline{K^0}\eta'$	$1.9 \pm 0.4 \pm 0.3$		
$\overline{K^0}\omega$	$2.2 \pm 0.7 \pm 0.6$	$3.4 \pm 0.9 \pm 1.0$	$3.2 \pm 1.3 \pm 0.8$
$\overline{K^{*0}}\eta'$	< 0.2 @ 90% CL		
$K^-\pi^+\pi^-\pi^+\pi^0$	$4.5 \pm 0.8 \pm 0.9$	$5.0 \pm 0.7^{+1.3}_{-1.0}$	
$\overline{K^{*0}}\omega$	$2.0 \pm 0.5 \pm 0.3$		
$\overline{K^{*0}}\eta$	< 2.9 @ 90% CL	$2.5 \pm 0.8^{+1.0}_{-1.3}$	

measurements are the first observation of a decay mode. These decays have final states with a K^0 or photons resulting from π^0, η, or η' decay. Nearly all these decays have branching fractions in the range 1-5%, consistent with theoretical expectations for Cabibbo-favored decays proceeding through internal spectator diagrams, where the W decay products interact with the spectator quarks.

There are also new measurements of Cabibbo-suppressed decays and resonance analyses for them from ARGUS and CLEO. These are summarized in Table 6. These modes have branching fractions

Table 6: Measurements of Cabibbo- suppressed D^0 branching fractions (in percent).

mode	ARGUS	CLEO
$K^0_s K^-\pi^+$	$.16 \pm .03 \pm .02$	
$K^{*+}K^-$	$.10 \pm .04 \pm .02$	
$K^0_s\overline{K^{*0}}$	$<.03$ @ 90% CL	
$\phi\pi^+\pi^-$		$.29 \pm .09 \pm .08$
$\phi\rho$		$.34 \pm .11$

in the range (0.1 to 0.4)%, consistent with expectation for Cabibbo-suppressed decays.

HADRONIC D_s^+ DECAYS

Absolute Branching Fraction to $\phi\pi$

It is of great interest to establish the absolute branching fraction for any decay mode of the D_s, the charm-strange meson. Such a measurement has been pursued with the Mark III experiment operating at a center of mass energy of 4.14 GeV. At this energy, the reaction

$$e^+e^- \to D_s^{*+}D_s^-, D_s^{*+} \to \gamma D_s^+$$

takes place. A "double tagged" analysis can be made in which both the D_s^+ and D_s^- are reconstructed. The masses of the D_s^+ and the D_s^- are constrained to be equal, but that value of mass is not fixed. All the Known D_s^+ channels which have resonances are used in the analysis. These include the modes $\phi\pi^+, \overline{K^0}K^+, f_0(975)\pi^+, \overline{K^{*0}}K^+, \overline{K^{*0}}K^{*+}, \phi\pi^+\pi^-\pi^+$ and $\phi\pi^+\pi^0$. The mass plot is shown in Figure 17. There are no candidate events in the mass range

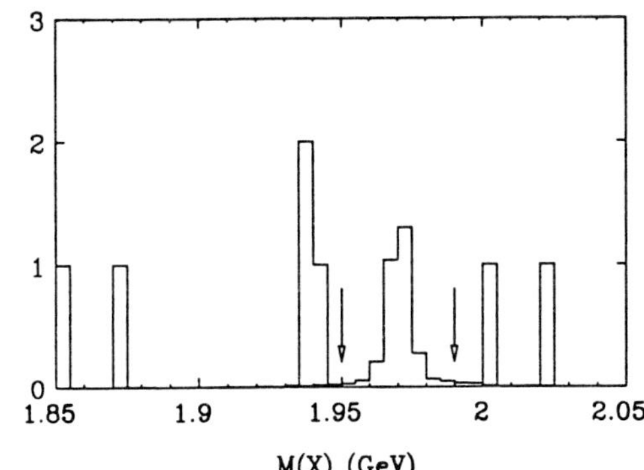

Figure 17: Masses of double-tagged $D_s^+ D_s^-$ candidates from the Mark III experiment. Entries between the arrows show the expected signal (Monte Carlo) assuming $B(D_s^+ \to \phi\pi^+) = 4.1\%$.

encompassing the D_s. The events shown in this range are from a Monte Carlo simulation assuming $B(D_s^+ \to \phi\pi^+) = 4.1\%$. A likelihood analysis gives an upper limit for the branching fraction of 4.1% at 90% confidence level. This result is consistent with the theoretical expectation of about 3%. The goal of measuring the absolute branching fraction for a D_s channel remains for the future.

$D_s^+ \to \eta\pi^+$ and $\eta'\pi^+$

There has been much interest recently in D_s^+ decays to η and η'. These decays should occur at a rate comparable to $\phi\pi$ since they all proceed through spectator decays to a state with $s\bar{s}$ quark content, as shown in Figure 18, Theoretically [31], there shouldn't be anything unusual about these decays, although they are more difficult to measure than $\phi\pi$. The existing and new measurements are summarized in Table 7. There is a measurement

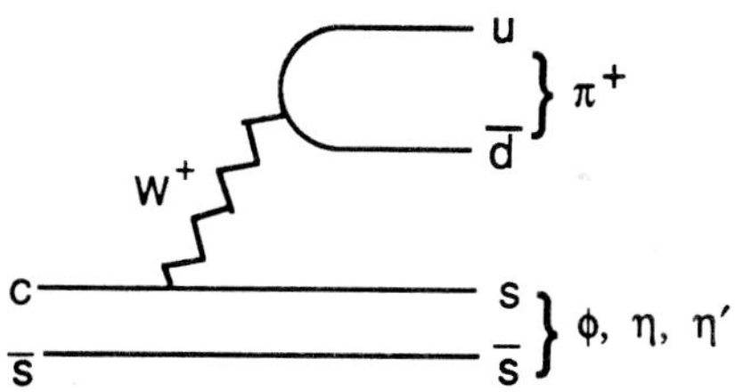

Figure 18: Diagram for D_s^+ spectator decay.

from Mark II [39] of

$$\frac{B(D_s^+ \to \eta\pi^+)}{B(D_s^+ \to \phi\pi^+)} = 3.0 \pm 1.3. \qquad (8)$$

Although the error is large, a value of 3 would be hard to understand. Indeed, there is more $s\bar{s}$ content in the ϕ than in the η or η', so the latter shouldn't occur at a higher rate. There have been some new results. E691 recently published [7] an upper limit of 1.5 at 90% C.L. for the ratio in Equation (8). Mark III submitted an updated measurement [32,4] to this conference of a limit of 2.5, also at 90% C.L. Thus, with increasing statistics the trend is to a lower $\eta\pi$ rate than initial indications. There is no obvious discrepancy at this time with the notion that the decay rates of D_s^+ to $\eta\pi^+$ and $\phi\pi^+$ are comparable.

To illustrate how these analyses are done, consider the the Mark III measurement of the $\eta\pi^+$ decay mode. The η is detected through its decay into two photons, measured with a calorimeter. The photons must have high enough energy to be measured well in the calorimeter. To further reduce background, a cut is made on the $D_s^+ \to \eta\pi^+$ decay angular distribution. Also, constraints are made on the η and D_s^+ masses. The resulting $\eta\ \pi$ invariant mass plot is shown in Figure 19. The

Table 7: Measurements of the relative branching fraction R = $B(D_s^+ \to \eta\pi^+)$ / $B(D_s^+ \to \phi\pi^+)$.

expt.	η mode	ratio R
Mark II	$\gamma\gamma$	3.0 ± 1.3
Mark III	$\gamma\gamma$	<2.5 @ 90% CL
	$\pi^+\pi^-\pi^0$	
E691	$\pi^+\pi^-\pi^0$	<1.5 @ 90% CL

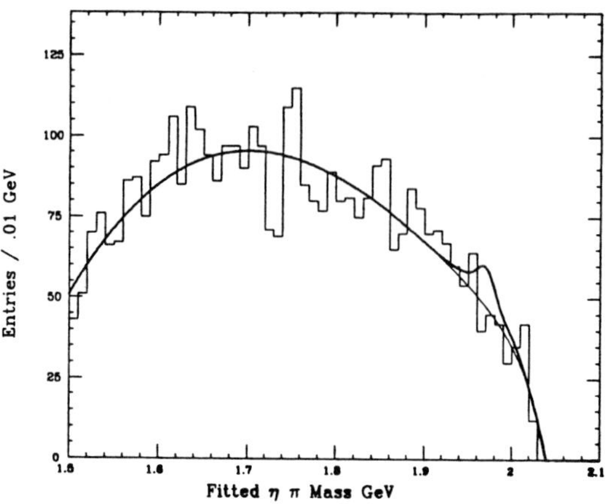

Figure 19: Search for $D_s^+ \to \eta\ \pi^+$ in the Mark III experiment, where $\eta \to \gamma\gamma$.

position of the expected signal is indicated with the small bump. Clearly the data are consistent with no signal. The η was also detected through its decay to $\pi^+\pi^-\pi^0$, the same mode studied by E691. This mode has lower background than for two photon decay. The 4π invariant mass plot is shown in Figure 20. There is an excess of events (16.6 ± 6.1) at the expected mass, but when this data is combined with that using $\eta \to \gamma\gamma$ there is no significant signal for $D_s^+ \to \eta\pi^+$.

The story for $D_s^+ \to \eta'\pi^+$ is similar to that for $D_s^+ \to \eta\pi^+$. There is less $s\bar{s}$ content in the η' than in the ϕ so the decay rate of D_s^+ to $\eta'\pi^+$ should be less than that to $\phi\pi^+$. The existing

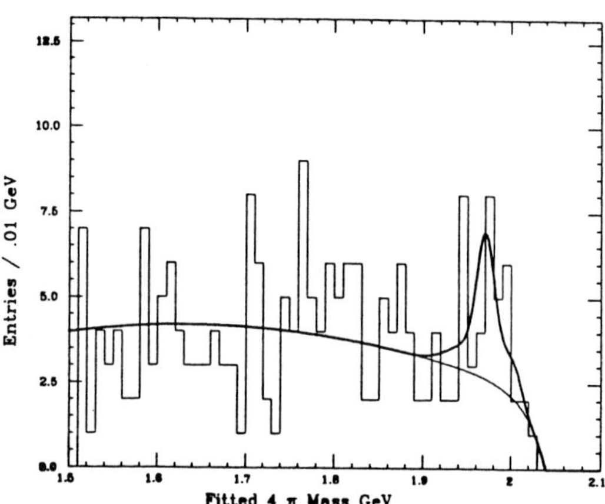

Figure 20: Search for $D_s^+ \to \eta\ \pi^+$ in the Mark III experiment, where $\eta \to \pi^+\pi^-\pi^0$.

and new measurements are summarized in Table 8. There are existing measurements [40] from Mark II and NA14′, a photo-production experiment at CERN. These experiments measured surprisingly high rates, although with rather large errors. The relative branching fraction with respect to $\phi\pi^+$ is about 7 from NA14′, and about 5 from Mark II.

Table 8: Measurements of the relative branching fraction R = $B(D_s^+ \to \eta'\pi^+)$ / $B(D_s^+ \to \phi\pi^+)$.

Expt.	η' mode	ratio R in %
Mark II	$\eta\pi^+\pi^-$ $\eta \to \gamma\gamma$	4.8 ± 2.1
NA14'	$\rho\gamma$ $\rho \to \pi^+\pi^-$	$6.9 \pm 2.4 \pm 1.4$
Mark III	$\eta\pi^+\pi^-$ $\eta \to \gamma\gamma$	<1.9 @ 90% CL
E691	$\eta\pi^+\pi^-$ $\eta \to \pi^+\pi^-\pi^0$	<1.7 @ 90% CL

There is a new result submitted to this conference from E691 of an upper limit of 1.7 with respect to $\phi\pi^+$. There is also an updated measurement from Mark III of an upper limit of 1.9. Thus, the trend of the recent data is towards a lower rate than previously indicated. The existing data is consistent with the expectation of comparable decay rates of D_s^+ to $\eta'\pi^+$ and $\phi\pi^+$.

As an example of the the analysis for $D_s^+ \to \eta'\pi^+$ consider the E691 measurement [12] where the π^0 is not measured. In Figure 21a, Monte Carlo data are displayed in a scatter plot of the invariant mass of the 5 charged pions against the invariant mass of the 4 charged pions from the η' decay. Since the π^0 is not measured, the 5-pion mass is always less than $m(D_s^+) - m(\pi^0)$. Similarly, the 4-pion mass is always less than $m(\eta') - m(\pi^0)$. The distribution of events is concentrated near those limits since the missing π^0 carries off, on average, a small fraction of the parent momentum. The data are shown in Figure 21b displayed in the same manner as the Monte Carlo events in Figure 21a. There is clearly no structure in the data like that in the Monte Carlo simulation. For comparison of the data and Monte Carlo scatter plots, note that the number of Monte Carlo events corresponds to the branching fraction $B(D_s^+ \to \eta'\pi^+) = 5 \times B(D_s^+ \to \phi\pi^+)$. A maximum likelihood fit gives a signal estimate of 5.8 ± 3.9 events for $D_s^+ \to \eta'\pi^+$ and 7.0 ± 4.5 events for $D_s^+ \to \eta'\pi^+$.

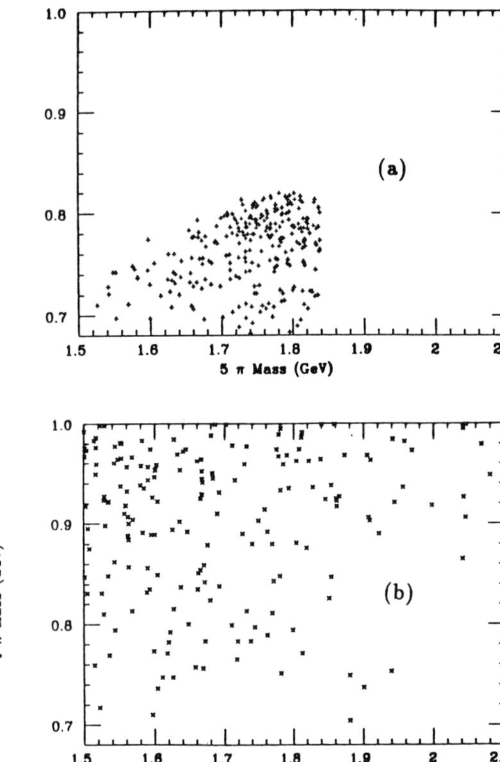

Figure 21: E691 analysis of $D_s^+ \to \eta'\pi^+$, $\eta' \to \eta\pi^+\pi^-$, $\eta \to \pi^+\pi^-\pi^0$. The invariant mass of the 5 charged pions is plotted against the invariant mass of the 4 charged pions from the η' decay (a) Monte Carlo events. (b) Data.

Miscellaneous New Measurements

There is an existing E691 measurement [8] of $D_s^+ \to f_0(975)\pi^+$. The $f_0(975)$ is a hadronic resonance that decays to pions, even though its quark content is mainly $s\bar{s}$. The $f_0(975)$ can be produced from spectator decay of the D_s^+. The existence of the decay $D_s^+ \to f_0(975)\pi^+$ was recently confirmed by Mark III [4]. They measure a relative branching fraction of $.58 \pm .21 \pm .28$ with respect to $\phi\pi^+$. This is consistent with the E691 measurement of $.28 \pm .10 \pm .03$.

A recent result from CLEO [24] is the first measurement of $D_s^+ \to K^{*+}\overline{K^0}$, $K^{*+} \to K^0\pi^+$. Two K_s^0's are detected. Because of the $s\bar{s}$ content, the decay rate is expected to be comparable to that for $\phi\pi^+$. The CLEO result is,

$$\frac{B(D_s^+ \to K^{*+}\overline{K^0})}{B(D_s^+ \to \phi\pi^+)} = 1.20 \pm .21 \pm .13,$$

consistent with the theoretical expectation.

COMMENTS ON D^+ DECAYS

There is a new measurement from E691 [12] of the Cabibbo-suppressed decay $D^+ \to \overline{K^0}K^+$. The relative branching fraction with respect to $\overline{K^0}\pi^+$ is $.271 \pm .065 \pm .039$ which confirms the measurement from Mark III [2] of $.317 \pm .086 \pm .048$.

Mark III has a new measurement and resonance analysis [4] of $D^+ \to K_s^0\pi^-\pi^+\pi^+$. In contrast to the decay $D^0 \to K^-\pi^+\pi^-\pi^+$ discussed previously, the D^+ decay discussed here does not have a possible vector-vector channel. The results for the D^+ decay are summarized in Table 9.

Table 9: Mark III measurement of branching fractions for the sub-channels of $D^+ \to K_s^0\pi^-\pi^+\pi^+$.

mode	branching fraction
$\overline{K^0}\pi^-\pi^+\pi^+$	$.014 \pm .005 \pm .007$
$K^0a_1(1260)^+$	$.077 \pm .020 \pm .020$
$K_1(1270)^0\pi^+$	$.007 \pm .003 \pm .003$
$K_1(1400)^0\pi^+$	$.019 \pm .007 \pm .015$

COMMENTS ON CHARMED BARYONS

The baryon states in the quark model that contain at least one charmed quark [35] are listed in Table 10. The charmed baryons include states containing charm and strange, two charmed quarks, and in the case of the Ω_{ccc}^{++}, 3 charmed quarks - a formidable experimental challenge!

None of the states with two or more charmed quarks have been established. For all the single-charmed states there is at least some evidence. Examples of the charmed baryon decay modes that have been observed are shown in Table 11, which is a fairly complete compilation. Note that for the Λ_c^+ there are a number of different decay modes observed, although for some only a few events have been seen in either bubble chamber or counter experiments. Stars next to some of the decay modes indicate new measurements [11] submitted to this

Table 10: Charmed baryon states expected in the quark model.

name	quark content	(I, I_3)	J^+
Λ_c^+	cud	(0,0)	$1/2^+$
Σ_c^{++}	cuu	(1,1)	$1/2^+$
Σ_c^+	cud	(1,0)	$1/2^+$
Σ_c^0	cdd	(1,-1)	$1/2^+$
Ξ_c^+	csu	$(\frac{1}{2}, \frac{1}{2})$	$1/2^+$
Ξ_c^0	csd	$(\frac{1}{2}, -\frac{1}{2})$	$1/2^+$
Ω_c^0	css	(0,0)	$1/2^+$
Ξ_{cc}^{++}	ccu	$(\frac{1}{2}, \frac{1}{2})$	$1/2^+$
Ξ_{cc}^+	ccd	$(\frac{1}{2}, -\frac{1}{2})$	$1/2^+$
Ω_{cc}^+	ccs	(0,0)	$1/2^+$
Ω_{ccc}^{++}	ccc	(0,0)	$3/2^+$

Table 11: Charmed baryon decay modes. New results were submitted to this Symposium by E691 for the modes indicated with (*).

$\Lambda_c^+ \to$	$PK^-\pi^+$, $P\overline{K^{*0}}$, $\Delta^{++}K^-$
(*)	$P\overline{K^0}$
(*)	$\Lambda^0\pi^+$
	$P\phi$
(*)	$\Lambda^0\pi^+\pi^-\pi^+$
(*)	$P\overline{K^0}\pi^+\pi^-$
	$\Xi^-K^+\pi^+$
	$PK^-\pi^+\pi^+\pi^-$
	$\Sigma^+\pi^+\pi^-$
	$\Sigma^+\pi^+\pi^-\pi^+\pi^-$
	$\Sigma^+\phi$
	$\Sigma^+K^-\pi^+$
$\Xi_c^+ \to$	$\Sigma^+K^-\pi^+$
	$\Xi^-\pi^+\pi^+$
	$\Lambda K^-\pi^+\pi^+$
	$\Sigma^0K^-\pi^+\pi^+$
$\Xi_c^0 \to$	$\Xi^-\pi^+$
$\Sigma_c^0 \to$	$\Lambda_c^+\pi^-$
$\Sigma_c^{++} \to$	$\Lambda_c^+\pi^+$

conference by E691 which confirm previous measurements.

For the Λ_c^+ there are substantial measurements of lifetime and decay modes, including resonance analysis for $PK^-\pi^+$. None of the other states have accurate lifetime measurements. The NA32 collaboration at CERN, using a hadron beam on a fixed target, recently measured [15] the Λ_c^+ lifetime to be $1.96^{+.23}_{-.20} \times 10^{-13}$ seconds. This is consistent with previous measurements and confirms that the Λ_c^+ lifetime is shorter than those of the charmed mesons. The D_s^+ and D^0 lifetimes are approximately 0.4 picoseconds and the D^+ has a lifetime of about 1.1 psec. The shorter Λ_c^+ lifetime is expected due to the exchange diagram, which is less suppressed in charmed baryon decay compared to charmed meson decay.

Even for some of the more exotic charm baryons, like the Ξ_c^+, there are a number of decay modes that have been measured. A recent observation by NA32 of 6 Ξ_c^+ decays provided a lifetime measurement [14,23] of $3.0^{+1.6}_{-0.9} \times 10^{-13}$ seconds. This result is consistent with the picture that the charmed baryon lifetimes are shorter than those of the charmed mesons.

As an example of the high quality of charmed baryon measurements other than those of the Λ_c, consider the recent data from CLEO [5] for $\Xi_c^+ \to \Xi^-\pi^+\pi^+$, $\Xi^- \to \Lambda^0\pi^-$, $\Lambda^0 \to P\pi^-$. In the $\Xi 2\pi$ mass plot, shown in Figure 22, there is a clean signal of 23 ± 6 events at a mass of 2467 ± 3 MeV/c^2. A cut on $x_p > 0.5$ has been applied, where x_p is the fraction of maximum momentum carried by the Ξ_c^0.

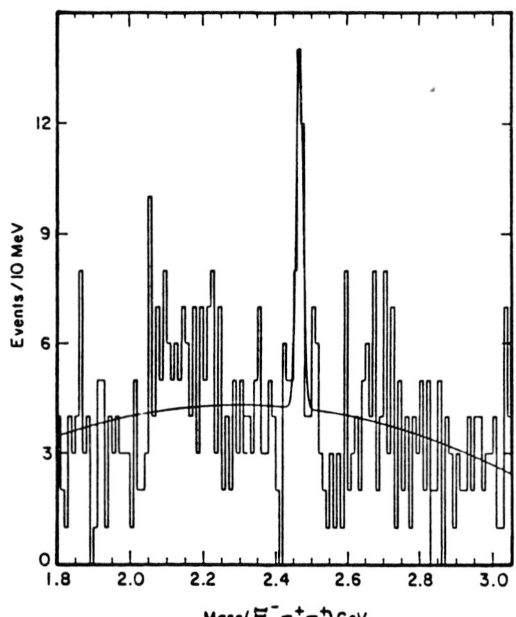

Figure 22: CLEO signal for $\Xi_c^0 \to \Xi^-\pi^+\pi^+$.

The mass plot for $\Xi_c^+ \to \Xi\pi^+$, shown in Figure 23, has a signal of 19 ± 5 events at a mass of 2472 ± 3 Mev/c^2. In addition to the x_p cut, there is a cut on the decay angle of $cos(\theta) > -0.8$.

Charmed baryon spectroscopy and decays are now being studied at statistical levels well beyond the discovery stage.

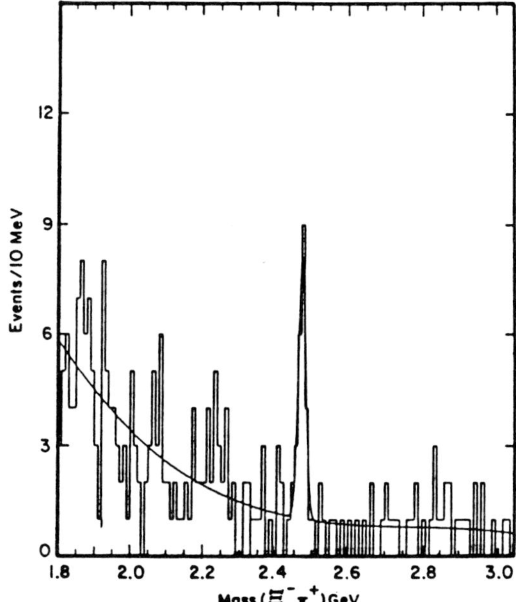

Figure 23: CLEO signal for $\Xi_c^0 \to \Xi\pi^+$.

CONCLUDING REMARKS

The semi-leptonic decays $D^0 \to K^-e^+\nu_e$ and $\pi^-e^+\nu_e$ are fairly well understood. Greater statistical precision is required for stringent tests of the CKM model or QCD-based predictions for the form factors. For $D^+ \to \overline{K^{*0}}e^+\nu_e$, first measurements have been made, but extraction of the three form factors will require both more precise data and more detailed analysis.

There is now a wealth of Knowledge on D^0 and D^+ hadronic decays, which can generally be understood in the quark diagram picture. The corrections due to final state interactions, which break SU(3) flavor symmetry, seem well understood now that the ratio $\Gamma(D^0 \to K^-K^+)$ / $\Gamma(D^0 \to \pi^-\pi^+)$ has been better measured.

For the D_s^+ decays, there is a need to establish more of the decay modes and to measure the absolute branching fraction for one of them. The current measurements of hadronic D_s^+ decays are consistent with the basic understanding of charmed meson decays.

For the charmed baryons, although there is now a lot of information on the Λ_c^+, study of the other states is just beginning. This is a rich field for future study which will some day culminate with the observation and detailed measurement of the quintessential charm state, the Ω_{ccc}^{++}.

ACKNOWLEDGEMENTS

I am indebted to Thomas Browder, Rollin Morrison, and Michael Witherell for helpful discussions. My thanks to Vincent Chabaud (NA32), David Hitlin (Mark III), David MacFarlane (ARGUS) and Sheldon Stone (CLEO) for transmitting contributions from their collaborations. I thank the conference organizers from SLAC, especially Peter Kim, the scientific secretary for this talk, and Vera Luth for their assistance.

For much effort in preparing this document and for her encouragement, I am grateful to my wife, Barbara Feitler-Karchin.

My work has been supported in part by a Yale University Junior Faculty Fellowship and by the United States Department of Energy.

References

[1] G.S. Abrams et. al., *Phys. Rev. Lett.* **43** (1979) 481.

[2] J. Adler et al., *Phys. Rev. Lett.*, **60** (1988) 89, and references therein.

[3] J. Adler et al., *Phys. Rev. Lett.*, **62** (1989) 1821.

[4] J. Adler et al., "Recent Results on Hadronic D_s^+ and D Meson Decays from the Mark III", paper No. 303 contributed to this Symposium.

[5] M.S. Alam et al., Cornell Preprint CLNS 89/21, CLEO 89-9.

[6] H. Albrecht et al., "A Study of Cabibbo-Suppressed D^0 Decays", paper No. 265 contributed to this Symposium.

[7] J.C. Anjos et al, *Phys. Lett.* **B 223** (1989) 267.

[8] J.C. Anjos et al, *Phys. Rev. Lett.* **62** (1989) 125.

[9] J.C. Anjos et al., *Phys. Rev. Lett.* **62** (1989) 722.

[10] J.C. Anjos et al., *Phys. Rev. Lett.* **62** (1989) 1587.

[11] J.C. Anjos et al., "A Study of Decays of the Λ_c^+", paper No. 85 contributed to this Symposium.

[12] Anjos et al., (E691), results submitted to the speaker.

[13] R.M. Baltrusaitis et al., *Phys. Rev. Lett.* **55** (1985).

[14] S. Barlag et al., CERN preprint EP/89-43, 1989.

[15] S. Barlag et al., *Phys. Lett.* **B 218** (1989) 374.

[16] M. Bauer, B. Stech and M. Wirbel, *Z. Phys.* **C34** (1987) 103.

[17] I. Bigi, to appear in proceedings of *1989 International Symposium on Heavy Quark Physics*, Cornell University.

[18] Buras et al., *Phys. Lett.* **B268** (1986) 16.

[19] L.L. Chau and H-Y. Cheng, *Phys. Rev. Lett.* **56** (1986) 1655.

[20] L.L. Chau and H-Y. Cheng, *Phys. Rev.* **D 36** (1987) 137.

[21] L.L. Chau and H-Y. Cheng, Preprint UCD-88-12, IP-ASTP-04, June, 1988.

[22] L.L. Chau and H. Y. Cheng, Preprint IP-ASTP-01-89, March, 1989, paper No. 75 contributed to this Symposium.

[23] V. Chabaud, to appear in proceedings of *Twelfth International Workshop on Weak Interactions and Neutrinos*, Ginosar, Israel, April, 1989.

[24] W.Y. Chen et al, Preprint CLNS 89/920, CLEO 89-8.

[25] CLEO Collaboration, "Study of Rare D^0 Decays", contributed paper to this Symposium.

[26] CLEO Preprint CBX 89-27, preliminary version submitted to the speaker.

[27] J.P. Cumulat et al., *Phys. Lett.* **B 210** (1988) 253.

[28] F. DeJongh, to appear in proceedings of *1989 International Symposium on Heavy Quark Physics*, Cornell University.

[29] C.A. Dominguez and N. Paver, *Phys. Lett.* **207B** (1988) 199.

[30] G. Gladding, to appear in proceedings of *1989 International Symposium on Heavy Quark Physics*, Cornell University.

[31] A.N. Kamal, N. Sinha and R. Sinha, University of Alberta Preprint, Thy- 7-88, 1988.

[32] P. Kim, to appear in proceedings of *1989 International Symposium on Heavy Quark Physics*, Cornell University.

[33] G. Martinelli, in these proceedings.

[34] X. Y. Pham, *Phys. Lett.* **B 193**, (1987) 331.

[35] S.P.K. Tavernier, *Rep. Prog. Phys.* **50** (1987) 1439.

[36] K. Terasaki and S. Oneda, *Phys. Rev.* **D38** (1988) 132.

[37] M. Wirbel, B. Stech, and M. Bauer, *Z. Phys.* **C29** (1985) 637.

[38] M. Witherell, to appear in proceedings of *1989 International Symposium on Heavy Quark Physics*, Cornell University.

[39] G. Wormser et al, *Phys. Rev. Lett.* **61** (1988) 1057.

[40] G. Wormser, to appear in proceedings of *1989 International Symposium on Heavy Quark Physics*, Cornell University.

DISCUSSION

A. Ali, DESY: You did not mention rare decays of charmed mesons, like doubly Cabbibo-suppressed decays $D^0 \to K^+\pi^-$ and $D^0 \to \mu^+\mu^-$ or e^+e^-. Would you like to comment on the experimental situation in this sector?

P. Karchin: No new results on rare charm decays were submitted to this conference, and I am not aware of any such recent results. For the doubly Cabibbo-suppressed decays, I think the best limit on the branching fraction is about 0.5% from E691. Also, there are a few candidate events from Mark III. For rare leptonic decays including lepton-number violating decays, I'm not aware of any new results. It's a very interesting field, though.

J.D. Jackson, University of California, Berkeley: First a comment and then a question. You mentioned in the quasi-three body decay $D^+ \to K^*e^+\nu$ that there was an "interesting" longitudinal polarization. To the extent that the invariant mass is small, the (V–A) structure of the leptonic current forces the leptons to have opposite helicities. When the $\nu - e$ invariant mass is small, the lepton momenta are nearly parallel. The projection of the leptonic angular momentum on the K^* momentum direction is then zero, and conservation of angular momentum requires the K^* to be longitudinally polarized.

Now the question. What is the evidence from the data on the $\nu - e$ invariant mass distribution in those decays?

P. Karchin: There is analysis in progress to extract some of those distributions. We'll look at the decay angle that I called θ_e.

P. Singer, Technion: You did not mention anything about $D^* \to D\gamma$ decays. There is a puzzle there, too, the reported $(D^{*+} \to D^+\gamma)$ / $(D^{*0} \to D^0\gamma)$ being quite different from the theoretically expected ratio, by nearly an order of magnitude. Are there any new results on this ratio?

P. Karchin: I'm not aware of this issue, and no results were submitted to this conference. It is an area that would be covered by Mark III. Would someone from Mark III care to comment? [No answer.]

M. Samuels, Oklahoma State University: You mentioned the charmed baryons. Is there any hope of measuring their magnetic moments, or is their lifetime too short?

P. Karchin: There's always hope. Such a measurement would require high statistics in an experiment where the charmed baryon decays in a magnetic field. It might be possible in a fixed-target experiment with a high-intensity hadron beam, or at a high luminosity e^+e^- machine. It would be a very interesting measurement.

STATUS AND PHYSICS PROGRAM OF BEPC

Ye Minghan and Zheng Zhipeng
China Center for Advanced Science and Technology (World Lab), and Institute of High Energy Physics, Beijing, China

ABSTRACT

BEPC, the first high-energy particle accelerator of the People's Republic of China, is an e^+e^- collider of 1.5–2.8 GeV per beam. The machine is built for charm and tau physics and synchrotron radiation applications. The current status of BEPC and BES is described. The checkout of BES has begun and the J/ψ resonance peak has been observed. The future BEPC physics program is discussed.

INTRODUCTION

The Beijing Electron-Positron Collider (BEPC for short) Project[1,2,3] consists of three main subsystems: a collider (BEPC) of 1.5–2.8 GeV per beam, a magnetic spectrometer called Beijing Spectrometer (BES for short) for high energy physics experiments, and synchrotron radiation facilities. There are two main purposes for the BEPC project. The first is to carry out research on particle physics; the center-of-mass energy attainable at BEPC, 3.0–5.6 GeV, is a region for charm and tau lepton physics. The second purpose is to provide synchrotron radiation in the VUV and x-ray regions for research and applications in sciences other than particle physics.

The BEPC project was officially approved on April 24, 1984, and groundbreaking took place on October 7 of the same year. After four years, on October 16, 1988, the first e^+e^- collision was realized in BEPC, and the first cosmic ray event was obtained by BES on October 23—both of which occurred more than two months ahead of schedule.

The site of BEPC is in the campus of the Institute of High Energy Physics, which is located to the west of Beijing, about 12 km from the center of the city. The total construction budget was 240 million Chinese Yuan (RMB), which was equivalent to 80 million US dollars in 1984. Table 1 shows the amount of money spent through May 1989. The expenditures have been well controlled, and the budget will be balanced, which is very rare for a scientific project in China.

Table 1: Financial summary of BEPC and BES.[a]

Item	Money spent (millions) Chinese Yuan	US Dollar[b]
Land and civil engineering	48.11	16.04
Injector	31.67	10.56
Storage ring and transportation line	43.22	14.41
Instrumentation and control	9.33	3.11
Detector BES and its electronics	38.74	12.91
Synchrotron radiation facilities	35.15	11.72
Computing center	13.00	4.34
Others	13.04	4.35
Total	232.26	77.44

(a) Money spent for hardware and installation, not including salaries and power for commissioning.
(b) According to the exchange rate in 1984, \$1 US = 3 Chinese Yuan (RMB).

BEPC

Figure 1 shows the layout of BEPC. The injector is a 202 m linear accelerator. The first part of it is a 30 MeV pre-injector. A movable tungsten target for positron production is located at the 150 MeV section. What follows is the main part of the injector, which accelerates electrons or positrons to 1.1–1.4 GeV. The injector operates in $2\pi/3$ mode at a frequency of 2856 MHz. Its disk-loaded waveguide accelerating sections are of the constant-gradient type with a shunt impedance of approximate 53 MΩ/cm. Each accelerating section is 305 cm long, and there is a total of 56 sections. A pulse-compression scheme based on the design of SLAC, called SLAC energy doubler (SLED), is employed, which can enhance the power delivered by a klystron to the accelerating section by about 50 percent. Table 2 shows the parameters and the present status of the injector; most of the

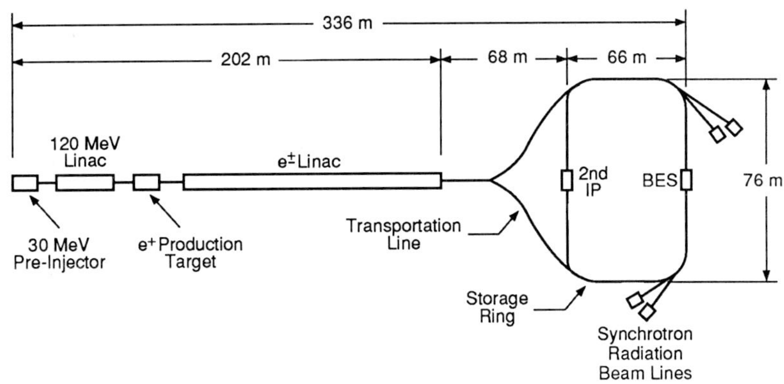

Figure 1: Layout of BEPC.

Table 2: Main parameters of the injector.

Parameter	Design Value	Actual Value
$e^{\pm}$ energy (GeV)	1.1–1.4	1.1–1.55 (e^-) 1.2 (e^+)
Electron gun output (A)	5	5
Maximum e^- beam current (mA)	200	1200 (1.1 GeV)
Bombarding energy for e^+ production (MeV)	150	150
e^+ production rate (e^+/e^-)	0.003	0.00375
Maximum e^+ beam current (mA)	4	3–6.4 (1.2 GeV)
$\Delta E/E$ of e^-	1%	$\sim$ 0.8%
$\Delta E/E$ of e^+	1%	$\sim$ 1%
Emittance of e^- ($\pi\cdot$ MeV/c $\cdot$ cm)	0.02	$\sim$ 0.015
Emittance of e^+ ($\pi\cdot$ MeV/c $\cdot$ cm)	0.22	0.16
Beam pulse width (ns)	2.5	2.5
Pulse repetition rate (Hz)	50	12.5
Working frequency of klystrons (MHz)	2856	2856
Average power output of klystrons (MW)	16	$\sim$ 17
Average multiplication factor of energy doubler	1.5	$\sim$ 1.4

experimental values obtained are in accord with or better than the design values. The injector has been running for more than 8000 hours with no klystron failures.

The storage ring has a circumference of 240.4 m. It is shaped like a racetrack with two long straight sections of 27.4 m and two approximately semicircular arcs. In the middle of each long straight section there

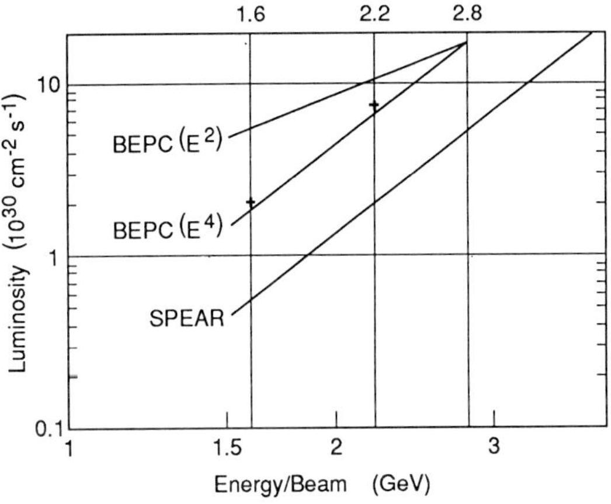

Figure 2: Luminosity of BEPC.

is an interaction region, where there is a 5 m straight section for the installation of the detector. The lattice is of separated-function FODO structure. The cross section of the vacuum beampipe is 56 mm high and 120 mm wide.

The luminosity is optimized at 2.8 GeV per beam. The design peak luminosity is 1.7×10^{31} cm^{-2} s^{-1}. Between 1.5 and 2.8 GeV, the peak luminosity is expected to be proportional to E^2. This behavior should be obtained by means of a system of emittance-control wiggler (ECW) magnets to keep the horizontal emittance constant. Figure 2 shows the expected luminosity for two cases: one is proportional to E^2; the other is proportional to E^4, if ECW magnets are not used.

Table 3 shows the parameters and present status of the storage ring, while Table 4 gives the single-

Table 3: Main parameters of the storage ring.

Parameter	Design Value	Actual Value
Energy per beam (GeV)	1.55–2.8	1.1–2.7
Circumference (m)	240.4	
# of IP	2	
Peak luminosity (10^{30} cm^{-2} s^{-1}) (without ECW magnets)	1.8 (1.6 GeV) 6.5 (2.2 GeV) 17 (2.8 GeV)	2.0 (1.6 GeV) 7.0 (2.2 GeV)
Circulation current per beam (mA)	65 (2.8 GeV)	30 (2.2 GeV)
Particle revolution frequency (MHz)	1.247	
Horizontal tune	6.18	5.8
Vertical tune	7.12	6.8
Horizontal emittance (mm mrad)	0.66	
Total beam lifetime (hrs)	5–6	5–6 (1.6 GeV) 3–4 (2.2 GeV)
Bending radius (m)	10.34	
Maximum strength of magnet field (G)	9028	
Horizontal beta at IP (m)	1.3	
Vertical beta at IP (m)	0.1	
RF frequency (MHz)	199.53	
RF power (KW)	200	
# of bunches/beam	1	
RMS energy spread (MeV)	2.67×10^{-7} E^2 (MeV)*	
RMS half bunch length (cm)	5.2 (2.8 GeV)	

* E is energy per beam in MeV.

Table 4: Maximum single-beam intensity.

Energy	1.1 GeV		1.6 GeV	
Current	Design Value	Actual Value	Design Value	Actual Value
Single bunch	e^- 27 mA e^+ 27 mA	84 mA 54 mA	e^- 37 mA	79 mA
Multiple bunch	e^- 150 mA	200 mA	e^- 150 mA	140 mA

beam intensity. The progress of the collider performance has been better than anticipated. The luminosity at 2.2 GeV per beam has reached the design value, 7×10^{30} cm^{-2} s^{-1}. The luminosity obtained without the ECW magnets is very encouraging; it is proportional to E^4. The single-beam intensity at 1.1 GeV is much higher than the theoretical design value. The ECW magnets will be installed in late 1990. The present value of the single-beam intensity is so good that we are optimistic the design luminosity proportional to E^2 can be obtained.

BES

The Beijing Spectrometer (BES), a general-purpose magnetic spectrometer,[3,4] is shown schematically in Fig. 3. Table 5 presents the parameters of its major components. The design of BES was modelled on the Mark III detector; however, some major improvements were made:

(a) Endcap time-of-flight counters (TOF) with 20 percent more solid angle coverage were added. Their time resolution is 250 ps.

(b) The main drift chamber has ten layers of sense wires, each layer composed of cells with four sense wires apiece. Each wire gives both timing and dE/dx signals; the sampling number is 40. The z coordinate is determined using axial and stereo wires.

(c) The self-quenching streamer (SQS) mode gas discharge has been chosen instead of the proportional mode for the shower counter, greatly increasing the output amplitude, so that no preamplifier is needed. The structure of 560 cells per layer provides a better space resolution, σ_ϕ = 4.5 mrad.

(d) Three layers of muon detectors, which should provide better muon identification with a lower momentum cut (500 MeV/c).

On October 23, 1988, the first cosmic-ray event was obtained by BES. In April 1989, BES was moved into the interaction point (IP) of BEPC and Bhabha scattering events were registered soon thereafter. On June 22, 1989, the J/ψ resonance peak was observed for the first time. From the shape of the resonance curve, we can infer that σ_E per beam is about 0.64 MeV. The calibration of BES will be continued in the next running period of BEPC (September 1989).

SYNCHROTRON RADIATION FACILITIES

Facilities are being built for applications of synchrotron radiation in various disciplines. In the first stage of construction, three beam ports, five beam lines and eight experimental stations will be built. The experimental stations are for x-ray topography, EXAFS, small-angle scattering, diffraction, diffuse scattering, photoemission spectroscopy, medical research and lithography. The front ends for beam

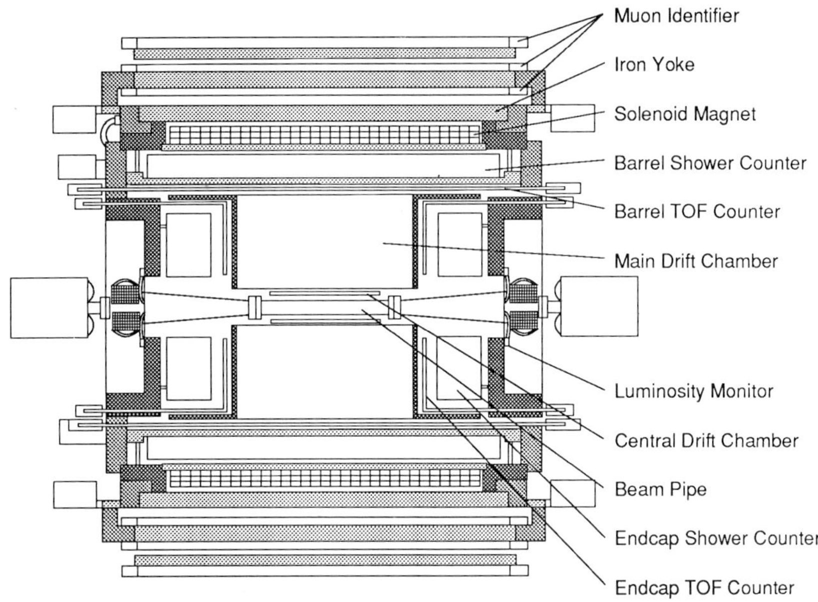

Figure 3: Cross-sectional diagram of BES.

ports and two beam lines were completed in April 1989. Synchrotron radiation has passed through the two beam lines and experiments are in preparation. Other beam lines and experimental stations will be completed at the end of 1989 or in 1990.

PHYSICS PROGRAM

There are many interesting physics problems in the center-of-mass energy range of 3.0–5.6 GeV.[4,5] That region has been scanned and studied by SPEAR, DORIS, DCI and other laboratories, and many important discoveries have been made by Mark III, DM2, Mark II, Mark I, Crystal Ball and other experimental groups. But there remain so many unanswered questions that BEPC and BES still can make important contributions. Since 1987 there have been several proposals for building a tau–charm factory[6–8] with a peak luminosity of 10^{33} cm^{-2} s^{-1}, which confirms that the choice of the energy range for BEPC is indeed correct. Some of the most important physics problems in this energy range are as follows:

(a) The search for glueballs and hybrids is a unique test of QCD. There are already a number of states observed in J/ψ decays that are regarded as glueball candidates, such as the $\iota/\eta(1460)$ and $\theta/f_2(1720)$. Furthermore, a controversy still exists about the very existence of the $\xi(2230)$. If it does indeed exist, what kind of particle is it? Radiative J/ψ decay provides the best window on glueballs and hybrids. With high statistics, the quantum numbers and decay properties of such candidates can be investigated in great detail and with unprecedented precision by BES. We hope that such studies will help determine their exact nature.

(b) The decay of the charmed mesons (D^0, D^+, D^-) is one of the major subjects to be studied by BES; for example, leptonic decay constants of D^0 and $D^{\pm}$, nonleptonic decays, and Cabibbo-favored and suppressed channels. At BEPC, a comparatively low-energy collider, there are possibilities for physics discoveries that would point beyond the Standard Model. One example at the moment is $D^0 - \overline{D}^0$ mixing; this quantity is small in the Standard Model, which makes it difficult to accumulate enough data to make a significant measurement, although present experiments are approaching the appropriate level of accuracy. Another example is a measurement of $c \to de\nu$ as compared to $c \to se\nu$ decays. Such a measurement, if done accurately and combined with theoretical analysis, will permit the extraction of the ratio of Kobayashi–Maskawa matrix elements V_{cd}/V_{cs}.

(c) So far the study of the properties of the D_s meson is only in its early stages. Not many hadronic D_s decay events have been obtained, and no semileptonic or Cabibbo-suppressed D_s decays have yet been observed. Further study strongly depends on how many D_s events can be accumulated. The

Table 5: Main parameters of BES.

Magnet	4.5 KG conventional aluminum coil, magnetic inhomogeneity in tracking area $< 3\%$; ID 3.48 m, OD 4.14 m, length 3.60 m.
Beam pipe	Aluminum 0.3 mm thick, reinforced by 2-mm-thick carbon fiber (1.04×10^{-2} rl); OD 15.4 cm.
Central drift chamber	ID 18.4 cm, OD 30.2 cm, length 110 cm; 4 layers of cells, 48 x 1-sense wire cells per layer. Resolution: $\sigma_x = 150\ \mu$m, $\sigma_z = 1$ cm (by charge division); solid angle 98% of 4π.
Main drift chamber	ID 31.0 cm, OD 230.0 cm, length 220.0 cm; 10 layers of cells, 5 axial, 5 stereo; # of cells per layer: 48, 48, 60, 48, 58, 68, 78, 88, 98, 108 (counting from innermost layer); 4 sense wires per cell, total 2808 sense wires; both time and dE/dx signal from each sense wire. Gas mixture: 89% Ar, 10% CO_2, 1% CH_4. Resolution: $\sigma_x = 200\ \mu$m, $\sigma_z = 4$ mm; $(\Delta p/p)^2 = (0.7\%p)^2 + (1.3\%)^2$; dE/dx: 8.2% FWHM; solid angle 96% of 4π (second layer).
Barrel TOF counter	48 15 cm x 5 cm x 284 cm NE110 scintillation counters forming a cylindrical array around the main drift chamber; XP2020 photomultiplier on each end of the scintillator. Time resolution: $\sigma_t = 200$ ps; solid angle 76% of 4π.
Endcap TOF counter	2 x 24 pieces 2.5 cm thick NE102A scintillator forming a disk of ID 75 cm and OD 211 cm; XP2020 photomultiplier on one end of the scintillator. Time resolution: $\sigma_t = 250$ ps; solid angle 20% of 4π.
Barrel shower counter	ID 247.0 cm, OD 338.2 cm, length 385.0 cm; 24 layers of Pb and gas chamber sandwich (12 rl total) forming a cylindrical array of 1.3 cm x (1.42–1.87) cm x 385 cm cells, 560 cells per layer, read out at both ends with charge division; gas chamber working in self-quenching streamer mode. Gas mixture: 33% Ar, 67% CO_2 bubbling through 0°C n-pentane. Resolution: $\sigma_\phi = 4.5$ mrad, $\sigma_z = 2$ cm, $\sigma_E/E = 16\%/\sqrt{E\ (\mathrm{GeV})}$; solid angle 80% of 4π.
Endcap shower counter	2 endcaps, each of ID 74.6 cm, OD 192.0 cm, length 41.0 cm; 24 layers of Pb and gas counter sandwich (12 rl total) forming an array of square cross section 1.28 cm x 1.28 cm aluminum tubes glued between 1/2 rl layers of Pb, read out at both ends with charge division; gas counter working in self-quenching steamer mode. Gas mixture: 33% Ar, 67% CO_2 bubbling through 0°C n-pentane. Resolution: $\sigma_x = 0.7$ cm, $\sigma_y = 13\%\ L$ (length of tube, cm), $\sigma_E/E = 16\%/\sqrt{E}$ (GeV); solid angle 14% of 4π.
Muon detector	3 double layers of 5.08 cm x 6.09 cm cross section, 4.2–4.7 m long proportional counter tubes mounted outside 12, 14, and 14 cm thick Fe flux return/absorber, respectively, read out at both ends with charge division. Gas mixture: 90% Ar, 10% CH_4. Resolution: $\sigma_z = 4.5$ cm, $\sigma_{\gamma\phi} = 3$ cm; solid angle 68% of 4π (at innermost layer).
Luminosity monitor	4 scintillator + shower counter telescopes.

best region in which to find the D_s in e^+e^- annihilation is believed to be in the energy range 4.03–4.16 GeV. Measurements of D_s at BES will provide insights into the dominant decay mechanisms and other properties of charmed mesons. BES also has a good chance to measure the mass of the D_s meson more accurately.

(d) There exists a problem in understanding the 1-charged particle decay modes of the tau; the measured inclusive branching fraction for 1-charged prong decays is larger than the sum of all the exclusive modes. New measurements with substantially lower systematic and statistical errors are required to resolve this issue. For the tau neutrino mass limit, the present value is $m_{\nu_\tau} < 35$ MeV (ARGUS 1988). Compared to the limits on the electron neutrino and muon neutrino masses, this limit is poorly determined, and a more precise measurement is needed. Some theories predict a mass hierarchy for the neutrinos; if they are not massless, it should be $m_e^2/m_{\nu e} = m_\tau^2/m_{\nu_\tau}$. A value of m_{ν_τ} equal to 10 MeV would therefore translate into a value for m_{ν_e} of 0.8 eV. We need to accumulate high-integrated luminosity in order to provide more tau events so that its 3-prong and 5-prong decay modes can be used to place more stringent limits on the tau neutrino mass and to help search for unusual tau decay modes.

(e) BES can study precisely the charmonium spectrum, especially to search for the charmonium state of $1\ ^1P_1(J=1)$.

(f) We plan to determine the decay channels of the charmed baryons (Λ_c, Ξ_c, etc.), measurements that are now only at their beginning stages. Some preliminary results on these charmed baryons were announced recently; most of them need to be investigated experimentally in more detail.

(g) BES will study the behavior of the ratio

$$R = \sigma(e^+e^- \to hadrons)/\sigma(e^+e^- \to \mu\mu) \quad ,$$

especially in the energy range 4.0–4.5 GeV where clarification of the charmonium spectrum could be useful.

With so many interesting physics projects to pursue, BES can be expected to make some important contributions to particle physics. There are several center-of-mass energies to be chosen. The anticipated sequence of physics programs is as follows:

First, we plan to study the decay of the J/ψ at a center-of-mass energy of 3097 MeV. We should collect 2×10^7 J/ψ events, which will be greater than the sum of all the J/ψ events observed so far by the Mark III (5.8×10^6), DM2 (8.6×10^6), Crystal Ball (2.2×10^6), Mark II (1.3×10^6) and Mark I (1.5×10^5) collaborations. The number of events to be collected can be estimated using the formula

$$N = \overline{L}\overline{\sigma}T\eta \quad ,$$

where $\overline{L}$ and $\overline{\sigma}$ are the average luminosity and cross section, T is the duration of data taking, and η is the detection efficiency. $\overline{L} = \epsilon L$, where L is the peak luminosity and ϵ is a factor related to the efficiency of injection and other factors; usually, we assume $\epsilon = 0.25$. If the energy spread ΔE of each beam of BEPC at 1.55 GeV is 1.50 MeV, then

$$\overline{\sigma} = \sigma\Gamma/\sqrt{2}\Delta E = 3.3 \times 10^{-30}\ \mathrm{cm}^2 \quad ,$$

where $\sigma = 1.05 \times 10^{-28}\ \mathrm{cm}^2$ and $\Gamma = 68$ keV. Assuming η to be 0.7, then

$$N = 0.25 \times 0.7 \times 3.3 \times 10^{-30}\ LT \quad .$$

Now we already have $L = 2 \times 10^{30}\ \mathrm{cm}^{-2}\ \mathrm{s}^{-1}$, so we should obtain 2×10^7 J/ψ events in the first year. If the results of Mark III concerning the $\xi(2230)$ are correct, then about 240 ξ events will be obtained by BES. We also have a good chance to study the $\iota/eta(1460)$, the $\theta/f_2(1720)$, and others.

After that, the machine energy will be moved to the $\psi''(3770)$ region. With the ECW magnets in operation by late 1990, we anticipate that the peak luminosity can reach $7.7 \times 10^{30}\ \mathrm{cm}^{-2}\ \mathrm{s}^{-1}$. Again assuming $\overline{L} = 0.25\ L$, we would require ten months to accumulate 50 pb^{-1} integrated luminosity. The running time of BES is about six months per year, so we need about two years to accumulate this much data. We shall do our best to improve the running efficiency of BEPC and BES in order to make ϵ as large as possible; the best value attained at SPEAR is about 0.5.

One reason for doing D physics after 1991 is the anticipated competition from fixed-target experiments and other e^+e^- experiments. Another reason is that our computing power at the end of 1990 will be sufficient for analyzing only about 2×10^7 J/ψ events per year, when our CPU power is nearly equivalent to that of twenty VAX 780's. We plan to develop software to use a gang of work stations for data filtering, track reconstruction and other purposes. Nowadays, such work stations with a UNIX operating system are very attractive with regard to low costs and high CPU power.

The $\psi(4050)$ region will be chosen next for studying decays of the D_s meson. BEPC will operate here for a year to accumulate 50 pb^{-1} integrated luminosity. The peak luminosity should be $8.9 \times 10^{30}\ \mathrm{cm}^{-2}\ \mathrm{s}^{-1}$; if $\overline{L} = 0.25\ L$, this phase of data taking will require about nine months. Again, we should be able to improve the running efficiency to get ϵ to be near 0.5.

When the center-of-mass energy of colliding beams increases above 3.57 GeV, studies of the tau lepton will begin. We expect to collect at least 10^5 tau pairs.

The machine energy will return to the J/ψ region. By this time, the peak luminosity should be $5.2 \times 10^{30}\ \mathrm{cm}^{-2}\ \mathrm{s}^{-1}$; we should be able to collect 10^8 J/ψ events within two years and might answer some questions about glueballs, hybrids and other puzzles. Later, $\psi'(3685)$ events will be collected. When we start to collect 10^8 J/ψ events will depend upon the outcome of the proposed tau–charm factory. We should have a large number of J/ψ events before its completion.

In the region approaching the maximum energy of our collider, which will be chosen later, we want to study the Λ_c, Ξ_c and other charmed baryons.

Right now we are calibrating the BES in the hope that serious data taking for the first sample of

2×10^7 J/ψ events will start near the end of this year. The running schedule of BEPC is as follows:

Running: September–December, March–June;

Shutdown: January–February, July–August.

There should be about six months per year of machine time available for high energy physics, with about two months for machine studies and other requirements.

At this time BEPC is the highest luminosity collider operating in the center-of-mass energy range 3.0–5.6 GeV. There will be about five years during which we can enjoy this superiority. After the completion of a tau–charm factory with a luminosity of 10^{33} cm^{-2} s^{-1}, what we can do best may be in the region above 5 GeV. The next five years will be crucial for us to make a contribution to particle physics. We also should try to increase the luminosity of BEPC above its design value.

ACKNOWLEDGMENTS

We greatly appreciate all the help received from the world high energy physics community. The US/PRC Joint Committee for High Energy Physics has made a major contribution to the progress of the project. We would like to thank, on the US side: ANL, BNL, FNAL, LBL, and SLAC. Special thanks must go to SLAC for all the information about SPEAR and the Mark III and for the very valuable help with which we have successfully built our two similar devices. The conceptual design of BEPC and BES was finalized at SLAC; the control system of BEPC, the on-line computer system of BES and a large part of the readout electronics were developed partly at SLAC. These were crucial to the timely completion of BEPC and BES.

We would like also to express our gratitude to CERN, DESY, KEK, RAL, and TRIUMF. KEK kindly provided the beams for the tests of our prototype drift chamber, shower counter and luminosity monitor. The insertion quadrupole magnets and the timing systems of BEPC were built in Japan with the help of KEK.

We are very grateful to Professor T. D. Lee, without whose painstaking efforts this project could not have happened, as well as to Professor W. K. H. Panofsky, who has given us much important help and advice.

REFERENCES

1. Xie Jialin *et al.,* "Summary of the Preliminary Design of Beijing 2.2/2:8 GeV Electron Position Collider" (IHEP, Beijing, December 1982).
2. Xie Jialin, *IEEE Trans. Nucl. Sci.* **NS–32** (1985), 3080.
3. Ye Minghan and Zheng Zhipeng, *Int. Journal of Modern Physics* **A2** (1987), 1707.
4. Ye Minghan and Zheng Zhipeng, *CCAST (World Laboratory) Symposium/Workshop Proceedings,* Vol. 2, (Beijing, 1987), p. 475.
5. Tao Huang, *CCAST (World Laboratory) Symposium/Workshop Proceedings,* Vol. 2, p. 503.
6. J. Kirkby, "A Tau–Charm Factory at CERN," CERN EP/87–210.
7. S. Kamada, "KEK Tau–Charm Facility Design," *Proceedings of the Tau–Charm Factory Workshop,* SLAC Report No. 343.
8. B. Barish *et al.,* "The Tau–Charm Factory: Physics and Design Status," SLAC–PUB–5033.

DISCUSSION

J. Haissinski, LAL Orsay: When you run your machine at 2.2 GeV per beam, what limits the luminosity? Is it the linear tune shift? And if so, what is the maximum linear tune shift that you achieve?

Y. Minghan: Probably the luminosity at 2.2 GeV was limited by the beam-beam tune shift. The tune measurement apparatus has not functioned properly yet, but we estimated that the maximum tune shift is about 0.035.

M. Samuel, Oklahoma State: We have many Chinese graduate students studying in the United States. Will there be jobs in China for them to return to?

Y. Minghan: Yes, there are many research positions and teaching positions in both universities and the Chinese Academy of Sciences. In the Institute of High-Energy Physics, we lack young doctors to participate in research, and there are post-doctoral and research associate positions.

B PHYSICS FROM CLEO*

DAVID L. KREINICK
Wilson Synchrotron Laboratory, Cornell University
Ithaca, NY 14853-8001, USA

ABSTRACT

New results on the charged and neutral B meson masses and branching ratios are presented. A measurement of the semileptonic branching ratio of the neutral B meson has been made using partially reconstructed $B \to D^* \pi$ decays as a tag. The neutral B decay rate is the same, within experimental errors, as the average B semileptonic decay rate. The consequences of this and other new semileptonic decay rate measurements on the mixing ratio r are explored, and a new value of $r = 0.14 \pm 0.05$ is obtained. Finally recently discovered evidence of charmless b quark decays in the lepton momentum spectrum from B decays is discussed.

SEPARATING $B^{\pm}$ FROM $\overline{B}^0$ MESONS

The decays of $\Upsilon(4S)$ lead to very simple final states, either $B^0\overline{B}^0$ or B^+B^-. Despite this, it has turned out to be difficult to separate the properties of the neutral B's from those of the charged. I would like to spend the first part of this paper discussing recent progress.

The data were taken with the CLEO detector[1] , which has rather good momentum resolution $\delta p/p = \sqrt{(0.23\% \times p)^2 + (0.7\%)^2}$, p in GeV/c, and a 6.5% dE/dx resolution within the central tracking chambers. The resolution in photon energy is less satisfactory, about $21\%/\sqrt{E}$, E in GeV. The data were gathered in 1987 at two energies. At the energy of the $\Upsilon(4S)$ resonance, 10.58 GeV, we accumulated 212 pb^{-1}, about 242,000 $B\overline{B}$ pairs. At 10.52 GeV, we accumulated 100 pb^{-1} for the purpose of estimating the continuum contribution to the event sample at the $\Upsilon(4S)$. A normalization factor of 2.079 ± 0.008 is required for the continuum subtraction.

The Masses of B^- *and* $\overline{B}^0$

Figure 1 shows all the cleanly reconstructed $B^{\pm}$ and B^0 decay modes that CLEO has analyzed in the 1987 data sample. Only about 66 neutral B's and 47 charged B's are seen above background levels out of nearly half a million B's produced. The plots in Fig. 1 are aligned so that you can see that there is rather little difference in the charged and neutral B masses, $m_{\overline{B}^0} - m_{B^-} = +0.4 \pm 0.6$ MeV.

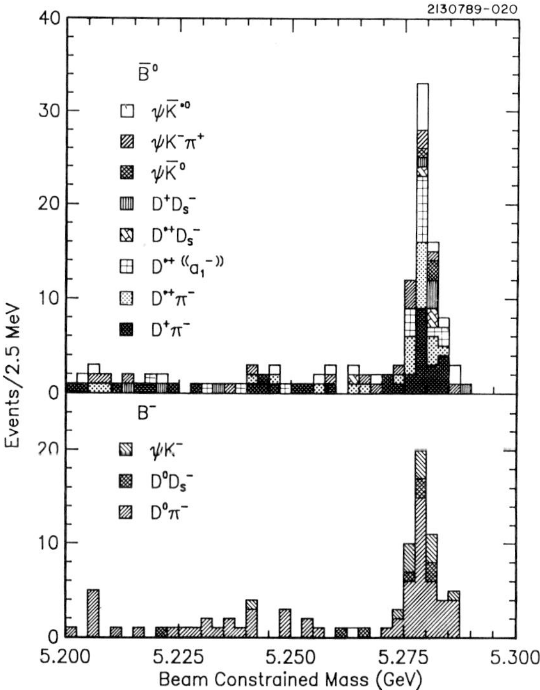

Figure 1: Missing masses of all fully reconstructed neutral B mesons (top) and charged B mesons (bottom).

This represents a shift from older CLEO [2] measurements (2.0 ± 1.1 MeV) and ARGUS [3] measurements (2.0 ± 1.4 MeV), though within errors all measurements agree. Averaging the two CLEO measurements, we obtain $m_{\overline{B}^0} - m_{B^-} = +0.8 \pm 0.5$ MeV as the best CLEO estimate of this quantity.

The mass difference is important for obtaining the production ratio of neutral to charged B's from

* Work supported in part by the National Science Foundation

$\Upsilon(4S)$ decay, which is in turn necessary for obtaining branching ratios. We assume that $\Upsilon(4S)$ decays into neutral B's a fraction α of the time and $(1-\alpha)$ into B^+B^-. One way to estimate the value of α is to assume that the branching ratio is determined by the phase space available in this $L = 1$ decay, which scales like the cube of the available momentum. However, this simple assumption has been criticized by Ono [4] , who predicts a much weaker dependence on the B momentum, and hence a ratio α very near to 0.5. Since the observed mass difference is very small anyway, I will assume in calculating branching ratios that α is 0.5, i.e. that equal numbers of charged and neutral $B\overline{B}$ pairs are produced at the $\Upsilon(4S)$.

B^- and $\overline{B}^0$ branching ratios

Table 1 lists CLEO's new B^- and $\overline{B}^0$ branching ratios [5] along with older CLEO[2] measurements and recent ARGUS[3] measurements. As you can see, quite a few modes are newly measured. Despite this fact, CLEO has only measured about 5% of the charged B and 15% of the neutral B hadronic modes to date. A lot remains to be done!

Table 1: B^- and $\overline{B}^0$ branching ratios in percent

Mode	CLEO 1987	CLEO 1985 †	ARGUS †	Bauer, et al. Model
$B^- \to D^0\pi^-$	$0.30 \pm 0.06 \pm 0.04$	$0.62 \pm 0.20 \pm 0.13$	$0.23 \pm 0.12 \pm 0.07$	$0.48(a_1 + 0.75a_2)^2$
$B^- \to D^{*+}\pi^-\pi^-$	< 0.4	$0.27 \pm 0.19 \pm 0.09$	$0.8 \pm 0.4 \pm 0.5$	
$B^- \to \psi K^-$	$0.08 \pm 0.02 \pm 0.02$	0.12 ± 0.08	0.09 ± 0.05	$1.01a_2^2$
$B^- \to \psi K^{*-}$	$0.13 \pm 0.09 \pm 0.03$			$4.33a_2^2$
$B^- \to \psi K^-\pi^+\pi^-$	$0.12 \pm 0.06 \pm 0.03$		0.13 ± 0.09	
$B^- \to \psi' K^-$	< 0.05		0.27 ± 0.21	
$B^- \to \psi' K^{*-}$	< 0.35			
$B^- \to D^0 D_S^-$	3.7 ± 1.9			$0.73a_1^2$
$\overline{B}^0 \to D^+\pi^-$	$0.27 \pm 0.08 \pm 0.05$	$0.45 \pm 0.23 \pm 0.11$	$0.25 \pm 0.11 \pm 0.08$	$0.48a_1^2$
$\overline{B}^0 \to D^{*+}\pi^-$	$0.33 \pm 0.09 \pm 0.06$	$0.23 \pm 0.11 \pm 0.07$	$0.26 \pm 0.14 \pm 0.10$	$0.37a_1^2$
$\overline{B}^0 \to D^{*+}\rho^-$	$1.9 \pm 0.9 \pm 1.3$			$1.18a_1^2$
$\overline{B}^0 \to D^{*+}\pi^-\pi^+\pi^-$	$1.5 \pm 0.4 \pm 1.0$	< 3.5	$3.2 \pm 0.9 \pm 1.5$	
$\overline{B}^0 \to D^{*+}a_1^-$	$1.8 \pm 0.5 \pm 1.2$			$1.63a_1^2$
$\overline{B}^0 \to \psi \overline{K}^0$	$0.06 \pm 0.03 \pm 0.02$			$1.02a_2^2$
$\overline{B}^0 \to \psi \overline{K}^{*0}$	$0.11 \pm 0.05 \pm 0.03$	0.31 ± 0.15	0.27 ± 0.15	$4.36a_2^2$
$\overline{B}^0 \to \psi K^-\pi^+$	$0.10 \pm 0.04 \pm 0.03$			
$\overline{B}^0 \to \psi' \overline{K}^0$	< 0.15			
$\overline{B}^0 \to \psi' \overline{K}^{*0}$	$0.14 \pm 0.08 \pm 0.04$			
$\overline{B}^0 \to D^+ D_S^-$	2.1 ± 1.1			$0.67a_1^2$
$\overline{B}^0 \to D^{*+} D_S^-$	3.2 ± 1.7			$0.30a_1^2$

†The ARGUS and previous CLEO results have been renormalized for equal charged and neutral B production on the $\Upsilon(4S)$.

There is room for only a few comments about this table. CLEO finds that all its $\overline{B}^0 \to D^{*+}\pi^-\pi^+\pi^-$ events are consistent with $\overline{B}^0 \to D^{*+}a_1^-$, where $a_1^- \to \rho^0\pi^-$. The detection efficiency differs between the two body decay $D^{*+}a_1^-$ and the four body decay mode, which accounts for the two different branching ratios listed.

The observation of $B^- \to D^0D_s^-$, $\overline{B}^0 \to D^+D_s^-$, and $\overline{B}^0 \to D^{*+}D_s^-$ represent the first observations of B decays to two charmed mesons. (The branching ratios listed in Table 1 assume $D_s \to \phi\pi$ 1.5% of the time.) The decays are expected in a simple spectator model, when the virtual W^- materializes as a $\bar{c}s$.

The rate of production, however, is somewhat surprising. The Bauer, Stech, and Wirbel [6] model predictions for relative branching ratios in terms of the parameters a_1 and a_2 are listed in the last column of Table 1. The model gives a fair description of the single-charm B decays within the large experimental errors, but predicts a total of only about two double-charm decays. We see eleven. This bad news for BSW may be good news for experimentalists seeking channels for observing CP violation in B decays. The large rate for $B \to D^+D_s^-$ implies a perhaps measurable rate for the Cabbibo suppressed decay $B \to D^+D^-$, a CP eigenstate. This could supplement the ψK_S^0 decay, for which a total of only three events have now been seen at CLEO.

$\overline{B}^0$ SEMILEPTONIC BRANCHING RATIO

Several years ago, experimentalists were surprised to discover a large difference in the D^0 and $D^\pm$ lifetimes, now known [7] to differ by a factor of about 2.5. Many possible explanations have been offered, including non-spectator diagrams and final state interactions, and a reasonable description appears to be in hand [8] . It would be useful, however, to check the ideas in an independent system like the B's.

Since the semileptonic decay widths of B^0 and B^- are expected to be equal (only the spectator diagram seems likely to contribute to semileptonic B decay), any difference in the $\overline{B}^0$ and B^- lifetimes should show up as a difference in their semileptonic decay rates. To date only the average B semileptonic rate has been measured. All four of the experiments at DORIS and CESR have measured it to be about 10 to 11%, with typical errors of less than 1%. (The new CLEO branching ratio is awaiting a careful calculation of the radiative corrections, which are more substantial than formerly assumed.)

The neutral and charged semileptonic decay rates can be measured separately in several ways. If the rates were very different, this fact could be determined by comparing dilepton with single lepton production rates, because if one charge mode's rate were very high, the partner B would also have a high semileptonic decay rate, leading to an unexpectedly large dilepton rate. This is not seen, and the ratio of charged to neutral semileptonic rates has for some time now [9] been limited at 90% confidence level to be between 0.44 and 2.05. This technique becomes insensitive when the rates agree within a factor of two, so a better technique is needed.

One such technique is to measure an exclusive semileptonic decay rate like $\overline{B}^0 \to D^{*+}\ell^-\bar{\nu}$, which is easy to calculate theoretically. CLEO [10] seeks events containing a lepton and a D^*, assumes that the particles come from the decay of a motionless B, and calculates the missing mass of the neutrino. In fact, a peak is observed at zero missing mass, indicating the presence of this decay. Both CLEO and ARGUS have beautiful, but similar results, and because Danilov has indicated he wishes to discuss them, I will leave the description to him.

Another technique is to tag $\overline{B}^0$'s by the decay of their partner B^0. Unfortunately, there are only about 65 fully reconstructed B^0's, too few to do the job well. However, $B^0 \to D^*\ell^-\bar{\nu}$ decays, reconstructed as described in the preceding paragraph, do yield a usable tag. Seven leptons are seen amongst about 120 B^0 tags. Here I will concentrate on yet another tag, the one with the largest number of events [11] , partially reconstructed $B^0 \to D^*\pi$.

Partial Reconstruction of $\overline{B}^0 \to D^{*+}\pi^-$

The basic idea is not new: you take advantage of the low Q value in the decay $D^{*+} \to D^0\pi^-$, the absence of a kinematically allowed corresponding $D^{*0} \to D^+\pi^-$ decay, and the slowness of the B motion to identify $B \to D^*\pi$ without observing the daughter D^0 from the D^{*+} decay. Observed are the fast π^- from the $\overline{B}^0$ decay and the π^+ from the D^{*+}decay. This π^+ has nearly the same vector velocity as the D^{*+} and hence is moving almost directly opposite the fast π^- and with a relatively low momentum. The kinematic constraints from the D^* and D^0 masses suffice to calculate all the four-vectors except for the ϕ angle between the plane defined by the initial $\overline{B}^0$ momentum and the fast π^- momentum, and the plane of the D^{*+} decay (in the $\overline{B}^0$ rest frame). Since this angle does not enter strongly into the calculation of the B^0 mass, we arbitrarily assume the two planes coincide, which maximizes the calculated B mass. Figure 2 shows the mass determined from the fast π^- and slow π^+ momenta in this way for 3000 simulated events. Demanding a pseudomass larger than the actual B^0 mass (5.280 in this simulation) keeps about 75% of the events.

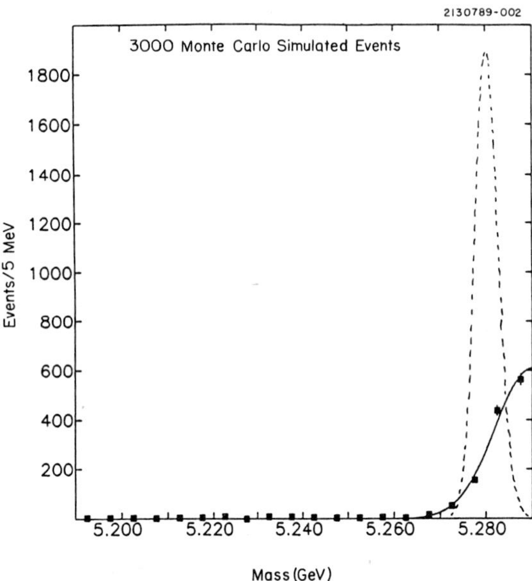

Figure 2: The results of the partial reconstruction technique applied to 3000 Monte Carlo $\overline{B}^0 \to D^{*+}\pi^-$ decays. The dotted curve is what would have resulted had the mass been calculated by fully reconstructing using all the tracks in the decay.

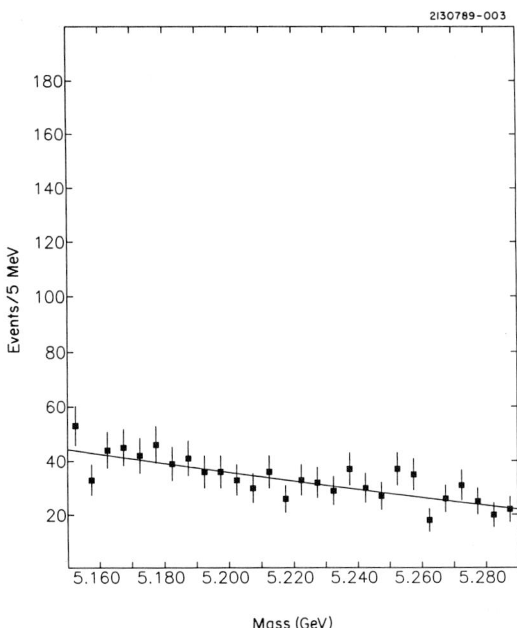

Figure 3: Contamination of the $\overline{B}^0$ tag from known $B^\pm$ decays as represented by the standard CLEO Monte Carlo.

Figures 3 and 4 represent a Monte Carlo estimate of the contamination of $B^\pm$ decays which could pass the criteria for a B^0 tag. Two separate contributions are simulated. Figure 3 shows the mass spectrum expected from a simulation which describes all B decays as well as we know them. Fewer than 18 events at 90% confidence level will be wrongfully classed as B^0's. However, as yet unobserved $B^\pm$ decays could also yield tags. For example, the decay $B^- \to D^{**0}(2420)\ \pi^-$ can produce a high mass as illustrated in Fig 4 if a low energy π from the D^{**0} is neglected. No evidence exists for this $B^\pm$ decay mode, and in fact the CLEO upper limit and the broadened shape of the expected spectrum permit us to set an upper limit of 21 false tags from such processes. Thus there are fewer than 39 $B^\pm$ decays contaminating our tagged B^0's.

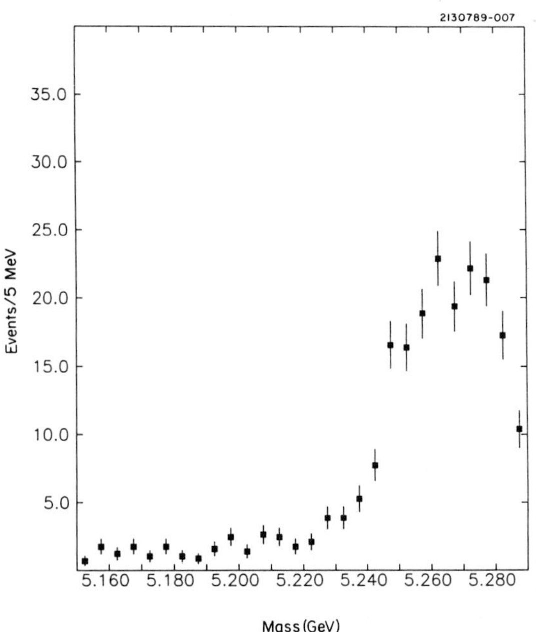

Figure 4: Contamination of the $\overline{B}^0$ tag from possible $B^- \to D^{**0}\pi^-$

Figure 5 shows the mass spectrum actually obtained. A clear peaking is observed at the high mass end; 289 ± 29 ± 26 events lie in the signal region above 5.280 GeV. As a check that the partial reconstruction technique is working, these events can be used to calculate the branching ratio for $\overline{B}^0 \rightarrow D^{*+}\pi^-$. The result, $(0.42 \pm 0.05\ ^{+0.04}_{-0.08})\%$, is in agreement with the result from fully reconstructed events where the D^0 is observed in its $K\pi$ or $K3\pi$ decay modes, (0.33 ±0.09 ±0.06)%.

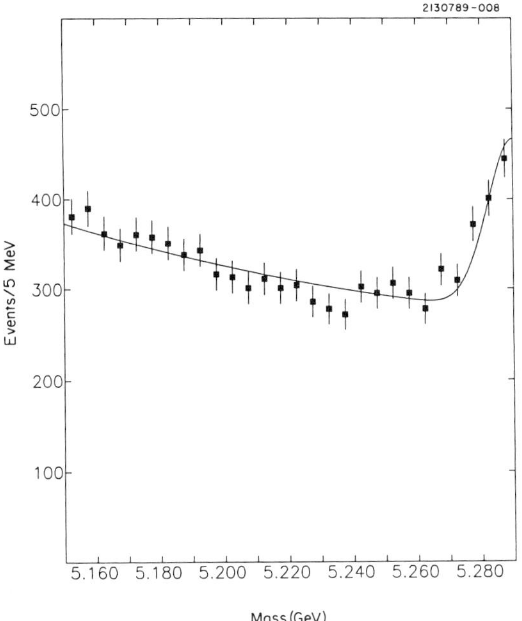

Figure 5: B mass spectrum obtained using the partial reconstruction technique on real data.

The background in Fig. 5 seems disturbingly high. However, the vast majority of the background events can be attributed to continuum processes which are less pernicious, since they are relatively lepton poor and since they can be estimated well.

Knowing the flavor of the B^0 or $\overline{B}^0$ tag, we know which charge of lepton to expect, as long as there is no $B^0\overline{B}^0$ mixing. Table 2 summarizes the right-sign leptons observed in the tagged events, together with the estimated backgrounds. Combining the electron and muon counts and the detection efficiency, we find the semileptonic branching ratio to right-sign leptons is (9.3 ± 3.1 ± 1.1)%.

Table 2: Right-sign leptons in tagged $\overline{B}^0$ or B^0 events, and estimated backgrounds

number of right-sign leptons	electrons	muons
Observed with tagged B^0	12.0 ±3.5	11.0 ±3.3
Background (sideband)	3.3 ±0.7	3.1 ±0.7
Cascade bkg, same B	1.2 ±0.1	0.4 ±0.1
Cascade bkg, other B	0.1 ±0.0	0.0 ±0.0
Fake leptons	0.1 ±0.0	0.1 ±0.0
Net signal	7.3 ±3.5	7.4 ±3.4

But of course B's do mix. One way to account for mixing is to count wrong-sign tagged leptons as well. Unfortunately, the backgrounds remain the same, while the signal is reduced by a factor of four, since mixing occurs only about 20% of the time. Thus only about three events are expected, while (2.0 ± 2.9) are observed. Adding in mixed B's this way would dilute the statistical accuracy of the result.

A more satisfactory technique is to correct for mixing using the additional measurement of the rate of like-sign to unlike-sign observed dileptons. The ratio $f = \frac{(N_{--}+N_{++})}{(N_{+-})}$, where N_{+-} is the total number of opposite-sign dileptons and $(N_{--} + N_{++})$ is the number of same-sign dileptons, is measured to be 0.076 ± 0.020 ± 0.022. This would be the mixing ratio r were it not for the fact that the denominator contains leptons from charged as well as neutral B decays. In order to account for this contamination, we need to know the semileptonic branching ratio of $B^{\pm}$ and B^0 separately, and also the rate of production of B^0 relative to that of $B^{\pm}$. However, we know [12] the average $B^{\pm}$ and B^0 semileptonic branching ratio is (10.1 ± 0.3 ± 0.7)%, and can reasonably assume that the production rates of charged and neutral B pairs from the $\Upsilon(4S)$ are equal. Thus the $\overline{B}^0$ semileptonic rate is both what we are solving for and the only remaining unknown for the denominator correction. It is only a matter of algebra to extract it from the equations to find (10.6 ± 2.8 ± 1.1)%. The $\overline{B}^0$ semileptonic decay rate is the same, within errors, as the average B semileptonic rate.

THE MIXING RATIO r

It should be clear that we can also solve the above equations to extract the mixing ratio r. This is defined as the number of B^0's which decay like $\overline{B}^0$'s divided by the number which decay as they were born,

$$r = \frac{\Gamma(B^0 \to \overline{B^0} \to \overline{X^0})}{\Gamma(B^0 \to X^0)} = \frac{(N_{--} + N_{++})(1+\lambda)}{N_{+-} - (N_{--} + N_{++})\lambda},$$

where $\lambda = \frac{(1-\alpha)}{\alpha} \times (b^-_{SL}/b^0_{SL})^2$ is an "ignorance parameter," which until now has been assumed [13,14] to be 1.2. In the above equations α is again the fraction of $\Upsilon(4S)$ decays yielding neutral B pairs, and b^-_{SL} and b^0_{SL} are the semileptonic branching ratios of charged and neutral B mesons, respectively.

Now, with the measurement of b^0_{SL}, it is possible to determine r without resorting to ad hoc assumptions about λ, so that the error on r is determined experimentally. As you saw in the previous section, the evaluation of r and b^0_{SL} are interdependent. Hence the procedure we use is to fit all available relevant information regarding r, α, b^0_{SL}, and b^-_{SL}. The following measurements are used.

- the average B semileptonic decay rate;
- the number of right-sign leptons per tagged B^0 using partially reconstructed $B^0 \to D^*\pi$ as tags: $(9.3 \pm 3.1 \pm 1.1)\%$;
- the same quantity using $B^0 \to D^*\ell\overline{\nu}$ as tags: $(9.0 \pm 3.4 \pm 0.7)\%$;
- the lifetime ratio $\frac{(1-\alpha)}{\alpha} \times (b^-_{SL}/b^0_{SL}) = 0.85 \pm 0.20\ ^{+0.22}_{-0.16}$. This is obtained from the measurement of the branching ratios for $B^- \to D^{*0}\ell^-\overline{\nu} = (3.9 \pm 0.8\ ^{+1.1}_{-0.8})\%$ and $\overline{B}^0 \to D^{*+}\ell^-\overline{\nu} = (4.6 \pm 0.5 \pm 0.7)\%$, along with the assumption of isospin invariance;
- the like-sign to opposite-sign dilepton ratio $\frac{(N_{--}+N_{++})}{(N_{+-})} = 0.076 \pm 0.020 \pm 0.022$.

Table 3 lists the results of the fitting. The two columns describe fits for which two different assumptions were made about the production rate of neutral and charged B mesons. In the first column, equal rates are taken, and in the second the rate is assumed to be governed by a phase space factor proportional to p^3_B. By comparing the two columns you can see the value $r = 0.14 \pm 0.05$ is insensitive to this assumption. This measurement is free of the systematic error previously caused by the need to assume a value 1.2 for the ignorance parameter λ.

Table 3: Optimal values of the parameters in the fit for the mixing ratio r, listed for two assumptions about the relative rates for charged and neutral B meson production.

Fit parameter	$\overline{B}^0B^0 = B^+B^-$	p^3 weighting
$\overline{B}^0B^0/B^+B^-$	$\equiv 1$	0.89 ± 0.08
b^0_{SL}	$(10.7 \pm 1.4)\%$	$(11.1 \pm 1.5)\%$
$\tau^-/\tau^0 = b^-_{SL}/b^0_{SL}$	0.88 ± 0.24	0.82 ± 0.21
λ	0.77 ± 0.39	0.75 ± 0.37
r	0.14 ± 0.05	0.14 ± 0.05

SEARCH FOR $b \to u$ TRANSITIONS

The standard model can explain CP violation only if all of the CKM matrix elements are nonzero. It is thus an important test of the model to demonstrate that V_{ub} is not zero. The value of V_{ub} is expected to lie in the range $0.04 \leq |V_{ub}/V_{cb}| \leq 0.16$, where the upper limit is obtained from model-dependent interpretation of experimental data, and the lower limit is where it becomes difficult to reconcile the high mass of the top quark, the size of $B^0\overline{B}^0$ mixing, and CP violation in the kaon system. Experiments measure $|V_{ub}|^2$, so we expect charmless transitions to be a few percent effect at best.

Several techniques have been employed to seek charmless b decays.

- One can seek a deficit in the number of B decays to charmed mesons. The most recent CLEO measurements use "reasonable" assumptions about such unmeasureable quantities as the the $D_s \to \phi\pi$ branching ratio to estimate (1.10 ± 0.15) charmed mesons per B decay[5]. Imprecise arguments based on spectator decay models predict roughly 1.15 charmed mesons per B decay. Clearly it is masochistic to seek a 2% effect this way.
- Exclusive decays like $\overline{B}^0 \to \pi^+\pi^-$ and $\overline{B}^0 \to \rho^+\ell^-\overline{\nu}$ which can be calculated rather reliably, can be used to seek a $b \to u$ transition [15] . While there is no reason in principle why this shouldn't work, the weaknesses of the CLEO-I detector and the small branching ratios to easily observed final states conspire to give rather unsatisfying upper limits.
- The most sensitive technique CLEO has is to analyze the shape of the lepton spectrum in B semileptonic decays. This technique has the advantage of summing over many exclusive modes at the price of having trickier backgrounds to handle.

The Endpoint of the Lepton Spectrum

Figure 6 shows the expected contributions of charmed and charmless semileptonic B decays to the lepton momentum spectrum. I have assumed that 2% of the decays are charmless, which corresponds to $|V_{ub}/V_{cb}| = 0.1$.

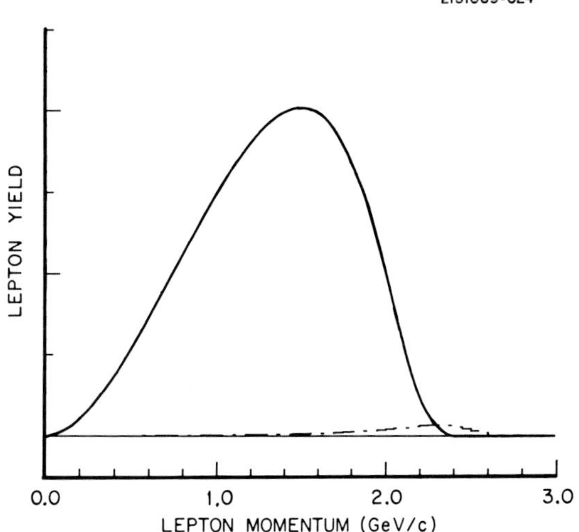

Figure 6: Expected lepton yields from $b{\to}c$ semileptonic decays (solid curve) and $b{\to}u$ decays (dotted), roughly in the expected proportions

Because the lightest charmed meson has mass 1.86 GeV, the $b{\to}c$ spectrum ends at lepton momentum about 2.4 GeV/c, while the $b{\to}u$ spectrum continues to about 2.6 GeV/c. Thus the momentum region 2.4 to 2.6 GeV/c represents a clean fiducial region for charmless B decays. The fiducial region can be pushed to lower momenta depending on the degree of confidence in the estimate of the $b{\to}c$ contribution. Clearly, the lower you attempt to go in lepton momentum, the more the systematic uncertainty in the charm contamination with which you have to deal. We have chosen the region 2.2 to 2.6 GeV/c.

The fully corrected and subtracted yield of electrons and muons is shown in Fig. 7. The high momentum tail of the $b{\to}c$ events can be seen falling rapidly toward its kinematic limit of 2.45 GeV/c. Above 2.7 GeV/c no particles can be produced in B decays at the $\Upsilon(4S)$ cm energy, and Fig. 7 shows no net events above this momentum, indicating that the continuum subtraction is working well. In the region 2.4 to 2.6 GeV/c, where charmless B decays are expected to be the dominant net source of leptons, there is a distinct lepton excess.

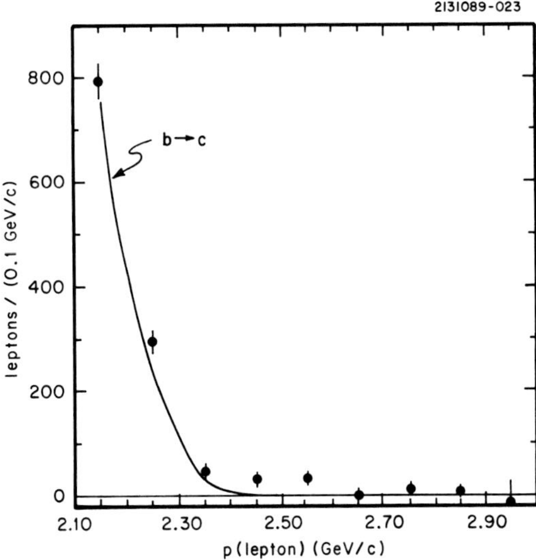

Figure 7: The yield of leptons in the endpoint region after all subtractions. The expected $b{\to}c$ contribution is indicated as the smooth curve.

Background Rejection

The backgrounds for this measurement can be read off the unsubtracted momentum spectrum, illustrated for electrons in Fig. 8.

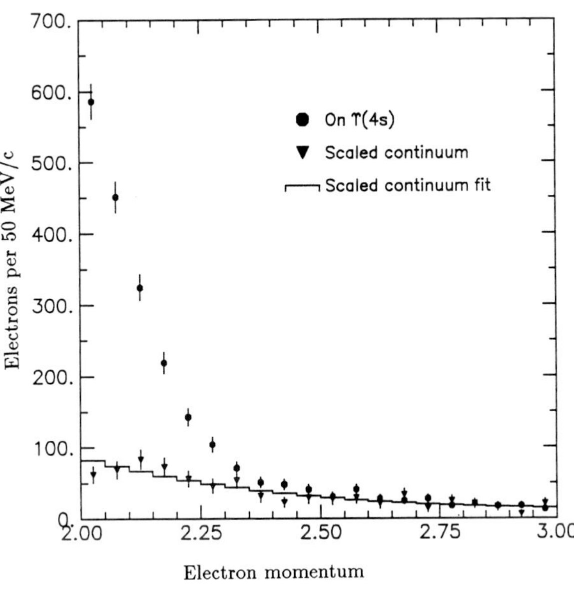

Figure 8: The electron yield before continuum subtraction.

Clearly, the majority of leptons in the momentum range of interest come from continuum processes. We suppress these with an event shape cut [16], demanding $R_2 = H_2/H_0 < 0.4$. This criterion rejects 70% of continuum events, as verified in off-4S running, and accepts 92% of the $\Upsilon(4S)$ events with semileptonic B decays, according to a Monte Carlo simulation. The remaining continuum events can be subtracted in two ways, a bin-by-bin subtraction of the off-4S running, scaled by the factor 2.079, and a smooth fit to the scaled continuum. We must keep careful control of the systematics of the subtraction, so we run in a pattern of two days on resonance alternating with one day off, so as to average over detector and CESR performance. The relative luminosity is measured to within 0.4% using wide angle Bhabhas. After the subtraction is performed, it is checked using tracks of momentum above 3 GeV/c. The agreement is good.

The rate at which hadrons are misidentified as leptons ("fakes") is calibrated using pions from $\Lambda \rightarrow p\pi$ and kaons from ϕ decays. These results are checked with $\Upsilon(1S)$ decays, which are depleted in leptons. The faking probabilities for electrons are $\sim 0.3\%$ in the 47% of 4π solid angle where the barrel shower counters and dE/dx chambers can be used, and $\sim 0.8\%$ where only the dE/dx of the tracking chambers is usable (32% of 4π). Muon fakes and random coincidences are 1.1% in the momentum region of interest.

As is clear from Fig. 8, if 2.2 GeV/c leptons from $B \rightarrow D\ell\bar{\nu}$ decays are mismeasured as having 0.1 to 0.2 GeV/c more momentum than they really have, an artificial excess of leptons in the fiducial region for charmless decays can be generated. Using $e^+e^- \rightarrow \mu^+\mu^-$ events we have verified that no non-Gaussian tails can be seen on the momentum distribution, once careful track quality cuts are applied to reject cases where the track finder fails (these cuts are 98% efficient). However, proving that mismeasurement is not a problem in the more complex hadronic events requires rather more effort. We extract well-measured electron tracks from radiative Bhabha events, and embed them hit by hit into hadronic events. If two hits arrive at the same wire, the time of the earlier hit is taken. We then run the trackfinder on the merged events. Fig. 9 shows the result; only three of 1727 tracks are seriously mismeasured, which indicates that less than one $b \rightarrow c$ lepton track will be boosted into the $b \rightarrow u$ fiducial region.

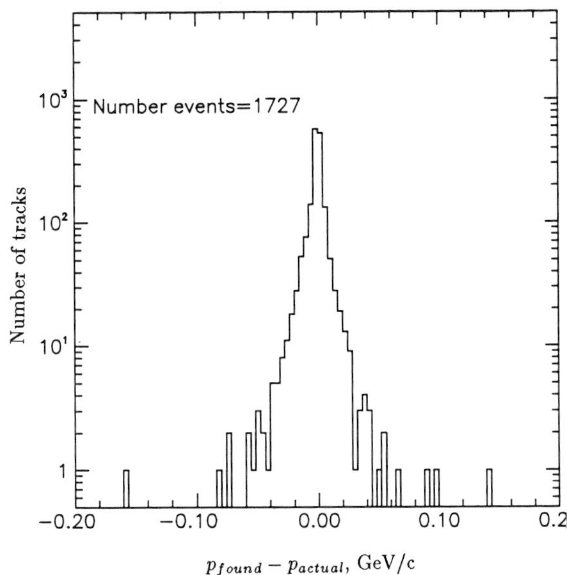

Figure 9: Histogram of the difference between the true momentum of electrons embedded into hadron events and the momentum found by the trackfinder. Note the log scale.

Finally, we need to estimate the lepton rate as a function of momentum for charmed semileptonic B decays, so that these can be subtracted. We have used shapes from the ISGW [17], ACCMM [18], and WSB [19] models, suitably smeared for our detector resolution, and corrected for QED effects. These are each normalized to the observed spectrum in the momentum region 1.5 to 2.2 GeV/c and extrapolated to higher momentum. The different models agree on the estimate to within 10%.

The Lepton Excess and V_{ub}

Table 4 summarizes the observed rates and backgrounds subtracted for electrons and muons in two momentum bins, 2.2 to 2.4 GeV/c and 2.4 to 2.6 GeV/c. One expects the muon and electron number to be nearly equal, and this is the case. Also, one expects 1.5 to 2 times as many leptons from charmless B decays in the lower momentum bin than the higher; this expectation does not appear to be met

Two reasonable hypotheses explain the data: this could be a roughly 3-standard-deviation statistical fluctuation, or this is first evidence for charmless B decays. If we assume that the latter is the case, we can calculate V_{ub} from the observed lepton cross section between 2.2 and 2.6 GeV/c, $\sigma_{ub} = 0.76 \pm 0.2$ pb using the equation

Table 4: The lepton yield and estimated backgrounds in two momentum bins near the endpoint for B semileptonic decay

2.4–2.6 GeV/c		
Lepton Source	Electrons	Muons
$B\overline{B}$ Events	$51 \pm 15 \pm 6$	$30 \pm 14 \pm 9$
Fakes	$1.2 \pm 0.6 \pm 0.3$	$4.0 \pm 2.0 \pm 1.0$
$B \to \psi X$	$1.0 \pm 0.5 \pm 0.5$	$0.8 \pm 0.4 \pm 0.4$
$B \to \psi' X$	$1.0 \pm 0.5 \pm 0.5$	$1.0 \pm 0.5 \pm 0.5$
$b \to c\ell\nu$	$0.4 \pm 0.1 \pm 0.3$	$0.6 \pm 0.1 \pm 0.3$
$b \to u\ell\nu$	$47 \pm 15 \pm 6$	$24 \pm 14 \pm 9$
2.2–2.4 GeV/c		
Lepton Source	Electrons	Muons
$B\overline{B}$ Events	$197 \pm 21 \pm 7$	$207 \pm 21 \pm 13$
Fakes	$15.0 \pm 2.0 \pm 2.7$	$28.8 \pm 2.1 \pm 7.2$
$B \to \psi X$	$2.7 \pm 0.9 \pm 0.5$	$1.7 \pm 0.6 \pm 0.4$
$B \to \psi' X$	$1.7 \pm 0.6 \pm 0.6$	$1.7 \pm 0.6 \pm 0.6$
$b \to c\ell\nu$	$145 \pm 2 \pm 10$	$146 \pm 2 \pm 10$
$b \to u\ell\nu$	$33 \pm 21 \pm 13$	$29 \pm 22 \pm 17$

$$\frac{|V_{ub}|^2}{|V_{cb}|^2} = \frac{\sigma_{ub}(p)}{\sigma_{cb}} \times \frac{1}{d(p)}.$$

Here, d(p) is a model-dependent factor accounting for the fraction of the entire spectrum contained in the momentum range p under consideration and for formfactors, phase space, and wave functions assumed in the particular model. $\sigma_{cb} = 236 \pm 14$ pb is the average cross section for lepton production from charmed B semileptonic decays. Table 5 lists the d(p) factor and relative V_{ub} size for the ISGW[17], ACCMM[18], WSB[19], and KS [20] models. Since the errors on V_{ub} are non-Gaussian because of the square root taken on the experimentally determined numbers, I prefer to quote 90% confidence level upper and lower limits, which are also given in the table.

Table 5: Values for d(p) and $|V_{ub}|/|V_{cb}|$ obtained in various models of B semileptonic decay

MODEL	d(p)	V_{ub}/V_{cb}	90% C.L. limits
ISGW	0.19	0.13 ±.02	0.09 to 0.16
WBS	0.31	0.10 ±.02	0.07 to 0.12
ACCMM	0.46	0.08 ±.01	0.05 to 0.10
KS	0.37	0.09 ±.02	0.06 to 0.11

SUMMARY

CLEO has fully reconstructed approximately 110 B decays in the 1987 data sample. A neutral-charged B mass difference of 0.8 ± 0.5 MeV is found. A number of B decay modes have been newly observed, most notably decays to two charmed particles.

Tagging of $\overline{B}^0$ by partial reconstruction of $\overline{B}^0 \to D^{*+}\pi^-$ permits measuring the $\overline{B}^0$ semileptonic branching ratio, $(10.6 \pm 2.8 \pm 1.1)\%$.

This result, combined with various other measurements on exclusive semileptonic B decays leads to an evaluation of the mixing parameter $r = 0.14 \pm 0.05$, without the need to resort to ad hoc assumptions about the "ignorance factor" λ.

CLEO has found evidence for charmless B decays in the lepton spectrum from B decays near the endpoint. The evidence is not statistically overwhelming, but if assumed valid, a value of $|V_{ub}|/|V_{cb}|$ in the range 0.06 to 0.17 is implied, where much of the spread is due to theoretical uncertainties in the models applied.

ACKNOWLEDGEMENTS

I would like to thank all my CLEO colleagues, but especially Mike Procario, Sheldon Stone, and Dave Cassel for their assistance in the preparation of this talk. Thanks are also due Steve Wagner for his able and efficient work as my scientific secretary.

REFERENCES

1. D. Cassel, *et al.*, *Nucl. Inst. Methods* **211** (1986), 325.
2. C. Bebek, *et al.*, *Phys. Rev.* **D36** (1987), 1289.
3. H. Albrecht, *et al.*, *Phys. Lett.* **185B** (1987), 218.
4. S. Ono, *Acta Phys. Pol.* **B15** (1984), 15.
5. W-Y Chen, *et al.* "B Decays to Charm: Exclusive Decays and Inclusive Rates", paper No. 283 contributed to this Symposium.
6. M. Bauer, B. Stech, and M. Wirbel, *Z. Phys.* **C34** (1987), 103.
7. Particle Data Group, G. P. Yost, *et al.*, *Phys. Lett.* **204B** (1988), 1.

8. R. Schindler, "Recent Results on the Charm Sector," in *High Energy Electon-Positron Physics,* ed. A. Ali and P. Söding, World Scientific, Singapore (1988) p234.
9. A. Bean, *et al., Phys. Rev. Lett.* **58** (1987), 183.
10. D. Bortoletto, *et al., Phys. Rev. Lett.* **63** (1989), 1667.
11. M. S. Alam, *et al.* "A Measurement of the B^0 Semileptonic Branching Ratio", paper No. 280 contributed to this Symposium.
12. R. V. Kowalewski, "Semileptonic B Meson Decay," Ph. D. Thesis, Cornell Univ., unpublished UMI 88-21246-m (microfiche) (1988).
13. H. Albrecht, *et al., Phys. Lett.* **192B** (1987), 245.
14. M. Artuso, *et al., Phys. Rev. Lett.* **62** (1989), 2233.
15. D. Bortoletto, *et al., Phys. Rev. Lett.* **62** (1989), 2436.
16. G. C. Fox and S. Wolfram, *Phys. Rev. Lett.* **41** (1978), 1581.
17. N. Isgur, D. Scora, B. Grinstein, and M. Wise, *Phys. Rev.* **D39** (1989), 799.
18. G. Altarelli, *et al., Nucl. Phys.* **B208** (1982), 365.
19. M. Wirbel, B. Stech, and M. Bauer, *Z. Phys.* **C29** (1985), 637.
20. J. G. Körner and G. A. Schuler, *Z. Phys.* **C38** (1988), 511

DISCUSSION

V. Khoze, Leningrad Nuclear Physics Institute: What limit on $|V_{bu}/V_{bc}|$ could be extracted from the charm counting data?

D. Kreinick: That is a really hard way to do it. There are decays, for example, into χ states, and we don't know the branching ratio for $D_s \to \phi\pi^+$. All those things come in. So the statistical errors, which I think I quoted as 0.05, are dwarfed by the systematic errors. I would guess that you can't do much better than a 20% limit, maybe 15% .

F. Sculli, Columbia University: Can one characterize with a number the theoretical error on $|V_{bu}/V_{bc}|$?

D. Kreinick: I think the best I can do is to give you various models, which is what I tried to do here. The spread among those models, which is a factor of 2, would be an indication of the theoretical error.

F. Sculli: Can it be improved?

D. Kreinick: I believe there are at least four sets of theorists who are working very hard on trying to improve it. I certainly wish them every good fortune. Would one of them like to comment? You can also do this measurement in a different way. When we have better statistics and we have our better detector running, we should be able to do this in some of the exclusive modes. That may be the right way to measure this quantity.

M. Peskin, SLAC: The theoretical models of $b \to u$ semileptonic decays differ not only in their predictions for the total rate but also in their predictions for the lepton momentum spectrum. So eventually it will be possible to tell experimentally which model is correct.

P. Minkowski, University of Bern: Are these latest numbers on V_{bu} critically dependent on the ratio of lifetimes on B^0 and B^+?

D. Kreinick: No, I don't believe so.

M. Danilov, ITEP, Moscow: Before *[Ed. note: at the Heavy Quark Symposium, June 1989]* there was a high-energy tail in your momentum resolution. Why did it disappear?

D. Kreinick: There was a long Gaussian tail before we did the cuts to clean up the track-finding procedure. There were four cuts: demanding hits on the track near the outer radius of the drift chamber; demanding hits in the vertex detector; requiring a small average residual; and requiring that the track point back to the beam spot within a couple of millimeters in the $x-y$ plane. If we don't to those cuts, then we get lots of junk. We did a lot of work on the track finder; in fact, we had to revise it three times before we got it in good enough shape to get rid of those tails. But once you preform those cuts, it's very clean.

RECENT ARGUS RESULTS ON B MESON DECAYS

Michael V. Danilov
Institute of Theoretical and Experimental Physics
117259, Moscow, USSR

ABSTRACT

Measurements of semileptonic B decays to D^{*-} and D^- mesons lead to consistent values for the Cabibbo-Kobayashi-Maskawa matrix element $|V_{cb}|$ of 0.046 ± 0.009 and 0.042 ± 0.008. A comparable result, $|V_{cb}| = 0.046 \pm 0.05$, is obtained from the study of the inclusive semileptonic decays. The lifetime ratio of B^+ and B^0 mesons is found to be $1.00 \pm 0.23 \pm 0.14$. A detailed study of the lepton momentum spectrum in $\Upsilon(4S)$ decays has been made. In the region from 2.3 to 2.6 GeV/c, which is above the endpoint for contributions from B decays via $b \to c$ transitions, 32 ± 10 events are observed in excess of known backgrounds. If these events are interpreted as a signal for $b \to u$ transitions, a model dependent value of 0.10 ± 0.02 for the ratio of CKM matrix elements $|V_{ub}|/|V_{cb}|$ is obtained. An update of the $B^0\overline{B}^0$ mixing measurement, with 70% more data, yields a value of the mixing parameter $r = 0.21 \pm 0.06$. Upper limits are obtained for many exclusive charmless B meson decays which can originate from $b \to u$ and $b \to s$ transitions.

1 Introduction

Of fundamental importance for our understanding of the weak interaction is a determination of the strength of the couplings between the third and first or second generation of quarks. The relationships among the quark-W boson couplings is expressed by the Cabibbo-Kobayashi-Maskawa (CKM) mixing matrix [1,2], with its four free parameters which have to be determined experimentally. Two of these elements are directly accessible through a study of B meson decays, namely V_{ub} and V_{cb}, which determine the strength of $b \to u$ and $b \to c$ transitions respectively. The latter process is known to dominate, as demonstrated by prolific charmed hadron production in B meson decays [3]. Searches for charmless decays, either in semileptonic or hadronic decay modes [3,4], have led to rather stringent limits on V_{ub}. Measurement of this coupling constant would greatly constrain our picture of weak-interaction physics. In particular, a non-zero value is essential for the Kobayashi-Maskawa explanation of the origin of CP violation [2].

The observation of an unexpectedly large $B^0\overline{B}^0$ mixing [5] demonstrated that V_{td} is nonzero. More precise measurements of the $B^0\overline{B}^0$ mixing rate can provide a stringent constraint on the free parameters of the CKM matrix. Finally, information about V_{ts} can be obtained from a study of loop induced $b \to s\gamma$ and $b \to s$ *gluon* transitions.

This review will summarize the recent ARGUS results relevant to the determination of the CKM matrix elements. Most of these were obtained using a data sample of about 170 pb^{-1} on the $\Upsilon(4S)$ and about 70 pb^{-1} in the nearby continuum at centre-of-mass energies below the open beauty threshold. There are about 150000 B meson pairs in this data sample.

2 The Transition $b \to c$

2.1 *The Decay* $B^0 \to D^{*-}\ell^+\nu$

A value for the V_{cb} matrix element can be extracted from measurements of the decay $B^0 \to D^{*-}\ell^+\nu$ provided its rich dynamics described by three substantial form factors is well understood (references in this paper to a specific charged state are to be interpreted as implying the charged-conjugate state as well). The decay $B^0 \to D^{*-}\ell^+\nu$ has been studied in detail by the ARGUS group [6, 7]. B^0 mesons are produced almost at rest in $\Upsilon(4S)$ decays. Their energy coincides with the beam energy and their small momenta can be neglected. The decay $B^0 \to D^{*-}\ell^+\nu$ is therefore seen as a peak in the spectrum of recoil mass squared spectrum at $M^2_{rec} = 0$, where M^2_{rec} is given by

$$M^2_{rec} = [E_{Beam} - (E_{D^{*-}} + E_{\ell^+})]^2 - [\vec{p}_{D^{*-}} + \vec{p}_{\ell^+}]^2.$$

In order to calculate the branching ratio one must know a fraction of B^0 mesons in $\Upsilon(4S)$ decays. New CLEO measurements of the B^0 and B^+ meson masses show that the mass difference is very close to zero $M_{B^0} - M_{B^+} = 0.8 \pm 0.9\ MeV/c^2$ [8]. Therefore, it is reasonable to expect equal branching ratios for $\Upsilon(4S)$ decays into charged and neutral B mesons. Until a more definitive measurement is made, the ARGUS group will assume equal fractions of B^0 and B^+ mesons in $\Upsilon(4S)$ decays: $f_0/f_+ = 0.5/0.5$. This will change slightly the published values of B meson branching ratios because the previous assumption was $f_0/f_+ = 0.45/0.55$.

Scaling the published ARGUS result $BR(B^0 \to D^{*-}l^+\nu) = (7.0 \pm 1.2 \pm 1.9)\%$ [6] to the new value of $BR(D^{*-} \to \overline{D}^0\pi^-) = (57 \pm 4 \pm 4)\%$ [9], and assuming $f_0/f_+ = 0.5/0.5$, one obtains

$$BR(B^0 \to D^{*-}l^+\nu) = (5.4 \pm 0.9 \pm 1.3)\% .$$

This is the dominant semileptonic decay mode since the inclusive semileptonic branching ratio is about 10% (see chapter 2.4).

The average D^{*-} polarisation in this decay was studied using the strong two-body decay $D^{*-} \to \pi^-\overline{D}^0$ as a polarisation analyser by measuring the angle of the slow π^-, $\theta^*_{\pi^-}$, in the rest frame of the $D^{*-}$with respect to the D^{*-}boost direction. The distribution of $\theta^*_{\pi^-}$ can be parametrised as

$$N(\theta^*_{\pi^-}) \sim 1 + \alpha \cdot \cos^2\theta^*_{\pi^-} ,$$
$$\alpha = \frac{2\Gamma_L - \Gamma_T}{\Gamma_T} ,$$

where Γ_L and Γ_T are the longitudinal and transverse contributions to the decay width. The measured pion angular distribution is almost flat which corresponds to comparable values of Γ_L and Γ_T in this decay:

$$\Gamma_L/\Gamma_T = 0.85 \pm 0.45 .$$

The new CLEO measurements confirm the ARGUS results. The CLEO group obtained [10]:

$$BR(B^0 \to D^{*-}l^+\nu) = (4.6 \pm 0.5 \pm 0.7)\%$$

and

$$\Gamma_L/\Gamma_T = 0.83 \pm 0.33 \pm 0.13$$

for lepton momenta larger than 1.4 GeV/c.

All recent theoretical calculations [11,12,13] are in a good agreement with the experimental results. This gives some confidence in the values of V_{cb} extracted from the data although the results are still model dependent. A knowledge of Γ_L/Γ_T permits one to express $\Gamma(B^0 \to D^{*-}l^+\nu)$ independent of the theoretically most uncertain form factor [14] :

$$\Gamma(B^0 \to D^{*-}l^+\nu) = |V_{cb}|^2\tilde{\Gamma}^{(T)}(1 + \frac{\Gamma_L}{\Gamma_T}),$$

where $\tilde{\Gamma}^{(T)}$= $1.2 \cdot 10^{13}\mathrm{sec}^{-1}$ can be reliably calculated [14]. Inserting the measured numbers for the branching ratio and Γ_L/Γ_T and assuming that the B^0 lifetime is equal to the average lifetime of beauty particles τ_B=$(1.15 \pm 0.14)ps$ [3], one obtains:

$$|V_{cb}| = 0.046 \pm 0.009 ,$$

where only experimental errors are included.

2.2 *The observation of the decay* $B^0 \to D^-\ell^+\nu$

Since the matrix element for reaction $B^0 \to D^-\ell^+\nu$ contains only one substantial form factor, a measurement of $BR(B^0 \to D^-\ell^+\nu)$ allows a nearly model independent determination of the CKM matrix element $|V_{cb}|$. However, it is more difficult to measure this decay because of a large combinatorial background and feed-down from the dominant decay $B^0 \to D^{*-}\ell^+\nu$ where $D^{*-} \to D^-\pi^0/\gamma$. Nevertheless, this channel was observed recently again using a missing mass technique [15]. The backgrounds were reduced considerably by requiring the D^- momentum to be larger than 1.5 GeV/c. Only 20% of D^- mesons from $B^0 \to D^-\ell^+\nu$ have momentum lower than 1.5 GeV/c, while the momentum spectrum of D^- mesons from the D^{*-} cascade decay $B^0 \to D^{*-}\ l^+\ \nu$ with $D^{*-} \to D^-\pi^0/\gamma$ is considerably softer [11]. Thus, the cut on the D^- momentum reduces the contribution from cascade decays, which is the main source of physical background. In addition this cut strongly reduces the combinatorial background.

Figure 1 shows the $K^+\pi^-\pi^-$ invariant mass spectra for events which contain a positive lepton in different intervals of M^2_{rec}. One observes a D^- peak at $M^2_{rec} \approx 0$ and no signal in the M^2_{rec} sidebands. The number of D^- mesons in each interval of M^2_{rec} was obtained from fits to these spectra.

Figure 2 shows the M^2_{rec} distribution for D^-l^+ combinations after subtraction of the backgrounds from continuum, uncorrelated $D^-\ell^+$, and fake leptons. The prominent peak at $M^2_{rec} \approx 0$ includes the decay $B^0 \to D^-\ell^+\nu$ as well as a contribution from D^{*-} cascade decays. The latter has been determined from the data using D^{*-} decays to $\overline{D}^0\pi^-$. For $\overline{D}^0\ l^+$ combinations the same kinematic cuts were applied as for D^- and l^+. After background subtraction 30 ± 7 $\overline{D}^0 l^+$ candidates are retained.

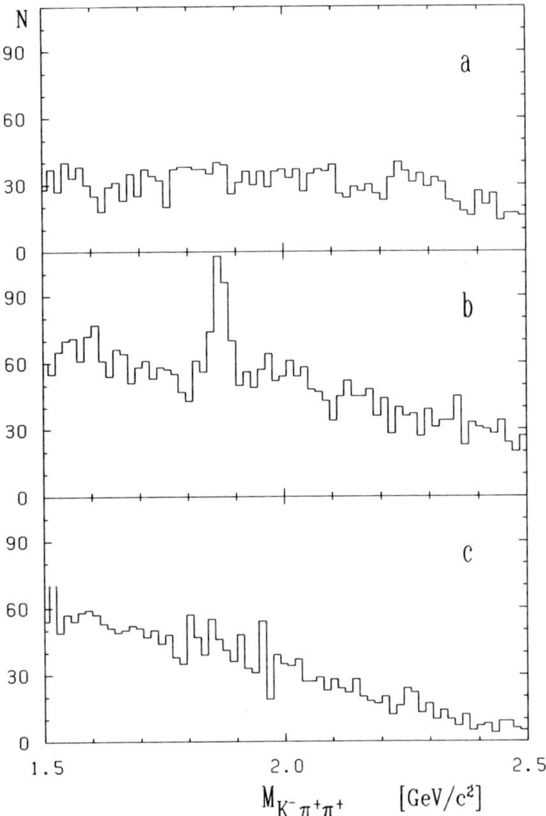

Figure 1: Invariant mass spectrum of $K^+\pi^-\pi^-$ combinations for different intervals of M^2_{rec} :
a) $-1.5 < M^2_{rec} < -0.5\ GeV^2/c^4$,
b) $-0.5 < M^2_{rec} < 0.5\ GeV^2/c^4$,
c) $0.5 < M^2_{rec} < 1.5\ GeV^2/c^4$.

A sum of two gaussians, representing the shapes of the direct and cascade decays, was fitted to the M^2_{rec} spectrum (Fig. 2). The contribution from D^{*-} cascade decays was constrained to lie within the errors of the measured number of $\overline{D}^0 l^-$ combinations multiplied by the factor 2.3±0.5, which accounts for the relevant branching ratios [9,16] and efficiencies. By this procedure ARGUS obtains a signal of 82±27 $B^0 \to D^-\ell^+\nu$ decays.

The dashed curve in Fig. 2 shows the fit result for the contribution (55 ± 18 events) from D^{*-} cascade decays. The contribution from other possible cascade decays has been tested by including the reflection from $\overline{D}_J(2420)^*$ in the fit. The fit result for this contribution is consistent with zero and does not change the number of $B^0 \to D^-\ell^+\nu$ decays.

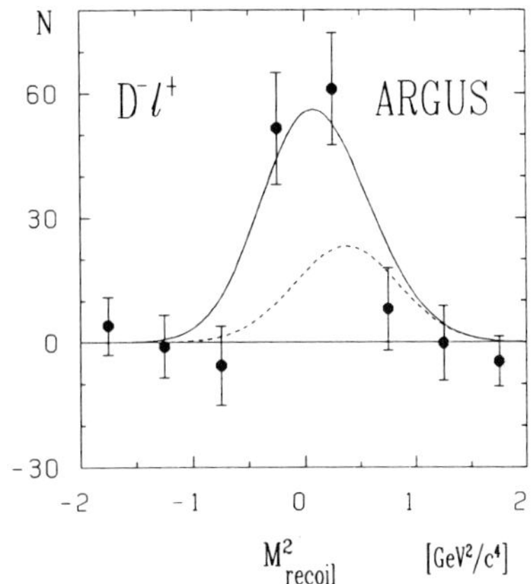

Figure 2: Distribution of M^2_{rec} against the D^-l^+ system. The dashed curve shows the contribution from the D^{*-} cascade decays.

Assuming $BR(\Upsilon(4S) \to B^0\overline{B^0}) = 50\%$ and electron-muon universality, ARGUS obtains

$$BR(B^0 \to D^-\ell^+\nu) = (1.7 \pm 0.6 \pm 0.4)\% .$$

Using this branching ratio, the WBS model [11], and $\tau_B = (1.15 \pm 0.14)ps$ [3] one obtains

$$|V_{cb}| = 0.042 \pm 0.008 .$$

Values for $|V_{cb}|$ obtained using other theoretical models [12,13,17,18,19] differ only by 10%. Thus the model dependence of the determination of $|V_{cb}|$ is considerably smaller than the experimental errors.

The ratio of branching ratios for semileptonic B decays into the vector and pseudoscalar D mesons is close to the expectations from naive spin counting:

$$R = \frac{BR(B^0 \to D^{*-}l^+\nu)}{BR(B^0 \to D^-l^+\nu)} = 3.3 \pm {3.7 \atop 1.1} ,$$

although the errors are still large. Model predictions for R vary from 1.2 to 9.6 [18,17].

2.3 *A Determination of* $\tau(B^+)/\tau(B^0)$

The lifetimes of B_u^+ and B_d^0 mesons have not yet been measured separately. Decay-in-flight measurements of the b hadron lifetime at PEP and PETRA only yield the average over an unknown mixture of b-flavoured hadrons. The difference in the lifetimes depends on the role of non-spectator processes in weak hadronic decays. In B decays these effects are assumed to be small. Hence, the lifetime ratio should be close to one [20], although some models predict higher values [21]. The B^+ and B^0 lifetimes enter into every evaluation of decay widths from measured branching ratios and thereby also into the determination of the CKM matrix elements. Currently, one has to assume the lifetimes to be equal. It is valuable to justify this experimentally. Previously, only weak bounds have been obtained from the single lepton and dilepton rates in $\Upsilon(4S)$ decays: $0.43 < \tau(B^+)/\tau(B^0) < 2.3$ at 90% C.L. [22].

Due to the absence of final state interactions, the partial widths of semileptonic B^+ and B^0 decays can be assumed to be equal. Therefore $\tau(B^+)/\tau(B^0)$ should be equal to the ratio of the B^+ and B^0 meson semileptonic branching fractions. This ratio can be extracted from a reconstruction of semileptonic B decays with a $\bar{D}^0$ or D^- meson detected in the final state, since B^+ decays mainly to $\overline{D}^0$ while B^0 decays mainly to D^-. Thus the charge of the D meson tags the charge of the parent B meson.

Decays into the lowest-lying charmed pseudoscalar and vector mesons dominate the total semileptonic decay rate. Therefore the production of excited charmed states in semileptonic decays can be neglected, with small corrections as discussed below. Since $\bar{D}^{*0}$ decays do not produce D^- mesons [9,23], all D^- mesons in semileptonic decays originate from B^0 decays, while semileptonic B^+ decays always yield $\bar{D}^0$ mesons. Thus only $\bar{D}^0$ mesons produced in the chain $B^0 \to D^{*-}\ell^+\nu$, $D^{*-} \to \bar{D}^0\pi^-$ have a wrong charge. The branching ratio for this decay has, however, been measured [6,10]. Taking this cross-talk into account, one has

$$\frac{f_+}{f_0}\frac{\tau(B^+)}{\tau(B^0)} = \frac{N(\bar{D}^0\ell^+) - r^*N(D^{*-}\ell^+)}{r^-N(D^-\ell^+) + r^*N(D^{*-}\ell^+)}, \tag{1}$$

where $N(\bar{D}^0\ell^+)$ $[N(D^-\ell^+)]$ denotes the observed number of $\bar{D}^0$ $[D^-]$ lepton pairs and $N(D^{*-}\ell^+)$ the number of D^{*-}- lepton combinations where the D^{*-} has been reconstructed in the decay channel $D^{*-} \to \bar{D}^0\pi^-$. The coefficients r^* and r^- account for the reconstruction efficiencies of $\bar{D}^0$ decays relative to D^{*-} and D^- decays, respectively.

After background subtraction, ARGUS obtains $325 \pm 28 \pm 9$, $183 \pm 37 \pm 12$ and $58 \pm 9 \pm 3$ events from the decays $B \to \bar{D}^0\ell^+X$, $B \to D^-\ell^+X$ and $B \to D^{*-}\ell^+X$ respectively. Inserting these numbers into (1), ARGUS obtains [24]

$$\frac{\tau(B^+)}{\tau(B^0)} = 1.00 \pm 0.23 \pm 0.14\ . \tag{2}$$

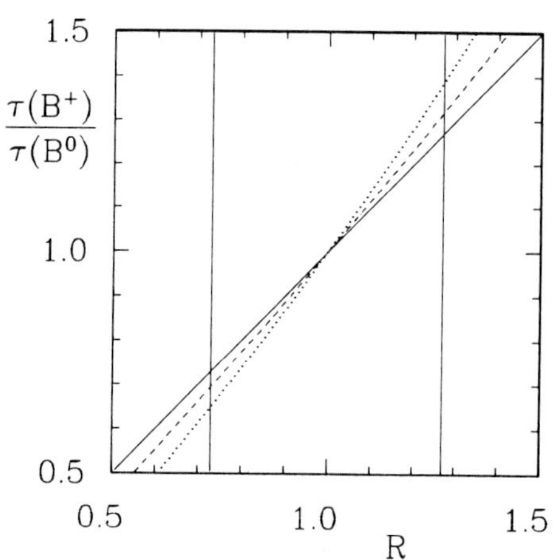

Figure 3: The lifetime ratio $\tau(B^+)/\tau(B^0)$ as a function of the measured ratio R for various contributions δ from decays $B \to \bar{D}_J^*\ell^+\nu$. Solid line: $\delta = 0$, dashed: $\delta = 0.1$, dotted: $\delta = 0.2$.

Semileptonic B decays to excited charmed meson states would partially violate the B meson tagging with the D meson charge. However, this effect can be taken into account assuming isospin conservation. Denoting by R the righthand side of equation (1), one finds

$$\frac{f_+}{f_0}\frac{\tau(B^+)}{\tau(B^0)} = \frac{R - \frac{2}{3}\delta(R+1)}{1 - \frac{2}{3}\delta(R+1)}, \tag{3}$$

where δ is a relative contribution of the excited charm states to the inclusive semileptonic B meson decay width. Figure 3 shows $\tau(B^+)/\tau(B^0)$ as a function of R for various values of δ. As can be infered from Fig.3, the possible contribution from excited states does not change the value obtained for $\tau(B^+)/\tau(B^0)$ and increases its error only slightly.

2.4 *The B Meson Semileptonic Branching Ratio*

The inclusive B meson semileptonic decay width depends on both V_{cb} and V_{ub}:

$$\Gamma(B \to l^+ \nu X) = \frac{G_F^2 m_b^5}{192\pi^3}(f_c|V_{cb}|^2 + f_u|V_{ub}|^2) ,$$

where the coefficients f_q are a product of a phase space factor and the lowest-order QCD correction. Since $|V_{ub}|$ is much smaller than $|V_{cb}|$ (see section 3) it can be neglected in this expression. Thus a value of $|V_{cb}|$ can be extracted from the measurements of the inclusive semileptonic branching ratio and the lifetime. Figure 4 shows the electron momentum spectrum on the $\Upsilon(4S)$ after continuum subtraction. The distribution is well described by contributions from $b \to cl^-\overline{\nu}$ transitions (dot-dashed curve) and secondary charm decays (dashed curve).

The electron and muon semileptonic branching ratios are obtained by fitting theoretical models to the experimental momentum distribution above 1.4 GeV/c where the contribution from the cascade decays is small. Using the ACM model [25] to extrapolate to the full spectrum ARGUS obtained [26]

$$BR(B \to \ell\nu X) = (10.3 \pm 0.7 \pm 0.2)\% ,$$

and

$$|V_{cb}| = 0.046 \pm 0.005 .$$

The CLEO and CUSB groups obtained similar branching ratios of $(10.1 \pm 0.3 \pm 0.7)\%$ and $(11.1 \pm 0.6)\%$ respectively [8]. The obtained value of $|V_{cb}|$ is in a good agreement with the results from the exclusive semileptonic decays although the sources of uncertainties in these approaches are very different. In the inclusive measurement, the main uncertainty is due to the very strong dependence of the semileptonic width on the not well known masses of b and c quarks. The agreement between the different methods of determining $|V_{cb}|$ provides some confidence that the model dependence of the obtained values is not too large.

However, it is worthwhile to mention that a somewhat larger value of $|V_{cb}| = 0.06 \pm 0.01$ was obtained using the parton model [27] for the analysis of the lepton spectrum measured by Crystal Ball [28]. The increase in $|V_{cb}|$ is only partially explained by the larger $BR(B \to \ell\nu X) = (12.0 \pm 0.5)\%$ obtained in this experiment.

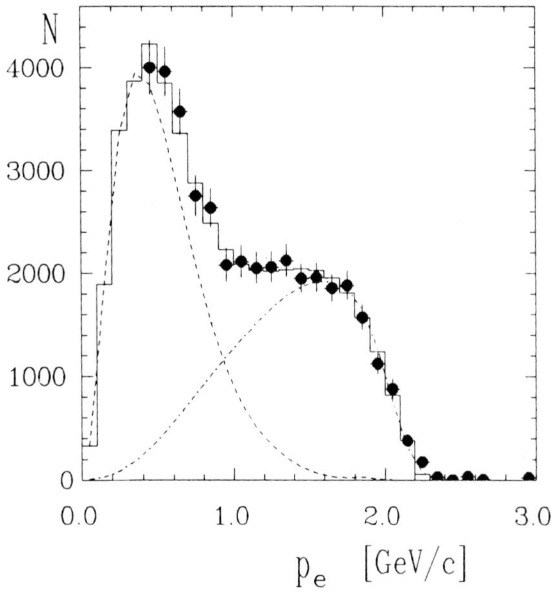

Figure 4: Electron spectrum at $\Upsilon(4S)$ after continuum subtraction.

The observed semileptonic branching ratio is smaller than the 12-15% predicted by the spectator model[29]. One possible explanation could be a substantial contribution from semileptonic decays into baryon antibaryon pairs such as $B \to \Lambda_c \overline{N} \ell \nu$

Leptons in these decays are soft because of the large baryon masses and would not influence a fit restricted to the high momentum range. No evidence for semileptonic B decays with antiprotons in the final state was found by ARGUS leading to the 90% CL upper limit $BR(B \to \overline{p}e^{\pm}X) < 0.36\%$. Protons from hyperon decays are included into this limit. The neutron yield is not expected to be larger than the proton yield. Therefore semileptonic B decays to baryons can not explain the difference between experiment and theory.

3 The $b \to u$ Transition

3.1 *A Study of the Lepton Spectrum Near the Endpoint*

The signature for a decay via a $b \to u$ transition would be seen as an excess of leptons with momenta above the kinematic limit of about 2.3 GeV/c for $b \to c$ decays. And there is some excess in the $\Upsilon(4S)$ data over the scaled continuum in the momentum interval from 2.3 to 2.6 GeV/c (see Fig.5).

The spectra in Fig.5 were obtained by selecting events with no particles above the kinematic limit for B decays (except identified leptons) and with not too small charged ($n_{ch} > 4$) and total ($n_{ch} + n_\gamma/2 > 5$) multiplicity.

The polar angle of electrons was restricted to the region $|cos\theta_e| < 0.85$ to ensure a good momentum resolution. Muons were used only in the barrel region where the fake rate is twice as small.

After bin-by-bin subtraction of the continuum, 73 ± 39 electrons and 131 ± 53 muons are left in the interval 2.3 to 2.6 GeV/c. The errors include the uncertainties in the continuum scaling factors for electrons and muons, which were taken to be the ratio of the number of leptons detected on the $\Upsilon(4S)$ and in the continuum, above the kinematic limit for B decays.

There are a number of backgrounds from $\Upsilon(4S)$ events themselves. Perhaps the most critical of these is the high momentum tail of the contribution from $b \to c$ decays. This depends on the experimental resolution function and on the model used for the semileptonic $b \to c$ decays.

The resolution function for muons has been checked by studying QED μ-pair events. The muon momentum distribution has practically no high energy tail and it is perfectly described by the detector Monte Carlo. The extrapolation to the lower momenta of the $b \to u$ analysis should be quite reliable, since this is still a regime where drift chamber resolution dominates the precision of the momentum determination. The same holds true for estimates of the electron response function which was studied using Bhabha events. Using the available models for $b \to c$ decays [11,12,13], and varying the D/D^* ratio within experimental limits, the uncertainty in the background from $b \to c$ decays in the signal region is estimated to be 25%.

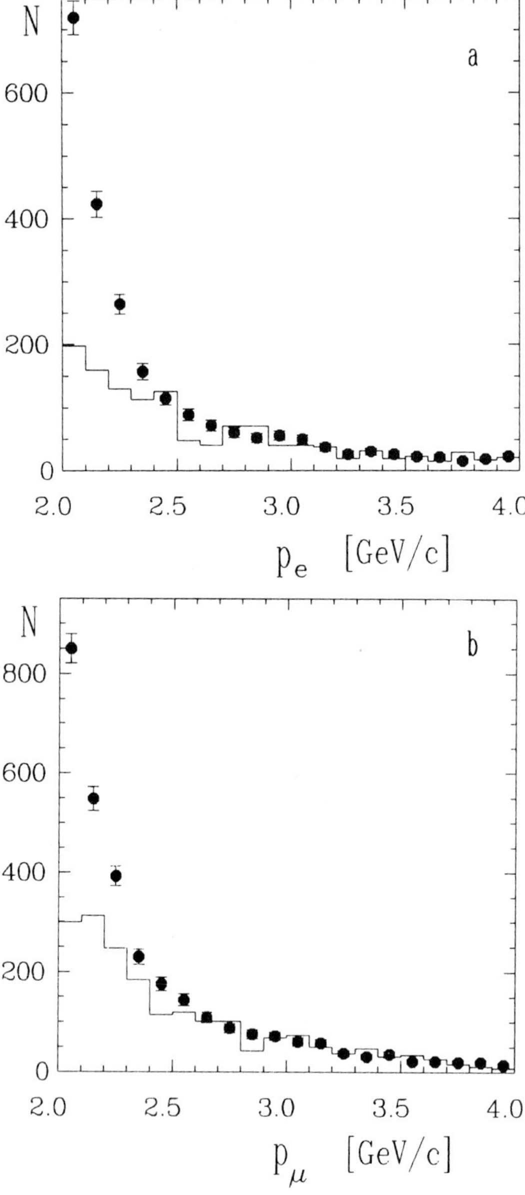

Figure 5: Momentum spectra for (a) electrons and (b) muons from $\Upsilon(4S)$ (points) and scaled continuum (histogram) data.

After subtracting continuum and backgrounds from $b \to c$ transitions, fake leptons and J/ψ decays, the number of signal events was found to be 39 ± 40 and 81 ± 54 for electrons and muons respectively. In the normalization region from 2.0 to 2.3 GeV/c, populated mainly by $b \to c$ transitions, the corresponding numbers were 864 and 852 events.

Clearly the sensitivity of the inclusive approach is limited by the subtraction of a large amount of continuum, using a scaling factor which by itself has large uncertainties. In proceeding further, the approach has been to use additional requirements which strongly suppress the continuum, but which have a reasonable efficiency for $b \to u$ decays. These requirements exploit the more spherical shape of $\Upsilon(4S)$ decays in comparison with continuum events and the existence of energetic leptons (including neutrinos) in B meson decays. The data were divided into two independent samples: one consisting of those events containing exactly one lepton and the other containing those events with exactly two. The requirement of an additional fast lepton ($1.2 < p_\ell < 2.3\ GeV/c$) already suppresses strongly the continuum contribution in the second sample. In the single lepton sample the continuum contribution was reduced by requiring a nonjet topology of the events. Finally, the presence of an energetic neutrino which manifests itself as a large missing momentum was required in both samples.

Events with a two-jet topology in the single lepton sample were removed by a requirement of a large transverse energy with respect to the lepton. Fig.6 shows the scalar sum of transverse momenta with respect to the lepton ($\sum p_T$) of all particles which have an angle with the lepton between 60° and 120°. There is an obvious difference between the $\Upsilon(4S)$ and continuum data. The requirement that $\sum p_T$ exceed 2 GeV/c reduces the continuum by a factor of 13, while its efficiency for $\Upsilon(4S)$ decays with $2 < p_\ell < 2.3 GeV/c$ is still 46%. The efficiency for the $b \to u$ transitions with $2.3 < p_\ell < 2.6 GeV/c$ was estimated using the WBS model [11] to be 41%.

Semileptonic decays produce a neutrino along with the charged lepton, resulting in missing momentum (p_{miss}) for the event. As noted previously, many of the continuum leptons, particularly in the muon sample, are misidentified hadrons, and so these events should have small missing momentum. A requirement of large p_{miss} also suppresses purely hadronic decays of the $\Upsilon(4S)$. This is of particular importance for reducing backgrounds from $B \to J/\psi X$, for example.

For semileptonic decays the direction of the neutrino ($\mathbf{p}_{miss}$) and the lepton momentum are strongly correlated, particularly for leptons in the endpoint region. The opening angle, $\cos\beta$, between the lepton direction and the direction of missing momentum should peak at $\cos\beta = -1$ for signal events. The actual strength of the correlation depends on the quality of the detector, since the resolution for $\cos\beta$ is a function of the hermiticity of the experiment.

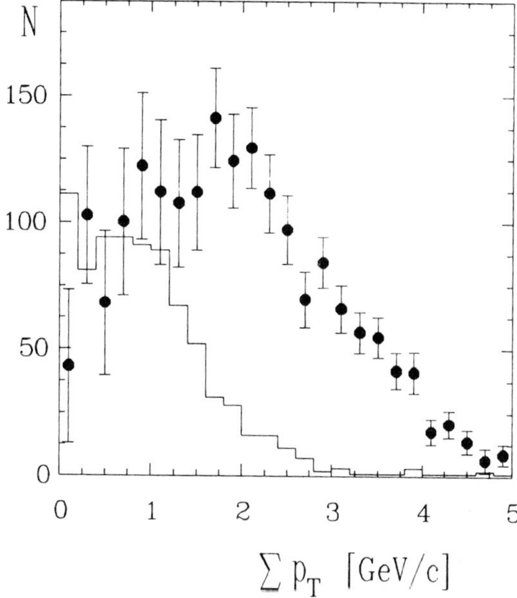

Figure 6: Distribution of $\sum p_T$ for continuum (histogram) and $\Upsilon(4S)$(points) events, for lepton momentum in the range 2.0 to 2.3 GeV/c.

Shown in Figure 7 and 8 are comparisons between Monte Carlo predictions and direct $\Upsilon(4S)$ decays for the distribution of p_{miss} and $\cos\beta$ in the region $2.0 < p_\ell < 2.3$ GeV/c populated mainly by $b \to c$ decays. Both the magnitude and angular correlation with $\mathbf{p}_{miss}$ are well described by the Monte Carlo. For comparison, the predicted p_{miss} and $\cos\beta$ distributions for $b \to u$ transitions are also shown. These were obtained by Monte Carlo calculation using the WBS model [11] for semileptonic B decays to π and ρ mesons.

In order to further demonstrate that the observed peaking of $\cos\beta$ near -1 is due to the neutrino and not an artifact, the opening angle between the known neutrino direction and $\mathbf{p}_{miss}$ for the exclusive sample of $B^0 \to D^{*-}\ell^+\nu$ was examined.

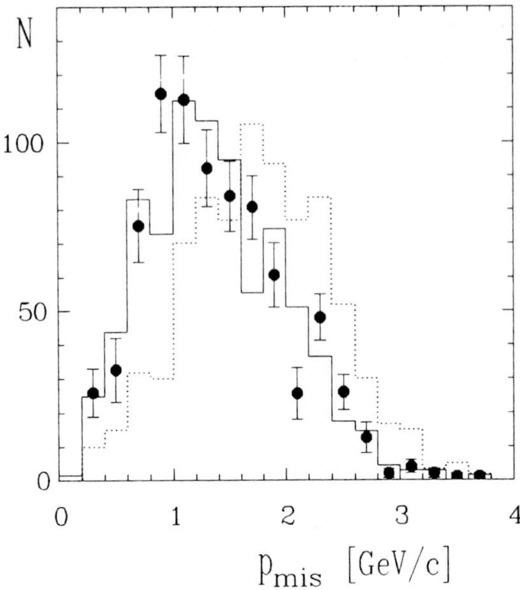

Figure 7: Distribution of p_{miss} for $\Upsilon(4S)$ data with leptons in the $b \to c$ range (circles), Monte Carlo prediction for $b \to c$ (open histogram) and $b \to u$ (dotted histogram) transitions.

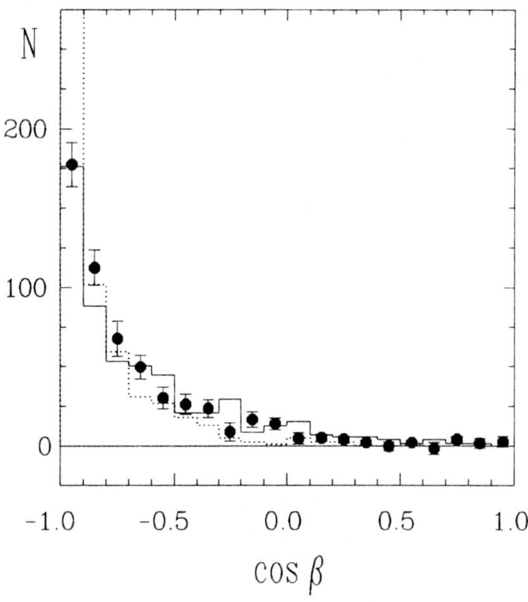

Figure 8: Distribution of $\cos\beta$. Notations as in Fig.7.

The result is shown in Figure 9, where $\mathbf{p}_{miss}$ can be seen to coincide with $\mathbf{p}_\nu$ with approxi-

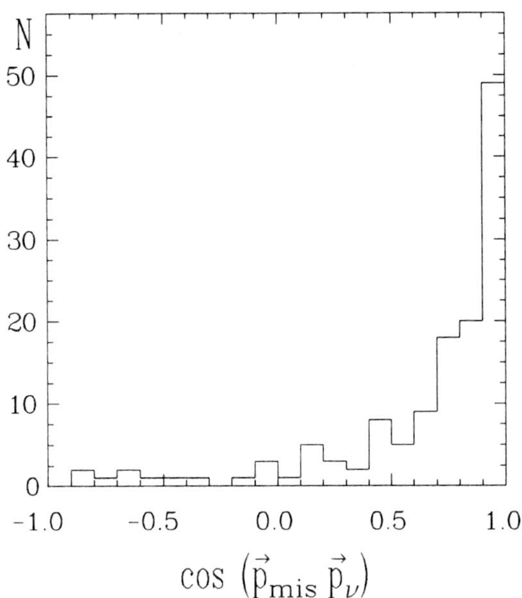

Figure 9: Opening angle distribution between $\mathbf{p}_{miss}$ and the known neutrino direction for $B^0 \to D^{*-}\ell^+\nu$ events.

mately the same precision as seen in Figure 8. For the single-lepton analysis, the requirement was made that $1.0 < p_{miss} < 3.5$ GeV/c and $\cos\beta < -0.5$. The efficiency of the cut has been estimated to be 81% for $b \to u$ decays, while at the same time the background is reduced by a factor of 3.

In total, the requirements discussed above on $\sum p_T$ and $\mathbf{p}_{miss}$ reduce the continuum contribution by the factor of more than 40 while keeping a reasonably high efficiency for a $b \to u$ signal of 0.33.

Finally, the contribution from J/Ψ decays was further suppressed by requiring the invariant mass of the lepton with any other oppositely charged track consistent with the lepton hypothesis be more than 100 MeV/c^2 away from the J/Ψ mass.

The momentum distributions for electrons and muons after application of these cuts are shown in Figure 10 for both $\Upsilon(4S)$ and continuum data. The continuum background can be seen to have been dramatically reduced. In the signal region for $b \to u$ decays, 40 leptons are observed with the estimated background of 18.4±6.5 events. Table 1 gives a detailed description of the background sources. Thus, there is an excess of 22 ± 9 leptons in the momentum region $2.3 < p_\ell < 2.6$ GeV/c. The normalization region for $b \to c$ decays was taken, as before, to be the interval $2.0 < p_\ell < 2.3$ GeV/c, where 206 electron and 201 muon candidates are observed after background subtraction.

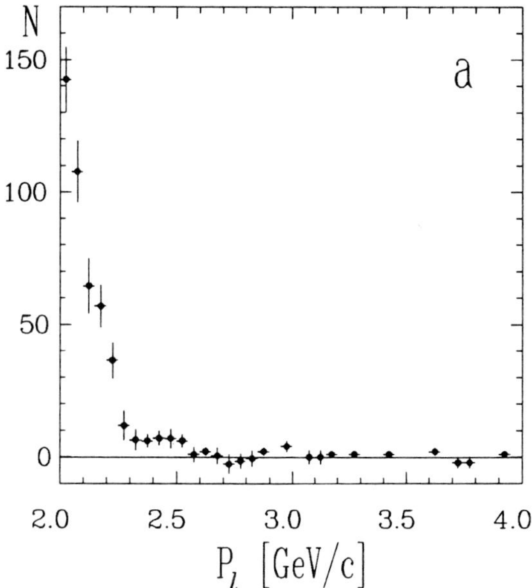

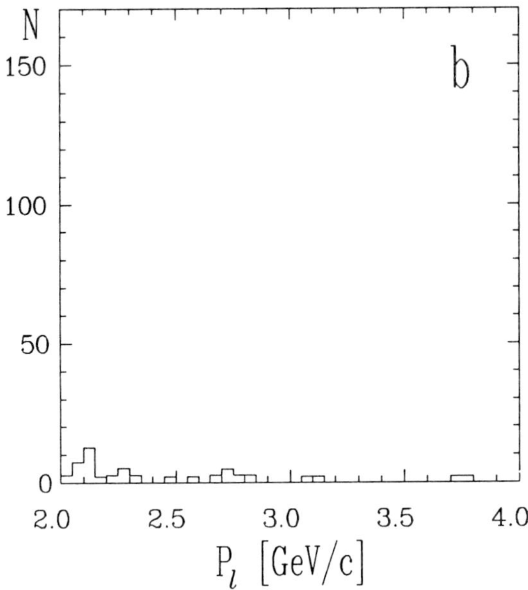

Figure 10: Lepton momentum spectra for (a) $\Upsilon(4S)$ data after continuum subtraction and (b) scaled continuum.

In the dilepton sample the small continuum contamination was further suppressed by requiring that the opening angle, $\theta_{\ell^+\ell^-}$, between the two leptons satisfy the requirement $\cos\theta_{\ell^+\ell^-} > -0.8$. Converted photons were removed by the restriction that $\cos\theta_{\ell^+\ell^-}$ be less than 0.95 for electron pairs. The J/ψ decays were removed as in the first sample. Finally, the missing momentum in the event was again required to lie in the interval $1.0 < p_{miss} < 3.5$ GeV/c.

The resulting momentum spectrum is shown in Figure 11, where the four events observed in the continuum are indicated by the histogram. There is no continuum event in the $b \to u$ signal region. However, we subtract the average level of continuum in the 2 to 3 GeV/c range. The $b \to u$ momentum interval is populated in the $\Upsilon(4S)$ data by 15 events, with an estimated background of 5.8 ± 1.6. There are 6 $e\mu$, 4 ee and 5 $\mu\mu$ candidates, a reasonable division of the signal. Table 1 gives a detailed breakdown of the known background sources. After background subtraction there are 59 electrons and 59 muons in the 2.0 to 2.3 GeV/c range.

Table 1. Observed single lepton and dilepton events in the momentum interval $2.3 < p_\ell < 2.6$ GeV/c and estimated backgrounds.

	Single Leptons		Dileptons	
	e	μ	e	μ
$\Upsilon(4S)$	24	16	7	8
Continuum	4.2	2.6	0.7	0.7
$b \to c$	4.1	4.6	1.2	1.3
J/ψ	0.5	0.3	0.2	0.1
Fakes	0.7	1.4	0.5	1.1
Sum	9.5 $\pm$4.4	8.9 $\pm$4.8	2.6 $\pm$0.8	3.2 $\pm$0.9
Signal	14.5 $\pm$6.6	7.1 $\pm$6.2	4.4 $\pm$2.8	4.8 $\pm$3.0

Figure 12 shows the combined lepton spectrum for both single and double lepton samples. Altogether there are 55 leptons between 2.3 and 2.6 GeV/c with the background estimated to be only 23 ± 7 events. The uncertainty of the background is dominated by the Poisson error on the continuum contribution. There are a total of 3 continuum events and the average continuum scaling factor is 2.35. Therefore the statistical significance of the signal corresponds to 3.3 standard deviations. In the momentum interval 2.4 to 2.6 GeV/c, where the $b \to c$ background vanishes completely, 32 leptons remain with an estimated background of 7.2 ± 7.

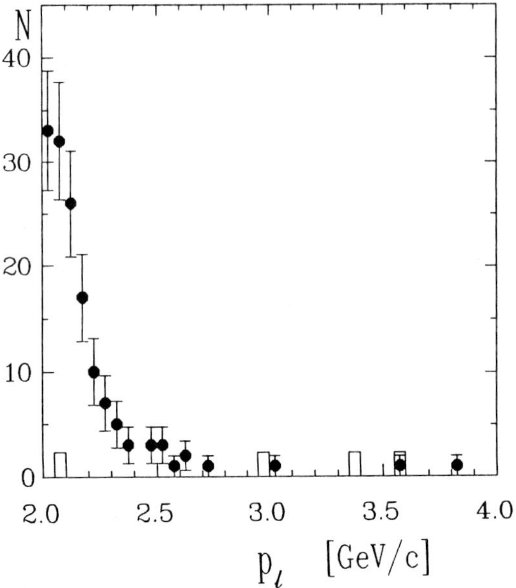

Figure 11: Momentum spectra in dilepton sample for $\Upsilon(4S)$ (points) and scaled continuum (histogram) data.

Assuming that $\Upsilon(4S)$ decays only to $B\overline{B}$ pairs, ARGUS interprets the observed excess of leptons beyond the endpoint for $b \to c$ decays as a signal for $b \to u$ transitions. Taking into account that the efficiency for $b \to u$ transitions is 1.3 times larger than the efficiency for $b \to c$ transitions with $2.0 < p_\ell < 2.3$ GeV/c, one obtains

$$\frac{BR_{sl}(2.3-2.6)}{BR_{sl}(2.0-2.3)} = 4.5 \pm 1.4\% \ .$$

The errors do not include the uncertainty due to the model dependence of the efficiency calculations for $b \to u$ transitions. Main uncertainty comes from the requirements on $\mathbf{p}_{miss}$. However this uncertainty can not lead to a substantial decrease of the obtained ratio of branching ratios since the used estimation of the corresponding efficiency is close to one (81%).

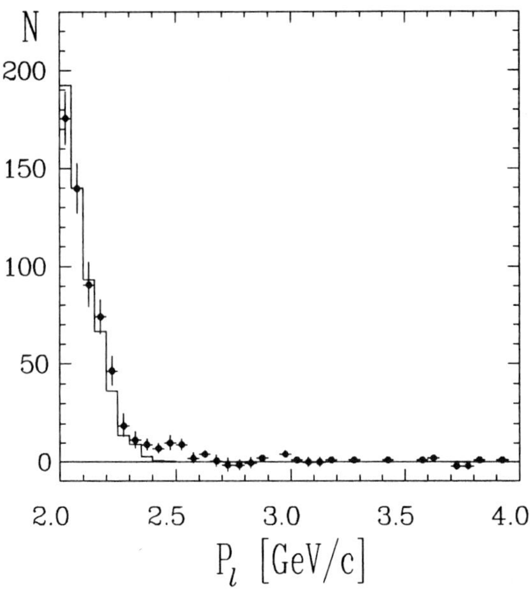

Figure 12: Combined lepton spectrum after continuum subtraction; histogram is a $b \to c$ contribution normalized in the region 2.0 to 2.3 GeV/c.

In order to calculate $|V_{ub}|/|V_{cb}|$ one must know the fractions of lepton spectra for $b \to u$ and $b \to c$ transitions in the selected momentum intervals. These fractions are different for electrons and muons because of electron bremsstrahlung. They are also model dependent, because of the very large theoretical uncertainties in the predictions for semileptonic B decays via $b \to u$ transitions. Using the ACM model [25] and neglecting the $b \to u$ contribution below 2.3 GeV/c one obtains a model dependent result

$$|V_{ub}|/|V_{cb}| = 0.10 \pm 0.02 \ .$$

It should be stressed that the errors for $|V_{ub}|/|V_{cb}|$ are not Gaussian and the statistical significance of the signal is 3.3 standard deviations.

3.2 *A Search for Exclusive Charmless Decays*

A search for exclusive decays $B^+ \to \rho^0 \ell^+ \nu$ and $B^0 \to \pi^- \ell^+ \nu$ was performed for leptons with $1.5 < p_\ell < 2.7 GeV/c$ [31] using the usual missing mass technique. No signal was observed. However, the background level is high and limits are model dependent and not very restrictive. The highest upper limit comes from the GISW model [13]:

$$|V_{ub}|/|V_{cb}| < 0.3 \ .$$

In 1987 ARGUS observed a signal of 25$\pm$8 events in the decays $B \rightarrow p\bar{p}\pi^+(\pi^-)$ [32]. The CLEO group, however, does not see a corresponding excess in the B mass region [33,4]. A preliminary analysis of new ARGUS data, corresponding to 70% of the previous sample, also do not show the expected signal. New data, which ARGUS is taking in 1989, will help to clarify the situation.

Theory predicts quite substantial branching ratios for B decays to multipion final states [34]. ARGUS performed a search for such decays with up to 6 pions. No signals were observed. The obtained upper limits are summarized in Table 2 together with the theoretical predictions [34] and CLEO results [4]. The backgrounds in channels with large expected branching ratios are also large and the upper limits are still much higher than the theoretical predictions for $|V_{ub}|/|V_{cb}|$=0.1.

Table 2. Limits on charmless B decays

Decay	ARGUS 90%CL	CLEO 90%CL	Theory $\lvert V_{bu}/V_{bc}\rvert$=0.1
$\pi^\pm\pi^0$	$5.0\cdot10^{-4}$	$2.6\cdot10^{-3}$	$0.6\cdot10^{-5}$
$\pi^+\pi^-$	$1.9\cdot10^{-4}$	$0.8\cdot10^{-4}$	$2.0\cdot10^{-5}$
$\pi^\pm\pi^+\pi^-$	$8.0\cdot10^{-4}$	$1.9\cdot10^{-4}$	$6\cdot10^{-5}$
$\rho^0\pi^\pm$	$1.9\cdot10^{-4}$	$1.7\cdot10^{-4}$	$2\cdot10^{-6}$
$\pi^+\pi^-\pi^0$	$1.8\cdot10^{-3}$	—	$2\cdot10^{-4}$
$\rho^0\pi^0$	$4.3\cdot10^{-4}$	—	$2\cdot10^{-6}$
$\pi^+\pi^+\pi^-\pi^-$	$1.0\cdot10^{-3}$	—	$1\cdot10^{-4}$
$\pi^\pm\pi^+\pi^-\pi^0$	$5.4\cdot10^{-3}$	—	$4\cdot10^{-4}$
$\pi^+\pi^-\pi^0\pi^0$	$5.7\cdot10^{-3}$	—	$5\cdot10^{-4}$
$\rho^+\rho^-$	$4.2\cdot10^{-3}$	—	$5\cdot10^{-5}$
$\pi^\pm2\pi^+2\pi^-$	$1.2\cdot10^{-3}$	—	$2\cdot10^{-4}$
$3\pi^+3\pi^-$	$3.3\cdot10^{-3}$	—	$2\cdot10^{-4}$

4 Update on $B^0\overline{B}^0$ Mixing

Large $B^0\overline{B}^0$ mixing is well established [5][35]. In this section an update on the $B^0\overline{B}^0$ mixing study [5] is presented with the 70% more data collected in 1988 (altogether 172pb^{-1} on the $\Upsilon(4S)$ and 55pb^{-1} in continuum).

The mass difference ΔM between the CP eigenstates in the B system is given by [36]

$$\Delta M = \frac{G_F^2}{6\pi^2} B_B f_B^2 m_b \left|V_{tb}^* V_{td}\right|^2 m_t^2 F(\frac{m_t^2}{M_W^2})\eta_{QCD} \ .$$

All parameters in this expression, except of V_{td} and m_t, are either known or can be calculated (although the accuracy in $B_B f_B^2$ is poor). Thus the knowledge of ΔM constrains the product of V_{td} and m_t. ΔM can be obtained from the measurements of the experimentally accessible mixing parameter r which is connected to ΔM by

$$r = \frac{Prob(B^0 \rightarrow \overline{B}^0)}{Prob(B^0 \rightarrow B^0)} = \frac{(\Delta M \cdot \tau_b)^2}{2 + (\Delta M \cdot \tau_b)^2} \ .$$

The most accurate method for measuring the mixing rate is by tagging B^0 and $\overline{B}^0$ mesons with the charge of the lepton from the semileptonic decays. B^0 mesons decay only to ℓ^+ while $\overline{B}^0$ produce only ℓ^-. Thus $B^0\overline{B}^0$ mixing should lead to like-sign lepton pairs. Using cuts similar to those in the previous study [5] ARGUS observes 70 like-sign and 403 unlike-sign lepton pairs. After background subtraction a mixing signal of 34.7 like-sign lepton pairs is observed as well as 381.4 unlike-sign leptons from $B^0, \overline{B}^0$ and B^+, B^- decays. Table 3 gives a detailed breakdown of the background sources.

Table 3. Dilepton rates

	$N(\ell^\pm\ell^\pm)$	$N(\ell^+\ell^-)$
$\Upsilon(4S)$ + Continuum	70	403
Continuum	2	7
$\Upsilon(4S)$ direct	64.2	382.4
Corrected for J/ψ cut	64.2	413.6
	$\pm$10.2	$\pm$23.3
Background		
Fakes	12.5	23.8
Conversion	0.6	0.6
Secondary decays	14.7	6.1
J/ψ decays	1.7	1.7
Signal	34.7	381.4
	$\pm$10.2	$\pm$23.3

For $\Upsilon(4S)$ decays the mixing parameter r is given by:

$$r = \frac{N_{\ell^\pm \ell^\pm}(1+\lambda)}{N_{\ell^+\ell^-} - N_{\ell^\pm\ell^\pm}\cdot\lambda}$$

where the parameter

$$\lambda = \frac{f^+}{f^0}\left(\frac{\tau_{B^+}}{\tau_{B^0}}\right)^2$$

accounts for unlike-sign dileptons which come from $\Upsilon(4S) \to B^+B^-$ decays. As discussed above (see chapters 2.1, 2.3), one expects $f_+/f_0 \approx 1$ and $\tau_{B^+}/\tau_{B^0} \approx 1$.Therefore $\lambda = 1$ is assumed for the determination of r (previously $\lambda = 1.2$ was assumed by ARGUS [5] and CLEO [35]). Using $\lambda = 1$ one obtains:

$$r = 0.20 \pm 0.06 \pm 0.05 \ ,$$

where the systematic error accounts for the uncertainties in the background estimates.

A second method is to reconstruct one B^0 meson using the decay $B^0 \to D^{*-}\ell^+\nu$ while the second B is tagged again with the lepton charge. Table 4 gives the observed number of events and background estimates. Using these numbers one finds:

$$r = \frac{N(D^{*+}\ell^-\ell^-)}{N(D^{*+}\ell^-\ell^+)} = 0.24 \pm 0.12 \ .$$

Since one B^0 meson is reconstructed the result does not depend on λ.

Table 4. Dilepton events with D^{*+} mesons

	$N(D^{*+}\ell^-\ell^-)$	$N(D^{*+}\ell^-\ell^+)$
At Υ(4S)	8	27
Background	1.9 ± 0.5	1.7 ± 0.5
Signal	6.1 ± 2.8	25.3 ± 5.9

A third independent method, assumes that D^{*+} mesons are only produced in the decay of $\overline{B}^0$ mesons and thus can serve as a tag. The other B^0 or $\overline{B}^0$ meson is again tagged with the charge of the lepton. A requirement that $M^2_{rec} < -2.5 GeV^2/c^4$ ensures that the D^{*+} and the $\ell^\pm$ come from different B mesons. The obtained M^2_{rec} spectra are shown in Fig.13 and the numbers are given in Table 5.

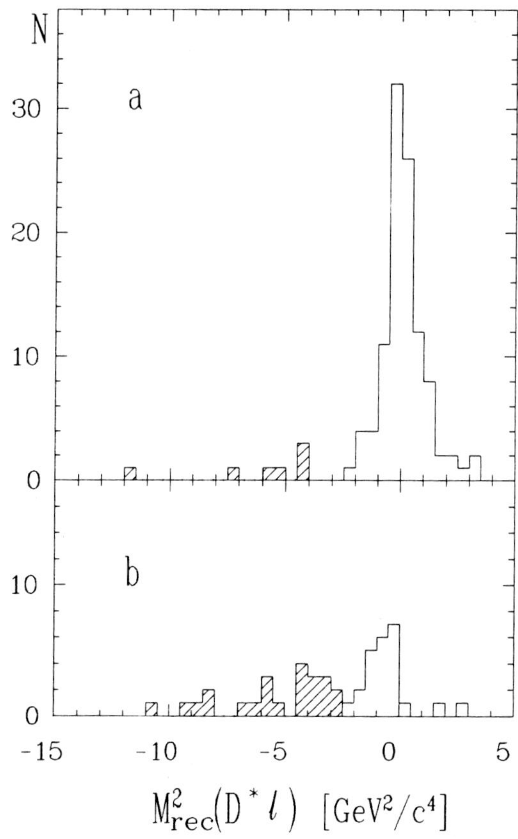

Figure 13: M^2_{rec} distribution for (a) $D^{*-}\ell^+$ and (b) $D^{*-}\ell^-$ combinations.

Table 5. $D^{*+}\ell^\pm$ events

	$N(D^{*+}\ell^-)$	$N(D^{*+}\ell^+)$
At Υ(4S)	7	23
Background	2.0 ± 0.5	2.3 ± 0.5
Signal	5.0 ± 3.1	20.7 ± 5.5

For the mixing parameter r, ARGUS obtains, under the above assumption,

$$r = \frac{N_{D^{*+}\ell^-}}{N_{D^{*+}\ell^+}} = 0.24 \pm 0.16$$

This result is, within large errors, close to the results obtained by the other methods, indicating that indeed most of the D^{*+} mesons are decay products of $\overline{B}^0$ mesons. Possible decays of charged B mesons into D^{*+} mesons would increase the value for r obtained with this method.

Combining all results on r ARGUS finds

$$r = 0.21 \pm 0.06$$

in perfect agreement with the original observation.

Using this result together with the estimate $B_B^{\frac{1}{2}} f_B = (140 \pm 40) MeV$ [37], $m_t > 78 GeV/c^2$ from direct searches [38] and $m_t < 190 GeV/c^2$ from the analysis of electroweak radiative corrections to M_W/M_Z [40], one obtains

$$0.007 \leq | V_{td} | < 0.04 \ .$$

The upper limit is less restrictive than the limit $|V_{td}| < 0.018$ infered by invoking unitarity of the CKM matrix [39], but it does not depend on the model dependent upper limit on $|V_{ub}|/|V_{cb}|$. Using $|V_{td}| < 0.018$ one can obtain a lower limit on the t quark mass of $m_t > 60 GeV/c^2$.

5 Searches for $b \to s$ transition

Penguin decays of B mesons with a radiated photon or gluon probe the electroweak interaction at the one-loop level and provide a possible window on physics beyond the directly accessible mass scale. Searches for $b \to s$ transitions were performed through the reconstruction of exclusive final states. No signal is seen, leading to the limits shown in Table 6 [41].

Table 6. Limits on "penguin-type" B decays

Decay mode	N(90%CL)	BR(90%CL)
$B^0 \to K^+\pi^-$	< 9.4	$< 1.8 \cdot 10^{-4}$
$B^+ \to K_S^0\pi^+$	< 3.2	$< 1.0 \cdot 10^{-4}$
$B^0 \to K^{*+}(892)\pi^-$	< 2.3	$< 6.2 \cdot 10^{-4}$
$B^+ \to K^{*0}(892)\pi^+$	< 3.6	$< 1.7 \cdot 10^{-4}$
$B^0 \to K_S^0\rho^0$	< 2.3	$< 1.6 \cdot 10^{-4}$
$B^+ \to K^+\rho^0$	< 6.4	$< 1.8 \cdot 10^{-4}$
$B^0 \to K^{*0}(892)\rho^0$	< 7.7	$< 4.6 \cdot 10^{-4}$
$B^+ \to K^{*+}(892)\rho^0$	< 3.4	$< 9.0 \cdot 10^{-4}$
$B^0 \to K_S^0\phi$	< 2.3	$< 3.6 \cdot 10^{-4}$
$B^+ \to K^+\phi$	< 2.9	$< 1.8 \cdot 10^{-4}$
$B^0 \to K^{*0}(892)\phi$	< 2.3	$< 3.2 \cdot 10^{-4}$
$B^+ \to K^{*+}(892)\phi$	< 2.3	$< 1.3 \cdot 10^{-3}$
$B^0 \to K^{*0}(892)\gamma$	< 7.9	$< 4.2 \cdot 10^{-4}$
$B^+ \to K^{*+}(892)\gamma$	< 2.3	$< 5.2 \cdot 10^{-4}$

These limits, as well as the limits obtained by CLEO [42], are close to the Standard Model predictions [43]. They start to constrain models involving more exotic phenomena.

Conclusions

The investigations of the exclusive and inclusive semileptonic B decays lead to the consistent results for the CKM matrix element V_{cb}. The $|V_{cb}|$ values of

$$|V_{cb}| = 0.046 \pm 0.009 \ ,$$
$$|V_{cb}| = 0.042 \pm 0.008 \ ,$$
$$|V_{cb}| = 0.046 \pm 0.005 \ ,$$

are obtained from the studies of the $B^0 \to D^{*-}\ell^+\nu$, $B^0 \to D^-\ell^+\nu$, and $B \to \ell\nu X$ decays respectively.

An excess of 32±10 leptons above known backgrounds is observed beyond the endpoint of the momentum spectrum for B decays via $b \to c$ transitions. If interpreted as an evidence for the $b \to u$ transitions this 3.3 σ signal leads to a model dependent value of 0.10±0.02 for the ratio of the CKM matrix elements $|V_{ub}|/|V_{cb}|$ (the value is given for the ACM model[25]). This corresponds to the 95% CL lower limit of

$$|V_{ub}|/|V_{cb}| > 0.06 \ .$$

The upper limit on $|V_{ub}|/|V_{cb}|$ of 0.16 [26] is also model dependent.

A study of $B^0\overline{B}^0$ mixing was performed using an additional 70% more data. The value obtained for the mixing parameter r

$$r = 0.21 \pm 0.06$$

agrees perfectly with the original observation [5]. This result, the limit $m_t < 190 GeV/c^2$ and information about other CKM matrix elements restrict V_{td} to the interval

$$0.007 < V_{td} < 0.018$$

in the Standard Model with 3 generations.

Searches for the loop-induced $b \to s$ transitions are not yet sensitive to the values of V_{ts} expected in the Standard Model.

These results can be summarized in a graphical form. The unitarity of the CKM matrix requires

$$V_{ub}^* V_{ud} + V_{cb}^* V_{cd} + V_{tb}^* V_{td} = 0 \ .$$

This expression can be approximated by

$$V_{ub}^* + V_{td} \approx s_{12} V_{cb}$$

because $V_{ud} \approx 1$, $V_{tb} \approx 1$ and $V_{cd} \approx -s_{12}$, where s_{12}=0.22 is the sine of the well known Cabibbo angle. The three elements of the last equation form a triangle in the complex plane shown in Fig.14.

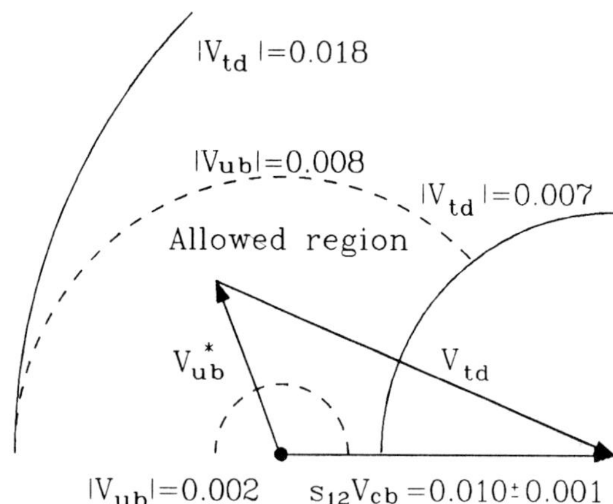

Figure 14: Unitarity triangle.

One side of the triangle is relatively well known: $s_{12}V_{cb}$=0.01 ± 0.001. The experimental constraints on V_{ub} and V_{td} are represented in the Fig. 14 by ellipses.

Recent studies of B mesons, especially the observation of the $B^0\overline{B}^0$ mixing and the evidence for the $b \to u$ transitions, put severe constraints on the parameters of the quark sector of the Standard Model. There is almost no flexibility left and new experiments have a good chance to reveal its incompleteness or provide important confirmations.

Aknowledgements

It is a pleasure to thank all members of the ARGUS collaboration who have contributed to the success of the experiment. I would like to express my gratitude to the organizers of the Symposium especially to R.E.Taylor, V.Lüth and F.Gilman.

References

[1] N.Cabibbo, Phys.Rev.Lett. **10** (1963) 531.

[2] M.Kobayashi and T.Maskawa, Prog.Theor.Phys. **49** (1973) 652.

[3] H. Schröder, Proc. of the 24th International Conference on High Energy Physics, Munich (1988).

[4] D.Bertoletto *et al.* (CLEO) Preprint CLNS 89/887.

[5] H.Albrecht et al (ARGUS), Phys.Lett. **192B** 245 (1987)

[6] H.Albrecht et al (ARGUS), Phys.Lett. **197B** 452 (1987)

[7] H.Albrecht et al (ARGUS), Phys.Lett. **219B** 121 (1989)

[8] D.L.Kreinick, These Proceedings

[9] J. Adler *et al.* (MARK III), Phys. Lett. **208B**, 152 (1988).

[10] D. Bortoletto *et al.* (CLEO), CLNS 89/922 (1989).

[11] M.Wirbel, B.Stech and M.Bauer, Z.Phys. **C29** (1985) 637.

[12] J.G.Körner and G.A.Schuler, Z.Phys. **C38** (1988) 511.

[13] N.Isgur, D.Scora, B.Grinstein and M.B.Wise, Phys.Rev. **D39** (1989) 799.

[14] T.Altomari and L.Wolfenstein, Phys.Rev. **D37**, 681 (1988)

[15] H. Albrecht *et al.* (ARGUS), DESY 89-082 (1989).

[16] J. Adler *et al.* (MARK III), Phys. Rev. Lett. **60**, 89 (1988).

[17] F. Schöberl and H. Pietschmann, Europhys. Lett. **2**, 583 (1986).

[18] A.A. Ovchinnikov and V.A. Slobodenyuk, IHEP 89-10 (1989).

[19] J.M. Cline, W.F. Palmer and G. Kramer, DESY 89-029 (1989).

[20] M.A. Shifman, Proceedings of the 1987 International Symposium on Lepton and Photon Interactions at High Energies, Hamburg, p. 289 (1987).

[21] A. Soni, Phys. Rev. Lett. **53**, 1407 (1984).

[22] A. Bean *et al.* (CLEO), Phys. Rev. Lett. **58**, 183 (1987);
E.H. Thorndike and R.A. Poling, Phys. Rep. **157**, 183 (1988).

[23] S. Abachi *et al.* (HRS), Phys. Lett. **212B**, 533 (1988).

[24] H. Albrecht *et al.* (ARGUS), DESY 89-117 (1989).

[25] G.Altarelli *et al.*, Nucl.Phys. **B208** (1982) 365.

[26] J.C.Gabriel, Dr.rer.nat.Thesis, Heidelberg University,(1988);
J.Spengler, Proc. of the 24th International Conference on High Energy Physics, Munich (1988);
M.V.Danilov, Preprint ITEP 88-180, Moscow (1988).

[27] A.Bareiss and E.A.Paschos, Preprint DO-TH 89/1,(1989).

[28] Crystal Ball Collab., K.Wachs *et al.*, Z.Phys. **C42** (1989) 33.

[29] R.Rückl, Habilitationsschrift,Universität München, Munich (1984).

[30] P.Haas *et al.*,(CLEO) Phys.Rev.Lett. **55** (1985) 395;
H.Albrecht *et al.*,(ARGUS) Phys.Lett. **B162** (1985) 395;
H.Albrecht *et al.*,(ARGUS) Phys.Lett. **B199** (1987) 451.

[31] T.Ruf,Dr.rer.nat.Thesis,Karlsruhe Univ. 1989.

[32] H.Albrecht *et al.*,(ARGUS) Phys.Lett. **B209** (1988) 119.

[33] D.L.Kreinick, Proc. of the 24th International Conference on High Energy Physics, Munich (1988).

[34] A.V.Dobrovolskaya et al. Preprint ITEP 108-89, Moscow (1989).

[35] M.Artuso et al (CLEO), Phys.Rev.Lett. **63** 2233 (1989)

[36] M.K.Gaillard and B.W.Lee, Phys.Rev. **D10** (1974) 897;
J.S.Hagelin, Phys.Rev. **D20** 2893 (1979)

[37] G.Altarelli, Proc. of the Int. Europhysics Conference on High Energy Physics, Uppsala (1987) p.1002.

[38] P.Sinervo (CDF) These Proceedings.

[39] K.Kleinknecht and B.Renk,Z.Phys. **C34** (1987) 209.

[40] G.Costa *et al.* Nucl.Phys. **B297** (1988) 244.

[41] H. Albrecht *et al* (ARGUS), DESY 89-086 (1989);
H. Albrecht *et al* (ARGUS), DESY 89-096 (1989).

[42] P.Avery *et al.* (CLEO), Preprints CLEO 89-2, CLEO 89-4 (1989).

[43] M.B.Gavela *et al.*, Phys.Lett. **154B** 425 (1985);
L.-L.Chau and H.Y.Cheng, Phys.Rev.Lett. **59** 958 (1987);
N.G.Deshpande *et al.*, Phys.Rev.Lett. **59** 183 (1987).

DISCUSSION

G. Farrar, Rutgers University: Can you tell us the prospects for ARGUS getting branching ratios for the exclusive charmless decays in the regime predicted by theory?

M. Danilov: We are considering the possibility of combining many exclusive channels; that may be feasible. Measuring individual channels is very hard.

E. Thorndike, University of Rochester: What is the factor by which the measured continuum background must be scaled, to allow for the difference in integrated luminosity?

M. Danilov: That factor is 2.3. So observing no events in the continuum running corresponds to an upper limit of $2.3 \times 2.3 = 5.3$ events in the background.

D. Kreinick, Cornell University: How do you know that your track finder does not get confused in complex hadronic events?

M. Danilov: This was studied carefully with the Monte Carlo. In addition, all events with leptons between 2.3 and 2.6 GeV were scanned visually.

V. Khoze, Leningrad Institute for Nuclear Physics: I would like to emphasize the importance of $|V_{bu}/V_{bc}|$ measurements for the prospects of CP-violation measurements in B mesons. If the $|V_{bu}/V_{bc}|$ results are confirmed, it will permit us to make rather stable theoretical predictions for CP-violating parameters.

A MEASUREMENT OF ε'/ε BY E731

B.Winstein
Enrico Fermi Institute and the Department of Physics
The University of Chicago, Chicago, Illinois 60637

The technique used by Fermilab E731 for determining the CP violation parameter ratio ε'/ε is described. Systematic uncertainties are treated and a result based upon an analysis of a part of the data taken is presented.

INTRODUCTION

This is a report from the E731 collaboration whose current members are: A. Barker, R. Briere, L. Gibbons, G. Makoff, V. Papadimitriou, J. R. Patterson, S. Somalwar, Y. Wah, B. Winstein, R. Winston, and H. Yamamoto, The University of Chicago; E. Swallow, Elmhurst College; G. Bock, R. Coleman, J. Enagonio, B. Hsiung, R. Stefanski, K. Stanfield, and T. Yamanaka, Fermilab; M. Karlsson, R. Tschirhart, and G. Gollin, Princeton University; P. Debu, B. Peyaud, R. Turlay, and B. Vallage, Saclay.

In order to observe direct CP violation in the decay of the neutral kaon to two pions, it is necessary to precisely determine the following four decay rates:

$$K_S \rightarrow \pi^0\pi^0, \quad (1)$$

$$K_L \rightarrow \pi^0\pi^0, \quad (2)$$

$$K_S \rightarrow \pi^+\pi^-, \quad (3)$$

$$K_L \rightarrow \pi^+\pi^-. \quad (4)$$

Then, by forming the double ratio:

$$R \equiv (1)/(2)/[(3)/(4)] \approx 1 + 6\,\mathrm{Re}(\varepsilon'/\varepsilon) \quad (5)$$

ε'/ε can be extracted. Given enough statistics, the major challenge is in designing the experiment and its subsequent analysis to minimize the resulting systematic uncertainty; this in practice involves the simultaneous collection of more than one of the above modes. At present, the CERN NA31 collaboration[1] has reported a 2.0% departure of R from unity, providing the first evidence for direct CP violation; this was accomplished by separately measuring the K_S and the K_L modes. This paper will report upon a set of data in which all four modes were simultaneously collected.

THE DETECTOR AND EXPERIMENTAL TECHNIQUE

The E731 detector[2] is shown schematically in Figure 1. Two nearly parallel K_L beams enter the detector region, one passing through a thick B_4C regenerator to provide K_S decays. In this manner, K_S and K_L decays are viewed simultaneously by the same detector. The regenerator alternates between the beams on a pulse-by-pulse basis (roughly every minute) so that any small difference in beam intensity or detector acceptance is essentially eliminated. There are different triggers for charged and neutral decays; however, the trigger in no way distinguishes between K_S and K_L. Thus the effect of the inevitable changes in chamber gas gain or drifts in calorimeter response over the course of a run will be reduced to high order. The same is the case for "noisy electronics" and similar effects. Moreover, changes in accelerator performance which affect the instantaneous rates in detector elements have little consequence with this technique.

With these advantages comes one principal disadvantage: the acceptance of the detector must be precisely known as a function of decay vertex because the lifetimes of the K_S and K_L are quite different.

The rates in the vacuum (R_V) and regenerated (R_R) beams are given (for example for charged decays) by:

$$R_V \propto |\eta_{+-}|^2 \tag{6}$$

and,

$$R_R \propto |\rho e^{-t/2\tau_s + i\Delta m t} + \eta_{+-}|^2 \tag{7}$$

where ρ is the regeneration amplitude for the material. Since $|\rho| \approx 10|\eta|$, the ratio of the integrated numbers of regenerated to vacuum decays (defined as R_{+-}) is given approximately by:

$$R_{+-} \propto |\rho/\eta_{+-}|^2. \tag{8}$$

Similarly,

$$R_{00} \propto |\rho/\eta_{00}|^2. \tag{9}$$

and, thus

$$R \approx R_{00}/R_{+-}. \tag{10}$$

The time dependence of the decay rate downstream of the regenerator, (7) above, can also be used to fit for τ_S, the K_S lifetime, Δm, the K_L-K_S mass difference, and Φ_η , the phase of the CP violating decay amplitudes (relative to the phase of ρ).

The charged decays are reconstructed by means of magnetic analysis using the four precision drift chambers shown in Figure 1. A large lead-glass array serves to detect the neutral decays and permit off-line rejection of the $\pi e\nu$ decay mode; $\pi\mu\nu$ decays are eliminated with the muon filter. A background in the neutral mode arises from $K_L \rightarrow 3\pi^0$ decays where photons miss the lead-glass; these are largely eliminated by means of many planes of "photon vetoes" situated outside the solid angle of the lead-glass array.

DATA COLLECTION AND TRIGGERS

The experiment took data from August of 1987 to February of 1988. About 75% of the data was taken with either the charged or neutral triggers in operation. For neutral mode running, a thin lead sheet was placed at the "trigger plane" in order to convert (usually) one photon. For the other 25%, all four modes were collected simultaneously (with no lead sheet) and it is from the bulk of this latter data set that we are now presenting a result.

The charged trigger required a two-track topology in the scintillator hodoscopes with no muon signal. In addition to the two pion decay, $\pi e\nu$ and $\pi^+\pi^-\pi^0$ events useful for calibration, alignment, and detector studies were collected.

The neutral trigger required[3] either four or six clusters and greater than about 30 GeV energy deposit in the lead glass; $3\pi^0$ decays were thus also collected.

The data set that we are reporting on consisted of approximately 20,000 spills (over an 18 day period) at the Fermilab Tevatron; each spill lasted for about 22 sec and the repetition rate was about one per minute. During each spill, a light flasher which illuminated each phototube of the lead-glass array was pulsed every second, determining the tube gains to about 1% per flash. Pedestal events were taken as frequently and an accidental trigger was taken about 70 times per spill to provide an unbiased sample of the underlying activity in the detector. Care was taken so that both triggers were "loose" and unbiased with respect to K_L vs K_S decays.

For this analysis, an event was retained only if its reconstructed z position lay between 120m and 137m and its reconstructed momentum lay between 40 GeV and 150 GeV; this selection was made to reduce systematic uncertainties, arising from acceptance and lead-glass energy scale uncertainties.

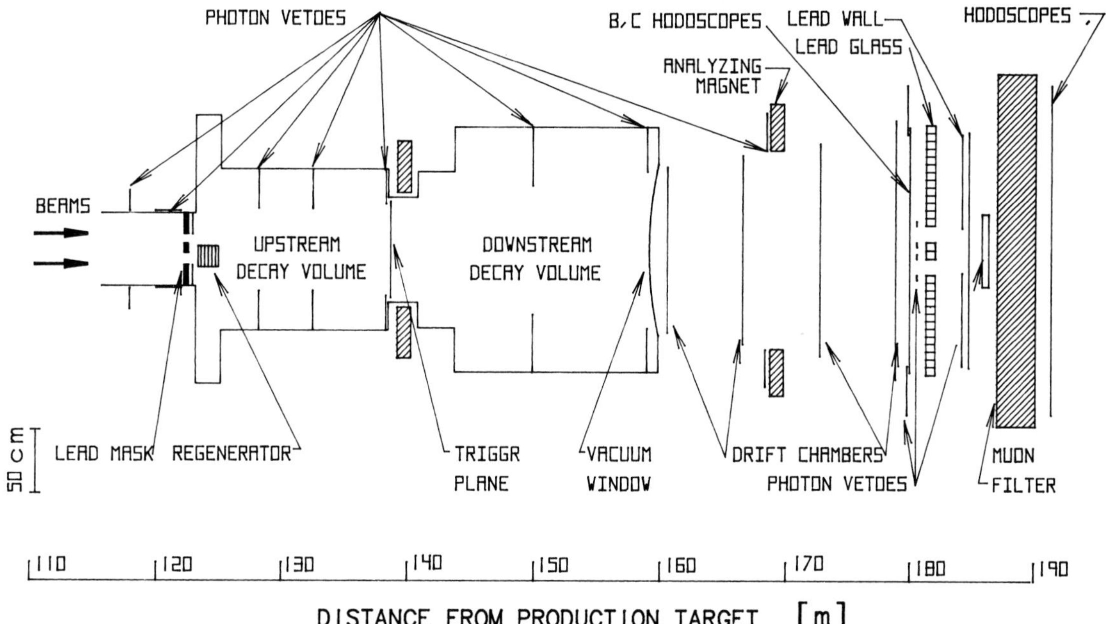

Fig. 1. Schematic of the E731 Detector.

MONTE CARLO SIMULATION

The simulation of the detector response is important. The philosophy is to understand as much as is possible about the detector and the beam to insure confidence in the eventual result. The difference in detector acceptance for K_L vs K_S decays is needed and for this a monte carlo is used and is extensively checked with the very high statistics modes. The simulation:

a) incorporates an extensive photon cluster library using data and EGS-simulated photon showers,

b) treats the initial pair production process of a photon in the lead-glass,

c) includes (different) light attenuation lengths in the lead-glass blocks,

d) treats shower leakage (transverse and longitudinal),

e) includes dead-time in the drift chamber electronics,

f) includes the few dead or inefficient drift chamber wires,

g) can overlay accidental activity when necessary, and,

h) includes the small variations in energy and flux across the beam.

Elements that are not included from "first principles," those that need to be tuned to the data, are the beam shape (which is sensitive to collimator misalignments) and the "dilution factor," the ratio of K^0 and $\overline{K}^0$ production at the target. In addition, the momentum spectrum[4] needs to be adjusted. Tuning these elements gives better agreement in comparisons of data and monte-carlo distributions; however, the resulting value of ε'/ε does not vary significantly.

RECONSTRUCTION OF $K_{L,S}\to\pi^+\pi^-$ CANDIDATES

In this section, we will treat the reconstruction of the charged decays. Each of the four drift chambers had two x and two y planes and the plane resolution was about 100μ. The transverse momentum of the analysis magnet was approximately 200 MeV/c and the resulting track momentum resolution was about 1%.

Pattern recognition and fiducial and kinematic cuts were all done without regard to whether an event was a candidate for a K_L or a K_S decay. The reconstruction efficiency, after all cuts, was about 90% for both modes. An event was classified as either K_L or K_S depending upon where its momentum vector extrapolated to the plane of the regenerator. Figure 2 shows the invariant mass distribution for candidate K_S decays; on the low-mass side, the data is broader as a result of radiative $\pi\pi\gamma$ decays. Figure 3 shows the reconstructed transverse momentum-squared (P_T^2) distribution for the K_S candidates (now with a mass selection between 484 and 512 MeV/c^2) showing a very sharp coherent peak above a background, consisting in diffractive and inelastic regeneration, of 0.13%. Figure 4 shows agreement between the data and monte carlo z-vertex distributions for these decays (with $P_T^2 < 250$ (MeV/c)^2). The mass and z-vertex distributions for K_L decays are shown in Figures 5 and 6 respectively; the $\pi\pi\gamma$ radiative tail is again present in addition to a residual background (0.31%) from semi-leptonic decays. The z-vertex distributions agree very well, and this coupled with the fact that the mean decay positions for K_L and K_S decays differ by only about 1.5m allows us to rule out a significant error in the determined ratio of K_L to K_S decays (see Figure 6b).

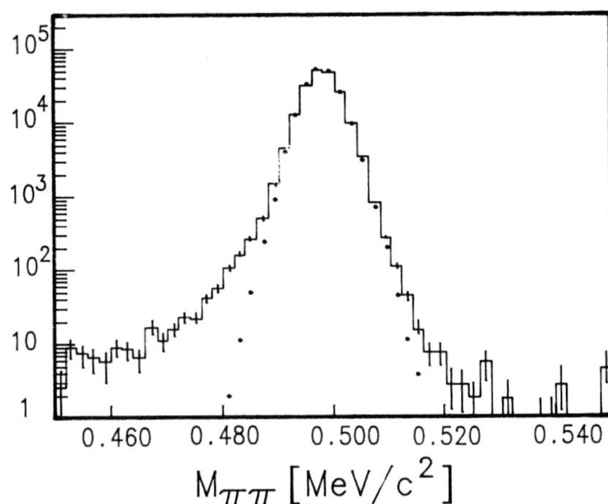

Fig. 2. Invariant mass distribution [events/2MeV/c^2] for $K_S \to \pi^+\pi^-$. The histogram is the data and the dots are the monte-carlo.

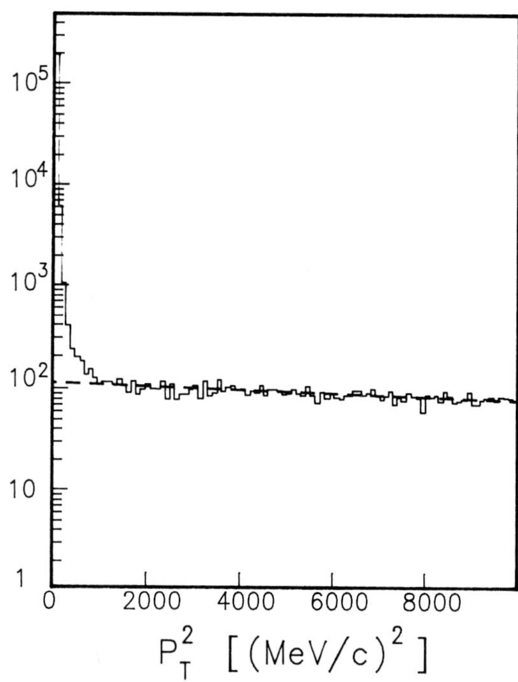

Fig. 3. P_T^2 distribution [events/(100MeV/c)^2] for $K_S \to \pi^+\pi^-$. The dashed line extrapolates the background under the coherent peak.

RECONSTRUCTION OF $K_{L,S} \to \pi^0\pi^0$ CANDIDATES

We turn now to the neutral mode. The lead-glass calibration[5] was important; it was accomplished by means of frequent special runs using electron positron pairs; this gave a very clean "beam" with negligible background and the sample used for this data set consisted of over 1 million electrons, after all cuts. The non-linearity of the blocks was determined from this sample; the energy E in a block was approximately related to the ADC pulse-height P for that block by the following expression:

$$E = g(P)^{\alpha} \qquad (11)$$

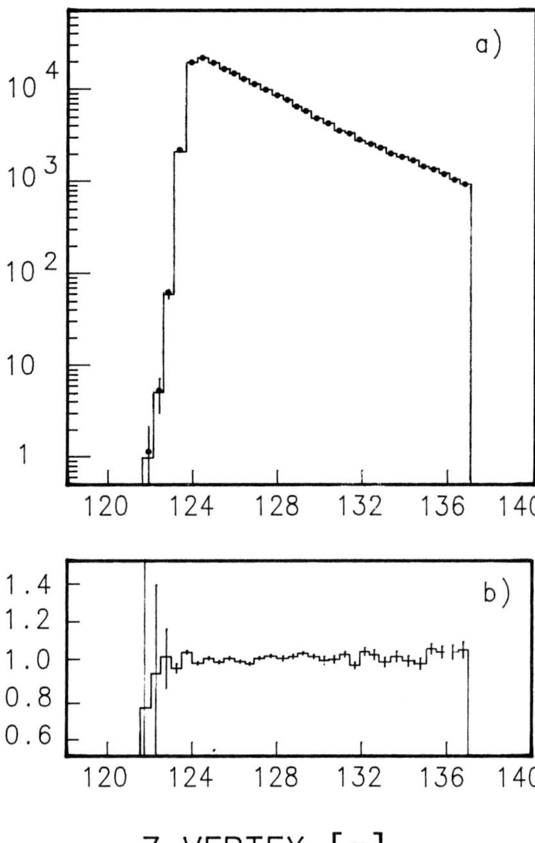

Fig. 4. a) Distribution in z-vertex [events/50cm] for $K_S \to \pi^+\pi^-$ events. The line (dot) is for data (monte-carlo); b)Ratio of data to monte-carlo.

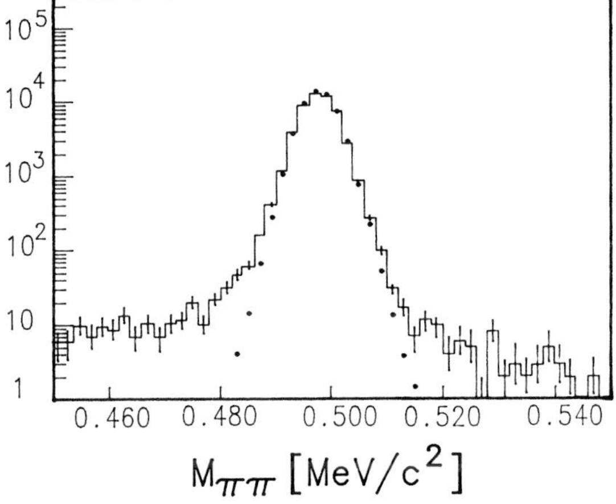

Fig. 5. Invariant mass distribution [events/2MeV/c^2] for $K_L \to \pi^+\pi^-$.The histogram is the data and the dots are the monte-carlo.

where g is the tube gain and α was typically 0.977; the departure of α from unity results from light absorption in the lead-glass. This calibration procedure was extensively checked with several million electrons from $\pi e\nu$ decays and several hundred thousand π^0's from $\pi^+\pi^-\pi^0$ decays.

The kaon vertex is determined by finding the best pairing for the $2\pi^0$ decay and using the constraint of the π^0 mass; then the $2\pi^0$ invariant mass is calculated.

Figure 7 shows the reconstructed center-of-energy distribution at the lead-glass of all 4-cluster events having invariant mass within 18 MeV/c^2 of the kaon mass. For this figure, the event coordinates have been "flipped" according to the position of the regenerator. The two modes (coherently regenerated K_S to $2\pi^0$ and vacuum K_L to $2\pi^0$) are distinct, and one can see the presence of incoherent regeneration in the small event rate outside of the two beam regions. Figure 8 displays the invariant mass distribution for the K_L candidates; overlaid is the distribution, absolutely normalized, of $3\pi^0$ background (0.37%) as predicted by monte-carlo. To subtract the background from incoherent regeneration, a plot of the event density in concentric rings away from each beam is made. This is shown for the vacuum beam in Figure 9; it can be seen that there is a flat distribution of "cross-over" events. The residual contribution under the peak can be estimated from the data alone, or it can be predicted absolutely by means of the P_T^2 distribution measured in the charged mode (Figure 3). This is our largest background (4.66%) but it is well understood. We remark that such an absolute prediction can be made since all four modes were simultaneously collected; otherwise, the rejection efficiency for inelastically regenerated events, which contain other particles in the final state, might depend upon the precise running conditions. However, since the pre-

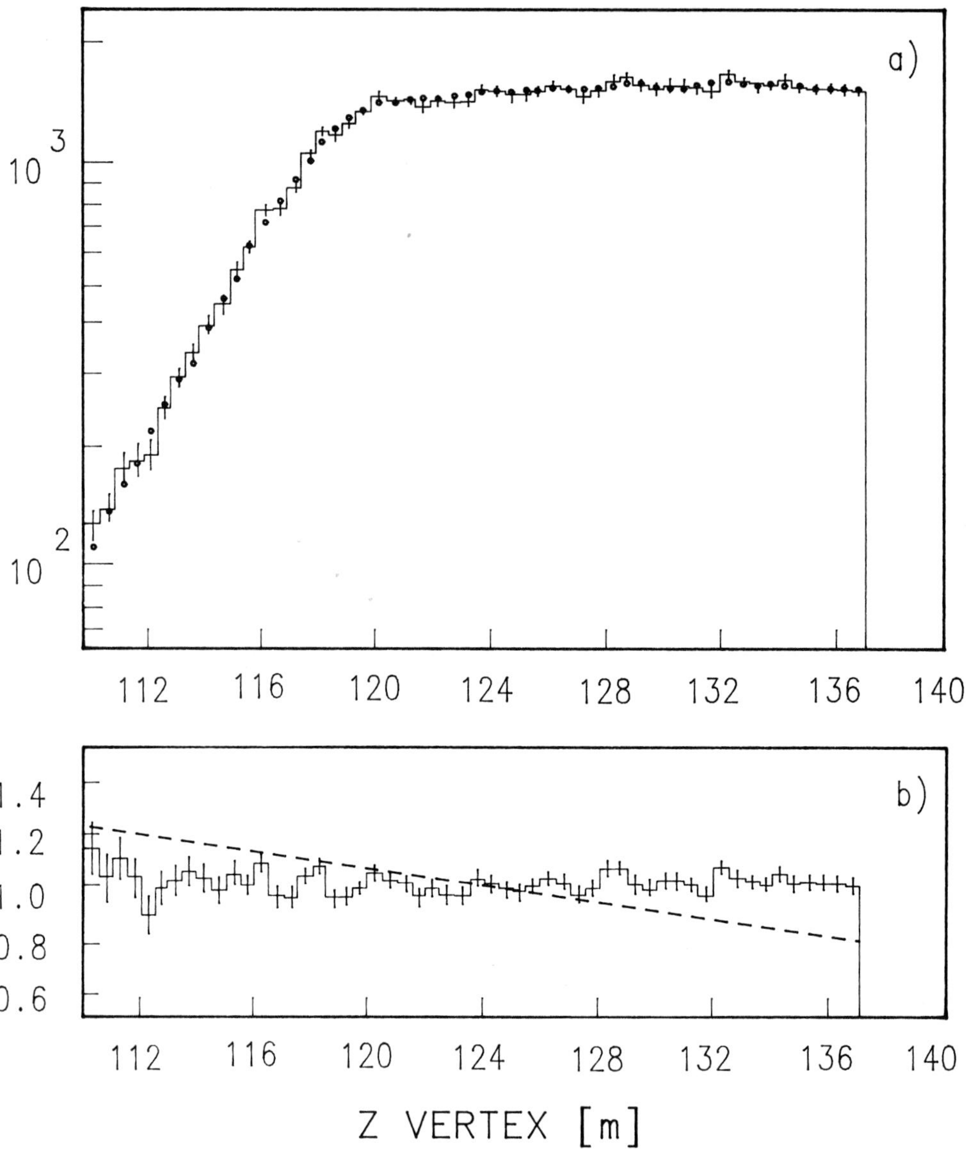

Fig. 6. a) Distribution in decay vertex [events/50cm] for $K_L \rightarrow \pi^+\pi^-$ events. The line (dot) is for data (monte-carlo); b) Ratio of data to monte-carlo. This ratio would follow the dashed line were there an acceptance error large enough to cause a 2% shift in the K_S to K_L ratio.

dicted distribution in ring number can easily be normalized to the data, we believe that the systematic uncertainty coming from this subtraction will not be appreciably larger for the remainder of the data set where charged and neutral data were separately taken.

Figure 10 shows the z-vertex distribution for $K_L \rightarrow 2\pi^0$ events: the agreement with monte-carlo is excellent.

EXTRACTION OF ε'/ε

Before fitting for the result, the backgrounds are subtracted and acceptance corrections are made.

The background summary and the estimated uncertainties are given in Table 1 below.

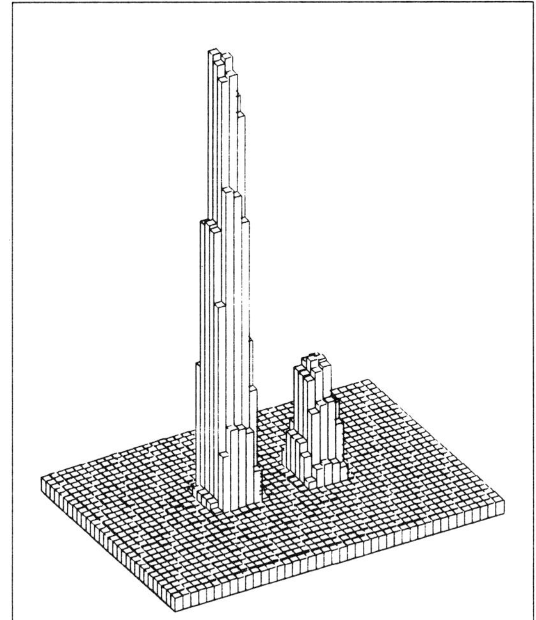

Fig. 7. The center-of-energy distribution at the lead-glass for $K_{L,S}\to 2\pi^0$ candidates. The vacuum beam is the smaller of the two.

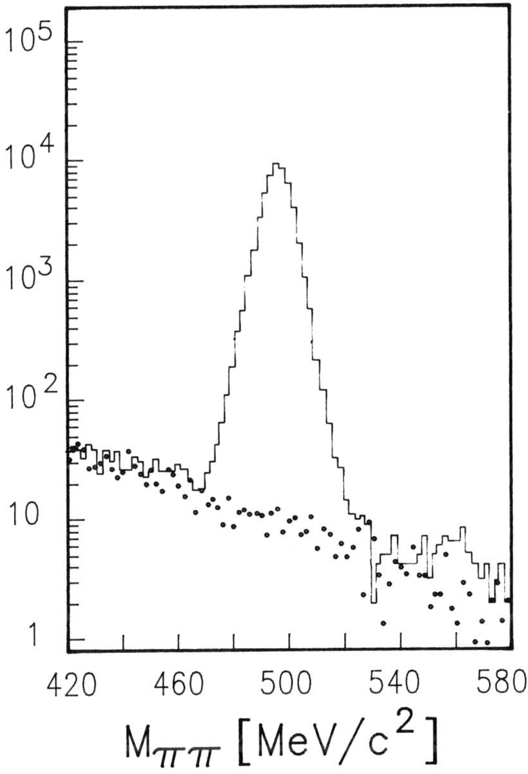

Fig. 8. Distribution in invariant mass [events/2MeV/c^2] for $K_L\to 2\pi^0$ events. The histogram is the data; the dots are the monte-carlo for the residual $K_L\to 3\pi^0$ background, absolutely normalized.

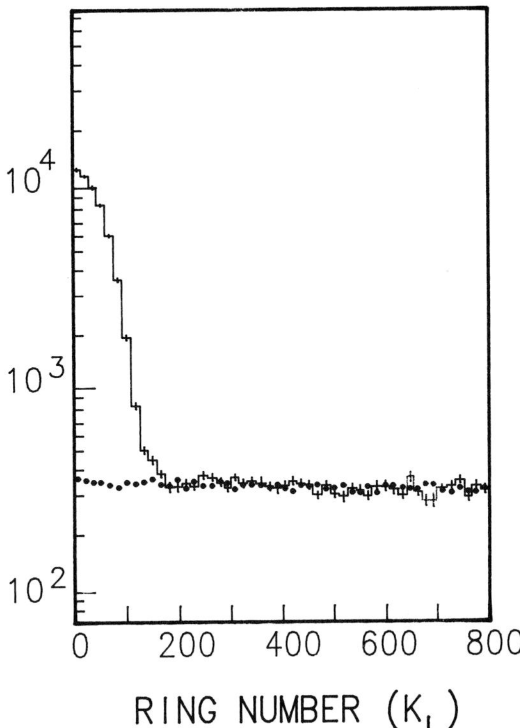

Fig. 9. The distribution in "ring number" around the vacuum beam for $2\pi^0$ events. The histogram is data; the dots are the (absolutely normalized) distribution for incoherent events as determined from $\pi^+\pi^-$.

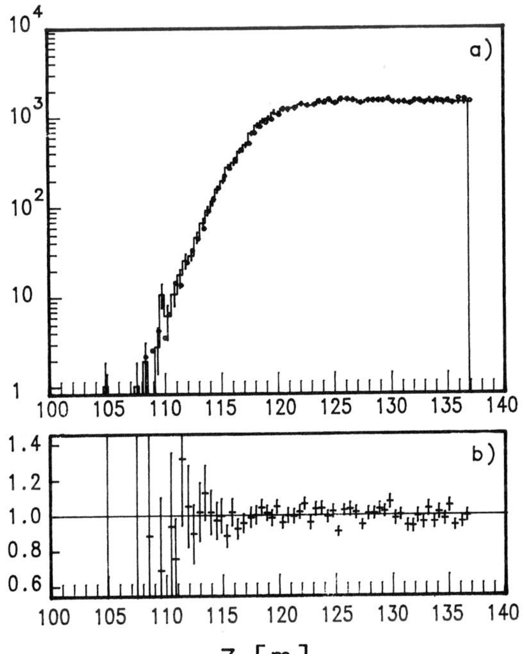

Fig. 10. a) Distribution in z-vertex [events/50cm] for $K_L\to\pi^0\pi^0$. The line (dot) is for data (monte-carlo); b) The ratio of data to monte-carlo.

Table 1. Summary of Backgrounds

Mode	Process	Correction [%]	Error [%]
$K_L \to \pi^+\pi^-$	$\pi e \nu$	0.31	0.06
$K_S \to \pi^+\pi^-$	incoherent regeneration	0.13	0.01
$K_L \to 2\pi^0$	"cross-over"	4.66	0.14
$K_L \to 2\pi^0$	$3\pi^0$	0.37	0.07
$K_S \to 2\pi^0$	incoherent regeneration	2.58	0.07

The magnitude of the acceptance and the event totals are given in the following Table.

Table 2. Event Totals and Corrections

	Neutral	Charged	R [from Eq. (5)]
Uncorrected Event Totals			
Vacuum	52,226	43,357	
Regen.	201,332	178,803	1.0698(78)
Background Fractions			
Vacuum	0.0515	0.0042	
Regen.	0.0257	0.0012	1.0445
Acceptance			
Vacuum	0.1885	0.5041	
Regen.	0.1813	0.5064	1.0001

A few comments are in order. First, the indicated background fractions include a correction for the presence of residual K_S decays in the K_L beam; this is on the order of 0.1% and is largely common to the two modes. Second, the acceptances are seen to be about the same for K_S and K_L decays; this follows from the similarity in the decay distributions and the fiducial cuts. Third, we see that the raw double ratio is about 7% from unity and that after background and acceptance corrections, it is consistent with zero.

The extraction of ε'/ε is then done by fitting (the integrated) decay distributions to the functional forms in (6) and (7). The regeneration amplitude ρ is proportional to the difference in forward scattering amplitudes of K^0 and $\bar{K}^0$ from the regenerator material:

$$\rho \propto (f - \bar{f})/k \qquad (12)$$

and this amplitude difference has a power-law dependence[6] upon momentum:

$$(f - \bar{f})/k \propto p^{-\alpha}. \qquad (13)$$

Fits were done for each mode separately, assuming that $\varepsilon' = 0$; the amplitude difference at 70 GeV/c and the power α were fit. The results of the fits are given in the following Table and they are displayed in Figure 11.

Table 3. Results of Fits for the B_4C Regeneration Amplitude

	Amplitude Difference at 70 GeV [mb]	Power-Law (α)	χ^2/ Degrees of Freedom
Charged	5.972(18)	0.602(10)	11.5/9
Neutral	5.963(18)	0.605(10)	10.3/9

The power laws are seen to be consistent with each other; they are in addition consistent with previous determinations[6,7,8] using light elements. The charged and neutral extracted amplitude differences are also consistent which points to a small value of ε'/ε, as did the indication from the double ratio.

A grand fit is next done to both modes simultaneously for the value of ε'/ε, again allowing the regeneration parameters to float. The value obtained is

$$\varepsilon'/\varepsilon = -0.0005 \pm 0.0014 \qquad (14)$$

where the error shown is statistical. We will now treat the sources of systematic uncertainty.

SYSTEMATIC ERROR DETERMINATION

The possible sources of systematic error involve uncertainties in the backgrounds (which have already been given), the stability of the results as a function of time and inten-

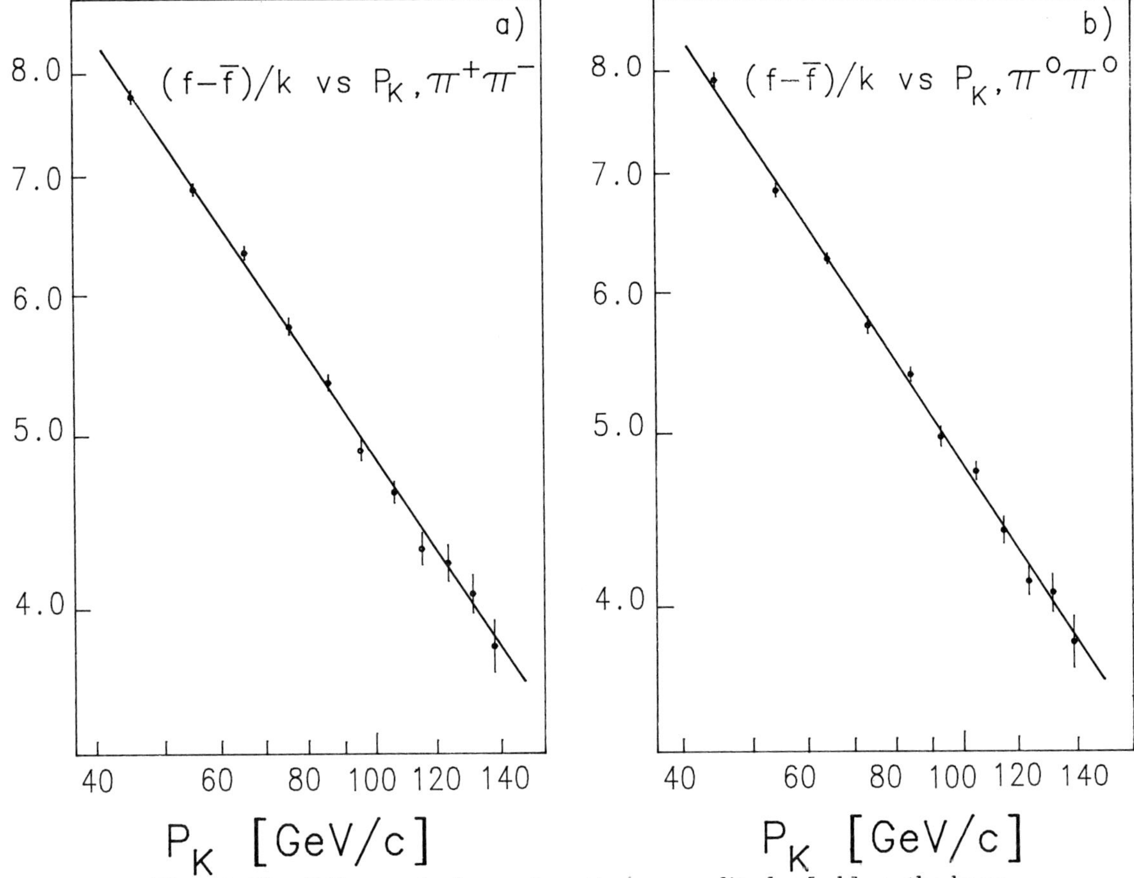

Fig. 11. The difference in forward scattering amplitudes [mb] vs. the kaon momentum. a) Charged mode; b) Neutral mode.

sity, the sensitivity to the energy scale and non-linearities in the lead-glass detector, sensitivity to accidental activity in the detector, and uncertainties in the detector acceptance.

Time Stability

In principle, because of drifts in the detector acceptance or in the beam position on the target, the yields could change with time. Indeed they do but with the method of recording long and short decays simultaneously, the effect of such drifts are greatly reduced. Alternating the regenerator every pulse eliminates effects which might change the relative acceptance between the two beams. In Figure 12 we show the ratios R_{+-} and R_{00} vs time into the run. Both ratios are time-independent even though the intensity varied.

Sensitivity to Energy Scale

The so called "energy scale" in the charged mode is easily determined: one simply adjusts the overall magnetic field so that the K^0 and Λ masses agree with the accepted values. It is important that the non-uniformities in the field be well understood. The neutral energy scale must be accurate as well: the double ratio R is formed in momentum bins and the decay rates, which are inversely proportional to energy, must be compared at identical energies for charged and neutral decays. The lead-glass is calibrated using momentum-analyzed electrons (using the same magnetic field map as the charged mode analysis) so that in principle the energy scales will coincide. In practice, this is only valid to

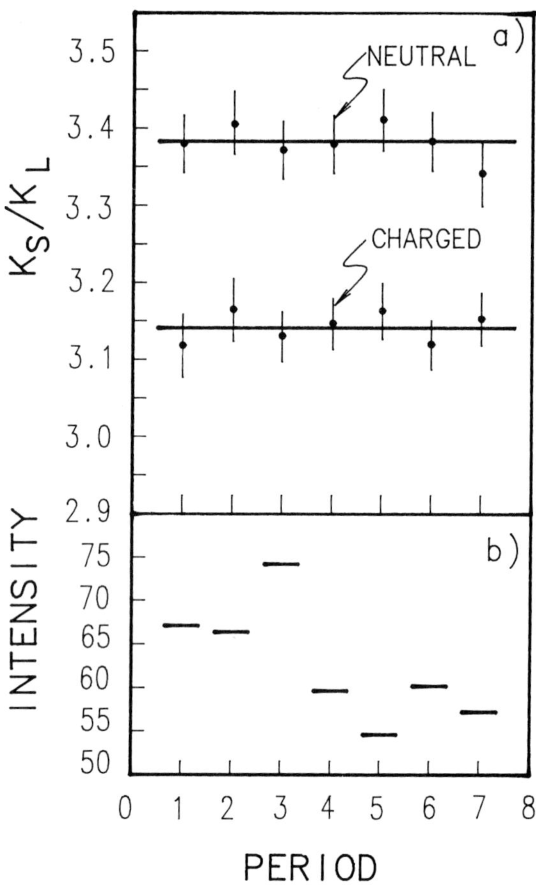

Fig. 12. a) The ratios of K_S to K_L decays and, b) the intensity, vs. the time into the run. (The selection is different from that described in the text.)

within a few tenths of one percent as a result of the difference in response of the lead-glass for photons and electrons. The overall neutral energy scale can be adjusted by aligning the observed distributions, but there remains a residual uncertainty on the order of 0.1%. The effects of this uncertainty of the ratio R_{00} are largely eliminated by the choice of the fiducial region (in z-vertex): the region, which includes essentially all of the K_S events, is chosen so that roughly the same number of K_L events leave (enter) at the downstream boundary as enter (leave) at the upstream boundary when the gamma energies are decreased (increased). In Figure 13 we see that R_{00} changes very little with lead-glass energy scale. However the energy resolution is not

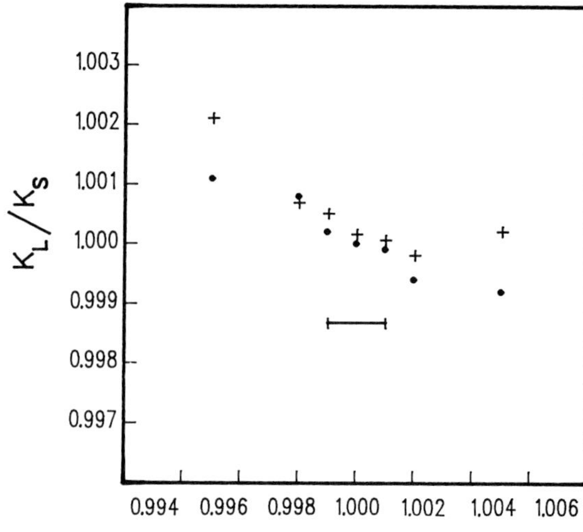

Fig. 13. The factor by which the ratio of K_L to K_S for $\pi^0\pi^0$ events changes vs. the neutral energy scale. The pluses (dots) are the data (monte-carlo). Also shown is the estimated residual uncertainty in the energy scale.

perfectly understood and as a result we estimate a possible 0.2% uncertainty in R_{00} .

Accidental Activity

In principle, the presence of accidental activity in the detector at the time of the event could result in a bias between K_S and K_L decays: the accidental activity tends to cluster near the vacuum beam and this could cause a systematic lowering of the efficiency for decays from that beam relative to decays from the other beam. Such a bias would obviously not cancel upon alternation of the regenerator position. The accidental activity, however, is at a low level: events have an extra cluster in the lead glass only 2.7% of the time (and only those which land on top of other clusters can cause a bias) and they have only about 8.5 extra chamber hits throughout the spectrometer (again only those which land on top of the true hits can cause a bias). We have searched for such biases, primarily using the high statistics modes; indeed, we have seen that distributions in track quality have a very slight broadening in the vacuum beam, but our cuts are sufficiently loose that this is inconsequential. We have used the technique of overlaying accidental events upon monte carlo events to investigate this possible bias. This procedure correctly reproduces a 3% loss in charged reconstruction efficiency (common to both modes) over the intensity range of the data (a factor of 4). There is, however, no asymmetry in K_L vs K_S for either mode within 0.07% so that the combined possible bias from accidental activity is taken to be 0.10%.

Acceptance

Possible uncertainties in the acceptance corrections are the most important source of systematic error for the measurement as we have chosen to perform it. The issue is of course the acceptance vs z-vertex; we have already seen that the agreement between data and monte-carlo is quite good and that, for there to be an effect at the percent level, the disagreements would be rather obvious (see Figure 6). In order to estimate the possible systematic error, we have relied upon the high statistics modes.

The z-vertex distribution for $K_L \to \pi e \nu$ decays is shown in Figure 14; the agreement over the fiducial decay volume is excellent, but with the high level of statistics (0.6 x 10^6 monte-carlo events, 1.3 x 10^6 data events) a small disagreement can be seen further upstream. The similar distribution for $3\pi^0$ decays is shown in Figure 15 for a portion of the decay region showing excellent agreement. The acceptance is easier to determine for the neutral mode: photons travel in straight lines and there is essentially only one aperture to model, the lead-glass array.

The monte-carlo includes the decay of the K_L beam along the decay region and neglecting this gives poorer agreement. For our mean momentum of about 70 GeV/c, the decay rate should decrease by about 0.05% per meter; using a sample of $3\pi^0$ decays, we determine that $c\tau_L = 9.5\text{m} \pm 3.1\text{m}$ where the accepted[9] value is 15.5m. The error is statistical and the result gives confidence in the neutral mode acceptance determination.

For the charged mode where the z resolution is about 15cm, our result is consistent when fitting in 2m z-vertex bins, thereby nearly eliminating the dependence upon acceptance. (This is less reliable in the neutral mode where the z resolution is about 1m.) We have also systematically determined the effect that the acceptance disagreement observed in the $\pi e \nu$ decays would have on the result: we would have a −0.05% shift in R_{+-}. From this information and the stability of R_{+-} and R_{00} when apertures, beam shapes, cuts, and detector efficiencies are varied, we assign a systematic uncertainty of $\leq 0.18\%$ for each mode.

A summary of the systematic uncertainties is given in Table 4 below.

Table 4. Systematic Error Summary

Source of Error	Uncertainty [%]
Accidental Effects	0.10
Backgrounds	0.18
Energy Scale Resolution/ Nonlinearity	0.20
Acceptance	0.25
TOTAL:	0.38

THE RESULT AND CONCLUSIONS

Before giving the result, we would like to present the outcome of additional fits to the data for other physical parameters. These may be regarded as internal consistency checks, and they provide interesting additional information. We have fit the data in 2m bins in z-vertex to determine the K_S lifetime, τ_S, and the K_L-K_S mass difference, Δm, for

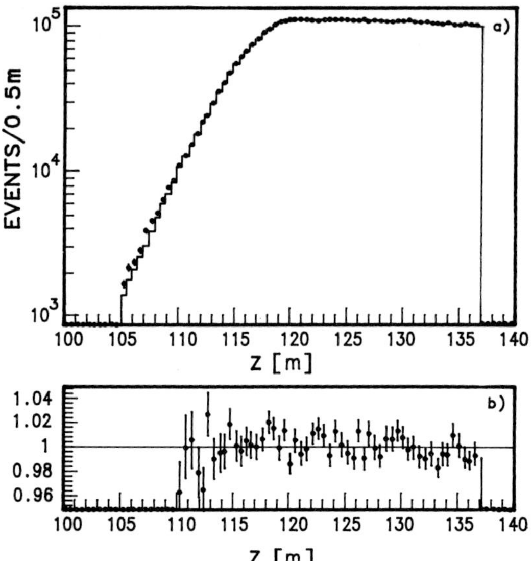

Fig. 14. a) Distribution in z-vertex [events/50cm] for $K_L \to \pi e \nu$. The line (dot) is for data (monte-carlo); b) Ratio of data to monte-carlo.

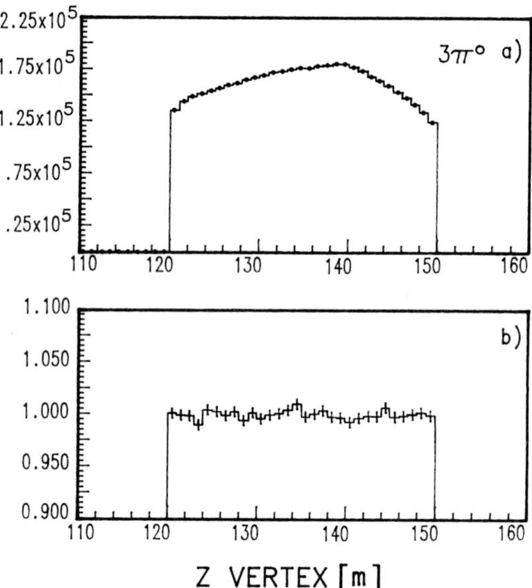

Fig. 15. a) Distribution in decay vertex [events/m] for $K_L \to 3\pi^0$. The line (dot) is for data (monte-carlo); b) Ratio of data to monte-carlo.

each mode. We have determined the phase difference between η_{+-} and η_{00}, $\Delta\Phi$. Finally, we give a value for the phase of η_{+-} itself which can be determined with the assumption of analyticity[10] for the regeneration amplitude.

Table 5. Fits for Other Physical Parameters

	Charged Mode	Neutral Mode	PDG Value (ref. 9)
τ_S [10^{-10} sec]	0.8909(63)	0.8940(61)	0.8923(22)
Δm[$10^{10}\,\hbar$ /sec]	0.524(18)	0.524(18)	0.535(2)
$\Delta\Phi$ [degrees]	2.3 ± 3.0		≤0.1 (by CPT)
Φ_{+-} [degrees]	47.5 ± 3.4	-	44.6 ± 1.2

The errors shown are statistical and these dominate. The results are obtained using the identical event selection and background subtractions as for the ε'/ε analysis; i.e., no optimization for the individual quantities has been made. The agreement of these results with the expected values gives confidence in the soundness of the data and its analysis.

Using (5) and (14) and Table 4, we then find:

$$\mathrm{Re}(\varepsilon'/\varepsilon) = (-5 \pm 14 \pm 6) \times 10^{-4} \qquad (15)$$

where the first error is statistical and the second is systematic. Combining the errors in quadrature then yields:

$$\mathrm{Re}(\varepsilon'/\varepsilon) = (-5 \pm 15) \times 10^{-4}. \qquad (16)$$

We show in Table 6 and in Figure 16 this result in relation to other recent measurements.

Table 6. Recent Measurements of $\mathrm{Re}(\varepsilon'/\varepsilon)$

Group	Result	Statis'l Error	Syst'c Error	Total Error
BNL-Yale[11]	+17	72	43	83
Chicago -Saclay[7]	−46	53	24	58
E731 Test[8]	+32	28	12	30
NA31[1]	+33	7	8	11
This Result	−5	14	6	15

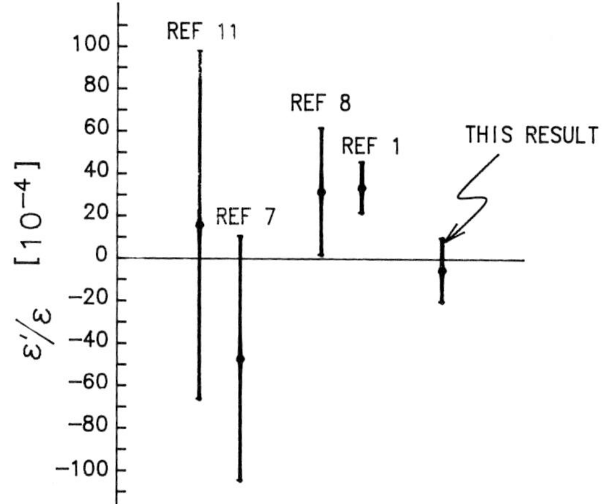

Fig. 16. This result with other recent determinations of Re(ε′/ε).

We can draw the following conclusions:

a) the result is consistent with zero and the systematic uncertainty has been significantly reduced;

b) we have simultaneously made precision determinations of an additional eight physical quantities (τ_S, Δm, and the regeneration slope for both modes; $\Delta\Phi$, and Φ_{+-});

c) the result does not confirm the NA31 claim of the observation of direct CP violation - the discrepancy is at the two standard deviation level;

d) the result is not inconsistent with the "Standard Model"[12] although it would favor a rather high top quark mass; it is consistent with the superweak[13] model.

We believe that the situation will be better clarified when our full data set is processed and a result obtained. We have every reason to believe that the resulting error will be significantly reduced over what is here presented.

ACKNOWLEDGEMENT

I would like to thank the organizers of this symposium for what was a very stimulating meeting. The high degree of concern for the well-being of the attendees was much appreciated.

I would also like to thank my colleague Jon Rosner for a careful reading of the manuscript.

REFERENCES

1. H. Burkhart *et al.*, Phys. Lett. **B206**, 169 (1988).
2. See M. Woods Ph.D. Thesis, University of Chicago, 1988 (unpublished), for a full description of most of the detector elements.
3. H. Sanders *et al.,* IEEE Trans. Nucl. Sci. **36**, 358 (1988).
4. A. J. Malensek, Fermilab FN–341 (1981) gives an excellent starting spectrum.
5. The calibration techniques will be described fully in the Ph.D. Thesis of J. R. Patterson, University of Chicago, 1990.
6. This follows from a single Regge exchange; in this case, the ω trajectory dominates. See J. Roehrig *et al.,* Phys. Rev. Lett. **38**, 1116 (1977).
7. R. H. Bernstein *et al.,* Phys. Rev. Lett. **54**, 1631 (1985).
8. M. Woods *et al.,* Phys. Rev. Lett. **60**, 1695 (1988).
9. Particle Data Group, G. P. Yost *et al.,* Phys. Lett. **B204**, 1 (1988).
10. The power α and the phase of π are connected by a dispersion relation.
11. J. K. Black *et al.,* Phys. Rev. Lett. **54**, 1628 (1984).
12. See, for example, F. Gilman, *Proc. of the Symp. on the 25th Anniversary of CP Violation,* Blois, France, 1989, to be published.
13. L. Wolfenstein, Phys. Rev. Lett. **13**, 562 (1964).

DISCUSSION

H. Wahl, CERN: How did you obtain the experimental value for the phase of η_{+-} independently from the phase of the regeneration amplitude?

B. Winstein: The phase of regeneration is linked (by analyticity) to the power-law behavior; thus by measuring the energy dependence of regeneration, the phase of the regeneration and hence that of η_{+-} can be extracted. As the regeneration may not obey an exact power-law, there will be an additional systematic error.

NA31 RESULTS ON CP VIOLATION IN K DECAYS, AND A TEST OF CPT

D. Fournier
Laboratoire de l'Accélérateur Lineaire, Orsay, France
and
CERN, Geneva, Switzerland

ABSTRACT

The NA31 experiment, dedicated to the study of CP violation in K decays, is reviewed. New, preliminary results on ϕ_{00} and ϕ_{+-} are presented. Some features of the most recent data taken for ϵ'/ϵ are analysed.

At the last lepton–photon conference, which took place in Hamburg, the NA31 Collaboration announced for the first time that they had found evidence for direct CP violation in K^0 decay [1]. Since then, the analysis has been finalized and the result, $\epsilon'/\epsilon = (3.3 \pm 1.1) \times 10^{-3}$, published [2]. Recently, preliminary results on part of the statistics accumulated by the E731 Collaboration working at Fermilab have also been presented [3].

The NA31 Collaboration has now taken new data, in 1987 for the phase of η_{00} and η_{+-}, and in 1988 again for ϵ'/ϵ. Preliminary results from these data are presented in Sections 2 and 3. In Section 1 the basic principles of the experiment and the method used to derive ϵ'/ϵ are described.

1. THE NA31 EXPERIMENT

In K^0 decays, and when CPT conservation is assumed (see Section 3), the parameter ϵ describes CP violation induced by mixing between the CP-even $[K_1^0 = (K^0 + \bar{K}^0)/\sqrt{2}]$ and CP-odd $[K_2^0 = (K^0 - \bar{K}^0)/\sqrt{2}]$ states. Direct CP violation may also occur in the decay of the CP-odd state (K_2^0) into two pions. This is parametrized, in the Wu–Yang convention [4] and with the usual approximations, by [5]

$$\epsilon' = \frac{1}{\sqrt{2}} \, \mathrm{Im}\left(\frac{A_2}{A_0}\right) \, \exp i(\pi/2 + \delta_2 - \delta_0) \, ,$$

where A_0 and A_2 are the amplitudes for the decay into isospin 0 and 2 two-pion states, and δ_0 and δ_2 are the corresponding strong phase-shifts at the kaon mass. With these definitions, the ratios of K_L and K_S decay amplitudes into $2\pi^0$ and $\pi^+\pi^-$ are

$$\eta_{00} = \epsilon - 2\epsilon' \, ,$$
$$\eta_{+-} = \epsilon + \epsilon' \, .$$

From the measured values of δ_0 and δ_2 the phase of ϵ' is expected to be close to the phase of ϵ (almost 45°, see Section 3) and therefore, to a good approximation,

$$\frac{\epsilon'}{\epsilon} = \frac{1}{6}\left[1 - \frac{|\eta_{00}|^2}{|\eta_{+-}|^2}\right] .$$

As is well known, the interest of the measurement of ϵ'/ϵ is to decide whether CP violation can be described by the Standard Model with three families of leptons and quarks [6, 7], or whether it needs physics beyond this Model.

In NA31, the above relation is used to determine ϵ'/ϵ from the double ratio R of K_s and K_L decay rates into charged and neutral pions:

$$R = \frac{|\eta_{00}|^2}{|\eta_{+-}|^2} = \frac{\Gamma(K_L \to \pi^0\pi^0)/\Gamma(K_L \to \pi^+\pi^-)}{\Gamma(K_S \to \pi^0\pi^0)/\Gamma(K_S \to \pi^+\pi^-)} .$$

The basic feature of the experiment [8] consists in detecting neutral and charged modes concurrently (this eliminates the overall normalization, the dead-time, accidental vetoes, etc.), but alternately in K_L and K_S beams. The alternating beams are collinear, thus illuminating the detector in the same way. Also, using a K_S train (see below) allows almost exact cancellation of acceptances.

Conversely, not all rates in all detectors can be made equal from K_L to K_S. Slow drifts of the apparatus have also to be monitored carefully. In order to reduce their possible effects, data are taken in 'miniperiods' of about 10 hours K_S and 30 hours K_L.

The K_L are produced by 450 GeV protons ($\leqslant 10^{11}$ per pulse) incident on a target located 120 m upstream of the

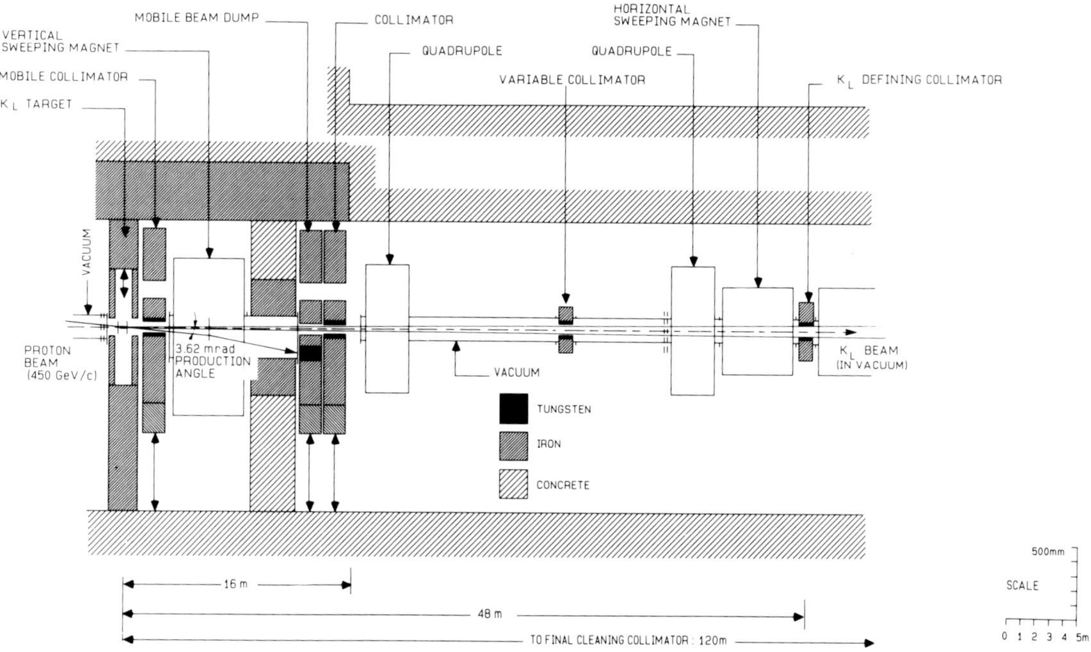

Fig. 1 The K_L beam line (1986).

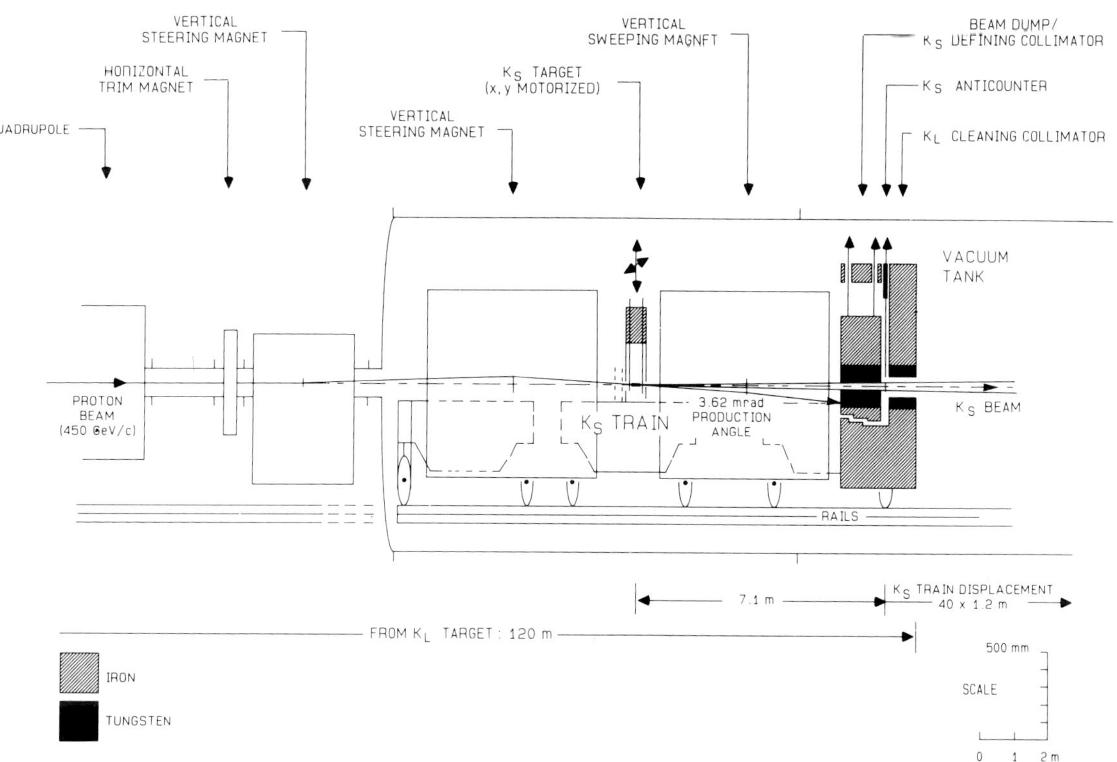

Fig. 2 The K_S movable beam and the K_L cleaning collimator.

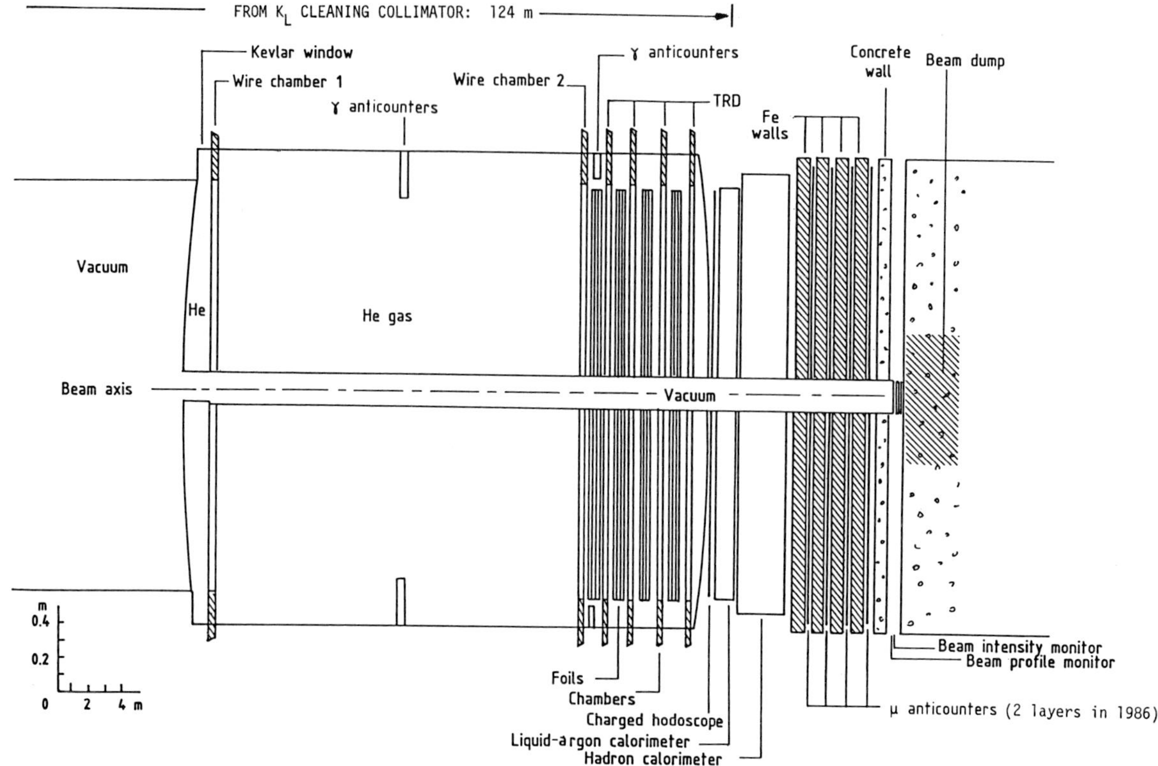

Fig. 3 The NA31 spectrometer. (The TRD was included in 1988.)

beginning of the decay region (Figs. 1, 2, and 3). In 1986, a double collimation scheme was used to reduce the beam halo. Since some regeneration from the K_L cleaning collimator was, however, observed in the upstream part of the decay volume, this was changed in 1988 to a three-stage system (see Section 2).

Alternatively, $\sim 10^7$ protons per pulse are brought onto a second target, from which the K_S are selected by collimation after 7 m. Mounted on a train, the beam elements can be moved over the 48 m of the K_L decay region.

The K_S data are taken with the beam train displaced in 1.2 m steps so that the K_S vertex distribution can be made effectively uniform, as in K_L (see Section 2).

An important feature of the K_S beam is an anticounter, preceded by 7 mm of lead, just downstream from the collimator. This counter vetoes decays in the collimator, defines the upstream edge of the decay region independently of resolution effects from the detector, and permits the relative calibration of the $2\pi^0$ and $\pi^+\pi^-$ energy scales to better than $\pm 10^{-3}$.

The detector does not use a magnet. The direction of the charged pions is measured by two sets of wire chambers, 25 m apart in an helium tank separated from the decay volume by a thin Kevlar window (3×10^{-3} radiation lengths). Photons from π^0 decays are measured in position and energy by a fine-grain lead/liquid-argon calorimeter (LAC). The resolution obtained for photons is 0.5 mm in position and $7.5\%/\sqrt{E}$ in energy. To measure charged-pion energies, the liquid-argon calorimeter is followed by an iron-scintillator hadron calorimeter (HAC). The overall resolution is $65\%/\sqrt{E}$. Both calorimeters are segmented longitudinally into two equally thick layers. The readout is by orthogonal strips organized in quadrants. The strip width is 1.25 cm in LAC and 12 cm in HAC. Two scintillator hodoscopes are used for pretriggering. The first one, in front of LAC, is used to trigger on charged events. The second one, inside LAC in between the two layers, is used to trigger on neutral events. Behind the HAC, two planes of hodoscopes after (each) 80 cm of iron and ring counters around the decay volume are used to veto, at the pretrigger level, $K \to \pi\mu\nu$ and other three-body decays.

Single rates in the first hodoscope are of the order of 100 kHz.

A trigger signal is accepted after cuts on the total energy, on the ratio of the energy between the front half of LAC and the rest of the calorimeter (electron rejection), on the number of hits in the first chamber, and on the number of photons in LAC. After digitization of calorimeter and chamber information, three-body decays are further rejected using on-line processors. Typically, 1000 events per burst are

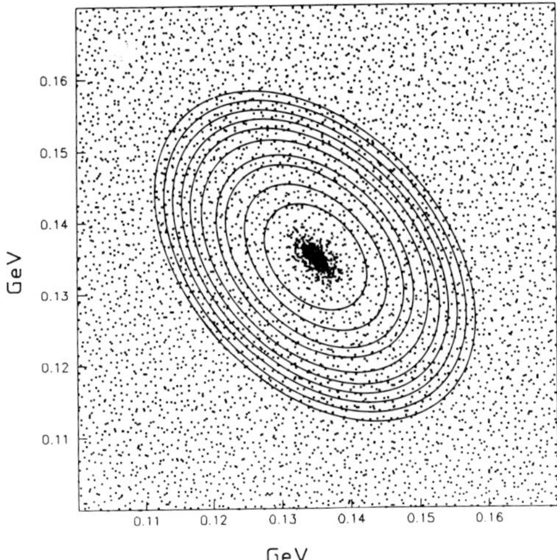

Fig. 4 Best combination of the two $\gamma\gamma$ masses (K_L data).

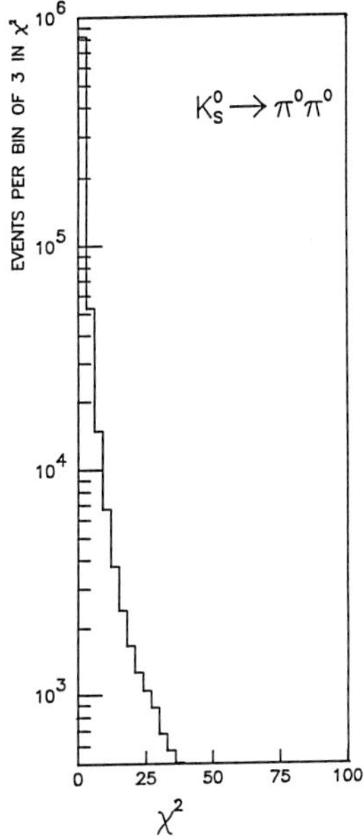

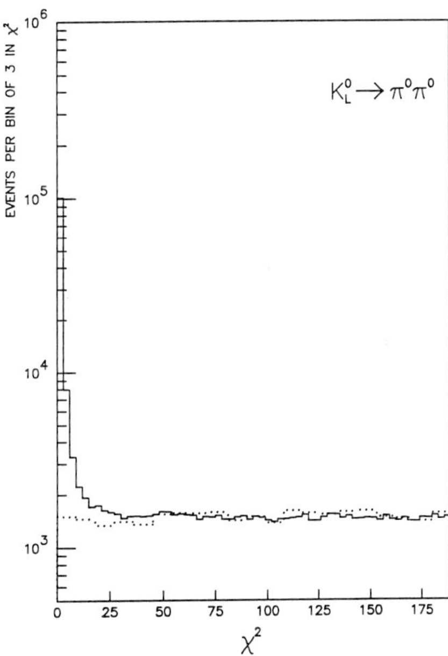

Fig. 5 Number of accepted 4γ events as a function of χ^2, for $K_S \rightarrow 2\pi^0$ and $K_L \rightarrow 2\pi^0$ data, and a Monte Carlo calculation for background originating from $K_L \rightarrow 3\pi^0$ decays. The signal region is taken as $\chi^2 < 9$ (3σ).

recorded in the K_L beam.

Neutral decays are reconstructed from events with four photons, each with at least 5 GeV (and no other shower of more than 2.5 GeV) and an energy centre of gravity in the beam profile. The longitudinal vertex position is calculated using the kaon mass as a constraint, and only then is the two-π^0 mass constraint applied. In K_S there is essentially no background, whilst in K_L the signal sits on a background dominated by $K_L \rightarrow 3\pi^0$ with undetected photons (Fig. 4). The signal region is defined as a 3σ contour*). Counting events in equal-area contours shows that the background distribution is flat. This is confirmed by a high-statistics Monte Carlo simulation of $K_L \rightarrow 3\pi^0$ decays in the apparatus (Fig. 5). In the signal region, the background is 4%, with a systematic error (from the extrapolation) of $\pm 0.2\%$.

For charged events, space points in each chamber (at least one hit in three planes out of four) are reconstructed, from which the vertex position and the opening angle θ are derived. The total energy is calculated from θ and the ratio ϱ of the two pion energies measured in the calorimeter. This measurement has a small error (1%) provided the decay is not too asymmetric. For this purpose, and to reduce the background of Λ decay (in K_S) to a negligible level, a cut on ϱ at 2.5 is applied. For each track, an energy ratio cut (between the front of LAC and HAC signals) is also used to reject electrons.

Events with isolated photons (such as those coming from $\pi^+\pi^-\pi^0$ or accidental photons) are also rejected, in a similar way to that for the $2\pi^0$ decay. This procedure minimizes the accidental correction from extra photons (see below). Events (charged or neutral) with an additional space point in the first chamber are rejected in the same way.

*) The peculiar shape of these contours (ellipses at 45° from the axes) reflects the K^0 mass constraint. The (unconstrained) error on $\gamma\gamma$ masses is about 2.5 MeV.

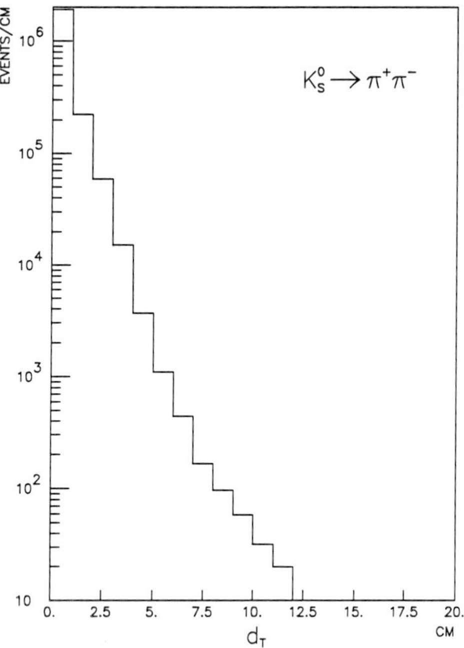

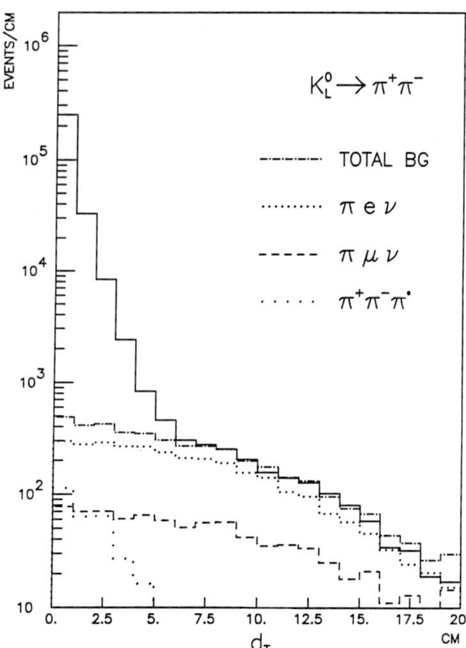

Fig. 6 Event distribution for charged decays as a function of distance d_T between the decay plane and the production target, for K_S and K_L decays and various background components.

Using calorimeter energies, the $\pi\pi$ invariant mass is then calculated (precision ~ 25 MeV) and a cut at 2.2σ is applied. The above-mentioned cuts reject about 50% of the $K^0 \to \pi^+\pi^-$ signals. Since most cuts are on energy ratios, the instability of the calorimeters enters only from non-linear terms. The stability is, however, carefully monitored. In the 1986 data the position of the K^0 peak has shown that HAC is stable to better than ±0.5%. Since LAC was stable to better than ±0.1%, in total, we estimated that the cuts differed from K_L to K_S by less than 0.1%.

After all cuts, the distribution of d_T—the distance between the decay plane and the target—is used to study the three-body decay background that remains in K_L (see Fig. 6, where the K_S distribution has been geometrically scaled for comparison).

The analysis of this background, using especially the shower widths in the calorimeters, has shown that it is dominated by $\pi e\nu$ and $\pi\mu\nu$. The dominant mechanism causing the $\pi\mu\nu$ background is a radiation of the muon in the calorimeter, hard enough for the mass constraint to be satisfied and for the muon to range out before the last veto plane. For $\pi e\nu$, similar contributions come from the production of hadrons in the electron shower, and from the superimposition of an accidental hadron on the electron.

The distributions shown correspond to a vertex position between 10.5 and 48.9 m. The first 10.5 m have been cut out to reduce the correction due to inelastic regeneration of K_L in the cleaning collimator (see Section 2).

In the control region (d_T between 7 and 12 cm), the background fraction was $(3.6 \pm 0.1) \times 10^{-3}$. After extrapolating the $\pi e\nu$ and $\pi\mu\nu$ contributions separately (using selected data samples and Monte Carlo events, respectively), and allowing for a certain fraction of $\pi^+\pi^-\pi^0$ events with unseen photons, the background in the signal region was estimated to be $(6.5 \pm 2.0) \times 10^{-3}$.

Table 1

Background composition for $K_L \to \pi^+\pi^-$ decays

Background	Control region ($\times 10^{-3}$)	Signal region ($\times 10^{-3}$)
$K^0 \to \pi^+\pi^-\pi^0$	0.1 ± 0.1	1.0 ± 1.0
$K^0 \to \pi e\nu$	2.8 ± 0.2	4.4 ± 0.3
$K^0 \to \pi\mu\nu$	0.5 ± 0.2	0.7 ± 0.3
Regenerated K_S	0.2 ± 0.10	0.4 ± 0.2
Total	3.6 ± 0.1	6.5 ± 2.0

Table 2

Event statistics and corrections

	Signal events ($\times$ 1000)	Background (%)	Scattering (%)	Inefficiencies		Accidental losses (%)
				Pretrigger (%)	Trigger (%)	
$K_L \to 2\pi^0$	109	4.0	< 0.1	0.06 ± 0.06	0.20 ± 0.10	2.6 ± 0.07
$K_L \to \pi^+\pi^-$	295	0.6		0.37 ± 0.07	0.05 ± 0.06	2.6 ± 0.05
$K_S \to 2\pi^0$	932	< 0.1	0.3	0.04 ± 0.02	0.12 ± 0.03	2.5 ± 0.05
$K_S \to \pi^+\pi^-$	2300	< 0.1		0.48 ± 0.03	0.01 ± 0.01	2.8 ± 0.05
Effects on R			0.3	-0.12 ± 0.10	-0.03 ± 0.12	-0.34 ± 0.10

This systematic error—one of the largest from the 1986 run (see Table 1)—led the Collaboration to improve the apparatus by adding more muon veto planes, and most important, by installing a transition radiation detector, to identify electrons before they start to radiate.

Table 2 shows the available statistics obtained from the 1986 run, after background subtraction, in the different modes. It also shows the pretrigger and trigger efficiencies, which are equal, within errors, for K_S and K_L. Although the rates were adjusted to be similar for K_S and K_L, and the cuts were designed to have similar effects in the presence of accidentals, for the charged and neutral modes, a correction is still necessary. This was estimated by overlaying on good events (however, before final cuts) the events taken at random at a rate proportional to the beam intensity. The net losses—around 2.5% for all decay modes—are given in Table 2.

To estimate the double ratio R, a fiducial cut was applied to the longitudinal vertex position z (10.5–48.9 m) and to the total energy E (70–170 GeV). For the charged events, both E and z depend only on the chamber geometry, which is known to high precision. For neutral events, both E and z depend on the LAC geometry and energy scale. This scale is adjusted to the charged energy scale by fits of the vertex distributions to the position of the anticounter in the K_S beam (Fig. 7). By studying the charged-to-neutral relative calibration as a function of the energy range used and for different K_S target stations, a systematic uncertainty of $\pm 10^{-3}$ was derived. To this uncertainty there corresponds an uncertainty on R (using the binning procedure, see below) of $\pm 3 \times 10^{-3}$. This 'amplification' factor of 3 reflects the difference that exists between K_S and K_L spectra (see Fig. 8) owing to the loss of low-energy K_S in the 7.2 m long K_S beam. The reduction

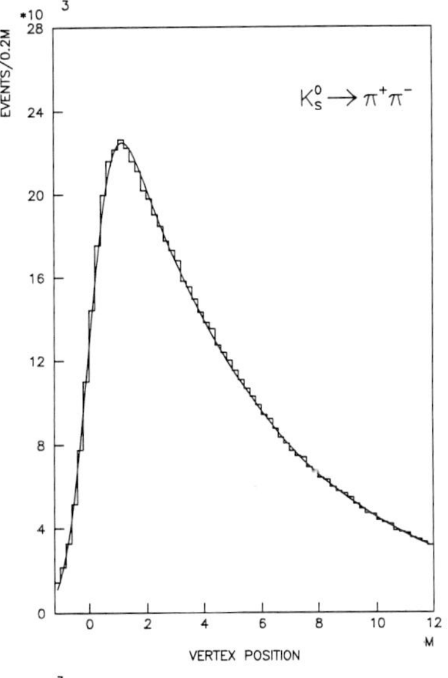

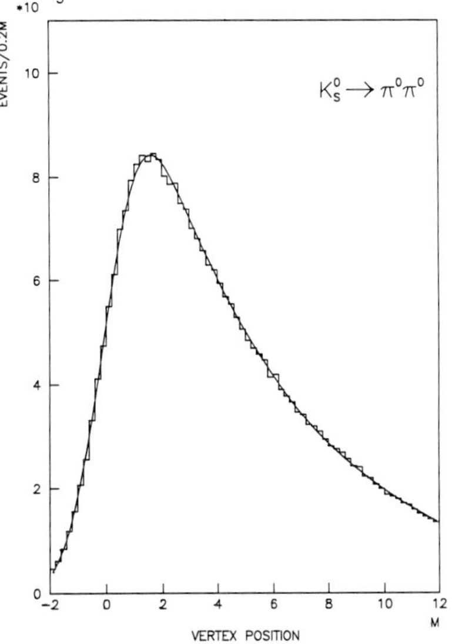

Fig. 7 $K_S \to 2\pi^0$ and $K_S \to 2\pi^+\pi^-$ event distributions as a function of distance from the anticounter in the K_S beam. The continuous lines show the best fits to the data.

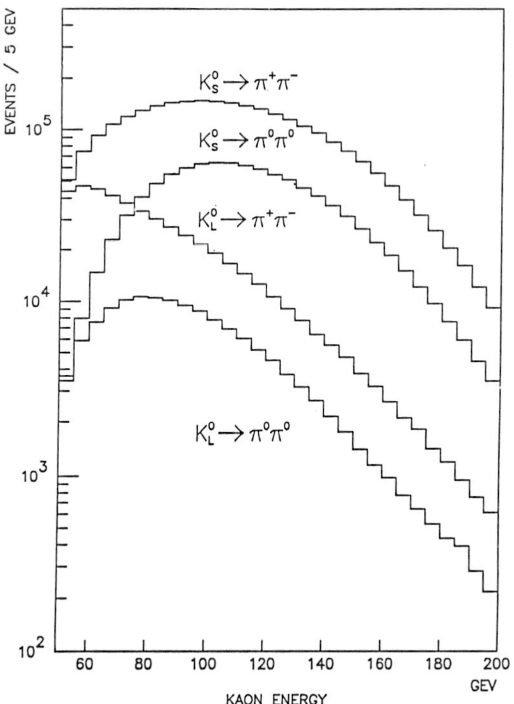

Fig. 8 Energy spectra for K_S and K_L decays in two pions (1986).

of this factor by running the K_S under different conditions was the other major improvement of the 1988 set-up (see Section 2).

The double ratio R was evaluated in 10×32 bins in E and z for each of the mini-periods. The weighted average was found to be $R = 0.977 \pm 0.004$ (statistical error).

Table 3

Systematic uncertainties on the double ratio R (in %)

Source	
Background subtraction for $K_L \to 2\pi^0$	0.2
Background subtraction for $K_L \to \pi^+\pi^-$	0.2
$2\pi^0/\pi^+\pi^-$ difference in energy scale	0.3
Regeneration in the K_L beam	< 0.1
Scattering in the K_S beam	0.1
K_S anticounter inefficiency	< 0.1
Difference in K_S/K_L beam divergence	0.1
Calorimeter instability	< 0.1
Monte Carlo acceptance	0.1
Gains and losses by accidentals	0.2
Pretrigger and trigger inefficiency	0.1
Total systematic uncertainty	± 0.5%

The residual differences in acceptance were estimated using a Monte Carlo simulation. This includes the differences in beam divergence, the scattering in the K_S collimator and anticounter, and the effects of resolution and finite bin size. The net effect found was 0.3% on R, leading finally to

$$R = 0.980 \pm 0.004 \text{ (stat.)} \pm 0.005 \text{ (syst.)} .$$

The dominant systematic uncertainties on R are summarized in Table 3. Converting to ϵ'/ϵ, and adding statistical and systematic errors in quadrature, gave as a final result from the 1986 data

$$\epsilon'/\epsilon = (3.3 \pm 1.1) \times 10^{-3} .$$

2. SOME FEATURES OF THE IMPROVED SET-UP

As mentioned in the preceding section, the main improvements realized in the NA31 set-up for a new run on ϵ'/ϵ, which started in spring 1988, are the following [9]:

- better K_L collimation,
- the addition of a transition radiation detector,
- the operation of the K_S beam under different conditions (360 GeV protons).

The data accumulated in 1988 correspond to 67,000 $K_L \to 2\pi^0$ events, i.e. within the same cuts, about two thirds of the 1986 data sample. At the time of the conference, the analysis of these data was still in progress, and in the following one only describes how the implemented modifications have performed. For this purpose, two main points are considered: the charged background in the K_L mode, and the sensitivity to a difference in energy scale between charged and neutral events.

In Fig. 9, the amount of charged background in the control region of d_T (7 to 12 cm), relative to signal, is plotted as a function of the vertex position of $K_L \to \pi^+\pi^-$ decays. In the fiducial region used in 1986 (10.5 to 48.9 m) this background is, on the average, 3.6×10^{-3}, but at low z it shows a large increase corresponding to some inelastic regeneration, in the cleaning collimator, of K^0 scattered from the defining collimator ($z = -72$ m). In 1988 the defining collimator was moved slightly backwards ($z = -78$ m), and a new collimator was added at $z = -20$ m. Under these conditions, the unwanted phenomenon is totally eliminated.

The 10 m of fiducial volume gained in this way are extremely valuable, being in the region where the $K_L \to 2\pi^0$ reaction has minimum background. The shape of the z distribution (see below and Fig. 12) is also such that the use of this 10 m upsteam region tends to reduce the sensitivity of the double ratio R to the energy-scale difference between charged and neutral decays.

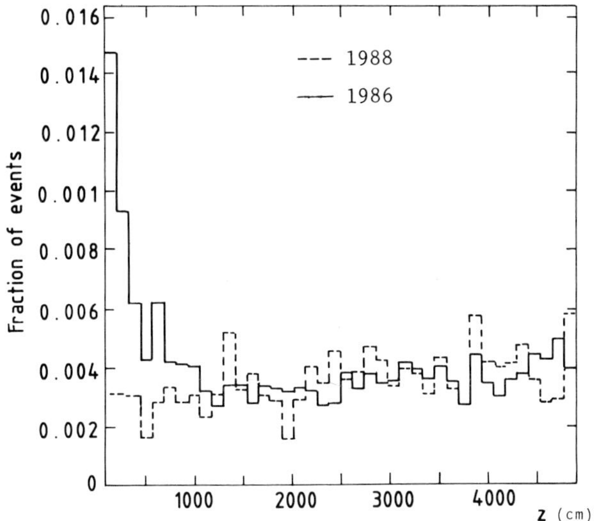

Fig. 9 Background for $K_L \to \pi^+\pi^-$ as a function of the longitudinal decay vertex position z (control region of d_T).

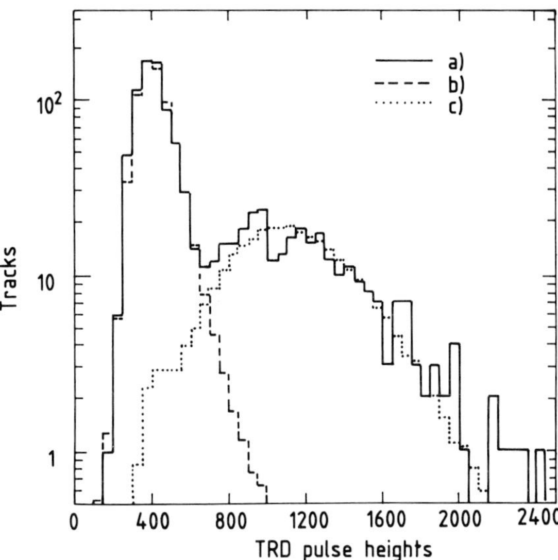

Fig. 10 TRD signal (average of 3 smaller signals out of 4).
a) Large d_T K_L events.
b) Pions from K_S.
c) Electrons from $\pi e \nu$.

To sort out the main $\pi e \nu$ component of the background in charged K_L decays, a transition radiation detector was placed downstream from the second wire chamber. This detector consists of four sets of radiators and chambers. Each radiator is a stack of 340 polypropylene foils, 19 μm thick and 600 μm apart. The chambers are filled with a xenon-helium mixture having the same density as that of the surrounding gas (CO_2). Signals from the wires are charge-integrated. The chamber gain is stabilized over time by controlling the applied high voltage in such a way that the signals from ^{109}Cd radioactive sources stay constant. The truncated mean of the three smallest signals out of four is used as an estimate of the detector response. The performances of this system have been evaluated using beams of electrons, muons, and pions, as well as samples of $K_S \to \pi^+\pi^-$ and $K_L \to \pi e \nu$ decays [10, 11].

Applying the same procedure as the one used in 1986 to the 1988 data, a similar fraction of background events was found in the high-d_T region (see Fig. 9). This background was then analysed by plotting, for each track, the signal from the transition radiation detector (Fig. 10). The electron component, associated with high TRD signals, shows up in a clear way. From a superimposition of signals from pions (K_S data) and electrons (selected $\pi e \nu$) on this plot, it is deduced that the $\pi e \nu$ mode represents $(64 \pm 3)\%$ of the background in the control region. This can then be extrapolated to the signal region, as was done in 1986. The TRD information however allows a more direct determination. Taking events in which one track has a TRD estimator above 1000 (see Fig. 10), one selects a known fraction (independently of d_T) of $\pi e \nu$ events, and obtains a d_T distribution which only has a small fraction of 2π events peaking at low d_T. This 2π contamination can be estimated from the K_S data, thus giving the full $\pi e \nu$ distribution down to $d_T = 0$. Applying this procedure to the 1988 data gives a $\pi e \nu$ fraction of $(4.2 \pm 0.3) \times 10^{-3}$ in the signal region. This is similar to the one reported in 1986. The statistical error is not smaller, but this time there is no significant systematic uncertainty in the procedure.

To reduce, in a definitive way, the 0.2% systematic error on R associated with the charged background in K_L, it still remains to complete the improved analysis of the non-$\pi e \nu$ component.

The other major step accomplished for the new run concerns the sensitivity to a difference in energy scale between charged and neutral events. As described above, the rather large sensitivity observed is a result of the K_S momentum spectrum being harder than the K_L one. This can be seen from Fig. 8, or perhaps even more clearly from Fig. 11 in which the ratio of the number of charged events, observed in K_S and K_L, is plotted as a function of energy, normalized arbitrarily to 1 at 105 GeV. To have more similar spectra, it was decided to operate the K_S beam, in the new run, with lower-energy protons (360 GeV) and at a somewhat larger angle (4.8 mrad). The K_L were also taken at a slightly smaller angle (2.8 mrad). The new spectra are now more similar (Fig. 11), although not yet identical. To estimate the actual sensitivity to a difference in energy scale between charged and neutral events, a full preliminary analysis in E and z bins has been performed. The z and energy spectra of the events used are shown in Figs. 12 and 13. A weight—which is a function of only the K_S station number—was applied when too many (or too few) data were taken at a given station. Under these new conditions, an uncertainty in the relative

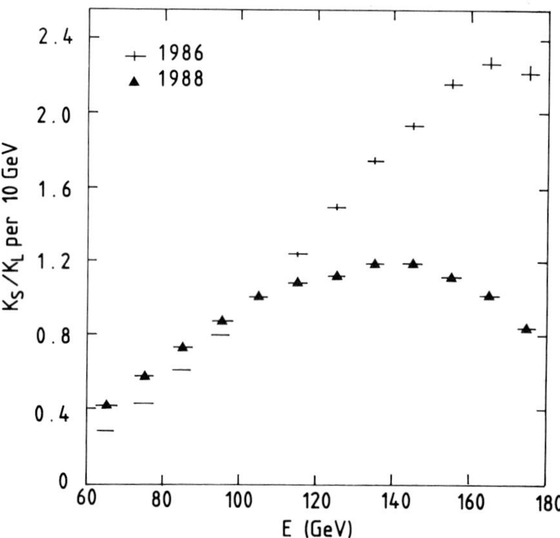

Fig. 11 Ratio of K_S to K_L energy spectra, arbitrarily normalized to 1 at 105 GeV.

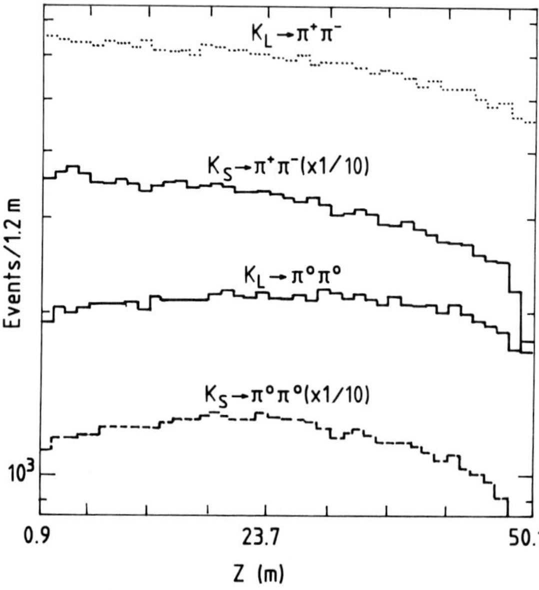

Fig. 12 Decay vertex distribution of the four categories of events (1988).

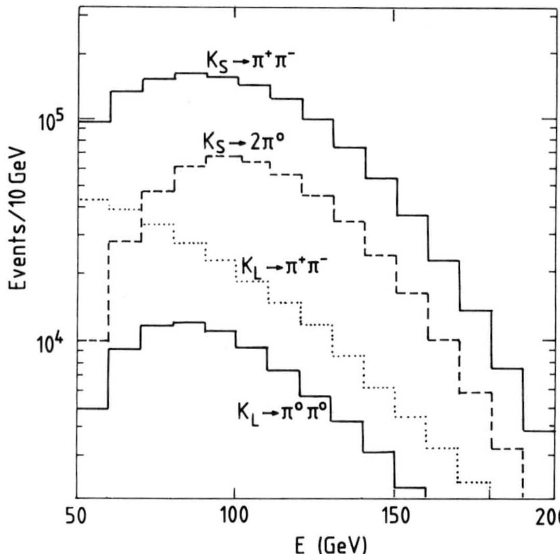

Fig. 13 Energy distribution of the four categories of events (1988).

energy scales of 10^{-3} is reflected in an uncertainty in R that is also of 10^{-3}. The improvement achieved is thus a factor of 3 on the error that was the largest in 1986.

To take advantage of this, large statistics are desirable. The NA31 Collaboration is at present accumulating these, with the aim of more than doubling the 1986 sample.

3. THE MEASUREMENT OF ϕ_{00}, ϕ_{+-}, AND THEIR DIFFERENCE

The amplitude ratios η_{00} and η_{+-} are physical quantities of exceptional richness. Whilst an accurate enough comparison of their moduli uncovers direct CP violation, their phases constitute the best presently known tool for a test of CPT. Therefore, a natural complement of the ϵ'/ϵ programme was to use the NA31 equipment [11] to measure these phases—especially ϕ_{00} [12]—more accurately than has been done so far (see Fig. 14).

The method used is based on the time dependence of the decay rate, in vacuum, of K^0 to two pions. The kaon beam is an incoherent superposition of K^0 and $\overline{K}^0$ produced by interactions of protons (450 GeV) on a target. It employs no regenerator. As a function of proper time t, the decay rate is

$$I(t) = S(E)\left\{\exp\left(-\frac{t}{\tau_S}\right) + |\eta|^2 \exp\left(-\frac{t}{\tau_L}\right) + 2D(E)|\eta| \exp\left[-\frac{t}{2}\left(\frac{1}{\tau_S}+\frac{1}{\tau_L}\right)\right]\cos(\Delta m t - \Phi)\right\} ,$$

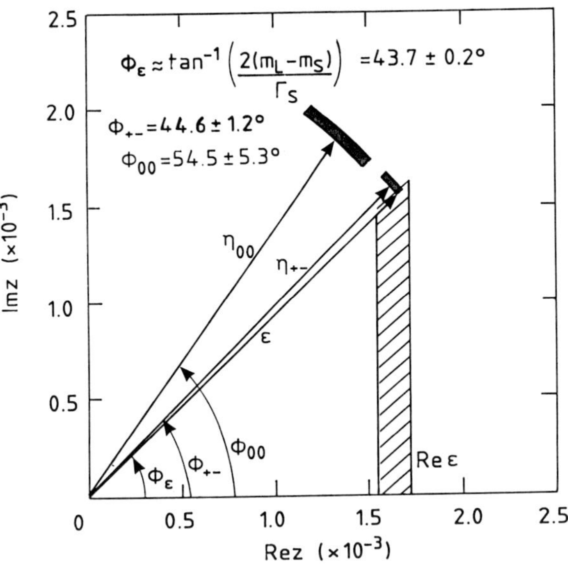

Fig. 14 ϵ, η_{00} and η_{+-} in the complex plane, prior to the present work.

where $S(E) = N_{K^0}(E) + N_{\overline{K^0}}(E)$ is the beam spectrum as a function of energy,

$D(E) = [N_{K^0}(E) - N_{\overline{K^0}}(E)]/S(E)$ is the dilution factor of the incoherent mixture,

τ_L (τ_S) is the K_L (K_S) lifetime,

$\Delta m = m_L - m_S$ is the mass difference.

This rate is most sensitive to the phase in the interference region, around 12 τ_S, where the K_L and K_S rates are similar. To match the NA31 acceptance to this would require a target about 40 m upstream of the K_L cleaning collimator (see Fig. 2). Also, for each energy, the information regarding the phase comes from the z dependence of the decay rate and therefore depends strongly on the knowledge of the acceptance. A way to avoid this constraint is to analyse a ratio of decay rates from targets at difference distances; in which case, for a given E and z bin, the acceptance cancels (to first order). The maximum sensitivity under these conditions is obtained when the interference patterns from the two targets are displaced by 90°. At 100 GeV this corresponds to a distance of about 15 m.

Two target stations were thus installed: one, K_N (near), at $Z = -33.6$ m; the other, K_F (far), 14.4 m further upstream (see Fig. 15). Both beams use a beryllium target, 40 cm long and 2 mm in diameter, from which the

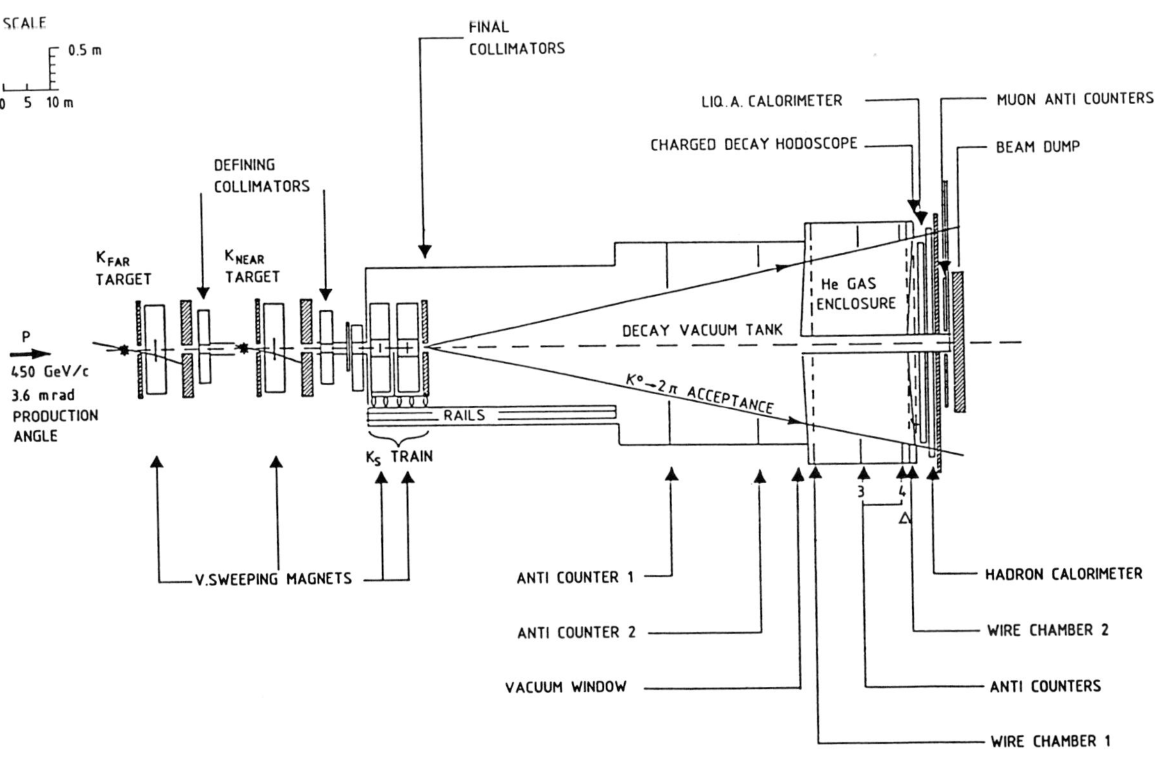

Fig. 15 The beam and detector for the phase measurement.

neutral beams are derived (at 3.6 mrad) using 5.6 mm diameter collimators at 10 m. Mobile collimators (diameter 2.8 cm for K_N, 4.2 cm for K_F) clean up the beam at the entrance to the large decay tube. The typical proton intensities used were 1.4×10^{10} (K_N) and 2.0×10^{10} (K_F) protons per pulse.

The detector elements and performances were the same as for the 1986 ϵ'/ϵ data-taking period (see Section 1).

The trigger and the on-line filtering were also essentially the same. A specific feature concerns the lifetime. This was calculated, for both charged and neutral events, from the energy and the vertex position available from on-line processors, and events that decayed before 7 K_S lifetimes were down-scaled. The down-scaling factors were chosen to give about the same number of events per K_S lifetime.

Data were taken over a period of 70 days in the summer of 1987. The data from K_N and K_F targets were interleaved as for ϵ'/ϵ, but in this case with a periodicity of about 16 hours. The time devoted to each beam was such that, on the average, equal fluxes in the K_L region were recorded from both beams. At the end of each cycle, pure K_S data, with the train (see Section 1) at the upstream end, and in the centre of the decay region, were also taken for the purpose of energy calibration. A total of about 150 million triggers were written to tape at a rate of typically 1000 events per burst.

The off-line reconstruction and the background subtraction procedure were also the same as for the 1986 ϵ'/ϵ data. Since in the present case the rate is dominated by the observed ($\gtrsim 5\tau_S$) K_S part, the background fraction is, on the average, very small. Its z (or τ) dependence is, however, strong: although in the K_S part it is totally negligible, in the K_L part it reaches levels similar to those found in pure K_L data. This is illustrated in Fig. 16 for the neutral background. Nevertheless, as is shown below, the backgrounds do not, on the whole, affect the phase result.

The data samples retained for the final analysis correspond to events with 1.2 m < Z < 49.2 m and 70 GeV < E < 170 GeV (see Table 4). The accuracy of these data is illustrated in Fig. 17 where the rate of decays to $\pi^+\pi^-$ and $\pi^0\pi^0$, combined from both beams and summed over energy, are plotted, after Monte Carlo correction, as a function of the lifetime. The interference term (shown in insets) has been extracted by subtracting the fitted lifetime distribution without interference from the data in energy bins, and then averaged. The cosine term is well visible—for the first time, in the neutral mode—over a whole period of the oscillation.

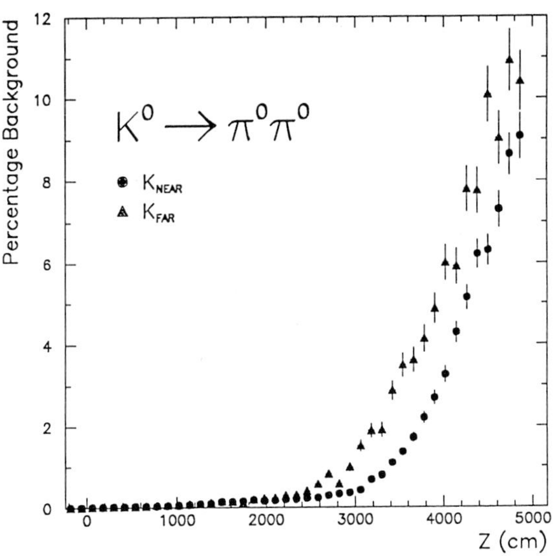

Fig. 16 Background in the neutral mode as a function of decay vertex position.

Table 4

Event statistics

	$K^0 \to \pi^0\pi^0$ ($\times 10^6$)	$K^0 \to \pi^+\pi^-$ ($\times 10^6$)
K_{near}	1.81	2.24
	(2.24) [a]	(5.99)
K_{far}	0.31	0.57
	(0.31)	(0.84)

a) The numbers in brackets include the lifetime downscaling weights.

To reduce to a minimum any systematic uncertainty from the acceptance correction, the final analysis was not, however, made using these distributions. Instead, we took the ratio of events observed from the two targets.

This quantity, analysed in bins of 'lifetime' computed from a fixed point (chosen to be at an equal distance from the K_N and K_F targets) has the behaviour illustrated in Fig. 18. The first plateau*) (visible only at 120 GeV in the figure) corresponds to a region where decays from both targets are in the K_S regime. This is followed by the interference region, and finally by a second plateau where decays from both targets are in the K_L regime.

*) The height of the first plateau is approximately exp (λ_T/λ_S), where λ_T is the distance between the two targets, and λ_S is the K_S mean path for the energy considered.

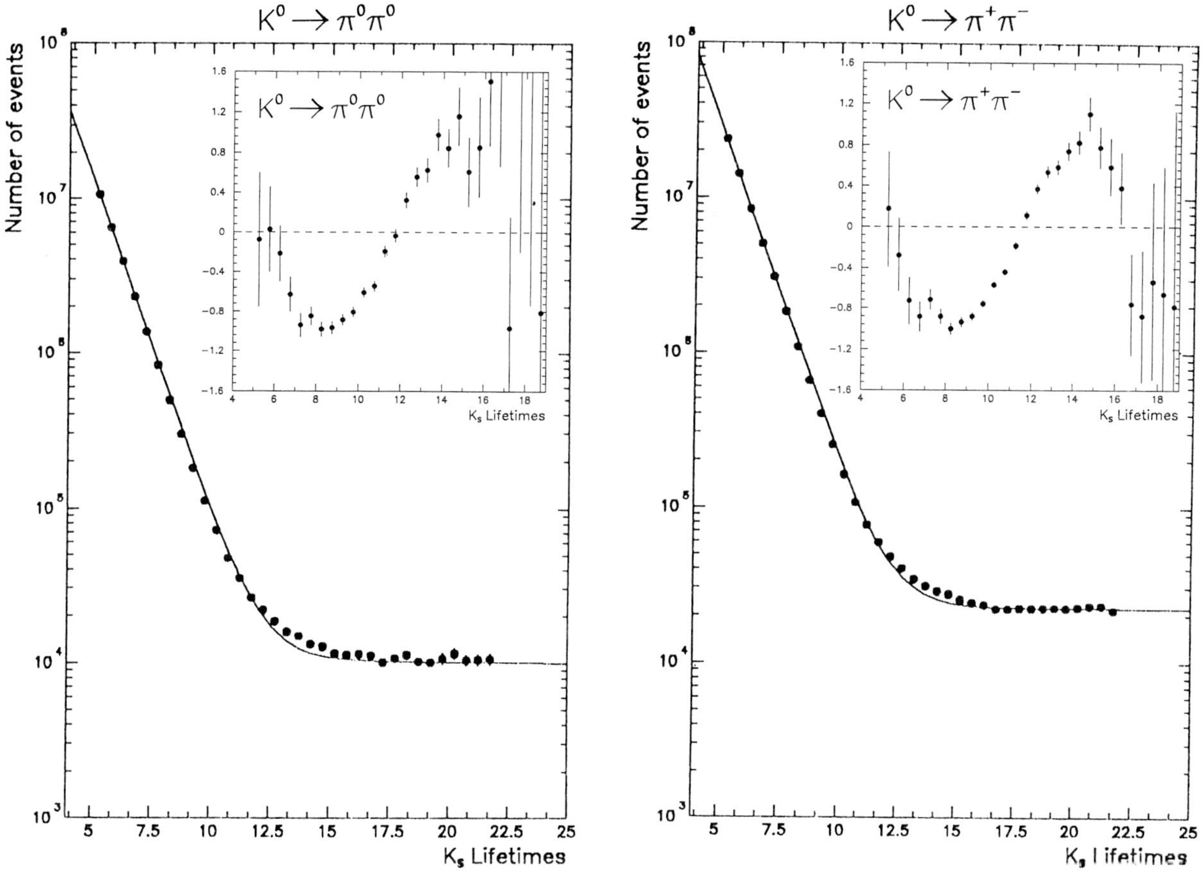

Fig. 17 Lifetime distribution, acceptance corrected. The line is a fit without the interference term. Shown in insets is the difference between this fit and the data, averaged over energy.

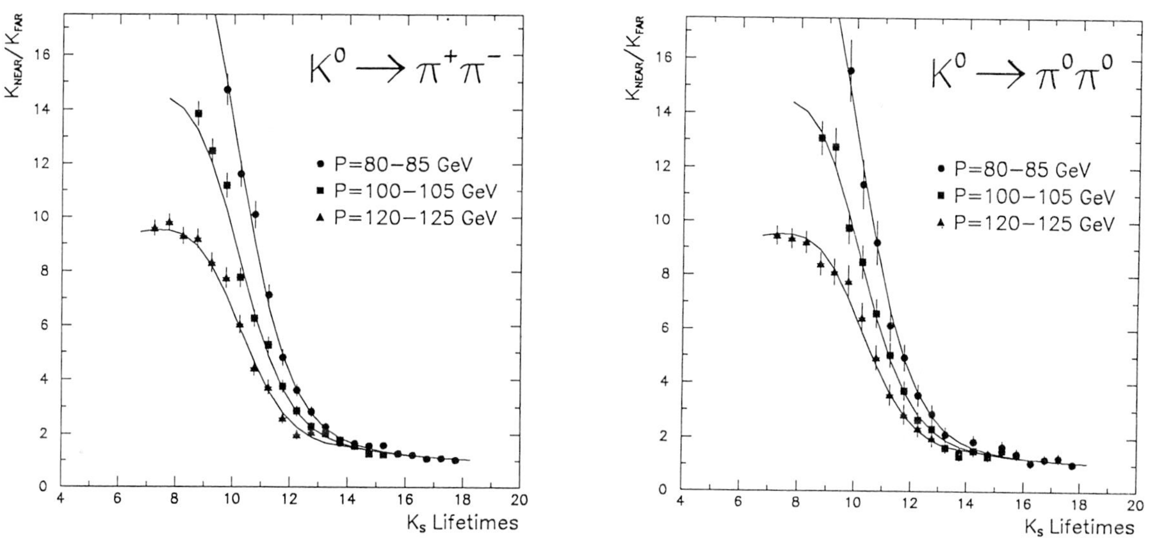

Fig. 18 Ratio of 'lifetime' distributions from the K_N and K_F targets.

A simultaneous fit to the charged and neutral ratios was made, using bins of 5 GeV and half a K_S lifetime. The phases ϕ_{+-} and ϕ_{00}, and a common unparametrized dilution factor in 10 GeV intervals, were varied in the fit. In addition, a charged and a neutral normalization K_N/K_F were allowed to vary as a function of momentum. The K_S and K_L lifetimes, and $|\eta_{+-}|$, $|\eta_{00}|$, and Δm, were assumed to be known and fixed to the central values given in Table 5. The dilution factor found rises smoothly (from ~ 0.2 to ~ 0.35) with energy. The results obtained for the phases (still preliminary) are (in degrees) [13, 14]

$$\phi_{+-} = 46.9 \pm 1.4 \quad \text{(stat.)} ,$$
$$\phi_{00} = 47.1 \pm 2.1 \quad \text{(stat.)} .$$

The χ^2 obtained was 796 for 768 degrees of freedom. The theoretical distribution calculated with these values is shown in Fig. 19, superimposed on the data scaled to 100 GeV.

Systematic uncertainties are given in Table 6. As mentioned earlier, the systematic uncertainty connected with background subtraction is small. In fact, if the presence of background is completely ignored, both phases change by less than 0.6°

The ratio method employed is characterized by the Monte Carlo acceptance correction being small, less than 1°, thus leading to a small systematic uncertainty.

The only significant systematic error connected with the NA31 set-up is in fact due to the energy scale. Statistical and systematic uncertainties in the K_S anticounter position fits (see Section 1) and anticounter inefficiency give an energy-scale uncertainty of $\leqslant 10^{-3}$, which in turn leads to a 0.8° (neutral mode) uncertainty.

Table 5

External parameters used in the fit

Parameter	Mean value	References
$\lvert\eta_{+-}\rvert$	$(2.272 \pm 0.021) \times 10^{-3}$	15
$\lvert\eta_{00}\vert\eta_{+-}\rvert$ a)	0.990 ± 0.003	2, 16
τ_S	$(0.8922 \pm 0.0020) \times 10^{-10}$ s	17
τ_L	$(5.18 \pm 0.04) \times 10^{-8}$ s	17
Δm	$(0.5351 \pm 0.0024) \times 10^{10}\ \hbar s^{-1}$	18

a) From ϵ'/ϵ and η_{+-}.

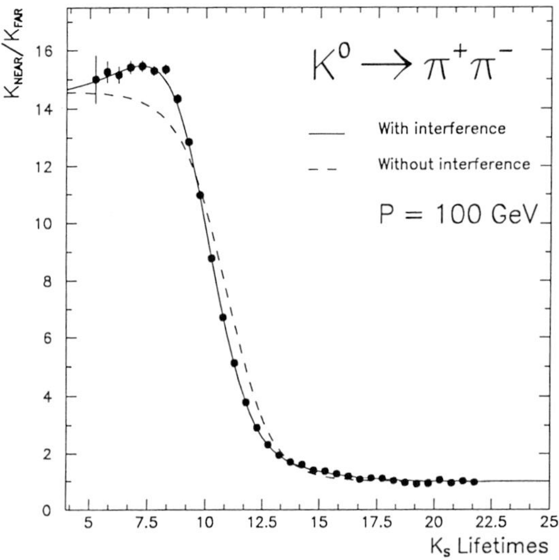

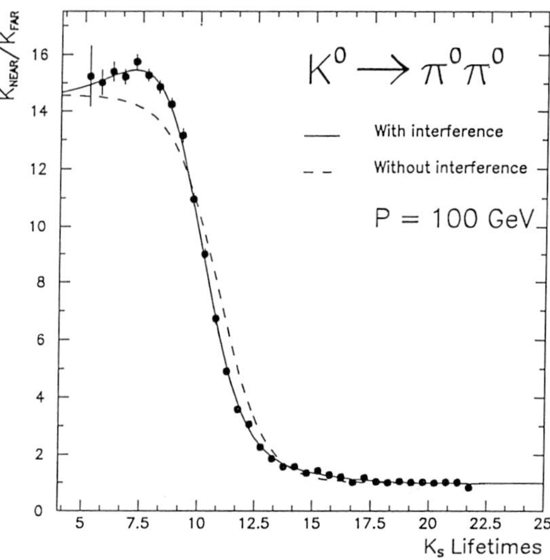

Fig. 19 Ratio of 'lifetime' distributions scaled to 100 GeV, and best fit to the data.

Table 6

Systematic uncertainties (in degrees)

Contributions	ϕ_{+-}	ϕ_{00}	$\Delta\phi$
Energy scale	0.6	0.8	1.0
Non-linearities	0.1	< 0.5	< 0.5
Background	0.1	0.1	0.1
Regeneration	0.1	0.1	-
MC acceptance	0.1	0.1	0.1
Resolution	Negligible	0.2	0.2
Mean production point	< 0.2	< 0.2	-

Systematic uncertainties on each phase come also from an imperfect knowledge of external parameters, mainly τ_S and Δm. Around the central values of Table 5 the sensitivity is $+2.68°$ $(+3.12°)$ for a percent increase in τ_S (Δm) for ϕ_{+-}, and $+2.23$ $(+3.12)$ for ϕ_{00}. Taking into account the present knowledge of these parameters [15–18] gives, as preliminary results from this experiment,

$$\phi_{+-} = [46.9 \pm 1.4(\text{stat.}) \pm 0.7(\text{syst.}) \pm 0.6 \pm 1.4(\text{ext.})]°,$$

$$\phi_{00} = [47.1 \pm 2.1(\text{stat.}) \pm 1.0(\text{syst.}) \pm 0.5 \pm 1.4(\text{ext.})]°,$$

$$\Delta\phi = [0.2 \pm 2.6(\text{stat.}) \pm 1.2(\text{syst.})]°.$$

A comparison with previous experiments (Fig. 20) shows [13] the improvement obtained in the neutral phase and in $\Delta\phi$.

To quantify the relevance of these results as a test of CPT, one should go back to the evolution of a $K^0-\bar{K}^0$ system, and its 2π decay, without the requirement of CPT invariance [5, 19].

Mass eigenstates of the mass matrix $(M_{ij} - i\Gamma_{ij})$ are written as

$$K_S = (K_1^0 + \epsilon_S K_2^0)/\sqrt{1+|\epsilon_S|^2} \ ,$$

$$K_L = (K_2^0 + \epsilon_L K_1^0)/\sqrt{1+|\epsilon_L|^2} \ .$$

Using $\epsilon = (\epsilon_S + \epsilon_L)/2$ and $\delta = (\epsilon_S - \epsilon_L)/2$, the diagonalization procedure gives

$$\epsilon = \frac{(\Gamma_{12} - \Gamma_{12}^*) + i(M_{12} - M_{12}^*)}{(\gamma_S - \gamma_L) - 2i\,\Delta m} \ ,$$

$$\delta = \frac{(\Gamma_{11} - \Gamma_{22}) + i(M_{11} - M_{22})}{(\gamma_S - \gamma_L) - 2i\,\Delta m}$$

From the equality of the mass and lifetime of particles and antiparticles, CPT invariance requires $\delta = 0$.

The amplitude for K^0 $[\bar{K}^0]$ decay to two pions in isospin state I is written $(A_I + B_I)e^{i\delta_I}$ $[(A_I^* - iB_I^*)e^{i\delta_I}]$; B_I is CPT-violating; Im A_I is CP-violating (direct) but CPT-conserving. In the Wu–Yang convention the dominant A_0 term is chosen to be real. In this case, $|\text{Im}\, M_{12}|$ dominates $|\text{Im}\, \Gamma_{12}|$ in the numerator of ϵ, and the phase of ϵ is $\phi_\epsilon = \text{A}\tan(2\Delta m/\gamma_S) = (43.7 \pm 0.2)°$.

As in Section 1, η_{00} and η_{+-} are written as a function of ϵ, $\epsilon' = (1/\sqrt{2})\, \text{Im}(A_2/A_0)\exp\, i[(\pi/2) + \delta_2 - \delta_0]$ (nearly parallel to ϵ), and an eventual CPT violating term, $\epsilon'_\perp = (1/\sqrt{2})$ (Re B_2/A_0) exp $i(\delta_2 - \delta_0)$. The compatibility of $\Delta\phi$ with zero thus indicates that with the present accuracy ($|\epsilon'_\perp/\epsilon| \leqslant 2\%$), no such direct CPT-violating term is needed.

Coming back to the mass matrix, one gets from the time evolution of $\langle K_S^0|K_L^0\rangle$, in the presence of the CPT violating term δ [19],

$$(1 + i\tan\phi_\epsilon)(\text{Re}\,\epsilon - i\,\text{Im}\,\delta) = (\tfrac{2}{3}\eta_{+-} + \tfrac{1}{3}\eta_{00}) + \alpha \ .$$

In this relation α is a correction term which accounts in the unitarity relation for decay channels (common to K_S and K_L) other than the dominant 2π, $I = 0$. Introducing $\delta_\perp$ $(\eta_\perp)$, the component of $\delta(\tfrac{2}{3}\eta_{+-} + \tfrac{1}{3}\eta_{00})$ orthogonal to ϵ in the complex plane, gives then

$$\delta_\perp = -\eta_\perp - \text{Im}\,\alpha\cos\phi \ .$$

Provided Im α can be neglected in this relation*), one sees that the difference between the phase of $(\tfrac{2}{3}\eta_{+-} + \tfrac{1}{3}\eta_{00})$

*) The correction term α, evaluated by Barmin et al. from experimental data, is $[(-0.006 \pm 0.068) - i(0.026 \pm 0.120)] \times 10^{-3}$, i.e. not that small compared with $\eta_\perp$. With some theoretical input, however, the value quoted is $(0.010 \pm 0.003) + i(0.001 \pm 0.003)] \times 10^{-3}$.

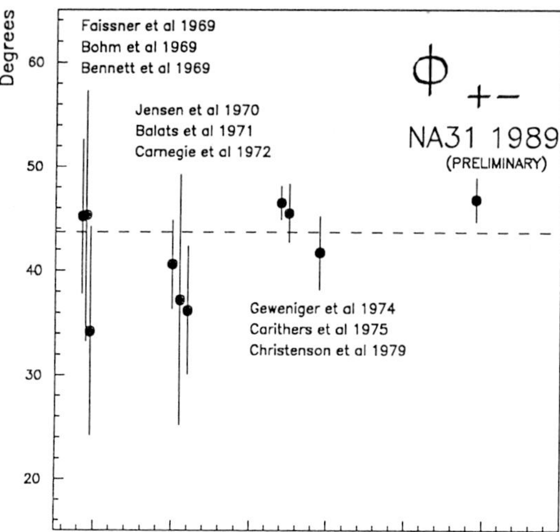

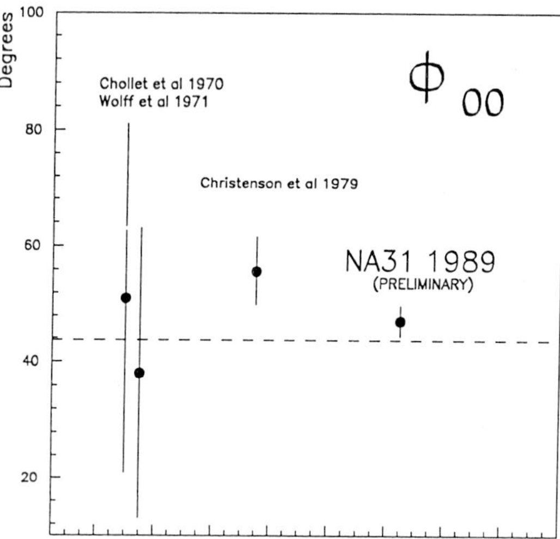

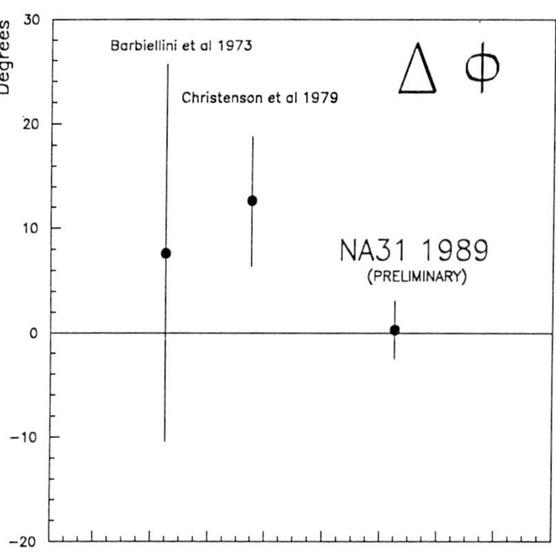

Fig. 20 Recent results on ϕ_{+-}, ϕ_{00} and $\Delta\phi$.

and ϕ_ϵ directly measures $\delta_\perp$, i.e. the mass difference term in δ.

The values of ϕ_{+-} and ϕ_{00} obtained in this experiment (including external errors) correspond to $\delta_\perp = -(1.3 \pm 0.8) \times 10^{-4}$. This is compatible with zero, and at the 90% confidence level $|\delta_\perp| < 2.6 \times 10^{-4}$, or conversely

$$\left| \frac{m_{K^0} - m_{\bar{K}^0}}{m_{K^0}} \right| \lesssim 5 \times 10^{-18} .$$

This is the best test of CPT conservation from a single experiment.

4. CONCLUSIONS

Using the NA31 set-up, the Collaboration has measured ϕ_{00} with an accuracy similar to the one achieved before on ϕ_{+-}. The charged phase has also been measured, and both values agree $[\Delta\phi = (0.2 \pm 2.6 \pm 1.2)°]$. The common charged and neutral phase is slightly larger than ϕ_ϵ. Allowing for the maximum possible deviations within errors this result makes a test of CPT at the 5×10^{-18} level.

New data, taken in 1988 for an improved measurement of ϵ'/ϵ, have had their new features analysed. Better performances were observed with respect to the charged background in K_L and the sensitivity to the difference in charged and neutral energy scales. This, together with the large statistics being accumulated at present in the 1989 run, gives good hope of obtaining a new result with significantly reduced errors.

Acknowledgements

I thank the organizers of the conference for their invitation to present the NA31 results. It is a pleasure to acknowledge all those who, in the laboratories of CERN, Edinburgh, Mainz, Orsay, Pisa, and Siegen contribute to the experiment.

I am indebted to D. Cundy for his careful checking of the manuscript and to the CERN Text Processing service for their competence and patience in preparing the final version of this paper.

* * *

REFERENCES

1. I. Mannelli, Proc. *23rd Int. Symp. on Lepton and Photon Interactions at High Energies,* Hamburg, 1987 (North-Holland, Amsterdam, 1988), p. 367.
2. H. Burkhardt et al., *Phys. Lett.* **B206** (1988) 169.
3. B. Winstein, talk given at the *25th Anniversary of the Discovery of CP Violation,* Blois, France, 1989, and at this conference.
4. T.T. Wu and C.N. Yang, *Phys. Rev. Lett.* **13** (1964) 380.
5. T.D. Lee and C.S. Wu, *Annu. Rev. Nucl. Sci.* **16** (1966) 511.
6. M. Kobayashi and K. Maskawa, *Prog. Theor. Phys.* **49** (1973) 652.
7. J. Ellis, M.K. Gaillard and D.V. Nanopoulos, *Nucl. Phys.* **B109** (1976) 213.
   F.J. Gilman and M.B. Wise, *Phys. Lett.* **B83** (1979) 83. *Phys. Rev.* **D20** (1979) 2392.
   B. Guberina and R. Peccei, *Nucl. Phys.* **B163** (1980) 289.
   L. Wolfenstein, *Annu. Rev. Nucl. Sci.* **36** (1986) 137, and references therein.
8. H. Burkhardt et al., *Nucl. Instrum. Methods* **A268** (1988) 116.
9. D. Cundy et al., CERN/SPSC/87-48 (1987).
10. A. Kreutz, Diplomarbeit, Universität Siegen (1989).
    I. Harrus, thèse Université Paris Sud (1989).
11. P. Clarke et al., CERN/SPSC/86-6 (1986).
12. J.H. Christenson et al., *Phys. Rev. Lett.* **43** (1979) 1209.
13. V. Gibson (NA31 Collaboration), An experimental test of CPT invariance, talk given at the *25th Anniversary of the Discovery of CP Violation,* Blois, France, 1989.
14. G.D. Barr et al. (NA31 Collaboration), A measurement of the phases of the CP-violating amplitudes in $K^0 \to 2\pi$ decays and a test of CPT invariance, CERN preprint in preparation.
15. C. Geweniger et al., *Phys. Lett.* **48B** (1974) 483.
    R. Messner et al., *Phys. Rev. Lett.* **30** (1973) 876.
    R. Devoe et al., *Phys. Rev.* **D16** (1977) 565.
    D.P. Coupal et al., *Phys. Rev. Lett.* **55** (1985) 566.
16. M. Woods et al., *Phys. Rev. Lett.* **60** (1988) 1695.
17. Particle Data Group, *Phys. Lett.* **B204** (1988) 1.
18. M. Cullen et al., *Phys. Lett.* **B32** (1970) 523.
    S. Gjesdal et al., *Phys. Lett.* **B52** (1974) 113.
    C. Geweniger et al., *Phys. Lett.* **B52** (1974) 108.
19. V.V. Barmin et al., *Nucl. Phys.* **B247** (1984) 293.

DISCUSSION

G. Barbiellini, CERN and INFN: On ϵ'/ϵ, if one is allowed to take a kind of weighted average, the final result is about 2×10^{-3}, averaging NA31 and E731. This is 1 standard deviation away from the CERN result and around 1.5 from Fermilab. I think both experiments are consistent and very well done. Do you agree with my analysis?

D. Fournier: There are two different kinds of error — the statistical error and the systematic error. If you make ten experiments, the statistical error will be divided by the square root of ten. On the systematic error, however, if your experiment is wrong by 0.5 you can do it ten times, and it will still be wrong by 0.5. It may be dangerous to the take such an average.

Rare Kaon Decays

Laurence S. Littenberg

Physics Department
Brookhaven National Laboratory
Upton, N.Y. 11973 USA

ABSTRACT

The results of the current generation of rare kaon decay experiments are reviewed. The present status of and future plans for such experiments are discussed.

Introduction

After more than a decade of relative inactivity, rare K decay is once again a flourishing sub-field of particle physics. Table I summarizes the current or recent experiments of this type, and their status. The first group are the dedicated rare decay experiments. These have been primarily performed at the Brookhaven AGS. Next are the two searches for direct CP-violation in $K^0 \to 2\pi$ discussed earlier at this conference. These experiments have also produced a number of rare decay results as by-products. The third group are experiments which have been approved but have not yet run.

The renaissance of this type of experiment is due largely to their physics reach, which is considerable and quite competitive with that of other modalities. However to be honest, a number of other factors have played roles in this, such as the opportunity to make very large advances in sensitivity at an early stage of the work, and the possibility of making such advances while working in relatively small groups. To illustrate the physics reach, one can compare the observable effects of a horizontal interaction[1](as might arise in attempts to explain the repetition of generations) in K decay (see Fig. 1a) with the effects of such an interaction in pp reactions at the SSC (Fig. 1b and c). Assuming a V-A interaction, the branching ratio of $K_L \to \mu e$ may be related to that of $K^+ \to \mu^+\nu$ via[1]

$$B(K_L \to \mu e) = B(K^+ \to \mu^+\nu) \times \tau_L/\tau_+ \times [\frac{M_W^2 g_H^2}{M_H^2 g_W^2 sin\theta_C}]^2 \quad (1)$$

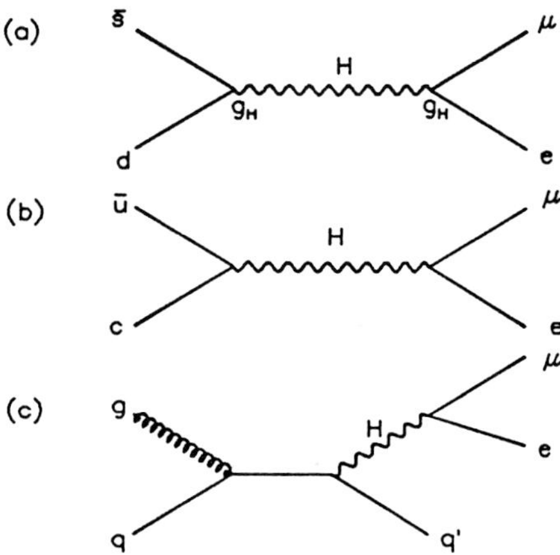

Figure 1: (a) $K_L \to \mu e$ via a horizontal gauge boson (HGB); (b) $\bar{u}c \to \mu e$ via an s-channel HGB; (c) Gluonic production of μe via an HGB.

For $g_H = g_W$, the current upper limit[2] of 2.2×10^{-10} on $B(K_L \to \mu e)$ corresponds to $M_H > 57$ TeV which is a formidably large scale. By contrast, for similar coupling strength, the effects observable at the SSC correspond to much smaller scales[3]. Fig. 2, taken from Ref. 3, gives the cross section of the processes shown in Fig. 1b and c for various assumptions on g_H. For $g_H = g_W$, assuming that 10^{-3}pb represents the limit of observability, masses beyond 10 TeVare not accessible. Thus in this regard, K decay can already probe beyond what will be possible at the SSC.

Table I: Rare Kaon Decay Experiments

EXPERIMENT	MODES	STATUS
KEK-137	$K_L \to \mu e,\ ee,\ \mu\mu$	continuing
AGS-780	$K_L \to \mu e,\ ee,\ \mu\mu,\ \pi^0 e^+ e^-, ...$	finished
AGS-845*	$K_L \to \pi^0 e^+ e^-,\ \gamma e^+ e^-,\ e^+ e^- e^+ e^-$	finished, analyzing
AGS-777	$K^+ \to \pi^+ e^- \mu^+,\ \pi^+ e^+ e^-, ...$	finished
AGS-851*	$K^+ \to \pi^+ e^+ e^-,\ \pi^0 \to e^+ e^-$	finished, analyzing
AGS-791	$K_L \to \mu e,\ ee,\ \mu\mu, ...$	continuing
AGS-787	$K^+ \to \pi^+ \nu\bar{\nu},\ \pi^+ \mu^+ \mu^-, ...$	continuing
NA-31†	$K_L \to \pi^0 e^+ e^-,\ \pi^0 \gamma\gamma, ...$	continuing
FNAL-731†	$K_L \to \pi^0 e^+ e^-,\ \pi^0 \gamma\gamma, ...$	finished, analyzing
FNAL-799*	$K_L \to \pi^0 e^+ e^-,\ \pi^0 \gamma\gamma,\ \gamma e^+ e^-, ...$	approved
KEK-162	$K_L \to \pi^0 e^+ e^-,\ \gamma e^+ e^-, ...$	approved

† Primarily CP-violation in $K^0 \to \pi\pi$ experiment.

⋆ This experiment is a 'descendent' of the experiment listed above it.

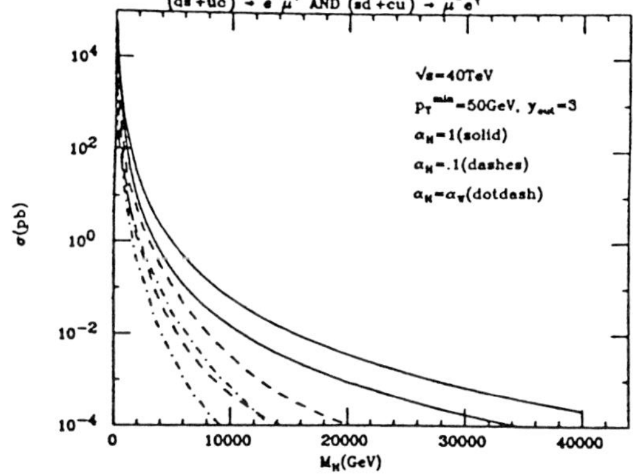

Figure 2: The cross section for HGB-mediated production of μe in a pp collider.

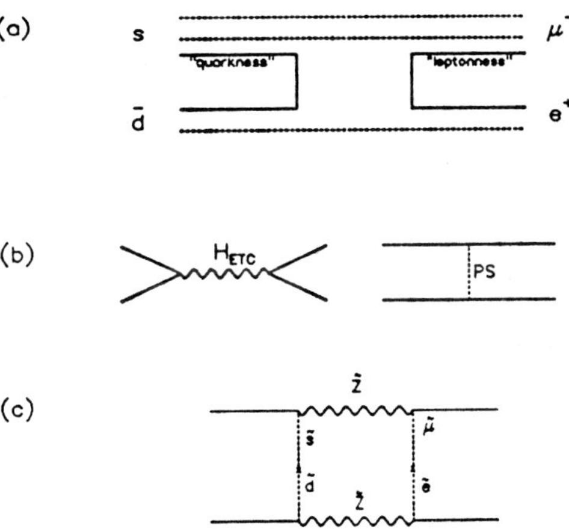

Figure 3: A selection of models which predict $K_L \to \mu e$: (a) constituent interchange, (b) technicolor, (c) supersymmetry.

Lepton flavor violation experiments

Lepton flavor-violating (LFV) decays such as $K_L \to \mu e$ are of particular interest because although they are strictly forbidden in the minimal Standard Model (SM), many of the proposed extensions to that Model predict such decays at accessible levels[4]. Fig. 3 shows a small selection of the large variety of proposed mechanisms for $K_L \to \mu e$. It is remarkable that many of these mechanisms were not introduced with the problem of generations in mind, yet when their consequences are explored, they turn out to predict LFV without any special tweaking.

The current generation of LFV experiments includes searches for both $K_L \to \mu e$ and $K^+ \to \pi\mu e$. Two of the former, KEK-137 and AGS-791, have new data. It's notable in this age of mega-experiments that the masthead of the KEK-137 preprint[5] submitted to this conference contains only 15 names. The apparatus built by this doughty group (KEK-Tokyo-Kyoto) is shown in Fig. 4. Typical of this type of experiment, it consists of a double arm drift-chamber spectrometer outfitted with e and μ identification devices.

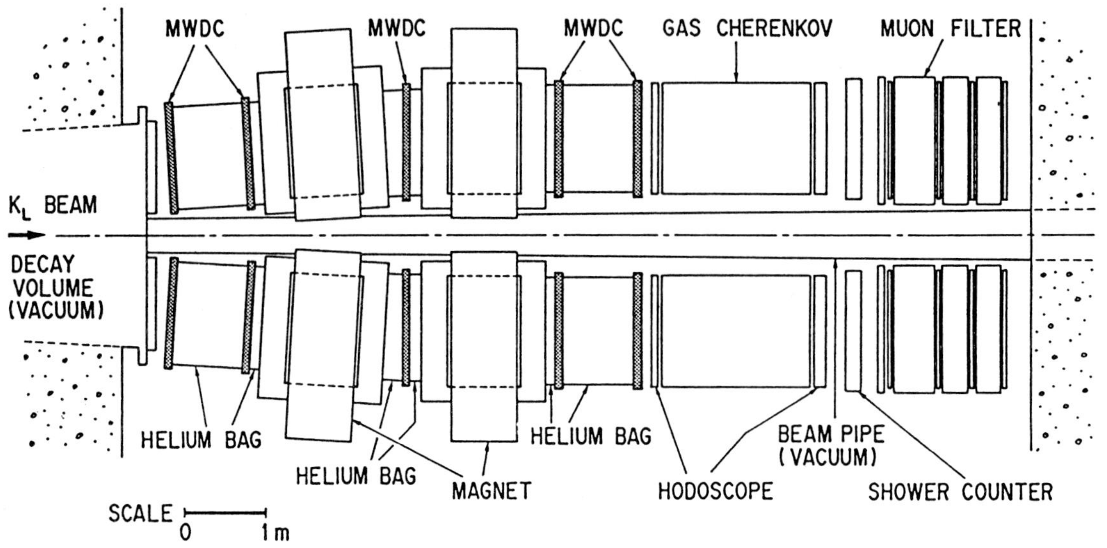

Figure 4: A plan view of KEK-137.

In principle the kinematically over-constrained, all-leptonic final states, $\mu\mu$, μe, and ee, are readily separable from the more copious K_L decay modes. However the sensitivity goal of the current round of searches ($< 10^{-10}$) is ambitious enough to move them into the technically challenging regime. For example, it is necessary to maintain extremely good resolution (e.g. $\sigma m_{\pi\pi} = 1.3\text{MeV}/c^2$ for calibration $K_L \to \pi\pi$ events) in the face of chamber rates in the tens of MHz. The neutral beam impinging on this apparatus contains some 15 million K_L and perhaps 10^9 neutrons/pulse. Roughly 7% of the K_L decay in a 10m vacuum decay tank. The acceptance of the spectrometer for 2-body K_L decays with $p_K \approx 2 - 8\text{GeV}/c$ is about 2%. Lead-scintillator plate shower counters and atmospheric Čerenkovs are used to identify electrons; an array of range-measuring hodoscopes serves to identify muons.

The logic of these experiments goes roughly as follows. CP-violating $K_L \to \pi^+\pi^-$ decays ($BR = 0.2\%$) provide spectrometer calibration and overall normalization. $K\ell 3$ decays are used as sources of leptons with which to determine the particle identification acceptance and rejection. Observation of the highly GIM-suppressed decay $K_L \to \mu^+\mu^-$, whose branching ratio is known to be $\lesssim 10^{-8}$, serves to demonstrate the overall ability of the experiment to detect rare decays of this general type. The measured value of this branching ratio is of considerable interest in itself, as will be discussed below. Another physics target is the related decay $K_L \to e^+e^-$. This is further suppressed relative to $K_L \to \mu\mu$ by helicity conservation to the *few* $\times 10^{-12}$ level, leaving a large window for the observation of possible new phenomena (intermediate scalars are of particular interest here).

Fig. 5 shows the latest $K_L \to \mu e, ee$, and $\mu\mu$ results from this experiment. In each case the effective 2-lepton mass is plotted against the square of the line-up angle. The latter quantity is the angle between the direction of the incoming K_L and that of the vector sum of the final state momenta. In the case of the μe and ee final states, there are no events within the fiducial region, leading to 90% c.l. upper limits of $B(K_L \to \mu e) < 4.3 \times 10^{-10}$ and $B(K_L \to ee) < 5.6 \times 10^{-10}$ respectively. These represent approximately an order of magnitude advances in sensitivity with respect to the Particle Data Book values[6]. In the case of $K_L \to \mu\mu$, there are 54 events in the fiducial region, corresponding to $B(K_L \to \mu\mu) = (8.4 \pm 1.1) \times 10^{-9}$. This is consistent with, but on the low side of the PDB value[6] for $B(K_L \to \mu\mu)$. All these results represent work in progress, and the experimenters expect to improve on them by significant factors by the time their final run (in 1990) is analyzed.

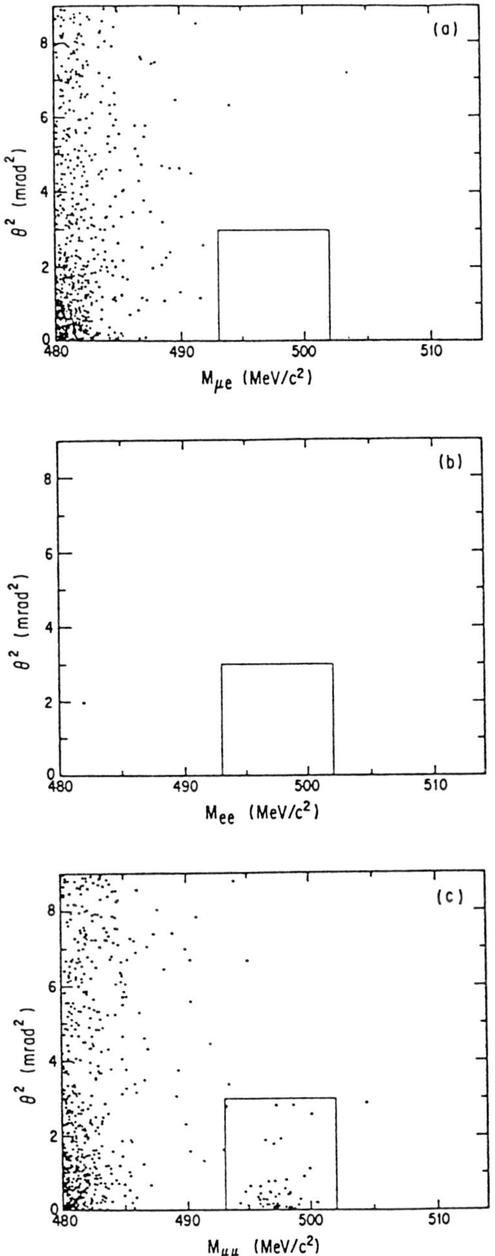

Figure 5: KEK-137 results:(a) μe effective mass vs. θ^2; (b) ee effective mass vs. θ^2; (c) $\mu\mu$ effective mass vs. θ^2. The rectangle in each figure represents the fiducial region.

The competition for KEK-137 is AGS-791, a UC Irvine, UCLA, LANL, U of Pennsylvania, Stanford, Temple, and William & Mary collaboration, whose apparatus is diagrammed in Fig. 6.

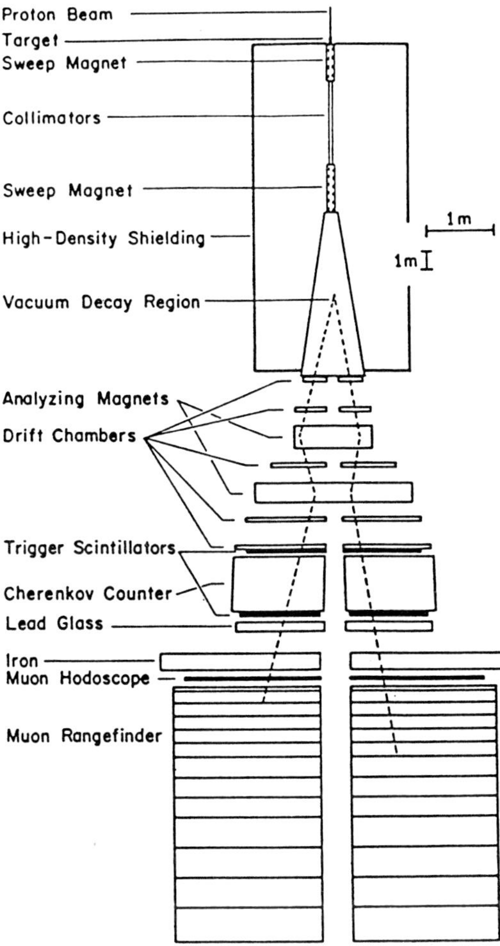

Figure 6: A plan view of AGS-791.

This experiment takes a similar rate of K_L/pulse to KEK-137, has a shorter decay region (in terms of K_L lifetimes), but a larger geometric acceptance. As in KEK-137, the AGS-791 detector is a double-arm, double-measuring drift chamber spectrometer. However in this case the two magnets, each of which spans both arms, have equal but opposite bending powers. This restores the original angles of the final state particles, a phenomenon of which use can be made in the trigger. Another difference is the use of lead glass rather than lead-scintillator shower counters. There are also differences driven by the disparity in beam energies: Čerenkov gases, muon identifier thickness, etc.

Fig. 7 shows the effective mass and lineup angle for calibration $K_L \to \pi^+\pi^-$ events. The

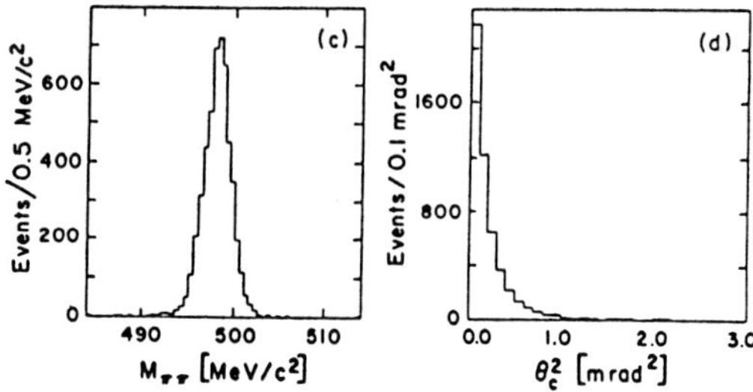

Figure 7: AGS-791 data: (a) $\pi\pi$ effective mass; (b) Square of the lineup angle.

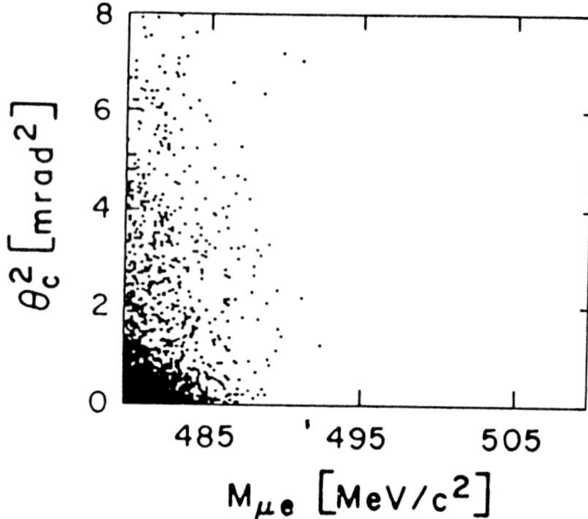

Figure 8: μe effective mass vs. θ^2 from AGS-791.

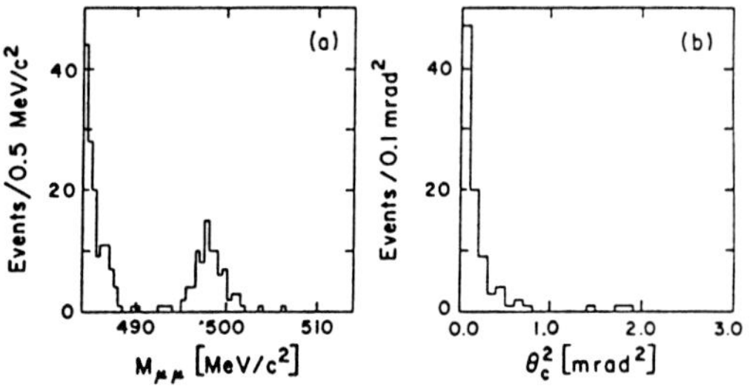

Figure 9: AGS-791 data: (a) $\mu\mu$ effective mass; (b) Square of the lineup angle.

resolutions in these quantities are excellent, as they must be in order to reject the $K_L \to \pi e\nu$ background that is the bane of this kind of experiment. Fig. 8 shows the two-body effective mass vs the square of the lineup angle for $K_L \to \mu e$ candidates[2]. As in the case of KEK-137, the signal region is devoid of events. The 90% c.l. upper limit derived from this is $B(K_L \to \mu e) < 2.2 \times 10^{-10}$. As mentioned above, this corresponds to a lower limit of 57 TeV on the mass of a horizontal gauge boson with couplings equal to g_W. In the case of $K_L \to ee$, the plot corresponding to Fig. 8 is completely empty. The branching ratio limit implied by this null result is $B(K_L \to ee) < 3.1 \times 10^{-10}$. It is notable that these new limits on the branching ratios of $K_L \to \mu e$ and $K_L \to ee$ represent 27,000-fold and 650-fold improvements respectively over the limits in force at the time the current round of K decay experiments was begun. Data sufficient to double or triple these sensitivities is already on tape. Running scheduled for the upcoming year is expected to yield a further factor of two or three.

$K_L \to \mu^+\mu^-$

Fig. 9 shows the effective mass and lineup angle of $K_L \to \mu^+\mu^-$ candidates from AGS-791[8]. A clear signal of some 87 events is seen, which corresponds to $B(K_L \to \mu^+\mu^-) = (5.8 \pm 0.6(stat.) \pm 0.4(syst.)) \times 10^{-9}$. This is a rather surprising result. It is not only significantly lower than previous results, it is lower (by a little more than one standard deviation) than the so-called "unitarity bound". The diagram responsible for this contribution to $K_L \to \mu\mu$ is shown in Fig. 10a.

The contribution of this diagram to the imaginary part of the amplitude for $K_L \to \mu\mu$ has been known for a long time[7]. This gives a lower limit for $B(K_L \to \mu\mu)$ in terms of the measured branching ratio for $K_L \to \gamma\gamma$.

$$B(K_L \to \mu\mu)|_{\gamma\gamma} = 1.195 \times 10^{-5} \times B(K_L \to \gamma\gamma) \qquad (2)$$

The first experiment[9] with sufficient sensitivity to test this limit found that it was apparently grossly violated! This set off both an explosion of theoretical effort[10] and a wave of rare K experiments[11]. The theorists were unable to find any significant contributions to the imaginary part of the amplitude that could interfere destructively with the 2γ piece. Of course any additional contributions to the *real* part of the amplitude could only make the discrepancy worse. It was left to the experimentalists to solve the mystery,

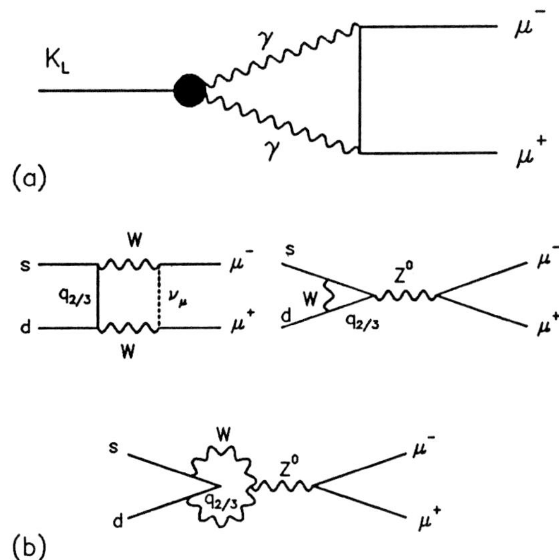

Figure 10: Contributions to $K_L \to \mu^+\mu^-$: (a) 2γ intermediate state; (b) bosonic loops.

by finding a total of some 27 events. This yielded $B(K_L \to \mu\mu) = (9.1 \pm 1.8) \times 10^{-9}$, a value safely above the unitarity limit ($\sim 6 \times 10^{-9}$). This emboldened some workers to attempt to extract constraints on Standard Model parameters from the excess of this branching ratio over the unitarity limit[12]. If this excess is identified with the short distance contribution to the branching ratio, one can use the theoretical expression[13]:

$$B(K_L \to \mu\mu)|_{sd} = B(K^+ \to \mu\nu)\frac{\tau_{K_L}}{\tau_{K^+}} \times \frac{\alpha^2}{4\pi^2 sin^4\theta_W} |\,\mathrm{Re} \sum_{j=c,t} V_{js}^* V_{jd} C(x_j)|^2 / V_{us}^2 \qquad (3a)$$

where

$$x_j \equiv m_j^2 / m_W^2 \qquad (3b)$$

and

$$C(x) \equiv \frac{3}{4}(\frac{x}{x-1})^2 \ln x + \frac{x}{4} - \frac{3}{4}\frac{x}{x-1} \qquad (3c)$$

Given the level of accuracy of the branching ratio measurements and the theoretical uncertainties, this did not really prove competitive with other bounds on the SM parameters. However with the advent of the more accurate measurements reported to this conference, one might want to reexamine this idea. Averaging the new and old data, yields $B(K_L \to \mu^+\mu^-) = (6.84 \pm 0.57) \times 10^{-9}$. Taking[6] $B(K_L \to \mu\mu)_{\gamma\gamma} = (6.81 \pm 0.27) \times 10^{-9}$ gives $B(K_L \to \mu\mu) - B(K_L \to \mu\mu)_{\gamma\gamma} < 0.8 \times 10^{-9}$ at 90% c.l. From this and Eq. (3) one obtains

$$|\,\mathrm{Re} \sum_{j=c,t} V_{js}^* V_{jd} C(x_j)|^2 / V_{us}^2 < 1.05 \times 10^{-5} \qquad (4)$$

This constraint couples K-M parameters and m_t, and can be used in a number of different ways. For purposes of illustration, assume m_t is large so that only one term is involved, and that $\mathrm{Re}[V_{ts}^* V_{td}]$ is at least 5×10^{-4}. One then would extract $m_t \lesssim 142\mathrm{GeV}$. Including the charmed quark term improves the limit to $\sim 132\mathrm{GeV}$. However even if we knew the exact value of $\mathrm{Re}[V_{td}^* V_{td}]$, limits on m_t obtained from Eq. (4) would be suspect. This is because there are potentially significant long-distance contributions to the real part of the amplitude for $K_L \to \mu\mu$, that could cancel the electroweak term. In fact there is at least one claim in the literature[14] that the relative signs of these contributions are indeed opposite. To get an idea of the size of such dispersive contributions, one can turn to the analogous decays of the η. Here the short distance contribution is negligible, and a comparison of $\Gamma(\eta \to \mu\mu)$ with the unitarity limit, gives[6] $|\,\mathrm{Re}\, A(\eta \to \mu\mu)/\,\mathrm{Im}\, A(\eta \to \mu\mu)| = 0.74^{+.28}_{-.51}$. Accommodating the possible influence of such a large fractional dispersive contribution raises the limit of Eq. (4) to $m_t \lesssim 280\mathrm{GeV}$. In order to reduce the uncertainty of this procedure, one would need both a more accurate measurement of $B(\eta \to \mu\mu)$, which is currently known to $\sim 30\%$, and also a reliable theoretical treatment of the long distance contributions to both $K_L \to \mu\mu$ and $\eta \to \mu\mu$. Neither of these seems to be forthcoming. There is, however, a much more direct way of obtaining equivalent information, with virtually no theoretical uncertainty. This is through measurements of the process $K^+ \to \pi^+\nu\bar{\nu}$, a subject which will be discussed below.

Lepton flavor violation experiments II

Complementing the $K_L \to \mu e$ results discussed above, there is also a new result on the closely related process $K^+ \to \pi^+\mu e$. Although three-body phase space and the fact that K^+ live only about a quarter as long as K_L make this

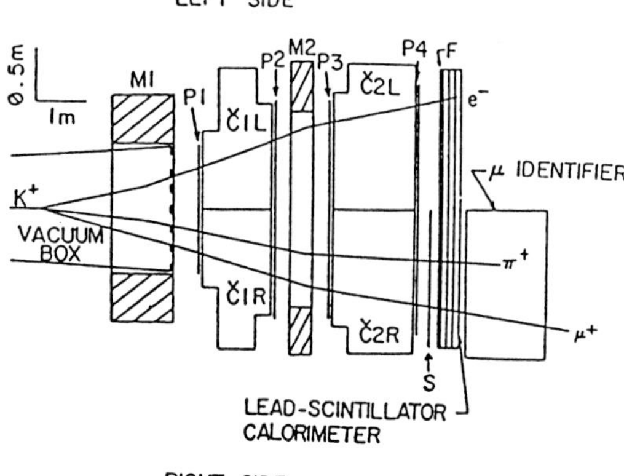

Figure 11: Plan view of AGS-777.

process intrinsically somewhat less sensitive than $K_L \to \mu e$ to generic LFV interactions, there are cases, such as that of a purely vector interaction, that cannot contribute to the K_L decay. Therefore it is essential to probe both processes. The apparatus of AGS-777, a BNL, FNAL, PSI, Washington, Yale collaboration, is shown in Fig. 11.

A 6 GeV/c positive beam containing $\sim 50M$ K^+ and roughly 20 times more π^+ and p per spill impinges on a vacuum decay tank. An upstream dipole separates the K^+ decay products by sign and kicks them out of the hot beam region. Their momenta are measured by an MWPC spectrometer and their identities are determined by two gas Č counters, a lead-scintillator shower counter, and a steel-plate-proportional tube muon range array. In order to exploit the triggering advantage implicit in demanding the appearance of an e^- in a K^+ beam, the apparatus was configured to detect only the $\pi^+\mu^+e^-$ charge combination. The most dangerous backgrounds for this process stem from daughter particle misidentification and/or decay in the processes $K^+ \to \pi^+\pi^+\pi^-$ and $K^+ \to \pi^+\pi^0$; $\pi^0 \to e^+e^-\gamma$. These processes are not all bad: they also serve to calibrate and normalize the experiment. The demands on kinematical reconstruction and particle identification power necessary to get down to the 10^{-10} level in this decay are considerable. Three-body effective mass resolution of $\sim$ 5MeV, and wrong particle rejection $\geq 10^6$ were achieved.

Fig. 12a shows the distribution of vertex miss distance versus three body effective mass for calibration $K^+ \to \pi^+\pi^+\pi^-$. The rectangular box shows the 3σ acceptance region for these events. Fig. 12b shows the corresponding distribution for for $K^+ \to \pi^+\mu^+e^-$ candidates. The increase in Q-value over $K^+ \to \pi^+\pi^+\pi^-$ mandates a larger signal region. There are no accepted events, allowing a 90% c.l. upper limit $B(K^+ \to \pi^+\mu^+e^-) < 2.1 \times 10^{-10}$ to be set[15]. This represents a 20-fold improvement over the results of previous experiments. This data can also be used to search for the LFV decay $\pi^0 \to \mu^+e^-$. Events consistent with the decay sequence $K^+ \to \pi^+\pi^0$; $\pi^0 \to \mu^+e^-$ are not found, allowing a 90% c.l. limit $B(\pi^0 \to \mu^+e^-) < 1.6 \times 10^{-8}$ to be set. This represents about a fourfold improvement of our knowledge of this process[16].

The successful prediction[17] of the rate of the GIM-suppressed decay $K^+ \to \pi^+e^+e^-$ by Gaillard and Lee was one of the early triumphs of the Standard Model. However long distance effects are important in this process and detailed predictions remain a challenge to theorists[18], so that more data is very welcome. AGS-777 has made a large contribution to the study of this decay, having increased the number of observed examples from tens[19] to hundreds of events [20]. They have also used this decay to search[21] for the process $K^+ \to \pi^+X^0$; $X^0 \to e^+e^-$ where X^0 is a new light particle (e.g. axion, Higgs).

The collaboration recently reoptimized their detector for $K^+ \to \pi^+e^+e^-$ and collected a sample of this decay that is expected to number in the thousands (AGS-851). This data can also be used to search for the process $\pi^0 \to e^+e^-$. One 'tags' the π^0 from the copious decay $K^+ \to \pi^+\pi^0$ (the so-called $K\pi2$ decay). Previous measurements of $B(\pi^0 \to e^+e^-)$ have been provocative and confusing. The earliest experiments[22] indicated a rate too high with respect to the unitarity contribution (the opposite of the situation in $K_L \to \mu\mu$) while later results[23] call this into question. A definitive experiment would be very welcome.

$K^+ \to \pi^+\nu\bar{\nu}$

Measurement of $K^+ \to \pi^+\nu\bar{\nu}$ offers an important test of the induced flavor-changing neutral current structure of the Standard Model. Fig. 13 shows the leading short distance contributions to this process. These are closely related to

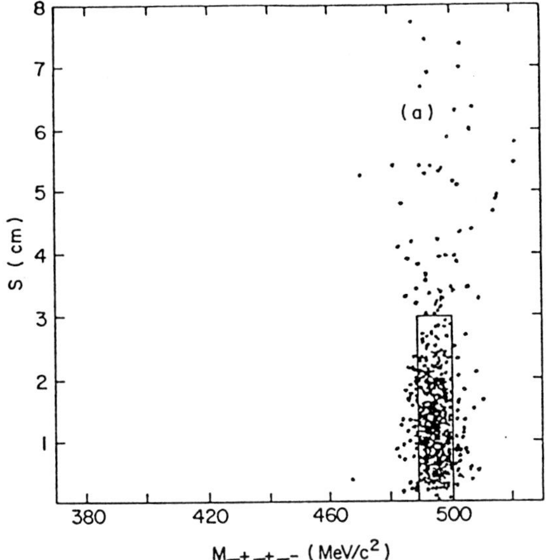

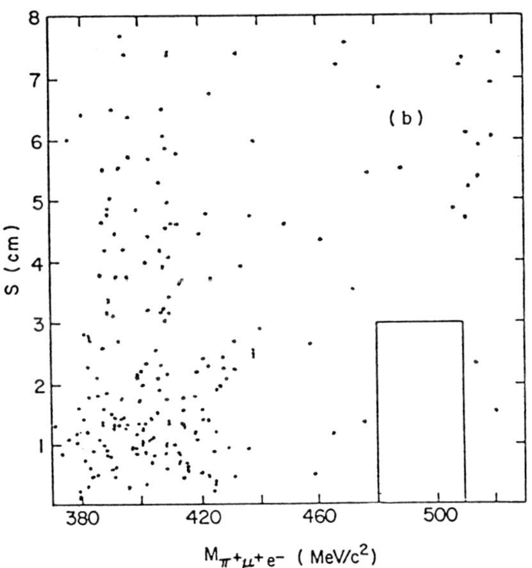

Figure 12: Vertex miss distance, S, versus effective mass for (a) $K^+ \to \pi^+\pi^+\pi^-$ and (b) $K^+ \to \pi^+\mu^+e^-$ candidates in AGS-777.

the diagrams of Fig. 10b, with internal and external lepton lines reversed. The resulting branching ratio (for a single flavor of neutrino) is given by[13]:

$$B(K^+ \to \pi^+\nu_i\bar{\nu}_i) = B(K^+ \to \pi^0 e^+ \nu)\times \frac{\alpha^2}{8\pi^2 sin^4\theta_W}|\sum_{j=c,t} V^*_{js}V_{jd}D(x_j)|^2/V^2_{us} \quad (5a)$$

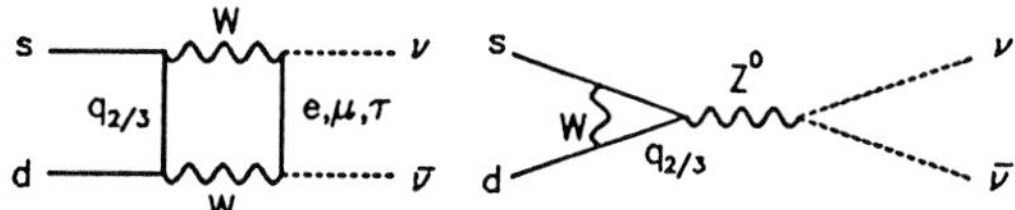
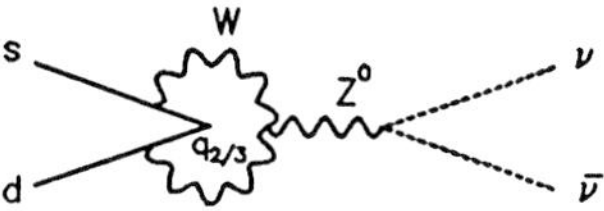

Figure 13: Short distance contributions to $K^+ \to \pi^+\nu\bar{\nu}$

where

$$D(x) \equiv \frac{1}{8}[1 + \frac{3}{(1-x)^2} - \frac{(4-x)^2}{(1-x)^2}]x\ln x + \frac{x}{4} - \frac{3}{4}\frac{x}{x-1} \quad (5b)$$

Long distance contributions to this decay are very small[4,24]. QCD corrections have been calculated by two different groups[25,26] who found them small and reasonably unambiguous, once the parameters of the theory are known. A rough idea of their influence on the size of $B(K^+ \to \pi^+\nu\bar{\nu})$ can be obtained by multiplying the charm term in Eq. (5) by 0.7. Estimates of the branching ratio using recent constraints on the KM angles and m_t[4,27] give a range of $\sim 1 - 7 \times 10^{-10}$ for three ν generations. It should be emphasized that this process offers the prospect of SM parameter constraints that are virtually free of 'hadronic engineering' and other theoretical uncertainties. For example, if m_t is known, an unambiguous value of $|V_{td}|$ can be extracted from such a measurement. However, since, as we will see, the experimental limit is not yet near the predicted range, there is a large window for possible new physics. Numerous candidates for this non-Standard physics have been proposed. These come in at least four categories:

- processes where the final state is indeed $\pi^+\nu\bar{\nu}$ but there are non-SM intermediate states, e.g. SUSYons replace their normal partners in the loops of Fig. 13
- the final state is $\pi^+\nu\nu'$, e.g. $\nu_e\bar{\nu}_\mu$. Thus virtually all LFV models will contribute here as in $K^+ \to \pi^+\mu e$

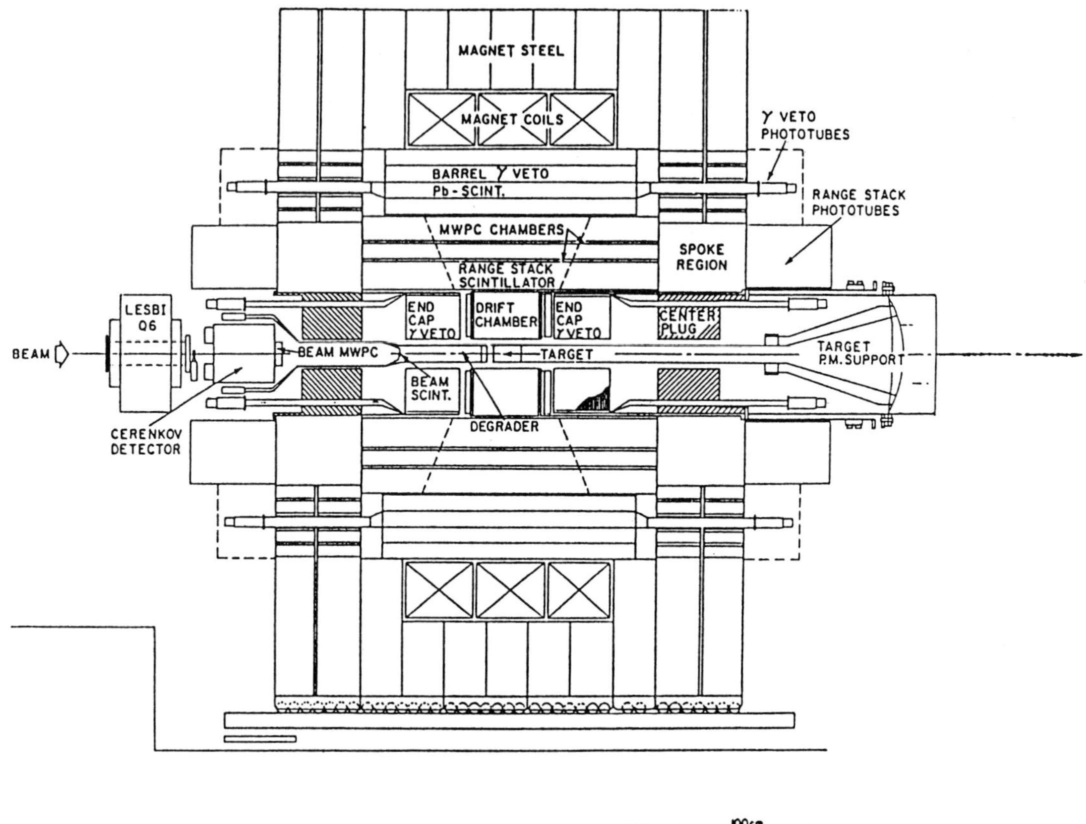

Figure 14: The AGS-787 detector

- the final state is $\pi^+ X^0 X^{0'}$ where $X^0 \neq \nu$. Candidates for X^0 include majorons[28], lightest supersymmetric particle[4], etc.
- the final state is actually $\pi^+ X^0$. This final state is experimentally distinguishable from $\pi^+ \nu\bar{\nu}$ and its observation at any level is *prima facie* evidence for physics beyond the Standard Model. There have been many candidates proposed for this over the years including axions[29], light Higgs, familons[30], hyperphotons[31], etc.

Previous to the current round of experiments, the measured upper limit on $B(K^+ \to \pi^+ \nu\bar{\nu})$ was[32] 1.4×10^{-7}, a number more than 100 times the SM prediction. This was obtained by the last in a series of small acceptance experiments[33] placed at the end of stopping K^+ beams. To make substantial progress in the exploration of the process $K^+ \to \pi^+ \nu\bar{\nu}$, with its complete lack of kinematic constraints, inimical backgrounds, and other experimental challenges, a large new initiative was required. This was provided by the AGS-787 collaboration[34] of BNL, Princeton, and TRIUMF.

Fig. 14 shows the large aperture solenoidal stopping K^+ detector built by this group. A beam of $\sim 10^6$ K^+ (and $\sim 2 \times 10^6$ π^+) impinges on a BeO cylinder which 'degrades' the K^+ momentum to $\sim 300\text{MeV}/c$ and delivers about 20% of the K^+ to a scintillating fiber stopping target. After a 2 nsec delay to assure c.m. kinematics, a gate opens for π^+ emerging transversely from the target. Fig. 15 shows an end view of such an event in in the stopping target. This target consists of 2269 2mm fibers grouped by sixes into triangles, each of which is read out by a 1 cm pmt. The momentum of pions is measured in a small cylindrical drift chamber upon which

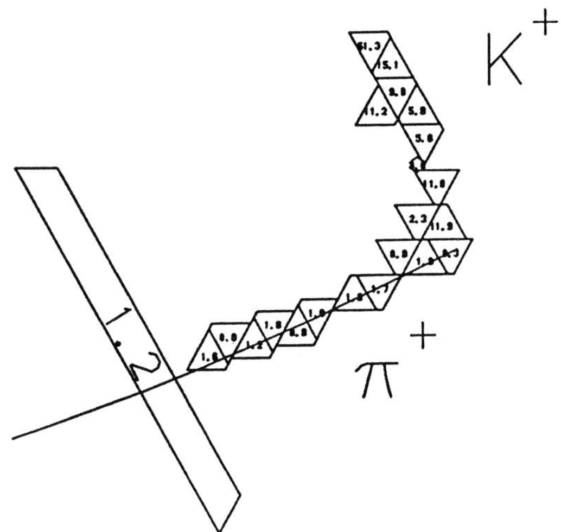

Figure 15: $K\pi2$ background event in the AGS-787 stopping target. Highly ionizing K^+ travels perpendicular to the plane of the picture. Lightly ionizing π^+ emerges transversely.

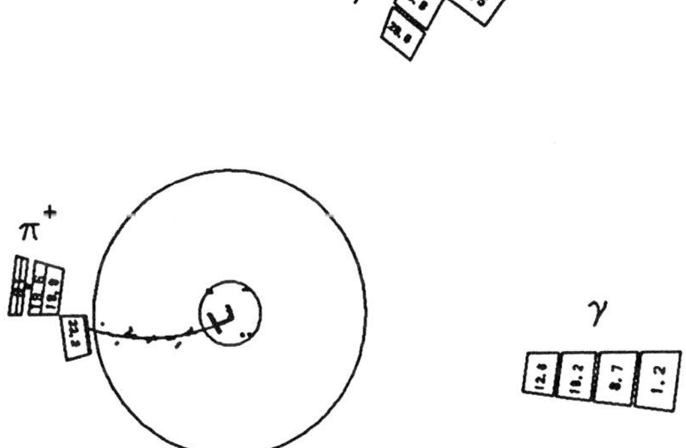

Figure 16: $K\pi2$ background event in the AGS-787 detector.

a 1 T field is imposed. The drift chamber surrounds the target and is itself surrounded by by a 24-sector cylindrical array of scintillator plates and PWCs referred to as the "range stack". Each sector consists of a 0.6cm triggering counter and 20 2cm thick scintillators interspersed with two layers of chambers. This serves to complete the measurement of the π^+ energy and range, and to record its decay sequence ($\pi \to \mu \to e$). Four layers of lead-scintillator shower counter (the "barrel veto") surround the range stack. Similar vetoes plug the upstream and downstream faces of the drift chamber, completing a nearly hermetic photon veto system ($\geq 15r.l.$ in almost all directions).

The rejection of π^0s from $K\pi2$ is roughly $10^6 : 1$. Fig. 16 shows a $K\pi2$ event in which both γs convert in the barrel veto. This device has sufficient resolution and granularity not only to serve as a veto but to reconstruct γs for physics analysis.

The most conspicuous backgrounds to $K^+ \to \pi^+\nu\bar{\nu}$ are the copious $K\pi2$ and $K\mu2$ ($K^+ \to \mu^+\nu$) decays. Each of these can be rejected by kinematics (the c.m. charged track momentum spectrum is a delta function). The ability of the detector to do so is illustrated by Fig. 17 which shows the resolution functions in momentum (p), kinetic energy (T), and range (R) for $K\pi2$ decays. A cut placed some 10MeV/c (in equivalent momentum for T and R) above the peak yields a rejection factor of several thousand to one. The kinematic rejection of $K\mu2$ is even better. $K\mu2$ can also be rejected via particle identification. The latter has two aspects, the first of which is kinematical - the relations between R, T, p, dE/dx, etc. are quite different for pions and muons in this energy range. Pions can also be distinguished from muons via their decay sequence $\pi \to \mu$ ($26nsec$) $\to e$ ($2.2\mu sec$), versus $\mu \to e$ ($2.2\mu sec$). Fig. 18 shows the pmt signals due to a pion in a sequence of range stack counters. The 4MeV muon has a range of only about 1mm in scintillator, so that all three pulses can be required to emanate from the same region of the counter. Custom transient digitizers were developed to record decay sequences in a deadtimeless fashion[35]. Other potential backgrounds, such as $K^+ \to \mu^+\nu\gamma$, are suppressed by the above methods and by their small branching ratios. In addition, only the high momentum end of the π^+ spectrum is used ($p > p_{K\pi2}$), a strategem which costs about a factor five in sensitivity, but which confines the search region to momenta above the physical range for π^+ from any significant K^+ branch.

Results of a two week engineering run are available at this time[36]. Fig. 19 displays the R versus T distribution for the final sample of $K^+ \to \pi^+\nu\bar{\nu}$ candidates. The rectangular box drawn in the figure indicates the fiducial region above the $K\pi2$ and below the $K\mu2$. In the absence of measuring errors, signal events would lie along a trajectory approximating the diagonal of the box. The residual events below and to the right of the box are $K\pi2$ events in which the π^0 has been missed. On this assumption, the π^0 inefficiency is $\sim 2 \times 10^{-6}$, a figure in reasonable agreement with Monte Carlo

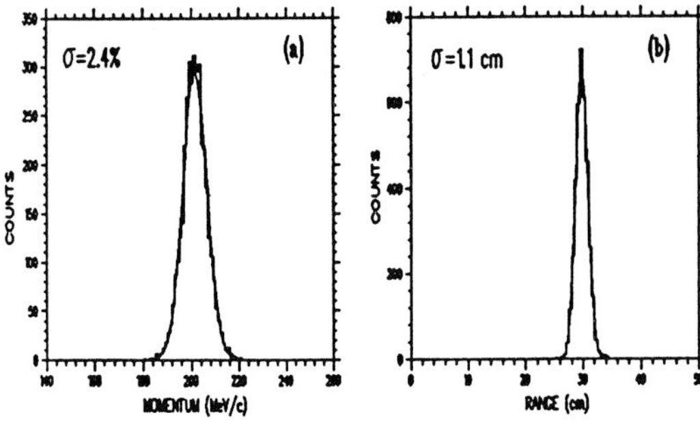

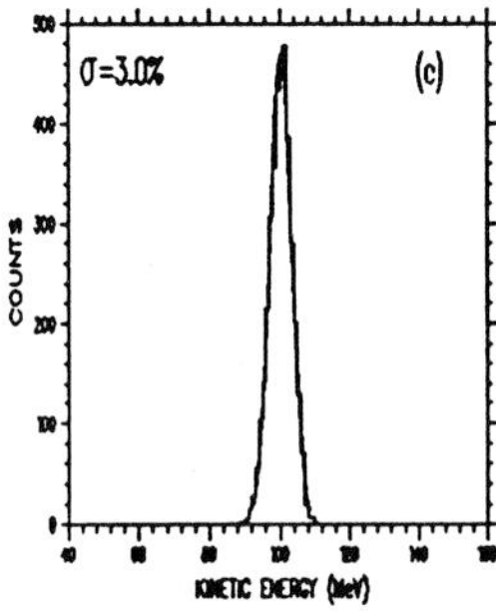

Figure 17: (a) Momentum, (b) range, and (c) kinetic energy of the π^+ from $K\pi 2$ decay.

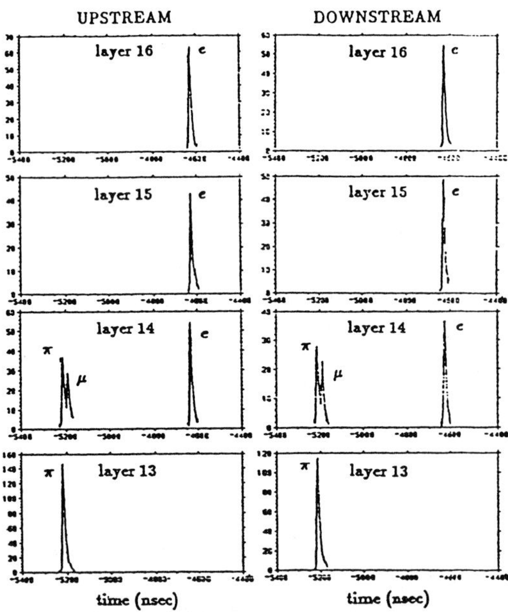

Figure 18: Digitized phototube pulses in a succession of range stack counters in AGS-787. Left and right columns are respectively upstream and downstream readouts.

estimates. The 90% c.l. upper limit extracted from this data is $B(K^+ \to \pi^+\nu\bar{\nu}) < 3.4 \times 10^{-8}$. The corresponding limit for a single, massless unseen recoil is considerably better, since in that case no penalty is paid for phase space cuts. One obtains $B(K^+ \to \pi^+X^0) < 6.4 \times 10^{-9}$. These represent approximately a fivefold improvement over previous data[32]. Fig. 20 shows the limits extracted for $K^+ \to \pi^+X^0$ when the assumption of $m_X = 0$ and $\tau_X = \infty$ are lifted. The dotted curve gives the limit for a very light Standard Model Higgs.

If, instead of attributing the residual $K\pi 2$ peak of Fig. 19 to γ veto inefficiency, it is attributed to the process $K^+ \to \pi^+\pi^0$; $\pi^0 \to X^0X^{0'}$, a limit on the branching ratio of π^0 to unseen particles can be obtained. Since this measurement is not statistics limited, it pays to tighten the γ veto cuts considerably beyond those used in the $K^+ \to \pi^+\nu\bar{\nu}$ analysis. The resulting 90% c.l. upper limit for this process[37], which includes $\pi^0 \to \nu\bar{\nu}$, is $B(\pi^0 \to X^0X^{0'}) < 7.7 \times 10^{-7}$. This represents a factor 30 improvement on the previous direct limit on this decay[38], and a factor 10 improvement over a recent indirect limit[39]. It is still a factor 400 short of the SM prediction for $\pi^0 \to \nu\bar{\nu}$ with $m_\nu = 35$MeV, however.

AGS-787 is also sensitive to a number of other K^+ decays. Analyses are in progress for $K^+ \to \pi^+\mu^+\mu^-$ and $K^+ \to \pi^+\gamma\gamma$. The former decay, closely related to the process $K^+ \to \pi^+e^+e^-$ discussed above, is expected[18] at the $few \times 10^{-8}$ level, but has never been observed. AGS-787 has three candidates for this decay[40]. Conservatively attributing these to background, a 90% c.l. upper limit of 2.3×10^{-7} can be put on this decay. This is an improvement of a factor 10 over previous data[41]. Note that several tens of $K^+ \to \pi^+\mu^+\mu^-$ events are expected to already be on tape as a result of the AGS-787 run earlier this year. Such events are also a hunting ground for light Higgs in the accessible mass range. For m_H significantly greater than $2m_\mu$, the 2μ branching ratio will completely dominate the $2e$. Fig. 21 shows the 90% c.l. upper limit attained for $K^+ \to \pi^+H$ as a function of m_H. According to recent theoretical work[42], this most probably rules out the possibility of a Standard Model Higgs in the mass range shown. Of course non-SM scalars are possible and experimenters will keep seeking them. However it will be increasingly difficult to do so in this mode, as the SM 'background' from $K^+ \to \pi^+\mu^+\mu^-$ comes into view. A cleaner environment could be found in the closely related process $K_L \to \pi^0\mu^+\mu^-$, where CP-conservation

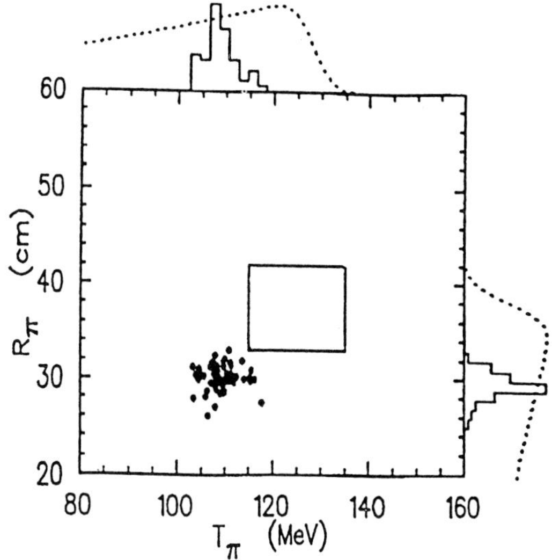

Figure 19: Range versus kinetic energy for residual $K^+ \to \pi^+\nu\bar{\nu}$ candidates in AGS-787. The box outlines the search region. The dotted curves on the projection axes show the expected spectrum.

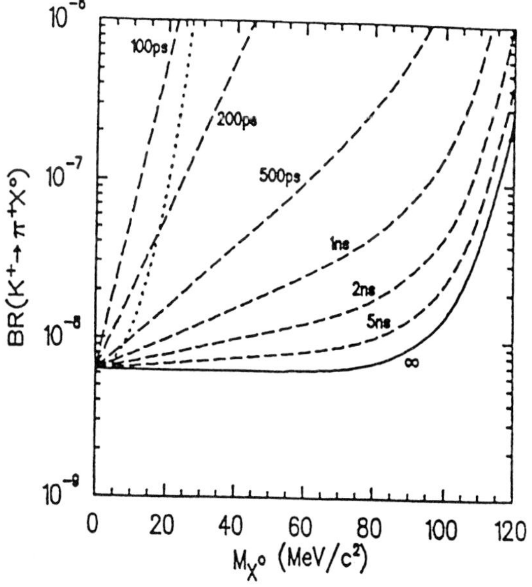

Figure 20: The solid curve is the 90% c.l. upper limit on $B(K^+ \to \pi^+ X^0)$ as a function of m_X. The dashed curves give the corresponding limits where X^0 has a finite lifetime. The dotted curve shows the limit on $B(K^+ \to \pi^+ H)$.

suppresses this 'background' by four orders of magnitude[43] (see below).

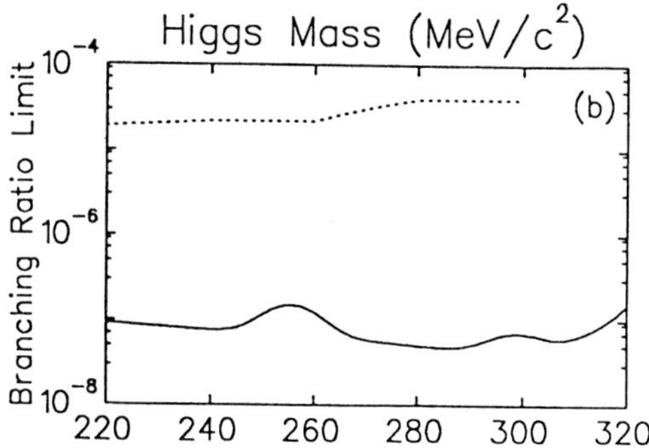

Figure 21: Solid line shows the 90% c.l. upper limit of the branching ratio for the decay $K^+ \to \pi^+ H, H \to \mu^+\mu^-$ as a function of m_H (AGS-787). Also shown (dashed line) is the results of an inclusive search for $K^+ \to \pi^+ X^0$ (Ref. 44).

AGS-787 uses the same three-body data set to search for the SM-allowed process $K^+ \to \mu^+\nu\mu^+\mu^-$, which is expected[45] at $\sim 3 \times 10^{-9}$. The 90% c.l. upper limit found, $B(K^+ \to \mu^+\nu\mu^+\mu^-) < 4.1 \times 10^{-7}$, is the first in the literature.

Finally, an analysis was done seeking $K^+ \to \pi^+\gamma\gamma$. This process is a proving ground for modern theoretical treatments[46,47] of the long distance effects in K decays[48]. Assuming a phase space matrix element (not theoretically fashionable), at 90% c.l., $B(K^+ \to \pi^+\gamma\gamma) < 10^{-6}$ was attained. This represents an eightfold improvement in sensitivity over previous data[49] which was analyzed under the same assumptions. The expected branching ratio is not much smaller[46,47], but it is concentrated in a region of phase space not accessed by this new data. When this is properly taken into account, the effective branching ratio probed is a factor ~ 1000 short of the theoretical estimate[50]. It may be possible for AGS-787 to access this region, which corresponds to relatively large $m_{\gamma\gamma}$, in future running.

As mentioned above, all AGS-787 results discussed here stem from a two week engineering run during 1988. In 1989, a run with ten times this sensitivity was completed. In 1990, yet another factor of three increment in exposure is expected, leading to a $K^+ \to \pi^+\nu\bar{\nu}$ branching ratio sensitivity $\sim 10^{-9}$ if all goes well. At that point, virtually all the 'window' for new physics in this decay will have been explored, and further advances (which the AGS-787 collaborators fully intend to pursue as the AGS Booster comes

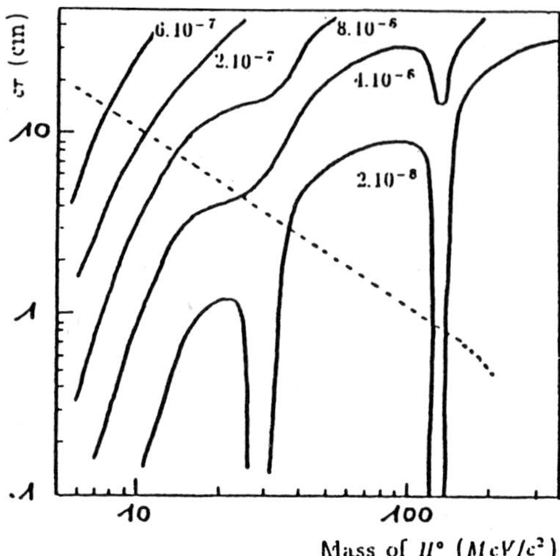

Figure 22: 90 % c.l. upper limit for $B(K_L \to \pi^0 X^0) \times B(X^0 \to e^+e^-)$ as a function of m_X and $c\tau_X$. The dotted line shows $c\tau_{Higgs}$ as a function of mass.

on line) will take us into the realm of detailed Standard Model measurement.

$K_L \to \pi^0 \ell\bar{\ell}$ and related processes

In the last few months the two CP-violation experiments that reported to this Conference on $K^0 \to 2\pi$ have each published data on the rare decay $K_L \to \pi^0 e^+ e^-$. The two results were very similar: at 90% c.l. NA-31[51] found $B(K_L \to \pi^0 e^+ e^-) < 4 \times 10^{-8}$ and FNAL-731[52], $B(K_L \to \pi^0 e^+ e^-) < 4.2 \times 10^{-8}$. The latter experiment also produced a limit on the corresponding K_S decay: $B(K_S \to \pi^0 e^+ e^-) < 4.5 \times 10^{-5}$. In the case of K_L, this represents about a two order of magnitude advance on the PDB level[6]. Since the SM prediction for this mode[53,54] is $\lesssim 10^{-10}$, a signal at anywhere near the current sensitivity would be a clear sign of new physics. The one exception to this is the possibility of observing the SM Higgs if it should be very light. The Higgs would be seen through the decay chain $K_L \to \pi^0 H$; $H \to e^+ e^-$. The extraction of this limit from a $K_L \to \pi^0 e^+ e^-$ exposure is not trivial, since for these very light masses the Higgs has a non-negligible lifetime. Nonetheless, NA-31 was able to produce such a limit[55], shown in Fig. 22. This is sufficient to rule out the existence of the Higgs[56] in the mass range 15 – 200MeV, barring an extremely fortuitous cancelation of amplitudes. The pursuit of non-Standard scalars in this process will no doubt continue as there is at least a two order of magnitude window before continuum 'background' becomes an impediment.

This SM 'background' is so small because in addition to the GIM suppression due a neutral flavor-changing decay, in leading order $K_L \to \pi^0 e^+ e^-$ violates CP. The diagrams governing the short distance contribution to $K_L \to \pi^0 \ell^+ \ell^-$ are essentially[57] those of Fig. 10b, with the addition of the electromagnetic penguin of Fig. 23a. The latter contribution makes the Standard Model calculation[53] less straightforward and unambiguous than those of $K_L \to \mu^+\mu^-$ or $K^+ \to \pi^+ \nu\bar{\nu}$. There is also an 'indirect' CP-violating contribution arising from the small admixture of K_1 in K_L. This is difficult to calculate *ab initio*, but can readily be determined once the (CP-conserving) process $K_S \to \pi^0 e^+ e^-$ has been measured. An estimate can be obtained by equating $\Gamma(K_S \to \pi^0 e^+ e^-)$ to $\Gamma(K^+ \to \pi^+ e^+ e^-)$ (a measured number[19]). Under this assumption, the branching ratio from indirect CP-violation alone can be determined[58]:

$$\begin{aligned} B(K_L \to \pi^0 e^+ e^-)|_\epsilon &= |\epsilon|^2 \frac{\tau_{K_L}}{\tau_{K^+}} \times \\ & B(K^+ \to \pi^+ e^+ e^-) \qquad (6) \\ &= 0.6 \times 10^{-11} \end{aligned}$$

This is comparable to the expected magnitude of the direct CP-violating contribution discussed above. In addition there is a long distance contribution mediated by the process $K_L \to \pi^0 \gamma\gamma$ (Fig. 23b). This is analogous to the contribution of Fig. 10a to $K_L \to \mu^+\mu^-$. Unlike the other terms, this contribution is CP-conserving. Until recently the size of this contribution was controversial, with some workers declaring it negligible with respect to the CP-violating terms[46,59], and others asserting the two to be at least comparable[60,61]. It now appears that the CP-conserving term cannot be ignored[62].

As intimated above, the immediate prospects for this decay are the search for non-SM physics including new scalars, and novel CP-violating currents[63,64]. Data taken earlier this year by a dedicated BNL experiment[65], AGS-845, should allow $B(K_L \to \pi^0 e^+ e^-)$ to be probed to at least 10^{-9}. There are also approved experiments at KEK[66](KEK-162) and Fermilab[67](FNAL-799), which promise to continue the search down past 10^{-10}, to the point where the SM contribution is expected to make its appearance.

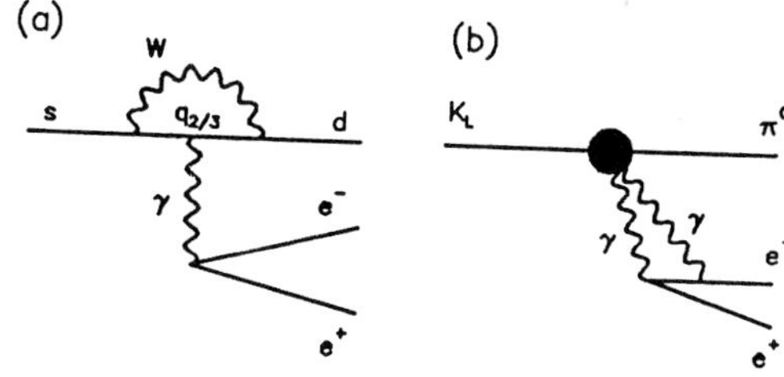

Figure 23: Additional contributions to $K_L \to \pi^0 e^+ e^-$: (a) electromagnetic Penguin, (b) two-photon intermediate state.

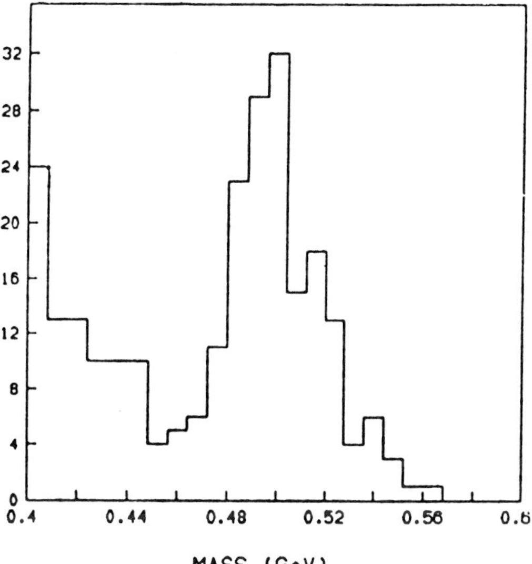

Figure 24: Preliminary data based on 7% of AGS-845 exposure. The three body effective mass is plotted for $ee\gamma$ candidates with lineup angle $\lesssim 4.5mr$.

These experiments are also likely to revolutionize our knowledge of the other decay modes such as $K_L \to e^+e^-\gamma$ and $K_L \to e^+e^-e^+e^-$. Fig. 24 shows the distribution of $m_{ee\gamma}$ for $K_L \to e^+e^-\gamma$ candidates from a small fraction of the data of AGS-845. There is a clear signal of ~ 150 $K_L \to e^+e^-\gamma$ events. The PDB value for this process is based on only 4 published events[68]. Thus these experiments will yield many benefits, but the detailed use of $K_L \to \pi^0\ell^+\ell^-$ to probe CP-violation awaits yet a further generation of experiments.

The incentive for such a generation, already considerable, seems to be growing even stronger. We have had for some time the NA-31 result[69] of a significant finite value for ϵ'/ϵ. As the first positive indication of a CP-violating effect that cannot be explained by the Superweak[70] model, if confirmed, this would certainly stand as one of the most important experimental results of the decade. The size of ϵ'/ϵ given by NA-31 is certainly within the allowed range of SM (Kobayashi-Maskawa[71]) CP-violation. However, at this Conference[72] we have been told that the FNAL-731 result, while statistically marginally inconsistent with that of NA-31, is quite consistent with $\epsilon'/\epsilon = 0$ (i.e. the Superweak value). What is more, recent theoretical work suggests[73] that for very large m_t the predicted SM value for ϵ'/ϵ might be smaller than previously suspected, perhaps too small for experiment to establish in the foreseeable future. Thus if the FNAL result[72] is confirmed and evidence arises that m_t is significantly larger than current limits, experiments on the $K^0 \to 2\pi$ system, into which so much recent effort has gone, might be demoted to the status of 'long-shots', wherein one could hope to find only non-SM sources of CP-violation. Other experiments in this category are the search for T-violating polarization in $K\mu3$, energy asymmetry in $K^0 \to \pi^+\pi^-\gamma$ and $Ke4$, etc. The same perceptual shift would tend to promote experiments on the $K_L \to \pi^0\ell\bar{\ell}$ system in importance. Although the predicted branching ratios are very small (typically 10^{-11}), once the requisite sensitivity level is reached, the *relative* direct CP-violating effects are extremely large. For example, in $K^0 \to 2\pi$ the ratio of direct/indirect CP-violation is $\sim 1/300$, whereas in $K_L \to \pi^0 e^+e^-$, this ratio is of order 1. What is more, many of the issues that go under the sobriquet "hadronic engineering" and which bedevil the interpretation of the $K^0 \to 2\pi$ experiments[74,75], are effectively absent in the latter case.

The presence of the CP-conserving 2γ contribution does complicate the analysis of $K_L \to \pi^0e^+e^-$, making precise determinations of CP-violating amplitudes dependent on acquiring reasonable statistics over a wide range of τ_K[76], but the job of merely establishing that there is *some* direct-CP violation is much simpler, and will be well worth the effort required, if further work on $K^0 \to 2\pi$ indicates that the first direct CP-violating signal is still to be found. Even if the finite ϵ'/ϵ is firmly established, the $K_L \to \pi^0e^+e^-$ measurement would still be worth pursuing, because of the difficulty of calculating ϵ'/ϵ in terms

Table II: Recent Progress in Rare Kaon Decay

Mode	result	Gain wrt previous data	(Gain with data now on tape)	Comments
$K_L \to \mu e$	$< 2.2 \times 10^{-10}$	27000	(81000)	$M_H > 57\ TeV$ c.f. 10 TeV for SSC
$K^+ \to \pi^+ \mu e$	$< 2.1 \times 10^{-10}$	23		complementary to $K_L \to \mu e$
$K^+ \to \pi^+ X^0$	$< 6.4 \times 10^{-9}$	6	(60)	new particle search; future: find/rule out familon
$K^+ \to \pi^+ \nu\bar{\nu}$	$< 3.4 \times 10^{-8}$	5	(50)	constrains new physics; future: KM angles, m_T
$K_L \to \pi^0 ee$	$< 4 \times 10^{-8}$	58	($\gtrsim 230$)	search for new scalars, non-S.M. CP-violation; future: S.M. CP-violation
$K_L \to e^+e^-$	$< 3.1 \times 10^{-10}$	600	(1800)	new scalar interactions?
$K^+ \to \pi^+ ee$	400 events	10	(50)	search for new scalars, S.M. study
$K^+ \to \pi^+ \mu\mu$	$< 2.3 \times 10^{-7}$	10	(100)	should be discovered in the current data sample; S.M. study
$K^+ \to \mu^+ \nu\mu\mu$	$< 4.1 \times 10^{-7}$	2.4M	(24M)	Higgs hunting ground; no previous limit
$K_L \to \mu^+\mu^-$	140 events	5	(12)	provocative result (BR seems low)
$K^+ \to \pi^+ \gamma\gamma$	$< 10^{-6}$	8	(20)	Long distance effects
$K_L \to \pi^0 \gamma\gamma$	$< 2.7 \times 10^{-6}$	90	(?)	Important for CP-violation in $K_L \to \pi^0 e^+ e^-$
$K_L \to e^+e^-\gamma$	~ 150 events	~ 40	(500)	Long distance effects
$\pi^0 \to \mu^+ e^-$	$< 1.6 \times 10^{-8}$	4		LFV in π decay
$\pi^0 \to \nu\bar{\nu}$	$< 10^{-6}$	8	($\gtrsim 20$)	Also search for new light particles

of the fundamental SM parameters and because a number of measurements in different systems will still be needed to in order to pin down the origin of CP-violation.

Experimentally the $K_L \to \pi^0 e^+ e^-$ signal is rather good, and to the sensitivity probed so far, it has proved rather clean. To reach the level needed for CP-violation studies, however, advances in experimental technique, beam design, and better sources of K^0s will probably be needed. As mentioned above, a measurement of $K_S \to \pi^0 e^+ e^-$ some five orders of magnitude beyond the current sensitivity[52] will be needed. Measurements in the $K_S - K_L$ interference region would also be highly desirable[76]. In addition a measurement of $B(K_L \to \pi^0\gamma\gamma)$ would be very useful in determining the size of the CP-conserving contribution to $K_L \to \pi^0 e^+ e^-$. Experiments are closing in on this process, which is predicted to be $\lesssim 10^{-6}$ in branching ratio[46,60,7-]. FNAL-731 has recently improved the sensitivity to this decay by almost two orders of magnitude[7-]. NA-31 is also potentially sensitive to this process, and by the time $\sim 10^{-11}$ is reached in $K_L \to \pi^0 e^+ e^-$, $K_L \to \pi^0\gamma\gamma$ should be well measured.

There are two processes related to $K_L \to \pi^0 e^+ e^-$ that are intrinsically of equal or even greater interest to the question of CP-violation, but which have experimental disadvantages with respect to the latter process. The first, $K_L \to \pi^0 \mu^+ \mu^-$, is also of particular interest to the search for new scalars. Although SM Higgs seem to have been ruled out in the accessible mass range, any new scalar that shares the Higgs' propensity to couple to mass will prefer the $\mu\mu$ channel to the ee. The SM branching ratio for $K_L \to \pi^0 \mu^+ \mu^-$ is predicted[46,60] to be a few times smaller than that of $K_L \to \pi^0 e^+ e^-$. The expected contributions, direct and indirect CP-violation, 2γ intermediate state, etc., are parallel to those of $K_L \to \pi^0 e^+ e^-$, but the differences are sufficient that valuable insights could be gained through experiments in this channel. Recent experiments don't seem to have made use of their potential sensitivity to this channel, so that the branching ratio upper limit stands at the decade-old value[68] of 1.2×10^{-6}.

Potentially even more interesting, but experimentally much more challenging is the all-neutral version[79] of $K_L \to \pi^0 \ell\bar{\ell}$, $K_L \to \pi^0 \nu\bar{\nu}$. The absence of charged daughters rules out contributions such as those of Fig. 23, leaving to a very good approximation only the one-loop contributions of Fig. 13. It is remarkable that for large m_t the imaginary part[80] of the amplitude given by these diagrams is O(10%) of the real part. As a result, the indirect CP-violating contribution to $K_L \to \pi^0 \nu\bar{\nu}$ is suppressed to $\sim 10^{-3}$ of the direct (versus $\sim 1:1$ for $K_L \to \pi^0 e^+ e^-$ and $\sim 300:1$ for $K^0 \to 2\pi$). QCD corrections to this process are also very small[27], so that a measurement would give an unambiguous determination of the CP-violating product of KM angles, $s_2 s_3 s_\delta$, once m_t is known[81]. One could probably make the case that a measurement of $B(K_L \to \pi^0 \nu\bar{\nu})$ would be the single most useful measurement in the entire field of CP-violation. One can certainly make the case that at the requisite 10^{-11} level, it is one of the most difficult. No experiment has purposely sought this decay, but limits in the range $\lesssim 10^{-2}$ or a little lower can be extracted[79] from old 'single-arm' $K^0 \to \pi^0 \pi^0$ experiments. The next generation of $K_L \to \pi^0 e^+ e^-$ experiments will probably be able to reach $\sim 10^{-8}$ in the neutral mode, if they choose to look for it. Further progress will then depend on the problems they discover.

Summary and prospects

Table II summarizes the impressive progress made in the present round of rare K decay experiments. In some cases our knowledge has improved by more than four orders of magnitude. The mass scales being probed by LFV searches are already far beyond what can be accessed at the SSC for similar interactions. Data taken within a year should push the mass reach of such K decay experiments to nearly 100 TeV.

The other important area of progress has been in one-loop processes ($K_L \to \mu\mu, K^+ \to \pi^+ \nu\bar{\nu}, K_L \to \pi^0 ee$, etc.). Plans are in place to continue the search for these processes until the window for new physics above the predicted SM level has been eliminated (or new phenomena discovered). This should be largely accomplished over the next two or three years by extant and proposed experiments at the present facilities. There is much enthusiasm for pushing these searches well into the SM regime. To complete such a program will probably require upgraded facilities. The Booster presently under construction at the AGS should increase the K flux available at that accelerator by a factor four or more. A further factor two or three would be provided by the proposed Stretcher for the AGS ring. The proposed new Main Injector at FNAL would allow comparable opportunities if fully exploited for K physics. Beyond these, the TRIUMF KAON proposal promises yet a further significant increase in K flux.

Continuing a vigorous program should allow unambiguous determinations of currently elusive quantities like the modulus and phase of V_{td}, and keep open the possibility of profound surprises like those K decay has provided in the past.

Acknowledgements

I wish to thank G. Barr, R. Cousins, S. Dawson, F. Gilman, J. Hagelin, T. Inagaki, Y. Kuno, P. Meyers, W. Marciano, W. Molzon, W. Morse, D. Pitman, N. Sasao, Y. Wah, B. Winstein, and M. Zeller for fruitful conversations, for providing data in advance of publication, and for other assistance with this paper. This work was supported by the U.S. Department of Energy under contract DE-AC02-76CH00016.

References

1. R.N. Cahn and H. Harari, Nucl. Phys. B176, 135 (1980).
2. C. Mathiazhagan et al., Phys. Rev. Lett. 63, 2181 (1989).
3. N.G. Deshpande et al., *Proceedings of the 1986 Summer Study on the Physics of the Superconducting Supercollider*, p. 250; C. Albright et al., *Proceedings of the 1984 Snowmass Workshop*, p.144.
4. J.S. Hagelin and L.S. Littenberg, Prog. Part. Nucl. Phys. 23, 1 (1989) Progress in Nuclear and Particle Physics, September 1989.
5. T. Inagaki et al., Phys. Rev. D40, 1712 (1989).
6. G.P. Yost et al., Phys. Lett. B204, 1 (1988).
7. L.M. Sehgal, Phys. Rev. 183, 1511 (1969); B.R. Martin, E. de Rafael, and J. Smith, Phys. Rev. D2, 179 (1970).
8. C. Mathiazhagen et al., Phys. Rev. Lett. 63, 2185 (1989).
9. A. Clark et al., Phys. Rev. Lett. 26, 1667 (1971).
10. H. Stern and M. Gaillard, Ann. Phys.(N.Y.) 76, 580 (1973).
11. W.C. Carithers et al., Phys. Rev. Lett. 30, 1336 (1973);W.C. Carithers et al., Phys. Rev. Lett. 31, 1025 (1973); Y. Fukushimia et al., Phys. Rev. Lett. 36, 348 (1976); M.J. Shochet et al., Phys. Rev. D40, 1712 (1989).
12. See e.g. R.E. Shrock and M.B. Voloshin, Phys. Lett. 87B, 375 (1979).
13. T. Inami and C.S. Lim, Prog. Theor. Phys. 65, 297 (1981).
14. V. Barger et al., Phys. Rev. D25, 1860 (1982).
15. A. Lee, et al., submitted to Phys. Rev. Lett.
16. D. Bryman, Phys. Rev. D26, 2538 (1982); P. Herczeg and C.M. Hoffman, Phys. Rev. D29, 1954 (1984).
17. M.K. Gaillard and B.W. Lee, Phys. Rev. D10, 897 (1974).
18. G. Ecker, A. Pick and E. deRafael, Nucl. Phys. B291, 692 (1987).
19. P. Bloch et al., Phys. Lett. 56B, 201 (1975).
20. M.E. Zeller,*Rare Decay Symposium*, Vancouver Canada, p. 138 (1988).
21. N.J. Baker, et al., Phys. Rev. Lett. 59, 2832 (1987).
22. J. Fischer, et al., Phys. Lett. 73B, 364 (1978); J.S. Frank, et al., Phys. Rev. D28, 423 (1983).
23. R. Engfer, reported in *Third Conference on the Intersections Between Particle and Nuclear Physics*, Rockport, Maine (1988).
24. L.M. Sehgal, Phys. Rev., D39, 3325 (1989).
25. J. Ellis and J.S. Hagelin,Nucl. Phys. B217, 189 (1983).
26. C.O. Dib,I. Dunietz, and F.J. Gilman, SLAC preprint, SLAC-PUB-4818, December 1988.
27. C.O. Dib,I. Dunietz, and F.J. Gilman, SLAC preprint, SLAC-PUB-4840, March 1989.
28. S. Bertolini and A. Santamaria, Nucl. Phys. B315, 558 (1989).
29. S. Weinberg, Phys. Rev. Lett. 40, 223 (1978); F. Wilczek, Phys. Rev. Lett. 40, 279 (1978).
30. F. Wilczek, Phys. Rev. Lett. 49, 1549 (1982).
31. E. Fischbach, et al., Phys. Rev. Lett. 56, 3 (1986); S.H. Aronson, et al., Phys. Rev. Lett. 56, 1342 (1986).
32. Y. Asano, et al., Phys. Lett. 107B, 159 (1981).
33. J.H. Klems, R.H. Hildebrand, and R. Stiening, Phys. Rev. D4, 66 (1971); G.D. Cable, et al.,Phys. Rev. D8 3807 (1973).
34. I.-H. Chiang, et al., "A Study of the Decay $K^+ \to \pi^+\nu\bar{\nu}$" BNL, Princeton, TRIUMF Proposal to the AGS (1983). Later joined by LASL.
35. M. Atiya, et al., Nucl. Instr. Meth. A279, 180 (1989).
36. M. Atiya, et al., BNL preprint # 43468, submitted to Phys. Rev. Lett.
37. P.D. Meyers, Moriond talk, March 1989.
38. P. Herczeg and C.M. Hoffman, Phys. Lett. 100B, 347 (1988). These authors extracted a limit from the data of Ref 33.

39. C.M. Hoffman, Phys. Lett. 208B, 149 (1988).

40. M.S. Atiya, et al., Phys. Rev. Lett. 63, 2177 (1989).

41. V. Bisi, et al., Phys. Lett. 25B, 572 (1967).

42. H.-Y. Cheng and H.-L. Yu, Taiwan Institute of Physics Preprint IP-ASTP-02-89-R, May 1989.

43. For the process $K \to \pi X$; $X \to \ell^+\ell^-$ where X is a new *pseudoscalar*, this CP-suppression turns into a relative disadvantage since $K_L \to \pi^0$ + *pseudoscalar* is also CP-violating.

44. T. Yamazaki, et al., Phys. Rev. Lett. 52, 1089 (1984).

45. S. Dawson, Phys. Lett. B222, 143 (1989).

46. G. Ecker, A. Pich, and E. de Rafael, Nucl. Phys. B303, 665 (1988).

47. H.-Y. Cheng, BNL preprint, October 1989.

48. K decay continues to be a fertile field for the development of methods for dealing with low energy hadronic phenomena, e.g. current algebra and more recently chiral perturbation theory.

49. Y. Asano, et al., Phys. Lett. 113B, 195 (1982).

50. Y. Kuno, private communication.

51. G.D. Barr et al., Phys. Lett. B214, 303 (1988).

52. L.K. Gibbons et al., Phys. Rev. Lett. 61, 2661 (1988).

53. F.J. Gilman, and M.B. Wise, Phys. Rev. D21, 3150 (1980).

54. C.O. Dib,I. Dunietz, and F.J. Gilman, Phys. Lett., B218, 487 (1989); C.O. Dib,I. Dunietz, and F.J. Gilman, SLAC preprint, SLAC-PUB-4818, December 1988.

55. E. Auge, LAL89-26, August 1989.

56. J.F. Gunion, et al.,*The Higgs Hunter's Guide*, BNL-41644, June 1989.

57. Although basically the same one-loop diagrams are involved, in the case of $K_L \to \mu\mu$ the real part of the corresponding amplitude is relevant, whereas in $K_L \to \pi^0\ell\bar{\ell}$ it is the imaginary part.

58. This is the branching ratio from indirect CP-violation in the event that there are no other contributions. This construct is a bit artificial since in reality the indirect contribution will interfere with the others.

59. J.F. Donohue,B.R. Holstein, and G. Valencia, Phys. Rev., D35, 2769 (1987).

60. L.M. Sehgal, Phys. Rev., D38, 808 (1988).

61. J.M. Flynn and L. Randall, Phys. Lett. 216, 221 (1988).

62. J.F. Donohue, private communication.

63. X.-G. He, B.H.J. McKellar, and N.E. Tupper, University of Melbourne preprint UM-P-88/40 (1988).

64. L.J. Hall and L.J. Randall Nucl. Phys. B274, 157 (1986).

65. R.K. Adair, et al., "A Search for the Rare Decay $K^0 \to \pi^0 e^+ e^-$" BNL, Yale, Vassar proposal, Jan 1988.

66. K. Miyake et al., "Measurement of CP-violating direct amplitude in $K_L \to \pi^0 e^+ e^-$ decay", Kyoto-KEK proposal, summary Aug 1988.

67. T. Barker, et al., "Search for the Rare Kaon Decay Mode $K_L \to \pi^0 e^+ e^-$", Chicago, Elmhurst College, FNAL, Princeton proposal, Dec 1988.

68. A.S. Carroll, et al., Phys. Rev. Lett. 44, 525 (1980).

69. H. Burkhardt et al., Phys. Lett. 206B, 169 (1988).

70. L. Wolfenstein, Phys. Rev. Lett. 13, 562 (1964).

71. M. Kobayashi and T. Maskawa, Prog. Theor. Phys. 49, 652 (1973).

72. B. Winstein, these Proceedings.

73. J.M. Flynn and L. Randall, Phys. Lett. B224, 221 (1989).

74. A.J. Buras, "Present Status of CP Violation in the Standard Model", *Rare Decay Symposium*, Vancouver Canada 30 Nov - 3 Dec 1988, p 249.

75. G. Martinelli, these Proceedings.

76. L.S. Littenberg,"CP violation in $K_L \to \pi^0 e^+ e^-$", *Proceedings of the Workshop on*

CP Violation at KAON Factory, December 3, 1988, ed. J.N. Ng, p. 19.

77. L.M. Sehgal,(Aachen. Tech. Hochsch. III Phys. Inst.), PITHA-89/16, Jan 1989.
78. V. Papadimitriou, et al., Phys. Rev. Lett. 63, 28 (1989).
79. L.S. Littenberg,Phys. Rev. D39, 3322 (1989).
80. The imaginary part gives the direct CP-violating decay $K_2 \to \pi^0\nu\bar{\nu}$ while the real gives the CP-conserving decay $K_1 \to \pi^0\nu\bar{\nu}$. It is the latter which generates the indirect CP-violating contribution through the small admixture of K_1 in K_L. Since $(Re/Im) \times \epsilon$ is small, the indirect CP-violation is small.
81. The dependence of this branching ratio on m_t is very close to that given by the appropriate term of Eqn. 5.

DISCUSSION

N. Cabibbo, INFN, Rome: In the future, will the experiments be limited by the intensities of the beams?

L. Littenberg: It varies with the type of experiment. By and large the lepton flavor-violation experiments are limited by trigger and chamber rates, but not by off-line background. This can be addressed by increasing the acceptance of future detectors, since much of the rates are due to K decays that are too low in momentum to be accepted. The $K^+ \to \pi^+\nu\bar{\nu}$ experiment has not been running long enough to encounter or identify any ultimate limit. At the moment the experiment appears to be flux limited.

People are also talking about new facilities. Fermilab, if it gets a new main injector, wants to do kaon physics there. The Canadians want to build a new facility with 100 times more current than the AGS. So people are optimistic about what one can do.

N. Shumeiko, INP, Byelorussian State University: Is there any news on four-lepton decays of kaons?

L. Littenberg: E787 has the first-ever limit on $K^+ \to \mu^+\nu\mu^+\mu^-$. The dedicated $K_L^0 \to \pi^0e^+e^-$ experiments will be sensitive to $K_L^0 \to e^+e^-e^+e^-$. E845 probably has many such events on tape already.

Experiments on Electroweak Interactions

Experiments on Electroweak Interactions

Thursday, 10 August 1989

Results from TRISTAN 203

Speaker: A. Maki

Chairperson: W. Paul

Scientific Secretary: D. Muller

Measurement of the Z Boson Resonance Parameters 225

Speaker: G. J. Feldman

Chairperson: W. Paul

Scientific Secretary: M. Woods

Results from the Mark II at SLC on Decays of the Z° 236

Speaker: A. J. Weinstein

Chairperson: T. Fujii

Scientific Secretary: T. Barklow

LEP Report 249

Speaker: F. Dydak

Chairperson: T. Fujii

Scientific Secretary: B. Mours

Neutrino-Electron Scattering 258

Speaker: J. Panman

Chairperson: B. Wiik

Scientific Secretary: L. Labarga

W and Z Production and Decay Results from UA2 266

Speaker: A. R. Weidberg

Chairperson: B. Wiik

Scientific Secretary: D. Williams

Electroweak Results from CDF 274

Speaker: M. Campbell

Chairperson: B. Wiik

Scientific Secretary: D. Williams

Theory of Precision Electroweak Experiments 286

Speaker: G. Altarelli

Chairperson: M. K. Gaillard

Scientific Secretary: J. Louis

RESULTS FROM TRISTAN

Akihiro MAKI

KEK National Laboratory for High Energy Physics
Tsukuba, Ibaraki 305, Japan

ABSTRACT

Since the start of physics run in May 1987, TRISTAN has gradually increased its beam energy from 25 GeV to 30.7 GeV. During the last two and a half years, four experiments have acquired data of about 30 pb^{-1} for each experiment. One of the experiments, SHIP, had completed the data taking and presents its final result. Recent results from TRISTAN experiments on searches for heavy flavors and other new particles and QCD studies are discussed.

1. HADRONIC CROSS SECTION

Three experiments, AMY [1], TOPAZ [2] and VENUS [3] measured the total cross section for e^+e^- annihilation into hadrons at various center-of-mass energies, $\sqrt{s}$, from 50 to 61.4 GeV and presented the results in terms of the ratio to the lowest order electromagnetic cross section for $e^+e^- \to \mu^+\mu^-$;

$$R_{had} \equiv \frac{\sigma(e^+e^- \to \text{hadrons})}{\sigma(e^+e^- \to \mu^+\mu^-)_{QED}} \quad (1.1)$$

$$= 3\Sigma Q_q^2(1 + QCD_{corr})(1 + EW_{corr}) \quad \text{(by Standard Model)} \quad (1.2)$$

$$= \frac{N_{ev} - N_{bg}}{\epsilon(1+\delta)L\sigma_0} \quad \text{(by Experiment)} \quad (1.3)$$

At TRISTAN energies, the QCD and electroweak corrections amount to about 5% and 30% respectively. Full production of a top quark (Q=+2/3e) and a fourth generation b′-quark (Q=-1/3e) would result in an increase in R of about 1.5 and 0.5 respectively.

The three TRISTAN experiments have all made careful studies of the backgrounds and systematic errors in the measurements and quoted the systematic errors as 3.5% (AMY), 5.4% (TOPAZ) and 4.2%(VENUS). The details of systematic errors are given in table 1. The largest item among them is the uncertainty from luminosity measurement. For the estimation of the radiative correction factor, 1+δ, they used the calculation program developed by Fujimoto and Shimizu [4] which calculates all the diagrams up to the third order of the electroweak interactions. This full EW correction turned out to be larger by about 3% around $\sqrt{s}$=60 GeV compared to that with the code developed by Berends, Kleiss and Jadach [5] which most of previous experiments used (Figure 1). At the same time they found only negligible difference between the two codes at the lower energies of PEP and PE-

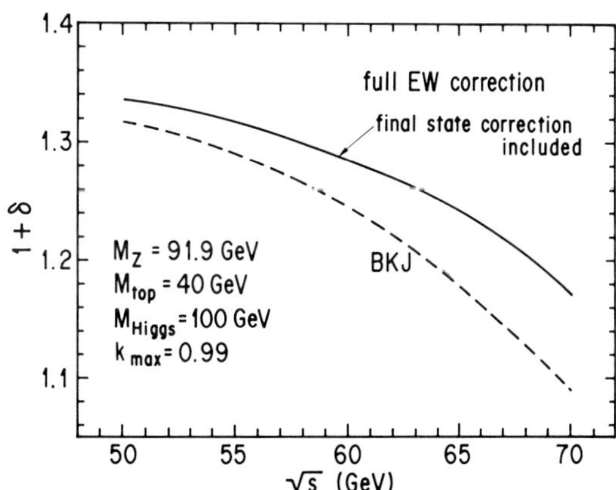

Fig. 1. Radiative correction factors for the total hadronic cross section calculated with codes by FS [4] and BKJ [5].

Table 1. Estimation of systematic errors by AMY, TOPAZ and VENUS

Systematic Errors (%)	AMY	TOPAZ	VENUS
1) Luminosity	2.5	<4.0	2.6
2) Radiative correction	1.3	1.5	2.1
3) Acceptance	1.3	1.0	1.8
4) Cuts, Backgrounds	1.7	2.3	1.7
Overall Norm. Error	3.5	5.4	4.2

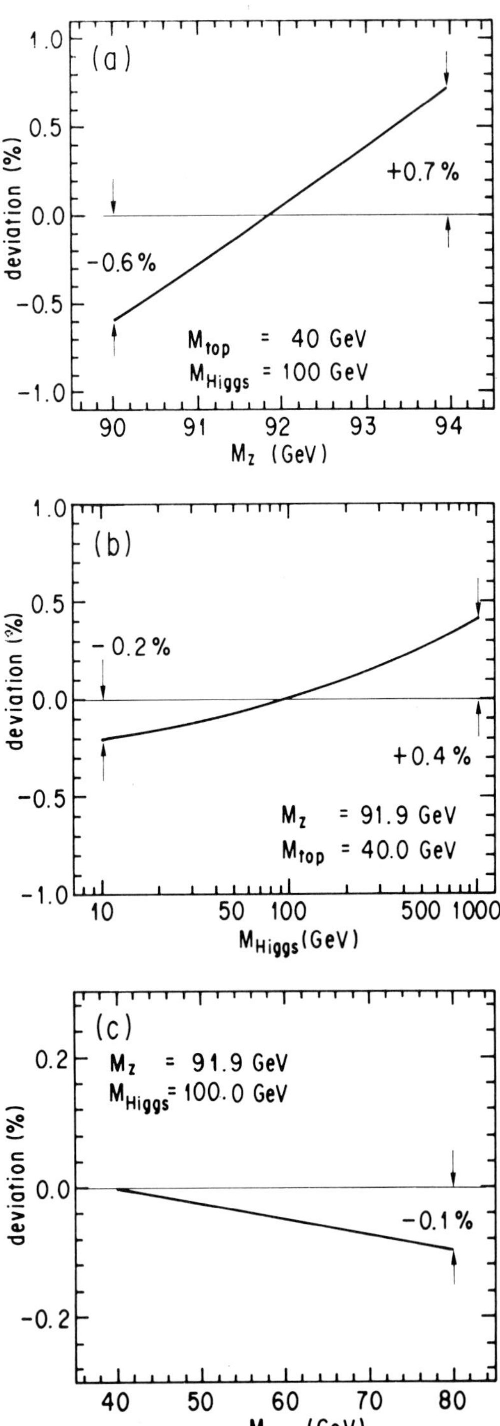

Fig. 2. Relative variations of EW radiative correction factor with (a) M_Z, (b) M_{Higgs} and (c) M_{top}.

TRA. The effects of Z^0 mass, top quark mass and Higgs mass on the radiative correction factor were studied and found to be relatively small (0.7 and 0.4% for top quark and Higgs masses respectively) (Figure 2). The $Z°$-mass dependence introduces a negligible effect because of the recent measurements of this mass by Mark II and CDF [6,7]. The largest uncertainty in the radiative correction factor comes from that of the maximum energy cut on radiated photons (1.2%).

The R measurements from the three experiments are shown in Figure 3. The error bars in the figure represent total statistical and systematic errors added in quadrature. The solid line is the prediction of the five-quark standard model with M_z=91 GeV, $\sin^2\theta_W$=0.23 and $\Lambda_{\overline{MS}} = 0.15$ GeV. The QCD corrections were calculated up to the third order, $O(\alpha_s^3)$ [8]. Also shown in the figure are the predictions of R from the Standard Model in the case of full production of a top quark (dotted line) and a fourth generation b′ quark (broken line). Most of the measured values lie higher than the solid line in the energy range above 56 GeV. Although the open top production is clearly ruled out, the possibility of b′ quark production can not be ruled out using R alone. In order to compare them with smaller error bars, the combined data of the three groups are given in Figure 4. In Figure 5, the measured values of each experiment are individually combined into two energy regions, 50 to 55 GeV and 56 to 61.4 GeV. Here only statistical errors are given. By taking the ratios between two R values at different energies, most of the systematic errors are expected to cancel out. The deviations from the five-quark standard model prediction at the highest energy are estimated to be more than two standard deviations for the AMY and TOPAZ data, but about one standard deviation for the VENUS data.

2.SEARCHES FOR HEAVY FLAVORS

Event shape parameters

Since simple total cross section (R_{had}) measurements are not very sensitive for heavy flavor production, all three groups have searched for them by various methods. They examined many different variables of event shape. In Figure 6, some examples are given. Figures 6 (a), (b) and (c) are thrust, sphericity and aplanarity distributions by AMY, VENUS and TOPAZ, respectively, at $\sqrt{s} = 61.4$ GeV. TOPAZ also gave an energy dependence of the cross section for high aplanarity events

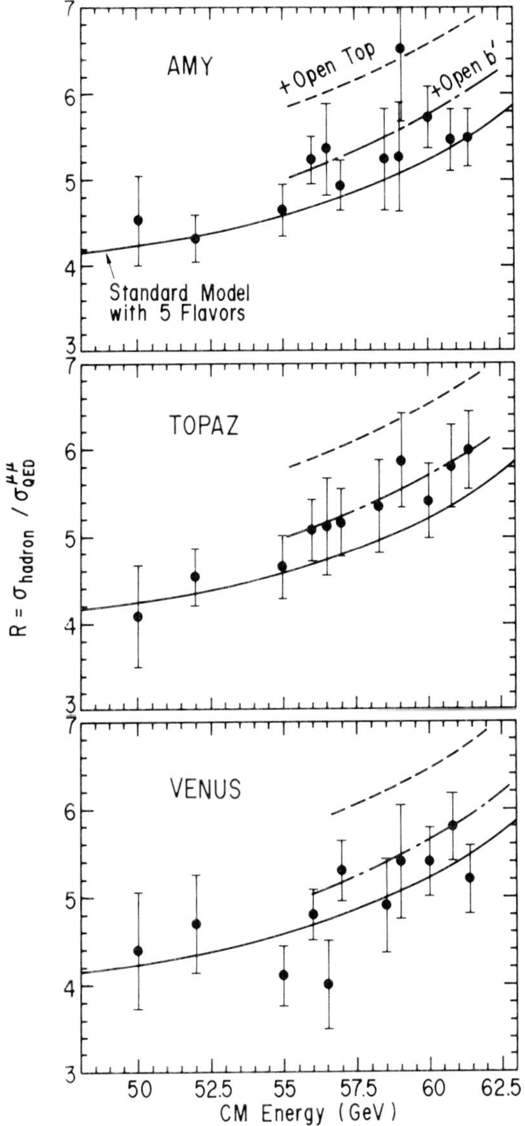

Fig. 3. R_{had} from AMY, TOPAZ and VENUS experiments. Lines are theoretical predictions for five-quark standard model with $M_Z = 91$ GeV, $\sin^2\theta_W$=0.23 and $\Lambda_{\overline{MS}}$=0.15 (solid), full top quark production (dashed) and full b′ quark production (dot-dash).

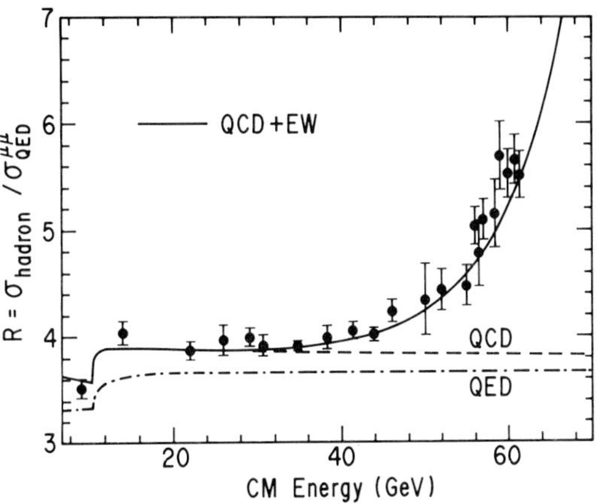

Fig. 4. Combined R_{had} from PEP, PETRA and TRISTAN.

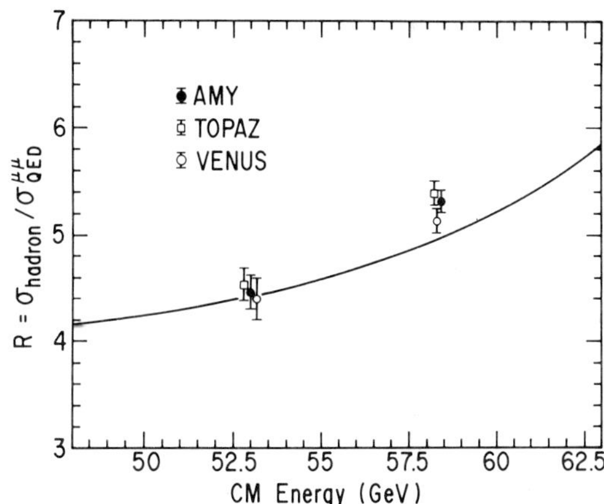

Fig. 5. R_{had} from TRISTAN experiments combined into two energy regions.

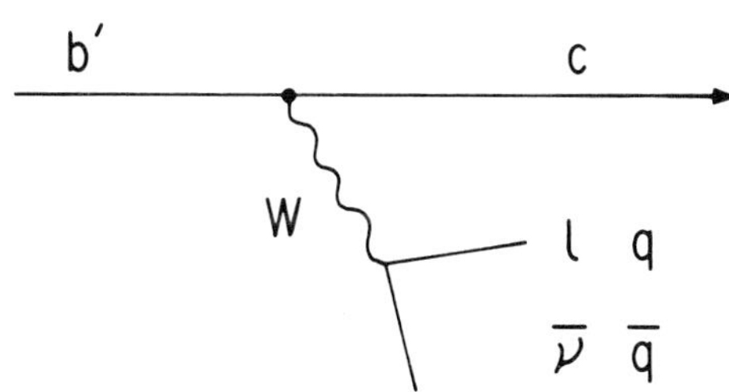

Fig. 7. Charged current decay of b′ quarks.

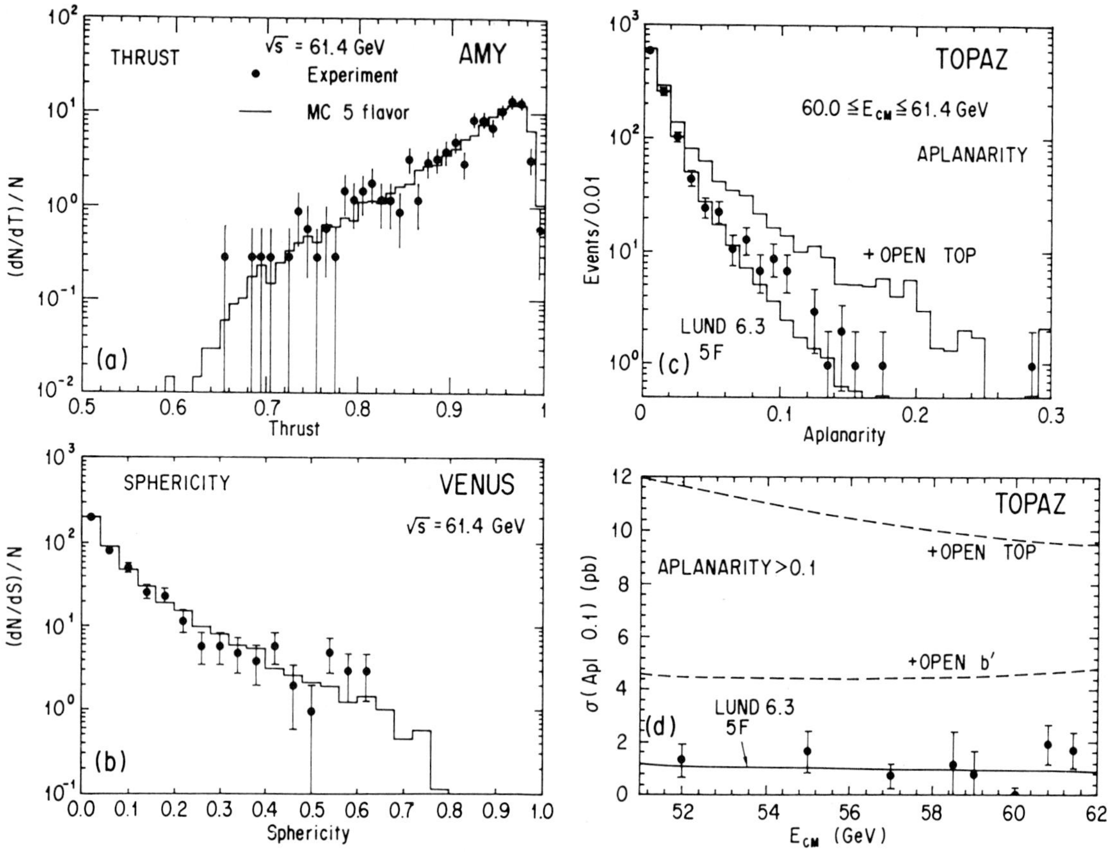

Fig. 6. Distributions at $\sqrt{s}$=61.8 GeV of (a) thrust from AMY, (b) sphericity from VENUS and (c) aplanarity from TOPAZ. (d) is the cross section as a function of $\sqrt{s}$ for events with high aplanarity (aplanarity>0.1).

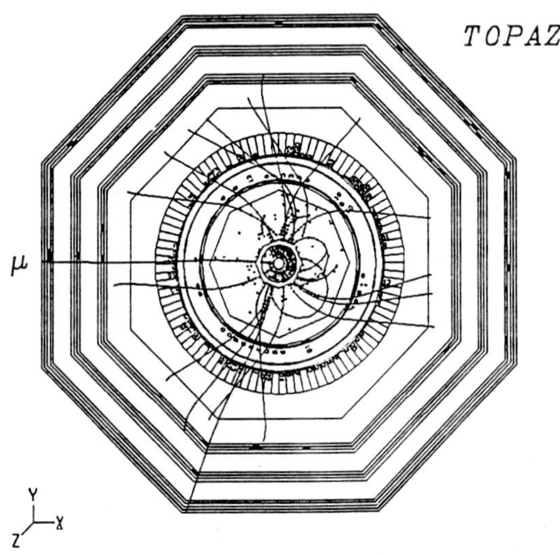

Fig. 8. Isolated muon event detected in the TOPAZ detector.

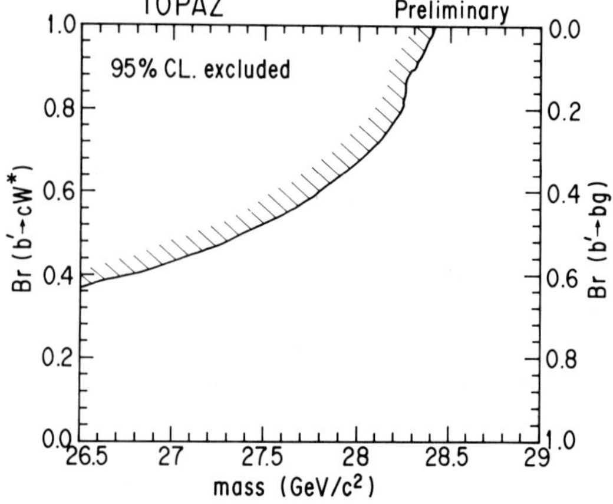

Fig. 9. mass limit of b′ quark set by an isolated lepton analysis of the TOPAZ experiment

(aplanarity >0.1) in Figure 6 (d) where the solid line is the expected cross section for the five flavor production estimated with the Lund Monte Carlo. The two broken lines are cross sections expected for the top quark production (top line) and for the b′ quark production (bottom line). All those data are consistent with the five flavor production except slight excesses at the highest values of sphericity and aplanarity in Figures 6 (b) and (c), respectively.

Isolated leptons

Quarks carrying a new flavor quantum number decay semi-leptonically into three-body final states with branching fraction of 1/9 for each of the e, μ and τ channels (Figure 7). Due to the kinematics of the semileptonic decay process, the final state lepton are usually well separated from the associated hadrons. This isolated lepton gives a clean signature of new quarks. Previously the VENUS experiment [9] has reported an anomalous yield of events with isolated leptons for center-of-mass energies of 60 GeV and above. They found 3 and 2 isolated muon and electron events respectively at $\sqrt{s}$= 60 and 60.8 GeV while they expect only 0.26 muon and 0.6 electron events for the five-quark production. On the other hand, they observed no such event at lower energies (56 to 59.5 GeV) with the expectation of 0.6 muon and 1.2 electron events for the five quark production. Since then, they have accumulated more data at 60.8 and 61.4 GeV, increasing the total amount of data above 60 GeV from 5.9 pb^{-1} to 12.8 pb^{-1}. Accordingly the expected numbers of events have doubled to 0.56 muon and 1.3 electron events. However they have observed no additional events. There still exists an excess but with lower statistical significance.

The TOPAZ [10] and AMY [11] groups also made similar studies. TOPAZ found only one isolated muon event. The TOPAZ event is totally consistent with their expectation of 1.43±0.27 for the five-quark production although it has large aplanarity of 0.17 (Figure 8). Based on the one event observed, TOPAZ gives b′ quark mass limits as a function of the charged current decay branching fraction (Figure 9). This means that the branching fraction of the charged current decay must be below 68% in order that b′ quarks are the source of the high cross section for $e^+e^- \rightarrow$ hadrons above 56 GeV. The same data was used to give the mass limit of top quarks to be 29.9 GeV. Since VENUS [9], Mark-J [12] and JADE [13] have claimed excesses in isolated lepton events, AMY analyzed their data with the same cuts as those groups in addition to their own cuts and found no excess at all. With their own cuts, AMY sets low mass limits for top and b′ quarks to be 30.4 GeV/c^2 and 28.9 GeV/c^2 respectively by assuming that these quarks decay through charged current interaction 100% of the time.

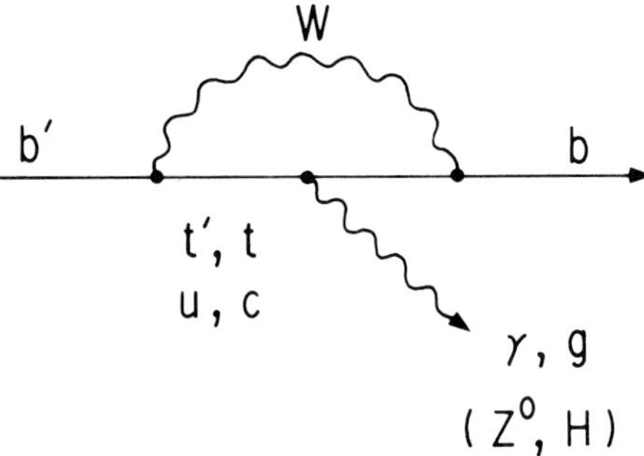

Fig. 10. Neutral decay of b′ quarks.

Neutral decay of b′ *quarks*

In the case that top quark is heavier than b′ quark, the tree-diagram decay of b′ quark (Figure 7) involves a change of two generations and is strongly "K-M suppressed". In this case, it was pointed out that the one-loop-diagram decay (Figure 10), involving a change of only one generation, is likely to be of greater importance [14]. The cleanest signature of b′ quark is then an isolated photon, instead of a lepton. A b′ quark can also decay to a bottom quark by emitting a gluon rather than a photon.

The VENUS group [15] has analyzed their data of 21.1 pb^{-1} up to 60.8 GeV by requiring 1) the energy flow within 30^0 around the photon is less than 1 GeV and 2) the photon energy is between 30 and 70% of the beam energy. They found 8 candidates out of 2702 hadronic events which is consistent with that expected from ordinary sources. From this analysis they give an excluded region in the mass of b′ quark and the branching fraction into a photon and a b quark (Figure 11 (a)). They ruled out b′ quark mass between 11.4 and 27.3 GeV/c^2 in the case of $\rho = \Gamma_{FCNC}/\Gamma_{tot}$=1 and $\xi = \Gamma(b' \rightarrow \gamma b)/\Gamma_{FCNC}$=0.1 where Γ_{FCNC} is neutral width. Combined with the the previous analysis of isolated leptons, they excluded b′ quark masses between 11.4 and 25.6 GeV/c^2 for ξ = 0.1 independent of the ρ value (Figure 11(b)).

The TOPAZ group [16] has made a 4-jet analysis using the "minimum invariant mass" jet algorithm introduced by JADE [17], in addition to an isolated photon analysis with similar cuts to those of VENUS. Their results are given in Figure 11 (c) where excluded mass regions are given for various branching ratios of photon channel together with that from the isolated muon analysis.

For a search of b′ quarks including neutral decays, AMY [18] made a different analysis. They do not require any isolation of photons and use only event shape parameters, thrust, acoplanarity and lesser biwidth. The last parameter is defined by

$$LBW \quad = \quad min\frac{\Sigma_i \mid P_t^i \mid}{E_{vis}} \qquad (2.1)$$

where i runs over all charged and neutral particles within each hemisphere separated by a plane perpendicular to the thrust axis of the event (Figure 12). They used looser cuts when they detected a hard photon in the event, but did not require any isolation of the photon from other activities in the event. The AMY results are given in Figure 11 (d).

From those results, the b′ mass is excluded up to 30 GeV/c^2 for the case where b′ quarks de-

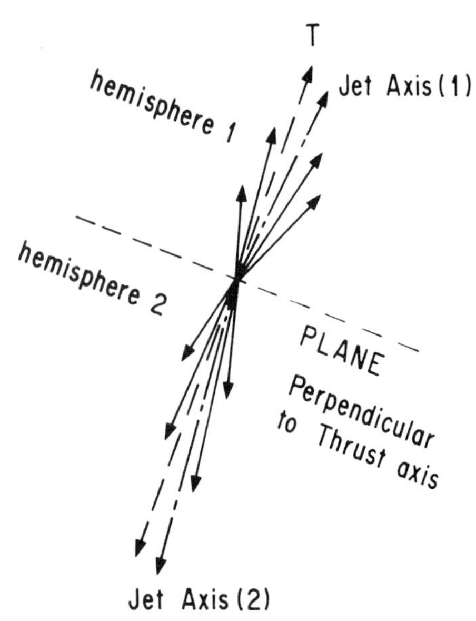

Fig. 12. Definition of lesser bi-width (LBW) variable.

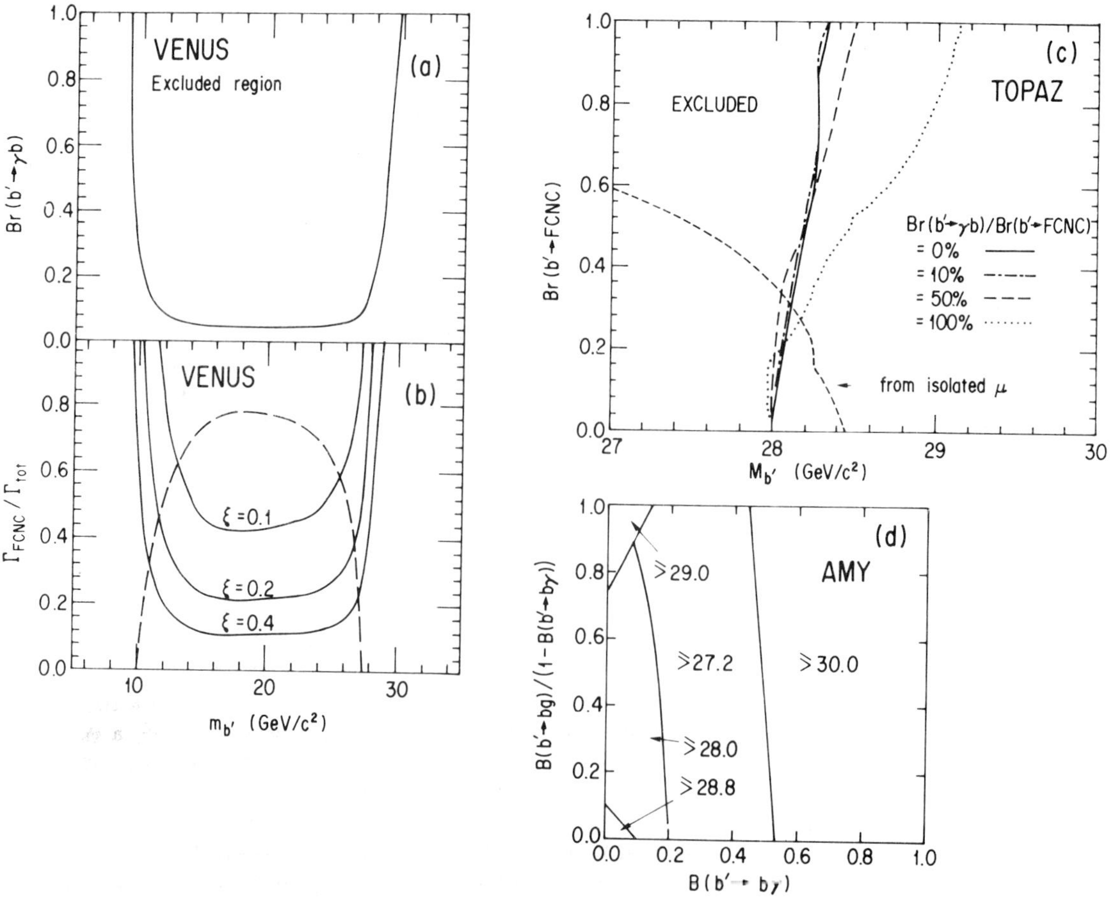

Fig. 11. Mass limits for b' quarks set by VENUS (a and b), TOPAZ (c) and AMY (d).

cay predominantly by emitting a lepton or a hard photon. However, if b′ quarks decay without emitting either a lepton or photon, the limits become lower, somewhere between 26 and 28 GeV/c^2. This means that b′ quarks are not completely ruled out as a possible source of the high cross section for $e^+e^- \to$ hadrons above 56 GeV, which all TRISTAN experiments are seeing.

Number of light neutrinos

The VENUS group [19] has contributed interesting information about the existence of the fourth generation. With the advantage of the super hermeticity of the VENUS detector, they have looked for events with a photon and nothing else. By reducing the veto area from 5^0 to 15^0 in the analysis, they obtained a photon energy spectrum as shown in Figure 13 and understood it with a Monte Carlo simulation as contributions from the $e^+e^- \to e^+e^-\gamma$ and $\gamma\gamma\gamma$. For the veto angle of 5^0, they obtained a photon spectrum given in the inset of Figure 13. With the cut at $x_t \equiv E_{photon}/E_{beam} = 0.18$, no event is seen while 0.79 events are expected from the process $e^+e^- \to \gamma\nu\bar{\nu}$. This gives limits on the number of different light neutrinos N_ν as,

$$\begin{aligned} N_\nu &< 11 \quad \text{(at 90\% C.L.)} \\ &< 15 \quad \text{(at 95\% C.L.).} \end{aligned}$$

Combined with previous e^+e^- results from ASP, CELLO, MAC and MARK-J [20], they reported limits to the number of the light neutrinos as

$$\begin{aligned} N_\nu &< 3.9 \quad \text{(at 90\% C.L.)} \\ &< 4.8 \quad \text{(at 95\% C.L.).} \end{aligned}$$

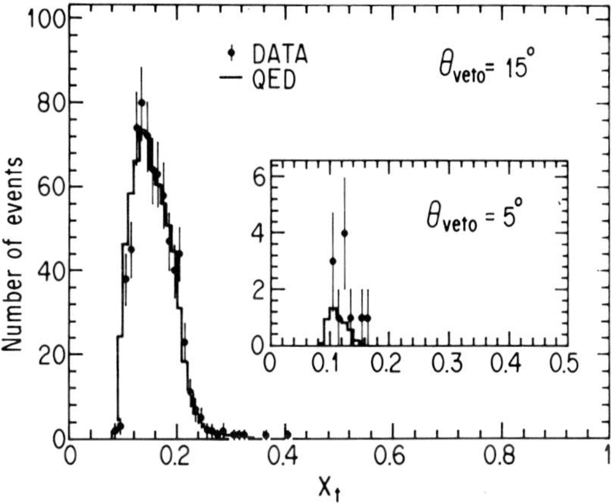

Fig. 13. Single phton energy spectrum by the VENUS experiment.

3. ELECTROWEAK ASYMMETRIES

Since all the particles involved in the reactions $e^+e^- \to \mu^+\mu^-$ and $e^+e^- \to \tau^+\tau^-$ are point-like, the reactions provide sensitive and unambiguous tests of the standard theory. Although the cross section is still small due to the small value of the vector coupling constant, v_l, the forward-backward asymmetry, A_{ll}, is sensitive to the axial vector coupling, $a_e a_l$, and is expected to be large at TRISTAN energies. Corresponding measurements for quark pair production are complementary for full test of the standard model.

Asymmetries in lepton pair production

All the three experiments [21,22,9] have measured these quantities based on about 30 pb^{-1} data between 50 to 60.8 GeV. The muon pair production has the simplest topology and the identification was done with the muon detection system consisting of several layers of drift chambers outside the thick iron plates except for TOPAZ, which did not use the muon detector at all. A time-of-flight technique was used with TOF counters (TOPAZ and VENUS) or muon counters (AMY) to single out muon pair signatures from cosmic ray backgrounds. Two track trigger efficiency was monitored during the data taking by using Bhabha events which were triggered with total energy deposit in the shower counter, as well as cosmic rays. A typical value is 97.0 ± 0.6% given by the VENUS experiment.

The backgrounds for the muon pair production from cosmic rays, Bhabha channel, tau pair production and two photon interactions were estimated to be as small as 0.2%. The largest among various systematic errors was the uncertainty in the luminosity measurement (2.6%), and the total systematic error was estimated to be 2.8% for the cross section. For the asymmetry, most of errors are canceled out and the total systematic error becomes negligible (~ 0.5%) compared to the current level of the statistical error (~ 20%).

For the tau pair production, one prong to n ($1 \le n \le 5$) prong topology was measured. Since no redundant trigger was available, the trigger efficiency was estimated with a Monte Carlo simulation to be 99.0 ± 0.2%. The largest backgrounds to this channel were two photon processes (2.2 ± 0.6%) and radiative muon pair production (1.1 ± 0.2%). For the cross section, uncertainties of detection efficiency (3.2%) and decay branching (2.0%) are major contributions to systematic error in addition to that of luminosity measurement. The systematic error in the asymmetry measurement is estimated to be 1.5%, larger than that of

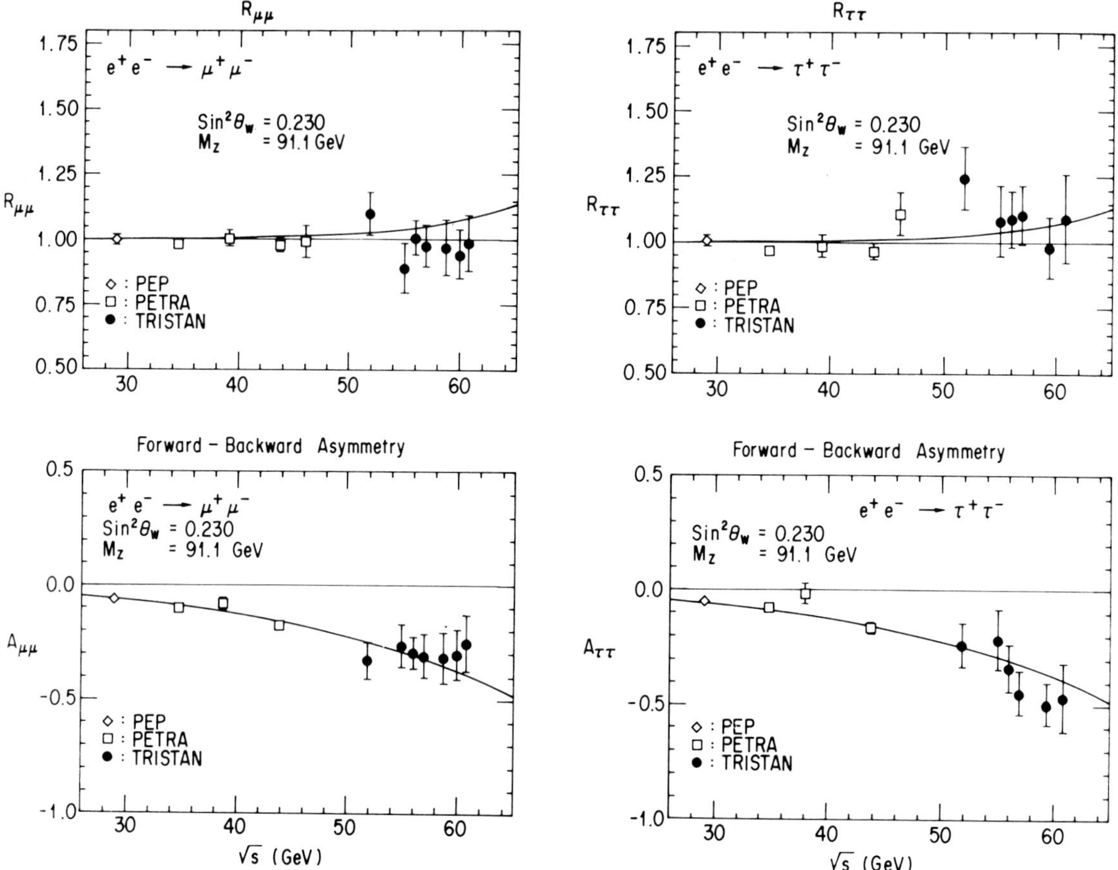

Fig. 14. Combined cross sections and asymmetries for $e^+e^- \to \mu^+\mu^-$, and $\tau^+\tau^-$ from PEP, PETRA and TRISTAN. Solid line is the standard model calculations with M_Z=91.1 GeV and $\sin^2\theta_W$=0.230.

the muon pair because of poorer angular measurements of tau particles.

The total cross sections and asymmetries for $e^+e^- \to \mu^+\mu^-$ and $\tau^+\tau^-$combined for the three experiments are shown in Figure 14. Generally the data follow the theoretical predictions except for the muon pair cross section ($R_{\mu\mu}$) which is consistently lower than the prediction.

In view of previous reports on the slight deviation from the lepton universality [23,24], it is interesting to check this with new TRISTAN data included. A 2.3σ deviation was reported on the Fermi coupling constants of the muon and tau [23] as

$$\frac{G_\tau}{G_\mu} = 0.957 \pm 0.019$$

based on a world average of tau lifetime measurements τ_τ=0.303± 0.008ps, by using the relation

$$\tau_\tau = (\frac{m_\tau}{m_\mu})^{-5}(\frac{G_\tau}{G_\mu})^2 BR(\tau^- \to \nu_\tau e^- \bar{\nu}_e) \cdot \tau_\mu \quad (3.1)$$

and a world average value of the tau branching ratio $BR(\tau^- \to \nu_\tau e^- \bar{\nu}_e) = 17.3 \pm 0.5\%$ [25]. For the axial vector couplings, a fit to the asymmetries from PEP and PETRA experiments [24] gave a ratio,

$$\frac{a_\tau}{a_\mu} = 0.79 \pm 0.09 \,.$$

That is again a 2.3σ deviation.

Fitting all data on muon and tau pair production from PEP, PETRA, and TRISTAN with M_Z = 91.1 GeV, $\sin^2\theta_W$ = 0.230 and a_e = -1 fixed, we obtained [26]

$$a_\tau = -0.949 \pm 0.075$$
$$a_\mu = -1.050 \pm 0.051 \,,$$

and then

$$\frac{a_\tau}{a_\mu} = 0.904 \pm 0.084 \,.$$

The deviation has become less significant. If we assume the lepton universality ($a_\tau = a_\mu$), $a_l = -1.018 \pm 0.042$.

Asymmetries in the quark sector

AMY and VENUS have measured not only lepton pair asymmetries but also quark pair asymmetries. The AMY group [27] identifies $b\bar{b}$ production by tagging muons with high transverse momentum. They got 123 hadronic events with at least one muon. From those events, they subtracted background events coming from $c\bar{c}$ production ($\simeq$ 30%) and hadron fakes ($\simeq$ 20%) with a Monte Carlo technique. Results are,

$$A_{b\bar{b}} = -0.72 \pm 0.23 \pm 0.13$$
$$R_{b\bar{b}} = 0.57 \pm 0.16 \pm 0.10$$

at an average center-of-mass energy of 55.2 GeV. The second and the third terms are respectively statistical and systematic errors. The measured $A_{b\bar{b}}$ is shown in Figure 15 together with the previous experiments at low energies [28] and the standard model prediction. The standard model values for the asymmetry at this energy is -0.56 without corrections for $B\bar{B}$ mixing. If the mixing effect measured by the ARGUS and CLEO experiments [29] is taken into account, the prediction changes to be in the range between -0.20 and -0.50. All experimental measurements in Figure 15 are not corrected for $B\bar{B}$ mixing. The AMY value of $A_{b\bar{b}}$ is different from zero by 2.3σ favoring iso-doublet structure of the third generation quarks (t, b).

VENUS [30] and AMY [31] have measured the forward-backward charge asymmetry of quark pair production. They selected two jet events by requiring thrust T > 0.85 (0.8 for AMY) and identified the quark charge with slightly different formula, i.e.,

$$Q_j = \Sigma Q_i (P_i/E_{beam})^{0.4} \qquad \text{(VENUS)} \quad (3.2)$$
$$Q_j = \Sigma Q_i \eta_i / \Sigma \eta_i \qquad \text{(AMY)} \quad (3.3)$$
$$\text{where } \eta : \ \textit{rapidity}.$$

They obtained similar results at average energies of 57.6 (VENUS) and 55.2 GeV (AMY);

$$A_{q\bar{q}} = 0.114 \pm 0.022 \pm 0.021 \quad \text{(VENUS)}$$
$$= 0.08 \pm 0.03 . \quad \text{(AMY)}$$

By fitting the differential cross section with the assumption of quark flavor independence, VENUS gave the axial vector coupling constant for quarks as $< a_e a_q > = -1.03 \pm 0.10$, consistent with the standard model value.

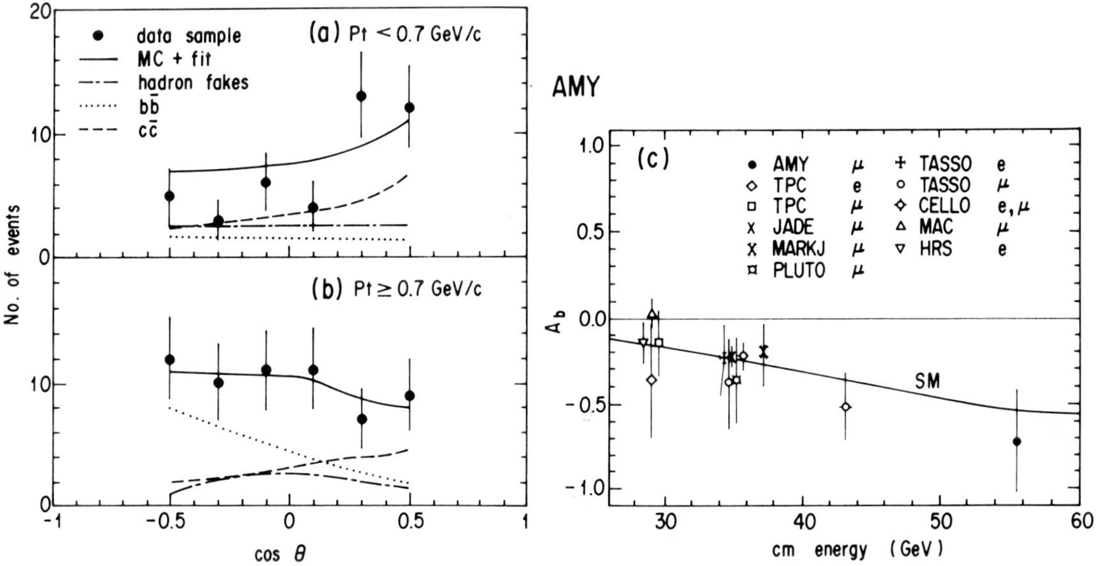

Fig. 15. Angular distributions of hadronic events with at least one muons of p_t <0.7 GeV/c (a) and p_t ≥0.7 GeV/c (b). (c) is the asymmetry of $b\bar{b}$ quark pair production measured by AMY in comparison with previous experiments.

4. QUANTUM CHROMODYNAMICS

Basic properties of quantum chromodynamics (QCD) were tested at TRISTAN. They are the non-Abelian nature of QCD, the renormalization scale and differences of the gluon and quark fragmentation. The non-Abelian nature was tested by measuring the QCD coupling strength of quarks and gluons from the asymmetry of energy-energy correlation and three-jet fractions over a wide energy range, and also by measuring distributions of the angular correlation in four-jet events.

Energy-energy correlation

Energy-energy correlation is defined as the energy weighted angular correlation between two particles in a multi-hadron final state [32],

$$EEC(\chi) \quad = \quad \frac{1}{N}\Sigma_k^N \Sigma_i \Sigma_j \frac{E_i E_j}{E_{cm}^2}\delta(\chi - \chi_{ij}), \quad (4.1)$$

where i and j run over all the particle combinations in an event. χ_{ij} is the angle between particles i and j. E_i is the energy of particle i, which was assumed to be a pion for a charged particle and a photon for a neutral. N is the number of hadronic events used for the analysis. If all events are perfect two jet events, two peaks should be seen only around 0° and 180°. The intermediate angle region is filled by the emission of hard gluons, and thus reflects the strength of the $q\bar{q}g$ vertex (α_s). Since this quantity tends to be affected by the uncertainty of the fragmentation, usually the asymmetry of the energy-energy correlation is used to extract the coupling strength α_s. It is defined as,

$$AEEC(\chi) = EEC(\pi - \chi) - EEC(\chi). \quad (4.2)$$

To extract α_s, the TOPAZ group [33] used a Monte Carlo program of the second order matrix element calculation by Gottshalk and Shatz (GS) [34] combined with the Lund string fragmentation model [35]. They ran it with 13 different $\alpha_s(Q^2)$'s and compared with their data. In order to compare with other experiments, they extracted α_s also using the matrix element calculation by Gutbrod, Kramer and Shierholz (GKS) [36]. The results are shown in Figure 16 together with the previous measurements [37]. Only the results with the matrix element calculations by GKS and by GS are shown there. The TOPAZ results obtained with the GS calculation are,

$$\begin{aligned} \alpha_s &= 0.129 \pm 0.007 \pm 0.010 \quad \text{at } 53.3 \text{ GeV} \\ &= 0.122 \pm 0.008 \pm 0.010 \quad \text{at } 59.5 \text{ GeV}. \end{aligned}$$

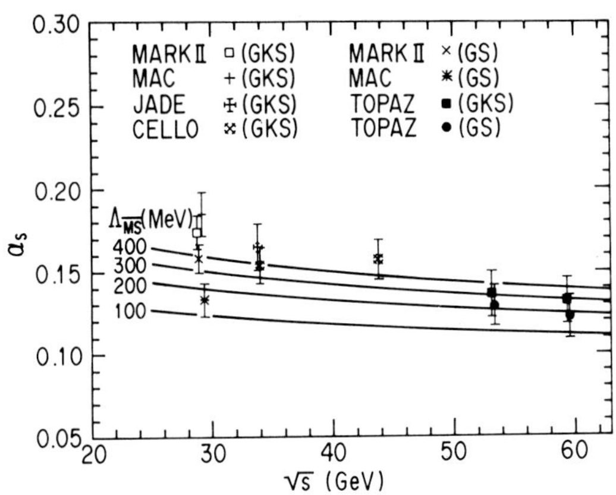

Fig. 16. α_s obtained from asymmetry of energy-energy correlation.

Using the two-loop β-function in the definition of $\alpha_s(Q^2)$ with $Q^2 = s$ [38],

$$\alpha_s(s) = \frac{12\pi}{b_0 ln(s/\Lambda_{\overline{MS}}^2)}[1 - \frac{b_1}{b_0^2}\frac{ln(ln(s/\Lambda_{\overline{MS}}^2))}{ln(s/\Lambda_{\overline{MS}}^2)}], (4.3)$$

where $b_0 = (33 - 2N_q)$ and $b_1 = 3(306 - 38N_q)$ and N_q is the number of quark flavors produced, they obtained $\Lambda_{\overline{MS}}^{(5)} = 209^{+104}_{-78}$ MeV from the above results.

Three-jet fraction

The running nature of the QCD coupling constant α_s is also tested by a comparison of a three-jet fraction with those from lower energy experiments at PEP and PETRA [39]. The jet multiplicity was identified by means of the jet-cluster algorithm introduced by the JADE group [17]. In the second order QCD [40], R_3 is given by

$$\begin{aligned} R_3 &= \frac{N_{3jet}}{N_{total}} && (4.4) \\ &= C_1(y_{min})\alpha_s + C_2(y_{min})\alpha_s^2 && (4.5) \end{aligned}$$

where y_{min} is the lower limit of the integration over the scaled invariant mass y_{ij} of the ij^{th}-parton pair. The correspondence between the theoretical y_{min} and the experimental y_{cut} was verified for $\sqrt{s} \geq 2$[illegible] GeV using several Monte Carlo event generators. Since both C_1 and C_2 are independent of Q^2 and the underlying Q^2 value is expected to scale with $\sqrt{s}$, the $\sqrt{s}$ dependence of R_3 with y_{min} fixed, if any, reflects the Q^2 dependence of the coupling constant α_s.

The three TRISTAN experiments measured R_3 at average s between 2800 and 3500 GeV^2 [41]. The results with $y_{cut} = 0.08$ are given in Figure 17 where the low energy experiments from PEP and PETRA [39] are included to see the s dependence. The solid line is the second order matrix element calculation by Kramer and Lampe (KL) [40] fitted to the data points (the VENUS and the TOPAZ points were not included in the fit) with $\Lambda_{\overline{MS}}$ as only parameter. The χ^2 of the fit is 6.1 for 6 d.o.f., while that is 23.3 for the case of constant α_s (confidence level is 0.1%).

Triple gluon coupling

Evidence for the triple gluon coupling (Figure 18(a)) was first shown by the AMY group [41] by measuring the angular correlation in four-jet events. To observe the effect of this diagram, they compared non-Abelian QCD to an Abelian version in which the color factor of the quarks is retained but the color of the gluon is turned off. AMY found that the relative contributions of these diagrams are (a) 66%, (b) 4% and (c) 30% for the non-Abelian case, while (a) 0% (by definition), (b) 51% and (c) 49% for the Abelian case. Bremsstrahlung gluons are polarized in the $q\bar{q}g$ plane. A pair of final gluons coupled to these gluons, as in diagram 18 (a), tend to be in the direction of this polarization due to the helicity conservation. For the same reason, a pair of quarks tends to be perpendicular to the gluon polarization when they couple to these gluons. Several observables had been proposed to see this effect [42].

The AMY group measured two different angles. The first angle is the one between two planes defined by the two largest jets and by the two smallest jets (Figure 19). The two largest jets and the two smallest jets are supposed to be most probably the pair produced quark and anti-quark and the pair of final gluons, respectively. The result is shown in Figure 20 (a) where two Monte Carlo simulations are also given. The data clearly prefer the non-Abelian case. The χ^2's are 0.3 and 6.5 with d.o.f.=1 for the non-Abelian and Abelian cases respectively.

The above angle has the disadvantage of reduced statistics, because additional cuts are needed on the angles between the jets that define the planes. This requirement eliminates about 50% of the four-jet events. The second angle proposed to study is the angle between the vector difference of the two largest jets and that of the two smallest jets. The result is shown in Figure 20 (b) together with the Monte Carlo predictions. Again the data favor the non-Abelian case with the $\chi^2 = 0.5$ (non-Abelian) against 7.3 (Abelian) for d.o.f.=3.

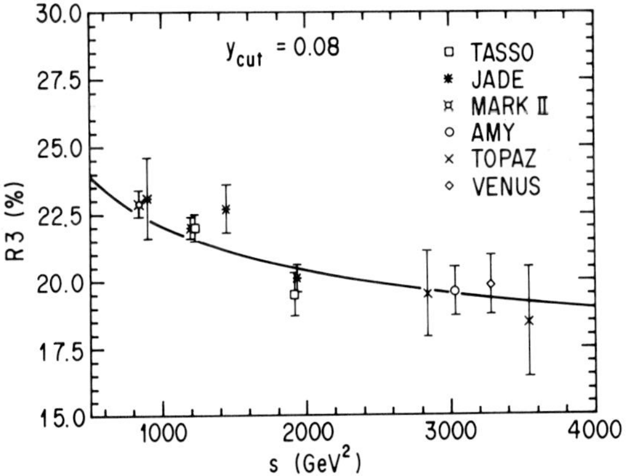

Fig. 17. Three jet fraction (R_3) in comparison with the O(α_s^2) QCD calculation [40].

Renormalization scale

Another contribution from AMY [43] is on the renormalization scale of QCD. It is well known that the second order perturbative QCD does not accurately describe the global shapes of multihadronic events. The AMY group made fits to multi-jet frac-

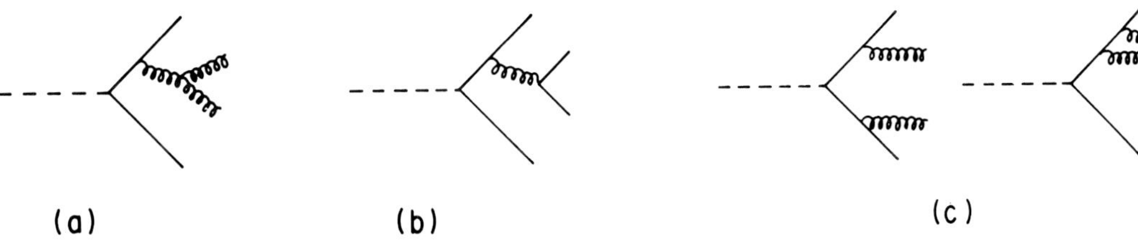

Fig. 18. Four jet production diagrams; (a) triple gluon coupling, (b) $q\bar{q}g$ coupling and (c) double gluon Bremsstrahlung.

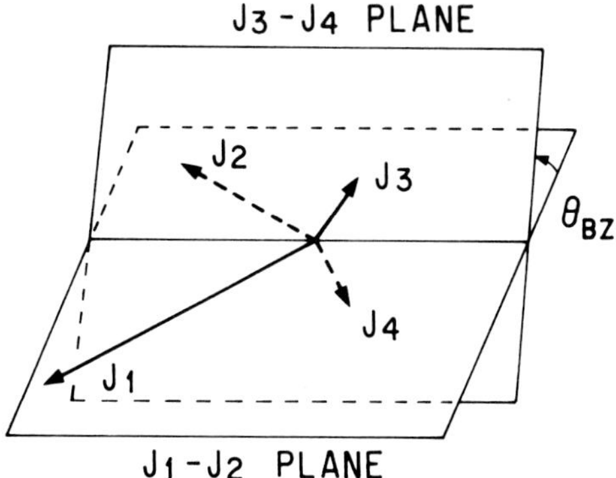

Fig. 19. θ_{BZ} angle in four-jet events.

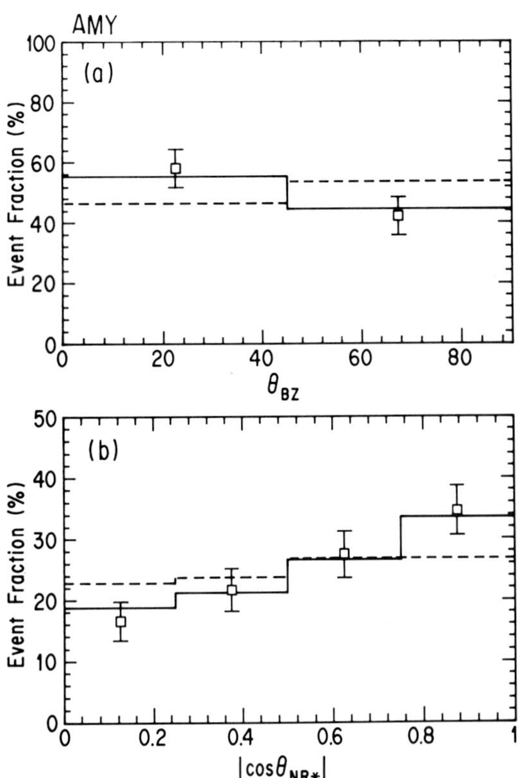

Fig. 20. Distributions of θ_{BZ} and $|\cos\,\theta_{NR*}|$ angles from the AMY experiment. Solid (dashed) curves are Monte Carlo predictions for the case of non-Abelian (Abelian) QCD.

tions with the second order QCD calculation by regarding the energy scale as a parameter. Multi-jet fractions are described in the second order QCD by,

$$\begin{aligned} R_2(f,\Lambda,y_{cut}) &= 1 + \quad C_{21}(y_{cut})\alpha_s(\mu^2) \\ &\qquad C_{22}(f,y_{cut})\alpha_s^2(\mu^2) \quad (4.6) \\ R_3(f,\Lambda,y_{cut}) &= \qquad C_{31}(y_{cut})\alpha_s(\mu^2) \\ &\quad + \quad C_{32}(f,y_{cut})\alpha_s^2(\mu^2) \quad (4.7) \\ R_4(f,\Lambda,y_{cut}) &= \qquad C_{42}(y_{cut})\alpha_s^2(\mu^2) \quad (4.8) \end{aligned}$$

where $f = \mu^2/s$ is the renormalization scale. Two parameters, f and $\Lambda_{\overline{MS}}$, were determined in the fit. Three different data sets were used for fits. They are (a) R_3 and R_4, (b) R_2 and R_4 and (c) R_2, R_3 and R_4 with the unitarity condition $R_4 = 1 - R_2 - R_3$. The results are quite stable for these different data sets and also for different y_{cut} values between 0.04 and 0.08. They gave the values,

$$\begin{aligned} f &= 0.0021^{+0.0015}_{-0.0006} \\ \Lambda_{\overline{MS}} &= 173 \pm 22 \pm 55 \text{ MeV}. \end{aligned}$$

This f value corresponds to $\mu = 2.5$ GeV. Then this $\Lambda_{\overline{MS}}$ is, by definition [38], $\Lambda^{(4)}_{\overline{MS}}$. In Figure 21 (a), the multi-jet fraction data are compared with this result (f=0.0021). In the figure, the calculation with f=1, i.e., $Q^2 = s$, is also shown. It is very clear that this fit reproduces the data very well. The obtained $\Lambda^{(4)}_{\overline{MS}}$ is compared with the $\Lambda^{(4)}_{\overline{MS}}$'s extracted from other experiments on different processes [44] in Figure 21 (b).

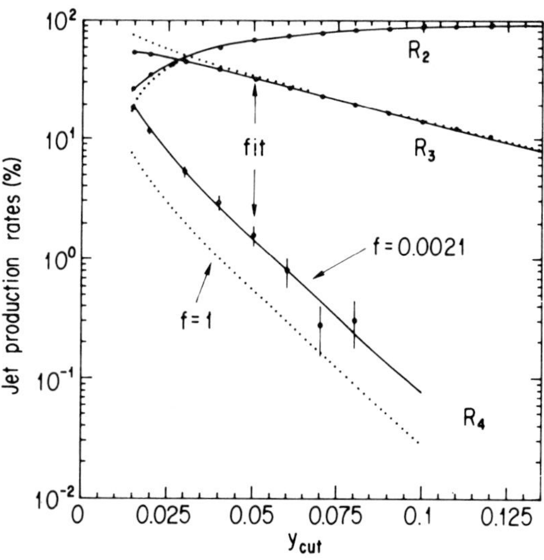

Fig. 21. (a) Fits to multijet fractions with the renormalization scale f=0.0021 in comparison with f=1 i.e., Q^2=s.

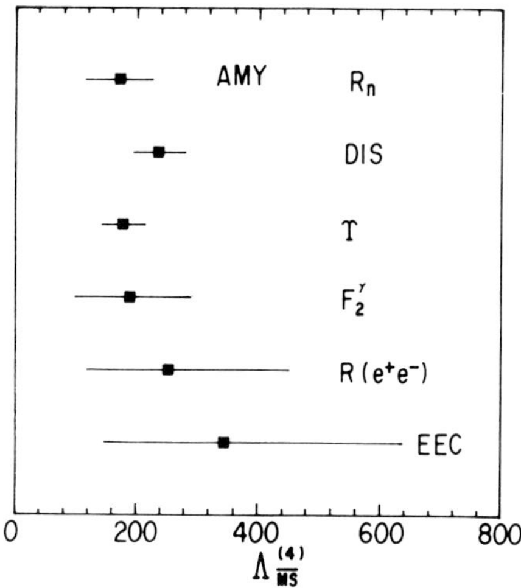

Fig. 21. (b) Comparison of $\Lambda_{\overline{MS}}$ from various experiments.

This AMY work was motivated by a similar work by Bethke for Mark II data at $\sqrt{s} = 29$ GeV [45]. He obtained f=0.0017 and $\Lambda^{(5)}_{\overline{MS}} = 95 \pm 30$. This can be translated into $\Lambda^{(4)}_{\overline{MS}} = 170$ MeV if we use the Marciano formula with consideration of the well known fact that the α_s formula used in ref.45,

$$\alpha_s(s) = \frac{12\pi}{b_0 ln(s/\Lambda^2_{\overline{MS}}) + \frac{b_1}{b_0} ln(ln(s/\Lambda^2_{\overline{MS}}))} \quad (4.9)$$

where the notations are same as equation (4.3), gives a smaller $\Lambda_{\overline{MS}}$ by 15% than that of formula (4.3) which AMY used. The two results are in good agreement in view of the difference of a factor 4 in Q^2.

Gluon and quark fragmentation

Gluons are expected to fragment into more hadrons with softer momentum distribution compared with quarks due to their larger color charges. To date two different methods have been used by several experiments to see this difference. HRS [46], Mark-II [47] and TASSO [48] used symmetric three jet events and two jet events at $\sqrt{s}$=2/3 of that of three jet data. Then all the jets to be compared have almost the same jet energy. The observables associated with gluons are extracted as,

$$\begin{aligned} O(g) &= O(3-\text{jet};\ \text{at}\ \sqrt{s} = 3E_{jet}) \\ &\quad - O(2-\text{jet};\ \text{at}\ \sqrt{s} = 2E_{jet}) \end{aligned} \quad (4.10)$$

This method is generally said to be the least biased. It is probably true when relative contributions from the electromagnetic and weak interactions are not too much different at the two energies. A disadvantage of this method is statistics. In addition to the symmetric three-jet requirement, which reduces the useful number of events substantially, one has to make a subtraction between the large fractions (3-2) to extract the relatively small fraction of gluon contribution (1). Since the difference between the two fragmentations are expected to be small in the present energy region, this method is rather insensitive. None of the above experiments have seen the difference except Mark II saw some indication.

The second method is to identify quark and gluon jets in three jets. Three jet events are produced by Bremsstrahlung of a gluon from one of the pair produced quarks. Therefore the gluon is expected to have the lowest energy among the three jets. The AMY group [49] has analyzed three jet events by using the JADE jet algorithm [17] and determined the jet energies from the measured angles of the jets and the energy momentum conservation. They used only events which passed the cut, $2/3 \leq (E^{jet}_{vis}/E^{jet}_{cal}) \leq 4/3$. With Monte Carlo, they found a high efficiency ($\simeq 67\%$) of identifying gluons due to the high reaction energy compared with that of JADE (51%) [50].

Their results are shown in Figure 22. Figure 22 (a) is mean core energy fraction, ξ, as a function of the jet energy. The core energy fraction was defined as the energy flow within the half cone angle, $60^0/\sqrt{E^{jet}_{cal}/E^{jet}_{vis}}$ (E in GeV). The solid circle is the lowest energy jet (jet-3; most likely gluon jet) and the open circle is the two largest energy jets (jet-1 and jet-2; most likely quark jet). The cross given at the highest jet energy is taken from two jet events where both of them are to be considered as quark jets. In the figure, Monte Carlo simulations are also given. The solid line is the independent fragmentation model where quarks and gluons are assumed to fragment in the same way. The dashed line is the leading log approximation model. The high energy side is for jet-1 and jet-2 and the low energy side is for jet-3. Figure 22 (c) and (d) are the distributions of the core energy fraction of the jet-1 and jet-2 combined (c) and the jet-3 (d) at the mean jet energy of 16 GeV. The notations are the same as those in Figure 22 (a). In Figure 22(b), the mean rapidity of the leading particle in the jet is shown. Those figures clearly show the difference between the fragmentations. It is unlikely that the observed differences are caused by the jet algorithm they used in the analysis because the Monte Carlo with the independent fragmentation model does not show any difference for the quark and gluon jets.

5. New Particles beyond Standard Model

Extra gauge boson

Several authers [51,52,53] have recently investigated constraints on extra Z bosons expected from E_6 grand unified theories. They gave lower mass limits of these bosons typically around 100 - 300 GeV (90% C.L.). All TRISTAN experiments report slightly high cross section for $e^+e^- \to$ hadrons and low cross section for $e^+e^- \to \mu^+\mu^-$. This suggests the possibility of extra Z bosons playing roles at TRISTAN energies. K. Hagiwara et al [54] made an analysis using available e^+e^- data from PEP, PETRA and TRISTAN.

In the model with breakdown $E_6 \to G \times U(1)_\beta$, where G contains the standard model group $SU(3) \times SU(2) \times U(1)_Y$, extra U(1) charge can be generally parameterized as,

$$Z_\beta = Z_\psi cos\beta + Z_\chi sin\beta \ . \qquad (5.1)$$

Angles β=0, $\pi/2$, $tan^{-1}\sqrt{3/5}$ and $tan^{-1}(-\sqrt{1/15})$ correspond to charges for Z_ψ, Z_χ, Z_η and Z_ν. The physical mass eigen states are expressed as a mixture of current eigenstates (Z_1^0, $Z_2^0 \equiv Z_\beta$) with a new mixing angle θ_E,

$$\begin{aligned} Z_1 &= Z_1^0 cos\theta_E + Z_2^0 sin\theta_E & (5.2) \\ Z_2 &= -Z_1^0 sin\theta_E + Z_2^0 cos\theta_E \ , & (5.3) \end{aligned}$$

where Z_1 and Z_2 are the ordinary Z and extra Z' bosons, respectively. The vector coupling of these bosons to ordinary leptons and down-type quarks are of the same magnitude with opposite signs, and that to up-type quarks is zero. The overall scale of the coupling $\lambda = g_E/g_Y$ is an arbitrary constant. However, renormalization group analyses suggest $\lambda = 1$ for the superstring-inspired models such as Z_χ and Z_η. In their analysis, λ=1 was assumed also for the general case, Z_β. QCD corrections were done up to the third order.

Chi-square fits were done to all available data of $R_{\mu\mu}$, $R_{\tau\tau}$, R_{had}, $A_{\mu\mu}$, $A_{\tau\tau}$ and A_{cc} in the energy range of $\sqrt{s}$ = 14 $\sim$ 57 GeV with M_2 and θ_E as the only free parameters. Other quantities were

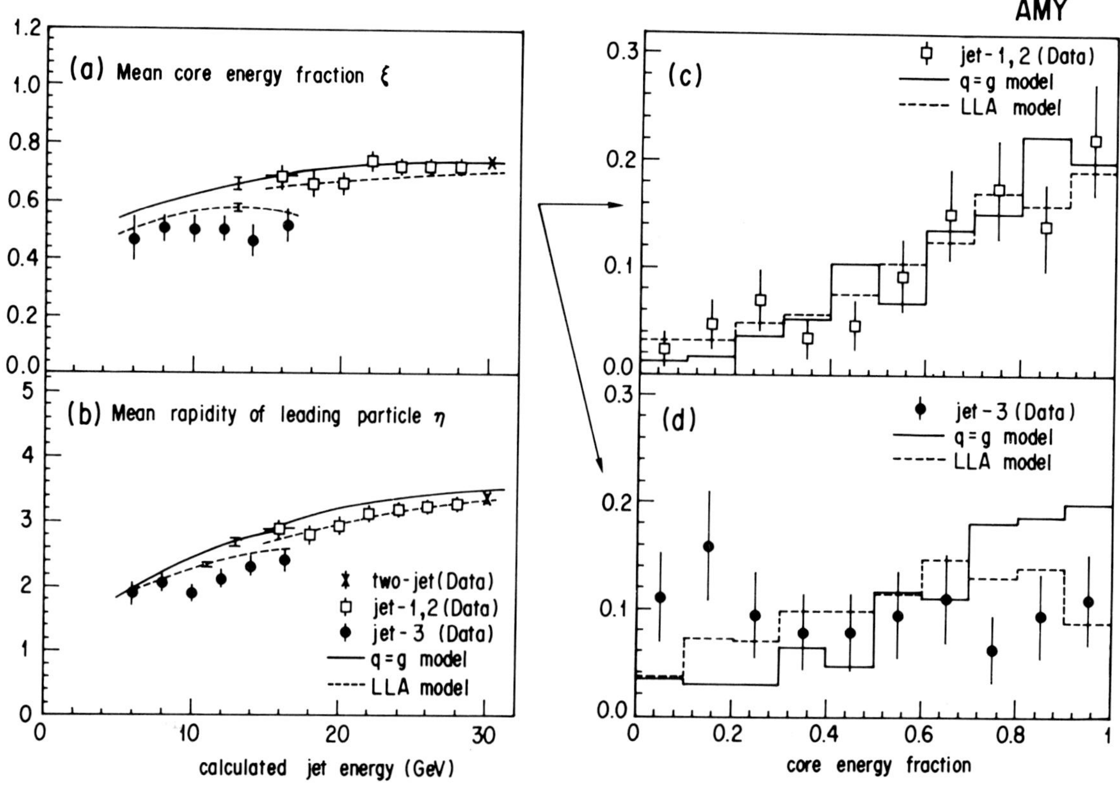

Fig. 22. Comparison between quark and gluon fragmentations in mean core energy fraction (a) and mean rapidity of leading particle (b). Distributions of core energy fraction for quark (c) and gluon (d) enriched sample .

fixed to or constrained within the experimental errors. $\Lambda_{\overline{MS}}$ was fixed to 0.15 GeV and $\sin^2\theta_W$ to the three values, 0.223, 0.229 and 0.235. M_W and M_Z were constrained to 80.9±1.4 GeV/c^2 and 91.9±1.8 GeV/c^2.

The results of the fit are given in Figure 23. The fit prefers β around $\pi/2$. The allowed regions for M_2 and θ_E are given in Figures 23 (a) and (b). The three lines at each boundary are for different values of $\sin^2\theta_W$. The lower mass limits from the analysis by Amaldi et al [52] are also shown in the figure. This analysis gives also upper limit for $\pi/4 \leq \beta \leq 2\pi/3$ which is still consistent with the previous lower limits. Figures 24 (a) and (b) show the allowed region in M_2 - θ_E plane for Z_χ and Z_η models. Numerically their mass limits at 90% confidence level are,

$$
\begin{aligned}
150\text{GeV} &\leq M_\chi \leq 363\text{GeV} \\
100\text{GeV} &\leq M_\eta \\
136\text{GeV} &\leq M_\psi \,.
\end{aligned}
$$

The best fit values are compared with the experimental data in Figure 25. Experimental data above 58 GeV are new TRISTAN points which were not included in the fit. The dashed line is this result and the others (solid and dotted) are the standard model calculations with different M_Z values ($M_Z = 91.9$ GeV for solid and $M_Z = 89.0$ GeV for

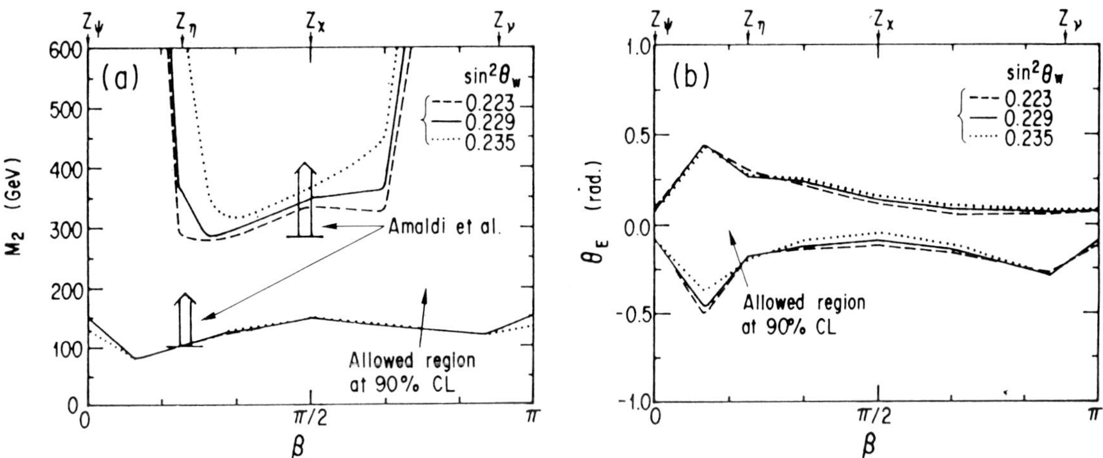

Fig. 23. Allowed region for extra Z′ mass and θ_E as a function of mixing angle β. Previous results of Amaldi et al [52] are given for comparison.

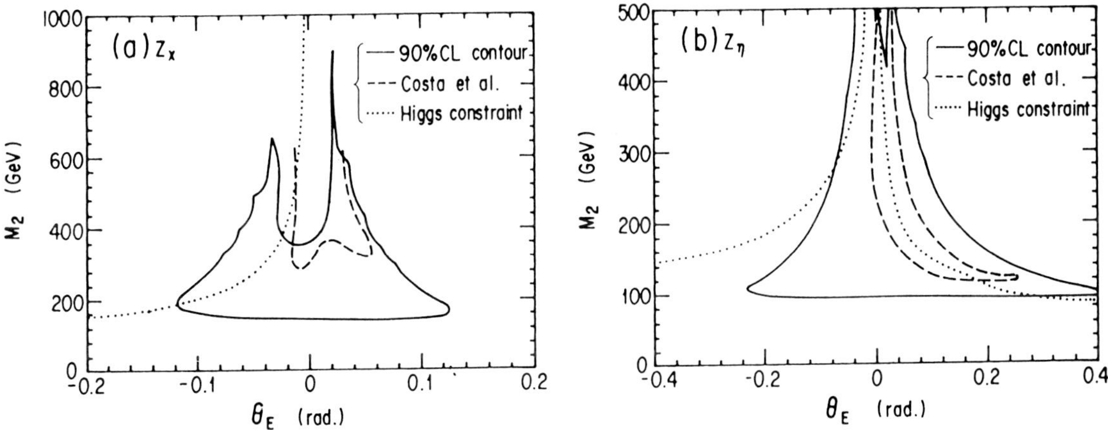

Fig. 24. Allowed mass regions for extra Z′ bosons for Z_χ and Z_η as a function of θ_E in comparison with the results of Costa et al. [53].

dot). The data must be further improved to a per cent level of accuracy in order to make a definite conclusion.

Sequential heavy leptons

A comprehensive analysis on sequential heavy leptons, both charged and neutral, was reported to the conference by the AMY experiment [55]. For charged leptons, low visible energy, imbalance in p_t and no correlation between the activities in the two hemispheres are good signatues for hadronic decays of heavy leptons because of undetected neutrinos as shown in Figure 26. They used cuts on thrust, acoplanarity angle, visible energy and transverse momentum, and found 3 events passed, while 3.8 ± 1.9 and 10.8 ± 0.8 events are expected for all known processes and the heavy lepton production of mass 29 GeV/c^2, respectively. With smaller branching fraction, charged heavy leptons decay into leptons. This gives a cleaner signatue. They looked for events with isolated electrons or muons and with small visible energy ($M_h \leq \sqrt{s}/2$). Two events were found which were consistent with expected backgrounds (3.4±1.7).

These searches are insensitive when the associated neutrino is as heavy as the charged lepton. For this case, they found out that with high probability, charged leptons decay into one prong, thus resulting in two prong events. A total of 7 events passed all cuts while 6.7 events are expected from $e^+e^- \rightarrow e^+e^-l^+l^-$ and 0.3 events from $\tau^+\tau^-$ processes. If charged leptons are stable and light enough, events look like the muon pair production. From the measured cross section of the muon pair production, they concluded the lepton mass must be greater than 21.8 GeV/c^2. On the other

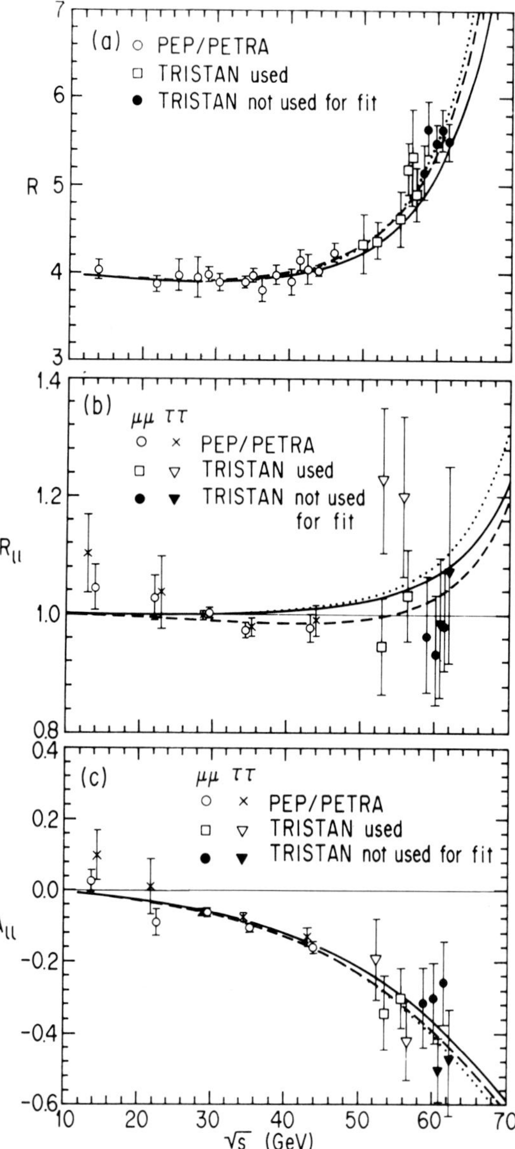

Fig. 25. Best fit value (dashed line) of the extra Z′ analysis [23] compared with experiments. Solid and dotted lines are the standard model calculations with M_Z=91.9 (solid) and M_Z=89.0 (dotted).

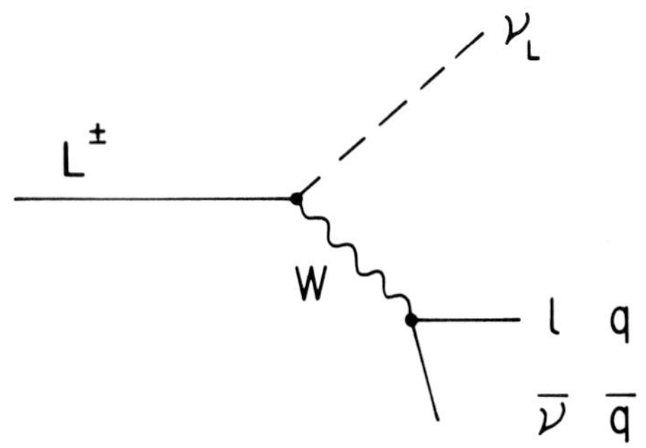

Fig. 26. Decay of sequential charged heavy leptons.

hand, events look like colinear non-muon and non-electron pairs if charged leptons are so heavy that they do not penetrate the muon absorber. They did not find such an event. Masses between 19.6 and 30.1 GeV/c^2 were thus excluded. Combined with the low mass case, M $\geq$ 30.1 GeV/c^2. Those are summarized in Figure 27 (a).

A pair of heavy neutral leptons (L^0) can be produced in the electron-positron annihilation. If leptons mix between generations like quarks do, neutral leptons can decay into an ordinary lepton and a pair of leptons or quarks through charged current interaction. Another possibility is to decay into a neutrino and a pair of ordinary leptons through a flavor-changing neutral current interaction. For the former case, the signature is two isolated leptons (electrons or muons). The lifetime of leptons could be long dependeng upon the mixing strength $\mid U_{lL} \mid$. In the analysis, the decay vertex was required to be within 5 cm from the interaction point to keep the track finding efficiency reasonably high. This limits the sensitivity in the mixing strength $\mid U_{lL} \mid$. Figure 27 (b) and (c) are for Dirac and Majorana L^0's, respectively. In both figures, two cases are shown where a neutral lepton exclusively couples to an electron or a muon. In the case of equal coupling to an electron and a muon, the results are given in Figure 27 (d) for Dirac and Majorana L^0's.

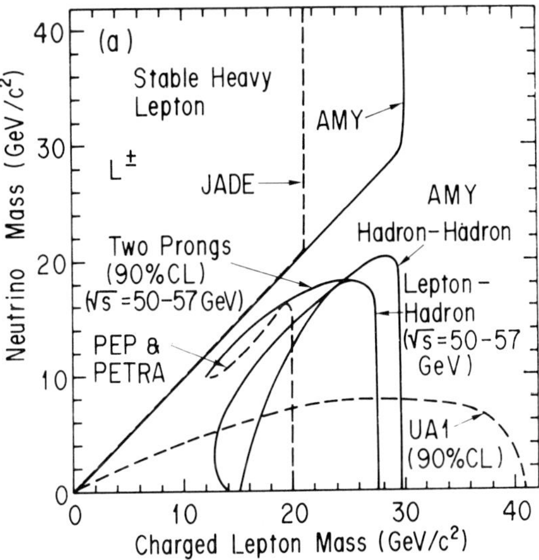

Fig. 27. Mass limits set by AMY for sequential charged heavy leptons (a) and neutral heavy leptons in case of Dirac (b) and Majorana (c) types which couple exclusively to either electrons or muons. (d) is the case where neutral leptons equally couple to both electrons and muons.

For the latter channel, the cleanest signature with large branching fraction ($\sim$ 28%) is expected to be a monojet where one neutral lepton decays into three neutrinos and the other into a neutrino and $q\bar{q}$ pair . However, no candidate event survives the selection while 6.4 events are expected for Dirac L^0's of 15 GeV/c^2. From this, masses between 6.9 and 20.1 GeV/c^2 are excluded for Dirac L^0's.

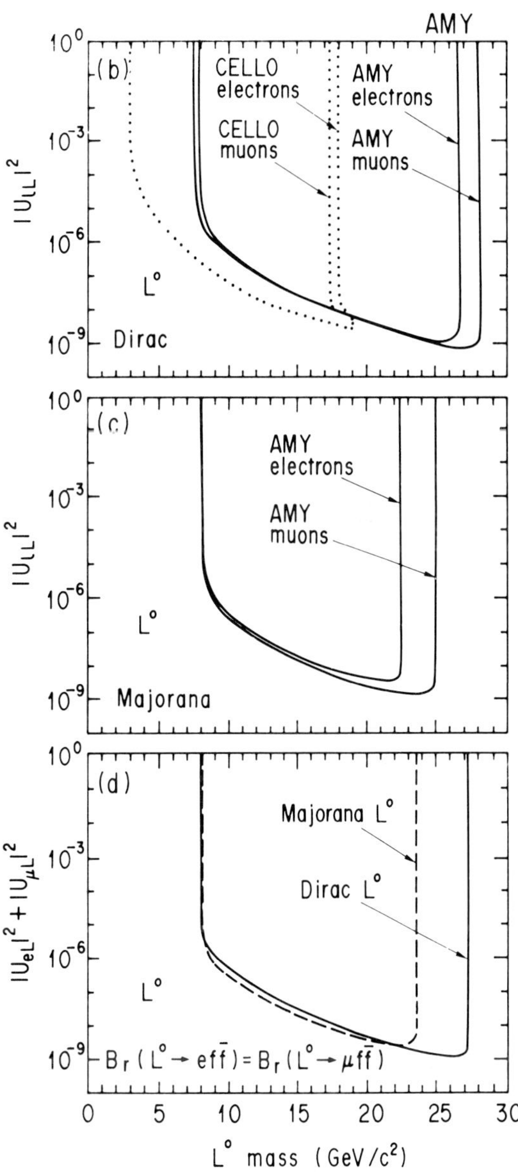

SUSY particles

Supersymmetric (SUSY) theories provide a natural solution to many open questions such as the naturalness problem which the standard model of elecroweak and strong interactions can not answer in spite of its extreme success. SUSY theories predict a rich spectrum of particles, requiring a set of SUSY partner particles to all known particles with spins differing by 1/2. Searches for a variety of SUSY particles have been made at TRISTAN [9, 55, 56]. In these searches, the following assumptions are made:
(1) R-parity conservation
(2) Lightest SUSY particle is the photino, $\tilde{\gamma}$, which is stable and undetectable
(3) All target SUSY particles promptly decay into $\tilde{\gamma}$ and ordinary particles.
Then the signature to look for is a large missing p_t and a large acoplanarity angle due to the undetectable photino. Limits for sleptons ($\tilde{e}, \tilde{\mu}$ and $\tilde{\tau}$) and squarks ($\tilde{q}$) are given in Figure 28 for four different cases per SUSY particle: two extreme cases of mass relationship between the left-handed and right-handed particles, and two different cases of the photino mass. The shaded ranges are excluded masses extended by TRISTAN experiments. Only for $\tilde{e}$, TRISTAN limits are lower than those of the ASP experiment [57].

Monopoles

The SHIP experiment [58] is the only experiment at TRISTAN which is dedicated to searches of heavily ionizing particles. They placed a number of CR-39 sheets around the interaction point with the coverage of about 90% of 4π steradian. The luminosity was measured with lead glass counters located at small and large angles. In the data of 30.2 pb^{-1} accumulated in the energy range from 50 to 60.8 GeV, they found no candidate track and calculated the limit of the cross section as,

$$\sigma \quad < \quad \frac{3.0}{\epsilon \int Ldt} \quad \equiv \quad \sigma_{lim} \quad \text{(at 95\% C.L.)}. \tag{5.4}$$

If one assumes a single-photon production process for Dirac monopoles, the amplitude for pair production is proportional to the magnetic charge. Then one can formulate a pair production cross section for monopoles of mass m as,

$$\sigma_D(m) \quad = \quad (\frac{g_D}{e})^2 \sigma_{\mu\mu}(> 2m)(1 - \frac{4m^2}{s}), \tag{5.5}$$

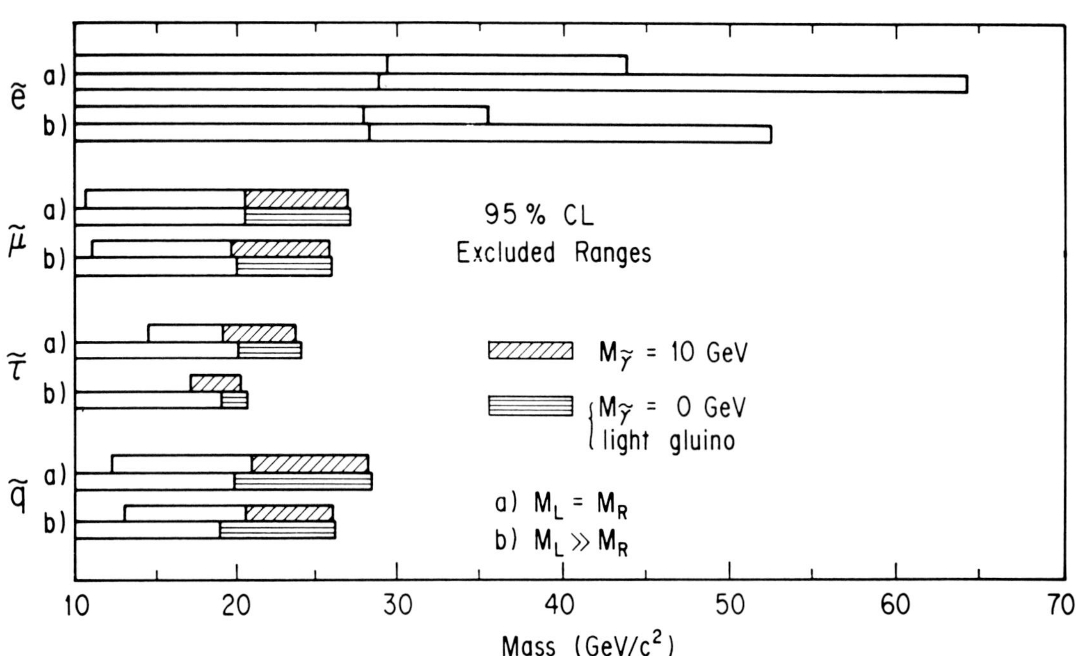

Fig. 28. Lower mass limits for SUSY particles. Shaded ranges are extended by TRISTAN experiments.

ignoring higher order effects. g_D and $\sigma_{\mu\mu}(> 2m)$ are charge of monopole and the cross section for production of a $\mu^+\mu^-$ pair with invariant mass greater than 2m. The quantity $R_D \equiv \sigma(m)/\sigma_D(m)$ would be expected to be of order unity for pointlike Dirac monopoles with magnetic charge g_D (and ~4 for charge $2g_D$) at energies above threshold. The limit on R_D is shown in Figure 29 in comparison with those from previous searches [59]. Pointlike Dirac monopoles with mass below 28.8 GeV/c^2 are ruled out.

If monopoles are not pointlike, the above cross section could be reduced by many orders and it is possible that a weaker interaction to which the monopole couples in a pointlike manner dominates the process [60]. If this interaction is due to annihilation of electron and positron into a neutral vector or scalar gauge boson with mass M_B, cross section is given by

$$\sigma_B \sim \sigma_{\mu\mu}(> 2m)(1 - \frac{4m^2}{s})(\frac{s}{s - M_B^2}). \quad (5.6)$$

Figure 30 shows a lower mass limit of such an interaction as a function of monopole mass.

6. CONCLUSIONS

Each TRISTAN experiment has acquired data of about 30 pb^{-1} in the energy range from 50 to 61.4 GeV. Three experiments observe slightly high total cross section for $e^+e^- \to$ hadrons which is inconsistent with full production of top quarks. Analysises on isolated leptons and photons, and multi-jet events set mass limits for top and b′ quarks to be respectively 30.4 GeV/c^2 and in the range from 27 to 30 GeV/c^2, depending on the unknown decay branching ratios of b′ quarks.

Basic properties of QCD associated with non-Abelian gauge theory, such as a running coupling constant and triple gluon coupling have been tested and favored to more than 99% C.L. Specially evidences of the triple gluon coupling were given for the first time. Differences between quark and gluon fragmentation were clearly shown by identifying the gluon jet in three-jet events.

None of the new particles expected from theories beyond the standard model have been found. A fit to $R_{had}, R_{l\bar{l}}, A_{l\bar{l}}$ and $A_{c\bar{c}}$ has shown, by the introduction of an extra gauge boson Z′, a possibility of explaining the high R_{had} and low $A_{l\bar{l}}$ that TRISTAN experiments are observing. However, it is necessary to make substantial improvements in the measurements of these quantities before definite conclusions can be reached.

Acknowledgement:

I am grateful to my colleges from all TRISTAN experiments for supplying materials of this talk and for numerous discussions.

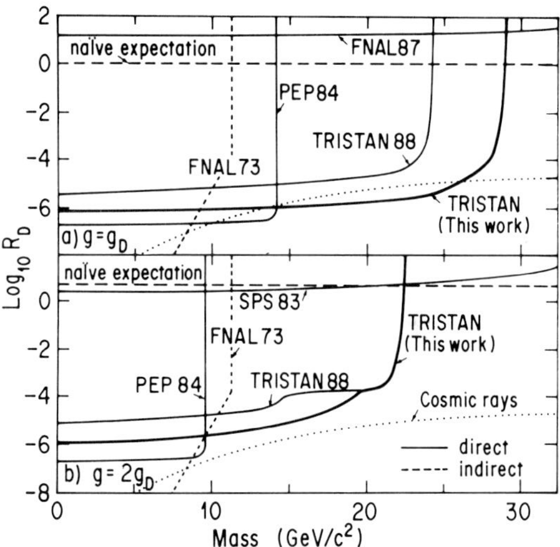

Fig. 29. Mass and cross section limts set by SHIP for monopoles with magnetic charge g=g_D and g=$2g_D$ in comparison with previous experiments.

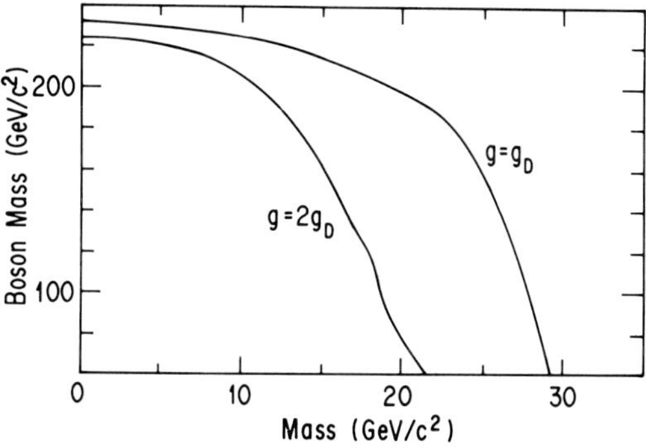

Fig. 30. Mass limits set by SHIP for new neutral gauge bosons as a function of monopole mass.

REFERENCES

1. H.Sagawa et al.(AMY), *Phys. Rev. Lett.* **60** (1988) 93; T.Mori et al.(AMY), *Phys. Lett.* **218** (1989) 499.
2. I.Adachi et al.(TOPAZ), *Phys. Rev. Lett.* **60** (1988) 97; and paper No.160 submitted to this symposium.
3. H.Yoshida et al.(VENUS), *Phys. Lett.* **B198** (1987) 570.
4. J.Fujimoto and Y.Shimizu, Mod. Phys. Lett. **A3** (1988) 581.
5. F.A.Berends and R.Kleiss, *Nucl. Phys.* **B178** (1981) 141; F.A.Berends, R.Kleiss and S.Jadach, *Nucl. Phys.* **B202** (1982) 63.
6. G.S.Abrams et al.(Mark II), *Phys. Rev. Lett.* **63** (1989) 724.
7. F.Abe et al.(CDF), *Phys. Rev. Lett.* **63** (1989) 720.
8. S.G.Gorishny, A.L.Kataev and S.A.Larin, *Phys. Lett.* **B212** (1988) 238.
9. K.Ogawa, KEK preprint 89-51, talk presented at KEK Topical Conference on e^+e^- Collision Physics, May 17-19, 1989, Tsukuba, Japan.
10. I.Adachi et al.(TOPAZ), KEK preprint 89-75, paper No.155 submitted to this symposium.
11. S.S.Myung et al.(AMY), KEK preprint 89-62, paper No.74 submitted to this symposium.
12. A.Adeva et al.(MARK J), *Phys. Rev.* **D34** (1986) 681.
13. W.Bartel et al.(JADE), *Z. Phys.* **C36** (1987) 15.
14. W.Hou and R.Stuart *Phys. Rev. Lett.* **62** (1989) 617.
15. K.Abe et al.(VENUS), KEK Preprint 89-39.
16. I.Adachi et al.(TOPAZ), paper No.151 submitted to this symposium.
17. W.Bartel et al.(JADE), *Z. Phys.* **C33** (1986) 23.
18. S.Eno et al.(AMY), KEK Preprint 89-46, paper No.105 submitted to this symposium.
19. K.Abe et al.(VENUS), KEK Preprint 89-68.
20. C.Hearty etal.(ASP), *Phys. Rev.* **D39** (1989) 3207; H.-J.Behrend et al.(CELLO), *Phys. Lett.* **B215** (1988) 186; W.T.Ford et al.(MAC), *Phys. Rev.* **D33** (1986) 3472; H.Wu, Ph.D. Thesis Univ. Hamburg, 1986.
21. A.Bacala et al.(AMY), *Phys. Lett.* **B218** (1989) 112.
22. I.Adachi et al.(TOPAZ), *Phys. Lett.* **B208** (1988) 319.
23. D.Muller in Proceedings of the 1988 International Conference on High Energy Physics, Munich, 1988, edited by R.Kotthaus and J.H.Kuehn, Springer-Verlag, 1989, p.884.
24. R.Felst in Proceedings of the 1987 Physics in Collision, Tsukuba,1987, edited by T.Kondo and K.Takahashi, World Scientific, 1988, p.87.
25. D. Hitlin in Proceedings of the 1987 International Symposium on Lepton and Photon Interactions at High Energies, Hamburg, 1987, edited by W.Bartel and R.Rueckl, N. Holland, 1988, p.179.
26. T.Tauchi, private communication.
27. H.Sagawa et al.(AMY), KEK Preprint 89-38, paper No.101 submitted to this symposium.
28. W.Bartel et al.(JADE), *Phys. Lett.* bf B146 (1984) 437; M.Althoff et al.(TASSO), *Phys. Lett.* **B146** (1984) 443; C.Kiesling, Talk at the 29th Moriond Conference (March, 1989); F.Ould-Saada, DESY report 88-177; H.R.Band et al.(MAC), *Phys. Lett.* **B218** (1989) 369; C.R.Ng et al.(HRS), ANL-HEP-PR-88-11.
29. H.Albrecht et al.(ARGUS), *Phys. Lett.* **B189** (1987) 245; M.Artuso et al.(CLEO), *Phys. Rev. Lett.* **62** (1989) 2233.
30. K.Abe et al.(VENUS), KEK preprint 89-89.
31. D.Stuart et al.(AMY), paper No.102 submitted to this symposium.
32. C.Basham, L.Brown, S.Ellis and S.Love, *Phys. Rev. Lett.* **41** (1978) 1585, *Phys. Rev.* **D19** (1979) 2018, *Phys. Rev.* **D24** (1981) 2383.
33. I.Adachi et al.(TOPAZ), KEK preprint 89-37, papaer No.147 submitted to this symposium.

34. D.Gottschalk and M.P.Shatz, *Phys. Lett.* **B150** (1985) 451, CALT-68-1172 and CALT-68-1173.

35. T.Sjostrand, Phys. Phys. Commum. **39** (1986) 347, *ibid* **43** (1987) 367.

36. F. Gutbrod, G. Kramer and G. Schierholz, *Z. Phys.* **C21** (1984) 235.

37. D.R.Wood et al.(Mark II), *Phys. Rev.* **D37** (1988) 3091; E.Fernandez et al.(MAC), *Phys. Rev.* **D31** (1985) 2724; W.T.Ford et al.(MAC), SLAC-PUB-4348 (1989); W.Bartel et al.(JADE), *Z. Phys.* **C25** (1984) 231; S.L.Wu, *Nucl. Phys.* **B** (proc. suppl.) 3 (1988), Proceedings of 1987 International Symposium on Lepton and Photon Interactions at High Energies, Hamburg, p.39.

38. G.P.Yost et al.(PDG), *Phys. Lett.* **B204** (1988) 96.

39. S.Bethke et al.(JADE), *Phys. Lett.* **B213** (1988) 235; W.Braunschweig et al.(TASSO), *Phys. Lett.* **B214** (1988) 286.

40. J.Ellis, M.K.Gaillard and G.G.Ross, *Nucl. Phys.* **B111** (1976) 253, *ibid.* **B130** (1977) 516; G.Kramer and K.Lampe, DESY-86-119 (unpublished).

41. I.H.Park et al.(AMY), *Phys. Rev. Lett.* **62** (1989) 1713; TOPAZ and VENUS, private communication.

42. M.Bengtsson and P.M.Zerwas, *Phys. Lett.* **B20** (1988) 306; M.Bengtsson, Aachen preprint PITHA 88/12 (1988) (unpublished); J.G.Körner, G.Shierholz and J.Willrodt, *Nucl. Phys.* **B185** (1981) 365; O.Nachtman and A.Reiter, *Z. Phys.* **C16** (1982) 45.

43. I.H.Park et al.(AMY), KEK preprint 89-53, paper No.103 submitted to this symposium.

44. Ref.36 and references given in it (p.100-101).

45. S.Bethke, *Z. Phys.* **C43** (1989) 331.

46. M.Derrick et al.(HRS), *Phys. Lett.* **B165** (1985) 449.

47. A.Peterson et al.(Mark II), *Phys. Rev. Lett.* **55** (1985) 1954.

48. W.Braunschweig et al.(TASSO), WIS-89/7/89-PH (unpublished).

49. Y.K.Kim et al.(AMY), KEK preprint 89-44, paper No.100 submitted to this symposium.

50. W.Bartel et al.(JADE), *Phys. Lett.* **B123** (1983) 460.

51. V.Barger, N.G.Deshpande and K.Whisnant, *Phys. Rev. Lett.* **56** (1986) 30; *Phys. Rev.* **D35** (1987) 1005; L.S.Durkin and P.Langacker, *Phys. Lett.* **B166** (1986) 436; F.del Aguila, G.Blair, M.Daniel and G.G.Ross, *Nucl. Phys.* **B283** (1987) 50; T.Matsuoka, H.Mino, D.Suematsu and S.Watanabe, *Prog. Theor. Phys.* **76** (1986) 915.

52. U.Amaldi et al., *Phys. Rev.* **D36** (1987) 1385.

53. G.Costa et al., *Nucl. Phys.* **B297** (1988) 244.

54. K.Hagiwara, R.Najima, M.Sakuda and N.Terunuma, KEK preprint 89-57.

55. Y.Sakai et al.(AMY), KEK preprint 89-42, paper No.132 submitted to this symposium; G.N.Km et al.(AMY), *Phys. Rev. Lett.* **61** (1988) 911.

56. I.Adachi et al.(TOPAZ), KEK preprint 89-31, paper No.152 submitted to this symposium.

57. C.Hearty et al.(ASP), *Phys. Rev. Lett.* **58** (1987) 1711.

58. K.Kinoshita et al.(SHIP), *Phys. Rev. Lett.* **60** (1988) 1610; *Phys. Lett.* **B228** (1989) 543.

59. R.A.Carrigan,Jr. et al., *Phys. Rev.* **D8** (1973) 3717; D.Fryberger et al., *Phys. Rev.* **D29** (1984) 1524; P.B.Price et al., *Phys. Rev. Lett.* **59** (1987) 2523; R.R.Ross et al., *Phys. Rev.* **D8** (1973) 698; B.Aubert et al., *Phys. Lett.* **B120** (1983) 465.

60. A.K.Drukier and S.Nussinov, *Phys. Rev. Lett.* **49** (1982) 102.

DISCUSSION

D. Haidt, DESY: Is R_μ in your plot shown with systematic errors? What fraction of the systematic error in the $\sqrt{s}$ dependence is common to AMY, TOPAZ and VENUS?

A. Maki: Yes, it includes systematic errors too; however, the statistical error still dominates. Out of $2.5 \sim 3.0\%$ of the systematics, only the 0.3% that comes from the radiative correction could be common to all experiments, because they are using the same full electroweak $O(\alpha_s)$ calculation code. However I do not think this has a large energy dependence.

V. Khoze, Leningrad Nuclear Physics Institute: I have a question connected with the quark-gluon comparison. How did you take into account that a large fraction of $q\bar{q}g$ events are connected with c-quarks that have specific features in their fragmentation?

A. Maki: We did not separate out the c-quark contribution. It was included in the quark-jet sample. When we estimated the jet identification efficiency, we considered any jet whose angle matches the direction of the jet that contains an original quark as a quark jet, and those which do not as gluon jets. We still do not know whether there is any difference between u- or d-quark and c-quark fragmentation.

P. Drell, Cornell University: What are the mass limits on sequential leptons?

A. Maki: I did not have time to mention this topic in my talk. However, the AMY group has made a comprehensive study of many different cases — hadron-hadron and lepton-hadron modes, stable cases of charged leptons, and Dirac and Majorana neutral leptons that couple to electrons or muons. Results are given in Figure 27 of the published version of my talk. Charged leptons are excluded up to 30 GeV.

M. Peskin, SLAC: At TRISTAN energies, the two-photon cross section is larger then the annihilation cross section and overlaps it. What cuts do you use to eliminate hard two-photon events? Are you sure that the extrapolation beyond this cut is not producing the measured excess in R_{had}?

A. Maki: For our selection of hadronic events, we required the visible energy to be greater than the beam energy and the longitudinal momentum balance to be less than 40% of the beam energy. These cuts take care of the two-photon background. Since we had plenty of two-photon events, we could tune the Monte Carlo and estimated this background to be about 0.5%. It is very hard to explain the 10% excess in R that we observed by the two-photon contribution, unless we have made an enormous error in this estimation.

MEASUREMENT OF THE Z BOSON RESONANCE PARAMETERS*

G. J. FELDMAN
Stanford Linear Accelerator Center,
Stanford University, Stanford, CA 94309

ABSTRACT

Using the Mark II detector at the SLC, we measure the Z mass and width to be 91.17 ± 0.18 GeV/c^2 and $1.95^{+0.40}_{-0.30}$ GeV, respectively. From a fit in which the visible Z width is constrained to its Standard Model value, the number of neutrino species is determined to be 3.0 ± 0.9 or < 4.4 at the 95% confidence level.

INTRODUCTION

There will be two presentations from the Mark II Collaboration[1] today. The division of labor is rather simple. I will talk about the production of Z bosons and Professor Weinstein will talk about their decay.

Three weeks ago, we submitted our initial measurements of the Z resonance parameters for publication in *Physical Review Letters.*[2] The results were

$$m = 91.11 \pm 0.23 \text{ GeV/c}^2 \ , \quad (1)$$

$$\Gamma = 1.61^{+0.60}_{-0.43} \text{ GeV} \ , \quad \text{and} \quad (2)$$

$$N_\nu = 3.8 \pm 1.4 \ . \quad (3)$$

Today, we will update these measurements based on two major improvements:

1. a doubling of the data from 106 events to 233 events, and
2. the use of the MiniSAM for point-to-point normalization.

This will result in a substantial improvement in the precision of these measurements.

We want to determine the Z boson resonance parameters by comparing the rate of Z formation in e^+e^- annihilation (Fig. 1) as a function of the center-of-mass energy, E, with that for a process with a known cross section, Bhabha (e^+e^-) scattering (Fig. 2). To accomplish this, we have to do four things:

1. measure E,
2. count Z's,
3. count Bhabha scatters, and
4. fit the ratios to obtain the Z parameters.

The above list will serve as an outline for this talk. I will simply explain how we do each of these things.

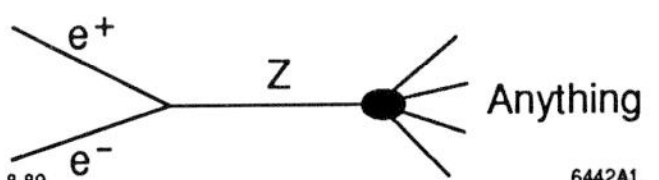

Figure 1: Z formation in e^+e^- annihilation.

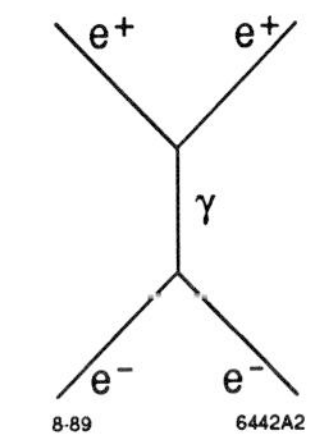

Figure 2. Small-angle Bhabha scattering.

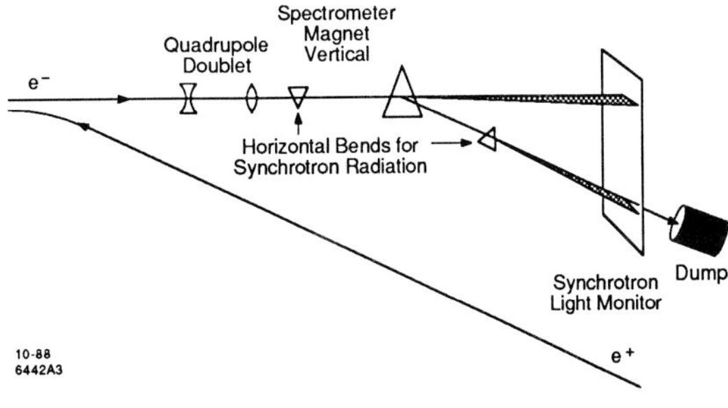

Figure 3. Schematic of one of the energy spectrometers.

⋆ Work supported by Department of Energy contract DE–AC03–76SF00515.

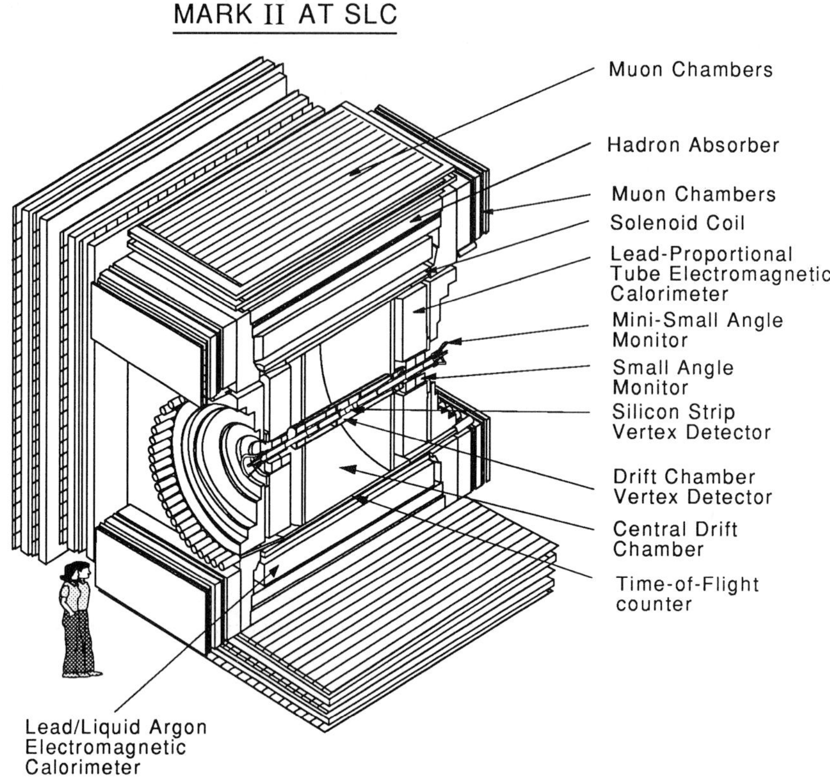

Figure 4. The Mark II detector.

Absolute Energy Measurement

We have built spectrometers of a novel design to measure the absolute energies of both beams to high accuracy.[3] Figure 3 shows a schematic drawing of one of the spectrometers. The electron beam first passes through a horizontal bend and emits a horizontal swath of synchrotron radiation in the initial electron beam direction. It then passes through an accurately-measured spectrometer magnet which bends it down. Finally, it traverses a second horizontal bend to give another swath of synchrotron radiation in the direction of the outgoing beam. The two swaths of synchrotron radiation are intercepted by a phosphorescent screen. It is clear that the mean energy of beam can be measured from the knowledge of three quantities:

1. the magnetic-field integral of the spectrometer magnet,
2. the distance between the center of the spectrometer magnet and the screen, and
3. the distance between the two synchrotron radiation swaths on the screen.

The spectrometer magnet field integral has been calibrated to a few parts in 10^5 by two independent techniques and is constantly measured by a rotating coil. The distance between the magnet center and the screen is determined to high precision by surveying techniques, and the distances on the screen are calibrated by accurately placed fiducial wires.

The systematic uncertainties in the measurement of each beam (itemized in Table 1) total to 20 MeV. Allowing for a possible correlation between beam dispersion and offset, the total systematic uncertainly in the measurement of E is 40 MeV.

Table 1: Systematic uncertainties in the energy measurement of each beam.

Item	Uncertainty (MeV)
Magnetic measurement	5
Detector resolution	10
Magnet rotation	16
Survey	5
Total	20

Table 2: Mark II triggers.

Purpose	Trigger	Requirements
Z decays	Charged	≥ 2 charged tracks with $p_t > 150$ MeV/c and $\lvert\cos\theta\rvert < 0.75$.
	Neutral	A single deposition of ≥ 2.2 GeV in the endcap calorimeter or ≥ 3.3 GeV in the barrel calorimeter.
Luminosity	SAM	≥ 6 GeV in both detectors.
	MiniSAM	≥ 15 GeV in both detectors.
Diagnostic	Random	Random beam crossings.
	Cosmic	Taken between beam crossings.

The energy spread of each beam is also measured to about 30% accuracy by the increased dispersion caused by the spectrometer magnet. The mean energy and energy spread are measured on every SLC pulse and are read by the Mark II on every trigger.

MARK II DETECTOR

A drawing of the Mark II detector[4] is shown in Fig. 4. The principal components which we will be interested in today will be the drift chamber, the calorimeters, and the luminosity monitors.

The drift chamber is a 72-layer, mini-jet cell, cylindrical chamber[5] immersed in 4.75 kG solenoidal magnetic field. It tracks charged particles in the region $\lvert\cos\theta\rvert < 0.92$, but the efficiency and momentum resolution begin to deteriorate at $\lvert\cos\theta\rvert = 0.82$. Without a vertex constraint, the momentum resolution is about $0.5\%p$ (p in GeV/c).

There are two sets of electromagnetic calorimeters, which, together, detect photons in the region $\lvert\cos\theta\rvert < 0.96$. The central calorimeters are lead-liquid argon sandwich ionization chambers[6] with an energy resolution of about $14\%/\sqrt{E}$ (E in GeV). The forward and backward calorimeters are composed of lead-gas proportional tube sandwiches with energy resolution of about $22\%/\sqrt{E}$. Both calorimeters have a strip geometry with three or four strip directions for stereographic reconstruction.

Figure 5 shows a close-up of the region around the beam-line. Note the two luminosity monitors at small angles, the Small Angle Monitor (SAM), followed at smaller angles by the MiniSAM.

TRIGGERS

The six main Mark II triggers are listed in Table 2. Monte Carlo simulations indicate that the charged and neutral triggers are 97% and 95% efficient for Z hadronic decays, respectively. In addition to being highly redundant, they are complementary in that the charged trigger is more efficient in the central region, while the neutral trigger is more efficient in the forward and backward regions. Together, they are calculated to be 99.8% efficient. (This number is irrelevant because, as we will see shortly, we will not be able to identify all of these decays.) Of the 215 hadronic events used in the present analysis, 211 satisfied both the charged and neutral trigger.

The random trigger is used to correct for beam-induced backgrounds. For example, in all Mark II SLC analyses, randomly triggered events are combined with Monte Carlo simulations of physical processes to give a complete simulation of both the physics and the backgrounds. In the next section we will see an important application of this technique.

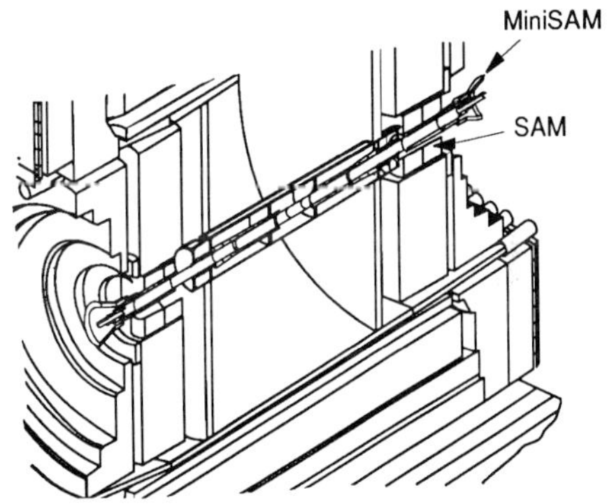

Figure 5. Detail around the beam line of the Mark II detector.

LUMINOSITY MEASUREMENTS

To obtain the optimum absolute and relative luminosity measurements we use a well-defined fiducial region of the SAM to measure the absolute luminosity, while we use the total SAM and the MiniSAM to determine the relative, or point-to-point, luminosity. The geometrical acceptance of these detectors is illustrated in Fig. 6.

Absolute Luminosity Measurement

A drawing of the SAM is shown in Fig. 7. Each SAM consists of nine layers of drift tubes for tracking and a six-layer lead-proportional-tube sandwich

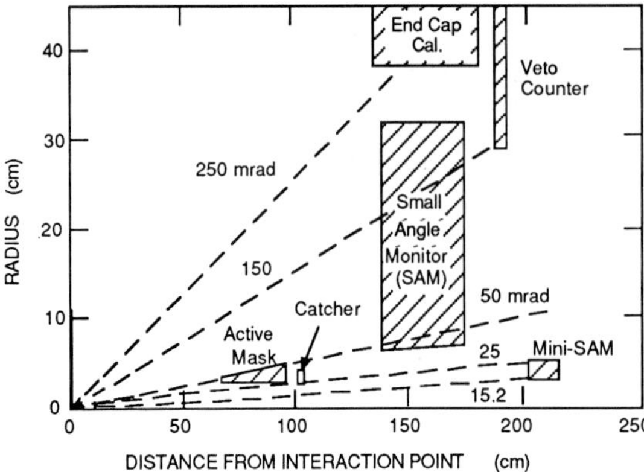

Figure 6. Geometrical acceptances of the SAM and MiniSAM.

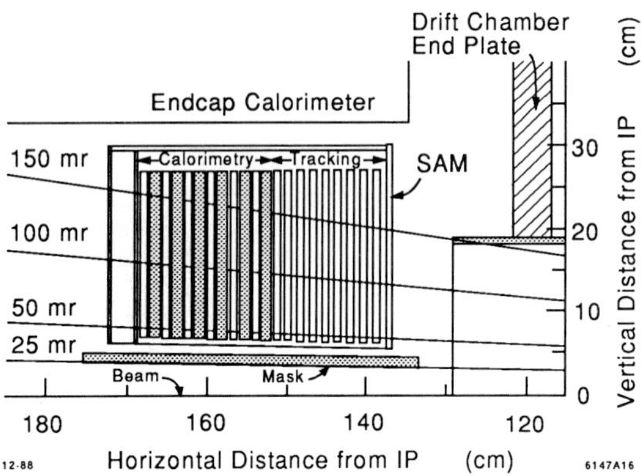

Figure 7. The Small Angle Monitor (SAM).

for measuring the electron energy and position. A typical event is shown in Fig. 8. The tracking information is not always available due to backgrounds, but the calorimetric reconstruction of the electron pulse is unmistakable and background free. The angular resolution from the shower reconstruction is about a milliradian.

Figures 9 and 10 show some of the results from the reconstructed shower measurements in the SAMs. Figure 9 shows the acollinearity angle compared with Monte Carlo calculations. Figure 10 shows the distribution in θ, the angle between the incident and scattered electron versus the number of events divided by θ^3. This quantity should be a horizontal line for full acceptance. The effect of the acceptance-defining mask can be clearly seen, and the region used for the precise luminosity measurement is shown.

The technique for determining the absolute luminosity was to count events with both the electron

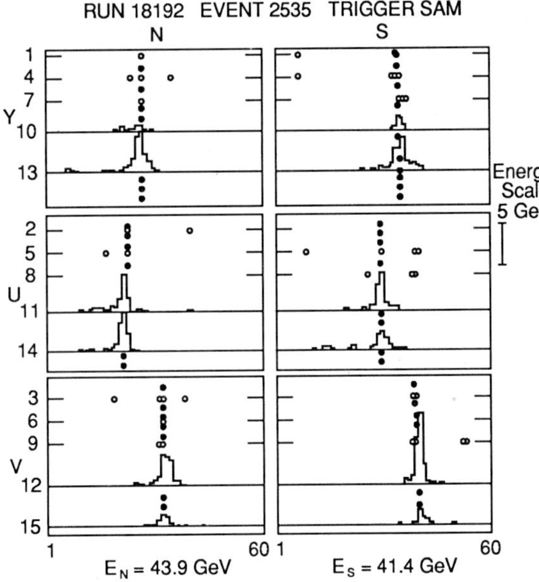

Figure 8. A typical SAM event.

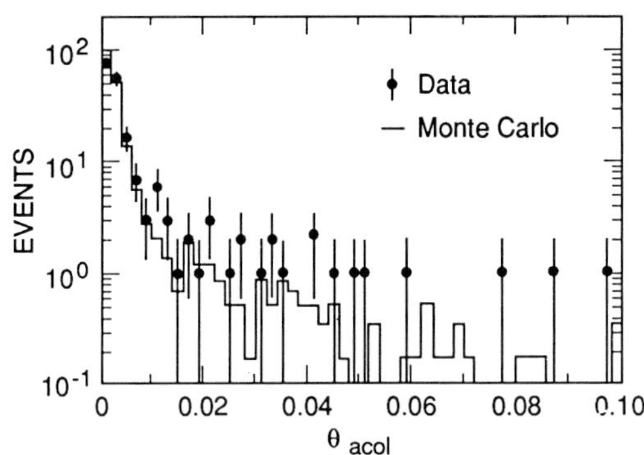

Figure 9. Acollinearity angle for Bhabha scattering events in the SAMs.

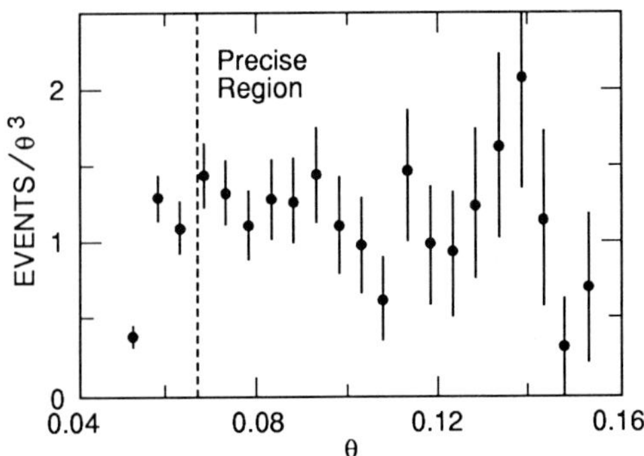

Figure 10. Angular distribution of Bhabha scattering events in the SAMs divided by θ^3.

and positron tracks in the angular region $65 < \theta < 160$ mrad with unit weight and events with only one track in this with region with half weight. This is a standard technique to reduce the sensitivity of the measurement to possible misalignments and detector resolution.

There were 236 events with both tracks and 21 events with only one track in the precise region. The cross section corresponding to the precise region was calculated to be 24.9 nb at 91 GeV.[7]

The systematic uncertainties in the absolute luminosity measurement total 3.0% and are equally divided between unknown higher-order radiative corrections and the effect of detector resolution on the SAM precise region acceptance.

Relative Luminosity Measurement

The most important part of the point-to-point luminosity measurement are the MiniSAMs, a drawing of which is shown in Fig. 11. These are simple tungsten-scintillator sandwiches divided into four quadrants which are separately read out.

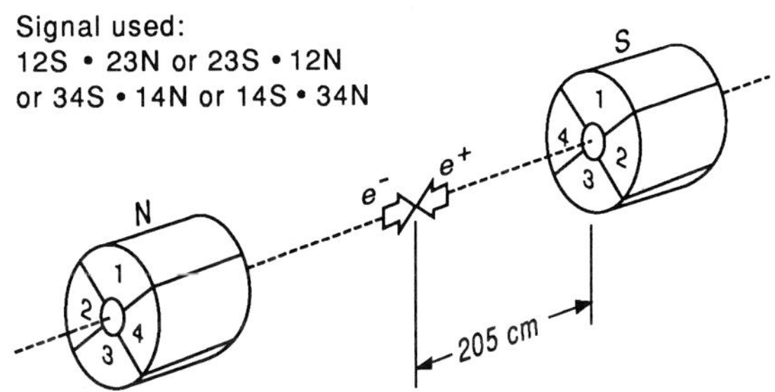

Figure 11. The MiniSAMs.

The requirement for detecting a Bhabha scattering event in the MiniSAMs is

1. back-to-back quadrants each with more than 15 GeV,
2. no nonadjacent quadrant with more than 15 GeV, and
3. time-of-flight measurement in all quadrants with more than 15 GeV consistent with Bhabha scattering.

There are two major sources of systematic error associated with the MiniSAMs. First the beam positions and angles can change from point to point. We track these changes closely and have concluded that they lead to a negligible error.

The second source is potentially more serious. Beam-related backgrounds can prevent the

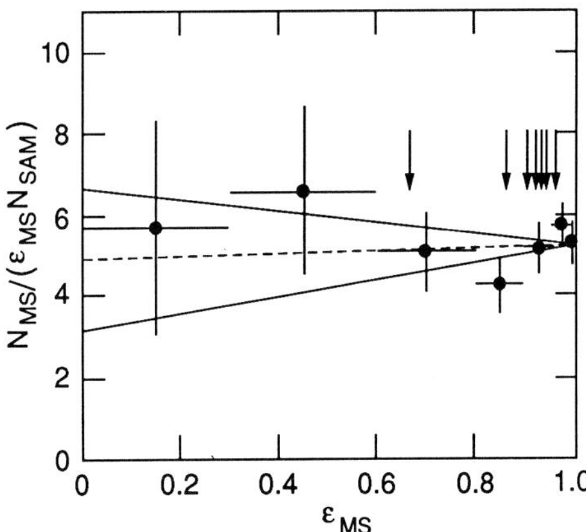

Figure 12. The ratio of the efficiency-corrected MiniSAM rate to the SAM rate as a function of the MiniSAM efficiency. The dashed line indicates a best fit, and the solid lines indicate the 1σ limits on the fit. The efficiencies of the seven scan points are shown by arrows.

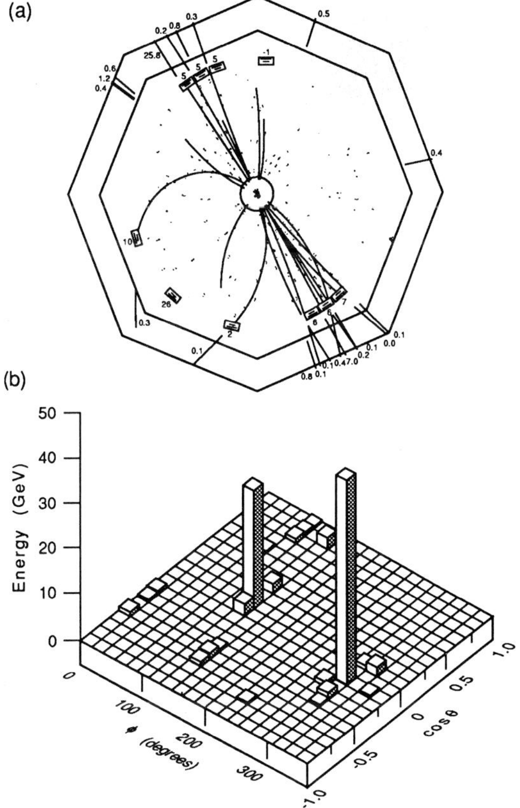

Figure 13. (a) A computer reconstruction of a typical hadronic Z decay viewed along the beam axis. (b) A plot of the detected energy for this event as a function of the azimuthal angle and the cosine of the polar angle.

MiniSAMs from meeting the above requirements for an otherwise valid Bhabha scattering event. To determine the magnitude of this effect, the raw data from Bhabha scattering events created by Monte Carlo simulations are added to randomly triggered events to determine if the events would have been counted if they had occurred. In this way, a MiniSAM efficiency is calculated for each run. The average efficiency was 90%, but it varied from 65% to 96% on different scan points.

Figure 12 shows the ratio of the efficiency-corrected MiniSAM rate to the SAM rate as a function of the MiniSAM efficiency. There is no indication that the efficiency calculation is biased, but we parameterize the uncertainty in the fit, shown by solid lines, as a possible systematic error.

Z DECAY EVENT SELECTION

Figure 13(a) shows a typical hadronic Z decay. The two-jet structure, shown graphically in the Lego plot of Fig. 13(b), and the charged multiplicity of about 20 tracks are typical of these events. About 7/8 of visible Z decays are into hadronic modes. The remainder are split among e, μ, and τ pairs. A τ pair decay is shown in Fig. 14, in which one of the τs decays into a 16 GeV/c muon and the other decays into a 17 GeV/c electron.

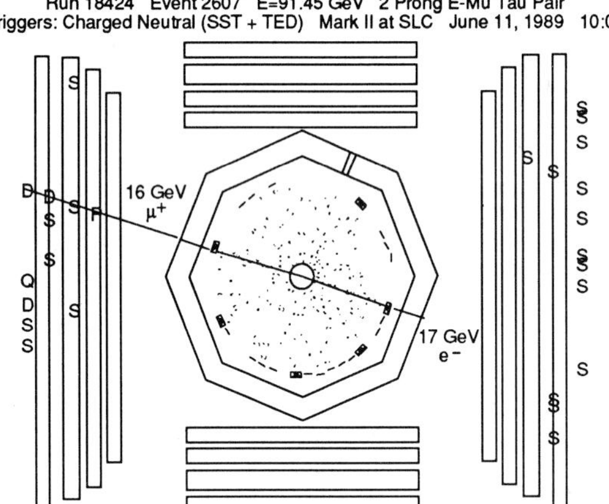

Figure 14. A computer reconstruction of a Z decay into τ pairs.

Z production is the dominant annihilation process, so the selection criteria can be quite loose. The only possible backgrounds come from beam-gas interactions and γ–γ interactions. Both of these process leave a large amount of energy in at most one of the forward-backward hemispheres.

Accordingly, the criteria for hadronic Z decays are:

1. ≥ 3 charged tracks from a cylindrical volume around the interaction point with a radius of 1 cm and a half-length of 3 cm, and
2. at least 0.05 E visible in both the forward and backward hemispheres.

These criteria give an efficiency of 94.5 ± 0.5%.

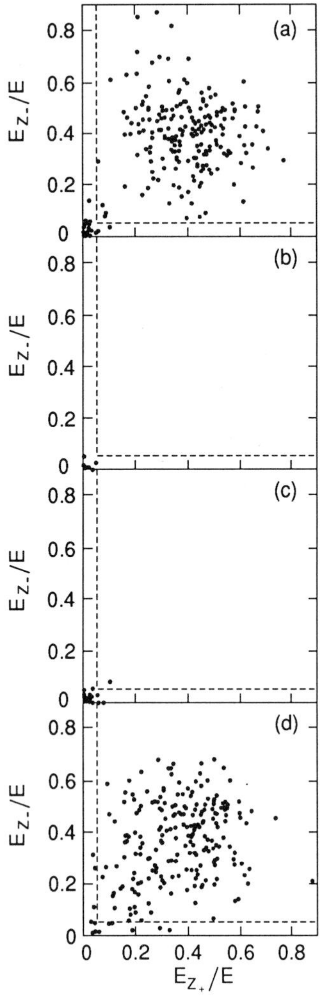

Figure 15. Fractional energies in the forward versus backward hemispheres for events with three or more charged tracks from the fiducial region for (a) events from the region of the interaction point, (b) events displaced along the beam line by ±10 and ±16 cm (four times the fiducial volume), (c) a Monte Carlo simulation of γ–γ interactions with 40 times the integrated luminosity, and (d) a Monte Carlo simulation of Z hadronic decays. The lines indicate the region for acceptable events.

Figure 15 shows the possible level of backgrounds. Each event with 3 or more charged tracks from the fiducial region is shown on a scatter plot of the amount of energy in each hemisphere expressed as a fraction of E. Figure 15(a) shows the data from the region of the interaction point. There is a diffuse cluster of events which meet the acceptance criteria and a another cluster near zero energy in both hemispheres. Figure 15(b) shows the distribution of beam-gas interactions. It is constructed by taking the fiducial regions displaced along the beam line by ±10 and ±16 cm, giving four times the fiducial volume. No event is within the acceptable region.

Figure 15(c) shows the results of a Monte Carlo simulation of γ–γ interactions corresponding to forty times the present luminosity. One event passes the acceptance criteria. Finally, Fig. 15(d) shows the result of a Monte Carlo simulation of Z hadronic decays. These plots demonstrate that the Z hadronic decays are essentially background-free.

To increase our statistical precision slightly, we also accept those leptonic Z decays for which the efficiency is high and the identification and interpretation is clear, namely, μ and τ pairs in the angular region $|\cos\theta| < 0.65$. To avoid backgrounds from γ–γ interactions, we require a minimum of 0.10 E visible energy for τ pairs. The efficiencies for detecting μ and τ pairs within the fiducial angular region are 99 ± 1% and 96 ± 1%, respectively.

The data are shown in Table 3 and Fig. 16. Note that we plot an unusual quantity, but onc that is closely related to what we actually measure, the cross sections for all hadronic decays and $\mu^+\mu^-$ and $\tau^+\tau^-$ with $|\cos\theta| < 0.65$. The rest of this talk discusses how we get information on the Z resonance parameters from these data.

Table 3: Summary of the data. (a) The errors do not include an overall 6.7% normalization uncertainty. (b) The total of 233 events is composed of 215 hadronic decays, 7 $\mu^+\mu^-$, and 11 $\tau^+\tau^-$. (c) Cross sections are for all hadronic decays and $\mu^+\mu^-$ and $\tau^+\tau^-$ with $|\cos\theta| < 0.65$. The errors are for 68.3% confidence level integrals. See Ref. [8].

$< E >$ (GeV)	$\int \mathcal{L}dt^{(a)}$ (nb^{-1})	$N_Z{}^{(b)}$	$\sigma_Z{}^{(a,c)}$ (nb)
89.24	0.67 ± 0.05	3	$4.6^{+4.6}_{-2.5}$
89.98	0.82 ± 0.06	10	$13.1^{+5.8}_{-4.2}$
90.74	1.31 ± 0.07	36	$30.5^{+6.4}_{-5.3}$
91.06	4.25 ± 0.13	120	$31.2^{+3.3}_{-3.0}$
91.50	1.33 ± 0.08	39	$32.4^{+6.5}_{-5.5}$
92.16	0.58 ± 0.05	11	$19.8^{+8.4}_{-6.1}$
92.96	1.10 ± 0.08	14	$13.0^{+4.7}_{-3.6}$
Totals	9.89 ± 0.21	233	

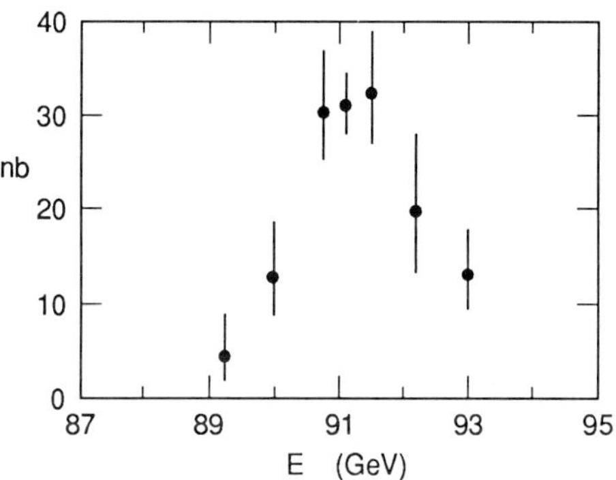

Figure 16. The cross sections for all hadronic decays and $\mu^+\mu^-$ and $\tau^+\tau^-$ with $|\cos\theta| < 0.65$.

FITS TO THE DATA

We perform maximum-likelihood fits using Poisson statistics to a relativistic Breit–Wigner line shape

$$\sigma(E) = \frac{12\pi}{m^2}\frac{s\Gamma_{ee}(\Gamma - \Gamma_{inv})}{(s-m^2)^2 + s^2\Gamma^2/m^2}[1+\delta(E)] \quad , \quad (4)$$

where Γ is the total width and Γ_{inv} is the partial width into invisible decay modes, *i.e.,* into neutrinos and neutrino-like particles. Large effects due to initial state radiation, represented in Eq. (4) by $[1 + \delta(E)]$, are calculated by an analytic form due to Cahn.[9] Alexander *et al.* have shown that this form has more than sufficient accuracy for our purposes.[10]

A Breit–Wigner shape has three parameters, a position, a width, and a height. We can fit for these three parameters as m, Γ, and Γ_{inv}, or equivalently, the number of neutrino species, N_ν.

The mass and width clearly determine the position and width of the resonance. The height is most sensitive to the third parameter, Γ_{inv}. This comes about because a Breit–Wigner is proportional to the partial width to the initial state times the partial width to the final state. The partial width to the initial state, e^+e^-, is well determined in the Standard Model. The final state can be taken to be all of the final states that we can see, in principle, in our detector, *i.e.,* all states except those into neutrino pairs (or pairs of neutrino-like objects).

Another way of viewing this is the following: If we could detect all of the final states, and if we integrated the resonance over energy, then we would find that the integral only depends on the width to the initial state. This is a statement that we pro-

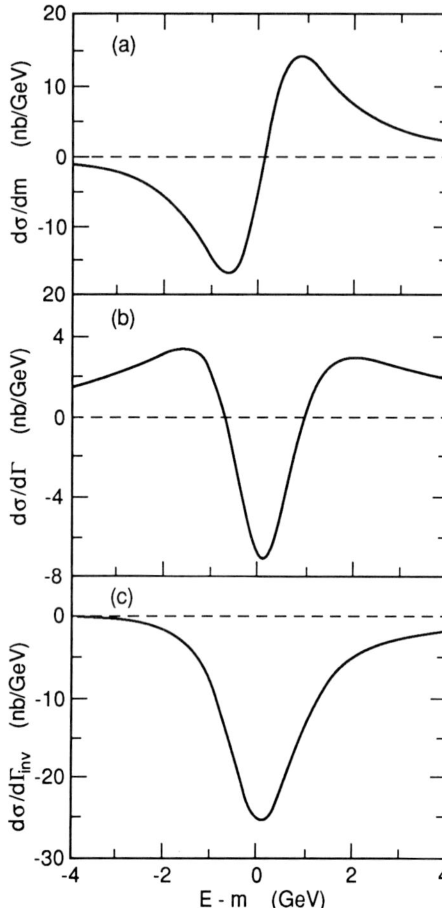

Figure 17. The derivatives of the Breit–Wigner shape, Eq. (4), with respect to the three fit parameters: (a) $d\sigma/dm$, (b) $d\sigma/d\Gamma$, and (c) $d\sigma/d\Gamma_{inv}$.

duce a Z and that it subsequently decays with unit probability. What we do not detect, then, must be those decays into neutrino pairs.

The sensitivities of the fits for the various parameters can be seen from the derivatives of Eq. (4), which are displayed in Fig. 17. The derivative $d\sigma/dm$ is an odd function, while both $d\sigma/d\Gamma$ and $d\sigma/d\Gamma_{inv}$ are even functions. This means that the determination of the mass is independent of the determination of the widths for scans which are symmetric with respect to the peak position. The determinations of Γ and Γ_{inv} differ in that the former is independent of the normalization (its derivative is bipolar), while the latter is strongly dependent on the normalization (its derivative is unipolar).

We perform three fits which differ in their reliance on the Standard Model, and thus address different questions that one may wish to ask.

1. In the "Standard Model Fit," m is the only parameter that is varied. The widths are taken to have their Standard Model values corresponding to the decays into five quarks and three charged and neutral leptons.
2. In the "Free ν Fit," both m and Γ_{inv} are allowed to vary. The visible width is constrained to its Standard Model value. The rationale for this is twofold:
   (a) New particle production in the quark-lepton sector might be expected to show up first with the lightest of particles, which, from the three examples we have seen so far, are the neutrinos.
   (b) Visible new particle production would probably show up first in the observation of distinctive decays.
3. Finally, the "Unconstrained Fit" allows all three parameters to be varied.

These fits are displayed in Fig. 18 and the results of the fits are displayed in Table 4. The mass values from the three fits are almost identical because of the orthogonality of the mass to the widths, as we just discussed. We choose the value from the Free ν Fit simply because it has the largest error. (I am often asked why the error on mass from the fit which has the most free parameters is the smallest. The reason is simple. First, the additional width parameters do not affect the mass determination because of the orthogonality of the functions. Second, the only scale of energy is the width. Since the unconstrained width is less than the Standard Model width, the error on the mass is smaller by roughly the same proportion.) All of the systematic errors in the mass determination are small, but the two largest sources of systematic uncertainty, included in the quoted errors, are 50 MeV for the MiniSAM efficiency and 40 MeV for the absolute energy determination.

The total width Γ is only determined by the Unconstrained Fit. The value of $1.95^{+0.40}_{-0.30}$ GeV should be compared to the Standard Model value of 2.46 GeV.[11] Two comments are in order:

1. The errors on the width are large because a good measurement of the width requires substantial data at ± 2 GeV from the peak, as can be seen from Fig. 17(b). We did not take very much data that far from the peak because the rate was just too low with our luminosity.
2. Although the width appears to be 1.3σ low, this is not a particularly meaningful statement be-

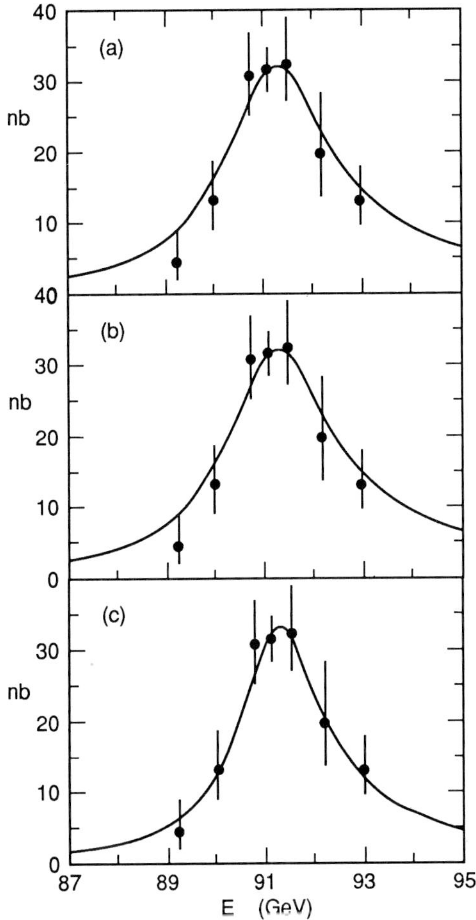

Figure 18. Fits to the cross section versus E. (a) the Standard Model Fit: m is free. (b) the Free ν Fit: m and Γ_{inv} are free. (c) the Unconstrained Fit: m, Γ, and Γ_{inv} are free.

Table 4: Results of the fits. The preferred values are shown in boldface. See text for explanation.

	Fit		
Variable (units)	Standard Model	Free ν	Uncon-strained
m (GeV/c^2)	91.17 $\pm$0.18	**91.17** **$\pm$0.18**	91.16 $\pm$0.16
N_ν	–	**3.0 ± 0.9**	3.5 ± 0.8
Γ (GeV)	–	–	$\mathbf{1.95^{+0.40}_{-0.30}}$

cause the likelihood function, shown in Fig. 19, is not Gaussian. A Monte Carlo simulation shows that if the true width were the Standard Model value, 11% of experiments with our integrated luminosity and scanning strategy would get a value for the width equal to or less than the one we obtained.

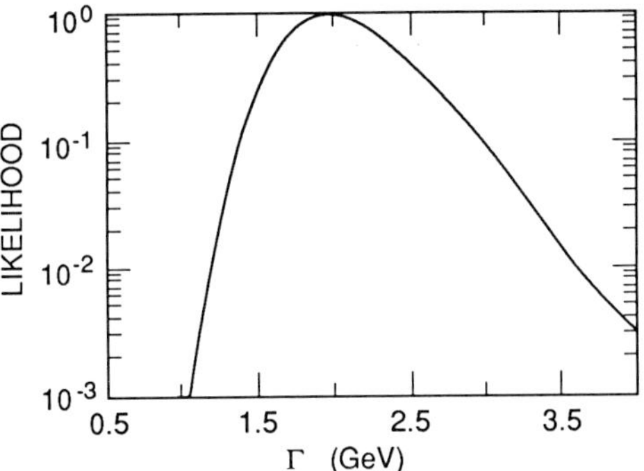

Figure 19. The relative likelihood function for Γ for the Unconstrained Fit.

Again all the systematic errors are small compared to the statistical error. The most significant contributions are 90 MeV from the uncertainty in the MiniSAM efficiency and 30 MeV from point-to-point energy uncertainties.

We take the value of $N_\nu = 3.0 \pm 0.9$ (or < 4.4 at the 95% confidence level) from the Free ν Fit as the preferred value. The value derived from the Unconstrained Fit cannot be interpreted as giving a consistent measurement of N_ν, since the fit has $\Gamma < \Gamma_{SM}$. This cannot happen in the Standard Model without a modification to the underlying gauge structure. However, in that case, Γ_{ee} and Γ_ν would not have Standard Model values, which we have assumed in calculating N_ν. The quoted errors include a contribution of 0.5 statistical and 0.25 systematic uncertainty coming from the overall absolute luminosity determination.

RELATIONSHIP BETWEEN THE MASS AND $\sin^2\theta_W$

The electroweak mixing angle, θ_W, can be expressed in terms of the Z mass by the relation

$$\sin 2\theta_W = \left(\frac{4\pi\alpha}{\sqrt{2}G_F m_Z^2(1-\Delta r)}\right)^{1/2} \quad , \tag{5}$$

where Δr represents weak radiative corrections. These corrections arise from loops and are sensitive to the masses of high mass particles.

The most common definition of $\sin^2\theta_W$ is the Sirlin form,[12] which is defined as

$$\sin^2\theta_W \equiv 1 - \frac{m_W^2}{m_Z^2} \quad . \tag{6}$$

For specific values of the two unknown masses in the Standard Model,

$$m_t = m_H = 100 \text{ GeV/c}^2 \quad , \tag{7}$$

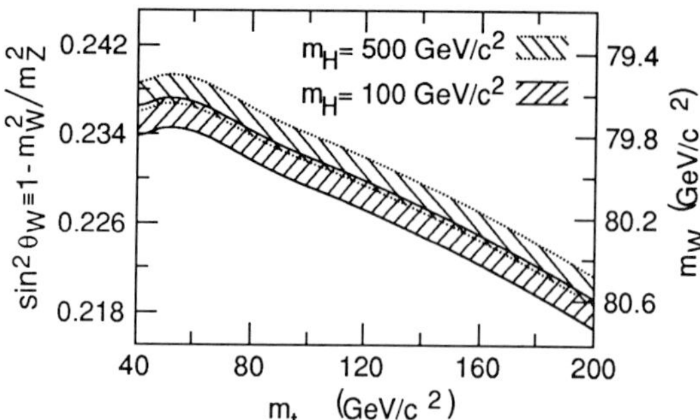

Figure 20. $\sin^2\theta_W$ as a function of the mass of the top quark for two values of the Higgs Boson mass. The bands represent ± 1 standard deviation about the values derived from Eq. (5).

our measured m_Z of 91.17 $\pm$ 0.18 GeV/c^2 implies

$$\sin^2\theta_W = 0.2307 \pm 0.0013 \quad . \tag{8}$$

The dependence of $\sin^2\theta_W$ on these two masses is shown in Fig. 20.

POSTSCRIPT

For the purpose of the historical record, the results given above were those reported at the Symposium. Between the time of the Symposium and the completion of this written version (mid-November 1989), the Mark II Collaboration doubled the data, from 233 events to 480 events, and published the following updated results[13]:

$$m = 91.14 \pm 0.12 \text{ GeV/c}^2 \quad , \tag{9}$$

$$\Gamma = 2.42^{+0.45}_{-0.35} \text{ GeV} \quad , \quad \text{and} \tag{10}$$

$$N_\nu = 2.8 \pm 0.6 \quad . \tag{11}$$

The last result translates into the upper limit

$$N_\nu < 3.9 \text{ at } 95\% \text{ C.L.} \quad , \tag{12}$$

which provides strong evidence that the number of light neutrino species is limited to the three that we have already discovered.

REFERENCES

1. The Mark II Collaboration consists of approximately 130 physicists from nine institutions: the California Institute of Technology, the University of California at Santa Cruz, the University of Colorado, the University of Hawaii, Indiana University, Johns Hopkins University, Lawrence Berkeley Laboratory, the University of Michigan, and the Stanford Linear Accelerator Center. The present members of the Collaboration are: G. S. Abrams, C. E. Adolphsen, R. Aleksan, J. P. Alexander, D. Averill, J. Ballam, B. C. Barish, T. Barklow, B. A. Barnett, J. Bartelt, S. Bethke, D. Blockus, W. de Boer, G. Bonvicini, A. Boyarski, B. Brabson, A. Breakstone, F. Bulos, P. R. Burchat, D. L. Burke, R. J. Cence, J. Chapman, M. Chmeissani, D. Cords, D. P. Coupal, P. Dauncey, H. C. DeStaebler, D. E. Dorfan, J. M. Dorfan, D. C. Drewer, R. Elia, G. J. Feldman, D. Fernandes, R. C. Field, W. T. Ford, C. Fordham, R. Frey, D. Fujino, K. K. Gan, C. Gatto, E. Gero, G. Gidal, T. Glanzman, G. Goldhaber, J. J. Gomez Cadenas, G. Gratta, G. Grindhammer, P. Grosse-Wiesmann, G. Hanson, R. Harr, B. Harral, F. A. Harris, C. M. Hawkes, K. Hayes, C. Hearty, C. A. Heusch, M. D. Hildreth, T. Himel, D. A. Hinshaw, S. J. Hong, D. Hutchinson, J. Hylen, W. R. Innes, R. G. Jacobsen, J. A. Jaros, C. K. Jung, J. A. Kadyk, J. Kent, M. King, S. R. Klein, D. S. Koetke, S. Komamiya, W. Koska, L. A. Kowalski, W. Kozanecki, J. F. Kral, M. Kuhlen, L. Labarga, A. J. Lankford, R. R. Larsen, F. Le Diberder, M. E. Levi, A. M. Litke, X. C. Lou, V. Lüth, G. R. Lynch, J. A. McKenna, J. A. J. Matthews, T. Mattison, B. D. Milliken, K. C. Moffeit, C. T. Munger, W. N. Murray, J. Nash, H. Ogren, K. F. O'Shaughnessy, S. I. Parker, C. Peck, M. L. Perl, F. Perrier, M. Petradza, R. Pitthan, F. C. Porter, P. Rankin, K. Riles, F. R. Rouse, D. R. Rust, H. F. W. Sadrozinski, M. W. Schaad, B. A. Schumm, A. Seiden, J. G. Smith, A. Snyder, E. Soderstrom, D. P. Stoker, R. Stroynowski, M. Swartz, R. Thun, G. H. Trilling, R. Van Kooten, P. Voruganti, S. R. Wagner, S. Watson, P. Weber, A. Weigend, A. J. Weinstein, A. J. Weir, E. Wicklund, M. Woods, G. Wormser, D. Y. Wu, M. Yurko, C. Zaccardelli, and C. von Zanthier.
2. G. S. Abrams *et al., Phys. Rev. Lett.* **63**, 724 (1989).
3. J. Kent *et al.,* SLAC–PUB–4922 (1989); M. Levi, J. Nash, and S. Watson, SLAC–PUB–4654 (1989); and M. Levi *et al.,* SLAC–PUB–4921 (1989).
4. G. S. Abrams *et al., Nucl. Instrum. Methods* **A281**, 55 (1984).
5. G. G. Hanson, *Nucl. Instrum. Methods* **A252**, 343 (1986).

6. G. S. Abrams *et al.*, *IEEE Trans. Nucl. Sci.* **NS–25**, 309 (1978) and **NS–27**, 59 (1980).
7. F. A. Berends, R. Kleiss, and W. Hollik, *Nucl. Phys.* **B304**, 712 (1988); S. Jadach and B. F. L. Ward, University of Tennessee report UTHEP–88–11–01 (1988).
8. F. James and M. Roos, *Nucl. Phys.* **B172**, 475 (1980).
9. R. N. Cahn, *Phys. Rev.* **D36**, 2666 (1987), Eqs. (4.4) and (3.1).
10. J. Alexander *et al.*, *Phys. Rev.* **D37**, 56 (1988).
11. Calculated using the program EXPOSTAR, assuming $m_t = m_H = 100$ GeV/c^2. D. C. Kennedy *et al.*, *Nucl. Phys.* **B321**, 83 (1989).
12. A. Sirlin, *Phys. Rev.* **D22**, 2695 (1980).
13. G. S. Abrams *et al.*, *Phys. Rev. Lett.* **63**, 2173 (1989).

DISCUSSION

R. Hofstadter, Stanford University: Do you need to include a correction for the Bhabha cross section because the e^+e^- scattering occurs in a high magnetic field?

G. Feldman: No, the magnetic fields from the SLC beams are far too weak to require such a correction.

Results from the Mark II at SLC on Decays of the Z^0

Alan J. Weinstein
California Institute of Technology
Pasadena, California, 91125
Representing the Mark II Collaboration[1]

ABSTRACT

New results from the Mark II at SLC on decays of the Z^0 are presented. The topics covered are leptonic decays, properties of hadronic decays (event shapes and inclusive distributions), and searches for heavy quarks and leptons. The leptonic branching ratios are consistent with Standard Model predictions, and the hadronic decays behave in accordance with QCD. No evidence is found for production of the top quark, a fourth-generation b' quark, or neutral heavy leptons.

INTRODUCTION

The Mark II has recently completed its first run at the SLC, accumulating approximately 350 Z^0 decays at eight cms energies near the peak of the resonance. In addition to extracting the resonance parameters, we have studied various aspects of the Z^0 decays. The subjects discussed in this report are the leptonic decay branching ratios, the properties of the hadronic decays (inclusive distributions, event shapes, and jet production), and searches for massive quarks and leptons.

Note that this paper includes results from the full data sample as of September 1, 1989, approximately three weeks after the Lepton Photon Symposium. The conclusions were not changed significantly by the addition of approximately 125 events, except that the lepton branching ratios (see below) are now in better agreement with the Standard Model predictions.

EVENT SELECTION

Events are selected based on charged tracks reconstructed in the central drift chamber, and showers reconstructed in the barrel and endcap calorimeters. We consider only charged tracks which

1. pass through a cylinder of radius 1 cm, half-length 3 cm along the beam axis, centered around the measured e^+e^- interaction point
2. have momentum transverse to the beam $p_T >$ 0.15 GeV/c
3. satisfy $|\cos\theta| < 0.82$ where θ is the angle with respect to the beam axis. This ensures efficient tracking, good modelling by the Monte Carlo, and good momentum measurement.

We consider all showers in the calorimeter which

1. have shower energy $E_{sh} > 1$ GeV
2. fall within fiducial volumes of barrel or endcap calorimeters ($|\cos\theta| < 0.96$)
3. are not consistent with being produced by muons from upstream collimators, based on longitudinal distribution of shower energy.
4. In isolated track searches, we avoid "double counting" of showers that have a charged track pointing to them by requiring $E_{sh} > 2 \cdot p_{track}$.

We then select events based on the total multiplicity N_{ch}, the total visible energy E_{vis} (charged plus neutral, assuming pion masses for charged tracks), and energy balance along the beam direction.

Simulation of machine backgrounds

Our events are contaminated with machine–related backgrounds:

1. muons, produced in upstream collimators, travelling approximately parallel to the beam axis;
2. synchrotron photons produced in the final quads, scattering into the drift chamber, leaving "salt–and–pepper" hits;
3. electromagnetic debris in the endcap calorimeter and outer muon chambers.

These backgrounds degrade the performance of the charged particle tracking, and produce fake showers in the calorimeters. We simulate these backgrounds in the Monte Carlo by "mixing" Monte Carlo Z^0 decay events with real events, triggered on random beam crossings near in time to our Z^0 decay events.

This procedure allows us to understand and correct for the effect of these backgrounds on our reconstruction efficiencies and resolutions. Note that these backgrounds decreased dramatically after the installation (near the beginning of our run) of carefully designed masks, collimators, and "muon-sweeper" toroids in the beam line.

LEPTONIC DECAYS OF THE Z^0

We choose to measure the ratio of the leptonic widths of the Z^0 to its hadronic width

$$\frac{\Gamma_{ll}}{\Gamma_{had}} = \frac{BR(Z^0 \to l^+l^-)}{BR(Z^0 \to \text{hadrons})} \tag{1}$$

For $m_Z = 91.1$ GeV/c^2 and $\sin^2\theta_W = 0.23$, the Standard Model predicts

$$\frac{\Gamma_{ll}}{\Gamma_{had}} = \frac{v_l^2 + a_l^2}{3\Sigma(v_q^2 + a_q^2)\left(1 + \frac{\alpha_s}{\pi}\right)} = 0.048. \tag{2}$$

Experimentally, we measure

$$\frac{\Gamma_{ll}}{\Gamma_{had}} = \frac{N_{ll}^{meas}}{N_{had}^{meas}} = \frac{\epsilon_{had}^{had}(N_{ll}^{obs} - N_{ll}^{BG} - \Sigma_x \epsilon_{ll}^x N_x^{prod})}{\epsilon_{ll}^{ll}(N_{had}^{obs} - N_{had}^{BG} - \Sigma_x \epsilon_{had}^x N_x^{prod})}, \tag{3}$$

where ϵ_{had}^{had} is the efficiency for reconstructing hadronic events, and ϵ_{ll}^{ll} is the same for leptonic events; N_{had}^{obs} is the observed number of hadronic events and N_{ll}^{obs} is the same for leptonic events; N_{had}^{BG} and N_{ll}^{BG} are the estimated number of beam-gas and cosmic ray events contaminating the hadronic and leptonic event sample; and the remaining terms represent production rates and efficiencies for feedthrough from one collision process to another ($x = e^+e^-, \mu^+\mu^-, \tau^+\tau^-, q\bar{q}, \gamma\gamma$). We estimate the beam-gas and cosmic ray background by performing the same event selection (described below) but requiring the collision point to be displaced along the beam line. The feedthrough from one collision process to another is estimated using the Monte Carlo.

The hadronic event selection requires $N_{ch} \geq 7$ and $E_{vis} \geq 15\% E_{cm}$. The efficiency for these cuts is estimated by Monte Carlo to be $(85 \pm 2)\%$. The leptonic event selection (not yet distinguishing between $e^+e^-, \mu^+\mu^-, \tau^+\tau^-$) requires

1. N_{ch} between 2 and 6;
2. no other charged tracks with $|\cos\theta| > 0.82$;
3. $E_{vis} > 15\% E_{cm}$;
4. $E_{ch} > 5\% E_{cm}$ where E_{ch} is the total energy observed in charged particles, assuming pion masses;
5. Thrust (computed using charged tracks only) > 0.99.

We then divide the candidate leptonic event into two thrust hemispheres. Each hemisphere must contain at least one charged track, and hemispheres with more than one charged track must have $m_{ch} \leq 2.2$ GeV/c^2, where m_{ch} is the invariant mass of the charged tracks.

We distinguish e^+e^- from $\mu^+\mu^-, \tau^+\tau^-$ on the basis of total energy deposited in the electromagnetic calorimeters, E_{cal}. Figure 1 shows the Monte Carlo predictions for the ratio E_{cal}/E_{cm} for $e^+e^-, \mu^+\mu^-, \tau^+\tau^-$ separately. A cut at $E_{cal} \geq 0.8E_{cm}$ selects our e^+e^- events.

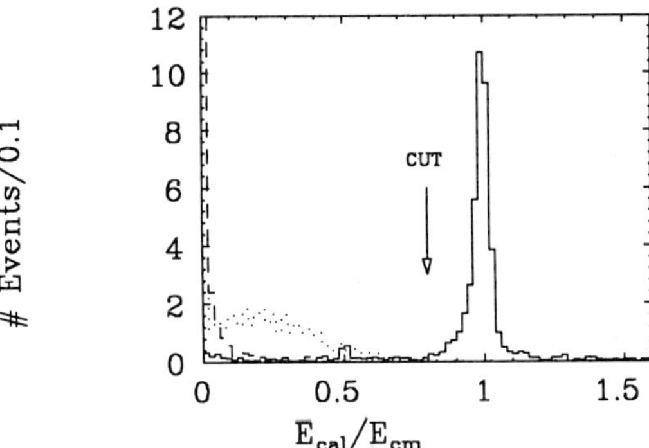

Fig. 1. Monte Carlo predictions of total energy deposited in the electromagnetic calorimeter from e^+e^- (solid histogram), $\mu^+\mu^-$ (dashed), and $\tau^+\tau^-$ (dotted). The cut for selecting e^+e^- is shown.

The remaining events are classified by hemisphere topology. Topologies that are one-prong versus three-prong (1-3) or (3-3) are classified as $\tau^+\tau^-$. Events in the (1-1) topology are classified as $\mu^+\mu^-$ or $\tau^+\tau^-$ based on the momentum of the lower-momentum track of the two in the event (p_{min}). Figure 2 shows the Monte Carlo predictions of p_{min} for e^+e^- which fail the E_{cal} cut, $\mu^+\mu^-$, and $\tau^+\tau^-$ separately. A cut at $p_{min} \geq 0.3E_{cm} \equiv 0.6p_{beam}$ separates our $\mu^+\mu^-$ events sample from our $\tau^+\tau^-$ event sample.

A summary of our data is given in Table 1. We measure

$$\Gamma_{\mu\mu}/\Gamma_{had} = 0.050^{+0.024}_{-0.017} \pm 0.002, \tag{4}$$

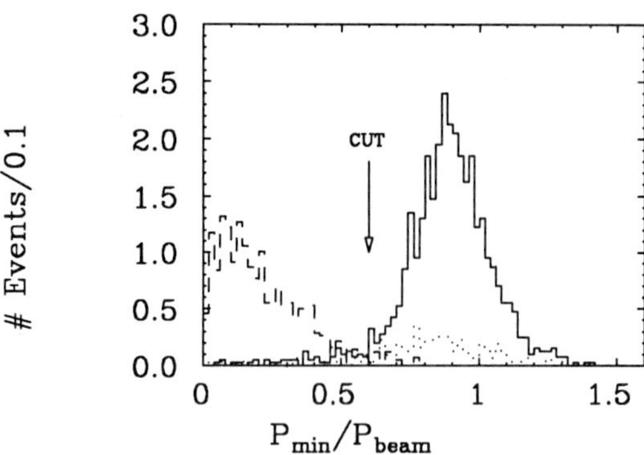

Fig. 2. Monte Carlo predictions of p_{min}/p_{beam}, the momentum of the lower momentum track in the event normalized to the beam energy, from $\mu^+\mu^-$ (solid histogram), e^+e^- which fail the E_{cal} cut (dotted), and $\tau^+\tau^-$ (dashed). The cut for selecting $\mu^+\mu^-$ is shown.

TABLE 1

Summary of the lepton branching ratio analysis. (Results for e^+e^- are for $|\cos\theta| < 0.82$.)

		e^+e^-	$\mu^+\mu^-$	$\tau^+\tau^-$	hadrons
N_{obs}		13	9	17	276
N_{bg}	ee	–	–	0.13	–
	$\mu\mu$	–	–	0.33	–
	$\tau\tau$	0.11	0.05	–	0.03
	$q\bar{q}$	–	–	0.42	–
	cosmics	–	0.12	0.12	<0.18
	$\gamma\gamma$	<0.01	<0.01	<0.01	<0.01
N_{bg}	sum	0.11	0.17	1.0	0.03
ϵ		0.93	0.54	0.65	0.85
N_{corr}		13.9	16.3	24.6	324.7
N_{SM}		15.8	15.6	15.6	

$$\Gamma_{\tau\tau}/\Gamma_{had} = 0.075^{+0.025}_{-0.020} \pm 0.005, \qquad (5)$$

with systematic errors due to uncertainties in the estimate of cosmic ray and beam–gas backgrounds, the calorimeter and drift chamber resolution simulation, and the hadronic model used (Lund parton shower or matrix element) to estimate ϵ_{had}. The $\mu^+\mu^-$ ratio agrees well with the Standard Model prediction. At the time of the Lepton Photon Symposium, the ratio $\Gamma_{\tau\tau}/\Gamma_{had}$ was higher than the standard model prediction by $\sim 1.9\sigma$ (probability of consistency with the Standard Model: 2%) now the discrepancy is $\sim 1.4\sigma$.

We also fit the angular distribution of the e^+e^- events, normalized using the luminosity measured with small angle Bhabhas, and taking into account the contribution from the pure QED and interference diagrams (which vary as a function of E_{cm}).[2] We extract

$$\Gamma_{ee}/\Gamma_{ee}^{SM} = 0.91 \pm 0.16. \qquad (6)$$

Again, at the time of the Lepton Photon Symposium, this ratio was observed to be $\sim 2\sigma$ lower than the Standard Model prediction; (probability of consistency with the Standard Model: 4%) now we see good agreement with the Standard Model.

HADRONIC DECAY PROPERTIES

In the Standard Model, partonic final states in $e^+e^- \rightarrow Z^0 \rightarrow q\bar{q}$ are kinematically identical to $e^+e^- \rightarrow \gamma \rightarrow q\bar{q}$. The only difference is in the mix of final state flavors (due to the different couplings of the Z^0 and the photon to the quarks) and in the (small) forward–backward asymmetry.

We can test this prediction by comparing event shape distributions and charged particle inclusive distributions from Mark II data at the SLC, $\sqrt{s} \simeq 91$ GeV, with Mark II data at PEP, $\sqrt{s} = 29$ GeV and other low energy experiments, and also with predictions from Monte Carlo fragmentation models tuned at PEP/PETRA energies.

The Monte Carlos that we use for comparison are:

1. the Lund version 6.3 parton phower model with string fragmentation (Lund 6.3 shower);[3]
2. the Webber–Marchesini version 4.1 parton shower model with cluster fragmentation (Webber 4.1);[4]
3. the Gottschalk and Morris parton shower model with string/cluster fragmentation (Caltech–II);[5]
4. the Lund model based on the second order QCD matrix element of Gottschalk and Shatz (Lund 6.3 ME).[3,6]

The parameters of each of these models was optimized at $\sqrt{s} = 29$ GeV using Mark II data at PEP.[7]

Using the standard selection of charged tracks and calorimeter showers described above, we select events with $N_{ch} \geq 7$ and $E_{vis} \geq 0.5E_{cm}$. A total of 185 events pass these cuts, with a Monte Carlo determined efficiency of $0.78 \pm .02$, where the error is the systematic uncertainty estimated by comparing the different Monte Carlo models. The events removed by this selection tend to have large undetected energy close to the beam line, and their removal does not bias the Z decay characteristics presented here. The background from beam–gas collisions, $\gamma\gamma$ collisions, and lepton pairs is estimated to be < 0.5 events.

Each of the distributions is corrected for detector effects by applying a bin-by-bin correction factor determined using the Lund shower Monte Carlo; the correction factors are defined as:

$$f_i = \frac{n_{gen}(i)/N_{gen}}{n_{det}(i)/N_{det}}, \qquad (7)$$

where N_{gen} and N_{det} are the generated and detected number of events and $n_{gen}(i)$ and $n_{det}(i)$ are the generated and detected number of events in the specified i–th bin. The effects included in the correction are decays of $\pi^{\pm}, K^{\pm}$, detector acceptance and resolution, and machine backgrounds. Initial state radiation effects are negligible. The correction factors varied from 0.7 to 1.2, and there is minimal bin–to–bin migration.

We apply these correction factors to the inclusive charged particle distributions $x_p \equiv 2pc/E_{cm}$, p_T^{in} and p_T^{out} (where *in* and *out* refer to the event plane defined by a sphericity analysis); the event shape parameters[7] sphericity (S), aplanarity (A), and thrust (T); and the total charged particle multiplicity (the statistics were insufficient for a full unfold on the multiplicity distribution, so we present the uncorrected distribution).

Inclusive Momentum Distribution and Scaling

The corrected x_p distribution for hadronic events at the SLC is shown in Figure 3.

Also included is the prediction from the Lund shower Monte Carlo tuned on the PEP data,[7] and the same distribution from the Mark II at PEP (E_{cm} = 29 GeV). We see good agreement between the data and the Monte Carlo prediction, and we see also that the 91 GeV and 29 GeV data are very similar, except at the lowest x_p and the highest x_p. The integral of this distribution is simply the mean charged multiplicity, and the expected increase in multiplicity at higher E_{cm} appears in the lowest x_p bins. Figure 4 shows the mean value of

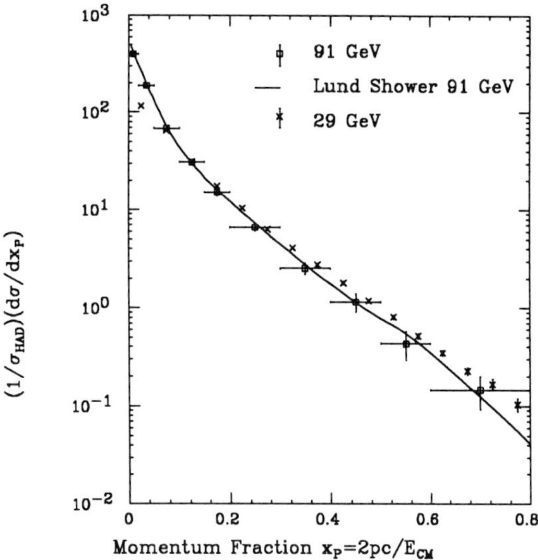

Fig. 3. The corrected distribution of $x_p = 2pc/E_{cm}$ for charged particles in hadronic events, at E_{cm} = 91 and 29 GeV, along with the prediction from the Lund shower Monte Carlo at 91 GeV.

$1/\sigma \cdot d\sigma(s)/dx_p$ in different x_p bins, for the SLC data and data at lower e^+e^- energies. The dependence on E_{cm} for each x_p bin is indicated by the solid lines, to guide the eye.

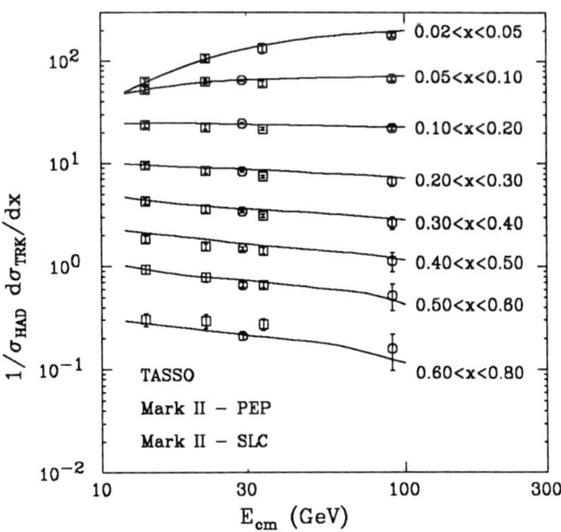

Fig. 4. The average value of $1/\sigma \cdot d\sigma(s)/dx_p$ in several bins of x_p, as a function of $s = E_{cm}^2$. Mark II data are at 91 and 29 GeV.

The agreement between the SLC and PEP datasets at intermediate x_p, and the disagreement at the highest x_p, is the consequence of scaling and scaling violations. This can be understood by comparing with the (perhaps more familiar) case of deep inelastic electron scattering (DIES). In that case, a spacelike photon probes the x, Q^2 distribution of the proton, $F(x, Q^2)$(Figure 5a). The observation that $F(x, Q^2)$ is independent of Q^2 (scaling) is taken as evidence for partons, since $x_{DIES} \simeq p_q/p_{proton}$. The later observation that $F(x, Q^2)$ increases with Q^2 at low x (scaling violations) is taken as evidence for gluon bremsstrahlung predicted by QCD.

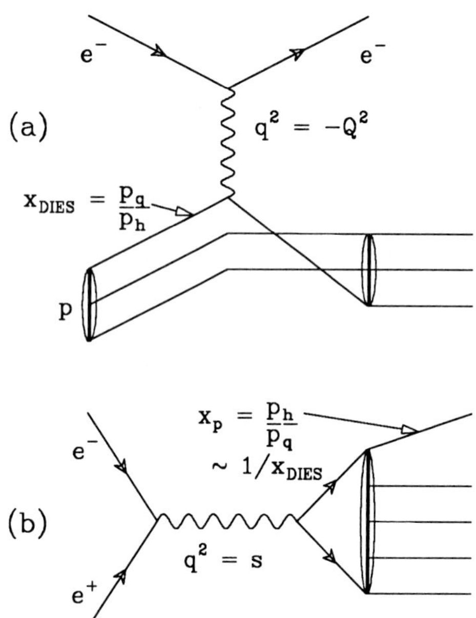

Fig. 5. Feynman diagrams for (a) Deep inelastic electron scattering; (b) $e^+e^- \rightarrow$ *hadrons*.

In e^+e^- collisions, a timelike photon or Z^0 probes the hadronization of a pair of quarks. in this case, $x_p = p_{hadron}/p_{quark} \simeq 1/x_{DIES}$. Scaling is observed when $1/\sigma \cdot d\sigma(s)/dx_p$ is independent of $s = -Q^2$. Scaling violations are observed when $1/\sigma \cdot d\sigma(s)/dx_p$ decreases with s at high x_p, again due to gluon bremsstrahlung. Scaling, and scaling violations, are as distinct in e^+e^- collisions as they are in DIES.

Jet Production and Hadronization

Let me remind you of some of the salient features of jet production, as described by QCD and QCD–inspired models.

1*a*. The distribution of the transverse momentum with respect to the jet axis of particles *within* a jet (p_T^{jet}) is limited ($\simeq \exp(-(p_T^{jet})^2/2\sigma(p_T)^2)$, with $\sigma(p_T)$ independent of the energy of the jet. Thus, we expect that the distribution of transverse momentum *out* of the event plane, $1/\sigma \cdot d\sigma/dp_T^{out}$, to be independent of $\sqrt{s}$, at least at low p_T^{out}. This can be seen when we compare distributions from hadronic events at $\sqrt{s} \approx 91$ GeV with data taken by the Mark II at PEP, at $\sqrt{s} = 29$ GeV. Indeed, Figure 6 indicates rough agreement between the two datasets at low p_T^{out}. The SLC dataset also agrees well with Monte Carlo predictions tuned on the PEP data.[7]

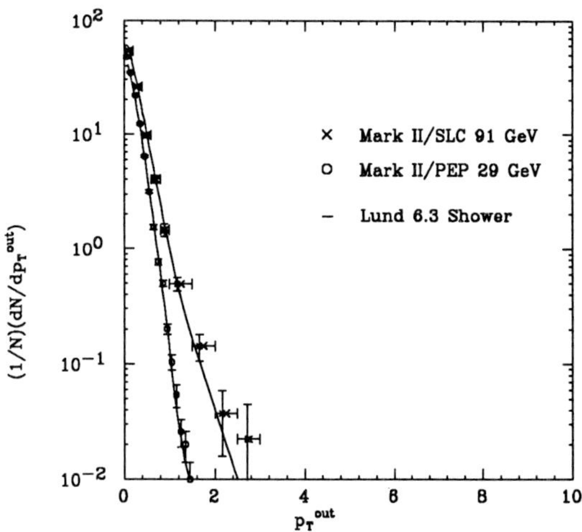

Fig. 6. Distributions of p_T^{out} from Mark II data at SLC and at PEP, compared with Monte Carlo predictions.

1*b*. Even though the p_T^{out}distributions are roughly invariant, the extra energy must go somewhere. We expect the mean charged multiplicity of each jet to rise with available energy E_{jet}. The hadronic event charged multiplicity distribution observed at the SLC is shown in Figure 7. We see good agreement with the predictions from the Monte Carlos. The mean multiplicity, compared with measurements at lower $\sqrt{s}$, is shown in Figure 8. We see a dramatic growth in multiplicity, as expected.

2*a*. As $\sqrt{s}$ grows, three–body phase space will increase. Single hard gluon bremsstrahlung will broaden the distribution of momentum transverse to the sphericity axis *in* the event plane (p_T^{in}); and this effect will tend to increase with $\sqrt{s}$. The p_T^{in} distribution seen at SLC is compared with data from PEP in Figure 9. As

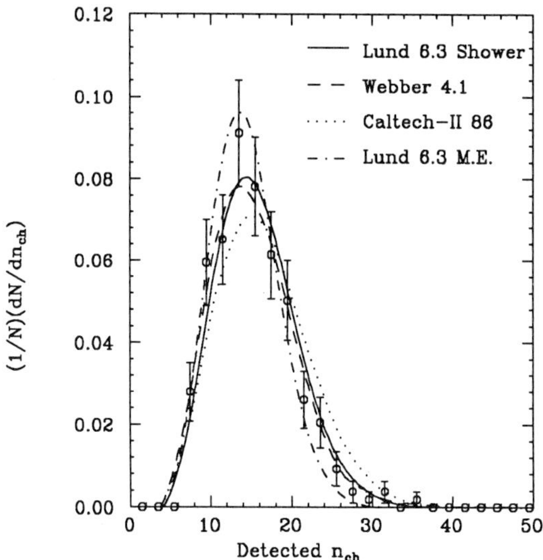

Fig. 7. Distribution of charged particle multiplicity (not corrected) from hadronic events at the SLC, compared with Monte Carlo predictions.

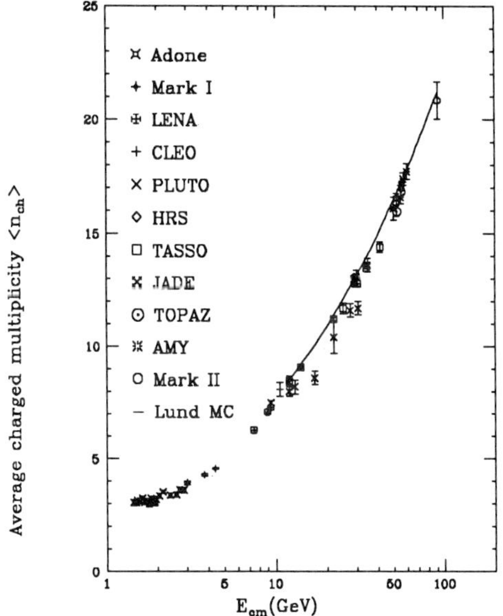

Fig. 8. Mean corrected charged multiplicity versus $\sqrt{s}$.

expected, we see a broadening of the distribution at high p_T^{in}, relative to the lower energy data. We have verified that this difference is not due to b quark production or the effect of radiative corrections.

2b. Still, at higher energies, two–body phase space increases faster than three–body; thus the event thrust will tend to grow with $\sqrt{s}$. The

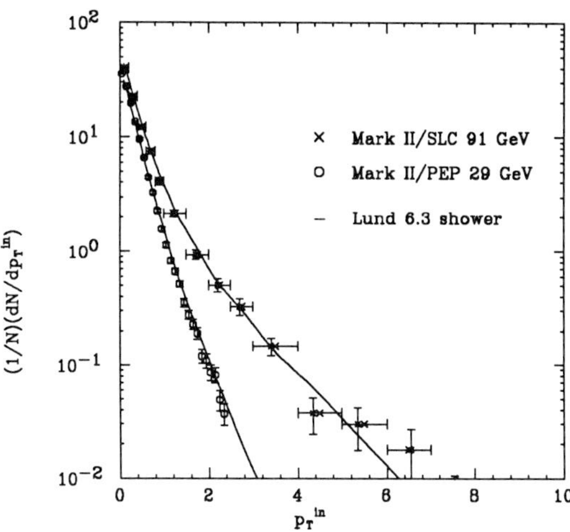

Fig. 9. Distributions of p_T^{in} from Mark II data at SLC and at PEP, compared with Monte Carlo predictions.

Thrust distribution observed at SLC is shown in Figure 10, compared with Monte Carlo predictions. The mean thrust, $\langle T \rangle$, is plotted in Figure 11, compared with data at lower energies. We see the growth in $\langle T \rangle$, as expected.

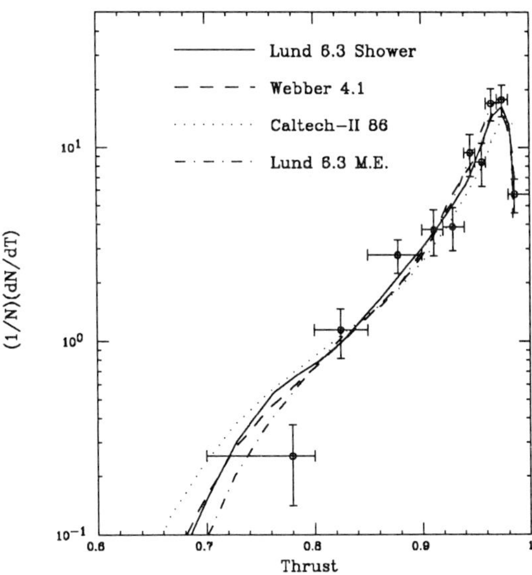

Fig. 10. Distribution of hadronic event thrust at SLC, compared with Monte Carlo predictions.

3a. As $\sqrt{s}$ grows, four–body phase space will increase. Double hard gluon bremsstrahlung will broaden the distribution of momentum transverse to the sphericity axis, *out* of the event plane (p_T^{out}); and this effect will tend to

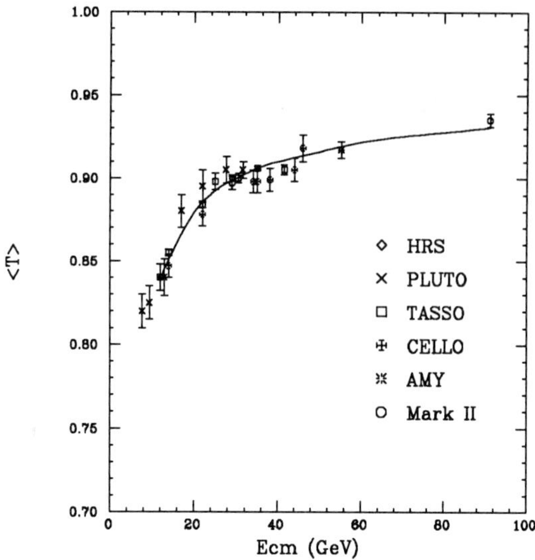

Fig. 11. Mean thrust of hadronic events versus $\sqrt{s}$, compared with Monte Carlo predictions.

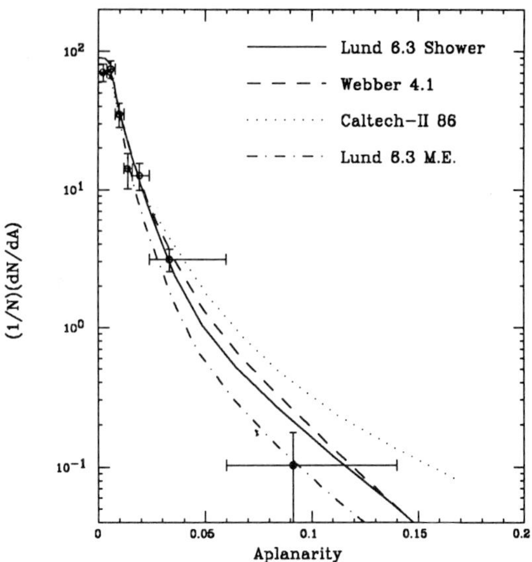

Fig. 12. Distribution of hadronic event aplanarity at SLC, compared with Monte Carlo predictions.

increase with $\sqrt{s}$. The p_T^{out} distribution seen at SLC is compared with data from PEP in Figure 6. As expected, we see a broadening of the distribution at high p_T^{out}, relative to the lower energy data.

3*b*. Still, at higher energies, two– and three–body phase space increases faster than four-body; thus the event aplanarity will tend to decrease with $\sqrt{s}$. The aplanarity distribution observed at SLC is shown in Figure 12, compared with Monte Carlo predictions. The mean aplanarity, $\langle A\rangle$, is plotted in Figure 13, compared with data at lower energies. We see the decrease in $\langle A\rangle$, as expected.

4*a*. Since the number of jets grows more slowly than the energy per jet as $\sqrt{s}$ increases, it should be easier to find jets. To do so, we partition particles into clusters with invariant mass no larger than m_{cut} ("jet resolution").[8] For fixed m_{cut}, the number of jets found will increase rapidly with energy, due to increasing phase–space for multi–jet states.

4*b*. If we define jets using a *scaled* "jet resolution" parameter $y_{cut} = m_{cut}^2 c^4/E_{vis}^2$, this will take out the effect of increasing phase–space (for fixed y_{cut}, *e.g.* $y_{cut} = 0.08$). In that case, the fraction of, *e.g.*, 3–jet events found will vary with $\sqrt{s}$ only due to the "running" of

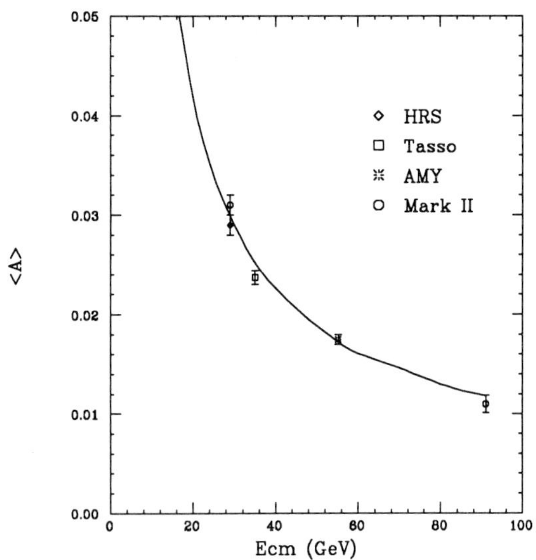

Fig. 13. Mean aplanarity of hadronic events versus $\sqrt{s}$, compared with Monte Carlo predictions.

the strong coupling constant;[9] *e.g.*,

$$R_3 \equiv \text{3–jet fraction} \simeq \alpha_s(\sqrt{s}). \qquad (8)$$

Figure 14 shows the value of R_3 measured at the SLC as a function of y_{cut}. The curves are predictions from QCD to order α_s^2.[9] Note that the data points are *not* statistically independent; each event shows up in class R_2, R_3, or R_4 for each value of y_{cut}.

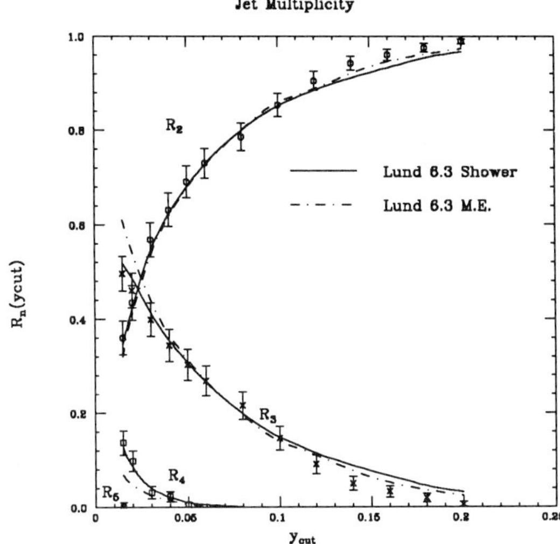

Fig. 14. Fraction of 2–, 3–, 4–, and 5-jet events in hadronic decays at SLC, versus y_{cut}. Curves are order α_s^2 QCD predictions.

Figure 15 shows the value of R_3 measured at the SLC and at lower energies for $y_{cut} = 0.08$. The curve is the prediction from QCD to order α_s^2.[9] The three–jet fraction R_3 is observed to decrease with $\sqrt{s}$, due to the running of α_s, in agreement with the prediction of QCD.

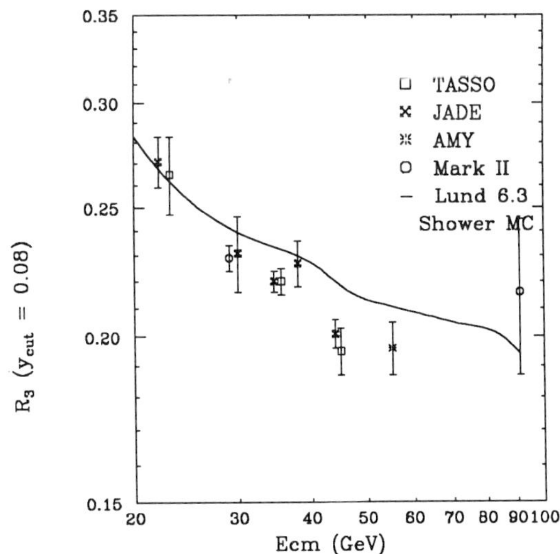

Fig. 15. Three–jet fraction, for different values of y_{cut}, versus $\sqrt{s}$, with order α_s^2 QCD predictions.

All the features of jet production discussed above are built into the QCD Monte Carlos (tuned at PEP/PETRA energies) and are well borne out by the data at the highest e^+e^- energies currently available ($\sqrt{s} \approx 91$ GeV).

SEARCHES FOR MASSIVE FERMIONS

We have searched for the pair–production of fundamental fermions with Standard Model couplings in our sample of 310 hadronic Z^0 decays. Figure 16 shows the prediction of the number of heavy fermions of each species produced in our sample, as a function of mass hypothesis.

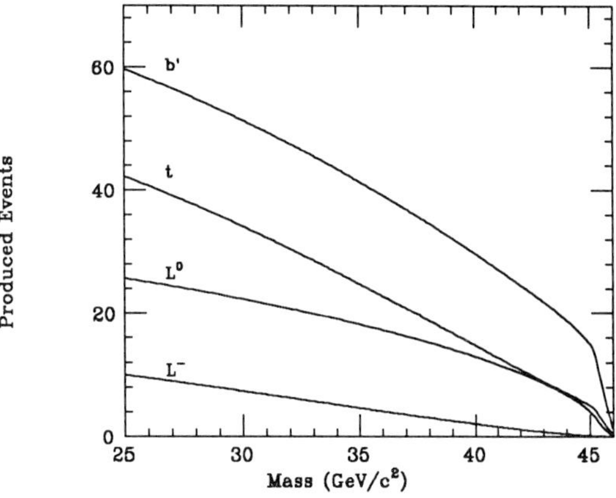

Fig. 16. Predicted number of heavy fermions produced in the Mark II data sample, as a function of mass hypothesis. Normalized to 310 observed hadronic decays.

The fourth generation b' decays via the weak charged current (CC) to a charmed quark plus a virtual W ($b' \to cW^*$). It may be that the 4×4 quark mixing matrix element governing the decay $V_{b'c}$ is very small since the transition is across 2 generations. If this is the case, the b' will be relatively long lived, and the dominant decay modes of the b' may be through flavor changing neutral current (FCNC) penguin diagrams,[10] $b' \to \gamma b, gb$, or $H^0 b$, where H^0 is a neutral Higgs boson which decays to a pair or heavy partons. The relative branching ratios for these modes is unknown.

Three types of searches were performed to detect different topologies in the decays of heavy fermions. We searched for isolated charged tracks from charged current semileptonic or leptonic decays of the type

$$\begin{array}{lll} b' \to W^* c, & t \to W^* b, & \text{or } L^0 \to W^* l^- \\ \quad \hookrightarrow l^- \bar{\nu} & \quad \hookrightarrow l^+ \nu & \quad \hookrightarrow l^+ \nu, q\bar{q}' \end{array} \tag{9}$$

The L^0 is a massive fourth–generation neutral Dirac

lepton. The neutrino weak eigenstates ν_l and mass eigenstates L_i^0 are assumed to mix in analogy with the quark sector:

$$\nu_l = \sum_{i=1}^{4} U_{li} L_i^0 \tag{10}$$

to permit the decay $L^0 \to l + W^*, (l = e, \mu, \tau)$.

The second search topology is for isolated photons from *e.g.*, the penguin decay of the $b' \to \gamma b$. The third is for hadronic decays of heavy particles into 4 or more jets, producing a non–planar topology, signalled by a large flow of momentum out of the event plane (M_{out}). Examples of such decays, in addition to the conventional charged current decays mentioned above, include

$$\begin{gathered} t \to bH^+ \to b(c\bar{s}), \quad b' \to cH^- \to c(s\bar{c}), \\ b' \to bH^0, \quad b' \to bg. \end{gathered} \tag{11}$$

Isolated Track Search

We select hadronic events using the standard track cuts described above, and require $N_{ch} \geq 6, E_{vis} \geq 0.5E_{cm}, |\cos\theta_{thr}| < 0.8$, where θ_{thr} is the angle between the event thrust axis and the beam, and Thrust < 0.9. For each selected event we determine an isolation parameter ρ_{max}, as follows. We perform the LUND cluster algorithm[11] on all tracks in the event excluding a candidate track. The cluster parameter d_{join}[11] is set to 0.5, corresponding to a jet mass resolution $m_{jet} \simeq 2$ GeV. The value of the parameter d_{join} was derived from Monte Carlo studies designed to maximize the signal (efficiency for detecting heavy quark decays) over the QCD backgrounds (due for fragmentation fluctuations). For each jet j we form the quantity

$$\rho \equiv \min_j \left\{ \sqrt{2E_l(1 - \cos\theta_{lj})} \right\}, \tag{12}$$

where E_l is the track energy in GeV and θ_{lj} is the angle between the track and jet j, and the minimum of this quantity over all jets is taken. The maximum value of ρ for all charged tracks in an event is called the isolation parameter ρ_{max}. Note that no lepton identification is used, in order to avoid efficiency losses.

Figure 17 shows the distribution of the isolation parameter ρ_{max} for charged tracks in the data, superimposed on the prediction from the Lund 6.3 shower Monte Carlo, with *udscb* quarks.

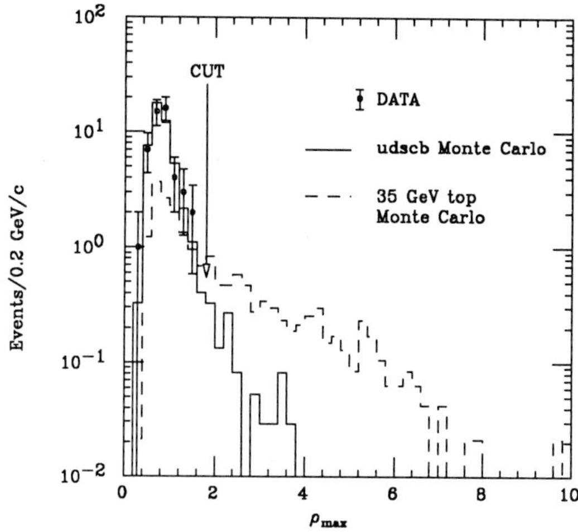

Fig. 17. Distribution of the isolation parameter ρ_{max} for charged tracks in the data (points); and predictions from the Lund 6.3 shower *udscb* Monte Carlo (solid histogram) and for a 35 GeV t (dashed histogram).

We require $\rho_{max} > 1.8$ GeV$^{1/2}$, and no events pass this cut. We expect 0.7 to 1.4 events from the *udscb* background, depending on the Monte Carlo model used. The dashed histogram in Figure 17 depicts the ρ_{max} distribution for a 35 GeV/c^2 top quark alone, normalized to the hadronic event sample. A b' quark of the same mass would be even more dramatic.

Figure 18 shows the distribution of the isolation parameter ρ_{max} for photon showers in the data, superimposed on the prediction from the Lund 6.3 shower Monte Carlo, with *udscb* quarks.

To search for decays of the type $b' \to \gamma b$, we require $\rho_{max} > 3.0$ GeV$^{1/2}$, and no events pass this cut. The cut is higher for photons than for charged tracks because the double–track separation for π^0's in the Mark II is considerably worse than for $\pi^\pm$, so we must require the photons to be more isolated. We expect 0.2 to 0.6 events from the *udscb* background. If a 35 GeV/c^2 b' quark decaying 100% of the time to γb were present in the data, the Lund Monte Carlo predicts the distribution shown as the dashed histogram in Figure 18.

No events with isolated charged tracks or photon showers are seen. To extract limits on the masses of heavy fermions, we compare the 95% confidence level upper limit on signal events with the number of

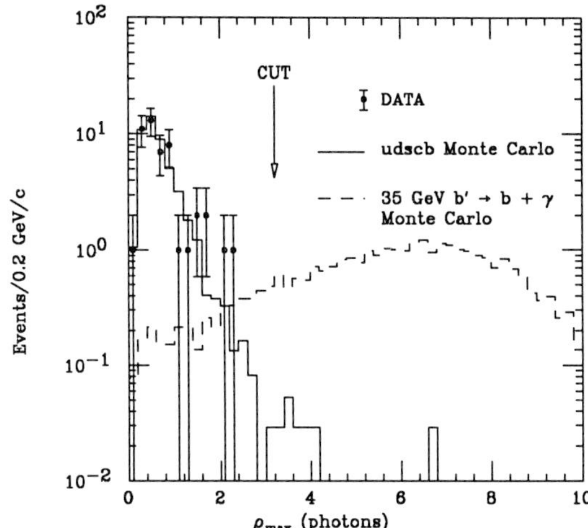

Fig. 18. Distribution of the isolation parameter ρ_{max} for photon showers in the data (points); and predictions from the Lund 6.3 shower *udscb* Monte Carlo (solid histogram) and for a 35 GeV b' decaying 100% of the time to γb (dashed histogram).

events expected, normalizing to the total hadronic event sample:

$$\text{expected events} = \frac{\epsilon_\rho \Gamma_x(m_x) N_h}{\epsilon_q \Gamma_q + \epsilon_x \Gamma_x(m_x)}, \tag{13}$$

where m_x is the mass of the heavy fermion, N_h is the number of hadronic events in the sample, $\epsilon_q = (94 \pm 1)\%$ is the efficiency for *udscb* events passing the event selection criteria; ϵ_x is the same for heavy fermion events (95 – 97% for b', t, and $\simeq$85% for L^0, depending on mass and mixing matrix element $U_{L^0 l}$); $\Gamma_q = 1.74$ GeV is the partial width of the Z^0 to *udscb* quarks; Γ_x is the partial width of the Z^0 to the heavy fermion; and ϵ_ρ is the efficiency for heavy fermion events to pass all cuts including the ρ_{max} cut. First order QCD corrections[12] are used when calculating Γ_q and Γ_x. We set a 95% confidence level upper limit on the number of signal events, based on 0 observed events, of 3.0.

To account for systematic errors, we reduce the number of expected signal events in equation (13) for each of the following effects, added in quadrature: higher order corrections to the order α_s QCD correction to Γ_x ($\leq 25\%$); normalization statistics (N_h); uncertainties in the semileptonic branching ratios of the heavy quarks (2%); and uncertainties in the efficiencies ($\epsilon_x, \epsilon_\rho$) due to detector simulation effects and different fragmentation models (including Lund Symmetric fragmentation and Petersen fragmentation) ($\leq 12\%$). In Figure 19, equation (13) is plotted as a function of the mass of the t or b' quark. The dashed curves are the predictions minus the systematic error.

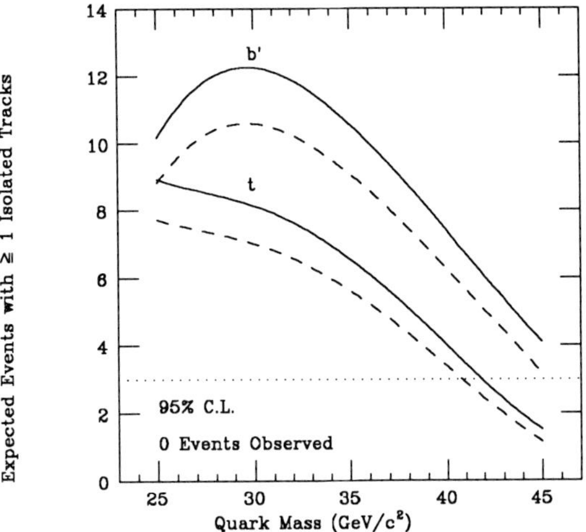

Fig. 19. Expected number of $b'\bar{b}'$ (lower pair of curves) or $t\bar{t}$ (upper pair) events passing the ρ_{max} cut, as a function of the quark mass. The dashed curves include the effect of systematic errors added in quadrature. The 95% confidence level upper limit on the number of signal events is also shown.

From this we conclude that $m_t \geq 40.7$ GeV/c^2 at 95% confidence level, and $m_b' \geq 45.1$ GeV/c^2 if the b' decays 100% of the time to some combination of cW^* and γb. Limits on L^0 production are sensitive to the mixing parameter $U_{L^0 l}$. If the sum $\Sigma|U_{L^0 l}|^2$ for $l = e, \mu, \tau$ is small enough, then the lifetime[13] of the L^0 will be so long that it will decay outside our fiducial region. This will produce spectacular signatures (which are not seen), but we have not yet set limits on decays of this type. We therefore set the limits shown in Figure 20, which depend somewhat on the species of light lepton that is dominant in the mixing (worst for mixing with ν_τ, best for mixing with ν_μ.)

M_{out} Analysis

To search for non–planar events containing four or more jets, we require $M_{out} > 18$GeV/c^2 where

$$M_{out} \equiv \frac{E_{cm}}{E_{vis}} \frac{1}{c} \Sigma |p_T^{out}|. \tag{14}$$

p_T^{out} is the momentum component of a charged track or neutral shower out of the event plane defined

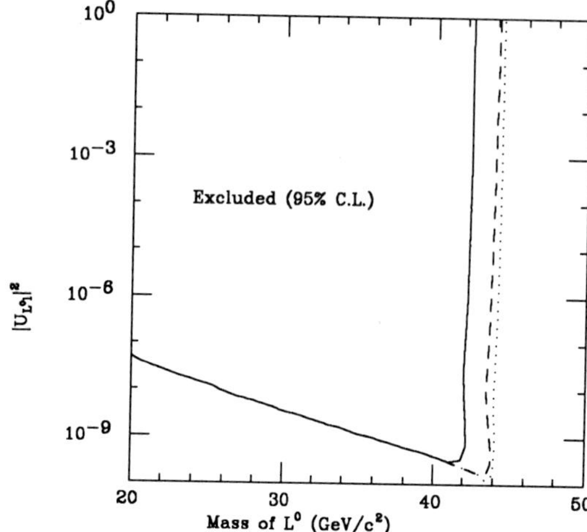

Fig. 20. 95% C.L. mass limits for an unstable neutral heavy lepton L^0 as a function of mass and mixing matrix element $|U_{L^0 l}|^2$ for BR($L^0 \to \tau W^*$) = 100% (solid); BR($L^0 \to eW^*$) = 100% (dashed); and BR($L^0 \to \mu W^*$) = 100% (dotted).

by the sphericity tensor, and the sum is over all charged tracks and neutral showers. The distribution of M_{out} is shown in Fig. 21 for the data sample, for a five-flavor QCD Monte Carlo, and for the process $b' \to cH^- \to c\bar{c}s$ for a 35 GeV/c^2 b'.

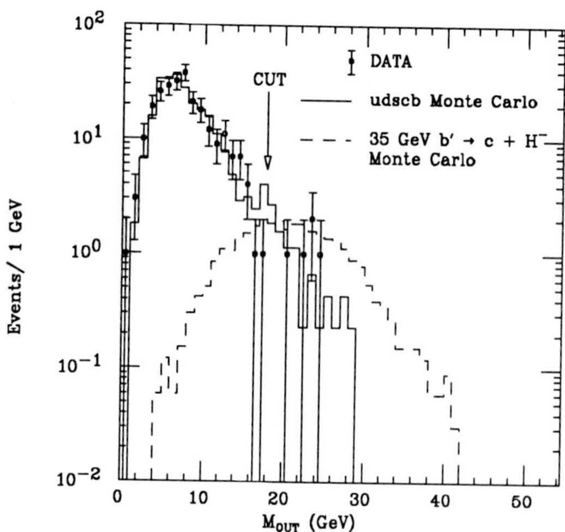

Fig. 21. M_{out} (equation (14)) for data (circles, with statistical errors), *udscb* QCD Monte Carlo (solid line), and for a 35 GeV/c^2 b' quark decaying into cH^- (hatched area), with M_{H^-} = 25 GeV/c^2 and the H^- decaying 100% into $\bar{c}s$.

Five events are observed in the data with $M_{\rm out} > 18$ GeV/c^2, while 3.8 events (Lund ME) to 9.3 events (Webber 4.1) are expected from QCD *udscb* processes. The tail of the QCD M_{out} distribution is very model dependent because of the different ways in which multiple-hard-gluon radiation is handled. To be conservative, background subtraction is performed using the smallest value (3.8 events) expected. We find the upper limit[14] at 95% C.L. to be 7.1 events for 5 observed events and 3.8 expected background events.

The above observation allows us to set the following limits. If t and b' decay 100% via the CC processes $t \to bW^*, b' \to cW^*$, then $M_t > 39.1$ GeV/c^2 and $M_{b'} > 43.3$ GeV/c^2 at 95% C.L. Note that these limits are more sensitive to the hadronic decays of t and b' (difficult to detect at hadronic colliders), while the isolated track analysis limits are sensitive to the semi-leptonic decays. If b' decays 100% into b+ gluon, then $M_b' > 40.6$ GeV/c^2 at 95% C.L. If t and b' decay through a charged Higgs boson of mass 25 GeV/c^2 which in turn decays 100% hadronically into $c\bar{s}$, then $M_t > 41.4$ GeV/c^2 and $M_b' > 45.1$ GeV/c^2 at 95% C.L. The case of the H^- decaying partially into $\tau\bar{\nu}$ is found to weaken the above limits, but if B.R.($H^- \to \tau\bar{\nu}$) < 50% both limits remain over 40 GeV/c^2.

The various mass limits obtained with the three event topologies are summarized in Table 2. Finally, the analyses of isolated tracks and photons

TABLE 2

Summary of the mass limits set by searching for the three event topologies described in the text. For $t \to bH^+$ and $b' \to cH^-$ limits, $M_{H^\pm}$ = 25 GeV/c^2. The L^0 limits are for decays within the vertex fiducial region (i.e. decay length less than ≈ 1 cm).

Heavy Particle	Decay Products (B.R. 100%)	Topology	Mass Limit (95% C.L.) (GeV/c^2)
top	bW^*	Isolated Track	40.7
	bW^*	M_{out}	39.1
	bH^+	M_{out}	41.4
b'	cW^*	Isolated Track	45.1
	cW^*	M_{out}	43.3
	cH^-	M_{out}	45.1
	b+ gluon	M_{out}	40.6
	$b\gamma$, BR≥ 40%	Isolated Photon	45.4
L^0	eW^*	Isolated Track	44.1
	μW^*	Isolated Track	44.4
	τW^*	Isolated Track	42.4

can be used to give mass limits on b' as a function of branching ratio into the CC process, assuming that the remaining decays are only through the FCNC decays $b+$ gluon and $b\gamma$ (we assume a Higgs particle is not kinematically accessible). In Figure 22, the dashed curve shows the region excluded by our isolated track analysis, as a function of b' mass and CC branching ratio. The remaining decays are assumed to be via the FCNC, to differing mixtures of $b+$ gluon and $b\gamma$. The isolated photon analysis allows us to rule out b' mass values when the FCNC branching ratio is large. The excluded region (solid line) depends on the ratio of $b' \to b\gamma$ decays to all FCNC decays. A substantial fraction of possible mass values of the b' are therefore excluded for a wide range of possible decay branching ratios.

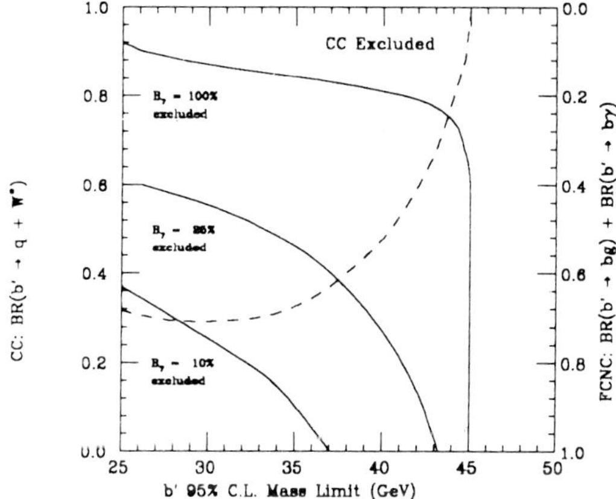

Fig. 22. 95% C.L. limit exclusion of b' quark production versus mass and FCNC branching ratio. The dashed curve shows the region excluded by the isolated track search, while the solid curves show the regions excluded by the isolated photon search, for different values of $B_\gamma = BR(b' \to \gamma b)/$ all FCNC decays.

CONCLUSIONS

The Mark II detector at the SLC has determined many of the properties of Z^0 decays with a relatively small sample of events. In addition to measuring the Z^0 resonance parameters (see G. Feldman's talk, these proceedings), we have observed hadronic decays of the Z^0 for the first time. The leptonic decay branching ratios are consistent with Standard Model predictions, and the hadronic decays behave in accordance with predictions based on QCD models. There is no evidence for the production of the top quark, or fourth-generation quarks or heavy neutral leptons.

ACKNOWLEDGEMENTS

The Mark II collaboration extends its most sincere gratitude to the SLAC accelerator staff for their tireless and heroic efforts at making the SLC a successful scientific instrument.

This work was supported in part by Department of Energy contracts DE-AC03-81ER40050 (CIT), DE-AM03-76SF00010 (UCSC), DE-AC02-86ER40253 (Colorado), DE-AC03-83ER40103 (Hawaii), DE-AC02-84ER40125 (Indiana), DE-AC03-76SF00098 (LBL), DE-AC02-76ER01112 (Michigan), and DE-AC03-76SF00515 (SLAC), and by the National Science Foundation (Johns Hopkins).

REFERENCES

1. The Mark II collaboration:
G. S. Abrams, C. E. Adolphsen, D. Averill, J. Ballam, B. C. Barish, T. Barklow, B. A. Barnett, J. Bartelt, S. Bethke, D. Blockus, G. Bonvicini, A. Boyarski, B. Brabson, A. Breakstone, F. Bulos, P. R. Burchat, D. L. Burke, R. J. Cence, J. Chapman, M. Chmeissani, D. Cords, D. P. Coupal, P. Dauncey, H. C. DeStaebler, D. E. Dorfan, J. M. Dorfan, D. C. Drewer, R. Elia, G. J. Feldman, D. Fernandes, R. C. Field, W. T. Ford, C. Fordham, R. Frey, D. Fujino, K. K. Gan, C. Gatto, E. Gero, G. Gidal, T. Glanzman, G. Goldhaber, J. J. Gomex-Cadenas, G. Gratta, G. Grindhammer, P. Grosse-Wiesmann, G. Hanson, C. A. Heusch, M. D. Hildreth, T. Himel, D. A. Hinshaw, S. J. Hong, D. Hutchinson, J. Hylen, W. R. Innes, R. G. Jacobsen, J. A. Jaros, C. K. Jung, J. A. Kadyk, J. Kent, M. King, S. R. Klein, D. S. Koetke, S. Komamiya, W. Koska, L. A. Kowalski, W. Kozanecki, J. F. Kral, M. Kuhlen, L. Labarga, A. J. Lankford, R. R. Larsen, F. Le Diberder, M. E. Levi, A. M. Litke, X. C. Lou, V. Lüth, J. A. McKenna, J. A. J. Matthews, T. Mattison, B. D. Milliken, K. C. Moffeit, C. T. Munger, W. N. Murray, J. Nash, H. Ogren, K. F. O'Shaughnessy, S. I. Parker, C. Peck, M. L. Perl, F. Perrier, M. Petradza, R. Pitthan, F. C. Porter, P. Rankin, K. Riles, F. R. Rouse, D. R. Rust, H. F. W. Sadrozinski,

M. W. Schaad, B. A. Schumm, A. Seiden, J. G. Smith, A. Snyder, E. Soderstrom, D. P. Stoker, R. Stroynowski, M. Swartz, R. Thun, G. H. Trilling, R. Van Kooten, P. Voruganti, S. R. Wagner, S. Watson, P. Weber, A. Weigend, A. J. Weinstein, A. J. Weir, E. Wicklund, M. Woods, D. Y. Wu, M. Yurko, C. Zaccardelli, and C. von Zanthier.

2. F. A. Berends, W. Hollik, R. Kleiss, *Nucl. Phys.* **B304** (1988) 712.
3. T. Sjöstrand, *Comp. Phys. Comm.* **39B** (1986) 347; T. Sjöstrand and M. Bengtsson, *Comp. Phys. Comm.* **43B** (1987) 367; M. Bengtsson and T. Sjöstrand, *Nucl. Phys.* **B289** (1987) 810.
4. G. Marchesini and B. R. Webber, *Nucl. Phys.* **B238** (1984) 1; B. R. Webber, *Nucl. Phys.* **B238** (1984) 492.
5. T. D. Gottschalk and D. Morris, *Nucl. Phys.* **B288** (1987) 729.
6. T. D. Gottschalk and M. P. Shatz, *Phys. Lett.* **150B** (1985) 451; T. D. Gottschalk and M. P. Shatz, Caltech preprints **68–1172, –1173, –1199** (1985) unpublished: T. D. Gottschalk, private communication.
7. A. Petersen *et al.*, *Phys. Rev.* **D37** (1988) 1.
8. W. Bartel *et al.*, *Z. Phys.* **C33** (1986) 23; S. Bethke *et al.*, *Z. Phys.* **C43** (1986) 325.
9. G. Kramer and B. Lampe, *Z. Phys.* **C39** (1988) 101.
10. V. Barger *et al.*, *Phys. Rev.* **D30** (1984) 947; V. Barger *et al.*, *Phys. Rev. Lett.* **57** (1986) 1519; W. Hou and R. G. Stuart, *Phys. Rev. Lett.* **62** (1989) 617.
11. T. Sjöstrand, *Comp. Phys. Comm.* **28B** (1983) 229.
12. J. H. Kühn, A. Reiter, and P. M. Zerwas, *Nucl. Phys.* **B272** (1986) 560.
13. F. Gilman, *Comments Nucl. Part. Phys.* **16** (1986) 231.
14. Particle Data Group, *Phys. Lett.* **204B** (1988) 81; O. Helene, *Nucl. Inst. Meth.* **C212** (1983) 319.

DISCUSSION

S. Olsen, University of Rochester: Do you see any radiative decays of the Z^0 such as $Z^0 \to \gamma\mu^+\mu^-$ or $Z^0 \to \gamma\tau^+\tau^-$?

A. Weinstein: No, we do not.

P. Sinervo, University of Pennsylvania: I think your limits on the top and b' masses are all significant — not just the charged Higgs mode. There must be some model of top quark fragmentation used in your Monte Carlo. How sensitive is your lepton isolation cut to this model? How large a variation did you consider in your model of the Peterson fragmentation?

A. Weinstein: We used both the Lund Symmetric fragmentation scheme and the Peterson fragmentation scheme, and we varied the parameters of the latter over some reasonable range. The uncertainty due to these models is $\leq$ 15%, including the uncertainty in the semileptonic branching ratios (assuming charged current decays).

V. Khoze, Leningrad Nuclear Physics Institute: I am slightly confused with your statement about higher-order QCD corrections. For example, near threshold, the higher-order corrections are of 100% importance. They change drastically the threshold behavior of the total cross section. Thus the total difference from the famous Schwinger correction near the threshold is a factor of two.

A. Weinstein: We use the order α_s prescription of Kühn, Reiter, and Zerwas. This is believed to give an underestimate of the cross section near threshold; thus we are being conservative. We also conservatively propagate higher-order uncertainties by reducing the order α_s estimate by 30%. Very close to threshold, resonance behavior may affect the cross section in either direction. Thus we do not quote limits any higher than about 0.4 GeV below $m_Z/2$.

R. Wilson, Boston University: What is the Standard Model probability of the combined

$$\Gamma_{ee}/\Gamma_{ee}^{SM} \oplus \Gamma_{\mu\mu}/\Gamma_{\mu\mu}^{SM} \oplus \Gamma_{\tau\tau}/\Gamma_{\tau\tau}^{SM}\ ?$$

A. Weinstein: There is no well defined answer to that question, unless you specify the region of compatibility in a three-dimensional space. The presence of correlations among the three measurements further complicates the issue, but we think we can tell the difference between ee, $\mu\mu$, $\tau\tau$ rather well. However, our extraction of Γ_{ee} from the data is preliminary, and so we assign a large systematic error to it.

E. Thorndike, University of Rochester: Is there a region of branching ratios of $b' \to b\gamma$, $b' \to bg$, $b' \to cW^*$ that you do not rule out?

A. Weinstein: Yes, for small $b' \to b\gamma$, large $b' \to bg$, small $b' \to cW^*$ we have no limit (see Fig. 22).

LEP REPORT

F. DYDAK
CERN, Geneva, Switzerland

ABSTRACT

We describe the status of the LEP machine and of the four LEP detectors, up to and including the "LEP pilot run" which took place from August 13 to 18, 1989.

1. INTRODUCTION

Before I start my report on LEP and the LEP detectors, I should like to convey my congratulations, and express my sincere appreciation for the results which have been achieved at the SLC, and which undoubtedly were the highlight of the 1989 Lepton-Photon Symposium: not only were the members of the MARK II collaboration deeply involved in the commissioning of the SLC, they had also made sure that their detector was ready, and they were able to turn the first events almost instantaneously into significant physics results.

In the light of this achievement, it became all the more interesting to learn where the people at CERN stood with their preparations, and what their status of readiness for physics data taking was. I will try to give a status report both on LEP, from the point of view of a user rather than a machine expert (which I am not), and on the four LEP detectors ALEPH, DELPHI, L3 and OPAL. My report will cover the time up to and including the "LEP pilot run", although this first attempt at producing Z events just missed the symposium: it started less than 24 hours after the end of the symposium, and lasted five days. The report on the four LEP detectors focuses on their state of readiness, and the upgrade plans which are known at present.

2. THE LEP INJECTOR COMPLEX

Besides the LEP main ring, a complex accelerator system must be operational for the injection of positrons and electrons (see Fig. 1):

- a 200 MeV linac which delivers electrons on a converter target for positron production.
- a 600 MeV linac which accelerates positrons or electrons.
- a 600 MeV storage ring.
- the CERN PS which operates at 3.5 GeV.
- the CERN SPS which operates at 20 GeV.

LEP is situated on the average some 40 m lower than the SPS. Two transfer tunnels of 400 and 650 m length respectively allow the injection of positrons and electrons into LEP.

The injector complex is designed in such a way as to allow fixed-target proton operation in parallel to LEP operation. Within one SPS "super-cycle" of 14.4 s length (see Fig. 2), a proton acceleration of up to 450 GeV is followed by two positron and another two electron acceleration cycles. The loss of proton intensity as compared with the stand-alone mode is minimal. Experience has shown that fixed-target operation is unperturbed by the simultaneous operation of LEP.

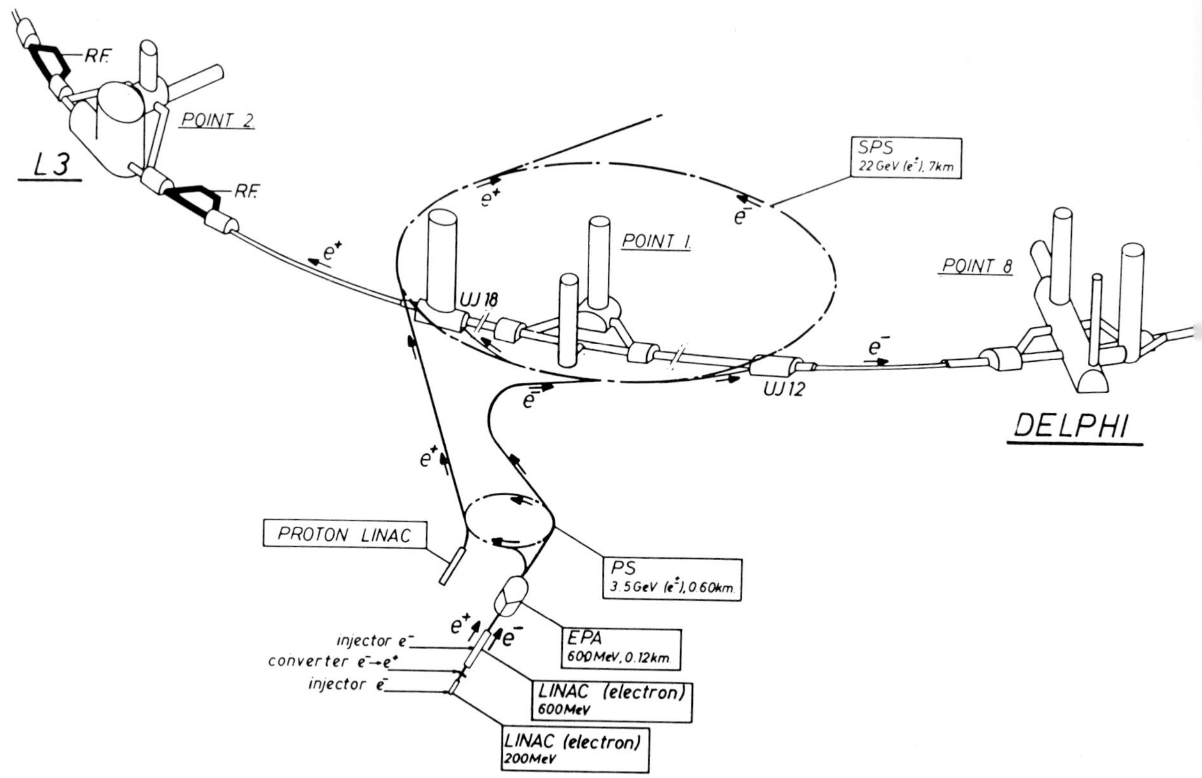

Fig. 1: The LEP Injector Complex

Protons: 450 GeV
e+e-: 4 x 20 GeV

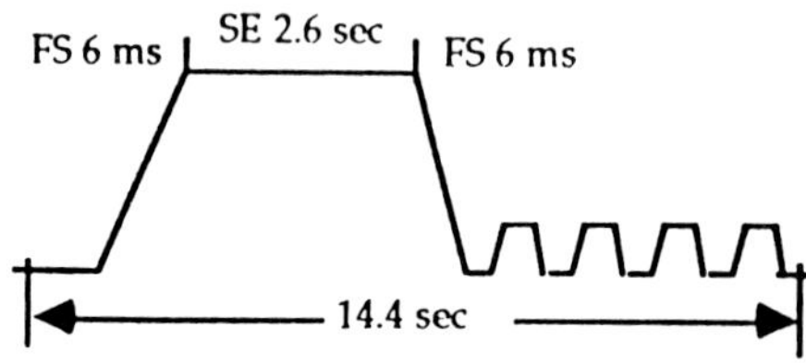

Fig. 2: The SPS "super-cycle"

3. LEP COMMISSIONING

The commissioning of LEP was an impressive exercise in that the planned schedule, which foresaw some 30 days between the first injection into the main ring and the first useful electron-positron collisions, was adhered to within a slip of schedule of only 10 days.

The highlights in the commissioning of LEP were:

- July 14 ("Bastille day"): positrons made their first full turn, less than one hour after the first injection into LEP.
- July 22: positrons were trapped by the RF.
- July 25: electrons were injected.
- July 30: 250 μA of positrons were stored in four bunches.
- August 3: 100 μA of electrons were stored.
- August 4: 700 μA of positrons were stored, and 56 μA of positrons ramped to 47.5 GeV.
- August 7: 150 μA of positrons and 30 μA of electrons simultaneously ramped to 45.5 GeV.
- August 10: ramping of positrons and electrons with the experiments solenoids powered, vertical beta squeeze to 42 cm.
- August 11: setting up of the collimator system.

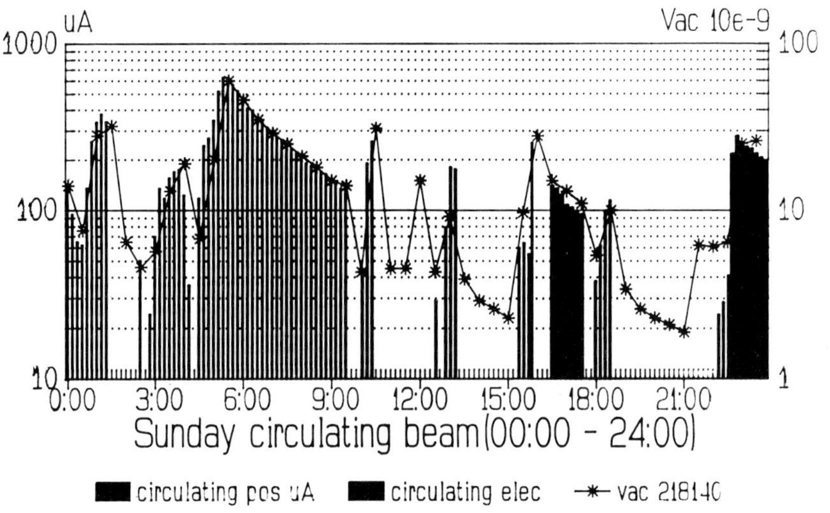

Fig. 3: LEP beam current and vacuum readings as a function of time

- August 13: start of the LEP pilot run. Recording of the first Z event by the OPAL detector.

Of course there were not only successes during the LEP commissioning. During the first days, progress was hampered by a delay in the commissioning of the beam orbit measurement system. The recovery system after controlled access periods had teething symptoms. With some regularity, the usual problems with vacuum leaks, cooling water, program bugs etc. plagued the commissioning crew. Nevertheless, the morale was such that problems of that kind were overcome as rapidly as they arose.

Fig. 3 shows the circulating current in the machine as a function of time (data from August 6, 1989). The currents were in the range of several 100 μA, and had a lifetime of several hours. Also shown are the vacuum readings which are closely related to the beam intensity: the higher the beam intensity the higher the gas pressure because of the outgassing caused by synchrotron radiation.

During this commissioning phase, the average "beam on" efficiency was 60 %.

There were also surprises in store: early during the commissioning it was noticed that there was an unexpectedly large x-y coupling in the betatron oscillations, which had a tendency to decrease in inverse proportion to the beam energy, and an unexpectedly large vertical momentum dispersion. While the theoretical integer betatron Q-values were $Q_x = 70$ and $Q_y = 78$, it was eventually decided to operate the machine with $Q_x = 71$ and $Q_y = 78$, which proved experimentally to be a rather tolerant mode of machine operation. The x-y coupling could thus be circumvented to a degree that did not affect the luminosity negatively during the LEP pilot run.

The LEP machine parameters at $E_b = 45$ GeV read as follows (wherever applicable, the values achieved up to and including the LEP pilot run are given as well):

	Planned	Achieved
Injection energy	20 GeV	20 GeV
Bunches per species	4	4
Filling time	12 min	O(1 hour)
Ramping time	80 s	O(10 mins)
Current per species	3 mA	O(500 μA)
Energy spread	0.7 10^{-3}	
Beam size @ I.P.	Hor: 300 μm Ver: 12 μm	
Betatron amplitude @ I.P.	Hor: 1.75 m Ver: 0.07 m	Ver: 0.32 m
Luminosity ($cm^{-2}s^{-1}$)	1.7 10^{31}	2 10^{28}

4. WHAT NEXT ?

Up to and including the LEP pilot run, only the standard quadrupoles were used to perform the beta squeeze at the interaction points. They will be replaced as soon as possible by superconducting quadrupoles which allow a reduction of the vertical beta function by a factor of three, with a corresponding increase of luminosity.

After the LEP pilot run, some 70 days of physics data taking are scheduled, with LEP and the SPS fixed-target programme working in parallel.

As for the likely luminosity, one can realistically expect a factor of three from the superconducting low-beta insertion, then quite some factor from an increase of the beam currents, and from an improvement of the overall reliability. Perhaps an average luminosity of 10 % of the design luminosity,

$$\langle L \rangle = 1\ 10^{30}\ cm^{-2}s^{-1},$$

i.e. an overall improvement by a factor of 50 with respect to what was achieved during the LEP pilot run, may be within reach. Assuming 70 days of running with 70 % machine efficiency and 70 % data taking efficiency, this would give an integrated luminosity of

$$3\ pb^{-1} \text{ or } 90000 \text{ Z events}$$

to each of the four experiments. Taking into account that some time will be spent scanning the Z resonance, a few times 10000 Z events per experiment, by the end of 1989, is a quite realistic expectation.

The machine physicists guarantee the absolute energy setting of the beam to be precise to 5 10^{-4}, from the uncertainty of the magnetic field integral around the ring, which permits a Z-mass determination with a precision of ± 50 MeV. This uncertainty may be further reduced at a later stage, when transversely polarized beams become available, by measuring the frequency of resonant spin precession. Thereby an accuracy level of ± 20 MeV may be achieved which is determined by the relative stability and reproducibility of the magnetic field of the machine.

5. THE LONGER PERSPECTIVE

During 1990 and 1991, the emphasis will undoubtedly be on optimizing the LEP luminosity and exploiting the Z resonance, with some 10^7 Z events collected per experiment by the end of 1991.

For the years thereafter, three improvements of LEP are envisaged:

- an energy upgrade to 100 GeV per beam, by the use of superconducting RF cavities. The completion of this task is dictated by the availability of financial resources, and could be accelerated with respect to the original schedule with more funds becoming available (completion in 1993/94).
- a luminosity upgrade at the Z resonance, by using the additional RF power which will have been installed for the energy upgrade, for working with many more bunches (36 rather than 4), thus permitting an increase of the luminosity by an order of magnitude.
- longitudinal polarization of the positron and electron bunches.

6. THE ALEPH DETECTOR

The components of the ALEPH detector, together with their performance, are the following:

Beam-pipe	0.5 mm Al r = 8 cm
Micro-vertex detector	Silicon strip detector (two layers)
Inner tracking chamber	Drift chamber $\sigma_{r\phi} = 100\ \mu m$
Central tracking chamber	TPC (r = 1.8 m) 1 atm pressure $\Delta p/p^2 = 0.13\ \%$ $\sigma(dE/dx) = 4.5\ \%$
Electromagnetic calorimeter	Lead / gas sandwich $\Delta E/E = 18\ \% / \sqrt{E}$
Magnet coil	1.5 T, cold
Hadron calorimeter	Iron / streamer tube sandwich $\Delta E/E = 80\ \% / \sqrt{E}$
Muon chambers	Two layers of streamer tube planes
Luminosity monitor	Lead / gas sandwich
Very small angle tagger	Tungsten / silicon sandwich

At the time of the LEP pilot run, the Micro-vertex detector was missing (scheduled for installation in 1990), as well as the outer muon layer (1990).

The ALEPH collaboration are discussing the replacement of the current beampipe by one with a somewhat smaller diameter, and the addition of a new luminosity monitor with smaller radius, built as a Tungsten / Silicon sandwich calorimeter.

Fig. 4 shows a Z event recorded by the ALEPH detector during the LEP pilot run.

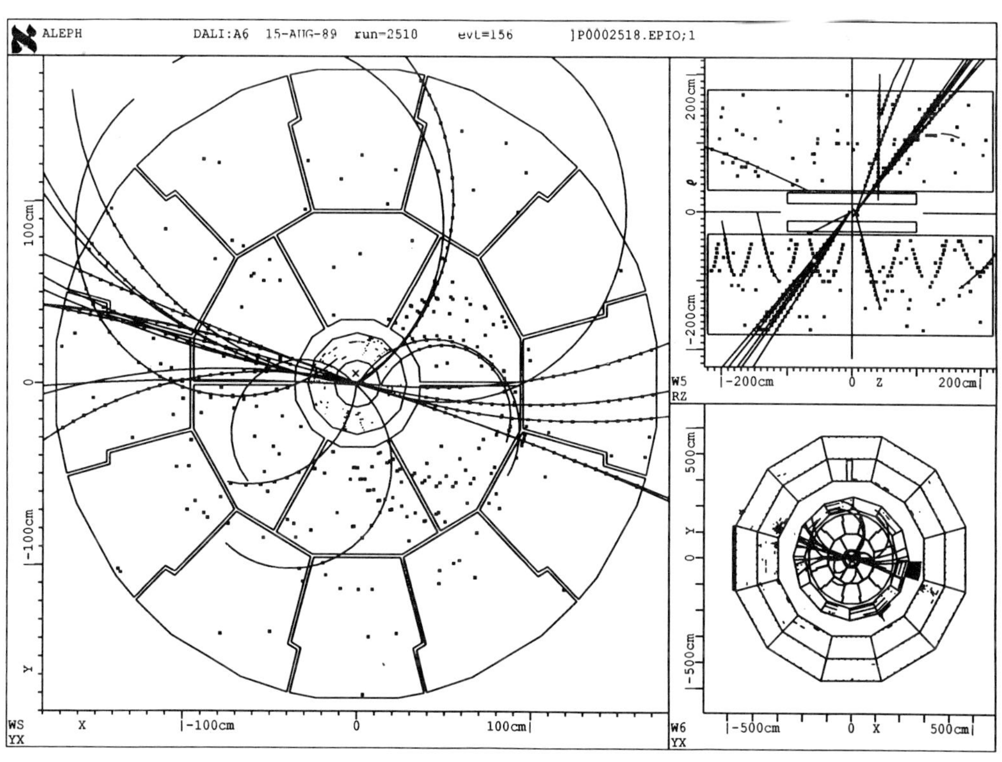

Fig. 4: Z event recorded by the ALEPH detector

7. THE DELPHI DETECTOR

The components of the DELPHI detector, together with their performance, are the following:

Beam-pipe	0.5 mm Al r = 8 cm
Micro-vertex detector	Silicon strip detector (two layers)
Inner tracking chamber	Drift chamber $\sigma_{r\phi}$ = 100 μm
Central tracking chamber	TPC (r = 1.2 m) 1 atm pressure $\Delta p/p^2$ = 0.15 % (incl. outer chamber) σ(dE/dx) = 5.5 %
RICH	Liquid and gas radiators; p/π/K separation
Outer chamber	Drift chamber $\sigma_{r\phi}$ = 150 μm
Electromagnetic calorimeter (Barrel)	HPC (Lead / gas) $\Delta E/E = 19\ \% / \sqrt{E}$
Electromagnetic calorimeter (Endcaps)	Leadglass $\Delta E/E = 8\ \% / \sqrt{E}$
Magnet coil	1.2 T, cold
TOF	Scintillator layer
Hadron calorimeter	Iron / streamer tube sandwich $\Delta E/E = 80\ \% / \sqrt{E}$
Muon chambers	Three layers of streamer tubes
Luminosity monitor	Lead / scintillator sandwich
Very forward monitor	Tungsten / silicon sandwich

At the time of the LEP pilot run, the Micro-vertex detector (scheduled for installation at the end of 1989), half of the Barrel RICH (begin of 1990) and the Forward RICH (1991) were missing.

The DELPHI collaboration are discussing the replacement of the current beampipe by one with a somewhat smaller diameter, and the addition of a third layer to their Micro-vertex detector.

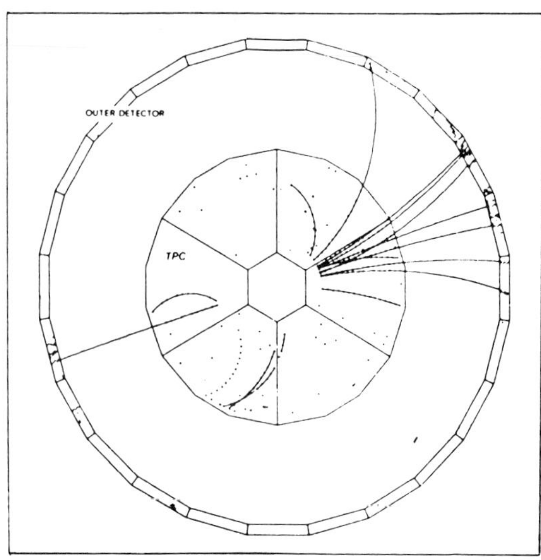

Fig 5. Z event recorded by the DELPHI detector.

Fig. 5 shows a Z event recorded by the DELPHI detector during the LEP pilot run.

8. THE L3 DETECTOR

The components of the L3 detector, together with their performance, are the following:

Beam-pipe	0.4 mm Be r = 8 cm
Central tracking chamber	TEC 1 atm pressure $\sigma_{r\phi}$ = 35 μm @ 1 cm $\Delta p/p^2$ = 0.7 %
Electromagnetic calorimeter	BGO $\Delta E/E \leq 1$ % for E > 1 GeV
Hadron calorimeter	Uranium / gas sandwich $\Delta E/E = 55\ \% / \sqrt{E}$
Muon chambers	Three layers of drift-chambers $\Delta p/p^2$ = 0.03 %
Magnet coil	0.5 T, warm
Luminosity monitor	BGO

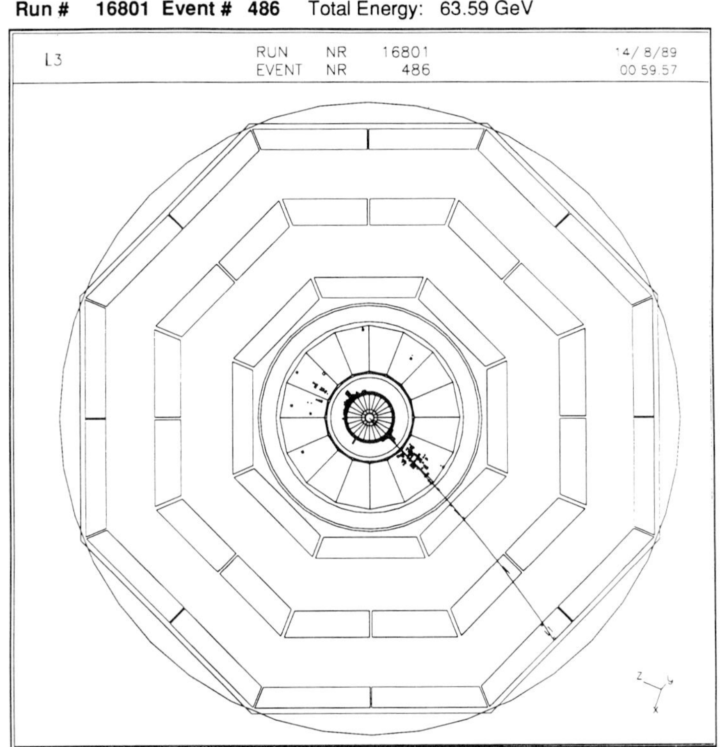

Fig. 6: Z event recorded by the L3 detector

At the time of the LEP pilot run, the endcaps of the electromagnetic calorimeter (scheduled for installation at the end of 1990) were missing.

As the L3 detector has in its current form no coverage of muons in the forward-backward direction, the L3 collaboration are studying the installation of a set of muon chambers, similar in design to those in the barrel part of the detector.

Fig. 6 shows a Z event recorded by the L3 detector during the LEP pilot run.

9. THE OPAL DETECTOR

The components of the OPAL detector, together with their performance, are the following:

Beam-pipe	0.1 mm Al plus 1.4 mm carbon fibre r = 8 cm
Inner tracking chamber	Drift chamber $\sigma_{r\phi} = 55\ \mu m$
Central tracking chamber	JET chamber pressure 4 atm $\Delta p/p^2 = 0.15\ \%$ $\sigma(dE/dx) = 3.5\ \%$
z - chamber	Drift chamber
Magnet coil	0.4 T, warm
TOF	Scintillator plane
Pre-sampler (Barrel)	Drift chambers
Presampler (Endcaps)	Streamer tubes
Electromagnetic calorimeter	Leadglass $\Delta E/E = 7\ \% / \sqrt{E}$
Hadron calorimeter	Iron / gas sandwich $\Delta E/E = 80\ \% / \sqrt{E}$
Muon chambers (Barrel)	Drift chambers

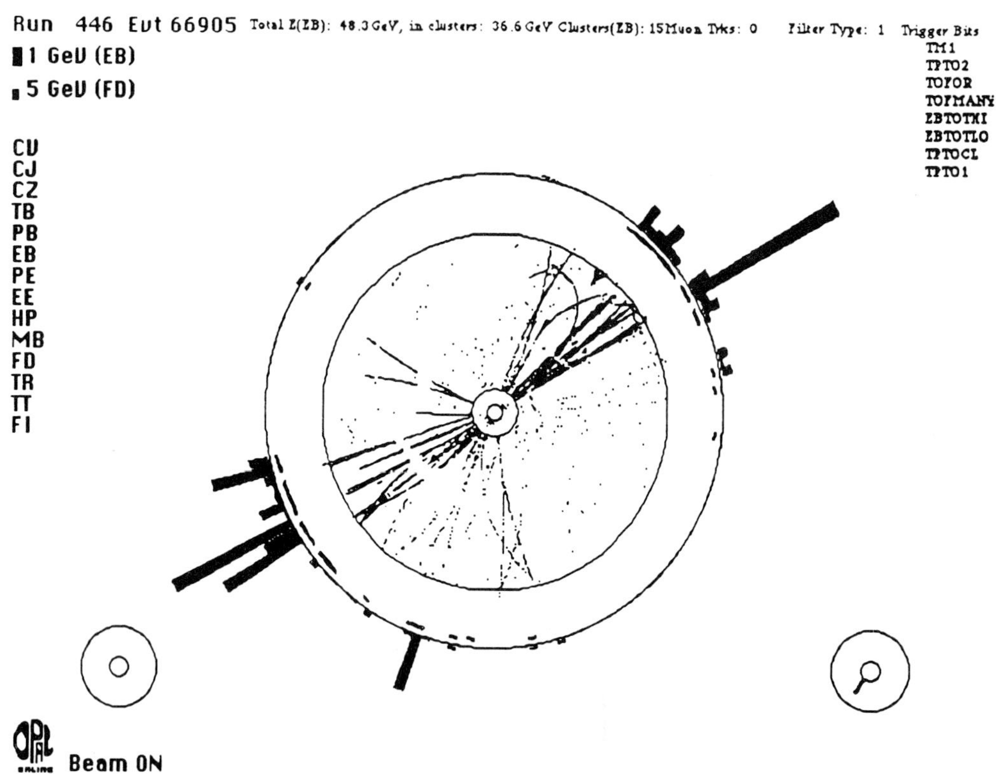

Fig. 7: Z event recorded by the OPAL detector

Muon chambers (Endcaps)	Streamer tubes
Luminosity monitor	Lead / scintillator sandwich
Very forward monitor	Lead / scintillator sandwich

At the time of the LEP pilot run, the OPAL detector was complete, the only one among the four LEP detectors to be so.

The OPAL collaboration are discussing the replacement of the current beam-pipe by one with a smaller diameter, and the installation of a Silicon Micro-vertex detector.

Fig. 7 shows a Z event recorded by the OPAL detector during the LEP pilot run.

10. SOME OBSERVATIONS

All four LEP collaborations devoted as much effort to their software and analysis preparations as to the strenuous task of detector installation and commissioning: the software chains had been thoroughly checked with Monte Carlo events to the extent that "further progress could only be made with real events". As a consequence, physics from LEP will not be significantly delayed by missing analysis software.

There were also snags: all four LEP collaborations underestimated the expenditure for material and manpower (by some 10 to 20 %), the efforts to be invested in cabling, and last but not least, the on-line system integration. With perhaps OPAL as the only exception, the collaborations had some troubles until the readout of complete events was mastered.

The overriding observation is of course that the physics community actually achieved the seemingly impossible: to record real events with all four LEP detectors at the same time when the machine was switched on for physics for the very first time.

11. THE LEP PILOT RUN

On Sunday, August 13, 1989, one hour before midnight, the LEP pilot run was started. Only minutes later, the OPAL collaboration announced that they had recorded the first Z event.

Over five days, from August 13 to 18, with many interruptions to fix machine problems, and to improve the data taking conditions (optimization of the collimator settings), a total of 15 hours of useful collisions was achieved. Typical beam currents were 300 μA, the typical beam life-time was 2 hours. All four LEP detectors were able to record Z and Bhabha events. The final score of Z events was:

15 in ALEPH
6 in DELPHI
16 in L3
21 in OPAL.

The collimator system proved effective in reducing the background by many orders of magnitude, such that the central tracking chambers of all four detectors were amazingly clean. Also, the machine itself proved very stable and reproducible in the orbit position from fill to fill, such that there was no need to re-optimize the position of the collimator jaws.

ACKNOWLEDGEMENTS

Rarely has it been such a pleasure to thank all those who made it possible for the above status report to be given: all the machine people who built and commissioned LEP, and all the experimental physicists and engineers who devoted six years of their lives to building and commissioning the LEP detectors.

Also, I wish to thank U. Amaldi, D. Plane, J. A. Rubio and D. Schlatter for useful comments, and G. Goggi for a careful reading of the manuscript.

NEUTRINO ELECTRON SCATTERING

JAAP PANMAN
CERN, 1211 Geneva 23
Switzerland

ABSTRACT

Recent results from neutrino electron scattering experiments are reported. The experiments are briefly described, and some future options are mentioned.

INTRODUCTION

The study of neutrino-electron scattering can provide sensitive tests of the Standard Model [1]. Improved precision on the measurements of neutral current processes can reveal small deviations from the predictions of the Standard Model, and can give information on unknown parameters in the model.

The theoretical understanding of neutrino electron scattering, being a purely leptonic process, is not affected by uncertainties such as those that limit the accuracy on $\sin^2\Theta_W$ obtained with neutrino nucleon scattering, which is limited by our understanding of the charm production threshold. A comparison of all measurements of $\sin^2\Theta_W$, including those obtained with neutrino electron scattering, and neutrino nucleon scattering [2] and from the masses of the intermediate vector bosons provides a test of the Standard Model beyond the tree level. Radiative corrections have been carefully considered [3].

A study of electron-neutrino electron scattering can give a handle on the interference of the neutral- and charged-current, which both contribute to this process.

The cross section of neutrino electron scattering processes can be written in terms of the vector and axial vector coupling constants of the electron, g_V and g_A,

$$\sigma(\nu_\mu e \to \nu_\mu e) = K \left\{(g_V + g_A)^2 + (1-y)^2 (g_V - g_A)^2\right\}$$

$$\sigma(\bar{\nu}_\mu e \to \bar{\nu}_\mu e) = K \left\{(g_V - g_A)^2 + (1-y)^2 (g_V + g_A)^2\right\}$$

$$\sigma(\nu_e e \to \nu_e e) = K \left\{(g_V + g_A + 2)^2 + (1-y)^2 (g_V - g_A)^2\right\}$$

$$\sigma(\bar{\nu}_e e \to \bar{\nu}_e e) = K \left\{(g_V - g_A)^2 + (1-y)^2 (g_V + g_A + 2)^2\right\}$$

where $y = E_e/E_\nu$ and $K = G^2 m_e E_\nu / 2\pi$. We have introduced here the energy of the neutrino, E_ν, and of the recoiling electron, E_e, its mass, m_e, and the Fermi constant G. It should be noted that these expressions are g_V-g_A symmetric. In the case of muon-neutrinos the cross sections are also invariant under sign changes. This is not the case for the electron-neutrino cross sections which contain a neutral-current charged-current interference term, and can therefore determine the relative sign of the couplings. New results have been reported on $\nu_\mu e$ and $\bar{\nu}_\mu e$ as well as a result on $\nu_e e$ scattering.

In the Standard Model g_A is predicted to be -1 /2 and $g_V = -1/2 + 2\sin^2\Theta_W$ for electrons.

Although the total cross section of $\nu_\mu e$ scattering is sensitive to $\sin^2\Theta_W$, a measurement of the ratio of the $\nu_\mu e$ and $\bar{\nu}_\mu e$ cross sections has experimental advantages, since it does not require knowledge on the absolute normalization of the neutrino flux and on the experimental efficiency to select $\overset{(-)}{\nu}_\mu e$ events [4].

The cross section ratio depends on the value of $\sin^2\Theta_W$ through the relation

$$R = \frac{\sigma(\nu_\mu e)}{\sigma(\bar{\nu}_\mu e)} = 3\,\frac{1-4\sin^2\Theta_W + (16/3)\sin^4\Theta_W}{1-4\sin^2\Theta_W + 16\,\sin^4\Theta_W}\ .$$

QED radiative corrections cancel to a large extent in the ratio and can be neglected [5]. Electroweak radiative corrections, however, change the effective value of $\sin^2\Theta_W$ by an amount which depends on the unknown mass of the top quark m_t and of the Higgs boson m_H [3, 5]. R is not directly accessible experimentally, since it would require pure ν_μ and $\bar{\nu}_\mu$ beams and a measurement with full kinematical acceptance.

Experimentally, the number of neutrino electron scattering events in the ν-beam, N(νe), and in the $\bar{\nu}$-beam, N($\bar{\nu}$e), are measured with recoil energies in a finite range. N(νe) and N($\bar{\nu}$e) contain events induced by the main $\overset{(-)}{\nu}_\mu$ helicity components of the ν and $\bar{\nu}$ beams, and events from interactions of wrong-helicity muon neutrinos, electron neutrinos and electron antineutrinos.

The distribution of y, the ratio of the electron recoil energy to the incoming neutrino energy, is also sensitive to the value of $\sin^2\Theta_W$, as can be seen in the expressions for the total cross sections. Thus, a measurement of the y-distribution and equivalently a measurement of the distribution of $E_e\Theta_e^2$, which is related to y by

$$E_e\Theta_e^2 = 2(1-y)m_e \qquad (\text{for } E_\nu >> m_e)\ ,$$

can give valuable information on $\sin^2\Theta_W$.

From the previous equation an upper limit can be derived for the quantity $E_e\Theta_e^2$,

$$E_e\Theta_e^2 < 2m_e\ ,$$

making use of the boundaries $0 < y < 1$. This relation shows that the electrons recoil within a narrow forward cone with respect to the neutrino direction, a fact which is used experimentally to distinguish νe scattering from background processes, which show much broader distributions.

The main experimental difficulty in studying νe scattering arises from the small cross section of these processes compared to neutrino nucleon scattering, e.g. $\sigma(\nu_\mu e) \simeq 10^{-4}\ \sigma(\nu_\mu N)$. Therefore, intense neutrino beams and targets with large mass and yet good segmentation are needed.

ν_ee SCATTERING

The Irvine - LANL - Maryland Collaboration has studied ν_ee scattering in an experiment performed at LAMPF [6]. A neutrino beam emerging from a proton beam dump from the decays of pions and subsequent decays of muons was used.

From a total of 1492 beam-related events, 295 ± 35(stat) ± 3(syst) neutrino-electron scattering events were obtained after subtraction of the $\nu_e C^{12}$ and $\nu_e C^{13}$ and neutron-induced background (see Fig. 1).

The number of ν_ee events is obtained after subtraction of $\overset{(-)}{\nu}_\mu$e induced interactions, and corresponds to 234 ± 35(stat) ± 8(syst) events.

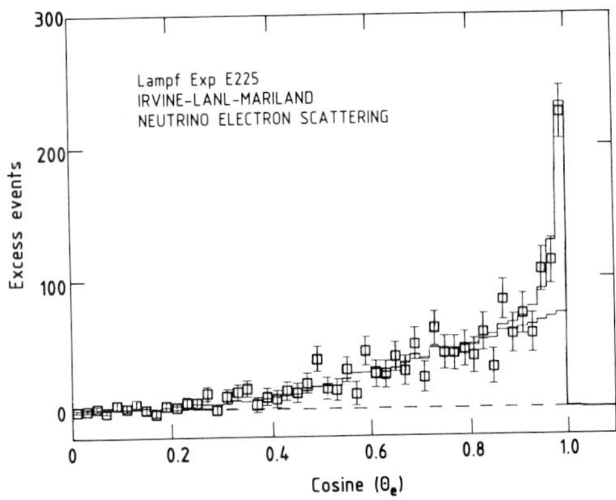

Fig. 1 : Irvine - LANL - Maryland experiment : The $\cos\Theta_e$ distribution of beam-related events. The dashed line shows the background which includes $\nu_e C^{12}$ and $\nu_e C^{13}$ and neutron induced background.

From this result the interference parameter, I, was measured to be $I = -1.03 \pm 0.17(\text{stat}) \pm 0.10$ (syst), to compare with the Standard Model prediction $I_{SM} = -1.08$ for $\sin^2\Theta_W = 0.23$. This result rules out non-interference by 5 standard deviations. In the g_V-g_A plane this result is shown in Fig. 2, together with older measurements from $\overset{(-)}{\nu}_\mu e$ scattering.

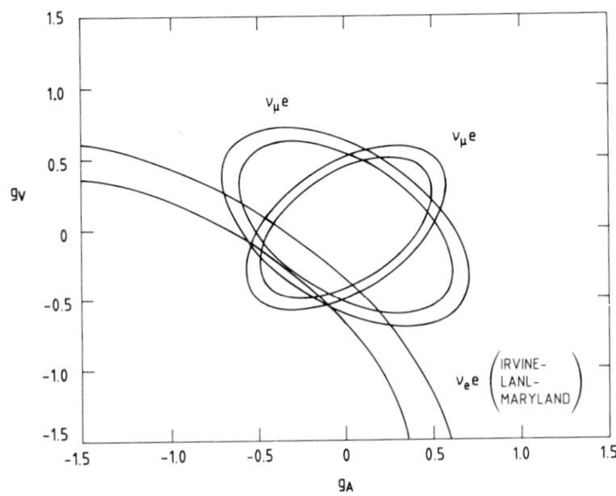

Fig. 2 : Irvine - LANL - Maryland experiment : g_V- g_A contours allowed by published $\overset{(-)}{\nu}_\mu e$ data and the region allowed by the new ν_ee-scattering result.

$\bar{\nu}_e$e SCATTERING

A preliminary result was presented on $\bar{\nu}_e$e-scattering [7] at this conference.

The experiment used a 103 kg. C_6F_6-scintillator target at a nuclear reactor. The experiment is still collecting data. A preliminary value

$$\sigma(\bar{\nu}_e e) = (7.5 \pm 3.5) \cdot 10^{-46} \text{ cm}^2/\text{fission event}$$

was found, corresponding to

$$\sin^2\Theta_W = 0.29 \pm 0.07 .$$

ν_μe SCATTERING (E734 AT BNL)

The US-Japan Collaboration has reported new results on the measurement of $d\sigma/dy$ of $\nu_\mu e$ and $\bar{\nu}_\mu e$ scattering [8]. A large fraction of the data had been published previously [9], where an analysis was presented using the ratio $R = \nu_\mu e/\bar{\nu}_\mu e$ to extract $\sin^2\Theta_W$. The data were taken in the wide-band horn-focussed neutrino beam of the BNL-AGS, in three periods (1981, 1983, 1986). The mean $\nu_\mu(\bar{\nu}_\mu)$ energy was 1.27(1.23) GeV/c. The neutrino detector was a 170 ton, fully active, highly segmented device. Electrons were measured with an angular resolution given by $\Delta\Theta_{x,y} = (13 \pm 1)$ mrad/$\sqrt{E/\text{GeV}}$.

In this experiment, samples of $160 \pm 17(\text{stat}) \pm 4$ $\nu_\mu e$ and $97 \pm 13(\text{stat}) \pm 5(\text{syst})$ $\bar{\nu}_\mu e$ events were obtained. The Θ_e distributions are shown in Fig. 3. The strong forward peak was fitted simultaneously for the ν and $\bar{\nu}$ data with the two y-dependent components present in the expression for the cross sections.

A likelihood fit was made to the data in Figs. 3a and b, making use of the normalized $d\sigma/dy$ distributions, varying as free parameters g_V, g_A and N_b, the number of background events. The present result is not independent of previously published analyses, but uses the additional information given by the shape of the Θ_e^2 distribution on top of the measurements of the total number of events. The results can be summarized as follows

$$g_V = -0.107 \; {}^{+0.035}_{-0.036} \text{ (stat)} \; {}^{+0.029}_{-0.028} \text{ (syst)}$$

$$g_A = -0.514 \; {}^{+0.023}_{-0.023} \text{ (stat)} \; {}^{+0.029}_{-0.027} \text{ (syst)}$$

when the only solution out of four possible ones is chosen compatible with $\bar{\nu}_e$e and e^+e^- data. In a fit allowing only $\sin^2\Theta_W$ as a free parameter

$$\sin^2\Theta_W = 0.195 \pm 0.018(\text{stat}) \pm 0.013(\text{syst})$$

was deduced.

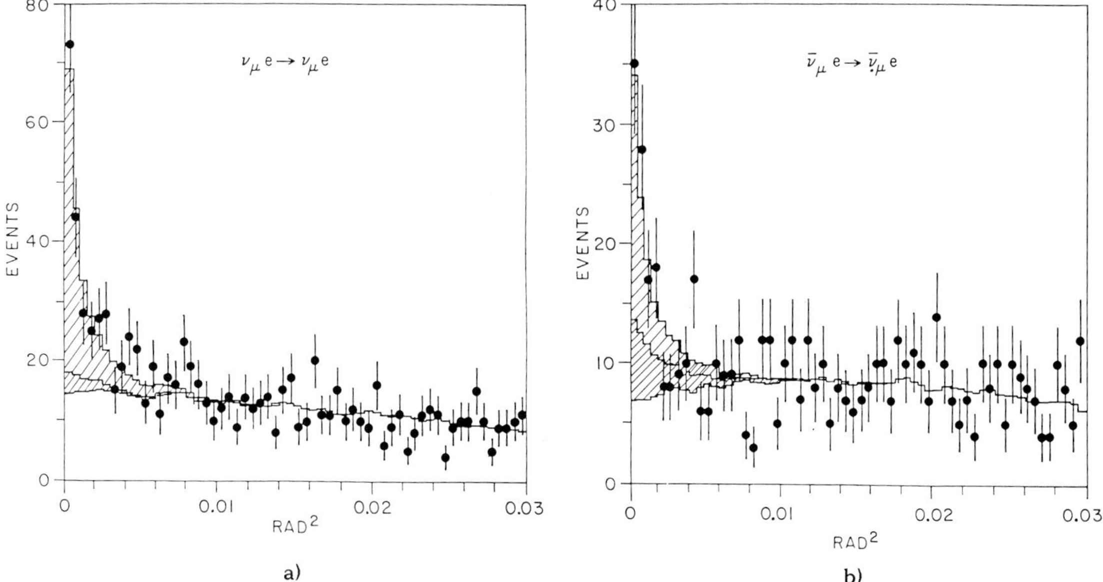

Fig. 3: BNL E734 experiment : a) differential distributions in Θ_e^2 for the neutrino data, b) differential distributions for the anti-neutrino data. Data are points with error bars; y-independent term is light- shaded; $(1-y)^2$ term is dark-shaded and background is unshaded.

$\nu_\mu e$ SCATTERING (CHARM-II EXPERIMENT)

The CHARM-II Collaboration reported data obtained with their new massive fine-grain and low-density detector in 1987 and 1988 [10]. This detector consists of a target calorimeter of 692 tonnes of glass followed by a muon spectrometer. The calorimeter is used to detect neutrino interactions and to measure the energy and direction of particle showers. It is instrumented with plastic streamer tubes. The resolution of the direction and the energy of electron showers is $\sigma(\Theta_e) = 22\,\mathrm{mrad}/\sqrt{E/\mathrm{GeV}}$ and $\sigma(E)/E = 0.3/\sqrt{E/\mathrm{GeV}}$, as determined in a test beam. The average energy of electrons selected in the analysis is 11 GeV.

The spectrometer is used to determine the momenta of muons produced in charged-current reactions of $\nu_\mu(\bar{\nu}_\mu)$ in the target. The momentum resolution is ±14% for 20 GeV/c muons.

This detector was exposed to specially designed horn focussed wide-band neutrino and antineutrino beams.

To select neutrino electron scattering candidates, events were selected with electromagnetic showers characterized by their small width in contrast to hadronic showers. Showers were selected with a high density of energy deposition starting with one hit in the first streamer tube plane following the vertex; showers with additional tracks or with backscattering were rejected. The combined efficiency of the selection criterion was determined to be 0.76.

The remaining background is composed of semileptonic interactions with an electromagnetic shower in the final state, mainly due to coherent and diffractive π^0 production by neutral current (NC) interactions and quasielastic scattering of a component of electron neutrinos and antineutrinos from K_{e3} - decays.

The $E\Theta^2$ distributions of events selected in the neutrino beam (a) and in the antineutrino beam (b) are shown in Fig. 5. The pronounced peak of $E\Theta^2 < 3$ MeV is attributed to neutrino electron scattering. To evaluate the number of events due to these reactions the expected distributions in $E\Theta^2$ for the background processes have been modelled using Monte Carlo techniques. The distribution for $\overset{(-)}{\nu}_e N \rightarrow e^\pm N'$ was constructed starting from observed quasielastic $\overset{(-)}{\nu}_\mu N \rightarrow \mu^\pm N'$ events. The muon track in these events was replaced by an electron shower of the same direction and energy. Events due to coherent as well as diffractive neutral-current production of a single π^0 were simulated following the theoretical description of Rein and Seghal [11] and Bel'kov and Kopeliovich [12]. The resulting distributions in $E\Theta^2$ and E were then fitted to the observed spectrum for $5 < E\Theta^2 < 72$ MeV and $3 < E < 35$ GeV to determine the relative contributions of the background reactions.

The background composition was independently determined for the subsample of events having their vertex in a glass plate followed by a scintillator plane. The energy loss distribution

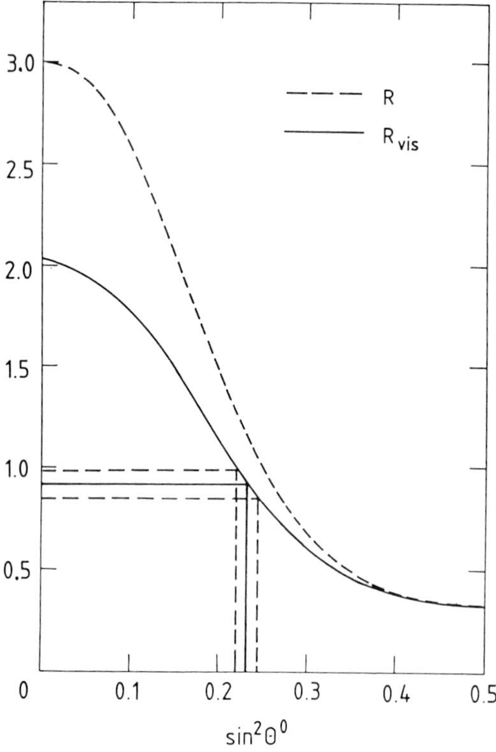

Fig. 4: Ratio of $\sigma(\nu_\mu e)$ and $\sigma(\bar{\nu}_\mu e)$ as a function of $\sin^2\Theta^o$ (without radiative corrections). The dashed curve (R) represents the case of full energy acceptance and pure ν_μ and $\bar{\nu}_\mu$ beams. The full curve (R_{vis}) is valid for the electron energy range 3-24 GeV and takes into account the fractions of wrong-helicity muon neutrinos and of ν_e and $\bar{\nu}_e$. The value of R_{vis} measured in this experiment and the corresponding value of $\sin^2\Theta^o$ are shown.

in the early stage of the electromagnetic shower corresponds to one single charged particle in the case of electron showers, whereas π^0 induced showers start with e^+e^- pairs. The relative abundances of these two components were determined from this energy loss distribution in the background region. The results from the two methods agree.

After subtraction of the background they found $N(\nu e) = 762 \pm 43$(stat) events in the neutrino data sample and $N(\bar{\nu}e) = 1017 \pm 51$(stat) events in the antineutrino data sample, with $E\Theta^2 < 3$ MeV and $3 < E < 24$ GeV. The observed shape of the $E\Theta^2$ distributions of these events is in good agreement with that expected from the experimental resolution and kinematics. This is shown in Fig. 6.

To extract $\sin^2\Theta_W$, the quantity R_{vis} is defined as the ratio of the number of neutrino electron scattering events with 3 GeV $< E_e <$ 24 GeV in the neutrino and the antineutrino beam multiplied by the energy-weighted antineutrino to neutrino flux ratio. The relation between R_{vis} and $\sin^2\Theta_W$, shown in Fig. 4, was determined taking the beam composition and finite energy window into account.

The flux ratio was determined in four independent ways: from the number of events recorded in the neutrino and antineutrino beams induced by inclusive NC and CC processes on nucleons and the known cross section ratio, from the number of NC produced π^0 events recorded in the range $5 < E\Theta^2 < 72$ MeV, from the number of quasielastic CC processes recorded in the range $0.05 < Q^2 < 0.15$ GeV^2 and from the integrated muon flux recorded in the shield for neutrino and antineutrino runs.

The four normalization factors agree within their errors, their weighted mean value $F = 1.215 \pm 0.028$ was used in determining R_{vis}.

Since the ratio of the mean energies does not only enter into the flux normalization, its uncertainty was taken into account separately in the determination of $\sin^2\Theta_W$. The value of R_{vis} was determined to be

$$R_{vis} = 0.910 \pm 0.068(\text{stat}) \pm 0.045(\text{syst})$$

which leads to

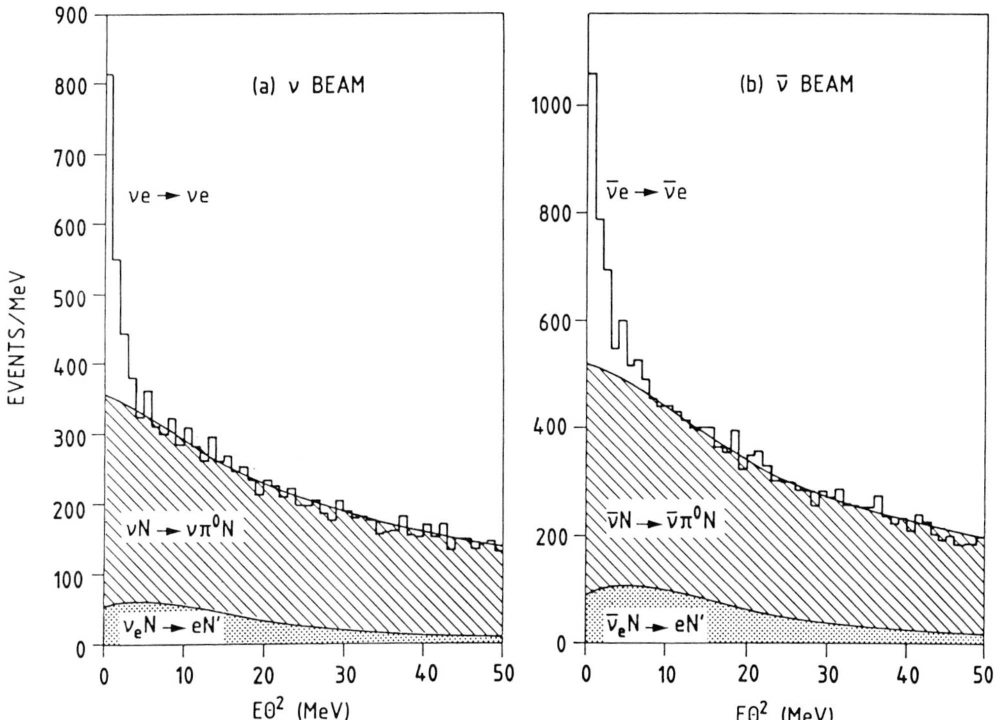

Fig. 5: Distribution of selected events as a function of $E\Theta^2$, a) in the neutrino, b) in the antineutrino beam. The background reactions $\nu_e N \to eN'$ and $\nu N \to \nu\pi^0 N$ are shown separately.

$$\sin^2\Theta_W = 0.233 \pm 0.012(\text{stat}) \pm 0.008(\text{syst}).$$

Applying electroweak radiative corrections with m_t = 100 GeV and m_H = 100 GeV gives

$$\sin^2\Theta_W(\text{corrected}) = 0.232 \pm 0.012(\text{stat}) \pm 0.008(\text{syst}).$$

The systematic error on $\sin^2\Theta_W$ contains contributions from uncertainties of the flux normalization (±0.003), of the abundancy of wrong-helicity muon neutrinos (±0.001) and of electron-neutrinos (±0.002), of the mean energies of the different beam components (±0.002), from the background subtraction (±0.007) and from the energy calibration of the detector (±0.002); they were combined quadratically.

OUTLOOK IN νe SCATTERING

The CHARM-II experiment is in progress and more data is being collected with the aim to more than double the final data sample. The analysis is being refined, in particular improvements in the understanding of the background are to be expected. The final precision will improve by a factor 2 to 2.5.

A proposal for a new experiment has been submitted at LAMPF [13], with the aim to measure $\sin^2\Theta_W$ with a 1% precision. The proposed method is different from that used in previous precision determinations of $\sin^2\Theta_W$ with neutrino electron scattering. The neutrino beam is obtained from pion and muon decays at rest in a proton beam dump. The scatterings of electrons by different neutrino species are separated by their different time of arrival with respect to the pulsed proton beam. The prompt ν_μ component is produced from π-decays while the $\bar{\nu}_\mu$ and ν_e component comes from μ-decays. The measurement of $\sin^2\Theta_W$ is based on the ratio

$$N(\nu_\mu e)/\left(N(\bar{\nu}_\mu e) + N(\nu_e e)\right).$$

The major experimental difficulties are the knowledge of the energy thresholds, of the efficiency of the timing separation and the calculation of the backgrounds.

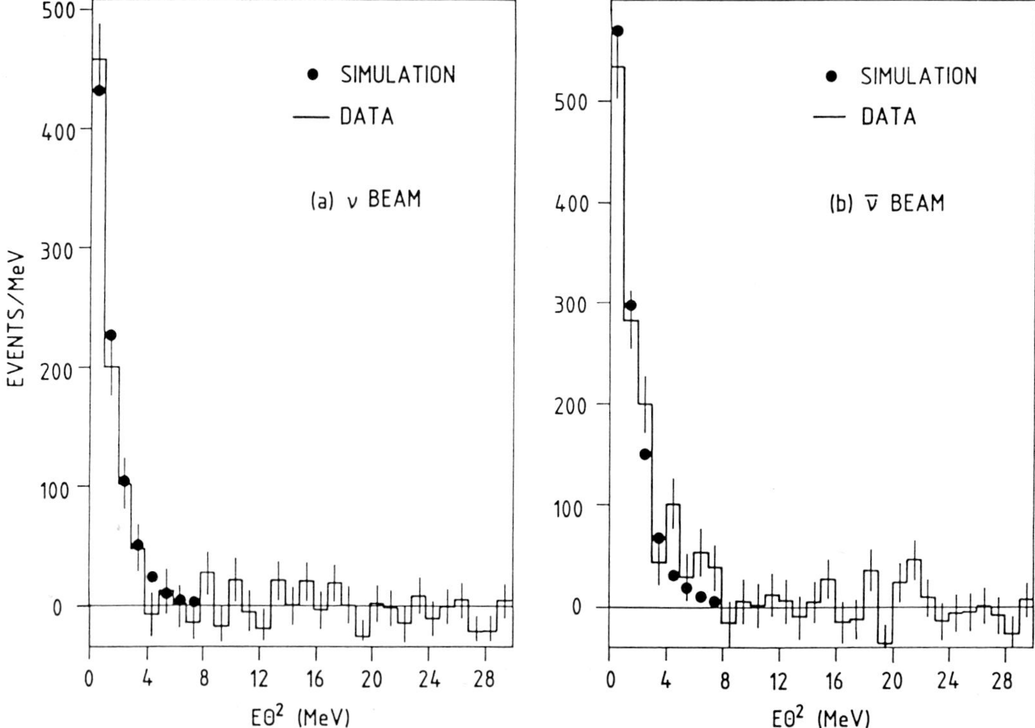

Fig. 6: Distribution of neutrino electron scattering events as a function of $E\Theta^2$, a) in the neutrino b) in the aintineutrino beam, after subtraction of the background. The points represent the expected distributions.

NEUTRINO NUCLEON SCATTERING

At present, the most precise measurements of $\sin^2\Theta_W$ with neutrino scattering have been obtained using the neutral current (NC) to charged current (CC) ratio of deep-inelastic neutrino nucleon scattering. The average value including the two most precise experiments is [2]

$$\sin^2\Theta_W = 0.230 \pm 0.004(\text{exp}) \pm 0.005(\text{theo})$$

including radiative corrections for $m_t = m_H$ = 100 GeV.

The largest part of the theoretical error is due to the uncertainty in the correction for the charm production threshold; the central value is quoted for m_c = 1.5 GeV, while $\Delta m_c = \pm 0.3$ GeV is assumed for the theoretical uncertainty.

A preliminary result from FMMF was reported, giving $R^\nu = \sigma^\nu(NC)/\sigma^\nu(CC) = 0.303 \pm 0.006$, measured in a Tevatron beam [14]. The interpretation in terms of $\sin^2\Theta_W$ is still in progress; the central value is expected to be $\sin^2\Theta_W$ = 0.237 - 0.241.

Several options for new neutrino beams are being proposed to improve the precision of $\sin^2\Theta_W$ in νN scattering [15]. These attempts are based on the benefits derived from a tagged neutrino beam. However, no scheme has emerged which can be demonstrated to improve the experimental and theoretical uncertainties significantly.

CONCLUSION

The new CHARM-II result can be compared with other determinations of $\sin^2\Theta_W$ derived from neutrino electron scattering based on the cross section ratio. The CHARM Collaboration [4] and the US-Japan Collaboration (E734 at BNL) have obtained a combined result of $\sin^2\Theta_W$ = 0.203 ± 0.020, which is 0.03 ± 0.02 lower than the result from neutrino nucleon scattering. The new result, although compatible with these older measurements, is in excellent agreement with the average value obtained with neutrino nucleon scattering provided m_t is lower than 250 GeV.

The comparison of $\sin^2\Theta_W$ derived from neutrino-electron scattering and from the new precise measurement of the Z^0 mass [16, 17, 18] shows good agreement with the prediction of the Standard Model. This result is nearly independent of the value of m_t.

REFERENCES

1. W.J. Marciano and A. Sirlin, *Physics Rev.,* D29(1984)945.

2. J.V. Allaby et al., CHARM Collaboration, *Z Phys.,* C36(1 987)611, gives $\sin^2\Theta_W$ = 0.235±0.005(exp)±0.005(theor) for $m_t = m_H$ = 1 00 GeV,
H. Abramowicz et al., CDHSW Collaboration, *Phys. Rev. Letters,* 57(1986)298. The final analysis of the CDHSW Collaboration gives $\sin^2\Theta_W = 0.226 \pm 0.005$ (exp)± 0.005(theor) for $m_t = m_H$ = 100 GeV (CERN/EP 89-101).

3. S. Sarantakos, A. Sirlin and W.J. Marciano, *Nucl. Phys.,* B217(1983)84,
M.J. Marciano and S. Sirlin, *Phys. Rev.,* D22(1980)2695, D29(1984)945 , D31 (1985)21 3E.

4. J. Dorenbosch et al., CHARM Collaboration, *Z Phys.,* C41 (1989)567.

5. D.Y. Bardin and O.M. Dokuchaeva, *Nucl. Phys.,* B246(1984)221, and preprint JINR-E2-86-260(1986).

6. R.C. Allen et al., *Phys. Rev. Lett.,* 55(1985) 2401, the final data : private communication X.-Q. Lu.

7. G.S. Vidyakin et al., paper No. 292 contributed to this Symposium.

8. Paper No. 047 contributed to this Symposium.

9. K. Abe et al., *Phys. Rev. Lett.* 58(1987)636.

10. D. Geiregat et al., paper No. 149 contributed to this Symposium.

11. D. Rein and L.M. Sehgal, *Nucl. Phys.,* B223(1983)29.

12. A.A. Bel'kov and B.Z. Kopeliovich, *Sov. J. Nucl. Phys.,* 46(1987)499.

13. R.C. Allen et al., A Proposal for a Precision Test of the Standard Model by Neutrino-Electron Scattering (Large Cerenkov Detector Project), LANL Proposal, LA - 11 300-P, 1988.

14. A. Abolins, R. Brock, private communication.

15. R.H. Bernstein, Neutrino Physics in a Tagged-Neutrino-Beam, FERMILAB-conf-89/34.
V.V. Amosov et al., Experiments with Tagged-Neutrino-Beam at IHEP Accelerator, Proposal SERP-E-152.
A.S. Vovenko et al., On the use of the Berlin-Budapest-Dubna-Serpukhov Neutrino Detector in the UNK Neutrino Beam.
I.A. Savin, Proceedings of the 13th International Conference on Neutrino Physics and Astrophysics, Boston (Medford) 1988, page 421, eds. Schneps, T. Kafka, W.A. Mann and P. Nath, *World Scientific,* Singapore.

16. F. Abe et al., CDF Collaboration, *Phys. Rev. Lett.,* 63(1989)720.

17. UA2 Collaboration, presented by A. Weidberg at this Symposium.

18. G.S. Abrams et al. MARK II Collaboration, *Phys. Rev. Lett.,* 63(1989)724 and more recent results reported by G. Feldman at this Symposium.

W AND Z PRODUCTION AND DECAY RESULTS FROM UA2

The UA2 Collaboration
Bern - Cambridge - CERN - Heidelberg - Milano - Orsay (LAL)
Pavia - Perugia - Pisa - Saclay (CEN) Collaboration

presented by
A. R. Weidberg
CERN, Geneva, Switzerland

ABSTRACT

Preliminary results on the production and decay of W and Z bosons produced at the CERN $\bar{p}p$ Collider are presented.The event sample consists of ~ 1200 $W \rightarrow e\nu$ decays and 85 $Z \rightarrow e^+e^-$ decays. The corresponding integrated luminosity is 7.1 pb^{-1}.

1. INTRODUCTION

From data taken at the CERN $\bar{p}p$ Collider in the period 1981-1985 results on the production and decay properties of the W and Z bosons were published by UA1 [1] and UA2 [2].The successful operation of the upgraded CERN $\bar{p}p$ complex in the period 1988-1989 allowed the UA2 detector to accumulate a sample of W and Z events which is a factor of eight times larger than the previously published one [2]. In this report results are presented on three measurements which were largely limited by statistical errors in the old results and which can therefore be significantly improved by an analysis of the present data sample.

1. The cross section ratio $R = \sigma_W/\sigma_Z$.
2. The distribution of the transverse momentum of the W, $P_T{}^W$ at high values of $P_T{}^W$.
3. The ratio of the mass values M_W/M_Z.

The value of R is sensitive to the number of light neutrino generations and to the mass of the top quark if it is lighter than the W. The distribution of $P_T{}^W$ is an important test of perturbative QCD and any significant discrepancies between the data and theory at high $P_T{}^W$ would be a signal for new physics beyond the standard model (SM). The errors on M_W and M_Z measured in this experiment are dominated by the systematic errors in the energy scale. Therefore the most precise tests of the M can be made using the ratio M_W/M_Z (see reference [3] for a more detailed discussion).

In Section 2 a brief review of the upgraded UA2 detector is given.In Section 3 results on σ_W, σ_Z and R are given and compared with theoretical predictions. In Section 4 results are given on the distribution of $P_T{}^W$ and comparisons are made with QCD predictions. In Section 5 results are presented for the mass values M_W and M_Z and the mass ratio M_W/M_Z. Note that the results presented are preliminary and may change slightly after further analysis of test beam data.

2. THE UA2 APPARATUS

The UA2 detector was upgraded during the period 1985 to 1987. Details of the construction and performance of the various detector elements can be found in the references given below. Only the main features relevant to this analysis will be summarised here.

2.1 Calorimetry

An overall view of the UA2 calorimeters is shown in Fig. 1. The UA2 central calorimeter [4] was retained

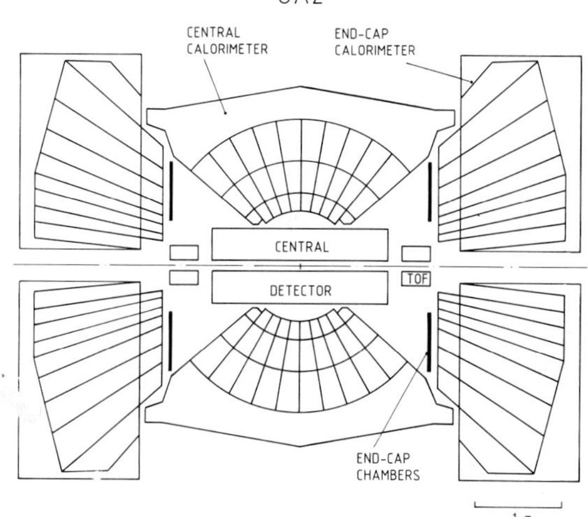

Fig. 1 Layout of the UA2 calorimeters.

with minor modifications. It covers the full azimuthal range, $0^o < \phi < 360^o$ and the polar angles $40^o < \theta < 140^o$. Each of the 240 electromagnetic and hadronic cells subtends 10° in θ and 15° in ϕ.

The electromagnetic part is a multi-layer sandwich of lead and scintillator, 17 radiation lengths thick, while the hadronic part is an iron-scintillator sandwich, giving a thickness of 4.5 absorption lengths including the electromagnetic cells. For the upgrade all the scintillator plates of the two hadronic compartments were replaced and, in order to increase the radial space available for the new central detector, the thickness of the edge cell electromagnetic compartments was reduced.

The end cap calorimeters (end-caps) [5] cover the pseudorapidity region $1 \leq |\eta| < 3$. Each end-cap consists of 12 modules and each module is segmented into 16 cells. In a given module the two cells closest to the beam axis ($2.5 \leq |\eta| \leq 3.0$ and $2.2 \leq |\eta| \leq 2.5$) cover 30° in azimuth. The other cells have a constant segmentation of $\Delta\phi = 15^o$, $\Delta\eta = 0.2$. All the cells in the pseudorapidity interval $1.0 \leq |\eta| \leq 2.5$ have one electromagnetic and one hadronic compartment. The electromagnetic compartment is a multi-layer sandwich of lead (3 mm thick) and acrylic scintillator (4 mm thick), with a total thickness varying from 17.1 to 24.4 radiation lengths depending on the polar angle. The hadronic compartment is a multi - layer sandwich of iron (25 mm thick) and scintillator (4 mm thick) corresponding to ~ 6.5 absorption lengths, including the electromagnetic cells. The cells nearest to the beam have only a hadronic compartment. In addition, cells with only a hadronic compartment cover the pseudorapidity interval $0.9 \leq |\eta| < 1.0$ to measure particles escaping from the interface between the end-cap and the central calorimeters (see Fig. 1). Each compartment is read out via two wave-length shifting plates placed on the opposite sides of each cell, introducing a dead space between adjacent cells of 7 mm for the electromagnetic compartments and of 13 mm for the hadronic ones. To minimize the effect of these dead spaces each module is rotated by 50 mr around the symmetry axis normal to the beam.

Clusters of deposited energy were formed in the calorimeters by joining all cells with an energy greater than 400 MeV sharing a common edge. Clusters with a small lateral size and a leakage into the hadronic compartments consistent with a shower from a single electron were marked as electromagnetic. Since the response of the calorimeter to showers depends on the fraction of the energy carried by hadrons, a weight was defined for each compartment of the calorimeter, which was applied to the observed energies in hadronic showers in order to compensate for the difference in response. The calorimeter weights applied to the electromagnetic cells were 1.18 in the central calorimeter and 1.20 in the end caps. The efficiency of finding an electromagnetic cluster from an electron candidate was measured from test beam data to be $\varepsilon_{cal} = 98.8 \pm 0.4\%$, in the fiducial region of the central calorimeter, integrating over all the allowed impact directions for electrons, and accounting for losses due to energy from other particles spoiling the electron signature. The fiducial region excludes 5 mm zones near the inter-cell boundaries and the shortened electromagnetic cells at the edge of the central calorimeter.

2.2 The Central Detector

The layout of the central detector is shown in Fig. 2. Around the beam-pipe, at radii of 3.5 cm (inner) and 14.5 cm (outer), are two arrays of silicon counters used for tracking and ionisation measurements [6]. Between the two is a cylindrical drift chamber with

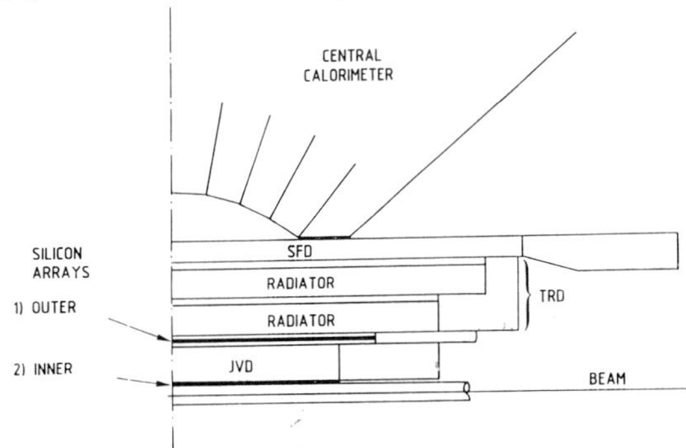

Fig. 2 La ıt of the UA2 central detectors.

jet geometry the Jet Vertex Detector or JVD) [7]. After the inner tracking detectors is the Transition Radiation Detector (TRD) [8] consisting of two sets of radiators and proportional chambers, able to distinguish electron tracks from those of hadrons. The particles are tracked onto the calorimeter surface by the Scintillating Fibre Detector (SFD) [9] which consists of fibres arranged on cylinders into 6 stereo triplets followed by a 1.5 radiation length thick lead converter covering the central calorimeter, and a further 2 stereo triplets used as a preshower detector. Electromagnetic showers initiated in the lead converter were detected using the SFD preshower detector.

The position of the event vertex and the charged tracks were reconstructed using the SFD in conjunction with the Silicon Hodoscopes and the JVD. The vertex finding efficiency was measured to be 100% for vertices on the beam line within 250 mm of the detector centre. The fraction of vertices inside this cut was measured to be $\varepsilon_V = 94.0 \pm 0.5\%$. The tracking efficiency for isolated high energy tracks was measured to be $\varepsilon_{trk} = 90.6 \pm 1.1\%$, using a sample of electrons produced in the decay of W bosons ("W electrons").

The tracking and preshower sections of the SFD were used to match the impact point of candidate electron tracks with the position of electromagnetic showers with a resolution of $\sigma_{r\phi} = 0.4$ mm in the r-ϕ plane (perpendicular to the beam axis) and $\sigma_z = 1.1$ mm along the beam direction. The quality of a track-preshower match was defined by the variable $d_\sigma^2 = (\Delta_{r\phi}/\sigma_{r\phi})^2 + (\Delta_z/\sigma_z)^2$ where $\Delta_{r\phi}$, Δ_z are the displacements between the track and shower positions. Accidental overlaps between photon showers and charged tracks give large values of d_σ^2, while candidate electrons were required to have $d_\sigma^2 < 25$. Preshower clusters for electron candidates were required to have a charge, detected in each of the three stereo views of the preshower detector, of at least twice that expected from a minimum ionising particle. The efficiency of the track-preshower matching with the above cuts was measured to be $\varepsilon_{ps} = 89.9 \pm 1.1\%$ using W electrons.

2.3 Forward Tracking

In front of the forward calorimeters in the pseudorapidity range $1.1 < |\eta| < 1.6$ tracking and preshower information is provided by the End-Cap Proportional Tubes (ECPT). The detector consists of 16 modules ($\Delta\phi = 45°$ each) of proportional tubes. In each module a stereo triplet is placed after a ~ 2 r.l. thick converter acting as a preshower, while two triplets in front of the converter act as tracking chambers.

2.4 The Trigger System

The trigger system consisted of 3 levels [10] based on calorimeter information and signals from the Time of Flight counters (TOF, see Fig. 1). These counters were used to generate a minimum bias trigger signal and to calculate the integrated luminosity. The first level electron triggers used analogue sums of the signals from the photomultipliers of the electromagnetic calorimeter cells up to $|\eta| = 2$. At the second level, electron and jet clusters were reconstructed in a fast processor using information from a fast digitization of the calorimeter cell signals. A full calorimeter reconstruction was performed in the third level processors using a complete set of calibration constants.

Two principal data samples were used in this analysis. The first sample (the "W sample") required an event to contain an electromagnetic cluster with a transverse energy above 15 GeV passing the third level cuts as well as requiring a missing transverse momentum $p_T^{miss\ raw}$ greater than 15 GeV/c reconstructed online :

$$p_T^{miss\ raw} = |\Sigma E^T_{cell} \times \vec{u}_{cell}|,$$

where E_{cell} is the sum of the electromagnetic and hadronic energies measured in each cell, weighted to compensate for the response of the electromagnetic compartments to hadron showers, $\vec{u}_{cell}$ is a unit vector in the transverse plane from the cell centre and the sum extends over all calorimeter cells. The second sample (the "Z sample") consisted of events containing two electromagnetic clusters with $E_T > 5$ GeV both passing the third level electron cuts. In addition the invariant mass of the pair of electromagnetic clusters was required to be above 25 GeV/c^2. Data were also taken with a less restrictive single electron trigger which was used for estimating the background contamination in the W sample.

2.5 Electron Identification

In the central region ($|\eta| < 1$), electron candidates were selected by searching for a track and preshower signal, matching within a tolerance of $d^2_\sigma < 25$, facing an electromagnetic cluster. The lateral and longitudinal profiles of the shower were required to be consistent with those expected for a single isolated electron incident along the track direction, using a quality factor $P(\chi^2)$ defined using extensive test beam measurements. Candidates with $P(\chi^2) < 10^{-4}$ were rejected as well as electrons hitting the shortened edge cells. The efficiency of this cut was measured to be $\varepsilon_{P(\chi2)} = 96.9 \pm 0.5\%$

for W electrons in the fiducial regions of the calorimeter (i.e. excluding an ~ 5 mm region near inter-cell boundaries). The electron energy was corrected for the impact point dependence of the calorimeter response as determined from test beam data. Only candidates with a transverse momentum $p_T^e > 20$ GeV/c were retained. In the forward regions ($1 < |\eta| < 1.6$) an equivalent selection was made using the ECPT tracking and preshower information.

2.6 Neutrino Identification

The neutrino transverse momentum was estimated offline by

$$\vec{p}_T^{\,\nu} = -\,\vec{p}_T^{\,e} - \Sigma\, E^T_{cell} \times \vec{u}_{cell}$$

where $\vec{u}_{cell}$ is now a unit vector in the transverse plane from the interaction vertex to the cell centre and E^T_{cell} has been corrected for the position of the event vertex. The sum over cells excludes the electron "core" cells and the electron p_T is corrected downward by ~ 1% to take into account the electron energy that is deposited outside the core cells.

3. W AND Z CROSS SECTIONS

In this section preliminary results are given for the cross-sections σ_W, σ_Z and the ratio $R_{exp} = \sigma_W/\sigma_Z$.

3.1 W sample

The W sample in the central region was defined by the following kinematic selection on the sample of events containing an electron candidate with $|\eta| < 1$.

1. $p_T^e > 20$ GeV/c
2. $p_T^\nu > 20$ GeV/c

To avoid regions of the calorimeter with low electron identification efficiency events with electron impact points on the calorimeter within ~ 5 mm of the cell boundary were rejected. The remaining sample of 1266 events shows clean Jacobian peaks in p_T^e and p_T^ν (see Fig. 3). From a partial sample of 2.7 pb^{-1} processed without any requirement on p_T^ν, the QCD background contamination was estimated to be ~ 1%. In the forward regions, a similar selection gave a sample of 361 events. The electron efficiencies were measured by applying tighter kinematic cuts to obtain a background free region and then relaxing the cuts one at a time. The acceptance was estimated using a simple Monte Carlo model for the detector response. The results are summarised in Table 1 below.

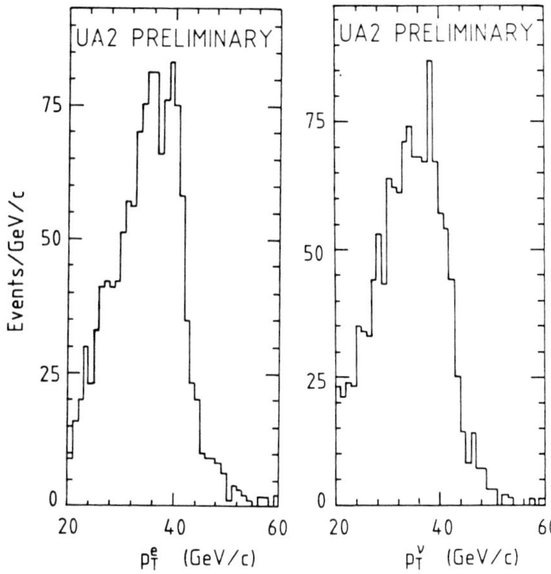

Fig. 3 Distribution of p_T^e (a) and p_T^ν (b) for the 1266 central-region W candidates.

TABLE 1 : The efficiency and acceptance values for the W sample

	Central	Forward
Number of events	1266	361
Electron selection criteria efficiency	(78.1±1.2±1.6)%	(81.5±2.7±1.5)%
Acceptance	(38.3±0.8)%	(9.9±0.2)%

From the numbers given in Table 1 and the value of the integrated luminosity of 7.1 ± 0.5 pb^{-1} plus small additional correction for trigger efficiencies and background subtractions the cross-section is measured to be

$$\sigma_W = 630 \pm 20\ (\text{stat}) \pm 50\ (\text{syst})\ \text{pb}$$

3.2 Z sample

The Z sample was obtained from a filter of the electron pair trigger sample requiring that the mass of two electromagnetic clusters $M_{e^+e^-} > 35$ GeV/c^2. Both clusters were required to satisfy the electron identification requirements (see Section 3.1). The mass spectrum of the resulting sample shows a clear peak at the Z mass above a small background (see Fig. 4).

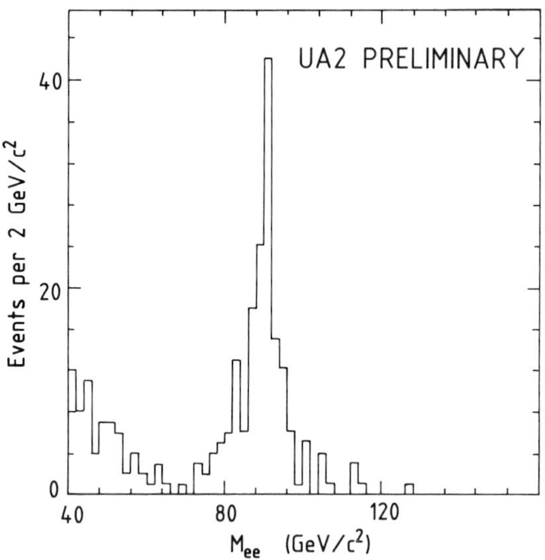

Fig. 4 The e^+e^- mass spectrum for the full sample.

There are 168 events with $Me^+e^- > 76$ GeV/c^2. For the present preliminary analysis of the cross-section both electrons are required to be in the fiducial regions of the calorimeters where it is easier to evaluate the efficiencies. The mass spectrum of the reduced sample is shown in Fig. 5 and contains 86 events with $Me^+e^- >$ 76 GeV/c^2. Correcting for the efficiencies, background

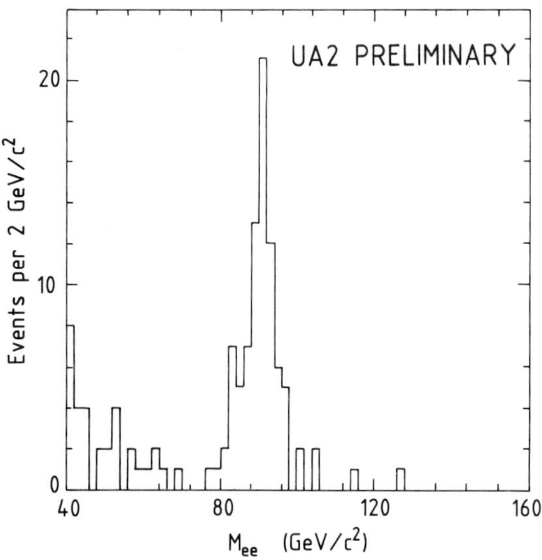

Fig. 5 The e^+e^- mass spectrum for the Z sample with both electrons in the fiducial volume.

subtraction and acceptance in a similar way as for the W, the resulting cross-section is

$$\sigma_Z = 61 \pm 7 \text{ (stat)} \pm 5 \text{ (syst) pb}$$

These values of σ_W and σ_Z are in agreement with the recent $O(\alpha(s))$ QCD predictions [11].

The cross secion ratio R can be compared to QCD predictions assuming three light neutrino generations as implied by the recent results from SLC [12] and LEP [13]. Most of the theoretical and experimental uncertainties cancel in the ratio R and the measurement value is sensitive to the mass of the top quark M_{top} because the value of M_{top} affects the branching ratio $B(W \to e\nu)$ if $M_{top} < M_W$ and $B(Z \to e^+e^-)$ if $M_{top} < M_Z/2$. The preliminary measured value is

$$R = 10.35 \,^{+1.2}_{-1.1} \text{ (stat)} \pm 0.3 \text{ (syst).}$$

This value is in agreement with the predicted value if $M_{top} > M_W$ but the errors are too large to conclude that this results excludes a light top quark. In the future analysis it is planned to use a larger Z sample to reduce significantly the statistical error on R.

4. $P_T{}^W$ DISTRIBUTION

The distribution of the W transverse momentum, $P_T{}^W$ can be calculated in perturbative QCD. For $P_T{}^W < 20$ GeV/c the $O(\alpha(s))$ calculations were used [11] and for $P_T{}^W > 20$ GeV/c the $O(\alpha(s)^2)$ calculations were used [14]. Experimentally $P_T{}^W$ is measured using the recoil hadrons as measured in the calorimeters. The theoretical predictions have been convoluted with a simple model for the detector response. The measured distributions are compared with the QCD predictions in Fig.6. The measured distributions are in good agreement with the QCD predictions up to large values of $P_T{}^W$ taking into account the experimental uncertainties on the detector response and the theoretical uncertainties.

5. MASS MEASUREMENTS

Although the most precise values for M_Z are now obtained from e^+e^- colliders [12], [13] hadron colliders still provide the only measurement of the W mass. Since the uncertainties in the mass measurements are dominated by the energy scale error, it will be

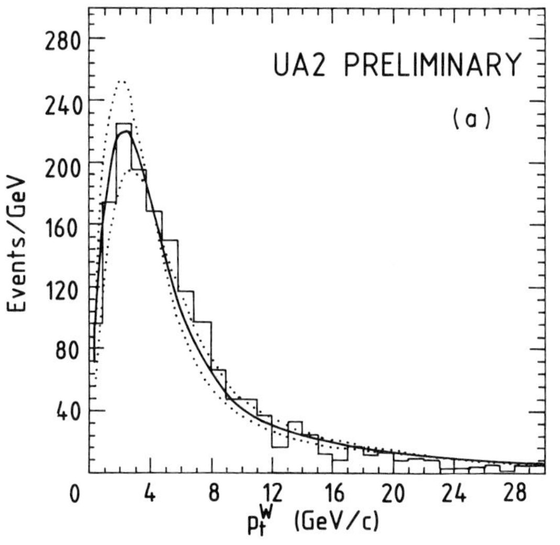

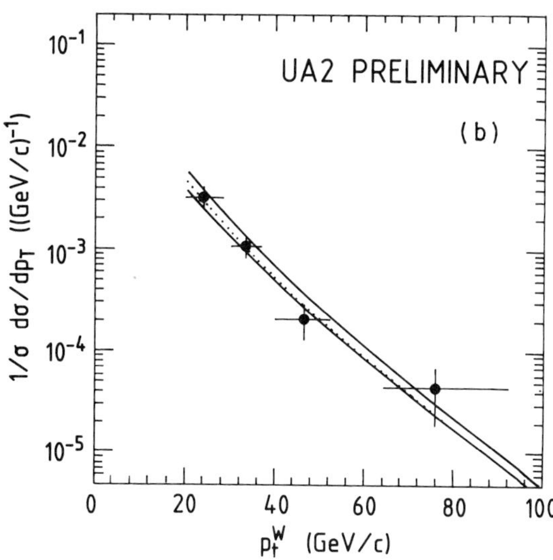

Fig 6 a) Comparison of the observed p_t^W distribution with theoretical predictions convoluted with detector response for $p_T^W < 30 GeV/c$. The solid curve uses the standard detector response and the dotted curves show the effect of varying the model for the detector response.
b) Same as (a) for $p_T^W > 20 GeV/c$. The solid curves use the results of Ref. [14] with $Q^2 = M_W^2$. The dotted curve shows the effect of choosing $Q^2 = (0.5 p_T^W)^2$.

interesting to calculate the mass ratio M_W/M_Z in which the energy scale uncertainty cancels. The Z mass can be determined in a straightforward way by fitting the Me^+e^- spectrum. The result of the fit (see Fig. 7) gives

$$M_Z = 90.53^{+0.46}_{-0.47} \text{ (stat) GeV/c}^2.$$

Since the value of M_W will be determined only from events with electrons in the Central Calorimeter it is also necessary to measure the Z mass for the sample of events with both electrons in the central calorimeter. The result of the fit to this sample (see Fig. 8) gives

$$M_Z = 90.19^{+0.59}_{-0.56} \text{ (stat)} \pm 1.4 \text{ (syst) GeV/c}^2$$

Fig. 7 Fit to the e^+e^- mass distribution for central-central and central-forward events.

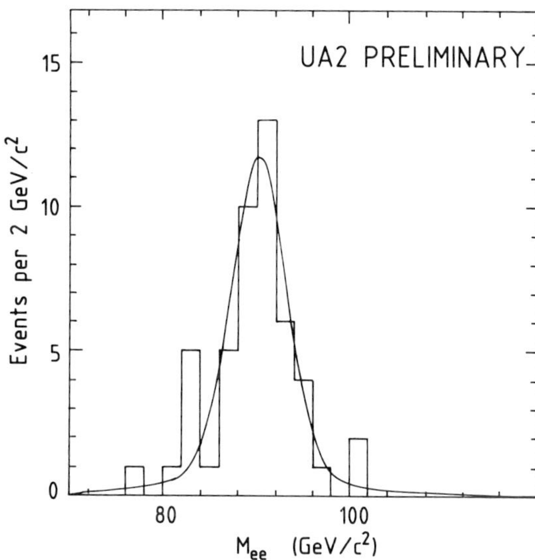

Fig. 8 Fit to the e^+e^- mass distribution for central-central events only.

which is in reasonable agreement with the recent measurements from SLC [12], and LEP [13].

5.1 M_W

The fit to the W mass is more involved since the longitudinal momentum of the neutrino cannot be measured and therefore the W mass can only be determined by fits to transverse variables. This means that for purely kinematic reasons the forward electrons carry little information for the determination of M_W and have not been used in the present analysis. Three kinematic variables have been used to fit M_W.

1. the electron transverse momentum p_T^e
2. the neutrino transverse momentum p_T^ν
3. the transverse mass, defined as

$$M_T = \sqrt{2 p_T^e \, p_T^\nu \, (1 - \cos\Delta\phi_{e\nu})}$$

where $\Delta\phi_{e\nu}$ is the azimuthal angle between the p_T^e and p_T^ν vectors.

In measuring the mass ratio M_W/M_Z, the energy scale error will cancel and the main systematic error comes from the M_W fit procedure. The three fits have the following main sources of systematic error.

1. p_T^e : the uncertainty on the theoretical distribution of p_T^W,
2. M_T : the uncertainty on the detector response for the measurement of p_T^ν,
3. p_T^ν : as for (1) and (2).

Therefore by fitting all three distributions, important cross-checks can be made on the analysis. The fits to p_T^e and M_T are shown in Fig. 9 and the results of all three fits are summarised in table 2 below.

TABLE 2 : W Mass fit Summary

Fit	M_W GeV/c2	Fit Syst. Error
p_T^e	79.98±0.36	0.3
M_T	$79.86\ ^{+0.30}_{-0.32}$	0.3
p_T^ν	$79.17\ ^{+0.40}_{-0.39}$	0.5

The agreement of the p_T^ν fit with the p_T^e fit is rather poor. Further studies of electron test beam data have shown that the discrepancy will be largely removed by using a more refined algorithm for the p_T^ν measurement. For the present preliminary analysis only the p_T^e fit will be used and the systematic error on the fit procedures has been increaed to 400 MeV/c^2 to account for the above problems. Therefore the best estimate of the W mass is

$$M_W = 80.0 \pm 0.4(\text{stat}) \pm 0.4(\text{syst1}) \pm 1.2(\text{syst2})\ \text{GeV/c}^2$$

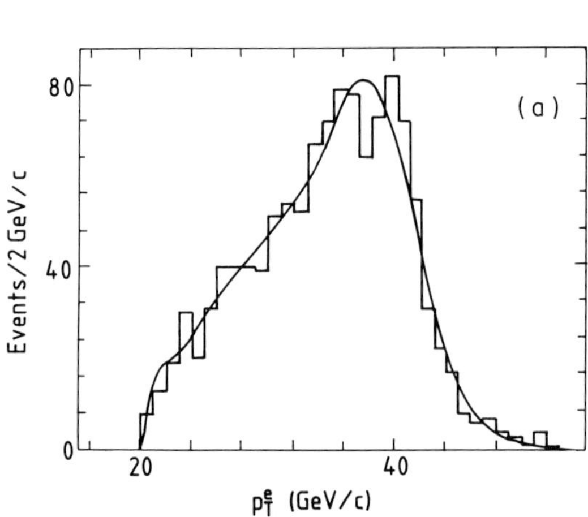

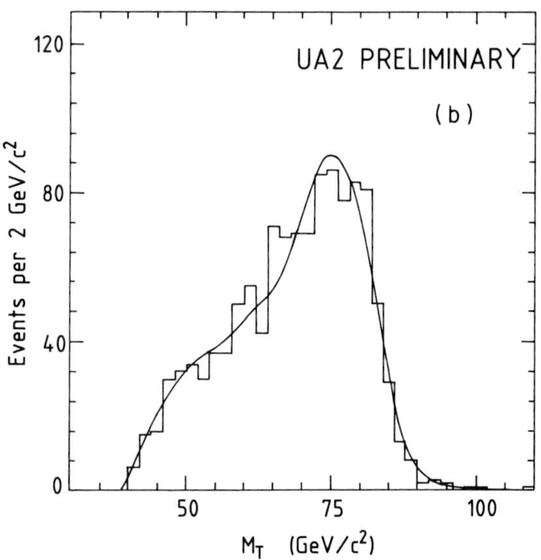

Fig. 9 a) Fit to the p_T^e distribution for the determination of M_W. b) Fit to the M_T distribution for the determination of M_W.

where the first systematic error comes from the fit procedure and the second comes from the uncertainty on the energy scale.

Therefore combining the values of M_W and M_Z obtained using central region electrons one obtains

$$M_W/M_Z = 0.887\pm0.007 \text{ (stat)}\pm0.004 \text{ (syst)}$$

or equivalently defining $\sin^2(\theta_W) = 1-(M_W/M_Z)^2$ gives

$$\sin^2(\theta_W) = 0.213\pm0.013\text{(stat)}\pm0.008\text{(syst)}.$$

In the future after further analysis of test beam data and using an extended sample of Z events it is hoped to reduce the overall error on $\sin^2(\theta_W)$ to $\pm$ 0.011 .

6. CONCLUSIONS

Preliminary results have been presented for W and Z cross-sections and masses based on an integrated luminosity of 7.1 pb^{-1}. Results were also presented on the $p_T{}^W$ distribution showing good agreement with perturbative QCD calculations.

REFERENCES

[1] C. Albajar et al., CERN EP/88-168.

[2] R. Ansari et al., Phys. Lett 186B (1987) 440; R. Ansari et al., Phys. Lett 194B (1987) 158.

[3] K. Einsweiler and A. Weidberg, Proceedings of the 7th Topical Worskshop on Proton-Antiproton Collider Physics, Fermilab, Batavia, USA 20-24 June 1988, p.30, World Scientific. and CERN-EP/88-152.

[4] A. Beer et al., Nuclear Instruments and Methods 224 (1984)360.

[5] P. Bagnaia et. al., The Electron, Jet and Missing Transverse Energy Calorimetry of the upgraded UA2 experiment at the CERN $\bar{p}p$ Collider, in preparation for Nucl. Instr. Meth. .

[6] R. Ansari et al., Nucl. Instr. Meth. A279(1989)388.

[7] F. Bosi et al., CERN-EP/89-82 (1989).

[8] R. Ansari et. al., Nucl. Instr. Meth. A263 (1988) 51.

[9] R.E. Ansorge et al., Nuclear Instr. Meth. A265 (1988) 33 and A273 (1988) 826.

[10] G. Blaylock et al., The Multi-Level Trigger and Data Acquisition System of the Upgraded UA2 Experiment at the CERN $\bar{p}p$ Collider in preparation for Nucl. Instr. Meth. .

[11] G. Altarelli et. al., Z. Phys. C27 (1985) 617.

[12] G. Abrams et. al. SLAC-PUB-5113.

[13] D. Decamp et. al., CERN-EP/89-132;
P. Aarnio et. al., CERN-EP/89-134;
M.Z. Aakrawy et. al., CERN-EP/89-133;
B. Adeva et. al., L3 Preprint #001.

[14] Arnold and Reno, FNAL PUB-88/168-T and 88/59-T.

DISCUSSION

G. Salvini, University of Rome: Considering the present knowledge on the lower limit for the top mass, we can perhaps compare the number of neutrino types as measured by UA2 with the result of $N_\nu = 3.0 \pm 0.9$ measured by the Mark II collaboration on SLC. Can you give your result?

A. Weidberg: We don not quote a limit on N_ν yet because we wish to reduce the statistical error (by increasing the size of the Z sample) first.

ELECTROWEAK RESULTS FROM CDF

The CDF Collaboration[1]

Reported by Myron Campbell

The University of Michigan

ABSTRACT

We summarize the electroweak results from 4.7 pb^{-1} of $p\bar{p}$ collisions at $\sqrt{s} = 1.8$ TeV. Results on the Z mass, W mass, W-Z cross section ratio and Z decay asymmetry are presented.

1. Introduction

The standard model of electroweak interactions[2] has three free parameters excluding the fermion masses and mixings and Higgs mass. One set of the parameters is α, the fine structure constant, G_F, the Fermi constant, and $\sin^2\theta_W$ defined as the ratio of the W and Z masses[3]

$$\sin^2\theta_W = 1 - \frac{M_W^2}{M_Z^2} \tag{1}$$

A measurement of the masses of the W and Z then gives the value of $\sin^2\theta$. By using

$$M_W^2 = \frac{(\pi\alpha/\sqrt{2}G_F)}{\sin^2\theta_W(1-\Delta r)} \tag{2}$$

limits on the size of the radiative correction can be placed which can be translated into limits on the mass of the top.

In this paper the measurements of the mass of the W and Z, the W-Z cross section ratio, and the Z-electron decay asymmetry will be discussed.

2. The CDF Detector

The CDF Detector is a solenoidal magnetic spectrometer used for the study of $\bar{p}p$ collisions at $\sqrt{s} = 1.8$ TeV. A detailed description of the detector has previously been published[4] and only a brief description will be given here.

2.1 Calorimeter

The calorimeter covers the pseudorapidity (η) range from -4 to $+4$ and has full azimuthal coverage. (FIG. 1) The calorimetry is divided into electromagnetic and hadronic compartments and is further divided into a central region ($|\eta| < 1.1$) plug region ($1.1 < |\eta| < 2.4$) and forward region ($2.3 < |\eta| < 4.2$). The central electromagnetic detector (CEM) consists of lead-scintillator layers with a segmentation into projective towers of $\Delta\eta = 0.1 \times \Delta\varphi = 15°$. Two phototubes view each tower. The central hadronic calorimeter (CHA) is steel-scintillator plates again with two phototubes for each $\Delta\eta = 0.1 \times\Delta\varphi = 15°$ tower. Imbedded in the electromagnetic calorimeter at 6 radiation lengths is a strip chamber consisting of wires parallel to the beam direction and cathode strips normal to the wires. Part of the central hadronic calorimeter is mounted outside the solenoidal region as end walls.

The plug calorimeter consists of lead plates with MWPC readouts. The hadronic part of the plug is steel plates with MWPC readout. The forward and backward calorimeters are also lead-MWPC and steel-MWPC layers.

2.2 Tracking

The Detector tracks and momentum analyzes charged particles. The magnetic field is generated by a superconducting solenoid 3 meters in diameter and 5 meters in length. The field strength is 1.5 Tesla. The central tracking chamber (CTC) measures charged tracks between $40° < \theta < 140°$ with 84 layers of sense wires. The resolution is $\delta p_T/p_T^2 = 0.0011$ $(\mathrm{GeV/c})^{-1}$.

Inside the CTC a Vertex Time Projection Chamber (VTPC) is used to measure tracks in the $R-Z$ plane. This chamber is used for measuring the vertex position of interactions.

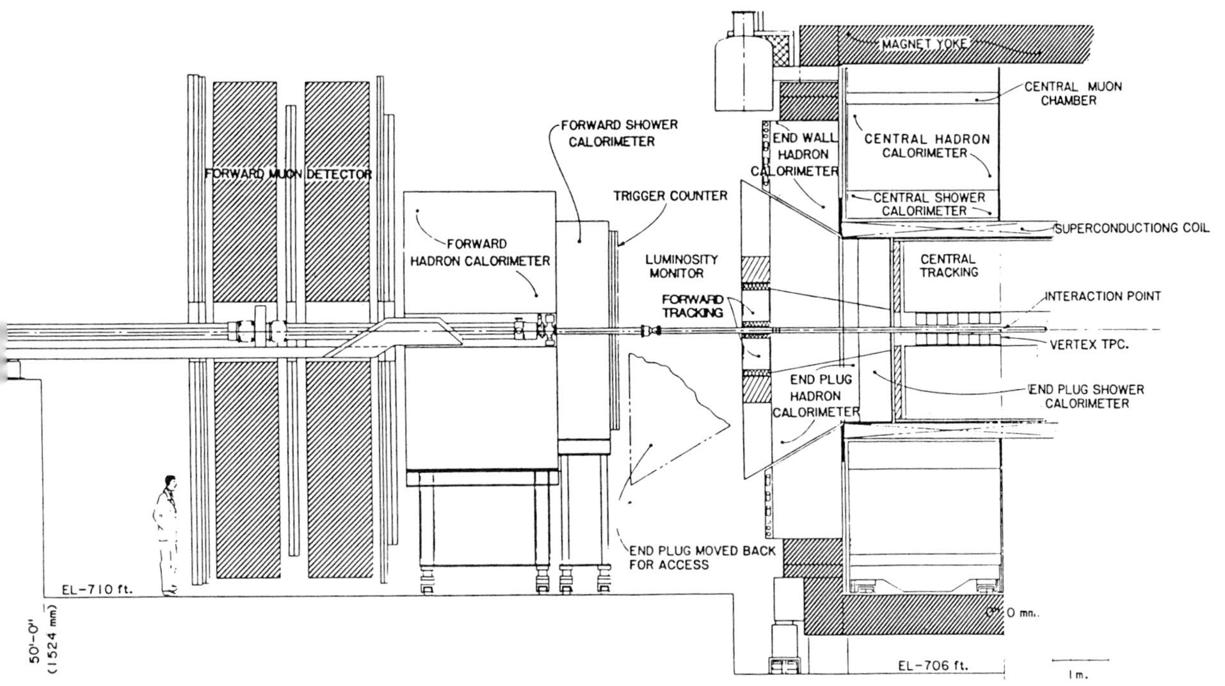

FIG.1 Cross section through a vertical plane of one half the CDF detector. The detector is symmetric about the midplane and roughly symmetric about the beam axis.

Drift chambers located at the outer edge of the central hadronic calorimeter are used to identify muons. There are four layers of chambers with alternate layers slightly staggered in order to resolve the left-right ambiguity. The muon chambers cover the range $56° < \theta < 124°$. Forward muon toroids cover the range $3° < \theta < 16°$ and $164° < \theta < 177°$.

2.3 Tracking Calibration

The calibration of the detector was crucial for obtaining accurate measurements of the W and Z masses. The method used was to understand the alignment of the CTC and then to carry the calibration of the CTC to the CEM calorimeter.

During each data run, which lasts several hours, minimum bias events were used to determine the t_o (drift time offset) and drift time. The cells of the CTC are tilted at the E $\times$ B angle so each track gave a measure of the drift time offset and drift time constant. (FIG. 2)

A sample of 1000 well identified central electrons from $W \rightarrow e\nu$ was used to determine the overall alignment of the CTC. Distortions could occur because of end plate deflections from wire tension. The overall relative rotations of the 84 layers of sense wires was determined by requiring that the mean of the positive electrons E/P distribution match the mean of the negative electrons E/P distribution. This alignment was then checked by using cosmic ray muons which passed through the detector. These muons would be reconstructed as one positive and one negative track; the momentum of the two tracks should be the same. The beam constrained resolution is 0.11% P_t after alignment.

The momentum can be derived from a curvature measurement and the magnetic field. The solenoidal field has been mapped[5] and the field is known to $\pm 0.05\%$ with the uncertainty coming from the fact that the field was mapped with the solenoid at 5000 amps while during the 1988-1989 run it was at 4650 amps.

The curvature measurement can be expressed as an expansion around C=0, *i.e.* an infinite momentum track, with both charge dependent and charge independent terms.

$$\begin{aligned} C_M = & C + [C_f + a_1 C + a_2 C^2 + \cdots] \\ & + (C/|C|)[b_o + b_1 C + b_2 C^2 + \cdots] \end{aligned} \tag{3}$$

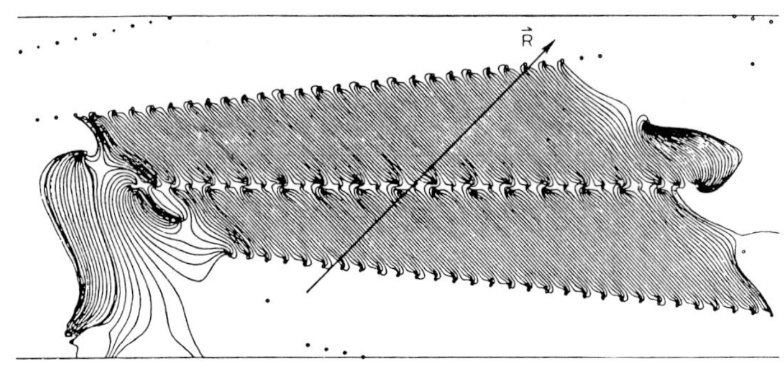

FIG. 2 Drift trajectories in a 15kg magnetic field. The radial direction is indicated by the arrow.

where C_M is the measured curvature, C_f a false curvature, and C the true curvature.

If we consider that infinite momentum tracks of both signs must have the same measured curvature then b_o must be zero, *i.e.* $C_{M+} = C_{M-} = C_f + b_o = C_f - b_o$.

The invariant mass for a particle which decays to two oppositely-charged particle is inversely proportional to the square root of the product of the measured curvatures.

$$M \sim \sqrt{E_{T+}E_{T-}} \sim \frac{1}{\sqrt{C_{M+}C_{M-}}} \tag{4}$$

Then using (3)

$$\begin{aligned} C_{M+}C_{M-} &= C_+C_-[(1+a_1)^2 - b_1^2] \\ &+ C_f(1+a_1)(C_+ + C_-) \\ &+ C_f b_1(C_+ - C_-) + C_f^2 + \mathcal{O}(C^3) \end{aligned} \tag{5}$$

The mass scale is set by the term $[(1+a_1)^2 - b_1^2]$. The invariant mass will be wrong if this term is different from 1. The term $(C_+ + C_-)$ will vanish for balanced decays.

The false curvature, C_f, was removed by the wire layer rotations discussed above. Since the $\mathcal{O}$ (C^3) terms get smaller as the curvature decreases extrapolating from measurements at low mass to high momentum will be valid.

From these considerations it is clear that the tracking mass scale can be checked by looking at the mass peaks of particles of known mass. This has been done for $J/\psi \to \mu^+\mu^-$ (FIG. 3) and $\Upsilon \to \mu^+\mu^-$ (FIG. 4) which resulted in measured masses of 3.097 ± 0.001 and 9.469 ± 0.010 GeV/c^2. The J/ψ mass was corrected for the muon dE/dx losses as they traversed the CTC. These masses agree well with accepted values[6] of 3.0969 and 9.4603 GeV/c^2. The mass scale uncertainty is set at 0.2% from these measurements.

2.4 CEM Calibration

The central calorimeter was calibrated in a test beam and the response as a function of position was mapped. The calibration for each tower was tracked with embedded Cs^{137} sources.

After the 1988-1989 data run 17,000 high quality (well identified, well isolated) electrons were used to recalibrate the calorimeters. The relative calibration scale between towers was adjusted by using the E_t/P_T mean value. The response maps from the test beam were still used. This calibration accuracy was limited by statistics from the number

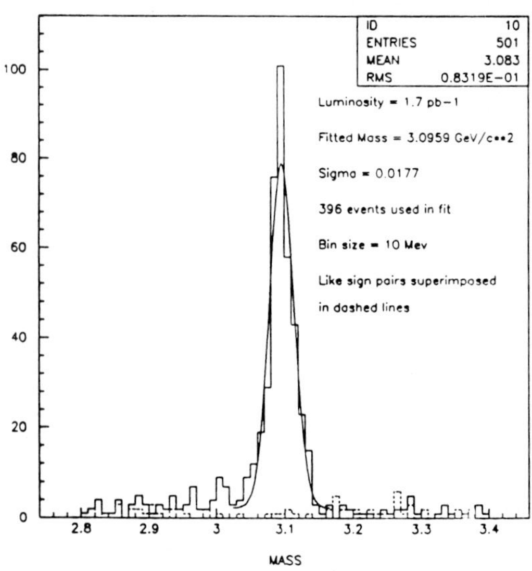

FIG. 3 J/ψ invariant mass plot. 1 MeV correction is required due to dE/dx.

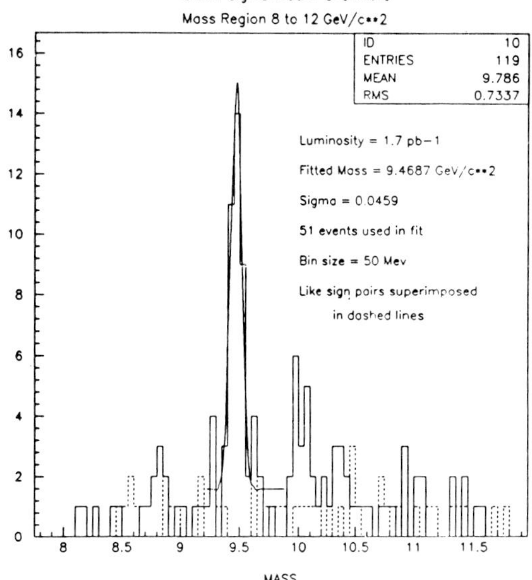

FIG 4. Υ invariant mass plot.

of electrons in each tower to about 1.7%. This factor was introduced as a constant term in the energy resolution

$$\left(\frac{\sigma_E}{E}\right)^2 = \left(\frac{13.5\%}{\sqrt{E\sin\theta}}\right)^2 + (1.7\%)^2 \qquad (6)$$

The average change from the test beam calibration was 2.7%

The absolute energy scale of the CEM was set by using 1000 well measured electrons from $W \rightarrow e\nu$ decays. The means of the measured and predicted E_t/P_T distributions (figure 5) were matched by the overall scale factor. The statistical limit on the accuracy of this calibration was 0.42%.

2.5 Data Sample

A nine month run during 1988 and 1989 with the full detector and trigger system resulted in a data set with integrated luminosity of about 4.7 pb^{-1}.

After the data was written to tape the events were analyzed and all events with a CEM electron candidate, or large missing E_t, or two or more leptons were selected for further processing. These events then had the full CDF analysis, including track reconstruction, applied. This analysis is the basis for the results presented here. A more complete analysis of the entire data set will follow.

3. Z° Mass Measurement

The Z° mass is measured using the $Z \rightarrow e^+e^-$ decay where the electron energy is measured in the calorimeter and $Z \rightarrow \mu^+\mu^-$ decay where the muon momentum is measured by the tracking system. Measuring $Z \rightarrow e^+e^-$ decay using the tracking system resulted in a large uncertainty due to radiative corrections.

3.1 $Z \rightarrow \mu^+\mu^-$

Dimuon events were selected by requiring the first muon to have

1. Track with $P_t > 20$ GeV/c.
2. A match in φ between the muon stub in the muon chamber and the track in the CTC.
3. Energy deposition in the calorimeter consistent with minimum ionizing, less than 2.0 GeV in CEM and less then 6 GeV for CHA.
4. No jets with $E_t > 15$ GeV within 10° of the muon track.

The second muon could either be another muon which went through the muon chambers, in which case it has the same selection criteria as the first muon, or it could be a muon with only a stiff track and minimum ionizing energy deposition in the calorimeter. Allowing the second type of muon increased the acceptance by increasing the available rapidity coverage to $|\eta| < 1.2$. Cosmic rays were removed by eliminating events with $|\Delta\varphi - 180°| < 1.5°$ and $|\Delta\eta| < 0.1$. Events having these criteria and with invariant mass between 50 and 150 GeV/c^2 were accepted. The sample consists of 132 events.

The effects of radiative corrections were calculated by a Monte Carlo event generator which used exact matrix elements to order α^2. External radiation (Bremsstrahlung) was studied by analyzing these generated events with a detailed simulation of the CDF detector.

The mass distribution for the dimuons is shown in figure 6a. This distribution was fit using a maximum likelihood fit with a signal modeled by a relativistic Breit-Wigner convoluted with a gaussian resolution in $1/P_T$. There are 123 events in the mass range 75 to 105 GeV/c included in the fit. The results of the fit are 90.7± 0.4 (stat) ± 0.2 (scale) GeV/c^2. The effects of radiative corrections, different structure functions, and the finite mass window are included in the uncertainties listed in Table 1.

TABLE I. Corrections and uncertainties in the Z^0 mass. All units are in GeV/c^2. The first uncertainty is statistical and the second is systematic.

	$Z^0 \to \mu^+\mu^-$ (Tracking)		$Z^0 \to e^+e^-$ (Tracking)		$Z^0 \to e^+e^-$ (Calorimeter)	
Number of events used in fit	123		58		65	
Observed fitted mass	90.41 ± 0.40		89.27 ± 0.80		90.93 ± 0.34	
Radiative corrections	+0.22	± 0.03	+2.19	± 0.30	+0.11	± 0.03
Structure functions	+0.08	± 0.03	+0.08	± 0.03	+0.08	± 0.03
E/P calibration	...	...	...	...		± 0.38
Mass scale		± 0.20		± 0.20		± 0.20
Corrected mass	90.7 ± 0.4 ± 0.2		91.5 ± 0.8 ± 0.4		91.1 ± 0.3 ± 0.4	

3.2 $Z \to e^+e^-$

The electron sample was selected by requiring two electrons, each with the following criteria:

1. The electron is several cm away from all calorimeter edges so the shower is completely contained.
2. The ratio of hadronic (HAD) to electromagnetic (CEM) energy be less than 0.10.
3. The position as measured by the strip chamber embedded in the CEM match the extrapolated track position to 3 cm in the Z direction and 1.5cm in the φ direction.
4. The ratio of calorimeter energy to track momentum, E/P, be less than 1.4.
5. The shower profile measured in the strip chambers be consistent with an electron shower using a χ^2 test in both dimensions.

A total of 73 events pass the above criteria and have an invariant mass between 50 and 150 GeV/c^2 (FIG. 6b). All of the events had opposite sign electrons.

The same mass-fitting procedure was used on the $Z \to e^+e^-$ sample as was used on the $Z \to \mu^+\mu^-$ sample. 64 of the 73 events had both tracks pass the high quality requirements needed to make beam constrained fits. Since the radiative effects for electrons are larger than for muons the best measurement of the Z° mass is from the calorimeter information. (see Table 1).

The mass of the Z° peak from the calorimeter measurement (FIG. 7) was fit using the maximum likelyhood method including the calorimeter resolution of equation 6. There were 65 events included in the fitting range of 80 to 100 GeV/c^2 and the result was 91.1 ± 0.3 (stat) ± 0.4 (sys)GeV/c^2. The uncertainty include estimates for effects of the choice of mass window, choice of structure functions and variations in detector resolution.

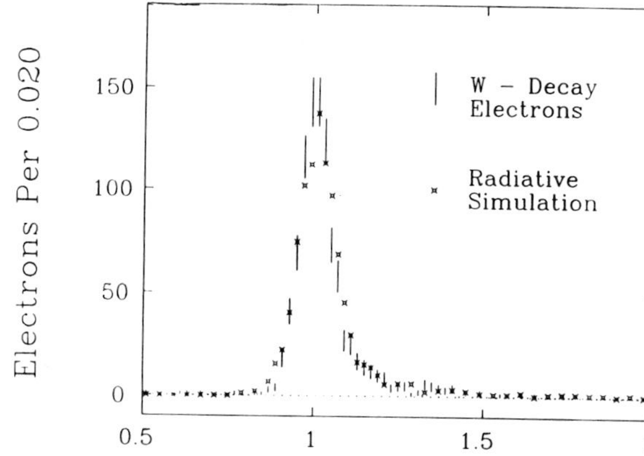

FIG. 5 E/P After Calibrations

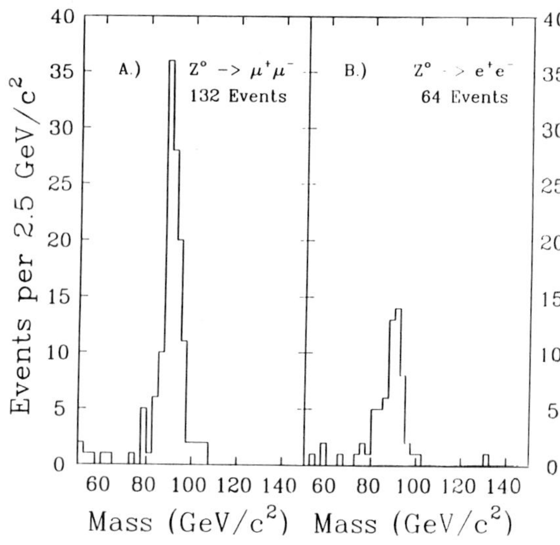

FIG. 6 Z invariant mass from muon and electron tracks

The best value for the Z° mass is a weighted mean of the tracking measurement of $Z \rightarrow \mu^{+}\mu^{-}$ and the calorimeter measurement of $Z \rightarrow e^{+}e^{-}$. Since the calorimeter was calibrated from the tracking chamber the mass scale uncertainty is common to both. The measurements were combined by taking the statistical and systematic errors in quadrature excluding the mass scale uncertainty. The result is the Z° mass is 90.9 ± 0.3 (stat + syst) $\pm$ 0.2 (scale) GeV/c^2.

4. W Mass Measurement

The W mass measurement is technically more difficult than the Z mass measurement. The decay $W \rightarrow e\nu$ was chosen to make the preliminary mass determination of the W. This is obviously different from the $Z \rightarrow e^{+}e^{-}$ by the substitution of a neutrino for one of the electrons. This complicates the analysis in two ways. First the neutrino momentum has to be inferred from summing the momentum of all particles in the event and measuring the missing momentum. Second, the longitudinal momentum of all the decay products cannot be measured because of the large amount of momentum carried by beam fragments at very low angles.

The missing transverse energy, $\not{E}t$, is defined as

$$\not{E}_t = -\overrightarrow{Et} = \Sigma \overrightarrow{E_{t_i}} \tag{7}$$

where the transverse energy is summed over all calorimetry towers.

In order to make this measurement as accurately as possible a correction for the non-linearity of the calorimeters at low energy must be made. This has been measured by using isolated single tracks in the CTC. The average E/P for low energy charged hadrons is 0.61 ± 0.04. The fraction of energy carried by photons in jets has been measured[7] to be $0.268 \pm 0.0003 \pm 0.045$. Assuming that this same fraction applies to minimum bias events and assuming that low energy photons are properly measured, then the average response to low energy particles in the underlying event is 0.71 $\pm$ 0.03. To correct for this non-linearity the total E_t summed over all towers not including the electron is multiplied by 1.4 ± 0.07.

The $\not{E}_t$ is a measure of two of the components of the neutrinos momentum. Since the third is not measured a transverse mass rather than invariant mass of the W is calculated

$$M_T^2 = 2E_t\not{E}_t(1 - \cos(\varphi_e - \varphi_\nu)) \tag{8}$$

where E_t is the electron E_t, $\not{E}_t$ is the total missing transverse energy and φ_e and φ_ν are the azimuthal angles of E_t and $\not{E}_T$ respectively.

Events for the W mass measurement are selected with the following criteria:

1. The $\not{E}_t$ must be greater than 25 GeV.
2. The ratio $\not{E}_t/\sqrt{\Sigma E_t}$ must be greater than 2.4.
3. The E_t of the electron must be greater than 25 GeV.
4. The ratio of energy to track momentum, E/P, of the electron must be less than 2.
5. The electromagnetic fraction of energy for the electron must be greater than 85%.
6. $|\eta|$ of the electron is less than 1.0.
7. Only one track is near the electron calorimeter cluster.
8. No jets with energy E_t greater than 5 GeV can be opposite the electron.
9. No additional cluster with E_t greater than 7 GeV.

The transverse mass plot of the W sample is shown in figure 8. The shape of the fit depends on many factors. The dominant factor is the W transverse momentum. The requirement of no clusters with energy greater than 7 GeV is an attempt to select a sample with low transverse momentum.

The natural width of the W also affects the shape of the transverse mass plot. We have chosen to fix the width at the standard model value of

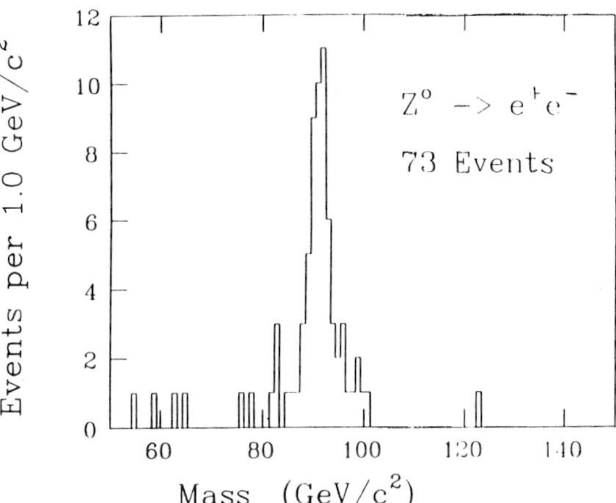

FIG. 7 Z invariant mass from electron calorimeter measurement.

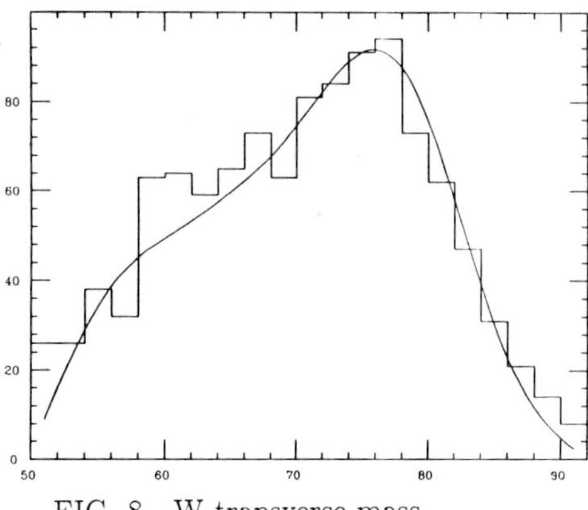

FIG. 8 W transverse mass

2.5 GeV. Allowing the width to be a free parameter in the fit increases the statistical uncertainty from 200 to 380 MeV. The energy scale is uncertain to 0.4% as mentioned in section 2.4. Different choices of models of the proton structure functions lead to different shapes of the fit which results in shifts of up to 300 MeV on the fitted mass. Finally by fitting different samples of the W's from Monte Carlo and fitting the Z sample a 250 MeV uncertainty due to the fitting procedure has been assigned. The results of the uncertainties are tabulated in Table 2.

The result of the fit to the W transverse mass is

$$80.0 \pm 0.2 \text{ (STAT)} \pm 0.3\text{(SCALE)} \pm 0.5\text{(SYST)} \quad (9)$$

PRELIMINARY

where the systematic and scale errors are quoted separately.

Table 2. Uncertainties in fitting the W mass

UNCERTAINTY	SIZE MeV
Statistical	200
Energy Scale	320
Structure Function	300
Resolution	400
Background	50
Fitting	250
Total Systematics	**620**
Total Uncertainty	**650**

5. W Z Cross Section Ratio

The ratio of cross sections for W and Z production with subsequent decay into $e\nu$ or ee is given by

$$R \equiv \frac{\sigma(W \to e\nu)}{\sigma(Z \to ee)} = \frac{\sigma(W)\Gamma(W \to e\nu)\Gamma(Z)}{\sigma(Z)\Gamma(Z \to ee)\Gamma(W)} \quad (10)$$

where $\sigma(W \to e\nu)$ and $\sigma(Z \to ee)$ are the measured cross sections times branching ratio, $\sigma(W), \sigma(Z)$ are the total cross sections, $\Gamma(W \to e\nu), \Gamma(Z \to ee)$ are the partial widths and $\Gamma(W)$, $\Gamma(Z)$ are the total widths.

The total widths are of interest because they are functions of well known parameters, the Weinberg angle and the W and Z masses. They are also functions of unknown quantities, the top mass, M_t, and the number of neutrino generations, N_ν. As the top mass increases the total width of the W decreases until $M_t > M_W$ when the channel W $\to$ tb becomes closed. The width of the Z increases with each additional light neutrino generation.

The advantage of measuring the ratio of cross sections is that many of the experimental uncertainties cancel in first order. The acceptance of the central electron should be nearly the same for $W \to e\nu$ and $Z \to ee$. The integrated luminosity will cancel if a common data set is used for both the W and Z. The remaining systematic errors are smaller than they would be if the two cross sections were measured independently.

The disadvantage of measuring the ratio is that the top mass limit and number of neutrino generations are no longer independent quantities, but instead there is a relationship between the two of them.

Equation 10 can be rewritten as

$$R = \frac{N(W)A_Z}{N(Z)A_W}\epsilon = \frac{\sigma(W)\Gamma(W \to e\nu)\Gamma(Z)}{\sigma(Z)\Gamma(Z \to ee)\Gamma(W)} \quad (11)$$

where $N(W)$, $N(Z)$ are the measured number of W's and Z's, A_W, A_Z are the acceptance of W's and Z's, and ϵ results from efficiency terms which don't cancel.

5.1 W-Z Ratio event selection.

The method of event selection was chosen to reduce the systematic error introduced by differences in efficiencies. A central electron with stringent criteria was selected for both the W and Z candidates. The efficiency for this electron cancels except for Z's where both electrons are in the central region. The second electron for the Z and the neutrino for the W were then selected with loose criteria in order to keep the efficiencies high.

The dominant background to W and Z events is from jets. In order to reduce this background for both bosons any event with a jet of energy greater than 10 GeV, excluding the electron, was vetoed.

The first electron was chosen with the following criteria:

1. E_t greater than 20 GeV.
2. Ratio of HAD/EM less than 5.5%.
3. E/P between 0.5 and 2.0.
4. The extrapolated central track and the strip chamber position must match to within 2.5 cm in φ and 3.0cm in Z.
5. The strip chamber profile must be consistent with an electron.

The W's were then selected by requiring in addition to the electron above:

1. $\not{E}_t$ greater than 20 GeV.
2. z position of the event vertex within 60 cm of nominal position.
3. No jets in the event greater than 10 GeV.

The Z's were selected by requiring the first electron as described above and a second electron with:

1. E_t greater than 10 GeV.
2. Ratio HAD/EM less than 10%.
3. E/P between 0.5 and 2.0 if the electron is in the central region.
4. Invariant mass of the pair between 65 and 115 GeV/c.
5. z position of event vertex within 60 cm of nominal position.

This leaves a sample of 192 Z candidates and 1828 W candidates

5.2 W and Z backgrounds

There are several background subtractions which must be made to the W and Z candidates. The decay $Z \to ee$ could be a background to $W \to e\nu$ if one of the electrons were lost. This background was estimated to be 12 ± 5 from looking for W events with an extra stiff track and from Monte Carlo. The decay $Z \to \tau\tau$ could mimic $W \to e\nu$ if one of the τ's decays to an electron. We expect this background to be 4 ± 1 events. The QCD background to W events was studied by using the isolation distribution of the electrons in W events and a sample of background events. 18 ± 9 events are expected as QCD background. The decay $W \to \tau\nu$ followed by $\tau \to e\nu\nu$ is a background to $W \to e\nu$. The ISAJET Monte Carlo was used to generate $W \to \tau\nu$ which were then propagated through the detector with the expected resolution. The background expected from this decay was found to be 67 ± 6 events.

The possible sources of background to $Z \to ee$ are $Z \to \tau\tau$, W + electromagnetic jet and two jet events. The $Z \to \tau\tau$ background was studied in the same way as this background to $W \to e\nu$. No events were found with invariant mass above 50 GeV/c^2, so this background is negligible. The W + jet background was estimated from Monte Carlo to be 1 ± 1 event. The Z background from QCD jets was estimated by looking at histograms of electron isolation as a function of invariant mass from the data. This procedure gave an estimate of 5 ± 3 events.

The resulting number of Z's and W's (Table 3) are 186 ± 14 ±3 and 1727 ± 43 ± 12.

Table 3: Summary of W-Z ratio results

	W Events	**Z Events**
Events	**1828**	**192**
Background		
W→ $\tau\nu$	67 ± 6	
Z→ ee	12 ± 5	
Z→ $\tau\tau$	4 ± 1	0.0
W + jet		1 ± 1
QCD	18 ± 9	5 ± 3
Total background	101 ± 12	6 ± 3
N(W), N(Z)	1727 ± 43 ± 12	186 ± 14 ± 3
A_W, A_Z	0.351	0.374
A_W/ A_Z	0.939 ± .026	
f_{cc}		.39
f_{cp}		.47
f_{cf}		.14
C_1	0.86 ± 0.03	0.86 ± 0.03
C_2		0.96 ± 0.02
p		0.96 ± 0.03
f		0.897 ± 0.03
ϵ_ν	0.965 ± 0.005	
ϵ_W, ϵ_Z	0.83 ± 0.03	0.86 ± 0.04
$\epsilon_Z, /\epsilon_W$	1.036 ± 0.026	
No jet corr.	0.992 ± 0.005	
Drell-Yan corr.	1.005	
R(exp)	10.21 ± 0.81 ± 0.42	

5.3 W-Z acceptance and efficiencies

The acceptance of the W and Z bosons will be different because of the differences in efficiencies for the neutrino and second electron. The calculation of the acceptance depends mostly on the choice of the proton structure functions to the extent the structure functions determine the P_t and P_l distributions of the bosons. Several different sets of structure functions were used which resulted in a 2.5% variation in the ratios of acceptances. From Monte Carlo studies acceptances of 35 ± 0.1% for the W and 37 ± 0.1% for the Z are assigned (see Table 3). The ratio of acceptances is 0.939 ± 0.026.

The efficiency term in equation (11) can be written as:

$$\mathcal{E} = \frac{\mathrm{f_{cc}} \cdot \mathrm{C_1(2C_2 - C_1) + f_{cp}C_1p + f_{cf}C_1f}}{\mathrm{C_1}\epsilon_\nu} \quad (12)$$

where f_{cc}, f_{cp}, f_{cf} are the fraction of central-central, central-plug, and central-forward Z's, and:

ϵ_ν = $\not{E}_t$ efficiency
C_1 = efficiency for first electron
C_2 = efficiency for second electron
p = efficiency for plug electron
f = efficiency for forward electron

Note that the term C_1 nearly cancels, only the term ($2C_2$-C_1) is left which is from Z's where either electron could satisfy the first electron criteria. The terms f_{cc}, f_{cp}, and f_{cf} are the measured fractions. The term ϵ_ν is studied with the same Monte Carlo used to determine the acceptance of the W and is set at 0.965 ± 0.005.

The central electron efficiencies were studied by using a $\not{E}_t$ selected W sample and by using the second electron from the Z sample. The efficiency of each criterion on the electron was measured and the overall efficiency was calculated to be 86 ± 3% for C_1 and 96 ± 2% for C_2. The plug and forward efficiencies were measured by using W candidates selected from $\not{E}_t$ and Z candidates with very loose cuts on second electron. The effect of the cuts used in the final event selection were found to give p=96 ± 3% and f=97 ± 3%. These results are tabulated in Table 3.

5.4 W-Z ratio results

From the total number of W's and Z's a ratio of 10.21 ± 0.81 (stat) ± 0.42 (syst) is obtained. Combining the two errors in quadrature yields 10.21 ± 0.91. The 90% and 95% confidence limits are then

$$9.05 < \mathrm{R} < 11.37 \quad (90\%\mathrm{CL})(\mathrm{PRELIMINARY})$$

$$8.71 < \mathrm{R} < 11.71 \quad (95\%\mathrm{CL})(\mathrm{PRELIMINARY})$$

The 90% CL limits are plotted along with the theoretical predictions[8] of the ratio as a function of top mass and number of neutrino generations in figure 8.

From this we can set the following limits:

$M_t > 48$GeV (90%CL) PRELIMINARY

$N_\nu < 4.5$ (90%CL) PRELIMINARY

$N_\nu < 5.0$ (95%CL) PRELIMINARY

The limit on the mass of the top is independent of the possible decay modes of the top quark.

6. Z→ e^+e^- Asymmetry

A preliminary measurement of the charge asymmetry in $p\bar{p} \rightarrow e^+e^-$ has been made. the electroweak cross section for this process has electromagnetic, weak, and interference terms:

$$\frac{d\sigma}{d\cos\hat{\theta}}(p\bar{p} \rightarrow e^+e^-) = \gamma^2 + \gamma Z + Z^2 \quad (13)$$

In this expression γ and Z represent the electromagnetic and weak interactions respectively and $\hat{\theta}$ is the emission angle of the lepton with respect to the incoming quark direction. Both the interference term and the Z^2 term have a dependence on cos$\hat{\theta}$. The distribution of electron pair candidates is fit to:

$$\frac{d\sigma}{d\cos\hat{\theta}} = A(1 + \cos^2\hat{\theta}) + B\cos\hat{\theta} \quad (14)$$

where A and B are functions of $\sin^2\theta_w$ the Weinberg

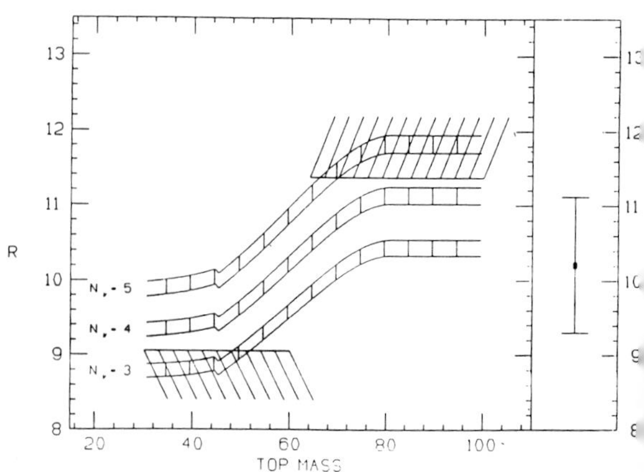

FIG. 9 W-Z cross section ratio as a function of top mass

angle. Since the term B vanishes when $\sin^2\theta_w = 1/4$ this is a measure of the deviation of $\sin^2\theta_w$ from 0.250.

The direction of the quark defines $\cos\hat{\theta}$, however in high P_t Z events the direction of the proton and the partons will be different. Since the quark direction is then not well defined $\cos\hat{\theta}$ can not be determined. The method of Collins and Soper[9] will be used to define $\cos\hat{\theta}$ in a P_t dependent way.

Interactions where the electron pair is produced by a sea quark from the proton or antiproton rather than a valence quark from the proton will introduce a symmetric dilution. Furthermore since all terms are included in the fit Drell-Yan production is not a background.

6.1 Asymmetry event selection

Events for this measurement are selected where both electrons can be identified and at least one electron can have its sign determined. This translates into a requirement that one electron have $|\eta| < 1.0$ and the second electron have $|\eta| < 3.8$. Note that if the detector and trigger acceptances are charge independent then the acceptance will be symmetric with respect to $\cos\hat{\theta}$.

Electrons were selected if the E_t was greater than 15 GeV and the E/P was less than 2 for central electrons. The extrapolated track and strip chamber positions were required to match and the shower profile was required to be consistent which an electron. All events were hand scanned.

The resulting acceptance corrected $\cos\hat{\theta}$ distribution is shown in Figure 10. The fit was a negative log likelihood to $d\sigma/d\cos\hat{\theta}$. If the acceptance is symmetric in $\cos\hat{\theta}$ then this fit is independent of acceptance. The preliminary results of the fit is:

$$\sin^2\theta_w = 0.216 \pm 0.015(\text{stat}) \pm 0.010(\text{syst}) \quad (15)$$

Electroweak radiative corrections have not been applied to this result. In addition to the acceptance independent fit, fits were made to the acceptance corrected data dropping the outer bins. The results are:

20 BINS: 0.217 ± 0.016
18 BINS: 0.216 ± 0.017
16 BINS: 0.220 ± 0.018

7. Conclusions

We have made the following measurement.

Z MASS	$90.9 \pm 0.3 \pm 0.2\ GeV/c^2$
W MASS	$80.0 \pm 0.2 \pm 0.3 \pm 0.5\ GeV/c^2$ *PRELIMINARY*
W − Z RATIO	$10.2 \pm 0.8 \pm 0.4$ *PRELIMINARY*
ASYMMETRY	$\sin^2\theta_w = 0.216 \pm 0.015 \pm 0.010$ *PRELIMINARY*

The boson masses can be used to determine $\sin^2\theta_w$ and through the radiative calculations, place

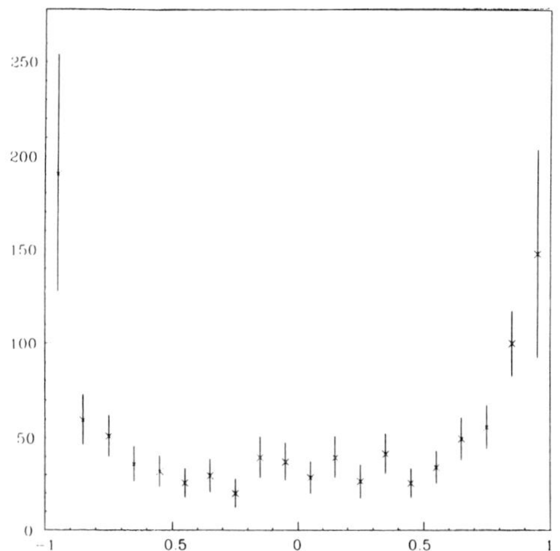

FIG. 10 Acceptance corrected $\cos\hat{\theta}$ distribution

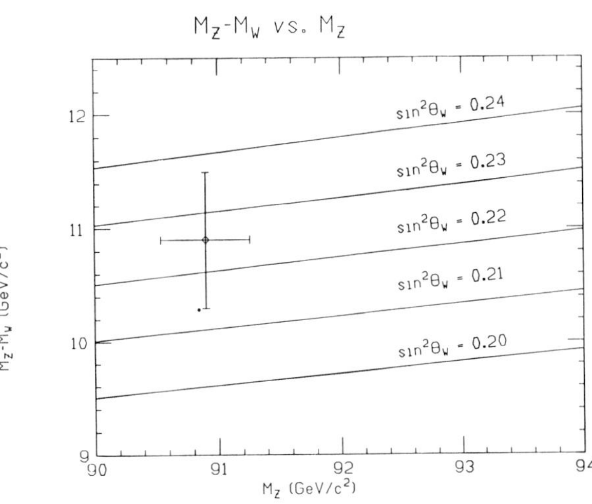

FIG. 11 $\sin^2\theta_w$ as a function of M_W and M_Z

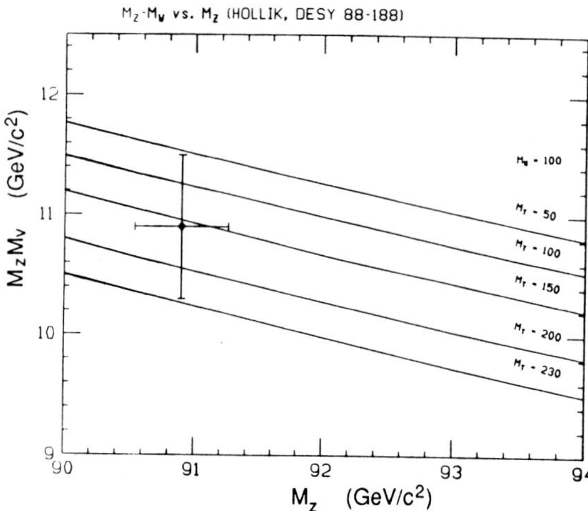

FIG. 12 Top mass contours as a function of M_W and M_Z

limits on the mass of the top quark. These are shown in Figure 11 and 12 against theoretical curves from reference 10.

8. Acknowledgements

We wish to thank the Fermilab Accelerator Division for the exceptional performance of the accelerator and antiproton source. This work was supported by the Department of Energy, the National Science Foundation, Instituto Nazionale di Fisica Nucleare, the Ministry of Science, Culture, and Education of Japan, and the A.P. Sloan Foundation.

9. References

1. The CDF collaboration is comprised of research groups from: Argonne National laboratory, Brandeis University, University of Chica- Fermi National Accelerator Laboratory, Laboratori Nasionali di Frascati, Harvard University, University of Illinois, National Laboratory for High Energy Physics (KEK), Lawrence Berkeley Laboratory, University of Pennsylvania, Instituto Nazionale di Fisics Nucleare, Purdue University, Rockefeller University, Rutgers University, Texas A&M University, University of Tsukuba, Tufts University and University of Wisconsin.
2. S. Weinberg, Phys. Rev. Lett. **19**, 1264 (1967); A. Salam in Elementary Particle Theory, ed N. Svarthalm (Almquist and Wiksells, Stockholm, 1969) p. 367; and S.L. Glashow, J. Illiopoulos, and L. Maigni, Phys. Rev. D2, 1285 (1970).
3. A. Sirlin, Phys. Rev. **D22**, 971 (1980).
4. F. Abe et al. (CDF Collaboration) Nucl. Instr. and Meth. **A271**, 387 (1988).
5. C. Newman-Holmes, E.E. Schmidt, and R. Yamada, Nucl. Instr. and Meth. **A274**, 443 (1989).
6. "Review of Particle Properties," Phys. Lett. B204 (April 1988).
7. Behrend et al. (CELLO Collaboration) A. Phys. C **14**, 189 (1982).
8. A.D. Martin, R.G. Roberts, W.J. Stirling, Phys. Lett. **B228**, 149 (1989).
9. J.C. Collins and D.E. Soper, Phys. Rev. D **16**, 2219 (1977).
10. W.F.L. Hollik, Radiative Corrections in the Standard Model and their Role for Precision Tests of the Electroweak Theory. DESY 88-188 (1988).

DISCUSSION

M. Samuel, Oklahoma State University: Can you say anything about your $W\gamma$ events and your radiative W decay events, and how they put a bound on the magnetic moment of the W boson?

M. Campbell: No, we do not have nearly enough events for that measurement.

P. Langacker, University of Pennsylvania: You quoted a value of the ρ parameter from measurements of M_W and M_Z. However, M_W and M_Z do not contain enough information to determine ρ unless additional input, such as a value of the top quark mass, is included. What did you do?

M. Campbell: To obtain ρ, you need to include another measure of $\sin^2\theta_W$. We used the world average in our analysis.

U. Nauenberg, University of Colorado: Can you show us a plot that shows the energy and angular distributions of electrons from your $J/\psi, \Upsilon$ and Z° decay samples?

M. Campbell: No, I do not have those plots with me. The distributions are very skewed by the fact that the trigger thresholds for these particles were very high. It would take a great deal of work to extract a physics result from them; the J/ψ data was used only for calibration purposes.

G. Barbiellini, CERN: Since the momentum of an electron from J/ψ decay must be around 3–5 GeV/c, how can you extrapolate your calibration of the energy scale to particles with momenta near 40 GeV/c?

M. Campbell: Please consult section 2.3 of the published version of my talk, which addresses this question in detail.

S. Olsen, University of Rochester: The last speaker reported an e^+e^- event with $M_{ee} = 280$ GeV. Does the CDF Collaboration see any dilepton events at high invariant masses?

M. Campbell: We are not reporting on that question at the present time.

W. Kozanecki, SLAC: Did the error you quoted on $\sin^2\theta_W$ (or, equivalently, on M_W/M_Z) take into account the cancellation of the common energy-scale error?

M. Campbell: No, we just added all errors in quadrature, as if they were independent. We are working on a better value for $sin^2\theta_W$ that takes this cancellation into account.

THEORY OF PRECISION ELECTROWEAK EXPERIMENTS

G. Altarelli
Theory Division, CERN
1211 Geneva 23, Switzerland

1. BASIC RELATIONS

In the standard electroweak theory [1], there is a number of basic relations that one wants to experimentally verify as precisely as possible. The same quantity $\sin^2\theta_W$ appears in all of these relations.

First, $\sin^2\theta_W$ can be measured from the value of m_W. Starting from the tree level relations $\sin^2\theta_W = e^2/g_2^2$ and $G_F/\sqrt{2} = g_2^2/8m_W^2$ [where $e^2 = 4\pi\alpha$ is the electron charge squared, g_2 is the weak SU(2) coupling and G_F is the Fermi coupling constant], one obtains

$$\sin^2\theta_W = \left(\frac{\pi\alpha}{\sqrt{2}\,G_F}\right)\frac{1}{m_W^2}\,\frac{1}{1-\Delta r} = \frac{(37.2802\ \mathrm{GeV})^2}{m_W^2(1-\Delta r)} \tag{1}$$

where $\Delta r \neq 0$ due to the effect of radiative corrections.

$\sin^2\theta_W$ is also related to the ratio of the vector boson masses. At tree level

$$\sin^2\theta_W = 1 - \frac{m_W^2}{\rho_{tree} m_Z^2} \tag{2}$$

where $\rho_{tree} = 1$ in the standard model with only doublets of Higgs bosons. In general, this relation is also modified by radiative corrections

$$\sin^2\theta_W = 1 - \frac{m_W^2}{\rho_{mass} m_Z^2} \tag{3}$$

with $\rho_{mass} = \rho_{tree}(1+\delta\rho_{mass})$. However, as we shall discuss in detail later, Eq. (2) with $\rho_{tree} = 1$ is often adopted as a definition of $\sin^2\theta_W$ at all orders. Clearly in this case, $\rho_{mass} = 1$ by definition.

Finally $\sin^2\theta_W$ can be obtained from neutral current couplings. At tree level, the four-fermion interaction from Z exchange is given by

$$M_{if} = \frac{\sqrt{2}\,G_F m_Z^2}{D(s)}\,\rho_{tree}\,(J_3^i - 2\sin^2\theta_W J_{em}^i)\cdot(J_3^f - 2\sin^2\theta_W J_{em}^f) \tag{4}$$

where D(s) is the Z propagator and J_3^f, J_{em}^f are the weak isospin and electromagnetic currents for the fermion f. Excluding pure QED corrections, electroweak radiative corrections modify M_{if} according to

$$M_{if} = \frac{\sqrt{2}\,G_F m_Z^2}{D(s)}\,\rho_{if}\,(J_3^i - 2k_i\sin^2\theta_W J_{em}^i)\cdot(J_3^f - 2k_f\sin^2\theta_W J_{em}^f) + \ldots \tag{5}$$

where $\rho_{if} = \rho_{tree}(1+\delta\rho_{if})$, $k_a = 1+\delta k_a$ $(a = i,f)$ are in general different for different fermions and depend on the scheme adopted (for example δk_a depend on the definition of $\sin^2\theta_W$). The ellipsis indicates possible additional non-factorizable terms.

2. INPUT PARAMETERS

For LEP1/SLC physics*) a self-imposing set of input parameters is given by α, α_s, G_F, m_Z, m_f and m_H. In fact m_Z has now been (and even more will be) precisely measured. Clearly, $G_F = 1.166389(22) \times 10^{-5}$ GeV^{-2} [3] is conceptually less simple than $\alpha_{weak} = g_2^2/4\pi$ (which would more naturally accompany $\alpha = 1/137.036$ and α_s) or $\sin^2\theta_W$ or m_W, but is known with practically absolute accuracy. Among the quark and lepton masses m_f the main unknown is the top quark mass m_t. Our ignorance of m_t is at present a serious limitation for precise tests of the electroweak theory because the radiative corrections depend strongly on m_t. The light quark masses are not very accurately known, but their effective values in all the relevant cases [for instance, in the evaluation of Δr in Eq. (1)] can be obtained from low and intermediate energy data. The Higgs mass m_H is almost totally unknown. The sensitivity of the radiative corrections to m_H in the standard model with doublet Higgses is small. It fixes the required level of accuracy as the experimental exploration of the symmetry breaking sector of the theory is of the utmost importance. The best value of α_s at the Z mass is at present given by $\alpha_s(m_Z) = 0.11\pm0.01$ [4]. The QCD corrections to processes involving quarks are typically of order of α_s/π, so that the present error on α_s leads to a few o/oo relative uncertainty on the corresponding predictions.

3. LARGE CONTRIBUTIONS TO RADIATIVE CORRECTIONS

The most important quantitative contributions to the radiative corrections arise from large logarithms [e.g., terms of the form $(\alpha/\pi \, \ell n(m_Z/(m_{f_\ell}))^n$ where f_ℓ is a light fermion] and from quadratic terms in m_t, i.e., terms proportional to $G_F m_t^2$.

The sequences of leading and close to leading logarithms are fixed [5] by well-known and consolidated techniques (β functions, anomalous dimensions, penguin-like diagrams, etc.). For example, large logarithms dominate the running of α from m_ℓ, the electron mass, up to m_Z, with the result [3]

$$\frac{\alpha(m_Z)}{\alpha} = \frac{1}{1-\delta\alpha} \tag{6}$$

At present, the best value of $\delta\alpha$, obtained by extracting the relevant effective light quark masses from the data on $e^+e^- \to$ hadrons, is given by [3]:

$$\delta\alpha = 0.0601 + \frac{40}{9}\frac{\alpha}{\pi}\ell n\left(\frac{m_Z}{91\text{GeV}}\right) \pm 0.0009 \tag{7}$$

Large logarithms of the form $(\alpha/\pi \, \ell n(m_Z/\mu))^n$ also enter, for example, in the relation between $\sin^2\theta_W$ at the scales m_Z (LEP,SLC) and μ (e.g., the scale of low energy neutral current experiments).

The quadratic dependence on m_t [6] (and on other possible widely broken isospin multiplets from new physics) arises because in spontaneously broken gauge theories heavy loops do not decouple. On the contrary, in QED or QCD the running of α and α_s at a scale Q is not affected by heavy quarks with mass $M >> Q$. According to an intuitive decoupling theorem [7], diagrams with heavy virtual particles of mass M can be ignored at $Q << M$ provided that the couplings do not grow with M and the theory with no heavy particles is still renormalizable. In spontaneously broken gauge theories, one important difference is in the longitudinal modes of weak gauge bosons. These modes are generated by the Higgs mechanism and their couplings grow with masses (as is also the case for the physical Higgs couplings). The upper limit on m_t from radiative corrections arises from this phenomenon. Other subtler sources of non-decoupling are related, for example, to the presence of chiral anomalies [8] (which may be not completely cancelled if heavy particles are removed). Another very important conse-

---

*) A thorough and updated review of physics at the Z pole in e^+e^- annihilation can be found in the final report [2] of the Workshop on Z Physics at LEP1 held at CERN during the last months. I will make large use in the following of the material collected in the various chapters of this book in three volumes.

quence is that precision tests of the electroweak theory are sensitive to new physics even if the new particles are too heavy for their direct production. With the value of m_t being continuously pushed up by experiment, the quantitative importance of the terms of order of $G_F m_t^2$ is increasingly large. Both the large logarithms and the $G_F m_t^2$ terms have a simple structure and are to a large extent universal, i.e., common to a wide class of processes. Their study is important for an understanding of the pattern of radiative corrections. One can also derive approximate formulae (e.g., improved Born approximations - see Section 5) which can be useful in cases where a limited precision may be adequate.

4. DETERMINATION OF $SIN^2\theta_W$ FROM m_Z

Once α, α_s, G_F, m_Z, m_f and m_H have been chosen as input parameters $\sin^2\theta_W$ is a derived quantity. We now consider the calculation of $\sin^2\theta_W$ beyond the tree approximation. A precise definition of $\sin^2\theta_W$ must be specified before its value can be computed. An infinite number of definitions are possible and many have been actually proposed and studied. Physical results are of course independent of the definition adopted for $\sin^2\theta_W$. Differences in physical results obtained from different schemes can only occur by terms of higher order, because the perturbative series is truncated at a given order. $\sin^2\theta_W$ is only a useful reference quantity: what really matters for precision tests is the prediction of the different measured quantities P in terms of the input parameters, or in practice, for fixed m_Z, the values $P = P(m_t, m_H)$ of each given physical quantity in the allowed range of m_t and m_H.

At tree level the relation between $\sin^2\theta_W$ and m_Z is obtained from Eqs. (1) and (2) (with $\Delta r = 0$):

$$\sin^2\theta_W \cos^2\theta_W = \frac{\pi\alpha}{\sqrt{2}\,G_F}\,\frac{1}{\rho_{tree} m_Z^2} \qquad (8)$$

Beyond the tree level the quantity Δr [9] is introduced by radiative corrections:

$$\sin^2\theta_W \cos^2\theta_W = \frac{\pi\alpha}{\sqrt{2}\,G_F}\,\frac{1}{\rho_{tree} m_Z^2}\,\frac{1}{1-\Delta r} \qquad (9)$$

where $\Delta r \equiv \Delta r(\alpha, \alpha_s, G_F, m_Z, m_f, m_H)$ is of course different for different definitions of $\sin^2\theta_W$. In the following, some particularly interesting definitions of $\sin^2\theta_W$ will be discussed. We now set $\rho_{tree} = 1$. Let us first consider the usual definition [10]

$$\sin^2\theta_W \equiv s_W^2 = 1 - \frac{m_W^2}{m_Z^2} \qquad (10)$$

From now on the symbol s_W^2 will always refer to this specific definition of $\sin^2\theta_W$. Clearly given m_Z from SLC or LEP, s_W^2 is directly equivalent to m_W. In this case Δr specifies the relation between m_Z and m_W:

$$\left(1 - \frac{m_W^2}{m_Z^2}\right) m_W^2 = \left(\frac{\pi\alpha}{\sqrt{2}\,G_F}\right) \frac{1}{1-\Delta r} \qquad (11)$$

The value of Δr as a function of the input parameters has been studied in great detail (Fig. 1). In particular one obtains the following results [3,9,11]:

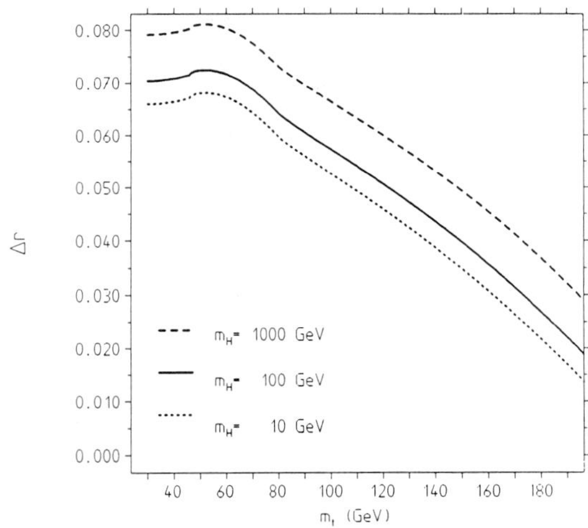

Fig. 1: Δr as a function of the top mass for various m_H (m_Z = 91 GeV) [9].

$$\frac{1}{1-\Delta r} = \frac{\alpha(m_Z)}{\alpha} \frac{1}{1+\frac{c_W^2}{s_W^2}\delta\rho} + \text{"small"} = $$

$$= \frac{1}{1-\delta\alpha} \frac{1}{1+\frac{c_W^2}{s_W^2}\delta\rho} + \text{"small"} \tag{12}$$

where $c_W^2 = 1-s_W^2$, $\delta\alpha$ was defined in Eqs. (6) and (7) and $\delta\rho \to \delta\rho_t$ for large m_t with $\delta\rho_t$ given by [6,12]

$$\delta\rho_t = \frac{3G_F m_t^2}{8\pi^2\sqrt{2}} + \left(\frac{3G_F m_t^2}{8\pi^2\sqrt{2}}\right)^2 \frac{19-2\pi^2}{3} + + o[(G_F m_t^2)^3] \tag{13}$$

In Eq. (12) by "small" we mean terms (which at one loop accuracy are known) without large logs and/or leading powers of $G_F m_t^2$. $\delta\rho_t$ is the dominant term for large m_t of the famous ρ-parameter first studied in Refs. [6]. The two-loop term was obtained in Ref. [12] and the geometric series resummation was recently advocated in Ref. [11].

Going back to the Z-exchange amplitude near the resonance and the parameters ρ_{if}, k_i and k_f defined in Eq. (5) with the definition of $\sin^2\theta_W \equiv s_W^2$, one obtains [3] that

$$\rho_{if} = 1+\delta\rho + \text{"small"} \tag{14}$$

and

$$k_f = 1 + \frac{c_W^2}{s_W^2}\delta\rho + \text{"small"} \tag{15}$$

(here $f \neq b$. The b quark will be reconsidered in the following) where the "small" terms are non-universal (i.e., process dependent). Note that k_f contains additional "large" terms with respect to those included in Δr and ρ_{if}. As these "large" terms are also universal, this suggests that ks_W^2 could be a better effective $\sin^2\theta_W$ than s_W^2 for physics at the Z pole. Before going into this matter, we add some comments on $\delta\rho$. For any weak isospin fermion doublet (e.g., new heavy quarks or leptons), the quantity "$3m_t^2$" in $\delta\rho_t$ (at one loop) becomes [6,9]:

$$\text{"}3m_t^2\text{"} = N_c\left[m_u^2 + m_d^2 - \frac{2m_u^2 m_d^2}{m_u^2-m_d^2} \ln \frac{m_u^2}{m_d^2}\right] \tag{16}$$

where N_c is the number of colours. For negligible $m_{d,u}$ "$3m_t^2$" $\to N_c m_{u,d}^2$, while for $m_u = m_d+\varepsilon$: "$3m_t^2$" $\to N_c 4/3\varepsilon^2$ to leading order in ε (the result vanishes for unsplit doublets). Similarly, many more kinds of broken weak isospin multiplets can contribute to "$3m_t^2$" [9,13] (squarks and sleptons [14], charged Higgses [15], etc.). The upper limit on "m_t" $\lesssim$ 210 GeV (see Section 6) together with the direct lower limit on m_t, $m_t \gtrsim$ 77 GeV leave little space for heavy multiplets except for nearly degenerate ones.

The contribution to $\delta\rho$ of the neutral Higgs mass is only logarithmic at one loop:

$$\delta\rho_{Higgs} \simeq - \frac{11G_F m_W^2}{24\pi^2\sqrt{2}} \, tg^2\theta_W \, \ln \frac{m_H^2}{m_W^2} + \ldots \tag{17}$$

There are no m_H^2 terms but only logs because the "custodial" SU(2) symmetry is not broken in the Higgs sector [16]. Power terms only appear at two loops [17], but their effect is only sizeable for $m_H \gtrsim$ 1 TeV.

We now consider a different class of definitions of $\sin^2\theta_W$. Assume that we fix $k_f = 1$ [defined by Eq. (5)] for one given Z vertex (e.g., $k_f = 1$ in $Z \to e^+e^-$ for on-shell Z). Then it follows from the previous discussion that δk is "small" for all neutral current processes (for $f \neq b$) near the Z pole. Let us introduce the notation

$$\sin^2\theta_W \text{ (from } Z \to e^+e^-) \equiv \bar{s}_W^2 \tag{18}$$

In this case Eq. (5) becomes:

$$M_{fi} = \frac{\sqrt{2}\, G_F m_Z^2}{D(s)} \rho_{if}\left(J_3^i-2\bar{s}_W^2 J_{em}^i\right)\left(J_3^f-2\bar{s}_W^2 J_{em}^f\right) + \text{"small"} \tag{19}$$

where ρ_{if} is given by Eq. (14) in terms of $\delta\rho$ which is specified in Eqs. (13) and (17). In the present case

$$\bar{s}_W^2 \bar{c}_W^2 = \frac{\pi\alpha}{\sqrt{2}\, G_F} \frac{1}{m_Z^2} \frac{1}{1-\Delta\bar{r}} \tag{20}$$

and one finds [3]

$$\bar{s}_W^2 \simeq s_W^2 + c_W^2 \delta\rho + \text{"small"} \tag{21}$$

In other words, apart from "small" terms one has:

$$\bar{s}_W^2 \simeq 1 - \frac{m_W^2}{\rho m_Z^2} \tag{22}$$

with $\rho \simeq 1+\delta\rho$. It is interesting to note that $\bar{s}_W^2$ is less dependent on m_t^2 than s_W^2. In fact [3,9]:

$$\Delta r \simeq \delta\alpha - \frac{c_W^2}{s_W^2}\delta\rho \tag{23}$$

$$\Delta\bar{r} \simeq \delta\alpha - \delta\rho \tag{24}$$

so that the amplifying factor c_W^2/s_W^2 in front of $\delta\rho$ is missing in $\Delta\bar{r}$. Note that given m_Z, $\bar{s}_W^2$ is known with an accuracy of about ±0.002 when m_t varies between 60 and 210 GeV (see Fig. 2). There is a class of definitions of $\sin^2\theta_W$ in the literature that correspond to $\bar{s}_W^2$ if "small" terms are neglected: $\bar{s}_W^2$ of Hollik [18], $s^{*2}(m_Z^2)$ of Lynn and Kennedy [19], $(\sin^2\theta_W)_{\overline{MS}}\,(m_Z^2)$ [20,21], etc.

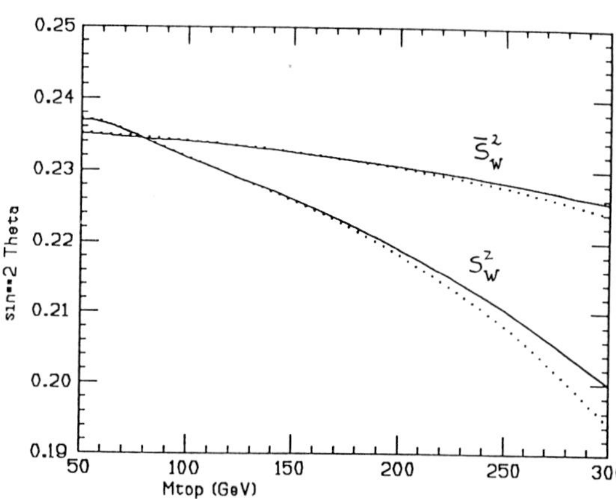

Fig. 2: The behaviour of s_W^2 and $\bar{s}_W^2$ [defined by Eqs. (10) and (18) respectively] as functions of m_t for m_Z = 91 GeV, m_H = 100 GeV. The solid (dashed) lines are computed to $O(\alpha^2)$ $[O(\alpha)]$. From Ref. [3].

5. IMPROVED BORN APPROXIMATION

For precision tests of the electroweak theory, the complete one-loop radiative corrections (plus higher order/exponentiated purely photonic corrections) are mandatory and are indeed available for the processes of practical relevance. However, in many cases a less accurate estimate can be enough. For this purpose, it is useful to know [3] that formulae as simple as those of the Born approximation can be written down in a way that takes all "large" corrections into account. For $e^+e^- \to f\bar{f}$, with $f \neq e,b$ the amplitude for γ and Z exchange near the resonance can be written down in the form:

$$M_{f\bar{f}} = Q_e Q_f \frac{4\pi\alpha(m_Z)}{s} J^e_{em} J^f_{em} + \frac{\sqrt{2} G_F \rho m_Z^2}{s - m_Z^2 + is\frac{\Gamma_Z}{m_Z}} \bar{J}^e \bar{J}^f \tag{25}$$

where $\rho = 1+\delta\rho$, $\delta\rho$ being given by Eqs. (13) and (17) and

$$J^f_{em} = \gamma_\mu$$
$$\bar{J}^f = \gamma_\mu[I_3^f(1-\gamma_5) - 2Q_f\bar{s}_W^2] \tag{26}$$

with Q_f and I_3^f being the electric charge and 3rd component of the weak isospin (e.g., for $f = \mu^-$, $Q = -1$, $I_3 = -\frac{1}{2}$). The important features are the replacement of α-fixed with α-running, the inclusion of the s dependence for the total width Γ_Z in the resonant denominator (see Ref. [7]), the presence of the factor of ρ multiplying G_F and the use in the Z couplings of the effective $\sin^2\theta_W = \bar{s}_W^2$, introduced in the previous section. Clearly for f = e, i.e., for Bhabha scattering, the t-channel exchange is also to be included. The improved Born approximations include the real parts of self-energies (sometimes [19] called "oblique" corrections). All large logs and all $G_F m_t^2$ terms are included. What are left out are "small" corrections from imaginary parts of self-energies, vertices and boxes.

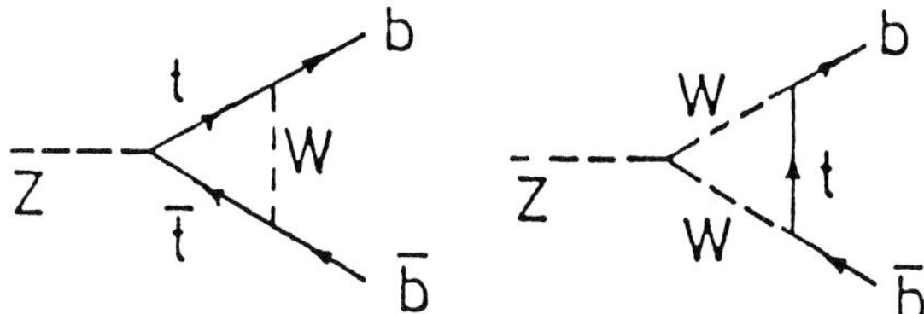

Fig. 3: Vertex corrections which contain large $G_F m_t^2$ terms.

For f = b there are additional large terms from the vertex corrections [18,22] of the type in Fig. 3. The longitudinal W modes are also in this case responsible for the presence of quadratic mass terms. One can simply modify the improved Born approximation in order to include these terms as well. The recipe [3,23] is as follows: in Eqs. (25) and (26) replace

$$\rho \to \sqrt{\rho\rho_b} \quad \rho_b \equiv \rho(1 - \frac{4}{3}\delta\rho) \tag{27}$$

$$\bar{s}_W^2 \to \bar{s}_W^2(1 + \frac{2}{3}\delta\rho) \equiv \bar{s}_W^2 k_b \tag{28}$$

Note that while many sorts of heavy particles can contribute to $\delta\rho$, only the t quark (or a new t') contributes to $\delta\rho_b$, because the W predominantly turns a b quark into a t (or a t') quark.

As a final example, the improved Born approximation for the inclusive widths $\Gamma[Z \to f\bar{f}(\gamma,g)]$ (including photons and gluons in the final state), is given by:

$$\Gamma\big(Z \to f\bar{f}(\gamma,g)\big) = N_c \frac{G_F \rho m_Z^3}{24\pi\sqrt{2}} \big(1-(1-4|Q_F|\bar{s}_W^2)^2\big) \tag{29}$$

where

$$N_c = \begin{cases} 1 \cdot (1+\frac{3\alpha}{4\pi} Q_f) & \text{(leptons)} \\ 3 \cdot (1+\frac{3\alpha}{4\pi} Q_f)\,(1+\frac{\alpha_s(m_Z)}{\pi}) & \text{(quarks)} \end{cases} \tag{30}$$

In the same approximation the total width is simply given by

$$\Gamma_Z = \Sigma_f\, \Gamma\big(Z \to f\bar{f}(\gamma,g)\big) + \text{rare decays} \tag{31}$$

where the rare decays [24] (e.g., $Z \to H\mu^+\mu^-$ for a light Higgs) are numerically unimportant in the standard model (for $m_H > 10$ GeV) [25]. For $\Gamma(Z \to b\bar{b})$ replace [3,23] ρ by $\sqrt{\rho_b}$ and $\bar{s}^2$ by $\bar{s}^2 k_b$ where ρ_b and k_b are given in Eqs. (27) and (28).

6. STATUS OF PRECISION TESTS OF THE ELECTROWEAK THEORY

After a period of stagnation, a big step forward in precision tests of the electroweak theory has been accomplished with the data presented at this Symposium. This makes my rapporteur task especially exciting. My predecessors in this rôle used to start by presenting a table of results on s_W^2 extracted from different processes with radiative corrections computed assuming a given value for m_t and m_H. Over the years the assumed value of m_t started from 15 GeV, then was set at 30 GeV, then again increased up to 45 GeV and more recently up to 60 GeV. This was a sensible procedure because for m_t not too large, the radiative corrections are not dominated by m_t. A moderate change of m_t in a range around and below 60 GeV does not very much affect the radiative corrections. This is, for example, evident from Fig. 1 where the value of Δr is plotted as a function of m_t. We now know from new data from CDF [26], UA1 [27] and UA2 [28] that probably $m_t > 70$-80 GeV. For large m_t the quadratic terms of order $G_F m_t^2$ become a driving term for the radiative corrections, which, as a result, strongly depend on m_t. It seems to me that a better strategy is to discuss the experimental results and the associated radiative corrections needed to extract $\sin^2\theta_W$ as functions of m_t, with m_t varying in the whole range of allowed values. The requirement of consistency of the electroweak theory with the data significantly restricts the allowed range of m_t. We shall see that not only an upper bound on m_t is obtained (which is an old result), but also a lower bound on m_t can now be obtained. This limit [see Eq. (38)] is interesting in itself although the limits from production experiments are more directly reliable (clearly a lower limit on m_t from radiative corrections can be evaded if heavy new particles contribute to vacuum polarization diagrams, while the upper limit is a fortiori valid).

Table 1

The underlined numbers refer to errors on the energy scale. Most of these errors cancel when taking the ratio m_W/m_Z.

	m_W (GeV)	m_Z (GeV)
	OLD DATA	
UA1	82.7 ± 1.0 ± <u>2.7</u>	93.1 ± 1.0 ± <u>3.1</u>
UA2	80.2 ± 0.6 ± <u>1.4</u>	91.5 ± 1.2 ± <u>1.7</u>
CDF	80.0 ± 3.3 ± <u>2.4</u>	
	NEW DATA	
UA2 [29]	80.0 ± 0.4 ± 0.4 ± <u>1.2</u>	90.2 ± 0.6 ± <u>1.4</u>
CDF [30]	80.0 ± 0.2 ± 0.5 ± <u>0.3</u>	90.9 ± 0.3 ± <u>0.2</u>
MARK II [31]		91.17 ± 0.18

The most important results are the new measurements of the weak gauge boson masses. The old and new data on m_W and m_Z are summarized in Table 1. The precise determination of m_Z by MARK II [31] (and CDF [30]) are among the highlights of this conference. The CDF result is quite impressive if one considers the difficulty of such a measurement at hadron colliders. The MARK II determination of m_Z is a great achievement not only in itself but also as a promise for the future of linear e^+e^- linear colliders.

By combining the CDF and MARK II results, one obtains:

$$m_Z = 91.12 \pm 0.16 \text{ GeV} \tag{32}$$

There is no theoretical error associated with the measurement of m_Z in e^+e^- annihilation and also the derivation of s_W^2 (or $\bar{s}_W^2$) from m_Z is very clear, i.e., the theory of Δr (or $\overline{\Delta r}$) has no ambiguities in the minimal standard model other than our ignorance of m_t and m_H. However, the value of s_W^2 obtained from m_Z strongly depends on m_t through Δr. This is seen in Fig. 4 where the value of s_W^2 corresponding to a given value of m_Z is plotted versus m_t. As discussed in Section 4 the value of $\bar{s}_W^2$

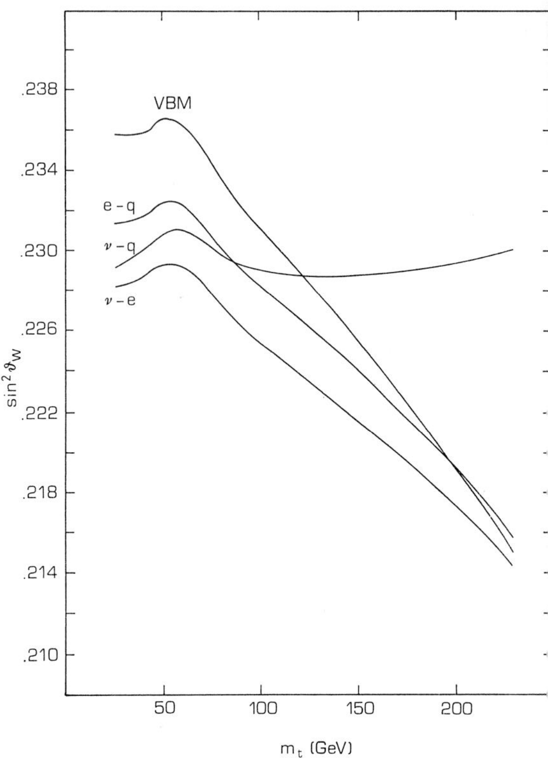

Fig. 4: Dependence on m_t of s_W^2 [defined in Eq. (10)] extracted from $m_{W,Z}$ (VBM) and from the neutral current e-q, ν-q or ν-e sectors (Ref. [32]).

extracted from m_Z is less sensitive to m_t. We shall come back later on the determination of $\bar{s}_W^2$.

The measurement of m_W/m_Z at $p\bar{p}$ colliders is also quite clean, although not as much as m_Z. The error on the energy scale drops in taking the ratio. The theoretical error is confined to some small uncertainties on the W transverse momentum distribution (e.g., structure functions, Λ_{QCD} and related problems) which somewhat affect the determination of m_W from the Jacobian peak in the electron distribution. The values obtained by CDF [30] and UA2 [29] are in excellent mutual agreement:

$$\begin{aligned} m_W/m_Z &= 0.887 \pm 0.009 \text{ (UA2)} \\ &= 0.880 \pm 0.007 \text{ (CDF)} \end{aligned} \tag{33}$$

In this case clearly there are no problems in obtaining s_W^2: the ratio m_W/m_Z directly determines s_W^2, by definition. The above results correspond to

$$[s_W^2]_{m_W/m_Z} = 0.213 \pm 0.015 \text{ (UA2)}$$
$$= 0.225 \pm 0.012 \text{ (CDF)} \quad (34)$$

The resulting combined value $s_W^2 = 0.220 \pm 0.009$ is reported in Fig. 5. Clearly, the relatively low value of s_W^2 obtained from m_W/m_Z when compared with m_Z, pushes the value of m_t toward large values (within the limited accuracy of the m_W/m_Z measurement). It is important to note that an upper limit on m_t around 260 GeV is already implied by the vector boson masses alone.

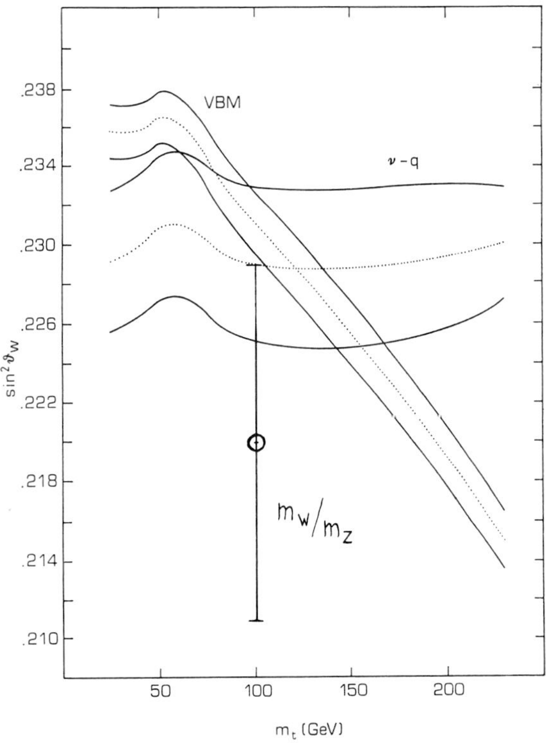

Fig. 5: s_W^2 [defined in Eq. (10)] obtained from m_Z, m_W/m_Z and $R_\nu = \sigma_{\nu N}^{NC}/\sigma_{\nu N}^{CC}$ as a function of m_t. The error shown for R_ν is only a part of the total: it corresponds to a fixed charm mass $m_c = 1.45$ GeV and $m_H = 100$ GeV (courtesy of G.L. Fogli).

We now consider neutral-current data. The most precise determination of s_W^2 is obtained from the ratio R_ν of neutral to charged current deep inelastic neutrino scattering. R_ν imposes an important constraint on the electroweak parameters when compared with m_W, m_Z because s_W^2 derived from R_ν is practically independent of m_t in the range of interest [33]. This near independence is apparent in Figs. 4,5. It arises because of a largely accidental cancellation of the m_t dependence introduced by defining s_W^2 from m_W/m_Z and that induced by the ρ parameter appearing in the neutral to charged current ratio. The most precise measurements of s_W^2 are due to the CHARM [34] and CDHS [35] collaborations at CERN:

$$[s_W^2]_{\nu N} = 0.236 \pm 0.007 \text{ (CHARM)}$$
$$= 0.228 \pm 0.007 \text{ (CDHS)} \quad (35)$$

Note that both $[s_W^2]_{\nu N}$ and $[s_W^2]_{m_W/m_Z}$ are independent (or nearly so for $[s_W^2]_{\nu N}$) on m_t. As a consequence, the direct comparison of Eq. (35) with Eq. (34) provides a quite remarkable and successful consistency check, independent of m_t, for the electroweak theory. Thus R_ν is a good measure of m_W/m_Z (as recently stressed again in Ref. [36]). Once R_ν is fixed, if m_Z is decreased, m_W has also to be decreased which means that Δr is decreased and finally that m_t is increased. The fact that the measured value of m_Z is below the previous world average of about 92 GeV pushes m_t toward the large values.

The values of s_W^2 obtained from m_Z, m_W/m_Z and R_ν as functions of m_t with the corresponding experimental errors are displayed in Fig. 5. Clearly, a lower and an upper limit on m_t are implied. For a quantitative determination of the allowed range of m_t and s_W^2, it must be stressed that the measurement of R_ν in neutrino deep inelastic scattering and the extraction of s_W^2 from R_ν are obviously affected by much larger theoretical uncertainties than those connected with m_Z and m_W/m_Z. In νN scattering, we must cope with uncertainties connected with the validity of the parton model (i.e., the importance of higher twist and other pre- asymptotic corrections), our relative ignorance of sea densities, the effects of the onset of the charm threshold and the validity of the slow rescaling procedure (the related uncertainties are roughly parametrized in terms of an effective "charmed mass" parameter m_c) the modelling of the effect of experimental cuts and so on. The errors stated by the experiments and quoted in Eq. (35) also include an estimate of theoretical errors. The error 0.007 arises from adding in quadrature the experimental and theoretical errors according to $\delta s_W^2 = 0.012(m_c-1.5) \pm 0.005$ (exp) ± 0.003 (theor) with $m_c \sim 1.2 - 1.8$ GeV. It is however to be kept in mind that the determination of s_W^2 from neutrino-nucleon scattering is quite involved and the evaluation of the related theoretical error is debatable to some extent. With the new data on m_W/m_Z and especially on Z physics from SLC/LEP1 R_ν will increasingly become the weak link in the chain.

There are new important results on other neutral current processes. First, I want to recall the recent measurement of parity violation in atomic cesium done in Boulder [37] (by a team of only three physicists!). They quote the remarkably accurate value

$$[s_W^2]_{Cs} = 0.219 \pm 0.019 \tag{36}$$

obtained for m_t light ($m_t \lesssim 60$ GeV). For m_t larger, the value of $[s_W^2]_{Cs}$ rapidly decreases as seen from the curve labelled "e-q" in Fig. 4. The value in Eq. (36) is in agreement with previous less precise results on Cs obtained by the Paris group [38]. A decade of continuous progress on atomic "table-top" experiments has allowed to establish the validity of the electroweak theory at very low energies with a precision not far from that of experiments at higher energies performed with large detectors at big accelerators.

New data on $\overset{(-)}{\nu}_\mu e$ scattering have been presented for the first time at this Symposium by the CHARM 2 collaboration [39]. As shown in Table 2 (where the numbers of observed events are also displayed) the CHARM 2 values are considerably more precise than in previous experiments. Also, the central value of s_W^2 is now in better agreement with the results from other processes. Including radiative corrections with $m_t = 100$ GeV (the behaviour at larger values of m_t is shown in Fig. 4) the CHARM 2 result is given by

Table 2

Results on ν-e scattering

	$\nu_\mu e$	$\bar{\nu}_\mu e$	$\sin^2\theta_W$ (no rad. cor.)
CHARM 1	83	112	0.211 ± 0.035 ± 0.011
E734	160	97	0.195 ± 0.018 ± 0.013
CHARM 2	762	1017	0.233 ± 0.012 ± 0.008

$$[s_W^2]_{\overset{(-)}{\nu}_\mu e} = 0.232 \pm 0.014 \tag{37}$$

The experiment is still running and the present stage of the analysis is far from the final aim of reducing the error on s_W^2 down to $\delta s_W^2 = \pm(6-7)10^{-3}$.

Going back to Fig. 4 showing the dependence on m_t of s_W^2 obtained from different processes, one sees that on one hand ν-q is similar to m_W/m_Z (leading to a value of s_W^2 nearly independent on m_t), while on the other hand m_Z, e-q and $\overset{(-)}{\nu}_\mu e$ lead to s_W^2 with a common trend in m_t. Figure 6 differs from

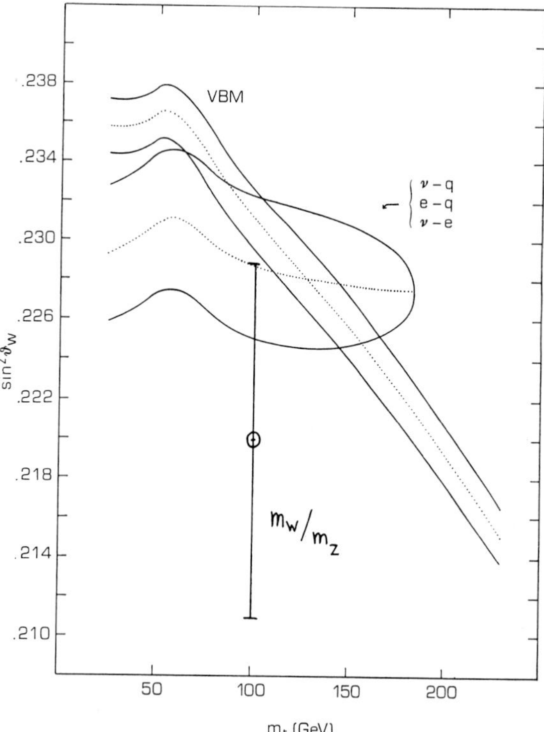

Fig. 6: Same as Fig. 5 but for the addition of the data on the e-q and ν-e sectors (courtesy of G.L. Fogli).

Fig. 5 because the constraints from e-q and ν-e have been added to those from ν-q in order to complete the information from neutral current processes.

Taking all the information together, I conclude that the ranges of m_t and s_W^2 indicated by the present data on the electroweak theory are

$$m_t = 130 \pm 50 \text{ GeV} \tag{38}$$

$$s_W^2 \equiv 1 - \frac{m_W^2}{m_Z^2} = 0.227 \pm 0.006 \tag{39}$$

These values are in agreement with the results of a number of complete detailed analyses of the data [32,40,41,42], (see for example Fig. 7 taken from Ref. [41]),

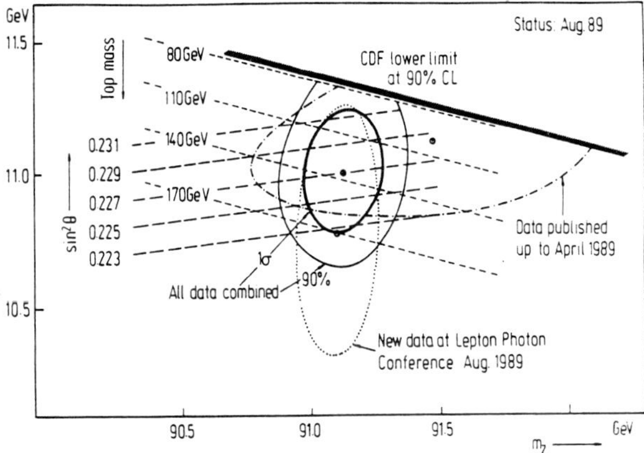

Fig. 7: Summary of present status of electroweak tests. Given m_Z, $m_Z - m_W = m_Z(1-m_W/m_Z)$ is equivalent to m_W/m_Z or s_W^2 (courtesy of D. Haidt).

although the errors I quote are in some cases more conservative. In particular, the upper (lower) limit on the top mass from Eq. (36) is about $m_t \lesssim 210$ GeV ($m_t > 50$ GeV) at 90% cl. Note that the uncertainty on s_W^2 is actually much smaller now than before this Symposium. In fact, the present error is for all allowed values of m_t and m_H (for m_H we take the range 10 GeV-1 TeV), while in the past the quoted errors referred to assumed fixed values of these masses.

A different question is what is at present the best estimate of $\bar{s}_W^2$, i.e., the effective $\sin^2\theta_W$ for the on-shell Z couplings introduced in Section 4, Eqs. (18)-(22). We recall that the value of $\bar{s}_W^2$ obtained from m_Z is less dependent on m_t and m_H than s_W^2. If we take for m_Z the value and the error given in Eq. (32), for m_t the range in Eq. (38) and for m_H 10 GeV < m_H < 1 TeV one obtains [21,36]

$$\bar{s}_W^2 = 0.233 \pm 0.002 \pm 0.0074\ \delta m_Z\ (\text{GeV}) \tag{40}$$
$$= 0.233 \pm 0.002 \pm 0.0012$$

We see that indeed the Z couplings can be predicted with more accuracy from $\bar{s}_W^2$ than one would naively obtain going through s_W^2. Note that as $\bar{s}_W^2 \sim (\sin^2\theta_W)_{\overline{MS}}$ (i.e., apart from "small" terms), this is also the relevant quantity to compare with GUT predictions. The value in Eq. (40) is in disagreement with naive SU(5) but in agreement with the minimal supersymmetric extension of SU(5) [43,44].

Before closing this section, I add some more comments on the results on the Z line shape and the total and partial widths presented by MARKII [31,45]. Quite interesting is the result on the Z width into invisible channels, expressed in terms of the number of standard neutrino species. The equivalent number of neutrinos, obtained from the value of the cross-section at the peak, is found to be $N_\nu = 3.0 \pm 0.9$. Additional indications consistent with $N_\nu = 3$ were presented by CDF [30]. These results leave little space for additional families of fermions (with $m_\nu < m_{Z/2}$). A fourth family is not yet excluded, but we are now getting close to this very important conclusion. Further data from SLC and LEP will decide the issue very soon. Knowing that there are precisely three families implies a formidable step forward in the predictivity of the standard model.

The total width presented by MARKII is a bit too narrow, $\Gamma_{exp} \simeq 1.95^{+0.40}_{-0.30}$ GeV, about 1.3 standard deviations away from the predicted value $\Gamma_{SM}(m_Z) \simeq 2.5$ GeV. Note that the CDF value for Γ [30] is $\Gamma_{exp} \simeq 3.8 \pm 1.5$ GeV. Similarly the partial width Γ_{ee} for $Z \to e^+e^-$ is 2 standard deviations below the standard value: $\Gamma_{ee}/\Gamma_{SM} = 0.52 \pm 0.23$. $\Gamma_{\tau\tau}/\Gamma_{q\bar{q}}$ is found at 2 standard deviations above the prediction (while $\Gamma_{\mu\mu}/\Gamma_{q\bar{q}}$ is in agreement with expectations within errors). Clearly there is still no reason to be excited. I just note that the precision of the data at the moment is such that it would only allow to detect rather large departures from the standard model which are practically unconceivable, given the constraints coming from other experiments. For example, smaller total and partial widths Γ and Γ_{ee} can indeed be obtained if a new heavy Z' exists (with appropriate couplings and mass). But given the existing limits, only a moderate narrowing of Γ and Γ_{ee} is possible. In particular, it is essentially impossible (i.e., without massive doses of fine-tuning) to obtain a decrease of Γ by more than 100 or 200 MeV at most and at the same time be compatible with existing data. Thus the best attitude at the moment is to wait for more data.

7. THE Z LINE SHAPE

The forthcoming experiments at SLC and especially at LEP will certainly mark a big step forward in the domain of precision tests of the standard theory. The strategy at LEP1/SLC is first to study the line shape and measure m_Z, Γ, the partial widths, perform the counting of neutrinos and look for possible surprises in the final state. This is the part of the programme that has already started at SLC and led to the important results that have been presented at this Symposium. At LEP1 the same measurements will presumably be completed very soon with considerably more precision*. The exploratory phase will be followed by a long period of running at the peak. The goals are to measure as precisely as possible a number of asymmetries that allow the most stringent tests of the theory and to search for interesting rare Z decays, the most important being those involving the Higgs boson (i.e., $Z \to H\mu^+\mu^-$ and $Z \to H\gamma$). A thorough and updated review of Z physics in e^+e^- annihilation can be found in Ref. [2]. In the following I will very briefly discuss some of the highlights of the physics programme outlined above.

A considerable amount of work has deservedly been devoted to the theoretical study of the Z line shape [25]. The experimental accuracy on m_Z which is planned at LEP is $\delta m_Z = \pm 50$ MeV. The error on m_Z can eventually go down to $\delta m_Z = \pm 20$ MeV if the transverse polarization of the beams measured by adequate polarimeters will be used in order to calibrate the energy by spin-resonance methods [46]. Similarly a measurement of the total width to an accuracy $\delta\Gamma = \pm 30$ MeV is now considered feasible. The prediction of the Z line shape in the standard model to such an accuracy has posed a formidable challenge to theory which has been successfully met. For the inclusive process $e^+e^- \to f\bar{f}X$, with $f \neq e$ (for simplicity, we leave Bhabha scattering aside) and X including γ's and gluons, the physical cross-section can be written in the form of a convolution [25]:

$$\sigma(s) = \int_{z_0}^{1} dz\ \hat{\sigma}(zs)\ G(z,s) \tag{41}$$

where $\hat{\sigma}$ is the reduced cross-section and $G(z,s)$ is the radiator function which describes the effect of initial state radiation. $\hat{\sigma}$ includes the purely weak corrections, the effect of final state radiation (of both γ's and gluons) and also non-factorizable terms (initial and final state radiation interferences, boxes, etc.) that being small can be treated in lowest order and effectively absorbed in a modified $\hat{\sigma}$. The radiator $G(z,s)$ has an expansion of the form [25]:

$$\begin{aligned} G(z,s) = {} & \delta(1-z) + \frac{\alpha}{\pi}(a_{11}L+a_{10}) + \\ & + \left(\frac{\alpha}{\pi}\right)^2 (a_{22}L^2+a_{21}L+a_{20}) + \\ & + \dots + \left(\frac{\alpha}{\pi}\right)^n \Sigma_{i=0}^{n} a_{ni}L^i \end{aligned} \tag{42}$$

where $L = \ln s/m_e^2 \simeq 24.2$ for $\sqrt{s} \simeq m_Z$. All first and second order terms are exactly known. The sequence of leading and next-to-

---

*This has indeed been the case. At the moment of this write-up, after three weeks of running of LEP, the combined values from the four LEP experiments are:
$m_Z = 91.10 \pm 0.033 \pm 0.045$ GeV
$= 91.10 \pm 0.06$ GeV
$\Gamma_Z = 2.584 \pm 0.075$ GeV
$N_\nu = 3.17 \pm 0.21$

leading logs can be exponentiated (closely following [47] the formalism of structure functions in QCD). For $m_Z \sim 91$ GeV the convolution displaces the peak by +110 MeV, and reduces it by a factor of about 0.74. The exponentiation is important in that it amounts to a shift of about 14 MeV in the peak position.

A model independent analysis [48] of the reduced cross-section $\hat{\sigma}$ leads to the following general expression (here $m \equiv m_Z$):

$$\hat{\sigma}(s) = \frac{12\pi\Gamma_e\Gamma_f}{|D(s)|^2}\left[\frac{s}{m^2} + R_f\frac{s-m^2}{m^2} + \frac{\Gamma}{m}I_f + + \ldots\right] + \frac{4\pi\alpha^2(m^2)Q_f^2N_c}{3s} \tag{43}$$

with N_c given by Eq. (35) and

$$D(s) = s - m^2 + im\Gamma\left[\frac{s}{m^2} + \varepsilon\frac{s-m^2}{m^2}\right] + \ldots \tag{44}$$

This form of the resonant term was obtained by starting from a general renormalizable field theory (the standard model being a particular case). Near the resonance $(s-m^2)/m^2 \approx \Gamma/m$ is of order α (or α_W). As only a perturbative calculation in α or α_W of $\hat{\sigma}$ is possible, at the same level of accuracy one can expand vertices, propagators, etc., in $(s-m^2)/m^2$ near the resonance. For example, for the inverse propagator one can write down the expansion:

$$\begin{aligned} D(s) &= s - m^2 + \Pi(s) = s - m^2 + \mathrm{Re}\Pi(m^2) \\ &+ (s-m^2)\mathrm{Re}\Pi'(m^2) + i\mathrm{Im}\Pi(m^2) + \\ &i(s-m^2)\mathrm{Im}\Pi'(m^2) + \ldots \\ &= [1+\mathrm{Re}\Pi'(m^2)]\left\{s-m^2+im\Gamma\left(\frac{s}{m^2} + \varepsilon\frac{s-m^2}{m^2} + \ldots\right)\right\} \end{aligned} \tag{45}$$

with

$$m\Gamma = \frac{\mathrm{Im}\Pi(m^2)}{1+\mathrm{Re}\Pi'(m^2)} \tag{46}$$

and $\mathrm{Re}\Pi(m^2) = 0$ because of the specific definition of $m \equiv m_Z$ which is adopted. The overall factor $1+\mathrm{Re}\Pi'(m^2)$ is reabsorbed in the numerator. The parameter ε measures the deviation from the scaling behaviour of the s-dependent width. ε is noticeably different from zero only if in the final state of Z decays there are important channels with massive particles. For example, for $Z \to A\bar{A}$, $\varepsilon_A \sim (4m_A^2/m^2)B(Z \to A\bar{A})$. The following identity is valid:

$$\begin{aligned} D(s) &= s-m^2 + im\Gamma\left(\frac{s}{m^2} + \varepsilon\frac{s-m^2}{m^2} + \ldots\right) = \\ &= [1+i\gamma(1+\varepsilon)](s-\bar{m}^2+i\bar{m}\bar{\Gamma}) \end{aligned} \tag{47}$$

with $\gamma = \Gamma/m$ and

$$\bar{m} = m\left(1 - \frac{\gamma^2}{2}(1+\varepsilon) + \ldots\right) \tag{48}$$

$$\bar{\Gamma} = \Gamma\left(1 - \frac{\gamma^2}{2}(1+3\varepsilon) + \ldots\right) \tag{49}$$

As the factor $1+i\gamma(1+\varepsilon) \simeq \exp i\gamma(1+\varepsilon)$ cannot be observed from the absolute square of $D(s)$, we see that, on one hand, at $\varepsilon = 0$ the replacement of the constant Γ with the s-dependent width $s\Gamma/m^2$ leads to a variation of the apparent mass by

$$\delta m = -\tfrac{1}{2}\gamma^2 m \simeq -34 \text{ MeV} \tag{50}$$

which is clearly an important effect [49,50]. On the other hand, the additional effect from ε:

$$\delta m_\varepsilon = -\tfrac{1}{2}\gamma^2\varepsilon m \simeq -34\varepsilon \text{ MeV} \tag{51}$$

is certainly small because ε cannot exceed 10% or so at most. Note that the effect of ε cannot be disentangled from m in the fit so that it leads to an ambiguity in the determination of m. In conclusion ε can be safely neglected at the price of allowing an error of a few MeV on m and Γ.

Of the two parameters R_f and I_f which appear in Eq. (43) for $\hat{\sigma}$ (in addition to the inclusive partial widths Γ_e and Γ_f), I_f like ε, depends only on the spectrum of particles below the Z. In fact, I_f is determined by absorptive parts of vertices and boxes. Thus large deviations from the standard model value cannot occur for I_f. I_f is very small in the standard model (for $\mu^+\mu^-$, $I_\mu \simeq (-(1\div2)10^{-2})$. As it is multiplied by Γ/m, it can be neglected. The main contribution to R_f is already present at the Born level and arises from γ-Z interference. Higher order corrections are relatively important especially for muons. New physics, for example a heavy Z', can modify R_f because the non-resonating background is changed. In the standard model R_f is small (for muons $R_\mu \simeq 4.5\div6\ 10^{-2}$, for hadrons $R_h \simeq (0.75\div1)10^{-1}$). Determining R_f from a fit is difficult. In first approximation, it can be fixed at its standard model value. Realistic deviations from the standard model would not appreciably affect the determination of m and Γ.

In conclusion, a model-independent analysis of the reduced cross-section proves that a modified Breit-Wigner with s-dependent width plus photon exchange and interference is a perfectly adequate basis for the experimental study of the line shape. While in the minimal standard model all the parameters Γ, Γ_e, Γ_f, R_f and I_f can be obtained from m_Z given m_t and m_H, the general expression of $\hat{\sigma}$, in principle, allows a model independent measurement of the Z parameters. Starting from Eq. (43) for $\hat{\sigma}$, approximate analytic solutions of the convolution integral can be found [48,51-53]. The resulting compact analytic expressions are sufficiently precise for most applications.

An indicative set of theoretical predictions on the line shape for $m_Z \simeq 91$ GeV is reported in Tables 3-5 taken from Ref. [25]. The agreement obtained on the line shape by different calculations using slightly different procedures or schemes is better than 0.1% for $\mu^+\mu^-$ corresponding to a few MeV error on m_Z. For hadrons, the uncertainty on Γ introduces by $\alpha_s(m_Z) = 0.11 \pm 0.01$ [4] is $\delta\Gamma = \pm6$ MeV. We see from Tables 4 and 5 that the visible cross-section at the peak for $N_\nu = 3$ is expected to be about 35nb (for reasonable cuts on Bhabha events near the forward direction). This implies that the number of visible Z is expected to be:

Table 3

Partial and total Z widths for $m_Z = 91$ GeV and $\alpha_s = 0.12$ obtained from Ref. [25].

m_t (GeV)	M_H (GeV)	$\sin^2\theta_W$	Γ_Z (MeV)	$\Gamma_{Z\to\nu\bar{\nu}}$ (MeV)	$\Gamma_{Z\to e^+e^-}$ (MeV)	$\Gamma_{Z\to u\bar{u}}$ (MeV)	$\Gamma_{Z\to d\bar{d}}$ (MeV)	$\Gamma_{Z\to}$ (MeV
60	100	0.2366	2460	165.0	82.7	292.7	378.3	376
90	10	0.2312	2463	164.9	82.7	293.4	379.0	375
90	100	0.2328	2466	165.3	82.9	293.6	379.3	376
90	1000	0.2360	2458	165.0	82.7	292.3	377.9	374
150	100	0.2257	2479	166.0	83.3	295.9	382.0	375
200	100	0.2181	2493	166.9	83.7	298.6	385.2	374
230	10	0.2106	2501	167.2	83.8	300.4	387.2	373
230	100	0.2123	2504	167.6	84.0	300.6	387.6	373
230	1000	0.2158	2497	167.3	83.8	299.4	386.2	372

Table 4

The total cross-section including all corrections for $e^+e^- \to \mu^+\mu^-$, using a Z mass of 91 GeV and a minimum for $\sqrt{zs}$ of 0.2 GeV. The results are obtained from Ref. [25].

m_t (GeV)	M_H (GeV)	Γ_Z (MeV)	σmax (nb)	$\sqrt{s}$max (GeV)	$\sqrt{s_-}$ (GeV)	$\sqrt{s_+}$ (GeV)
60	100	2460	1.492	91.090	89.802	92.656
90	10	2463	1.490	91.091	89.801	92.657
90	100	2466	1.492	91.091	89.800	92.659
90	1000	2458	1.493	91.090	89.804	92.654
150	100	2479	1.493	91.091	89.794	92.668
200	100	2493	1.497	91.091	89.786	92.678
230	10	2501	1.499	91.092	89.782	92.682
230	100	2504	1.500	91.092	89.781	92.685
230	1000	2497	1.500	91.091	89.785	92.680

Table 5

The total cross-section including all corrections for $e^+e^- \to$ hadrons, using a Z mass of 91 GeV and a minimum for $\sqrt{zs}$ of 10 GeV. The results are obtained from Ref. [25].

m_t (GeV)	M_H (GeV)	Γ_Z (MeV)	σmax (nb)	$\sqrt{s}$max (GeV)	$\sqrt{s_-}$ (GeV)	$\sqrt{s_+}$ (GeV)
60	100	2460	30.652	91.091	89.820	92.637
90	10	2463	30.653	91.092	89.819	92.639
90	100	2466	30.660	91.092	89.817	92.641
90	1000	2458	30.658	91.091	89.821	92.635
150	100	2479	30.699	91.092	89.811	92.649
200	100	2493	30.761	91.093	89.805	92.659
230	10	2501	30.818	91.093	89.801	92.665
230	100	2504	30.817	91.093	89.800	92.667
230	1000	2497	30.796	91.093	89.803	92.662

$$N_Z = \sigma^{peak}_{visible} \int Ldt = 3.5\cdot10^6 \frac{\int Ldt}{10^{38}cm^{-2}} \qquad (52)$$

At the project luminosity of LEP, $L = 1.7\cdot10^{31}cm^{-2}s^{-1}$, in a "year" of 10^7 s, with an overall efficiency $\varepsilon \sim \frac{1}{2}$ one expects for each experiment

$$N_Z/\text{year} \simeq 3.5\cdot10^6\cdot1.7\cdot0.5 \simeq 3\cdot10^6 \quad (53)$$

At LEP, if everything works well, ν-counting [54] can be completed very quickly. The fastest procedure is to extract N_ν from the peak value of the cross-section assuming the standard model. This method, already followed by MARKII [31], and similar ones based on the partial widths would allow to measure N_ν in the next few months, the expected accuracy being $\delta N_\nu = \pm0.2\text{-}0.3$ with $3pb^{-1}$. A similar precision $N_\nu = 0.25\text{-}0.35$, can also be obtained by collecting a few pb^{-1} a number of GeV above the resonance and looking for $e^+e^- \to \gamma$+nothing.

8. FUTURE PERSPECTIVE ON PRECISION TESTS

The goals of LEP and SLC at longer term are to measure several asymmetries which provide the optimal precision tests at the Z peak. With no longitudinal polarization of the beam, the interesting quantities are the forward-backward asymmetries. A^f_{FB} for $e^+e^- \to f\bar{f}$ with $f = \mu,b,c,\ldots$ and the τ-polarization asymmetry A^τ_{pol}, measured in $e^+e^- \to \tau^+\tau^-$ by reconstructing the τ helicity from the $\tau \to \pi\nu$ decay. At LEP with $\gtrsim 100\ pb^{-1}$ of integrated luminosity, one expects the accuracies reported in Table 6 [55-57]. Note that the quoted error on $\sin^2\theta_W$ actually refers to the quantity $\bar{s}^2_W$ defined in terms of effective couplings at the Z. A more significant representation of the accuracy planned is in terms of plots like in Figs. 8 and 9, where the predicted values of A^μ_{FB} and

Table 6

Experimental errors on asymmetries and $\bar{s}^2_W$ expected at LEP [55] with >100 pb^{-1} of integrated luminosity.

Measurement	Error	$\delta\bar{s}^2_W$
A^μ_{FB}	0.0035	0.0017
A^s_{FB}	0.007	0.0012
A^c_{FB}	0.007	0.0015
A^b_{FB}	0.005	0.0009
A^u_{FB}	0.010	0.0020
A^τ_{pol}	0.011	0.0014

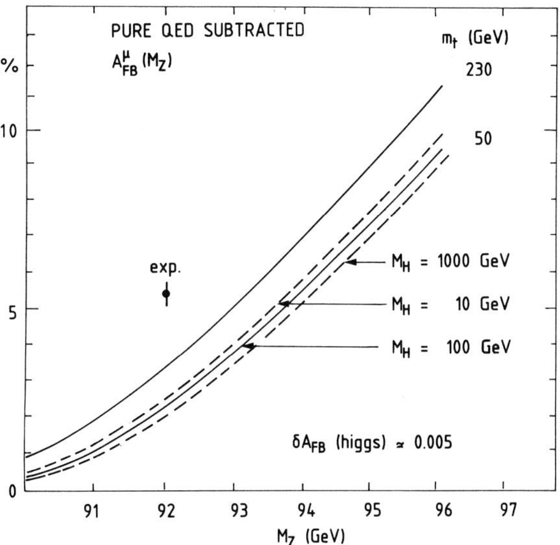

Fig. 8: A^μ_{FB} versus m_Z for given m_t, m_H (from Ref. [56]). The experimental error expected at LEP is also shown for comparison.

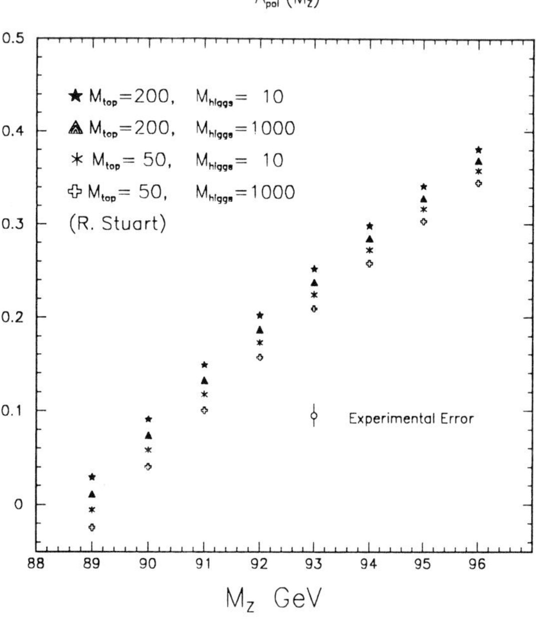

Fig. 9: A^τ_{pol} versus m_Z for given m_t, m_H (from Ref. [57]). The experimental error expected at LEP is also shown for comparison.

A^{τ}_{pol} are shown as functions of m_Z for different m_t and m_H. The curves are obtained [56,57] by using the most updated and complete sets of electroweak radiative corrections. In fact, approximations like the improved Born formulae of Section 5 are not sufficient for precision tests at the level of these asymmetries. The pure QED corrections, being detector dependent, have to be subtracted away [the asymmetries shown are computed from the analogues of the reduced cross-section $\hat{\sigma}$ in Eq. (41)]. The size of the experimental error is also shown. One sees that the expected accuracy is of the order of the effect corresponding to a variation of m_H in the range 10 GeV $\lesssim m_H \lesssim$ 1 TeV. This means that the measurements are indeed probing the fine structure of radiative corrections.

Going back to Table 6, we note that the forward-backward asymmetry for the b quark is really promising. The error shown in Table 6 for A^b_{FB} (which requires $\sim 10^7$ Z events) also includes the uncertainty associated with the correction [23] demanded by B^0-$\bar{B}^0$ mixing. In particular, the mixing parameter averaged over all produced b particles, has to be precisely measured at LEP (e.g., from the ratio of equal to opposite sign dileptons) in order to use this information for extracting A^b_{FB}. An updated analysis can be found in Ref. [23]. Note that A^b_{FB} is also important in another respect. The other asymmetries are all essentially measuring the same vacuum polarization effects. For example, if one plots A^{τ}_{pol} versus A^{μ}_{FB} the resulting curve is practically independent on m_t, as can be seen from Fig. 10. On the contrary A^b_{FB} is also strongly dependent on large corrections from vertex diagrams, as was discussed in Section 5.

It is well known that implementing longitudinally polarized beams is a great advantage for precision tests of the electroweak theory. This is because the left-right asymmetry A_{LR} has the same sensitivity of A^{τ}_{pol} to weak effects but is much easier to measure [55]: one simply measures the difference of the totally inclusive cross-sections from left- and right-handed electrons and then divides by the sum. The

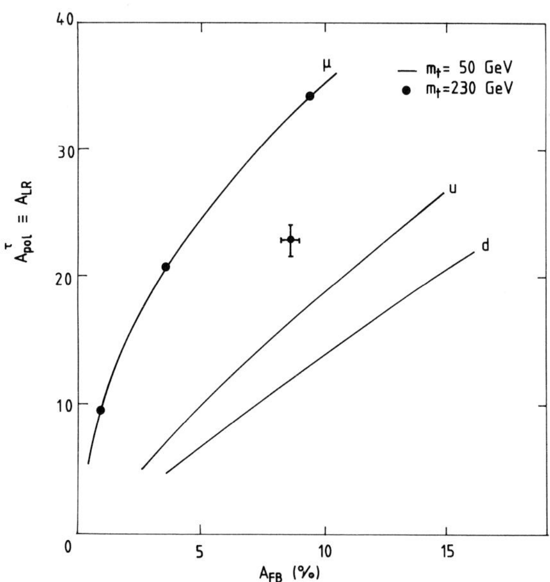

Fig. 10: A^{τ}_{pol} versus A_{FB} (for μ, u, d final states). The curves for m_t = 50 GeV and 230 GeV are practically coincident (from Ref. [56]).

structure of radiative corrections is also much simpler. At LEP with 40pb^{-1} of integrated luminosity and polarization P ≃ 0.5 fixed with accuracy δP ≃ (1%), one can obtain [55] $\delta A_{LR} \simeq \pm 0.003$, corresponding to $\delta \bar{s}^2_W = \pm 0.0004$. This is a very exciting possibility and longitudinal polarization is actually under serious consideration at CERN. But implementing longitudinally polarized beams at LEP is difficult (because of the long time needed to build up the polarization in comparison with the beam lifetime, of depolarizing effects due to the spin time and the energy spread, etc.) hence costly. Also the necessary polarization rotators have to be provided to all four experiments. On the contrary, polarized beams are simpler to be obtained at SLC. Thus the most obvious development of the experimental programme at SLC with the SLD detector is to implement polarization and measure A_{LR}. Even with only 10^4Z, one can expect [55] $\delta A_{LR} \simeq \pm 0.025$ corresponding to $\delta \bar{s}^2 = \pm 0.003$, which is a quite respectable result.

At LEP2, the measurement of m_W with precision $\delta m_W = \pm 100$ MeV will be possible [58]. Hadron colliders can hardly approach a comparable accuracy. In any case the improvements needed in terms of integrated luminosity and/or detectors are such that it would more or less take the same time than waiting for LEP2. A precision $\delta m_W \simeq \pm 100$ MeV is equivalent, given m_Z, to $\delta s_W^2 = \pm 0.002$. Not only is this accuracy of the same order of that obtained from the measurement of asymmetries, but it has the additional advantage of carrying information of a very different qualitative nature.

The search for the Higgs boson [59] is one of the most important goals of LEP. There is no better accelerator than LEP for discovering a Higgs with mass $m_H \lesssim 40\text{-}50$ GeV at LEP1 and $m_H \lesssim 70\text{-}90$ GeV at LEP2. Actually, the Higgs search is a reference problem. More in general, one looks for a solution of the problem of the origin of the Fermi scale of mass $\sim G_F^{-\frac{1}{2}}$. Indeed LEP2 is a formidable discovery machine and I am confident that the low-lying fringes of the rich spectroscopy associated with all conceivable scenarios for new physics will already be observed at LEP2.

In conclusion a very exciting period for phenomenology is in front us. We all hope that the wealth of data which are about to be produced at SLC, LEP, $Sp\bar{p}S$, Tevatron, HERA, etc. will lead to new important breakthroughs in our understanding of particle physics.

REFERENCES

[1] S.L. Glashow, Nucl. Phys. 22 (1961) 579;
S. Weinberg, Phys. Rev. Lett. 19 (1967) 1264;
A. Salam, Proc. 8th Nobel Symposium, Aspenäsgarden, ed. N. Svartholm (Almqvist and Wiksell, Stockholm, 1968), P. 367.

[2] Z Physics at LEP1, eds. G. Altarelli, R. Kleiss and C. Verzegnassi, CERN report 89-08 (1989).

[3] M. Consoli, W. Hollik and F. Jegerlehner, "Electroweak radiative corrections for Z physics", Ref. [2], Vol. 1, p. 7.

[4] G. Altarelli, CERN preprint TH.5290 (1989), Annual Review of Nuclear and Particle Science, Vol. 39 (1989).

[5] See for example: F. Antonelli and L. Maiani, Nucl. Phys. B186 (1981) 269;
S. Bellucci, M. Lusignoli and L. Maiani, Nucl. Phys. B189 (1981) 329.
A recent discussion of "large" terms can be found in G. Gounaris and D. Schildknecht, Z. Phys. C42 (1989) 107.

[6] M. Veltman, Nucl. Phys. B123 (1977) 89;
M.S. Chanowitz, M.A. Furman and I. Hinchliffe, Phys. Lett. 78B (1978) 285.

[7] T. Appelquist and J. Carazzone, Phys. Rev. D11 (1975) 2856.

[8] S.L. Adler, Phys. Rev. 177 (1969) 2426;
J.S. Bell and R. Jackiw, Nuov. Cim. 51 (1969) 47;
see also C. Bouchiat, J. Iliopoulos and Ph. Meyer, Phys. Lett. 38B (1972) 519;
D. Gross and R. Jackiw, Phys. Rev. D6 (1972) 477.

[9] G. Burgers, F. Jegerlehner et al., "Δr or the relation between the electroweak couplings and the weak vector boson masses", Ref. [2], Vol. 1, p. 55.

[10] A. Sirlin, Phys. Rev. D22 (1980) 971;
W.J. Marciano and A. Sirlin, Phys. Rev. D22 (1980) 2695; Phys. Rev. D29 (1984) 75, 945.

[11] M. Consoli, W. Hollik and F. Jegerlehner, CERN preprint TH.5395/-89 (1989).

[12] J.J. van der Bij and F. Hoogeeven, Nucl. Phys. B283 (1987) 477.

[13] B.W. Lynn, M. Peskin and R.G. Stuart, in "Physics at LEP", eds. J. Ellis and R. Peccei, CERN report 86-02 (1986), p. 90.

[14] R. Barbieri and L. Maiani, Nucl. Phys. B224 (1983) 32;
L. Alvarez-Gaumé, J. Polchinski and M. Wise, Nucl. Phys. B221 (1983) 495.

[15] W. Hollik, Z,. Phys. C37 (1988) 569.

[16] M. Veltman, Acta Phys. Pol. B8 (1977) 475.

[17] J. van der Bij and M. Veltman, Nucl. Phys. B231 (1984) 205.

[18] W. Hollik, DESY preprint DESy 88-188 (1988).

[19] D.C. Kennedy and B.W. Lynn, Nucl. Phys. B322 (1989) 1;
D.C. Kennedy et al., Nucl. Phys. B321 (1989) 83;
B.W. Lynn, SU-ITP-867 (1989).

[20] W.J. Marciano and A. Sirlin, Phys. Rev. Lett. 46 (1981) 163
J. Sarantakos, W.J. Marciano and A. Sirlin, Nucl. Phys. B217 (1983) 84.

[21] A. Sirlin, CERN preprint TH.5506/89 (1989).

[22] W. Beenakker and W. Hollik, Z. Phys. C40 (1988) 141.

[23] J.H. Kühn, P.M. Zerwas et al., "Heavy Flavours", Ref. [2], Vol. 1, p. 267.

[24] E.W.N. Glover, J.J. van der Bij et al., "Rare Z decays", Ref. [2], Vol. 2, p. 1.

[25] F.A. Berends et al., "Z Line Shape", Ref. [2], Vol. 1, p. 89.

[26] P. Sinervo, these proceedings.

[27] K. Eggert, these proceedings.

[28] L. Di Lella, these proceedings.

[29] A. Weidberg, these proceedings

[30] M.K. Campbell, these proceedings.

[31] G. Feldman, these proceedings.

[32] J. Ellis and G.L. Fogli, CERN preprints TH.5457/89 and TH.5511/89 (1989).

[33] R.G. Stuart, Z. Phys. C34 (1987) 445.

[34] J.V. Allaby et al., Phys. Lett. 177B (1986) 446; Z. Phys. C36 (1987) 611.

[35] H. Abramowicz et al., Phys. Rev. Lett. 57 (1986) 298;
A. Blondel et al., CERN preprint EP 89-101 (1989).

[36] A. Blondel, CERN preprint EP-89-84 (1989).

[37] M.C. Noecker, B.P. Masterson and C.E. Wieman, Phys. Rev. Lett. 61 (1988) 310.

[38] M.A. Bouchiat et al., J. Phys. 47 (1986) 1709.

[39] J. Panman, these proceedings.

[40] P. Langacker, Univ. of Pennsylvania preprint UPT-0400T (1989).

[41] D. Haidt, DESY preprint, DESY 89-073 (1989) (updated: private communication).

[42] Z. Hioki, Tokushima Univ. preprint 89-05 (1989).

[43] U. Amaldi et al., Phys. Rev. D36 (1987) 1385;
G. Costa et al., Nucl. Phys. B297 (1988) 244.

[44] W.J. Marciano, Brookhaven report BNL-41498 (1988).

[45] A. Weinstein, these proceedings.

[46] See, for example, the articles in "Polarization at LEP", eds. G. Alexander et al., CERN report 88-06 (1988).

[47] E.A. Kuraev and V.S. Fadin, Sov. J. Nucl. Phys. 41 (1985) 466;
G. Altarelli and G. Martinelli, Physics at LEP, eds. J. Ellis and R. Peccei, CERN report 86-02 (1986), Vol. 1, p. 47.

[48] A. Borrelli et al., CERN preprint TH.5441/89 (1989).

[49] W. Wetzel, Nucl. Phys. B227 (1983) 1 and Physics at LEP, eds. J. Ellis and R. Peccei, CERN report 86-02 (1986) Vol. 1, p. 40.

[50] D.Y. Bardin et al., Phys. Lett.B206 (1988) 539.

[51] R.N. Cahn, Phys. Rev. D36 (1987) 2666.

[52] F. Aversa and M. Greco, Frascati preprint LNF-89/025 (PT) (1989).

[53] S.N. Ganguli, A. Gurtu, K. Mazumdar, Tata preprints, TIFR-EHEP/89/2 and 89/3.

[54] L. Trentadue et al., "Neutrino Counting", Ref. [2], Vol. 1, p. 129.

[55] D. Treille, "Polarization at LEP", eds. G. Alexander et al., CERN report 88-06 (1988), vol. 1, p. 265.

[56] M. Böhm and W. Hollik, "Forward-backward asymmetries", Ref. [2], Vol. 1, p. 203.

[57] S. Jadach and Z. Was, "The τ polarization measurement", Ref. [2], Vol. 1, p. 235.

[58] P. Roudeau et al., ECFA workshop on LEP200, eds. A. Böhm and W. Hoogland, CERN report 87-08 (1987), Vol. 1, p. 49.

[59] P.J. Franzini and P. Taxil, "Higgs search", Ref. [2], Vol. 2, p. 58.

DISCUSSION

N. Shumeiko, Byelorussian State University, Minsk: What is the general status of alternative electroweak models — for example, the left-right symmetric models — and similar ideas?

G. Altarelli: At the moment these models are on "stand-by." We do not need them, because the Standard Model is working well. The limits on additional gauge bosons are currently being updated for each model. Maybe we will need these models in the future (for example, if the partial widths of the $Z°$ do not turn out to be in agreement with the Standard Model).

M. Gaillard, University of California and LBL: You did not make any comparisons with predictions of minimal GUTs, SUSY GUTs, etc. Can you comment on the present status of $\sin^2\theta_W$ in this respect?

G. Altarelli: The value of $\sin^2\theta_W$ decreased a bit at this symposium. Thus it is closer to, but still far above, the minimal SU(5) value, and moves slightly away from the minimal supersymmetric value, which (if I remember correctly) is around 0.235.

M. Peskin, SLAC: I think the answer to the previous question was precisely wrong. Grand-unified theories predict the value of $(\sin^2\theta_W)_{\overline{MS}}$, which is roughly equal to $(\sin^2\theta_W)_{Sirlin}$ for low values of the top quark mass. Thus the new values of the $Z°$ mass confirm that the value of $\sin^2\theta_W$ is much higher than that coming from SU(5) grand unification. In fact, even supersymmetric GUTs have trouble attaining such a high value (according to, for example, the work of Kennedy and Lynn).

G. Altarelli: Do you agree with me that $\sin^2\theta_W$ is far away from the predictions of SU(5)?

M. Peskin: Yes, precisely.

D. Schildknecht, University of Bielefeld: I would like to make one comment. You showed various plots with the well-known very weak Higgs-mass dependence of radiative effects together with the accuracy to be obtained in future measurements. Precision measurments are great; however, from the plots it is clear that the bosonic vacuum polarization, which is close to the very heart of the renormalizability of the theory, cannot be quantitatively measured. Even precision data will thus be unable to discriminate the Standard Model from a conceivable (even though unknown) scenario that is different from the standard one on the TeV scale and may also lead to small bosonic vacuum polarization effects. So experimentalists can still hope for deviations from the Standard Model when going on to LEP200 and other new machines on the TeV scale.

S. Brodsky, SLAC: I would like to suggest experiments at LEP where one searches for $e^+e^- \to \gamma Z°$ events at total energies slightly above the $Z°$ mass. The interference with the initial state radiation can lead to sensitive measures of the $Z°$ parameters, real part, axial-vector separation, etc. Has this been studied?

G. Altarelli: For purposes of radiative ν counting, it is planned to run above the resonance and look for $e^+e^- \to \gamma$ + nothing. Thus, if one replaces "nothing" by "something," one can study $e^+e^- \to \gamma + Z°$ as well. However, the radiative ν counting is a fast measurement, so not very high statistics will be collected, while you may want a lot of events.

The Standard Model
at the High Energy Frontier

The Standard Model at the High Energy Frontier

Friday, 11 August 1989

Status of the UA1 Top Search 305

Speaker: K. Eggert

Chairperson: S. P. Denisov

Scientific Secretary: U. Schneekloth

Search for New Particles with the UA2 Detector 318

Speaker: L. DiLella

Chairperson: S. P. Denisov

Scientific Secretary: U. Schneekloth

New Particle Searches at CDF 328

Speaker: P. K. Sinervo

Chairperson: S. P. Denisov

Scientific Secretary: C. Adolphsen

The Production of High P_T Photons and Lepton Pairs 348

Speaker: J. Huston

Chairperson: P. Söding

Scientific Secretary: A. Breakstone

QCD and Jets 368

Speaker: J. Huth

Chairperson: P. Söding

Scientific Secretary: P. Burrows

The Physics of QCD Jets 387

Speaker: V. A. Khoze

Chairperson: R. Peccei

Scientific Secretary: R. Brooks

STATUS OF THE UA1 TOP SEARCH

Karsten Eggert
CERN, Geneva, Switzerland
Representing the UA1 Collaboration

Aachen–Amsterdam (NIKHEF)–Annecy (LAPP)–Birmingham–Boston–CERN–Helsinki–Kiel–Imperial College, London–Queen Mary College, London–Madrid (CIEMAT)–MIT–Padua–Paris (College de France)–Rome–Rutherford Appleton Lab–Saclay (CEN)–UCLA–Vienna

ABSTRACT

The status of the search for new heavy quarks in the UA1 experiment is reported, including the preliminary analysis of the new 4.9 pb^{-1} of data collected in the 1988-89 runs. The search is conducted in a variety of channels, primarily isolated muons accompanied by jets, and dimuons with jets. The properties of these events are consistent with the expectations from the standard physics processes, and do not show a signal for a new heavy quark. Combining all the UA1 data, and all the available channels, the lower limit on the mass of the top quark is 61 GeV/c^2 (95% CL); and a fourth-generation quark (b') is excluded below 43 GeV/c^2 (95% CL). The previous UA1 B^0 $\bar{B}^0$-mixing result has also been confirmed with the order of magnitude increase in statistics; our preliminary result is $\chi = 0.17 \pm 0.05$.

INTRODUCTION

In the framework of the Standard Model, the top quark should exist as the weak Isospin partner of the beauty quark. For example, it is needed to suppress flavour-changing neutral currents involving the beauty quark (GIM mechanism) and to explain the large B^0_d mixing [1]. Furthermore, the electroweak data are sensitive to the top mass due to radiative corrections involving heavy quark loops. The masses of quarks and leptons are not yet theoretically understood and hence no prediction for the top quark mass exists. A global fit to all electroweak data favours a top-mass above the W-mass and below 170 GeV/c^2 [2]. Direct lower limits on the top mass come from electron-positron and proton-antiproton - colliders. The successful running of the SLC allowed the search for pair-produced top quarks in hadronic Z^0 decays. From these data Mark II quotes a lower top mass limit of about 40 GeV/c^2 at 95% C.L., including a possible decay of the top into a charged Higgs [3].

The UA1 Collaboration has made a systematic study of heavy-flavour production in proton-antiproton collisions using muons. The topology of standard heavy flavour (b- and c- quarks) events has been studied in detail and has led to a test of QCD [4], the first observation of B^0 mixing [5], and a measurement of the b-quark production cross-section [6]. The knowledge of standard heavy-flavour topologies allows us to estimate their contribution to the background for a top signal. Using both the electron and the muon channels for various topologies that are characteristic of the semileptonic decay of a top quark, an upper limit for the production cross-section was given as a function of the top mass [7]. Taking into account the contribution from the W decays into top (derived from our measurement of the W - cross-section) and using the calculation [8] of the top cross-section to order α_s^3, a lower limit of m_t = 41 GeV/c^2 (95% C.L.) was obtained [9].

We present here an update on our top search, using a sample which includes the recent 1988/89 data (total integrated luminosity 5.6 pb^{-1}) and which is ten times larger in the muon channel than the one previously used.

THE 1988/89 COLLIDER RUNS WITH UA1

The commissioning of the Antiproton Collector (ACOL) brought about an order of magnitude increase in the luminosity at the CERN Collider. The peak luminosity exceeded 2 x 10^{30} cm^{-2} sec^{-1}, and UA1 collected 1.4 pb^{-1} (3.5 pb^{-1}) of data at $\sqrt{s}$ = 630 GeV in the 1988 (89) runs. The UA1 detector had been modified as follows:

The central electromagnetic calorimeters and the forward calorimeters had been removed in order to prepare for the installation of the new Uranium - TMP calorimeter [10].

The muon detection capability has been improved by the installation of an additional 820 t of iron shielding in the forward region. The muon chambers allow the detection of muons in a pseudorapidity range $|\eta| \leq 2.3$. Muon trigger processors, using the precise drift-time information in a second level and requiring tracks pointing back to the interaction region, enable an inclusive trigger for muons with $|\eta| \leq 1.5$. The muon trigger acceptance is well matched to the detection of heavy quarks, since 80% of all muons from top quark semileptonic decays ($m_t > 40$ GeV/c^2) are in the above rapidity region.

The Central Detector was operated in a new mode in order to cope with the higher currents induced by the increased Collider luminosities. The wire gain was lowered by a factor of 4, which was compensated by a factor of 3 increase in the electronic gain. This resulted in a degradation of the charge-division resolution by about 50%, whereas the resolution in the drift-time measurement remained essentially unchanged at 300 μ.

A powerful new first-level calorimeter trigger processor [11] designed for the upgraded calorimeter was installed and used with the existing hadron calorimeters. At high luminosities the forward muon trigger rate was reduced by requiring a jet of transverse energy (E_t) greater than 10 GeV in coincidence with the muon trigger. In addition, the trigger processor was used to collect a data sample for background calculations by triggering on jets with $E_t > 10$ GeV.

The calorimeter measurements now rely on the hadron calorimeters alone. A detailed study has shown that the present calorimetry is still adequate for measuring missing transverse energy (E_t^{miss}) and jet topologies that are relevant to the search for new heavy quarks. However, the identification of electrons and the measurement of total energy are no longer possible. A large amount of effort was invested in calibrating the central barrel and the end-cap hadron calorimeters using ^{106}Ru source measurements, cosmic muons, minimum bias data, laser calibration, and finally, by studying a complete module in SPS and PS test beams. The conclusions are that the E_t^{miss} resolution is now deteriorated by 15% compared with the previous UA1 calorimeter configuration, and that the estimated systematic error on the jet energy scale is degraded from 8% to 10%; the hadronic energy resolution is essentially unchanged at $\Delta E/E = 0.8 / \sqrt{E}$. The comparison of jet profiles and of jet azimuthal-angle distributions between Monte Carlo and minimum bias data shows that the simulation of jets ($E_t^{jet} > 7$ GeV) in the present UA1 detector is well understood.

PRODUCTION OF HEAVY FLAVOURS

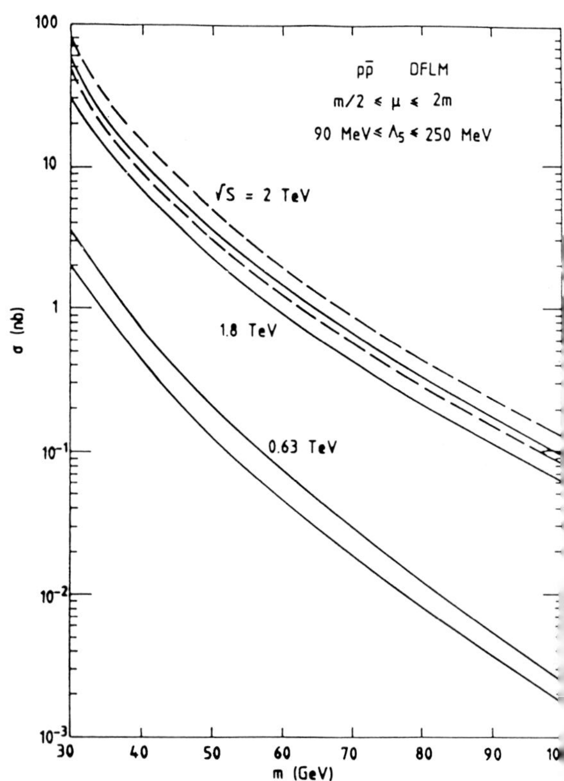

Fig.1 Predicted strong top cross-section fc CERN and FNAL energies versus top mass (taken fro [9])

A precise determination of the top productio cross-section is manditory in order to convert cross-section limit into a mass limit. Calculations of th top and beauty cross-sections in hadronic collisions order α_s^3 are now available [8]. The dominar uncertainties in the cross-sections arise from the choic of the scale, the value of the strong coupling consta

α_s, and the choice of the structure functions. Fig.1 shows the top quark production cross-section for the CERN and FNAL energies using the DFLM structure functions [12]. The bands include the above mentioned uncertainties. However, in the top mass region which we are sensitive to the weak top production via W - decays is dominant (Fig.2) , and we are therefore less sensitive to the uncertainties in the theoretical prediction.

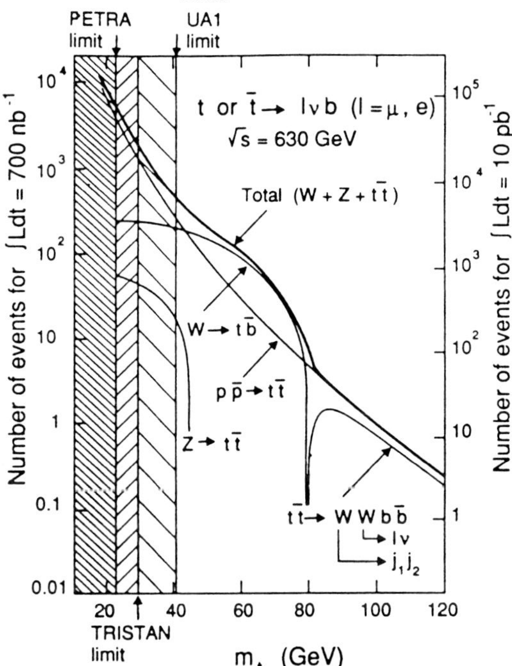

Fig. 2 Number of semileptonic top decay events, as a function of the top mass at √s=0.63 TeV. No experimental cuts or efficiencies are included.

In the UA1 top search beauty quarks constitute one of the main backgrounds. Understanding strong beauty production and the topology of these events is therefore of the highest importance. Recent calculations [8] demonstrated that higher order contributions (α_s^3) like gluon splitting and flavour excitation dominate the strong beauty production. These topologies are similar to the top topologies. In Fig. 3 the transverse momentum dependance of the beauty cross-section, as measured by UA1 [6], is compared to the calculations. Note the excellent agreement . The lowest order beauty production alone only contributes 40% to the total cross-section.

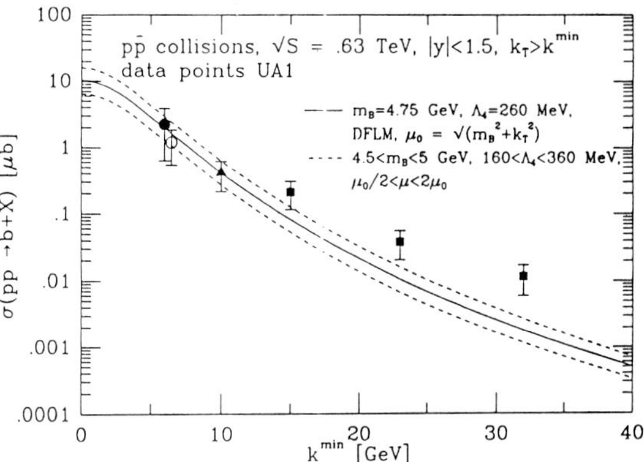

Fig. 3 Transverse momentum distribution of beauty quarks. The UA1 measurements [6] (points) are compared to higher order calculations [8] (line).

TOP SEARCH IN THE SINGLE MUON + JET(S) CHANNEL

In view of the large background of events containing high p_t jets from QCD processes, the best way of identifying a top quark is through its semileptonic decay modes with typical branching ratios of 11%. The inclusive muon p_t spectrum for the initial event sample with p_t> 6 GeV/c is shown after background subtraction in fig. 4. The Monte Carlo prediction for the different known muon sources reproduces both the shape and the normalization of the distribution. The predicted contribution for a top quark of 50 GeV/c^2 is small everywhere (signal to background ratio of ~ 10^{-2}). However, as a consequence of its large mass, top decays have a characteristic topology, which will allow one to discriminate against the other muon sources. The top search is made by selecting events which contain an isolated muon, well seperated from the remaining beauty fragments, accompanied by at least two jets.

The following variables are best suited to differentiate between top and background processes :

1) The muon isolation is measured in the calorimeter and in the central detector in a cone ($\Delta R \leq 0.7$) around the muon. The isolation variable I combines the two measurements and is defined as :

$$I^2 = (\Sigma E_t / 3)^2 + (\Sigma p_t / 2)^2$$

$$\text{in } \Delta R = \sqrt{\Delta\eta^2 + \Delta\phi^2} = 0.7$$

2) The transverse momenta of the jets and the neutrino E_t^{jet1}, E_t^{jet2}, E_t^{ν}.

3) The angle Θ^*_{jet2} between jet2 and the beam direction in the rest system of the muon, jet1, jet2, and the missing energy

4) The azimuthal angular difference between the muon and the fastest jet $\Delta\phi_{\mu,jet1}$

The above variables are shown in fig. 5 which sketches a typical top decay topology.

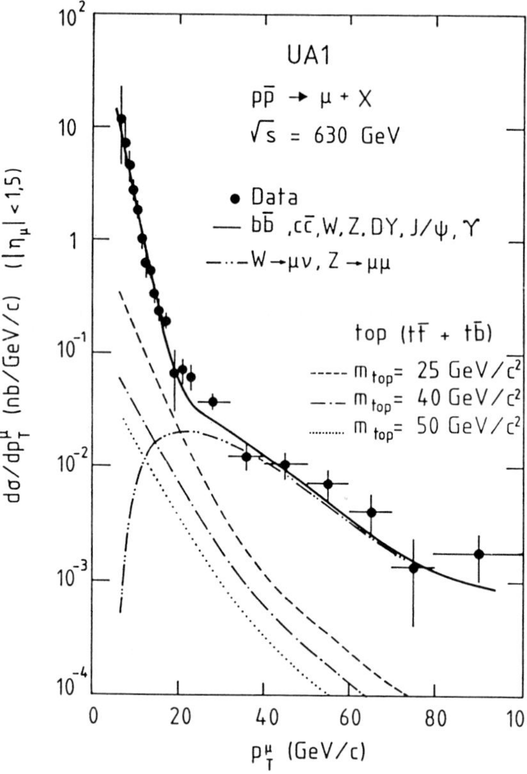

Fig. 4 The inclusive muon p_t spectrum, background subtracted. The solid line is the sum of all known muon sources. Also indicated is the contribution from top quarks.

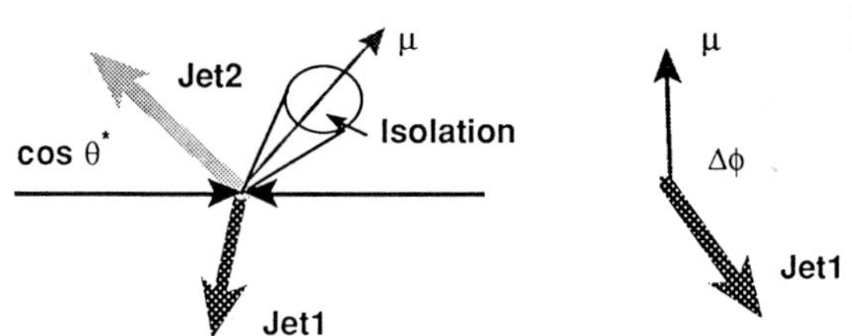

• additional jets are counted, if $E_t^{jet} > 7$ GeV

Fig. 5 Typical event with a semileptonic top decay, e.g. W → tb.

The first step in the search for the top quark is the extensive study of the background which is made with a control data sample dominated by beauty quark production. We selected events which contained one muon with $12 \leq p_t^{\mu} \leq 15$ GeV/c in $|\eta| \leq 1.5$ and at least one jet with $E_t \geq 12$ GeV in $|\eta| \leq 2.5$. The transverse momentum interval of the muons was chosen such that the background contributions from fake muons due to kaon and pion decays and from $W \to \mu\nu$ decays are small.

With the ISAJET Monte Carlo events were generated for all known physics processes contributing to large-p_t muon production. The simulated raw data were then processed through our standard reconstruction program and were compared with our experimental data. As can be concluded from fig. 6 the data are in good agreement with the sum of the absolutely predicted pion and kaon decay backgrounds and the standard QCD processes normalized to the data. The key variable with the best discrimination power in the top search is the isolation I. As expected for charm and beauty production most of the muons are produced inside jets as shown by the distribution of the isolation I

Our analysis suggests that the best sensitivity to a possible top quark signal is obtained by requiring events which contain an isolated muon $I \leq 2$, $p_t^{\mu} > 12$ GeV/c, and at least two accompanying jets with $E_t^{jet1} > 15$ GeV and $E_t^{jet2} > 7$ GeV. We also cut on the transverse mass of the muon-neutrino system ($M_t^{\mu\nu} < 60$ GeV/c^2) to reject $W \to \mu\nu$ background.

Table 1 shows the number of top quark events expected in our data after applying these criteria

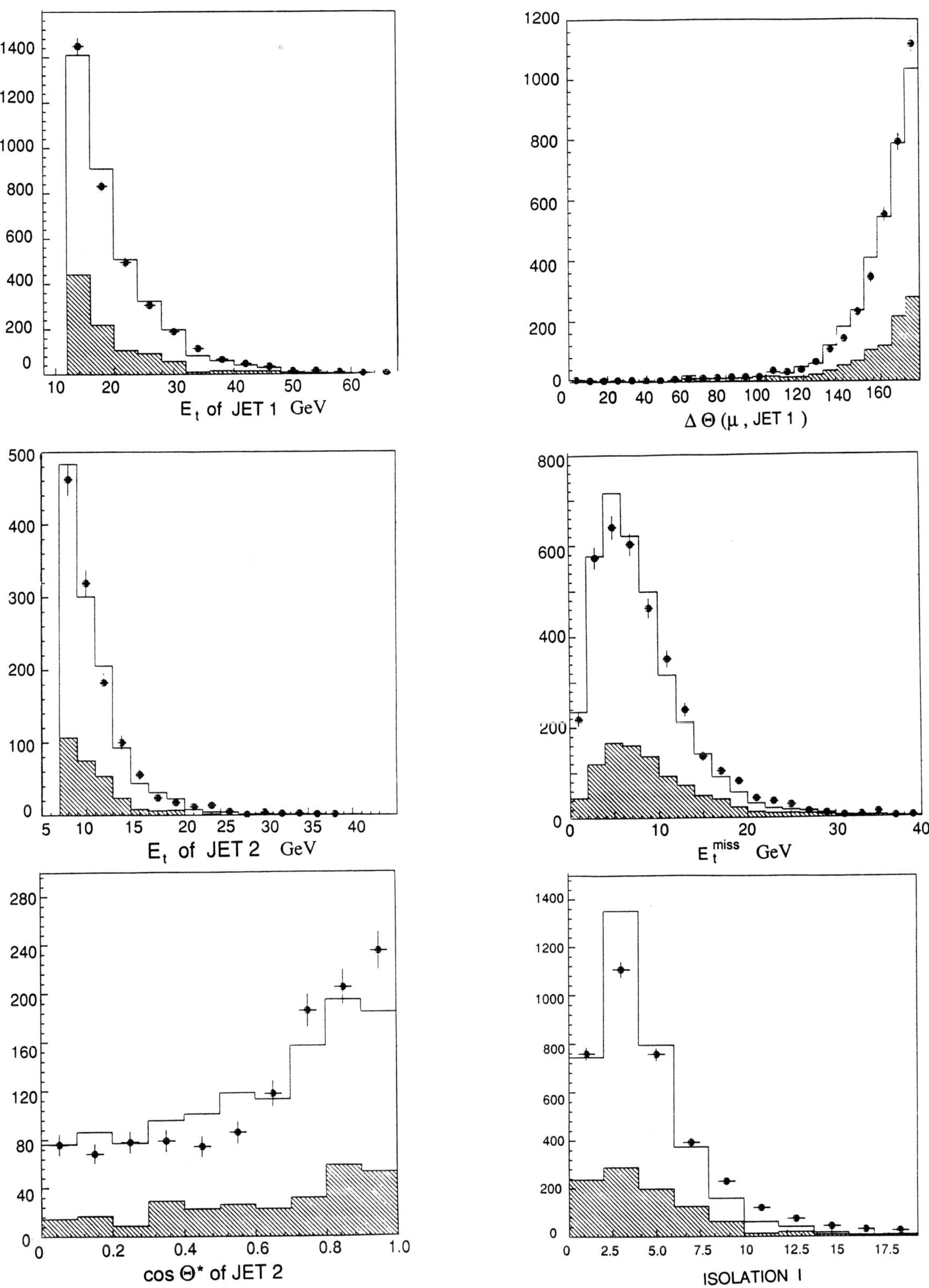

Fig. 6 Comparison of the data (points) in the control sample with the ISAJET calculation (solid line) which includes the relevant physics processes, the decay background is hatched.

Table 1 Number of top quark events expected for various t-masses

TOP MASS (GeV/c²)	40	50	60	70
$t\bar{b}$ $t\bar{t}$	21 18	20 9	10 4	4 1
TOTAL	39 ± 10	29 ± 7	14 ± 4	5 ± 1

Table 2 Comparison between data and an absolute prediction of standard processes

ISOLATED μ + ≥2 JETS

K/π Decay	22 ± 5
W / Z	6 ± 2
D. Y , J/ψ,γ	6 ± 3
$b\bar{b}$ + $c\bar{c}$	44 ± 8
TOTAL	77 ± 10
Data	76

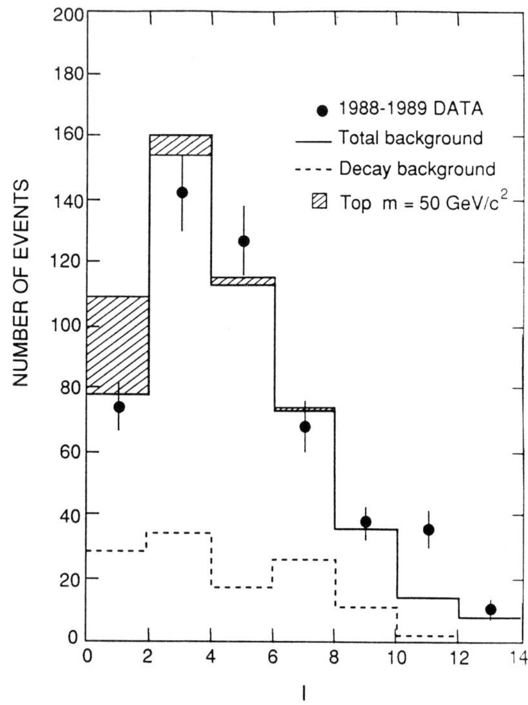

Fig. 7 Isolation distribution for muons which passed the top selection cuts

for various top masses. In table 2 the data are compared with the expectations from background processes.

Fig. 7 shows the isolation distribution for muons which have passed our selection. The data agree well with the absolute prediction of the background. The whole distribution is well reproduced without the need for a contribution from a new heavy quark. For a top mass of 50 GeV/c^2 the signal to background ratio is 0.38 in the isolation bin $I < 2$. In order to reduce the large systematic uncertainty on the contribution from beauty and charm, the background distribution is normalized to the data in the control region $I > 2$, where the above background dominates. Using the large Monte Carlo statistics (50 pb^{-1}) we then extrapolate to determine the background contribution in the signal region $I < 2$. With this procedure, we estimate a total background of 77±10 events in the signal region, where we observe 76 events.

In order to calculate the upper limit on the expected number of top events, we assume Gaussian errors on the predicted number of top and of background events. Table 3 summarizes the individual contributions to these two errors. The main uncertainties in the background come from the limited Monte Carlo statistics and the shape of the isolation distribution, whereas the signal is largely influenced by the uncertainty in the energy scale of the calorimeters. Using Poisson statistics with Gaussian distributions of the number of predicted background events and expected top events we determine the upper limit on the top cross-section, which is plotted in fig. 8 as a function of the top mass. Note that a top mass limit cannot simply be read off

Table 3

SYSTEMATIC ERRORS			
	Δ n = ERROR ON THE PREDICTED NUMBER OF TOP EVENTS		
	Δ b = ERROR ON THE PREDICTED NUMBER OF BACKGROUND EVENTS		
	ERROR	Δ n	Δ b
• ENERGY SCALE	± 10%	± 18	~0*)
• LUMINOSITY	± 8%	± 8%	–
• CROSS SECTIONS			
$W \rightarrow t\bar{b}$	± 5 ± 10%	± 11%	–
$t\bar{t}$	± 30%		–
$W \rightarrow \mu$ + JET	–	–	± 4%
D-Y J/ψ, Y	± 50%	–	
• SELECTION EFFICIENCY	± 8%	8%	–
• DECAY BACKGROUND	± 20%	–	± 5%
• BACKGROUND STAT.	±10%	–	± 10%
• TOP EVENTS STAT.	± 8%	8%	–
• ISOLATION SHAPE	± 20%	–	8%
• GLOBAL	–	25%	14%
*$b\bar{b}$, $c\bar{c}$ NORMALIZED TO THE DATA			

Table 3 The systematic errors which contribute to the top search in the single muon channel for the predicted number of top events Δn and for the background events Δb.

from this graph since it does not include the error on the theoretical top cross-section. However, including the theoretical uncertainty on the prediction of the top cross-section [8] and [9], we obtain the lower mass limit of :

$$\mathbf{m_{top} > 53\ GeV/c^2\ at\ 95\ \%\ C.L.}$$

This limit will slightly improve in the future when a likelihood fit will be made of the individual distributions which will further discriminate top from the background.

TOP SEARCH IN THE DIMUON CHANNEL

In the semileptonic cascade decays of top and beauty quarks multimuons can easily be produced. The following analysis is optimized for $W \rightarrow tb$ decays since this weak top production is dominant for top masses between 40 and 75 GeV/c^2 at $\sqrt{s}$ = 630 GeV. In general, the muon coming from the top quark is expected to have the largest transverse momentum and be isolated, whilst the muon from the recoiling beauty quark should not be isolated. If the two muons both would come directly from the first generation top and beauty decays, they should have the same electrical charge. In fact, this is one of the best signatures for top.

Here we combine all the data collected between 1983 and 1989 corresponding to an integrated luminosity of 5.6 pb^{-1}. We made the following selection:

i) $p_t^{\mu 1} > 8$ GeV/c in $|\eta| \leq 1.6$
ii) $p_t^{\mu 2} > 3$ GeV/c
iii) mass $(\mu 1, \mu 2) > 4$ GeV/c^2
iv) Isolation $I(\mu 1) < 6$, $I(\mu 2) > 2$
v) at least one jet with $E_t > 10$ GeV

The mass cut and the isolation cut of the second muon suppresses substantially the background from J/ψ and Drell-Yan. With this selection, 102 events were found in good agreement with the expected background (tab.4) whereas only 10 top events are expected for a mass m_{top} = 50 GeV/c^2. Hence this selection provides little sensitivity to the top quark production. In order to increase the sensitivity we make use of all the event properties which can differentiate between top and "non-top" events. The variables chosen are the isolation $I(\mu 1)$ of the highest p_t muon ($\mu 1$) in a cone $\Delta R < 0.4$ around the muon, the transverse momentum of muon $\mu 1$, and the difference in the azimuthal angle between the two muons. As can be seen from fig. 9 the data and the background agree with each other in all the three variables. The distributions for a top quark (mass of 50 GeV/c^2) differ considerably. To combine the discrimination power of these three distributions we define a likelihood variable L :

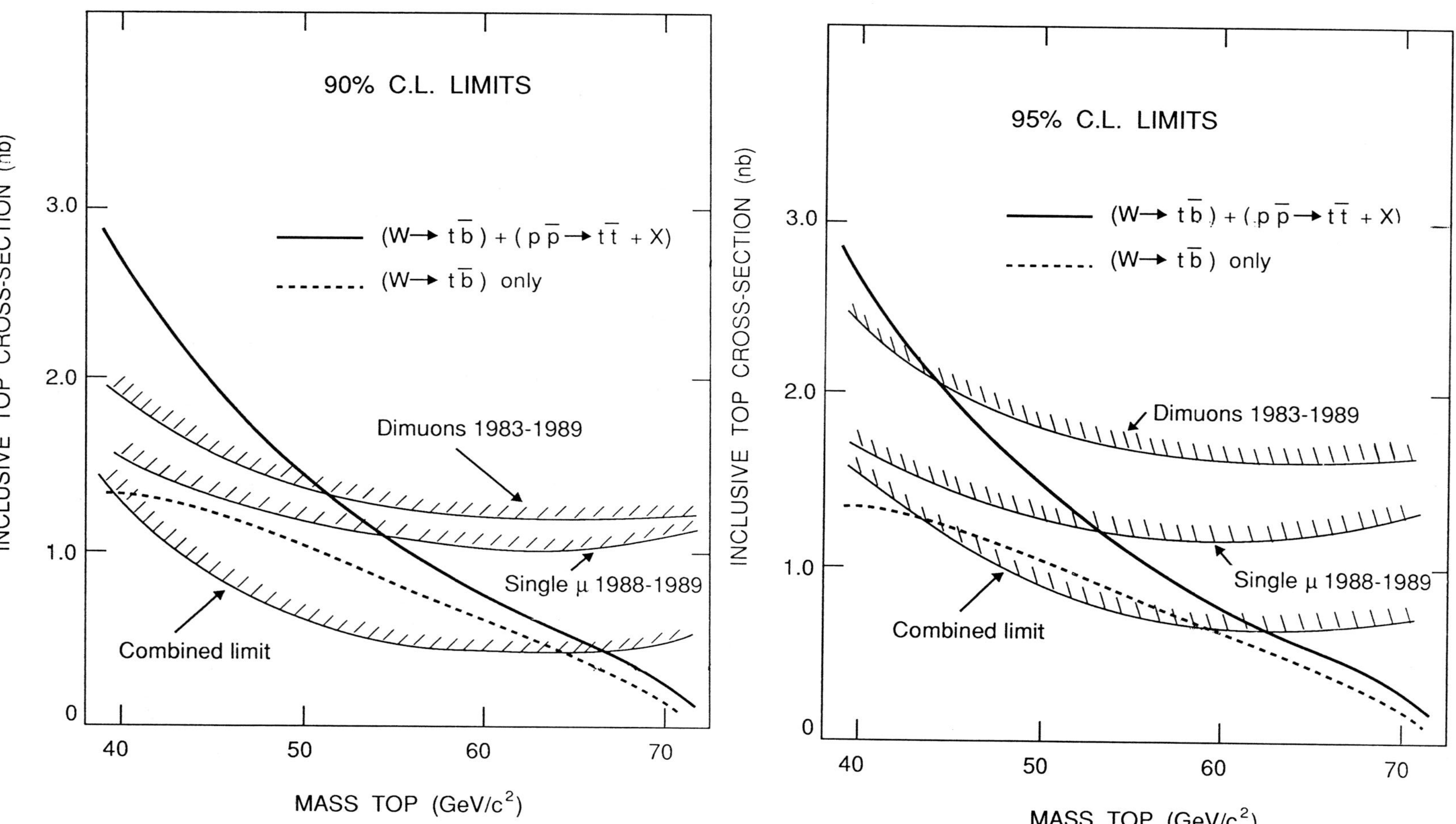

Fig. 8 Upper limit on the top cross-section (90% and 95% C.L.) from the single muon - and the dimuon - data. Also given is the combined limit of all UA1 data from 1983 onwards.

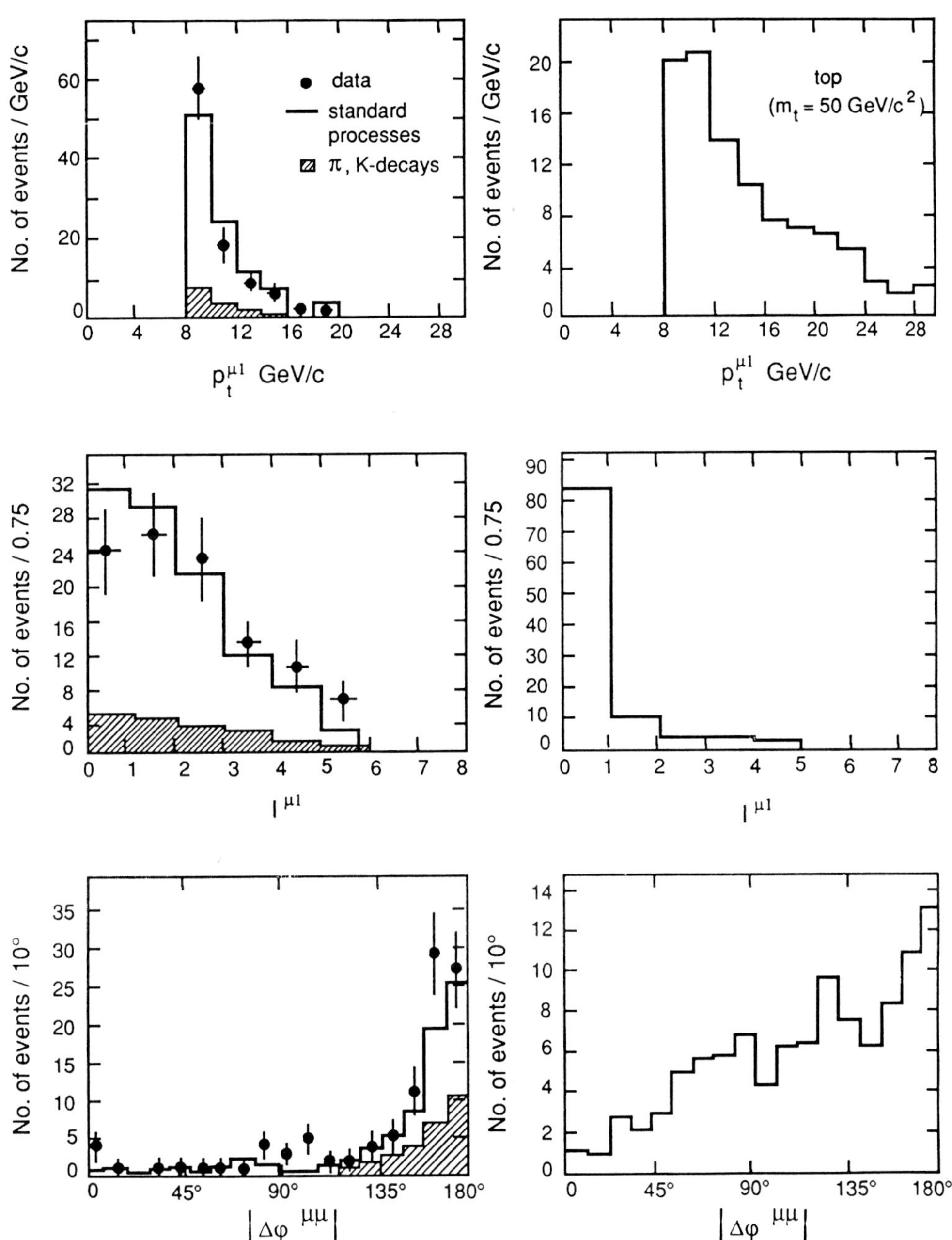

Fig. 9 The distributions of the transverse momentum and the isolation of the fast muon, and the difference of the azimuthal angle of the two muons.

Left side : data and standard backgrounds (decay background is hatched)

Right side : expectations for a top with a mass of 50 GeV/c^2

$$L = \Pi_i P_{top}(x_i) / P_{beauty}(x_i),$$

where P_{top} is the probability density function for the variable x_i for top quark events and P_{beauty} for beauty events. Π indicates the product for all variables of the index i . Fig. 10 shows the distribution of the logarithm of the likelihood L for the data, the predicted background,and the prediction from tb and tt production (multiplied by a factor 10). The background prediction is normalized to the number of events for $\ln L < 1$ by adjusting the beauty cross-section by only 5%. The data do not show a signal for top quark production. If we select events in the signal region $\ln L > 1$, we find 10 events in the data, while we predict 8.9 events from the background and would expect 7.1 events for a top quark with a mass of 50 GeV/c^2. This result can be converted into a lower bound on the top quark mass of:

$$\mathbf{m_{top} > 46\ GeV/c^2\ at\ 95\%\ C.L.}$$

The cross-section limit from this analysis is also given in fig. 8.

COMBINATION OF ALL UA1 TOP LIMITS

Tab. 5 summarizes all the UA1 results on the top quark search for the different channels used. The backgrounds quoted are obtained by normalizing the calculated background to the data in the control region, and then extrapolating it into the signal region for the top. The renormalization factors for the background were always close to 1. The systematic errors listed are the errors from the extrapolation method.

The limits from the old data (1983 -88) have been re-evaluated using the cross-sections from [8] and [9] and combining the statistical and systematic errors in quadrature as for the new data.

All these channels, which do not overlap, are combined to obtain an overall mass limit for the top quark. Most of the systematic errors are correlated between the channels : for instance, the error on the intgrated luminosity and the error on the theoretical production cross-section.We take the most unfavourable

	TOTAL	$\ln L > 1$
π , K-DECAY	16.0 ± 3.3	1.9 ± 0.4
Drell-Yan Y, J/ψ, $W^{\pm}$, Z°	2.3 ±0.6	1.0 ± 0.6
$b\bar{b}$, $c\bar{c}$ (normalized)	85.5 ± 9.5	8.9 ± 2.1
TOTAL	103.8	11.8 ± 2.4
Data (83-89)	102	10
Top m_t = 40 GeV/c^2	14.4	9.0 ± 1.7
50	10.4	7.1 ± 1.3
60	4.6	3.2 ± 0.6
70	1.7	1.3 ± 0.2

Tab. 4 Contributions of the different background sources and expectations for a top signal in the dimuon data sample.

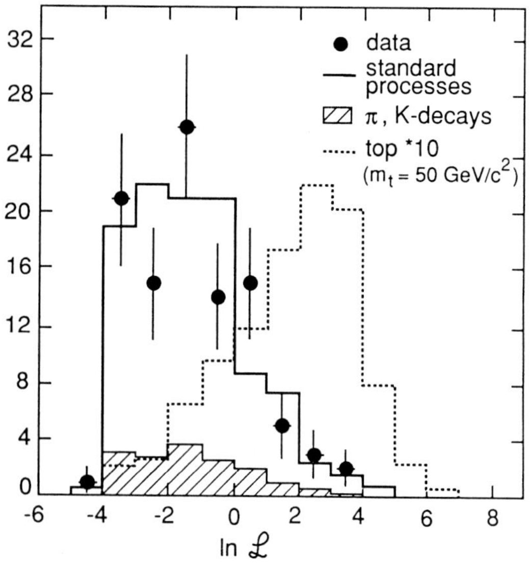

Fig. 10 Distribution of the likelihood for the dimuon events

case where we assume that all errors are correlated. This will give a conservative mass limit.

The limit quoted below is obtained by combining the likelihood distributions from all the independent channels. For each of these channels we define a likelihood which takes into account the statistics in the

CHANNEL	DATA	BACKGROUND + SYST. ERRORS	TOP EXPECTATION $m_t = 50$	LIMIT m_t 95% C.L.
e + JET (85)	26	26.0 + 2.8	8.5	41
μ + JET (85)	10	11.4 +1.2	4.7	40
μ + e (85)	0	1.6 + 0.2	1.2	25
μ + JET (88 + 89)	76	77 + 11	29.0	53
μ μ (83 - 89)	10	11.8 + 2.3	7.1	46
COMBINED LIMIT				61

Tab. 5 Top limits for the different UA1 data samples and the combined limit. The numbers of data - , background - and top - events for $m_{top} = 50$ GeV/c^2 are also given.

signal and in the background regions, and two types of systematic errors: those affecting the signal and those affecting the background. Fig.8 shows the 90% and 95% C.L. top production cross-section limits as a function of the mass. The new UA1 mass limits derived from the cross-section of [9] and our measured W-cross-section are:

$$m_{top} > 61 \text{ GeV}/c^2 \text{ at } 95\% \text{ C.L.}$$

$$m_{top} > 66 \text{ GeV}/c^2 \text{ at } 90\% \text{ C.L.}$$

If we only consider top production via W - decays we obtain the following limit:

$$m_{top} > 58 \text{ GeV}/c^2 \text{ at } 95\% \text{ C.L.}$$

MASS LIMITS ON THE b' QUARK

The single muon + jets data of 5.6 pb^{-1} collected during the 1983-1989 physics runs also affords the opportunity to search for the b'. The same analysis methods and cuts were applied to the data sample as those used in the top search, with the exception that the b' is considered to be produced via direct QCD production only and not via weak decays from the W. The number of expected events for a b' of mass 40 GeV/c^2 is 29, compared to 39 events that were expected for the top. The number of background and data events are the same as in the top analysis. The limits thus obtained are:

$$m_{b'} > 43 \text{ GeV}/c^2 \quad \text{at } 95\% \text{ CL}$$
$$m_{b'} > 45 \text{ GeV}/c^2 \quad \text{at } 90\% \text{ CL}$$

These limits are made with the assumptions that the mass of the b' is less than the mass of the top and that the b' decays only via the charged current.

However, to address the possibility of a flavour changing neutral current component to the b' decay, one must consider different percentages of the decays. This analysis is not efficient for muons from neutral current decays. Assuming that both b' quarks in the event must decay via charged currents the limit on $m_{b'}$ as a function

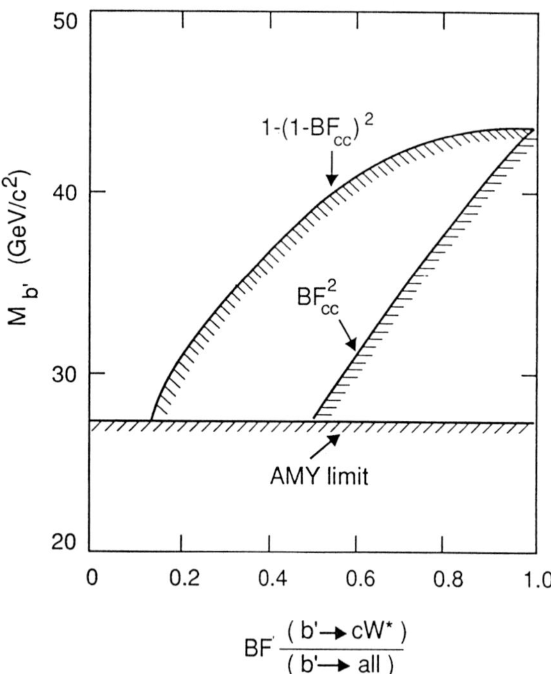

Fig. 11 Excluded b' masses as a function of the branching fraction of charged current decays.

of BF(b' --> cW* / b' --> all) is shown by the curve labelled by BF_{cc}^2 in Fig. 11. However, if we assume the jet topologies for charged and neutral currents to be similar, then our analysis is also sensitive to events with at least one charged current decay in the event. This limit is shown by the upper curve. Also indicated is the mass limit from the search for the b' by the Amy collaboration [13] at KEK.

B_0 - $\overline{B}_0$ MIXING

Time-averaged B_0- $\overline{B}_0$ oscillations have been first observed by the UA1 Collaboration in 1985 [5],[14], using dimuon events with $p_t^{\mu} > 3$ GeV/c. These events mainly originate from pair-produced beauty particles, since muons from charm are strongly suppressed by the softer fragmentation of the charm quark. The small contributions from Drell-Yan and Υ decays can be easily identified, as the muons are isolated. All the above processes lead to unlike-sign dimuons. The experimental signature for beauty mixing is an excess of like-sign dimuon events over the number of like-sign events from secondary charm decays and background. The best measurement of beauty mixing is obtained by fitting the number of like- and unlike-sign muon pairs, taking into account the transverse momenta of the two muons. Second-generation decays generally give one high-p_t muon from the beauty decay with a lower-p_t muon from the charm decay. In contrast, first-generation decays give two muons of approximately equal transverse momenta. Also, the fractional background from decays-in-flight falls rapidly with the transverse momentum.

The mixing is usually expressed by the variable χ :

$$\chi = P(b \rightarrow \overline{b} \rightarrow \mu^+) / P(b \rightarrow \mu^{\pm}) = \frac{N(\overline{B}^0)}{N(\overline{B}^0) + N(B^0)}$$

A likelihood fit used to determine χ gave the following result for the 1985 data:

$$\chi = \mathbf{0.158 \pm 0.059}$$

We performed a preliminary mixing analysis with the same technique using a fraction of the 1988/89 data (3.0 pb^{-1}). Two muons are required with $p_t > 3$ GeV/c and with an invariant mass $m_{\mu\mu} > 6$ GeV/c^2. The latter requirement ensures that the two muons do not originate from the same beauty quark. Furthermore, the muons should not be isolated. We define an isolation variable :

$$S = [\Sigma E_t(\mu_1)]^2 + [\Sigma E_t(\mu_2)]^2 ,$$

where $\Sigma E_t(\mu)$ is the total transverse energy contained in a cone of $\Delta R = 0.7$. Events with $S < 9$ GeV^2 come mainly from Drell-Yan and Υ. We therefore consider only the non-isolated events with $S > 9$, for which beauty decays are the main source. The contributions to the non-isolated dimuon sample is listed in table 6. The

	Data	$\pi/K \rightarrow \mu$	DY, Υ	After subtraction
± ±	482	267 ± 68	–	215
+ –	725	267 ± 68	48 ± 17	410

Tab. 6 Contributions to the non-isolated dimuons

number of Drell-Yan and Υ events is calculated from the observed number of isolated events, allowing for the 18% probability that the events have S > 9. The background from decays-in-flight of pions and kaons is estimated, using single-muon data, by calculating the probability for hadrons in the single-muon events to decay and be reconstructed as high-p_t muons. With a likelihood fit taking into account the transverse momenta of the two muons we obtain from the 1988/89 data:

$$\chi = \mathbf{0.18 \pm 0.08}$$

Combining the results from the old and the new data yields: $\chi = \mathbf{0.17 \pm 0.05}$

Note that the analysis of the 1989 data is preliminary and that we expect to reduce significantly the errors on the measurement. The results are shown in fig. 12, together with measurements of χ from subsamples of the 1989 data using higher cuts on the muon p_t. The results at higher p_t where the background is much reduced are consistent with the lower-p_t results.

UA1 measures a mixing parameter which is averaged over B_d and B_s. Assuming for the probability P_d (P_s) to produce a B_d (B_s) the values 0.36 (0.18) we can present our result in the χ_d - χ_s plane together with the results of ARGUS [1] and CLEO [15] (fig. 13).

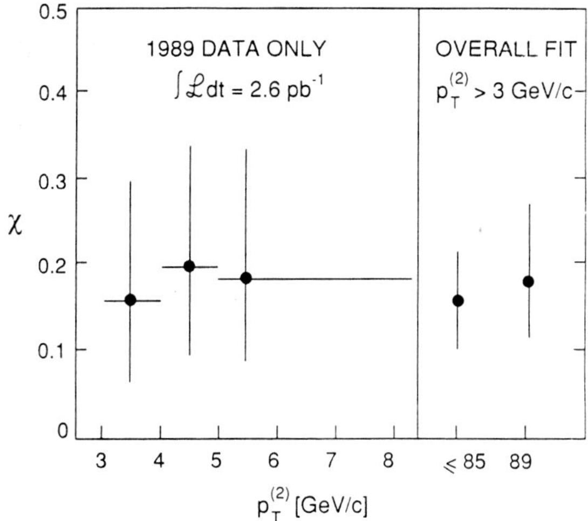

Fig. 12 The mixing variable χ versus the muon p_t

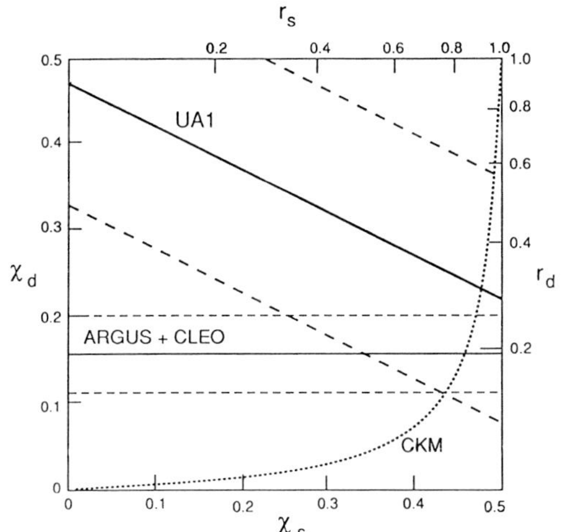

Fig. 13 The mixing result in the χ_d - χ_s plane

REFERENCES

1. ARGUS Collaboration, H. Albrecht et al., Phys. Lett. **192 B** (1987) 245
2. J. Ellis and G.L. Fogli, CERN-TH. 5511/89 D. Haidt, DESY 89-073
3. Mark II Collaboration, A.J. Weinstein, these proceedings
4. UA1 Collaboration, C. Albajar et al., Z. Phys. **C37** (1988) 489 C. Albajar et al., Phys. Lett. **200 B** (1988) 380 C. Albajar et al., Phys. Lett. **186 B** (1987) 237
5. UA1 Collaboration, C. Albajar et al., Phys. Lett. **186 B** (1987) 247
6. UA1 Collaboration, C. Albajar et al., Phys. Lett. **213 B** (1988) 405
7. UA1 Collaboration, C. Albajar et al., Z. Phys. **C 37** (1988) 505
8. P. Nason, S. Dawson, R.K. Ellis, Nucl. Phys. **B 303** (1988) 607
9. G.Altarelli, M.Diemoz,G.Martinelli and P.Nason Nucl. Phys. **B 308** (1988) 724
10. UA1 Collab. CERN UA1 TN/86 - 112 (1986) M. Albrow et al., NIM **A 265** (1988) 303 A. Gonidec et al., CERN - EP/88-36
11. N. Bains et al., report RAL 88-026 (1988)
12. M. Diemoz et al., Z. Phys. **C 39** (1988) 31
13. AMY collaboration S. Eno et al., KEK 89 - 46
14. K. Eggert, Proc. Conf. on New Particles, Madison, (1985), p. 207, ed. by V. Barger

15 CLEO Collaboration M. Artuso et al., Phys. Rev. Lett. **62** (1989) 2233

SEARCH FOR NEW PARTICLES WITH THE UA2 DETECTOR

THE UA2 COLLABORATION

Berne–Cambridge–CERN–Heidelberg–Orsay (LAL)–Milan–Pavia–Perugia–Pisa–Saclay (CEN)

presented by

L. DILELLA

CERN, Geneva, Switzerland

ABSTRACT

The upgraded UA2 detector has collected data corresponding to a total integrated luminosity of 7.5 pb^{-1} from $p\bar{p}$ collisions at a centre-of-mass energy of 630 GeV during 1988 and 1989. A search has been performed for the production and decay of top quarks (t) or a member of a hypothetical fourth family (b′). No evidence has been found for such processes. Using the expected rates for production and decay branching ratios from the Standard Model, this implies that the t-quark mass is greater than 69 (71) GeV/c^2, and that the b′ mass is greater than 54 (57) GeV/c^2, at 95% (90%) confidence.

1. INTRODUCTION

The UA2 detector has been used to search for new particles in $p\bar{p}$ collisions at a centre-of-mass energy $\sqrt{s}$ = 630 GeV. A data sample of integrated luminosity 7.5 pb^{-1} was collected during the 1988 and 1989 runs of the CERN $p\bar{p}$ Collider thanks to the successful construction and operation of the new Antiproton Accumulator Complex (AAC) which enabled the Collider to run at peak luminosities of $\approx 3 \times 10^{30}$ cm^{-2} s^{-1}. These data have been used to search for the signature of t-quark production and semileptonic decay, which is the main subject of this paper. This study provides a lower limit on the mass of the t-quark, m_t, and also on the mass of a possible b′ quark belonging to an additional, as yet undiscovered quark family.

At the $p\bar{p}$ Collider, the t-quark can be produced from two dominant processes, either mediated by the weak interaction ($t\bar{b}$):

$$p\bar{p} \rightarrow W + X\,;\, W \rightarrow t\bar{b} \qquad (1)$$

or by the strong interaction ($t\bar{t}$):

$$p\bar{p} \rightarrow t\bar{t} + X\ . \qquad (2)$$

The cross-section for reaction (1) can be deduced from the cross-section times branching ratio measured in the same experiment [1]. The number of t-quarks produced by such a mechanism, N_t, is then:

$$N_t = 3 \int L\, dt\, \sigma(p\bar{p} \rightarrow W \rightarrow e\nu_e)\, PS(m_t)\, F_{QCD}\ ,$$

where the factor 3 relates the production of coloured quarks to leptons, $\int L\, dt$ is the integrated luminosity, $PS(m_t)$ is a phase-space factor depending on m_t, and F_{QCD} is a correction for higher-order QCD processes which becomes important if this mass is close to that of the W boson, m_W. For example, for m_W = 80.2 GeV/c^2 and m_t = 65 GeV/c^2, F_{QCD} = 1.5 [2]. Since this correction is determined from an incomplete theoretical calculation, it was set equal to the conservative value of 1.0 in this analysis, and only the well understood lowest-order calculation was used to estimate the rate of t-quark production. The dependence of the width of the W on m_t was taken into account. The cross-section (including the phase-space factor) is shown in Fig. 1.

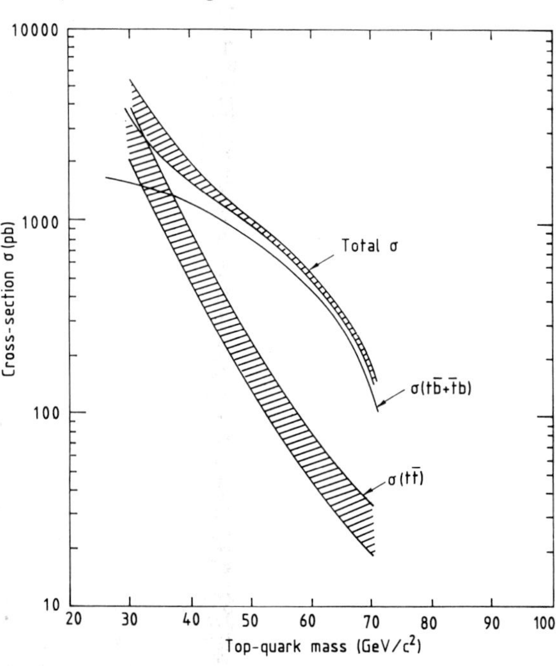

Figure 1 Cross-sections for t-quark production in $p\bar{p}$ interactions at $\sqrt{s}$ = 630 GeV. The shaded regions represent theoretical uncertainties.

The cross-section for reaction (2) has been evaluated [3] using the full next-to-leading order calculation of Ref. [4]. The result is shown in Fig. 1 as a band indicating the theoretical uncertainties.

Decays of the t-quark into final states containing only hadronic jets are very difficult to distinguish from the large background due to QCD processes. The search was therefore performed using the decay mode

$$t \to b e \nu_e ,$$

which has a branching ratio of 1/9 in the Standard Model. In this case, the signature of t-quark production consists of events containing an electron, one or more hadronic jets from the associated t- or b-quarks, and a transverse momentum imbalance (missing p_T) due to the neutrino produced in the t-quark decay. Because of the high m_t considered here, the electron and jets will tend to be produced in the central region. Accordingly, only electron candidates detected in the central calorimeter were considered in this analysis.

2. THE UA2 APPARATUS

The UA2 detector was substantially upgraded between 1985 and 1987. Details of the construction and performance of the various detector elements can be found in the references given below. Only the main features relevant to this analysis will be summarized here.

Calorimetry

An overall view of the UA2 apparatus is shown in Fig. 2. The UA2 central calorimeter [5] was retained with minor modifications. It covers the full azimuthal range, $0° < \phi < 360°$ and the polar angles $40° < \theta < 140°$. Each of the 240 electromagnetic (e.m.) and hadronic cells subtends 10° in θ and 15° in ϕ. The e.m. part is a multilayer sandwich of lead and scintillator, 17 radiation lengths (r.l.) thick, while the hadronic part is an iron–scintillator sandwich, giving a thickness of 4.5 absorption lengths including the e.m. cells.

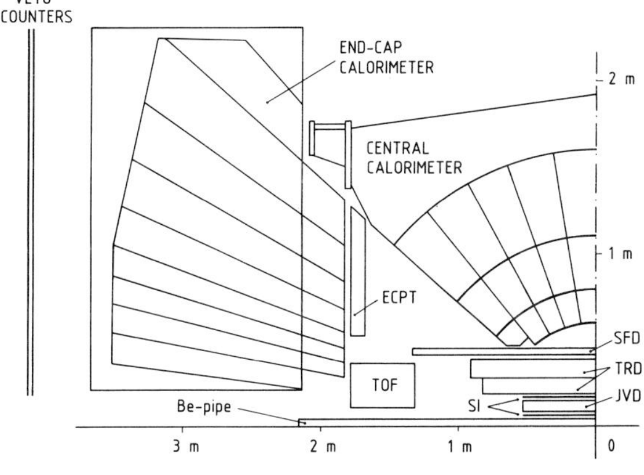

Figure 2 Layout of the UA2 detector.

The end-cap calorimeters (end-caps) [6] cover the pseudorapidity region $1 \leq |\eta| \leq 3$. Each end-cap consists of 12 modules and each module is segmented into 16 cells. In a given module the two cells closest to the beam axis ($2.5 \leq |\eta| \leq 3.0$ and $2.2 \leq |\eta| \leq 2.5$) cover 30° in azimuth. The other cells have a constant segmentation of $\Delta\phi = 15°$, $\Delta\eta = 0.2$. All the cells in the pseudorapidity interval $1.0 \leq |\eta| \leq 2.5$ have one e.m. and one hadronic compartment. The e.m. compartment is a multilayer sandwich of lead (3 mm thick) and acrylic scintillator (4 mm thick), with a total thickness varying from 17.1 to 24.4 r.l. depending on the polar angle. The hadronic compartment is a multilayer sandwich of iron (25 mm thick) and scintillator (4 mm thick) corresponding to ~ 6.5 absorption lengths, including the electromagnetic cells. The cells nearest to the beam have only a hadronic compartment, as well as those covering the pseudorapidity interval $0.9 \leq |\eta| \leq 1.0$ to measure particles escaping from the interface between the end-cap and the central calorimeters (see Fig. 2). Each compartment is read out via two wavelength shifting plates placed on the opposite sides of each cell, introducing a dead space between adjacent cells of 7 mm for the electromagnetic compartments and of 13 mm for the hadronic ones. To minimize the effect of these dead spaces each module is rotated by 50 mrad around the symmetry axis normal to the beam.

Clusters of deposited energy were formed in the calorimeters by joining all cells that have a transverse energy $E_T > 400$ MeV and share a common edge. Clusters with a small lateral size and a leakage into the hadronic compartments consistent with a shower from a single electron were marked as electromagnetic. Since the response of the calorimeter to showers depends on the fraction of the energy carried by hadrons, the observed energy depositions in each calorimeter compartment were multiplied by appropriate factors (of the order of 1.2 for the e.m. compartments), in order to compensate for the difference in response. The efficiency of finding an e.m. cluster from an electron candidate with $E_T > 12$ GeV in the central calorimeter, was measured from test-beam data to be $\epsilon_{cal} = (90 \pm 1)\%$, integrating over all the allowed impact directions for electrons, and accounting for losses due to energy from other particles spoiling the electron signature. The main loss was for electrons which struck an inter-cell boundary, or the shortened e.m. cells at the edge of the calorimeter (see Fig. 2), giving a large hadronic leakage.

The Central Detector

The layout of the central detector is shown in Fig. 2. Around the beam pipe, at radii of 3.5 cm and 14.5 cm, are two arrays of silicon counters (SI) used for tracking and ionization measurements [7]. Between the two is a cylindrical drift chamber, the Jet Vertex Detector or JVD [8]. After the outer silicon array is the Transition Radiation Detector (TRD) [9], consisting of two sets of radiators and proportional chambers, able to distinguish electron tracks from those of hadrons. The particles are tracked onto the calorimeter surface by the Scintillating Fibre Detector (SFD) [10] which consists of fibres arranged on cylinders into six stereo triplets followed by a 1.5 r.l. thick lead converter covering the central calorimeter, and a further two stereo triplets used as a preshower detector.

The position of the event vertex and the charged tracks are reconstructed using the SFD in conjunction with the silicon hodoscopes and the JVD. The fraction of reconstructed vertices within ± 30 cm from the detector centre is measured to be $\epsilon_v = (98 \pm 1)\%$. The tracking efficiency for isolated high-energy electrons is measured to be $\epsilon_{trk} = (90.6 \pm 1.1)\%$, using a sample of electrons produced in the decay of W bosons ('W electrons').

The outer silicon detector is used to reduce the large background from electron pairs arising from photon conversions in the material closer to the beam pipe, and from Dalitz decays. The candidate electron tracks are required to give a signal in an outer silicon counter, with measured charge between 0.6 and 1.6 times that expected from a minimum ionizing particle. The efficiency of this cut is measured to be $\epsilon_{sil} = (73.6 \pm 1.1)\%$, using W electrons.

The tracking and preshower sections of the SFD are used to match the impact point of candidate electron tracks with the position of e.m. showers, with a resolution of $\sigma_{r\phi} = 0.4$ mm in the r–ϕ plane (perpendicular to the beam axis) and $\sigma_z = 1.1$ mm along the beam direction. The quality of a track–preshower match is defined by the variable $d_\sigma^2 = (\Delta_{r\phi}/\sigma_{r\phi})^2 + (\Delta_z/\sigma_z)^2$, where $\Delta_{r\phi}$ and Δ_z are the displacements between the track and shower positions. Accidental overlaps between photon showers and charged tracks give, in general, large values of d_σ^2, while candidate electrons are required to have $d_\sigma^2 < 25$. Preshower clusters for electron candidates are required to have a charge, detected in each of the fibre layers of the preshower detector, of at least twice that expected from a minimum ionizing particle. The efficiency of the track–preshower matching with the above cuts is measured to be $\epsilon_{ps} = (89.9 \pm 1.1)\%$ using W electrons.

The Trigger System

The trigger system [11], based on calorimeter information and signals from the time-of-flight counters (TOF, see Fig. 2), consists of three levels. The first-level trigger uses analogue sums of the signals from the photomultipliers of the e.m. calorimeter cells up to $|\eta| = 2$. At the second level, electron and jet clusters are reconstructed in a fast processor using information from a fast digitization of the calorimeter cell signals. A full calorimeter reconstruction is performed in the third-level processors using a complete set of calibration constants.

Two data samples are used in the analysis. The first is taken from a total of 2.7 pb^{-1} (1.2×10^6 triggers) of data collected during the 1988 run, and consists of all events containing an e.m. cluster with $E_T > 12$ GeV passing the third-level cuts. This gives a trigger rate of ≈ 1 Hz at a luminosity of 2×10^{30}. The second sample is a subset of 3×10^5 triggers taken from 4.4 pb^{-1} of data collected during the 1989 run with a trigger using the above cut, as well as requiring a hadronic cluster with $E_T > 6$ GeV and $p_T^{miss} > 9.5$ GeV/c. At the trigger level, p_T^{miss} is calculated on-line from the relation

$$p_T^{miss} = |\Sigma\, E_{cell}\, \vec{u}_{cell}^{T}| \ ,$$

where E_{cell} is the sum of the e.m. and hadronic transverse energies measured in each cell, $\vec{u}_{cell}^{T}$ is the transverse projection of a unit vector from the centre of the detector to that of the cell, and the sum extends over all calorimeter cells. These additional cuts reduce the third-level trigger rates to ≈ 0.1 Hz.

3. SELECTION OF THE TOP CANDIDATES

Electron Identification

Electron candidates are selected by searching for a track and preshower signal, matching within a tolerance of $d_\sigma^2 < 25$, facing an e.m. cluster. The lateral and longitudinal profiles of the shower are required to be consistent with those expected for a single isolated electron incident along the track direction, using a quality factor $P(\chi^2)$ defined using extensive test-beam measurements. Candidates with $P(\chi^2) < 0.01$, or with an energy greater than 1 GeV in the second hadronic compartment, are rejected. The efficiency of this cut is measured to be $\epsilon_{P(\chi 2)} = (88.7 \pm 0.6)\%$ for W electrons. The electron energy is taken as the sum of the e.m. and first hadronic compartments of those cells expected to contain more than 0.5% of the shower energy from genuine electrons. Only candidates with $p_T^e > 12$ GeV/c are retained.

Multiplying all the quoted efficiencies together, we obtain the overall efficiency to find a W electron with these cuts:

$$\epsilon_e^W = (47.6 \pm 1.6)\% \ .$$

Efficiency of the Electron Selection Cuts in t-Quark Events

Top-quark events are expected to have more complex topologies and lower energy electrons than the W events which are used to determine the efficiencies of the electron cuts (see Section 2). This results in a possible loss of efficiency, which is evaluated as follows.

i) The loss of efficiency of the cut on the shower quality factor $P(\chi^2)$ is studied by overlaying the energy pattern of W electrons onto Monte Carlo simulations of t-quark events, and is found to be $(3 \pm 1)\%$.

ii) The loss of efficiency of the matching between tracking and preshower detectors and the calorimeter is studied using W events with underlying events characterized by either high total transverse energy or high charged-track multiplicity. The loss is found to be $(3 \pm 3)\%$.

iii) A loss of $(2 \pm 1)\%$ associated with a drop in the response of the preshower detector for lower energy electrons is estimated using test-beam data.

Taking these losses into account, the efficiency of finding electrons in semi-electronic t-quark decays is found to be

$$\epsilon_e^t = (43.2 \pm 3.0)\% \ .$$

Neutrino Identification

After reconstructing the event vertex and identifying the electron candidate, p_T^{miss} is recomputed as

$$p_T^{miss} = |-\vec{p}_T^{\,e} - \Sigma E_{cell} \vec{v}_{cell}^{\,T}| \ ,$$

where $\vec{p}_T^{\,e}$ is the reconstructed electron p_T vector, $\vec{v}_{cell}^{\,T}$ is the transverse projection of a unit vector from the event vertex to the cell centre, and the sum extends over all cells not used in the electron definition.

A useful variable for discriminating between various classes of events is the transverse mass of the electron-p_T^{miss} system

$$M_T = \sqrt{2p_T^e \, p_T^{miss} (1 - \cos \Delta\phi_{e\nu})}$$

where $\Delta\phi_{e\nu}$ is the azimuthal angle between the p_T^e and p_T^{miss} vectors. The distribution of p_T^{miss} as a function of M_T is shown in Fig. 3 for the 1988 sample, which is selected by

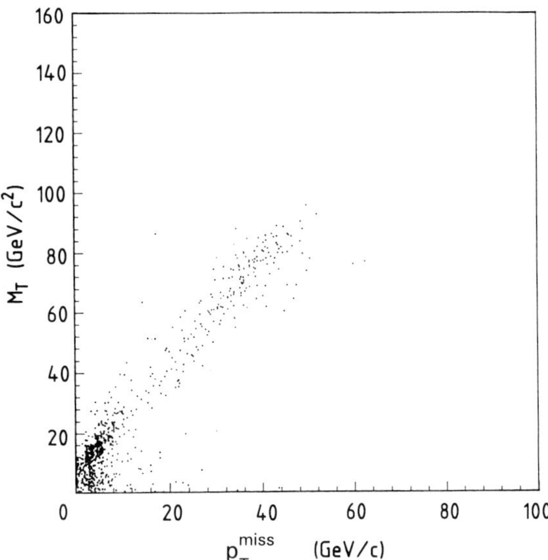

Figure 3 M_T plotted against p_T^{miss} for events with an electron candidate (1988 data only).

applying no cut on p_T^{miss}. The events at low p_T^{miss} and low M_T are dominated by background processes in which a hadronic jet fakes the electron signature and the small value of p_T^{miss} results from detector resolution effects, or particles escaping the acceptance. The p_T^{miss} distribution of these events, integrated over all values of M_T, is shown in Fig. 4. Events consistent with the emission of a high-energy neutrino are selected by requiring $p_T^{miss} >$ 15 GeV/c for both samples.

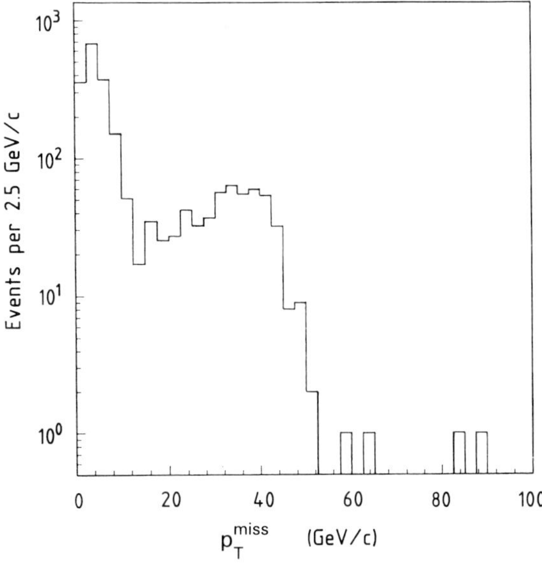

Figure 4 Missing transverse momentum distribution in events with an electron candidate (1988 data only).

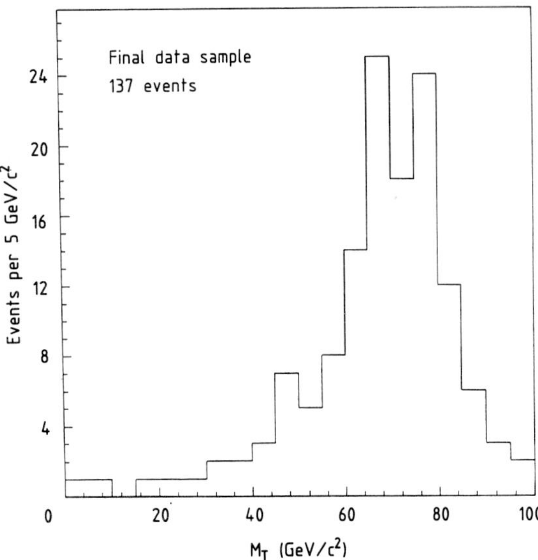

Figure 5 Transverse-mass distribution for the final sample.

Jet Identification

Any cluster failing to pass the electron cuts is considered to be a jet and its energy is defined as the sum of the energies of all cells in the cluster. In this analysis only jets with $|\eta| \leq 2.2$ are retained in order to reduce the background from QCD processes for which the jet angular distribution peaks at large pseudorapidity. Events in which the highest-E_T jet (jet 1) is below 10 GeV are rejected. At this stage the data sample still contains two-jet events in which an electromagnetic jet fakes the electron signature. Such events are expected to show an azimuthal difference $\Delta\phi(e - \text{jet}\,1)$ close to 180° between the electron and the highest-energy jet. Events such that

$$160° < \Delta\phi(e - \text{jet}\,1) < 200°$$

are, therefore, removed. After all the above cuts a total of 58 events are selected from the 1988 data sample, and 79 events from 1989. The M_T distribution for the combined sample of 137 events is shown in Fig. 5.

4. ESTIMATE OF THE EXPECTED SIGNAL

Monte Carlo Simulation of t-Quark Events

The acceptance for t-quark events and their expected M_T distribution are obtained using the EUROJET Monte Carlo [12], which contains the correct matrix elements for higher-order tree-level processes in heavy-quark production (order of α_s for $t\bar{b}$ and order of α_s^3 for $t\bar{t}$).

The presence of these higher-order processes increases the acceptance for t-quark events, since they lead to final states with additional jets. The fact that the theoretical calculation is not complete to all orders should result in an underestimate of the acceptance for events containing an electron and a jet in the final state. The calculation is regularized using a cut-off of 5 GeV on the E_T of additional partons in the higher-order terms. The acceptance is not significantly changed if this cut-off is varied between 2 and 7 GeV.

The t-quark decay in EUROJET is simulated after hadronization into a t-meson or baryon, with branching fractions as expected in the Standard Model for a free quark decay. The fraction z of the t-quark momentum carried by the t-hadron is drawn from the parametrization [13]:

$$f(z)\,dz = N\frac{dz}{z}\left[1 - \frac{1}{z} - \frac{\epsilon_Q}{(1-z)}\right]^2 ,$$

where N is a normalization factor and $\epsilon_Q = (m_Q/m_t)^2$, with m_Q being the mass of the light quark in the t-hadron. Given the large values of m_t considered, the exact value of ϵ_Q does not affect the results.

The b and c hadron decays are generated using extrapolations from known exclusive branching ratios, and the simulations are found to be insensitive to the exact values used. After reconstruction in the calorimeter, jets from hadronic b decays are very similar to those from gluons of the same initial parton energy, and only slightly broader than those from light quarks.

Gluons are fragmented into light-quark pairs, each with an average p_T, $\langle p_T \rangle = 0.4$ GeV/c, relative to the gluon direction, and light-quark fragmentation follows the parametrization of Field and Feynman [14]

$$f(z) = 1 - a + 3a(1-z)^2 ,$$

where, at each step of the fragmentation, z is the fractional longitudinal momentum carried by the generated hadron, which has a p_T distributed with an average value $\langle p_T \rangle = 0.4$ GeV/c. The value of the parameter $a = 0.89$ is chosen to agree with UA1 data on jet fragmentation [15].

Finally, a full simulation is performed of the calorimeter response to all the generated particles, using extensive test-beam measurements with hadron and electron beams, over an energy range from 300 MeV to 150 GeV. The Monte Carlo events are then analysed in the same way as the data.

Systematic Errors in the Acceptance

Several sources of systematic error in the acceptance are considered.

i) The underlying event generated by the EUROJET Monte Carlo is replaced by the energy pattern of minimum-bias events measured in UA2. It is found that a reasonable simulation of the transverse energy of W events is obtained by using the superposition of two minimum-bias events. The systematic error due to this procedure is estimated by comparing the results obtained using one or three minimum-bias events as the underlying event for the generated top signal, and is found to be $\pm 4\%$ for $t\bar{b}$ and $\pm 2\%$ for $t\bar{t}$, for m_t = 65 GeV/c^2.

i) The calorimeter response to hadron jets is very sensitive to the response to low-energy hadrons (< 1 GeV). The measured response curve is adjusted to give the lowest response consistent with the test-beam data, thus reducing the acceptance for events having at least one jet with E_T > 10 GeV. The uncertainty in the absolute energy scale of the calorimeter ($\pm 1\%$ in the electromagnetic and $\pm 2\%$ in the hadronic compartments) is also taken into account by adjusting the response downwards. In the worst case the loss in acceptance is 10% for $t\bar{b}$ and 2% for $t\bar{t}$, for m_t = 65 GeV/c^2, the difference being due to the higher jet multiplicity in the $t\bar{t}$ final state.

ii) The parameters a and $\langle p_T \rangle$ used in the fragmentation functions are varied within limits consistent with the observed energy flow in jets with $E_T \approx$ 10 GeV measured in UA2. In the worst case the loss in acceptance is 2% for both production processes.

All of the above effects are then combined to derive a lower limit for the acceptance. The errors on the integrated luminosity (7.1 ± 0.5 pb^{-1}), on the observed number of $W \rightarrow e\nu_e$ decays, on the electron detection efficiency, and on the Monte Carlo statistics are treated as independent Gaussian errors in the extraction of the m_t limit (see Section 6).

Production Cross-Sections and Acceptances

Table 1 gives the production cross-sections ($\sigma_{t\bar{b}}$ and $\sigma_{t\bar{t}}$) used (see Section 1, Fig. 1). For $\sigma_{t\bar{b}}$ the lower limit in brackets is obtained assuming m_W = 79.8 GeV/c^2, one standard error lower than the value of 80.2 ± 0.4 GeV/c^2, obtained by combining the best value of the Z-boson mass [16] with the average of low-energy measurements of $\sin^2\theta_w$, $\sin^2\theta_w$ = 0.231 ± 0.006 [17]. Because of the small available phase space, this change has a significant effect on $\sigma_{t\bar{b}}$ for t-quark masses above 60 GeV/c^2. We note that most systematic uncertainties on $\sigma_{t\bar{b}}$ have no effect on the estimate of the expected signal, since $\sigma_{t\bar{b}}$ is measured in this same experiment [1].

In the case of $\sigma_{t\bar{t}}$ the lower limit in brackets is taken from Ref. [3], corresponding to about 70% of the central value.

Also shown in Table 1 are the acceptances obtained for each process, defined as the fraction of generated

Table 1

Estimated Acceptance and Signal Rates for Various t-Quark Masses

	Cross-sections (pb)		Acceptance (%)		Expected events	
m_t	$t\bar{b}$	$t\bar{t}$	$t\bar{b}$	$t\bar{t}$	All M_T values	M_T between 15 and 50 GeV/c^2
30	1522 (1522)	3040 (2128)	1.8 (1.5)	1.9 (1.7)	39.2 (26.2)	33.4 (22.4)
40	1211 (1211)	643 (450)	4.1 (3.5)	7.1 (6.4)	39.2 (28.3)	34.9 (25.4)
50	845 (820)	188 (132)	8.1 (7.1)	16.8 (15.3)	38.2 (28.9)	34.4 (26.2)
60	459 (436)	66.9 (46.8)	12.3 (11.0)	25.2 (22.9)	26.7 (20.8)	22.6 (17.6)
63	349 (328)	50.3 (35.2)	12.1 (10.8)	29.6 (27.0)	21.3 (16.3)	17.0 (13.3)
65	283 (263)	40.9 (28.6)	12.4 (11.1)	29.5 (26.9)	17.6 (13.4)	13.6 (10.6)
67	218 (200)	34.4 (24.1)	12.5 (11.1)	32.4 (29.5)	14.5 (10.8)	11.0 (8.2)
70	136 (122)	26.7 (18.7)	15.3 (13.6)	35.5 (32.3)	11.5 (8.5)	8.3 (6.2)

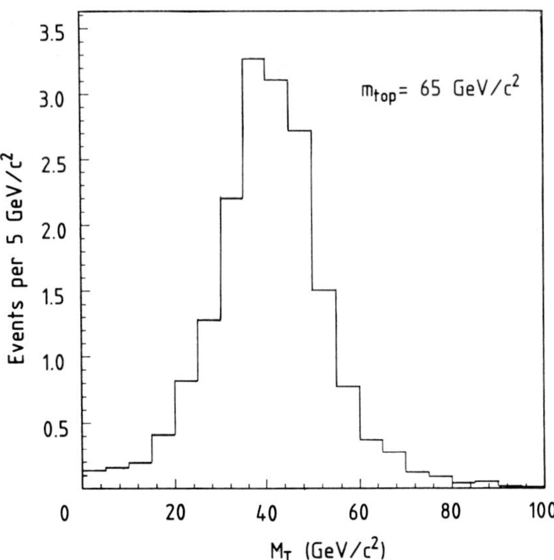

Figure 6 Transverse-mass distribution from t-quark decays with m_t = 65 GeV/c^2.

t-quark events with a semi-electronic t-quark decay which pass the kinematic cuts defined in Section 3. In each case the central value is given, followed (in brackets) by the lowest value consistent with the systematic errors, as discussed above. Because the E_T spectrum of jets from t-quark decays is soft, we obtain a significantly larger acceptance by requiring only one jet to pass the E_T threshold.

Table 1 also gives the number of events expected from both processes after taking into account the electron detection efficiency and the semileptonic branching ratio. The lower limit (in brackets) on the number of events uses the lowest production cross-sections and acceptances. The expected number of events is also given for the transverse mass range $15 < M_T < 50$ GeV/c^2, where most of the signal is expected. As an example the M_T distribution predicted for m_t = 65 GeV/c^2 is shown in Fig. 6.

5. BACKGROUND PROCESSES

$W \to e\nu_e$ and $W \to \tau\nu_\tau$, $\tau \to e\bar{\nu}_e\nu_\tau$

The main source of associated high-energy electrons and neutrinos in the Standard Model is W-boson production and decay via:

$$W \to e\nu_e \quad \text{or} \quad W \to \tau\nu_\tau,\ \tau \to e\bar{\nu}_e\nu_\tau \ .$$

These events ('W events') will enter the sample if the W boson is produced in association with a high-E_T jet. The Jacobian peak expected from W events can clearly be seen in the p_T^{miss} distribution of the data (Fig. 4). The M_T distribution expected for such events is modelled using the EKS Monte Carlo [18], which includes a full order α_s^3 tree-level calculation. Since this variable depends only on p_T^e and p_T^{miss}, it is insensitive to details of the event associated with the W, such as the number of jets or the jet fragmentation model used. However, the total number of events depends on higher-order QCD corrections, fragmentation functions, and the details of the simulation of the detector response to jets. The M_T distribution is therefore normalized to the 105 events observed in the region $M_T > 60$ GeV/c^2, where little t-quark signal is expected, giving an expectation of 148 ± 14 events in the full M_T range — in agreement, within errors, with the EKS prediction of 112 ± 40 events.

$Z \to ee$

Decays of the Z boson to e^+e^- can simulate a large p_T^{miss} if one of the electrons is misidentified as a jet. The e.m. part of its energy is then multiplied by the correction factor described in Section 2, thus generating a spurious momentum imbalance. This process is simulated using the EKS Monte Carlo, which predicts a total of 1.7 ± 0.3 events passing the cuts. Since the electron, 'jet', and neutrino are all due to the e^+e^- pair in the final state, this estimate does not depend on QCD corrections or jet simulation. In order to check this estimate, the cut on $\Delta\phi$(e – jet 1) is removed and all the events are inspected for Z candidates. A total of 14 candidates is found compared with the Monte Carlo prediction of 15 ± 3 events.

$Z \to \tau\bar{\tau}$, $\tau \to e\bar{\nu}_e\nu_\tau$, $\tau \to \bar{\nu}_\tau X$

This process can also give rise to the required signature although the p_T spectrum of the electron is softer than in the previous case. The EKS Monte Carlo gives an estimate of 0.8 ± 0.2 events for the total background from this source.

$p\bar{p} \to b\bar{b} + X$, $b \to e\bar{\nu}_e c$

This process produces electrons by semileptonic b decay in association with a c-quark which fragments into hadrons. Because of the relatively small b mass, the electron will generally fail to pass the initial trigger cuts on the e.m. shower profile, or the cut on $P(\chi^2)$ which is calibrated using isolated electrons. The efficiency for electrons from b decays is estimated to be (16 ± 3)% of that for electrons from t-quark decays, for events passing the topological cuts. In addition, the recoiling $\bar{b}$ jet generally fails to pass the cut on $\Delta\phi$(e – jet 1). Although neutrino and muons produced in the decay chains of the heavy quarks will penetrate the calorimeter and give rise to p_T^{miss}, the cut at $p_T^{miss} > 15$ GeV/c is rarely satisfied. A full simulation is performed using the EUROJET Monte Carlo [12] normalized to the results of the UA1 experiment [19]. The total estimated background is found to be 1.0 ± 0.6 events.

Jets Misidentified as Electrons

Most of the electron candidates with $p_T^{miss} < 10$ GeV/c are misidentified hadronic jets. In order to estimate the contribution of this background to the final sample ($p_T^{miss} > 15$ GeV/c and $E_T^{jet} > 10$ GeV), events are selected which fail to pass the electron cuts. The events with $p_T^{miss} < 10$ GeV/c from the unbiased 1988 sample are used to normalize the samples relative to each other. The number of QCD events (N_{QCD}) in the final sample with $p_T^{miss} >$ 15 GeV/c and $E_T^{jet} > 10$ GeV is then obtained by using the same normalization factor. The M_T distributions of the signal and background samples with $10 < p_T^{miss} <$ 15 GeV/c are found to agree well with each other. The estimate of the QCD background is

$$N_{QCD} = 2.4 \pm 1.5 \text{ events},$$

where the error includes systematic uncertainties. The QCD background for $M_T > 50$ GeV/c^2 is estimated to be negligible.

Summary of the Background Estimates

The various background contributions to the event sample are given in Table 2. Since the t-quark signal is expected to concentrate at intermediate values of M_T the values are also given for the region $15 < M_T <$ 50 GeV/c^2.

Table 2

Summary of the Event Sample and Expected Background

	All M_T values	M_T between 15 and 50 GeV/c^2
$Z \rightarrow e^+e^-, \tau\tau$	2.5 ± 0.6	1.6 ± 0.5
$b\bar{b}$	1.0 ± 0.6	0.5 ± 0.3
QCD	2.4 ± 1.5	2.1 ± 1.5
Total of above backgrounds	5.9 ± 1.7	4.2 ± 1.6
W events	148.5 ± 14.5	22.0 ± 3.0
Total backgrounds	154.5 ± 14.6	26.2 ± 3.4
Observed events	137	17

6. LIMITS ON THE t-QUARK MASS

The number of events observed as well as that expected from background sources is given in Table 2 for the full sample and also for the range $15 < M_T < 50$ GeV/c^2 where the top signal is expected to be concentrated (see Table 1). No excess of events is observed, from which it is concluded that there is no indication for the existence of new quarks in the sample. For $m_t = 65$ GeV/c^2, 13.4 events would be expected in the most pessimistic case (Table 1). Using a simple calculation based on Poisson statistics, accounting for the error on the background estimate [20], this hypothesis is excluded at the 99% confidence level.

Limits on m_t are obtained by comparing the M_T distribution of the observed events with that expected from background sources alone, or in the presence of a t-quark signal of given mass. A likelihood fit is performed with two free parameters giving the fraction of the event sample due to t-quark decays and to background sources other than W events. The exact shape of the M_T distribution for background sources other than $W \rightarrow e\nu_e$ or $\tau\nu_\tau$ decays is rather uncertain, mainly because of the small number of events found in each background sample. However the results are found to be insensitive to the exact shape of the background distribution. The expected signal distribution is taken from the EUROJET Monte Carlo using the appropriate t-quark mass (see Fig. 6 for the distribution relative to $m_t = 65$ GeV/c^2). No normalization is imposed on the number of W events, but for the other background sources the estimate of Table 2, with its error, is used as a constraint. The 90% and 95% confidence limits on the number of t-quark events in the sample are obtained by integrating the likelihood distribution over all possible values of the signal.

For each value of m_t the fitted signal is found consistent with no t-quark production. Figure 7 shows the best fit to the data with no t-quark contribution. The expected contribution from top for $m_t = 65$ GeV/c^2 (13.4 events in the most pessimistic case, see Table 1) is superimposed on the result of this fit. The fit excludes this hypothesis at the 99.0% confidence level.

Figure 8 shows the total expected cross-section for t-quark production, as a function of m_t, using the lower cross-section limits quoted in Table 1. Also shown are the 90% and 95% confidence level cross-sections excluded by

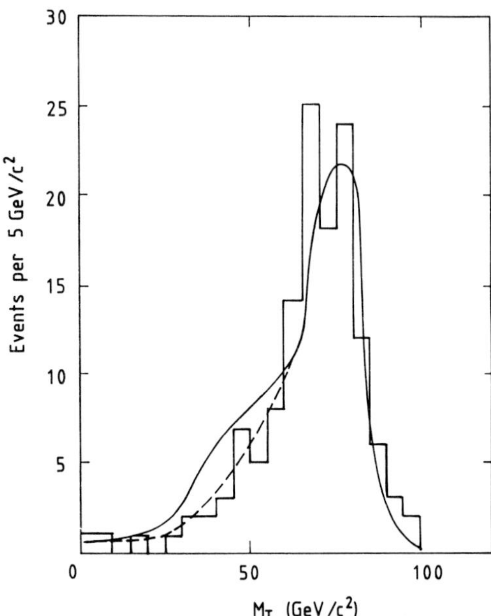

Figure 7 Fits to the experimental M_T distribution (histogram) with no t-quark signal (dashed line) and including t-quark decays with $m_t = 65\ GeV/c^2$ (full line).

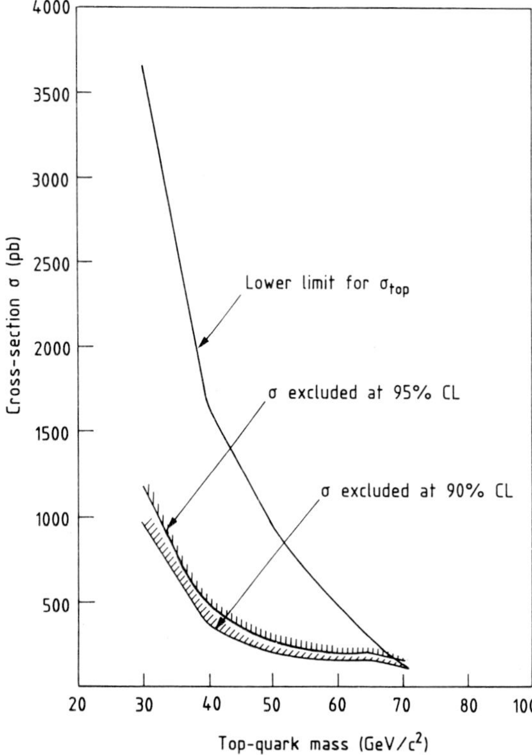

Figure 8 Estimated lower limit for the total cross-section for t-quark production as a function of m_t (full line). Also shown (shaded areas) are the regions excluded at the 90% and 95% confidence levels.

the fit. Top-quark masses smaller than 69 (71) GeV/c^2 are excluded with 95% (90%) confidence. Using the central values for the acceptance and production cross-sections would increase these limits by approximately 3 GeV/c^2.

The fit procedure takes into account the uncertainty on the background estimate. The conservative hypothesis that the total background is equal to the observed number of events would reduce the limits by approximately 4 GeV/c^2.

Since the acceptance is different for top production via reaction (1) or (2), the limits of Fig. 8 are only valid for the cross-sections listed in Table 1. If the t-quark were produced through electroweak processes alone [$\sigma(t\bar{t}) = 0$], we would exclude the region $35 < m_t < 66\ GeV/c^2$ using the lower limits of $\sigma(W \to t\bar{b})$ given in Table 1.

Our results are not very sensitive to the exact value of $\sigma(t\bar{t})$. In particular, the 95% confidence level limit on m_t increases from 69 to only 70 GeV/c^2 if the central prediction for $\sigma(t\bar{t})$ is used instead of the lowest theoretical estimate.

7. LIMITS ON THE b′ MASS

The analysis described above can be used to extract a mass limit on a hypothetical member of a new quark family (b′), assuming that its partner is too heavy to allow production mediated by the weak interaction, that its decay matrix element is identical to that of a t-quark, and that the branching ratio to a c-quark and a virtual W boson is 100%. The excluded region for $\sigma(W \to t\bar{b}) = 0$ gives $m_{b'} > 54$ (57) GeV/c^2 with 95% (90%) confidence. Using the correct matrix element for the b′ decay does not change these limits significantly.

8. CONCLUSIONS

A search has been performed for evidence for production and decays of t- or b′-quarks with the upgraded UA2 detector at the CERN $p\bar{p}$ Collider. No evidence was found for such processes, in agreement with recent preliminary results presented by other experiments [21], providing the following lower bounds on the t- and b′-quark masses:

$$m_t > 69\ GeV/c^2, \quad m_{b'} > 54\ GeV/c^2$$

at the 95% confidence level

and

$$m_t > 71\ GeV/c^2, \quad m_{b'} > 57\ GeV/c^2$$

at the 90% confidence level.

REFERENCES

1. A.R. Weidberg, these Proceedings.
2. L.J. Reinders et al., *Phys. Rep.* **127** (1985) 1.
   T. Alvarez et al., *Nucl. Phys.* **B301** (1988) 1.
3. G. Altarelli et al., *Nucl. Phys.* **308B** (1988) 724.
4. P. Nason, S. Dawson and R.K. Ellis, *Nucl. Phys.* **B303** (1988) 607.
5. A. Beer et al., *Nucl. Instrum. Methods* **224** (1984) 360.
6. P. Bagnaia et al., *The electron, jet and missing transverse energy calorimeter of the upgraded UA2 Experiment at the CERN $p\bar{p}$ Collider* (in preparation).
7. R. Ansari et al., *Nucl. Instrum. Methods* **A279** (1989) 388.
8. F. Bosi et al., preprint CERN-EP/89-82 (1989).
9. B. Merkel, *Proc. 7th Topical Workshop on Proton-Antiproton Collider Physics,* Batavia, Ill., 1988 (World Scientific, Singapore, 1989), p. 182.
10. R.E. Ansorge et al., *Nucl. Instrum Methods* **A273** (1988) 826.
11. K. Einsweiler, *Proc. Int. Conf. on the Impact of Digital Microelectronics and Microprocessors on Particle Physics,* Trieste, 1988 (World Scientific, Singapore, 1988), p. 247.
    S. Stapnes, ibid., p. 254.
12. A. Ali et al., *Nucl. Phys.* **B292** (1987) 1.
13. C. Peterson et al., *Phys. Rev.* **D27** (1983) 105.
14. R.P. Feynman and R.D. Field, *Phys. Rev.* **D18** (1978)3320.
15. G. Arnison et al., *Nucl. Phys.* **B276** (1986) 253.
16. G. Feldman, these Proceedings.
17. D. Haidt, *Status of the electroweak standard model,* DESY 89-073 (1989).
18. S.D. Ellis, R. Kleiss and W.J. Stirling, *Phys. Lett.* **154B** (1985) 435.
19. C. Albajar et al., *Phys. Lett.* **200B** (1988) 380 and **213B** (1988) 405; Z. Phys. **C37** (1988) 489.
20. Particle Data Group, *Phys. Lett.* **204B** (1988) 81.
21. K. Eggert, P. Sinervo, these Proceedings.

NEW PARTICLE SEARCHES AT CDF

The CDF Collaboration[1]

presented by Pekka K. Sinervo
Physics Department, University of Pennsylvania
Philadelphia, PA 19104

Abstract

Preliminary results of searches for new particles at a center of mass energy of 1.8 TeV using the CDF detector are reported. A study of events with missing energy and several jets using $\sim$ 1 pb^{-1} of data suggests that the mass of the squark in a naive supersymmetry model is greater than $\sim$120 GeV/c^2 if the gluino is heavier than the squark. An analysis of 4.4 pb^{-1} of data selecting events with an energetic electron, missing energy, and two or more jets excludes a Standard Model top quark with $40 < M_{top} < 77$ GeV/c^2 at 95% confidence level (CL). A search for events with an energetic electron ($E_T > 15$ GeV) and muon ($P_T > 15$ GeV/c) excludes a top quark with $28 < M_{top} < 72$ GeV/c^2 at 95% CL.

1. Introduction

The Tevatron collider, which produces $p\overline{p}$ collisions at a center of mass energy $\sqrt{s} = 1.8$ TeV, is an excellent place to search for new particles with masses of order 100 GeV/c^2 and greater. At these energies, the Standard Model predicts copious production of the undiscovered top quark, the SU(2) partner to the bottom quark. In addition, extensions to the Standard Model such as Supersymmetry (SUSY) require the existence of a SUSY companion to every one of the known elementary fermions and bosons. These SUSY particles, such as the squark and gluino, are predicted to be pair-produced with strong cross sections in $\overline{p}p$ collisions.

I will summarize the current status of the SUSY particle and top quark searches in the 1988-89 Tevatron collider data collected by the Collider Detector at Fermilab (CDF) collaboration.

2. The CDF Experiment

2.1 The Detector

The CDF detector is a general purpose magnetic spectrometer.[2] It consists of a solenoid magnet of radius 1.5 m and length 5 m that achieves a field strength of 1.41 T, tracking devices that instrument the magnet volume, and calorimeters and muon systems outside the magnet. A vertex time projection chamber (VTPC) provides charged particle tracking out to radius of 0.22 m and covers an angular region of pseudorapidity $|\eta| \lesssim 3$ ($\eta \equiv -\ln\tan\theta/2$ with θ defined as the polar angle relative to the proton beam axis). A central tracking chamber (CTC) consisting of an 84-layer drift chamber fills the rest of the magnet volume. This device, which has both axial and stereo wire layers, provides three-dimensional charged particle tracking and yields a momentum measurement of $\sigma_p = 0.0011p^2$ for particles with $|\eta| \lesssim 1.0$.

A central electromagnetic calorimeter (CEM) is located outside the solenoid magnet and consists of 18 radiation lengths of a lead plate-scintillator sandwich. A wire proportional chamber (WPC) is located at a depth of six radiation lengths and consists of axial wires oriented along the beam (Z) direction and cathode strips placed normal to the wires. This "strip chamber" localizes an electromagnetic shower to an accuracy of 0.2 and 0.5 cm in R-ϕ and Z views, respectively, and measures the shower profile independently in each view. A central hadronic calorimeter (CHA) is located outside the CEM and consists of 5 interaction lengths of a steel plate-scintillator sandwich. The CEM and CHA are segmented into a projective tower geometry with a tower size of $\Delta\eta \times \Delta\phi \sim 0.1 \times 15°$ (ϕ is the azimuthal angle about the proton beam axis). A central muon detector (CMU), located outside the CHA, consists of four layers of drift chambers that identify particles that penetrate through the CEM and CHA. This system provides muon identification in the region $|\eta| \lesssim 0.7$.

The intermediate pseudo-rapidity region ($1.1 < |\eta| < 2.4$) is instrumented with an electromagnetic (EM) calorimeter consisting of lead plates and WPC readout, and a hadronic calorimeter consisting of steel plates. The forward-backward region ($2.3 < |\eta| < 4.2$) is covered with additional EM and hadronic calorimeters constructed of lead and steel plates with WPC readout. Toroid magnets instrumented with six planes of drift chambers provide muon identification in the region $|\eta| \gtrsim 2.0$. A set of scintillator hodoscopes cover the region $3.0 \lesssim |\eta| \lesssim 6.0$ and are used for triggering purposes.

2.2 Triggering

At typical Tevatron luminosities of 10^{30} cm^{-2}s^{-1}, the 50 kHz interaction rate is reduced to 1-3 Hz of recorded events by four levels of trigger decisions. The Level 0 trigger requires either several hits in the BBC counters or several hits in the inner two layers of the CTC. The Level 1 trigger decision is based on an analogue sum of the transverse energy ($E_T \equiv E \sin\theta$ where E is the energy deposited in a calorimeter tower at polar angle θ) deposited in each of the calorimeter "trigger" towers above a set threshold (trigger towers are defined as sums of calorimeter towers that yield an effective tower size of $\Delta\eta \times \Delta\phi \sim 0.2 \times 15°$).

The events passing the Level 1 criteria are examined with the Level 2 system, which consists of several special purpose processors that search for stiff tracks in the CTC, E_T clusters in the calorimeter, and muon stubs in the CMU. The different data are correlated and used to identify electron and muon candidates, and the amount of missing transverse energy ($\not\!\!E_T$) in the event is computed. Various trigger criteria are then used to determine if the event is to be accepted and passed to the Level 3 trigger. This final trigger level is implemented as a set of $\sim$ 55 Motorola M68020 microprocessors that execute a FORTRAN program to analyze the calorimeter and tracking information and make more stringent requirements on the quality of the lepton and jet candidates in the event.

The primary triggers for the detector consist of inclusive electron, muon, jet, and $\not\!\!E_T$ requirements. Several additional triggers require various combinations of these objects (such as the existence of two electrons with somewhat looser requirements than those imposed by the inclusive electron trigger). For example, the CEM electron trigger requires: i) a trigger tower with 3 GeV or more of electromagnetic energy at Level 1; ii) a CEM cluster with $E_T >$ 12 GeV, a track with $P_T \gtrsim$ 6 GeV/c pointing at the cluster, and the ratio of hadronic-to-electromagnetic energy (had/em) in the cluster < 0.125 at Level 2; and iii) the distribution of transverse energy in the cluster to be consistent with that of an electron, and a stiff track with $P_T >$ 6 GeV/c in the event at Level 3.

The efficiencies of the electron and $\not\!\!E_T$ triggers have been determined using W boson events, which typically satisfy both of these two independent triggers. Additional studies have been performed using minimum bias data (*i.e.*, events triggered only by the beam-beam counters) and jet data. For example, the electron trigger is $98 \pm 0.5\%$ efficient for inclusive electrons with $E_T >$ 15 GeV, and the $\not\!\!E_T$ trigger is $> 90\%$ efficient for events with $\not\!\!E_T >$ 25 GeV.

2.3 The 1988-89 Collider Data Run

The CDF collaboration performed an initial engineering run in 1987 where $\sim$ 30 nb^{-1} of data was acquired using only the Level 1 trigger. A year long run during 1988-89 using the full trigger system gathered a data set with an integrated luminosity of $\sim$ 4.4 pb^{-1}. The analyses I will discuss are based on studies of this latter data set.

These data were passed through a special software filter that identified events with either a CEM electron candidate satisfying very loose cuts, or one of several additional criteria, such as a large value of $\not\!\!E_T$, two or more lepton candidates satisfying a loose selection, or a high E_T photon candidate. These events were selected and processed with the full CDF offline reconstruction algorithm, and form the basis of the studies described here. This procedure provided access to the fully reconstructed event samples needed for these analyses within one month after completion of the run.

3. The Supersymmetry Search

3.1 Introduction

Supersymmetry theories postulate the existence of additional fermions and bosons beyond those in the Standard Model, which carry a new quantum number that is conserved in strong and weak interactions. These particles would interact strongly and would be pair-produced in $\bar{p}p$ collisions with large cross sections. Supersymmetry predicts these cross sections to fall rapidly as the mass of the SUSY particle increases. The decays of such particles have rather striking event topologies, with the most interesting being those where the lightest SUSY particle carries off a significant amount of missing energy and where several jets are produced.[3]

Previous searches for such decays have set lower limits on the mass of the squark ($\tilde{q}$) and gluino ($\tilde{g}$).[4,5] An analysis of CDF data taken during the 1987 run yielded even stronger $\tilde{q}$ and $\tilde{g}$ limits of $M_{\tilde{q}} > 73$ GeV/c^2 and $M_{\tilde{g}} > 74$ GeV/c^2.[6] These studies have assumed that the photino ($\tilde{\gamma}$) is the lightest SUSY particle and that the $\tilde{q}$ and $\tilde{g}$ decay primarily in the modes listed in Table 1. The common signature in these decays is missing transverse energy carried off by the non-interacting $\tilde{\gamma}$, and one or more energetic jets from the gluons or quarks. These assumptions are reasonable for light squarks and gluinos, but it has been pointed out that other decay channels are possible when the SUSY particle masses are of order 100 GeV/c^2 and greater.[7,8] These additional decays reduce the sensitivity of searches for large missing energy and jets by a factor that depends on the fraction of the total decay rate into the modes listed in Table 1.[9] In the results quoted below, we have assumed the decay model in Table 1. Further studies taking into account the effects of these "cascade" decay modes is underway.

3.2 SUSY Event Selection

Events are selected by requiring a significant amount of missing transverse energy, $\not{E}_T$, which we define as the magnitude of the vector sum of the transverse energy deposited in the calorimeter towers in the region $|\eta| < 3.6$. Due to shower fluctuations and calorimeter response, the $\not{E}_T$ resolution is $\sim 0.8\sqrt{\sum E_T}$, where $\sum E_T$ is the scalar sum of the transverse energy deposited in the entire calorimeter. Hence, we define the $\not{E}_T$ "significance,"

Table 1. The assumed decay modes of the squark and gluino.

	$M_{\tilde{q}}<M_{\tilde{g}}$	$M_{\tilde{q}}>M_{\tilde{g}}$
$\tilde{q}$ Decay	$\tilde{q}\to q+\tilde{\gamma}$	$\tilde{q}\to q+q+\bar{q}+\tilde{\gamma}$
$\tilde{g}$ Decay	$\tilde{g}\to q+\bar{q}+\tilde{\gamma}$	$\tilde{g}\to q+\bar{q}+\tilde{\gamma}$

$$S \equiv \not{E}_T/\sqrt{\sum E_T}, \qquad (1)$$

and require this to be greater than 2.8.

Jets are identified as clusters of energy in the calorimeter using a cone clustering algorithm with a cone radius $R = \sqrt{(\Delta\eta)^2 + (\Delta\phi)^2}$ of 0.7.[10] We require the measured E_T of each cluster to be greater than 15 GeV and to be within the interval $|\eta| < 3.5$; these requirements are efficient for events from heavy SUSY particle production. We require at least two jets that satisfy these criteria. We also require that at least one of these jets is detected in the central calorimeter (*i.e.* $|\eta| < 1.0$) and that it has an electromagnetic energy fraction (EMF) within the range 0.05 to 0.95; jets with small EMF have high probability of being due to calorimeter defects, while jets with large EMF are most likely electrons or photons.

Cosmic ray events are vetoed by requiring that less than 6 GeV of detected hadronic energy is out-of-time in the event. To further reduce cosmic ray backgrounds, at least 20% of the central jet E_T is required to be observed in the CTC as charged particles in the cone of the jet. We remove the dijet events where mismeasurement of one of the jets has produced a spurious $\not{E}_T$ signature by rejecting any event that has a jet cluster with $E_T > 5$ GeV within 30° of being back-to-back in azimuth with the most energetic jet in the event. We also identify muon candidates and high P_T minimum-ionizing particles and correct the $\not{E}_T$ measured in the calorimeter to take into account the momentum of the muon

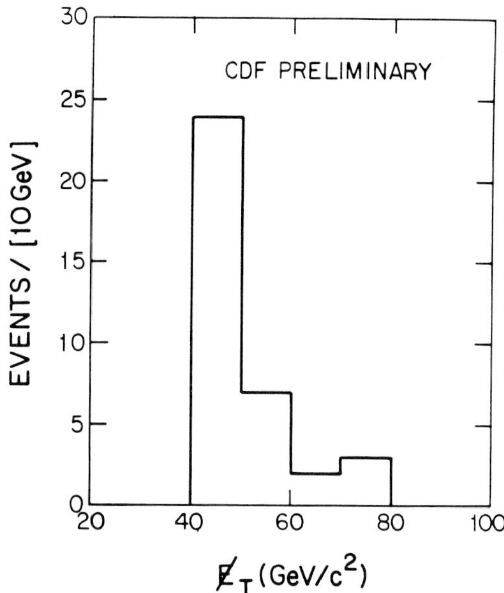

Figure 1. The $\not{E}_T$ distribution for the $\not{E}_T + \geq 2$ jet sample. The events were required to satisfy the $\not{E}_T$ significance cut and to have two or more jets passing the cuts described in the text.

candidate. The $\not{E}_T$ distribution for this $\not{E}_T + \geq 2$ jet sample is shown in Fig. 1. There are 36 events in the sample with $\not{E}_T > 40$ GeV and 5 events with $\not{E}_T > 60$ GeV.

Although these events have the signatures expected of SUSY production, there are also Standard Model sources of background events with similar characteristics. The largest background sources are from the processes listed in Table 2. We have estimated the rate of events from W + jet production where either the electron, μ or τ from the W boson decay has not been detected by measuring the rate of W + jet production in the data and the probability that a lepton from a W decay is undetected. We have estimate the maximum possible rate of events from c and b quark production by generating heavy quarks with the ISAJET Monte Carlo program, processing these events through the full CDF detector simulation and analyzing the resulting sample in the same manner as the data for various top quark masses. The estimated rate of events from all Standard Model sources in the $\not{E}_T + \geq 2$ jet sample is 34 ± 8 events, whereas we see 36 data events. Hence, we conclude that the observed events can be accommodated entirely by Standard Model

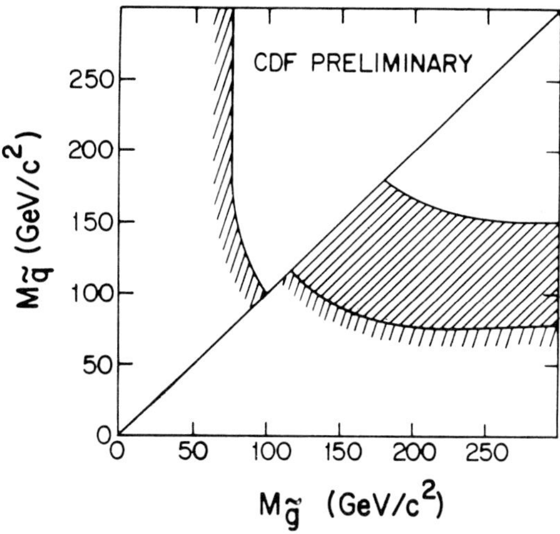

Figure 2. The region of $M_{\tilde{q}}$-$M_{\tilde{g}}$ excluded by the present analysis. The dashed boundary represents the previous CDF lower limits. SUSY particles with masses in the shaded region are considered unlikely.

processes.

To determine an upper limit on the number of events in the data sample that can be attributed to SUSY, we need to pursue several studies. Our background estimates are based on a partial data set and so we are currently analyzing the full sample to make more precise estimates of the background rates. In addition, we are studying the degradation of the $\not{E}_T$ signal due to cascade decays. The shaded region in Fig. 2 is the region in the $M_{\tilde{q}}$-$M_{\tilde{g}}$ plane where our analysis suggests the existence of SUSY particles is unlikely. A complete analysis is in progress.

Table 2. The rate of background events from Standard Model sources.

Process	$\not{E}_T > 40$ GeV	$\not{E}_T > 60$ GeV
$W \rightarrow e\nu_e$	5 ± 3	1 ± 1
$W \rightarrow \mu\nu_\mu$	4 ± 2	0
$W \rightarrow \tau\nu_\tau$	8 ± 4	0
$Z \rightarrow \nu\bar{\nu}$	8 ± 4	4 ± 3
Heavy Quarks	9 ± 4	3 ± 1.4
Total	34 ± 8	8 ± 3

For the case of $M_{\tilde{q}} > M_{\tilde{g}}$, the predominant source of SUSY particles will be the process

$$\bar{p}p \longrightarrow \tilde{g}\bar{\tilde{g}} \quad \hookrightarrow q\bar{q}q\bar{q}\tilde{\gamma}\tilde{\gamma}. \tag{2}$$

The three-body decays of the $\tilde{g}$ typically result in events with less $\not\!\!E_T$ and so the analysis described above is rather insensitive to this scenario. Studies are underway to recover sensitivity in this region.

4. The Top Quark Search

4.1 Introduction

The top quark is an essential ingredient of the Standard Model. Perhaps the strongest experimental evidence for the top quark comes from the observation of a forward-backward asymmetry in $e^+e^- \rightarrow b\bar{b}$.[11,12] In the Standard Model, this asymmetry is a natural consequence of the b quark being in an SU(2) doublet. Other phenomena also suggest the existence of the top quark; the list includes i) the rate of B°-$\bar{B}^\circ$ mixing,[13] ii) the size of the radiative corrections when comparing the values of $\sin^2\theta_W$ measured at different energy scales,[14] and iii) the ratio of the W and Z boson production cross sections.[15]

Standard Model calculations[16] predict that $t\bar{t}$ pair production is the dominant source of top quarks in $\bar{p}p$ collisions at $\sqrt{s} = 1.8$ TeV, in contrast to the situation at the CERN collider energies of $\sqrt{s} = 630$ GeV, where W production followed by the decay $W \rightarrow t\bar{b}$ is the largest source of top quarks. This is illustrated in Fig. 3, where we compare the $t\bar{t}$ cross section calculated to order α_S^3 with the rate of top quarks from W production as a function of top quark mass, M_{top}.[17,18] These calculations predict that for a top quark mass of 80 GeV/c^2, $t\bar{t}$ production would yield

$$4.4\,\text{pb}^{-1} \times 285\,\text{pb} = 1250 \text{ events} \tag{3}$$

in the 1989 CDF data sample.

The Standard Model predicts that the top quark decays via the charged-current interaction, *i.e.* into a W boson and b quark 100% of the time.

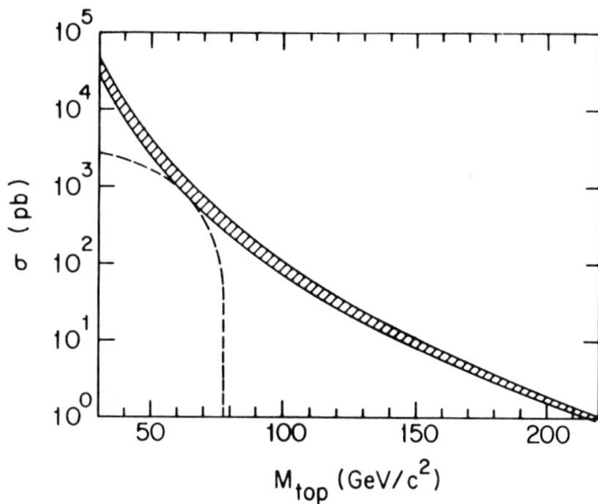

Figure 3. The cross section for top quark production at the Fermilab collider. The shaded band shows the theoretical lower and upper bounds of the rate of top quarks from $t\bar{t}$, while the dashed line represents the rate from W$\rightarrow t\bar{b}$.

In the case where both W's in a $t\bar{t}$ event decay hadronically, the final state consists only of jets. However the rate of such events from top quark production is a small fraction of that expected from the QCD multijet processes, making a search for this decay mode difficult. In the cases where one W decays leptonically, the final state is a lepton, neutrino, and four quarks, while in the case of both W's decaying leptonically, one is left with a final state with two energetic leptons, two neutrinos, and two b quark jets. The presence of the leptons provide a powerful signature as the backgrounds from most of the QCD induced final states are avoided. Of these decay modes, the lepton + jets final states are the most copious as each lepton species corresponds to $\sim$ 15% of the $t\bar{t}$ final states. However, other Standard Model processes such as b quark and W + jet production also contribute to the lepton + jets channel, and these may create formidable signal-to-noise problems in a top quark search.

The dilepton decay modes have the clearest experimental signature, but also have the lowest yields. From rate considerations alone, the $e^\pm\mu^\mp$ final state is the most promising di-lepton mode, as it corresponds to $\sim$ 2.4% of the total $t\bar{t}$ rate, whereas e^+e^- and $\mu^+\mu^-$ individually are only half of this. In addition, the $e^\pm\mu^\mp$ final state does not face contamination from other sources of dileptons such as Drell-Yan production. Although it has only 1/6 of the overall rate of the inclusive electron mode, it's cleanliness makes it an attractive channel for a top quark search.

We have performed a search for the top quark in both the electron + jets and $e^{\pm}\mu^{\mp}$ final states, and we present the preliminary results of these analyses here.

4.2 Electron + Jets Analysis

4.2.1 Introduction

The CDF detector is well suited to search for top quarks decaying into the electron + jets final state, due to the excellent electron identification afforded by the CEM and the good $\not{E}_T$ resolution of the device. However, as we show below, this channel also has a large background resulting from the QCD production of W bosons with associated jets. To overcome this problem, we have used the transverse mass variable to efficiently discriminate between W + jet and $t\bar{t}$ events..

We first describe the data selection used to define the electron + jets data sample, and then discuss the transverse mass analysis.

4.2.2 Electron Selection

The data sample for the electron + jets analysis comes from events that satisfy the inclusive electron trigger described earlier. We select these events by requiring an electron candidate in the CEM, which we define as an electromagnetic cluster with $E_T >$ 15 GeV, where a cluster is 1 to 3 CEM towers adjacent in η (the structure of the CEM prevents electron showers from spreading across towers adjacent in azimuth). We further require that i) the ratio of hadronic-to-electromagnetic energy, had/em, in the cluster be less than 0.05 and the distribution of transverse energy in the cluster be consistent with that of an electron; ii) a CTC track point at the cluster and that the ratio of calorimeter E_T to the track transverse momentum P_T (E/P) be less than 1.5; iii) the track position extrapolated into the CEM match the position of the electromagnetic shower observed in the CEM strip chamber to within 1.5 cm in R-ϕ and 3.0 cm in the beam direction and that the strip chamber

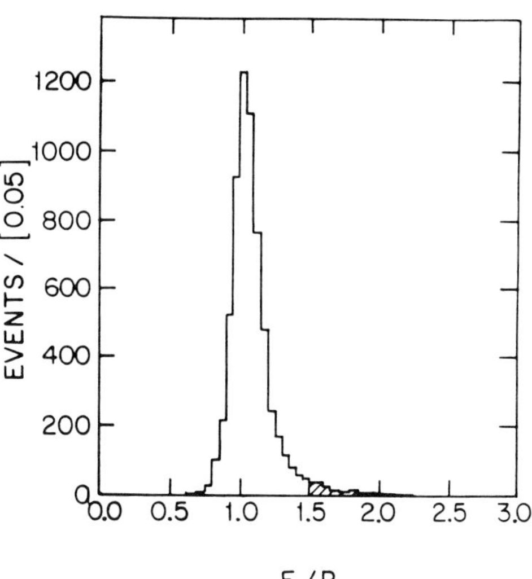

Figure 4. The E/P distribution for the electron candidates with $E_T > 15$ GeV. The shaded region is excluded by the cut on this variable. The final calibration corrections to the measured energy and momentum have not been applied.

shower profiles be consistent with an electron; and iv) the position of the electromagnetic shower be well away from the edge of the calorimeter towers to ensure that its energy is well measured.

This electron selection results in a data sample of 17,200 electron candidates, corresponding to an observed cross section of $\sim$ 4 nb. The E/P and had/em distributions prior to cutting on these two variables are shown in Fig. 4 and Fig. 5, respectively. The small tail of events with E/P > 1.5 indicates that the background of fake electrons from $\pi^{\pm}\pi^{0}$ overlap is virtually non-existent, as this background has an approximately uniform distribution for E/P $\gtrsim$ 1.0. We can use the had/em distribution to estimate the amount of background in the sample arising from high P_T charged pions that interact in the CEM to produce an electromagnetic shower, as such background events have a flat had/em distribution, whereas test beam measurements show that only 5% of real electrons have $had/em > 0.05$. Using this approach, we estimate the size of the π interaction background in the inclusive electron sample to be $8 \pm 3\%$.

We have measured the efficiency of this selection for inclusive electrons with E_T > 20

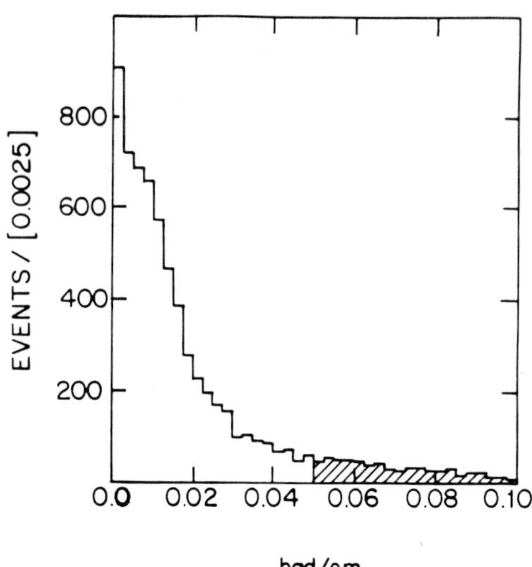

Figure 5. The *had/em* distribution for the electron candidates with $E_T > 15$ GeV. The shaded region is excluded by the electron selection.

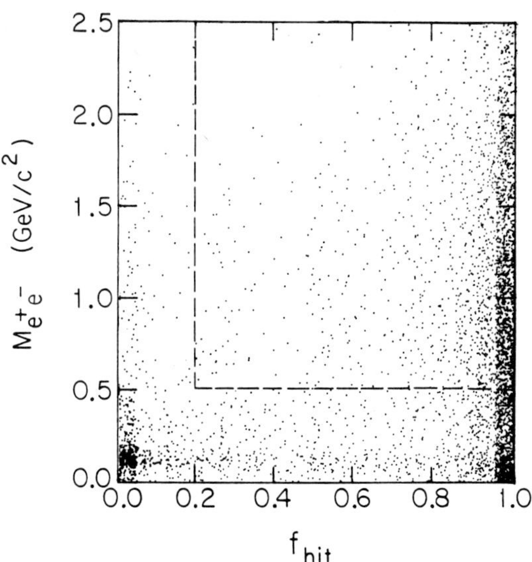

Figure 6. The distribution of effective electron pair mass versus VTPC hit fraction for inclusive electrons with $E_T > 15$ GeV. The dashed line shows the cuts used to identify electrons from conversions and Dalitz decays.

using electrons from W boson decays (where the requirement of a large amount of $\not\!\!E_T$ is sufficient to select a clean sample of $W \rightarrow e\overline{\nu_e}$ decays) and electrons from Z° boson decays. The selection efficiency is $77 \pm 3\%$.

Electrons from photon conversions in the inner and outer layers of the VTPC and from π° Dalitz decays form a significant background to "prompt" electrons. We identify conversions in the outer layer by requiring the existence of a reasonable fraction of the expected hits in the VTPC, f_{hit}, and identify conversions in the VTPC inner layer and Dalitz decays by searching for a second track that forms an effective e^+e^- mass, M_{ee}, less than 0.5 GeV/c^2. We show the distribution of M_{ee} versus f_{hit} in Fig. 6 for the electron candidates where a nearby oppositely-charged CTC track has been found. This distribution shows two clusters of events at low M_{ee}; one cluster is at large f_{hit} and consists of electrons from Dalitz decays and conversions in the beam pipe and inner wall of the VTPC, and the other is at small f_{hit} and consists of electrons from conversions at the outer wall of the VTPC. The diffuse band of events at $f_{hit}\sim 1$ and larger M_{ee} is from "prompt" electrons. We define a conversion electron sample by requiring either $f_{hit} < 0.2$ or $M_{ee} < 0.5$ GeV/c^2, and reject these events. The cut on M_{ee} identifies $\sim 75\%$ of the conversions and Dalitz decays, whereas the cut on f_{hit} is essentially 100% efficient at identifying conversions that occur at the outer wall of the VTPC. This yields an overall conversion finding efficiency of $87 \pm 2\%$ with an associated inefficiency for real electrons of $7 \pm 1\%$. This cut rejects $\sim$ 30% of the inclusive electrons, and we estimate from these measured efficiencies that the resulting inclusive electron sample has a remaining "unseen" conversion background of $\sim$ 9%. There are $\sim$11,700 events that survive this cut.

A cut requiring the event interaction vertex along the beam direction to be within 60 cm of the nominal collision point reduces the event sample to $\sim$11,100 events. This cut ensures that the events are fully contained within the detector.

The production of Z° bosons are an additional source of electrons in our event sample. We remove them by rejecting those events that contain a second electromagnetic cluster forming an effective electron pair mass with the electron candidate greater than 70 GeV/c^2. Figure 7 shows this effective mass distribution before and after the Z° removal. This cut removes 480 events, leaving an inclusive electron sample of 10,644 events.

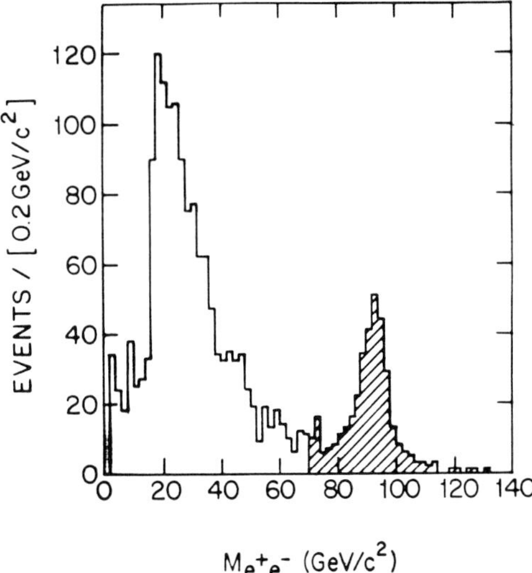

Figure 7. The M_{ee} effective mass distribution for events with a second electromagnetic cluster. The events in the shaded region are removed to reject backgrounds to the top quark search from Z°+jet production.

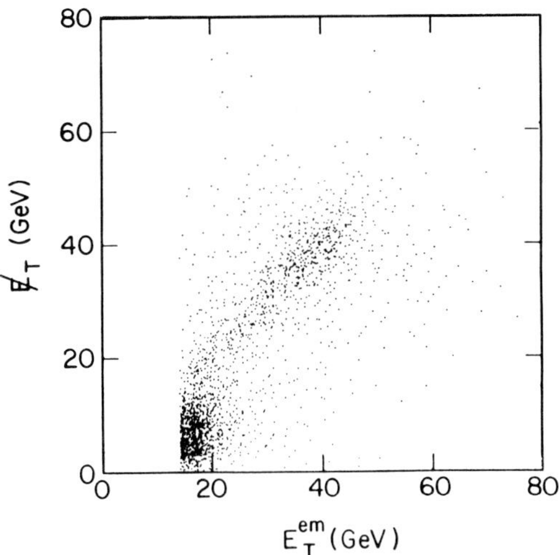

Figure 8. The distribution of $\not{E}_T$ versus the E_T of the electron (E_T^{em}) for electrons with $E_T > 15$ GeV. Conversion and Z° candidates have been removed.

4.2.3 *Properties of the Inclusive Electron Sample*

We show in Fig. 8 the distribution of $\not{E}_T$ versus E_T for the inclusive electron sample after the conversion and Z° removal. There is a broad distribution of events at large $\not{E}_T$ and E_T, the signature expected from W boson production. There is a much larger concentration of events at low $\not{E}_T$ and low electron E_T, which is where events from b and c quark semileptonic decay are expected to appear. An electron from W decay is known to be well isolated from other energy deposition,[19] whereas the electrons from c and b quark decay tend to be surrounded by additional energy due to the low P_T of the hadronic decay fragments relative to the electron axis. A measure of this isolation is the sum of the transverse energy deposited in the calorimeter towers adjacent to the electromagnetic cluster:

$$E_T^{iso} \equiv \sum_{\text{border}} E_T. \tag{4}$$

This sum is over from eight to twelve towers, depending on the size of the electron shower; typically, electrons deposit all of their energy in one calorimeter tower so that in most cases there are only eight adjacent towers in this sum. The distribution of this quantity for the electrons with $E_T > 20$ GeV and $\not{E}_T > 20$ GeV is shown in Fig. 9 where it is compared with the E_T^{iso} distribution for the electrons with $E_T < 20$ GeV or $\not{E}_T < 20$ GeV. The lower energy electrons are clearly less isolated than the electrons at higher E_T and $\not{E}_T$. An

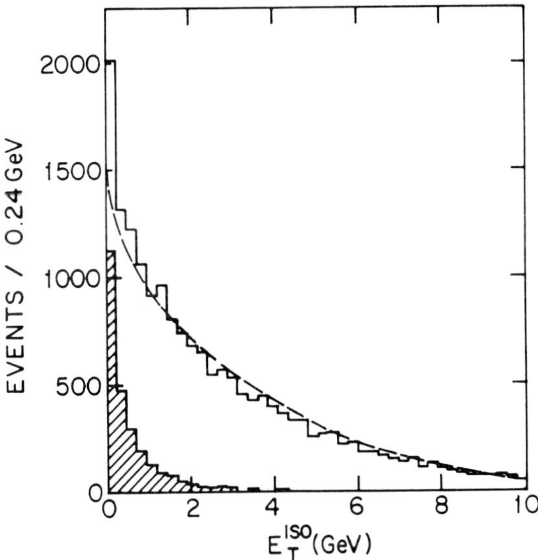

Figure 9. The distribution of the isolation variable E_T^{iso}. The solid histogram is for events with either $E_T < 20$ or $\not{E}_T < 20$ GeV, while the shaded histogram is for events with both $E_T > 20$ and $\not{E}_T > 20$ GeV.

ISAJET Monte Carlo calculation of inclusive b and c quark production predicts the E_T^{iso} distribution shown as the curve in Fig. 9. The agreement with the E_T^{iso} distribution for the low energy electrons is excellent.

Other features of the low E_T electrons are well modeled by the ISAJET calculations for b and c quark production. Such quantities as the angle between the electron and leading jet, the number of jet clusters in the event, and the E_T distribution of the electrons and jets are reproduced by this Monte Carlo. In addition, the absolute cross section predicted by the Monte Carlo agrees with the observed inclusive electron cross section to within 20%.

This Monte Carlo calculation predicts that the low energy inclusive electrons in the sample come primarily from b quark semi-leptonic decays. A unique signature for this is the presence of a charmed hadron resulting from the $b \to c$ transition. We have searched for this signature using the channel

$$\begin{aligned} b &\longrightarrow \overline{B} + X \\ &\quad \hookrightarrow D^\circ e^- \overline{\nu_e} X' \\ &\qquad \hookrightarrow K^- \pi^+, \end{aligned} \tag{5}$$

and its charge conjugate mode. This process has two very specific signatures; the final state $K\pi$ system must reconstruct to a D° meson and the electron and kaon must have the same charge. We have searched for D° and $\overline{D^\circ}$ signals by forming the $K^+\pi^-$ and $K^-\pi^+$ effective masses for all combinations of pairs of oppositely-charged tracks within an $\eta - \phi$ cone of radius 1.0 around the electron or positron candidate for a subsample of the data. We show the resulting $K\pi$ effective mass combinations in Fig. 10, where we plot the "right-sign" effective mass combinations (*i.e.* the $K^-\pi+$ effective mass for electrons and the $K^+\pi-$ effective mass for positrons) in (a) and the wrong-sign combinations in (b). A clear D° peak is seen in the right-sign distribution with no evidence of a signal in the wrong-sign data. We fit these two distributions to a smooth background and a resonance and obtain a signal of 62 ± 17 events.

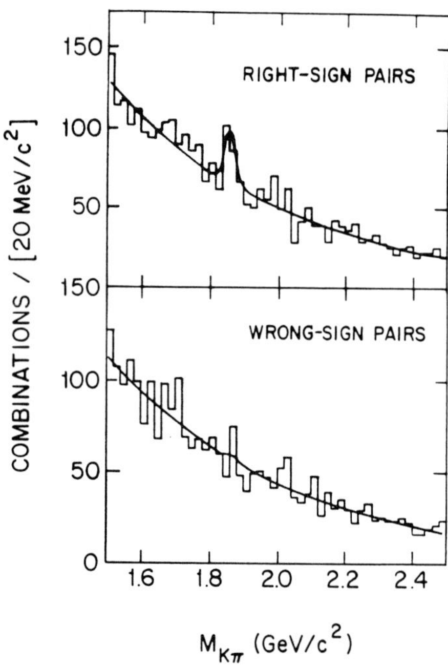

Figure 10. The $K\pi$ effective mass combinations for charged tracks in a cone around the electron candidate. The "right-sign" combinations are shown in (a), while the "wrong-sign" combinations are shown in (b) (see text).

We estimate the expected rate of D° events from b quark decay by assuming that 80% of the low energy inclusive electrons come from b's, which takes into account an estimated non-electron background of $\sim$ 10% and the unseen conversion background of $\sim$ 10%. Using the latest CLEO measurement[20] that 68±15% of the semi-leptonic B decays yield a D° and the world average value[21] for $BR(D^\circ \to K\pi)$ of $3.8 \pm 0.4\%$, we expect to find 87 ± 19 events in the correct-sign distribution, in good agreement with the observed rate. We conclude that the primary contribution to the low energy inclusive electron sample is b quark production.

ISAJET Monte Carlo calculations for $t\bar{t}$ production show that the electrons from top quark decay are well isolated, typical of real W decay. Hence, we require the electron candidates to have $E_T^{iso} < 2$ GeV in order to reduce the rate of events from b quark production. The efficiency of this cut for $t\bar{t}$ events is $\sim$ 80% for top quark masses in the range of 40-80 GeV/c^2, whereas only about 45% of the inclusive electrons from b quark decay are expected to survive this cut. This leaves a sample of 6070 events.

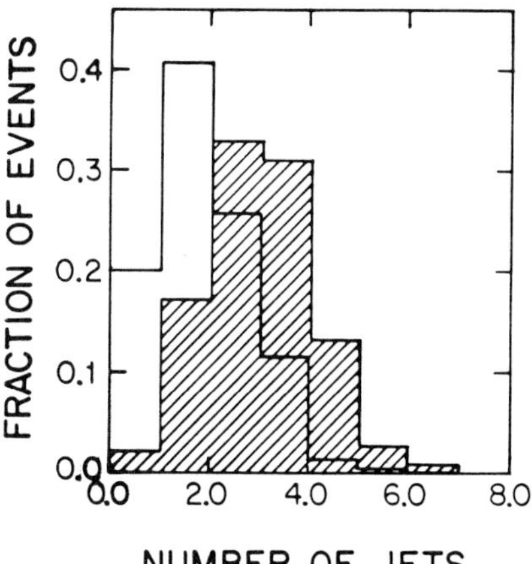

Figure 11. Jet multiplicities for $t\bar{t}$ events with electron $E_T > 20$ GeV and $\not{E}_T > 20$ GeV.The histograms show the fraction of events as a function of jet multiplicity for M_{top} of 40 (line) and 80 (shaded) GeV/c^2.

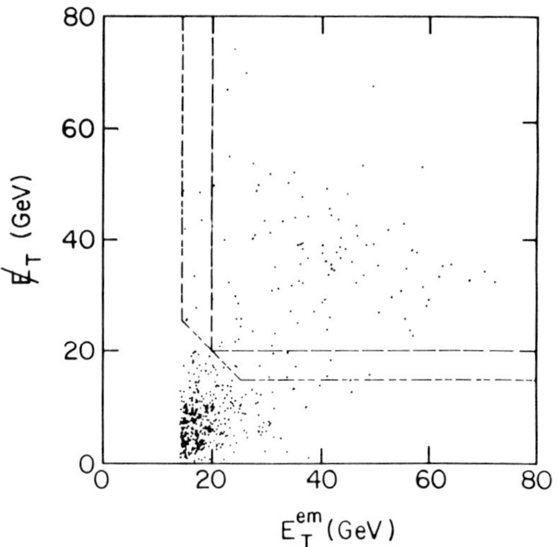

Figure 12. The distribution of $\not{E}_T$ versus E_T for electron + $\geq$ 2 jet events. The dashed (dot-dashed) lines represent the "tight" ("loose") selection.

4.2.4 The Jet and $\not{E}_T$ Cuts

A $t\bar{t}$ event has at least four quarks in the final state, which would appear as hadronic jets in the detector. Because some of the daughters from a top quark decay will not be very energetic, we do not expect to observe all four quarks as hadronic jets. We therefore require that the events have at least two jets with measured $E_T > 10$ and be within the pseudorapidity interval $|\eta| < 2.0$. There are 512 events that survive this cut. Monte Carlo predictions for the multiplicity of jets in a $t\bar{t}$ event satisfying these criteria are shown in Fig. 11 for $t\bar{t}$ events with $M_T^{e\nu}$= 40 and 80 GeV/c^2. Because the two highest energy jets in a $t\bar{t}$ event have transverse energies that are well above the jet E_T cut, the $t\bar{t}$ acceptance is not very sensitive to the choice of the jet E_T cut, especially at higher top quark masses. For example, the $t\bar{t}$ acceptance decreases by only 5% if we raise the jet E_T cut from 10 to 12 GeV for a top quark mass of 80 GeV/c^2.

The $\not{E}_T$ versus E_T distribution for the electron + $\geq$ 2 jet events is shown in Fig. 12. There is still an enhancement at high $\not{E}_T$ and E_T from W production and a cluster of events at low E_T or $\not{E}_T$, which are the remnant of the non-isolated electrons that survive the border tower cut. We require $\not{E}_T$ > 20 GeV and $E_T > 20$ GeV to further reduce the background from the non-isolated electrons. This results in a sample of 104 events. This $\not{E}_T$ and E_T selection is efficient for top masses above $\sim$ 50 GeV/c^2, but is increasingly inefficient as the top quark mass decreases. In order to be sensitive to a lighter top quark, we also consider the selection requiring $\not{E}_T > 15$ GeV, $E_T > 15$ GeV and $\not{E}_T +$ $E_T > 40$ GeV, which forms the dashed boundary shown in Fig. 12. This "loose" selection, which yields 123 events, is about twice as efficient for a 40 GeV top quark, but also introduces more background.

The $\not{E}_T$ versus E_T distribution of the data surviving the "tight" selection has been compared with that expected for W + jet production. The W + 2 jet process has been calculated using a QCD tree-level calculation exact to all orders[22] as implemented in the PAPAGENO Monte Carlo program.[23] We processed the W + 2 jet events resulting from this calculation through the CDF detector simulation and analysis program.

The properties of the W + 2 jet events agree well with those of the electron + $\geq$ 2 jet events. To illustrate the level of agreement, we show the data and MC distributions of the dijet invariant mass in Fig. 13 for the two leading jets, where we have corrected the observed jet energies to take into

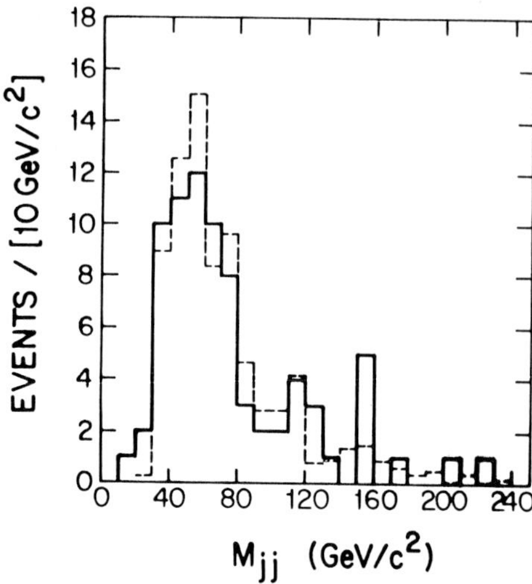

Figure 13. The dijet invariant mass for the two leading jets. The prediction from the PAPAGENO $W + 2$ jet calculation is shown as the dashed histogram. The observed energies of the jets have been corrected to take into account the calorimeter response and jet fragmentation.

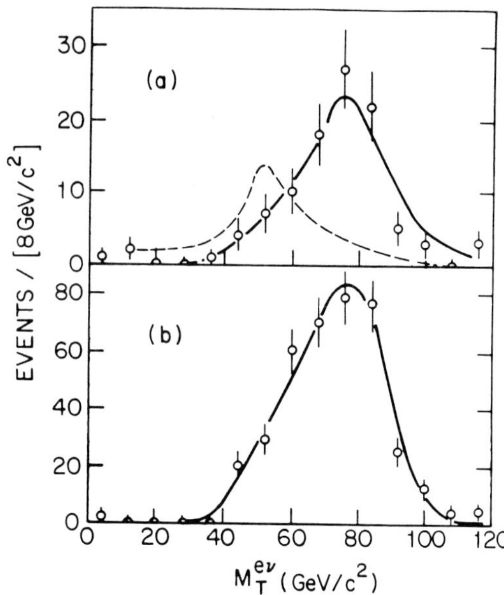

Figure 14. The $M_T^{e\nu}$ distribution for the electron $+ \geq 2$ jet data satisfying the "tight" cuts is shown in (a), and compared with the PAPAGENO prediction (solid curve). The dashed curve shows the distribution predicted for $t\bar{t}$ production for a top quark mass of 70 GeV/c^2. The electron $+ \geq 1$ jet $M_T^{e\nu}$ distribution is shown in (b) and compared with the PAPAGENO $W + 1$ jet calculation.

account the detailed calorimeter response to jets (the ratio of corrected to observed jet energies is typically 1.5). The observed electron $+ \geq 2$ jet rate of ~ 24 pb is 30% higher than the rate predicted by the PAPAGENO calculation, but this excess is well within the theoretical uncertainties in the W jet cross section calculation, which are estimated to be 30-50%.[24]

The backgrounds in the electron $+ \geq 2$ jet sample from b and c quark production, unseen conversions, and other QCD processes give rise to non-isolated electrons. We have compared the E_T^{iso} distribution of the event sample to the E_T^{iso} distributions observed for W bosons and low-energy inclusive electrons to determine the size of the non-isolated electron backgrounds remaining in the event sample. This results in estimated rates of background from non-isolated electrons of 12% and 20% in the electron $+ \geq 2$ jet samples satisfying the tight and loose selection, respectively.

4.2.5 *Transverse Mass Analysis*

We use the momentum vectors of the electron and unobserved neutrino, represented by the $\not{E}_T$ vector, to define the transverse mass

$$M_T^{e\nu} = \sqrt{2E_T \not{E}_T(1 - \cos\phi^{e\nu})}, \qquad (6)$$

where $\phi^{e\nu}$ is the difference in the azimuthal angles of the electron and neutrino momentum vectors. We show the $M_T^{e\nu}$ distribution for both the data and $W + 2$ jet MC in Fig. 14(a), and for comparison we show the $M_T^{e\nu}$ distribution expected for $t\bar{t}$ production for M_{top}= 70 GeV/c^2.

We use the difference between the $W + 2$ jet and $t\bar{t}$ transverse mass distributions[25] to estimate the number of observed electron $+ \geq 2$ jet events that one can attribute to these two sources. We fit the observed $M_T^{e\nu}$ distribution to the form

$$\frac{dN}{dM_T^{e\nu}} = \alpha T(M_T^{e\nu}) + \beta W(M_T^{e\nu}), \qquad (7)$$

where $T(M_T^{e\nu})$ is the predicted $M_T^{e\nu}$ distribution for $t\bar{t}$ events and $W(M_T^{e\nu})$ is the $M_T^{e\nu}$ distribution for W +2 jet events. The function $T(M_T^{e\nu})$ depends on the top quark mass and is calculated using the ISAJET Monte Carlo program and the CDF detector simulation, while the function $W(M_T^{e\nu})$ is calculated using the PAPAGENO Monte Carlo and detector simulation. Both T and W are normalized so that $\alpha = 1.0$ and $\beta = 1.0$ if the predicted amounts of the $t\bar{t}$ and and $W + 2$ jet processes are required in the fit, taking into account detector acceptance and integrated luminosity. We have used the $t\bar{t}$ cross sections in reference 18 and the $W + 2$ jet cross section from the PAPAGENO calculation.

This method will only be valid for top quark masses at least several GeV/c^2 below the mass of the $W + b$ quark threshold since above that mass, a real W is emitted in the top quark decay and the $M_T^{e\nu}$ distributions for $t\bar{t}$ and $W + 2$ jets would be similar. This method is also sensitive to the detector $M_T^{e\nu}$ resolution for electron + jet events. We have checked the $M_T^{e\nu}$ resolution by studying inclusive electron events, Z° events, and the electron + ≥ 1 jet events. The latter event sample provides a particularly relevant test of our understanding of the detector; it is dominated by $W + 1$ jet production with any $t\bar{t}$ contribution being less than 10-20%, and the $M_T^{e\nu}$ resolution of the detector for the electron + ≥ 1 jet events is comparable to that of the electron + ≥ 2 jet events. We have used the PAPAGENO Monte Carlo to generate the $W + 1$ jet process and have then processed the events through the detector simulation and analysis. The comparison of the electron + ≥ 1 jet $M_T^{e\nu}$ distribution with the $W + 1$ jet prediction is shown in Fig. 14(b). The excellent agreement is confirmation that the $M_T^{e\nu}$ resolution is well understood.

The transverse mass distribution was fit to the form in equation 7 using a binned maximum likelihood method. The results of the fit are shown in Table 3, where we also list the expected number of observed electron + ≥ 2 jet events from $t\bar{t}$. The fits for $M_{top} > 65$ GeV/c^2 were made on the data satisfying the tight selection, while the fits to the lower top masses were made on the data satisfying the loose selection. The fit requires essentially no $t\bar{t}$ contribution (*i.e.* $\alpha \sim 0$), but the statistical uncertainty in α increases with M_{top} such that for $M_{top} = 80$ GeV/c^2, the statistical 95% confidence level (CL) upper limit on α is ~ 0.4. The fitted values for β are typically 1.28±0.15. The fitted $M_T^{e\nu}$ distribution is in good agreement with the data, with the binned χ^2 per degree of freedom being ~ 1.0.

4.2.6 Systematic Uncertainties

The sources of systematic uncertainties in this estimate of the $t\bar{t}$ fraction fall into two categories: The first category consists of those uncertainties that effect both the predicted $M_T^{e\nu}$ distributions and potentially the $t\bar{t}$ acceptance. These are properly considered as uncertainties on the fitted value of α, and include i) jet energy scales; ii) contributions from the "underlying" event (i.e. the energy deposited in the calorimeter that is not related to the hard scattering process); and iii) the choice of $M_T^{e\nu}$ interval included in the fit. The second category of systematic uncertainties are those that effect only the $t\bar{t}$ acceptance or the normalization of the experiment. These include i) uncertainties in the model used for top quark fragmentation; ii) variations in the produced top quark kinematics; iii) uncertainties in the amount of initial state radiation; and iv) uncertainties in the electron selection efficiency and the integrated luminosity of the data sample. Other sources of systematic uncertainty have been studied and found to have no effect on the results of the analysis.

The calorimeter energy response to high P_T particles was calibrated using test beam data, and this response function was extended to low P_T particles by using minimum-ionizing particles

Table 3. The number of predicted $t\bar{t}$ events, $n_{t\bar{t}}$, the fitted $t\bar{t}$ contribution to the electron + ≥ 2 jet rate and the 95% CL upper limits on the acceptance corrected $t\bar{t}$ cross sections.

M_{top} (GeV/c^2)	$n_{t\bar{t}}$ predicted	$\alpha \pm (stat) \pm (syst)$	$\sigma_{t\bar{t}}$ (pb)
40	130 ± 44	$0.07 \pm 0.05 \pm 0.02$	< 2410
50	123 ± 31	$0.06 \pm 0.05 \pm 0.03$	< 648
60	101 ± 22	$0.11 \pm 0.08 \pm 0.04$	< 408
70	43 ± 8	$0.00^{+0.12}_{-0.00} \pm 0.11$	< 266
80	32 ± 5	$0.00^{+0.27}_{-0.00} \pm 0.17$	< 281

observed *in situ* in minimum-bias events. This response function was then incorporated into the CDF detector simulation, and the jet fragmentation function tuned using dijet events.[10] We then used jets recoiling against isolated high P_T photons to verify this jet energy scale; such events are primarily produced by direct photon production with a single gluon or quark recoiling against the photon, so that the photon's E_T is strongly correlated with the E_T of the recoiling jet system. We have also verified that the properties of jets recoiling against electrons in the inclusive electron sample agree with the results of the Monte Carlo calculations for inclusive b quark production. From these studies, we conservatively estimate that the jet energy scale is uncertain to within 20% for 10 GeV jets.

The uncertainty in the amount of transverse energy coming from the "underlying" event in the Monte Carlo calculations has been estimated to be ±20% from comparisons of the underlying events in W candidate events and Monte Carlo predictions. We have also varied the $M_T^{e\nu}$ interval over which the fit is performed and assign an additional systematic uncertainty due to this.

We have estimated the effect of these uncertainties on the fitted values of α by introducing a one standard deviation change in each of the quantities in the Monte Carlo calculations, recalculating the T and W distributions, and refitting the data. We assign the systematic uncertainty on α to be the resulting variation in the fitted value. We list the resulting systematic uncertainties in Table 4. Since there are several uncorrelated uncertainties, we are justified in convoluting them together quadratically

Table 4. The systematic uncertainties in estimating α.

M_{top} GeV/c^2	Jet Scale	Underlying Event	$M_T^{e\nu}$ Interval	$\Delta\alpha$
40	0.02	0.00	0.02	**0.028**
50	0.02	0.01	0.03	**0.036**
60	0.03	0.01	0.03	**0.044**
70	0.10	0.04	0.04	**0.115**
75	0.05	0.01	0.08	**0.095**
80	0.14	0.05	0.10	**0.179**

to obtain the total systematic uncertainty on α, $\Delta\alpha$, listed in Table 4, and to treat it as a Gaussian statistic.

The second category of systematic effects are uncertainties in the overall $t\bar{t}$ acceptance and normalization. The top quark fragmentation function, $D(z)$, which is the probability that the top meson carries a fraction z of the produced top quark energy, is parametrized in the acceptance calculations by the Peterson formula[26]

$$D(z) \propto \left[1 - \frac{1}{z} - \frac{\epsilon}{(1-z)}\right]^{-1}. \qquad (8)$$

In this model, the ϵ parameter is expected to be proportional to $(1/m_q)^2$, where m_q is the mass of the heavy quark, and the differences in the measured c and b quark fragmentation functions support this ansatz.[27] We have assumed that this ϵ parameter is uncertain to a factor of three for top quark decays. We have estimated the uncertainty in the model for the production of the top quark by comparing the results of several Monte Carlo calculations. The only significant difference in the $t\bar{t}$ acceptance appears at top quark masses ≤ 40 GeV/c^2; for M_{top} of 40 GeV/c^2, we assign an additional uncertainty of 17%.

We have studied the rate and E_T spectra of low energy jets in the inclusive electron data, and have verified that the ISAJET prediction for b quark production reproduce these distributions. Since a significant fraction of the jets in these data come from initial state radiation, this checks that the Monte Carlo calculation is reliably predicting the rate of these jets. To be conservative, we assume that the rate of jets coming from initial state radiation in the Monte Carlo $t\bar{t}$ events is uncertain by ±50% and have propagated this into an uncertainty on the $t\bar{t}$ acceptance. With increasing top quark mass, the fraction of observed jets in an event resulting from initial state radiation decreases. Hence, the size of the uncertainty in the electron + ≥ 2 jet acceptance decreases.

Our studies of the electron efficiencies in inclusive W and Z° decays have been compared with the efficiencies obtained from Monte Carlo calculations with full detector simulation, and we have found the Monte Carlo calculations to reproduce the measured efficiencies. We conservatively assign a $\pm 5\%$ uncertainty in the electron selection efficiency in $t\bar{t}$ events as estimated by the Monte Carlo calculations. The integrated luminosity is known to within 15%. We add these systematic uncertainties in quadrature to obtain a total systematic uncertainty on the acceptance and normalization. These uncertainties are summarized in Table 5 and are reflected in the uncertainty in the number of predicted $t\bar{t}$ events in Table 3.

We use the likelihood function for α determined from the fit as the estimate of the statistical uncertainties, and convolute this together with the systematic uncertainties on α and the acceptance to obtain a second likelihood function that now incorporates the statistical and systematic uncertainties. The 95% confidence level (CL) upper limit on the $t\bar{t}$ production cross section is then calculated from this likelihood function for each top quark mass. These upper imits are listed in Table 3 and are plotted in Fig. 15. We compare our 95% CL upper limit with the lower bound on the $t\bar{t}$ cross section calculated in reference 18, which is also shown in Fig. 15, and find that we exclude a top quark mass less than 77 GeV/c^2 at 95% CL. We

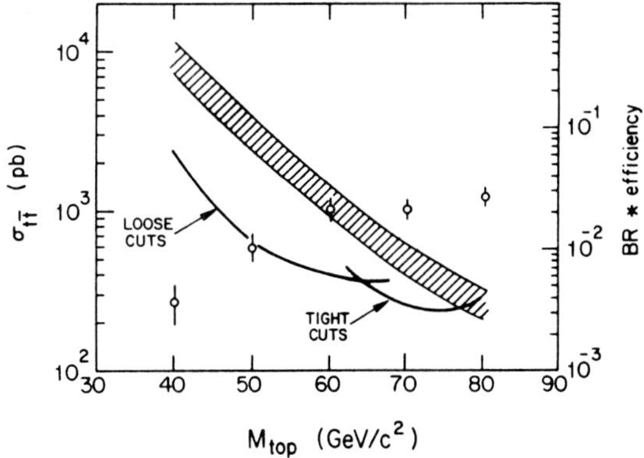

Figure 15. The 95% CL upper limits on $t\bar{t}$ cross section from the electron + $\geq$ 2 jets analysis. The curves represent the upper limits from the loose and tight selection, while the band shows the upper and lower bounds on the predicted $t\bar{t}$ cross section. The plotted points show the acceptance of the electron + $\geq$ 2 jet selection for $t\bar{t}$ events (right hand scale).

choose to quote the excluded mass range

$$40 < M_{top} < 77 \text{ GeV/c}^2 \tag{9}$$

as the systematic uncertainties grow rapidly for top masses below 40 GeV/c^2.

Table 5. The systematic uncertainties on normalization and acceptance, expressed as fractional uncertainties. The uncertainties in the electron efficiency (0.05) and the integrated luminosity (0.15) are not listed in the table but are included in the total uncertainty. An additional uncertainty of 0.17 due to the top quark production model is also included in the total uncertainty for M_{top} of 40 GeV/c^2.

M_{top} GeV/c^2	Fragmentation Model	Initial State Radiation	$\Delta\epsilon$
40	0.130	0.210	**0.33**
50	0.103	0.170	**0.25**
60	0.076	0.125	**0.22**
70	0.050	0.090	**0.19**
75	0.050	0.070	**0.18**
80	0.050	0.045	**0.17**

4.3 THE $e^{\pm}\mu^{\mp}$ ANALYSIS

4.3.1 *Introduction*

The production of $t\bar{t}$ pairs followed by the semi-leptonic decay of both top quarks yields a final state with two energetic leptons, some missing energy arising from the undetected neutrinos and two b-quark jets. A lepton pair requirement using the CDF central electron and muon systems yields an event sample free of fake-lepton contaminations and background from W + jet production. In the case of the $e\mu$ mode, the additional requirement of having the leptons come from different families avoids the problems associated with a large Drell-Yan background.

We have performed a top quark search in this channel by studying an event sample containing a central electron candidate and a muon candidate observed in the central region of the detector. The following section describes this analysis and its conclusions.

4.3.2 *The eμ Event Selection*

We define the event sample by requiring both a CEM electron candidate and a central muon candidate. The electron candidate is selected with the same requirements used in the electron + jets analysis. An isolation requirement on the electron candidate is not applied in this case as the backgrounds from sources of non-isolated electrons are small.

In the region $|\eta| \lesssim 0.65$, which is covered by the central muon chambers (CMU), we define a μ candidate by a track segment in the CMU and a CTC track satisfying the following requirements: i) the P_T of the CTC track is greater than 5 GeV/c; ii) the energies deposited in the CEM and CHA towers pierced by the muon candidate are less than 2 GeV and 6 GeV, respectively; and iii) the CMU and CTC tracks match to within 10 cm in the $R\phi$ direction. These requirements are efficient for real μ's as the matching cut takes into account the multiple scattering of the μ in the calorimeter, while the calorimeter cuts are equivalent to requiring that the particle be minimum-ionizing.

For muon $P_T > 10$ GeV/c, the fake muon backgrounds are negligible in the inclusive electron data so that a clean sample can be obtained without requiring the presence of CMU track. Hence, we extend the muon coverage over the region $|\eta| < 1.2$ by defining a muon candidate as a CTC track with $P_T > 10$ GeV/c that is minimum-ionizing in the calorimeter (with the cuts described in ii) above). We also require these muon candidates to satisfy a relatively loose isolation cut. We demand that there be less than 5 GeV of transverse energy deposited in the calorimeter cells that are in a cone of radius $\sqrt{(\Delta\eta)^2 + (\Delta\phi)^2} = 0.4$ around the CTC track. Since this alternative muon selection does not require a CMU track, it identifies all of the muons with P_T> 10 GeV/c that do not fall within the fiducial region of the CMU chambers. We only consider those events in which the electron and muon candidates are oppositely charged. The same-sign sample, interesting in its own right for other reasons, is not discussed here.

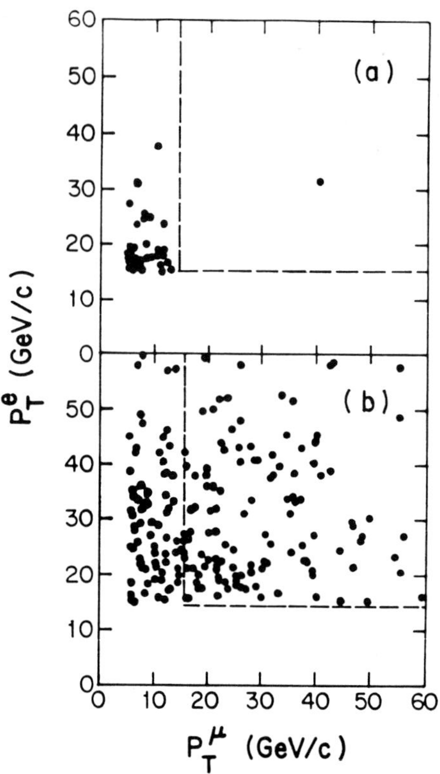

Figure 16. The $P_T{}^\mu$ vs $P_T{}^e$ distributions for the $e^\pm\mu^\mp$ data is shown in (a). Note that the acceptance for $P_T{}^\mu < 10$ GeV/c is lower because a CMU segment is required for such muon candidates. The distribution expected from $t\bar{t}$ production for M_{top} of 70 GeV/c² is shown in (b) for an integrated luminosity of 80 pb^{-1}.

The efficiency of the electron selection is $77 \pm 3\%$, as discussed earlier in the section describing the inclusive electron selection. The efficiency of the muon selection has been measured using Z° events decaying into $\mu^+\mu^-$. We have found that the selection criteria are $98 \pm 2\%$ efficient for muon candidates within the CMU fiducial volume, while they are $96 \pm 2\%$ efficient for muons outside this region.

The resulting distribution for the P_T of the electron ($P_T{}^e$) versus the P_T of the muon ($P_T{}^\mu$) is shown in Fig. 16(a). There are 45 events with $P_T{}^e > 15$ and $P_T{}^\mu > 5$ GeVc, and only 1 event has $P_T{}^e > 15$ GeV/c and $P_T{}^\mu > 15$ GeV/c. The corresponding distribution of $P_T{}^e$ versus $P_T{}^\mu$ for the expected signal from $t\bar{t}$ production with

M_{top}= 70 GeV/c^2 is shown in Fig. 16(b) for an integrated luminosity of ~80 pb^{-1}. The expected number of events for this top mass, taking into account the acceptance determined by an ISAJET calculation, the full detector simulation and the $t\bar{t}$ cross section[18] for M_{top} of 70 GeV/c^2 is 7.5 events.

We have studied the various sources of background to this event selection. The decay $Z^\circ \rightarrow \tau^+\tau^- \rightarrow e^\pm\mu^\mp X$ can produce high energy $e\mu$ pairs, and we estimate that we should see one event from this process, based on the observed rate of $Z^\circ \rightarrow e^+e^-$, the known branching ratios of the τ lepton, and the efficiency of the lepton selection for electrons and muons from τ decays. The pair production of b quarks is another potential background source, and estimates of its rate are very uncertain because of the uncertainty in the b quark production cross section. ISAJET Monte Carlo calculations for b quark production indicate that the observed rate of events with ${P_T}^e > 15$ GeV/c and lower P_T muons is consistent with that expected from b decays. The expected contribution to the event sample with ${P_T}^\mu > 15$ GeV/c is of order 1 event or less. Contributions from WW and WZ production are estimated to be 0.15 and 0.05 events, respectively. The rate of events from sources of fake lepton candidates are estimated to be negligible. We conclude that the one event in our sample is consistent with the expected rate from sources other than $t\bar{t}$ production.

4.3.3 The eμ Top Quark Mass Limit

The systematic uncertainties that affect the $t\bar{t}$ acceptance calculations have to be taken into account to convert the rate of 1 $e^\pm\mu^\mp$ event into an upper limit on the $t\bar{t}$ production cross section and a limit on the top quark mass. Most of these uncertainties also affect the electron + jets analysis and have been described above. These are: i) uncertainties in the model for top quark fragmentation; ii) uncertainties in the model for top quark production; iii) uncertainty in the integrated luminosity; and iv) the uncertainty in the electron selection efficiency and integrated luminosity. The only additional uncertainty comes from the muon selection efficiency; we estimate this to be ±3%. As

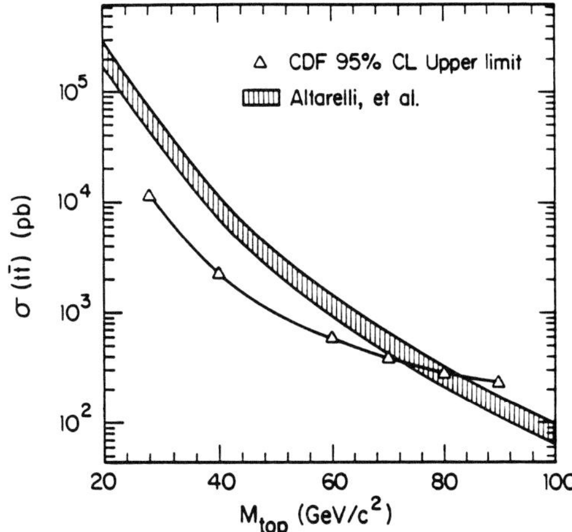

Figure 17. The upper limit on the $t\bar{t}$ production cross section from the $e\mu$ analysis. The shaded band shows the upper and lower limits of the theoretical prediction for the $t\bar{t}$ cross section as given by Altarelli *et al.*

in the electron + jets analysis, we convolute these systematic uncertainties in quadrature to obtain a total systematic uncertainty for the $e\mu$ acceptance. This uncertainty varies from 45% at a top mass of 28 GeV/c^2 to 19% for M_{top} of 80 GeV/c^2.

We determine the 95% CL upper limit on $t\bar{t}$ production by finding the mean of a parent Poisson distribution that, when convoluted with the Gaussian distribution representing the total systematic uncertainty on the $t\bar{t}$ acceptance, yields a probability of 5% for observing zero or one event in our sample. This results in the 95% CL upper limit on the $t\bar{t}$ production cross section shown in Fig. 17, where we compare it with the theoretical upper and lower limit for $t\bar{t}$ production.[18] A comparison of our upper limit with the lower bound on the theoretical cross section excludes a top quark mass less than 72 GeV/c^2 at 95% CL. The analysis excludes top quark masses lower than 28 GeV/c^2, but as in the electron + jets case, we choose to define the excluded range of top quark masses as

$$28 < M_{top} < 72 \text{ GeV/c}^2 \tag{10}$$

due to the increasing systematic uncertainties on the top quark acceptance at low top quark masses and because this bound overlaps sufficiently with lower limits on the top quark mass from other analyses.

This limit is also applicable to a fourth generation, charged 1/3, b' quark that decays predominantly via the charged current interaction to second generation quarks, and whose lifetime is sufficiently short that its decay products appear to come from the interaction vertex.

5. Conclusion

We have searched for evidence of the strong production and decay of the $\tilde{q}$ and $\tilde{g}$ using a simple decay model that excludes "cascade" decays of these particles. Preliminary results of a search for events with $\not{E}_T >$ 40 GeV and two or more jets indicates that the $\tilde{q}$ mass is above 120 GeV/c^2 if the $\tilde{g}$ is lighter than the $\tilde{q}$. An analysis of the transverse mass distribution of events with an electron candidate, missing transverse energy, and two or more jets excludes the production and decay of the Standard Model top quark for top quark masses between 40 and 77 GeV/c^2 at 95% CL. A search for events with an energetic electron and muon yields one such event. This excludes a top quark with masses between 28 and 72 GeV/c^2 at 95% CL.

6. Acknowledgements

We wish to thank the staff of the Fermilab Accelerator Division for their outstanding efforts and express our appreciation to the CDF technical and support staff. We also wish to acknowledge the International Business Machines Corporation for the 3909 supercomputer CPU time they made available to perform several of the Monte Carlo calculations. This work was supported by the Department of Energy, the National Science Foundation, Istituto Nazionale di Fiscia Nucleare, the Ministry of Science, Culture and Education of Japan, and the A. P. Sloan Foundation.

REFERENCES

1. The CDF collaboration is comprised of research groups from: Argonne National Laboratory, Brandeis University, University of Chicago, Fermi National Accelerator Laboratory, Laboratori Nazionali di Frascati, Harvard University, University of Illinois, National Laboratory for High Energy Physics (KEK), Lawrence Berkeley Laboratory, University of Pennsylvania, Istituto Nazionale di Fisica Nucleare, Purdue University, Rockefeller University, Rutgers University, Texas A&M University, University of Tsukuba, Tufts University, and University of Wisconsin.
2. F. Abe *et al.*, Nucl. Instr. Meth. **A271**, 387 (1988).
3. For reviews of SUSY, see H. P. Nilles, Phys. Rep. **110**, (1984), and
   H. E. Haber and G. L. Kane, Phys. Rep. **117**, 75 (1985).
4. R. Ansari *et al.*, Phys. Lett. **B195**, 613 (1987).
5. C. Aljabar *et al.*, Phys. Lett. **B198**, 261 (1987).
6. F. Abe *et al.*, Phys. Rev. Lett. **62**, 1825 (1989).
7. H. Baer, J. Ellis, G. Gelmini, D. Nanopoulos and X. Tata, Phys. Lett **B161**, 175 (1985).
8. G. Gamberini, Z. Phys. **C30**, 605 (1986).
9. H. Baer, X. Tata, and J. Woodside, Phys. Rev. Lett. **63**, 352 (1989).
10. F. Abe *et al.*, Phys. Rev. Lett. **62**, 613 (1989).
11. W. W. Ash *et al.*, Phys. Rev. Lett. **58**, 1080 (1987).
12. The JADE Collaboration, DESY preprint 88-154, (1988).
13. H. Albrecht *et al.*, Phys. Lett. **192B**, 245 (1987).
    M. Artuso *et al.*, Phys. Rev. Lett. **62**, 2233 (1989).
14. U. Amaldi *et al.*, Phys. Rev. **D36**, 1385 (1987).

15. See, for example, A. D. Martin, R. G. Roberts and W. J. Stirling, Phys. Lett. **B189**, 221 (1987).
16. R. K. Ellis, Proceedings of the Seventh Topical Workshop on Proton-Antiproton Collider Physics, edited by R. Raja, A. Tollestrup, and J. Yoh, World Scientific (1989), p. 639.
17. P. Nason, S. Dawson and R. K. Ellis, Nucl. Phys. **B303**, 607 (1988).
18. G. Altarelli, M. Diemoz, G. Martinelli and P. Nason, Nucl. Phys. **B308**, 724 (1988).
19. F. Abe *et al.*, Phys. Rev. Lett. **62**, 1005 (1989).
20. P. Tipton and E.H. Thorndike, CLEO internal report CBX 87-26; and E.H. Thorndike and R.A. Poling, Phys. Rep. **157**, 183 (1988).
21. The Particle Data Group, G.P. Yost *et al.*, Phys. Lett. **B204**, 1 (1988).
22. R. K. Ellis and R. J. Gonsalves, Proceedings of the Supercollider Physics Topical Conference, edited by D. Soper, World Scientific (1985), p. 287.
23. I. Hinchliffe, report in preparation. Version 2.71 of PAPAGENO was used for the W + multijet calculations.
24. M. Mangano and S. Parke, Fermilab Pub 89/106-T (July, 1989).
25. J. Rosner, Phys. Rev. **D39**, 3297 (1989). J. Rosner, Phys. Rev. **D40**, 1701 (1989). H. Baer, V. Barger, and R. J. N. Phillips, Phys. Lett. **B221**, 398 (1989).
26. C. Peterson, D. Schlatter, I. Schmitt and P. Z. Zerwas, Phys. Rev. **D27**, 105 (1983).
27. R. Cashmore, Oxford preprint NP-96/87, Sep. 1987.

DISCUSSION

L. Di Lella, CERN: You quote top mass limits at the 90% confidence level. We quoted 95%. If UA2 had quoted 90%, the top mass limit would have gone from 67 to 70 GeV. Also, when you summarized the CERN top mass results at the beginning, you quoted 69 GeV from UA1, which I missed in the talk of Eggert. I thought he had said 65 GeV. You gave them a bonus of 4 GeV.

P. Sinervo: Yes, our 95% confidence level limit is about 2 GeV lower than the 90% limit. With regard to the reference to the UA1 limit, my apologies. The 95% UA1 limit is 65 GeV; I confused it with their 90% confidence level limit.

M. Samuel, Oklahoma State University: Are you allowed to say anything about your $W\gamma$ events or radiative W decays?

P. Sinervo: I'm not prepared to say anything on that subject. You might question one of the following speakers, John Huth, as he will be covering the rest of the CDF results on W production.

G. Altarelli, CERN: Concerning the theoretical calculation of the $t\bar{t}$ cross section, I must say that I am embarrassed when one starts discussing a difference of a few GeV in the top mass bound, or the difference between 90% and 95% confidence levels, and all of these very detailed questions. Our calculation is, I believe, the best estimate of the cross section that one can make — given the status of perturbative QCD and the uncertainties in Λ, the structure functions, etc. But, I repeat, it cannot be taken as too sacred!

P. Sinervo: I believe that it is very useful if not necessary to have a well-defined cross section calculation, and I feel that your calculation has been done extremely well. As the UA1 and UA2 people have chosen to use it also, I believe that it is a reasonable standard. I think the important issue is that the data should be presented in such a way that if another cross section calculation is made that differs from yours, then a top mass limit using that calculation can be easily derived by laying it on top of the experimental data, which is an upper bound on the cross section and not the top mass.

M. Peskin, SLAC: I'd like to ask about the case in which the top quark mass is greater than the W mass. In this case, the top quark will decay to an on-shell W; then $t\bar{t}$ production is just another mechanism for W production and so does not show up in the $e\nu$ transverse mass distribution (except in its normalization, which is in any case quite uncertain). In this regime, you have to try to combine W's with jets to make a mass peak. Have you done this analysis?

P. Sinervo: Yes, we are working very hard on that. That is the thrust of the next phase of our search. However, one encounters problems with jet counting when the top mass is around the $W + b$ threshold ($\sim$ 90 GeV). The acceptance drops in this region because the bottom quark jets have very little P_T in the lab frame (due to the difference in the W and b quark masses and the Lorentz boost into the lab frame), making them difficult to reconstruct. We hope that we will be able to do something above that region, around 95 to 100 GeV. But it's effectively what you suggest; one is looking for di-boson production with associated hadronic activity.

M. Peskin: And well above the W mass, in the region of top masses around 100 GeV, have you anything to say yet?

P. Sinervo: No, we have nothing to report as yet. However, I hope that we'll have something to say on the timescale of six months.

B. Ward, University of Tennessee: So it appears then that you are beginning to set somewhat large limits on what is called the Standard Model top quark. Let me just emphasize or recall that there are models in which there exists what we call the charged Higgs particle, and in these scenarios there are regions of parameter space in which this Higgs has a mass less that of the top quark — sufficiently less, in fact, that the top quark can decay into it. I would like to know if you are paying any attention at all to this case in your analysis, or are you just concentrating on the simplest case?

P. Sinervo: We are aware of this issue, which is why we try to qualify the results of our search by emphasizing that it assumes a minimal form of the Standard Model that excludes the possibilities you mention. The case of a charged Higgs is problematic because it will prefer to decay to the heaviest possible fermions. This would substantially reduce the sensitivity of our event selection (as well as those of UA1 and UA2) and degrade the top mass limit. We have not had the time to devote to understanding all the ramifications of such scenarios, however. Certainly we would like to say more about it, but right now we haven't worked on it hard enough.

G. Feldman, SLAC: My question is just a few GeV less than Michael Peskin's. You showed us a transverse mass plot of what a 70 GeV top quark looked like in the $e + jet$ analysis compared to the $W + jet$ background. But you've set a limit at 78 GeV, and it's clear that peak will slide into the W peak. What is it that distinguishes the $W + jet$ events from the top events as you approach the W mass?

P. Sinervo: Even with an 80 GeV top, since it is decaying into a W plus a b-quark, the transverse mass distribution is still quite distinct from what you'd expect from the W alone, because the W from top decay is still substantially off-mass shell. If you look at the transverse mass distribution for an 80 GeV top quark, you see that it peaks at around 60 GeV, which is still 10-15 GeV below where the transverse mass distribution peaks for the $W + jets$ process. If top really is there with a mass of 80 GeV, its effect would then be to create a shoulder in the transverse mass distribution in the 40 to 60 GeV region below the peak caused by W production. It then becomes simply a question of Poisson statistics to determine how small the expected shoulder can be in the transverse mass distribution and still be consistent with the observed data.

In terms of how much higher the top mass can rise before the transverse mass distributions for top and W production are indistinct, my guess is that it can up increase up to a few W widths below the W plus b-quark mass (85-90 GeV). You can see that the discrimination of this analysis is already deteriorating by looking closely at our confidence level contour in the $e + jets$ cross section plot. It starts to turn up between 75 and 80 GeV. For example, we would not have been very comfortable quoting a mass limit at 85 GeV from this analysis, because that would have been too close to the point where one becomes much more sensitive to the details of the transverse mass distribution both for top and $W + jets$.

P. Minkowski, University of Bern: UA1 and UA2 saw some W and associated jet events with very large P_T and invariant masses. Do you see similar events?

P. Sinervo: John Huth, the next CDF speaker, will actually show the P_T distribution and make some comment on it. It agrees fairly well with the recent QCD calculations of Arnold and Reno.

J. Kuhn, Max Planck Institute, Munich: In your production cross section, you have taken account of $O(\alpha_s)$ QCD corrections. I would think that at the same time, you should consider the QCD corrections in the top quark decay that affect the resulting lepton spectrum. Have you done that? Have you any idea about the size of these corrections?

P. Sinervo: It depends on which corrections you refer to. There are "hadronic" effects when a top quark fragments into a hadron. Then there are other effects which are of much shorter range. We have tried to understand what a reasonable range of uncertainty is by comparing Monte Carlos such as PAPAGENO and ISAJET, which have different models in detail for $t\bar{t}$ production. We beat those two models against each other and have taken the difference between the two as one of the uncertainties in the transverse mass distribution that we fit.

J. Kuhn: But that's still in the production mechanism. The top decay should also receive QCD corrections and the spectrum will change.

P. Sinervo: I think we are talking about the same thing. When the top decays, it's a semileptonic decay and you have to ask how well do the Monte Carlos model that process, and what are the uncertainties in those models. I can't speak to that myself very much. When we have talked to theorists about this, they tend to believe that the semileptonic decay of the top quark is better understood than the fragmentation or the production process. Hence, we've focussed on estimating uncertainties in the latter two.

G. Salvini, University of Rome: I think that with the present limit from CDF and the others on the top mass, one can look more closely at the limits on the number of light neutrinos. If I recall the transparency shown yesterday by Myron Campbell from CDF on the dependence of the W and Z production cross section ratios with top mass and the number of neutrinos, one prefers three neutrinos, plus or minus something. Is that correct?

P. Sinervo: It is certainly one of the strongest constraints on the number of light neutrinos. As you saw, the W and Z production cross section ratio is sensitive to the top mass, but it is approximately independent of this once you are above our lower limit, as for such masses the decay $W \to tb$ is not possible. Hence, a more stringent top mass lower limit will not improve the limit on the number of neutrinos beyond what one can extract from the current data.

THE PRODUCTION OF HIGH PT PHOTONS AND LEPTON PAIRS

Joey Huston

Physics and Astronomy Department, Michigan State University
East Lansing, MI 48824

Recent experimental results and phenomenology are presented for the production of lepton pairs and for the production of photons at high transverse momentum. The concentration of this review will be on the subject of direct photons. Several topics related to direct photons, such as the hadroproduction of photon pairs and the fragmentation properties of jets opposite direct photons, will also be presented.

1. DIRECT PHOTONS

1.1 Motivation

The study of direct photons is a subject that has reached a certain degree of maturity, on both the experimental and theoretical sides. Several excellent recent reviews provide a summary.[1,2] The motivation for pursuing direct photon physics can be summarized as follows:

-simplicity: to first order, only two subprocesses contribute to direct photon production, q $\overline{q}$ annihilation and gluon Compton scattering (18 diagrams for 3 quark flavors). For comparison, in the calculation of the invariant cross section for single particle inclusive hadron production(using 3 quark flavors), there are 127 separate two-body scattering contributions. Also, a direct photon in the final state removes inherent ambiguities present in the case of jets, which can be either quarks or gluons.

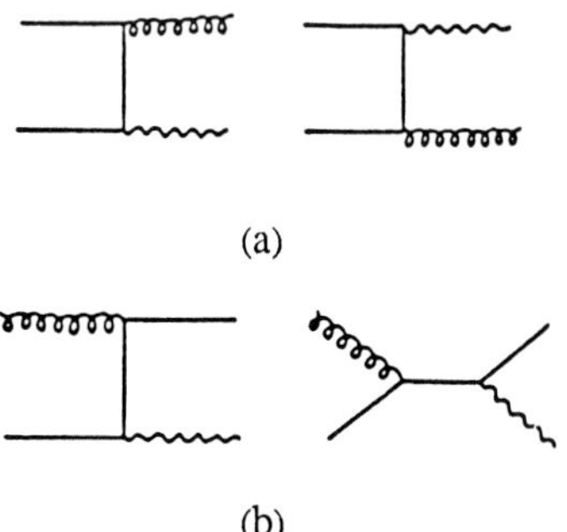

Fig. 1: (a) q $\overline{q}$ annihilation (b) gluon Compton scattering.

-gluon physics: gluons are involved in both of the diagrams shown in Fig. 1, namely in the final state for the annihilation graph and in the initial state for the Compton graph. An isolation of the Compton graph allows a measurement of the gluon structure function of the colliding hadrons; an isolation of the annihilation term allows a measurement of the gluon fragmentation function. As will be seen later in this review, this isolation can be accomplished by considering specific kinematic regimes or by comparing production by particles and antiparticles(for example, p and $\overline{p}$, π^- and π^+).

-QCD tests; detailed comparisons are possible between theory and experiment. The next-to-leading order (NLO) predictions have been available for several years.[3] On the experimental side, it is easier to reconstruct the 4-vector of a photon than of a jet. Because of this ease of reconstruction, direct photon measurements are possible over a wider kinematic range.

1.2 Experimental Review

One caveat must be mentioned, however, in regards to the point mentioned above. A direct photon measurement suffers from potentially large backgrounds, primarily from the decays $\pi^o \rightarrow \gamma\gamma$ and $\eta \rightarrow \gamma\gamma$, in which one of the photons is not detected. There are a number of different measurement strategies that are possible, each designed to attempt to minimize the backgrounds from these meson decays.

-reconstruction; this technique involves simply measuring the positions and energies of the two photons and requiring the resultant mass to be consistent with that of the π^o or η within experimental resolution. In practice, this technique is applicable mainly for fixed target experiments due to the requirements of a large separation from the interaction point to the calorimeter and/or fine lateral calorimeter sampling. (For a π^o of 50 GeV energy, the two photons have an opening angle of about 5.5 mrad. This corresponds to a two photon separation of about 1 cm at a distance of 2 m.) Losses are inevitable even if the reconstruction technique is possible. The energy asymmetry distribution for the two photons from a π^o or η is shown in Fig. 2. For perfect detection, this distribution would be flat from 0 to 1($A=\beta|\cos(\theta^*)|$ where θ^* is the decay angle in the π^0 rest frame; since the π^0 has spin 0, the decay distribution should be flat in $\cos(\theta^*)$.) An experimental measurement of this variable usually shows a "rolloff" at high asymmetry. This rolloff is due to losses of the soft photon, either because it's beyond the acceptance of the calorimeter or the energy is too low to be successfully reconstructed. In addition, for asymmetries near zero, the two photons have their minimum opening angle and there may be a loss of π^0's due to the two photons coalescing. These losses of π^0's (and η's) can cause a significant background to direct photons; however, it's a background that can be well-calculated given the experimental measurement of the $\pi^0(\eta)$ cross sections and asymmetry distributions.

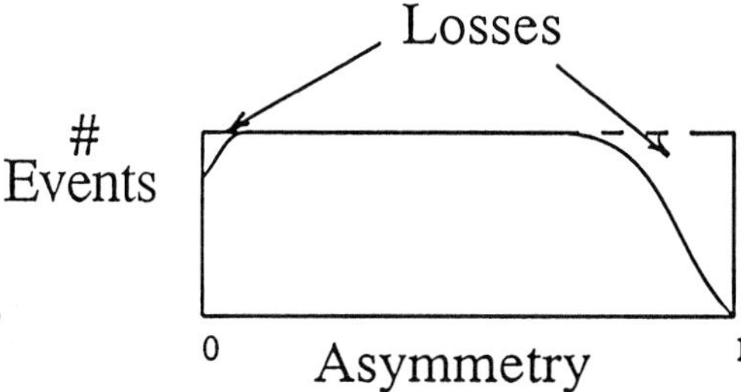

Fig.2: The π^0(or η) energy asymmetry distribution.

-conversion; the percentage of electromagnetic showers (due to direct photon candidates) that convert in the material between the interaction point and the calorimeter (typically 1-2 X_0) is measured. Showers originating from π^0 or η decays will have a conversion fraction roughly twice that of showers from direct photons. A calculation of the amount of material traversed by the photons and the observed conversion percentage allows an extraction of the direct photon fraction in the data sample. This technique works best if the direct photon fraction of the sample is at least of the same order as the π^0 and η background.

-profiles; even if the two photons cannot be resolved, a measurement of the lateral and/or longitudinal profile of the electromagnetic shower may allow a discrimination between direct photons and π^0's. Showers originating from π^0's appear broader due to the opening angle between the two photons. This technique loses effectiveness as the π^0 energy increases since the opening angle decreases as $1/E_{\pi^0}$. The longitudinal development of direct photon and π^0 showers will also differ as the average energy of a π^0 photon is half that of the direct photon. Since the longitudinal development of an electromagnetic shower varies only logarithmically with the photon energy, the differences may be subtle.

-isolation; this technique is essential in the collider regime and requires that the photon candidate be "unaccompanied" inside a cone of a certain radius R($R=\sqrt{(\Delta\eta^2+\Delta\phi^2)}$; typically R=0.5-1.0) centered on the photon direction. An application of this criterion discriminates strongly against π^0 events since the π^0 is usually accompanied by additional particles from the fragmentation of the jet. Direct photons due to the sources mentioned above ($q\bar{q}$ annihilation and gluon Compton scattering) are relatively unaffected since the photon is isolated. Photons originating from Bremsstrahlung processes, as shown in Fig.3, are also strongly discriminated against, again because of the presence of a nearby jet.

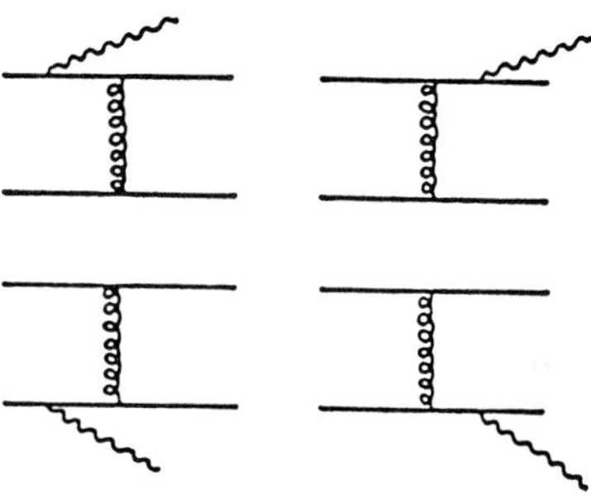

Fig. 3: Some of the diagrams responsible for Bremsstrahlung production of direct photons.

Note that the photon fragmentation functions are proportional to α/α_s. A convolution of two-body parton-parton subprocesses(of order α_s^2) with the fragmentation function of a parton to a photon results in a contribution of order $\alpha\alpha_s$ (for non-isolated photons), or the same as the annihilation and Compton scattering diagrams. Direct photon measurements at the colliders ususally involve the applicaton of an isolation cut in conjunction with one of the two techniques(conversion or profiles) mentioned above. Again, an isolation cut is usually unnecessary for fixed target experiments.

A tremendous amount of direct photon data is available, taken in a variety of kinematic regimes and with a variety of particle types.[2] Only the collider experiments and one fixed target experiment(E706) are still actively taking data. This review will concentrate on data presented in the last two years.

1.3 Phenomenology Review

As was mentioned in Section 2.1, the NLO (or equivalently the next-to-leading logarithm) calculations for inclusive direct photon production have been available for several years.[3] At the leading logarithm level, the magnitude of the photon cross section depends sensitively on the exact scales that are used in the structure functions and in α_s. (At the leading log level, there is an equivalence among all "large scales.") When the next-to-leading logarithm terms are included, some of this sensitivity is reduced since changes in the scales are partially compensated by sub-leading logarithms. This is demonstrated in Fig. 4. A cross section appropriate to CERN experiment WA-70 is plotted versus the relative scale Q/p_T (Q is used both in α_s and in the structure functions) for a leading log and next-to-leading log calculation. A "reasonable" variation of this scale is between the limits $Q/p_T=0.5$ and $Q/p_T=2.0$. The NLO calculation does have less sensistivity to the exact value of the scale. However, between the limits mentioned above, the cross section still varies by somewhat less than a factor of 2. (Note that cross sections at low x_T and large energies will have less of a sensitivity to the value of the scale used. As Q increases α_s will decrease. However,this will be partially compensated by increases of F(x,Q).)

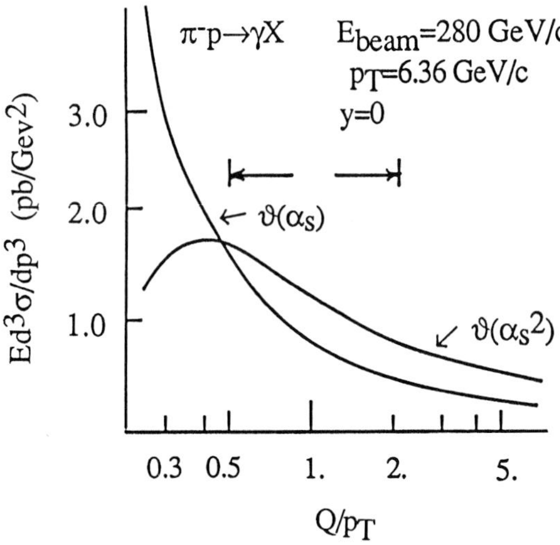

Fig. 4: A direct photon cross section (LO and NLO) plotted as a function of the relative scale Q/p_T.

One choice, by the authors of the NLO calculation has been to go beyond perturbation theory and to attempt to minimize the remaining sensitivity of the cross section to the choice of scales by choosing two new scales (μ,M) according to an optimization procedure.[4] (Here, μ is the renormalization scale, used in the argument of α_s; M is a factorization scale, used for evaluation of structure functions.) The justification of this procedure is that an all-orders calculation of the cross section would produce a result insensitive to the scales chosen. A calculation of that type is not possible but it is possible to impose on the truncated series the same stability condition as exists for the all-orders cross section, i.e. $\mu\partial\sigma/\partial\mu=0$ and $M\partial\sigma/\partial M=0$. This is accomplished by choosing the value of μ and M for which the above conditions are satisfied. This is known as the Principle of Minimal Sensitivity(PMS).[5] (Another optimization technique consists of choosing scale(s) such that the higher order terms in the cross section vanish. This is known as the technique of Fastest Apparent Convergence (FAC).[6]) The point given by the PMS conditions is simply the saddle point in the NLO cross section plotted as a function of μ and M, as shown in Fig. 5.

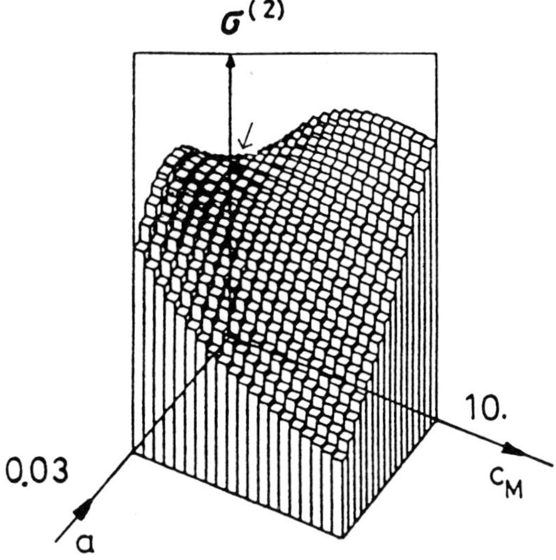

Fig. 5: The renormalization scheme and factorization scheme dependence of the second order QCD invariant cross section $\sigma^{(2)}$ for $pp\rightarrow\gamma X$ at $p_T=10$ GeV/c, $\sqrt{s}=63$ GeV. Here, $a=a(\mu)(=\alpha_s/\pi)$ and $c_M=M^2/(p_T{}^2(1-x_T))$. The saddle point is indicated by the arrow.

This choice has been somewhat controversial. The scales obtained by this procedure tend to be small, usually less than the "reasonable" limit of 0.5 p_T stated before. The factorization scales obtained can be parameterized by $M^2 = C_M\, p_T{}^2(1-x_T)$, where C_M is usually on the order of 0.1-0.4. The renormalization scale μ does not lend itself to such easy parameterization. (Note that the optimization procedure must be carried out at each point in phase space.) For WA-70, the scale μ can be parameterized as $\mu^2= 0.49p_T{}^2 -3.93p_T+8.28$, or from ~0.5 GeV^2 to 5 GeV^2 for $4<p_T<7$ GeV/c.[7] The smaller scales result in larger theoretical cross sections. Since, typically, the direct photon data lie above the theory, the optimized cross sections are in better agreement with experimental results. Another disturbing feature is that optimization is not possible in some kinematic regions (see for example the discussion on E706 in 1.41) where scales such as $\mu=M=p_T$ give reasonable results.

One nice feature of the optimization procedure is that for $\mu=\mu_{opt}$, the magnitude of the cross section is insensitive to the value of the factorization scale. This can be seen in Fig. 6 where the cross section shows little variation as M changes by several orders of magnitude.

One point of concern in perturbative QCD involves the convergence of the perturbation series. The relative size of the next-to-leading logarithm terms, in the direct photon calculation, with respect to the leading terms, is controlled, in part, by the choices

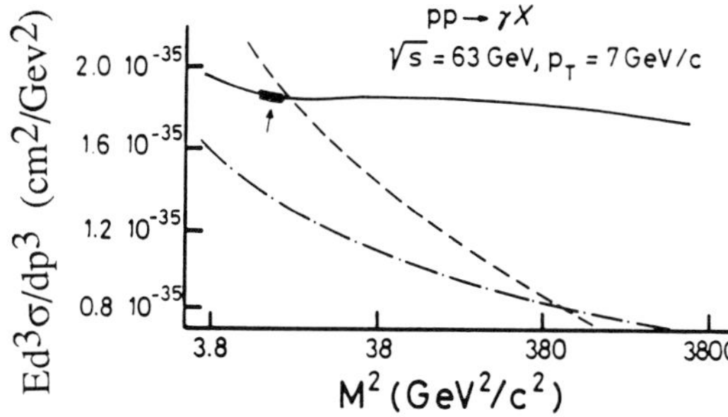

Fig. 6: The variation of the QCD direct photon cross section versus the factorization scale M^2 for (solid line) $a(\mu)=a_{OPT}$ and (dashed-dotted) $a(\mu)=a(p_T)$. For the optimized case, the saddle point is indicated by the black square. For comparison, the leading order prediction is shown by the dashed curves.

made for the factorization and renormalization scales. For optimized scales, the K factor (with K defined as (LO terms + NLO terms)/LO terms) is given by $K \cong 1 - \alpha_s/\pi$. In other words, the NLO corrections to the leading order cross section are small, as might be hoped, and negative. The effects of this choice of scales on the still higher order terms are unknown.

The procedure of optimization amounts, in some sense, to resumming the uncalculated higher order terms.[4] For example, optimization of the Drell-Yan cross section at order α_s roughly agrees with the order α_s^2 (next-to-next-to-leading log) results of Van Neerven et al., using Q as the scale.[8]

Another test of the validity of the optimization procedure is possible with the recent calculation to order α_s^3 of $R=\sigma_{tot}(e^+e^- \rightarrow hadrons)/\sigma(e^+e^- \rightarrow \mu^+\mu)$[9]; this extends by one order of α_s the existing calculations. The convergence of this ratio R $[(R^{(3)}-R^{(2)})/R^{(2)}]$ is found to be worse for optimized scales ,derived either from PMS or FAC(0.12), than for the conventional scale $\mu=M=p_T(0.09)$.[10] Note that the perturbation series does not seem to be well-behaved in this case. For the $\overline{MS}$ scheme, the third order term is 1.75 times as large as the second order term. For the optimized scheme, the third order term is the same size as the second order term(and negative).

1.4 New Results on Direct Photon Inclusive Cross Sections

1.41 E706

E706 is a fixed target direct photon experiment operating in the Meson West experimental area at Fermilab.[11] The experiment is designed to perform a large acceptance measurement of the production of direct photons and the awayside jets for $3 \leq p_T \leq 10$ GeV/c for a variety of incident particles. The first run for the experiment occured in 1987-88. This first run was primarily an engineering one, but still resulted in the accumulation of of ≈ 1 pb^{-1} of data at a beam momentum of 530 GeV/c. The data was roughly evenly divided between positive and negative running with results being available for π^-,K^-($\approx$2% of the negative beam), p and π^+($\approx$6% of the positive beam). Most data was taken with a beryllium target(with 10% of the data being on a copper target).

The key element of the experiment is the 3.0 m diameter electromagnetic calorimeter.[12] The calorimeter is lead/liquid argon with 2.0 mm lead plates and a 2.5 mm liquid argon gap. The lead plates are kept at a negative high voltage(2.5 kV) and copper-clad G-10 readout boards are kept approximately at ground. The readout boards have strips cut in them measuring, in alternate gaps, r positions and ϕ positions. The fine granularity ($\approx$5.5 mm in r) and long distance from the target (9.0 m) are necessary to resolve the two photons from π^0's of energies up to 300 GeV.

The two photon mass spectrum is shown in Fig. 7 for $p_{T\gamma\gamma}$>3.5 GeV/c. Clear π^0 and η peaks are evident. The asymmetry distribution for two photon events in the π^0 mass band(.11-.16 GeV/c^2) is shown in Fig. 8. Again, the rolloff at high asymmetry is due to losses of soft photons.

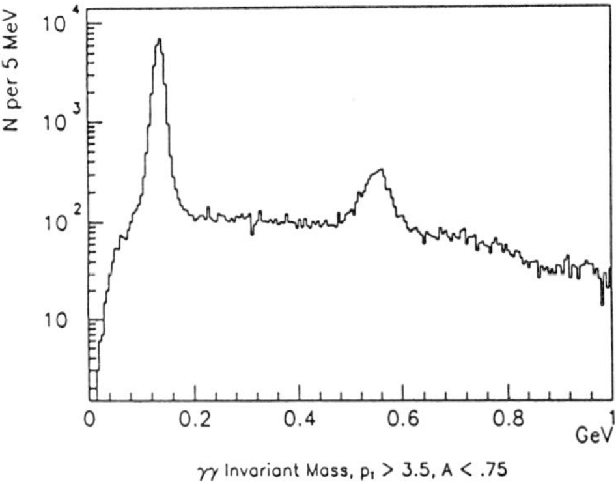

Fig. 7: The two photon mass spectrum from E706 for $p_{T\gamma\gamma}$>3.5 GeV/c.

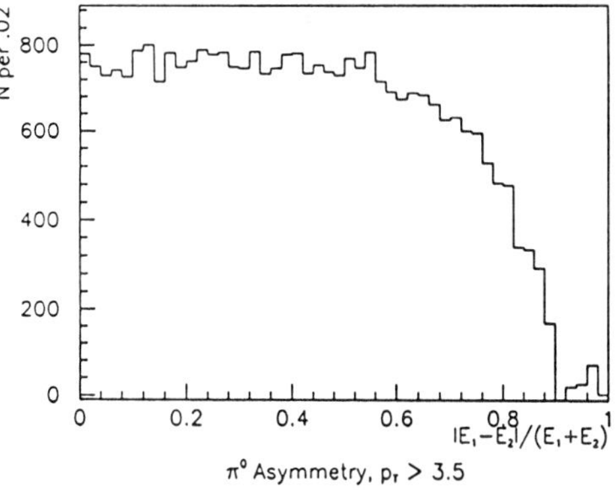

Fig. 8: The energy asymmetry distribution ($A=|E_1-E_2|/(E_1+E_2)$) for events in the π^0 mass band(.11-.16 GeV/c^2).

For the results presented here, photons are considered as direct photon candidates if, in combination with any other photon in the detector, they do not form a two photon mass either in the π^0 mass band or the η mass band with an asymmetry of less than 0.75. Backgrounds to the direct photon signal from π^0 and η decays are calculated via Monte Carlo using the experimental cross sections.

One traditional form in which to present direct photon results is in the ratio of the direct photons measured to π^0's measured as a function of transverse momentum. The raw ratio is shown in Fig. 9a for π^- Be and the subtracted ratio is shown in Fig. 9b. (Here, raw simply means that backgrounds from π^0 and η decays have not been subtracted.) In the subtracted distribution, the γ/π^0 ratio is small for low values of p_T, as expected. At the higher values of p_T, the ratio rises, exceeding unity for $p_T \geq 8$ GeV/c. (Some care must be taken ,however, in the interpretation of this direct photon data as the background calculations are still incomplete.) The rise with p_T is expected as a result of (1) the lack of a fragmentation function for the lowest order direct photon subprocesses(in contrast to π^0 production)and (2) the increasing dominance of the u $\bar{u} \rightarrow \gamma g$ diagram as x_T increases.

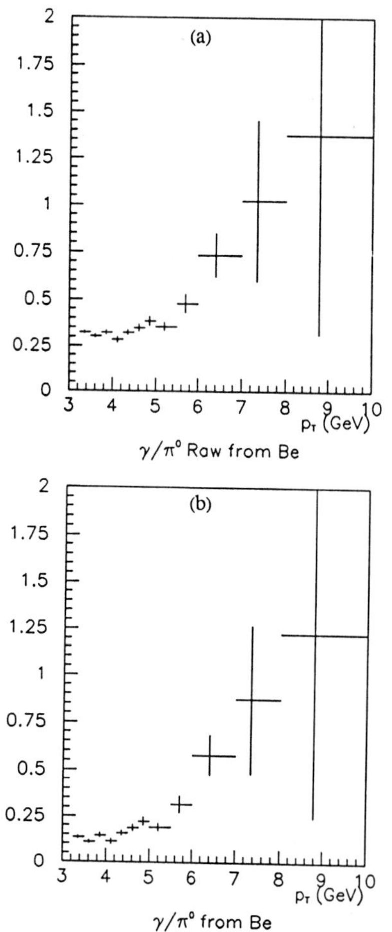

Fig. 9: (a) the raw (unsubtracted) γ/π^0 ratio for π^-Be (b) the subtracted γ/π^0 ratio for π^-Be

The photon cross section for π^- on Be is shown in Fig. 10(a). Some theoretical predictions using the calculations of Aurenche et al are also shown in Fig. 10(b).[13] (The data and theoretical predictions are plotted separately to emphasize the preliminary nature of the experimental cross sections.) One prediction uses optimized scales and a gluon distribution for the pion $\alpha(1-x)^{1.79}$ and for the proton $\alpha(1-x)^{4.0}$ [both at $Q_o^2=2$ GeV^2] The motivation for these gluon distributions will be discussed in the next section. The other two predictions use the same structure functions but scales of μ=M=p_T and μ=M=p_T/2. For the higher values of transverse momentum($p_T \geq 7$ GeV/c), the optimized predictions coincide with the predictions with μ=M=p_T/2. At lower values of p_T, the optimized slope is noticeablely steeper. Below transverse momenta of 4 GeV/c, the optimization criteria are not satisfied,i.e. there is no saddle point.

The subtracted γ/π^0 ratio is shown in Fig. 11 for protons on Be. (The same caution concerning the background subtraction applies.) The subtracted ratio is relatively flat, rising only to approximately 40% for a transverse momentum of 8 GeV/c. (With the absence of valence antiquarks, the cross section is dominated by gluon Compton scattering over the entire kinematic range.) The direct photon cross section is shown in Fig. 12(a). Again, the predictions of Aurenche et al.

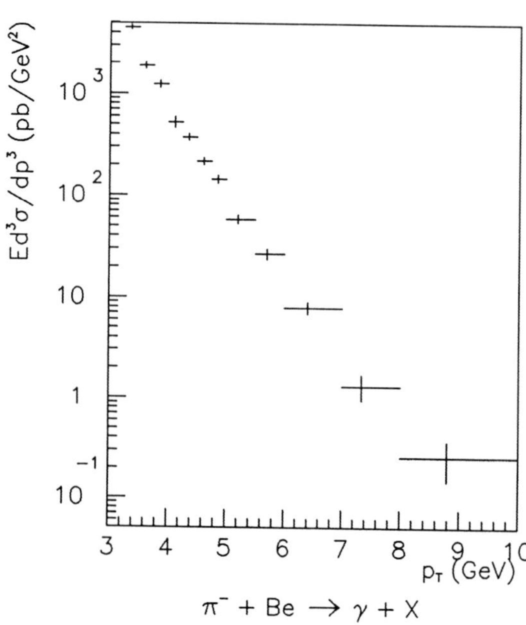

Fig. 10(a): The preliminary direct photon cross section for E706(π^-Be$\rightarrow\gamma$X).

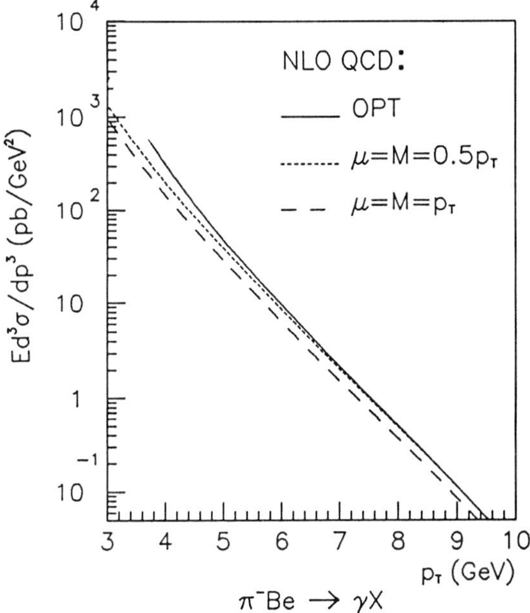

Fig. 10(b): NLO QCD predictions for π^-Be→γX at $\sqrt{s}$=31.5 GeV with optimized scales(solid line), μ=M=p_T/2 (short dashed lines), and μ=M=p_T (long dashed lines). Note that optimized cross sections are not possible for p_T<4 GeV/c.

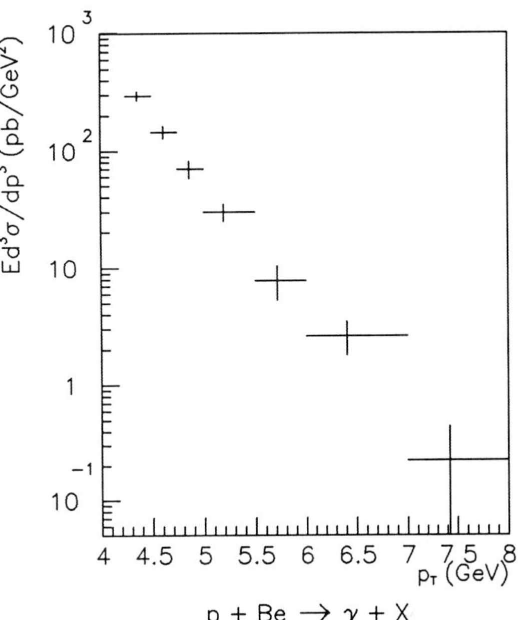

Fig. 12(a): The preliminary direct photon cross section for E706(pBe→γX).

are shown separately in Fig.12(b). The optimized cross sections are close to the cross sections with μ=M=p_T/2 over most of the p_T range with a slight increase in slope for p_T near 4 GeV/c.

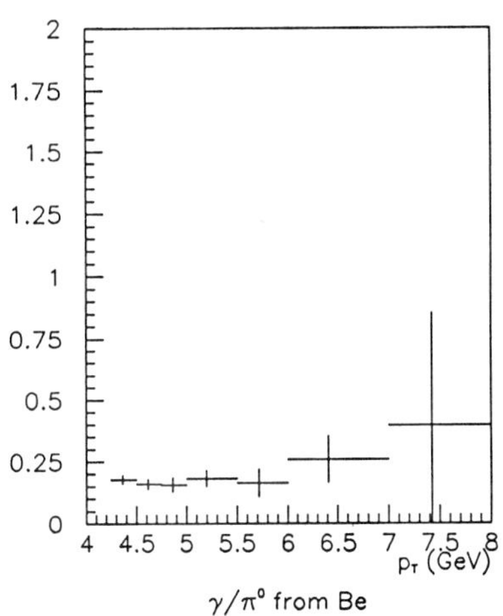

Fig. 11: The subtracted γ/π^0 ratio for pBe.

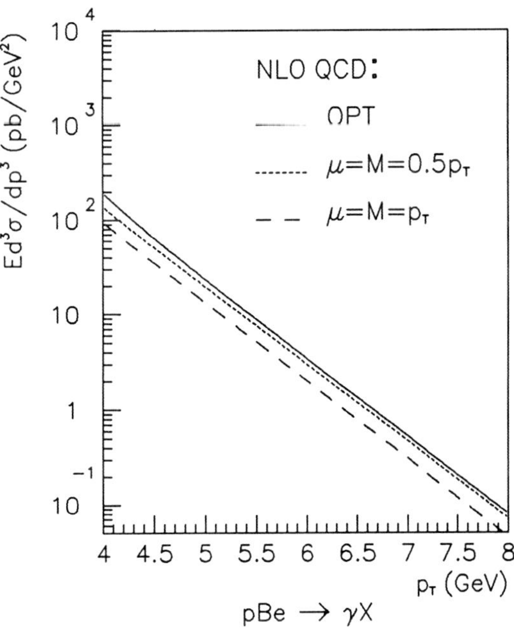

Fig. 12(b):NLO QCD predictions for pBe→γX at $\sqrt{s}$=31.5 GeV with optimized scales (solid line), μ=M=p_T/2(short dashed lines), and μ=M=p_T (long dashed lines).

1.42 CDF

The Collider Detector at Fermilab (CDF)[14] has performed a measurement of direct photon production at $\sqrt{s}$=1800 GeV for $p_T \geq$ 15 GeV/c and in the central rapidity region ($-0.8<\eta<0.8$). Analysis has been completed on the data sample from the 1987 run consisting of 27 nb^{-1}.[15] The distance from the interaction point to the electromagnetic calorimeter (1.75 m) and the granularity of the calorimeter towers (0.1 units in rapidity and 15° in azimuth) do not allow for separation of the two photons from a π^o decay. Electromagnetic showers originating from π^o's are distinguished from those originating from single photons by measuring the lateral shower profile with a single MWPC embedded in the calorimeter at approximately shower maximum, 6 radiation lengths. The sample size is 1.5 cm, compared to a typical photon shower FWHM of 0.6 cm.

An isolation cut is applied to each photon candidate by requiring that the total energy measured in the calorimeters inside a radius R ($=\sqrt{(\Delta\eta^2 + \Delta\phi^2)}$) of 0.7 be less than 1.15 times the E_T of the photon. As was mentioned earlier, this strongly discriminates against π^o events. It also discriminates against production of photons from bremsstrahlung processes(which form the bulk of the cross section in this x_T region), but the backgrounds that mimic this process are too large to allow a direct measurement.

For low E_T showers ($\leq$30 GeV/c) the shower profile measured in the MWPC for most π^o initiated showers will be measureably broader than for a corresponding shower initiated by a single photon. A χ^2 is calculated which indicates the level of agreement between a measured shower profile and that expected for a single electromagnetic shower. A significant fraction of showers from π^o's will have high χ^2 while those from single photons will not. The drawback to the technique is that, for high E_T, π^o showers start to become indistinguishable from photon showers. This can be seen in Fig. 13 where the efficiency for the

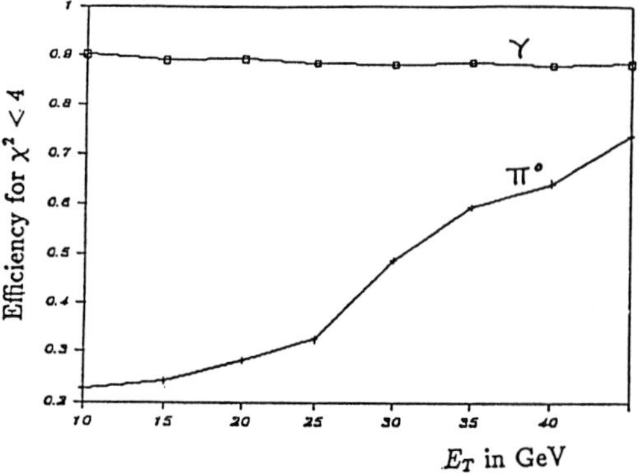

Fig. 13: The efficiency for the $\chi 2$ to be less than 4 for a fit to the lateral profile of the showers to a single shower shape.

shower χ^2 to be less than 4 is shown for both π^o's and single photons. The effective limit of this technique is 30-35 GeV/c. From fits to the observed χ^2 distributions, the direct photon cross section can be measured for $15<E_T<30$ GeV/c. The experimental cross section is shown in Fig.14 along with NLO predictions with optimized scales and with $\mu=M=p_T$[16]. The data are consistent with either choice of scale.

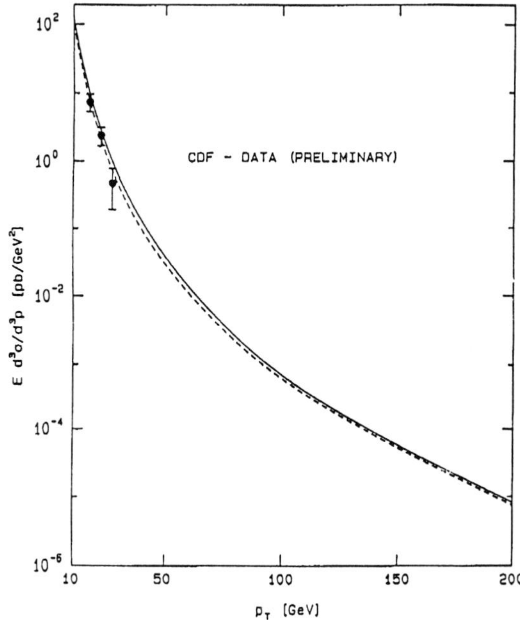

Fig. 14: The CDF direct photon cross section. Also shown are NLO predictions with optimized scales(solid line) and $\mu=M=p_T$(dashed line).

1.5 Structure Functions Fits and Comparisons to Data

1.51 Proton

Direct photon production, especially for pp scattering, is very sensitive to the strength of the gluon distribution, due to the dominance of the gluon Compton scattering term. A recent analysis[17] utilized the available high statistics direct photon data (primarily from WA-70[18]) in conjunction with DIS results (primarily from BCDMS[19]), to attempt to determine the gluon distribution in the proton. In principle, the gluon distribution can be determined solely from DIS data. However, as the gluon enters into DIS only as a second order process, the sensitivity is not as great as with direct photon production. This point will be discussed further below.

The following parameterization was used to describe the distribution functions in the proton (at $Q_o^2 = 2$ GeV^2):[20]

$$x(u^{val}(x)+d^{val}(x))=N(\eta_1,\eta_2,\gamma_{ud})x^{\eta_1}(1-x)^{\eta_2} *(1+\gamma_{ud}x) \quad (1)$$

$$xd^{val}(x) = N(\eta_3,\eta_4)x^{\eta_3}(1-x)^{\eta_4} \quad (2)$$

$$2x(\bar{u}(x) + \bar{d}(x) + \bar{s}(x)) = A_s(1-x)^{\eta_s} \quad (3)$$

$$xg(x) = A_g(1-x)^{\eta_g} \quad (4)$$

A fit to the BCDMS data alone (only hydrogen and deuterium data) yielded the following results:

$\eta_g=6.35\pm1.76$
$\Lambda_{\overline{MS}}=.209\pm.027$ GeV
$\eta_s=9.92\pm0.72$
$\chi^2=147.5/144$ DOF (statistical errors only)

The fit has a quite reasonable χ^2, yielding a fairly soft value for the gluon exponent of 6.35. However, as shown in Fig. 15,if the gluon exponent is varied between 3.5 and 8.5 the χ^2 only changes by 2. The fact that the χ^2 values are almost independent of of η_g is related to the above parameterization. Large variations of η_g are compensated by small changes of the parameters for the quark distributions(mainly sea-quark) and of $\Lambda_{\overline{MS}}$. (Note the correlation between η_g and the value of Λ.)

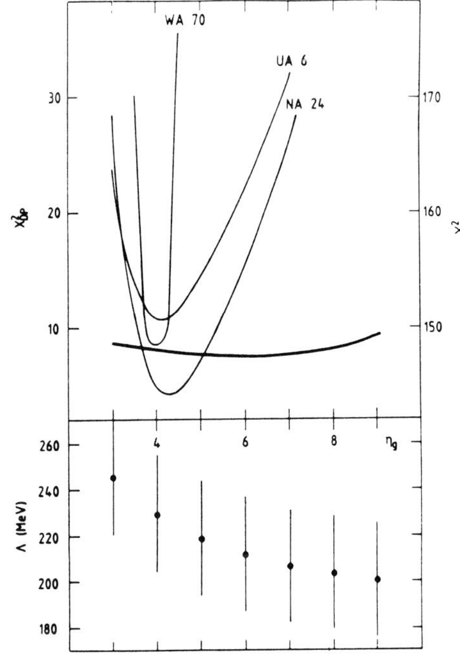

Fig. 15: The χ^2 of fits to DIS(BCDMS) and direct photon cross sections as a function of η_g. Variation of the fitted $\Lambda_{\overline{MS}}$ value.

A different approach is to take the fits to the BCDMS data for different values of η_g and calculate, beyond leading order, the direct photon cross section (using optimized scales). The χ^2 dependence for direct photon data is sharply peaked around a gluon exponent of approximately 4. This joint fit gives the following values:

$\eta_g= 4.0\pm0.11$(stat) (+0.8 -0.6) (sys)
$\Lambda_{\overline{MS}}= 230 \pm 17$ (stat) $\pm$ 50 MeV
$\chi^2_{WA-70}= 7.9$
$\chi^2_{TOT}=\chi^2_{BCDMS} + \chi^2_{WA-70} = 156$ (for 151 DOF)

The corresponding error ellipse is shown in Fig. 16 for $\Delta\chi^2_{TOT}=1$ and $\Delta\chi^2_{TOT}=4.61$(statistical errors only). The systematic errors quoted above for $\eta_g(\Lambda_{\overline{MS}})$ are due primarily to the data from WA-70(BCDMS). This gluon distribution is plotted in Fig. 17 along with Duke-Owens I.[21]

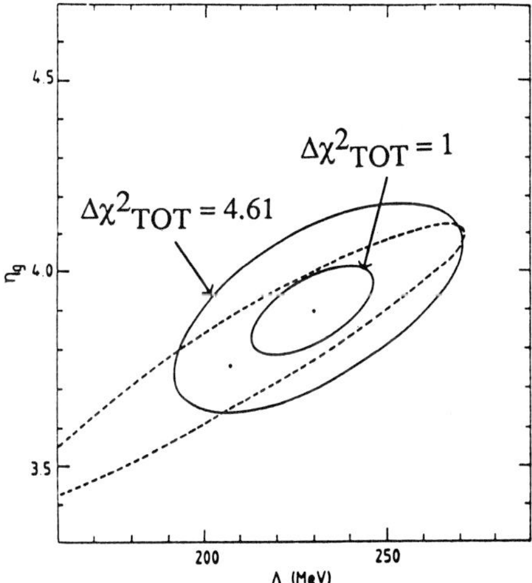

Fig. 16: Error ellipses corresponding with $\Delta\chi^2_{TOT}=1$ and $\Delta\chi^2_{TOT}=4.61$(90% CL). The dashed contour is the $\Delta\chi^2_{WA70}=1$ contour.

If the scales are chosen as $\mu=M=p_T$, the resulting χ^2 for WA-70 is 13.8 with $\Lambda_{\overline{MS}}=257$ MeV and $\eta_g=2.5$. The value of $\Lambda_{\overline{MS}}$ is largely constrained by the BCDMS data. The value of η_g decreases to 2.5 to fit the normalization of the WA-70 data. The flatter predicted p_T spectrum leads to a deterioration of the χ^2.

As might be expected, the gluon distribution given above provides a good description of the direct photon data from WA-70. It also provides a good description of the world's other pp direct photon data[18,22] as shown in Fig. 18 where the ratio

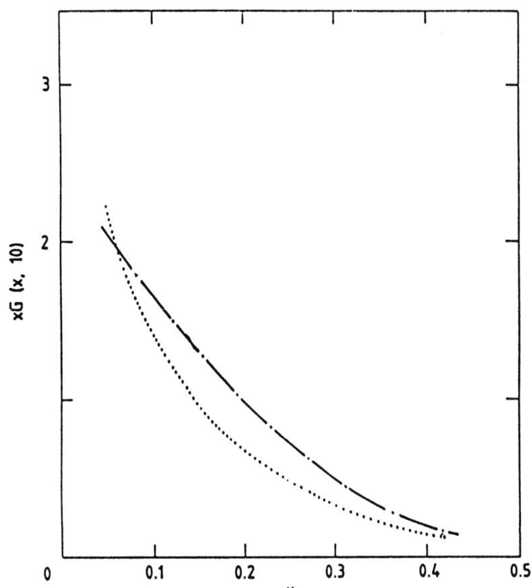

Fig. 17: A comparison, at Q^2=10 GeV^2, of the DOI gluon structure function(dash-dotted) and the gluon distribution discussed in the text $\alpha(1-x)^4$(dotted).

Data/Theory is plotted. (Theory =NLO calculation with optimized scales.) Good agreement, with statistical errors alone taken into account, is seen for most of the x_T range with the exception of some of the data from the ISR. Shifting the R806 cross section upwards within their systematic errors(by about 15%) leads to a much better agreement. An important check of this fitting procedure will be provided by experiment E706 whose data will overlap the x-ranges of the ISR and CERN fixed target experiments. Also shown is a comparison to the WA-70 data with the cross sections calculated with scales $\mu=M=p_T/2$. Scales chosen this way seem to give a description of the data similar to the optimized scales.

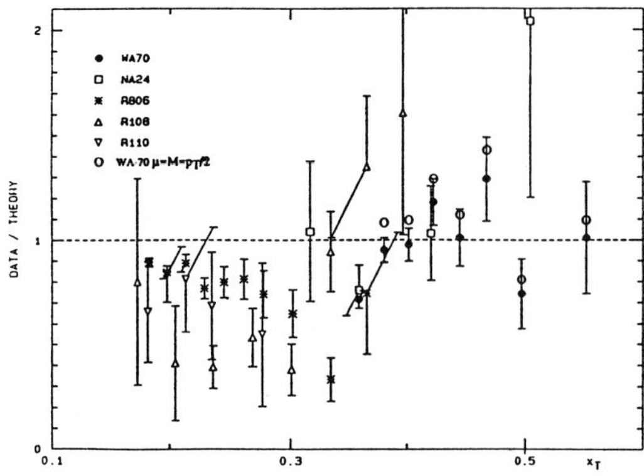

Fig. 18: Ratio Data/Theory for fixed target and ISR pp→γX experiments.

The same ratio (Data/Theory) is shown in Fig 19 for $\bar{p}p$ data[23], where the q $\bar{q}$ annhilation term is also present. For comparison purposes, the NLO cross sections for UA-6 are also shown with $\mu=M=p_T/2$. Again, the descriptions of the data are equally good.

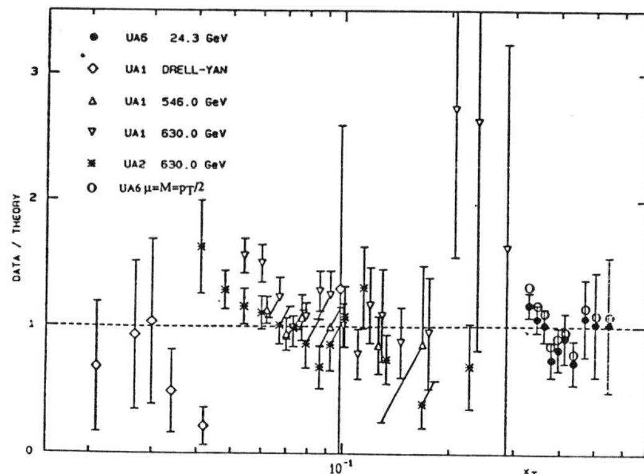

Fig. 19: Ratio Data/Theory for p $\bar{p}$→γX.

A NLO calculation[24] using a physical scale $\mu=M=p_T$ (and introducing a parton intrinsic k_T with a gaussian distribution and $<k_T>=0.7$ GeV/c) favors a gluon distribution of intermediate softness, between DO-I and DO-II. Allowing the scale (μ=M) to vary between $p_T/2$ and $2p_T$, in fact, allows either DO-I or DO-II to agree with the data.

1.52 Pions

A similar analysis for pion structure functions[25] was performed using direct photon[26] and Drell-Yan[27] data from $\pi^{\pm}$ beams. In addition to gluon Compton scattering, the q $\bar{q}$ annihilation term is also present for both π^+ and π^-. For π^- the annihilation term dominates at high x_T while for π^+, the Compton term dominates in all kinematic regions.

The proton structure functions used were those discussed in the last section. The pion structure functions, parameterized at $Q_0{}^2$ = 2 GeV^2, are given by the forms

$$xV(x)=N_V(\alpha,\beta)x^{\alpha}(1-x)^{\beta} \tag{5}$$

$$xS(x)=2x(\bar{u}(x)+\bar{d}(x)+\bar{s}(x))=N_S(1-x)^{\delta} \tag{6}$$

$$XG(x)=N_g(1-x)^{\eta} \tag{7}$$

for respectively, the valence, sea and gluon distributions. The parameter β (=0.93) was taken from the data of E615, which covers the largest x range. The valence contribution $s<xV(x)>=2\alpha/(1+\beta+\alpha)$ was left as a free parameter in the fit The sea distribution was taken from NA-3 with

δ=8.4±2.5 and <xS(x)>=0.19±0.06. The parameters were evolved from Q^2=25 GeV^2 down to the reference value of 2 GeV^2.

The NLO cross section calculations for direct photon production utilized scales either determined by the Principle of Minimal Sensitivity or given by $\mu=M=p_T$. The free parameters in the fit were Λ(in α_S), η and <xV(x)>. The data are well described when either PMS or conventional scales are used. However, for PMS scales the value of Λ in α_S (Λ=231 MeV in used in the structure functions for all fits) is 222 MeV while it turns out to be 693 MeV with conventional scales ($\mu=M=p_T$).

With PMS scales (and with Λ in α_S fixed to 231 MeV), the gluon structure function obtained is given by $xG(x)=1.1(1-x)^{\eta}$ with $\eta=1.70\pm0.19(stat)\pm 0.28(sys)$. This gluon distribution is shown in Fig. 20. It lies somewhat below previous determinations based on a leading log analysis of J/ψ production.[28] The valence distribution is given by 2<xV(x)> = 0.416±0.020(stat)±0.042(sys), consistent with most previous determinations. The NLO cross sections using these structure functions are shown in Fig. 21a(21b) for WA-70(NA-24).

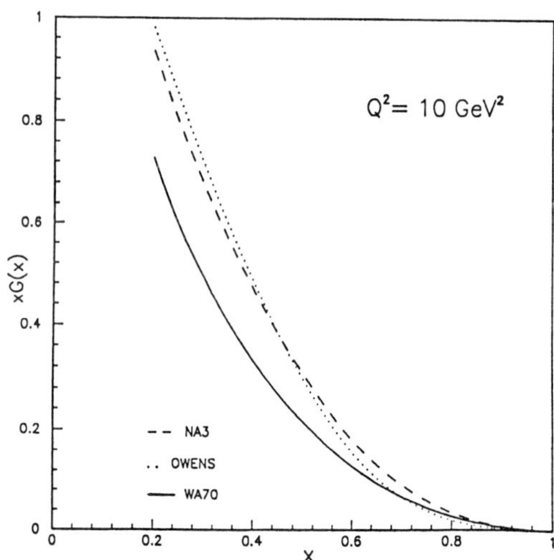

Fig. 20: The gluon structure function of the pion at Q^2=10 GeV^2 from the WA-70 experiment compared to previous determinations.

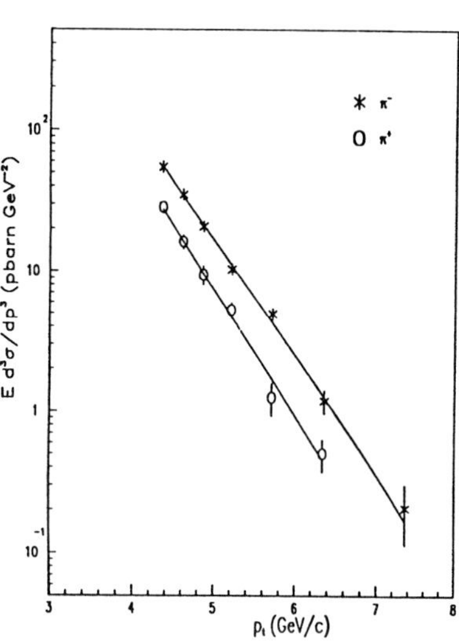

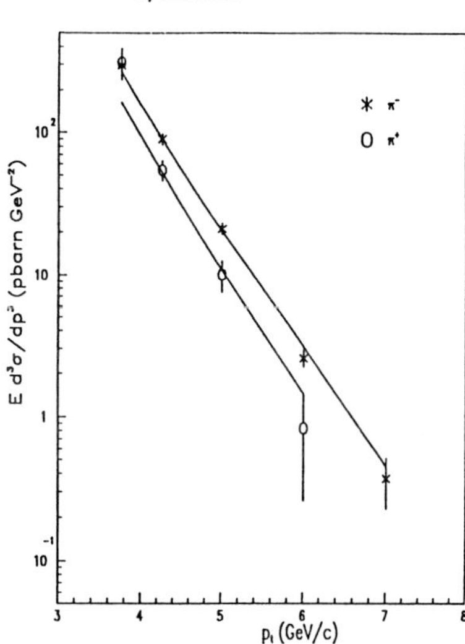

Fig. 21: The direct photon cross sections for $\pi^-p\rightarrow\gamma X$ and $\pi^+p\rightarrow\gamma X$ from (a) WA70 and (b) NA-24 and QCD predictions.

1.6 Event Structure in Direct Photon Events

1.61 WA-70

The structure of events associated with the production of direct photons in π^-p, π^+p and pp collisions has been studied using data from WA-70.[29] The integrated luminosities for π^-p,π^+p and pp were 10.8 pb^{-1}, 1.3 pb^{-1} and 5.2 pb^{-1} for an incident momentum of 280 GeV/c. Photons were reconstructed in a lead-liquid scintillator calorimeter and charged tracks were detected in the OMEGA spectrometer.

In this kinematic regime, jet structure is not as obvious as, for example,at the colliders. One advantage, however, lies in the presence of only one recoil jet in the final state(assuming 2→2 kinematics) with the photon taking the place of the other jet. For WA-70, the recoil jet was reconstructed using an algorithm making use of the fact that the fragments of the recoil jet should have a limited transverse momentum with respect to the direction of the jet. Due to losses of undetected neutral hadrons and inefficiencies in photon and charged track reconstruction, only about 70% of the jet's momentum is reconstructed, on average. The loss is corrected by assuming that the jet has the same transverse mometum as the direct photon. This procedure neglects the instrinsic transverse momentum of the partons and any transverse momentum acquired by the beam or target jets in the fragmentation process. A test of the algorithm using the LUND Monte Carlo[30] with parton showers and a Gaussian primordial k_T of the partons (σ=0.44 GeV) indicates that the corrected jet momentum is centered on the parent parton momentum with a σ of approximately 15%. For $p_T \geq 4$ GeV/c, the number of photon events with reconstructed jets were 18063,1425, and 4067 for π^-p,π^+p and pp(with π^0 backgrounds of 30%,44% and43%).

The fragmentatation distributions for direct photon events (with backgrounds subtracted) $D(z_{||})=(1/N_{jets})\ (dN/dz_{||})$ (where $z_{||}=p_{||}/p_{jet}$), are shown in Fig. 22. For high $z_{||}$, the proton data lies above the π^- data. Since 50% of the π^- events have a gluon jet in the final state (and almost all of the proton events have quark jets), this may be an indication of subtle differences between quark and gluon jets. The solid curve is a prediction from TASSO[31] for the subprocess center of mass energy appropriate for these measurements. A χ^2 comparison between the TASSO extrapolation and the direct photon data (assuming a systematic error of 15% added in quadrature to the statistical error) gives a $P(\chi^2)$=78% for quark fragmentation of pp data while the $P(\chi^2)$=6% for the π^-p data.

The ratio of positive to negative particles is shown in Fig. 23 a,b as a function of $z_{||}$. For the π^- beam, the ratio is approximately 1 over the entire range. For both π^0 and direct photons the recoil jet is almost equally likely to originate from a gluon, positively charged quark or negatively charged quark. For incident protons, the ratio rise to approximately 2 at high $z_{||}$ opposite π^0's (as expected from quark counting), while opposite direct photons the ratio rises to over 4. This indicates the dominance of the gluon Compton scattering off a u quark, $gu \rightarrow \gamma u$.

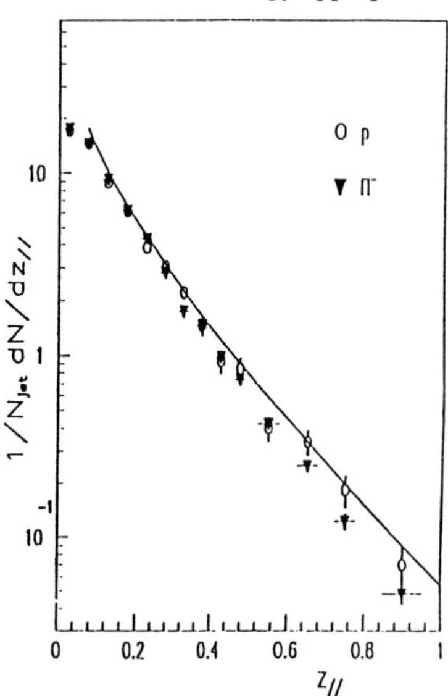

Fig. 22: Normalized $z_{||}$ distributions of pp (circles) and π^-p (triangles) for direct γ events after background subtraction. The errors are statistical only.

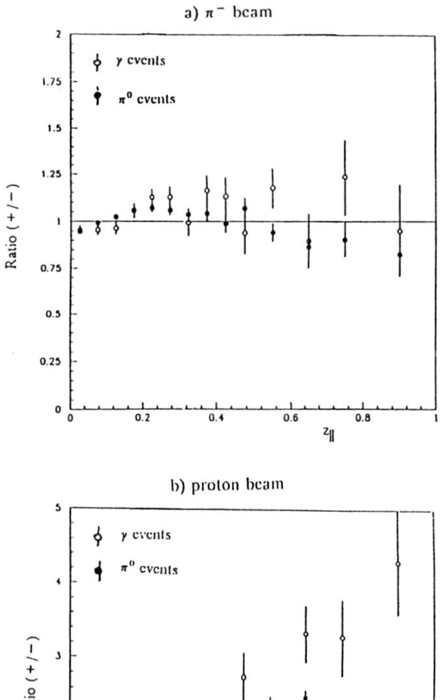

Fig. 23: Ratio of positive over negative hadrons as a function of $z_{||}$, comparison of direct γ (open circles) and π^0 events (full circles). The errors are statistical only. The background on direct γ events has been subtracted.

1.62 E706

E706 deals with jets in a slightly higher p_T range than WA-70 and as a result the jet structure is somewhat clearer. One event from the '87-'88 run is shown in Fig. 24. (The figure shows the particle distributions in ϕ with the length of the vector indicating the value of transverse momentum that the particle has. Charged particles are shown by solid lines, photons by dashed.) The event shown in Fig. 24 is interesting in that it shows that the $2\to 2$ assumption is not always valid at these energies. Opposite the direct photon candidate of approximately 7 GeV/c are two clear jets.

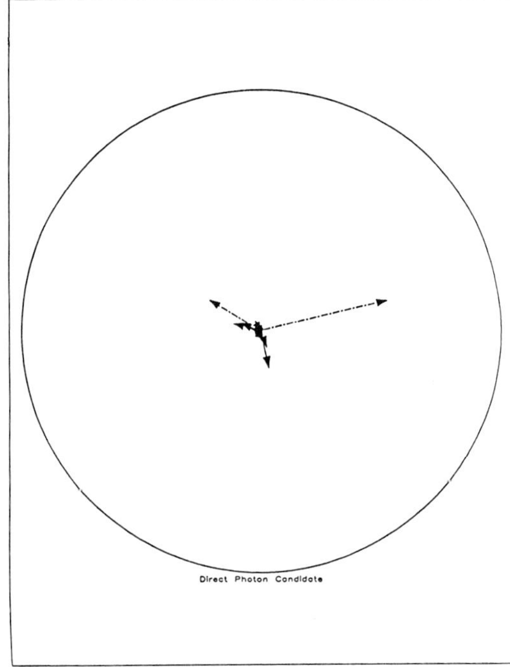

Fig. 24: A direct photon event from E706. The length of the vector indicates the size of the transverse momentum. Photons(charged particles) are dashed(solid).

The ϕ distributions of particles opposite π^0's and direct photons (background subtracted) are shown in Fig. 25. Opposite the π^0 or photon the distributions look very similar. On the same side, the π^0's clearly are accompanied by additional particles. (The x's indicate the level expected from minimum bias events.)[32] A challenging and very interesting measurement to perform is the percentage of Bremsstrahlung processes in direct photon production. This should allow the extraction of the fragmentation function of a quark to a photon. Bremsstrahlung photons unlike the $2\to 2$ subprocesses discussed earlier, will tend to be accompanied by same-side hadrons. Bremsstrahlung processes are not expected to play an important role in direct photon production at fixed target energies(approximately 20-30% of the total cross section for p_T=4 GeV/c at $\sqrt{s}$=31.5 GeV), but the inclusive nature of the measurement technique allows an extraction of the Bremsstrahlung component. Bremsstrahlung is much more important in the low x_T regions explored by the collider experiments, but the necessity of an isolation cut in the analysis precludes a measurement.

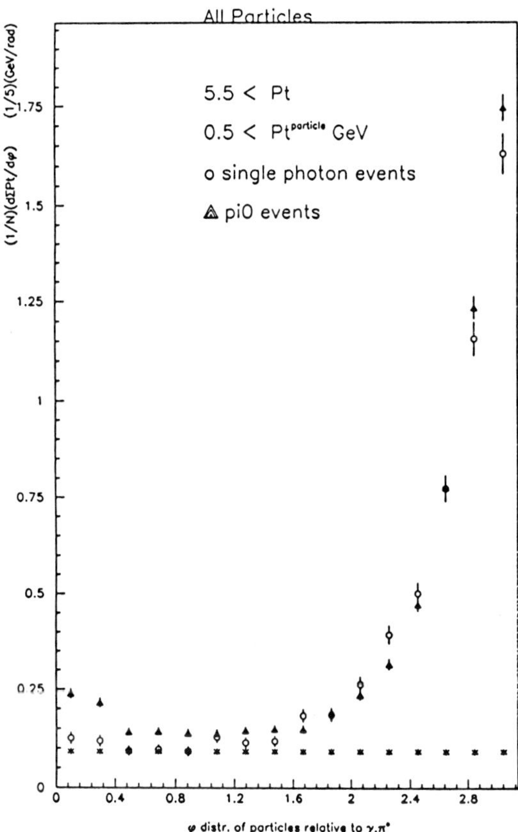

Fig.25: The ϕ distributions of particles opposite direct γ's (background subtracted) and π^0's (ϕ in radians).

1.63 UA-2

The dominant contribution to "isolated" direct photon production in the p_T range accessible to the collider experiments is gluon Compton scattering. For example, for $\sqrt{s}$=630 GeV, about 75% of the jets recoiling against a photon with $p_T\cong 16$ GeV/c are of quark origin. For comparison, only 36% of jets from two-jet events in the same p_T range are expected to be of quark origin. This allows a test for differing fragmentation properties of quark and gluon jets. The track density as a function of azimuthal distance from the trigger photon is shown in Fig. 26, from a study performed by UA-2.[33] For comparison, the same distribution is shown for events with a trigger "π^0"(actually a multi-photon jet failing some of the isolation criteria). At all azimuthal angles the track density in the π^0 sample is larger than that in the photon sample, which again is larger than the track density in events with a $W \rightarrow e\nu$ decay.

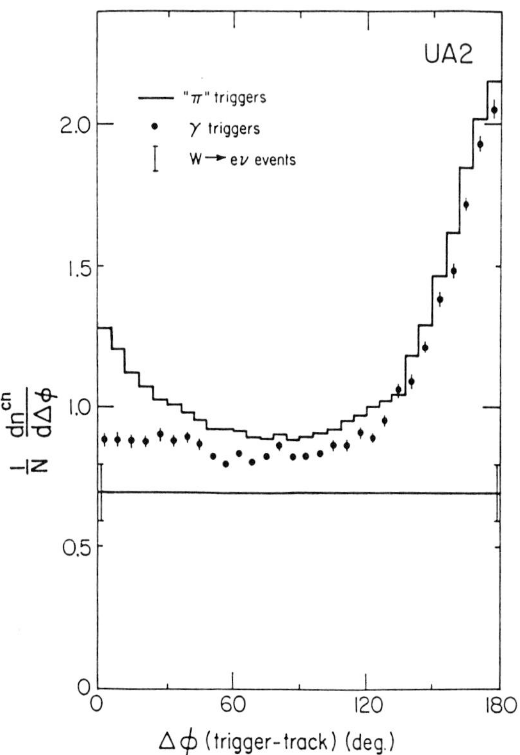

Fig. 26: The azimuthal distribution of charged tracks associated with a multiphoton jet("π^o"), a photon candidate or a $W \rightarrow e\nu$ decay. The distributions have not been corrected for background and inefficiencies.

The net result is a slight difference in the total charged multiplicity in the hemisphere opposite the trigger:

$$\bar{n}_{ch}(\text{opposite } \pi^o) - \bar{n}_{ch}(\text{opposite } \gamma) = 2.0 \pm 0.3(\text{stat}) \pm 0.7(\text{sys})$$

This difference is consistent with the expected difference between quark and gluon fragmentation using a QCD cascade model[34] but could also be explained by a different track density in the underlying event.

One interesting comparison between γ+jet and π^o+jet events that can be performed is that of the angular distribution in the parton-parton center of mass,$d\sigma/d\cos\theta^*$. Two jet production receives strong contributions from t-channel spin-1 exchanges. These exchanges are absent in direct photon production, at least at the Born level. As a result, the expectation is that direct photon production would be less peaked in the forward direction than jet production. The ratio between the scattering angle distributions of γ+jet and π^o+jet systems is shown in Fig. 27 for UA-2. Corrections have been made for background contamination and the ratio has been normalized to the ratio in the first bin. The result is consistent with QCD expectations.

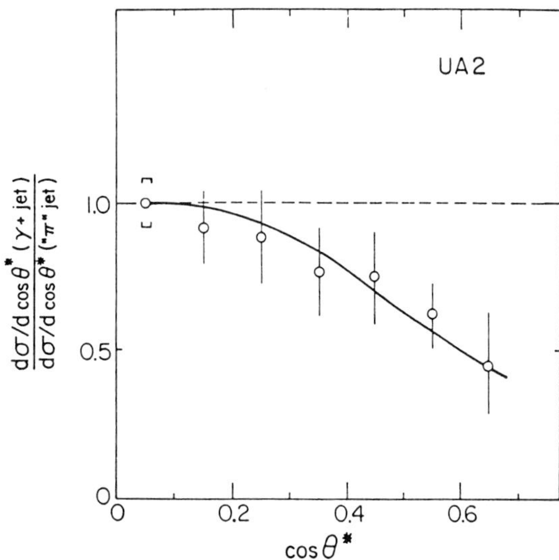

Fig. 27: Ratio between the scattering angle distributions of γ+jet and "π^o"+jet. The curve is the ratio between the subprocesses leading to γ+jet final states and two-jet final states. The error on the first bin represents the normalisation error of the points relative to the curve.

1.7 Two Photon Production

A process closely related to direct photons is the production of double prompt photons.[35] Double prompt photon production can proceed through the QED annihilation process $q\bar{q} \rightarrow \gamma\gamma$ or through gluon-gluon scattering $gg \rightarrow \gamma\gamma$(Fig. 28). The first subprocess mentioned above is similar to the $q\bar{q}$ annihilation term important in direct photon production but with the final state gluon replaced by a photon. The second subprocess can be important at low x_T since the gluon distribution is comparable to or larger than the quark distribution at low values of x.

For both of the above subprocesses, neglecting intrinsic transverse momentum effects, the photons have the same transverse momentum. (A measurement of the p_T distributions of the two photon system, in fact, can allow one of the cleanest measurements of the size of intrinsic k_T effects.) Two photon production can also proceed through the direct photon subprocesses $q\bar{q} \rightarrow \gamma g$ and $gq \rightarrow \gamma q$, where the second photon originates from Bremsstrahlung from the recoiling parton. In some kinematic regimes,this contribution can be comparable to the two

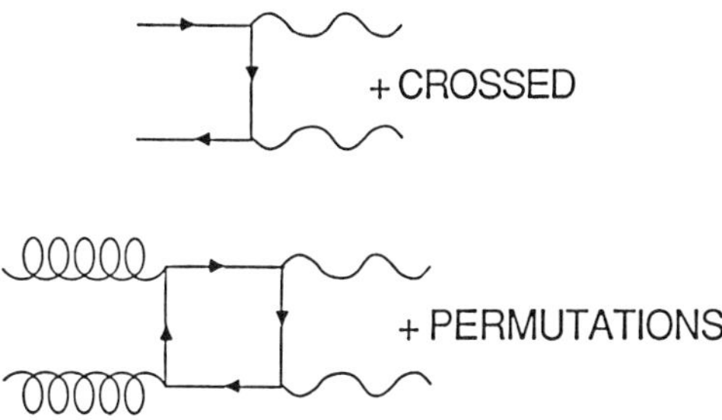

Fig. 28: Two of the subprocesses responsible for double prompt photon production.

subprocesses, $q\,\bar{q}\rightarrow\gamma\gamma$ and $gg\rightarrow\gamma\gamma$, discussed above. In general, the two photons will have unequal transverse momentum.

WA-70 has recently reported on a measurement of two photon production for π^-p at 280 GeV/c incident momentum.[7] The data sample in the study corresponded to an integrated luminosity of 7.7 pb^{-1}. The analysis required that one photon of the pair have a transverse momentum greater than 3.0 GeV/c and the other greater than 2.75 GeV/c. The event reconstruction and data selection followed the same procedure as that used for the analysis of high p_T π^0 and prompt photon events.[36] The measurement is a difficult one, as backgrounds can arise not only through subprocesses producing π^0 pairs but also through subprocesses producing direct photons where the recoiling parton fragments into a leading π^0(and the π^0 is seen as one photon).

The two photon signal obtained by WA-70 is statistically significant(6 standard deviations) after a background subtraction of 50%. The cross section, integrated for p_T>3.0 GeV/c(and corrected for the full production rapidity and ϕ ranges) is 69±11.5 pb. The data have been compared to a BLL(Beyond Leading Log) calculation using Duke-Owens I structure functions with Λ=200 MeV and $Q^2=p_T^2$[37]. (The sensitivity of the predictions to the scale is expected to be small for $\gamma\gamma$ production.) The calculation includes both the $q\,\bar{q}\rightarrow\gamma\gamma$ term with its order α_s corrections as well as the order α_s^2 box diagram. For this kinematic region the $q\,\bar{q}\rightarrow\gamma\gamma$ term is very dominant.

The BLL prediction is 57.8 pb while the leading log prediction from $q\,\bar{q}\rightarrow\gamma\gamma$ gives 39.4 pb. The differential cross section $d\sigma/dp_T$ is shown in Fig. 29 as a function of p_T(both γ's are included) along with the BLL and leading log predictions. The data agree with BLL QCD predictions both in magnitude and slope within the statistical errors.

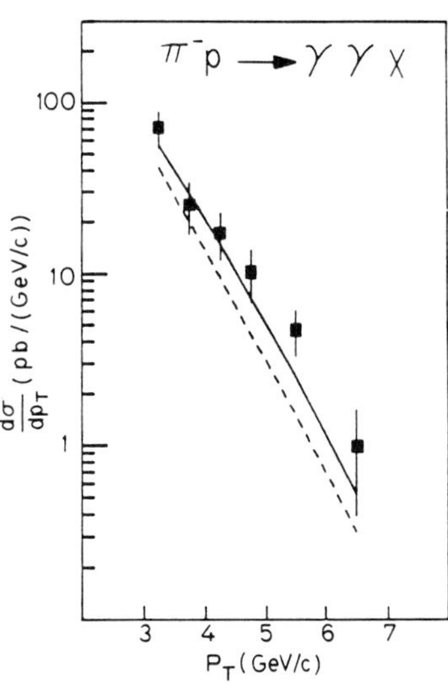

Fig. 29: The $\gamma\gamma$ cross section $d\sigma/dp_T$, where p_T is the transverse momentum of either of the two prompt photons. The dashed line is the prediction of the $q\,\bar{q}\rightarrow\gamma\gamma$ Born term; the full line is the prediction of the BLL calculation.

Photon pairs have also been observed in pp collisions at the CERN ISR with a two standard deviation signal[38] and in π^-C and pC collisions at 200 GeV/c.[39]

2. DRELL-YAN PAIRS

2.1 Low-Mass Lepton Pairs (UA-1)

A process clearly related to direct photons is that of lepton pair production. The process is somewhat complicated by the presence of another kinematic variable, the dilepton invariant mass. In the region where the dilepton mass is less than the p_T, $m^2_{\mu\mu} \ll p_T{}^2_{\mu\mu}$, the Drell-Yan process can be simply related to high p_T direct photon production by the electromagnetic coupling α multiplied by a factor that is essentially a measure of the virtuality of the intermediate photon.[40]

$$\int_{m^2_{min}}^{m^2_{max}} E\frac{d^4\sigma^{DY}}{dm^2d^3p} \cong C\,E\frac{d^3\sigma^{\gamma}}{d^3p}$$

$$C=\frac{\alpha}{3\pi}\ln\frac{m^2_{max}}{m^2_{min}} \qquad (8)$$

Application of this relationship can allow measurement of "direct photon" production at colliders in a lower p_T range than possible with direct photon techniques.

UA-1 has performed a study of low mass, high

p_T muon pairs from data at $\sqrt{s}$=630 GeV with a total integrated luminosity of 552 nb^{-1}.[41] Events were selected if they contained at least two muons with transverse momentum ($p_{T\mu}$) greater than 3 GeV/c for each muon and if the dimuon mass were less than 6 GeV/c^2. Isolation cuts were applied ($\Sigma E_T < 3$ GeV inside a cone of radius $R=\sqrt{(\Delta\phi^2 + \Delta\eta^2)}$ both to each muon and to the dimuon system to discriminate against backgrounds from heavy flavor decay.

The Drell-Yan contribution was extracted from a fit to the mass distribution for isolated unlike sign muon pairs (Fig. 30).

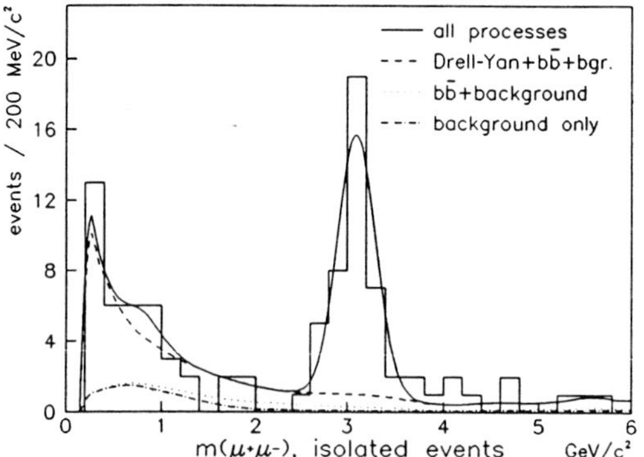

Fig. 30: The mass distribution for 93 isolated unlike-sign muon pairs. Also shown are contributions from individual processes. In addition to the processes mentioned explicitly, the curve "all processes" includes contributions from J/ψ decay and from the decays of ρ, ω, φ, η and η'.

An upper limit for the dimuon mass of 2.5 GeV/c^2 was chosen. This cut at 2.5 GeV/c^2 was chosen to eliminate the J/Ψcontribution. For this mass range ($m_{min}=2m_\mu$ and m_{max}=2.5 GeV/c^2) the factor C relating the Drell-Yan and direct photon cross sections is given by $C=\alpha/1.91$. The invariant Drell-Yan cross section scaled according to this relation is compared to direct photon data[42] and NLO predictions[4]in Fig. 31. The Drell-Yan data are in good agreement and allow direct photon measurements to be extended to lower values of p_T than would otherwise be possible.

2.2 Sea-to-Valence Ratio (E615)

Lepton pair production by π^+ and π^- beams on a nuclear target allows a direct measurement of the ratio of sea to valence quarks in the nucleon. At high energies, the dominant contribution to lepton pair production is quark-antiquark annihilation. For an isoscalar target, the lepton pair production ratio (π^+/π^-) should be 1/4, if only valence antiquarks contribute. Deviations from this result indicate the participation of antiquarks from the sea.

Experiment E615 at Fermilab has performed this measurement using dimuon data taken with π^+ and π^-

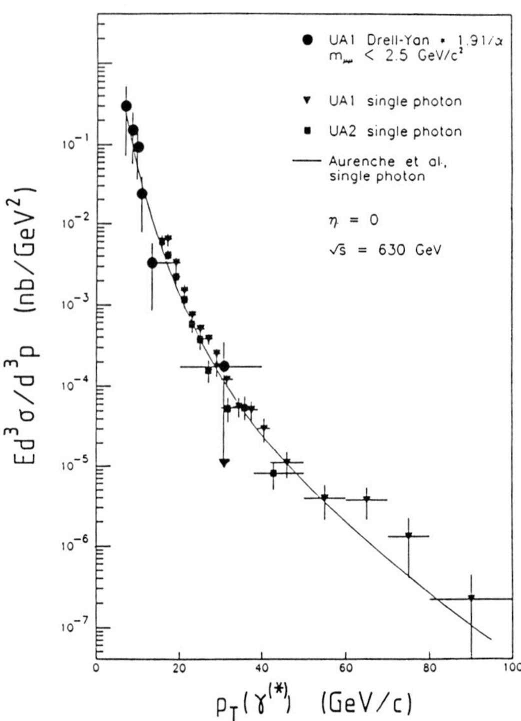

Fig. 31: Comparison of the Drell-Yan and single photon cross sections. The Drell-Yan cross section is scaled by a factor of 1.91/α.

at 250 GeV/c on a tungsten target.[43] The data sample used in this analysis consisted of ~5×10^3(~3×10^4) $\mu^+\mu^-$ pairs with $m_{\mu\mu}$<4.05 GeV/c^2 for the positive (negative) beam. A large number of J/ψ events were also present in the data. The J/ψ's were used to determine the proton contamination in the positive beam(~46%) and the relative positive to negative beam normalization. To eliminate muon pairs from Υ decays an upper limit on $m_{\mu\mu}$ of 8.55 GeV/c^2 was also used. To insure no significant contribution from the pion sea, the momentum fraction x_π of the annihilating quark or antiquark in the pion was required to be greater than 0.36. These criteria constrained the nucleon momentum fraction to be $0.04 < x_N < 0.36$.

The ratio $R=\sigma(\pi^+)/\sigma(\pi^-)$ was obtained for intervals of x_N, correcting the data in each interval for the proton contamination. Neglecting the pion sea, an expression can be obtained for the sea to valence ratio

$$\frac{S_p}{V_p^u+V_p^d} = \frac{2R-0.5+\varepsilon(4R+1)(V_p^u-V_p^d)/(V_p^u+V_p^d)}{5(1-R)} \quad (9)$$

where $V_p{}^u$ and $V_p{}^d$ are the proton u and d x_N distributions and S_p is the x_N distribution of the nucleon sea. $\varepsilon(=Z/A-1/2)$ is the deviation of the target from isoscaler. (For the tungsten target, $\varepsilon=-0.095$.)

The sea to valence ratio obtained from this procedure is shown in Fig. 32 together with the ratio derived from CDHS[44] and the parameterization from Duke and Owens[28] and EHLQ[45]. The results are in agreement with the previous CDHS results and are in best agreement with the Duke-Owens II parameterization. A parameterization of the sea to valence ratio in the form $(1-x_N)^{\omega}x_N^{-0.5}$ results in $0.120(1-x_N)^{7.5}x_N^{-0.5}$ with a χ^2 of 2 for 9 degrees of freedom.

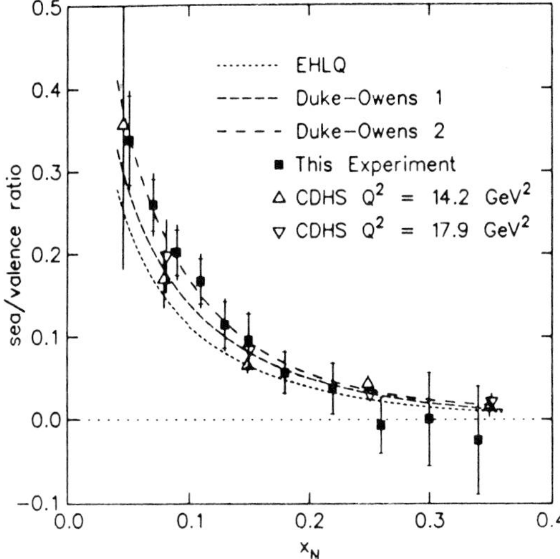

Fig. 32: The measured sea to valence ratio for $0.04<x_N<0.36$. The ticks on the error bars represent the statistical component alone and the total error bars represent statistical and systematic errors added in quadrature.

2.3 New Dimuon Results(E605)

Fermilab experiment E605 has recently reported a measurement of the mass spectrum from 6-18 GeV/c^2 of dimuons produced in 800 GeV proton-copper collisions.[46] The integrated luminosity for the measurement corresponded to 1.4×10^{42} cm^{-2}. A scaling form of the cross section,$d^2\sigma/d\sqrt{\tau}dy$ (evaluated at y=0.2) is plotted for the E605 data in Fig. 33 versus $\sqrt{\tau}$ ($\tau=m_{\mu\mu}/\sqrt{s}$). Also shown is the data from Fermilab experiment E288[47], at three different $\sqrt{s}$ values, and a next-to-leading order (order α_s) QCD calculation of Martin et al[48] for $\sqrt{s}$=19.4 GeV (dashed line) and $\sqrt{s}$=38.8 GeV (solid line). The lower yield predicted at 800 GeV (and qualitatively agreeing with the trend of the data) is a manifestation of QCD scale violation effects at high $\sqrt{\tau}$.

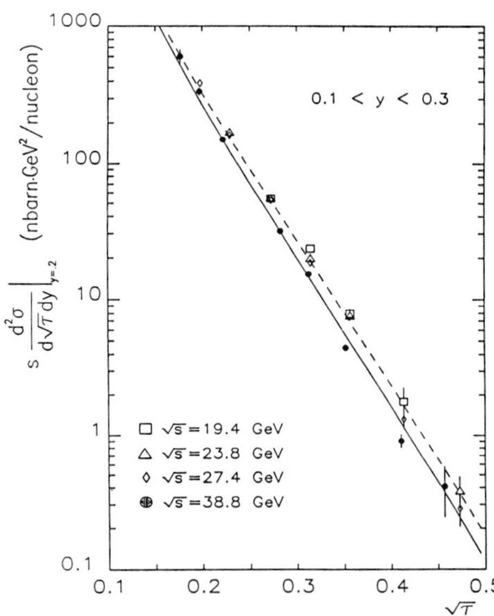

Fig. 33: Comparison of the data from E605 (solid circles) with data from E288. The dashed line corresponds to $\sqrt{s}$=19.4 GeV(solid line, $\sqrt{s}$=38.8 GeV) order α_s QCD predictions of Martin et al.

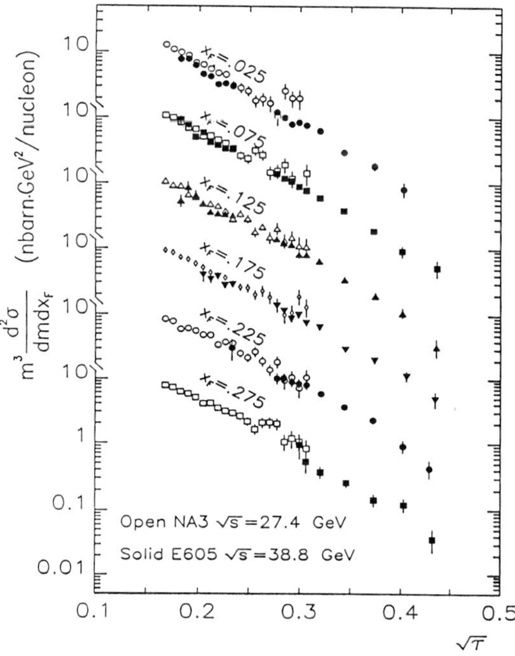

Fig. 34: Comparison of the data from E605 and NA-3 for the cross section versus $\sqrt{\tau}$ at constant x_F.

A comparison of the E605 data to that of NA-3[49], in the scaling form $m^3d^2\sigma/dmdx_F$, versus $\sqrt{\tau}$ for different x_F bins is shown in Fig. 34. The data appear consistent with the higher $\sqrt{s}$ E605 data again being slightly lower. The data sets shown in the above plots (E605, NA3, E288) are consistent within reported normalization uncertainties, possible dependence of structure functions on nuclear environment[50] and uncertainties in the perturbative calculation. The ensemble of these data sets should provide a tight constraint on hadron structure function analyses, especially the anti-quark content of the nucleon.

3. CONCLUSIONS

QCD has proven remarkably successful in describing the production of direct photons over a wide kinematic range. Advances on both the theoretical and experimental fronts make comparisons of data to theory on semi-log plots no longer adequate.

Application of optimization techniques allows greater "precision" but the technique remains controversial for the reasons stated in the paper. For most kinematic regions, optimization does not lead to a change of slope of the cross section (with respect to p_T), but rather a change in normalization. In a sense, it seems to provide the right "K-factor". A similar normalization can be obtained, in many cases, using scales of $\mu=M=p_T/2$, scales on the edge of "acceptability".

Direct photon data has proven very useful in the extraction of gluon structure functions. Many analyses use optimized scales but, as discussed above, optimized QCD predictions differ mainly by a normalization factor from predictions using "physical scales", so the shape of the extracted gluon structure function is not totally a function of the scales chosen.

To date, the NLO direct photon calculations have been for the inclusive cross section. Recently, NLO predictions have become available for γ+jet cross sections.[51] Measurements of double differential cross sections of this type will lead to new test of the precision of the theory and will allow an even greater sensitivity to the gluon structure function.

There seems to be a wealth of direct photon, Drell-Yan and DIS scattering available. Fits incorporating several different processes such as these should prove very successful in constraining both quark and gluon structure functions.

Several interesting new results should be available in the next two years. UA6 is currently working on an analysis of their pp data (at $\sqrt{s}$=24.3 GeV). A comparison of the pp data to their $\bar{p}p$ data should allow an independent measure of the value of Λ. The next fixed target run (1990) should allow E706 to increase their statistics by approximately a factor of 10(to ~10 pb^{-1}), allowing a measurement of direct photons (and awayside jets) over the p_T range 3-4 $<p_T<10$ GeV/c ($0.2<x<0.6$) for $\pi^{\pm}$,p ($K^{\pm}$). Comparison of results from beryllium and copper targets should allow a measurement of the A-dependence of direct photon production. UA2 is currently analyzing a data sample with approximately 8 times the statistics as that discussed here.[52] CDF has recently finished a run with ~4.6 pb^{-1}. The limitations of the lateral profile technique will limit the direct photon measurement to p_T<40 GeV/c, but the installation of a preconverter detector for the next run should allow the measurement of the direct photon cross section out to the highest p_T values accessible.[53] E615 is extending the analysis of their dimuon data down to masses of ≥3 GeV/c^2. It should be interesting to see if the higher twist effects observed at large x_F in their higher mass range become even more dominant.[54]

ACKNOWLEDGEMENTS

I would like to thank the conference organizers for a successful and enjoyable conference. I would also like to thank my collaborators on E706 and would like to acknowledge the following people for useful discussions: P. Aurenche,R. Baier, E. Berger, R. Blair, C. Bromberg, C. Brown,J. Collins, K.Eggert, T. Ferbel, K. Freudenreich, M. Kienzle-Foccaci,S. Mani,B. Pope,J. Rutherfoord,P. Slattery, S. Smith, W. K. Tung,M. Werlen,M. Zielinski.

REFERENCES

1. E.L. Berger, E. Braaten, R.D. Field, Nucl. Phys. B239(1984)52; J.F. Owens, Rev. Mod. Phys. 59(1987)465.

2. T. Ferbel, W.R. Molzon, Rev. Mod. Phys. 56(1984)18; J.P. Rutherfoord, Proceedings of 1985 International Symposium on Lepton and Photon Interactions at High Energies(Kyoto,1985);F. Richard, Nucl. Phys. B(Proc. Suppl)3(1988)639;T. Ferbel, Proceedings of Physics in Collision(Capri,1988)P. Strobin et al., eds.;see also the recent compilation of direct photon experiments by P. Aurenche, M. Whalley, DPDG189/04, RAL-89-106.

3. P. Aurenche, R. Baier, M. Fontannaz and D. Schiff, Phys. Lett. 140B(1984)87.

4. P. Aurenche, R. Baier, M. Fontannaz, and D. Schiff, Nucl. Phys. B297(1988)661.

5. P.M. Stevenson, Phys. Rev. D23(1981) 2916;P.M. Stevenson,Nucl. Phys. B203(1982)472; P.M. Stevenson, H.D. Politzer, Nucl. Phys. B277(1986)758.

6. G. Grunberg, Phys. Lett. 95B(1980)70;G. Grunberg, Phy. Rev. D29(1984)2315.

7. E. Bonvin et al.,Z. Phys. C41(1989)591.

8. W.L. Van Neerven, Phys. Lett. 147B (1984)175;T. Matsuma, S.C. van der March and W.L. van Neerven, Nucl. Phys. B319(1989)570;P. Aurenche, private communication.

9. S. Gorishny, A. Kataev and A. Larin, Phys. Lett. 212B(1988)238.

10. C.S. Maxwell and J.A. Nicholls, Phys. Lett. B213(1988)217; A.P. Contogouris and N. Mebarki, Phys. Rev. D39(1989)1464.

11. G. Alverson et al., XXIVth Conference on High Energy Physics, R. Kotthaus and J.H. Kuhn, eds.(1988)719.

12. F. Lobkowicz et al., Nucl. Instr. and Meth. A235(1985)332.

13. I would like to thank the authors of ref[3,4] for use of their computer program.

14. F. Abe et al., Nucl. Instr. and Meth. A271(1988)387.

15. R. Blair et al., 7th Topical Conference on Proton-Antiproton Collider Physics ed. R. Raja, A. Tollestrup and J. Yoh, World Scientific(1989)361.

16. P. Aurenche, private communication.

17. P. Aurenche, R. Baier, M. Fontannaz, J.F. Owens, M. Werlen ,Phys. Rev. D39(1989)3275.

18. M. Bonesini, et al., Z. Phys. C38(1988)371.

19. A.C. Benvenuti et al., Phys. Lett. B223(1989)485,490.

20. The $\overline{MS}$ scheme was adopted and the universal convention was used for the quark distributions. The higher order corrections were performed in the $\overline{MS}$ renormalization scheme, so Λ_{QCD} denotes $\Lambda_{\overline{MS}}$(with 4 flavors).

21. D.W. Duke and J.F. Owens, Phys. Rev.D30 (1984)149.

22. (NA-24)C. de Marzo et al., Phys. Rev. D36(1987)8;(R108)A.L.S. Angelis et al.,Phys. Lett. B94(1980)106,B98(1981)115;(R806)E. Anassontzis et al.,Z. Phys.C13(1982)277;(R110)C.W. Salgado-Galeazzi, Ph.D. Dissertation(1988)Michigan State University.

23. (UA6)A. Bernasconi et al., Phys. Lett. B206 (1988)163;(UA1)C. Albajar et al., Phys. Lett. B209(1988)385,397; (UA-2)R. Ansari et al.,CERN-EP/88-102.

24. A. P. Contogouris et al., paper No. 182 submitted to this Symposium.

25. P. Aurenche et al., paper No. 175 submitted to this Symposium.

26. M. Bonesini et al., Z. Phys. C37(1988)535;see also C. de Marzo in [22].

27. J.S. Conway et al., Phys. Rev. D39(1989)92;J. Badier et al., Phys. Rev. D36(1987)281.

28. J.F. Owens, Phys. Rev. D30(1984)943; J. Badier et al., Z. Phys. C20(1983)101.

29. Paper No. 178 submitted to this Symposium.

30. H.U. Bengtsson, G. Ingleman, T. Sjostrand: The Lund Monte Carlo for high p_T scattering, PYTHIA version 4.2, update of H.U. Bengtsson and G. Ingleman, Comp. Phys. Comm. 34(1985)251;T. Sjostrand, Jet fragmentation, JETSET version 6.2, update of T. Sjostrand, Comp. Phys. Comm. 39(1986)347.

31. M. Althof et al., Z. Phys. C22(1984)307; W. Braunschweig et al.,DESY 89-038(1989).

32. See Ph.D. theses of A. Sinanidis(Northeastern U., 1989),S. Easo(Pennsylvania State U., 1989),D. Brown(Michigan State U., 1989)

33. R. Ansari et al., Z. Phys. C41(1988)395.

34. B.R. Webber, Nucl. Phys. B238(1984)492.

35. S. Berman, J. Bjorken and J. Kogut, Phys. Rev. D4(1971)1918; C. Carimalo, M. Grozon, P. Kessler and J. Parisi, Phys. Lett. 98B(1981)105;A. P. Contogouris, L. Marleau, et al., Phys. Rev. D25(1982)2459. see also E. Berger et al., in [1].

36. See M. Bonesini et al., in refs [18,26].

37. P. Aurenche, A. Douri, R. Baier, M. Fontannaz, D. Schiff, Z. Phys. C29(1985)459.

38. T. Akesson et al., Europhysics Conference on High Energy Physics, Bari, 1985.

39. J. Badier et al., Phys. Lett. 164B(1985)184.

40. P. Aurenche, R. Baier, M. Fontannaz, D. Schiff, Phys. Lett. 209B(1988)375.

41. C. Albajar et al., Phys. Lett. B209(1988)397.

42. J.A. Appel et al., Phys. Lett. 176B(1986) 239;C. Albajar et al., Phys. Lett. 209B(1988)385.

43. J.G. Heinrich et al., Phys. Rev. Lett. 63(1989)357.

44. H. Abramowicz et al., Z. Phys. C17(1983)283.

45. E. Eichten, I. Hinchliffe, K. Lane and C. Quigg, Rev. Mod. Phys. 56(1984)579.

46. C.N Brown et al., submitted to Phys. Rev. Lett.

47. A.S. Eto et al., Phys. Rev. D23(1981)604.

48. A.D. Martin, R.G. Roberts, W.J. Stirling, Phys. Lett. 206B(1988)327.

49. J. Badier et al., Z. Phys. C26(1985)489.

50. All three experiments assume a linear A-dependence. The E288 data was taken with platinum and copper targets while NA-3 used platinum. The data from NA-3 and E288 are corrected for Fermi motion in the target while the data from E605 do not have that correction.

51. H. Baer, J. Ohnemus, and J.F. Owens, FSU-HEP-891010.

52. L. di Lella, private communication.

53. CDF memo 760.

54. K. McDonald, private communication.

DISCUSSION

G. Martinelli, University of Rome: I have two comments. The first is that if you try to combine different information coming from deep inelastic, Drell-Yan and direct-photon experiments to do a common fit, you unrealistically constrain the shape of the gluon distribution, because the systematics are very different in different experiments.

My second comment is on the optimization procedure. In the only case we know where the higher-order corrections can be computed, namely $R_{e^+e^-}$, the optimizations do not work. So why should one believe this kind of approach in direct-photon experiments? It does not have any theoretical basis, in my opinion.

J. Huston: My comments were just based on pragmatics. The approach seems to work very well in fitting the data, even though there is no theoretical basis for it.

E. Reya, University of Dortmund: There was a recent analysis of Contogouris *et al.* where they got a similar agreement without using this optimization procedure. So the agreement between theory and experiment does not depend on the optimization. If we use the $\overline{MS}$ scheme consistently for deep inelastic and for direct-photon data, then the procedure works well. At the moment, however, we know that it does not work, for instance, in $R_{e^+e^-}$.

J. Huston: I think that I showed a somewhat similar conclusion. If one uses a scale that's relatively small, say $p_t/2$, then there is good agreement between the data and theory. But I think that the Contogouris *et al.* conclusion was actually the opposite, and I don't know if I quite understand it. That is, they can't get a consistent gluon distribution if they use conventional scales of just p_t. I'd also like to mention that the rapidity distributions are somewhat harder to fit if you use one consistent scale rather than optimized scales.

S. Brodsky, SLAC: I would like to make two comments, one about the use of the PMS optimization procedure for scale-setting in QCD, and the other about higher-twist effects. The PMS optimization procedure is in conflict with the method for setting the scale in QED, where there is no ambiguity. In this case one sums the effects of light lepton pairs into the QED running coupling constant. This gives a sensible, physically reasonable scale μ for the argument of α_{QED} for the higher order calculations. If one adopts the same procedure for QCD, that is, sums the effects of light $q\overline{q}$ pairs into the QCD running coupling constant, then one obtains physically reasonable predictions. This is the method of "automatic scale-fixing" proposed by Lepage, Mackenzie, and myself a number of years ago. I urge that, at least as an alternative, you utilize this method. I expect that the effective scale for the high-p_t photon predictions will be somewhat less than $p_t/2$ in the $\overline{MS}$ scheme.

Secondly, in your discussion you mentioned a number of reasons why photon production at moderate to low p_t did not resemble the leading-twist predictions. I am reminded of the ISR and Chicago-Princeton data for hadron production $pp \to hX$ in the $p_t = 4GeV$ region, where anomalous power-law scaling was observed. The proton production data was particularly anomalous: the inclusive cross section scaled as P_t^{-12} at fixed θ_{cm} and x_t. This is a clear signature for higher-twist subprocesses. Thus it is unlikely that the low p_t region for photon production can be fit simply with leading-twist predictions. Experimentalists can look for explicit signals for "direct" higher-twist contributions, by identifying isolated hadron production. In general the various effects can be separated by using fixed-angle and fixed-x_t scaling-law parameterizations of data obtained over a range of incident energy.

J. Huston: It's a subject of some interest to me, but I was actually looking for this behavior at high x_t, just where the pion becomes very efficient at producing high-x_t particles because you just have one structure function, rather than dividing it between quarks and gluons. For the high-statistics running planned for next year, we actually expect to see that at the higher p_t values, some fraction of events can be clearly identified as due to higher-twist effects.

K. Freudenreich, Federal Institute of Technology, Zurich: Two remarks. First, you compared the pion valence structure function obtained from the direct-photon data to several from Drell-Yan experiments. There is one warning: the use of different external inputs may change, especially, the normalization. So if you have a discrepancy, it may just be due to different inputs. For a valid comparison, the analysis must be carried out with the *same* external inputs.

Second, you should add to your list of future results the analysis of NA10, which has by far the largest dimuon data sample at hand.

QCD and Jets

John Huth
Fermilab
Box 500
Batavia, IL 60510, USA

ABSTRACT

In this talk, recent progress in QCD measurements involving jet activity will be outlined. Inclusive jet production and high p_t W/Z production will be reviewed for Hadron Collider experiments. The running of α_s, n-jet production cross sections, and tests for gluon-gluon coupling in 4 jet events will be examined for e^+e^- experiments. Recent fragmentation measurements will also be reviewed.

1 Introduction

In this talk, recent progress in the testing of Quantum Chromodynamics will be outlined. An assesment of prospects for measurements in the near future will also be given. Results presented include work from both $\overline{p}p$ and e^+e^- colliders. From hadron colliders, recent work will be reviewed on the inclusive jet cross section and the p_t spectrum of W/Z^o bosons. From electron-positron colliders, work on jet multiplicities and the running of α_s will be presented. Finally, recent results from both $\overline{p}p$ and e^+e^- fragmentation measurements will also be shown.

2 Jet Definition

The common definition of a jet is a cluster of particles with some limited momentum transverse to a central axis. Operationally, jets are what a reconstruction algorithm in a particular experiment defines them to be. In $\overline{p}p$ experiments, final state energies are measured in calorimeter towers which project back to the origin. A common algorithm is the cone algorithm, which groups calorimeter towers together in a coordinate system segmented in pseudorapidity (η) and azimuth (ϕ). Calorimeter towers within some radius in this coordinate system, $\Delta R = \sqrt{\Delta\eta^2 + \Delta\phi^2}$ are grouped together to form a jet [1],[2]. ΔR can vary from 0.4 to 1.2 units, depending on the measurement. Figure 1 shows a typical jet as seen by the CDF calorimeter, which has a projective geometry with η-ϕ segmentation.

In e^+e^- experiments, where charged particles from tracking, rather than calorimetric measurements are used to define jets, a typical algorithm employed involves the invariant masses of all pairs of charged particles. This algorithm, introduced by the JADE collaboration [3] uses the quantity y_{cut} to define the effective cutoff on the invariant mass:

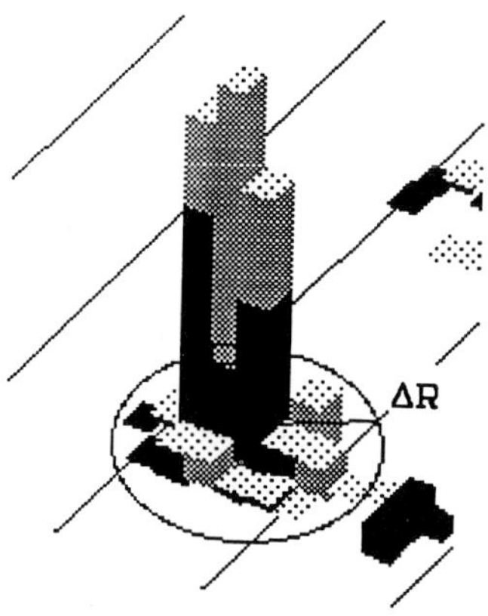

Figure 1: Typical jet cluster as seen in calorimeter cells.

$$y_{cut} = \frac{M_{ij}^2}{E_{cm}^2} \tag{1}$$

where M_{ij} is the invariant mass between any pair of charged particles.

In addition to the above two algorithms, experiments use global event measures, such as sphericity or triplicity to define explicit n jet final states [4]. UA2 uses a clustering algorithm based on contiguous calorimeter towers [5] to define clusters. Here the relevant clustering parameter is the minimum tower energy used to associate a tower to a

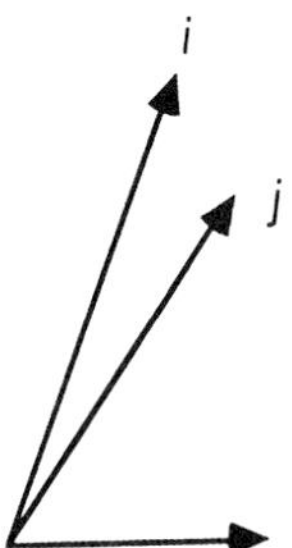

Figure 2: Jet pairing based on the invariant mass of pairs of charged particles.

jet, typically 400 MeV.

3 Hadron Collisions

In $\bar{p}p$ collisions, four fundamental processes which test perturbative QCD are:

1. Jet production
2. High p_t boson production
3. Direct photon production
4. Heavy quark production

Each of these processes now have cross sections evaluated at either order α_s^3 or (α_s^2, α). The latter two have already been covered at length in other talks at this conference [6],[7],[8],[9].

3.1 Inclusive Jet production

Inclusive jet production is defined as the cross section for the reaction $\bar{p}p \rightarrow$ JET $+X$. Although this process is dominated by QCD production, it explores the highest values of momentum transfer available in colliders and is sensitive to non-standard model processes which have large cross sections. Examples of such processes include axigluon production [10], and the possibility of quark compositeness [11]. Two jet events in the central region of the CDF detector have been observed with transverse energies in excess of 450 GeV/jet. At low values of jet E_t, gluon-gluon scattering dominates the cross sections, and at the highest energies, quark-quark scattering give the largest contribution. t channel diagrams like those in figure 3 are part of the graphs contributing at leading order to the cross section.

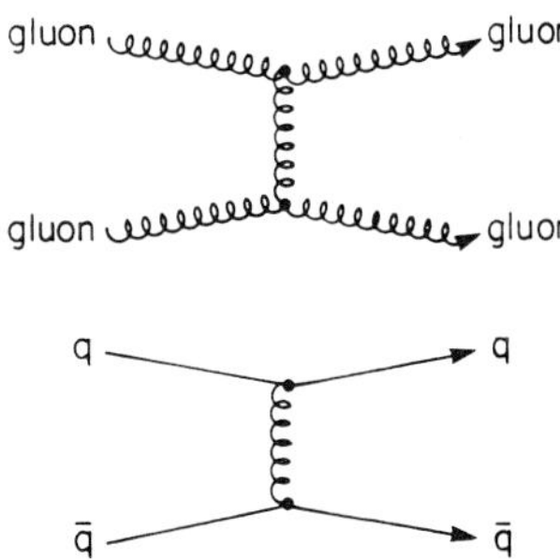

Figure 3: Diagrams contributing to the inclusive jet cross section at leading order.

The cross section for inclusive jet production is well defined to all orders of perturbation theory. This is to be distinguished from measurements such as the dijet invariant mass spectrum, which explicitly involves some definition of the type of final state to be considered, (*eg.* two back-to-back jets, with no third jet) Being more of an exclusive process, the invariant mass spectrum is harder to define uniquely at all orders of perturbation theory. The low energy production of jets is dominated by gluon initial states, and the high energy end tend to be dominated by valence quarks. At SSC energies, the gluon dominance will be apparent for most processes.

Figure 4 shows the inclusive production cross section taken at several values of $\sqrt{s}$ for different $\bar{p}p$ experiments. Also shown are the results of a leading order QCD calculation with absolute normalization [12][13][14] [15]. The perhaps surprising feature is that leading order QCD provides such a good description of the data, over 7 orders of magnitude of cross section, spanning transverse energies ranging from 6 to 400 GeV. Plots displayed on a logarithmic scale can hide multitudes of sins, therefore also displayed on a linear scale is the ratio of data/QCD for the three different center of mass energies (fig 5). As one can see, the data agree with QCD at about the 50 % level. The size of the theory error bar indicates the uncertainty associated with the leading order QCD prediction, and was obtained by varying the Q^2 scale used to evaluate both the proton structure functions and the strong coupling constant, α_s from $E_t^2/2$ to $2E_t^2$. The experimental error bar illustrates the uncertainty typical to all four experiments. The single

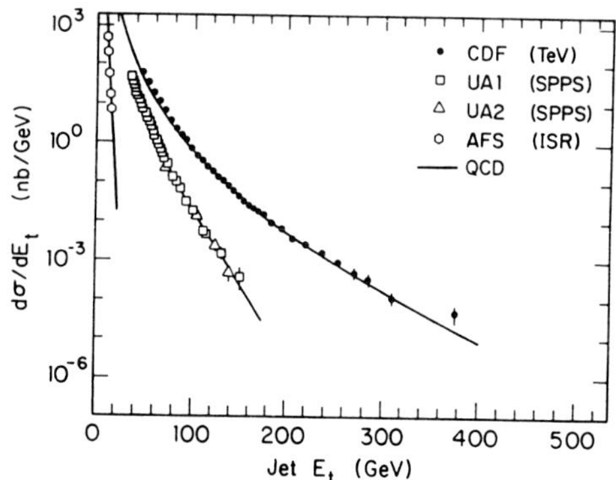

Figure 4: Inclusive jet E_t spectrum from the ISR, SPPS, and the Tevatron. Also shown is the prediction of a leading order QCD calculation made with the choice $Q^2 = E_t^2/2$ for all values of $\sqrt{s}$. The normalizations are absolute.

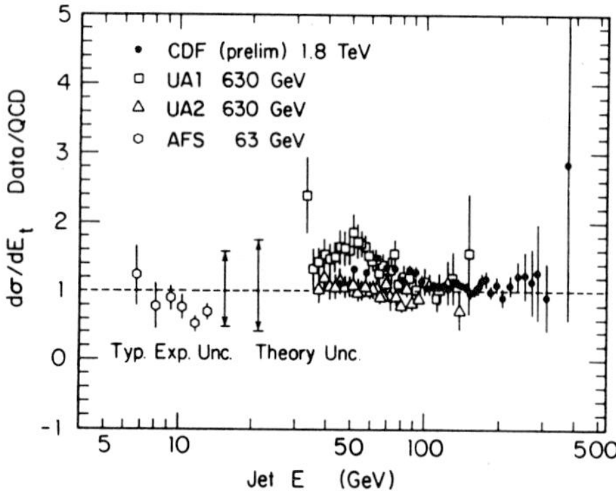

Figure 5: The ratio data/QCD for the data shown in figure 4. The error bars shown illustrate the typical uncertainty associated with both the theoretical calculations at leading order and the experimental uncertainties.

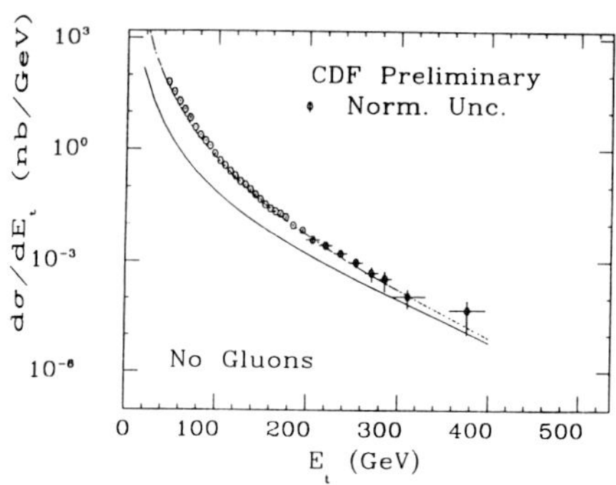

Figure 6: The effect of turning off the gluons in the structure function in performing the leading order calculation.

largest contribution results from the uncertainty of the energy scale for jets. The uncertainty also contains contributions from luminosity, and the effects of resolution smearing (ie. the convolution of finite jet energy resolution on a falling spectrum). Parts of the experimental uncertainty are totally correlated over the entire range of E_t; the luminosity uncertainty is an example. If one uses the shape, rather than the absolute cross section, a large fraction of the uncertainty is reduced.

Here, CDF data will be used to illustrate some features of the inclusive cross section. Gluons begin to play a significant role at the Tevatron energies, and will dominate the initial state partons at the SSC. As an illustration of this, in figure 6 the cross section is shown with the gluons turned off in the structure functions. The absence of gluons has the effect of reducing the cross section by about an order of magnitude for E_t below 100 GeV.

In figure 7 CDF data are displayed along with a range of possible leading order QCD predictions, taking the momentum transfer scale, Q^2, from E_t^2 to $2E_t^2$. The next-to-leading order calculation currently being performed [16] [17] [18] should substantially shrink these uncertainties in the future. As an example of how these will be reduced, figure 8 shows the dependence of the cross section on the scale chosen for the evaluation of α_s and the structure functions at both leading order (order α_s^2) and at the next to leading order (α_s^3), for a specific value of jet E_t. This is for the scattering of gluons only, but does illustrate the reduction in the uncertainty associated with the calculations at

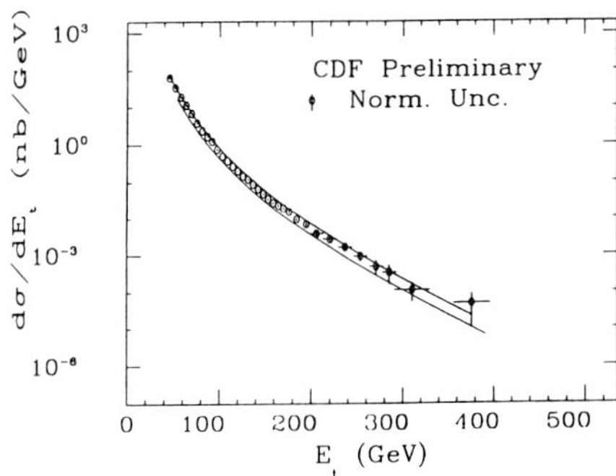

Figure 7: The variation in the inclusive jet cross section as a function of renormalization scale chosen for the evaluation of both the structure function and α_s.

order α_s^3.

The existence of a quark substructure should be apparent well below the relevant energy scale associated with the binding of the constituents. This low energy effective behavior can be approximated by the addition of a contact term to the QCD Lagrangian [19]. There are many different effective interactions that could be chosen, but one form in particular is in common use:

$$L_{int} = \frac{g^2}{4\pi}\frac{1}{\Lambda^*}(\bar{q}_l\gamma^\mu q_l)^2 \qquad (2)$$

where Λ^* is a constant with units of energy which is related to the constituent scale. The effect of this term is to raise the jet cross section above the QCD prediction at high E_t. From CDF, the inclusive cross section appears to be consistent with the shape of the leading order QCD cross section. If the QCD prediction is normalized by a multiplicative constant at low E_t ($E_t \leq 120$ GeV), then the shape of the cross section extrapolated to higher values of E_t can be used to set limits on possible contact terms. Currently, the most stringent published lower limit on Λ^* is 750 GeV [20]. Figure 9 shows the data with both the leading order QCD prediction and the effect of a contact term with $\Lambda^* = 800$ GeV on the cross section, which appears to be readily excluded by the data. It should be noted that the effective interaction term can take many forms, and that this particular form is used by most collider experiments as a basis for comparison.

1. Born Level

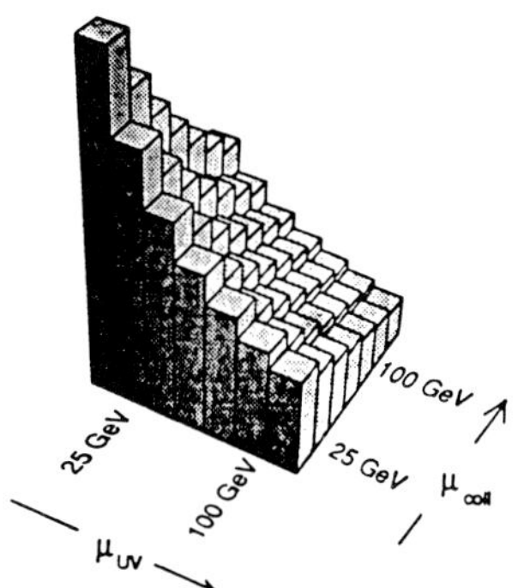

2. Full calculation

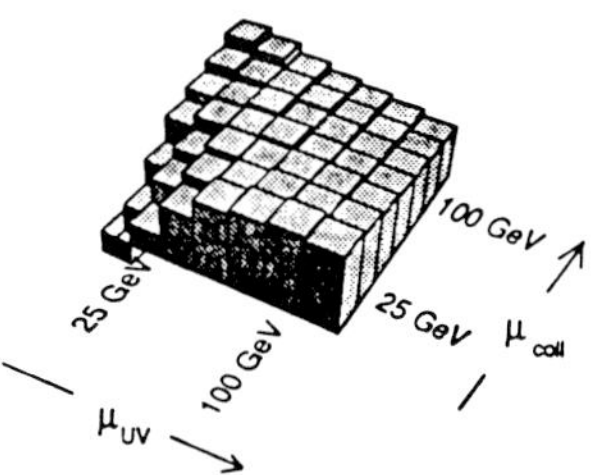

Figure 8: Variation of the jet cross section evaluated at an E_t of 100 GeV for both the leading order and next to leading order as a function of the renormalization scales used to evaluate both the strong coupling constant, and the structure functions [17].

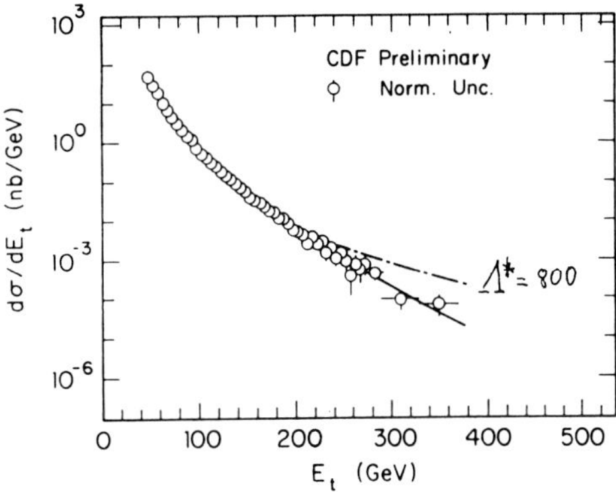

Figure 9: The effect of a contact term of the form described in the text, with $\Lambda^* = 800$ GeV.

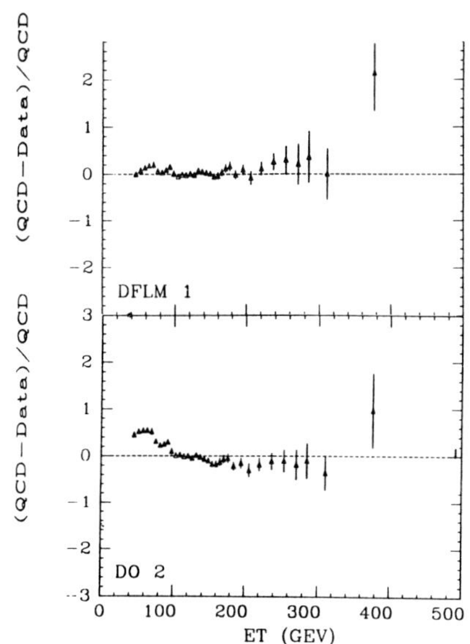

Figure 10: The ratio of data/QCD for two different sets of structure functions. The upper plot is for DFLM set 1, the lower is for Duke and Owens set 2.

Differences in the commonly used parameterizations of structure functions start to become apparent with a wide range of jet E_t. If the uncertainty in the theory can be shrunk, as well as the experimental systematics, then it may be possible to exclude certain parameterizations. As an example, one can plot on a linear scale the ratio of data/QCD for the inclusive jet cross section, with different choices of structure function. Figure 10 shows one such example where the DFLM (set 1) [21] and the Duke and Owen (set 2) [22] structure functions are used for the theory comparison. The particular Duke and Owens structure function is formulated with a hard gluon distribution $\approx (1-x)^4$ as an input at a $Q^2 = 4$ GeV^2.

The experimental uncertainties may shrink in the near future. One of the largest contributions to the experimental uncertainty results from the non-compensating nature of the calorimeters currently in use at the hadron colliders. Figure 11 illustrates the response of the CDF central calorimeter to incident pions of varying energies. At lower energies, there is a substantial deviation from linearity. Since most of the jet energy is carried by particles with momenta below 10 % of the total for the jet, uncertainties in the fragmentation spectra can result in uncertainties in the jet energy scale at the level of 8 % of the jet energy.

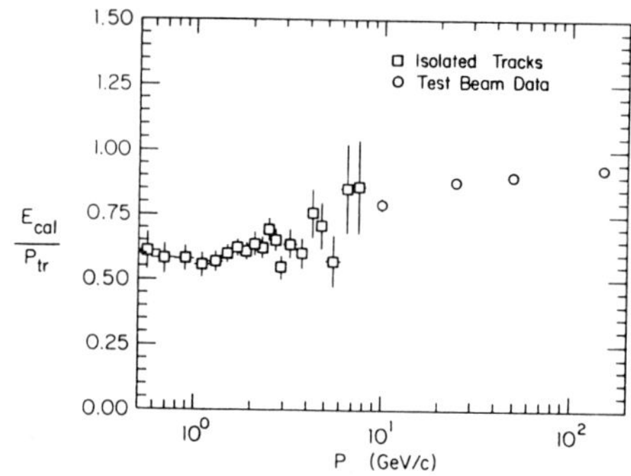

Figure 11: Response curve for the CDF central calorimeter to low energy pions. Ideal response (full compensation) would be at E_{cal}/p=1.

For the non-compensating calorimeters, some improvement could be gained by using tracking measurements to partially alleviate the energy losses from the nonlinear behavior. Calorimeters which are much closer to compensating should have reduced energy scale uncertainties, although there are other loss mechanisms such as cracks and neutrinos which are unavoidable. If this is possible, the next generation of measurements may keep pace with the theoretical improvements in calculation of jet cross sections.

3.2 p_t Spectrum of W and Z bosons

In addition to being relatively clean probes of the quark-gluon vertex, involving terms of order α_s at leading order, high p_t production can be a background for heavy top and higgs searches, and are thus important to understand. Experimentally they have the virtue of being very clean, and (particularly for the Z) more precisely measured than events with all jets. Figure 12 shows typical leading order graphs for high p_t W/Z production. In the very high energy limit, the behavior is very much like that for direct photon production. Currently there are theory calculations for high p_t ($\geq$ 15 GeV) to order (α_s^2, α) [23] [24]. At order (α_s, α), the perturbative result has also been matched onto the low p_t region where leading terms have to be resummed to obtain a cross section [26]- [28].

The $O(\alpha_s^2, \alpha)$ calculations show a reduction of the uncertainty associated with the renormalization scale used over the $O(\alpha_s, \alpha)$ result. To illustrate the uncertainties due to choice of scale, figure

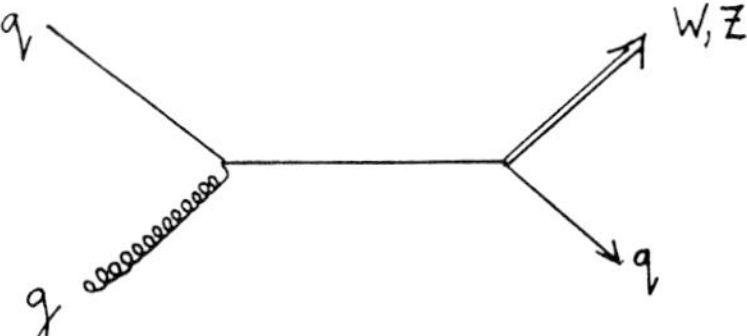

Figure 12: Typical leading order graph for high p_t production of W and Z bosons.

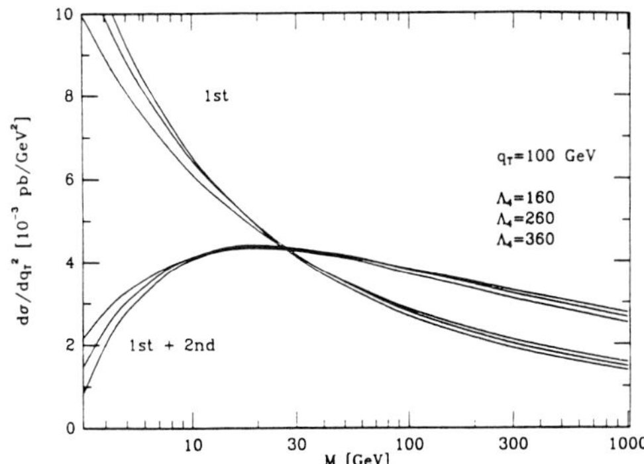

Figure 13: A comparison of the cross section for a W with a p_t of 100 GeV for $\sqrt{s} = 1.8$ TeV taken at leading and next to leading order [24] .

13 is included from reference [24] where the cross section for a 100 GeV p_t W is shown for different choices of renormalization scale. In addition, one can see that the cross section evaluated at this order is about 30% higher.

At present, extensive data exist on high p_t W and Z events from both the SPS and Tevatron colliders. Figure 14 plots CDF data [25], and figure 15 shows UA1/UA2 [24] data for the p_t spectrum of W's. The theory curves displayed for both sets of data are the calculations performed at order (α_s^2, α) by Arnold and Reno [24]. Theory and experiment show good agreement for all three data sets.

It is possible to examine more exclusively defined final states; requiring, for example, a specific number of identified jets. Calculations can be made for these final states, but must take into account detector acceptances, and even resolution effects. One example of such a comparison is the spectrum of jet multiplicities recoiling against W's

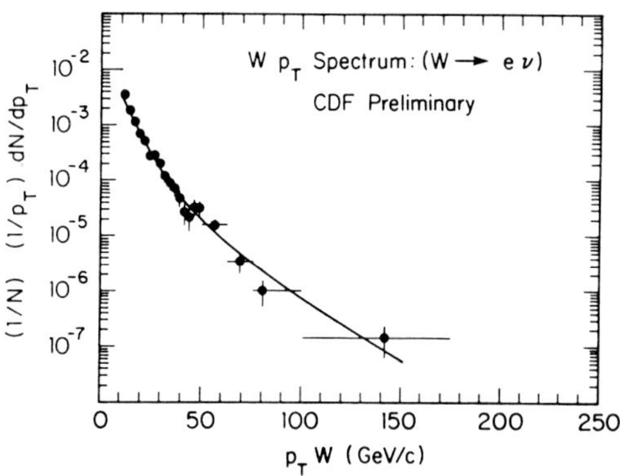

Figure 14: p_t spectrum for W bosons seen in the CDF detector, along with the predictions of an order (α_s^2, α) calculation by Stirling *et al* [23].

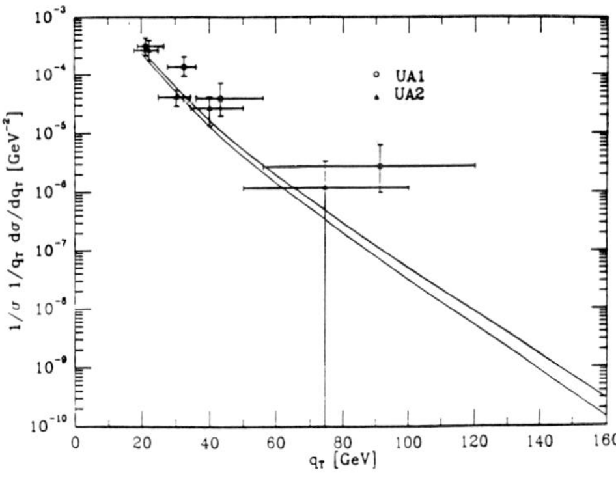

Figure 15: p_t spectrum for W bosons from UA1/UA2 along with the predictions of a order (α_s^2, α) calculation [24].

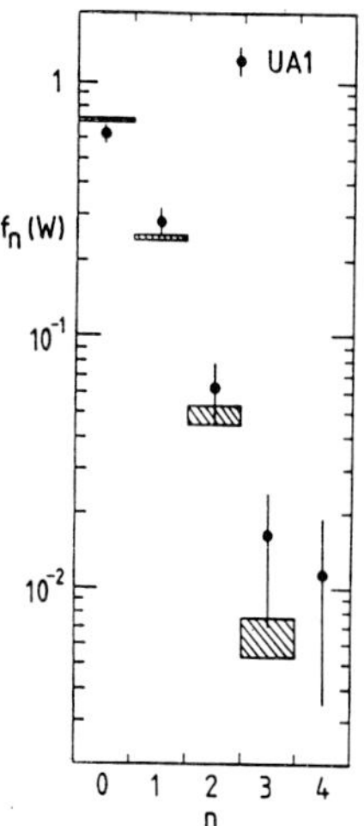

Figure 16: The multiplicity of jets found in W events for the UA1 experiment. The theory comparisons are for both the analytic calculation of Berends *et al* [29].

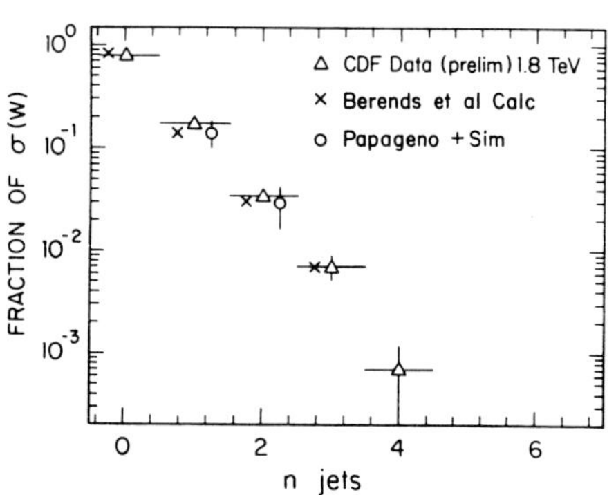

Figure 17: The multiplicity of jets found in W events for CDF. The theory comparisons are for both the analytic calculation of Berends *et al* [29] and the result of a detector simulation using the papageno event generator as an input [30].

and Z's. Berends, Giele *et al* have performed calculations of the jet multiplicities given the standard detector acceptances of both UA1 and CDF [29]. Figures 15 and 16 shows the fraction of N jet+W events as seen in the UA1 and CDF detectors, and compared to these predictions. There is already a good agreement seen between the data and the calculation, even though minute details of detector acceptance were not included. Also shown for the CDF data is the result of the PAPAGENO [30] event generator plus detector simulation. Reasonable agreement is seen in this comparison.

The primary reason for focusing on such specific final states is that QCD forms a background for top and Higgs production (or more broadly any W pair production mechanism), where the jet multiplicity in the final state is a tag. One test relevant for a heavy top is the p_t spectrum for 1, 2...N jet final states, where the signature of W pair production would coincide with an enhancement in the high end of the spectrum relative to QCD backgrounds. Figures 17 and 18 show the W p_t spectrum for 1 and 2 jet events in CDF data, and the comparison to the PAPAGENO simulation [30]. Note a reasonable, although not perfect, agreement with the simulation.

One of the problems of measuring high p_t W bosons is measurement of the neutrino energy via missing E_t in mixed events where one measures high energy electrons, jets, and the underlying event.

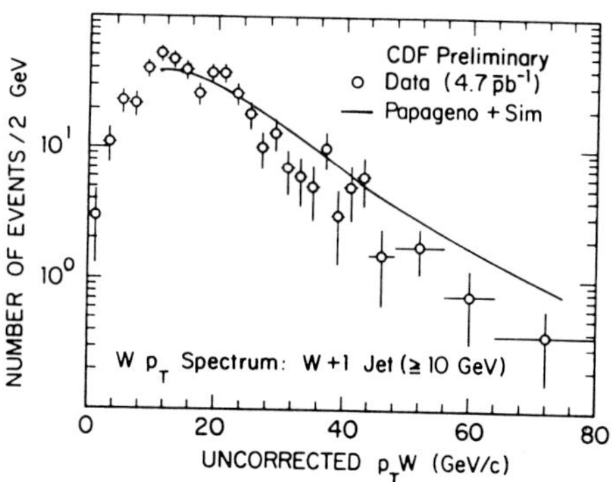

Figure 18: The p_t spectrum for CDF W events with 1 jet identified in the final state. Also shown are the results of a detector simulation and the PAPAGENO event generator.

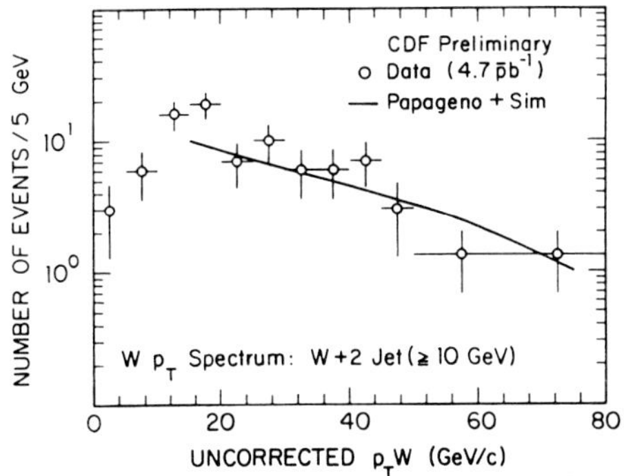

Figure 19: The p_t spectrum for CDF W events with 2 jets identified in the final state. The PAPAGENO event generator and a simulation of the CDF detector was used to derive these curves.

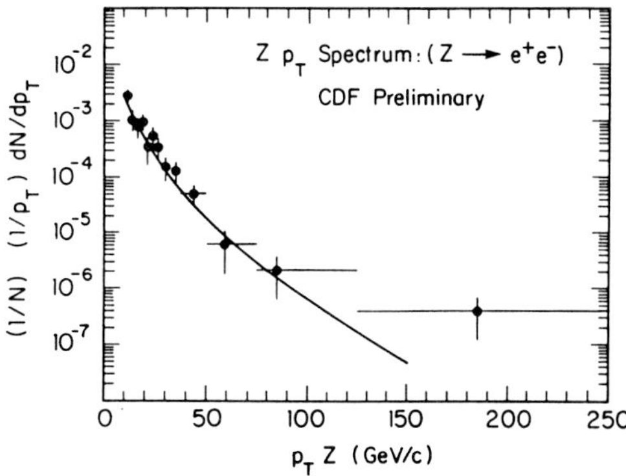

Figure 20: p_t spectrum of Z's seen in the CDF detector. The solid line is the prediction of QCD [23].

Because of nonlinear calorimeter effects (described above) , both the mean, and the spread of the W p_t can be biased. Care must be taken to correct for these, and to estimate the systematic uncertainties introduced in the process. The use of high p_t Z's has the advantage that the two electrons are measured on an equal footing (no missing energy measurement is necessary. With high statistics, the Z sample will provide a very precise measurement of the high p_t spectrum. Figure 20 shows the spectrum of high p_t Z's seen in the CDF detector. Also shown are the predictions of QCD at order (α_s^2, α), which exhibit good agreement with the data.

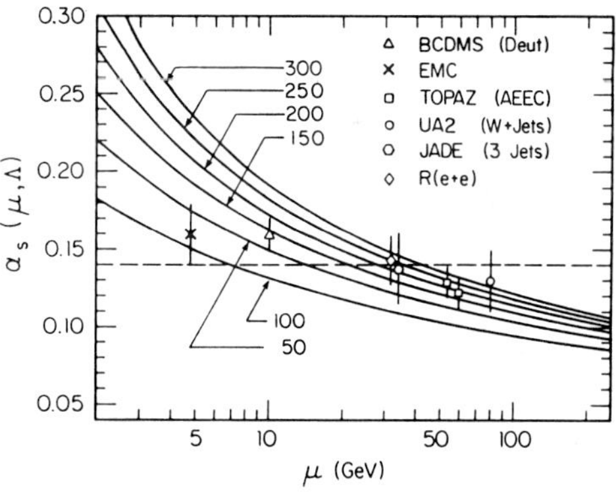

Figure 21: A representative set of measurements of α_s taken from different sources. Also shown are the predictions of QCD.

4 e^+e^- tests of QCD

Several topics of current interest in e^+e^- test perturbative aspects of QCD, and involve distinguishing the number or fraction of events with significant amounts of gluon emission from canonical two jet events.

4.1 Running of α_s

Figure 21 shows a compendium of measurements of α_s from different sources [31][32]. For reference also shown are possible trajectories of α_s given 5 flavors and several different values of $\Lambda_{\overline{MS}}$. These measurements are intended as a representative sampling, rather than an exhaustive listing. Here too is drawn a skeptic's line (dashed) which is meant to illustrate that one cannot yet rule out a fixed

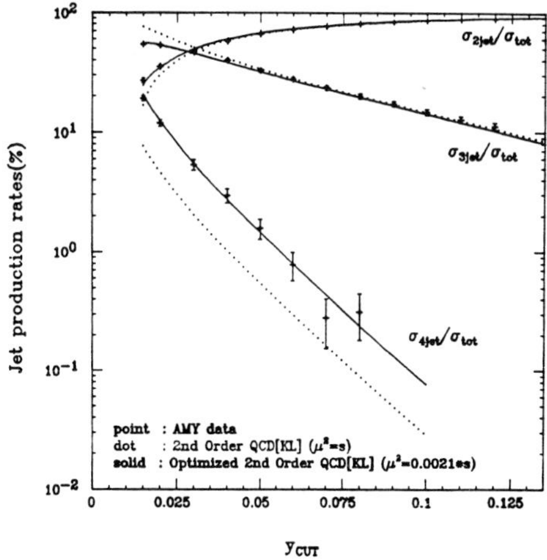

Figure 22: The fraction of 2,3 and 4 jet events as determined using the JADE clustering algorithm in the AMY detector. The number of jets is clearly a function of the cutoff imposed on the invariant mass of pairs of particles to be included in a cluster.

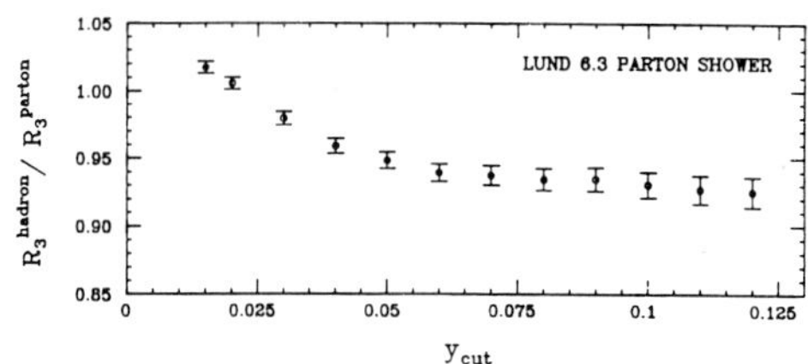

Figure 23: The fraction of the time that a 3 parton final state is identified as a three jet event in the AMY detector as a function of the Y_{cut} parameter.

α_s. There is not a lot of room left to draw such a line indicating the importance of high statistics measurements at LEP and SLC.

4.2 3 jet fraction

In this measurement, the fraction of 3 jet events, R_3 is determined using the JADE algorithm (described above). The number of jets found is a function of the invariant mass cutoff, Y_{cut}, as can be seen in figure 22 [33]. For this algorithm above some value of Y_{cut}, there is a good correspondence between the number of n parton final states in theory and those identified in the detector (figure 23).

If one plots R_3 events as a function of the e^+e^- center-of-mass energy, W [35][34], (figure 24) one finds that it drops as W increases. This is shown for a number of experiments, all with a fixed value of $Y_{cut} = 0.08$. Besides W, another relevant scaling parameter for R_3 is the associated invariant mass cut for jet determination, which will vary with W as $(W\sqrt{y_{cut}})$.

The TASSO collaboration [35] has studied extensively systematic effects associated with the R_3 variable. This study used the LUND [36] event generator to simulate jets. The matrix elements used in the generator can allow α_s to either be fixed, or run. In figure 24 the two models are compared. In the first model (dotted line) α_s is held fixed in simulation, and partons are allowed

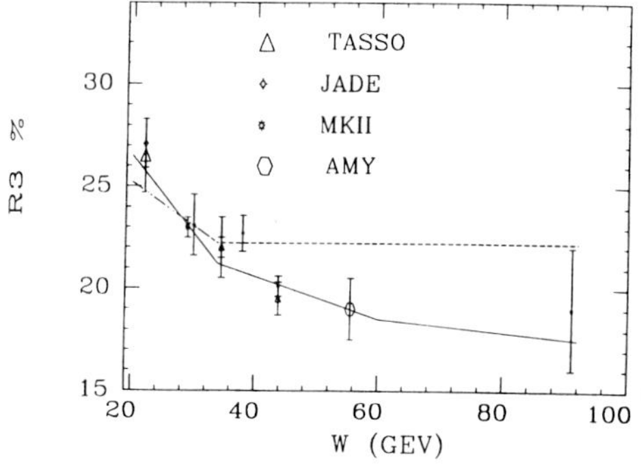

Figure 24: The fraction of three jet events, R_3, plotted a function of the e^+e^- center-of-mass energy. The solid line indicates what the effect of a running α_s in a detector simulation would produce, whereas the dashed line shows the scaling with a fixed α_s in the Monte Carlo.

to fragment. In the second, both fragmentation and a running α_s are allowed (solid line). One can see from this exercise that R_3 can me made to appear to run with the CMS energy just from fragmentation effects up to around 35 GeV. The scaling of R_3 is indicative of a running coupling constant above this energy. It is evident from this discussion that high statistics measurements of R_3 at both Tristan, SLC and LEP are important.

With the advent of order α_s^2 calculations of the QCD cross section for hadronic e^+e^- final states [37], it is possible to compare for 2,3 and 4 jet predictions with data. This is shown above in figure 22 for the 2,3 and 4 jet components. The solid and dashed lines refer to evaluations of the cross section at renormalization scales of 57 GeV (Tristan CMS energy) and at 3 GeV respectively. The data show some preference for the 3 GeV scale; but given that the Q^2 scale associated with the quark-gluon vertex is typically the gluon virtuality, this trend is not surprising. The relevance of results of actually fitting the renormalization scale, μ [33],[38] for the 2-3-4 jet data, however can be questioned because of the possibility of a large contribution to the 4 jet cross section coming from the next order in perturbation theory.

An alternative to jet counting to determine α_s is the use of global event measures, such as the energy-energy correlation. The energy-energy correlation is taken from the charged particle data in e^+e^- collisions, and is defined to be:

$$EEC \equiv \frac{1}{N_{ev}} \sum_{ev} \sum_{i} \sum_{j} \frac{E_i E_j}{E_{cm}^2} \delta(\chi - \chi_{ij}) \quad (3)$$

where i and j are particle indices, and χ refers to the angle between particles. The asymmetry in energy-energy correlations, defined as:

$$AEEC \equiv EEC(\pi - \chi) - EEC(\chi) \quad (4)$$

Figure 25 shows the AEEC plots from two different CMS energies from the TOPAZ detector [32]. The large angle region, where the effect of hard gluon *bremsstrahlung* is calculable is used to fit to a combination of an event generator, incorporating the Gottschalk-Schatz matrix elements and string fragmentation. α_s is varied in the Monte Carlo until a best fit is derived with the AEEC data. The result for TOPAZ data for the two different energies gives $\alpha_s = 0.129 \pm .007$ (stat)$\pm .010$(sys) GeV CMS energy, and $0.122 \pm .007 \pm .010$ at 59. GeV. The errors quoted are typical for AEEC measurements, which, when coupled with the slow variation of α_s with CMS energy, is one of the reasons it

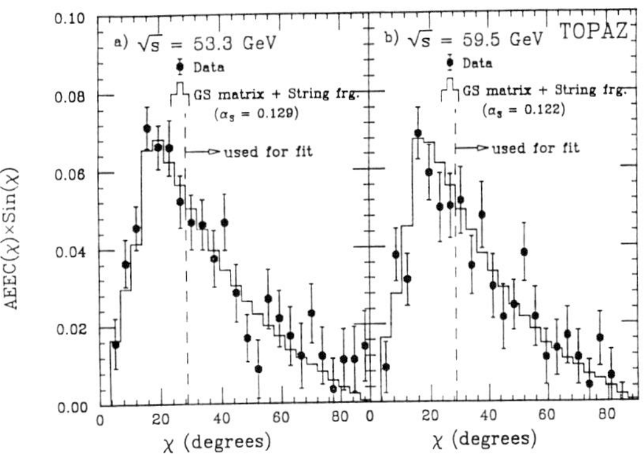

Figure 25: AEEC plots from the TOPAZ experiment [32] along with the fits obtained from a Monte Carlo incorporating the GS matrix elements [37] coupled with a string fragmentation model [36].

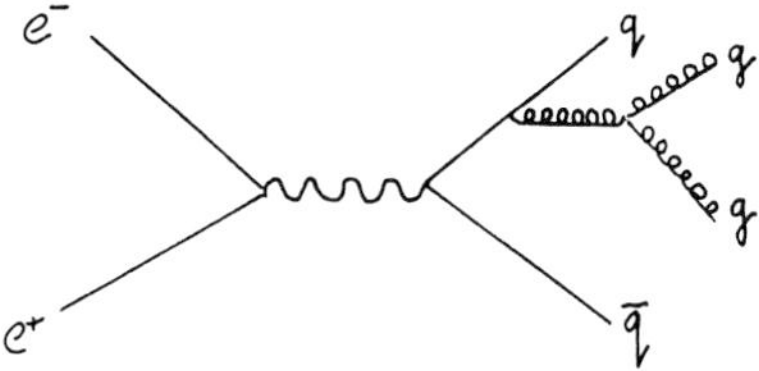

Figure 26: Typical graph at the lowest order in e^+e^- collisions in which the gluon-gluon coupling characteristic of QCD becomes evident.

is difficult to conclusively determine that α_s runs. Again, high statistics measurements at LEP and SLC now have sufficient lever arm to resolve this.

4.3 Non-Abelian Couplings of Gluons

Although the non-Abelian coupling of gluons is an essential ingredient in QCD in order to have asymptotic freedom, and a strong, confining regime direct evidence of this coupling is not trivial. In e^+e^- collisions, the lowest order at which the gluon-gluon coupling is apparent involves 4 jets (figure 26). Several variables have been devised to test this coupling [39]. In particular, one variable which should be sensitive to this coupling is the Bengtsson-Zerwas angle, θ_{BZ}, which describes the angle between two planes in 4 jet events [40]. One plane

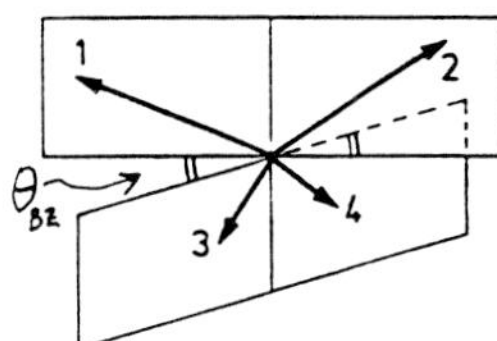

Figure 27: The Bengtsson-Zerwas angle, θ_{BZ} described by the angle between the plane containing the leading two jets, and the plane containing the remaining two jets in a four jet event.

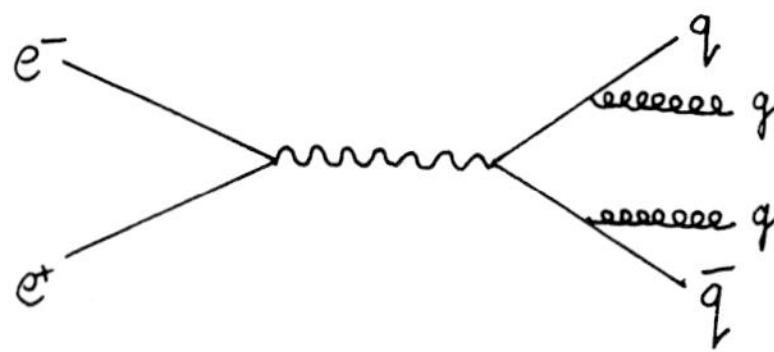

Figure 28: Example of a diagram producing four jets in the Abelian model.

contains the leading two jets, and the other contains the remaining jets (figure 4.3). It is difficult to obtain sufficient statistics to be able to measure the distribution of θ_{BZ} with any precision because of the penalty of α_s^2 in event rate for well defined four jet events.

In order to illuminate the distinction between a possible Abelian and QCD model, the distribution for an Abelian model was also generated, and shows a substantial difference from the QCD case. It should be noted that the Abelian model must be tinkered with in order to obtain the correct behavior for $R_{e^+e^-}$ [42]. The Abelian model includes graphs such as those seen in figure 28 but not graphs like the ones in figure 26. Figure 29 shows the result of a Monte Carlo study [42] where 100,000 events on the Z peak were generated, and θ_{BZ} was measured for the four jet events. The behavior of θ_{BZ} tends to be flat for QCD, implying more coplanar events for this than for the Abelian case.

The AMY collaboration has studied θ_{BZ} in four jet events observed at TRISTAN [41]. The distribution for θ_{BZ} is shown in figure 30. As one can see, the statistics are rather low, making a definitive statement difficult. This result was based on 139 events; a more concrete test will require a larger amount of data.

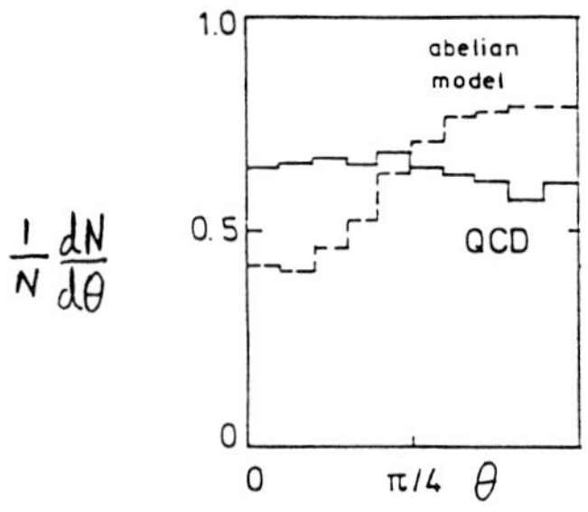

Figure 29: Result of a Monte Carlo study of the distribution of θ_{BZ} from 100,000 events generated on the Z_o peak [42]. The two curves show the difference between standard QCD, and the result from an Abelian model.

5 Fragmentation

To a large extent, predictions for the particle distributions and energy flow within jets are limited to various semi-empirical parton shower models, and contains much that is put in by hand, limiting their usefuleness as tests of QCD. On the other hand, there are several aspects of jet fragmentation that are directly related to perturbative QCD. These will be examined below.

5.1 Charm Production in $\overline{p}p$ collisions

The direct production of charm in $\overline{p}p$ collisions is predicted by QCD [43] to arise mostly from gluon splitting processes, such as from the graph in figure 31. Perturbation theory can be used to determine the ratio, $\rho_{\bar{c}c}$, of the number of charm pairs per gluon jet [43]. This can be approximated by:

$$\rho_{\bar{c}c} \approx \frac{1}{3\pi}\int_{4M_c^2}^{Q^2}\frac{dk^2}{k^2}\alpha(k)(1+\frac{2M_c^2}{k^2})(\frac{1}{4}-\frac{M_c^2}{k^2})^{\frac{1}{2}} \quad (5)$$

where M_c is the charm quark mass. The number of D^* s produced can be used as a rough guide to $\rho_{\bar{c}c}$. From spin arguments the number of D^* s produced per jet ρ_{D^*} should be approximately equal to $\frac{3}{4}$ of $\rho_{\bar{c}c}$. The UA1 collaboration has reported $\rho_{D^*} = 0.08 \pm 0.02$ (stat) ± 0.04 (syst) [44], substantially lower than an earlier report of $0.65 \pm 0.2 \pm 0.3$ [45].

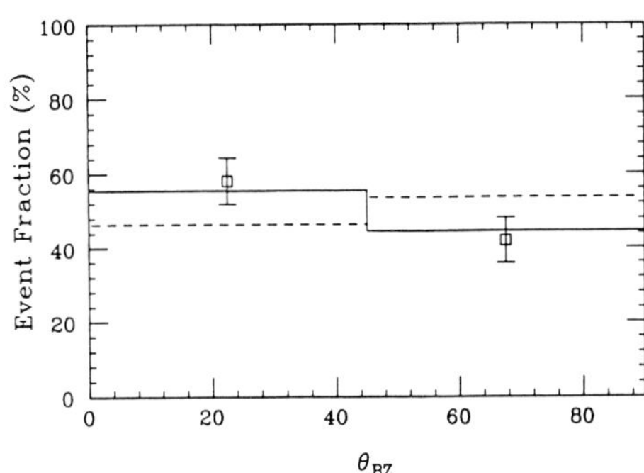

Figure 30: Distribution of θ_{BZ} reported by the AMY collaboration from four jet events. The two curves result from the Abelian model (dashed) and QCD (solid).

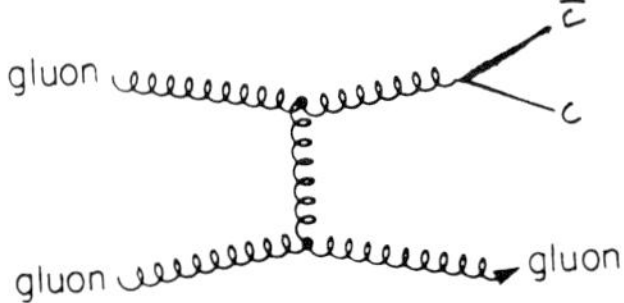

Figure 31: Diagram of a gluon splitting contribution to direct charm production in $\bar{p}p$ collisions.

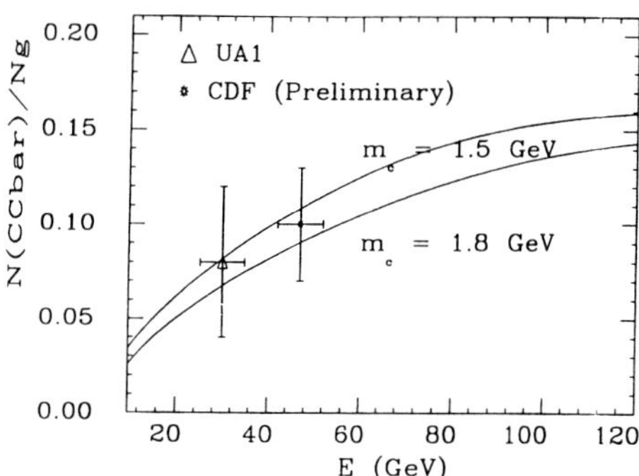

Figure 32: CDF and recent UA1 value of ρ_{D*} plotted against the theory calculation of Muller and Nason [43].

In reference [44] it is speculated that this earlier number may result from a sample biased by the trigger selection criteria.

CDF has a measurement of ρ_{D^*} of $0.1 \pm 0.03 \pm 0.03$, for D^* s with fragmentation z (defined below) greater than 0.1. Figure 32 shows these two values of ρ_{D^*} plotted versus the average jet energy. Also shown are the values predicted by Muller and Nason for $\rho_{\bar{c}c}$. Although the error bars are large, the CDF number is more in agreement with the more recent value reported by UA1, and more in accord with thc thcory predictions.

5.2 Quark and Gluon Fragmentation

The question of whether quarks fragment differently from gluons has occupied the literature for at least 12 years now. In the naive view, one might expect from counting arguments that there would be $\frac{9}{4}$ more particles in a quark jet than in a gluon jet [46]. This ratio can be washed out by the presence of heavy quarks in the sample, and by other effects. In principle, the differences between the jets produced from quarks versus gluons should become more apparent at larger energies.

There are two directions that these studies can go. One can try to use a process or kinematic constraint to obtain an enriched quark/gluon sample, and then examine the tagged samples for differences in fragmentation properties. The other direction, where one tags a gluon or quark sample via fragmentation variables, has not been demonstrated to work experimentally, and may be ex-

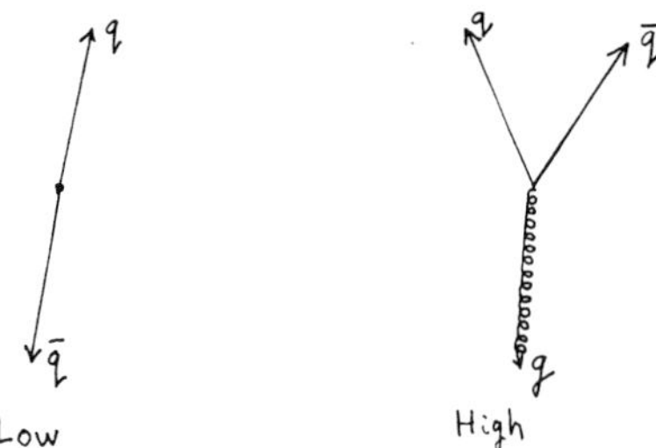

Figure 33: Events taken at two different CMS energies in e^+e^- events for two jet final states, and the symmetric three jet case can be used as a means of obtaining gluon and quark enriched samples at the same energies for the parton.

tremely difficult to implement.

In e^+e^- experiments, gluon enriched samples can be found by selecting the lowest energy jet in symmetric three jet (*ie* Mercedes Benz) events. This could be compared to quark enriched samples from two jet events. To be unbiased by scale breaking effects, one must compare jets of the same energy, which implies that the three jet sample must come from events taken at a higher CMS energy than the two jet events.

This approach has been tried by the Mark II [47], HRS [48], AMY [49], TASSO [50] and TPC [51] collaborations. The Mark II collaboration has used the variable $x_p \equiv \frac{p_{ch}}{E_{jet}}$ as a measure of the hardness or softness of the jets. Since a gluon would contain more particles, one expects the x_p distribution to be steeper, or softer than for quark tagged jets. The Mark II gluon enriched sample was obtained from 560 symmetric three jet events taken at a CMS energy of 29 GeV. Data did not exist at 20 GeV which would be needed to obtain the two jet sample with jets at the same energy as the symmetric, three jet sample. Instead, the x_p distribution for the quark enriched case was derived from an interpolation of distributions of x_p from two jet events taken at both lower and higher CMS energies. The result of the comparison of the x_p distributions is seen in figure 33. The average E_t of jets in the sample is 10 GeV. The gluon enriched sample appears to be softer than the quark sample.

The TASSO collaboration has recently performed a similar study 5.2, using data taken at different CMS energies, and without interpolation to do the comparison. The equivalent plot is shown in figure 35. The three jet sample consisted of 396 events.

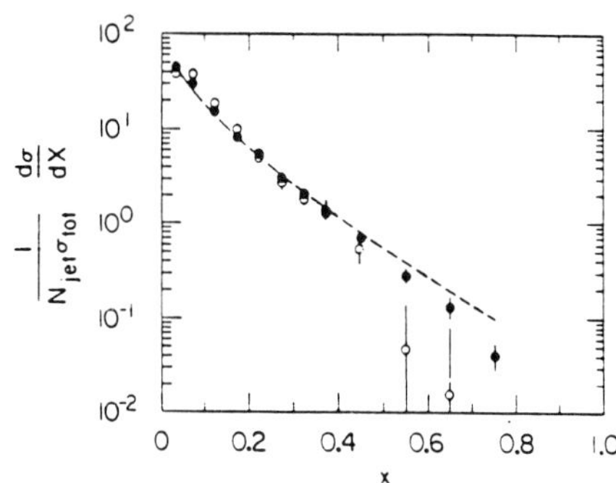

Figure 34: x_p distributions for a quark enriched sample (dots) and from a gluon enriched sample (circles), taken from Mark II data [47].

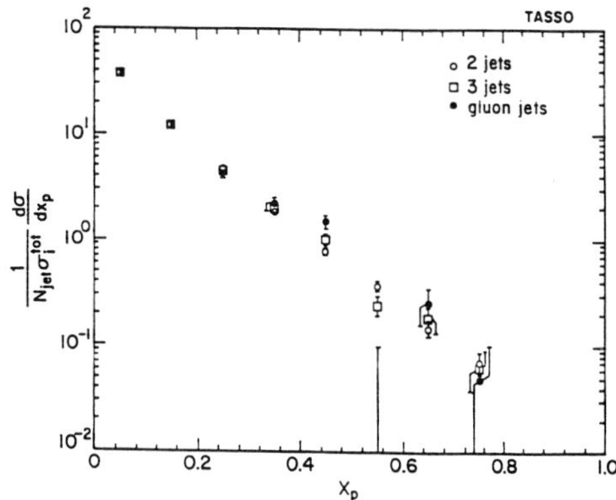

Figure 35: x_p distributions for a quark and and gluon enriched sample from TASSO data taken for three and two jet events.

These data indicate that there is no discernible effect between the three and two jet cases. In their analysis, the TASSO group claims to understand possible effects involving event selection that could explain the discrepency between the two findings.

The AMY collaboration has performed a similar analysis, but using instead the energy and charged particle multiplicity in the jet core as a means of exploring differences between quark and gluon enriched samples [49]. Jet energies in the three jet events are ordered as $E_1 \geq E_2 \geq E_3$. The jet core here is defined by the opening angle of a cone for particles to be included. The half angle of the cone is $\theta = 60^\circ/\sqrt{E_{jet}}$. For three jet events, the mean energy carried by charged particles in the core is compared for jets 1,2 and 3. The jet energies range from 4 to 30 GeV, but only at 15 GeV is a direct

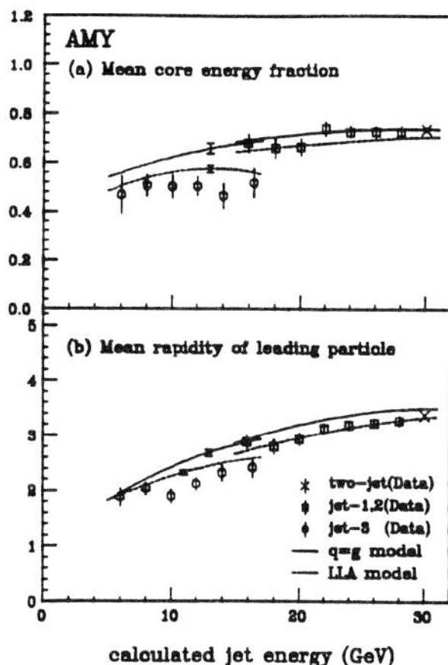

Figure 36: Comparison of fragmentation properties of jets 1,2 and 3 in three jet events by the AMY collaboration [49]. At 15 GeV (approximately $\frac{1}{3}$ of the CMS energy), a direct comparison is possible between jets 1, 2 and 3.

comparison between jets 1 and 3 really possible (figure 36). In this case, it appears that the third jet has less energy in the core than the other jets. The data are in more in agreement with a model featuring parton showering which is generated according to standard QCD leading-log prescriptions than with a model where quarks and gluons fragment in the same way.

In $\overline{p}p$ collisions, one can compare jet events produced with different boosts, but with approximately the same transverse energy. In data published by the UA1 collaboration [52] a small difference was reported for events separated by this method. Another method for obtaining enriched samples is to use direct photon events that explore a relatively low region of x_{bj} to tag quark enriched samples, and compare these to two jet events in the same kinematic range. Two jet events in this regime ($x_{bj} \leq 0.1$) should be dominated by gluons in the final state. This can be seen in figure 37 where events with direct photons are dominated by quark-gluon (as opposed to quark-antiquark) initial states. Likewise, two jet final states originating from partons at these values of x_{bj} are mostly from gluon-gluon scattering ($\geq 75\%$ enhancement).

Direct photon production from different reactions can also provide a means of tagging the final state partons as either quarks or gluons. The WA70 collaboration has compared final state jets in the reactions $\pi + p \rightarrow \gamma + jet$ and $p + p \rightarrow \gamma + jet$. In the first case, $\overline{q}q$ annihilation dominates, whereas in the latter case, q-glue in the

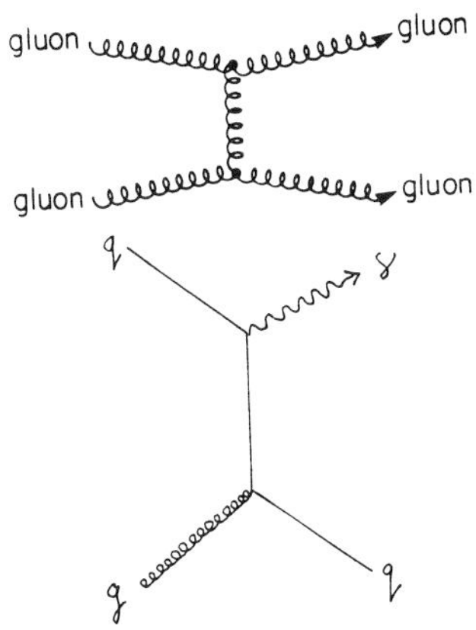

Figure 37: Diagrams with large contributions for direct photon and two jet production at moderate transverse energies in $\overline{p}p$ collisions. Events with direct photons can be used to obtain a quark enriched sample and compared to the predominately gluon jets in the two jet final state.

initial state is enhanced (figure 38) [54]. The fragmentation function $D(z) \equiv 1/N_{jet}dN_{ch}/dz$, where $z = p_{\|}/p_{jet}$, is sensitive to difference in quarks and gluon jets. Figure 39 shows the ratio of $D(z)$ for jets associated with direct photons coming from πp and pp scattering. There is a tendency for the quark enriched sample to have a harder fragmentation spectrum. This is not unexpected, although the estimate of the systematic errors from WA70 is too large to draw any definitive conclusions.

5.3 Jets in e^+e^- vs. $\overline{p}p$ events

One can compare jets in e^+e^- events to jets of the same energy in $\overline{p}p$ collisions. The effects of the

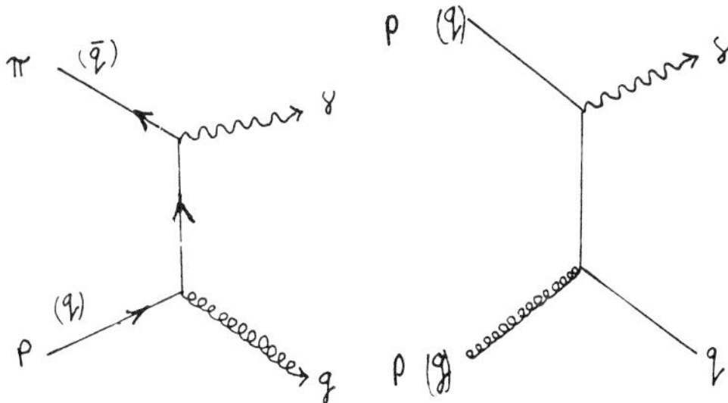

Figure 38: Different final states can be enhanced by choosing reactions for direct photon production. In the first case, $\pi + p$ scattering enhances final state gluons opposite the direct photon, in the second, $p + p$ enhances the quark final state.

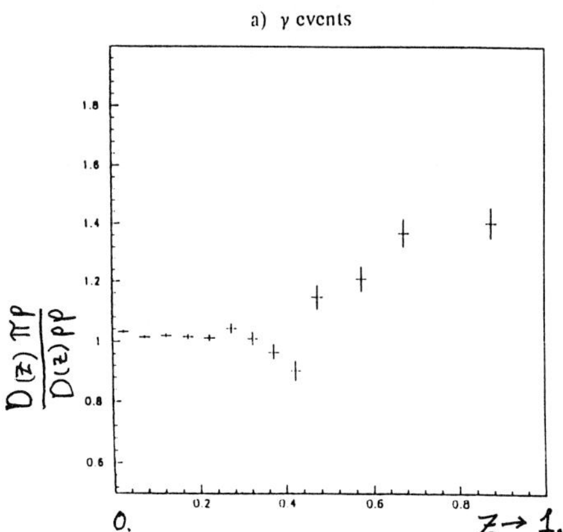

Figure 39: The ratio of the fragmentation function $D(z)$ for jets opposite direct photons in the reactions πp and pp. The jets from the πp reaction appear to have a harder spectrum than for those from pp reaction.

underlying event in $\bar{p}p$ collisions may make direct comparisons more difficult, nonetheless, universal aspects of jet behavior such as scale breaking effects, and multiplicities can be compared, particularly for the jet core where the similarities should be most striking.

Figure 40 shows the rapidity distribution of charged particles with respect to the jet axis for both $\bar{p}p$ and e^+e^- collisions [55],[54]. The distribution from $\bar{p}p$ collisions is subject to large uncertainties for rapidities below 2 units because of the influence of the underlying event. Above 2 units, however the distributions become quite similar. It is not clear which quantities are appropriate for the comparison of fragmentation properties of $\bar{p}p$ and e^+e^- collisions. One good guess would be that the CMS energy in e^+e^- and the dijet invariant mass (M_{jj}) form a reasonable basis for scaling comparisons. Figure 41 shows a comparison of the multiplicity of charged particles in the jet core as a function of these scaling variables. Here, the core is defined as the region of rapidity greater than 2 units. From the figure, it is apparent that in this region, the multiplicity has a more or less universal behavior.

Another universal component of fragmentation is the scale breaking effects associated with gluon bremsstrahlung. These can be observed in the fragmentation function scaling due to the emission of gluons inside a cone associate with the jet cone. this can be related directly to the scale breaking effects in structure functions. The relevant vari-

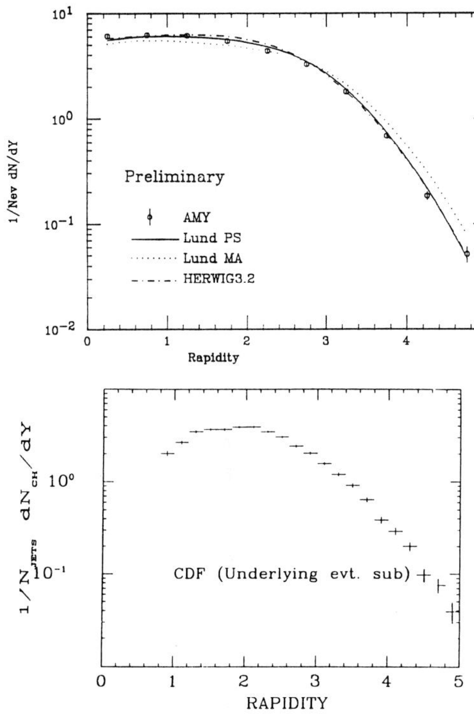

Figure 40: Distributions of charged particle rapidity with respect to the jet axis from the AMY and CDF collaborations. Note that only above 2 units is a comparison meaningful because of underlying event subtraction uncertainties associated with $\bar{p}p$ collisions.

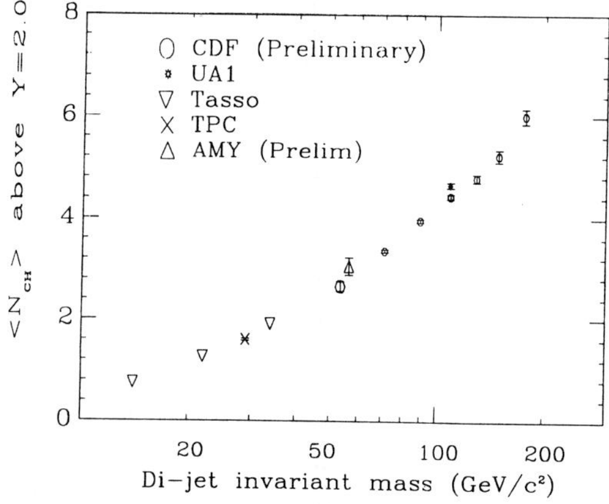

Figure 41: Jet core multiplicity as a function of M_{jj} for $\bar{p}p$ collider experiments, and as a function of CMS energy for e^+e^- experiments.

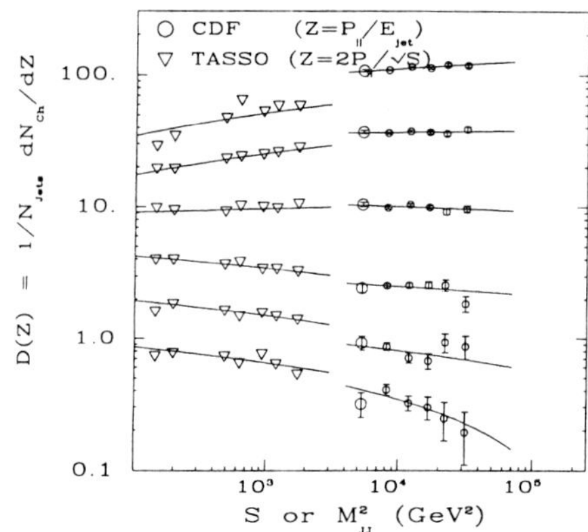

Figure 42: Evolution of the fragmentation function, $D(z)$, as a function of CMS energy for e^+e^- and as a function of M_{jj} for $\overline{p}p$ collisions.

able for fragmentation is z, defined above, rather than x_{bj}. Figure 42 shows the fragmentation function for both CDF and TASSO data sliced into different z bins, plotted as a function of a relevant q^2-like parameter. In the case of TASSO [56], it is $\sqrt{s}$, in the case of CDF, it is taken to be M_{jj}. As one can see, there is the appearance of scale breaking for both the e^+e^- and $\overline{p}p$ jets.

6 Conclusions

To conclude, much progress has been made in QCD studies recently, in both theory and experiment. With more detailed studies at the Tevatron, LEP, SLC and Tristan, along with more precise theoretical predictions, measurements are becoming more meaningful.

The inclusive jet cross section in $\overline{p}p$ collisions, now has been explored over 8 orders of magnitude of cross section, from 6 to 470 GeV. New theory calculations are expected by early 1990 with the prospect of more precise comparisons. These together help establish greater sensitivity to possible deviations from the standard model, such as quark substructure effects.

The p_t spectra of the $W^\pm$ and Z^o have been measured with higher precision, and, although no surprises have been seen, it is critical to be able to understand these spectra in detail in order to explore vector boson pair production. The new α_s^2, α calculations allow more precise comparisons.

The running of α_s as seen in e^+e^- is very close to being nailed down. High statistics measurements at LEP, SLC and Tristan are needed to confirm this most important feature of QCD. The jet multiplicity up to 4 jets appears to have a satisfactory theoretical comparison, although, again, the statistics are low. The study of gluon-gluon coupling remains to be thoroughly explored at these machines and awaits higher statistics.

In fragmentation studies, charm production in $\overline{p}p$ collisions appears to be well described by the perturbative description of gluon splitting, but awaits higher statistics. Attempts to find differences between quark and gluon enriched samples have shown at best only modest effects, and in at least one case, when examined closely, no effect. Similarities between e^+e^- and $\overline{p}p$ jets, particularly in the jet core are very striking, pointing to a similar fragmentation mechanism.

Acknowledgements

I would like to thank the organizers of the symposium, Phil Burrows, Steve Olsen, Steve Schnetzer, Keith Ellis, Paul Mackenzie and David Soper for all of their assistance in the preparation of this talk.

References

[1] F. Abe *et al*, CDF Collab., Phys. Rev. Lett. **62** (1989) 613.

[2] G. Arnison *et al*, UA1 Collab. Phys. Lett. **132B**, 214 (1983) and **172B**, 461 (1986).

[3] W. Bartel *et al*, JADE Collab., Z. Phys. **C33**(1986) 23.

[4] S. Wu, G. Zobernig, Z. Phys. **C2**(1979) 107.

[5] J. Appel, *et al* Z. Phys. **C30**(1986) 341.

[6] L. DiLella, these proceedings.

[7] K. Eggert, these proceedings.

[8] P. Sinervo, these proceedings.

[9] J. Huston, these proceedings.

[10] P. Frampton and S. Glashow, Phys. Lett. **190B** (1987) 157.

[11] E. Eichten, G. Kane and M. Peskin, Phys. Rev. Lett., **50** (1983) 811.

[12] T. Akesson et al., Phys. Lett. **123B** , 133, (1983).

[13] F. Abe *et. al.* , Phys. Rev. Lett., **62** 613 (1989); J. Huth, in Proc. of "Particle Physics in Prospective", La Thuile, Italy, 1989, ed. G. Belletini.

[14] G. Arnison et al., Phys. Lett. **172B** , 461, (1986).

[15] J. Appel et al., Phys. Lett. **160B** , 349, (1985)

[16] R. Ellis and J-C. Sexton, Nucl. PHys. **B269**, 445 (1986).

[17] S. Ellis, Z. Kunzst, and D. Soper, Oregon Preprint OITS 395 and 396 (1988).

[18] F. Aversa, *et al*, Annecy preprint LAPP-TH-255/89 (1989).

[19] E. Eichten, K. Lane and M. Peskin, Phys. Rev. Lett. **50,** 811 (1983).

[20] F. Abe *et al*, Phys. Rev. Lett., **62** (1989) 613.

[21] M. Diemoz, F. Ferroni, E. Longo, and G. Martinelli, Z. Phys.**C39** (1988) 21 (sets 1,2 and 3).

[22] D. W. Duke and J. F. Owens, Phys. Rev. **D30** (1984) 49 (sets 1 and 2).

[23] Bawa, A.C. and Stirling, W.J., Phys. Lett. B203, 172 (1988); Martin, A.D. *et al.*, Z.Phys. **C42**, (1989) 277.

[24] P. Arnold and H. Reno, Fermilab preprint FERMILAB-PUB-88/168-T (1988); ERRATA, FERMILAB-PUB-89/59-T (1989); to appear in Nucl. Phys. B.

[25] T. Kamon, CDF Collab., Proc. of 8th Topical Workshop on Proton-Antiproton Collider Physics, Castiglione, Italy, Sept. 1989.

[26] G. Altarelli *et al*, Nucl. Phys. **B157** , (1979) 461.

[27] G. Altarelli *et al*, Nucl. Phys. **B246** , (1984) 12.

[28] G. Altarelli *et al*, Z. Phys. **C27** , (1985) 617.

[29] F. Berends, *et al*, Phys. Lett. **B224** (1989), 237.

[30] Ian Hinchliffe, private communication.

[31] V. Ruhlmann, UA2 collaboration, in Proceedings of the 7th Topical Workshop on Proton-Antiproton Collider Physics, June 1988, Fermilab, Box 500, Batavia Il; ed. Raja, Tollestrup and Yoh, World Scientific (1989); A. Benvenuti *et al*, BCDMS collaboration, Paper 109, contributed to this symposium; S. Bethke *et al*, JADE collaboration, Phys. Lett., **B213** (1988) 235; J. Aubert*et al*, EMC collab., Nucl. Phys. **B259** (1985) 189; W. deBoer, Proc. of XXIV Int. Conf. on High Energy Physics, Munich, August 1988, Springer Verlag, R. Kotthaus, J. Kuhn eds. (1989).

[32] H. Aihara *et al* , TOPAZ collaboration, Paper 147 contributed to this symposium.

[33] The AMY collaboration, Paper 231 contributed to this symposium.

[34] A. Weinstein, Plenary talk, these proceedings.

[35] P. Burrows, D. Phil. Thesis, Oxford University (1988); W. Braunschweig *et al*, TASSO Collab., Phys. Lett. B214 (1988) 286.

[36] T. Sjostrand, Comput. Phys. Commun. **39** (1986) 347; **43** (1987) 367.

[37] G. Kramer and K. Lampe, DESY Report DESY-86-119;
T. Gottschalk and M. Shatz, Phys. Lett **B150** (1985) 451;
F. Gutbrod, G. Kramer and G. Schierholz, Z. Phys. **C33** (1984) 235.

[38] S. Bethke, LBL Report LBL-26958 (1989).

[39] J. Korner, *et al*, Nucl. Phys. **B206** (1981) 365; O. Nachtmann, A. Reiter, Z. Phys., **C16** (1982) 45.

[40] M. Bengtsson, P. Zerwas, Aachen preprint PITHA-88/04 (1988).

[41] M. Bengtsson, Z. Phys., **C42** (1989) 75.

[42] AMY Collaboration, contribution to this conference.

[43] A. Muller and P. Nason, Nucl. Phys., **B266** (1986) 265.

[44] M. Della Negra (UA1 collaboration), Proceedings for the 6th Topical Workshop on Proton-antiproton Physics, Aachen, W. Germany (1986).

[45] G. Arnison *et al* Phys. Lett. **147B** (1986) 265.

[46] S. Brodsky and J. Gunion, Phys. Rev. Lett. **37** (1976) 402.

[47] A. Petersen **et al**, Mark II Collab., Phys. Rev. Lett. **55** (1985) 1954.

[48] M. Derrick *et al*, HRS Collab., Phys. Lett. **165B** (1985) 449.

[49] The AMY Collaboration, contribution to this conference.

[50] W. Braunschweig *et al*, TASSO Collab, DESY preprint DESY 89-032 WIS-89/7/Feb-Ph (1989) contribution to this conference.

[51] R. Madaras, Proc. of Rencontre de Moriond on Strong Interaction and Guage Theories, Les Arcs (1986).

[52] P. Ghez, PhD Thesis, Annecy (1986) unpublished; G. Arnison *et al*, Nucl. Phys. B **276** 253 (1986).

[53] M. Bonesini *et al*, WA70 Collab., Paper 178 contributed to this symposium.

[54] Steve Schnetzer, private communication.

[55] B. Hubbard, Proc. of 7th Topical Workship on Proton-Antiproton Collider Physics, Fermilab, Batavia Il, June 1988, Ed. Raja, Tollestrup and Yoh, World Scientific (1989).

[56] M. Althoff *et al*, Z. Phys C **22**, 307 (1984).

DISCUSSION

S. Komamiya, SLAC: You showed the AMY plot of fractions for 2-jet, 3-jet and 4-jet events as a function of y_{cut} and concluded that you can fit the 4-jet data better by choosing smaller Q^2. This does not make sense because the 4-jet fraction is calculated only to leading order $(O(\alpha_s^2))$ and no higher-order corrections are made to eliminate Q^2 ambiguities. I think that a simultaneous fit of Q^2 and $\Lambda_{\overline{MS}}$ by using 2- and 3-jet fractions (or 3- and 4-jet fractions) has serious problems: the 4-jet fraction does not have higher-order calculations, and the 2- and 3-jet fractions are fully correlated.

J. Huth: My only comment on this result is that the AMY data indicate a renormalization scale somewhat less than the collision center of mass energy is favored. I avoided a discussion of the actual fit in part because of the problems you have just mentioned; at higher than second order, the meaning of the renormalization scale becomes murky.

S. Komamiya: I have a second comment on the natural Q^2 scale of qqg events. According to the method of Brodsky, Lepage and MacKenzie, which eliminates the Q^2 ambiguity for QED and QCD, the running of α_s comes only from vacuum polarization of gluons — at least for second-order QCD calculations. The natural Q^2 scale of a $q\bar{q}g$ event is not the gluon-quark invariant mass, but has something to do with the virtuality of the gluon. Therefore the natural scale can be very small, and it must depend on the event kinematics and the infrared cut-off (y_{min}) of the QCD calculation. For the case of hadron colliders, the gluon virtuality is much larger, so the natural Q^2 scale is much larger than it is for e^+e^- collisions.

J. Huth: If you could do the calculation to all orders, the dependence on the renormalization scale should in principle vanish; so it is difficult to arrive at a natural scale that works to all orders. The gluon-quark invariant mass is an example of one possible scale and was mentioned to illustrate that there are scales in the problem other than the collision center-of-mass energy.

V. Khoze, INP, Leningrad: I'd like to make a comment connected with this comparison of quark versus gluon fragmentation. I think one of the best places for real investigations of the gluon jet will be with heavy tagged quarks — for example, at the Z resonance in $c\bar{c}g$ production. Because of the possibility of tagging the quark jets, you can be sure that you are investigating the gluon jet.

J. Huth: Good point.

Unidentified questioner: Are there possibilities for polarization experiments in proton-antiproton collisions?

J. Huth: There was a workshop about a year ago at Fermilab on polarization schemes. There was discussion of the various possibilities for polarized proton-proton collisions. Although there were some interesting ideas, I believe that in practice polarized collisions may be difficult to realize with sufficient luminosity to do anything significant. But if an ingenious idea comes up, maybe it's possible.

M. Einhorn, University of Michigan: I like to question your determination of the compositeness scale, based on a lowest-order QCD fit. I'd guess that higher-order QCD corrections might be 40-50% and not necessarily give the same p_t dependence. So I wonder how to interpret your advertised lower limit on compositeness.

J. Huth: The limits quoted take into account possible variations in the gluon structure function, with the most conservative limit quoted. In particular, we are measuring deviations in the shape, not absolute magnitude of the cross section. I expect that the next-to-leading order calculations may only cause logarithmic deviations of the shape from the leading-order calculations, in which case the limits should still be approximately valid. In any event, two groups are in the process of calculating the next-to-leading order cross sections, and I expect these results by the end of the year, so we are in the fortunate situation of being able to check against these calculations.

M. Peskin, SLAC: I'd like to make a comment on Marty Einhorn's question. First, when you test for compositeness in proton-proton collisions, you can rely on the fact that the shape of the contribution from the contact interactions is different from QCD. Effectively the Q^2 dependence is different, so the curves turn up in the last few bins. So you can normalize the curve and slope, and then just look at the effect in the last few bins. That's a very strong procedure that's roughly independent of the gluon distribution. Now the only comment one should make on the other side is that in proton-proton collisions, there are (from complete ignorance) 17 parameters that you could write for the effective Lagrangian. Eichten, Lane, and I chose one completely arbitrarily, and so this should be taken as a figure of merit rather than as any definite limit on composite interactions.

J. Huth: Right. We use an isoscalar, color-singlet term, as quoted in your paper, with a coupling to left-handed quarks. This is the same term used by both the UA1 and UA2 collaborations in deriving compositeness limits, and thus is the most natural basis for comparison.

G. Martinelli, University of Rome: You talked about scalar-scalar production, what about vectors?

J. Huth: Again, this is just a specific interaction that we chose to put in. I just consider it something of an alchemist's touchstone that gives a basis for comparison. I could put in other terms and derive quite different limits.

THE PHYSICS OF QCD JETS

VALERY A. KHOZE
Leningrad Institute for Nuclear Physics,
188350 Gatchina, USSR

ABSTRACT

We review the analytical perturbative approach to QCD jet physics. The role of coherent phenomena, reflecting the collective character of multiple hadroproduction, is emphasized.

1. INTRODUCTION

This review summarizes recent advances in applications of the perturbative QCD approach (PA) to the description of jet-like final states in hard processes (HP). The interest in the detailed studies of jets is twofold: On the one hand, they are important for testing both perturbative and nonperturbative QCD, for design of experiments and the analysis of their data. On the other hand, the characteristic features of jet-like states could provide a valuable additional tool, helping to extract and to study manifestations of new physics. The main aim of this talk is to show where we are now and where we are going.

It would be impossible to cover the subject of QCD jets completely, as it is quite large. So I must apologize in advance for being selective in terms of topics discussed and references cited. In the talk, we shall discuss in detail only some specialized items, emphasizing the manifestations of color coherence in the jet-like final states in HP.[1] For recent excellent reviews, covering other topics in jet physics, see Refs. [2–9].

The outline of the talk is as follows:

1. Introduction.
2. The essence of the perturbative approach.
3. QCD portrait of an "individual jet."
4. Intrajet coherence and humpbacked QCD spectra.
5. Interjet coherence and radiophysics of particle flows.
6. Perturbative approach and QCD-inspired fragmentation schemes.
7. Experimental selection procedures.
8. Specific properties of jets generated by heavy quarks.
9. QCD coherence in deep inelastic scattering (DIS).
10. Prospects for future experiments.
11. Conclusions.

2. THE ESSENCE OF THE PERTURBATIVE APPROACH

The existence of hadron jets is one of the most striking successes of QCD. Since the first pioneering paper on jets in e^+e^- collisions,[10] written here at SLAC, jet physics in HP has reached a mature level of sophistication.[1–8,11–13] As has been discussed by the previous speaker,[9] hadronic jets are now intensively studied both at e^+e^- and hadronic machines. Jet dynamics will be one of the central problems of investigation for the colliders of the present and of the future.

The field of jet physics has gone through a qualitative transformation in the last few years:

i. At the partonic level, developments in QCD allow one to make testable quantitative predictions, with controllable accuracy, in terms of analytical perturbative calculations; see, e.g., Ref. [1]. The key idea is to invoke the parton shower picture where one views the evolution, say, of a jet as a sequence of parton branchings. Generally speaking, using a shower picture does not necessarily lead to a loss of accuracy. The main concept is to reorganize the perturbative expansion in such a way that its zero-order approximation is systematic and involves an arbitrary number of produced particles. This approximation can be achieved through an iteration of basic $A \to B+C$ branchings. In principle, it should be possible to include higher corrections in order to systematically improve the accuracy. This procedure[14] is closely related to a renormalization group approach[15] in which the branchings are not so visible and in which higher corrections can be systematically calculated.

The approximation mentioned above is the Modified Leading Logarithmic Approximation (MLLA), taking care of both double logarithmic and single logarithmic effects in a systematic way.[14,16]

In constructing the MLLA, one should pay special attention to maintaining the probabilistic interpretation of jet evolution. The existence of such an interpretation is far from trivial in the problems connected with description of soft particle distributions ($x \ll 1$). Here, interference contributions play an important role and prove to be unavoidable, unlike the case of the familiar problems dealt with near $x \sim 1$. Nevertheless, it appears to be possible, by choosing an appropriate evolution parameter (jet opening angle θ_0) and accounting for the specific angular dependence of soft emission probabilities to maintain probabilistic interpretation of the dynamics of soft partonic cascades; see, e.g., Refs. [1, 14]. This not only helps one's physical intuition but also provides the basis for Monte Carlo simulations of jet physics.

ii. The well-elaborated Monte Carlo algorithmic models[6–8,17–22] are becoming better and better at building in realistic fragmentation and proper QCD evolution. These schemes successfully describe the experimental data and prove to be very useful tools for experimentalists.

It is important to notice, however, that the description of multipartonic system development in terms of classical Markov chains is of limited value in principle. Of special interest here are $1/N_c^2$ corrections. Thus, the possibility to provide a probabilistic scheme of the partonic cascade evolution does not exist beyond the MLLA (the so-called "color monsters": $1/N_c^2$ suppressed soft contributions[14,23]). Another example is connected with the interjet collective phenomena that could be reproduced in a probabilistic way only in the large N_c limit.[23]

iii. A vast amount of experimental data on jets in HP has been accumulated over the last few years.[9,11–13] These data have demonstrated convincingly that at high energies, the dominant role in multiparticle production is played by the perturbative stage of jet evolution. In particular, nowadays it seems to be mandatory, for a good description of experiment, to use a low value of the formal boundary Q_0 between the perturbative and nonperturbative stages of jet development, typically, $Q_0^2 \simeq 1$ GeV^2; see Refs. [6, 17]. Moreover, the data indicate that the distributions of hadrons follow unexpectedly closely those of the fundamental partons of QCD, even in the cases of some subtle collective characteristics. The fact that nonperturbative effects do not radically rearrange a parton system at the confinement stage provides evidence in favor of locality of parton branching and hadronization processes in configuration space,[23] thus supporting the hypothesis of Local Parton-Hadron Duality (LPHD),[14,24] based on the "preconfinement"[25] properties of the parton cascades.

The LPHD approach attempts to describe the global features of the hadronic systems in HP, such as the mean multiplicities and multiplicity distributions, angular patterns of energy and multiplicity flow, inclusive energy spectra and correlations of particles, etc., without making any reference at all to a fragmentation scheme. The nonperturbative effects are reduced to normalizing constants relating hadronic characteristics to partonic ones (for various quantitative realizations of LPHD, see Ref. [26]). This makes predictions very restrictive. Since there are only a few parameters to vary in connecting PA results to experiment, these predictions are simply testable. Thus, the main equation of PA reads

$$PA = MLLA + LPHD \qquad (1)$$

We shall focus on the manifestation of coherent phenomena. The rediscovery of coherence in the context of QCD in the early eighties[27,28] has led to a dramatic revision of theoretical expectation about the structure of soft particle distributions (see Refs. [1, 6–8, 20, 29–33]).

Thus, the coherent effects in the intrajet partonic cascades, resulting, on average, in the angular ordering (AO) of sequential branching, gave rise to the humpbacked plateau—one of the most striking perturbative predictions.[14–16,24,29,30]

Due to the interjet coherence, which is responsible for the drag effects in multijet events,[20,23,31–33] a very important physical phenomenon can be experimentally verified, namely, the fact that it is the dynamics of the color which governs the particle production in accordance with the QCD radiophysics of hadron flows.

Surely, the main lesson of these observations is not the proof of coherence: it would be inexcusable to check quantum mechanics at modern accelerators. Of real importance is that the coherence has revealed itself in hadron spectra, i.e., confinement has not disturbed the perturbative picture of multiparticle production.

3. QCD PORTRAIT OF AN "INDIVIDUAL JET"

At sufficiently high energies, HP events should possess the clear geometry that reflects the topology of the participating partons. The space-energy portrait of events represents a natural partonmeter for registering the kinematics of the energetic partons. While the *hard component* of a hadron system (a few hadrons with the energy fraction $z \sim 1$) determines the partonic skeleton of an event, the *soft component* (the other hadrons with $z \ll 1$) forms the bulk of the multiplicity. This soft component is concentrated inside the bremsstrahlung cones of jets. The opening angle θ_0 of each cone is bounded by the nearest other jet, since at larger angles particles are emitted coherently by the overall color charge of both jets.[23,32] Even though the bremsstrahlung cones of the neighboring jets strongly overlap, the resulting total multiplicity can be presented as the additive sum of the contributions of the individual jets.

One can study the properties of an individual quark jet when measuring the different inclusive distributions in the process $e^+e^- \to$ *hadrons*. The decay into two gluons of the C-even ultra-heavy quarkonium states, $\chi_Q = Q\overline{Q}$, might define by analogy the individual gluon jets. In spite of the great importance of coherence phenomena, the notion of an isolated jet makes sense if one does not deal with the azimuthal effects, but considers only multiplicities, energy spectra and correlations, etc. In this case, all the influence of the jet ensemble on a given jet may be encoded in a single parameter θ_0, the jet "opening angle."

Jet characteristics prove to depend not on the jet energy E but on the hardness Q of the process producing the jet, $Q = E\theta_0$ at $\theta \ll 1$. At $Q_0 = \Lambda$, the multiplicity of gluons in a gluon jet, $N_g^g(Q,\Lambda)$ is[1]

$$\begin{aligned} N_g^g(Q,\Lambda) &= \Gamma(B)\left(\frac{X_1}{2}\right)^{1-B} \cdot I_{1+B}(X_1) \\ &\simeq \Gamma(B)\left(\frac{X_1}{2}\right)^{(1/2)-B} \frac{e^{X_1}}{\sqrt{2\pi}} \ , \quad \frac{1}{\sqrt{Y}} \ll 1 \ ; \end{aligned} \tag{2}$$

with I a modified Bessel function with

$$X_1 = \left[\frac{16N_c}{b}\,Y\right]^{1/2} \ , \quad Y = \ell n\,\frac{E\theta_0}{\Lambda} \ , \tag{3}$$

where

$$b = \frac{11}{3}\,N_c - \frac{2}{3}\,n_f \ , \quad B = \left(\frac{11}{3}\,N_c + \frac{2}{3}\,\frac{n_f}{N_c^2}\right)\frac{1}{b}$$

at $N_c = n_f = 3$, $b = 9$, and $B = 101/81$.

In the LPHD phenomenology[24] for hadron h production, one has

$$N_g^h(Q) = K^h\,N_g^g\,(Q,\Lambda) \ . \tag{4}$$

The ratio of particle multiplicities in gluon jets and quark jets with the accuracy to $O(\alpha_s)$ terms is given by[34]:

$$\frac{N_g}{N_q} = \frac{9}{4}\,(1 - 0.27\sqrt{\alpha_s} - 0.07\alpha_s + \cdots) \ . \tag{5}$$

The asymptotic factor $9/4 = N_c/C_F$ ($C_F = (N_c^2 - 1)/2N_c$) of PA appeared in the pioneering paper.[35]

In the LPHD picture, the multiplicity of charged particles in e^+e^- annihilation is[1,24]:

$$N_{e^+e^-}^{ch}\,(W) \simeq N_g^g(Q,\Lambda) \ , \quad W = 2Q \ . \tag{6}$$

Note that at high energies ($W \gtrsim 1$ TeV) there is a significant difference between the expectations of various Monte Carlo algorithms. The analytical PA predicts values that are close to the "dipole formulation" of the Lund Monte Carlo.[21,22]

The collimation of a QCD jet around the parent parton becomes stronger as the parton energy E increases. Moreover, the collimation of an energy flow grows much more rapidly compared to a multiplicity flow. Consider a jet with energy E and opening angle θ_0. The angular cone $\theta_{\mathcal{E}}$, in which an energy fraction $\mathcal{E}$ is deposited, is given by[23,36]:

$$\frac{\theta_{\mathcal{E}}}{\theta_0} = \left(\frac{E\theta_0}{\Lambda}\right)^{-\gamma_j(\mathcal{E})} \ , \tag{7}$$

where, for example,

$$\begin{aligned} \gamma_q(0.9) &\simeq 0.55 \ , \qquad \gamma_q(0.5) \simeq 0.83 \ , \\ \gamma_g(0.9) &\simeq 0.30 \ , \qquad \gamma_g(0.5) \simeq 0.54 \ . \end{aligned} \tag{8}$$

It follows from Eqs. (7) and (8) that the energy collimation in a quark jet is stronger than in a gluon one, and that the collimation grows rapidly as the energy increases.

The collimation of multiplicity inside a jet is described by[23,36]:

$$\frac{\theta_\delta}{\theta_0} \sim [N_j\,(E\theta_0)]^{-(b/8N_c)\,\ell n\,1/\delta} \quad , \qquad (9)$$

where θ_δ is the cone in which the jet multiplicity fraction δ is concentrated. Thus, $\theta_{1/2}$ decreases with the increase of jet hardness $Q = E\theta_0$ approximately as $N_j^{-1/4}(Q)$, i.e., parametrically much slower than in the case of the energy collimation in Eq. (7). Note that fixing a jet axis with the accuracy better than the natural angular width of the corresponding energy flux is unreasonable.

4. INTRAJET COHERENCE AND HUMPBACKED QCD SPECTRA

Roughly speaking, there are two types of coherence effects. The first manifestation of coherence is *AO* of the sequential parton decays. Coherence of the second type deals with the angular structure of particle flows when three or more partons are involved in HP. Here, the particle distributions depend on the geometry and color topology of the whole jet ensemble (radiophysics of particle flows).

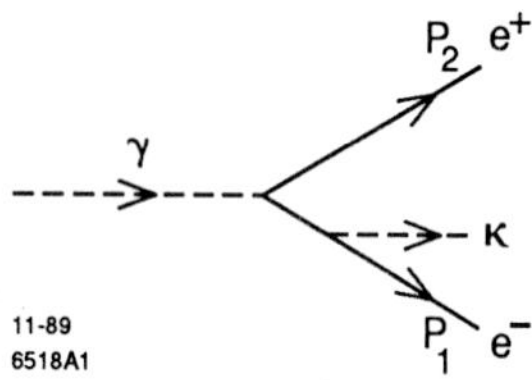

Figure 1: Emission of a soft photon k after e^+e^- pair production.

To elucidate the physical origin of *AO*, let us consider a simple mode of the jet cascade, namely, the radiation pattern of soft photons produced by a relativistic e^+e^- pair; see Fig. 1. The question is to what extent do the e^+ and e^- independently emit photons. To answer this question, one has to estimate the formation time needed for a photon to be radiated from, say, the e^- leg. Using the uncertainty relation, one finds

$$t_{form} = \frac{1}{k\cdot\theta_{\gamma e}^2} = \frac{k}{k_\perp^2} \quad , \qquad (10)$$

where $\theta_{\gamma e}$ is the angle between the emitted photon and the electron. Now $k\cdot\theta_{\gamma e} = k_\perp = \lambda_\perp^{-1}$, with $\lambda_\perp$ the transverse wavelength of the radiated photon. Thus,

$$t_{form} = \lambda_\perp/\theta_{\gamma e} \quad .$$

During this period of time, the e^+e^- pair transversely separate a distance

$$\rho_\perp^{e^+e^-} \approx \theta_{e^+e^-}\cdot t_{form} \approx \lambda_\perp \frac{\theta_{e^+e^-}}{\theta_{\gamma e}} \quad . \qquad (11)$$

One concludes that for large angles

$$\theta_{\gamma e^-} \approx \theta_{\gamma e^+} \gg \theta_{e^+e^-} \quad ,$$

the separation of e^+ and e^- is smaller than $\lambda_\perp$. In this case, the photon cannot resolve the internal structure of the e^+e^- pair and probes only its total electric charge, which is zero. Thus, for $\theta_{\gamma e} \gg \theta_{e^+e^-}$, we expect emission to be strongly suppressed. This is the Chudakov effect,[37] well known in the physics of the QED shower. The e^+ and e^- can emit photons independently only when $\rho_\perp^{e^+e^-} \gg \lambda_\perp$, that is, when $\theta_{\gamma e^+}$, $\theta_{\gamma e^-} < \theta_{e^+e^-}$.

The same discussion can be given for QCD cascades in which gluon radiation is governed by the conserved (color) currents. The only difference is that the coherent radiation of soft gluons by an unresolved pair of gluons or quarks is no longer zero, but the radiation acts as if it were emitted from the parent gluon imagined to be on shell, as is illustrated in Fig. 2(a). The remarkable fact is that one gets all double log and single logarithmic effects correctly, for angular averaged observables, by emitting the gluon independently off line 1 when $\theta_{1k} < \theta_{12}$, off line 2 when $\theta_{2k} < \theta_{12}$, and off the parent, line g, when $\theta_{kg} > \theta_{12}$; see Refs. [1, 14]. In the general case of large-angle soft gluon emission off the multiparton jet [see Fig. 2(b)] the intrajet partons p_i can be considered as being collimated. As a result, this gluon can be treated as a classical probe testing the total color charge of the jet.

The *AO* occurs not only for the time-like jet evolution, but also for the space-like partonic cascades.

An additional gluon can be emitted really only if its formation time [Eq. (10)] is smaller than the hadronization time

$$t_{form} < t_{had} \sim kR^2 \quad . \qquad (12)$$

Then one gets the requirement

$$k_\perp\, R > 1 \quad , \qquad (13)$$

which justifies the very opportunity of using quark-gluon language.

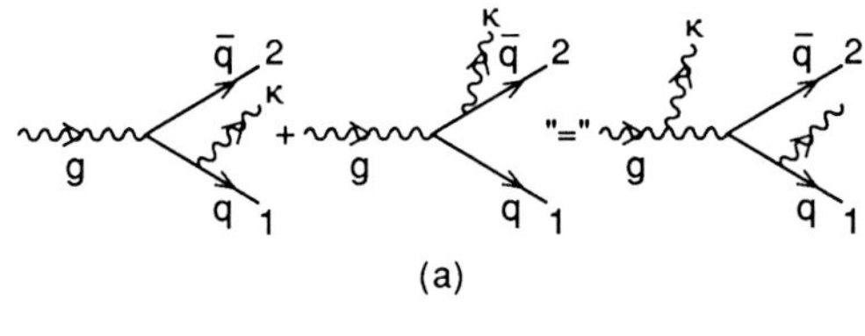

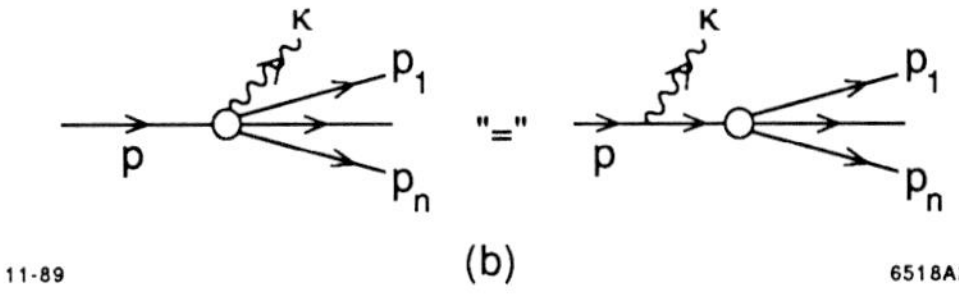

Figure 2: Illustrations of coherence where wide-angle emission of soft gluon, k, acts as if the emission came off the parent parton: (a) gluon conversion into the quark-antiquark pair; (b) parton jet emission off a hard parton p.

Let us discuss the origin of the depletion of soft-particle emission inside a jet (humpbacked plateau). The suppression of soft radiation follows from the AO of the partonic cascade and is a direct manifestation of color coherence. This behavior can be understood on kinematical grounds as the result of two conflicting tendencies: on the one hand, due to the restriction $k_\perp > 1/R$, a slow particle is "forced out" at large emission angle $\theta > 1/kR$; and on the other hand, the allowed decaying angle, after a few successive branchings, has shrunk to small values. For the purpose of finding clear manifestations of QCD coherence, the rapidity y is not a good variable; see Ref. [24] for details.

Not too far from the peak, the asymptotic shape of the single-particle energy distribution can be described by the MLLA expression,[1]

$$x\overline{\mathcal{D}}(x,Y) \sim \sqrt[4]{\frac{36N_c}{\pi^2 bY^3}} \times \exp\left[-\sqrt{\frac{36N_c}{b}}\left(\frac{\ell n\,\frac{1}{x} - \frac{Y}{2} - B\sqrt{\frac{b}{16N_c}Y}}{Y^{3/4}}\right)^2\right] \tag{14}$$

where $Y = \ell n\ E/\Lambda$, $x = E_p/E$, E is the jet energy and E_p is the energy of the measured particle. It has a broad Gaussian shape with a peak at $E_p = E_0 = x_0 E$,

$$\ell n\ 1/x_0 = 1/2 \cdot Y + B\sqrt{\frac{b}{16N_c}Y} \quad , \tag{15}$$

that grows rather slowly with the jet energy E:

$$\frac{E\ dE_0}{E_0\ dE} = 1/2 - \sqrt{\frac{b}{64N_cY}} \quad . \tag{16}$$

For example, Eq. (16) gives $d\ \ell n\ E_0/d\ \ell n\ E \approx 0.4$ at $E = 30$ GeV.

It is noteworthy that the recent data on the energy distribution of hadrons in e^+e^- annihilation[38–40] favor the existence of a humpbacked plateau, thus supporting the concept of LPHD. Unfortunately, the energy corresponding to x_0 in Eq. (15) is not large at present energies: $E_0 \simeq 0.6$ GeV in the PEP/ PETRA region. E_0 grows rather slowly with the jet energy E; even at $E = 1.5$ TeV, its value reaches only $E_0 \simeq 3$ GeV. Therefore, nowadays there is room for skepticism, since one should be able to disentangle the standard kinematical dip in the spectrum from the dynamical dip required by Eq. (14). So far, probably the most convincing evidence for the QCD-inspired humpbacked distribution is the growth of the energy $E_0 = E_{hump}$ at which the spectrum reaches its maximum as one increases the jet energy E.[40] Additional data for higher-energy jets at SLC and LEP should make this growth much clearer and perhaps allow one to check the relation [Eq. (16)].

To sharpen the influence of AO on the parton multiplication process, and in an attempt to find the dip in jets produced in hadronic collisions, it proves to be important to look at the spectra of particles restricted to lie within a particular opening angle with respect to the jet.[14] For example, one might consider the energy distribution of particles accompanying the production of an energetic particle and lying within an opening angle θ_0 about the direction of the trigger particle momentum.

Parton cascades in these situations will populate mainly the region

$$\frac{m_h}{\sin\theta_0/2} < E_h < E \quad ,$$

with E_h and m_h being the energy and the mass of the observed particle, respectively ($m_h \gtrsim \Lambda$). The maximum of the distribution, in E_h, is now forced to larger energies[41] [cf. Eq. (15)],

$$\ell n\ \frac{E_0}{m_h} = \frac{1}{2}\ \ell n\ \frac{E}{m_h\ (\sin\theta_0/2)} - B\left(\sqrt{\frac{b}{16N_c}}\ \ell n\ \frac{E(\sin\theta_0/2)}{\Lambda} - \sqrt{\frac{b}{16N_c}}\ \ell n\ \frac{m_h}{\Lambda}\right) . \tag{17}$$

Clearly, by choosing a small θ_0, one can move the peak of the distribution away from the phase-space limit of $E_h \sim m_h$.

Note that if one chooses θ_0 moderately small and varies E, coherence will give a moving peak in accordance with the relation [cf. Eq. (16)]:

$$\frac{E}{E_0}\frac{dE_0}{dE} = \frac{1}{2} - B\sqrt{\frac{b}{64N_cY(\theta_0)}} \quad , \tag{18}$$

with

$$Y(\theta_0) = \ell n \, \frac{E(\sin\theta_0/2)}{\Lambda} \quad .$$

The angular cut θ_0 is especially useful for jets produced in hadronic collisions, since one is able to eliminate much of the soft background.

Interesting prospects for studying particle distributions in different cones around the jet axis arise in the Z^0 peak, where one can easily fix a heavy quark direction. In Fig. 3, the expected charged-particle distributions are shown for jets with $E \simeq M_Z/2$.

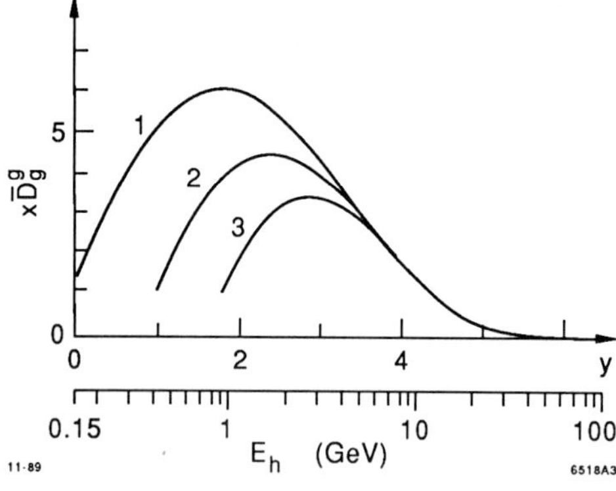

Figure 3: Energy distributions of charged hadrons (vs. $y = \ell n\ E_h/\Lambda$) in different cones around the jet axis in e^+e^- annihilation at $W = 2E = M_Z$: (1) for the whole opening angle $\theta = 180°$; (2) $\theta_0 = 45°$; and (3) $\theta_0 = 20°$. Numerically, these spectra coincide with appropriate parton spectra $x\overline{D}_g^g$ for an "isolated" gluon jet; see Eq. (6).

5. INTERJET COHERENCE AND THE RADIOPHYSICS OF PARTICLE FLOWS

In the framework of LPHD, the source of multiple hadroproduction in HP is gluon bremsstrahlung, so one should expect that all produced hadrons are the consequences of the color dynamics. Therefore, the properties of the partonic skeleton, such as the flow of color quantum numbers, should influence the distribution of color singlet hadrons in the final state. Such a phenomenon was first observed in experiments[42] studying the angular flows of hadrons in three-jet ($q\overline{q}g$) events from e^+e^- annihilation, the so-called string[43] (or drag[31]) effect. The data strongly support the predicted drag of the interjet particles in the direction of the gluon jet.

Detailed studies of the string-like phenomena are of importance for the high-energy HP. These effects are interesting not only in their own right as tests of QCD, but also as a valuable aid in distinguishing new physics signals from the conventional QCD backgrounds. So far, the most striking experimental test[38,44] of the interjet coherence was the comparison of hadron production in $q\overline{q}g$ events with that of $q\overline{q}\gamma$ events with the g and γ having similar kinematics.[31] In the plane of the jets, one finds a suppression of associated hadrons in the region between the q and $\overline{q}$ in $q\overline{q}g$ events as compared to $q\overline{q}\gamma$ events.

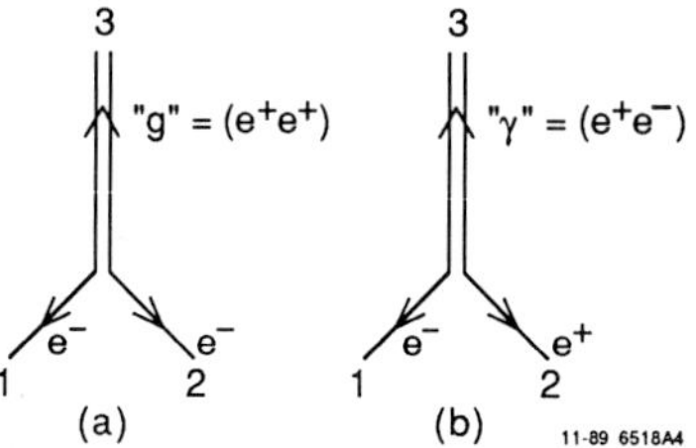

Figure 4: QED model for illustrating drag effect: (a) the gluon is represented as having double electric charge as compared to the electron; (b) the photon is replaced by a collinear e^+e^- pair.

To illustrate the physical origin of the destructive interference, one can use the simple QED model with the q and $\overline{q}$ replaced by electrons and the gluon by a collinear e^+e^+ pair; see Fig. 4(a). The $q\overline{q}\gamma$ event is illustrated in this model by Fig. 4(b). The corresponding radiation pattern is

$$dW_{q\overline{q}\gamma} = \frac{dk}{k}\, d\Omega_k\, \frac{\alpha}{2\pi^2}\, (\widehat{12}) \quad . \tag{19}$$

Here, the "antenna" $\widehat{ij}$ is

$$\begin{aligned} \widehat{ij} &= a_{ij}/a_i a_j \quad , \\ a_{ij} &= (1 - \vec{n}_i\vec{n}_j) \quad , \\ a_i &= (1 - \vec{n}\vec{n}_i) \quad , \end{aligned}$$

where n_i and n_j denote the directions of 'q' and '$\overline{q}$', respectively; and $\vec{n}$ is the direction of the emitted quantum. The soft radiation in the $q\overline{q}$'g' case is determined by the standard classical currents:

$$dW_{q\overline{q}\gamma} = \frac{dk}{k}\, d\Omega_k\, \frac{\alpha}{\pi^2}\, (\widehat{13} + \widehat{23} - 1/2 \cdot \widehat{12}) \quad . \tag{20}$$

Let us pay attention to the negative contribution of the antenna $\widehat{12}$, connected with the "repulsion" of electrons. Analogously in QCD, the opposite color charges of q and $\overline{q}$ in the $q\overline{q}\gamma$ are replaced by the effectively equal ones in the $q\overline{q}g$ case, which leads to destructive interference:

$$\frac{dN_{q\overline{q}g}/d\vec{n}_2}{dN_{q\overline{q}\gamma}/d\vec{n}_2} = \frac{N_c^2-2}{2(N_c^2-1)} = \frac{7}{16} \quad .$$

Due to the constructive interference, there is a surplus of radiation in the $q-g$ and $g-\overline{q}$ regions: for the symmetric configuration

$$\frac{dN_{(qg)}}{dN_{(q\overline{q})}} = \frac{22}{7} \simeq 3.14$$

(here, (ij) denotes the direction between partons i and j).

Note that the destructive interference proves to be so strong that the particle flow in the region opposite to g is smaller than that in the most kinematically "unfavorable" direction (normal to the event plane):

$$\frac{N_{<\perp>}}{N_{<q\overline{q}>}} = \frac{17}{14} = 1.2 \quad .$$

It is worthwhile to note that a clear proof of the QCD nature of the interjet drag phenomena will require future experiments at higher energies and/or with clearly identified q- and g-jets; see Secs. 8 and 10 and Ref. [7].

The rich diversity of the collective drag phenomena has been studied for high $P_\perp$ hadronic collisions.[20,23,32,33] Such effects are necessary for designing future experiments and analysis of their data. In particular, they could provide a valuable additional tool, helping to extract and to study new physics.

One of the simplest examples is the prompt γ-production at large $P_\perp$, as shown in Fig. 5; see Refs. [23, 41]. Here, the radiation pattern is evaluated for final pions projected onto the scattering plane. The particle production proves to be largest between the directions for the incoming gluon and the outgoing quark, but approximately 2.8 times smaller between the directions of the incoming quark and the outgoing quark.

An interesting manifestation of the QCD wave nature of hadronic flows arises from studying the double-inclusive correlations of the interjet flows in $e^+e^- \to q\overline{q}g$ events. The point is that here one faces

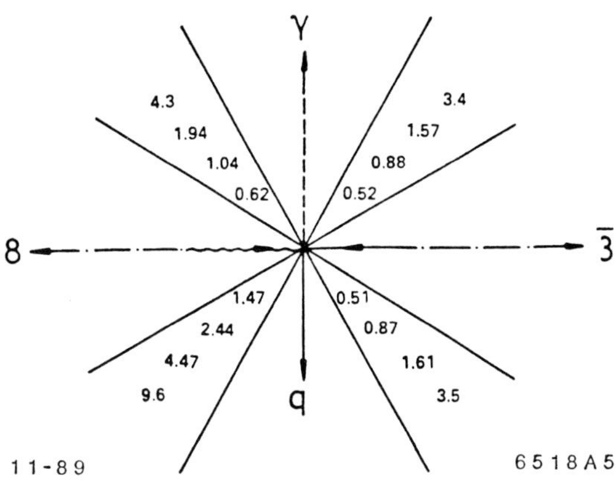

Figure 5: The energy rise of $\pi^\pm$ multiplicities in the interjet 30° sectors in large $P_\perp\gamma$ production: $E_\perp/\Lambda$ = 60, 200, 10^3, and 10^4.

such tiny effects as the mutual influence of different $q\overline{q}$ antennae. As a consequence:

$$\begin{aligned} r_2 &= \frac{d^2N}{d\Omega_{(1+)}d\Omega_{(1-)}} \Big/ \frac{d^2N}{d\Omega_{(+-)}d\Omega_{(1-)}} \\ < r_1 &= \frac{dN}{d\Omega_{(1+)}} \Big/ \frac{dN}{d\Omega_{(+-)}} \end{aligned} \tag{21}$$

(the elaborated analytical formulae may be found in Ref. [23]). To quantify the correlation effects one can compare the ratios of particle flows, projected onto the event plane, in the case of threefold symmetric events: $r_1 \simeq 2.42$, $r_2 \simeq 2.06$. Such color screening effects cannot be mimicked by the canonical Lund string model. Only the "dipole formulation" of the Lund Monte Carlo[21,22] may reproduce them.[45]

The azimuthal asymmetry of QCD jets seems to be of fundamental importance. Recall that the treatment of the structure of final states given by the string picture qualitatively reproduced the QCD radiation pattern only up to $O(1/N_c^2)$ corrections (the large N_c limit).[31] However, under specific conditions $O(1/N_c^2)$ terms become sizable or even dominant.

The simplest example is given by the azimuthal asymmetry of a quark jet in the events $e^+e^- \to q_+\overline{q}_-g_1$. The azimuthal distribution of particles produced inside a cone of the opening half-angle θ_0 may be characterized by an asymmetry parameter:

$$A(\theta_0) = \frac{N_{\to g}(\theta < \theta_0) - N_{\to \overline{q}}(\theta < \theta_0)}{N_{tot}(\theta < \theta_0)} = \frac{(\Delta N)_{as}}{N_{tot}} \quad . \tag{22}$$

For parametrically small θ_0,

$$A(\theta_0) \simeq \frac{2\theta_0}{\pi} G \sqrt{4N_c \frac{\alpha_s(E\theta_0)}{2\pi}} \; ,$$
$$G = \frac{N_c}{2C_F} \cot \frac{\theta_{+1}}{2} + \frac{1}{2N_cC_F} \cot \frac{\theta_{+-}}{2} \; . \tag{23}$$

The first color-favored term in Eq. (23) describes the asymmetry due to the "boosted string" connecting q- and g-directions. The corresponding asymmetry vanishes with increase of θ_{+1} as the string straightens. Here, however, the second term enters the game, forcing the asymmetry to increase anew as shown in Fig. 6. This behavior might be interpreted as an additional repulsion between particles from q- and $\overline{q}$-jets.

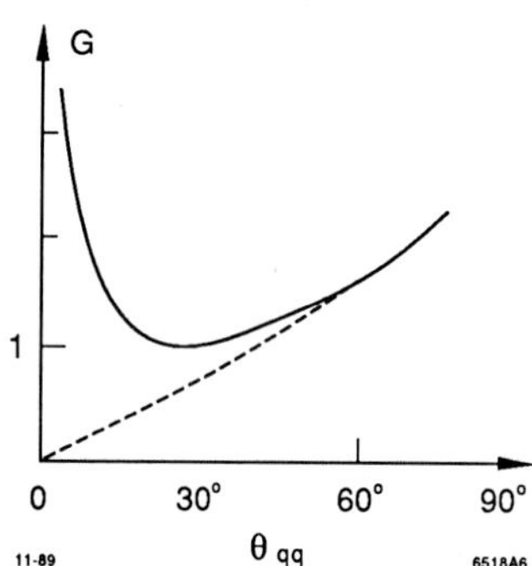

Figure 6: QCD (solid) vs. string (dashed) prediction for G factor; see Eq. (23).

For a symmetric configuration

$$\theta_{+1} = \theta_{-1} = \pi - \theta_{+-}/2 \quad ,$$

the color-suppressed term in Eq. (23) prevails when

$$\theta_{+-} \leq 2\arctan\,(1/N_c) \approx 37^\circ \quad . \tag{24}$$

To realize this effect, one has to select $q\overline{q}g$-events with special kinematics: when the hard gluon moves in the opposite direction to the quasi-collinear $q\overline{q}$ pair.

The asymmetry of jets is certainly not specific to 3-jet events. An analogous picture should be observed, e.g., in high $P_\perp$ processes. The qualitative difference between the predictions of QCD and its large N_c limit (string) proves to be the most spectacular in the case of the $p\overline{p}$ scattering with identification of the scattered q-jet.[23]

6. PERTURBATIVE APPROACH AND QCD-INSPIRED FRAGMENTATION SCHEMES

As is well known, the intertwining of perturbative and nonperturbative phases of jet evolution can so far only be formulated in algorithmic models. A wide selection of Monte Carlo simulation programs for jet physics has been presented on the market; see, e.g., Refs. [6–8, 17–22]. All of them are of a probabilistic and iterative nature. The models evolve steadily in time, in particular, by building in realistic fragmentation and proper QCD cascades. Modern elaborate QCD-inspired models describe very successfully the bulk of the existing data, based on the concept of the well-developed parton shower, and become very useful tools for experimentalists.

The model-builders aim to formulate the Monte Carlo algorithms to be as faithful as possible representatives of the QCD dynamics of jets. Over the last few years, one could observe an obvious convergence between the different classes of models and a growing tendency to reproduce the important ingredients of PA.

One of the main approaches for model improvement is the most adequate incorporation of the color coherence, reflecting the quantum mechanics of QCD. As the first important steps in this direction, one may consider the Marchesini–Webber recipe for the strict AO in partonic cascades[46] and Lund string scenario,[18] mimicking the interjet collective phenomena.

To our understanding, the modern algorithm, based on the Lund dipole Monte Carlo,[21,22] can reproduce the bulk of collective phenomena, even including some subtle effects.

As we have already emphasized, the very possibility of absorbing the interference effects into the local probabilistic scheme is far from obvious, and the use of Monte Carlo generators for the multipartonic system development proves to be of limited value in principle; for more details, see Refs. [6, 23].

For example, the collective phenomena in the multijet ensembles could be reproduced in classical probabilistic language only in the large N_c limit (see previous section). The problem is that nowadays we simply do not know how to handle $1/N_c^2$ terms in

Monte Carlo simulations. Normally, the neglect is not serious ($\lesssim 10\%$), but under special conditions (see previous section) these terms may even become dominant.

Let us make some additional remarks:

i. Up until now, different Monte Carlo models had coexisted more or less peacefully. An effective tool to discriminate between the various schemes proves to be the energy and particle flow study. As an instructive example, recall the celebrated string/drag effect. Originally, it was predicted[43] as a result of a Lorentz boost exerted by a gluon on the nonperturbative string stretched between quarks. But in the framework of PA, this effect is a direct consequence of interference among the gluon waves radiated from the $q\overline{q}g$ emitter.[31]

At PETRA/PEP energies, the observation of this phenomenon, in principle, could not be considered as a clear proof of its QCD wave nature. This should be true no longer at higher energies. For example, string effects in the ratio of particle flows are enhanced by selection of only particles with large P_{out} (momentum out of the event plane) or large m_h. If interpreted in terms of the canonical Lund string model, this increase will be of the same strength at all CM energies. By contrast, in the PA scenario at high enough energies, no enhancement is expected for the subsamples with large P_{out} or m_h. So, with increasing energy, the PA picture should win. For illustration, we present in Fig. 7 the energy evolution of the particle flow ratios in the ARIADNE program,[21,22] which contains both soft-gluon emission and string fragmentation. One can easily see that PA regime here becomes dominant at the energies of SLC/LEP.

ii. Now we come to the correlations of interjet particle flows. The canonical Lund scenario for $q\overline{q}g$ events, in which three independent partons are replaced by two fragmenting string segments, in principle, represented an important step forward. However, even such a picture incorporates the traditional ideas of independent fragmentation, in that the two string segments are assumed to act as two uncorrelated particle sources. As discussed in the previous section, the limitation of this approach to a basically quantum mechanical problem could be manifested, e.g., in double-inclusive correlations of particle flows. In the canonical string scenario, the effect of the additional detection of particle

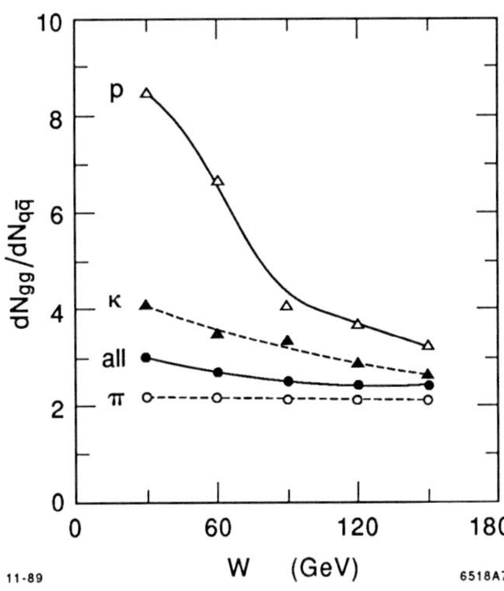

Figure 7: The ratio of particle flow between q and g jets to that between q and $\overline{q}$ jets, for symmetric $q\overline{q}g$ configurations (using 50° sectors). Results for all hadrons, and separately for $\pi^{\pm}$, for $K^{\pm}$ and for P, $\overline{P}$, as obtained from the ARIADNE program at different CM energies; see Refs. [6, 45].

flow cancels in the ratio of correlation functions, and one naturally expects $r_1 \simeq r_2$. On the other hand, PA predicts $r_1 > r_2$; see Eq. (21).

In concluding this section, let us emphasize that the relative smallness of the nonclassical effects does not diminish their fundamental importance. These consequences of QCD radiophysics are a serious warning against the traditional ideas on independently evolving partonic subsystems.

7. EXPERIMENTAL SELECTION PROCEDURES

Traditionally, the final state structure in a hard collision is interpreted in terms of jets of hadrons with kinematics corresponding to those of the energetic quarks and gluons. The standard procedures for jet finding, as well as the reconstruction of the jet parameters, differ in some details, but all of them are based on the idea of assigning each hadron in an event to a certain jet. This has been a very fruitful approach, especially regarding the first stage of QCD jet study; for details, see Refs. [11, 39] and references therein. However, the separation into a specific number of jets is inherently ambiguous, especially as one goes to higher energies:

i. In a particular part of an event, it may be equally correct to identify a set of particles as belonging to one jet, two jets or even more jets.

ii. Such a procedure ignores the collective nature of particle production. In particular, due to coherence, soft hadrons do not belong to any particular jet, but have emission properties dependent on a jet ensemble.

In particular, this phenomena is not the least of the factors explaining the ratio measured in the $q\bar{q}g$ events (see Refs. [38, 39] and references therein) of multiplicities in g- and q-jets, which turns out to be lower than the famous value 9/4[35]; see Sec. 3.

Attempting to force particles to belong to some jet in an event may cause difficulties. This leads, e.g., to the sizable uncertainties in the finding of a "jet axis," resulting in bias effects for the particle distribution relative to such an axis; for details, see Ref. [23].

Especially as higher energies are attained, a purely inclusive procedure for quantitatively dealing with HP is preferable to organizing the event according to a certain number of jets. In general, there is a direct correspondence between jets and energy correlations, so that any observable that can be described in terms of jets also can be described in terms of energy correlations.[23,41]

The simplest example is the angular distribution of the multiplicity flow in two-jet events. Its study is accessible through an (energy)^2 multiplicity correlation (E^2MC). The drag effect physics becomes accessible through a correlation E^3MC. In the general case, multiple correlations between the energy and multiplicity flows could be referred to as $E^\alpha M^\beta C$. This means that one should fix α jet directions from the energy flows, and then consider particle multiplicity correlations.

8. SPECIFIC PROPERTIES OF JETS GENERATED BY HEAVY QUARKS

Jets from heavy quarks (Q) are now extensively studied experimentally. The interest in this subject is connected not only with testing the fundamental aspects of QCD but also with its large practical importance for measurements of heavy particle properties: lifetimes, spatial oscillations of flavor, searching for CP-violating effects in their decays, and so on.

Let us briefly enumerate the main lessons from the application of PA to studies of Q-initiated jets.[20,32,47]

(a) As is well known, the main specific QCD properties of these jets are connected with the bremsstrahlung factor

$$dW_k = \frac{\alpha_s}{\pi} C_F \frac{k_\perp^2 dk_\perp^2}{\left[k_\perp^2 + \left(\frac{w m_Q}{E_Q}\right)^2\right]^2} \frac{dw}{w} , \tag{25}$$

which effectively leads to the depletion of primary gluon radiation at small angles

$$\frac{1}{wR} < \theta \simeq \frac{k_\perp}{w} \leq \frac{m_Q}{E_Q} . \tag{26}$$

Then, $k_\perp \geq m_Q$ for a hard gluon, and, e.g., the energy loss due to gluon emission before the nonperturbative jet development is governed by the standard factor[48]

$$\Delta\xi = \int_{m_Q^2}^{E_Q^2} \frac{dk^2}{k^2} \frac{\alpha_s(k^2)}{4\pi} = \frac{1}{b} \ell n \frac{\alpha_s(m_Q^2)}{\alpha_s(E_Q^2)} . \tag{27}$$

(b) The hard fragmentation of heavy quarks was first predicted in the framework of the parton model.[49] Now several forms of the primordial fragmentation functions W_H exist, which model the nonperturbative phase, $Q \to (Q\bar{q})+q$, of the jet development.[7]

Because of gluon radiation off a heavy quark, the average scaled energy of a heavy quark Q in HP with the characteristic hardness $E^2 \gg m_Q^2$ is[47]

$$\begin{aligned} < x_Q > &\simeq \exp\left(-\frac{8}{3} C_F \Delta\xi\right) \times \left(1 + \frac{11}{9} C_F \frac{\alpha_s}{\pi} + \cdots\right) \\ &= \left[\frac{\alpha_s(E^2)}{\alpha_s(m_Q^2)}\right]^{32/9b} \times \left(1 + \frac{11}{9} C_F \frac{\alpha_s}{\pi} + \cdots\right) , \\ x_Q &= \frac{E_Q}{E} . \end{aligned} \tag{28}$$

For PETRA/PEP energies, the average values $< x_c > \simeq 0.75$ and $< x_b > \simeq 0.85$ for charm and bottom quarks are numerically close to experiment.[38]

(c) In the realistic case, when $\Delta\xi$ is small and one can neglect the additional $Q\overline{Q}$ pair production, the Q-quark spectrum $D_Q(x_Q)$ is described by a simple interpolation formula; more elaborated analytical results may be found in Ref. [47].

$$D_Q(x_Q) \;=\; A\,\frac{1+x_Q^2}{2}\,(1-x_Q)^{-1+4C_F\Delta\xi}\;,$$

$$\text{where} \int dx_Q\, D_Q\,(x_Q) \;=\; 1$$

$$\text{and}\;\; 1-x_Q \;\gg\; \frac{1}{m_Q R} \qquad (29)$$

(A is the normalization factor).

The heavy hadron (H) distribution can be presented in the form

$$D_H(x) \;=\; \int_x^1 \frac{dx_Q}{x_Q}\, D_Q(x_Q) W_H(x/x_Q)\;, \qquad (30)$$

where $\int dz W_H(z) = 1$. Because the primordial functions $W_H(z)$ are sharply peaked, the behavior of $D_H(x)$ at $1-x \gg 1/m_Q R$ is determined by the quark spectrum D_Q (known as the "radiative tail" phenomenon). It is convenient to use the approximate formula

$$D_H(x) \;\simeq\; \frac{1}{<z>}\cdot D_Q\left(\frac{x}{<z>}\right)\;, \qquad (31)$$

where $<x> \;=\; <x_Q><z>$, and all the model dependence is concentrated in the quantity $<z>$ $(1-<z>\simeq 1/m_Q R)$.

(d) The string-like phenomena appear to be the same for light and heavy quarks. This behavior arises from the fact that the radiation off a quark at large angles, determining *interjet particle flows,* proves to be insensitive to the value of quark mass. This makes it possible to use the QCD analysis of the string effect for study of interjet particle flows with "tagged" heavy quarks, say, in the Z^0-peak.

(e) At $E \gg m_Q \gg \Lambda$, the spectra of associated light particles with $x \ll 1$ in the Q-initiated jets (within the MLLA accuracy) may be presented in the form[32]

$$x\overline{D}_Q^h(x,Y) \;=\; x\overline{D}(x,Y) - x\overline{D}(x,Y(m_Q))\;, \qquad (32)$$

where $Y(m_Q) = Y(E = m_Q)$ and $x\overline{D}(x,Y)$ is the standard inclusive spectrum in $e^+e^- \to q\overline{q}$; see Eq. (14).

This expression reflects the lack of energetic particles with

$$E/m_Q R \;<\; E_h \;<\; E\;\;.$$

For the multiplicity of light particles, one can obtain

$$N_Q^h(E) \;=\; N_q^h(E) - N_q^h(m_Q)\;\;. \qquad (33)$$

(f) An instructive lesson comes from the case of ultra-heavy top (m_t > 100 GeV).[50] Due to the decay $t \to W + b$, its lifetime τ_t becomes shorter than the hadronization time

$$\tau_t \;\sim\; 1\,fm\cdot\left(\frac{M_W}{m_t}\right)^3 \;<\; t_{hadr} \;\sim\; 1\,fm\;\;,$$

and all the bremsstrahlung processes prove to be under the jurisdiction of perturbative QCD. One can say that this quark in all aspects behaves as if it were a free colored object.

Let us emphasize that for an ultra-heavy unstable ($\Gamma_Q \gg R^{-1}$) quark, the inclusive distributions of final light particles prove to be the same as in the case of the light quark jet, and no traces of m_Q and Γ_Q can be seen in these spectra. It is of interest that even in the b case, the observed gross features of the jets are very similar to those from light quarks.[38]

9. QCD COHERENCE IN DEEP INELASTIC SCATTERING

The spectrum of particles associated with DIS has only recently been studied in detail in the framework of PA.[20,51,52] A hard lepton-hadron interaction with high $-q^2 = Q^2$ and fixed $x = Q^2/2Pq$ knocks the quark with longitudinal momentum $k = xP$ at virtuality level $k_\perp^2 \lesssim Q^2$ out of the initial partonic fluctuation which was prepared long before scattering.

The structure of the final state is governed by two main phenomena: disassociation of the initial parton system whose coherence was destroyed [target fragmentation (TF)], and evolution of the struck quark [current fragmentation (CF)].

Fragmentation regions are best separated kinematically in the Breit frame ($q_0 = 0,\; 2x\vec{P} = -\vec{q}$). Here, the process looks like abrupt spatial spreading of two color states 3 (the struck q) and $\overline{3}$ (the disturbed proton) moving in opposite directions.

The current jet should be identical to that in e^+e^- annihilation at $W^2 = Q^2$. One again finds a hump in the energy distribution with the dip occurring for particles with finite energies in the Breit frame. In the TF region, the situation proves to be much more complicated, especially for $x \ll 1$. The DIS occurs in this case on a sea quark from bremsstrahlung of soft $(q\bar{q})_g$ pairs in a color octet state (g exchange in t-channel). The dominant structure of the appropriate fluctuation can be characterized in terms of the multirung ladders of Fig. 8, which determine the small-x behavior of the structure function $D_p^q(x,Q^2)$. It is important to mention that sets of graphs which cancel in the structure function no longer cancel in the inclusive spectrum. In addition to the offspring of decaying subjets—remnants of the ordered "ladders" (structural contribution)—the collective coherent accompaniment arises, which is determined by the overall color topology of the partonic system (t-channel contribution). As in the time-like case, the PT analysis has proved that the AO of the soft gluons is associated with the space-like fluctuations.[20,51,52]

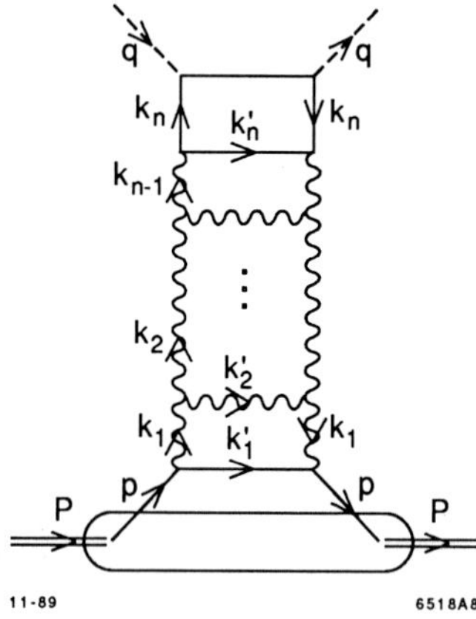

Figure 8: Ladder graphs, with transverse momentum ordering, dominating in structure functions at small x.

In the TF, there also arises the humpbacked particle distribution evolving with $\ell n\ Q^2$ and $\ell n\ 1/x$. The resulting spectrum of hadrons h in the TF can be represented as a sum of three terms as shown in Fig. 9. Contribution I is related to the upper quark cell in Fig. 8:

$$\left(\frac{d\sigma}{\sigma dy}\right)_I = \frac{C_F}{N_c}\left[\frac{\omega}{xP}\overline{D}_g^h\left(\frac{\omega}{xP}, \ell n\ \frac{Q}{\Lambda}\right)\right], \qquad (34)$$

where $y = \ell n\ \omega/\Lambda$; ω being the energy of hadron h. Contribution II accounts for effective emissions off the vertical gluon lines

$$\left(\frac{d\sigma}{\sigma dy}\right)_{II} = \frac{1}{D_p^q(x,Q^2)}$$

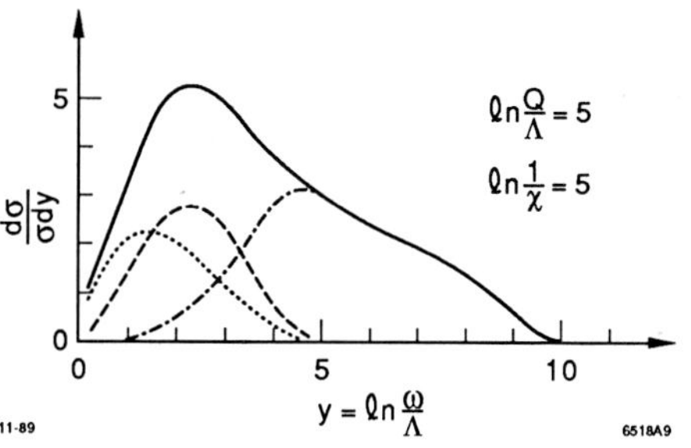

Figure 9: Contributions to the energy particle spectrum in DIS target fragmentation region at $\ell n\ Q/\Lambda = 5$, $\ell n\ 1/x = 5$. (a) Quark box contribution (dotted line); (b) coherent "t-channel" color radiation (dashed); (c) fragmentation of ladder rungs (dash-dotted); sum of (a), (b) and (c) (solid line).

$$\times \int_0^{\ell n\ Q/\Lambda} d\,\ell n\ \frac{k_\perp}{\Lambda}\ \frac{\partial}{\partial \ell n\ (k_\perp/\Lambda)}\ D_p^q(x,k_\perp^2)$$

$$\times \left[\frac{\omega}{xP}\overline{D}_g^h\left(\frac{\omega}{xP}, \ell n\ \frac{k_\perp}{\Lambda}\right)\right]. \qquad (35)$$

Contribution III combines relatively soft gluons ($xP < \ell < P$) off the lower part of the ladder with fragments of ladder rungs:

$$\left(\frac{d\sigma}{\sigma dy}\right)_{III} = \frac{1}{D_p^q(x,Q^2)} \int_{xP}^{P} \frac{d\ell}{\ell} \int^{\xi_Q} d\xi_k$$

$$\times D_P^g\left(\frac{\ell}{P}, k_\perp^2\right) \frac{\partial^2}{\partial \xi_k^2} D_g^q\left(\frac{xP}{\ell}, Q^2, k_\perp^2\right)$$

$$\times \int_\Lambda^{k_\perp} \frac{d\ell_\perp}{\ell_\perp} \frac{\alpha_s(\ell_\perp^2)}{2\pi} \left[\frac{\omega}{\ell}\overline{D}_g^h\left(\frac{\omega}{\ell}, \ell n\ \frac{\ell_\perp}{\Lambda}\right)\right], \qquad (36)$$

where $\xi_k = 1/b\ell n\ \ell n\ k_\perp/\Lambda$. Contributions II and III, which are negligible at $x \sim 1$, increase with decreasing x.

The evolution of the resulting hadron energy spectra with q^2 and x is shown in Fig. 10. The left wings correspond to the current fragmentation, the right ones to the hadrons from target fragmentation. As in the e^+e^- annihilation case, coherence in DIS stiffens the energy spectra. The yield of "slow" particles ($\ell n\ \omega/m \lesssim 1$) should be independent of Q^2.

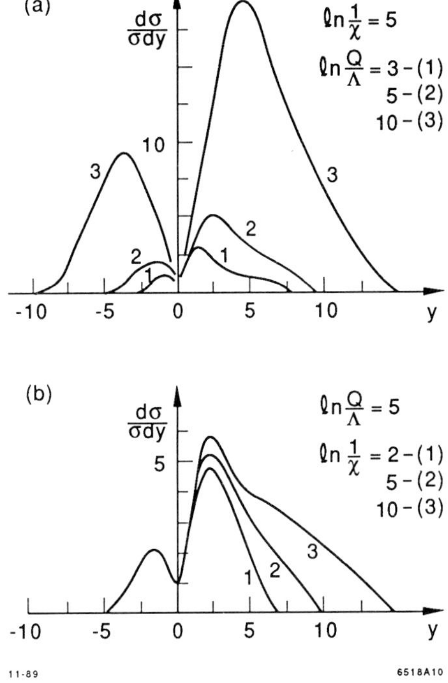

Figure 10: Evolution of DIS energy hadron distribution in the Breit system with (a) $Q^2 : \ell n\ Q/\Lambda = 3, 5, 10$ at $\ell n\ 1/x = 5$, and (b) $x : \ell n\ 1/x = 2, 5, 10$ at $\ell n\ Q/\Lambda = 5$.

10. PROSPECTS FOR FUTURE EXPERIMENTS

Taking into account the shopping lists of the newcomers SLC and LEP, we shall concentrate here mainly on the physics of particle flows in c and b events.[6,7,23] As is well known, the experimental studies of the rich body of subtle QCD collective phenomena, as discussed in Secs. 4 and 5, and their confrontations with the predictions of fragmentation algorithms are often hampered by the fact that quark and gluon jets, in principle, cannot be separated on an event-by-event basis. The new important prospects on this problem are connected with the large rates of charm and bottom decays of Z^0 bosons, in which the quark nature of the Q-initiated jet can be identified by some specific features of Q (hard prompt leptons, specific final states, vertex detection, etc.).[7,23] In the following, we shall briefly comment on some of the most interesting points in this context.

The detailed tests of color coherence effects and their discrimination against nonperturbative dynamics require comprehensive studies of the total three-dimensional pattern of particle flows in three-jet events. Comparison with analytical results (accounting for both interjet and intrajet coherence phenomena) should make it possible to distinguish reliably the PA predictions from fragmentation schemes. Of special interest here are the energy dependence of the multiplicity flow and its dependence on m_h and P_{out}.

With jet identification, one can study the double-inclusive correlations of the interjet flows (see Sec. 5), where the quantum mechanical collective nature of radiation manifests itself.

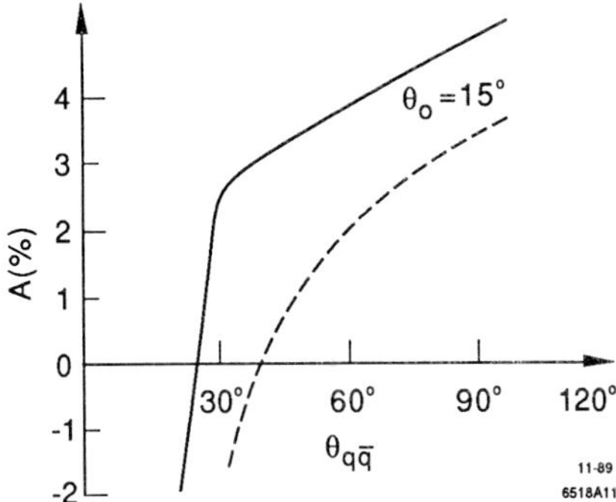

Figure 11: Asymmetry parameter A in symmetric $q\overline{q}g$ events ($\theta_{qg} = \theta_{\overline{q}g}$); QCD (solid) vs. large N_c limit predictions.

By studying the radiation pattern of a tagged quark jet in $q\overline{q}g$ events, one can measure the azimuthal asymmetry of a jet. An interesting vista on this problem is connected with the $Z^0 \to c\overline{c}g$ events. To investigate the role of $1/N_c^2$ interference terms, one has to select $c\overline{c}g$ events in which the hard gluon moves in the direction opposite to quasi-collinear $c\overline{c}$ pair, and to measure the asymmetry parameter $A(\theta_0)$ at $\theta_0 \gtrsim 3$–$4°$. Figure 11 illustrates the predicted asymmetry for $\theta_0 = 15°$ as a function of the relative angle $\theta_{q\overline{q}}$ between the q and $\overline{q}$ jets. The increase of A is of comfortable size at moderate $\theta_{q\overline{q}}$ angles. Of course, before the different behavior in QCD prediction and its large N_c limit can be studied, the asymmetry itself must be investigated experimentally.

Of special interest are the particle distributions inside jets initiated by heavy quarks. One of the prospective ways of fixing jet direction arises from the vertex detection. Note that the technique of tagging through the long decay length of the heavy hadron will be increasingly powerful at higher energies. In particular, one can study here distributions of particles inside fixed cones, energy and angular spectra. Of special interest is the observation of particle depletion in the forward region and the search for the manifestations of nonperturbative dynamics.

One of the interesting tasks for HERA seems to be the study of the coherent phenomena as discussed in Sec. 9.

11. CONCLUSIONS

1. There is a continually developing understanding of QCD jet physics. A theoretically substantiated scheme exists (PA = MLLA + LPHD) for making quantitative predictions of jet characteristics.
2. There is a continually improving evolution of Monte Carlo algorithms, which are coming closer and closer to a "true" description of the underlying physics.
3. Purely inclusive studies of jet characteristics (calorimetric and multiparticle $E^{\alpha}M^{\beta}$ correlations) are probably the best way to make sharp connections between theory and experiment.
4. Hadroproduction studies within the perturbative approach are far from being exhausted. Future experiments need PA activity.

ACKNOWLEDGMENTS

I would like to express my appreciation to the organizers of this excellent symposium. I wish to thank Yu. Dokshitzer, V. Fadin, A. Mueller and S. Troyan for a good collaboration. I am indebted to Ya. Azimov, S. Brodsky, F. Gilman, V. Gribov, W. Hofmann, L. Lönnblad and T. Sjöstrand for fruitful discussions. I am grateful to Yu. Dokshitzer, V. Gribov, J. Huth, A. Mueller and M. Peskin for valuable advice during preparation of my talk. I am very much obliged to my scientific secretary, R. Brooks, for all the help I received in preparing the talk as well as these proceedings.

REFERENCES

1. For reviews, see A. Bassetto, M. Ciafaloni, and G. Marchesini, *Phys. Rev.* **100** (1983) 201; Yu. L. Dokshitzer, V. A. Khoze, A. M. Mueller, and S. I. Troyan, *Rev. Mod. Phys.* **60** (1988) 373; Yu. L. Dokshitzer, V. A. Khoze, and S. I. Troyan, in *Perturbative QCD,* ed. A. H. Mueller (World Scientific, Singapore, 1989), p. 241.
2. R. K. Ellis, *Proc. 24th Int. Conf. on High Energy Physics,* eds. R. Kotthaus and J. Kühn (Springer–Verlag, Berlin, 1989), p. 48.
3. G. Altarelli, preprint CERN–TH.5290/89, to be published in *Ann. Rev. Nucl. Part. Sci.* **39** (1989).
4. A. Ali and F. Barreizo, preprint DESY 88–075.
5. P. Bagnaia and S. D. Ellis, *Ann. Rev. Nucl. Part. Sci.* **38** (1988) 659.
6. B. Bambah *et al.,* preprint CERN–TH.5466/89, to be published in *Proc. 1989 LEP Physics Workshop.*
7. *Report of the Heavy Flavor Working Group,* convenors H. Kühn and P. Zerwas, to be published in *Proc. 1989 LEP Physics Workshop.*
8. B. Webber, *Ann. Rev. Nucl. Part. Sci.* **36** (1986) 253.
9. J. Huth, plenary talks in these proceedings.
10. G. Hanson *et al., Phys. Rev. Lett.* **35** (1975) 1609.
11. For recent experimental reviews, see Ref. [9] and A. Maki and A. Weinstein, plenary talks in these proceedings; W. Hofmann, *Proc. Int. Symp. on Lepton and Photon Interactions at High Energies,* Hamburg, July 1987, eds. W. Bartel and R. Rückl (North Holland, 1988), p. 671; S. L. Wu, *op. cit.,* p. 39; M. J. Shochet, *Proc. 24th Int. Conf. on High Energy Physics,* eds. R. Kotthaus and J. Kühn (Springer–Verlag, Berlin, 1989), p. 18; T. Kamae, *op. cit.,* p. 156; P. Mättig, preprint DESY 88–125; K. Sugano, *Int. Journ. Mod. Phys.* **A3** (1988) 2249.
12. AMY Collaboration, papers 99 and 100 contributed to this symposium.
13. S. Bethke, paper 46 contributed to this symposium; preprint LBL–26958.
14. Yu. L. Dokshitzer and S. I. Troyan, *Proc. XIX Winter School of the LNPI,* V. I. 144 (1984); preprint LNPI–92 (1984).
15. A. H. Mueller, *Nucl. Phys.* **B228** (1984) 351.
16. A. H. Mueller, *Nucl. Phys.* **B213** (1983) 85 and erratum quoted in *Nucl. Phys.* **B241** (1984) 141.
17. T. Sjöstrand, *Int. Journ. Mod. Phys.* **A3** (1988) 751.
18. B. Andersson, B. Gustafson, G. Ingelman, and T. Sjöstrand, *Phys. Rev.* **97** (1988) 31.
19. B. Andersson, Lund preprint LU TP 88–2.
20. G. Marchesini and B. R. Webber, *Nucl. Phys.* **B310** (1988) 461; Cambridge preprint Cavendish–HEP–88/9.
21. G. Gustafson, *Phys. Lett.* **175B** (1986) 453; G. Gustafson and U. Pettersson, *Nucl. Phys.* **B306** (1988) 746.
22. L. Lönnblad and U. Pettersson, Lund preprint LU TP 88–15.
23. Yu. L. Dokshitzer, V. A. Khoze, and S. I. Troyan in Ref. [1].

24. Ya. I. Azimov *et al.*, *Z. Phys.* **C27** (1985) 65; *Z. Phys.* **C31** (1986) 213.

25. D. Amati and G. Veneziano, *Phys. Lett.* **83B** (1979) 87; G. Marchesini, L. Trentadue, and G. Veneziano, *Nucl. Phys.* **B181** (1981) 335.

26. G. Cohen-Tannoudji and W. Ochs, *Z. Phys.* **C39** (1988) 513; L. Van Hove and A. Giovannini, *Acta Phys. Pol.* **B19** (1988) 917 and 931; B. Andersson, P. Dahlquist, and G. Gustafson, Lund preprints LU TP 89–2 and 89–4.

27. B. I. Ermolayev and V. S. Fadin, *JETP Lett.* **33** (1981) 285.

28. A. H. Mueller, *Phys. Lett.* **104B** (1981) 161.

29. Yu. L. Dokshitzer, V. S. Fadin, and V. A. Khoze, *Phys. Lett.* **115B** (1982) 242; *Z. Phys.* **C15** (1982) 325; *Z. Phys.* C18 (1983) 37.

30. A. Bassetto *et al.*, *Nucl. Phys.* **B207** (1982) 189.

31. Ya. I. Azimov *et al.*, *Phys. Lett.* **165B** (1985) 147; *Yad. Fiz.* **43** (1986) 149.

32. Yu. L. Dokshitzer, V. A. Khoze, and S. I. Troyan, *Proc. 6th Int. Conf. on Physics in Collisions*, ed. M. Derrick (World Scientific, Singapore) 365 (1987).

33. R. K. Ellis, G. Marchesini, and B. R. Webber, *Nucl. Phys.* **B286** (1987) 643.

34. J. B. Gaffney and A. H. Mueller, *Nucl. Phys.* **B250** (1985) 109.

35. S. J. Brodsky and J. F. Gunion, *Phys. Rev. Lett.* **37** (1976) 402.

36. Yu. L. Dokshitzer, V. A. Khoze, and S. I. Troyan, *Yad. Fiz.* **47** (1988) 238.

37. A. E. Chudakov, *Isv. Akad. Nauk. SSSR, Ser. Fiz.* **19** (1955) 650.

38. T. Kamae, W. Hofmann, P. Mättig, and K. Sugano, in Ref. [11].

39. W. Hofmann, *Ann. Rev. Nucl. Part. Sci.* **38** (1988) 279.

40. TASSO Collaboration, M. Althoff *et al.*, *Z. Phys.* **C22** (1984) 307.

41. Yu. L. Dokshitzer, V. A. Khoze, A. M. Mueller, and S. I. Troyan, in Ref. [1].

42. JADE Collaboration, W. Bartel *et al.*, *Phys. Lett.* **101B** (1981) 129; *Phys. Lett.* **134B** (1984) 275; *Z. Phys.* **C21** (1983) 37; TPC Collaboration, H. Ahaiza *et al.*, *Z. Phys.* **C28** (1985) 31; TASSO Collaboration, M. Althoff *et al.*, *Z. Phys.* **C29** (1985) 29.

43. B. Andersson, G. Gustafson, and T. Sjöstrand, *Phys. Lett.* **94B** (1980) 211.

44. TPC Collaboration, H. Ahaiza *et al.*, *Phys. Rev. Lett.* **57** (1986) 945; MARK II Collaboration, P. Sheldon *et al.*, *Phys. Rev. Lett.* **57** (1986) 1398.

45. L. Lönnblad, private communication.

46. G. Marchesini and B. Webber, *Nucl. Phys.* **B238** (1984) 1; B. Webber, *op. cit.*, p. 492.

47. Ya. I. Azimov *et al.*, *Yad. Fiz.* **36** (1982) 1510; *Yad. Fiz.* **40** (1984) 777.

48. Yu. L. Dokshitzer, D. I. Dyakonov, and S. I. Troyan, *Phys. Rev.* **58** (1980) 270.

49. Ya. I. Azimov, L. L. Frankfurt, and V. A. Khoze, preprint LNPI–222 (1976); J. D. Bjorken, *Phys. Rev.* **D17** (1978) 171; M. Suzuki, *Phys. Lett.* **71B** (1977) 139.

50. I. I. Bigi *et al.*, *Phys. Lett.* **181B** (1986) 157.

51. L. V. Gribov *et al.*, *Phys. Lett.* **202B** (1988) 276; *JETP* **34** (1988) 12.

52. M. Ciafaloni, *Nucl. Phys.* **B296** (1987) 249.

DISCUSSION

B. F. L. Ward, University of Tennessee: From the standpoint of your perturbative approach to QCD processes, what can you say about the large $O(\alpha^3)$ correction to the total cross section for $e^+e^- \to X$ in perturbation theory?

V. Khoze: I agree with you that the perturbation series for the e^+e^- total cross section is an important part of physics. I cannot add more than the Moscow group, which has calculated the three-loop contribution. And I don't think there are any new ideas beyond their known result.

S. Brodsky, SLAC: Some of the results you showed have structure in the rapidity variable. Actually, you're not using a covariant rapidity variable, but a log of an energy fraction. If you really used a covariant variable, a frame-independent variable, then you could not have structures at fixed points in rapidity. Have you reanalyzed these results to show that you do have smooth structures if you use strictly covariant variables? And shouldn't these be the variables that experimentalists should use rather than frame-dependent ones?

V. Khoze: First of all, I think that was one of the reasons we used this variable. We wanted to enhance the effect, not smear it out. By using these natural variables that you suggest, the effect will be smeared out. But to answer the second part of your question, we have not done that. So maybe you are right. Perhaps it is necessary to use the light-cone variable and to see what happens.

The Sky Is the Limit

The Sky Is the Limit

Saturday, 12 August 1989

High Energy Astrophysics 403

Speaker: J. W. Cronin

Chairperson: V. Singh

Scientific Secretary: C. Munger

Accelerators for Very Energetic Gamma Rays from Neutron Star Systems 420

Speaker: M. Ruderman

Chairperson: V. Singh

Scientific Secretary: P. Rowson

Gauge Fields, Geometry, Strings, and Gravity 433

Speaker: J. A. Harvey

Chairperson: M. Green

Scientific Secretary: D. Kastor

The Future e^+e^- Colliders 442

Speaker: G. A. Voss

Chairperson: W. K. H. Panofsky

Scientific Secretary: N. Toge

Summary of the Conference 452

Speaker: F. Sciulli

Chairperson: W. K. H. Panofsky

Scientific Secretaries: D. Coupal, I. Dunietz

HIGH ENERGY ASTROPHYSICS

James W. Cronin
Enrico Fermi Institute and Department of Physics
The University of Chicago
Chicago, Illinois 60637, USA

INTRODUCTION

In recent years a new branch of experimental science has emerged which might be called particle astrophysics. Its scope is wide ranging. Among its many facets is included the detection of solar neutrinos, searches for dark matter, searches for monopoles, observation of supernovae in neutrinos, observation of 10^{12}-10^{18} eV particles from astrophysical objects, and the exploration of the highest energy cosmic rays ($\geq 10^{19}$ eV). These observations bear on both astronomy and particle physics in a manner that cannot be neatly separated. The detection techniques are those of particle physics and the field has attracted a substantial number of physicists who had formerly carried out experiments with high energy accelerators.

I will not attempt to survey all the aspects of particle astrophysics. Rather I will describe in some detail the search for high energy particles ($>10^{12}$ eV) emitted by astrophysical objects, pulsars and x-ray binaries. I will try to give a flavor of the field by specific examples, pointing out both weaknesses as well as successes. It is a field in which I am now personally involved.

The observation techniques use ground based equipment. The high energy particles are observed by the extensive air showers they produce in the upper atmosphere. These showers are detected either by the Cerenkov light produced in the upper atmosphere or by direct detection on the ground of the shower particles. Figure 1 shows schematically how the two methods work. The first method is the atmospheric Cerenkov technique (ACT). The optical photons produced by the Cerenkov effect of the shower particles in the upper atmosphere travel in a flat disk to the ground where they are detected by a single photomultiplier (PM) or by an array of PMs at the focus of a steerable telescope mirror. The actual shower particles need not reach the ground. The radius of the disk of optical photons is typically about 100 meters, so that a single detector can cover an area $>10^4 m^2$. For a shower induced by a 10^{12} eV hadron or photon, the disk contains about 50 optical photons/meter^2. Thus, a mirror of modest size can be very effective. The sensitivity of the ACT can be pushed as low as a few hundred GeV. Given that most sources are expected to have a spectrum which falls sharply with increasing energy, a low threshold is desirable. The angular resolution is related to the size of the PM and the focal

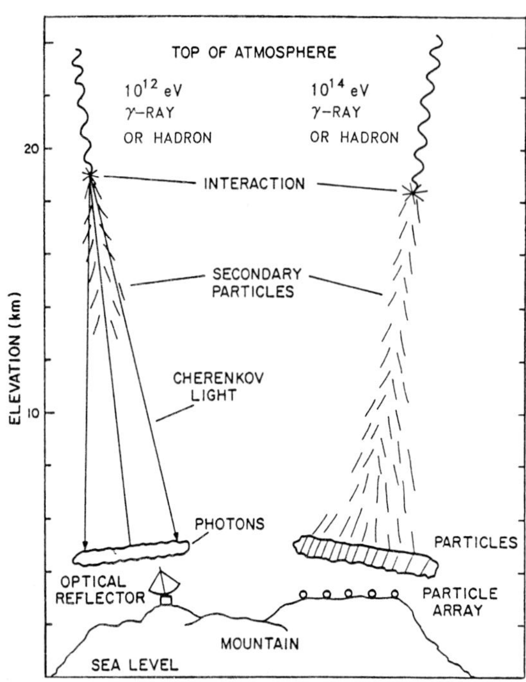

Fig. 1: Methods of air shower detection.

length of the mirror.

The technique has its limitations. It will only work on cloudless, moonless nights. The accessibility of any given source is seasonal, since a source cannot be seen when it is on the same side of the earth as the sun. For a given mirror, only one source can be followed at a time. Typically, when a major effort is directed to a particular source only a few hundred hours of exposure per year can be obtained. Nevertheless as we shall see in the course of this talk, the most convincing observations have been made with ACT.

The second method involves the direct detection of the shower particles (air shower technique, AST). The threshold energy for this method is typically 10^{14}eV when the detectors (normally scintillation counters) are placed at 1600m elevation where the atmospheric thickness is about 860 gm/cm^2. The shower particles arrive in a disk (of ~1m thickness), which travels with the speed of light. The time differences between successive detections of the shower particles by a two-dimensional array of scintillation detectors permits the direction of the shower to be measured. With an array spacing of 15m and a time resolution of 1 ns, the shower direction can be measured with a precision of ~1°. A 10^{14}eV particle at the zenith produces about 25,000 particles at a depth of 860 gms. These particles (principally electrons and positrons) have a lateral distribution which is shown in Fig. 2. The particle density at the core of the shower is ~40/m^2 droping to ~1/m^2 40 meters from the core. The core is the projected impact point of the particle which initiated the shower.

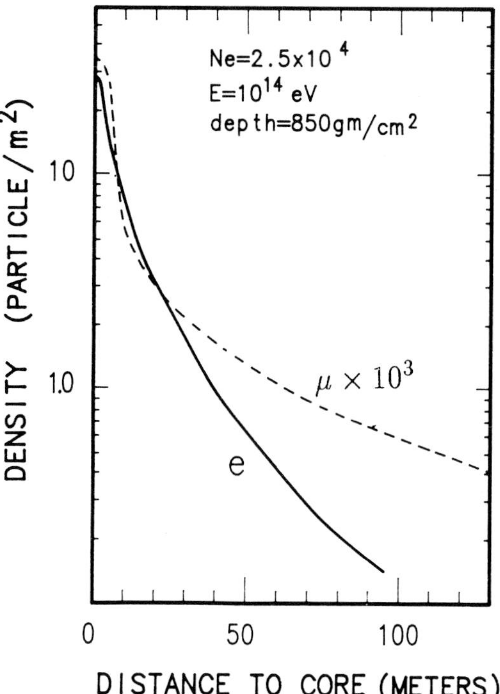

Fig. 2: Lateral density distribution of electrons and muons in an air shower.

While the AST has the disadvantage of a higher threshold, it has the advantage that it operates under all conditions. It observes without selection any source which passes overhead. The AST is complementary to the ACT in that it can search for radiation at higher energies from sources seen by ACT. At the conclusion we will discuss the possibilities of pushing the AST into the ACT range.

The energy resolution of either technique is very poor. One can imagine trying to measure the energy of a particle in an electromagnetic calorimeter with only one sampling plane at a depth of 23 radiation lengths, which is the situation for the AST. For the ACT, the lack of knowledge of the position of the core of the shower with respect to the mirror produces a large uncertainty in the relation between the number of optical photons and the energy of the particle which initiated the shower. Thus, measurement of the energy of an individual shower is poor, but the mean energy of an ensemble of events can be determined to better than 50% in both techniques.

If a point source is to be observed, the high energy particle it emits must be neutral. There is a galactic magnetic field of 10^{-6} gauss. In this field a singly charged particle of 10^{15} eV has a radius of curvature of 3×10^{16} m or 1 parsec (pc). Since the size of our galaxy is 10 kpc, charged particles less than 10^{20} eV cannot reveal their source. The only stable neutral particles within the framework of conventional particle physics are photons and neutrinos. Since neutrinos are weakly interacting, they do not produce air showers. Hence, the only particles which can point to a source and be observed by the two techniques described above are photons. Between 10^{15} and 10^{16} eV, absorption of the photons can be important. This is due to the process $\gamma + \gamma \rightarrow e^+e^-$ on the microwave background. At 3×10^{15} eV the mean free path for absorption is ~5 kpc, significant for sources at the edge of the galaxy. A neutron with 10^{18} eV has a mean decay length of 10 kpc. The observation of a point source at $>10^{18}$ eV could be either a neutron or a photon.

Ordinary cosmic rays produce a nearly overwhelming background in the search for point sources. These cosmic rays have a steeply falling spectrum. The integral spectrum at 10^{12} eV is 1.6×10^{-5} cm^{-2}-sec^{-1}-ster^{-1}, at 10^{14} eV is 8×10^{-9}, and at 10^{19} eV is about 0.7 km^{-2}-year^{-1}-ster^{-1}.

The cosmic rays are principally charged particles which strike the upper atmosphere isotropically and consist of protons and heavier nuclei. Measurements of composition in the range of energies considered here suggest that the principal component is protons but significant amounts of heavier nuclei (helium to iron) are present (20-50%). These cosmic rays initiate hadronic cascades in the atmosphere. Deep in the atmosphere the showers consist mostly of electromagnetic particles. There are about 10% muons and a few percent hadrons near the core. With the exception of the muons a hadronically initiated shower and a photon initiated shower hardly differ when detected on the surface.

In order to achieve good sensitivity for photon showers, characteristics that provide a separation between photon-induced showers and hadron-induced showers must be exploited. One obvious property of a detector which can accomplish this is angular resolution. If $\Delta\Omega$ is the solid angle required to accept a significant fraction of events from a particular point source, then the background rate from the isotropic cosmic rays is $F_B\Delta\Omega$ events/cm^2/sec, and the smaller $\Delta\Omega$ the lower the background.

An additional rejection of background for the AST involves the detection of muons which accompany a hadronically-produced air shower. Figure 2 also shows the lateral distribution of the muons. While the peak density is much lower, the lateral distribution is much broader. In a photon-induced shower, many fewer muons are produced because the cross section for photon production of hadrons, $\sigma_{\gamma,air}$ is ~2 mb (extrapolating from accelerator energies), while $\sigma_{p,air}$ ~270mb, so one expects many fewer muons to be present in a photon-initiated shower. Thus, if an air shower array on the surface is supplemented with sufficient muon detection, a large suppression factor for hadronically-induced showers can be obtained. Figure 3 shows a recent calculation of the relation between the number of electrons, N_e, and mean number of muons $\langle N_\mu \rangle$ expected for hadronic and photonic showers(1). The reduction factor for muons is greater than 50. With sufficient area of muon detection a rejection factor for hadronic showers greater than 100 can be obtained.

For the ACT, a different technique has been successfully employed(2). The major yield of optical photons comes from the shower maximum. For a hadronic shower the spread of angles of radiating particles is much broader than in a photonic shower. This is because transverse momenta in the hadronic interactions are typically 200 MeV/c, while in a purely electromagnetic shower, they are negligible. Thus, the image of the Cerenkov photons is much broader and more stochastic for a hadronic shower than a photon shower, and this effect can be used as a discrimination technique.

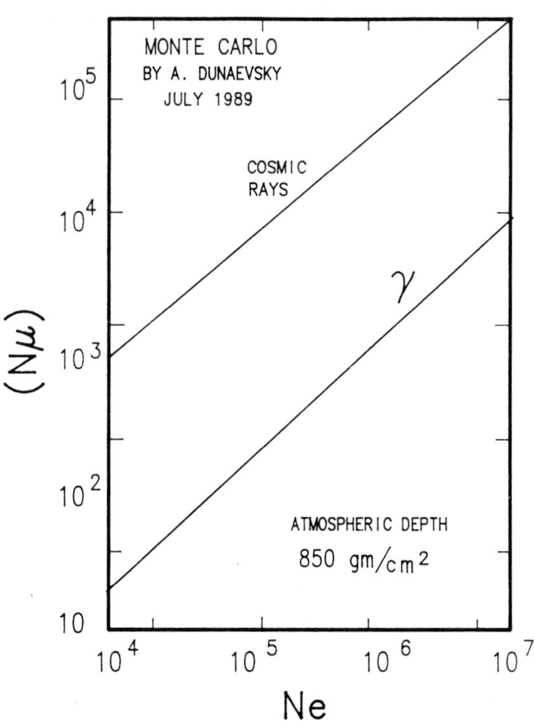

Fig. 3: Mean number of muons as a function of electron number for showers produced by cosmic rays and γ-rays.

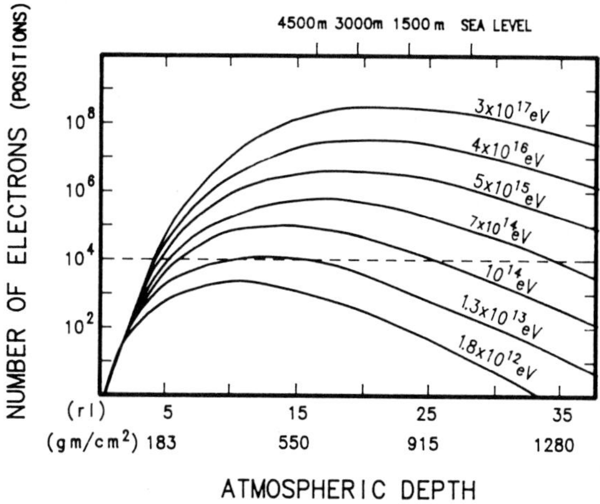

Fig. 4: Shower development as a function of energy and atmospheric depth. Figure adapted from K. Greisen, Progress in Cosmic Ray Physics 3 (1956) 1.

DESIGN CONSIDERATIONS FOR THE AIR SHOWER TECHNIQUE.

Figure 4 shows the development of photon-induced showers in air as a function of depth and initial photon energy. The depth is measured in gms/cm^2 or radiation lengths, and the corresponding elevations above sea level are indicated. The curves for hadron-induced showers are rather similar. These curves are average values and the fluctuations of individual showers are very large. With this curve and the "back of an envelope" one can roughly design an air shower array. Typically the shower is sampled with detectors which cover about 1% of the actual surface. In order to make a reasonable measurement of the shower direction about 100 particles are required. The dashed line corresponds to these conditions. The remaining variable is the atmospheric depth which is controlled by the altitude of the installation.

At sea level the threshold of such an array is about 2×10^{14}eV. At 1500 meters it is $\sim 0.8 \times 10^{14}$eV and at 3000 meters it is $\sim 0.5 \times 10^{14}$eV. It should be noted that these energies are for vertical showers. The median energies, folding in the expected falling spectrum and the passage overhead of a source, are about a factor 2 larger. If one can afford a continuous coverage rather than 1% then the threshold can be $\sim 10^{12}$eV, and the AST would overlap the ACT.

In Fig. 5 we show a typical response of an array to a source. Suppose we look at the Crab nebula which has a declination of about 20°N with respect to the earth's equator, and suppose the array is at a latitude of 40°N. The sensitivity of the array depends on the zenith angle of the source. Showers at larger zenith angles must have larger energies to reach the threshold of the array. Shown on the figure of declination and right ascension are contours of equal zenith angle. Typically

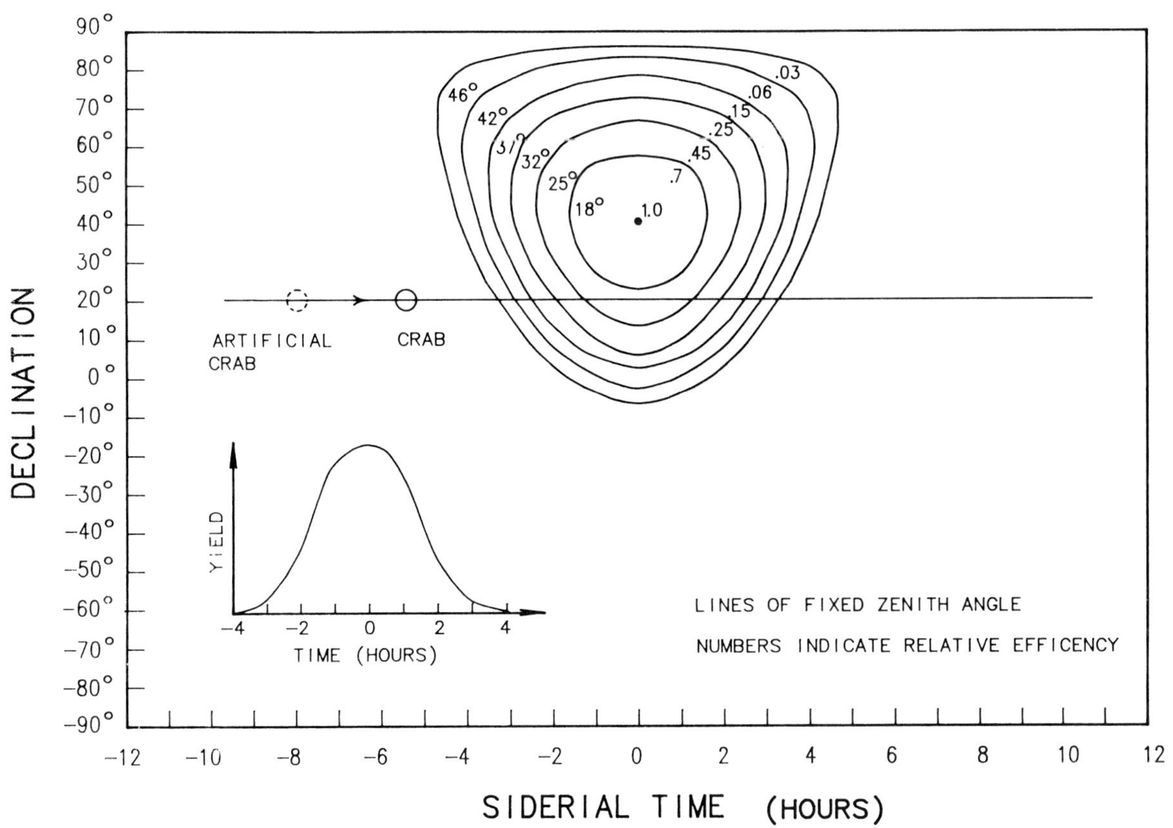

Fig. 5: Response of an air shower array at 40°N latitude to the passage of the Crab.

the response drops to less than 10% at zenith angles greater than 40°.

As the Crab passes each day the relative yield of the array will vary as indicated in the inset. Typically the integrated yield of a source as it passes close to overhead is equivalent to its peak rate for a period of four hours. As shown in the figure any number of spots in the sky can be used to evaluate backgrounds (artifical Crabs).

An air shower array samples the number of charged particles which strike the ground. With this information, the total number of charged particles, N_e, can be estimated. A crude estimate of the energy of the incident particle can be made using N_e and the angle of incidence.

A typical air shower array at an elevation of 1600 meters (860 gms/cm^2) with an area of $10^5 m^2$ and a sampling of 1% of the surface will have a threshold $N_e=10^4$ particles, and a resolution of 1°, giving $\Delta\Omega = 9 \times 10^{-4}$. If the flux of a source passing overhead is 3×10^{-13}/cm^2/sec, one will record 4.5 events/day within $\Delta\Omega$ from the source and 190 background events. The time required to observe a 6σ effect for a steady source of this strength is about 340 days.

The latter figure is derived from the formula $T = (N^2 F_B \Delta\Omega)/(F_S^2 AR)$ where T is the time required for an effect of N sigma, F_B the flux of background cosmic rays, $\Delta\Omega$ the angular resolution, F_S the signal flux, A the area of the array, and R a rejection factor for hadronic showers. To reduce the time for a given significance there is a premium on large area and good angular resolution. Note that the source strength appears squared in the denominator. Thus, if the source strength were ten times smaller than above, a prohibitive 100 years would be required for a result of similar significance.

For this reason the rejection factor R is important. Rejection factors >100 can be obtained in the AST by muon detection. Similar suppression factors can be obtained in ACT using the image shape in the focal plane of the Cerenkov mirror. As will be seen later, there is some evidence that point sources produce showers with abundant muons. If this is the case a powerful suppression factor is lost.

CHARACTERISTICS OF SOME SOURCES OBSERVED WITH $E>10^{12}$ eV

In the northern hemisphere three sources have each had a number of reported observations. More details of these sources and references to the original literature can be found in a number of excellent review articles(3). A feature of these sources which is often exploited in their identification is periodicity in emission which matches known orbital or pulsar periods.

Many observations of particles with $E>10^{12}$ eV have been observed from the Crab nebula. The most convincing have used the ACT. The Crab nebula contains SN1054, the famous supernova reported by Chinese astronomers. It contains a 33 msec pulsar at decl. 20°59′ and R.A. 5h32m. Its distance is ~2 kpc. The integral flux observed by the ACT is $I(E>10^{12} eV) \sim 7 \times 10^{-12} cm^{-2} sec^{-1}$. Recent observations of the Crab of great significance are discussed below. A few observations have been reported by the AST.

The most well known source as reported by the ACT and AST is Cygnus X-3. This is an x-ray binary observed prominently by its x-ray and radio emission. The orbital period is 4.79 hours. This object may contain a pulsar with a period of 12.6 msec, but the evidence is controversial. Its coordinates are decl. 40°47′ and R.A. 20h31m. Its distance is uncertain, but estimated to be ~10 kpc. It has been observed by both the ACT and AST. This is the first source to have been observed, in 1972 by the ACT by the Crimean observatory(4), and 1984 by the AST by the Kiel group(5). It is the latter observation that has stimulated the recent burst of activity in this field. On average, observations of Cygnus X-3 indicate a steady integral flux, $I(E>10^{14}eV) \sim 3 \times 10^{-13} cm^{-2} sec^{-1}$. Figure 6 shows a summary of reported integral fluxes. The integral spectrum has an energy dependence $\sim E^{-1}$. This spectrum is much harder than the cosmic ray spectrum. Cygnus X-3 has also been observed to exhibit bursts of radiation with $E>10^{12}eV$, which may be expected as it has outbursts of radio emission almost yearly. It is also interesting that in the few cases where muon detection is available, the showers from Cygnus X-3 show a normal muon content.

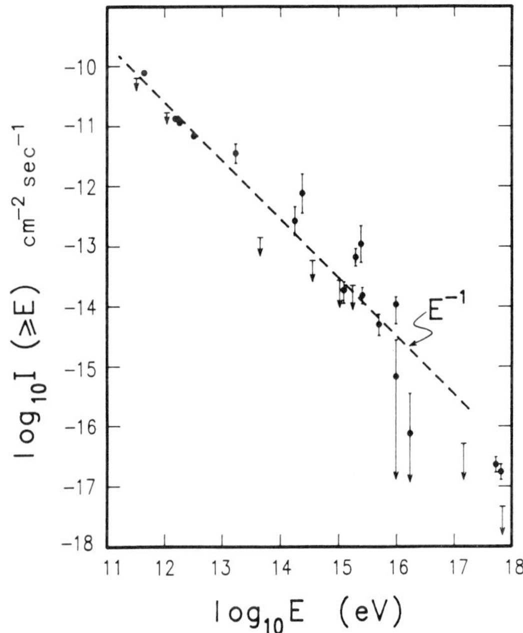

Fig. 6: Summary of measured fluxes of high energy γ-rays from Cygnus X-3.

A special note must be made of observations of Cygnus X-3 at $E>5 \times 10^{17}eV$ which have been observed by installations designed to explore the cosmic ray spectrum at the highest energies extending to nearly $10^{20}eV$. The Fly's Eye at Dugway, Utah, detects the weak fluorescence produced by ($>10^{17}eV$) shower particles in the nitrogen of the atmosphere. In the period 1983-1988 the Fly's Eye observed showers from the direction of Cygnus X-3 with $E>5 \times 10^{17}eV$(6). Very recently the 20 km^2 array at Akeno reported showers from the direction of Cygnus X-3 in the period September 1984 to July 1989(7). Both experiments report a flux $I(E>5 \times 10^{17}eV) \sim 2 \times 10^{-17} cm^{-2}\text{-}sec^{-1}$. However, the array at Haverah Park saw no signal and reported an upper limit $I(E>5 \times 10^{17}eV)$ to be less than $8 \times 10^{-18} cm^{-2} sec^{-1}$ with 90% confidence(8). The Akeno array observed a muon content which was the same as for normal cosmic rays. Neutrons of $5 \times 10^{17}eV$ have a mean decay length of 5 kpc so it is possible that these are showers induced by neutrons, for which a normal muon content is expected.

A third source is Hercules X-1. It is a well understood x-ray binary. It consists of a 1.2379 sec pulsar orbiting a companion star with a period of 1.7 days. There is also a 35-day precessional period. Its coordinates are decl. 35°25′ and R.A. 16h56m. Its distance is 4.5 kpc. It has been observed by both the ACT and AST but only in bursts. During the burst, integral fluxes as high as $I(E>4 \times 10^{11}eV) \sim 2 \times 10^{-8} cm^{-2} sec^{-1}$ and $I(E>10^{14}eV) \sim 2 \times 10^{-11} eV$ were observed for intervals of ~30 min.

A RECENT OBSERVATION OF HERCULES X-1

We discuss in some detail an experiment reported by the air shower array called "CYGNUS" located at Los Alamos(9). This is a good example of the kinds of observations from astrophysical sources which, at once, are fascinating and incredible in the true sense of the word.

The CYGNUS air shower array began operation in 1986 with ~60 detectors with a mean spacing of 14m. It was built at the end of the LAMPF accelerator at the Los Alamos National Laboratory to take advantage of the high altitude (800 gm/cm^2) and existing neutrino detectors which could serve as muon detectors. The area of the muon detector used in the experiment was 40m^2, which is too small to reject hadronic background, but quite adequate to decide if a point source is muon poor.

On July 24, 1986 an excess emission from the direction of Hercules X-1 was observed. This was on one day of a data set consisting of 340 days. Forty-six events were observed from the source where 24 were expected. A significant fraction of these events occurred in two bursts of about 30 min. each. In the first, seven events were found where 0.5 were expected. In the second, ten events were found where 2.6 were expected. The universal time of each event is recorded with a precision of better than 1 msec. A period analysis was made with a search range of ±0.3% of the known x-ray period. Periods of 1.23572 [40] and 1.23575 [30] sec were found for the two bursts. When the two bursts were analyzed together, it was found that the phase was maintained during the two hour interval between bursts and a common period of 1.23568 [25] was found which is 0.16% shorter than the x-ray period. Curiously ACT observations by the Whipple group on June 11, 1986(10), and the Haleakala group on May 13, 1986(11) found in bursts of Hercules X-1 periods also shifted from the x-ray period by the same amount. While these were not time coincident observations the agreement in period is remarkable. The shift in period needs to be understood. Other observations of Hercules X-1 by the ACT show periodicities more in agreement with the x-ray period.

One was astonished when it was reported that the mean number of muons for each event was 5.9 where 3.7 was expected, a result not anticipated if the showers were produced by γ-rays. If the showers were produced by a new strongly interacting neutral particle, its mass must be less than ~50 MeV to preserve the time correlation with the emitting pulsar. It has been suggested by Halzen(12) that above 10^{14}eV the photoproduction cross section increases sharply. We have investigated this possibility and find it unlikely. The geometric cross section of air is about 270mb. The equivalent cross section for pair production is 550mb. Thus, in the first interaction the γ-ray has a two-to-one chance to produce an e^+e^- pair. If the primary energy is in the few 10^{14}eV range, the energy of a subsequent γ-ray is usually well below 10^{14}eV where the photoproduction cross section is more constrained by measurement. It is not possible for a photon in this energy range to mimic a hadron.

Dunaevsky(1) has calculated, with a detailed hadronic Monte Carlo, two cases which are shown in Fig. 7. Case I is a photoproduction cross section following Halzen's prediction. Case II is more dramatic having the cross section increase to geometric by 10^{14}eV. Figure 8 shows the relation between the mean number of muons and the size of electron shower for 850 gms/cm^2, close to the Los Alamos depth. The mean size of the Los Alamos showers was ~10^5, so one would expect a factor 10 less muons for Case I and a factor 3 less for Case II. Both are difficult to reconcile with the data.

Thus, one is forced to believe in a new strongly interacting particle of very light mass that heretofore has not been observed. The CYGNUS group evaluates that the probability of these observations being produced by a chance fluctuation in background is less than 2×10^{-5}. Given what one is asked to believe, this result must be confirmed. The conclusion cannot yet be accepted as established fact.

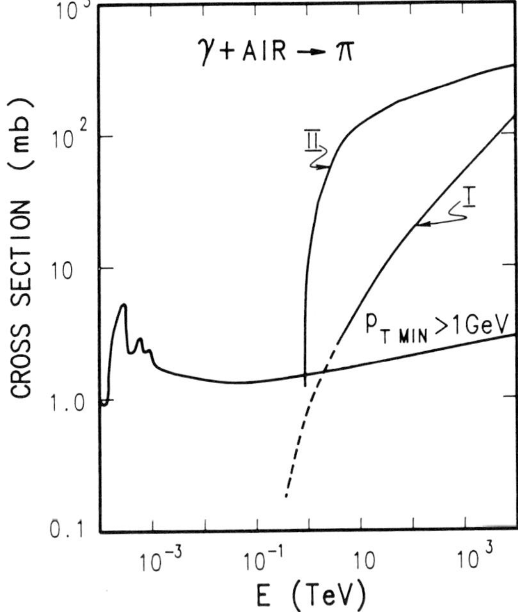

Fig. 7: Modifications of the photoproduction cross section. Case I from Ref. 12. Case II is an extreme example by the author.

OBSERVATION OF GAMMA RAYS FROM THE CRAB NEBULA

Recently the Whipple Observatory has convincingly observed showers from the direction of the Crab nebula(13). These showers demonstrate properties consistent with γ-rays. At the focus of the Whipple 10m telescope is an array of 109 PMs. A gamma ray shower in the upper atmosphere is essentially a line source. Hence the angles of Cerenkov photons lie in a plane containing the source, the core position on the ground and the telescope. We assume the

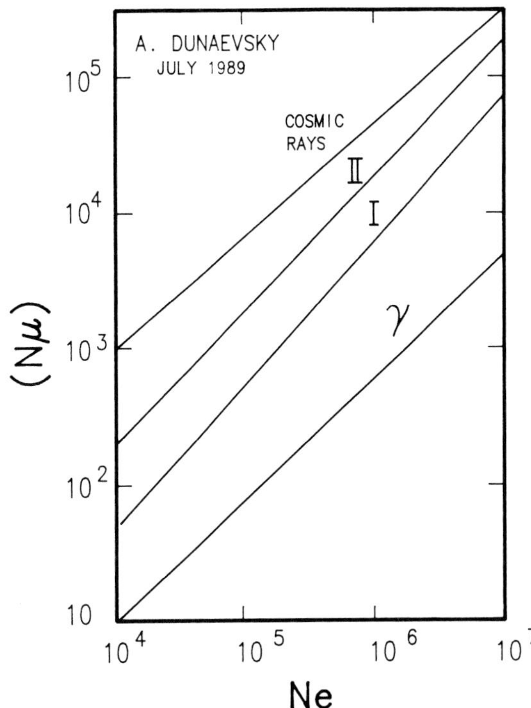

Fig. 8: $\langle N_\mu \rangle$ versus N_e for various assumptions for the γ-air cross section. The curve for normal cosmic rays is also shown.

telescope is pointed exactly at the source. If the core of the shower lands some distance from the telescope (50-100m), the image in the focal plane will be a line pointing toward the optical axis. The length of the line will be related to the distance along the shower path that produces detectable Cerenkov light. The width will be related to the angles of the radiating particles transverse to the plane defined above. This width is very small for a gamma-induced shower. A proton-induced shower will be much broader and will in general not point to the optic axis. In Fig. 9a is shown the angular distribution of detected photoelectrons for an event interpreted as hadronically-induced. Fig. 9b shows an observed event which is attributed to a gamma-ray initiated shower. The Whipple group has used a quantity azwidth, a single number which is small for gamma-induced showers and large for hadron-induced showers. Figure 10a shows how the azwidth is constructed from moments of the image. Figure 10b shows the separation in azwidth between gamma showers and hadron showers expected by Monte Carlo calculations.

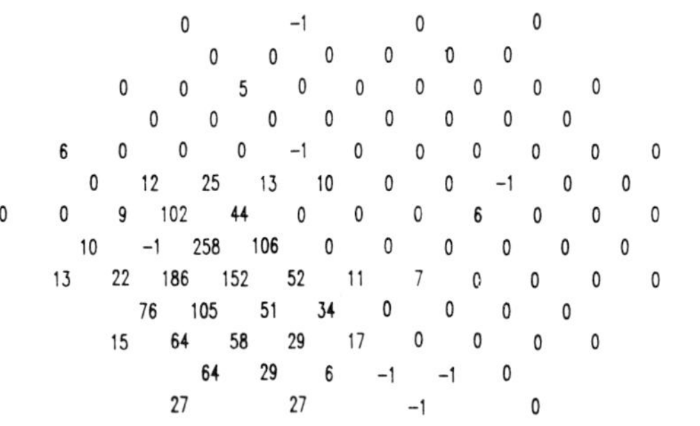

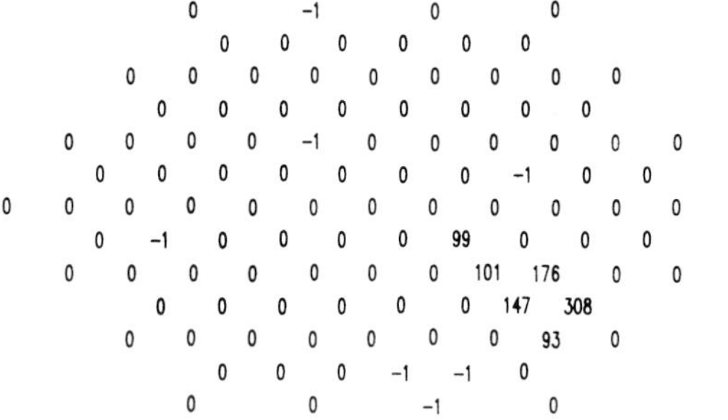

Fig. 9: Distribution of photoelectrons observed with the 109 PM detector of the Whipple telescope. (a) For hadron-induced shower, (b) for photon-induced shower. The entries indicated by "-1" are inoperative PMs.

In 1988 and 1989, 403379 events were observed on the Crab and with equal exposure 398563 events were observed slightly off the Crab. With an azwidth cut of 0.25, 12352 events remained on, and 10065 off, giving a signal of uncontestable significance. The cut reduces the background by a factor 40.

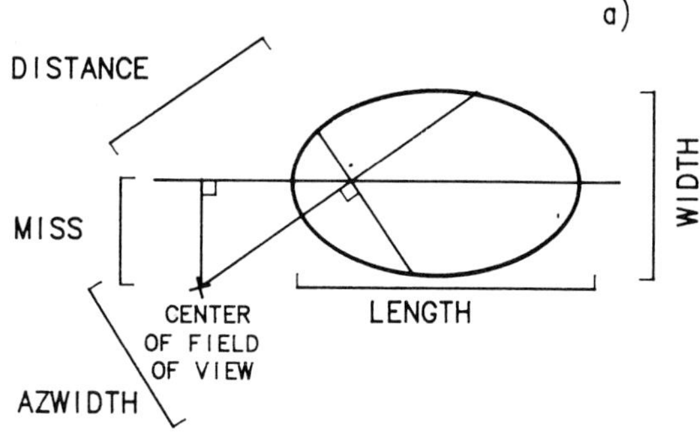

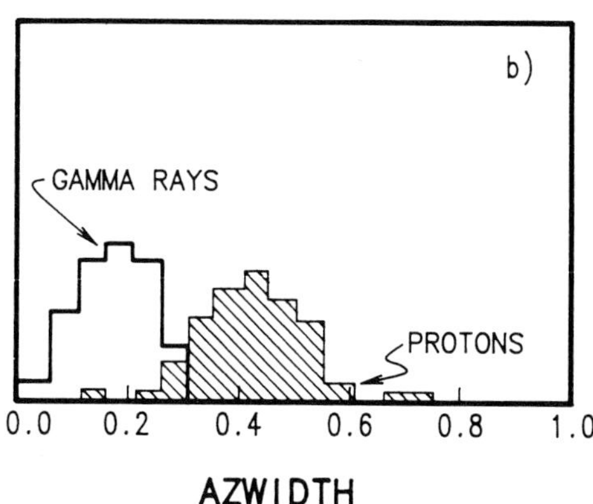

Fig. 10: (a) Construction of the azwidth parameter; (b) calculated distribution of the azwidth parameter for photon and proton showers.

Figure 11a shows the azwidth distribution for ON and OFF and Fig. 11b shows the significance of the difference in each azwidth bin. These results correspond to a steady (unmodulated at the Crab pulsar frequency) emission with an integral flux $I(E>0.6 \times 10^{12} eV) \sim 1.5 \times 10^{-11}\ cm^{-2} sec^{-1}$. A prior result using a 37 PM array at the focus gave a 9σ effect with $I(E>0.4 \times 10^{12} eV) \sim 7 \times 10^{-11} cm^{-2} sec^{-1}$ (2). These results have also been confirmed by groups from Michigan(14) and Riverside(15).

At $E \sim 10^{12}$ eV there appears to be a steady reliable astrophysical source of γ-rays from the Crab nebula. It is interesting to note that the imaging method which greatly increased the significance of the Crab signal completely eliminated the signal from Hercules X-1 which coincided in periodicity with the Los Alamos result discussed above(16).

As we shall see, it will be difficult to construct an air shower array which can observe these Crab γ-rays.

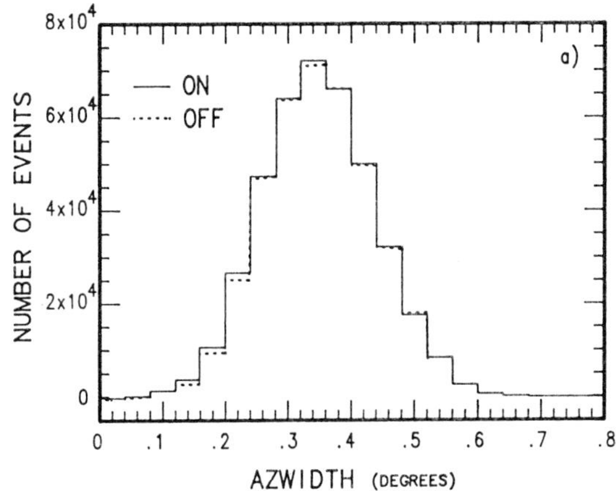

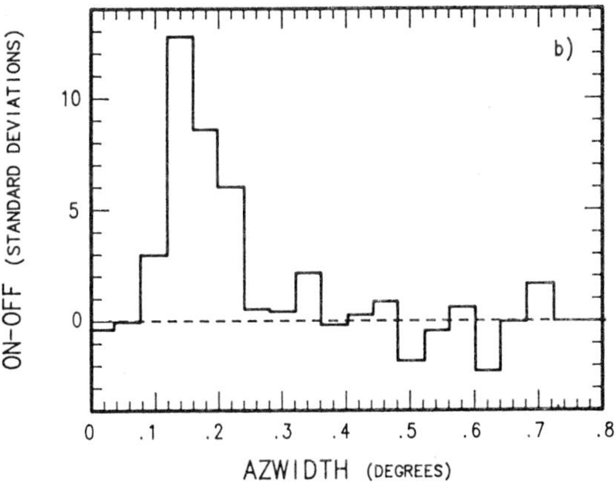

Fig. 11 (a) Azwidth distributions on and off source; (b) difference of on and off distribution expressed in standard deviations.

A SURFACE ARRAY WITH SIGNIFICANT MUON COVERAGE

At the site of the Fly's Eye at Dugway, Utah, the University of Michigan has installed $1250m^2$ of muon detectors. These have been operating since the beginning of 1988 in conjunction with an array of 33 surface counters of $1.5m^2$ built by the University of Utah(17). The layout of the detectors is shown in Fig. 12. The mean spacing of the surface counters is about 30 meters; threshold is several times 10^{14} eV. The muon detectors are arranged in eight patches of 64 scintillation counters each. The counters are buried about 3m below the surface in soil of density 2.2. This amounts to 20 radiation lengths of low Z material so the punch-through of the electromagnetic component of the shower is negligible.

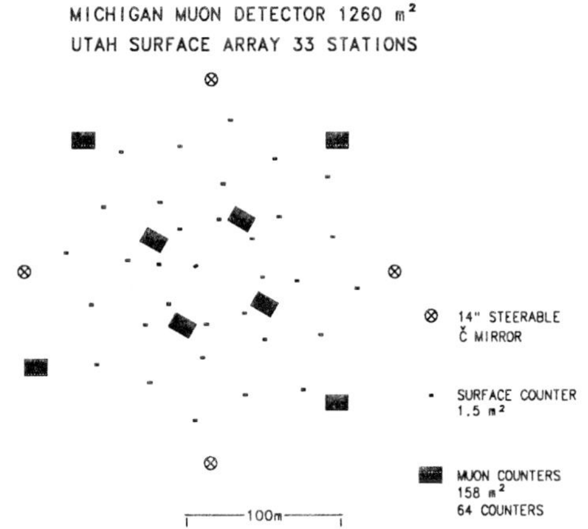

Fig. 12: The Michigan-Utah array.

The relation between the mean number of muons $\langle N_\mu \rangle$ and the number of electrons N_e has been measured to be $\langle \log_{10} N_\mu \rangle = -1 + .6\sec\theta + .8\log_{10}(N_e)$. In Fig. 13 is plotted the observed distributions of $N_\mu/\langle N_\mu \rangle$ for minimum N_e of 10^5 and 3×10^5. If one makes a cut at $\log (N_\mu/\langle N_\mu \rangle) = -1$, then the rejection factors for hadron-induced showers are 1/1100 and 1/4500, respectively. For $N_e > 10^4$ one expects from these data a rejection factor of 1/80. These rejection factors, large as they are, are still far from being able to detect a component of diffuse γ-rays in the normal cosmic rays which is expected to be a factor of 10^{-5} of normal cosmic rays.

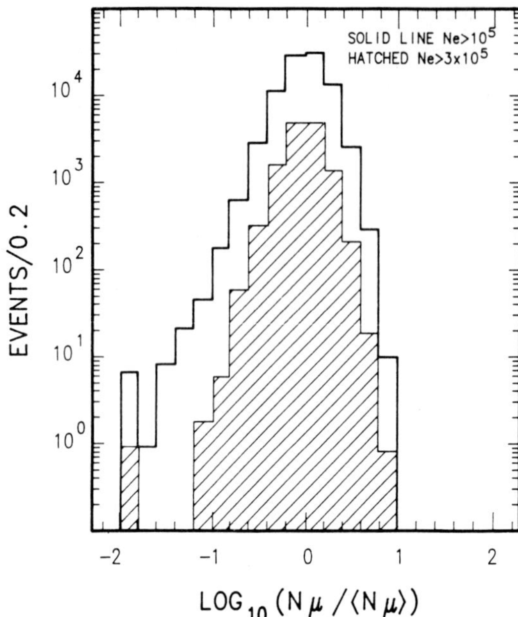

Fig. 13: Distribution of $N_\mu/\langle N_\mu\rangle$ measured by the Michigan-Utah array.

The Utah-Michigan group has searched for gamma rays from Cygnus X-3, Hercules X-1 and the Crab, all with negative results(18,19). Significant upper limits were obtained only with the application of muon rejection. If they do not use the muon rejection, the limits are about ten times larger. As an example, their results on Cygnus X-3 are shown in Fig. 14. The solid curve is the best estimate of the steady integral flux from Cygnus X-3 using all the results plotted in Fig. 6. The dip around 10^{15} eV corresponds to absorption by microwave radiation superimposed on a power law spectrum. The data represent about a year of operation. No significant results can be obtained with an array of this size if muon rejection cannot be applied.

NEW INSTRUMENTS

The discussion above demonstrates that there is a new field of astronomy developing. The convincing results from the Crab guarantee that there is an astronomy to be explored using photons of ~10^{12} eV. A second 10m telescope has been proposed and will likely be built at the Whipple Observatory(20).

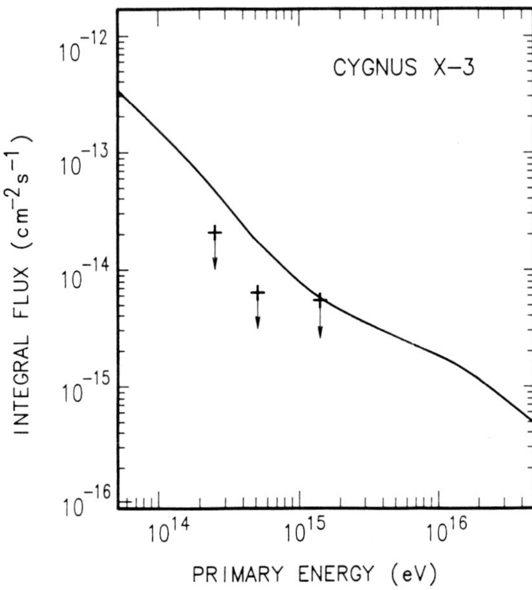

Fig. 14: Limits on the flux of high energy γ-rays from Cygnus X-3 by the Michigan-Utah array.

As we have pointed out, the incredible results indicated by the AST need to be confirmed, or laid to rest. This requires much larger instruments than have been used in the past. One such instrument is well along in its construction at Dugway, Utah, at the site of the Fly's Eye and the University of Michigan muon counters(21). This is the Chicago Air Shower Array (CASA) being built by a group from the University of Chicago. A layout of the array is shown in Fig. 15. When complete it will consist of 1089 detectors covering an area 0.5 km x 0.5 km on a rectangular grid of 15 meter spacing. It uses a combination of brute force (size) with modern electronics (analog electronics, data storage, and a microprocessor in each station) to be ensitive but relatively affordable.

Since December 1988 a prototype ngineering array of 49 detectors has been n operation. By December 1989, 529 etectors will be operating, and the array ill be completed by December 1990. It will un in coincidence with the Michigan muon ounters which were described above. By ecember 1990 Michigan will have installed nother eight muon patches making a total of 500 m^2 of muon counters in sixteen patches.

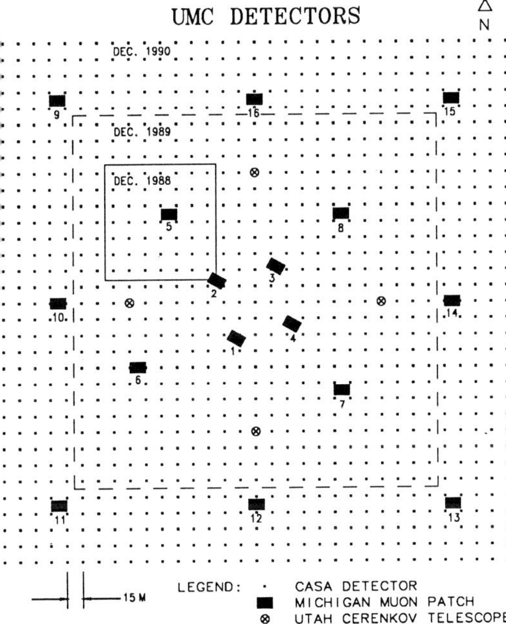

Fig. 15: Plan view of the Utah-Michigan-Chicago array now under construction at Dugway, Utah.

The "CYGNUS" array at Los Alamos is eing enlarged(22). To the original array, ow consisting of 108 counters with ~14 eter spacing, is being added an additional 00 detectors with a 30 meter spacing. dditional muon detection is being added, ufficient to ask whether point sources are uon rich or not, but insufficient to gain a ig suppression factor for hadronic showers. The design of the addition is predicated on the properties of the Hercules X-1 burst previously discussed. It is of great value that the Dugway and Los Alamos sites are within 15° in longitude. This will permit simultaneous viewing of the same source even if it bursts for a very short time.

The number of installations in the southern hemisphere, where the galactic center is visible, is much less than in the northern hemisphere. One impressive installation is a 14 element array being operating at the South Pole by groups from Leeds and Bartol(23). The South Pole, aside from the harsh climate, is an ideal location. The elevation is greater than 3000m and any object in the field of view remains at fixed zenith twenty-four hours per day. After plans were made for its installation, SN1987A, at a zenith angle of 23°, occurred. The Leeds-Bartol group intends to improve the array so that a lower threshold (>10^{13} eV) can be obtained. There are a number of groups who are proposing to install ACT detectors at the South Pole(24).

The Kiel group is building a >200 element detector at La Palma in the Canary Islands (elevation 2200 meters). Muon detectors are also being planned(25).

A very ambitious program is being planned by the Heidelberg group(26). Being experts on multiwire drift chambers (jet chambers), they plan to track the shower particles. Calculations suggest that far better angular resolution can be obtained by this method rather than by timing. At present, they are building prototypes as it must be demonstrated that the actual resolution agrees with the calculations. They have chosen a site in Venezuela, 9° north latitude, with an altitude of 3600 meters. The array would consist of two parts, an inner part of ~256 detectors, covering 6% of $10^4 m^2$. The dense array is in the center of a sparse array of ~256 identical detectors which will cover about 2.5 x 10^5 m^2. Because of its high altitude and fine angular resolution, this installation is by far the most sensitive one being considered.

New installations which emphasize the high end of the cosmic ray spectrum are being planned. In the prototype stage is a new Fly's Eye detector which will have a collecting aperture of 5000 km^2-ster(27). It can collect yearly ~10^5 events with E > 5 x 10^{17} eV. In the Soviet Union a surface array with an area of 1000 km^2 is being planned to be located near Alma-Ata(28). It would have denser arrays and Cerenkov detectors at its center, so gamma ray astronomy could be followed from ~10^{12} eV to 10^{19} eV.

We have given only a partial listing of new projects. There is an enormous activity in this field worldwide, so at the very least much more sensitive measurements will be forthcoming during the next ten years.

CONCLUSION

This is a conference on particle physics, so we ask if this work has a direct relevance to particle physics. To the uncritical observer, the answer must be affirmative since there are many reports of point sources producing showers with excessive muon content. Accepting this, one is forced to conclude some new particle physics is emerging, whether it is a new stable neutral particle, an anomolous rise in neutrino cross section(29), or possibly new photon interactions. I believe the latter can be excluded. These phenomena must be observed with more sensitive detectors and with a confidence which requires absolutely no fancy statistics other than the square root of N. The fact that these point sources show an excessive muon content may suggest that these observations are somehow fluctuations where the authors in some way have improperly estimated the chance probability that normal cosmic rays have produced the results. With respect to Cygnus X-3, an entire Physics Reports(30) article has been written on this issue!

If we are groping towards new physics with these techniques, one must credit many in the field of cosmic rays who have had to work for years with marginal apparatus and little funding.

The aspect of this work which is gamma ray astronomy is most promising, particularly the observations of the emissions from the Crab nebula at ~10^{12} eV. It is of great interest to push the threshold of the air shower technique down to 10^{12} eV. Then sources such as the Crab could be observed in all seasons, and one could pass from limited observations with the ACT to daily observations.

It is possible to achieve this goal with a body of water some 10^4 m^2 in area and 10 meters deep at an altitude of >3000 meters. Appropriately instrumented with large diameter photmultipliers, such a lake can detect separately the electromagnetic, and muon component of 10^{12} eV showers. In Fig. 16 we plot the sensitivity of various installations to spectra expected from the Crab and Cygnus X-3. The Crab spectrum when extrapolated as a power law would have only a slight chance of being detectable at 10^{14} eV. Instruments being built by Chicago-Michigan and Heidelberg can substantially improve the sensitivity for Cygnus X-3, even if muon suppression is not possible, but only a lake detector will have a good chance of observing the Crab.

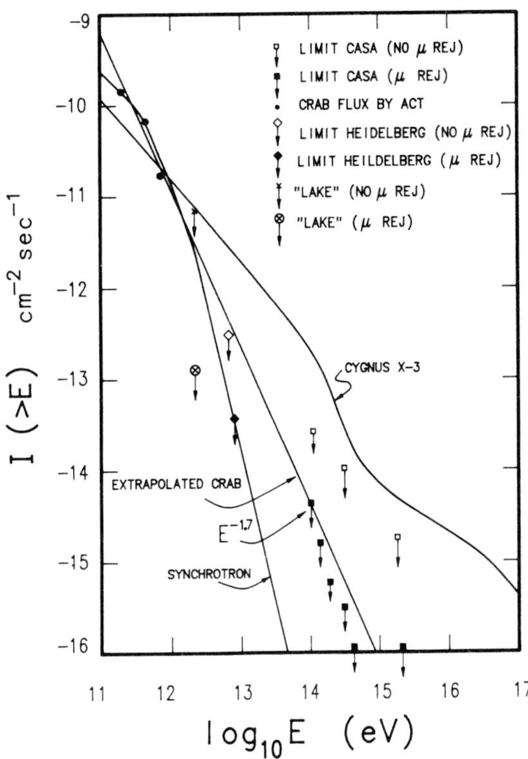

Fig. 16: Sensitivities of new detectors compared with γ-ray fluxes expected from Cygnus X-3 and the Crab nebula.

ACKNOWLEDGEMENTS

The author wishes to thank the many people who have given permission to present their data in this talk and who have helped to clarify the author's understanding of many points. Support of the National Science Foundation, Grant PHY87-02822 and PHY88-23033 is gratefully acknowledged.

REFERENCES

(1) A. Dunaevsky, P.N. Lebedev Institute, Moscow, 1989, private communication.

(2) T.C. Weekes et al. Astrophys. J. 342 (1989) 379.

(3) T.C. Weekes, Physics Reports 160, (1988); R.J. Protheroe, Proc. 20th Intl. Cosmic Ray Conference, Moscow, 8 (1987) 21; D.E. Nagle, T.K. Gaisser and R.J.Protheroe, Ann. Rev. Nucl. Sci. 38 (1988) 609.

(4) B.M. Vladimirsky, Yu. I. Neshpor, A.A. Stepanian and V.P. Fomin, Proc. 14th Intl. Cosmic Ray Conference, Munich, 1 (1975) 113.

(5) M. Samorski and W. Stamm, Astrophys. J. 268 (1983) L17.

(6) G.L. Cassiday et al., Phys. Rev. Lett. 62 (1989) 383.

(7) M. Teshima et al., Paper OG 4.1-23, submitted to 21st Intl. Cosmic Ray Conference, Adelaide, January 1990.

(8) M.A. Lawrence, D.C. Prosser and A.A. Watson, Phys. Rev. Lett. 63 (1989) 1121.

(9) B.L. Dingus et al., Phys. Rev. Lett. 61 (1988) 1906.

(10) R.C. Lamb et al., Astrophys. J. 328 (1988) L13.

(11) L.K. Resvanis et al., Astrophys. J. 328 (1988) L9.

(12) M. Drees and F. Halzen, Phys. Rev. Lett. 61 (1988) 275; M. Drees, F. Halzen and K. Hikasa, Phys. Rev. D39 (1989) 1310.

(13) M.J. Lang et al., Proc. Workshop Physics and Experimental Techniques of High Energy Neutrino and VHE and UHE Gamma Ray Particle Astrophysics, Little Rock, Arkansas (1989) to be published in Nucl. Phys. B.

(14) C. Akerlof et al., Proc. Gamma Ray Observatory Workshop, NASA, Goddard, (1989) to be published.

(15) O.T. Tumer, J.S. Hammond, A.D. Zych and C. MacCallum, University of California, Riverside Report UCR/IGPP 89/13 (1989) unpublished.

(16) T.C. Weekes, Preprint 2802, Harvard Smithsonian Center for Astrophysics (January 1989) unpublished.

(17) G.L. Cassiday et al., Paper HE 3.4-1 submitted to 21st Intl. Cosmic Ray Conference, Adelaide, January 1990.

(18) D. Ciampa et al., Papers OG 4.2-9, OG 4.1-12, submitted to 21st Intl. Cosmic Ray Conference, Adelaide, January 1990; D. Sinclair, Nucl. Instrum. Meth. A278 (1989) 583.

(19) S.C. Corbato et al., Paper OG 4.3-11 submitted to 21st Intl. Cosmic Ray Conference, Adelaide, January 1990.

(20) C.W. Akerlof et al., Preprint 2945, Harvard Smithsonian Center for Astrophysics (August 1989) unpublished.

(21) K.G. Gibbs, Nucl. Instrum. Meth. A264 (1988) 67; R.A. Ong, Proc. Workshop Physics and Experimental Techniques of High Energy Neutrino and VHE and UHE Gamma Ray Particle Astrophysics, Little Rock, Arkansas, (1989), to be published in Nucl. Phys. B.

(22) T.J. Haines, Proc. Workshop Physics and Experimental Techniques of High Energy Neutrino and VHE and UHE Gamma Ray Particle Astrophysics, Little Rock, Arkansas (1989), to be published in Nucl. Phys. B.

(23) N.J.T. Smith et al., Nucl. Instrum. Meth. A276 (1988) 622.

(24) K. Harris et al., Proc. Workshop Astrophysics in Antarctica, Bartol Research Institute (Newark, Delaware 1989).

(25) Hegra Proposal, M. Samorski, Kiel University (1989) private communication.

(26) B. Schmidt, Proc. Workshop Physics and Experimental Techniques of High Energy Neutrino and VHE and UHE Gamma Ray Particle Astrophysics, Little Rock, Arkansas (1989) to be published in Nucl. Phys. B.

(27) High Resolution Fly's Eye Proposal, University of Utah, Salt Lake City (1989),

(28) G.B. Khristiansen, Institute of Nuclear Physics, Moscow State University (1989) private communication.

(29) G. Domokos, B. Elliot, S. Kovesi-Domokos and S. Mrenna, Phys. Rev. Lett. 63 (1989) 844.

(30) J.M. Bonnet-Bidaud and G. Chardin, Physics Reports 170 (1988) 326.

DISCUSSION

M. Drees, CERN: I would like to point out that the calculation of the photoproduction cross section by Halzen *et al.* does not rely on fancy non-standard physics, but only on a minijet-type phenomenon as predicted by QCD, including effects due to the hadronic structure of the photon. Although there are large uncertainties in this calculation, I think it is fair to say that QCD tells us that $\sigma(\gamma N \longrightarrow \mu + X)$ at high energy should be higher than what one expects from a logarithmic extrapolation of existing photoproduction data.

J. Cronin: The point I was trying to make is that we are working only a factor of a hundred above where we know the cross section; when you follow the development of a shower, there's no way to increase the muon count to where it's comparable to ordinary hadrons. I'm not making a comment about the theory. I'm saying that even if you make it wilder than Francis Halzen, it still won't work.

G. Yodh, Univ. of California, Irvine: I would like to point out that, in the CYGNUS experiment's observation of the Hercules X-1 burst in UHE, the signal was sought without asking about the muon content of the showers. One may ask about the significance of detection, but not about muon content, which was unambiguously determined. What, in your opinion, is the achievable angular resolution in UHE experiments?

J. Cronin: With electronic techniques of timing I believe the best limit is about 0.5° for showers near the threshold of any particular detector. By the use of tracking detectors, as the Heidelberg group proposes, one may be able to achieve resolutions of 0.25°. As you know, however, it may be difficult to know the alignment of an array to a good enough precision that one can use such a high resolution. I believe we can continue to push other techniques to suppress further the hadron showers. One such technique may be to use the stochastic nature of the hadron induced showers.

V. Tsarev, Lebedev Physical Institute: What can you say about possible use of underwater Cerenkov detectors for γ-ray astronomy?

J.Cronin: The instrumentation of a lake or pond with photomultipliers in order to have a continuously sensitive detector for air showers is the most attractive way to lower the threshold of an air-shower array to overlap the capabilities of the atmospheric Cerenkov technique.

HIGH ENERGY ASTROPHYSICS: ACCELERATORS FOR VERY ENERGETIC GAMMA-RAYS FROM NEUTRON STAR SYSTEMS

M. Ruderman[1]

Department of Physics and
Columbia Astrophysics Laboratory
Columbia University
New York, New York 10027

ABSTRACT

Mechanisms are considered for the production of very energetic γ rays from neutron star systems. These include particle accelerators in certain radiopulsars, neutron stars in accreting binaries, and speculations about transient and unidentified sources. Questions still exist about the interpretation of the γ-ray data.

I. INTRODUCTION

From Jim Cronin's earlier discussion [1] of problems and paradoxes in identifying astrophysical sources of $10^{11} \sim 10^{16}$ eV γ-rays it is clear that our present view of the point-like γ-ray sources in our Galaxy is still rather poor and restricted. It is based upon a total of fewer than 10^7 collected γ-ray photons (about as many as detected in 0.1 s by an optical astronomer viewing a typical star). The average cost for each of these γ-ray photons is about $10 and there is certainly an extra impetus to refine and massage these γ-ray data until they suggest some possible models for the needed astrophysical accelerators and the beams they produce. In many cases the data are so sparse and even controversial that the models and conclusions they support are very fragile.

These conclusions are the following:

a) Among already identical spinning neutron stars in accreting binaries and rapidly rotating radiopulsars a signifcant fraction are (claimed to be) strong γ-ray sources.

b) Most γ-ray sources have not yet been associated with any particular astronomical object. Otherwise undetected neutron stars are the leading candidates for unidentified γ-ray sources.

c) Among the $10^8 - 10^9$ neutron stars in our Galaxy at least 10^5, and perhaps most, are probably strong transient sources of bursts of γ-rays.

d) A variety of different kinds of γ-ray sources seem to have qualitatively similar γ-ray spectra at high energies (Crab radiopulsar, Vela radiopulsar, most otherwise unidentified 100 MeV - GeV γ-ray sources, many transient burst sources).

From the observations three crucial questions have emerged:

1. Why do such a variety of neutron stars produce the very high energy particle accelerators needed to make energetic γ-rays?
2. Why are the neutron star accelerators so efficient as energetic γ-ray sources? The observed γ-ray luminosities are typically about 10^{-2} the total energy loss rate from the neutron star system. In many cases the γ-ray power may approach 100% of the total power which can be mobilized by the neutron star for an accelerator.
3. Are the highest energy γ-rays really γ-rays? Is there "new physics" in the TeV - PeV regime?

We consider first a brief summary of the claimed point sources of energetic γ-rays.

II. "POINT SOURCES" OF COSMIC GAMMA-RAYs [2,3,4]

Two point-like (for $\gtrsim 1°$ resolution) γ-ray source families were anticipated before any were detected. Cosmic ray protons within dense interstellar clouds produce π^0-mesons which decay into pairs of γ-rays. About 10 such sources have been observed. A second anticipated source was the Crab supernova nebula. Its optical light is synchrotron radiation from 10^{12} eV electrons (positrons). These same very energetic $e^-(e^+)$ will collide with the optical photons and exchange energy with them (inverse Compton scattering) to produce 10^{12} eV γ-rays. The crab nebula has indeed been confirmed as such a γ-ray source [5].

Other published reported sources include the following:

1. **Active galaxies including M 31, Cen A, and the quasi-stellar object 3C 273.** Claimed observaitons have been in the range 10^2 MeV – 1 GeV (HE), at 10^{12} eV (VHE), and at 10^{15} eV (UHE), but all have been rather marginal and unconfirmed [2].

[1] Research supported in part by NSF grant AST-86-02381.

2. **A variable source of 0.51 MeV $e^+ - e^-$ annihilation γ-rays from the Galactic Center Region.** This has been attributed to processes around a massive black hole and to an accreting neutron star in the close binary system GX1-4 [6,7].

3. **Radiopulsars.** Of the 400 observed radiopulsars, four have been reported as γ-ray sources. The Crab pulsar, the 935 year old neutron star which powers the Crab supernova remnant nebula, is a source of energetic quanta from optical frequencies up to 10^{12} eV [1,2,8]. It is most powerful in the hard X-ray regime. The 10^4 year old Vela pulsar has not been observed as a pulsed X-ray source but resembles the Crab pulsar in its emission above a few MeV. Its reported VHE emission at 10^{12} eV has not yet been confirmed. VHE emission also has been reported but not confirmed for the radiopulsars PSR 1953+29 and PSR 1801–23 [2]. All four of the claimed γ-ray radiopulsars have very short periods in the range $10^{-2} - 10^{-1}$ s.

4. **Unidentified Galactic sources** [9]. The Cos B γ-ray satellite detected about 10 distant HE sources (10^2 MeV – 1 GeV) in the Galactic disk which do not coincide with otherwise identified objects. Their γ-ray luminosities are in the range $5 \cdot 10^{36} > L_\gamma > 4 \cdot 10^{35}$ erg s^{-1} $\sim 10^2 L_\odot$, much brighter in γ-rays than the Crab or Vela radiopulsars. There are not enough neutron stars in the Galaxy younger than the Crab pulsar to provide plausible candidates for the unidentified Cos B sources from this population.

5. **X-ray pulsars in accreting binaries.** Of the approximately 40 neutron stars in accreting binaries with measured spin and/or orbital periods a half dozen have been claimed as sources of γ-rays with energy $\gtrsim 10^{12}$ eV: Cyg X-3, Her X-1, Vela X-1, Cen X-3, LMC X-4, and 4U0115+63 [2]. None are among the established Cos B sources at lower γ-ray energies. The two sources reported most often by different groups are Cyg X-3, and Her X-1 at both 10^{12} and 10^{15} eV. Each observation has been rather marginal but the total number is statistically impressive. The neutron star binary systems which have been reported as γ-ray sources do not form a special subset with respect to spin or orbital period, X-ray luminosity, or size of the secondary (c.f., however, Ref. [10]).

6. **Gamma-ray Burst Sources (GRB)** [11]. Several times a week transient γ-ray emission lasting $10^{-2} - 10^2$ s is observed from unknown sources. While they are observed these can be the strongest γ-ray sources in the sky. When they are dormant nothing is detected from them in X-rays or optically. There seem to be at least two different population of of GRB's. One appears to be almost isotropically distributed with a hard γ-ray spectrum extending to perhaps 10^2 MeV or beyond. These do not repeat on a timescale of 10 years or more. The other has much softer spectra with not much energy above an MeV and can repeat often but not periodically. There is now considerable evidence that strongly magnetized neutron stars constitute a major source (if not the only one) of GRB's. Cyclotron absorption features and submillisecond time structure indicate sources of dimension $\lesssim 10^7$ cm with magentic fields $\sim 3 \cdot 10^{12}$ G. If the GRB population is mainly in the Galactic disk the source population is $\gtrsim 10^5$. Neutron stars in close accreting binaries which are not detectable as accreting X-ray sources but in which the secondary is close enough to cause explosive γ-ray emission probably constitute too small a population to include most GRB candidates. Single neutron stars seem the most promising possibility.

We shall consider in the next sections, an interpretation of the part of the γ-ray source data which seems both least fragile and least likely to be greatly extended by new observations in the next several years. It is based upon the Crab and Vela pulsar observations. Models for them suggest speculations about some of the unknown sources such as the unknown Cos B sources of 4 and the GRB's of 6 above. However, the models and speculations are and must be very subjective since there is not yet a consensus in the community about just what is happening in these γ-ray stars.

III. ENERGETIC EMISSION FROM THE CRAB AND VELA PULSARS

About 10^{-3} of the total spin-down power of the Crab pulsar is emitted in hard X-rays and γ-rays (assuming the rotating emission beams are shaped much more like fans or large hollow cones than filled cones to account for the apparently high probability for observing them in both the Crab and Vela pulsars). The "light curves" at 10^2 MeV and 10^3 GeV are shown in Figure 1 from Ref. [8]. Here the average background counting rate is indicated by the horizontal dashed line and the arrows indicate the arrival time for the main pulse and subpulse at all lower energies. Two unambiguous pulses exist at the same phase as those at 10^2 MeV in the entire range from IR to GeV implying that all of this radiation is emitted from the same place near the neutron star. At 10^{12} eV the signal is only slightly above that from the cosmic ray background but additional support comes from a pulse phase exactly the same as that of all the lower photon energy pulses. The presence of the second pulse in these 10^{12} eV observations is much more problematic.

The Vela pulsar shows a similar double-pulsed structure with the same 0.4 phase separation as that for the Crab. This phase separation equality, however, appears to be coincidental. PSR 0540-693 is a radiopulsar similar in age, spin, and magnetic field to those of the Crab [12]. Because it is much further from us, only its optical and X-ray emissions have

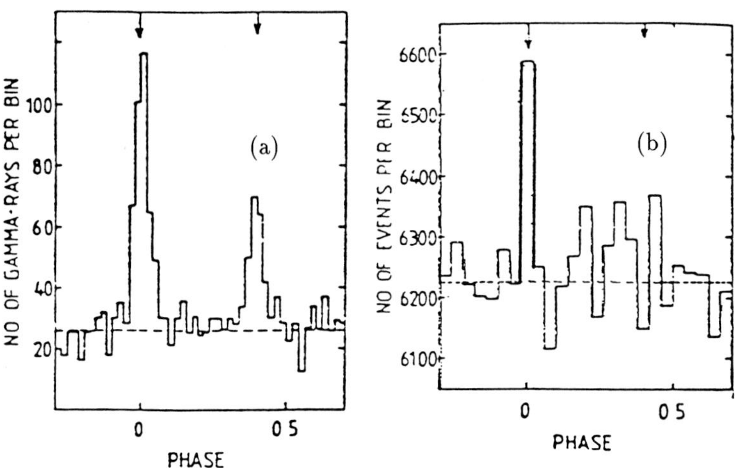

FIGURE 1. Light curves for the Crab pulsar at a) 10^2 MeV and b) 10^{12} ev (after Dowthwaite *et al.* [8]). The arrows point to the phases of the main pulse and interpulse observed at lower energies.

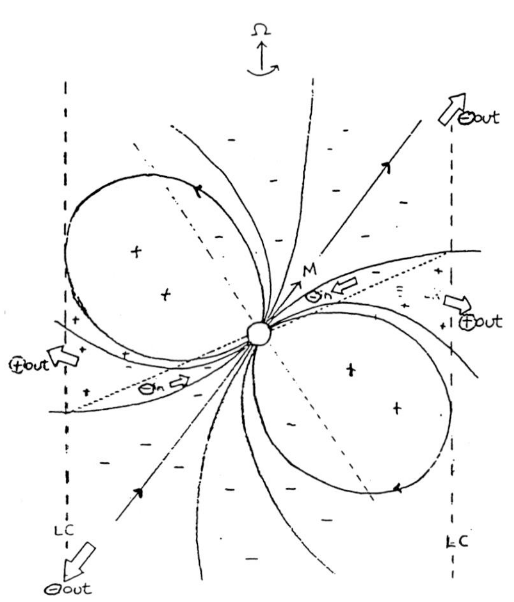

FIGURE 2. The corotating magnetosphere of a non-aligned neutron star with magnetic dipole moment $\vec{M}$ [10].

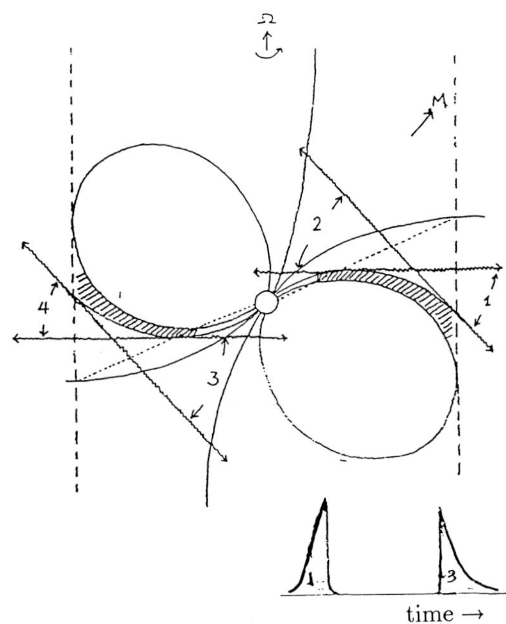

FIGURE 3. Outer magnetosphere "gap" accelerator regions from the current flow of Fig. 2 [10,14].

been observed. Its double pulse phase separation has narrowed to only 0.2.

A final clue to the mechanism of the γ-ray emission may be the very large $e^{\pm}$ flux ($\gtrsim 10^{39}$ s^{-1}) which seems to be injected by the Crab pulsar into its surrounding nebula [10,13].

From the above observations and considerations about the possible structure of the near environment around such rapidly spinning strongly magnetized (surface dipole $\sim 10^{12}$ G) neutron stars several features emerge about the properties of the source of their γ-ray emission: [10]

1. There must be a powerful accelerator near the star which can give e^-/e^+ energies exceeding 10^{12} eV. If the accelerator lies within the "light cylinder" of the star (where the corotation speed reaches c) the accelerator must be accomplished by a strong electric field ($\vec{E}$) directed *along* the magnetic field ($\vec{B}$). Ultrarelativistic acceleration perpendicular to $\vec{B}$ could be accomplished only by an $|\vec{E}| \geq |\vec{B}|$ which could not be achieved near the star by rotation.

2. The maximum net current flow through the Crab's accelerator is $\sim 10^{34}$ s^{-1} of e^- (or e^+ in the opposite direction). A larger current would itself give a larger $\vec{B}$ near the light cylinder than that of the stellar dipole and quench the accelerator.

3. The potential drop of the accelerator along $\vec{B}$ must exceed 10^{14} volts to supply the needed γ-ray power.

4. The location of the accelerator must be at least about 10^8 cm (10^2 stellar radii) away from the Crab neutron star. If it were closer, 10^{12} eV γ-rays could not escape before conversion to $e^{\pm}$ pairs in the strong stellar magnetic field. In addition the needed $\gtrsim 10^{14}$ volt potential drop would be quenched at much lower values by various $e^{\pm}$ production mechanisms in a much stronger $\vec{B}$.

5. A 10^{39} s^{-1} $e^{\pm}$ pair production could be a consequence of $\gamma+\gamma \rightarrow e^+ + e^+$ around the neutron star from the crossing γ-ray beams of the observed flux only if this interaction takes place within several $\times 10^8$ cm of the star. To be consistent with 4 this would indicate an accelerator about 10^8 cm away from the neutron star.

6. The success in observing the double pulses of the Crab and Vela pulsars and the small phase separation between those pulses in the Crab-like pulsar PSR 0540 are more consistent with a rotating double fan beam structure than with narrow radial cones from above the north and south magnetic poles.

We turn now to consider how a pulsar model might account for the existence and suggested properties of the needed accelerator.

IV. A MODEL FOR ENERGETIC EMISSION FROM THE CRAB PULSAR [10,14,15]

Within a spinning magnetized neutron star

$$\vec{E} = \frac{(\vec{\Omega} \times \vec{r})}{c} \times \vec{B}, \tag{1}$$

where $\vec{\Omega}$ is the star's angular velocity. It follows that

$$\vec{E} \cdot \vec{B} = 0 \tag{2}$$

(in both the laboratory and rotating frames). Equation (2) fails at the stellar surface and beyond if the star is in a vacuum. However, $(\vec{E} \cdot \vec{B})$ would then exceed 10^{12} Volts cm^{-1} outside the Crab and Vela surfaces so that charge would be pulled from those stars no matter what the surface structure. That charge flow would continue until Eq. (2) is satisfied outside the star as well as inside. Equation (1) would then also hold outside as well as inside and the outside plasma would corotate with the star ("corotating magnetosphere"). The plasma pulled outside the star would act like an extension of the star with a plasma charge density

$$\rho = \frac{\vec{\nabla} \cdot \vec{E}}{4\pi} \sim \frac{\vec{\Omega} \cdot \vec{B}}{2\pi c}. \tag{3}$$

This plasma would be "charge separated", consisting exclusively of negative or positive charge according to the sign of $\vec{\Omega} \cdot \vec{B}$ in different regions of the magnetosphere as indicated in Fig. 2. Here the sign of the charge density is indicated by + or − and the dotted lines are surfaces where $\vec{\Omega} \cdot \vec{B} = 0$ and the magnetosphere charge vanishes. The arrows indicate the assumed current flow of + and − on open field lines ($\oplus$ and $\ominus$) out through the light cylinder (LC) and into the star. Strong charge depletion is expected where $\ominus$ and $\oplus$ flow in opposite directions.

The difficult problem, on which there is not yet a consensus, is understanding in detail how to describe a corotating magnetosphere near, at, and beyond the light cylinder at $r = c\Omega^{-1}$ where corotation must fail. If a cylindrical insulating sheath of radius $\ll c\Omega^{-1}$ isolated the magnetosphere from the rest of the Universe, Eq. (2) could be achieved everywhere within the sheath and no accelerator or γ-ray emission from the magnetosphere would be expected. When such a sheath is removed particle flow along $\vec{B}$ out through the light cylinder should begin. Charge depletion in the outer magnetosphere in the volume between the surface where $\vec{\Omega} \cdot \vec{B} = 0$ and the light cylinder can result in the failure of Eq. (3) and consequently of Eq. (2).

Such a charge deficient region ("gap") in the outer magnetosphere is a common feature of a variety of models and is an expected consequence of the assumed magnetospheric current flow patterns. We shall use the model of Cheng, Ho, and Ruderman [10,13,14] for further analysis, but expect the general features of γ-ray pulsar evolution which are based upon it to be more robust than model details. Because of charge separated current flow a stable outer magnetosphere gap forms and expands in the region indicated in Fig. 3. When such a gap begins to grow, the accelerating $\vec{E} \cdot \vec{B}$ in it also increases until the resulting e^-/e^+ accelerator becomes strong enough to support so much $e^\pm$ production that the supply of pair plasma quenches further gap growth. For the Crab pulsar that is the result of a series of processes:

a) An $e^\pm$ pair produced in the gap is instantly separated by the large $\vec{E}\cdot\vec{B}$ there which accelerates the e^- and e^+ in opposite directions. Because of magnetic field line curvature each lepton radiates multi-GeV curvature γ-rays.

b) These are converted into $e^\pm$ pairs in collisions with KeV X-rays [from (d) below]. Pairs created in the gap repeat process (a).

c) Pairs created beyond the gap boundary lose their energy to synchrotron radiation (optical to MeV) and to higher energy γ-rays from Compton scattering on the same X-ray flux responsible for the pair creation.

d) The X-ray flux from the sychrotron radiation of c) is that which initially caused the curvature radiation γ-rays of (a) to materialize and the inverse Compton scattering of the pairs in (e). Since the entire series of processes is powered by the extremely energetic e^- and e^+ of (a) moving in opposite directions within the gap, all of the resulting fluxes of photons and $e^\pm$ pairs have a symmetric flux which is oppositely directed. The pairs and γ-rays moving in one direction then interact mainly with the X-rays moving oppositely.

e) A third generation of $e^\pm$ pairs comes from partial materialization of the crossed γ-ray beams from (c).

The above processes (a – d) "bootstrap" the creation of an $e^\pm$ plasma until enough is produced to form a gap boundary layer which quenches further growth. The charge depleted gaps where $\vec{E} \cdot \vec{B} \neq 0$ and e^-/e^+ are strongly accelerated along $\vec{B}$ are shown crosshatched in Fig. 3. Pair production from $\gamma + \gamma \rightarrow e^+ + e^-$ sufficiently fills all other regions of the outer magnetosphere to satisfy Eqs. 3 and 2 despite current flow. (Sustained current flow and gap formation are expected only on "open" $\vec{B}$ field lines – those that pierce the light cylinder. Most γ-rays from relativistically accelerated e^-/e^+ there are emitted almost tangentially to the local $\vec{B}$. Thus the γ-rays from the crosshatched accelerator region may supply $e^\pm$ to all other open field line regions of the outer magnetosphere but not vice-versa.) Most of the γ-ray emission which escapes the magnetosphere comes from synchrotron radiation and inverse Compton scattering along the border region between the accelerating gap and the rest of the open field lines of the outer magnetosphere. Because of the symmetry between the flow of e^- in one direction and e^+ in the other together with that between the predominantly dipolar $\vec{B}$ from both poles near the light cylinder, the γ-ray emission is always in the form of four fan beams as shown in Fig. 3, and any observer will see two of them (latitudinal width $\sim \pi/2$) with a phase separation (partly because of different travel times and partly from aberration) which depends on the tilt between the dipole ($\vec{M}$) and $\vec{\Omega}$ and on the direction to the observer. Crossed γ-ray beams fill the rest of the magnetosphere with pairs from $\gamma + \gamma \rightarrow e^+ + e^-$. A very small fraction of the e^- or e^+ are reversed by weak $\vec{E}\cdot\vec{B}$ to keep $\vec{E}\cdot\vec{B} \sim 0$ everywhere except in the accelerator gaps. The dotted line on which $\vec{\Omega}\cdot\vec{B} = 0$ vanishes separates the regions in which net current is carried by e^- moving in toward the star for those in which e^+ is moving out, the needed e^+/e^- source coming from local $e^\pm$ production. Photons of 10^{12} eV in beam 1 may reach an observer without absorption. Photons of this energy or higher in beam 3 whch must pass through a stronger $\vec{B}$ before leaving the magnetosphere may be converted first into $e^\pm$ pairs so that the subpulse peak in Fig. 1a may be absent in Fig. 1b.

The $e^\pm$ production from mechanism (e) above has been estimated to produce about 10^{39} $e^\pm$ per second whch suggests that these may be the source of those further accelerated and ultimately injected into the nebula in which the pulsar is imbedded. The calculated spectral shape from $1 - 10^9$ eV which results from mechanisms (a – d) approximately fits the Crab pulsar observations in Fig. 4. This shape depends mainly only on the width of the gap in Fig. 3, but is not very sensitive to it or other parameters [14,15]. The local B is $\sim 10^6$ G. The regimes where synchrotron (SYN) and inverse Compton scattering (ICS) occur are indicated. Some 10^{13} eV emission (not shown) is expected from inverse Compton scattering of gap e^-/e^+ on optical photons in the gap process (c) and from the synchrotron radiation by the $e^\pm$ pairs these ultra high energy γ-rays make beyond the gap boundary. (Radiation reaction limits gap e^-/e^+ to about 10^{13} eV. Higher energies would be given to protons which traverse the entire 10^{15} Volt gap potential drop. These 10^{15} eV protons may make π^0-mesons in collisions with the soft X-ray photons from the Crab pulsar, but 10^{15} eV γ-rays from π^0 decay will not survive if not made well beyond the pulsar light cylinder. Nebular matter is another possible target for π^0 production).

V. EVOLUTION FROM THE CRAB TO THE VELA PULSAR [16–18]

Observations of the Crab and Vela pulsars imply that these spinning neutron stars differ only in their spin periods: the Crab has a period $P \sim 3 \times 10^{-2}$ s while for Vela $P \sim 9 \times 10^{-2}$ s. Both pulsars appear to have the same dipole field strengths (surface dipole $B_s \sim 4 \times 10^{12}$ G) and an almost identical double subpulse structure which suggests very similar relative

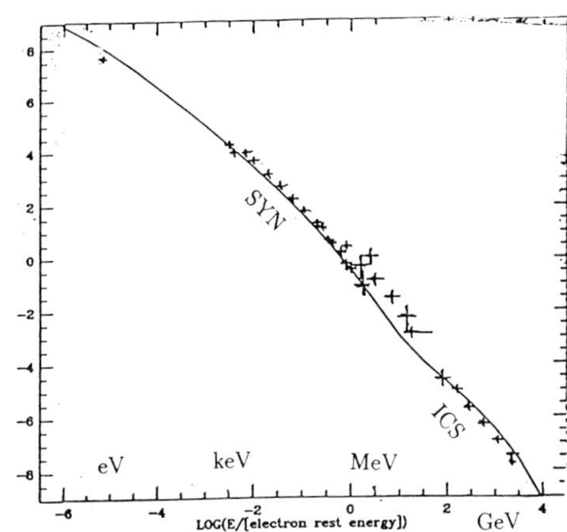

FIGURE 4. Calculated and observed spectra of the energetic pulsed photon emission from the Crab pulsar [14]. (The absolute magnitude is arbitrarily adjusted.)

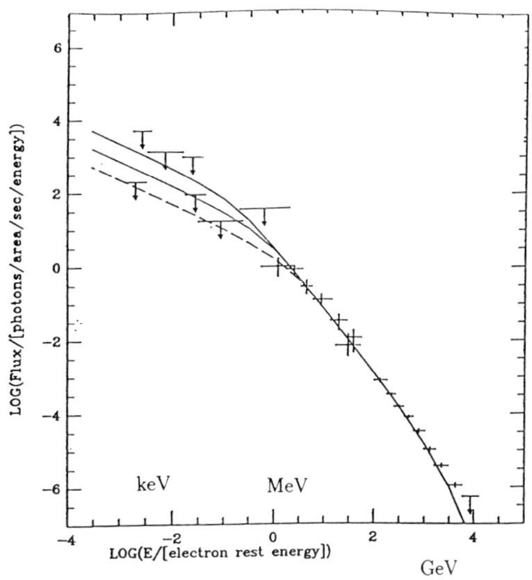

FIGURE 5. Calculated and observed spectra of energetic pulsed photon emission from the Vela radiopulsar [14]. (The absolute magnitude is arbitrary)

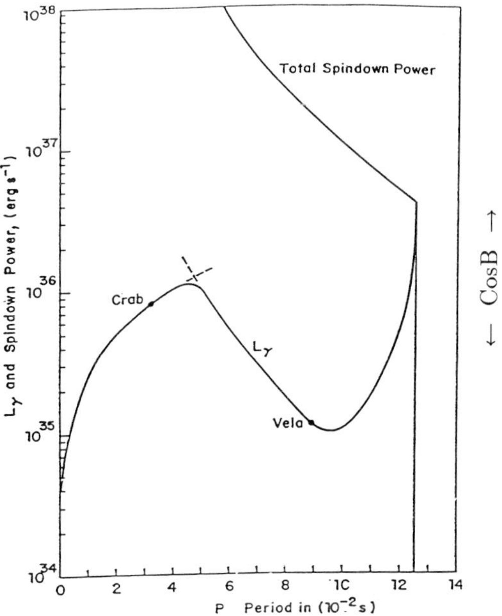

FIGURE 6. Calculated evolution of γ-ray luminosity (L_γ) and total spin-down power as a function of pulsar period for a pulsar with the magnetic dipole field of the Crab and Vela radiopulsars [18].

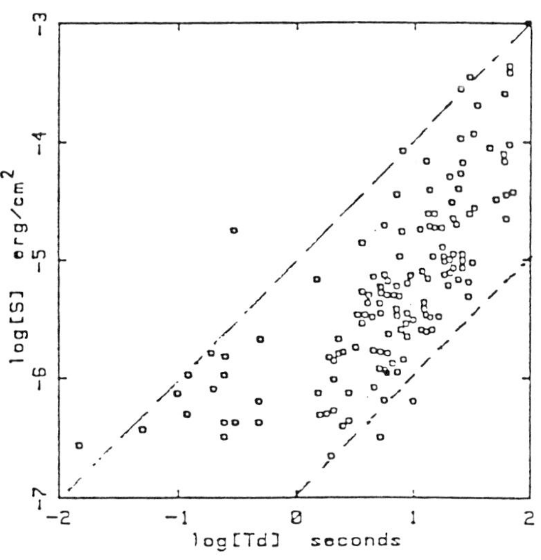

FIGURE 7. Gamma-ray burst fluence S (time-integrated intensity) as a function of burst duration time from KONUS [23] experimental data whch measured mainly X-rays of energy < 1 MeV [11]. The two diagonals give the fluence of a 10^{36} erg s^{-1} source at 10^2 and 10^3 lt-yrs.

orientations between their spins and magnetic dipoles and probably also with the line of sight to us. It seems very plausible, therefore, that in less than 10^4 years, when the Crab pulsar's spin will have slowed to that of Vela, its observed energetic radiation will have evolved into that seen today from Vela. This implies that certain large changes must occur as the neutron star spins slowly by a factor of three:

1. The Crab and Vela energetic radiation spectra are similar at photon energies above 1 MeV as shown in the spectra of Figs. 4 and 5. (In Fig. 5 the spectral break at three possible energies covers a range of possibilities for e^-/e^+ energies and local $\vec{B} \sim 5 \cdot 10^3$ G for all of which $e^\pm$ exit from the magnetosphere before strong synchrotron emission has reached the X-ray regime.) From the Crab the emitted power remains strong down to optical frequencies. In Vela's spectrum, however, there is a sharp break at around one MeV and a very great relative intensity suppression at all lower photon energies.
2. The fraction of the total spin-down power of the Crab which is radiated as energetic radiation ($\sim 10^{-3}$ if the radiated beams are much more greatly extended in latitude than in longitude as in Fig. 3) is 10^{-1} of that from the more slowly spinning Vela.

Both of these features have a natural explanation in the outer gap accelerator model.

The series of mechanisms which limit the growth of an outer magnetosphere gap in the Crab pulsar are not nearly so efficient in the Vela pulsar. The synchrotron X-rays of §IV, item c, necessary for the $e^\pm$ conversion of curvature γ-rays from the gap, are greatly suppressed in Vela. This is because the time it takes for a relativistic e^-/e^+ to synchrotron radiate down to a particular characteristic frequency is proportional to B^{-3}. Since Vela has about the same dipole moment as the Crab, but spins 3 times less rapidly, the local outer magnetosphere B is smaller by $3^3 = 27$. This is enough to suppress strong X-ray synchrotron radiation since the radiating pairs will leave the magnetosphere, either directly or after reflection if spiralling inward along converging magnetic field lines, before such radiation is strong. The Vela pulsar limits its outer magnetosphere gap growth by mechanisms rather different from those of the Crab, and the gap grows to almost 1/3 the available outer magnetosphere volume before this is accomplished. The calculated and observed spectra are compared in Fig. 5.

Because of the strong synchrotron X-ray emission from the Crab pulsar's outer magnetosphere, $e^\pm$ production there can utilize more efficient mechanisms than those available for Vela. For pulsars spinning more rapidly than the Crab but which are otherwise identical, L_γ should be less. This is because in the larger outer magnetosphere magnetic fields of faster spinning pulsars it becomes easier to make $e^\pm$ pairs. (If a residual pulsar of SN 1987A is in that family, it would therefore be expected to be much less bright in γ-rays than the Crab pulsar.) The calculated evolution of γ-ray luminosities as a function of pulsar periods up to that of Vela is shown in Fig. 6. As a Vela pulsar slows to a period somewhat greater than 9×10^{-2} s, calculation of its expected L_γ indicates a change from one which diminishes with the decreasing spin-down power to one whch increases sharply until almost all of the pulsar's spin-down power is being radiated away as energetic γ-rays. We turn next to that stage of the pulsar's evolution.

VI. POSSIBLE FUTURE EVOLUTION OF VELA TO A STRONGER COS B γ-RAY SOURCE [16–18]

Key mechanisms in the symbiotic complex which sustains the $e^\pm$ pair production needed for Vela's outer magnetospheric current flow include $\gamma+\gamma \to e^- + e^+$ by crossing radiation beams which contain the observed γ-ray flux. Pair production from such crossed beam photon collisions is largest for photon center of mass energies between the theshold $2mc^2 \sim 10^6$ eV and several MeV. As a young pulsar's spin slows to around that of Vela, L_γ remains a fixed fraction ($\sim 10^{-2}$) of the pulsar's total spin-down power (c.f. Fig. 6). However, as the spectral break exceeds several MeV, the number of γ-rays in the beams with energies in the most effective range to make $e^\pm$ pairs greatly decreases. Moreover the suppression of the lower end of the γ-ray spectrum leaves many fewer γ-ray photons in the beams. Because the spectral break rises so sharply with increasing pulsar period, a very much larger fraction of the pulsar's spin-down power must then be devoted to sustaining the needed outer magnetospheric pair production and thus to the γ-rays associated with it. In Vela the outer magnetosphere accelerator already occupies around 1/3 the total open field line volume. As P increases even very modestly, that fraction must grow very considerably in order to maintain required outer magnetosphere $e^\pm$ production until, finally, all of the available volume is used for the accelerator. Then most of the pulsar's spin-down power will be dissipated in γ-rays produced near the pulsar light cylinder. For $P \sim 10^{-1}$ s the consequent upper bound to the pulsar's γ-ray luminosity is

$$L_\gamma \sim 5 \times 10^{36} \text{ erg s}^{-1}. \tag{4}$$

(It has been estimated that this maximum L_γ will be achieved for Vela when its period reaches $P \sim 0.13$ s, as indicated in Fig. 6.) In the interval between $P = P(\text{vela}) \sim 9 \times 10^{-2}$ s and $P \sim 0.13$ s, the pulsar γ-ray luminosity could almost reach that of Eq. (4). The radiated hard γ-ray spectrum would be expected to remain that of Fig 5 which can be approximated by a flux spectrum $N(\omega) \sim \omega^{-2}$ at the higher energy end of its range. The number of such luminous "post-Vela" γ-ray sources is the number of Vela-like pulsars whose periods lie between that of Vela and one a factor 1.4 greater. For a nominal birthrate of pulsars in this family of one per 10^2 years the total Vela-like Galactic population in this period interval would be 40, a large fraction of which could have L_γ approaching that of Eq. (4). The population, the γ-ray luminosity and spectrum, and the Vela-like suppression of X-ray emission are consistent with those of the 10 "unidentified" Cos B sources. It is tempting to suggest that the latter may consist of such post-Vela neutron stars with rising L_γ and $P \sim 10^{-1}$ s.

We turn next to the evolution of these γ-ray pulsars when their outer magnetospheres can no longer sustain the large $e^\pm$ production rates needed to sustain strong magnetospheric flow.

VII. TERMINAL ALIGNMENT AND DEATH OF POST-VELA GAMMA-RAY PULSARS [16–18]

In the very luminous γ-ray emission phase of §VI the charge deficient open field line accelerator region of the outer magnetosphere had to grow with increasing P until it occupied such a large region that L_γ approached the entire spin-down power of the pulsar. That growth was necessary to sustain needed large $e^\pm$ creation rates there. When P lengthens still further the accelerator, already at its maximum relative volume, can no longer sustain the needed steady copious $e^\pm$ production. As a result we may expect that a very large part of the magnetosphere current flow between the polar cap and the light cylinder will also be quenched. A great suppression of current flow throug the polar cap ($\vec{J}$) would give a corresponding reduction its spin-down torque from $\vec{J} \times \vec{B}$ inside the neutron star beneath the cap. Then any large further spin-down of the neutron star would have to be accomplished through the Maxwell torque from the radiation of the star's spinning non-aligned magnetic dipole moment. If the neutron star is approximated by a rigid sphere or, more realistically, as a spinning fluid whose shape is always axially symmetric about its spin axis, then that torque would not only spin-down a Vela-like star in about 10^4 years, it would also, on that same time scale, align its magnetic moment with $\vec{\Omega}$ [19,20] and thus quench the Maxwell spin-down torque. Long lived solid crust deformations and neutron superfluid vortex pinning may complicate estimates of alignment timescale [21] but crustal relaxation mechanisms would ultimately allow alignment. If such relaxation is within 10^4 years, then the range of final periods for the family of aligning post-Vela γ-ray pulsars is $P \sim (1-2) \times 10^{-1}$ s if the current flow through the stellar magnetosphere is largely quenched. Because they no longer spin-down, the expected number of solitary aligned rapidly spinning former γ-ray pulsars in the Galaxy could be of order the birthrate of Vela-like pulsars ($\sim 10^{-2}$ yr) times the age of the Galaxy. (This proposed evolutionary scenario is certainly very different from that which describes the evolution of conventional radiopulsars. These pulsars have periods in the range $3 \times 10^{-1} < P < 3$ s, and are much older than 10^4 years. There is almost certainly considerable open-field-line current flow through their polar caps which is the source of RF radiation. Most significantly, typical 10^6 year old radiopulsars are not yet aligned and are spinning down canonically. How and why does their evolution differ so strikingly from that proposed here for Vela-like γ-ray pulsars? One possibility is, of course, frozen crustal deformations or internal coupling between $\vec{\Omega}$ and $\vec{B}$ which do not relax for 10^6 yrs in canonical pulsars. Another is variations in polar cap local magnetic field geometry which directs the flow of $e^\pm$ pairs created above the polar caps (in the very strong $\vec{B}$ there by $\gamma + \vec{B} \to e^- + e^+ + \vec{B}$) to the outer magnetosphere regions. There they may replace the otherwise needed local creation of $e^\pm$ pairs so that strong current flow there continues at spin periods where true pair creation is no longer sustainable.

A spinning neutron star with suppressed magnetosphere current flow and an aligned dipole would no longer spin down. Thus such a population of extinct Vela-like pulsars, despite an enormous rotational energy, is dead. But we are assured in the English Book of Common Prayer: "After death there is the sure and certain hope of ressurection" and we turn next to more detailed consideration of such a prediction.

VIII. EXCITATION OF TRANSIENT X-RAY AND γ-RAY EMISSION FROM LATENT ALIGNED POST-VELA PULSARS: GRB SOURCES? [16–18]

If the magnetic field of strongly magnetized neutron stars decays substantially after 10^7 years, then only 10^5 of the solitary aligned rapidly spinning neutron stars in the Galaxy are still Vela-like. If strong dipole fields survive longer this population would be increased proportionately. It has been proposed that members of this large population of aligned latent pulsars can be re-ignited during certain brief transient events (e.g., a modest flux of soft X-rays passing through their outer magnetospheres). Their temporary emissions would then resemble those of the Vela family except that the alignment of the pulsar dipole would greatly diminish strong modulation at the pulsar spin period. A possible 10^5 population is not inconsistent with the minimum number of distinct GRB sources of §II, item 6. Many GRB spectra resemble that of Vela [26]. A key feature of a Vela-like model is that the main source for L_γ is the outer manetosphere, far enough from the stellar surface that reprocessing of γ-rays into X-rays by stellar surface γ-ray absorption does not give $L_x/L_\gamma > 3 \times 10^{-2}$. A standard candle for energetic radiation up to 1 MeV of $L_{x,\gamma} \sim 10^{36}$ erg s^{-1} suggested by Eq. (4) is compared to observed GRB time integrated fluences in Fig. 7, where the two diagonals give the fluence of 10^{36} erg s^{-1} sources at 10^2 and 10^3 lt-yr distances [18]. These data do not conflict with the possibility that most GRB's are temporarily reignited Velas.

Among many claimed but controversial GRB spectral features is an occasional (10^{-1} of GRB's) bump at photon energies near 450 KeV which contains over 10^{-2} of $L_{\gamma,x}$ [23]. If this feature is interpreted as a 510 KeV γ-ray from $e^+ + e^- \rightarrow \gamma + \gamma$ gravitationally red-shifted on escaping from the neutron star surface, a 10^{36} erg s^{-1} GRB source must put about 10^{40} $e^\pm$ per second onto the stellar surface. The crab pulsar produces $e^\pm$ at a rate $\sim 10^{39}$ s^{-1} mainly from corssed beam $\gamma + \gamma \rightarrow e^- + e^+$ in its outer magnetosphere [15]. A similar estimate for reignited latent Vela-like pulsars with $L_\gamma \sim 10^{36}$ erg s^{-1} give $\dot{N}_\pm \sim 10^{40}$ s^{-1}. Those $e^\pm$ pairs which are created moving out of the magnetosphere along open field lines will escape. A comparable flux will be created moving inward along converging field lines. Most of these may be reflected and also escape. Some can synchrotron radiate rapidly enough to avoid such reflection and reach the neutron star surface where $e^\pm$ will annihilate with emission of a γ-ray pair. An unknown fraction would be created on closed field lines on which all e^+ would suffer the same fate. If, in either of the above ways, a substantial fraction of $\dot{N}_\pm$ reach the stellar surface, red-shifted e^+ annihilation γ-rays with the claimed observed intensity would be emitted.

The first observations of GRB's were very quickly followed by a large number of suggested models in which neutron stars were the sources. Examples included neutron stars in binaries with occasional nuclear explosions of accreted material or accretion instabilities. Models with solitary neutron star sources invoke cometary or asteroidal impacts, internal structure readjustments, surface magnetic field reconnection and flares, and explosions from interstellar matter accretion [11]. A scaled down version of any of these could serve simply as a "match" which injects enough soft X-rays or plasma into the partially charge depleted outer magnetosphere of a latent γ-ray pulsar to ignite, temporarily, outer magnetosphere production and the γ-ray/X-ray emission associated with it. The match need supply only a modest power, just enough to support closing of the outer magnetospheric current circuit where it was opened by a charge deficient void, and allow the former pulsar to radiate again at nearly its full spin-down power.

Quasiperiodic surface explosions of accreted interstellar matter may be the most plausible match candidate for igniting a solitary latent γ-ray pulsar. That accretion mass flow down onto the stellar polar cap however, should not continually spin-down the neutron star prior to the transient explosion. It has been estimated that a polar cap X-ray burst from runaway nuclear burning of accreted interstellar matter may be repeated at intervals of order 10^2 years [24]. The relatively weak match with $L_x \sim 10^{34}$ erg s^{-1} could stay lit up for 10^2 s and supply enough soft X-rays to keep a latent outer gap active with an $L_\gamma \sim 10^2 L_x$.

IX. EVOLUTION IN A MILLISECOND PULSAR FAMILY

The magnetospheres of low magnetic field millisecond period pulsars are qualitatively similar to the outer magnetosphere of Vela. The maximum potential drop from charge depletion which can be realized along $\vec{B}$, the maximum current flow, the maximum luminosity, and the position of the computed spectra break in the γ-ray/X-ray spectrum are comparable for Vela and the millisecond pulsar PSR 1937. A millisecond γ-ray pulsar family may also go through an evolution which passes through Cos B sources and ends as GRB sources. However, with the same spin-down power as a Vela-like pulsar the millisecond pulsar will stretch the duration of the Cos B evolutionary phase by the ratio of its spin energy to that of Vela, $\sim 4 \times 10^3$. Therefore the formation rate of millisecond pulsars need be only 3×10^{-4} that of Vela-like pulsars for them to be a significant part of, and perhaps even most of, the Cos B sources. Such a birthrate does not conflict with the observed population of millisecond radiopulsars. If millisecond pulsars are almost 10^{-1} of all radiopulsars [25] and each survives as a radiopulsar for $10^8 - 10^{10}$ years, then they could constitute $10^{-2} - 10^{-4}$ of all neutron stars. If their aligned latent millisecond pulsar cousins are similarly abundant they would have a birthrate $10^{-2} - 10^{-4}$ that of Vela-like pulsars and evolve into $10^4 - 10^6$ potential GRB sources.

X. ACCELERATORS IN ACCRETING NEUTRON STAR SYSTEMS [27,28]

For the accreting nutron stars in binaries cited in §II.5 as TeV/PeV γ-ray sources, neutron spin periods (1.2 s for Her X-1; 280 s for Vela X-1) are far too small to support significant particle accelerators if those same stars were isolated. Several different kinds of accelerators have been suggested which involve the special accretion environment around the star, e.g., magnetohydrodynamic turbulence below a standing accretion shock above the star or after a similar shock in a radiation driven jet from the neutron star polar cap [29–31]. We shall emphasize here, however, only accretors that are a consequence of the dynamo formed by the spinning star, the rotating Keplerian accretion disk around it, and the magnetic field which links them. These can be of two sorts, a "static" accelerator, possible only when the whole inner accretion disk rotates more rapidly than the star (and radial flow is neglected) and a "dynamic" accelerator which would replace it if this condition is not met or if sufficient current flows through an otherwise static accelerator.

If the accretion disk were modelled as a thin rigid conducting disk whose angular rotation rate ($\vec{\Omega}_d$) everywhere equaled the spin vector of the neutron star ($\vec{\Omega}_*$) which it surrounds, a corotating charge separated magnetosphere forms between the star and the disk as shown in Fig. 8a. The charge distribution and "null surface" $\vec{\Omega}_* \cdot \vec{B} = 0$ are just those of §IV. (More precisely, $\vec{E} \cdot \vec{B} = 0$ is replaced by $\vec{F} \cdot \vec{B} = 0$ where

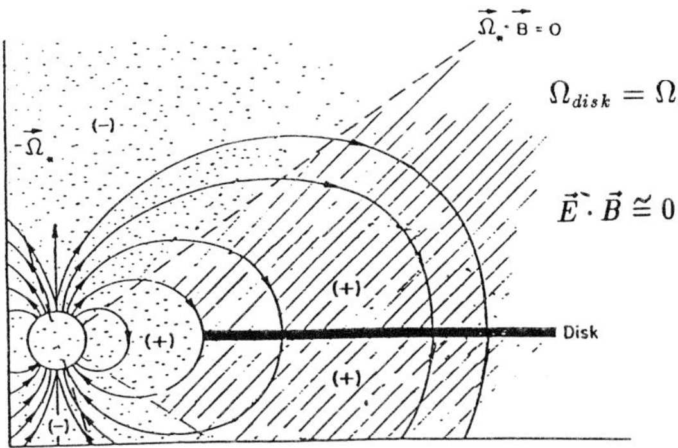

Figure 8a. The charge distribution around a spinning neutron star and an aligned corotating rigid conducting disk.

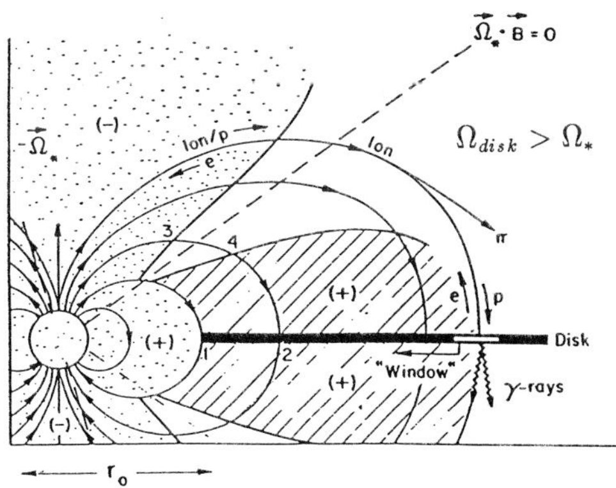

Figure 8b. Corotating magnetospheres around a spinning neutron star (spin Ω^*) with an aligned dipole and a more rapidly spinning conducting disk. The dotted region corotates with the star. In the hatched region, each charged particle corotates with that part of the disk to which it is attached by a magnetic field line. Between the star's magnetosphere and the disk there is an empty conelike "gap" around the "null-surface" $\vec{\Omega}_* \cdot \vec{B} = 0$. Where there is magnetosphere plasma, it is completely charge separated. Within the dotted region above the null surface, only negative charges remain. Below it are only positive charges.

$\vec{F}$ includes gravitational and centrifugal forces as well as the electric one.) When Ω_d varies radially and $\Omega_d > \Omega_*$, an empty "gap" in which $\vec{E} \cdot \vec{B} \neq 0$ forms around the null surface as shown in Fig. 8b. The potential across this gap is

$$\Delta V \cong r_0^2(\Omega_* - \Omega_d)B\, r_0/c \tag{5}$$

with r_0 the radius of the inner edge of the disk or, equivalently, the outer edge of the corotating magnetosphere around the star (c.f., ref. [32] for a different kind of disk dynamo model with a qualitatively similar ΔV and ref. [31] for related criticism.)

For Her X-1 $\Delta V \sim 5 \cdot 10^{14}$ Volts; for Vela X-1 $\Delta V \sim 10^{14}$ Volts. (In the case of Cyg X-3 where the system parameters are much more uncertain $\Delta V \lesssim 10^{17}$ Volts.) These ΔV are sufficient for accelerators which generate TeV γ-rays. For PeV γ-rays, however, a nucleon must be accelerated to of order 10^{16} eV. (Radiation reaction keeps e^-/e^+ energies from exceeding 10^{14} ev.) The nucleon could create π^0 in the disk and the decay γ-rays can often escape before pair creation.

The maximum current ($J_{\max}$) which could flow through the gap accelerator (before the local magnetic field from that current exceeds the initial local B and pushes apart the accelerator) gives a maximum power

$$\Delta V J_{\max} \sim \frac{GM\dot{M}}{r_0} \sim \frac{R_*}{r_0} L_x . \tag{6}$$

Here M is the neutron star mass, $\dot{M}$ its mass accretion rate, R_* the star radius, and L_x its X-ray luminosity powered by the gravitational energy loss of the accreted mass. The ratio $R_*/r_0 \sim 10^{-2}$ for typical X-ray pulsars in accreting binaries such as Her X-1 and Cen X-3. It may be smaller for Vela X-1 and much larger for Cyg X-3 if it is a weak magnetic field millisecond pulsar with $r_0 \sim R_*$.

The maximum power of Eq. (6) is compared below to some time averaged γ-ray luminosities (erg s^{-1}) from various X-ray binaries:

	L_x [erg/s]	$\bar{L}_\gamma$(TeV)	$\bar{L}_\gamma(PeV)$
Her X-1	$2 \cdot 10^{37}$	$3 \cdot 10^{35}$	$2 \cdot 10^{36}$
Vela X-1	$2 \cdot 10^{36}$	10^{34}	$2 \cdot 10^{34}$
Cen X-3	10^{37}	10^{35}	...
Cyg X-3	$0.2 - 2 \cdot 10^{38}$	$2 \cdot 10^{36}$	$6 \cdot 10^{36}$

With the above R_*/r_0 the static gap accelerator maximum power appears sufficient to be compatible with $\bar{L}_\gamma$, especially if there is some γ-ray beaming.

When $\Omega_d < \Omega_*$ no static gap is possible. A similar situation results if pair production within the gap from the mechanisms of §IV (a–e) leads to $\dot{J}(e^\pm) \rightarrow rBc(\Omega_d - \Omega_*)$. Then

$$\Delta V \sim \mathcal{L}\dot{J} \tag{7}$$

with $\mathcal{L} \sim r_0 c^{-2}$ the inductance of the current circuit between the star and the disk. The power from the dynamo is mainly diverted into increasing magnetic field energy through building up a large azimuthal magnetic field (B_φ). At that radius where the Keplerian disk has $\Omega_d(r) = \Omega_*$, B_φ reverses sign: the disk-star dynamo power pumps up adjacent oppositely directed toroidal B_φ in the magnetosphere. This may lead to reconnection above and below the disk when $B_\varphi \sim B$. (This is in addition to B_φ reconnection through the disk which involves a much greater plasma density and much smaller energy per particle [33].) The time averaged power is again given by Eq. (6) but instead of steadily accelerating particles in a static gap accelerator, the energy $\iiint B_\varphi^2/8\pi \, d^3r$ is (erratically) put into particle acceleration in the much shorter timescale $r_0 c^{-1} \sim 10^{-2}$ s. The number of particles accelerated per unit time is smaller and, instead of Eq. (5),

$$\Delta V \sim B_0 r \sim 10^{17} \text{ Volts.} \tag{8}$$

This could be sufficient to account for PeV as well as TeV sources.

XI. WHAT ARE TeV/PeV "GAMMA-RAYS"?

The main argument for interpreting the initiator of the detected atmospheric showers as γ-rays is the preservation of source period and direction in the received signal after flight times greatly exceeding 10^4 years. The emitted particles should, therefore, be neutral, almost massless relative to their energies, stable enough to survive the flight, and able to penetrate through more than 10^{-2} g cm^{-2} of interstellar matter without degradation. Among presently known particles only γ-rays and neutrinos are possibilities, but at 10^{12} eV (and probably even above 10^{15} eV) neutrinos would not cause the $e^\pm$ air showers in the atmosphere. Very disquieting, however, is the fact that the claimed signals are not μ-meson poor at 10^{15} eV as had been expected for γ-ray initiated showers. In addition, that discrimination against cosmic ray proton induced showers which worked so well in increasing the signal to background ratio in the Atmospheric Cerenkov Technique detection of 10^{12} eV γ-rays from the Crab nebula has not been similarly successful for Her X-1 observations. Thus instead of corroboration of certain X-ray pulsars in binaries as VHE and UHE γ-ray sources we have counterindications [1].

Three kinds of explanation involving increased $\pi \to \mu$ in PeV "γ-ray" showers have been suggested:

1. A large rise in the $\gamma + p \to \pi$ cross section from the measured 120 μb at $E_\gamma \sim 0.2$ TeV ($E_{cm} \sim 20$ GeV) to several mb at $E_\gamma \sim 10^{15}$ eV ($E_{cm} \sim 1.4$ Tev) [34–37].
2. Strong interactions of neutrinos above 10^{15} eV ($E_{cm} > (\sqrt{2}G_F)^{-1/2} \sim 0.25$ TeV) [38–40].
3. New light neutral quasistable particles (e.g., color singlet "cygnets") with parameters adjusted to give enough μ-mesons in the atmosphere, to avoid detection so far in laboratory beam dump experiments, and even to penertrate deeply enough to accomodate some of the reported underground fluxes from Soudan and Nusex which have been associated with Cyg X-3. (These are too high to be from incident particles with canonical hadronic or electromagnetic interactions) [41–45].

All of these proposals fail to explain the similarity of Her X-1 TeV γ-ray showers to those from collisions of background hadrons. Here laboratory experiments seem to exclude anomalous $\gamma + p$ or $\nu + p$ interactions. A hypothetical cygnet would need a mass $\lesssim 2$ MeV$/c^2$ just to preserve the 1.2 s pulse timing.

Ideally, one wishes an unambiguous TeV γ-ray detector which could measure and confirm incident γ-rays by directly observing the $e^\pm$ showers they originate. However, among those relatively few cases in which the $e^\pm$ shower of a TeV cosmic γ-ray has been followed there are examples which may make one uncomfortable about assuming that everything is understood clearly. In the 1955 Rochester Conference [46] Schein showed details of a very strange $\gamma - e^\pm$ shower. Very similar ones were reported by the Turin cosmic ray group, one of which is shown in Fig. 9. A VHE shower initiated by a cosmic ray particle was observed in a photographic emulsion carried to 80,000 ft. in a balloon. The mean distance for VHE Bremsstrahlung or pair production (λ) is about 3.7 cm. The incident shower develops in a very peculiar way. No charged particle entered the emulsion and no $e^\pm$ formation was observed for almost one radiation length, after which two $e^\pm$ pairs materialized. After another 2.7 cm gap an additional 21 $e^\pm$ pairs are created within only 0.54 λ. Estimated $e^\pm$ energies do not decline with distance down the shower but these energies may be underestimated because of multiple scattering contributions to pair opening angles. The minimum total observed energy in the shower is 0.3 TeV.

The original Schein event [48] begins with a single pair, followed by the materialization of 18 more pairs within the first radiation length. The minimum total shower energy is about 0.1 TeV. All pairs point toward a common origin 2.5 cm away in the Al cover surrounding the emulsion. (The distribution of pair origins around the shower axis is highly ellipsoidal instead of spherical indicating that the originator of this shower is very unlikely to be a single photon.)

These several events and two additional anomalous ones reported by Kaplon [46] seem to be initiated by large groups of almost parallel multi-GeV γ-rays rather than one or even a very few TeV one's. However, it is difficult to discount the remote possibility of unusual statistical fluctuations in the canonical development of a shower initiated solely by some π° γ-decays. A paper at the present conference on direct observation in an electromagnetic shower detector in an emulsion chamber gives 2 of the 29 reported cascades with anomalously large initial multipliers [49]. All of this may hint that we should not yet be completely comfortable about the origin and details of all cosmic TeV $e^{\pm}$ showers. Meanwhile, the putative TeV γ-rays from Her X-1 remain a puzzle.

It is a pleasure to thank Drs. K. S. Cheng, A. Szentgyorgyi and P. C. Rowson, and Prof. P. J. Ellis for helpful conversations.

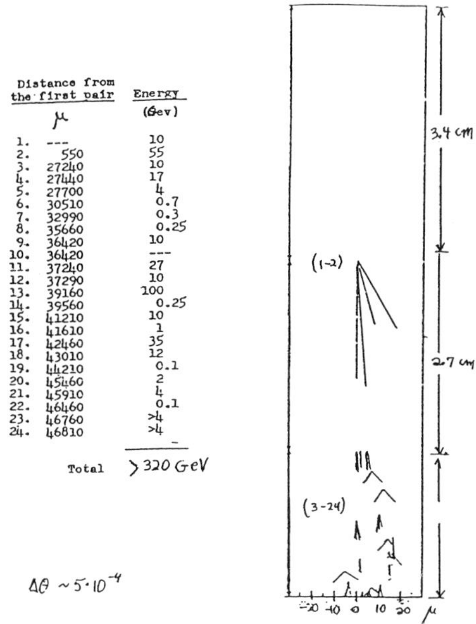

	Distance from the first pair μ	Energy (Gev)
1.	---	10
2.	550	55
3.	27240	10
4.	27440	17
5.	27700	4
6.	30510	0.7
7.	32990	0.3
8.	35660	0.25
9.	36420	10
10.	36420	---
11.	37240	27
12.	37290	10
13.	39160	100
14.	39560	0.25
15.	41210	10
16.	41610	1
17.	42460	35
18.	43010	12
19.	44210	0.1
20.	45460	2
21.	45910	4
22.	46460	0.1
23.	46760	>4
24.	46810	>4
	Total	> 320 GeV

Figure 9. A Schein type shower reported by the Turin cosmic ray group [47].

REFERENCES

[1] J. Cronin, these Proceedings. References for TeV/PeV observations are covered there and are not given in detail in the present paper.

[2] T. Weekes, in *Proc. of 14th Conference on Relative Astrophysics*, ed. E. Feneyes (World Scientific, Singapore, 1989) and references therein.

[3] R. Protheroes, in *Proceedings of 20th International Cosmic Ray Conference* (Moscow, 1987) and references therein.

[4] M. Ruderman in *Trends in Theoretical Physics*, Vol. 1, P.J. Ellis and Y.C. Tang eds. (Addison Wesley, Reading MA 1989). Sections II-VIII of this paper are mostly taken from this volume and reference 16.

[5] T. Weekes, M. Cawley, D. Fegan, K. Gibbs, A. Hillas, P. Kowk, R. Lamb, D. Lewis, D. Macomb, N. Porter, P. Reynolds and G. Vacanti, *Ap.J.* (1989), in press.

[6] M. Levinthal, in *13th Texas Symposium on Relativistic Astrophysics*, ed. M. Ulmer (World Scientific, Singapore, 1987) p. 382 and references therein.

[7] J. McClintok and M. Levinthal, *Ap.J.* (1989) in press.

[8] J. Dowthwaite, A. Harrison, I. Kirkman, H. McCrae, T. McComb, K. Oxford, K. Turner and M. Walmsley, *Ap.J.*, **286**, (1984), L35.

[9] G. Bignami, in *HIgh Energy Phenomena around Collapsed Stars*, ed. F. Pacini, NATO ASI; Cargese 1985 (Reidel, 1987).

[10] M. Ruderman, in *HIgh Energy Phenomena around Collapsed Stars*, ed. F. Pacini (Reidel, Dordrecht, 1987) p. 145 and references therein.

[11] *Gamma Ray Bursts*, ed. E. Liang and V. Petrosian (AIP Conf. Proc. **141**) and references therein.

[12] J. Middleditch, C. Pennypacker, and M. Burns, *Ap.J.*, **315**, (1987) 142 and *Nature*, **313**, (1985) 659.

[13] K. Cheng, C. Ho, and M. Ruderman, *Ap.J.*, **300**, (1986) 495.

[14] K. Cheng, C. Ho, and M. Ruderman, *Ap.J.*, **300**, (1986) 522.

[15] C. Ho, *Ap.J.*, (1989) in press.

[16] M. Ruderman, in *Timing Neutron Stars in Proceedings of NATO Advanced Study Institute*, Cesme, 1988, ed. H. Ogelman and E. Van den Heuvel (North Holland, Amsterdam, 1989).

[17] M. Ruderman, in *13th Texas Symposium on Relativistic Astrophysics, 1986*, ed. M. Ulmer (World Scientific, Singapore, 1987) p. 448.

[18] M. Ruderman and K. Cheng, *Ap.J.*, **355**, (1988) 306.

[19] L. Davis and M. Goldstein, *Ap.J.*, **5**, (1970) 21.

[20] F.C. Michel and H. Goldwire, *Ap.J.*, **5**, (1970) L21.

[21] P. Goldreich, *Ap.J.*, **160**, (1970) L11.

[23] E. Mazets and S. Golensetskii, *Astrophys. and Space Phys. Rev.*, **1**, (1981) 205.

[24] R. Taam, in *13th Texas Symposium on Relativistic Astrophysics, 1986*, ed. M. Ulmer (World Scientific, Singapore 1987) p. 546.

[25] J. Taylor, in *13th Texas Symposium on Relativistic Astrophysics, 1986*, ed. M. Ulmer (World Scientific, Singapore 1987) p. 546.

[26] G. Share, S. Matz, D. Messina, P. Nolan, E. Chupp, D. Forrest, and J. Cooper (1986) preprint.

[27] K.S. Cheng and M. Ruderman, *Ap.J. (Letters)*, **337**, (1989) L77.

[28] K.S. Cheng and M. Ruderman, to be published.

[29] D. Eichler and T. Vestrand, *Nature*, **307**, (1984) 767.

[30] D. Kazanas and D. Ellison, *Nature* , **319**, (1986) 380.

[31] P. Király and P. Meszaros, *Ap.J.*, **333**, (1988) 719.

[32] E. Chanmugam, G. and K. Brecher, *Nature*, **313**, (1985) 767.

[33] P. Ghosh and F. Lamb, *Ap.J.*, **234**, (1979) 296.
[34] W. Ochs and L. Stodolsky, *Phys Rev*, **D33**, (1986) 1247.
[35] M. Drees and F. Halzen, *Phys. Rev. Lett.*, **61**, (1988) 275.
[36] J. Rosner and L. Stodolsky, preprint (1989).
[37] M. Drees, F. Halzen, and K. Hikasa, *Phys. Rev.*, **D39**, (1989) 1310.
[38] G. Domokos, B. Elliott, S. Kovesi-Domokos and S. Mrenna (1989) preprint.
[39] G. Domokos and S. Kovesi-Domokos, *Phys. Rev.*, **D38**, (1988) 2833.
[40] G. Domokos and S. Nussinov, *Phys. Lett.*, **B 187**, (1987) 372.
[41] G. Baym, R. Jaffe, R. Kolb, L. McLerran and T. Walker, *Phys. Lett*, **B 160**, (1985) 181.
[42] V. Berezinsky, J. Ellis, and B. Ioffe, *Phys. Lett.*, **B 172**, (1986) 423.
[43] K. Ruddick, *Phys. Rev. Lett.*, **57**, (1986) 531.
[44] J. Collins and F. Olness, *Phys. Lett.*, **B 187**, (1987) 376.
[45] J. Collins, A. Kaidalov, A. Khodjamirian, and L. McLerran, *Phys. Rev.*, **D39**, (1989) 1318.
[46] *Proc. of Fifth Annual Rochester Conf.*, Jan. 1955, eds. H. Noyes, E. Hafner, E. Yekutieli and B. Raz.
[47] A. Debenedetti, C. Garelli, L. Tallone, and M. Vigone, *Il. Nuovo Cimento*, **II**, **2**, (1955) 220.
[48] M. Schein, D. Haskin, and M. Glasser, *Phys. Rev.*, **95**, (1954) 855.
[49] T. Ross, J. Iwai, J. Lord, S. Strausz, R. Wilkes and J. Woosley, these Proceedings.

DISCUSSION

R. Hofstadter, Stanford University: Years ago, I attempted to calculate the contribution of certain "colliding-beam"-like processes to account for the gamma rays, processes of the form $e^+e^- \rightarrow \gamma\gamma$ or $e^+e^- \rightarrow$ hadrons with subsequent γ production. Unfortunately, the resulting flux at the Earth's surface turns out to be a factor of 10^6 too small. Perhaps the cross sections are much larger. This is a problem I'm still thinking about.

M. Ruderman: For astrophysics, a factor of 10^6 is not too bad!

G. Yodh, University of California, Irvine: What scenario for acceleration in Cygnus X-3 could account for the recently reported observation of EeV signals by the Fly's Eye detector?

M. Ruderman: It would seem to require a non-static accelerator — for example, one in which energy is put into an azimuthal magnetic field which then reconnects or bursts out of the system when it gets to be comparable to the original local B-field.

Gauge Fields, Geometry, Strings, and Gravity

JEFFREY A. HARVEY

Jadwin Hall, Princeton University
Princeton, NJ 08544

ABSTRACT

Recent developments in and connections between gauge theory, geometry, and quantum gravity are discussed. The status of superstring theory is briefly reviewed followed by a discussion of the notion of fundamental length in string theory.

Gauge Theory and the Standard Model

The 1970's and 1980's have seen the development and triumph of gauge symmetry as the fundamental theoretical concept underlying the weak, strong, and electromagnetic interactions. This has culminated in the development of the standard model, which, as we have dramatically heard here, continues to be experimentally confirmed by increasingly precise tests.

As we try to move beyond the standard model, particularly to such ambitious areas as grand unified theories, quantum gravity, and string theory, it may be useful to remember some of the difficulties that were faced in developing the standard model. From a theoretical perspective there are three areas I would like to emphasize. They are:

(i) Technical developments, including the formulation and quantization of gauge theories, the proof of the renormalizability of gauge theory, and the formulation of lattice gauge theory. One might argue that we are still lacking a useful nonperturbative formulation of the weak interactions due to the problem of putting chiral fermions on the lattice.

(ii) Understanding the dynamics of gauge theories, particularly the possible phases of gauge theory and the ways that the symmetry can be realized. It is quite striking that the standard model exhibits such a rich spectrum of possibilities. The $U(1)$ symmetry of electromagnetism is manifest, and the fundamental particles exist as asymptotic states.[1] The $SU(2)$ symmetry of the weak interactions on the other hand is hidden as a consequence of the Higgs mechanism, and one must go to high energies to see the effects of the symmetry and produce the fundamental quanta. The $SU(3)$ symmetry of the strong interactions is even more hidden by the confinement of color and can only be inferred indirectly since the fundamental particles of the theory (quarks and gluons) do not appear as asymptotic states.

(iii) Model building. Choosing the correct gauge group, fermion representations, and symmetry breaking patterns, followed by calculation and comparison with experiment.

Of course at any one time all three of these were going on, but in describing progress in new areas it will be useful to keep these three stages in mind.

The problems with the standard model are familiar. They include the necessity of many input parameters which are not related to each other by theory and must be determined by experiment (fermion masses, mixing angles, etc.), and the presence of a symmetry breaking mechanism which is theoretically unsatisfactory and experimentally unconfirmed. On a more formal level it is disturbing that gravity is left out of this grand synthesis. There is also a strong possibility that the standard model does not exist as a fundamental quantum field theory since it involves interactions which are not asymptotically free. If this is true then the standard model can only be an approximation at low energies and must be modified in a fundamental way at some scale. Since quantum gravity introduces a fundamental length scale, the Planck scale, given by

$$l_P = \sqrt{\frac{G\hbar}{c^3}} \sim 10^{-33}\text{cm} \sim \frac{1}{10^{19}\text{GeV}}, \qquad (1)$$

it seems reasonable that the standard model must be modified at this length scale. There are theoretical reasons connected with symmetry breaking for

[1] Modulo technical problems involving infrared divergences.

expecting modifications at the TeV scale, perhaps in the form of new gauge interactions or supersymmetric particles. These possibilities have profound theoretical and experimental consequences, but presumably can be described in terms of field theory and do not require a fundamental revision of our ideas. Gravity is the most geometrical theory of nature we know. If it is to be unified with gauge theory at some scale then there should be some geometrical structure which unites the structure of gravity and gauge theory.

In many applications of gauge theories it is only necessary to perform perturbative calculations about simple background fields. These calculations do not reveal the fact that gauge theory has a beautiful geometrical structure. The topological and geometrical aspects of gauge theories become evident only when one studies non-perturbative phenomenon involving large gauge field configurations, in the same way that one has to travel large distances to appreciate the topology and geometry of the earth. Thus the study of the strong coupling QCD vacuum, or toy models of confinement such as gauge theories in 2+1 dimensions, often involve study of topologically non-trivial gauge fields. Another aspect of this is the possibility of having objects in gauge theory whose existence is predicted by topological arguments based on gauge symmetry.

Two familiar examples are magnetic monopoles and cosmic strings. It is well known in electromagnetism that we must choose a gauge to perform calculations, and that the physical answers should be independent of gauge. We usually pick a single gauge throughout all of space, but there is nothing in gauge theory that requires this. We are free to pick different gauges in different regions, as long as we can smoothly relate the different gauge choices along the boundaries of the regions. Thus we can consider a two-dimensional sphere surrounding some point in space, divide it into two hemispheres, and make different gauge choices on the two hemispheres, say

$$A_N^\mu = A_+^\mu + \frac{1}{e}\partial^\mu \Lambda_N \tag{2}$$

on the northern hemisphere and

$$A_\mu^S = A_-^\mu + \frac{1}{e}\partial^\mu \Lambda_S \tag{3}$$

on the southern hemisphere as shown below.

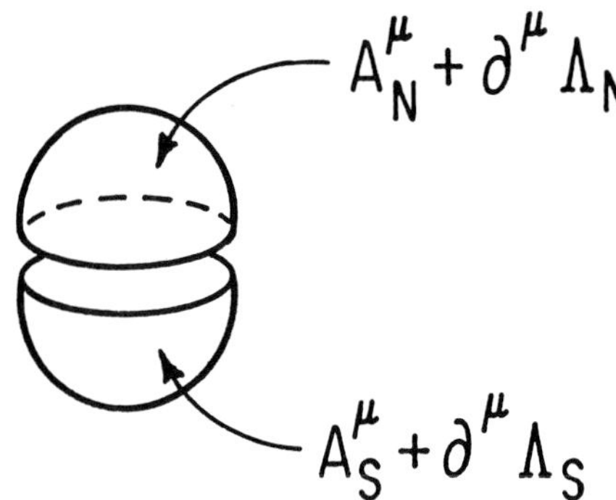

Figure 1. Patching together gauge fields to obtain a magnetic monopole.

Here A_+^μ and A_-^μ are smooth solutions of the free Maxwell equations. Along the equator we must be able to relate the two gauge choices by a gauge transformation. If we take

$$e^{i\Lambda^N} = e^{in\phi} e^{i\Lambda^S} \tag{4}$$

where ϕ is the azimuthal angle then by performing the integral

$$\int \vec{B}\cdot d\vec{S} = \int (\nabla_\times \vec{A})\cdot d\vec{S} = \frac{1}{e}\oint \partial(\Lambda^N - \Lambda^S)\cdot d\vec{l} \tag{5}$$

we find that the total magnetic flux emerging from the sphere is n times the charge of a Dirac monopole. Thus we discover the possibility of magnetic monopoles by patching together solutions of Maxwell's equations in a non-trivial way [1].

As another example consider a cosmic string. These can arise when there is a new $U(1)$ gauge interaction like QED, but spontaneously broken at some large scale so that its quanta have not yet been observed. Let the symmetry be broken by the vacuum expectation value of a complex scalar field, $\langle\phi\rangle = v$. The phase of v is arbitrary, it can be changed by a gauge transformation. Consider a situation in which the phase of v changes by 2π as we move around some fixed line, say the z axis as shown below.

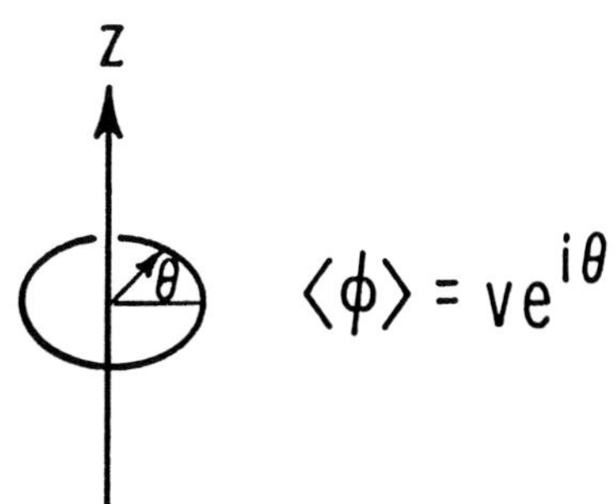

Figure 2. A cosmic string configuration.

We can still try to make the phase of v constant everywhere by a gauge transformation, but the gauge transformation will now be singular along the z axis. The singularity just corresponds to the fact that there must be an infinitely thin tube of flux running along the z axis.

In both of these examples we have field configurations which are locally trivial, but when we try to put things together globally, we find that we can do it only at the expense of a singularity with non-zero field strength. In both cases by embedding the theory into a larger theory and including massive fields one finds that the singularities are smoothed out and one is left with interesting non-singular configurations. What is important is that the possible existence of such objects can be inferred without knowing about the massive fields or details of the symmetry breaking and is closely related to the underlying gauge invariance.

From a theoretical point of view it is unfortunate that the standard model by itself does not require either of these possibilities. It allows magnetic monopoles, but to have these arise as actual solutions with definite properties requires embedding the standard model in a grand unified theory. Cosmic strings also require the addition of new interactions.

It has been realized recently that there are even deeper connections between gauge theories and geometry. In fact, gauge theories have become one of the central tools in the study of geometry in three and four dimensions. A detailed description of these connections would be beyond my competence, so my explanation of this work will be rather brief. The essence of the previous constructions was that patching together locally trivial configurations can lead to globally non-trivial configurations. In ordinary Yang-Mills theory the classical vacuum is given by zero field strength, but the classical equations allow other types of solutions and in formulating the theory on a curved space the metric on the space enters into the equations of motion. In recent developments it has been realized that one can use alternate formulations of gauge theory which carry only topological information and have no local dynamics. In these theories the vanishing of the field strength actually arises as the classical equation of motion and the quantization of the theory just quantizes the topological aspects without introducing any dynamical degrees of freedom. In addition, these theories can be formulated without the metric of the space ever occuring, so they encode purely topological information which is independent of the choice of metric.

An example of such a theory in $2+1$ dimensions is constructed by taking the action for a $U(1)$ gauge field A^μ to be (on a general curved manifold)

$$\int d^3x \epsilon^{\mu\nu\lambda} A_\mu \partial_\nu A_\lambda. \tag{6}$$

which gives the classical equation of motion $F_{\mu\nu} = 0$. Note that the action is independent of the metric since indices are contracted using only the epsilon symbol. The only way that the classical equations can have non-trivial solutions is if the manifold has some holes or handles through which we can pass imaginary flux tubes. Then, much like in the cosmic string case, we can have non-trivial solutions due to the global behavior of fields which are locally trivial. An example of such a possibility is shown below. If the manifold has a handle then by passing an imaginary magnetic flux tube through the handle we can have $\oint_C \vec{A}\cdot d\vec{l} \neq 0$ along a closed curve C even though the gauge field is locally trivial. In this way the theory captures information about the topology of the manifold it is defined on.

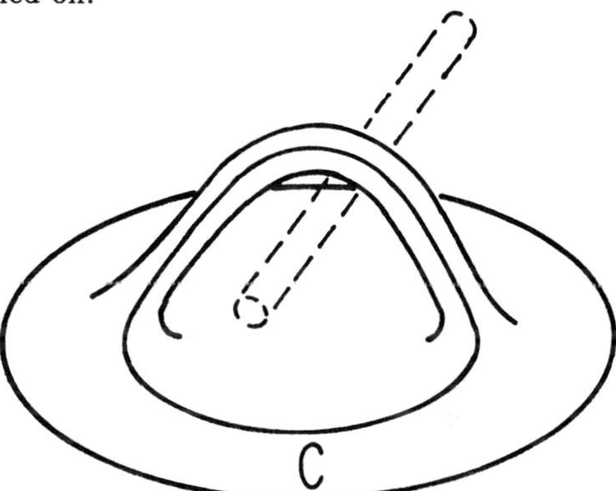

Figure 3. Inserting an imaginary flux tube yields a non-zero line integral for the vector potential around a closed loop.

Compare this with ordinary electromagnetism in $2+1$ dimensions. The action is then

$$\int d^3x \sqrt{g} g^{\mu\nu} g^{\lambda\rho} F_{\mu\lambda} F_{\nu\rho} \tag{7}$$

and clearly depends on the choice of metric g. In addition, the classical field equations now have non-trivial plane wave solutions and do not require that the field strength vanish.

Such theories and their generalizations have been used by mathematicians to prove the existence of so-called "exotic" versions of four-dimensional space and to construct new invariants of knots in three dimensions [2]. Whether there are more down to earth physical applications remains to be seen, but it is quite remarkable that gauge theories are such a powerful tool in studying geometry, and that, for reasons I have not explained, they are most powerful in four dimensions.

Quantum Gravity

The formulation of a quantum theory of gravity has long been one of the fundamental problems in theoretical physics. The successes of the standard model and our desire to probe beyond it coupled with the revival of string theory have renewed interest in this problem. Before describing some of the recent progress in quantum gravity I would like to make a few general comments.

1. It has been known for some time that there are many similarities between gravity and gauge theory with either general coordinate invariance or Lorentz invariance playing the role of the gauge symmetry depending on the precise formalism. However, there are also significant differences. For one thing, Einstein's action has no direct analog in gauge theory. Also, perturbative calculations in gravity have a much different flavor in quantum gravity than in gauge theory. In gravity we usually expand the metric about flat spacetime:

$$g_{\mu\nu} = \eta_{\mu\nu} + \kappa h_{\mu\nu} \tag{8}$$

with $\eta_{\mu\nu}$ the flat space metric and $\kappa = 1/M_P$. In gauge theories on the other hand we usually expand about zero gauge field, $A_\mu = 0$. The expansion about flat spacetime breaks general coordinate invariance (since $\eta_{\mu\nu}$ is not invariant) and in this regard is somewhat analogous to a Higgs phase in gauge theory. Quantum gravity also involves the dimensionfull coupling κ.

2. Because of the existence of a dimensionfull coupling with dimensions of inverse mass, quantum gravity is non-renormalizable. If we cut off the divergences of the theory at a scale Λ, then in the nth order of perturbation theory we expect to find amplitudes involving the dimensionless combination $(\Lambda\kappa)^n$ which indicates that the theory has an infinite number of possible divergences which cannot be renormalized by adjusting the parameters of the theory. This problem is less severe but still exists in supergravity theories. It seems to be solved only in string theory.

3. Einstein's theory of gravity is a theory of geometry. One of the conceptual problems in quantum gravity is to find a framework which incorporates and generalizes the geometrical structure in a way consistent with quantum mechanics. In particular, at short distances it is expected that spacetime itself should melt away into something else. Perhaps even the topology of spacetime is not fixed, an idea which has been popular recently under the name of wormholes.

4. A correct theory of gravity should explain why the cosmological constant is zero. String theory so far has not had much to say about this.

5. Quantum gravity is plagued with interpretational problems connected with the proper interpretation of the wave function of the universe. Worrying about these questions often makes one feel like Schrödinger's cat. It is quite possible that the framework of quantum mechanics will also have to be changed in order to incorporate quantum gravity.

Clearly quantum gravity is a difficult subject, and we have a long way to go before coming to grips with it, string theory not withstanding. I would like to discuss three recent developments which touch on different aspects of the problems of quantum gravity and the search for a unified theory. The first of these is the development of tools for studying the structure of quantum gravity in 2+1 spacetime dimensions. The second goes under the name of wormholes and is an attempt to study topology change in quantum gravity; the recent revival of interest in this subject is due to claims that such processes lead to a natural explanation for the vanishing of the cosmological constant. The third is string theory, which appears to solve the problem of the non-renormalizability of quantum gravity, but is still lacking a fundamental formulation.

Gravity in $2+1$ Dimensions

In D spacetime dimensions the graviton has

$$\frac{D(D-1)}{2} - D \tag{9}$$

physical degrees of freedom with the $D(D-1)/2$ given by the number of spatial components, g_{ij} and the $-D$ given by the gauge degrees of freedom corresponding to the D independent coordinate transformations. For $D = 4$ this gives two physical degrees of freedom corresponding to the two familiar polarizations of a graviton. In $D = 3$ this gives 0, so we conclude that in three spacetime dimensions there are no propagating gravitons. In spite of this fact it seems that quantum gravity in $2+1$ dimensions is non-renormalizable due to the dimensionfull coupling constant, as in $3+1$ dimensions. It would be rather peculiar for a theory that describes nothing to be non-renormalizable. As it turns out, the theory describes more than nothing and is not only renormalizable, but finite.

The basic observation, made earlier but utilized effectively by Witten [3], is that the analogy between gauge theory and gravity can be made quite precise in $2+1$ dimensions. In particular, the Einstein action $\int \sqrt{g}R$ is identical to an action of the topological type described earlier for a gauge theory with gauge group $SO(2,1)$, i.e. the Lorentz group in $2+1$ dimensions. Furthermore, when expanded not about flat space, but about $g_{\mu\nu} = 0$, the theory is seen to be finite. The choice of $g_{\mu\nu} = 0$ as the vacuum seems peculiar in a theory of geometry but is analogous to expanding about $A_\mu = 0$ in gauge theory. However expanding about $g_{\mu\nu} = 0$ means that there can be no real dynamics in the theory since it is not possible to define light cones or causality. As a purely gravitational theory this is fine, since the gravitons are not dynamical anyway. What the theory then describes is in effect the topology of spacetime without any local dynamics. Thus this theory allows one to calculate topology changing amplitudes in a well defined theory. The main drawback is that it is not possible to introduce dynamical matter into the theory in a consistent way.

This work emphasizes the importance of thinking about different phases or ways of realizing the basic symmetries in quantum gravity. It is tempting to think that there might two phases in quantum gravity. A short distance phase in which there is "nothing there" such as the phase of $2+1$ gravity above, and a large distance phase with a flat space metric and dynamical matter fields. A concrete realization of this idea even in $2+1$ dimensions would be an important step in our understanding of quantum gravity. There has also been recent progress in understanding the canonical formulation of gravity in $3+1$ dimensions which shows some interesting connections to $2+1$ dimensions [4].

Wormholes

The second topic I wish to mention goes under the name of wormholes. There seem to be two major motivations for studying these objects.

The first is the search for a solution to the cosmological constant problem. The most puzzling aspect of this problem is that it seems to require a connection between short distance physics (such as the vacuum fluctuations and symmetry breaking which give rise to the vacuum energy) and long distance physics (the present universe is so large compared to the Planck scale because the cosmological constant is so close to zero). We are used to the idea that one can integrate out short distance physics to find an effective theory of large distance physics, but there seems to be no conventional mechanism that can give rise to a zero cosmological constant after integrating out the short distance effects. Wormholes provide a sort of short-circuit of spacetime giving a direct connection between short distance and long distance physics.

The second is to explore in a general way the consequences of topology change in quantum gravity. There is no obvious reason why the topology of spacetime should not change in quantum gravity, and if it does it may be possible to say something about the consequences even without doing a full calculation in a consistent theory of gravity, such as some future second-quantized form of string theory.

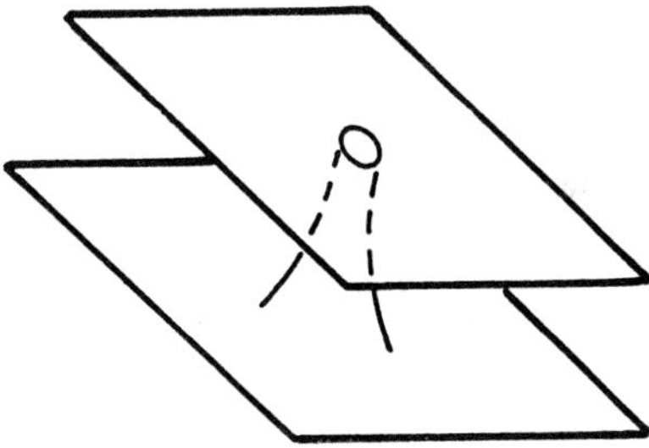

Figure 4. A wormhole connecting causually disconnected parts of spacetime.

The basic wormhole picture is shown in the figure above. The two mouths of the wormhole may connect different regions of spacetime separated by a macroscopic distance, or even different universes. This picture should not be taken too seriously since it makes it seem as if such topology change can occur in a completely smooth way. In fact, this picture only makes sense in a Euclidean spacetime with imaginary time. In ordinary spacetime such a configuration must be singular somewhere. Nonetheless, in a full theory of quantum gravity such configurations may well be allowed. At first it would seem that allowing configurations which connect macroscopically separated parts of spacetime would lead to blatant violations of causality since one part of spacetime could influence another outside its light cone by making use of the wormhole. The proponents of wormholes claim however that the only non-locality can be absorbed by saying that wormholes produce universal shifts in the fundamental coupling constants. The fact that coupling constants are the same throughout the universe is a sort of non-locality that we do not find disturbing and it is certainly not in conflict with experiment.

The claim that is more controversial is that the shift of coupling constants shifts one particular constant, the cosmological constant, to zero. Whether or not this is the case is still the subject of much debate. There seem to be two main problems. The first is that all the calculations are carried out in either non-existent or over simplified theories. The second is that there is no clear framework for interpreting the calculations, so that different people will draw different conclusions from the same calculation. Nonetheless, this has been the most promising attempt to date to solve the cosmological constant problem, and many of the issues raised will continue to motivate further work.

String Theory

It has been five years since the revival of string theory as a fundamental theory of all interactions. String theory not only seems to give a perturbatively consistent theory of quantum gravity, free of divergences, but also includes gauge interactions and chiral fermions in a way which is compatible with attempts to unify the gauge interactions of the standard model into a single gauge group. It predicts the number of spacetime dimensions (ten) and allows only a small number of different gauge groups ($E_8 \otimes E_8$, $SO(32)$, or $SO(16) \otimes SO(16)$ for tachyon free theories in ten dimensions). It seemed at first that there would be only a small number of ways of compactifying the six extra dimensions and breaking the gauge symmetry and that the chances were good that one of these might be close to the standard model. Unfortunately the theories were less restrictive than they seemed and there are now many compactification schemes as well as ways of of formulating string theories that make no explicit mention of extra dimensions.

To discuss the present situation in string theory it might be useful to compare our progress with the development of the standard model I outlined earlier. The first area was technical developments, learning how to formulate and quantize the theory. In string theory we have learned after much hard work to first quantize string theory about a given classical configuration and we have learned how to formulate a perturbative expansion for string interactions. The construction of explicit classical string solutions and the working out of perturbation theory about these solutions has required the development of many extremely technical tools. But we still lack a proper formulation of the theory that is background independent and does not rely on perturbation theory. In QCD we could discuss high energy scattering with just perturbation theory but we would not get far in understanding the structure of the vacuum and the spectrum of hadrons. Similarly we are probably missing many important phenomenon in string theory because of our lack of a proper formulation of the theory.

The second area had to do with understanding the possible phases or ways of realizing the symmetries of the theory. In string theory we are completely ignorant in this regard. It is assumed implicitly that the massless modes of the string correspond directly to the fundamental particles we observe, but there is really no justification for this assumption. We do not know whether there are other phases of string theory, or whether string interactions are weak or strong.

The third area was model building. In principle this should be trivial in string theory since there are only a few consistent string theories and they have no adjustable parameters. The $E_8 \otimes E_8$ heterotic string seems the most promising, all we have to do is understand what spectrum of particles it predicts. As I mentioned earlier, life is not so simple. We have to get down to four dimensions and that requires choosing a form for the six internal dimensions. If we think that string theory is weakly coupled (and we have no reason to think so other than lack of anything better to do), then the internal space should correspond to a classical solution of string theory. At present we only know how to look for solutions that preserve spacetime supersymmetry, so we also assume that supersymmetry is a good symmetry at energies small compared to the Planck scale. It still turns out that there are myriad multi-parameter solutions which satisfy this criterion and they yield an almost limitless number of possible gauge groups, fermion representations, and Higgs fields. Faced with this situation most people have either tried to select a small number of these solutions by requiring consistency with the standard model, e.g. three generations, acceptable proton decay rates, etc. or have abandoned the model building enterprise until we have a better understanding of string theory. It is still possible that the model builders will get lucky and find one solution which happens to be close to the standard model and is picked out by string theory for some as of yet unknown reason. It seems equally likely that there are many solutions satisfying various criterion which will make different "predictions" for unobserved phenomenon. String inspired predictions should probably be regarded with suspicion until we have a better understanding of how the correct vacuum is selected.

It should be clear that there are some fundamental gaps in our knowledge of string theory that are keeping us from really testing whether the theory makes correct predictions or not. Also, while there has been great technical progress, if string theory is the correct framework for a truly unified theory, there should be some beautiful and simple physical concepts which capture its basic principles. In the remainder of this talk I want to discuss some hints about what these concepts might be that can be explained in simple physical terms. They all have to do with the notion of a fundamental length in string theory.

Of course any theory of quantum gravity has a fundamental length scale in it given by the Planck length. There is no a priori reason however why we shouldn't have to worry about new phenomenon at distances much less than the Planck scale, no matter how remote they might seem. If string theory is fundamental it should not only contain a fundamental length, it should explain why we don't need to concern ourselves with distances less than that fundamental length. In quantum mechanics the introduction of Planck's constant allows us to postulate the commutation relation $[x,p] = i\hbar$ which implies the uncertainty relation $\Delta x \Delta p \geq \hbar/2$. Thus the notion of a cell in phase space with size smaller than $\hbar$ is not needed in our description of physics. In string theory we introduce a new constant of nature with dimensions of $(\text{length})^2$, namely α' or the inverse string tension (related to the Planck mass by $\alpha' = 16\pi G/g^2 = 16\pi/g^2 M_{Pl}^2$ with g the string coupling constant). We might think that in some sense we could now have $[x,x] = \alpha'$ so that distances below a certain scale are also an unnecessary concept. Of course this equation doesn't make any sense as it stands, but nonetheless there is evidence in string theory that distances less than $\sqrt{\alpha'}$ are unnecessary in our description of physics. We should look for an equation in string theory that is a correct version of the wrong equation $[x,x] = \alpha'$.

I will give three arguments in string theory that point towards a conclusion like this. None is very convincing. Again, we need to find one good argument to replace a variety of bad ones. The first argument requires a complicated formula to explain in detail, but the basic observation is quite simple [5]. It is tempting to think of a string as an infinite collection of particles corresponding to the different modes of the string. We know that the vacuum energy associated with a single particle is ultraviolet divergent, it takes the form

$$\frac{1}{2}\sum_k \hbar\omega_k = \int \frac{d^3k}{(2\pi)^3}\sqrt{k^2+m^2} = \int_0^\infty ds \frac{e^{-sm^2}}{s^3} \tag{10}$$

in three spatial dimensions. In the last form it is written as a integral over the proper time that the particle propagates. From this point of view the divergence arises when the particle propagates for very short times. In more dimensions this is more divergent, and if we sum this for an infinite number of particles corresponding to the modes of the string it is infinitely infinite. But in string theory something rather wonderful happens. In summing over all the intermediate states of the string there is a symmetry called modular symmetry that relates strings propagating for a very short proper time to those propagating for a very long proper time. This symmetry says that it is actually redundant and wrong to sum over configurations related by modular symmetry. So in the correct sum in string theory you never have to include configurations that propagate for very short times. This turns the infinitely infinite particle expression into a perfectly finite string expression.

The second argument has to do with a sort of string uncertainty principle. In quantum mechanics we know that we need energies $E \sim 1/\Delta x$ (setting $c = \hbar = 1$) in order to probe distances of order Δx. It has been realized since the early days of quantum gravity that putting too much energy in too small a space will lead to large curvature and thus may lead to an inability to probe too small a distance. On would expect a relation like

$$\Delta x \sim \frac{1}{E} + \frac{E}{M_{Pl}^2} \tag{11}$$

which would express the minimum distance we could probe with energy E consistent with quantum mechanics and without running into problems of strong gravity. This equation does not indicate that distances less than the Planck length are not possible, just that we are ignorant about how to deal with them. On the other hand, in string theory there seems to be a similar equation but with a different interpretation.

Since strings are extended objects with an an energy associated with increasing their length, an increase in the energy of a string can go either into making it move faster or into making it longer. It is plausible then that if we try to probe very short distances using strings as a probe that at some energy scale the size of the probe will be greater than the distance we are trying to probe. This would indicate a fundamental inability to probe distances below a certain scale. This idea is backed up by semi-classical calculations of high energy scattering in string theory[6]. By looking at the semi-classical trajectories the authors of [6]find a string uncertainty relation of the form

$$\Delta x \sim \frac{1}{E} + \frac{E}{g^2 M_{Pl}^2} \tag{12}$$

where g is the string coupling constant. If g is

sufficiently small than this indicates an inability to probe below a certain scale but without having to invoke problems of strong gravity. The limit is occuring in a regime where gravity is still weak. Thus although the two equations are very similar, there seems to be a fundamental difference in what they are telling us, with the string expression indicating that distances less than gM_{Pl} are a superfluous concept since even in principle they cannot be measured.

The third argument is connected with duality in string compactification. In higher dimensional field theories such as Kaluza-Klein theory we assume that spacetime has the form

$$M^{3,1} \otimes K \tag{13}$$

with $M^{3,1}$ four-dimensional Minkowski space and K some internal compactified space. In field theory if we let $R(K)$ be some sort of average radius of K then as we take $R(K) \rightarrow 0$ all the modes of the fields with momentum $p \sim \hbar/R(K)$ become infinitely massive and we are left with an effective four-dimensional theory.

In string theory this argument must break down, otherwise we could define string theory directly in four dimensions without any additional degrees of freedom. In many cases we can see precisely how the argument breaks down. In string theory in addition to momentum modes with energies of order $1/R(K)$ there are also winding modes with energies of order $R(K)$ as illustrated below.

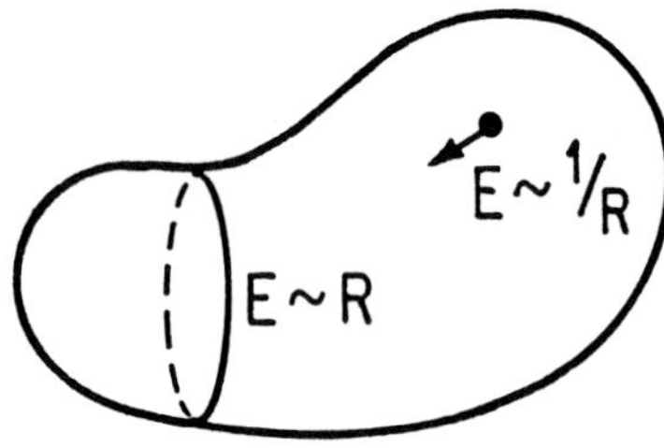

Figure 5. Momentum and winding modes of a string on a compactified space of average radius R.

As $R(K) \rightarrow 0$ these winding modes become very light, and we must include them in the low-energy effective theory. In many cases that have been looked at there is in fact an exact symmetry which relates two spaces K and K' with $R(K)$ related to $\alpha'/R(K')$. This is an exact string symmetry which says that the spectra and interactions on the two spaces K and K' are identical. If such a symmetry existed for all string compactifications (and this seems to be the case) then it would be unnecessary to ever talk about spaces with $R(K) \leq \sqrt{\alpha'}$ since they would always be equivalent to some space K' with $R(K') \geq \sqrt{\alpha'}$.

In all these cases the notion of distances or times less than some fundamental scale seem to be redundant in our description of string theory. There are other indications as well of such a phenomenon which I have not discussed having to to with the existence of other limiting quantities in string theory such as temperature. We are familiar with redundancy in gauge theories where the gauge degrees of freedom of the field are redundant but useful in formulating the theory. In string theory it seems that there ought to be some gauge-like symmetry which relates large and small distances, but again it is not clear exactly what is meant by this. It would be an important step forward to be able to derive these sort of arguments from a general principle rather than by specific calculations. This seems to be one more indication that our understanding of string theory is lacking some crucial ingredients.

In conclusion let me reiterate that although quantum gravity and string theory present very difficult problems, there has been progress and there are interesting new ideas in the air. Still, the recent revolution in string theory has been more technical than conceptual. The real revolution should change the way we think about space and time, may change our ideas of what we can hope to predict, and may even force us to alter the basic structure of quantum mechanics.

References

[1] T.T Wu and C.N. Yang, *Phys. Rev.* **D12** (1975) 3845.
[2] See E. Witten, *Comm. Math. Phys.* **121** (1989) 351 and references therein.
[3] E. Witten,*Nucl.Phys* **B311** (1989) 46.
[4] A. Ashtekar, *Phys.Rev.Lett.***57** (1986) 2244.
[5] J. Polchinski,*Comm.Math.Phys.***104** (1986) 37 and references therein.
[6] D. Gross and P. Mende, *Nucl.Phys.* **B103** (1988) 407.

DISCUSSION

G. Altarelli, CERN: You have indicated that the group $E_8 \otimes E_8$ is the most likely possibility for the gauge symmetry of string theories. However, at some stage we were told that, with the advent of four-dimensional string theories, a much larger class of gauge groups is possible. Can you clarify this point?

J. Harvey: To be more specific, I should have said that models based on the heterotic string (as compared to the type II closed superstring or type I open superstring) seem the most promising. There have been great advances in our ability to formulate string theories where the extra dimensions are replaced by some explicit conformal field theory. This can lead to a greater variety of gauge groups than was originally thought to be the case, and many of these seem at least as promising as the original suggestions based on an E_6 subgroup of one of the E_8 factors. These constructions raise the interesting question of whether or not there is always a generalization of six-dimensional geometry underlying the construction. To my knowledge, all the constructions are equivalent to particular Calabi-Yau spaces, or to orbifolds of tori or Calabi-Yau spaces (including asymmetric orbifolds).

M. Peskin, SLAC: I'd like to highlight two of the issues you raised as pointing to a conceptual problem for physics — that we may not in principle be able to predict the fundamental physical constants. We are familiar with the situation in the Standard Model in which we cannot predict the electron mass (or Yukawa coupling), but we hope that this can be remedied by a deeper theory. However, your talk raised two issues that make this difficult. First, you discussed the presence in string theory of many possible vacuum states that are almost exactly degenerate. This is already a problem in supersymmetric grand unified theories, but in string theory it is enormously multiplied. The laws of physics do not seem to choose one of these vacuum states. Second, you mentioned that wormholes, while they may shift the cosmological constant to zero, also introduce arbitrary shifts in all the other fundamental constants. It seems to me that we cannot recover it except through a very deep conceptual revolution.

J. Harvey: I agree with you that a conceptual revolution may well be needed if we are going to be able to make certain fundamental predictions, but I do not think the situation is quite as bleak as you have made it sound. The many degenerate vacua of string theory are degenerate only at the level of perturbation theory. In many cases we know from field theory arguments that the vacuum should be unstable, and will hopefully lead to supersymmetry breaking. The problem in string theory is that we do not know how to deal with supersymmetry breaking in a reliable way. If this problem is solved, at least most of the continuous degeneracies should disappear. This of course still leaves the problem of discrete degeneracies, such as vacua leading to different numbers of generations or even vacua with different numbers of spacetime dimensions. Here I think the solution will involve conceptual progress and understanding of quantum cosmology. As you pointed out, wormholes lead to unsettling questions about our ability to make fundamental predictions. However, our understanding of wormholes is so primitive that an optimist can still hope that they do not actually exist, or that if they do their effects are somehow suppressed.

V. Baluni, SLAC: Does string theory incorporate a description of the evolution of the universe? This problem seems to be intimately related to identifying the correct vacuum. Please comment on this point.

J. Harvey: Since string theory incorporates Einstein's theory of gravity, it must also incorporate the evolution of the universe. However with our present understanding of string theory, time dependent solutions of the theory are difficult to formulate. This is one of the centrol problems in classical string theory and its solution may well shed light on the problem of identifying the correct vacuum.

B. Ward, University of Tennessee: You started out by discussing the beautiful work in lower dimensions which suggest that the quantum gravitation theory should be expanded about $g_{\mu\nu} = 0$. You ended by discussing the recent work that shows the duality of string theories with distances below the Planck length to those with distances larger than it, so as to give a minimum distance to be considered. Are these ideas in contradiction with each other?

J. Harvey: They don't seem obviously in contradiction to me. The two ideas arise in rather different places. We don't have a model of string theory where the idea of expanding around $g_{\mu\nu} = 0$, or some string field equals zero, gives us something that looks at all like the real world. I think trying to extend the idea of expanding around $g_{\mu\nu} = 0$ in order to find a new phase of string theory is appealing, but I'm not aware of any attempts to implement it.

THE FUTURE e^+e^- COLLIDERS*

Gustav-Adolf Voss**

Stanford Linear Accelerator Center
Stanford University, Stanford, California 94309

ABSTRACT

At present, the highest energy e^+e^- colliders are the SLC and LEP. Their future improvement programs for increasing luminosity and/or energy, and the use of longitudinally polarized beams at the interaction point (IP) are discussed.

An e^+e^- collider in the SSC tunnel does not seem to be an attractive option, on both technical and economical grounds, and with LEP, circular colliders have reached the sensible limit of size and cost. Linear colliders which have, in principle, no high energy limit, must overcome a new set of technical problems having to do with beam power limitations, emittance control, superstrong focusing at the IP, strong bunch-bunch interactions at the IP and related backgrounds.

Present R&D activities in various laboratories are directed toward linear accelerator colliders using very high gradients (80 to 180 MeV/m) and operating at very short wavelengths (2.5 to 1 cm). This R&D is not yet far enough advanced to expect an upcoming proposal for a linear collider in the immediate future.

INTRODUCTION

During the past twenty years, positron–electron (e^+e^-) colliders have played an ever-increasing role in high energy physics research, and they are likely to continue to do so in the years to come. Hadron colliders, like the Tevatron or the SSC, will reach much higher energy regions than those accessible to e^+e^- colliders in the foreseeable future and will thereby scan the field for qualitatively new phenomena. But quantitative investigation of any new processes is more easily done with electron machines. Circular e^+e^- colliders may have reached their ultimate size with the LEP machine at CERN. Synchrotron radiation losses force the size and the costs of an optimized e^+e^- storage ring to increase quadratically with energy. Linear colliders should be more economical at collision energies just beyond those of LEP.

Linear colliders represent a new technology. Normally new technologies are tested in small prototypes, leading to small pilot machines, then to larger ones, eventually replacing the old concepts in the push for new frontiers. There is not that much time for the evolution of linear colliders. At least a substantial part of the physics community hopes and expects to see the machine which extends the LEP energy range by a significant factor before the end of this century.

Most of what we know about linear colliders is the result of the work on the SLC project. Because it uses magnetic arcs to direct electron and positron beams against each other, the SLC is only in a very limited sense a linear collider. Many of the problems of the SLC are atypical of future linear colliders. On the other hand, all of the problems of linear colliders also occur in the SLC and solutions have had to be developed.

In trying to look at future e^+e^- colliders, I will start by describing the future plans for LEP and the SLC—the only two e^+e^- colliders now, and for some years to come, which reach the Z^0 energy. I will

* Work supported by Department of Energy contract DE-AC03-76SF00515.

** On leave from Deutsches Elektronen Synchrotron (DESY), Hamburg, Germany.

then make a brief comment on the possibility of using the SSC tunnel for an even larger e^+e^- storage ring. Then I will describe some of the more principal problems of linear colliders, and the R&D work going on at various laboratories to find practical solutions to these problems. Since no formal proposal has yet been made for a linear collider, I feel free to speculate on what the next machine might look like.

LEP II AND SLC

The present LEP configuration (LEP I) works with accelerating cavities made out of copper which are powered with 16 MW of RF power. With this, LEP is supposed to reach a peak luminosity of 1.6×10^{31} cm^{-2} s^{-1} at a beam energy of about 50 GeV. The maximum energy should be around 60 GeV, where the luminosity drops to zero.

The program to upgrade the LEP energy relies on superconducting RF cavities operating at accelerating gradients of $5 \sim 7$ MV/m. A pilot series of eight 4-cell cavities will be installed in LEP soon after commissioning of the storage ring, i.e., in the latter half of 1989. After enough experience has been gathered with this installation, 32 superconducting 4-cell units will be installed which, by themselves, should sustain a 50 GeV operation for Z^0 production (completion 1992). As a third step, installation of up to 192 4-cell cavities is planned, leading to a peak energy of 95.2 GeV/beam.

With a beam current of 3 mA, a luminosity of 2.8×10^{31} cm^{-2} s^{-1} is anticipated. Depending on the availability of funds (293 MSFr.), this third step may be completed by 1995.

A program to increase the luminosity at the Z^0 peak is under discussion: With the help of horizontal electric fields the orbits for electrons and positrons can be separated such that multibunch operation becomes possible. With 36 bunches in each beam, luminosities in excess of 10^{32} cm^{-2} s^{-1} are expected. The increase of average current makes the use of superconducting cavities necessary. This means that the program presumably will not be in effect until 1992.

Polarization of beams in LEP is considered to be a possibility; it is by no means certain at this moment. A program to study a possible radiative buildup of polarization has been initiated. Conversion of the natural transverse polarization into a longitudinal one at the interaction point (IP) is possible in principle.

The SLC will go through an improvement program starting in October 1989. At this moment, the typical luminosity in the SLC is 1.4×10^{28} cm^{-2} s^{-1}, corresponding to the production of 1.5 Z^0 per hour. This number is lower than the original design by a factor of 400. Plans for upgrading the SLC starting in early 1990 call for an increase of the repetition frequency from 60 to 120 Hz, and an increase of the number of particles per bunch from 1.5×10^{10} to 4.5×10^{10}. An attempt will also be made to decrease the dilution factor of the invariant emittance. If these measures are successful, a luminosity larger than 2×10^{29} cm^{-2} s^{-1} is expected. But in order to be successful, the increase of bunch current has to be accompanied by a study and improvement of the background situation at the detector.

Another important area of machine development is to have electrons longitudinally polarized at the IP. Up to 40% polarization can be expected when the present linac gun is replaced by a gun using a gallium arsenide cathode in conjunction with a polarized laser. Preserving the polarization in the electron damping ring during the subsequent linac acceleration and through the magnetic arcs of the SLC should be relatively certain.

Long-range plans for the SLC might involve increasing the repetition frequency to 180 Hz, building a new Final Focus System leading to much smaller beam cross sections at the IP, and rebuilding the two injector damping rings. All these measures, if successful, could increase the luminosity beyond the original design numbers of 6×10^{30} cm^{-2} s^{-1} into the 10^{31} cm^{-2} s^{-1} range. However, only the implications of a repetition frequency increase have been worked out in technical detail. Rebuilding the Final Focus and the damping rings still requires technical studies before they can be proposed to obtain the necessary funding.

ELECTRON RING IN THE SSC TUNNEL

Before discussing the upcoming linear colliders, mention should be made of a conceivable proposal nobody has made yet, but which surely must be considered: Does it make sense to install an electron ring in the SSC tunnel? The SSC is more than three times as large as LEP. Applying the quadratic scaling law, the first rough estimate would indicate a machine with a peak energy of 1.75 times that of LEP or two times 170 GeV in the center-of-mass energy. If I take out of the LEP costs those parts which are already available at the SSC, the price tag would be \$1–2 billion. This is, of course, a gross oversimplification. An electron machine of this size needs a distributed RF accelerating system arranged in a large number (32?) of straight sections. This is not compatible with the existing SSC design. Extra civil engineering work would be required on a fairly large scale and even then the SSC tunnel could only be used in a rather inefficient way.

Putting a high-energy electron ring into the same tunnel with a superconducting proton ring could also significantly increase the cryogenic losses in the proton ring from hard photons from synchrotron radiation, which are absorbed in the cold steel of the superconducting magnets. Of course, one day e–p collisions at very high center-of-mass energies may become very interesting, and an electron ring in the SSC tunnel may be the way to do it. But this ring probably will have a lower energy than a dedicated e^+e^- machine of comparable size. The SSC project is not suited for accommodating a cost-effective e^+e^- machine.

LINEAR COLLIDERS

Linear colliders have been discussed during the last 20 years and have been the subject of at least 30 workshops. But most of what we know about linear colliders, I believe, is the result of the SLC project here at Stanford and the discussions it has stimulated worldwide. SLAC is committed to developing the new technologies necessary for linear colliders. SLAC is also the only place with a significant experience in this new field, and, because of its low emittance 50 GeV linear accelerator, is the only place where a number of these new technologies can reasonably be tested.

When discussions of linear colliders became fashionable 20 years ago, a large number of new ideas were promoted which ranged from the daring via the exotic to the absurd. Once people sat down and seriously looked at the currents and luminosities which they possibly could achieve with their new schemes, almost all of the new accelerator ideas evaporated. At the same time, there was precious little experimental work done on testing these new ideas. New concepts like the laser beat-wave and the wakefield accelerators have been tried experimentally and seem to work in principle, but they are at this time far from being serious contenders for the next linear collider proposal.

There are now four laboratories in which serious work toward a linear collider is being done and where such a machine might be proposed within the next decade: SLAC, CERN, KEK and Novosibirsk. The collider made by the Novosibirsk group would be built at the Serpukhov laboratory. There is basically a great similarity between the studies at these four laboratories: In all cases, electrons and positrons will be accelerated in standard linear accelerators, but at considerably higher gradients and higher frequencies than the ones used by the familiar S-band linacs. Low emittance beams are focused to unusually small spot sizes and directed against each other at the IP. Center-of-mass energies in these studies reach from 500 GeV to 2 TeV, and the luminosities are in the 10^{33} cm^{-2} s^{-1} range. None of these studies have, at the present time, matured enough to be even close to a proposal. A considerable amount of R&D work is still necessary to make a convincing argument that linear colliders with interesting energies and luminosities can be built at affordable costs. In the following discussion, I will compare various aspects of circular colliders with those of linear machines, point out the problems which have to be overcome in linear colliders, and discuss how close we are to a solution.

BEAM POWER

The electrical power consumption of any high-energy colliding beam machine clearly must be limited. In a circular machine like LEP II, energy loss from synchrotron radiation is of the order of 2% per turn.

In a linear collider, each bunch collides only once with its counterpart before it is dumped. The beam energy loss per collision is therefore at least 50 times as large as in circular machines. To remain within comparable power consumption levels, linear colliders must have a smaller bunch collision frequency (typically 100 to 1700 Hz, as compared to 11245 Hz in LEP) and/or smaller numbers of particles per bunch (a few $\times 10^{10}$ in linear colliders vs. 4×10^{11} in LEP). Therefore, to obtain comparable luminosities at the reduced bunch collision frequency and smaller bunch charge, linear colliders must have much smaller beam cross sections at the IP as compared to storage rings.

BEAM CROSS SECTIONS AT THE INTERACTION POINT

In storage rings, beam emittances are determined by the interplay of the quantum nature of the synchrotron radiation and its damping effect. Typical invariant transverse emittances (emittance multiplied by the Lorentz factor) are 5.6×10^{-3} m (LEP) as compared to 3.5×10^{-6} m in present linear collider studies. The emittance of a linear collider is fundamentally determined by the emittance of low-energy damping rings at injection, but it is important not to spoil this invariant emittance during acceleration. Only then can one hope to realize the small spot sizes at the IP. This is one of the less assured parameters: Today the SLC, the only operating linear collider, has horizontally 10 times and vertically 1000 times larger invariant emittances than those assumed for future linear colliders. There has been little attempt at the SLC so far—and also little reason—to improve particularly the vertical emittance. But there is also no experimental proof that it can be done easily.

The other factor entering the expression for the beam size at the IP is the amount of focusing that can be applied there. This can be expressed in terms of beta-functions β_x,β_y such that $\sigma_{x,y} = \sqrt{\epsilon_{x,y}\ \beta_{x,y}}$. Future linear colliders are assumed to have beta-functions at the IP smaller than 0.1 mm vertically, leading to vertical beam sizes as small as 2.5 nm, more than a factor 1000 smaller than those achieved in the SLC project. Such strong focusing is only possible if the bunch length is shorter than the beta-function. Linear colliders require bunch lengths of the order of 0.1 mm, as compared to 1.5 cm in a circular machine (LEP). The small bunch length and the extremely small transverse beam dimensions lead to very strong beam–beam interactions in linear colliders.

BEAM–BEAM INTERACTIONS

Colliding beams affect each other at the IP via their electromagnetic fields. In storage rings, such interactions lead to an increase of betatron frequencies and, if the interaction gets strong enough, they lead to a beam blowup and beam loss caused by nonlinear resonances excited by these fields. In linear colliders, beam–beam interactions may be larger by several orders of magnitude. At high interaction strength, they lead first to a pinch effect and then to a blowup of the emittance. The pinching is described by a disruption factor and a simultaneous luminosity enhancement, which can be as large as 5–10. Large disruption factors may also lead to unacceptable instabilities which must be avoided by a corresponding choice of interaction region parameters.

The strong beam–beam interaction forces in linear colliders, which are the result of trying to get high luminosities with moderately small currents, cause "beamstrahlung." This can be understood as synchrotron radiation in the electromagnetic field of the opposing bunch, leading to individual energy losses of the particles and a broadening of the energy spread in the center-of-mass system. This may be unacceptable for doing good physics experiments and a smaller strength of beam–beam interaction may be indicated. (In this respect, a high aspect ratio of the beam dimensions at the IP is desirable. High aspect ratios keep the beam–beam effects small without lowering luminosity).

Another highly undesirable effect of beam–beam interaction at very high energies was discovered recently: Photons created by beamstrahlung can, in the same bunch, produce electron–positron pairs. While one particle polarity of these pairs is kept close to the beam because of the focusing pinch effect, the other one is deflected straight into the detector, producing severe background problems. It is evident that the permissible strength of beam–beam interaction is somewhat limited (especially at very high center-of-mass energies).

BACKGROUNDS

In a storage ring such as LEP II, a beam lifetime of three hours corresponds to a loss of 2 W of high-energy radiation more or less distributed around the ring. A linear collider, on the other hand, loses all of its high-energy particles at each pulse. This loss rate corresponds to 1 MW, and all of this loss occurs not too far from the detector. Although this simple comparison does not allow an estimate of the background rates in the detector, it may give some feeling for the difficulty of controlling backgrounds in linear colliders.

THE LINACS

Overall power efficiency is most important in the choice of linac frequencies. The higher the frequency, the smaller the amount of stored energy per unit length for a given accelerating gradient. This would tend to favor very high frequencies, except that intense bunches interact with their environment and excite wakefields in the accelerating structure. These fields can act back on the tail of the bunch, causing transverse deflection (if the beam is not perfectly centered in the structure) and an additional energy spread. This energy spread increases quadratically with the frequency, if all other parameters are kept constant. Although an emittance growth from misalignment errors can be mitigated or avoided by strong focusing through quadrupole fields combined with a betatron frequency spread within the bunch (BNS damping), alignment tolerances of the RF structures become much more critical as the frequency increases.

Similar alignment problems arise from misalignment of quadrupoles. Such misalignments create orbit distortions and, at the same time, energy dispersion effects, i.e., particles with slightly different energies within the bunch move on different trajectories. This in turn may cause "filamentation," an effective emittance increase. There are other reasons for not choosing too high a frequency: In the effort to make linear colliders short so as to make them cheaper, very high accelerating gradients are required. This, in turn, requires RF tubes with very large peak RF power at very high frequencies (although with very short pulse lengths). To maintain high RF power as one increases frequency is difficult, because the dimensions of the tubes (cathode area, cathode–anode spacing, etc.) all get smaller at high frequencies.

To increase efficiency of power transfer from RF to the beam and to increase the overall luminosity, multibunch operation is a necessity. To avoid beam–beam interactions at places other than the IP, opposing beams will have to cross at the IP. To ensure good bunch–bunch overlap, this crossing angle must be small. Another way, which allows much larger crossing angles without luminosity reduction, may be the so-called "crab crossing." Here, RF deflecting fields are used to change the angle of the bunch axis with respect to its flight path from zero to the half-crossing angle. Perfect bunch–bunch overlap at the IP can be reached this way, while relatively large crossing angles between beams can be maintained.

OPTIMIZATION

From the above, it is evident that the design of a linear collider is a very complex problem: All of the effects mentioned above affect each other in an optimized design. Computer programs have been written to illuminate the interaction of various parameter choices and to help find an optimized design. The problem with such an approach is that most of the functional dependencies have only small ranges of validity, and that the financial implications are uncertain as well. Many of the new technologies in question have not been developed to the point where clear economical choices can be made.

Most linear collider studies contain a number of initial decisions, some of which appear to be rather arbitrary:

1. *Energy:* For the next generation of e^+e^- colliders, an energy of $2 \sim 10$ times that of LEP II seems to be desirable.

2. *Luminosity:* Since particle production cross sections in e^+e^- colliders scale like $1/E^2$, the luminosity should increase quadratically with E. Luminosity values are, therefore, around 10^{33} cm^{-2} s^{-1}.

3. *Power Consumption:* 50–200 MW are typical power consumption values for large high energy physics laboratories. At this level, power costs are a sizable portion of the operations budget.

4. *Length and Gradient:* The impact of length on costs is obvious. The impact of the choice of accelerating gradient on costs is less clear, particularly if one does not know how the large gradients are to be made. This has led to an unjustifiable bias toward high gradients, and correspondingly shorter overall lengths. Gradients of 80–190 MV/m have been chosen rather arbitrarily, resulting in lengths of 6–26 km.

5. *Frequency:* The fact that stored energies in linear accelerators scale like the inverse square of the accelerating frequency has led to a heavy bias towards very high frequencies. Higher breakdown fields at higher frequencies may have supported that trend. Neither argument for high frequencies may be important. The choice of wavelengths between 10 and 26 mm corresponding to frequencies of 30–11.4 GHz is, in my opinion, arbitrary, and can be subject to debate. But since there is no economical power source yet which is in that frequency range and which comes close to fulfilling the requirements, there is also no rational basis for any particular choice.

Table 1: Main parameters of machines under study and the SLC design (actual numbers).

	SLC	CLIC	ILC	JLC	TLC	VLEPP
Energy cms (GeV)	100	2000	500	1000	1000	2000
Linac length (km)	3	2 × 13	2 × 3	2 × 5	2 × 3	2 × 10
Av. acc. grad. (MV/m)	17	80	93	100	186	100
RF power (MW/m)	13	177	147	?	586	217
# of bunches/ pulse	1	5	10	15	10	1
# of particles/ bunch (10^{10})	7 (1.5)	0.5	0.7	0.4	1.4	10
$\gamma\epsilon_x(10^{-5}$ m)	3.5 (8)	0.1	0.3	0.3	0.5	2
$\gamma\epsilon_y(10^{-5}$ m)	3.5 (8)	0.1	0.003	0.01	0.006	0.02
RF wave-length (mm)	105	10	17.5	26	17.5	21.4
Luminosity (10^{30} cm^{-2} s^{-1})	4 (0.01)	1000	1700	1700	7900	1000

Table 1 shows a list of the main parameters of the various machines under study. Two of the key problems are immediately evident:

1. All machines use much higher accelerating gradients than the SLC so as to keep the overall length small. This leads to much higher RF power per meter, and at the more difficult higher frequencies to boot.

2. In order to reach the desired luminosity with relatively modest currents, the invariant emittances are much smaller than those in the SLC. This requires, if it can be done at all, unusually tight tolerances of accelerator alignment and beam stability.

To understand some of the technical problems, let us examine some more parameters of one of these machines. Table 2 shows the final focus parameters of the ILC. In order to reach the luminosity of 10^{33} cm^{-2} s^{-1}, not only do the emittances have to be very small, but the beta-functions at the IP lead to a vertical beam size of 2.6 nm—three orders of magnitude smaller than that at the SLC! A chromatically corrected Final Focus System has been worked out, but the stability requirements for the optical elements are somewhat mind-boggling. The final quadrupoles obviously may not move by more than the final beam size with respect to each other, i.e., relative alignment tolerance of the order of a few angstroms are required. This can only be accomplished by near-perfect separation from ground motion at high frequencies, a good mechanical feedback system, and beam steering at low frequencies.

Table 2: ILC parameters (see also Table 1).

Parameter	Value	Unit
Vertical beta function at the IP	0.1	mm
Crossing angle	4.2	mrad
Final quadrupole bore radius	0.17	mm
Free length to the IP	0.36	m
Vertical beam size at the IP	2.7	nm
Horizontal beam size at the IP	356	nm
Bunch length	0.07	mm

The final focusing system is complex and difficult to build. In order to reach the small spot size with a beam of a finite energy spread (±0.3%), it has to be chromatically corrected up to at least third order. It uses a series of small bending magnets, interspersed with quadrupoles and sextupoles. To reach the required short focal length, the final focusing quadrupoles have bore radii of 0.17 mm! The tune-up of such a system will require measurement of the final spot size, a problem for which no good technical solution exists at the moment. Alignment tolerances of the optical elements are in the micron range.

PRESENT ACTIVITIES

From the above, it is obvious that a large amount of R&D work is required before a serious proposal for a next-generation linear collider can be submitted. There are many areas where the required technology is not yet in hand. To name the most important areas:

1. If one wants to use the high frequencies of the

present collider studies, RF power tubes need to be developed at frequencies between 11 GHz and 30 GHz which are efficient and so economical that one can afford to have power inputs of hundreds of MW/m over a length of several kilometers.

2. Linear accelerator structures have to be developed:

   a. with very tight mechanical tolerances,

   b. which will not break down with accelerating gradients of 100–200 MV/m,

   c. which will strongly damp higher modes to make multibunch operation possible, and

   d. which can be economically mass-produced.

3. Final focusing systems need to be developed that can focus beams down to spot sizes in the nanometer range, together with all the required beam position and spot size monitors, and all of the technology to maintain the elements accurately in place.

All of these problems are simultaneously being attacked and much work is going on in the different laboratories which I have named above. I can only mention the highlights:

1. Four different approaches are being pursued at SLAC to solve the RF power problem.

   a. The relativistic klystron, developed with the Livermore Lab., uses high-power induction linacs to produce 1 MeV electron beams with currents of the order of 1000 A. Two hundred megawatt RF power has been produced in such a device at a frequency of 11.4 GHz, but the chances of this being the preferred route to an economically competitive power source are not considered to be very good.

   b. RF pulse compression through a system of directional couplers in combination with phase shifting the RF output of a standard 100 MW klystron could increase the peak power by a factor of 8 at the cost of pulse length. Low power tests of the compressor indicate an efficiency of 71%. Standard klystrons at a frequency of 11.4 GHz are presently being built for power levels of 100 MW. High power tests of the compressor will be done in the near future.

   c. Cross field amplifiers (phasable magnetrons) with voltages of 300 kV and currents of 10 kA should be able to produce up to 1 GW. They would have the great advantage of needing no expensive modulators but could run directly off a DC supply. A scaled-down version with an output of 120 MW is presently being built.

   d. A cluster klystron combining 42 small 30 MW tubes in one body is also being designed. Running many small tubes in parallel, each operating with a hollow beam, should overcome the space charge problems one normally faces in high-power tubes. The inventors hope that modulating anodes in this cluster klystron combined with a new way of powering the klystron may do away with the expensive modulators.

The Novosibirsk group developed 60 MW Gyrocons at 7 GHz and is also working on other concepts (separatron at 3 GHz).

The KEK laboratory is engaged in the development of 11.4 GHz klystrons at power levels of 30 MW, 60 MW and 120 MW. Work is also done on pulse compression schemes.

A radically different approach is being taken by the CERN group in their CLIC study: An intense 3 GeV beam of tightly bunched electrons is being accelerated in a 350 MHz superconducting linac. In special transfer structures, 10 nsec-long RF pulses of 100 MW are then generated at frequencies of 30 GHz. These pulses are used to power a 30 GHz linac, which runs parallel to the 350 MHz linac. Funding for building a prototype of this two beam arrangement is not available, and R&D work is limited to small aspects of this project.

Many more ideas on linear colliders are on the market. Because of time limits, I will only mention the most important ones.

2. Accelerating structures in the frequency range from 3–30 GHz have been built, or are being built, by the Novosibirsk group, at CERN, at SLAC, at KEK and at Orsay. One of the aims is a study of the fabrication methods; another is the test of break-down voltages and field emission currents at high gradients. For the multibunch operation, there is also the desire to damp higher modes very strongly to avoid multibunch instabilities. Several groups have attained accelerating gradients in excess of 100 MV/m. A SLAC/Livermore collaboration actually accelerated an electron beam in a 26 cm-long waveguide section at a frequency of 11.4 GHz and with gradients of 120 MV/m. Low-power measurements on "slotted" structures for very strong damping of higher modes are in progress. High-gradient tests on slotted structures have demonstrated that the attenuation of higher waveguide modes is quite compatible with high-field strengths in these structures.

3. For the experimental study of the Final Focus Systems, a collaboration of Novosibirsk, KEK, CERN and SLAC is being formed. The aim of this collaboration is the construction of one Final Focus Test Beam (FFTB) to be located at the end of the 50 GeV SLAC linac. This seems to be the only place in the world where an intense electron beam of sufficiently small emittance and sufficiently high energy is available. It is planned to focus the electron beam to a size 1 μm-wide and 65 nm high, to develop the beam position and beam size monitors necessary for tuning up the Final Focus, and to test all other techniques required to stabilize and survey the electron optical elements to the necessary levels of accuracy.

OUTLOOK

During the last half-hour, some of you may have developed a feeling of depression when you think about the conceivable schedule for an e^+e^- collider in the TeV range. Particularly in the area of RF power sources, it seems that we are still a long way from the goals described in the collider studies. Also, much of what I have said—nanometer-size beams, quadrupole systems with bore sizes of one-tenth of a millimeter, mechanical alignment to micron accuracy—may sound to some of you like science fiction. The most depressing aspect is how little of the necessary R&D work is presently going on—a situation which is determined to a large extent by the degree of funding. There seems to be a huge discrepancy between our aspirations and the efforts which can be mustered. Is it that going from the SLC to the TLC or only to the ILC is just too big a step in new technological developments?

A question that certainly must be addressed by the physics community is: What kind of e^+e^- collider is worth building? What is the minimum for doing good physics, physics which cannot be done in any other way? What might such a machine cost? It is customary to assume that e^+e^- colliders must have about one-tenth of the center-of-mass energy of a comparable hadron collider. A 4 TeV e^+e^-

collider, comparable to the SSC, cannot be built in this century!

But if we scale down our ambitions and ask, "What can be done today with only modest extrapolations of the existing SLC technology, and may be worth doing?" then things may not look so bleak.

At the beginning, I promised to speculate about what the next linear collider might look like. For example, a machine of the same length as the circumference of LEP, using the SLC gradient of 17 MV/m, would get us to a center-of-mass energy of 500 GeV—five times as large as SLC and more than twice as large as LEP would ever be able to reach. I would estimate the cost of the machine to be of the order of $2 billion, although I have heard experts in the field who insisted on some considerably smaller number. Such a machine would probably run with a much larger number of bunches për pulse, to make efficient, use of the available RF power. Depending on the invariant emittance one can maintain in such a machine, spot sizes may be considerably larger than those assumed in the ILC, resulting in correspondingly lower luminosities. An emittance dilution in the vertical plane of 100 would still leave us with a luminosity of 10^{32} cm^{-2} s^{-1}. If one is worried that such a machine, after having been built, turns out to have too small an energy because the SSC has just discovered some very important effects outside of this energy range, then one must remember that linear colliders, unlike storage rings, can be extended; more length or more RF power or both will always buy you more energy.

I believe that a machine with a modest energy increase over LEP II and one that uses only small extensions of existing technologies has a better chance of being built. Construction of such a project would also provide the urgently needed continuity in the evolution of the SLC into the next generation of linear colliders.

DISCUSSION

J. Haissinski, CERN: At the end of your talk you suggested that we should not take such a large step forward with the linear collider, can you say what you think of the possibility of combining a linear accelerator with a storage ring, as has been advocated by some people. It might allow us to investigate center of mass energy of a few hundred GeV.

G. Voss: You're talking about a linear accelerator in conjunction with LEP?

J. Haissinski, CERN: Yes, for instance.

G. Voss: I had not really thought of that. You would save half of the linac perhaps, but you would not get to the energies we have been talking about here.

B. Richter, SLAC: I want to comment on Haissinski's question. The linear collider and the storage ring really only works and produces decent luminosities with a superconducting linac. The problem is that high luminosity requires a match between the rotation frequency of the storage ring and the repetition rate of the linear accelerator. Conventional linacs have low repetition rates, but superconducting linacs can have high rates. If you wish to build a superconducting linac, the optimists think that you can get 5, 10, maybe 15 mega-volts/meter. They're very expensive at the moment and if you want to go in that direction, you really have to push superconducting cavity technology very far from where it is now.

SUMMARY OF THE CONFERENCE

Frank Sciulli

Physics Department and Nevis Laboratories
Columbia University
New York, NY 10027

This conference is being held on the threshold of a new decade; it seems appropriate to observe where we stand in an historical context. Twenty-five years ago, we were in the closing stages of a true "period of discovery": checking Quantum Electrodynamics (QED) with high precision experiments; completing a consistent framework for the weak interactions, especially the selection rules for hadronic quantum numbers in weak decay; and groping toward an understanding of the strong interactions. Though we had SU_3 commutation relations and the quark model as a structure in which to imbed hadron spectroscopy, we had not yet made the critical connection between these and strong interaction dynamics, which began a few years later with the discovery of Bjorken scaling, and the advent of Quantum Chromodynamics (QCD).

Today we are in (probably the closing stages of) a period of consolidation. We have a unified theory of the electromagnetic and weak interactions with the $SU_2 \times U_1$ Electroweak Theory. The QCD description of strong interactions, though not subject to the same kinds of precision tests, stands unchallenged as the description of strong interactions. The constituent base states seem to cleanly reside in three families; and there is no evidence for more families. (And, at this conference, there is evidence that there are few additional families!) All of this information is amalgamated neatly into the "standard model", which stands as a beautiful consolidation of our present understanding and for which there are no clear experimental discrepancies.

However, we are uncomfortable with this standard model for many well-known reasons. It is incomplete. The theorists need to find some ways out of certain internal inconsistencies; the experimentalists need to provide some clue or direction by providing data which are not consistent with it.

Years ago, it was recognized that there existed a unique opportunity for testing the standard model: e^+e^- collisions at the Z^0 mass. Two colliders to accomplish this have been built. At this conference, first results have been presented from the first operational Z^0 machine: the SLAC Linear Collider (SLC). These data also represent the first observation of the Z^0 boson decaying into hadronic decay modes. Measurements of the mass and width were presented here by Gary Feldman, obtained from 233 events taken at seven different center-of-mass energies. The results are

$$M_Z = 91.17 \pm 0.18 \text{ GeV}$$

$$\Gamma_Z = 1.95 \begin{array}{l} +0.40 \\ -0.30 \end{array} \text{ GeV}$$

The latter value, while smaller than the standard model prediction of 2.49 GeV, is consistent within experimental precision. More directly significant is a fit to the "invisible width", which should be equal to $.167N_\nu$ GeV, where N_ν is the number of low mass neutrinos. The best fit to the Mark II

TABLE 1: Summary of Z/W Measurements (Masses in GeV/c^2)

	Mark II	CDF	UA2	Average
M_Z	91.17±.18	90.9±.3±.2	90.2±.6±1.4	91.10±.16
Γ_Z	$1.953^{+0.40}_{-0.30}$	3.8±1.1±1.0		$2.04^{+.39}_{-.29}$
M_W		80.0±.2±.3±.5	80.0±.4±.4±1.2	80.0 ± .5
M_Z-M_W		10.9 ± 0.7	"10.2±.7±.4"	10.6 ± .6
R		10.3±.8±.5	$10.35^{+1.5}_{-1.0}$	10.3 ± 0.9

data gives

$$N_\nu = 3.0 \pm 0.9 \text{ neutrinos.}$$

At ninety-five percent confidence level, the data require less than 4.4 neutrino families.

The branching ratios of the leptonic and hadronic modes were presented by Alan Weinstein. While the numbers of individual electron and tau events are somewhat smaller and larger, respectively, than the standard model predictions, we are still dealing with small numbers (6 and 15 events, respectively). Even with this small preliminary sample, significant bounds can be placed on the masses of various hypothetical leptons and bosons, described in detail by Weinstein.

Results from Tristan were presented by Akihiro Maki; these addressed several important questions. First, can the newly measured Z^0 parameters from SLC be accommodated by the TRISTAN data within the context of the standard model? (It will be remembered that the TRISTAN data tended to be a bit higher than previous standard model parametrizations.) The AMY, TOPAZ, and VENUS data are consistent with the parametrization using the SLC values and the systematic errors of measurement. The second question is "Are any other measurements from TRISTAN inconsistent with the standard model?" The energy dependences of cross-sections to hadrons, to taus, to muons, as well as the muon and tau asymmetry parameters, are all consistent with prediction.

Important complementary measurements of Z^0 and $W^\pm$ boson properties were reported from hadron collider experiments. Results on the production and decay of the vector bosons by Pekka Sinervo (CDF), Luigi di Lella (UA2), and Karsten Eggert (UA1) were presented; these include directly measured values of the Z^0 mass from the e^+e^- and $\mu^+\mu^-$ decay modes, and of the $W^\pm$ transverse mass from the corresponding leptonic decays. The results, together with Z^0 values from Mark II, are shown in Table I.

In this table, R represents the ratio of production and decay for $W \to e + \nu$ relative to $Z \to e^+ + e^-$. This value depends on the number of neutrino families and on the top quark mass. For three neutrino families, masses of the top quark larger than 80 GeV are favored if $R \simeq 10$.

The value of M_Z-M_W in quotes is my calculation and error estimate obtained from the UA2 value for the mass ratio:

$$M_W/M_Z = .887 \pm .007 \pm .004,$$

combined with the UA2 value for the Z mass. Figure 1 shows the Z-mass plotted versus the difference between the boson masses. Curves (1) of constant $\sin^2\theta_W$ and constant top quark mass, M_t, are superimposed. Data published prior to April 1989 for this plot are dominated by existing deep inelastic measurements of $\sin^2\theta_W$; such measurements are relatively insensitive to the top quark mass. The earlier data give the point and the one standard deviation contour (dot-

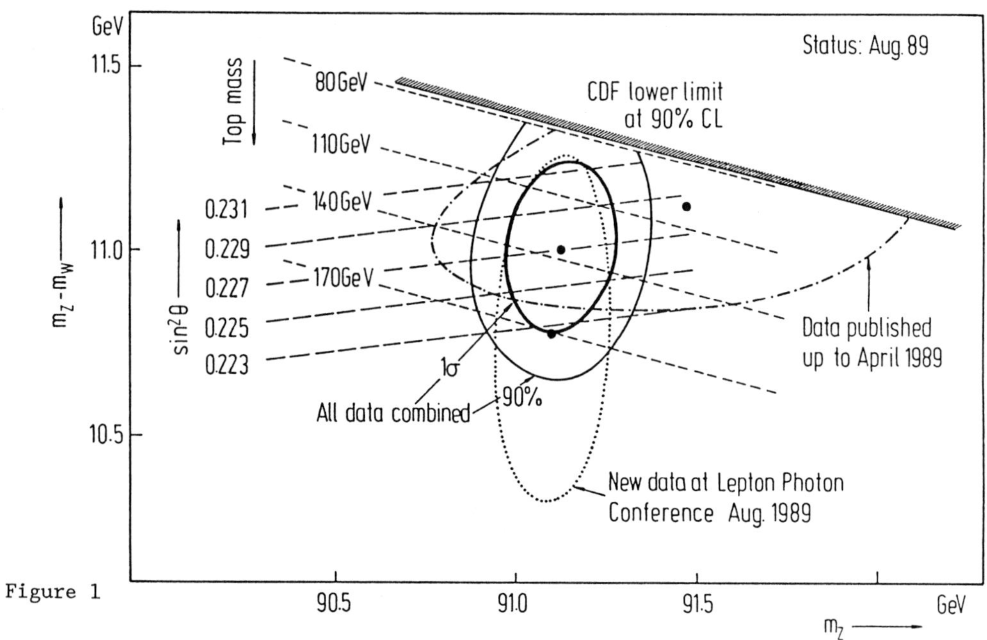

Figure 1

dash) near the direct experimental lower limit quoted by CDF; newer data in the above table correspond to the lower point and the dotted one standard deviation contour. These data are quite consistent among themselves, and together define the point and heavy one standard deviation contour shown. The ninety percent confidence limit from all combined data tends to indicate that high mass top quarks are favored:

$100 < M_t < 170$ GeV

in the context of present data and presently available calculations. A contemporary estimate (2), done independently on much the same data, gives a preferred mass of 132 GeV with a similar range. Though there are some calculational uncertainties, it is clear that newer data has pushed expectations for the top quark mass higher.

While such indirect limits indicate a high top quark mass, it should not deter experimenters from searching for it wherever possible. Limits presented at this conference include

TRISTAN	>30.4 GeV/c^2
Mark II	>37.5
UA2	>67
UA1	>69
CDF	>76 GeV/c^2.

It should be noted that these direct experimental limits on the mass of the top quark are approaching the mass of the weak boson. At that point, decay modes change and techniques for searching for the top quark will require revision.

One potentially difficult problem for the standard model was a low average value of the Weinberg angle obtained from the elastic scattering of neutrinos with electrons, relative to such angles obtained from the deep inelastic scattering of neutrinos with quarks. A new measurement from the purely leptonic process reported at the conference by Jaap Panman gives

$\sin^2\theta_W = .235 \pm .012 \pm .008$,

which is in good agreement with the deep-inelastic average value. The CHARM II experimental group expects the error on this quantity to be reduced ultimately to about .005, which will be competitive with the precision from deep-inelastic scattering.

The summary of the situation in the lepton sector by Martin Perl concluded that there are no outstanding large anomalies; limits of approximately 30 GeV/c^2 exist for masses of various kinds of hypothesized heavy leptons. These present limits principally arise from TRISTAN data.

Prior indications of neutrino oscillations have not been corroborated. This was pointed out in the summary by I. V. Kirpichnikov. The most sensitive experiments on the mass of the electron neutrino are consistent with zero. This is illustrate with the recent measurements from Los Alamos and INS (Tokyo), which give lower limits of 13.4 ev and 11.0 ev, respectively. A difficulty has existed for many years with the <u>other</u> standard model, the Solar standard model (SSM). This was first posed by the Davis experiment, and now this is corroborated by the Kamiokande experiment, which is sensitive at a higher threshold energy. The latter experiment, over several years of running, sees a high energy neutrino flux which is less than half the SSM. We of course do not know whether the origin of this effect is in the solar model, or in our understanding of the properties of the neutrino. Many new experiments in the pipeline, some with very low threshold energies, should help provide clues to the resolution of this conflict.

The talk by Guido Martinelli beautifully summarized the properties of the Cabibbo-Kobayashi-Maskawa (CKM) matrix. These new numbers describe the weak couplings among the quarks. In the conventional formalism, V_{ab} represents the coupling of quark type <u>a</u> to quark type <u>b</u>. Since the top quark has not yet been found, we know nothing (except through unitarity constraints) about the three V_{tx} values, where x represents the d, s, and b quarks. Of the other six values, two are close to unity, with measured values for V_{ud} (from β-decay) and V_{cs} (from D^0 decays). Three others are clearly non-zero: V_{us} (from K_{e3} and hyperon decays), V_{cd} (from neutrino production of charm, followed by muonic decay), and V_{cb} (from B lifetime measurements). The last member, V_{ub}, has had its ups and downs. The qualitative question has been whether the number is near to or identically zero.

This question of the finiteness of V_{ub} is extremely important within the formalism of the standard model. A body of opinion exists that the origin of the CP-violation (seen in K-decay) can be traced directly to a non-zero phase contained in the CKM matrix. If, on the other hand, V_{ub} (or any other element), were identically zero, the phase could not provide the source for CP-violation. At this conference, results were presented by two groups on the inclusive leptonic decays of B-mesons. Decays corresponding to $b \rightarrow u$ will include charged leptons of momentum higher than the endpoint for leptons from the more dominant $b \rightarrow c$ transitions. Observation of these high energy leptons signals the finiteness of the V_{ub} matrix element. The CLEO group showed data (presented by David Kreinick) indicating the presence of such events with a confidence of about 2.2 standard deviations; the ARGUS group data (presented by Michael Danilov) showed the presence of such events with a confidence of more than three standard deviations. While the precise translation of this signal into a value for V_{ub} (about 0.1 V_{bc}) contains some calculational uncertainty, the finiteness of the matrix element seems established.

The decays of charm mesons were reviewed by Paul Karchin. The semileptonic decays of the D^0 mesons into a single kaon or a single pion provide measurements of the form factor in the decay; this agrees generally with expectations. The semileptonic decay into the K^*, which involves three form factors, is used to demonstrate that we are dealing with highly polarized K^* mesons. The wealth of information available on the nonleptonic modes is generally compatible with models. Decays into strange baryons are beginning to be observed.

The decays of the B-mesons were discussed at length by Martinelli. The Argus and CLEO collaborations are beginning to see many individual decay modes, which should facilitate precision measurement of the B^0-B^+ mass difference and permit comparison with the intense theoretical work extant on interpreting the hadronic decays. Calculations of the weak amplitudes, particularly within the context of lattice gauge theories, was viewed with optimism. It was emphasized by both Martinelli and Altarelli that the large value of the top quark mass requires a new approach to the calculations of these amplitudes, and to amplitudes arising elsewhere, particularly the important CP-violating, $\Delta I=3/2$ amplitude (ϵ') present in K-decay.

Precision measurements of ϵ' and related parameters were important contributions to this conference. In such experiments, the aim is to measure the $\pi^+\pi^-$ and $\pi^0\pi^0$ decay modes of the K^0, and hence to extract the amplitude ratio of $K_L \rightarrow \pi^+\pi^-$ relative to $K_S \rightarrow \pi^+\pi^-$ (η^{+-}); and the corresponding ratio for the neutral mode, $\pi^0\pi^0$, (η^{00}). These amplitude ratios permit direct evaluation of the isospin amplitudes, ϵ and ϵ':

$$\eta^{+-} = \epsilon + \epsilon \qquad\qquad \eta^{00} = \epsilon - 2\epsilon'.$$

Each of these has a phase (ϕ^{+-} and ϕ^{00}, respectively); CPT requires that these phases be equal. We have known since 1964 that $|\epsilon| \simeq 2 \times 10^{-3}$; we have also known that ϵ' is much smaller. The question is ... how small? As pointed out by Guido Altarelli, standard model calculations performed to date predict that ϵ' should be finite, with values for ϵ'/ϵ estimated to be about 5×10^{-3} for top quark mass of 50 GeV, and decreasing monotonically as the top quark mass increases. It was also emphasized that these calculations need to be redone, avoiding approximations that assume that the top quark mass is smaller than the weak boson mass. Hence, the experimenters have the initiative in this area until the calculators resolve their questions.

Several years ago, the CERN NA31 experiment provided a precise measurement for this ratio which is three standard deviations from zero. At this conference, Daniel Fournier of that group presented results from a new run on the CPT-violating phase difference:

$$\Phi^{CPT} = \phi^{+-} - \phi^{00}.$$

New results on ϵ'/ϵ and Φ^{CPT} were presented here by Bruce Winstein from the Fermilab E731 experiment. Both experiments use a double ratio technique, measuring the four rates corresponding to K_L and K_S decaying into $\pi^+\pi^-$ and $\pi^0\pi^0$, respectively. This double ratio directly measures ϵ'/ϵ:

$$|\eta^{+-}/\eta^{00}|^2 \simeq 1 + 6\mathrm{Re}(\epsilon'/\epsilon).$$

Though the experiments are similar in concept and in statistical precision, there are some potentially important differences in execution. If the two experiments obtain inconsistent results in the end, it will be very important to examine those differences. Table 2 shows the values reported by the two groups.

Table 2

Group	Φ^{CPT}(degrees)	$\epsilon'/\epsilon \times 10^3$
NA31	+0.3±2.6±1.1	+3.3 ± 1.1
E731	2.3 ± 3.0	-0.5 ± 1.5

Both measurements are consistent with a CPT-violating phase of zero. The new measurement of ϵ'/ϵ by E731 is consistent with zero. Such a result, at least in the context of existing calculations (which predict that the value is finite and positive), is extremely interesting and contrasts with the three standard deviation difference from zero for the NA31 result. On the other hand, the two experiments differ from each other by only two standard deviations. This is not yet significant enough to cry, "experimental inconsistency". Clearly, we need more data and higher precision to reconcile the situation. The NA31 group has run during 1988/89 with improvements in beam and detector. The E731 group is analyzing data corresponding to five times that used for the above result. We can expect, therefore, to hear more in the near future from these two important experimental efforts.

Clues to breakdowns in the standard model are being sought in other rare decays of the K-meson. Laurence Littenberg presented a review of the subject at this conference. Recent experiments from KEK and BNL on the decays into $\mu^+\mu^-$ give finite results, generally consistent with expectations scaled from the decay into $\gamma\gamma$. Experiments are in the pipeline which should provide thousands of events for $K^+ \to \pi^+e^+e^-$ (BNL851) and $K_L \to \pi^0e^+e^-$ (BNL845). With only a few weeks of running, BNL experiment E787 has set limits on the branching ratio for $K^+ \to \pi^+\nu\nu$ at the level of 3×10^{-8}. Small limits, at the level of 10^{-11}, should soon become available on the forbidden process $K_L \to \mu e$ (KEK137,BNL791). In general, we should begin within the next few years to be sensitive at levels of standard model predictions to modes such as $K \to e^+e^-$, πee, $\pi\nu\nu$, and $\pi\mu\mu$; and to be sensitive to forbidden modes at levels below 10^{-10}.

The situation in nuclear double β-decay was summarized by I. V. Kirpichnikov. Observation of finite rates have now been claimed by two groups for decay of nuclei into two electrons and two neutrinos. A very important result would be the observation of neutrinoless double β-decay. For this to occur, the electron neutrino would need to be massive and Majorana, with a rate approximately proportional to the electron neutrino mass. Limits on the rate of decay for ^{76}Ge exist in excess of 10^{24} years from two groups (UCSB/LBL, ITEP/Verevan).

The field of two-photon physics was summarized by Robert Cahn. The spectroscopy of states formed by the photon-photon interaction has been eminently successful for the 0^{-+} and 2^{++} states, while the 0^{++} and 1^{++} states require further work. No clear evidence yet exists for existence of glueballs or radial excitations.

The nucleon structure functions of deep inelastic scattering were reviewed by Joel Feltesse. There is now reasonable agreement among all the various measurements of isoscalar structure functions in the values and general trends, though there is some ongoing concern over the consistency with predictions of QCD from some of the experiments. New measurements of the ratio of longitudinal to transverse cross-sections (R-parameter) performed at SLAC demonstrate the existence of non-perturbative contributions to R at low Q^2. New data from hydrogen and deuterium targets permit consistent extraction of u- and d-quark distribution functions, provided one uses global normalization corrections. One specific question, largely unrelated to normalization, is the beautiful agreement of the BCDMS data with the requirements of perturbative QCD, and the qualitative disagreement by several other groups with the same hypothesis. The argument is presented that such discrepancies can be a consequence of systematic errors of measurement; however, this explanation is not universally accepted. New data from the CCFR neutrino program and from the muon program at FNAL may help to resolve these issues.

A subject of substantial interest over the past year or so, the study of spin structure functions for the nucleon, was reviewed by Graham Ross. Measurements by the EMC group, using polarized muons and a polarized proton target, permit extraction of the proton spin structure function. Without comparable data from polarized neutrons, the Bjorken sumrule (a consequence of current algebra) cannot be directly tested. The Ellis-Jaffe sumrule, which assumes flavor SU_3 symmetry and ignores the effect of strange quarks, is not satisfied by the EMC data. If one assumes the Bjorken sumrule, the contribution of nonstrange quarks to the nucleon spin can be directly calculated, and this contribution is smaller than intuitively expected (at least by many). The conclusion seems inescapable that an appreciable fraction of the nucleon helicity is provided by strange quarks and/or gluons and/or orbital angular momentum.

The physics of Drell-Yan lepton pairs and direct photons was reviewed by Joey Huston. There is good agreement between QCD prediction, using a simple optimization procedure, and available photon data. The data can also be used to extract the gluon structure function (with sensitivity primarily at large x) which complements deep inelastic data (applicable at smaller x values). Considerable data, primarily from collider experiments at CERN and at Fermilab, as well as from fixed target experiments at Fermilab, are in the pipeline.

Comparisons of QCD predictions with data on large transverse momentum jets was discussed by John Huth. We saw measurements of jet cross-sections from ISR, CERN collider, and Tevatron encompassing nearly four orders of magnitude in cross section and transverse energies out to 400 GeV. These agree well with QCD evolved structure functions over this wide range. Work is proceeding on understanding the fragmentation of these jets, and also algorithms which would permit statistical separation of quark from gluon jets. A review of these and other theoretical aspects of QCD jets was presented by V. A. Khoze.

With all the beautiful data and with the explanations of that data presented at this conference, we find ourselves a bit frustrated. We cannot yet experimentally complete the standard model with the top quark that is essential to the model. We know that the behavior of known leptons, quarks, and bosons must become anomalous and/or there must be new, still undiscovered, particles. We find neither anomalies nor new particles; worse, we haven't a clue yet where they will emerge.

While this situation may <u>seem</u> discouraging, it seems to me to be just the opposite when we view it from an historical perspective. The two decades between 1950 and 1970, in retrospect, clearly formed an era of understanding. Many experimental and theoretical discoveries were made during this period in which we learned to ask the right questions, found the answers to those questions, and evolved the standard model. Between 1970 and 1990, we lived through the era of the standard model: a time of consolidation during which we corroborated many of the major predictions of the model and clearly understood its predictions and limitations. It seems likely that 1990 will begin a new era of discovery: we will begin to see the experimental limitations of the model; as we understand those limitations and explore their implications, we will develop an entirely new level of understanding for the microworld.

We have an array of tools poised to make the next discoveries. Those tools are reasonably balanced; this is important because we do not know from what source the next clues will come. We heard, at this conference, descriptions of the newly commissioned BEPC e^+e^- collider in Beijing, which will continue investigations of J/Ψ physics; of the SLC collider which will provide important new information about the Z^0 as the luminosity is improved and large beam polarization is exploited; of LEP, which has begun commissioning: four large collider experiments lie poised to exploit the physics of the Z^0 with high luminosity. Within the next two years, the HERA collider will provide the world's first beam-beam interactions between electrons and protons. The list of existing facilities which provided important data for this conference is long: TRISTAN, Tevatron collider and fixed target, SPS $p\bar{p}$ and fixed target, CESR, Brookhaven AGS, LAMPF, proton decay detectors, etc. Even this does not exhaust our resources: information flows from astrophysics and cosmic ray detectors, as discussed in the talks today by Jim Cronin and Mal Ruderman; from nuclear physics experiments, like double β-decay. Such diversity is healthy and must continue.

For the future, we foresee entirely new domains available for exploration. In a few years, the LEP II facility will permit direct observation of the production of weak vector boson pairs. A few years later, the SSC (and perhaps the LHC and UNK) will be commissioned to explore the spectroscopy of boson pairs at high rate with center-of-mass energies at which the standard model almost certainly must encounter difficulties.

We have heard mention of other possible new facilities at this conference. Machines devoted to systematic exploration of the charm-tau regime, a "factory' to explore the details (including possible CP violation) of the B-mesons, a modest energy "super linear collider" to develop e^+e^- colliders and explore their collisions so as to preserve this important component of our arsenal into the high energy regime. These suggestions speak to the need to maintain diversity in a field which requires diversity. We must continue to be open in our approach to physics, while we effectively utilize all available tools. The only limits to this diversity are the physical resources available to build and operate new facilities and the people to effectively design, construct, and exploit them.

We are in a field of great intellectual ferment and excitement. Moreover, that excitement can be fulfilled with a multitude of resources. We must be sure to communicate this excitement to others: to those who help to provide the resources and, most importantly, to the young generations of potential physicists who will be making the important discoveries of the future. Because, in the end, it is the physicists who make the physics. If we can create a next generation of physicists who continue the communication, the camaraderie, and the competitive spirit that we have seen at this conference, the field will flourish through the next decades with society and science enjoying the fruits of an era of discovery.

I would like to acknowledge the important help provided by the organizers of the conference. I would especially like to express my sincere gratitude to my scientific secretaries, Isi Dunietz and Dave Coupal, who were unstinting in their help and cooperation in preparing this talk on very short notice.

REFERENCES

(1) D. Haidt, "Status of the Electroweak Standard Model, DESY preprint 89-073 (June 1989); also private communication.

(2) John Ellis and G.L. Fogli, "The Implications of Recent Electroweak Data for m_t and M_H", CERN preprint TH.5511/89 (August 1989).

Contributed Papers

CONTRIBUTED PAPERS

There were a total of 319 contributed papers submitted to this Symposium. They are listed below in the order received, with name of author(s), title and abstract (where available) given for each. A reference to a contributed paper in these Proceedings usually specifies the paper number.

1. H. Inazawa (Shoin Women's U.), T. Morii (Kobe U.)
PRODUCTION OF ULTRAHEAVY LEPTONIUM AND/OR QUARKONIUM AT e^+e^- COLLIDERS IN TeV REGIONS.

The production cross section of the ultraheavy leptonium and quarkonium with the mass around 1 TeV, which have the contribution of the Higgs-boson exchange, is calculated in TeV regions at e^+e^- colliders. In some kinematical regions, we find that the Drell ratio for a bound state production shows a prominent resonance structure.

2. Zenro Hioki (Tokushima U.)
W BOSON MASS DETERMINATION FROM THE THRESHOLD BEHAVIOR OF $\sigma_{tot}(e^+e^- \to W^+W^-)$.

The total cross-section of $e^+e^- \to W^+W^-$ has been studied in relation to the W-boson mass determination in the framework of the standard $SU(2) \times U(1)$ electroweak theory. We use the cross-section $\sigma_{tot}(e^+e^- \to W^+W^-)$ including the leading logarithmic and heavy quark corrections, and also taking account of the finite W width effects in the final state. Taking the top quark mass and even α and M_Z as free parameters, we show that that the behavior of $\sigma_{tot}(e^+e^- \to W^+W^-)$ as a function of M_W in the threshold region is quite universal as expected from tree level analyses, and therefore adequate for the kinematical W-mass determination.

3. Zenro Hioki (Tokushima U.)
HEAVY FERMION CORRECTIONS TO HIGGS COUPLINGS REVISITED.

Heavy fermion-loop corrections for the ZZH, WWH and $f\bar{f}H$ couplings are calculated in the framework of the on-mass-shell renormalization scheme. By taking $\alpha^{\rm exp}$, $G_F^{\rm exp}$ and $M_Z^{\rm exp}$ as input data, explicit formulas of these corrections are presented which are more practical than those given previously by other authors especially for a fermion doublet whose mass difference is very large.

4. U. Baur (CERN), D. Zeppenfeld (Wisconsin U., Madison)
MEASURING THE $WW\gamma$ VERTEX IN SINGLE W PRODUCTION AT ep COLLIDERS.

A detailed analysis of single W production in ep collisions via $ep \to eW^{\pm}X$ is presented by using helicity amplitudes for general $WW\gamma$ couplings. Analytic expressions are given for the $\gamma q \to W^{\pm}q'$ helicity amplitudes which describe $ep \to eWX$ in Weizsäcker–Williams approximation and provide a semiquantitative understanding of the results obtained with the full matrix elements. Possibilities to test the gauge theory structure of the $WW\gamma$ vertex at HERA ($\sqrt{s} = 314$ GeV) and at an ep collider in the LEP tunnel ($\sqrt{s} \approx 1.4$ TeV) are explored. It is found that at HERA the $WW\gamma$ vertex can be measured with 30–50% accuracy after a few years of running.

5. Carla Goldman, Carlos O. Escobar (Sao Paulo U.)
THE $s \to d\gamma$ CONTRIBUTION TO HYPERON RADIATIVE DECAYS: A SUM RULE APPROACH.

We analyze, using the method of the QCD sum rules, the short distance contribution $s \to d\gamma$ to the hyperon radiative decays, $\Sigma^+ \to p\gamma$ and $\Xi^- \to \Sigma^-\gamma$. A more reliable estimate of this contribution is obtained, giving branching ratios that are larger than obtained before for the same process, although still small when compared with experimental data. The asymmetry parameter α, turns out to depend on the decay considered and surprisingly is equal to -1 for $\Sigma \to p\gamma$, for dynamical reasons which we discuss at length.

6. Yoshio Koide (Shizuoka U.)
CHARGED LEPTON MASS MATRIX WITH DEMOCRATIC FAMILY MIXING.

By starting from a mass matrix with family-independent mixing (the so-called 'democratic' family mixing) and by requiring some simple constraints on the additional mass terms, but without any small parameters, a mass relation which is in excellent agreement with experiment is derived for charged leptons.

7. Seitaro Nakamura (Tokai U.)
WEAK BOSONS.

A dynamical model of weak bosons is presented by taking an analogy of the diatomic molecule. We tentatively assume that the weak bosons now observed are the ground state of the vibrational levels, and expect the possible existence of the exotic weak bosons which correspond to the excited vibrational states, say 270 GeV, 450 GeV, ... The observabilities of these exotic weak bosons are discussed.

8. Seitaro Nakamura (Tokai U.)
HEAVY MESONS IN CHARGE SPIN MODEL.

In order to understand the dynamical reasons of the multiplet structures and the selection rules of the interactions of particles, we extend, instead of the standard quark model, the idea of 'global symmetry' in the 4-dimensional Euclidean space and introduced two charge spin operators, $\vec{\tau}$ (isospin) and $\vec{\zeta}$ (hypercharge spin), and classified leptons as well as baryons and mesons. Further, we introduce, in the same 4-dimensional Euclidean space, the second charge spins, $\vec{H}$ (isospin-like) and $\vec{U}$ (hypercharge spin-like) in order to include heavy mesons, J/ψ, D, Υ and B. By space reflection in this space, we have to interchange:

$$\begin{aligned} \tau &\rightleftharpoons \zeta \\ H &\rightleftharpoons U \\ J/\psi &\rightleftharpoons \Upsilon \\ D &\rightleftharpoons B \\ e &\rightleftharpoons \mu \end{aligned}$$

To heavy lepton tau (τ) we need a complementary partner, say sigma (σ), for the sake of space reflection symmetry in this space:

$$\tau \rightleftharpoons \sigma \, .$$

Selection rules and branching ratio problems for the weak interactions of heavy mesons are analyzed in this scheme and obtained the satisfactory results, which could not easily be obtained in the standard quark model.

9. Bing An Li, Keh-Fei Liu (Kentucky U.)
$K^*\bar{K}^*$ MESONIUM PRODUCTION IN $\gamma\gamma$ REACTIONS AND HADRONIC COLLISIONS.

Data of $\gamma\gamma \to K^{*+}K^{*-}$ in the 1.7–2.7 GeV region show a large cross section compared to the small cross section of $\gamma\gamma \to K^{*0}\bar{K}^{*0}$ and $\rho^0\phi$ in the same region. This pattern can be understood in terms of the mixing of the $Q^2\bar{Q}^2$ mesoniums. Small cross section of $\gamma\gamma \to \omega\phi$ and large cross sections of $K^{*0}\bar{K}^{*0}$ and $\omega\phi$ mesoniums in inclusive πp and pp productions are also expected.

10. H. Inazawa (Shoin Women's Univ.), T. Morii, S. Tanaka (Kobe U.)
ENHANCEMENT OF SCALAR QUARKONIUM $\widetilde{Q}\widetilde{Q}^*({}^1S_0)$ PRODUCTION VIA HIGGS BOSON EXCHANGE AT A MULTI TeV HADRON COLLIDER.

We calculate the production cross section of the bound state $\widetilde{Q}\widetilde{Q}^*({}^1S_0)$ composed of the scalar quark (squark) in hadron collisions through two-gluon fusion. With the very heavy mass of $\widetilde{Q}$, the production rate is largely enhanced due to the Higgs-boson exchange over the gluon exchange. In addition to the enhancement effect due to the strong coupling of the Higgs boson to $\widetilde{Q}$, the wave function at the origin is enhanced further by the ratio of vacuum expectation values of two Higgs fields.

11. H. Inazawa (Shoin Women's U.), T. Morii, S. Tanaka (Kobe U.)
PRODUCTION AND DECAY OF ULTRAHEAVY BOUND STATES.

We study the production and its subsequent decay of ultraheavy bound states $Q\overline{Q}$ composed of superheavy (4-th generation) quarks in a two-Higgs-doublet model. It is found that in such heavy quarkonia the Higgs-boson exchange dominates largely over the gluon exchange and the production rate of such quarkonia increases hugely in multi-TeV collider energies.

12. E. Gotsman (Tel Aviv U. & DESY), U. Maor (Tel Aviv U. & Illinois U., Urbana)
ESTIMATES OF THE GLUON-FUSION CONTRIBUTION TO THE PROTON STRUCTURE FUNCTION AT EXCEEDINGLY SMALL x.

Updated estimates are given for the contribution of gluon fusion with virtual photons to theproton structure function at very small x. The sensitivity of these estimates to the assumed input of parton distributions and gluon structure functions are examined in detail.

13. G.C. Branco, M.N. Rebelo (Lisbon, IFM), J.W.F. Valle (Valencia U.)
LEPTONIC CP VIOLATION WITH MASSLESS NEUTRINOS.

Leptonic CP Violation may arise in the $SU(2)\otimes U(1)$ theory even if the neutrinos are strictly massless and, as a result, it is potentially large. Theoretical scenarios include isosinglet neutral heavy leptons which are present in many extensions of the standard electroweak theory such as superstring inspired models. We discuss in detail the nature of this CP Violation and show that it can occur in a two-generation model. Possible effects are briefly discussed.

14. J.W.F. Valle (Valencia U.)
NEUTRINOS BEYOND THE STANDARD MODEL.

I review some basic aspects of neutrino physics beyond the Standard Model such as neutrino mixing and non-orthogonality, universality and CP violation in the lepton sector, total number and lepton flavor violation, etc. These may lead to neutrino decays and oscillations, exotic weak decay processes, neutrinoless double β decay, etc. Particle physics models are discussed where some of these processes can be sizable even in the absence of measurable neutrino masses. They may also substantially affect the propagation properties of solar and astrophysical neutrinos.

15. R.S. Fletcher, F. Halzen (Wisconsin U., Madison), R.W. Robinett (Penn State U.)
PROBING THE GLUON STRUCTURE OF THE PHOTON WITH HERA.

We study the production of ψ-mesons in ep collisions and argue that ψ's form the ideal experimental signature for directly identifying the 'anomalous' gluon structure of the photon. The observation is of relevance to studies of the nucleon structure functions using HERA and to high energy γ-ray astronomy.

16. J.W.F. Valle (Valencia U.)
THEORY AND IMPLICATIONS OF NEUTRINO MASS.

I briefly review the basic theory of neutrino mass from the point of view of modern gauge theories. Some of the implications of neutrino masses for particle physics, nuclear physics, cosmology and astrophysics are discussed.

17. A. Santamaria (Carnegie Mellon U.), J.W.F. Valle (Valencia U.)
SOLAR NEUTRINO OSCILLATION PARAMETERS AND THE BROKEN R-PARITY MAJORON.

Matter-enhanced neutrino-oscillation parameters can be probed in a variety of conventional experiments in supergravity models where the small neutrino mass arises from spontaneous R-parity violation. A combined analysis of astrophysical and laboratory limits tends to *exclude* regions of oscillation parameters where the high-energy neutrinos are *adiabatically* converted. This suggests the possibility of a large reduction in the pp and ^7Be neutrino flux even for a mildly reduced ^8B neutrino flux, thus stressing the importance of gallium experiments.

18. M. Dittmar (UC, Riverside & CERN), A. Santamaria (Carnegie Mellon U. & Valencia U.), M.C. Gonzalez-Garcia, J.W.F. Valle (Valencia U.)
PRODUCTION MECHANISMS AND SIGNATURES OF ISOSINGLET NEUTRAL HEAVY LEPTONS IN Z^0 DECAYS.

Neutral Heavy Leptons (NHL) arise in many extensions of the standard electroweak theory such as superstring inspired models. The possibility of gauge singlets NHLS is especially attractive because it gives an explanation to the observed smallness of the neutrino mass. Existing limits on the possible existence of such particles are still fairly poor. We have investigated isosinglet NHL production and decays within different models. The dominant production cross section is single production (*i.e.* $Z^0 \to N + \bar{\nu}$ or $Z^0 \to \overline{N} + \nu$) as a result of mixing with the standard doublet neutrinos. Subsequent NHL decays lead to striking signatures. Taking into account the expected luminosities and typical detector efficiencies of the different LEP/SLC experiments we conclude that these may discover isosinglet NHLS or else substantially improve and extend present limits on their mass and coupling strength.

19. M.C. Gonzalez-Garcia, J.W.F. Valle (CERN & Valencia U.)
FAST DECAYING NEUTRINOS AND OBSERVABLE FLAVOR VIOLATION IN A NEW CLASS OF MAJORON MODELS.

Neutrinos can have any mass (allowed by laboratory limits) without violating limits from cosmology, astrophysics or laboratory searches for lepton violation phenomena. We present a simple extension of the standard theory where neutrinos decay dominantly into invisible models involving a majoron associated with the spontaneous violation of B-L symmetry due to physics at or below the electroweak scale. Measurable branchings for lepton-flavor-violating processes such as $\mu \to e + \gamma$, and for non-standard Z decays *e.g.* $Z \to e + \bar{\tau}$ and $Z \to \mu + \bar{\tau}$ (plus their conjugates) at LEP are possible without unnatural fine tuning of the parameters. Lepton-number-violating effects such as neutrinoless $\beta\beta$ decay may also be present at a measurable level.

20. R.N. Mohapatra (Maryland U.), E. Rusjan (Virginia Tech), G. Senjanovic (CERN & Zagreb U.), A. Sokorac (Kidric Inst., Belgrade)
27^3 YUKAWA COUPLINGS IN THE THREE GENERATION SUPERSTRING MODEL.

We compute Yukawa couplings for various values of complex moduli on Tian-Yau manifold. For certain specific choices of the parameters, a maximally symmetric case being one of them, we find many couplings to vanish for geometric reasons.

21. K.S. Babu (Maryland U.), D. Eichler (Maryland U. & Ben Gurion U. of Negev), R.N. Mohapatra (Maryland U.)
RIGHT-HANDED NEUTRINO AS WEAKLY UNSTABLE DARK MATTER.

We suggest that the massive right-handed neutrino in a specific version of left-right symmetric model can have a lifetime $\simeq 10^{25}$ sec and therefore, cannot only act as a dark matter but also provide a source of anomalous positrons with energies higher than 10 GeV reported in cosmic ray experiments.

22. K.S. Babu, R.N. Mohapatra (Maryland U.)
MODEL FOR LARGE TRANSITION MAGNETIC MOMENT OF THE ELECTRON NEUTRINO.

We propose a simple extension of the standard model by adding an $SU(2)_H$ Horizontal gauge symmetry in the leptonic sector, which leads to a large transition magnetic moment of the electron neutrino while keeping the neutrino mass naturally small. This model can provide a solution to the solar neutrino puzzle while at the same time avoiding the SN1987A bound on the neutrino magnetic moment.

23. Rabindra N. Mohapatra (Maryland U.)
FERMION MASSES AND MIXINGS OUT OF RADIATIVE CORRECTIONS.

We present a scheme where the mass of the third generation of quarks and charged leptons appears at the tree level and the masses of the second and first generation fermions arise at one and two loop levels respectively. This provides a natural explanation of the observed mass hierarchy among fermions, without any fine tuning of parameters and without the need for any horizontal symmetry. The small mixing angles between the quarks and the tiny mass for neutrinos is also explained as they arise purely from higher loop effects.

24. K.S. Babu, Rabindra N. Mohapatra (Maryland U.)
CP VIOLATION IN SEE-SAW MODELS OF QUARK MASSES.

We study CP violation in the left-right symmetric models with 'see-saw' formula for charged fermions. CP-violation phenomenology closely parallels that of the usual left-right symmetric models with the additional advantage that it provides a natural solution to the strong CP problem. For the case where the third generation mixing parameter V_{ub} is extremely small, the neutral Higgs interactions leads to $\epsilon'/\epsilon \simeq 10^{-3}$ and the electric dipole moment of the neutron $d_n \simeq 10^{-25}\, e\, cm$. Smallness of the neutrino masses is understood as a two-loop effect.

25. B.S. Balakrishna, R.N. Mohapatra (Maryland U.)
RADIATIVE FERMION MASSES FROM NEW PHYSICS AT TeV SCALE.

We show how the smallness of neutrino masses and the observed mass hierarchy among charged leptons can be understood within a scheme recently proposed to understand the quark mass and mixing hierarchies. The scale of new physics in this model is in the TeV range and could therefore be experimentally accessible.

26. V. Barger, J.L. Hewett (Wisconsin U., Madison), T.G. Rizzo (Wisconsin U., Madison & Ames Lab. & Iowa State U.)
ρ PARAMETER CONSTRAINTS ON FOURTH-GENERATION QUARK MASSES.

Constraints on the masses of possible fourth generation quarks (a, v) are obtained from measurements of the ρ parameter and the elements of the quark mixing matrix. Stringent mass limits are found when the off-diagonal elements V_{tv} and V_{ab} are large. For example, with $m_t = 90$ GeV and $|V_{tv}| \simeq 0.5$ we find $M_{a,v} \leq 300$ GeV. Stronger constraints are obtained as m_t or $|V_{tv}|$ increase.

27. JoAnne L. Hewett (Wisconsin U., Madison), Thomas G. Rizzo (Ames Lab. & Iowa State U.)
LOW-ENERGY PHENOMENOLOGY OF SUPERSTRING-INSPIRED E_6 MODELS.

In this report we survey the low energy phenomenological implications of superstring-inspired E_6 models. The motivation for such models is reviewed. New particles including new gauge bosons, exotic fermions, Higgs bosons, and their superpartners are expected to exist in models of this type. We summarize the present experimental limits on these particles from both accelerator and non-accelerator data. Techniques for producing these new particles directly at existing and planned colliders as well as searching for their indirect effects are examined in detail. Other phenomenological implications of such models are also reviewed.

28. Hidezumi Terazawa (Tokyo U., INS)
SUPERSYMMETRY IN COMPOSITE MODELS AND THE MASS SPECTRUM OF QUARKS AND LEPTONS.

It is shown that supersymmetry turns out to be very useful and powerful enough to determine the mass spectrum of quarks and leptons in composite models. For quarks and leptons as either almost or quasi Nambu-Goldstone fermions, many mass formulas and sum rules are derived. Among others, the mass sum rules of $m_e^{1/2} - m_{\nu_e}^{1/2} = m_d^{1/2} - m_u^{1/2}$, $m_\mu^{1/2} - m_e^{1/2} = m_s^{1/2} - m_d^{1/2}$, $m_\mu^{1/2} - m_{\nu_e}^{1/2} = m_s^{1/2} - m_u^{1/2}$ and $m_\tau^{1/2} - m_{\nu_\mu}^{1/2} = m_b^{1/2} - m_c^{1/2}$ are satisfied remarkably well by the experimental values and estimates.

29. Hidezumi Terazawa (Tokyo U., INS)
NEUTRAL CURRENT EFFECT IN NUCLEAR BETA DECAYS.

It is pointed out that the electron energy spectrum of nuclear β-decays is affected by the weak neutral current interaction in nuclei to the order of several eV. The effect has become a serious obstacle to future measurements of the neutrino mass within the accuracy of a few eV but, once observed, it will become very useful for measurements of the weak mixing angle of the quark density in nuclei.

30. Hidezumi Terazawa (Tokyo U., INS)
RELATION BETWEEN THE AVERAGE MULTIPLICITY AND TRANSVERSE MOMENTUM OF PRODUCED PARTICLES IN HADRON-HADRON COLLISIONS AT VERY HIGH ENERGIES.

It is pointed out that the average charged multiplicity ($\langle n_{ch}\rangle$) and transverse momentum ($\langle p_T\rangle$) of produced particles in hadron-hadron collisions at very high energies ($\sqrt{s}$) have a simple relation of $\langle n_{ch}\rangle^2\langle p_T\rangle/\sqrt{s}$ = constant (= 0.70 ± 0.05) in the generalized Fermi-Landau statistical and hydrodynamical model. The relation is satisfied remarkably well by the experimental data up to the SPS $p - \bar{p}$ Collider energies and will soon be tested by the Tevatron Collider experiments.

31. Hidezumi Terazawa (Tokyo U., INS)
NEW PHYSICS AT THE TeV SCALE.

Many important predictions mostly in composite models which have already been checked or will be tested in the future experiments at TeV energy scale are summarized. They include predictions for 1) the gauge coupling constants (and the weak mixing angle), 2) the masses of the Higgs scalar and top quark, 3) the quark mixing parameters, 4) the excited quarks, leptons, gauge bosons and Higgs scalars and 5) the exotic particles, especially the 'color ball'. Various topics including 1) supersymmetry, 2) superstring phenomenology, 3) technicolor, 4) composite model, 5) new gauge interaction (5-th force, horizontal gauge model, etc.), 6) flavor mixing and symmetry violation ($B^0 - \overline{B^0}$ mixing, lepton mixing, CP violation search, etc.), 8) underground experiments and ν physics (solar ν problem, ν from supernova, etc.) and 7) physics perspectives in TRISTAN/SLC/LEP/HERA/TEVATRON/SSC experiments are also discussed.

32. V. Barger (Wisconsin U., Madison), W.Y. Keung (Illinois U., Chicago), T.G. Rizzo (Ames Lab. & Iowa State U.)
Z BOSON BREMSSTRAHLUNG IN HEAVY QUARK DECAY.

We calculate the rate for Z boson bremsstrahlung in the charged current decay $Q \to qWZ$ of a heavy fermion Q to a light fermion q. This process may be used to test the WWZ trilinear gauge boson coupling of the standard model. We evaluate the branching fraction for this decay mode of fourth generation and E_6 exotic fermions: the branching fraction increases dramatically with increasing m_Q and may reach 1% for the TeV fermion masses.

33. Thomas G. Rizzo (Wisconsin U., Madison & Ames Lab. & Iowa State U.)
SQUARK AND SQUARKONIUM PRODUCTION BY GAUGE BOSON FUSION AT TeV e^+e^- COLLIDERS.

We explore the production of squarkonium as well as squark and slepton pairs by gauge boson fusion at TeV e^+e^- colliders. Although squarkonium production rates are found to be too small to be observable, gauge boson fusion contributions to squark and slepton pair production can be sizeable for a reasonable range of model parameters.

34. Thomas G. Rizzo (Wisconsin U., Madison & Ames Lab. & Iowa State U.)
EXOTIC QUARK PRODUCTION AT e^+e^- COLLIDERS.

The prospects for discovering exotic quarks (h), present in E_6 models, at existing and future high energy e^+e^- colliders are examined. The flavor-changing decay mode, $h \to \text{jet} + \ell^+\ell^-$, clearly separates the production of h from that of new sequential quarks. Although this mode has a small branching fraction, reasonable event rates at each of the various colliders is expected.

35. Thomas G. Rizzo (Wisconsin U., Madison)
SINGLE $W_R^\pm$ PRODUCTION AT TeV e^+e^- COLLIDERS IN AN ALTERNATIVE LEFT-RIGHT SYMMETRIC MODEL.

We examine the cross section and p_T distribution for the production of a single, right-handed, charged gauge boson ($W_R^\pm$) at new TeV e^+e^- colliders within the context of the E_6 superstring-inspired Alternative Left-Right Symmetric Model proposed by Ma. The calculations are performed using the effective photon approximation. For $\sqrt{s} = 1(2)$ TeV, significant production rates are found for W_R masses as large as 0.8 (1.5) TeV with distinct signatures and p_T distributions.

36. Thomas G. Rizzo (Wisconsin U., Madison & Ames Lab. & Iowa State U.)
ORDINARY-EXOTIC QUARK MIXING.

We reconsider the possibility that a new $Q = -1/3$, vector-like, iso-singlet quark, h, which can appear in superstring-inspired E_6 models, may have appreciable mixing with the usual bottom (b) quark. Using the most recent data on the Kobayashi-Maskawa mixing matrix, ϵ'/ϵ, and $B - \overline{B}$ and $D - \overline{D}$ mixing, we analyze the influence of h on rare processes, e.g., $K \to \pi\nu\bar{\nu}$, $b \to s\gamma$, and $D \to \mu^+\mu^-$.

37. D.A. Dicus (Texas U.), J.L. Hewett (Wisconsin U., Madison), C. Kao (Texas U.), T.G. Rizzo (Wisconsin U., Madison & Ames Lab & Iowa State U.)
$W^\pm H^\mp$ PRODUCTION AT HADRON COLLIDERS.

We have examined the production of a charged Higgs boson in association with a W boson at high energy hadron colliders, i.e. $pp \to W^\pm H^\mp + X$ and find that the production rates can be large. The various subprocesses which contribute to this mechanism at the tree and one-loop level are compared, and we find that quark annihilation, $b\bar{b}, t\bar{t} \to W^\pm H^\mp$ is dominant. Production via gluon-gluon fusion, which proceeds through box and triangle diagrams is also found to yield a significant contribution. We also compare our results with those obtained in rank-5 E_6 models for this process.

38. Thomas G. Rizzo (Wisconsin U., Madison)
ADDENDUM TO 'PRODUCTION OF FLAVOR CHANGING GAUGE BOSONS FROM E_6 AT e^+e^- COLLIDERS': SINGLE PRODUCTION IN e^+e^- ANNIHILATION.

We extend our previous analysis of the production of neutral, non-Hermitian, flavor-changing gauge bosons (W_I) to include single, associated production in e^+e^- collisions. Such gauge degrees of freedom are associated with the $SU(2)_I$ subgroup of E_6 which can lead to new low-energy electroweak interactions and produce distinct decay signatures.

39. Thomas G. Rizzo (Wisconsin U., Madison & Ames Lab & Iowa State U.), Richard W. Robinett (Penn State U.)
TRIPLE GAUGE-BOSON DECAY OF NEW NEUTRAL GAUGE BOSONS.

The decay of new heavy neutral gauge bosons (Z') into three gauge boson final states (e.g. W^+W^-Z) is examined within the context of E_6 superstring-inspired models. We find that the ratio of decay rates $\Gamma(Z' \to W^+W^-Z)/\Gamma(Z' \to W^+W^-)$ grows quadratically with the Z' mass and can be as large $\simeq 0.43$ for a Z' mass of 3 TeV. Such final states should be observable above the continuum at both e^+e^- and hadron colliders.

40. Thomas G. Rizzo (Wisconsin U., Madison & Ames Lab. & Iowa State U.)
NEW QUARK AND LEPTON PRODUCTION AT HIGH-ENERGY e^+e^- COLLIDERS.

We compare and contrast production and decay signatures for several kinds of new quarks and leptons which may be produced at future high energy e^+e^- colliders. Mirror, iso-doublet and iso-singlet vector-like, and sequential fourth generation fermions are considered along with the iso-triplet lepton of Foot et al. We find that total cross sections, forward-backward, and left-right polarization asymmetries combined with particular final state decay signatures, can be used together to distinguish among these possible sets of new fermions at e^+e^- colliders.

41. N.P. Merenkov (Kharkov, FTI)
THE CALCULATION OF THE RADIATIVE CORRECTIONS TO THE ELECTRON NUCLEUS SCATTERING CROSS SECTION BY THE METHOD OF THE STRUCTURE FUNCTIONS.

No abstract included.

42. M.G. Bowler (Oxford U., NPL), P.N. Burrows, D.H. Saxon (Rutherford)
BARYON FRAGMENTATION FUNCTIONS AND
DIQUARK STRUCTURE.

We present a simple dynamical model of baryon production which predicts the form of baryon fragmentation functions to be softer than for mesons. These predictions are in good accord with recent measurements of the spectra of baryons produced in e^+e^- annihilation. The model relates the degree of softening of the baryon fragmentation function to the probability of producing mesons between baryon and anti-baryon.

43. Charles A. Nelson (SUNY, Binghamton)
SPIN CORRELATIONS IN Z^0 DECAYS.

In contrast with the elegant τ-polarization technique, measurement of the energy correlation function $I(E_A, E_B)$ for the decay sequence $Z^0 \to \tau^+\tau^- \to A^+B^-X$ will determine independently the fundamental parameters $\sin^2\theta_W$, ξ_A, and ξ_B. For 10 million Z^0 events, the ideal statistical percentage error for $\sin^2\theta_W$ is 0.3%, for the Michel parameter ξ it is 2.6%, and for the analogous chiral parameter ξ_π it is 1.4%. Analogously, $D^{*+}D^{*-}$ correlation functions in Z^0 decays can be used to study the spin dynamics of heavy quarks in QCD physics without instrumentation of longitudinal beam polarization.

44. Charles A. Nelson (SUNY, Binghamton)
TESTS FOR 'NEW PHYSICS' FROM TAU SPIN CORRELATION FUNCTIONS FOR $Z^0 \to \tau^+\tau^- \to A^+B^-X$.

The fundamental parameters ξ_A, ξ_B, and α_H can be independently determined by measurement of the energy correlation function $I(E_A, E_B)$ where ξ is the Michel polarization parameter for $\tau^- \to \ell^-\nu\bar{\nu}$, $\xi_A = (|g_L|^2 - |g_R|^2)/(|g_L|^2 + |g_R|^2)$ is the chiral polarization parameter for $\tau^- \to A^-\nu$, and $\alpha_H \simeq -2a_\tau v_\tau/(a_\tau^2 + v_\tau^2)$ where a_τ, v_τ describe $Z^0 \to \tau^+\tau^-$ at tree level. In contrast, measurement of $I(E_A)$ by the τ-polarization technique only determines $\xi_A\alpha_H$. For 10 million Z^0 events and for $\sin^2\theta_W = 0.23$, the ideal statistical percentage-errors in the determination of the Michel parameters are ξ [2.6%], δ [3.5%], ρ [1.0%], and of the chiral parameters ξ_π [1.4%], ξ_ρ [3.0%], ξ_{K^*} [18%]. For α_H, the ideal statistical error is $\sigma(\alpha_H) = 0.005$ [3.3%], equivalently, for $\sin^2\theta_W$ [0.3%]. Explicit formulas for the full sequential decay correlation function $I(E_{\bar{\mu}}, E_e, \cos\psi_{\bar{\mu}e})$ are given for arbitrary Z mass, and for $I(E_{\bar{\mu}}, E_B)$ for arbitrary Michel parameters. These results can be used for analyzing the decay of a Z' boson and for $q\bar{q}$ modes. The present world average for ρ only excludes more than 47% of a 'V+A' coupling at the 95% confidence level.

45. H. Baer (Florida State U.), V. Barger (Wisconsin U., Madison), R.J.N. Phillips (Rutherford)
SEARCH FOR TOP QUARK DECAYS TO REAL W BOSONS AT THE TEVATRON COLLIDER.

We propose ways in which eventual signals from heavy top quark decays into on-shell or nearly-on-shell W bosons may be discriminated more sharply from background at the Tevatron $\bar{p}p$ collider. For single-lepton $\bar{t}t$ signals, the principal background from direct W production cannot be eliminated by acceptance cuts alone, but the signal/background ratio can be greatly increased by requiring high jet multiplicity n. Top quark signals can be detected and m_t can be measured via enhancements in the single-lepton event rates at high n, or via a peak in the invariant mass of the three hardest jets at high n. A $\bar{t}t$ signal can be extracted up to $m_t = 150$ GeV (180 GeV) from selected data with $n \geq 3$ ($n \geq 4$) jets. For two-lepton $\bar{t}t$ signals, the principal standard model backgrounds from $\bar{b}b$, $\bar{c}c$ and Drell-Yan dilepton production can all be removed by cuts. The remaining backgrounds from direct W^+W^- and Drell-Yan $\bar{\tau}\tau$ production can be suppressed to a level well below the $\bar{t}t$ signals, for the whole of the mass range up to $m_t = 200$ GeV allowed for the standard model. Eventual $\bar{t}t$ signals can be confirmed by characteristic dynamical distributions. An integrated luminosity of 10 pb^{-1} (100 pb^{-1}) would be enough to detect top quarks up to $m_t = 120$ GeV (200 GeV).

46. Siegfried Bethke (LBL, Berkeley)
AN EXPERIMENTAL APPROACH TO OPTIMIZE AND TEST PERTURBATIVE QCD TO $\mathcal{O}(\alpha_s^2)$.

Production rates of multijet hadronic final states observed in e^+e^- annihilation are compared with recent QCD calculations using renormalization group improved perturbation theory to $\mathcal{O}(\alpha_s^2)$. The scale parameter $\Lambda_{\overline{MS}}$ and the renormalization scale μ^2 are adjusted in order to reproduce the observed 2-, 3- and 4-jet event production rates. Small scales of $\mu^2 \approx 0.002 \cdot E_{\text{cm}}^2$ provide a significantly better description of the overall 4-jet production rates and of 2- and 3-jet rates at small jet pair masses than calculations using $\mu^2 = E_{\text{cm}}^2$. The adjusted value of $\Lambda_{\overline{MS}}$ depends on the choice of μ^2. It decreases by a factor of two when going from $\mu^2 = E_{\text{cm}}^2$ to $\mu^2 = 0.002 \cdot E_{\text{cm}}^2$ and results in $\Lambda_{\overline{MS}} = 95$ MeV $\pm$ 30 MeV for $0.001 \leq \mu^2/E_{\text{cm}}^2 \leq 0.020$. The predictions of an Abelian vector theory complete to $\mathcal{O}(\alpha_A^2)$ are not compatible with the observed dynamics of jet production. The experimental evidence for the energy dependence of α_s, obtained from 3-jet event production rates observed in the center-of-between 22 GeV and 56 GeV, does not depend on the detailed choice of μ^2.

47. K. Abe, et al. (Brookhaven & Brown U. & Hiroshima U. & KEK, Tsukuba & Osaka U. & Penn U. & SUNY, Stony Brook)
DETERMINATION OF $\sin^2\theta_W$ FROM MEASUREMENTS OF DIFFERENTIAL CROSS SECTIONS FOR MUON NEUTRINO AND ANTINEUTRINO SCATTERING BY ELECTRONS.

Measurements of the combined differential cross sections $(d\sigma/dy)$ for ν_μ and $\bar{\nu}_\mu$ scattering by electrons yield the value $\sin^2\theta_W = 0.194 \pm 0.018 \pm 0.013$. This value, at $Q^2 \simeq 1 \times 10^{-3}(GeV/c)^2$, is more precise than and in good agreement with our determination from the ratio of the total cross sections for the same reaction. It is consistent with, but lower than, the value determined from deep inelastic neutrino scattering.

48. A. Bartl (Vienna U.), H. Fraas (Wurzburg U.), W. Majerotto, N. Oshimo (Vienna, OAW)
THE NEUTRALINO MASS MATRIX IN THE MINIMAL SUPERSYMMETRIC MODEL.

We study systematically the dependence of the neutralino mass eigenvalues and eigenstates on the parameters of the mixing matrix. Starting from analytical solutions for special values of the parameters we work out various practical approximation formulae for the masses. We examine their applicability in the different regions of parameters accessible with the next generation of accelerators.

49. Masako Bando (Aichi U.), Taichiro Kugo (Kyoto U.), Koichi Yamawaki (Nagoya U.)
NONLINEAR REALIZATION AND HIDDEN LOCAL SYMMETRIES.

The idea of dynamical gauge bosons of hidden local symmetries in nonlinear sigma models is reviewed. Starting with a fresh look at the Goldstone theorem and low energy theorems, we present a modern review of the general theory of nonlinear realization both in nonsupersymmetric and supersymmetric cases. We then show that any nonlinear sigma model based on the manifold G/H is gauge equivalent to a 'linear' model possessing a $G_{\text{global}} \times H_{\text{local}}$ symmetry, H_{local} being a hidden local symmetry. The corresponding supersymmetric formulation is also presented. The above gauge equivalence can be extended to a model having a larger symmetry $G_{\text{global}} \times G_{\text{local}}$. Also reviewed are dynamical calculations showing that in some two-, three- and four-dimensional models, the gauge bosons of the hidden local symmetries acquire the kinetic terms via quantum effects, thus becoming 'dynamical'. We suggest that such a dynamical gauge boson may be a rather common phenomenon realized in Nature. As a realistic example, we examine the QCD case where we identify the vector mesons $(\rho, \omega, \phi, K^*)$ with the dynamical gauge bosons of the hidden $U(3)_v$ local symmetry in the $U(3)_L \times U(3)_R/U(3)_v$ nonlinear sigma model. The totality of the vector meson phenomenology seems to support our basic idea. The axial-vector mesons are also incorporated into our framework. Also given is a brief sketch of some applications of this formalism to unified models beyond the standard model, such as technicolor, composite W/Z boson and supergravity models.

50. S.Y. Choi, Taeyeon Lee, H.S. Song (Seoul National U.)
DENSITY MATRIX FOR POLARIZATION OF HIGH SPIN PARTICLES.

The Bouchiat and Michel formula for the polarization of spin-1/2 particle is extended to higher spin cases. Explicit formulas for massive spin-1 and spin-3/2 particles are given and discussed in connection with density matrices.

51. V. Barger (Wisconsin U., Madison), R.J.N. Phillips (Rutherford)
SIGNALS OF TOP QUARKS WITH CHARGED HIGGS DECAY AT $p\bar{p}$ COLLIDERS.

If the top quark decays dominantly via a charged Higgs boson $t \to bH^+$, as occurs if $m_t < M_W + m_b$ and $m_{H^\pm} < m_t - m_b$, the conventional top-quark signals at $p\bar{p}$ colliders (based on $t \to bW^*$ charged-current decays) are suppressed. Distinctive new signals appear, based on $H^+ \to \bar{\tau}\nu$ decays: they include events with large missing p_T, events with isolated τ-jets or leptons plus hadronic jets, and events with two isolated τ-jets. We calculate typical cross sections and distributions for the Tevatron $p\bar{p}$ collider: the results are sensitive to the $H \to \tau\nu$ branching fraction, that may well be large.

52. H. Baer (Florida State U.), V. Barger (Wisconsin U., Madison), R.J.N. Phillips (Rutherford)
TOP QUARK DETECTION VIA W + n-JET MEASUREMENTS.

In $t\bar{t}$ and W production at $p\bar{p}$ colliders, top quark semileptonic decays $t \to b\ell\nu$ contribute to the same class of events as $W \to \ell\nu$ decays. The top contribution is particularly significant in $n \geq 2$ jet events and can be observed (a) as a distortion in the $\ell\nu$ transverse mass distribution below the $W \to \ell\nu$ Jacobian peak if $m_t < M_W$: (b) as a systematic increase in the jet multiplicity distribution at large n for any m_t: (c) as a shift in the apparent $(W + n\text{ jet})/(Z + n\text{ jet})$ event ratio at large n for any m_t. The observation of a top quark signal in these ways does not require prior elimination of the W background.

53. V. Barger (Wisconsin U., Madison), G.F. Giudice (Fermilab), T. Han (Wisconsin U., Madison)
SOME NEW ASPECTS OF SUPERSYMMETRY R-PARITY VIOLATING INTERACTIONS.

We examine R-parity breaking interactions with lepton number violation in the minimal supergravity model and obtain new stringent bounds on the coupling constants from low energy processes. We discuss the decay of the lightest supersymmetric particle and analyze possible signatures for R-parity violating interactions in W, Z and Higgs boson decays, and in e^+e^- collisions.

54. V. Barger, T. Han (Wisconsin U., Madison), J. Ohnemus (Florida State U.), D. Zeppenfeld (Wisconsin U., Madison)
PERTURBATIVE QCD CALCULATIONS OF WEAK BOSON PRODUCTION IN ASSOCIATION WITH JETS AT HADRON COLLIDERS.

We present perturbative QCD calculations of $W, Z+$ n-jet production ($n \leq 3$) in $p\bar{p}$ collisions at $\sqrt{s} =$ 0.63 TeV and 1.8 TeV and also extrapolate the production cross sections to supercollider energies. The dependence of the cross sections on experimental cuts and the theoretical ambiguities are discussed in detail. It is shown that W/Z ratios of transverse momentum distributions are quite insensitive to these uncertainties and can be predicted rather accurately. Neutrino counting and a top quark search at large p_T are addressed as applications. Our results are based on complete tree level calculations of $W, Z+(n+2)$-parton amplitudes.

55. V. Barger (Wisconsin U., Madison), R.J.N. Phillips (Rutherford)
HIDDEN TOP QUARK WITH CHARGED HIGGS DECAY.

In top quark decays, there may be competition between decays mediated by real or virtual W bosons and decays into a charged Higgs boson. We evaluate the branching fractions to delineate the regions where standard top signals would be suppressed. We also show that both the standard semileptonic decay signals $t \to bW^+ \to b\ell\nu$ and unconventional signatures such as $t \to bH^+ \to b\tau\nu$ may be simultaneously suppressed with suitable parameter choices in the two-Higgs-doublet model: in such circumstances the top quark would be very difficult to detect at hadron colliders.

56. Wei-Shu Hou (Munich, Max Planck Inst.), Robin G. Stuart (Mexico, IPN)
FLAVOR CHANGING NEUTRAL CURRENTS INVOLVING HEAVY FERMIONS: A GENERAL SURVEY.

We calculate, using computer algebra techniques, a general expression for the one-loop fermion-neutral boson coupling keeping all masses and momenta. The correctness of the result is guaranteed by the application of a number of severe analytic and numerical checks. The result is applied to the calculation of the flavor changing decays of the Z^0 in the 2 Higgs doublet model and to the flavor changing neutral current decays of the fourth generation b' quark. In the latter case it is found that this process can dominate charged current decays. It is likely that a b' quark decaying via flavour changing neutral modes could escape detection by conventional searches.

57. Stephen L. Adler (Princeton, Inst. Advanced Study)
A NEW ELECTROWEAK AND STRONG INTERACTION UNIFICATION SCHEME.

I explore the consequences of requiring the electroweak gauge group to contain an $SU(2)$ subgroup rotating the charged fermions to their antiparticles. This is implemented by coupling the leptons ℓ_L, ℓ_L^c, ν_L of each family as a 3_L representation of an electroweak $SU(3)$ gauge group, and coupling the corresponding quarks as a second 3_L and two 3_R's or 3_L^*'s. In the absence of strong interactions, the model predicts $\sin^2\theta_W = 1/4$ at the electroweak unification energy. Since the electroweak $SU(3)$ does not commute with the color $SU(3)$ group, the minimal consistent electroweak-strong gauge group is obtained from the closure of their commutator algebra. This has been determined to be the 151-generator semi-simple Lie algebra $SU(3)\times SU(12)$, and gives a corrected formula $\sin^2\theta_W = 1/4 + 1/3\,(\alpha/\alpha_s)$ at the $SU(3) \times SU(12)$ unification energy, which is computed to be $10^{10.8\pm0.8}$ GeV. Unification in the simple group $SU(15)$ is attained at an energy approximately a factor of 10 higher.

58. Chiara R. Nappi (Princeton, Inst. Advanced Study)
INTERACTING LAGRANGIANS FOR MASSIVE SPIN-TWO PARTICLES FROM STRINGS AND KALUZA-KLEIN THEORIES.

Working at the first order in the massive field we derive explicitly, both from bosonic strings and Kaluza-Klein theory, lagrangians for massive spin two interacting with a massless background. The purpose is to see explicitly how well-known inconsistencies and difficulties of higher spin interactions are circumvented in these theories.

59. Hoi Lai Yu, Ta Shin Lee, Wai Bong Yeung (Taiwan, Inst. Phys.)
MAGNETIC EFFECTS AND CHIRAL SYMMETRY BREAKING IN STRONGLY COUPLED QUANTUM ELECTRODYNAMICS.

We study analytically the effect of magnetic interactions on dynamical symmetry breaking in strongly coupled quantum electrodynamics. By solving the Schwinger-Dyson equation for the electron propagator in the asymptotic region with the inclusion of magnetic effects in the vertex function, we find a nonperturbative solution whose existence indicates a spontaneous breaking of chiral symmetry with a critical coupling constant depending on the magnetic-interaction strength.

60. P.C.M. Yock (Auckland U.)
AN ALTERNATIVE APPROACH TO ELEMENTARY PARTICLE PHYSICS.

A gauge theory of sixteen fermions and electrostrong interactions appears to be as systematic and natural as the quark model. Photon-nucleon collisions at cosmic ray energies may reveal the fundamental structure of the nucleon.

61. H. Inazawa (Shoin Women's U.), T. Morii (Kobe U.), J. Morishita (Dokkyo U., Soka)
SPECTROSCOPY OF ULTRAHEAVY QUARKONIA.

We study the spectroscopy of ultraheavy quarkonia which are composed of superheavy (4th generation) quarks. Owing to the strong Higgs-boson exchange force over the gluon exchange force, the situation of the mass levels of such quarkonia becomes drastically different from the ordinary quarkonia like $c\bar{c}$ or $b\bar{b}$. The hyperfine splitting becomes larger than the fine splitting.

62. C.L. Bilchak, D. Schildknecht, J.D. Stroughair (Bielefeld U.)
SHADOWING IN DEEP INELASTIC MUON SCATTERING ON HEAVY NUCLEI.

It is shown that the simple generalized vector dominance model developed a long time ago, consistently and quantitatively describes the shadowing effect recently detected in deep inelastic muon scattering on heavy nuclei.

63. G. Gounaris (CERN & Bielefeld U.), D. Schildknecht (Bielefeld U.)
ON THE RADIATIVE CORRECTIONS TO THE ELECTROWEAK PARAMETERS AND PRECISION TESTS OF THE ELECTROWEAK THEORY.

The radiative corrections to the electroweak parameters are reconsidered with an emphasis on analysing prospects for future tests of the as yet untested parts of the electroweak theory, in particular the 'new physics' of vector-boson self-interactions and the Higgs scalar. The vacuum polarization due to the light fermions is treated in the leading-log approximation, while the top-quark is taken into account exactly. A detailed analysis of the errors involved in our approximations and a comparison with the results of complete one-loop calculations shows that vacuum polarization due to bosons is negligible, if $m_H = 100$ GeV, while it may become visible in precision tests in e^+e^- annihilation, if $m_H \cong 1$ TeV. We also give detailed results (as a function of the top-quark mass) on the radiatively corrected parameters used in model-independent fits to neutrino scattering and in the interpretation of atomic-parity violation experiments. Technically, we diagonalize the $\gamma - Z$ propagator for any q^2, and we show, when treating the top-quark vacuum polarization exactly, that the intuitively appealing notion of running coupling constants can be used beyond the leading-log approximation.

64. F.M. Renard (Montpellier U.), D. Schildknecht (Bielefeld U.)
RELATING THE FUNDAMENTAL COUPLING CONSTANTS VIA A STABILITY PRINCIPLE.

We propose a dynamical stability principle related to the interplay of electromagnetic and weak interactions and predict $\sin^2\theta_W$ in terms of the lepton and quark quantum numbers in good agreement with its empirical value. The application of our stability principle to the interplay of electromagnetic and strong interactions yields a value for the $\gamma\rho^0$ transition strength (and thus implicitly for α/α_s), which is satisfactory, if one takes into account the crudeness of the approximations involved.

65. F.M. Renard (Montpellier U.), D. Schildknecht (Bielefeld U.)
THE STABILITY PRINCIPLE, $\sin^2\theta_W$ AND THE MASS OF THE TOP QUARK.

A global (integral) form of the recently proposed stability principle, which led to the successful prediction of $\sin^2\theta_W \cong 0.23$, is introduced. The global form allows one to take into account particle masses. Consistency with the empirical value of $\sin^2\theta_W \cong 0.23$ requires the decays $Z^0 \to t\bar{t}$ or $Z^0 \to b'\bar{b}'$ to occur, unless other hitherto unknown particles (e.g., supersymmetric partners of the known particles) are present in the Z^0 decay.

66. K.S. Babu, R.N. Mohapatra (Maryland U.)
GAUGE MODEL FOR THE EXCESS COSMIC MICROWAVE BACKGROUND.

We point out that the radiative neutrino decay mechanism proposed by Fukugita to explain the observed excess microwave background radiation can be naturally realized in a class of gauge models, where the primary decay mode of the heavy neutrino is $\nu_H \to 3\nu_\epsilon$ with a radiative branching ratio $B(\nu_H \to \nu_\epsilon + \gamma) \simeq 10^{-5}$. These models also provide a natural understanding of the smallness of the neutrino masses via the see-saw mechanism. Constraints from muon decay experiments require the existence of a light Higgs particle ($M_H \leq 50$ GeV) which couples dominantly to the neutrinos.

67. James M. Cline, William F. Palmer (Ohio State U.), G. Kramer (Hamburg U.)
EXCLUSIVE SEMILEPTONIC DECAYS OF B AND D MESONS.

Exclusive semileptonic decays of D and B mesons to one and two pseudoscalar mesons are studied with respect to form factor dependence, finite width effects in intermediate resonant states, and nonresonant background terms calculated from chiral Lagrangians. Branching ratios, electron decay spectra and asymmetry parameters for angular distributions are given for CKM (Cabibbo-Kobayashi-Maskawa) favored and unfavored transitions and compared to experimental data.

68. A. Feldmeier, H. Salecker, F.C. Simm (Munich U., Theor.Phys.)
SUBSTRUCTURE TESTS WITH POLARIZED $e^+e^- \to 2\gamma$.

It is shown that two-quantum pair annihilation with polarized beams is useful in the search for electron substructure at high-energy e^+e^- colliders. We consider modification models with excited electrons of spin 1/2 and 3/2. The test will be most sensitive for longitudinally polarized beams of equal handedness.

69. W.G. Bauer, H. Salecker (Munich U., Theor.Phys.)
PROPOSAL FOR MEASURING THE LONGITUDINAL POLARIZATION OF SCATTERED ELECTRONS AT HIGH ENERGIES.

A proposal is given for measuring the longitudinal polarization of electrons (or positrons) in the final state of experiments in high energy physics. The final-state electrons will be Compton-scattered within a very dense target of circularly polarized laser photons, thus producing hard backscattered γ-quanta. The energy spectrum of these particles will yield the information on the electron polarization. We discuss the efficiency and the resolution of the proposed scheme, as well as the possibility of its realization. It is shown that appropriate photon targets can be produced with the help of present-day high power lasers (peak power $\approx$ 1 TW, pulse duration $\approx$ 100 ps). The energy spectrum of the backscattered photons can be resolved by means of a conventional magnetic detector (e.g. the SLD). The repetition rate of existing high power lasers is still too low to enable reasonable counting rates. Higher repetition rates seem to be achievable with present-day technology.

70. Y.M. Cho (Seoul National U.)
UNIFIED COSMOLOGY AND MISSING MASS PROBLEM.

A unified cosmology is proposed as a higher-dimensional generalization of the standard model in which the Dirac's conjecture is naturally realized from the unification. It is argued that the theory has the potential ability to resolve the major problems of the standard cosmology without necessarily compromising its successes.

71. Z.G. Berezhiani, A.G. Doroshkevich, M. Yu. Khlopov, V.P. Yurov, M.I. Vysotsky (Moscow, ITEP)
COSMIC BACKGROUND RADIATION SPECTRAL DISTORTION AND RADIATIVE DECAYS OF RELIC NEUTRAL PARTICLES.

The recently observed excess of photons on a short wave-length side of the peak of a cosmic background radiation spectrum can be described by radiative decays of relic neutral particles. The lifetime and mass of a decaying particle must satisfy the following conditions: $2 \cdot 10^9 s < \tau \ll 3.6 \cdot 10^{14} s$, $0.4\ eV < m < 170\ eV$. The transitional magnetic moment must be of the order of $(3 \cdot 10^{-9} - 8 \cdot 10^{-8})\mu_B$, and the interaction of new particles with the usual matter must be rather strong. The generalization of the standard $SU(3) \times SU(2) \times U(1)$ model is presented which includes new particles with the desired properties.

72. ARGUS Collaboration (H. Albrecht, et al.)
LIFETIMES OF CHARMED MESONS.

Using the ARGUS detector at the e^+e^- storage ring DORIS II at DESY, we have measured the lifetimes of the D^0, D^+ and D_S^+ mesons. We find $\tau_{D^0} = (4.8 \pm 0.4 \pm 0.3) \times 10^{-13}\ s$, $\tau_{D^+} = (10.5 \pm 0.8 \pm 0.7) \times 10^{-13}\ s$ and $\tau_{D_S^+} = (5.6^{+1.3}_{-1.2} \pm 0.8) \times 10^{-13}\ s$.

73. A. Datta, J. Frohlich, E.A. Paschos (Dortmund U.)
QUANTUMCHROMODYNAMIC CORRECTIONS FOR $\Delta F = 2$ PROCESSES.

We calculate the QCD corrections to the weak $\Delta F = 2$ transition amplitudes for the cases $m_t \approx m_W$ and $m_t \gg m_W$. We find that for heavy top quark masses a whole class of diagrams decouples reducing the corresponding anomalous dimensions by a factor of two. We present explicit formulas for the corrections with numerical results summarized in tables and figures. We arrive at a QCD corrected effective Hamiltonian which holds for all ranges of m_t.

74. AMY Collaboration
SEARCH FOR e^+e^- ANNIHILATION TO CHARGED HIGGS PAIR AT TRISTAN.

We present a search for the production of charged scalar particles from e^+e^- annihilation at center-of-mass energies in the range of 50 to 60 GeV. In the general two-doublet model for the charged Higgs there is a parameter $\tan\beta$ which controls the branching ratio of the charged Higgs to decay into different weak-isospin doublets (quark-antiquark or lepton-neutrino pairs). Pair production with Higgs decaying into jets or $\tau\nu_\tau$ are examined to cover a wide range of $\tan\beta$ values.

75. Ling-Lie Chau (UC, Davis), Hai-Yang Cheng (Taiwan, Inst. Phys.)
ON THE NON-RESONANT THREE-BODY DECAYS OF CHARMED MESONS.

Non-resonant 3-body decays of charmed mesons are first studied in the approach of effective $SU(4) \times SU(4)$ chiral Lagrangians. It is pointed out that the predictions of the branching ratios in chiral perturbation theory are in general too small when compared with experiment. However, the experimental results are comprehensible in the general framework of the quark-diagram scheme. The existence of a sizable W-annihilation amplitude, which is evidences by the observation of $D_s^+ \to (\pi^+\pi^+\pi^-)_{NR}$, is the key towards an understanding of the 3-body non-resonant decays of D^+ and D_s^+. The measurement of $D^0 \to \overline{K}^0 K^+K^-$ and $D^0 \to \overline{K}^0\pi^+\pi^-$ indicates that color suppression is not effective in the 3-body decay. Based on the quark-diagram analysis, predictions for some other non-resonant modes are given.

76. D. Morgan (Rutherford), M.R. Pennington (Durham U.)
$I = 0$ SCALAR DYNAMICS IN THE $K\overline{K}$ THRESHOLD REGION: WHAT NEW EXPERIMENTAL RESULTS TELL US.

The $I = J = 0$ interaction in the $K\overline{K}$ threshold region is rich in structure with hints of states beyond the simple quark model. New experimental results on $J/\psi \to \phi\pi\pi$ and $\phi K\overline{K}$ from $DM_2, K\overline{K} \to K\overline{K}$ from LASS and $D_s \to \pi^\pm\pi^+\pi^-$ from Fermilab further explore this mass region and supplement previous information from classic meson scattering and production experiments. By parametrizing the scattering matrix in terms of the Jost function, thereby automatically satisfying coupled-channel unitarity, alternative hypotheses for the pole content of the $f_0(975)$ (the S^* effect) are confronted with this new, extended, data set.

77. D. Morgan (Rutherford), M.R. Pennington (Durham U.)
AMPLITUDE ANALYSIS OF $\gamma\gamma \to \pi\pi$ DATA BELOW 1.4 GeV.

A new amplitude analysis of presently available experimental results on $\gamma\gamma \to \pi^+\pi^-$ from Mark II and $\gamma\gamma \to \pi^0\pi^0$ from Crystal Ball is reported. Key constraints from the low energy theorem and from analyticity and unitarity that relate this process to other meson scattering reactions are built into the trial forms. Despite these restrictions, solutions with quite different proportions of S-wave and D-waves of both helicities 0 and 2 are admitted. This leads to a much wider range of possibilities for resonance couplings than is commonly supposed, even for the dominant $f_2(1270)$.

78. M. Gluck, E. Reya, A. Vogt (Dortmund U.)
RADIATIVELY GENERATED HADRONIC PHOTON STRUCTURE AT LOW Q^2.

Absolute predictions for the photon structure function $F_2^\gamma(x, Q^2)$ and for $\sigma_{\gamma\gamma}^{tot}(W)$ are presented for very low values of Q^2. These predictions are generated radiatively from the underlying assumption that at some low resolution scale the hadronic component of the photon consists only of its VMD induced valence quarks. These speculative predictions for $Q^2 \lesssim 1$ GeV^2 agree surprisingly well with presently available data and are actually not worse than the results of fitted VMD inspired models.

79. M. Gluck, E. Reya, W. Vogelsang (Dortmund U.)
POLARIZED PARTON DISTRIBUTIONS OF THE NUCLEON.

Spin-dependent (polarized) parton distributions are presented satisfying the available experimental and theoretical constraints. A prediction for the structure function $g_1^n(x, Q_0^2 \simeq 10\ \mathrm{GeV}^2)$ is given for the case of lowest-twist dominance.

80. M. Gluck, E. Reya (Dortmund U.)
POLARIZED LEPTOPRODUCTION AND DYNAMICALLY GENERATED SPIN-DEPENDENT PARTON DISTRIBUTIONS.

The first moment of the polarized parton distributions of the proton is evaluated dynamically using the assumption that at some low resolution scale the proton consists of valence quarks only. The spin carried by gluons, ΔG, at the EMC resolution scale $Q_0^2 = 10$ GeV is *predicted* to be $\Delta G/s_p \simeq 5$ with $s_p = 1/2$ denoting the spin of the proton. This result furthermore explains the dynamical origin of the recent EMC measurement of the first moment of the polarized structure function $g_1^p(x, Q^2)$.

81. B.K. Kerimov, A.M. Safin, S.M. Zeinalov (Moscow State U.)
ON THE ROLE OF THE ELECTROMAGNETIC MOMENTS OF THE NEUTRINO IN PROCESSES OF STELLAR NEUTRINO EMISSION.

Stellar neutrino luminosities due to neutrino emission, $Q_{\nu\nu}^{\mathrm{em}}$, in the processes of neutrino pair photoproduction $e^- + \gamma > e^- + \nu + \nu$, pair annihilation $e^- + e^+ > \nu + \bar{\nu}$, plasmon decay $\gamma^* \rightarrow \nu + \bar{\nu}$, and neutrino pair bremsstrahlung $e^- + Z \rightarrow c + Z + \nu + \bar{\nu}$ are calculated taking into account the neutrino magnetic and electric dipole moments. Our results are compared with those available from earlier calculations of $Q_{\nu\nu}^{\mathrm{weak}}$ in context of the Standard electroweak model ($\mu_\nu = d_\nu = 0$). It is shown that with $u_\nu, d_\nu \sim 10^{-10}\,\mu_B$ and $T - 10^9 - 10^{10} K$. $Q_{\nu\nu}^{\mathrm{em}}$ is of the same order of magnitude, or even greater than $Q_{\nu\nu}^{\mathrm{weak}}$. All the results obtained show that neutrino electromagnetic moments open up new channels for neutrino emission, thus could play an important role in stellar neutrino luminosity.

82. D. Haidt (DESY)
STATUS OF THE ELECTROWEAK STANDARD MODEL.

No abstract included.

83. C.D. Froggatt, H.J. Porter (Glasgow U.)
ON A NEW MECHANISM FOR THE ENHANCEMENT OF THE $\Delta I = 1/2$ WEAK INTERACTIONS.

We reanalyze and correct a 'superstring inspired' extension to the standard model due to Ma. The model contains new penguin diagrams which contribute to $\Delta I = 1/2$ processes but it has to be constrained to preserve the observed small $K_L - K_S$ mass difference. Preliminary results show that, for certain choices of the new particle masses, the extra contributions give a comparable rate to the standard model for $\Delta I = 1/2$ decays: but the model is unable to provide the necessary enhancement required to fully understand the rule.

84. Chikashi Iso (Tokyo Inst. Tech.), Kenju Mori (Saitama U.)
NEGATIVE BINOMIAL MULTIPLICITY DISTRIBUTION FROM BINOMIAL CLUSTER PRODUCTION.

Two step interpretation of negative binomial multiplicity distribution as a compound of binomial cluster production and negative binomial like cluster decay distribution is proposed. In this model we can expect the average multiplicity for the cluster production increases with increasing energy, different from a compound Poisson-Logarithmic distribution.

85. Fermilab E691 Collaboration (J.C. Anjos, et al.)
A STUDY OF DECAYS OF THE Λ_c^+.

We report measurements of the decays $\Lambda_c^+ \rightarrow p\overline{K^0}$, $\Lambda_c^+ \rightarrow p\overline{K^0}\pi^+\pi^-$, $\Lambda_c^+ \rightarrow \Lambda^0\pi^+$, and $\Lambda_c^+ \rightarrow \Lambda^0\pi^+\pi^+\pi^-$ from Fermilab photoproduction experiment E691. We have measured the relative branching ratios: $Br(\Lambda_c^+ \rightarrow p\overline{K^0})/Br(\Lambda_c^+ \rightarrow pK^-\pi^+) = 0.55 \pm 0.17 \pm 0.14$, $Br(\Lambda_c^+ \rightarrow \Lambda^0\pi^+\pi^+\pi^-)/Br(\Lambda_c^+ \rightarrow pK^-\pi^+) = 0.82 \pm 0.29 \pm 0.27$, $Br(\Lambda_c^+ \rightarrow \Lambda^0\pi^+)/Br(\Lambda_c^+ \rightarrow pK^-\pi^+) =\leq 0.33$ at 90% confidence level, and $Br(\Lambda_c^+ \rightarrow p\overline{K^0}\pi^+\pi^-)/Br(\Lambda_c^+ \rightarrow pK^-\pi^+) < 1.7$ at 90% confidence level.

86. Hans D. Dahmen (Siegen U.), Thomas Mannel, Panagiotis Manakos (Darmstadt, Tech. Hochsch.), Thorsten Ohl (Siegen U.)
THE PHOTON TWO-POINT FUNCTION IN UNITARY APPROXIMATION.

The photon two-point function is calculated using the exponential representation for the time-evolution operator taking into account terms up to second order in e in the exponent. The limiting case of vanishing electron mass is examined and it is shown that mass singularities cancel by means of unitarity. Near mass shell the causal Greens function exhibits in the limit of vanishing electron mass the expected cut behavior, off mass shell we recover the result of the usual perturbation theory.

87. A.A. Pankov (ICTP, Trieste & Trieste U., IFT), I.S. Satsunkevich (Minsk, Inst. Phys.)
ON THE POSSIBILITY OF STUDYING EXTRA GAUGE Z'-BOSON EFFECTS IN $e^+e^- \to \ell^+\ell^-$ AT TRISTAN.

A new possibility is suggested in order to study in the process $e^+e^- \to \ell^+\ell^-$ at energies below the standard Z-boson production threshold the effects coming from an extra neutral gauge boson Z' in superstring-inspired E_6 models. It is shown that at energies $\sqrt{s} \sim$ 50-60 GeV the cross-section for relatively light $Z'(M_{Z'} \lesssim 2.5\ M_Z)$ can have an interference dip absent in the standard model. The depth and energy coordinates of this dip depend on Z'-boson parameters. This information concerning the Z'-boson can be obtained by making use of TRISTAN, which is a high-luminosity set-up.

88. TASSO Collaboration (W. Braunschweig, et al.)
COMPARISON OF INCLUSIVE FRACTIONAL MOMENTUM DISTRIBUTIONS OF QUARKS AND GLUON JETS PRODUCED IN e^+e^- ANNIHILATION.

Inclusive charged particle production in e^+e^- annihilation into hadrons is studied in terms of the particle fractional momentum x_p. The x_p distribution for gluon jets is extracted by comparing two data samples measured in the TASSO detector: nearly symmetric three jet events at centre-of-mass energy $W \sim 35$ GeV and two jet events at $W \sim 22$ GeV, yielding quark and gluon jets of similar energies ($\sim$ 11.5 GeV). No significant difference is observed between quark and gluon jets. Monte Carlo models based on parton showers describe the trend and energy variation of the data better than a model with a second order matrix element in α_s.

89. TASSO Collaboration (W. Braunschweig, et al.)
A MEASUREMENT OF ELECTROWEAK EFFECTS IN THE REACTION $e^+e^- \to \tau^+\tau^-$ AT 35.0 AND 42.4 GeV.

We report on total cross section and forward backward charge asymmetry measurements of the reaction $e^+e^- \to \tau^+\tau^-$ at centre-of-mass energies of 35.0 GeV and 42.4 GeV using the TASSO detector. Including previous data an analysis in terms of electroweak parameters of the standard model is presented, and lower limits on mass scale parameters of residual contact interactions are given. A combined analysis of electroweak couplings using all our results on leptonic reactions $e^+e^- \to \ell^+\ell^-$ has been performed.

90. Ken-iti Matumoto (Toyama U.)
TECHNIAXION.

Techniaxion, a new kind of axion relating to the technicolor gauge theory, is investigated. On the general basis of the technicolored preon model, the mass ratio of the techniaxion to the ordinary QCD axion and the mass and the life time of the techniaxion are estimated.

91. S.I. Bilenkaya (Dubna, JINR), D.B. Stamenov (Sofia U.)
NEXT-TO-LEADING-ORDER QCD ANALYSIS OF EMC DEEP INELASTIC μp AND μd SCATTERING DATA.

A combined next-to-leading order QCD analysis of the European Muon Collaboration (EMC) μH_2 and μD_2 scattering data is presented. The nucleon structure functions are given in terms of parton distributions. The Buras-Gaemers method is used to solve the QCD equations for these distributions. Unlike most of the papers on this subject the cross section scattering data are fitted. The QCD mass scale parameter Λ is determined. Unlike the good non-singlet fit to the data ($x > 0.3$) the data fit in the full x range (a complete flavor singlet and non-singlet treatment) is not satisfactory.

92. Louis Lyons (Oxford U.), Alex J. Martin (Bristol U.), David H. Saxon (Rutherford)
ON THE DETERMINATION OF THE B-LIFETIME BY COMBINING THE RESULTS OF DIFFERENT EXPERIMENTS.

The current estimate of the B-lifetime is obtained by averaging the results of six experiments. We point out problems connected with the conventional approach to combining data, and discuss ways of overcoming them. Our recommended value is 1.13 ± 0.15 ps. Our discussion is relevant to a wide range of problems in which data are combined.

93. Pran Nath (Northeastern U.), R. Arnowitt (Texas A and M)
NEUTRINO MASSES IN THREE-GENERATION CALABI-YAU MODELS.

Neutrino masses for the three generation $CP^3 \times CP^3/Z_3$ Tien-Yau model are analyzed within the framework of matter-parity invariance. It is shown that mixing of the three light generation neutrinos with heavy $SU(5)$ single ν^c and $\bar{\nu}^c$ fields lead to seesaw masses for the three light generations of neutrinos. The seesaw mechanism involving the heavy ν^c, $\bar{\nu}^c$ fields is deduced by integrating out the heavy fields. A non-trivial dependence of the heavy sector remains in the seesaw masses. Neutrino masses in these models are found to be generally in the range $0(10^{-6} - 10^{-5})$ eV allowing for the possibility of neutrino oscillations in the solar neutrino flux.

94. R. Arnowitt (Texas A and M), Pran Nath (Northeastern U.)
MATTER PARITY CONSTRAINTS ON PARTICLE SPECTRUM IN THREE-GENERATION CALABI-YAU MANIFOLDS.

An analysis of the particle spectrum after intermediate scale breaking in the three generation Calabi-Yau manifold $CP^3 \times CP^3/Z_3$ is given. The spectrum is investigated within the framework of intermediate scale breaking which preserves matter parity and breaks the rank six gauge group $SU(3)_C \times SU(3)_L \times SU(3)_R$ to the Standard Model gauge group $SU(3)_C \times SU(2)_L \times U(1)_Y$. It is shown that a large number of particles and mirror particles associated with the extra families and mirror families on $CP^3 \times CP^3/Z_3$ can all be made superheavy through mass generation induced by non-renomalizable interactions that are present in the superpotential below the compactification scale. The final spectrum of the compactified superstring below the intermediate mass scale breaking is shown to be the spectrum of the Standard $N = 1$ SUSY theory and a few additional exotic leptons. Some of these exotic leptons are hybrid structures containing equal mixture of leptons and mirror leptons.

95. R. Arnowitt (Texas A and M), Pran Nath (Northeastern U.)
PROTON DECAY IN THREE-GENERATION MATTER PARITY INVARIANT SUPERSTRING MODELS.

Proton decay in three-generation superstring models with intermediate scale is investigated. It is shown that matter-parity (M_2) invariance alone does not stabilize the proton. However, models with M_2 invariance and Calabi-Yau manifolds such that the exotic C-odd quarks and leptons are superheavy will generally lead to proton lifetimes consistent with existing data.

96. Pran Nath (Northeastern U.), R. Arnowitt (Texas A and M)
MATTER PARITY, INTERMEDIATE SCALE BREAKING AND $\sin^2\theta_W$ IN CALABI-YAU SUPERSTRING MODELS.

Consequences of matter-parity invariance on intermediate scale breaking in three-generation Calabi-Yau superstring models are discussed. It is shown that values of $\sin^2\theta_W$ in conformity with the current experimental limits can arise in the Calabi-Yau models when the extra families and mirror families gain superheavy masses through nonrenormalizable interactions. A small number of exotic leptons with electroweak-scale masses remain and may be accessible at accelerator energies.

97. Pran Nath (Northeastern U.), R. Arnowitt (Texas A and M)
SYMMETRY BREAKING IN THREE-GENERATION CALABI-YAU MANIFOLDS.

The spontaneous breaking of $[SU(3)]^3$ symmetry to $SU(3) \times SU(2) \times U(1)$ due to nonrenormalizable interactions in the three-generation Calabi-Yau superstring model is considered. It is seen that these models lead naturally to intermediate scales $M_I \sim 10^{15}$ GeV with simultaneous formation on N and ν^c vacuum expectation values (VEV's). The lowest-lying extrema automatically preserve matter parity and hence protect the model against too rapid proton decay. Models with matter parity also protect against electroweak Higgs VEV's from forming at M_I. The intermediate mass scale produces some particles with mass $O(M_I)$. It also implies the existence of others with mass ~ 1 TeV, which are possible accessible to low-energy experiments.

98. AMY Collaboration
MEASUREMENTS OF R FOR $\sqrt{s}$ FROM 50 to 60.8 GeV.

Results from an experimental study of electron-positron annihilation into hadronic final states for center-of-mass energies from 50 to 60.8 GeV, measured by the AMY detector at the TRISTAN e^+e^- collider, are presented. A data sample corresponding to a total integrated luminósity in excess of 28 pb^{-1} is used. The ratios of the total cross section to the lowest order QED cross section for $e^+e^- \to \mu^+\mu^-$, R, are determined. Using our measurements for R combined with lower energy results from DORIS, CESR, PEP, PETRA and other TRISTAN experiments, we fit to the standard model prediction to determine the electroweak mixing angle $\sin^2\theta_W$ and $\Lambda_{\overline{MS}}$, the scale parameter in second-order QCD.

99. AMY Collaboration
MULTIHADRONIC EVENT PROPERTIES IN e^+e^- ANNIHILATION AT $\sqrt{s} = 52$ TO 57 GeV.

We present the general properties of hadronic final states produced by e^+e^- annihilation at center of mass energies from 52 GeV to 57 GeV in the AMY detector at TRISTAN. Corrected distributions from global shape analyses, inclusive charged particle distributions, and particle flows are presented. Our measurements are compared with QCD+fragmentation models which use either leading-logarithmic shower evolution or QCD matrix elements at the parton level, and either string or cluster fragmentation for hadronization.

100. AMY Collaboration
A COMPARISON OF QUARK AND GLUON JETS PRODUCED IN HIGH ENERGY e^+e^- ANNIHILATION.

Three-jet events produced in e^+e^- annihilations, measured in the AMY detector at the TRISTAN storage ring, are used to provide comparisons between quark and gluon jets. Clear differences between quark-induced and gluon-induced jets are observed. Quark jets tend to have a more tightly collimated structure than gluon jets. This is seen as a difference in the fraction of the total jet energy in the core of the jet. Also, the rapidity relative to the jet direction of the most energetic particle in gluon-jets is observed to be substantially lower than that of its counterpart in quark-jets.

101. AMY Collaboration
A MEASUREMENT OF $e^+e^- \rightarrow b\bar{b}$ FORWARD-BACKWARD CHARGE ASYMMETRY BETWEEN $\sqrt{s} = 52$ AND 57 GeV.

Using 123 multihadronic inclusive muon e^+e^- annihilation events at an average center-of-mass energy of 55.2 GeV, we extracted the forward-backward charge asymmetry of the $e^+e^- \rightarrow b\bar{b}$ process and the R ratio for $b\bar{b}$ production. We used an analysis method in which the behavior of the c quark and lighter quarks is assumed, with only that of the b quark left indeterminate. The results of $A_b = -0.72 \pm 0.28(\text{stat.}) \pm 0.13(\text{sys.})$ and $R_b = 0.57 \pm 0.16 \pm 0.10$ are consistent with the standard model, implying the existence of the t quark.

102. AMY Collaboration
FORWARD-BACKWARD CHARGE ASYMMETRY IN $e^+e^- \rightarrow$ HADRON JETS.

A measurement of the jet production angular distribution in $e^+e^- \rightarrow$ hadron jets, at center-of-mass energies from 52 to 60 GeV, is reported. Several methods of determining the sign of the charge of the underlying quarks are compared. The measured value of the forward-backward charge asymmetry is compared with the prediction of the Standard Model of the electroweak interaction and used to determine the weak axial-vector coupling constants of quarks.

103. AMY Collaboration
A DETERMINATION OF QCD PARAMETERS FROM A STUDY OF MULTI-JET EVENTS PRODUCED IN HIGH ENERGY e^+e^- ANNIHILATIONS.

We have fit the two, three, and four-jet fractions produced in e^+e^- annihilations to a perturbative QCD calculation of Kramer and Lampe in which the renormalization energy scale can be varied. This fit allows both the renormalization energy scale and $\Lambda_{\overline{MS}}$ to be determined. Using this fitted value of the renormalization scale, we find that, unlike the case in which the scale is chosen to be E_{CM}, the perturbative QCD calculation gives good agreement with the measured jet rates. In particular, the underprediction by the perturbative calculation of the four-jet rate by about a factor of two is cured. Furthermore, we find that $\Lambda_{\overline{MS}}$ is relatively insensitive to the value of the energy scale when the fitted renormalization scale is used.

104. TASSO Collaboration (W. Braunschweig, et al.)
PRODUCTION AND DECAY OF CHARMED MESONS IN e^+e^- ANNIHILATION AT $\sqrt{s} > 28$ GeV.

We report on a study of inclusive production of $D^{*\pm}$ mesons in e^+e^- annihilation at c.m. energies between 28 and 46.8 GeV using the TASSO detector at the PETRA storage ring. A hard $D^{*\pm}$ energy spectrum is measured with a maximum near $E_{D^{*\pm}} \simeq 0.6\ E_{\text{beam}}$. The measured cross section ratio $(\sigma_{D^{*+}} + \sigma_{D^{*-}})/\sigma_{\mu\mu} = 1.28 \pm 0.09 \pm 0.18$ indicates that D^* production accounts for a large fraction of the observed charm production. Two complementary methods have been used to determine the forward-backward asymmetry of charm pair production due to electroweak interference. Combining both measurements the product of the axial vector couplings of the electron and the charm quark to the weak neutral current was determined to be $g_A^c g_A^c = -(0.276 \pm 0.073)$, in agreement with the standard model prediction of -0.25. Using a sample of reconstructed $D^{*\pm}$ mesons, the relative strength of the strong interaction coupling of the c quark compared to that of an average of all flavors is measured as $\alpha_s(c)/\alpha_s(\text{all}) = 0.91 \pm 0.38 \pm 0.15$, consistent with the coupling constant being flavor independent. An update of our D^0 lifetime measurement is presented, based on considerable increase in statistics, the final result being $\tau_{D^0} = (4.8\ ^{+1.0+0.5}_{-0.9-0.7})\ 10^{-13}\ s$.

105. AMY Collaboration (S. Eno, et al.)
SEARCH FOR A FOURTH GENERATION CHARGE -1/3 QUARK .

By studying e^+e^- annihilations in the center-of-mass energy range between 50 and 60.8 GeV, we have established a 95% confidence level lower limit on the mass of a fourth-generation charge -1/3 quark, b', of 27.2 GeV. In contrast with all previous searches, this limit has been obtained through consideration of the decay processes $b' \rightarrow b\gamma$ and $b' \rightarrow bg$ as well as $b' \rightarrow cW^-$ For the cases where any one of the three decay modes dominates, we obtain higher mass limits.

106. AMY Collaboration
A STUDY OF e^+e^- ANNIHILATION INTO FOUR-LEPTON FINAL STATES.

We report results of a study of four-lepton final states produced by e^+e^- annihilation at center-of-mass energies from 50 to 60.8 GeV, measured in the AMY detector at TRISTAN. Cross sections for processes where two, three and four tracks are observed in the detector are used to test predictions of QED to order α^4, for large values of Q^2 and a wide range of masses for the produced lepton pairs.

107. AMY Collaboration
A MEASUREMENT OF THE PHOTON STRUCTURE FUNCTION F_2 AT AN AVERAGE Q^2 OF 67 $(\mathrm{GeV}/c)^2$.

The photon structure function F_2 has been measured in the Q^2 range from 30 to 110 $(\mathrm{GeV}/c)^2$ with the average value being 67 $(\mathrm{GeV}/c)^2$. The results are consistent with a calculation based on QCD.

108. V. Elias, R.R. Mendel (Western Ontario U.), M.D. Scadron (Arizona U.)
ON-SHELL RENORMALIZATION OF THE $\langle\bar{q}q\rangle$ CONTRIBUTION TO THE $\Delta I = 1/2$ s-d SELF-ENERGY TRANSITION.

The dimension-3 quark condensate contribution to the renormalized s-d self energy transition is evaluated within the framework of on-shell-renormalization with respect to the (Lagrangian) quark mass matrix. This transition is then related to $\Delta I = 1/2$ matrix elements for nonleptonic kaon decay rates. The contributions to such matrix elements from the $\langle\bar{q}q\rangle$-component of the s-d self-energy transition, when considered in isolation, are seen to generate rates of the same order of magnitude as the experimental values for $K_L \rightarrow \gamma\gamma$ and $K_S \rightarrow \pi\pi$ decays.

109. BCDMS Collaboration (A. Benvenuti, et al.)
A HIGH STATISTICS MEASUREMENT OF THE DEUTERON STRUCTURE FUNCTIONS $F_2(x,Q^2)$ AND R FROM DEEP INELASTIC MUON SCATTERING AT HIGH Q^2.

We present results on a high statistics study of the deuteron structure functions $F_2(x,Q^2)$ and $R = \sigma_L/\sigma_T$ measured in deep inelastic scattering of muons on a deuterium target. The data were recorded at beam energies of 120, 200 and 280 GeV and cover the kinematic range $0.06 \leq x \leq 0.80$ and $8\ GeV^2 \leq Q^2 \leq 160\ GeV^2$. Scaling violations observed in the data are compared to predictions of perturbative QCD and allow to determine the QCD mass scale parameter Λ.

110. BCDMS Collaboration (A. Benvenuti, et al.)
A COMPARISON OF THE STRUCTURE FUNCTIONS F_2 OF THE PROTON AND THE NEUTRON FROM DEEP INELASTIC MUON SCATTERING AT HIGH Q^2.

High statistics data on the structure functions $F_2(x,Q^2)$ of the proton and the deuteron, measured with the same apparatus, are used to study the ratio of structure functions of the neutron and the proton F_2^n/F_2^p and the structure function $F_2^p - F_2^n$. Both measurements are in good agreement with data from earlier experiments.

111. R.R. Volkas, R. Foot, X.G. He, G.C. Joshi (Melbourne U.)
STRONG INTERACTION NON UNIVERSALITY.

The universal QCD color theory is extended to an $SU(3)_1 \otimes SU(3)_2 \otimes SU(3)_3$ gauge theory, where quarks of the ith generation transform as triplets under $SU(3)_i$ and singlets under the other two factors. The usual color group is then identified with the diagonal subgroup, which remains exact after symmetry breaking. The gauge bosons associated with the sixteen broken generators then form two massive octets under ordinary color. The interactions between quarks and these heavy gluon-like particles are explicitly non-universal, and thus an exploration of their physical implications allows us to shed light on the fundamental issue of strong-interaction universality. Non-universality and weak flavour mixing are shown to generate heavy-gluon induced flavor changing neutral currents. The phenomenology of these processes is studied, as they provide the major experimental constraint on the extended theory. Three symmetry breaking scenarios are presented. The first has color breaking occurring at the weak scale, while the second and third divorce the two scales. The third model has the interesting feature of radiatively induced off-diagonal Kobayashi-Maskawa matrix elements.

112. R. Foot, H. Lew, G.C. Joshi (Melbourne U.)
IMPLICATIONS, CONSTRAINTS AND SIGNATURES OF A FOURTH-GENERATION LEPTON TRIPLET.

The possibility of a fourth generation weak $SU(2)_L$ triplet of leptons is further investigated. Constraints on the masses of the charged triplet of leptons are derived from their contribution to the ρ parameter. Partial wave considerations are then applied to the lepton-triplet-Higgs boson interactions, where it is argued that these interactions are likely to become strongly interacting at high energies. Finally, possible signatures of the lepton triplet are considered. These include the ratio Γ_Z/Γ_W, the partial width $\Gamma(Z \to f\bar{f})$, the cross section $\sigma(e^+e^- \to f\bar{f})$, and the forward-backward asymmetry parameter.

113. R. Foot, H. Lew, R.R. Volkas, G.C. Joshi (Melbourne U.)
THE STRUCTURE OF EXOTIC GENERATIONS.

Theoretical and phenomenological aspects of non-standard multiplets of anomaly-free fermions (exotic generations) are examined. The standard quark-lepton generation is shown to be the simplest non-trivial member of a class of fermion multiplet structures allowed by the gauge and Higgs structure of the minimal Standard Model. The next simplest member includes weak-isospin triplet quarks and leptons, whose structure and phenomenology is investigated. A feature of some of these models is the presence of small flavor-changing neutral currents at tree level, however these currents are shown to be within existing experimental bounds. It is shown that it is possible that some of the known members of the incomplete third generation (b-quark, τ and ν_τ may actually be part of such a triplet family, rather than a standard doublet family. Upper bounds on the masses of the new particles are obtained from data on the ρ-parameter.

114. X.G. He, G.C. Joshi, B.H.J. McKellar, R.R. Volkas (Melbourne U.)
CONNECTION BETWEEN GENERATION NUMBER AND ANOMALY CANCELLATION IN SUPERSYMMETRIC MODELS.

We construct supersymmetric theories in which the number of generations of quarks and leptons is related by gauge anomaly-cancellation to the spectrum of Higgs fields. Models yielding at least three generations with the minimal Higgs spectrum assumed are discussed in detail. This mechanism requires an extension of the Standard Model gauge group such that an ordinary quark-lepton generation is not anomaly-free. We discuss models of $SU(3)$ weak isospin, separate isospin groups for quarks and leptons, and chiral color.

115. R. Foot, H. Lew, X.G. He, G.C. Joshi (Melbourne U.)
SEE-SAW NEUTRINO MASSES INDUCED BY A TRIPLET OF LEPTONS.

Neutrinos can acquire mass through the see-saw mechanism by extending the lepton sector of the Standard Model. The most general $SU(2)_L \times U(1)_Y$ assignments allowed for the exotic leptons are the singlet and triplet representations. A model is constructed involving a triplet of exotic leptons. In this model lepton number is broken. Therefore lepton number violating processes, in particular those involving the lepton triplet, could provide distinctive experimental signatures.

116. R. Foot, H. Lew, R.R. Volkas, G.C. Joshi (Melbourne U.)
ANOMALY CANCELLATION IN CHIRAL COLOUR MODELS WITH A MINIMAL HIGGS SECTOR.

It has been proposed that the strong interactions arise from the chiral color group $SU(3)_L \otimes SU(3)_R$. In such models exotic fermions are required in order to cancel the anomalies. We propose a very simple extension of the fermion sector which involves no exotic color quantum numbers, but instead involves an exotic $SU(2)_L \otimes U(1)_Y$ generation. This exotic generation preserves the lepton-quark symmetry and requires no additional exotic Higgs bosons to give mass to the exotic fermions.

117. R. Foot, G.C. Joshi, H. Lew (Melbourne U.)
GAUGED BARYON AND LEPTON NUMBERS.

A possible extension of the Standard Model can be defined by gauging the global baryon and lepton number $U(1)$ symmetries. Gauging Baryon and Lepton numbers provide a natural framework for the see-saw mechanism in the lepton sector, and the Peccei-Quinn mechanism in the quark sector. Another consequence of this extension is that the usual three generations of fermions are not anomaly free. However we consider a wider framework involving the existence of generations with exotic $SU(2)_L \otimes U(1)_Y$ quantum numbers. This allows us to 'derive' a minimal spectrum of fermion which contain the known quarks and leptons.

118. R. Foot, G.C. Joshi, H. Lew, R.R. Volkas (Melbourne U.)
HORIZONTAL SU(2) SYMMETRY REVISITED.

One possible approach to understanding the pattern of masses and mixing angles in the quark sector is the concept of horizontal symmetry. We consider the $SU(2)$ horizontal symmetry candidate, and examine the possibility of assigning the three generations of fermions to the doublet and singlet representations of $SU(2)_H$. This assignment allows a natural explanation of the near masslessness of the first generation, and the smallness of the Kobayashi-Maskawa matrix elements connecting the third and first generation, and the second and third generation. We suggest that these masses and mixings may arise radiatively, and an example is given to illustrate how this may come about. In this example the inequality $m_u < m_d$ arises naturally.

119. R. Foot, G.C. Joshi (Melbourne U.)
ON THE SPACE-TIME SYMMETRIES OF THE SUPERSTRING AND THE JORDAN ALGEBRAS.

The sequence of Jordan algebras $\mathcal{M}_3^n$, whose elements are the 3×3 Hermitean matrices over the division algebras R,C, Q and O are considered. These algebras are naturally related to supersymmetric structures in space-time dimensions of 3, 4, 6 and 10, as the Lorentz groups in these dimensions can be expressed in a unified way as a subgroup of the structure group of the Jordan algebras $\mathcal{M}_3^n$. The generators of the complete structure group and the automorphism group can be separated into bosonic and fermionic generators depending on their transformation properties under the Lorentz subgroup. A peculiar connection between these fermionic generators and the supersymmetry generators of the superstring action is introduced and discussed.

120. R.R. Volkas, A.J. Davies, G.C. Joshi (Melbourne U.)
A STUDY OF ONE-LOOP RADIATIVE CORRECTIONS IN THE INVISIBLE AXION MODEL.

The 1-loop effective (Coleman-Weinberg) potential for the original invisible axion model is constructed. We discuss the finetuning that is necessary to establish the enormous hierarchy between the doublet and singlet Higgs vacuum expectation values. We then show that no further fine-tuning is necessary at the 1-loop level to retain the hierarchy. We also discuss the renormalization group evolution of the Higgs quartic couplings from the point of view of stability and triviality. The relevance of this analysis to invisible axion models featuring spontaneous CP violation is emphasized.

121. K.G. Chetyrkin (Moscow State U.)
TWO-LOOP CALCULATIONS IN THE LARGE QUARK MASS EXPANSION IN THE FRAMEWORK OF THE QCD SUM RULES METHOD.

The first seven moments of the coefficient function

$$C_{G^2}(t,\alpha_s) = \frac{\alpha_s}{\pi} \sum_{n=0}^{\infty} a_n \left(1 + \frac{\alpha_s}{\pi} b_n\right) t^n$$

(here $t = q^2/4m_c^2$, and a_n are one-loop moments) of the operator $G^2 = G^a_{\mu\nu} G^{a\mu\nu}$ in the large quark mass expansion of the correlator of two vector currents are calculated in the two-loop approximation. The obtained results show, in particular, that α_s^2-corrections to this coefficient function are reasonably small and hardly can reduce the disagreement between the values of the corresponding condensate $\langle G^2 \rangle_0$ found in different approaches. We describe a method of approximate recovering of the very coefficient function starting from the calculated moments and the asymptotic behavior in the large n limit. We also present the results of the calculations of the first seven moments for the case of scalar and pseudoscalar currents.

122. G.S. Atoyan, S.N. Gninenko, V.I. Razin, Y.V. Ryabov (Moscow State U.)
A SEARCH FOR PHOTONLESS ANNIHILATION OF ORTHOPOSITRONIUM: $e^+e^-(^3S_1) \rightarrow$ nothing.

We have searched for the photonless decay of orthopositronium $o-Ps \rightarrow$ nothing (nothing denotes the weakly-interacting particles which are not detected). The upper limit on the ratio of this decay rate to the 3γ decay rate obtained is $\Gamma(o-Ps \rightarrow$ nothing$/\Gamma(o-Ps \rightarrow 3\gamma) < 5.8 \times 10^{-4}$ (90% c.l.). This result excludes the hypothesis that this decay is responsible for the discrepancy between the theoretical and experimental values of the $o-Ps$ lifetime in vacuum.

123. V.N. Bolotov, et al. (Moscow, INR)
THE EXPERIMENTAL STUDY OF THE $\pi^- \rightarrow e^-\bar{\nu}\gamma$ DECAY IN FLIGHT.

We describe an experiment in which studied the radiative $\pi^- \rightarrow e^-\bar{\nu}\gamma$ decay. The measurements were carried out on a 17-GeV secondary meson beam of the IHEP machine on the ISTRA facility of the Institute for Nuclear Research. The high energy beam used in the experiment enabled us to investigate this decay in wide range of kinematic variables and to determine unambiguously the ratio of the axial-to-vector form factors $\gamma = 0.41 \pm 0.23$. The value for the vector form factor has been determined in model independent way $F_v = 0.014 \pm 0.009$.

124. S.B. Gerasimov (Dubna, JINR)
ELECTROMAGNETIC MOMENTS OF HADRONS AND QUARKS IN A HYBRID MODEL.

Magnetic moments of baryons are analyzed on the basis of general sum rules following from the theory of broken symmetries and quark models including the relativistic effects and hadronic corrections due to the meson exchange currents. A new sum rule is proposed for the hyperon magnetic moments, which is in accord with the most precise new data and also with a theory of the electromagnetic $\Lambda\Sigma^0$-mixing. Our results favor the latest experimental value $\mu(\Sigma^+) = 2.379 \pm 0.025$ n.m. for the Σ^+ hyperon magnetic moment, differing from latest measurements. The quark magnetic moment ratios $\mu(u)/\mu(d) = -1.80 \pm 0.02$, $\mu(s)/\mu(d) = 0.67 \pm 0.02$ derived from baryon data bring the ratio of the $K^{*\pm} \to K^{\pm}\gamma$ and $K^{*0} \to K^0\gamma$ decay widths close to experiment. The numerical values of the quark electromagnetic moments are obtained within a hybrid model treating the pion cloud effects through the local coupling of the pion field with the constituent (massive) quarks. Possible sensitivity of the weak neutral current 'magnetic' moments to violation of the OZI rule is emphasized and discussed.

125. BEBC WA21 Collaboration (G.T. Jones, et al.)
PRELIMINARY RESULTS ON THE RATIO $d_v/u_v(x)$ DETERMINED IN ν AND $\bar{\nu}$ SCATTERING ON HYDROGEN AND DEUTERIUM.

Preliminary results are presented on a determination of the quark momentum distribution function $xq(x)$ in the proton and on a measurement of the valence quark ratio $r_v = d_v/u_v(x)$. The analysis is based on data from neutrino and antineutrino scattering on hydrogen and deuterium, obtained with BEBC in the (anti)neutrino wideband beam of the CERN SPS. Using the ratios hydrogen/deuterium of the normalized x, y, and Q^2 distributions enables a precision measurement of r_v with the systematic errors expected to be small. It is found that r_v decreases with increasing x and drops considerably below the naive QPM expectation for $x \gtrsim 0.2$.

126. A.A. Pankov, I.S. Satsunkevich (Minsk, Inst. Phys.)
P-ODD EFFECTS OF EXTRA Z' BOSONS IN DILEPTON HADRONIC PRODUCTION.

Possible observation of P-odd effects produced by extra superstring Z' bosons in hadron-hadron collisions with inclusive production of a lepton-antilepton pair is studied for nonpolarized and longitudinally polarized incident hadron beams. The polarization asymmetry, the forward-backward polarization asymmetry, and the longitudinal polarization of the lepton are numerically estimated for various models which describe the superstring Z' boson. The numerical results are presented for the CERN $\bar{p}p$ collider and UNK pp collider.

127. A.A. Pankov, I.S. Satsunkevich (Minsk, Inst. Phys.)
P-ODD EFFECTS OF FERMION EXTRA WEAK NEUTRAL CURRENTS IN e^+e^- ANNIHILATION.

We study possibilities of observing at PEP, SLC, and LEP setups the effects produced by extra Z' bosons which are predicted in the unified theories on the basis of the gauge group E_6 in e^+e^- annihilation. The Z' boson effects can be detected by measuring the longitudinal polarization of final fermion states $(\tau, u, d, \ldots)$ at e^+e^- beam energy being less than, equal to, or higher than the production threshold of the standard Z boson.

128. OMEGA Photon Collaboration
SEARCH FOR HIGHER TWIST IN γp INTERACTIONS FOR $E_\gamma = 65 - 175$ GeV.

The reaction $\gamma p \to MX$, where M is $\pi^\pm$ or $K^\pm$, has been studied in order to isolate the QCD higher twist contribution $\gamma q \to Mq$. The study was restricted to mesons in the range $p_t \geq 2.0$ GeV $x_F \geq -0.2$. The γp data are assumed to be an incoherent superposition of contributions from the hadronlike component of the photon and from the minimum twist and higher twist pointlike interactions of the photon. The hadronlike component is obtained from $\pi^\pm$, $K^\pm$ beam data taken in this experiment by applying an appropriate Vector Meson Dominance (VMD) factor. Monte Carlo calculations show that the higher twist contribution can be enhanced relative to the other two contributions by using x_F and the topological variable d, the minimum 'distance' between the highest p_t track and any other particle. The higher twist cross section extracted by this method is compared with the QCD prediction.

129. OMEGA Photon Collaboration
COMPARISON OF PHOTO AND HADROPRODUCTION OF ρ^0 MESONS ON HYDROGEN AS A FUNCTION OF x_F AND p_t.

We compare inclusive production of ρ^0 mesons in γp and $h^\pm p$ collision with beam energies of $65 < E_\gamma < 175$ and $E_h = 80$ and 140 GeV, where h is π or K. The ratio $\sigma(\gamma p \to \rho^0 X)/\sigma(h^\pm p \to \rho^0 X)$ of inclusive ρ^0 production cross section is measured with respect to $x_F(-0.2 \leq x_F \leq 1)$ and $p_t(0 \leq p_t \leq 1.4$ GeV$)$. In a nondiffractive kinematical domain where no effects due to different quantum numbers of beamphotons and beamhadrons are expected we observe within the errors an x_F and p_t independent ratio as expected by Vector Meson Dominance. The forward part ($x_F \geq 0.8$) of inclusive ρ^0 photoproduction is quantified by a phenomenological calculation based on the triple Regge exchange mechanism and VMD. Nondiffractive photoproduction and hadroproduction of ρ^0's can be described in terms of quark antiquark fusion models based on the Drell-Yan mechanisms. With actual parametrisations of these structure functions extrapolated to the soft regime we obtain good agreement in a restricted x_F range of $0.2 \leq x_F \leq 0.8$ in either photon and hadron data. At smaller x_F the predictions exceed the measured results in either data set.

130. OMEGA Photon Collaboration
CHARGE ASYMMETRY IN LOW-p_t PHOTOPRODUCTION OF HADRONIC FINAL STATES IN FORWARD DIRECTION.

The charged particle asymmetry, $(n^+ - n^-)/(n^+ + n^-)$, in inclusive photoproduction of charged particles on protons has been studied as a function of x_F and p_t in the ranges $x_f > 0$ and $0 < p_t < 2$ GeV and for photon energies in the range $65 < E_\gamma < 175$ GeV. The asymmetry is found to be positive, increasing with x_F and p_t and decreasing with E_γ. Asymmetries of π's and K's in the final state have been obtained separately. These data have been compared with a parton recombination model which provides a qualitative explanation of parts of the effects observed.

131. OMEGA Photon Collaboration
ANALYSIS OF THE TOPOLOGICAL BEHAVIOUR OF THE POINTLIKE γ-INTERACTIONS IN γp REACTIONS IN THE ENERGY RANGE $E_\gamma = 65 - 175$ GeV.

The production of charged particles in γp, $\pi^\pm p$ and $K^\pm p$ reactions has been studied in terms of energy flow in the eventplane. For values of $\sum_i p_{tin}^{i2}$ above 2.0 GeV^2 in the plane, a depletion of the energy flow in the beam direction is observed for both photon and hadron beams. It is found that the photon induced data exhibit a more pronounced structure than the hadron beam data. The energy flow distribution for the *pointlike* photon interaction has been obtained using a subtraction technique analogous to that employed in our earlier study of inclusive charged particle distributions. This assumes that the hadronlike component of the photon can be obtained from the hadron beam data by applying an appropriate Vector Meson Dominance (VMD) factor. The subtracted energy flow distribution has been compared with first order QCD calculations for the pointlike photon implemented in the LUND program LUCIFeR.

132. AMY Collaboration (Y. Sakai, et al.)
A SEARCH FOR HEAVY LEPTONS AND NEW PARTICLES BEYOND THE STANDARD MODEL IN e^+e^- ANNIHILATIONS AT $\sqrt{s} = 50-60.8$ GeV.

Searches for production of heavy leptons and various kind of new particles beyond the standard model have been made in e^+e^- annihilations at center-of-mass energies of $\sqrt{s} = 50 - 60.8$ GeV using the AMY detector. Improved mass limits for charged and neutral heavy leptons are reported. We have searched for both Dirac and Majorana type neutral heavy leptons that decay via mixing with electron and muon generations through either charged or neutral currents. In addition, results of searches for SUSY particles, non-minimal Higgs, Leptoquarks, and colored leptons are reported.

133. D.G. Richards (Edinburgh U.)
A MODEL OF THE INTERNUCLEON EXCHANGE POTENTIAL IN LATTICE GAUGE THEORY .

The internucleon exchange potential between mesons consisting of a heavy quark and a light quark is investigated in a lattice gauge theory model. The range of the force is determined as a function of the t-channel quantum numbers, and the exchange potential is estimated for a range of light quark masses.

134. TASSO Collaboration (W. Braunschweig, et al.)
MEASUREMENT OF THE AVERAGE LIFETIME OF B HADRONS.

The average lifetime of B hadrons has been measured by the TASSO collaboration using 110 pb^{-1} of data collected in 1986 at a center-of-mass energy of 35 GeV. Measurements by three different methods are presented and a method of B-enrichment is discussed. The result of the Impact Parameter Method is $\tau_B = 1.36 \pm 0.13 \pm 0.28$ ps, that of the Vertex Method is $\tau_B = 1.30 \pm 0.10 \pm 0.28$ ps and that of the Dipole Method is $\tau_B = 1.47 \pm 0.14 \pm 0.30$ ps. These are combined to give the result $\tau_B = 1.35 \pm 0.10 \pm 0.24$ ps.

135. G. Shaw (Manchester U.)
SHADOWING IN THE SCALING REGION.

The approximate scaling behavior of the shadowing effect at small x, recently discovered in deep inelastic muon scattering experiments on nuclei, is shown to arise naturally in the context of hadron dominance models, formulated many years ago, which are dual to the parton model. The effect decreases very rapidly with increasing x, at a rate which is only weakly dependent on the atomic number A for reasonably large A.

136. Avinash Khare (Bhubaneswar, Inst. Phys.), Trilochan Pradhan (Utkal U.)
MAGNETOELECTRIC SUSCEPTIBILITY OF THE VACUUM.

A magnetoelectric medium is one which becomes magnetized when it is placed in an electric field and electrically polarized when placed in a magnetic field. The question of vacuum having under a property is examined in $2+1$ and $3+1$ QED and expressions for relevant susceptibilities are obtained. It is shown that whereas in $2+1$ dimensions, such an effect can arise from Chern-Simons action, in $3+1$ dimensions it can arise from the interaction of photon with spin-2 fields.

137. R.C.E. Devenish, et al. (Oxford U. and University College, London)
THE ZEUS CENTRAL TRACKING CHAMBER SECOND-LEVEL TRIGGER.

With the very brief interval of $96ns$ between bunch crossings, the HERA *ep* collider puts stringent requirements on the design of trigger and data acquisition systems. The ZEUS detector has a three stage trigger system, the first two levels are component based and the third level will be an ACP-II farm using full event data. The complete trigger system is designed to reduce the rate from the raw interaction rate of about 10^5 Hz to a few Hz at which rate the 60-100 kbytes of even data can be written to tape. The Central Tracking Detector (CTD) provides information on charged particle tracks and hence on the event vertex. The paper describes the propose and design of the CTD second level trigger (SLT) within the context of the ZEUS trigger as a whole. The SLT is based on a network of transputers closely integrated with the CTD data acquisition system. The network and algorithm have been designed to exploit the natural parallelism inherent in the CTD design. Some preliminary performance estimates are given and a comment made on the use of formal methods in specifying and implementing the SLT design.

138. R. Rodenberg (Aachen, Tech. Hochsch., III Phys. Inst.)
RENORMALIZATION GROUP ANALYSIS OF THE QUARTIC COUPLINGS AND BOUNDS ON THE HIGGS MASS SPECTRUM IN TWO HIGGS MODELS AT THE FERMI SCALE.

Using renormalization group techniques we investigate without any necessary explicit knowledge of the effective potential the Coleman-Weinberg symmetry breaking in the standard model with two Higgs doublets. From the requirement of vacuum stability and finiteness of the quartic couplings up to a unification scale bounds on the Higgs mass spectrum will be found and compared with previously obtained results derived from the explicitly known effective potential for Higgs-gauge boson and Higgs-fermion interactions only in case of two Higgs doublets.

139. J. Maalampi, J.T. Peltoniemi, M. Roos (Helsinki U.)
THE MAGNETIC MOMENT OF THE ELECTRON NEUTRINO IN MIRROR MODELS.

The electromagnetic properties of the electron neutrino are analyzed in theories involving mirror fermions. We find that the magnetic dipole moment of the neutrino can be as large as $10^{-10}\mu_B$, which may explain the observed solar neutrino deficit, as previously suggested. We show that this is not contradictory to other electroweak phenomena. To this end we calculate the weak correction to the anomalous magnetic moment of the electron, from which we get the upper limit for the electron-mirror electron mixing angle $\theta \lesssim 0.05$.

140. K. Enqvist, J. Maalampi (Helsinki U.), A. Masiero (Helsinki U. & INFN, Padua)
MAGNETIC MOMENTS IN THE SEE-SAW MODEL.

We study the transition magnetic moments of Majorana neutrinos in the two generation see-saw model. The stability of the neutrino mass matrix under the radiative corrections induced by the transition couplings gives an important constraint. We show that in the see-saw model there is an unexpected enhancement of the transition magnetic moment of the electron neutrino ν_e. It is, however, far too small to explain the solar neutrino deficiency problem by the helicity flip transition of ν_e in the magnetic field of the sun.

141. J. Maalampi, M. Roos (Helsinki U.)
NON-STANDARD FERMIONS AT LEP.

Many theoretical schemes predict the existence of unconventional quarks and leptons not fitting the standard pattern of fermion families. We consider two possible kinds of non-standard fermions, mirror fermions and exceptional fermions. Mirror fermions arise in a variety of models ranging from family unification to extended supersymmetry and Kaluza-Klein theories: exceptional fermions come along with the group E_6 which is believed to be the low energy gauge symmetry of the superstring theory. We discuss some physical properties of these non-standard particles relevant for the LEP.

142. J. Maalampi, M. Roos (Helsinki U.)
DISTORTION OF THE COSMIC BACKGROUND RADIATION BY RADIATIVE NEUTRINO DECAY IN A LEFT-RIGHT SYMMETRIC MODEL.

We show that the recently observed distortion of the cosmic background radiation in the submillimeter region can be explained in terms of radiative decay of right-handed neutrinos in a left-right symmetric model.

143. V. Gupta, K.V.L. Sarma (Tata Inst.)
ANALYSIS OF THE DATA IN HEAVY FLAVOR DECAYS.

The charmed and bottom hadron decays have been analyzed phenomenologically, keeping both the spectator and non-spectator contributions to the total decay rates. A sum rule relating the total decay widths of the four low-lying charmed baryons is given. It is pointed out that all the data which have become recently available can be understood in all cases with negligible Pauli interference effects but with important exchange-mechanism contributions. Predictions for the lifetimes of Ω_c^0 and Ξ_c^0 together with expected lifetime hierarchy for the bottom hadrons are given. The corresponding semileptonic branching ratios, where possible, are also predicted.

144. T.A. Koss, Jun Iwai, J.J. Lord, S. Strausz, R.J. Wilkes (Washington U., Seattle)
COSMIC RAY PHOTON INTERACTIONS AT 1–15 TeV.

Radiations from Cygnus X-3 produce electromagnetic showers in the earth's atmosphere and are associated with underground muons of energies in excess of 1 TeV. If the radiations are gamma rays, their muon production rates are about 100 times greater than expected (Berezinsky, Ellis, and Ioffe). A cosmic ray interaction was observed in which the visible (electromagnetic calorimeter) energy was estimated to be about 130 TeV, implying a primary energy $\sim$ 900 TeV. The vertex of the interaction was $>$ 100 meters from the balloon borne shower calorimeter in which it was detected. About 31 individual well separated showers were observed, most of which displayed the cascade development characteristics of gamma rays of energies from 1 to 15 TeV. However, in at least two cases, the shower development was neither in agreement with that to be expected of gamma rays or for hadronic processes due to neutrons or other strongly interacting particles. In the two anomalous showers, the first interaction occurred at an improbably shallow depth in the calorimeter. The initial vertices in the 1mm lead plates of the calorimeter resulted in interactions of multiplicity approximately 10. (All of the normal initial shower particles for the anomalous events were emitted at angles characteristic of hadroproduction, much larger than in the case of ordinary electromagnetic processes at TeV energies, and the development of the cascades was characteristic of hadron interactions. The energies of the anomalous events were 1.5 and 6 TeV (assuming them to be electromagnetic cascades). These events may represent the onset of a new character of the electromagnetic interaction for energies in excess of a few TeV as suggested by the work of F. Halzen.

145. T.A. Koss, J. Iwai, J.J. Lord, S. Strausz, R.J. Wilkes, J. Woosley (Washington U., Seattle)
INTERACTION OF COSMIC RAY GAMMA RAYS FROM 1 – 15 TeV.

A very high energy family of gamma rays having energies between 1 TeV and 15 TeV was detected in an electromagnetic shower detector in an emulsion chamber flown by balloon in the stratosphere. Motivated by the observations of muon rich showers due to radiations from Cygnus X-3 and Hercules X-1 suggesting anomalous interactions of gamma rays with energies above 50 TeV, a reexamination has been made of the interactions of gamma rays produced in cosmic ray interactions. To first order, the development of the resulting cascade showers was found to be in agreement with theory. The starting point distribution of the electromagnetic showers was, within statistical error, in agreement with the Landau-Pomeranchuk theory. While the cascade development of the gamma rays in this experiment was in agreement with theory, two additional interactions produced by uncharged particles or radiation were found which could not be explained. The two interactions occurred within 1.4 radiation lengths in the lead plates of the calorimeter but their shower development did not agree with that expected for gamma rays or by interactions produced by neutral hadrons. The results are compared with the speculation that at energies much greater than obtained in accelerators, the nucleus begins to act like a bag of gluons rather than a bag of quarks producing an increased photon-nuclear cross section for hadron production.

146. TOPAZ Collaboration (I. Adachi, et al.)
A STUDY OF PION PAIR PRODUCTION IN TWO-PHOTON PROCESS.

We studied $\gamma\gamma \rightarrow \pi^+\pi^-$ interaction with the untagged two photon sample obtained by the TOPAZ detector at the TRISTAN e^+e^- collider. The sample was taken in the energy region, $\sqrt{s} = 55.0 \sim 60.8$ GeV. One of the QED backgrounds, $\gamma\gamma \rightarrow e^+e^-$ is removed by measurement of dE/dx of the TOPAZ-TPC. The $\pi^+\pi^-$ invariant mass spectrum is well reproduced by the f_2 resonance, the Born term and their interference when $\gamma\gamma \rightarrow \mu^+\mu^-$ background is subtracted. From this, we determined $\Gamma_{f_2 \rightarrow \gamma\gamma}$ to be 2.25 ± 0.18 (stat) $\pm$ 0.25 (sys.) keV (preliminary).

147. TOPAZ Collaboration (I. Adachi, et al.)
MEASUREMENTS OF α_s IN e^+e^- ANNIHILATION AT $\sqrt{s} = 53.3$ GeV AND 59.5 GeV.

We studied the energy-energy correlation (EEC) and its asymmetry (AEEC) using e^+e^- hadronic annihilation events obtained at $\sqrt{s} = 53.3$ GeV and 59.5 GeV with the TOPAZ detector at the TRISTAN collider. We used a Monte Carlo simulation combined with the QCD matrix elements by Gottshalk and Shatz and the Lund string fragmentation model. By comparing the experimental data with simulated events, we determined the strong coupling constant α_s at both energies. The results are 0.129 ± 0.007 (stat) $\pm$ 0.010 (syst) at $\sqrt{s} = 53.3$ GeV and 0.122 ± 0.008 (stat) $\pm$ 0.010 (stat) at 59.5 GeV.

148. Harry F. Contopanagos, Martin B. Einhorn (Michigan U.)
IS THERE A RADIATIVE BACKGROUND TO THE SEARCH FOR RIGHT HANDED CHARGED CURRENTS?

The probability for an electron to emit a collinear, hard photon while undergoing helicity flip is finite in the limit that the electron mass tends to zero. We point out that this provides a background of order α to tests of the Standard Model, which may affect for the search for right-handed, charged currents in deeply inelastic scattering, especially at HERA. It would appear to be exceedingly difficult to discriminate against this background in practice and may make it difficult to improve substantially the mass limits on a right-handed charged vector boson. While our calculation is completely standard, we offer some theoretical conjectures which, if established, could obviate the surprising result.

149. CHARM-II Collaboration (D. Geiregat, et al.)
A NEW DETERMINATION OF THE ELECTROWEAK MIXING ANGLE FROM MUON-NEUTRINO ELECTRON SCATTERING.

The aim of the CHARM-II experiment is a precise determination of $\sin^2\theta_W$ from the ratio of $\nu_\mu e$ and $\bar{\nu}_\mu e$ scattering cross sections. The detector was exposed to the Wide-Band Neutrino Beam of the 450 GeV CERN SPS. An analysis of data in 1987 and 1988 will be presented and a value of the electroweak mixing angle will be given based on almost 1000 $\nu_\mu e$ and $\bar{\nu}_\mu e$ events.

150. F. Csikor, Z. Fodor, E. Lendvai, G. Pocsik (Eotvos U.)
$O(\alpha_s^2)$ PLANAR TRIPLE ENERGY CORRELATION IN QCD.

The exact $O(\alpha_s^2)$ QCD corrections to the planar triple energy correlation are studied. Both the two-dimensional distribution and the integrated correlation are considered. It is shown that planar triple energy correlation is infrared insensitive down to zero jet resolution.

151. TOPAZ Collaboration (I. Adachi, et al.)
SEARCH FOR TOP AND b' QUARKS AT TRISTAN.

The top and b' (charge $-1/3e$ quark of the 4th generation) quarks were searched for in e^+e^- annihilation up to $\sqrt{s} = 60.8$ GeV. In the b' search, the possible decay modes $b' \to \gamma b$ and $b' \to gb$ were considered in addition to the charged current decay. The search was done using the aplanarity distribution, the rate of four jet events, and the rate of isolated photon. We observed no signature for the heavy quarks, and set new limits to their masses.

152. TOPAZ Collaboration (I. Adachi, et al.)
SEARCH FOR NEW PARTICLES IN e^+e^- ANNIHILATION AT TRISTAN.

The TOPAZ detector has accumulated about 20 pb^{-1} of data in the energy range $\sqrt{s} = 52.0 - 60.8$ GeV. Using this data sample, we searched for new particles including supersymmetric particles, heavy leptons, and abnormal dE/dx particles. No significant signal was observed and new mass limits were obtained.

153. TOPAZ Collaboration (I. Adachi, et al.)
SEARCH FOR PAIR PRODUCTION OF HEAVY STABLE PARTICLES AT TRISTAN.

We searched for signatures of new stable particles in e^+e^- annihilation using the data taken by the TOPAZ detector between $\sqrt{s} = 52$ and 60 GeV at TRISTAN. The search made use of the particle identification by the Time Projection Chamber (TPC). We set new limits on the pair production of new stable charged particles with spin 0 or 1/2 and charge between 2/3 and 4/3. The limits on the production cross section was translated to the lower mass limit of hypothetical new particles, SUSY particles, free quarks and heavy leptons.

154. TASSO Collaboration (W. Braunschweig, et al.)
STUDY OF INTERMITTENCY IN ELECTRON-POSITRON ANNIHILATION INTO HADRONS.

Intermittency effects have been observed directly for the first time in e^+e^- annihilation, using 37,509 hadronic events at an average c.m. energy of $\langle\sqrt{s}\rangle =$ 35 GeV. The factorial moments F_2, F_3 and F_4 are given for the rapidity distribution and for the two dimensional distribution in rapidity and azimuthal angle. The effects of cuts in sphericity and particle momentum are large. Comparison with several fragmentation models are made: some models like the Lund model with $O(\alpha_s^2)$ matrix element give qualitative description of the phenomena. The importance of detector effects is demonstrated. The results are discussed in terms of various suggested interpretations of this effect.

155. TOPAZ Collaboration
A SEARCH FOR NEW HEAVY QUARKS USING INCLUSIVE MUONS IN e^+e^- ANNIHILATIONS AT $\sqrt{s} = 56.5$ TO 60.8 GeV.

We searched for a top and fourth generation charge $-1/3$ (b') quarks in e^+e^- annihilations at $\sqrt{s} = 56.5$ to 60.8 GeV using multi-hadron events accompanying muons by the TOPAZ detector at TRISTAN for a total luminosity of 12.31 pb^{-1}. Our observation of low thrust events with isolated muons agrees with the expectation of the standard model with five quark flavors. We set a lower limit for the top quark mass. For a b' quark, lighter than a top quark, we set a mass lower limit as a function of its branching ratio for charge current decay, $Br(b' \to c + w^*)$.

156. TOPAZ Collaboration
STUDY ON THE INCLUSIVE HADRON SPECTRA IN e^+e^- ANNIHILATION AT $\sqrt{s} = 52 - 60.8$ GeV.

We studied the inclusive cross sections and mean multiplicities for $\pi^\pm$, $K^\pm$ and $(p + \bar{p})$ using the TOPAZ detector. The particle species were identified using the energy loss measurements in the Time Projection Chamber. The differential cross sections were compared with various hadronization models.

157. TASSO Collaboration (W. Braunschweig, et al.)
STRANGE BARYON PRODUCTION IN e^+e^- ANNIHILATION.

The production of strange baryons in e^+e^- annihilation has been studied at centre of mass energies of 34.8 GeV and 42.1 GeV, using the TASSO detector at DESY. Inclusive cross-sections have been obtained for Λ^0 and Ξ^- production and an upper limit has been placed upon the production rate of $\Sigma^{*\pm}(1385)$. We measure the Λ^0 multiplicity per event to be $0.218^{+0.011}_{-0.011} \pm 0.021$ and $0.256^{+0.030}_{-0.029} \pm 0.025$ at $\sqrt{s} = 34.8$ and 42.1 GeV respectively. The Ξ^- multiplicity per event is found to be $0.014^{+0.003}_{-0.003} \pm 0.004$ at $\sqrt{s} = 34.8$ GeV. An investigation has been made of the extent to which Λ^0 are produced in pairs. The Λ^0 cross-section has been studied as a function of event sphericity.

158. TOPAZ Collaboration (I. Adachi, et al.)
TEST OF STANDARD MODEL OF ELECTROWEAK INTERACTION BY LEPTON PAIR PRODUCTION IN e^+e^- ANNIHILATION AT $52 < \sqrt{s} < 60.8$ GeV.

Three kinds of lepton pair production of $e^+e^- \rightarrow e^+e^-$, $\mu^+\mu^-$, $\tau^+\tau^-$ have been studied at center-of-mass energies of $52 < \sqrt{s} < 60.8$ GeV with the TOPAZ detector at TRISTAN. Two basic parameters of the standard model of electroweak interaction: mass of $Z(M_z)$ and Weinberg angle ($\sin^2\theta_W$), are determined by their total and differential cross sections. Our results are in good agreement with the world average values. Compositeness scales of the contact interaction are also obtained for these three leptons.

159. TASSO Collaboration
EXPERIMENTAL STUDY OF ORIENTED THREE-JET EVENTS IN e^+e^- ANNIHILATION AT PETRA.

No abstract included.

160. TOPAZ Collaboration (I. Adachi et al.)
MEASUREMENT OF THE TOTAL HADRONIC CROSS SECTION IN e^+e^- ANNIHILATION AND DETERMINATION OF THE STANDARD MODEL PARAMETERS.

We have measured the total hadronic cross sections, R, in e^+e^- annihilation at center-of-mass energies of 50.0 GeV $\leq \sqrt{s} \leq$ 60.8 GeV with the TOPAZ detector at TRISTAN. Systematic errors on R were studied in detail, especially full radiative corrections up to $O(\alpha^3)$ in electroweak theory were taken into account. It allowed us to derive the exact R values in the energy region where the effects due to Z^0 boson become sizable. We determined the standard model parameters: M_{Z^0}, $\sin^2\theta_W$ and $\Lambda_{\overline{MS}}$ by fitting our results and also including other e^+e^- experimental data.

161. TOPAZ Collaboration (I. Adachi, et al.)
SEARCH FOR CHARGED HIGGS PARTICLES AT TRISTAN.

Based on the data accumulated by the TOPAZ detector at the TRISTAN e^+e^- storage ring, the charged Higgs particles ($H^\pm$) were searched for. Taking into account the decay branching ratio, we studied three decay modes: $H^+H^- \rightarrow \tau\bar{\nu}_\tau\bar{\tau}\nu_\tau$, $\tau\bar{\nu}_\tau\bar{c}q$, and $c\bar{q}\bar{c}q$, $(q\ :\ s,b)$. Results obtained by combining these three analyses are reported.

162. CUSB Collaboration (D.M.J. Lovelock, et al.)
A DETAILED STUDY OF THE Υ'' INCLUSIVE PHOTON SPECTRUM.

A 120 pb^{-1} run, using the CUSB-II detector at CESR, on the Υ'' has been used to make a precision determination of the fine splitting of the χ'_b states and the electric dipole (E1) rates. The measurements of the fine structure were obtained from an analysis of the inclusive photon spectrum on the Υ''. They have been used to determine the Lorentz structure of the spin dependent interquark potential. There are 25 possible electromagnetic transitions that can contribute to monochromatic photon signals. Using a photon algorithm that has been optimised for photon efficiency, we have observed many of these photon lines. The fine structure is measured using the observed $3^3S_1 \to 2^3P_J$ photons which are found to have energies of $87.6 \pm 0.5 \pm 1.5$, $100.1 \pm 0.5 \pm 1.5$, $121.0 \pm 1.6 \pm 1.5$ MeV and branching ratios of $(11.4 \pm 1.0 \pm 0.6)\%$, $(11.3 \pm 1.1 \pm 0.6)\%$, $(4.2 \pm 0.8 \pm 0.2)\%$ for $J = 2, 1, 0$ respectively. When these results are combined with those obtained from an 'exclusive' analysis, the fine structure splitting is found to be $M(\chi'_{b2}) - M(\chi'_{b1}) = (13.9 \pm 0.5)$ MeV and $M(\chi'_{b1}) - M(\chi'_{b0}) = (20.7 \pm 1.2)$ MeV, leading to a ratio $r = 0.67 \pm 0.05$. This measured fine structure has been used to probe both the relative contributions of the spin-orbit vs tensor interactions ($a = 9.2 \pm 0.3$ MeV and $b = 1.9 \pm 0.2$ MeV respectively), as well as the Lorentz nature of the confining potential (i.e. scalar or vector). The agreement of the data with the predictions of Gupta, Radford, and Repko are usually taken as proof that the confining potential is 'scalar'. Other photon lines corresponding to the transitions $2^3P_{2,1,0} \to 2^3S_1$ at around 240 MeV, $3^3S_1 \to 1^3P_{2,1,0}$ at around 450 MeV, and $2^3P_{2,1,0} \to 1^3S_1$ at around 750 MeV have now been seen for the first time in the inclusive photon spectrum. They have been observed at the 3.1σ, 4.7σ, and 6.3σ levels respectively, and at the expected branching ratios. We have also searched for other Υ states, the η_b and the $1D$ state, in the inclusive photon spectrum.

163. CUSB Collaboration (P.M. Tuts, et al.)
HYPERFINE SPLITTING OF B-MESONS AND PRODUCTION OF STRANGE B's AT THE $\Upsilon(5S)$.

Data was taken recently at CESR with the CUSB-II detector mostly at the $\Upsilon(5S)$, at a center-of-mass energy of 10.87 GeV. The much improved energy resolution of the CUSB-II detector allows us to clearly observe a photon line from B^* decays and obtain the average mass difference. The observed line is however narrower than expected if due only to $B^*_{d,u}$ decays but is quite consistent with large production of B^*_s, as predicted by our previous coupled channel calculation which had successfully described the cross section above the free flavor threshold.

$$\begin{aligned} \langle E_\gamma \rangle &= 47.5 \pm 0.4\ MeV\ , \\ N(\gamma)/n(\Upsilon(5S)) &= 1.09 \pm 0.06 \\ \langle \beta \rangle &= 0.156 \pm 0.01\ . \end{aligned}$$

Additional information is obtained by satisfying all observed facts: γ yield, width and shape of the signal. Both coupled channel model decay amplitudes have been used, as well as just statistical spin factors and P-wave $B\bar{B}$ pair production. The fraction of B_s production, at the 1σ limit is found to be always larger than 30%. The shape of the observed photon spectrum also implies a very similar value for the hyperfine splitting of normal and strange B meson systems.

164. CUSB Collaboration (C. Yanagisawa, et al.)
B MESON SEMILEPTONIC DECAYS AT THE $\Upsilon(4S)$ AND THE $\Upsilon(5S)$.

B meson semileptonic decay spectra have been obtained from data taken at the $\Upsilon(4S)$ and at the $\Upsilon(5S)$ at the CESR collider with the CUSB detector. The branching ratio for $B \to e\nu X$ at the $\Upsilon(4S)$ is found to be $(11.1 \pm 0.6)\%$. The electron spectrum of B-mesons from $\Upsilon(5S)$ is observed for the first time and the average branching ratio for $B, B_s \to e\nu X$ is consistent with that for B's from the $\Upsilon(4S)$ decays. The shape of the electron spectrum at the $\Upsilon(5S)$ indicates production of B mesons which are heavier than $B_{u,d}$, presumably B_s, B^*_s, at the $\Upsilon(5S)$.

165. CUSB Collaboration (M. Narain, et al.)
SEARCH FOR THE NEUTRAL HIGGS IN RADIATIVE Υ AND Υ'' DECAYS.

CUSB has obtained new lower bounds based on searching for the Higgs in the radiative decay of the Υ. From the strength of the $q\bar{q}H$ vertex, $M_q/v = M_q\sqrt{\sqrt{2}G_F}$, we can compute the annihilation rate. The $|\Psi(0)|^2$ dependence is avoided by comparing to the decay $\Upsilon \to \mu\mu$. Specifically, for the minimum standard model we have

$$\frac{\Gamma(\Upsilon \to \gamma + H)}{\Gamma(\Upsilon \to \mu\mu)} = \frac{G_F M_b^2}{\sqrt{2}\pi\alpha}\left(1 - \frac{M_H^2}{M_\Upsilon^2}\right)$$

which for Υ decays to a very light Higgs gives $BR(\Upsilon \to \gamma + H) \sim 2.3 \times 10^{-4}$. There are however large QCD radiative corrections which strongly suppress $\Upsilon \to \gamma + H$. Depending on the method of calculation, combining the effect of radiative corrections to both the $\mu\mu$ the $H\gamma$ rates results in a reduction of the above BR of a factor of about one and a half to two. Our null result comes from a study of 800,000 Υ's and 600,000 Υ'' decays. Depending on which radiative correction prescription is used, at 90% CL, we exclude Higgs bosons of mass smaller than 5 GeV to 5.75 GeV. The same CUSB data can be also used to exclude gluinos of mass $M_{\text{gluino}} \leq 3.6$ GeV.

166. CUSB Collaboration (U. Heintz, et al.)
HADRONIC WIDTHS OF THE χ_b' STATES, OBSERVATION OF $\Upsilon'' \to (\chi_b')\chi_b\gamma \to (\Upsilon')\Upsilon\gamma\gamma \to \mu\mu(ee)\gamma\gamma$.

The CUSB-II detector at CESR (Cornell Electron Storage Ring) has been used to observe the decay $\Upsilon'' \to \chi_b'(\chi_b)\gamma \to \Upsilon'(\Upsilon)\gamma\gamma \to \mu\mu(ee)\gamma\gamma$. From a sample of 6×10^5 produced Υ'''s obtained from a run with an integrated luminosity of $\approx 140\ pb^{-1}$ we collected ≈ 200 of such events. We use them to derive the χ_b''s hadronic widths (because these widths are not directly accessible) by using the measured $BR(\chi_{bJ}' \to \Upsilon'(\Upsilon)\gamma)$'s combined with $E1$ partial rates calculated by potential models.

$$\text{Experiment}\quad \begin{aligned} \Gamma_{\text{had}}(\chi_{b2}') &= 74 \pm 11 \pm 11\ \text{keV} \\ \Gamma_{\text{had}}(\chi_{b1}') &= 63 \pm 9 \pm 8\ \text{keV} \\ \Gamma_{\text{had}}(\chi_{b0}') &= 360 \pm 157 \pm 39\ \text{keV} \end{aligned}$$

$$\text{Theory}\quad \begin{aligned} \Gamma_{\text{had}}(\chi_{b2}') &= 145(153)\ \text{keV} \\ \Gamma_{\text{had}}(\chi_{b1}') &= 51\ \text{keV} \\ \Gamma_{\text{had}}(\chi_{b0}') &= 542(866)\ \text{keV}\ . \end{aligned}$$

The first error is statistical the second one systematical. QCD to leading order (next-to-leading order) plus a potential model calculation of the wave function predicted instead the values quoted on the right. We see the first evidence for the suppressed transition: $\Upsilon'' \to \chi_b\gamma \to \Upsilon\gamma\gamma \to \ell\ell\gamma\gamma$, for which we determine $\sum BR(\Upsilon'' \to \chi_b\gamma) \times BR(\chi_b \to \Upsilon\gamma) = (0.9 \pm 0.5) \times 10^{-3}$. Comparison with three potential model predictions indicates that the experimental value lies in the middle of the wide range resulting from the different treatments of the relativistic corrections.

$\sum_{J=2,1} BR(\Upsilon'' \to \chi_{bJ}\gamma) \times BR(\chi_{bJ} \to \Upsilon\gamma), (\times 10^{-3})$

Experiment	EI	KR	GRR
0.9 ± 0.5	10.2 ± 1.5	0.47 ± 0.07	0.08 ± 0.01

167. CUSB Collaboration (T.Kaarsberg, et al.)
MEASUREMENT OF THE BRANCHING RATIO FOR DECAY OF Υ STATES TO $\mu^+\mu^-$.

Using the CUSB detector at CESR, we have measured $B_{\mu\mu}$ the branching fraction into muons, of the $1S$, $2S$ and $3S$ Υ mesons. These branching ratios allow the determination of total and partial widths, from which one can then obtain $\Gamma_{ggg}/\Gamma_{\mu\mu}$ and therefore α_s and $\Lambda_{\overline{MS}}$, the coupling constant and the scale parameter of QCD, where $\overline{MS}$ denotes the modified minimal-subtraction scheme. We obtain $B_{\mu\mu}(1S) = (2.61 \pm 0.09 \pm 0.11)\%$, $\Gamma_{\text{tot}}(1S) = 51.1 \pm 3.2$ keV, $B_{\mu\mu}(2S) = (1.38 \pm 0.25 \pm 0.15)\%$, $\Gamma_{\text{tot}}(2S) = 42.3 \pm 0.2$ keV, $B_{\mu\mu}(3S) = (1.73 \pm 0.15 \pm 0.11)\%$ and $\Gamma_{\text{tot}}(3S) = 27.7 \pm 3.7$ keV. The branching ratios are in good agreement with the values obtained under the assumption that all annihilation processes scale in the same ratio from any Υ to any other (factorization). We derive, from these results, $\alpha_s = 0.1736 \pm 0.0033 \pm 0.0173$ and $\Lambda_{\overline{MS}} = 157 \pm 12 \pm 60$ MeV, where the second error is due to the lack of theoretical guidance in how to fix the energy scale. This value for the QCD scale parameter is in good agreement with the values obtained using other methods.

168. TASSO Collaboration (W. Braunschweig, et al.)
CHARGED MULTIPLICITY DISTRIBUTIONS AND CORRELATIONS IN e^+e^- ANNIHILATION AT PETRA ENERGIES.

We report on an analysis of the multiplicity distributions of charged particles produced in e^+e^- annihilation into hadrons at c.m. energies between 14 and 40.8 GeV. The charged multiplicity distributions of the whole event and single hemisphere deviate significantly from the Poisson distribution but follow approximate KNO scaling. We have also studied the multiplicity distributions in various rapidity intervals and found that they can be well described by the negative binomial distribution only for small central intervals. We have also analyzed forward-backward multiplicity correlations for different energies and selections of particle charge and shown that they can be understood in terms of the fragmentation properties of the different quark flavors and by the production and decay of resonances. These correlations are well reproduced by the Lund string model.

169. K.S. Babu, R.N. Mohapatra (Maryland U.)
IS THERE A CONNECTION BETWEEN QUANTIZATION OF ELECTRIC CHARGE AND A MAJORANA NEUTRINO?

In the standard electroweak model with right-handed neutrino, as well as in the minimal left-right symmetric models, quantization of electric charge follows from the requirement of anomaly cancellation only if the neutrino is assumed to be a Majorana particle.

170. B. Strongin, et al. (MIT, Fermilab, Florida U. and Michigan State U.)
OPPOSITE-SIGN DIMUON PRODUCTION IN HIGH ENERGY NEUTRINO-NUCLEON SCATTERING.

The results of a study of opposite sign dimuons produced in the Lab C neutrino detector exposed to the Fermilab quadruplet neutrino beam are presented. The amount of strange sea in the nucleon, the semileptonic branching ratio of the D meson, the elements U_{cd} and U_{cs} of the Kobayashi-Maskawa matrix, and the mass of the charm quark are measured. Various kinematic properties of the standard charm model of opposite sign dimuons are compared with data. The amount of strange sea relative to the up and down sea is found to be $\kappa = 0.56 \pm 0.06 \pm 0.07$. The semileptonic branching ratio is $B = 0.084 \pm 0.03 \pm 0.014$. The matrix elements U_{cd} and U_{cs} are found to be $U_{cd} = 0.225 \pm 0.038 \pm 0.019$ and $U_{cs} = 0.973 \pm 0.061\ ^{+0.036}_{-0.019}$. The mass of the charm quark is found to be consistent with $M_c = 1.5$ GeV/c^2. No discrepancy between the data and the standard charm production model of opposite sign dimuons is found.

171. T.S. Mattison, et al. (MIT, Fermilab and Michigan State U.)
NUCLEON NEUTRAL CURRENT STRUCTURE FUNCTIONS.

The structure of the nucleon is studied by means of deep-inelastic neutrino-nucleon scattering through the weak neutral current at high energies. The neutrino-nucleon scattering events were observed in a 340 metric ton fine grained calorimeter exposed to a narrow-band (dichromatic) neutrino beam at Fermilab. The valence and sea quark distributions of the neutral current nucleon structure functions are presented and are compared to the predictions of the standard model.

172. T. Appelquist, et al. (Yale U., Michigan U. and Cincinnati U.)
HIGH ENERGY ISOSPIN BREAKING IN TECHNICOLOR THEORIES.

Previously, within technicolor models for the electroweak Higgs mechanism, we suggested that four-fermion interactions may play an important role in electroweak symmetry breaking and quark-lepton mass generation. In this paper, we explore the potential impact of strong four-technifermion interactions on such mass splittings via explicit breaking of the custodial $SU(2)_R$ symmetry. A consequence of assuming the four-fermion interactions are of current-current type and the ladder approximation for the fermion self-energies is that they cause no mixing between the two members of a fermion isodoublet, a seemingly necessary feature to produce large mass splittings. We discuss the generality to be expected of this result. The main phenomenological issue, assuming the four-fermion interactions do not lead to unacceptably large flavor-changing neutral currents, is whether the dynamics can be made simultaneously comparable with the apparently large b-t-quark mass splitting and the experimental limits on $\rho-1$ where $\rho \equiv (M_W/M_Z \cos\theta_W)^2$. While this cannot be answered precisely, our conclusion is rather optimistic that this mechanism could be consistent. We recall also that certain four-fermion interactions contribute linearly to ρ and therefore need not be positive, contrary to the lowest order Standard Model radiative corrections.

173. Gongru Lu, Bing-Lin Young (Ames Lab. & Iowa State U.), Dong-Sheng Du (Utah State U.)
UNIFIED MODEL OF METACOLOR INTERACTION.

Unified composite models in which the metacolor symmetry group and the standard model group are embedded in a single group of the $SU(n)$ type and unified above a certain energy scale are constructed. Such a unification scheme provides constraints to the composite model building. The proposed models are analyzed in terms of complementarity principle to show that they satisfy the physics Higgs phase criterion. In order to obtain a metacolor scale of the order of a few tens TeV, the rank of the unification group is restricted.

174. WA70 Collaboration (M. Bonesini, et al.)
THE STRUCTURE OF EVENTS TRIGGERED BY DIRECT PHOTONS IN π^-p, π^+p AND pp COLLISIONS AT 280 GeV/c.

The structure of events associated with the production of direct photons in π^-p, π^+p and pp reactions at 280 GeV/c has been studied using data from the WA70 experiment at the CERN SPS. Results are presented on the distributions of the fractional momenta of the colliding partons and on the fragmentation of the recoil jet and a comparison is made with predictions using the structure functions of Duke and Owens in the Lund Monte Carlo with string fragmentation.

175. P. Aurenche, et al. (Annecy, LAPP & Bielefeld U. & Orsay, LPTHE & Geneva U. & Lausanne U.)
THE GLUON CONTENT OF THE PION FROM HIGH p_t DIRECT PHOTON PRODUCTION.

A fit to the large p_t direct photon cross-sections for the reactions $\pi^{\pm}p \to \gamma X$, using complete beyond leading logarithm QCD expressions, allows the determination of the relative content of gluon and valence quark inside the pion and defines the shape of the gluon structure function for $x > 0.2$. The QCD scale parameter $\Lambda_{\overline{MS}}$ is also in good agreement with precise BCDMS determinations.

176. WA70 Collaboration (M. Bonesini, et al.)
A COMPARISON OF HADRONIC FINAL STATES IN HIGH p_t DIRECT PHOTON AND π^0 PRODUCTION WITH THOSE IN e^+e^- INTERACTIONS.

The characteristics of the jet opposite a high p_t π^0 or direct photon are studied for pp, π^+p and π^-p interactions at 280 GeV/c. The multiplicities of charged hadrons inside the jet and the $p_{||}$ and p_t distributions are compared with e^+e^- results.

177. WA70 Collaboration (E. Bonvin, et al.)
INTRINSIC TRANSVERSE MOMENTUM EFFECTS IN THE $\pi^-p \to \gamma\gamma X$ REACTION AT 280 GeV/c.

The correlation between the two high p_t gammas is studied in the reaction $\pi^-p \to \gamma\gamma X$, observed at 280 GeV/c when each photon has $p_t > 2.75$ GeV/c. QCD calculations beyond leading logarithms are not sufficient to explain the width of the distributions and the effects due to intrinsic transverse momentum of the incident partons are discussed.

178. WA70 Collaboration (M. Bonesini, et al.)
SPECTATOR JETS IN HIGH p_t DIRECT PHOTON AND π^0 PRODUCTION IN 280 GeV/c pp AND $\pi^{\pm}p$ INTERACTIONS.

The characteristics of charged hadrons produced in pp, π^+p and π^-p interactions at 280 GeV/c for high p_t π^0 and direct gamma events are studied. The charge multiplicities, charge ratios and fragmentation distributions in the forward and backward regions in the centre of mass are compared with the expectations of a parton picture.

179. WA21 Collaboration (G.T. Jones, et al.)
W^2 AND Q^2 DEPENDENCE OF CHARGED HADRON AND PION MULTIPLICITIES IN νp AND $\bar{\nu}p$ CHARGED CURRENT INTERACTIONS.

Using data on νp and $\bar{\nu}p$ charged current reactions, obtained in a bubble chamber experiment with BEBC at CERN, the average multiplicities of charged hadrons and pions are determined as functions of W^2 and Q^2. The analysis is based on $\sim$ 20000 events with incident ν and $\sim$ 10000 events with incident $\bar{\nu}$. In addition to the known dependence of the average multiplicity on $\ln W^2$ a weak dependence on $\ln Q^2$ for fixed intervals of W is observed. For $W > 2$ GeV and $Q^2 > 0.1$ GeV^2 the average multiplicity of charged hadrons is well described by $\langle n\rangle = a_1 + a_2 \ln(W^2/\mathrm{GeV}^2) + a_3 \ln(Q^2/\mathrm{GeV}^2)$ with $a_1 = 0.465 \pm 0.053$, $a_2 = 1.211 \pm 0.021$, $a_3 = 0.103 \pm 0.014$ for the νp and $a_1 = -0.372 \pm 0.073$, $a_2 = 1.245 \pm 0.028$, $a_3 = 0.093 \pm 0.015$ for the $\bar{\nu}p$ reaction.

180. BEBC WA59 Collaboration (W. Wittek, et al.)
PRODUCTION OF $\rho^{+,-,0}(770)$, $\eta(550)$, $\omega(783)$ AND $f_2(1270)$ MESONS IN $\bar{\nu}$ NEON AND ν NEON CHARGED CURRENT INTERACTIONS.

The production of the meson resonances $\rho(770)$ (all three charge states), $\eta(550)$, $\omega(783)$ and $f_2(1270)$ in $\bar{\nu}$Ne and νNe charged current interactions is investigated in a bubble chamber experiment with BEBC at CERN. Except for the f_2, the main features of resonance production are reasonably well described by the Lund model, although the average resonance multiplicities are overestimated by the model by $(67 \pm 30)\%$. The average multiplicities of all resonances, including the f_2, are well reproduced by a semi-empirical model, whose parameters were determined from hadron interaction data.

181. BEBC WA21 Collaboration (G.T. Jones, et al.)
EXPERIMENTAL TEST OF THE PCAC HYPOTHESIS IN THE REACTIONS $\nu_\mu p \to \mu^- p\pi^+$ AND $\bar{\nu}_\mu p \to \mu^+ p\pi^-$ IN THE $\Delta(1232)$ REGION.

Data on the reactions $\nu_\mu p \to \mu^- p\pi^+$ and $\bar{\nu}_\mu p \to \mu^+ p\pi^-$ in the $\Delta(1232)$ region are presented and a test of the PCAC hypothesis, using a modified version of the *Adler* model, is performed. The analysis is based on 1081 events in the neutrino and on 180 events in the antineutrino reaction, obtained in a bubble chamber experiment with BEBC at CERN. The experimental cross-sections for an invariant hadronic mass $W < 1.4$ GeV and an (anti-) neutrino energy $E_\nu^L > 10$ GeV are determined to be $(0.628 \pm 0.059) \cdot 10^{-38}$ cm^2 for the neutrino and $(0.168 \pm 0.023) \cdot 10^{-38}$ cm^2 for the antineutrino reaction. The Q^2 and W distributions, the density matrix elements of the Δ resonance, and moments of the pion angular distribution are discussed. The data are found to be in good agreement with the *Adler* model in the Q^2 region below 1 GeV^2. A maximum likelihood fit for the axial mass m_A in the axial-vector form factor yields a value of $m_A = 1.31 \pm 0.12$ GeV. At low Q^2 the data confirm the PCAC hypothesis and the discrepancy, formerly observed between the experimental and theoretical cross-sections for $\nu_\mu p \to \mu^- p\pi^+$ at low momentum transfers ($Q^2 \lesssim 0.2$ GeV^2), is understood as being due to inadequate pion 'off-mass-shell' corrections.

182. A.P. Contogouris, S. Papadopoulos, D. Atwood (McGill U.)
DIRECT PHOTON PRODUCTION WITH COMPLETE QCD CORRECTIONS AND PHYSICAL SCALES.

Recent data on large-p_T direct photon production at $\bar{p}p$ collider, ISR and fixed target energies are analyzed using complete next-to-leading ($O(\alpha_s^2)$) corrections. We employ physical scales $\mu = M = p_T$ and use three parton distribution sets differing essentially in the softness of the gluon. With $\mu = M = p_T$ we find that, on the whole, the data favor a distribution of intermediate softness. We discuss theoretical uncertainties related with the photon Bremsstrahlung contribution and with the variation of the scales ($p_T/2 \leq \mu = M \leq 2p_T$). We conclude that some ambiguity in the gluon distribution still remains.

183. European Muon Collaboration
STRUCTURE FUNCTIONS IN THE PRESCALING REGION .

Structure functions of a nucleon in deuterium, carbon and calcium have been measured at low x, $0.002 < x < 0.17$ and in the prescaling region, $0.2 < Q^2 < 8$ GeV^2, using a trigger designed especially to take data with a muon scattering angle down to 2 milliradians. F_2 was extracted from the inelastic scattering cross sections in the ν range $10 - 150$ GeV. The nucleon F_2 obtained from deuterium does not show any significant x dependence in any Q^2 interval for $Q^2 < 3$ GeV^2, as it is expected by a Regge type theory. $F_2(Q^2)$ dependence is logarithmic, i.e. the same as in the scaling region. For calcium, a depletion of F_2 is observed at low x by 30% as compared with the values at $x = 0.1$ where F_2(Ca) and F_2(D) are not significantly different. This depletion is attributed to shadowing. Also carbon structure function exhibits a similar but less pronounced x dependence. This behavior is quite independent of Q^2 which suggests that the signal is due to some parton mechanism rather than to shadowing caused by a vector meson structure in the virtual exchange photon. The present data add valuable information on the low x and low Q^2 behavior of the free and bound nucleon structure functions. Together with the results obtained with other charged leptons, both at the similar and higher values of x and Q^2 as well as with those obtained in a photoproduction experiment, they make a consistent picture of the nucleon structure functions over a wide kinematic interval. The (x, Q^2) region of our measurements will be a part of the one covered by the future HERA experiments and therefore our data are of interest for constraining the relevant predictions. Results obtained in this experiment were compared to different theoretical models.

184. Yumi S. Hirata (Kitasato U.), Hisakazu Minakata (Tokyo Metropolitan U.)
QUANTUM FIELD THEORIES AROUND A LARGE-Z NUCLEUS. II: METASTABLE STATES IN SUPERCRITICAL PHASE.

We show that new metastable charge-neutral excitations exist in the supercritical phase of QED around a hypothetical large-Z nucleus. They are the vibration modes of the electron cloud induced around the external charge, and therefore do not have their counterpart in the normal phase. To uncover their properties we utilize our bosonized formulation of QED with proper care of renormal-ordering. In this framework the states can be described as small fluctuations on the solitonic ground state. Their energy levels and the widths are estimated by using the WKB approximation. Some of them are found to be very narrow, of the order of 1-10 keV, and have energies around ~ 1.6 MeV. The nature of the states is clarified and their quantum numbers are determined: $J^{PC} = 0^{+-}$ and 1^{++}. Possible relevance of these states to the anomalous peak structure found in heavy ion collision experiments is briefly discussed and a new coincidence experiment is proposed to test this picture.

185. Hisakazu Minakata, Hiroshi Nunokawa (Tokyo Metropolitan U.)
HYBRID SOLUTION OF THE SOLAR NEUTRINO PROBLEM IN ANTICORRELATION WITH SUNSPOT ACTIVITY.

We describe a mechanism for the new solution of the solar neutrino problem with natural anticorrelation with the solar magnetic-field activity in the convection zone. Our mechanism requires the simultaneous presence of the neutrino flavor mixing and the magnetic (or transition) moment as large as $\mu B \sim 10^{-10}$ in units of Bohr magneton·kG.

186. A.S. Joshipura, G. Rajasekaran (Matscience, Madras), V. Gupta, K.V.L. Sarma (Tata Inst.)
INTERPRETATION OF THE RECENT KOLAR EVENTS.

We give plausible interpretations of the unusual events seen in the proton decay detector at Kolar Gold Fields indicating the existence of a massive ($\gtrsim$ 2 GeV) longlived ($10^{-8} - 10^{-9}$ sec.) particle. We show that it is possible to accommodate the particle in the standard model as a fourth generation neutrino, or in E_6 grand unified theory as a neutral fermion occuring in the $\underline{27}$ representation or in supersymmetric theory as a scalar neutrino. However, there is a difficulty in explaining the large production rate for the particle.

187. Nils A. Tornqvist (Helsinki U.)
THE PROTON SPIN, AXIAL VECTOR CURRENT AND MESONIC CLOUD.

We argue that the suppression of the nonsinglet axial vector couplings g_A/g_V from the naive quark model value can easily be understood as an effect of the mesonic cloud around the proton. However, it does not seem possible to generate a sufficiently large polarization of the $s\bar{s}$ sea needed to understand, within the naive parton model, the EMC data on the polarized structure function. Thus our results strengthen the arguments that the gluons in the proton contribute through the Adler-Bell-Jackiw anomaly to the EMC data.

188. Amitava Datta (Jadavpur U.), Dipankar Home (Bose Inst., Calcutta)
IS BELL-TYPE INEQUALITY VIOLATED FOR $K^0 - \overline{K}^0$, $B^0 - \overline{B}^0$,... SYSTEMS?

We extend the Clauser-Horne formulation of Bell-type inequality for the correlated pair of decaying and oscillating (neutral pseudo-scalar) meson-antimesons. The quantum mechanical predictions for the $K^0 - \overline{K}^0$ system do not seem to violate this inequality, whereas the predictions for the $B^0 - \overline{B}^0$ system do lead to such violations in certain situations.

189. ARGUS Collaboration (H. Albrecht, et al.)
SEARCH FOR D^0 DECAYS INTO LEPTON PAIRS.

We have searched for the lepton flavor violating decay $D^0 \to \mu^\pm e^\mp$ and for the rare decays $D^0 \to \mu^+\mu^-$, e^+e^- using the ARGUS detector at the e^+e^- storage ring DORIS II. No candidates were found, leading to the upper limits $\mathrm{BR}(D^0 \to \mu^+\mu^-) < 7{\cdot}10^{-5}$, $\mathrm{BR}(D^0 \to e^+e^-) < 1.7{\cdot}10^{-4}$, and $\mathrm{BR}(D^0 \to \mu^\pm e^\mp) < 1.0 \cdot 10^{-4}$ at the 90% confidence level.

190. John F. Donoghue, B.R. Holstein, Y.C. Lin (Massachusetts U., Amherst)
THE REACTION $\gamma\gamma \to \pi^0\pi^0$ AND CHIRAL LOOPS.

Two-photon production of a neutral-pion pair is uniquely predicted near threshold by the theory of chiral symmetry. The prediction vanishes at the tree level and is nonzero only at one-loop order, yielding a finite result without any unknown counterterms. In this paper we calculate the cross-section for both on-shell and off-shell photons, as both cases can both be studied at e^+e^- storage rings. This reaction is the most accessible process which directly probes the loop structure of chiral-$SU(2)$ symmetry.

191. Masako Bando (Aichi U.), Takanori Fujiwara (Ibaraki U.), Koichi Yamawaki (Nagoya U.)
GENERALIZED HIDDEN LOCAL SYMMETRY AND THE A_1 MESON.

We present a generalized formalism of the hidden local symmetry in which any nonlinear sigma model based on the manifold G/H is gauge equivalent to a model possessing $G_{\mathrm{global}} \times G_{\mathrm{local}}$ symmetry, which is a natural extension of the well-known gauge equivalence between a G/H nonlinear sigma model and a $G_{\mathrm{global}} \times H_{\mathrm{local}}$ 'linear' model. As an application of this formalism, we reexamine a possibility that the A_1 mesons as well as the ρ mesons and their $U(3)$ partners are dynamical gauge bosons of the hidden local symmetry, $[U(3)_L \times U(3)_R]_{\mathrm{local}}$, in the $U(3)_L \times U(3)_R/U(3)_V$ nonlinear sigma model.

192. Kei-ichi Kondo (Chiba U.), Hidetoshi Mino, Koichi Yamawaki (Nagoya U.)
CRITICAL LINE AND DILATON IN SCALE-INVARIANT QED.

We have found a novel spontaneous chiral symmetry breaking solution to the ladder Schwinger-Dyson equation for QED plus a chiral-invariant four-fermion interaction. The critical line is explicitly obtained in the plane of two coupling constants of gauge and four-fermion interactions. The existence of a dilaton pole has also been examined on the full critical line.

193. V.A. Miransky, K. Yamawaki (Nagoya U.)
ON GAUGE THEORIES WITH ADDITIONAL FOUR-FERMION INTERACTION.

In the ladder QED with four-fermion interaction, the anomalous dimension γ_m of the operator $\bar{\psi}\psi$ is calculated on the full critical line. It is shown that in the nonperturbative phase with the gauge coupling $\alpha^{(0)} < \alpha_c = \pi/3$, the dynamical dimension $d_{\bar{\psi}\psi} = 3 - \gamma_m$ is smaller than two ($\gamma_m > 1$) near the critical line. This suggests that gauge theories with four-fermion interaction may be renormalizable. The applications of these theories to the technicolor models are discussed.

194. V.A. Miransky, Masaharu Tanabashi, Koichi Yamawaki (Nagoya U.)
DYNAMICAL ELECTROWEAK SYMMETRY BREAKING WITH LARGE ANOMALOUS DIMENSION AND t QUARK CONDENSATE.

A new class of models with dynamical electroweak symmetry breaking is proposed. In these models the composite operator $\bar{\psi}\psi$ has a large anomalous dimension $\gamma_m \simeq 2$ and the t quark plays an important role in the generation of W and Z boson masses without destroying the relation $M_W^2/M_Z^2 \cos^2\theta_W \simeq 1$. The scenario naturally leads to a large value of the t quark mass.

195. Ken-Ichi Aoki (Kyoto U., RIFP), Masako Bando (Aichi U.), Hidetoshi Mino, Tatsuhiko Nonoyama (Nagoya U.), Hiroto So (Niigata U.), Koichi Yamawaki (Nagoya U.)
HOW DOES TECHNICOLOR SURVIVE THE FCNC SYNDROME?

A novel systematic numerical analysis of the chiral condensate in technicolor theories is carried out with a very careful treatment of cut-off dependence in the ladder Schwinger-Dyson equation. In accord with the previous analytical study, we find that the 'slowly running' coupling constant does not yield enough suppression of the flavor-changing neutral currents in sharp contrast to the scale-invariant model.

196. V.A. Miransky, Masaharu Tanabashi, Koichi Yamawaki (Nagoya U.)
IS THE t QUARK RESPONSIBLE FOR THE MASS OF W AND Z BOSONS?

Based on the dynamical model having a large anomalous dimension $\gamma_m \simeq 2$, we propose that the t quark condensate is responsible for the mass of W and Z bosons within the framework of the $SU(3)_C \times SU(2)_L \times U(1)_Y$ gauge theory (without Higgs) plus four-fermion interactions among quarks and leptons. This accounts for the largeness of the t quark mass m_t without violating the weak-isospin relation $\rho \equiv m_W^2/m_Z^2 \cos^2\theta_W \simeq 1$. Quark/lepton mass matrices are obtained through the four-fermion interactions, which yields no serious flavor-changing neutral currents. The model predicts a $(\bar{t}t)$ composite scalar meson with a mass $\simeq 2m_t$. A possibility for an unusual resolution of the strong CP problem is also suggested.

197. Koichi Yamawaki (Nagoya U.)
SCALE INVARIANT TECHNICOLOR AND A TECHNIDILATON.

I shall describe the scale-invariant technicolor model characterized by the ladder Schwinger-Dyson equation. It is found that the theory has a non-trivial ultraviolet fixed point with large anomalous dimension, which naturally suppresses the flavor-changing neutral currents (FCNC's) and at the same time raises the mass of the problematic light technipions. The model is expected to predict a dilaton (technidilaton), a light $J^{PC} = 0^{++}$ pseudo Nambu-Goldstone boson associated with the spontaneous breakdown of the scale invariance, whose dynamical issues are argued in detail. Our model is compared with a similar scenario of asymptotically free theories with the running coupling constant varied slowly in the ladder-like Schwinger-Dyson equation: it is found that enough suppression of FCNC's is only possible in the scale-invariant model.

198. Koichi Yamawaki (Nagoya U.)
RESURRECTING TECHNICOLOR — SCALE-INVARIANT TECHNICOLOR AND A TECHNIDILATON.

I shall explain how the technicolor is resurrected in the light of new trends in strong-coupling gauge theories. Basic idea and ingredients of technicolor are critically reviewed. Then I shall describe the scale-invariant technicolor model characterized by the ladder Schwinger-Dyson equation. It is found that the theory has a non-trivial ultraviolet fixed point with large anomalous dimension, which naturally suppresses the flavor-changing neutral currents and at the same time raises the mass of the problematic light technipions. The model is expected to predict a dilaton (technidilaton), a light $J^{PC} = 0^{++}$ pseudo Nambu-Goldstone boson associated with the spontaneous breakdown of the scale invariance, whose dynamical issues are argued in detail.

199. Tatsuhiko Nonoyama, Tsuyoshi B. Suzuki, Koichi Yamawaki (Nagoya U.)
PHASE STRUCTURE AND DILATON IN QED WITH FOUR-FERMION INTERACTION.

The chiral phase transition of QED plus four-fermion interaction is studied through the quenched planar (ladder) Schwinger-Dyson equation. We derive the critical line and the scaling relation near the critical line from the analytical solution which is valid for the whole coupling region including the pure four-fermion limit. Along the full critical line we find no dilaton in the fermion-antifermion scattering amplitude and also in the effective potential approach.

200. V.A. Miransky, Tatsuhiko Nonoyama, Koichi Yamawaki (Nagoya U.)
ON THE PHASE DIAGRAM OF ASYMPTOTICALLY FREE GAUGE THEORIES WITH ADDITIONAL FOUR-FERMION INTERACTION.

In the ladder asymptotically free gauge models with one-loop running coupling constant and additional four-fermion interaction, the structure of the phase diagram in coupling constants is determined. The anomalous dimension γ_m of the composite operator $\bar{\phi}\phi$ is calculated on the full critical line of the diagram. The possibilities of the examination of the phase diagram of such theories in lattice computer simulations are discussed.

201. A.L. Golendukhin, et al. (Yerevan Phys. Inst.)
INVESTIGATION OF POLARIZATION PARAMETERS Σ, P AND T FOR π^0-MESON PHOTOPRODUCTION IN THE RESONANCE ENERGY RANGE.

At the Yerevan synchrotron the polarization parameters Σ, T and P have been measured for the reaction $\gamma p \to \pi^0 p$ in the photon energy range $0.9 - 1.35$ GeV and c.m. angle of 125° in an experiment with a polarized target. The obtained results are compared with the existing theoretical predictions in the resonance region.

202. R.O. Avakian, et al. (Yerevan Phys. Inst.)
INVESTIGATION OF RECOIL PROTON POLARIZATION IN DOUBLE POLARIZATION EXPERIMENT OF THE REACTION $\gamma p \to p\pi^0$ IN THE RESONANCE PHOTON ENERGY REGION OF $(0.73 - 1.17)$ GeV.

Results on measurements of P_{xz}- and P_y-components of the vector of recoil proton polarization in the reaction $\gamma p \to p\pi^0$ by linearly polarized photons are given. The measurements were carried out for π^0-meson production at cms angles $\theta^*_{\pi^0} = 70°$ and 80° and photon energies $E_\gamma = 0.73$ to 1.17 GeV. The results of measurements are compared with predictions of various theoretical analyses.

203. N.E. Sekhposyan, N.L. Ter-Isaakyan (Yerevan Phys. Inst.)
ANALYSIS OF THE DEEP INELASTIC ELECTROPRODUCTION DATA ON NUCLEON.

Model independent description and analysis of the EMC data on deep inelastic μN-scattering (J. Aubert *et al.*) together with SLAC data (M. Mestauer *et al.*) and SLAC-MIT data (A. Bodek *et al.*) on eN-scattering are carried out. The structure functions were parametrized taking into account logarithmic and power contributions to describe the scaling violation. We find that EMC data on the deuteron are 1.2 times lower as compared to SLAC and SLAC-MIT data. The data of all other groups coincide within 5%. The power contributions corresponding to higher twists turned out to be high and make up about 24% for the proton and 18% for the deuteron at $x = 2/3$, $Q^2 = 20$ GeV^2. For leading twist structure function the data indicates the following physical results: a. The fraction of nucleon momenta carried by gluons slowly increases with decreasing Q^2 from 40% at $Q^2 = 100$ GeV^2 to 50% at $Q^2 = 1$ GeV^2. b. The fraction of momentum carried by $q\bar{q}$-pairs decreases from $\sim 50\%$ at $Q^2 = 100$ GeV^2 to a few percent at $Q^2 = 1$ GeV^2. The ratio $F^n/F^p \sim 0.6 + 0.1$ for $x = 0.5$ and points out no tendency to decrease when $x \to 1$.

204. I.G. Aznauryan, A.S. Bagdasaryan, S.V. Esaybegyan, N.L. Ter-Isaakyan (Yerevan Phys. Inst.)
HADRON FORM FACTORS IN RELATIVISTIC QUARK MODEL. BEHAVIOR IN LOW MOMENTUM TRANSFER REGION AND QCD ASYMPTOTICS.

It is shown that in the region of relatively low momentum transfer $0 \leq Q^2 \leq 5$ GeV^2 the pion form factor and nucleon electromagnetic and axial form factors are described well in the relativistic quark model by simple quark diagrams without gluon exchanges. The hard gluon exchange diagrams are about 50 times lower and their signs are opposite to those of the experiment. The low-energy pion and nucleon wave functions which determine these contributions were found to differ slightly from the asymptotic wave functions and their moments coincide within 10%, which is in contradiction to QCD dispersion sum rules.

205. TOPAZ Collaboration (I. Adachi, et al.)
A SEARCH FOR EXCITED LEPTONS IN THE ENERGY REGION $\sqrt{s} = 52 - 60.8$ GeV.

Excited leptons, e^*, μ^* and τ^*, have been searched for with the TOPAZ detector at TRISTAN e^+e^- collider in the energy region $\sqrt{s} = 52 - 60.8$ GeV. e^* and μ^* have been searched for in the l^* single production process, $e^+e^- \to l^+l^-\gamma$, and the l^*-pair production process, $e^+e^- \to l^+l^-\gamma\gamma$. e^* has been also searched for in the processes $e^+e^- \to e\gamma(e)$ (quasi-real Compton scattering) and $e^+e^- \to \gamma\gamma$. τ^* has been searched for in the τ^*-pair production process, $e^+e^- \to \tau^+\tau^-\gamma\gamma$. No evidence for the production of excited leptons has been found, and new limits on e^*, μ^* and τ^* production are set.

206. B.A. Kniehl, J.H. Kuhn (Munich, Max Planck Inst.)
QCD CORRECTIONS TO THE AXIAL PART OF THE Z DECAY RATE.

As a consequence of the large splitting between top and bottom quark masses QCD corrections to the axial part of the Z decay rate into b quarks differ from those to the vector part in order α_s^2. Analytic results for these two-loop corrections are presented.

207. TASSO Collaboration (W. Braunschweig, et al.)
EXPERIMENTAL STUDY OF JET MASSES IN e^+e^- ANNIHILATION AT C.M. ENERGIES BETWEEN 12 AND 43.5 GeV.

Data on jet masses, resulting from the decomposition of e^+e^- hadronic final states into two hemispheres, are presented at centre of mass energies between 12 and 43.5 GeV. Comparisons are made with bare $O(\alpha_s^2)$ QCD predictions as well as with QCD based fragmentation models. Values for α_s and $\Lambda_{\overline{MS}}$ are determined, both with and without hadronization effects included. Upper and lower limits for $\Lambda_{\overline{MS}}$ independent of fragmentation models have been determined to be 0.480 ± 0.025 GeV and 0.047 ± 0.007 GeV respectively.

208. TASSO Collaboration
STRANGE MESON PRODUCTION IN e^+e^- ANNIHILATION.

The full TASSO data have been used to study the inclusive production of strange mesons in e^+e^- annihilations. Differential and total cross sections have been measured in the centre of mass energy range 14 to 44 GeV for $K^0, \overline{K}^0$ and 34.5 to 44 GeV for $K^{*\pm}(892)$. We have investigated the strange meson production properties in jets by studying the rapidity and p_t^2 distributions as well as the evolution of the multiplicities as a function of the event sphericity. We find no evidence that the strange meson yields increase with increasing sphericity faster than the total charged multiplicities.

209. T. Inagaki, et al. (KEK, Tokyo U. and Kyoto U.)
SEARCH FOR $K_L^0 \to \mu e$ AND $K_L^0 \to ee$ DECAYS.

We report on the present status of a search for the rare decays $K_L^0 \to \mu e$ and $K_L^0 \to ee$ at the KEK 12 GeV Proton Synchrotron. We found no $K_L^0 \to \mu e$ or $K_L^0 \to ee$ event and found 54 $K_L^0 \to \mu\mu$ events. The upper limits of the $K_L^0 \to \mu e$ and $K_L^0 \to ee$ branching ratios are $B(K_L^0 \to \mu e) < 4.3 \times 10^{-10}$ and $B(K_L^0 \to ee) < 5.6 \times 10^{-10}$ at 90% confidence level. The 54 $K_L^0 \to \mu\mu$ events correspond to an absolute branching ratio of $(8.4 \pm 1.1) \times 10^{-9}$.

210. Johannes Blumlein (GDR Academy of Sciences)
LEADING LOG RADIATIVE CORRECTIONS AT HERA.

The $O(\alpha)$ radiative corrections for neutral and charged current deep inelastic ep-scattering at HERA are calculated in the leading log approximation. It is demonstrated that all light quark mass terms can be eliminated from the cross section. The numerical results agree within a few percent with those found in a complete $O(\alpha)$ calculation by Bardin *et al.*

211. Wei-Shu Hou, Robin G. Stuart (Max Planck Inst., Munich)
POSSIBILITY OF DISCOVERING THE NEXT CHARGE -1/3 QUARK THROUGH ITS FLAVOR-CHANGING NEUTRAL CURRENT DECAYS.

If a fourth fermion generation exists, it is possible that the next charge $-1/3$ quark (b') is lighter than the top. The leading charged-current transition $b' \to cW^*$ is expected to be strongly suppressed by mixing angles. This leads to the interesting and likely consequence that the b'-quark decay is completely dominated by loop-induced flavor-changing neutral currents, which have distinctive experimental signatures. The relative importance of $b' \to bg^*$, $b' \to b\gamma$, and $b' \to bZ$ transitions (as well as comparisons with $b' \to cW^*$) are explored for b' masses up to 130 GeV.

212. A.D. Martin, W.J. Stirling (Durham U.), R.G. Roberts (Rutherford)
CHARM QUARK CONTRIBUTION TO THE W AND Z CROSS-SECTIONS AT $p\bar{p}$ COLLIDERS.

The ratio R of the cross sections for $p\bar{p} \to W \to l\nu$ and $p\bar{p} \to Z \to l^+l^-$ is one of the most important precision measurements at present day $p\bar{p}$ colliders. In previous studies we have investigated some of the theoretical uncertainties in the calculation of this ratio. We further refine the analysis by considering the uncertainty coming from the charm quark content of the nucleon. We show that data on the charm quark contribution to deep inelastic structure function F_2 are already sufficiently precise to place limits on this uncertainty, and we conclude that the charm quark uncertainty in the ratio of W and Z cross sections does not pose a serious problem. We predict that $R = 10.4 \pm 0.1$ at $\sqrt{s} = 1.8$ TeV if $N_\nu = 3$ and $m_t > M_W$.

213. A.D. Martin (Durham U.), R.G. Roberts (Rutherford), W.J. Stirling (Durham U.)
IMPROVED PARTON DISTRIBUTIONS AND W, Z PRODUCTION AT $p\bar{p}$ COLLIDERS.

We obtain improved parton distributions from a structure function analysis which includes new data on deep inelastic scattering and on dilepton production. These lead to improved predictions for $\sigma(W)/\sigma(Z)$, which when combined with the forthcoming data from $p\bar{p}$ colliders, will give tighter constraints on the values of m_t and N_ν. We emphasize how new measurements of the $W^\pm$ charge asymmetry may resolve the EMC – BCDMS F_2 discrepancy and hence sharpen the prediction for $\sigma(W)/\sigma(Z)$ even further.

214. Jean-Marc Gerard, Wei-Shu Hou (Munich, Max Planck Inst.)
CP NONCONSERVATION AND CPT: A RE-ASSESSMENT OF LOOP EFFECTS IN CHARMLESS B DECAYS.

Absorptive parts in QCD-induced charmless B decays allow for a new mechanism of CP violation. Existing discussions in the literature do not satisfy the CPT theorem. Solving the problem, we find that, inclusively, a strong cancellation remains between the $b \to su\bar{u}$ mode CP rate asymmetries and the purely loop-induced $b \to sd\bar{d}$, $ss\bar{s}$, sgg modes. Exclusively, the asymmetries of purely loop-induced modes (such as $B \to K\phi$) are unaffected, but those related to $b \to su\bar{u}$ (such as $B \to K\pi$) change sign. The discussion for $b \to d$ modes is similar.

215. Sachio Komamiya (SLAC)
Q^2 SCALE OF α_s.

The Q^2 scale of the strong coupling α_s is discussed. On the one hand, the Q^2 is simply an 'unphysical' parameter, since it is an artifact of the renormalization procedure. On the other hand, the origin of the QCD running coupling is nothing but the 'physical' effects of vacuum polarization so that the Q^2 should have 'physical' meaning. This paper applies the methods of Stevenson and that of Brodsky, Lepage and Mackenzie to eliminate the Q^2 ambiguity in the study of multijet rates in high energy e^+e^- annihilation. The predictions of jet multiplicities for these methods are compared with the Mark-II data at $\sqrt{s} = 29$ GeV.

216. Yung Su Tsai (SLAC)
PRODUCTION OF NEUTRAL BOSONS BY AN ELECTRON BEAM.

We investigate the possible existence of neutral bosons which are coupled to leptons. The cross section for the process $e+p \to e+X+$ anything, where X is a neutral boson of spin-parity $0^\pm$ or $1^\pm$ emitted by the electron, is calculated and its energy-angle distribution discussed. Assuming X to decay predominantly into a lepton pair we investigate the characteristics of the background. It is pointed out that the signal to the background ratio can be greatly enhanced if one selects high $x = E_x/E_1$ and also uses the outgoing electron as a tag at a slightly non-forward angle with the X particle arranged in such a way that the momentum transfer to the target particle is near its minimum. This happens when the outgoing electron momentum $\vec{P}_2$ is parallel to $\vec{P}_1 - \vec{k}$, where $\vec{P}_1$ and $\vec{k}$ are the momenta of the incident electron and the X particle respectively.

217. Yung Su Tsai (SLAC)
EFFECTS OF CHARGED HIGGS IN τ DECAY.

An experiment to test the effect of charged Higgs exchange in τ decay is proposed. It is pointed out that in the decay $\tau^- \to \nu_\tau + \pi^- + \pi^0$ the effect due to Higgs exchange can be obtained by observing the angular distribution of the π^- in the rest frame of the $\pi^- + \pi^0$. For each value of invariant mass of the $\pi^- + \pi^0$ system, the π^- must have a unique p-wave angular distribution independent of the strong interaction if it is due to W^- exchange alone. Any s-wave interference with this p-wave angular distribution can be attributed to the scalar exchange.

218. Yung Su Tsai (SLAC)
EFFECTS OF W_R AND CHARGED HIGGS IN THE LEPTONIC DECAY OF THE τ.

An experimental test of the existence of the right-handed W boson and the charged Higgs particle is suggested. The experiment involves measurement of muon polarization from the decay of polarized τ's.

219. D.O. Caldwell, et al. (UC, Santa Barbara and LBL, Berkeley)
RECENT RESULTS FROM THE UCSB/LBL DOUBLE BETA DECAY EXPERIMENT.

A new limit has been set on neutrinoless double beta decay, a process which would require lepton number nonconservation, plus one other piece of new physics. That second item might be light Majorana electron neutrino mass, a very heavy Majorana neutrino, right-handed currents, or supersymmetric particles in theories with R-parity violation. Limits on all these quantities can be obtained from the result that the UCSB/LBL Ge multidetector system shows the half-life for ^{76}Ge $\to$ ^{76}Se $+ 2e^-$ to be $> 1.2 \times 10^{24}$y at the 90% confidence level or $> 2.4 \times 10^{24}$y at the 68% confidence level.

220. Wei-Shu Hou (Munich, Max Planck Inst.)
B' MIXING.

The possibilities of B' meson mixing is discussed. It is found that if $f_{B'}$ is not much smaller than f_π, B'_s (or B'_d) mixing is possible if $|V_{t's}V_{t'b'}/V_{cb'}|$ is large, but only for not too heavy b' masses, e.g. below LEP I energies. The B'_b meson will have large mixing for a broad range of parameters: however, it is hard to produce. The vector mesons (B'^*) may exhibit mixing phenomena as well. A discussion of our present understanding of meson decay constants of a heavy-light system (f_{Qq}) is given. Further efforts on this problem is needed, which will have an impact on our understanding of B mixing as well.

221. C.S. Kim, A.D. Martin (Durham U.)
ON THE DETERMINATION OF V_{ub} AND V_{cb} FROM SEMILEPTONIC B DECAYS.

All available data on the inclusive semileptonic decays of B mesons are analyzed and it is found that $|V_{ub}/V_{cb}| < 0.20$ at the 90% confidence level, and that $|V_{cb}| = 0.049 \pm 0.006$. The experimental and theoretical model uncertainties in these determinations are critically discussed.

222. V. Barger (Wisconsin U., Madison), C.S. Kim (Wisconsin U., Madison & Durham U.), R.J.N. Phillips (Rutherford)
FERMI MOTION IN THE SEMILEPTONIC DECAY OF B MESONS.

In the independent-quark decay model for inclusive semileptonic decays $B \to \ell\nu X$, the Fermi momentum distribution in the initial B meson is loosely constrained by the lepton momentum spectrum. We examine the additional constraints from the final hadronic invariant mass spectrum and show that a step-function Fermi distribution gives a somewhat better overall fit to data than the conventional gaussian shape.

223. K. Hagiwara, F. Halzen, C.S. Kim (Durham U. & Wisconsin U., Madison)
NEUTRINO COUNTING WITH W, Z AND WEAK BOSON PRODUCTION BY CHARM QUARKS.

Production of weak intermediate bosons by charm quarks in the nucleon is expected to occur at the few percent level at the energy of the Tevatron collider. As $c\bar{s} \to W$ could be much larger than $c\bar{c} \to Z$, production via charm has a strong influence on the ratio of W and Z events and therefore on the neutrino count. The range $c = 0$ and $c = s$ corresponds to a variation mimicking two neutrinos. We show that the charm structure function of the nucleon $c(x, Q^2 = m_W^2)$ can be extracted from $p\bar{p}$ collider measurements of $\sigma_W^\pm(y)$ and $\sigma_Z(y)$.

224. G. D'Agostini, W. de Boer, G. Grindhammer (INFN, Rome & Max Planck Inst., Munich & SLAC)
DETERMINATION OF α_s AND THE Z^0 MASS FROM MEASUREMENTS OF THE TOTAL HADRONIC CROSS SECTION IN e^+e^- ANNIHILATION.

We have made a fit to the measurements of the normalized cross section R for the process $e^+e^- \rightarrow$ *hadrons* at centre of mass energies between 7.0 and 57.0 GeV with α_s and M_Z as free parameters. At the highest TRISTAN energies, the increase in R from the tail of the Z^0 resonance allows a precise measurement of the Z^0 mass, while the lower energy data determine the value of the strong coupling constant. The result is: $\alpha_s(34^2\ \mathrm{GeV}^2) = 0.143 \pm 0.015$, if $O(\alpha_s^3)$ QCD corrections are taken into account, and $M_Z = 89.3 \pm 1.5$ GeV for a top mass of 60 GeV and fixed $\sin^2\theta_W$. The off shell determination of the Z^0 mass depends almost linearly on the top mass through the electroweak radiative corrections: it is lowered by 1.0 GeV for $M_t = 180$ GeV.

225. Xiao-Gang He (Melbourne U.), Wei-Shu Hou (Munich, Max Planck Inst.)
RELATING THE LONG B LIFETIME TO A VERY HEAVY TOP.

The long B lifetime is related to the heaviness of the top quark by a particular mass-mixing ansatz. The u-type quark mass matrix is of the Fritzsch form, while for the d-type it is diagonal except in the $d-s$ plane, which generates the Cabibbo rotation. We predict $m_t \gtrsim 200$ GeV from $V_{cb} \lesssim 0.06$. One gets 'maximal CP', and the relations $|V_{ub}/V_{cb}| = \sqrt{u/c}$, $|V_{td}/V_{cb}| = \sqrt{d/s}$ and $|V_{ts}| = |V_{cb}|$ are close to exact. An interesting Wolfenstein pattern emerges. We discuss the viability and implications of having such a heavy top (such as lower M_Z), and the possibility of a vanishing V_{ub} and its impact on CP violation.

226. Wei-Shu Hou (Munich, Max Planck Inst.), Robin G. Stuart (Mexico, IPN)
FLAVOR CHANGING DECAYS OF THE Z^0 BOSON TO MASSIVE QUARKS IN THE TWO HIGGS DOUBLET MODEL.

The rates for rare flavor chaning Z^0 decays ($Z^0 \rightarrow Q\bar{q}$) are calculated for one massless and one massive external quark in the 2 Higgs doublet extension of the Standard Model with three or four fermion generations. Upper bounds for the flavor changing branching ratios could increase by 2—3 orders of magnitude compared to those of the Standard Model, independent of Kobayashi-Maskawa matrix elements.

227. CELLO Collaboration (H.J. Behrend, et al.)
INCLUSIVE STRANGE AND CHARMED PARTICLE PRODUCTION IN e^+e^- ANNIHILATION.

We present an analysis of strange and charmed particle production from e^+e^- annihilation into multihadronic final states. The experiment was performed with the CELLO detector at the PETRA storage ring at DESY. The data was taken at a centre of mass energy of 35 GeV with an integrated luminosity of 86 pb^{-1}. The particles K_S^0, $K^{*\pm}$, Λ and $D^{*\pm}$ have been identified by their characteristic decays, and differential cross sections for their production have been obtained. From a comparison of K_S^0 and $K^{*\pm}$ rates the vector meson suppression parameter $V/(V+P)_s$ has been determined to be $0.59^{+0.20}_{-0.10}(\mathrm{stat.})^{+0.10}_{-0.05}(\mathrm{syst.})$. The $D^{*\pm}$ cross section is used to extract the Peterson fragmentation parameter ε_c for charm quark fragmentation, where $\varepsilon_c = 0.26 \pm 0.17(\mathrm{stat.})$ is found.

228. CELLO Collaboration (H.J. Behrend, et al.)
EXCLUSIVE τ DECAYS WITH THE CELLO DETECTOR AT PETRA.

Precise measurements of all major exclusive leptonic and semihadronic decay channels of the τ lepton are presented. The results are based on 6064 τ decays observed at $\sqrt{s} = 35$ GeV, using the CELLO detector at PETRA. The sum of the measured branching ratios saturates the total τ decay rate within a statistical error of 2.6% and a systematic uncertainty of 1.2%. Comparing with the topological branching ratios measured recently by CELLO, the sums of the respective exclusive branching ratios are found to saturate the inclusive fractions. Our results are in good general agreement with existing measurements. However, the decay fractions for $\tau^- \rightarrow \pi^-\pi^+\pi^-\nu_\tau$ and $\tau^- \rightarrow \pi^-\pi^0\pi^0\nu_\tau$ are larger than the present world averages. All measured branching ratios are in good agreement with the expectations from a standard τ lepton.

229. CELLO Collaboration (H.J. Behrend, et al.)
$a_2(1320)$ AND $\pi_2(1670)$ FORMATION IN THE REACTION $\gamma\gamma \rightarrow \pi^+\pi^-\pi^0$.

The reaction $e^+e^- \rightarrow e^+e^-\pi^+\pi^-\pi^0$ in the untagged mode has been measured with the CELLO detector at PETRA. We present preliminary results. The cross section is dominated by exclusive $a_2(1320)$ production. Whose radiative width is determined to be $\Gamma_{\gamma\gamma} = 1.00 \pm 0.07$ (stat) $\pm$ 0.19 (syst) keV. An angular correlation analysis indicates pure helicity 2 with an upper limit on the helicity 0 contribution of 5.8% (at 95% c.l.). Excitation of the pseudotensor meson $\pi_2^0(1670)$ is observed for the first time in this reaction, confirming the preliminary Crystal Ball result in the $3\pi^0$ decay mode. The π_2 radiative width and its strong dependence on interference effects are studied.

230. Jon Pumplin (Michigan State U.)
NONPERTURBATIVE EFFECTS IN TAU DECAY.

It has been suggested that the lifetime of the τ lepton can be calculated using perturbative QCD, with sufficient accuracy that comparison with the measured lifetime can be used to determine the QCD scale parameter Λ. Contrary to this suggestion, non-perturbative effects are shown here to induce uncertainties in the calculated lifetime which are so large as to make the determination of Λ unreliable. These non-perturbative effects are associated with hadronic thresholds and resonances, and are much larger than previous estimates based on QCD sum rules. Independent experimental tests, which make use of the hadronic mass spectrum in τ decay, are proposed.

231. AMY Collaboration (I.H. Park, et al.)
DETERMINATION OF THE QCD RENORMALIZATION SCALE AND $\Lambda_{\overline{MS}}$ FROM MULTI-JET EVENTS PRODUCED IN ELECTRON-POSITRON COLLISIONS.

We have fit the multi-jet fractions from e^+e^- annihilations at center-of-mass energies from 50 to 57 GeV to an $O(\alpha_s^2)$, perturbative QCD calculation in which the renormalization energy scale is a free parameter. Using the fit values for the optimal renormalization scale and for $\Lambda_{\overline{MS}}$, the perturbative QCD calculation reproduces well the measured multi-jet fractions. The determined value of $\Lambda_{\overline{MS}}$ is 173 ± 60 MeV.

232. TASSO Collaboration (W. Braunschweig, et al.)
STUDY OF CHARMED MESON PRODUCTION IN $\gamma\gamma$ INTERACTIONS.

Charm production in $\gamma\gamma$-interactions at PETRA energies has been observed in the TASSO detector. Both inclusive $D^{*\pm}$ and $D^0\overline{D}^0$ production seem to have similar cross-sections, which are larger than expected by some simple models.

233. ARGUS Collaboration (H. Albrecht, et al.)
ARGUS: A UNIVERSAL DETECTOR AT DORIS-II.

The detector ARGUS has been designed as a universal tool to investigate final states from e^+e^- annihilation processes in the energy range of the Υ resonances. ARGUS started operation in October 1982 and has since successfully taken data at the $\Upsilon(1S)$, $\Upsilon(2S)$ and $\Upsilon(4S)$ energies, and in the nearby continuum. The detector combines excellent charged particle identification and good photon energy resolution over more than 90% of the full solid angle. A particle originating from the interaction vertex and leaving the beam tube traverses the following components: the vertex drift chamber, the main drift chamber which determines its momentum and specific ionization, the time-of-flight system through which its velocity is determined, and the electromagnetic calorimeter. Muons pass through the magnet coils and the flux return yoke and finally hit the muon chamber system which surrounds the detector. The momentum resolution of ARGUS is $\sigma(p_T)/p_T = \sqrt{0.01^2 + (0.009 \cdot p_T[GeV/c])^2}$, and the photon energy resolution in the barrel shower counters is $\sigma(E)/E = \sqrt{0.072^2 + 0.065^2/E[GeV]}$. Combining the information from all particle identification devices, more than 80% of all charged hadrons can be recognized unambiguously. The electron-hadron and muon-hadron rejection rates are 1:200 and 1:50 respectively.

234. ARGUS Collaboration (H. Albrecht, et al.)
INCLUSIVE ϕ-MESON PRODUCTION IN ELECTRON-POSITRON INTERACTIONS IN THE ENERGY REGION OF THE Υ-RESONANCES.

We report on a high precision measurement of ϕ-meson production in continuum events and in direct decays of the $\Upsilon(1S)$- and $\Upsilon(2S)$-mesons. The ratio of the total production rate of the ϕ-mesons in direct $\Upsilon(1S)$- and $\Upsilon(2S)$- decays over that in continuum events is $1.32 \pm 0.08 \pm 0.09$ and $1.07 \pm 0.13 \pm 0.11$ respectively. This is compatible with the corresponding ratio obtained for lighter mesons, but is appreciably smaller than the relative baryon production rate.

235. ARGUS Collaboration (H. Albrecht, et al.)
OBSERVATION OF THE ORBITALLY EXCITED $\Lambda(1520)$ BARYON IN e^+e^- ANNIHILATION.

We report the first observation of an orbitally excited baryon, the $\Lambda(1520)$, in quark and gluon fragmentation. The production rate is found to be $(1.15 \pm 0.21 \pm 0.16) \times 10^{-2}$ and $(0.80 \pm 0.17\ ^{+0.10}_{-0.13}) \times 10^{-2}$ $\Lambda(1520)$ hyperons per event in direct Υ decays and in the continuum, respectively. In contrast to the observed situation for ground state baryons, the production rate of the $\Lambda(1520)$ in direct Υ decays shows little or no enhancement with respect to continuum production.

236. ARGUS Collaboration (H. Albrecht, et al.)
OBSERVATION OF THE CHARMED BARYON Λ_c IN e^+e^- ANNIHILATION AT 10 GeV.

Using the ARGUS detector at DORIS II, we have studied the production of the charmed baryon Λ_c in e^+e^- annihilation at centre-of-mass energies near 10 GeV. The Λ_c^+ was seen in the three decay modes $pK^-\pi^+$, $\Lambda\pi^+\pi^-\pi^+$ and $\overline{K}^0 p$, with products of normalized cross-section times branching ratio $[R \cdot Br]$ of $(10.8 \pm 1.4 \pm 1.2) \times 10^{-3}$, $(6.6 \pm 1.5 \pm 0.9) \times 10^{-3}$ and $(6.7 \pm 1.4 \pm 0.8) \times 10^{-3}$ respectively. The measured mass for the Λ_c was $(2283.1 \pm 1.7 \pm 2.0)$ MeV/c^2. A limit on the decay rates to $\Lambda\pi^+$ is reported. The fragmentation function of the Λ_c was measured.

237. ARGUS Collaboration (H. Albrecht, et al.)
OBSERVATION OF INCLUSIVE B MESON DECAYS INTO Λ_c^+ BARYONS.

We report the first direct observation of B meson decays into Λ_c^+ baryons using the decay channel $\Lambda_c^+ \to pK^-\pi^+$. The product of branching ratios $Br(B \to \Lambda_c^+ X) \cdot Br(\Lambda_c^+ \to pK^-\pi^+) = (0.30 \pm 0.12 \pm 0.06)\%$ is derived from an observed signal of 208 ± 89 events. Using previous measurements of inclusive baryon rates we find a branching ratio for $\Lambda_c^+ \to pK^-\pi^+$ of $(4.1 \pm 2.4)\%$. The measured Λ_c^+ momentum spectrum indicates that the multiparticle final states dominate the decays $B \to \Lambda_c^+ X$.

238. ARGUS Collaboration (H. Albrecht, et al.)
A SEARCH FOR $\omega\phi$ AND $\phi\phi$ PRODUCTION IN THE REACTIONS $\gamma\gamma \to K^+K^-\pi^+\pi^-\pi^0$ AND $\gamma\gamma \to 2K^+2K^-$.

The reaction $\gamma\gamma \to K^+K^-\pi^+\pi^-\pi^0$ has been observed for the first time, using the ARGUS detector at the e^+e^- storage ring DORIS II at DESY. The cross section shows an enhancement for $W_{\gamma\gamma}$ close to 3 GeV/c^2. Searches for $\gamma\gamma \to \omega\phi$ and for $\gamma\gamma \to \phi\phi$ leading to this final state, as well as for $\gamma\gamma \to \phi\phi \to 2K^+2K^-$, have been performed. The derived upper limits for $\omega\phi$ and $\phi\phi$ production are compatible with $q\bar{q}q\bar{q}$ model predictions.

239. ARGUS Collaboration (H. Albrecht, et al.)
OBSERVATION OF THE CHARMLESS B MESON DECAYS.

Using the ARGUS detector at the e^+e^- storage ring DORIS II, we have observed charmless decays of B mesons into the final states $p\bar{p}\pi^\pm$ and $p\bar{p}\pi^+\pi^-$. The significance of the signal corresponds to more than five standard deviations. The branching ratios are $(5.2 \pm 1.4 \pm 1.9) \cdot 10^{-4}$ for the three-body and $(6.0 \pm 2.0 \pm 2.2) \cdot 10^{-4}$ for the four-body final state. These decays cannot proceed via the dominant $b \to c$ transitions, and we show that they are not the result of penguin-type processes. Thus, the observed decays must represent $b \to u$ quark transitions. Consequently, the Kobayashi-Maskawa matrix element V_{ub} is non-zero.

240. ARGUS Collaboration (H. Albrecht, et al.)
OBSERVATION OF THE CHARMED BARYON Σ_c IN e^+e^- ANNIHILATIONS.

Using the ARGUS detector at the DORIS II storage ring, we have observed the charmed baryons Σ_c^{++} and Σ_c^0, through their decays to $\Lambda_c^+\pi^\pm$. We have measured the mean $\Sigma_c - \Lambda_c^+$ mass difference as $(167.6 \pm 0.3 \pm 1.6)$ MeV/c^2. The isospin mass splitting between the Σ_c^{++} and the Σ_c^0 was found to be $(1.2 \pm 0.7 \pm 0.3)$ MeV/c^2. The rate of Λ_c^+ production from Σ_c decays was found to be $(36 \pm 12. \pm 11)\%$ of the total rate of Λ_c^+ production. The Σ_c x_p-spectrum was observed to be similar to that of the Λ_c^+, with a Peterson function parameter ϵ of (0.29 ± 0.06).

241. ARGUS Collaboration (H. Albrecht, et al.)
SEARCH FOR THE DECAY $B \to K^*\gamma$.

Using the ARGUS detector at the e^+e^- storage ring DORIS II at DESY, a search for the decays $B \to K^*\gamma$ has been made. The following upper limits were derived at 90% CL: $BR(B \to K^*(892)\gamma) < 2.4 \cdot 10^{-4}$, $BR(B \to K_1(1400)\gamma) < 4.1 \cdot 10^{-4}$, $BR(B \to K_2^*(1430)\gamma) < 8.3 \cdot 10^{-4}$ and $BR(B \to K_3^*(1780)\gamma) < 3.0 \cdot 10^{-3}$.

242. ARGUS Collaboration (H. Albrecht, et al.)
INCLUSIVE π^0 AND η MESON PRODUCTION IN ELECTRON-POSITRON INTERACTIONS AT $\sqrt{s} = 10$ GeV.

We report on a high statistics study of π^0 and η production in continuum events and in direct decays of the $\Upsilon(1S)$ and $\Upsilon(2S)$ resonances. The measured production rates per event are $\langle n_{\pi^0} \rangle = 3.22 \pm 0.07 \pm 0.31$ $(3.97 \pm 0.23 \pm 0.38)$ and $\langle n_\eta \rangle = 0.19 \pm 0.04 \pm 0.04$ $(0.40 \pm 0.14 \pm 0.09)$ for continuum events (direct $\Upsilon(1S)$ decays).

243. ARGUS Collaboration (H. Albrecht, et al.)
OBSERVATION OF $\Delta^{++}(1232)$ PRODUCTION IN e^+e^- ANNIHILATIONS AROUND 10 GeV.

We report on the first observation of $\Delta(1232)^{++}$ and $\overline{\Delta}(1232)^{++}$ baryons in e^+c^- annihilation at energies around 10 GeV, using the ARGUS detector at DORIS II. The sum of the rates of Δ^{++} and $\overline{\Delta}^{++}$ per hadronic event in the continuum is measured to be $0.040 \pm 0.008 \pm 0.006$, and the rate in direct $\Upsilon(1S)$ decays is $0.124 \pm 0.016 \pm 0.015$. The momentum spectrum of Δ^{++} baryons in direct $\Upsilon(1S)$ decays has been measured.

244. ARGUS Collaboration (H. Albrecht, et al.)
INCLUSIVE PRODUCTION OF CHARGED PIONS, CHARGED AND NEUTRAL KAONS AND ANTI-PROTONS IN e^+e^- ANNIHILATION AT 10 GeV AND IN DIRECT UPSILON DECAYS.

Using the ARGUS detector at the e^+e^- storage ring DORIS II, we have investigated inclusive production of $\pi^\pm$, $K^\pm$, K_s^0 and $\bar{p}$ in multihadron events at 9.98 GeV and in direct decays of the $\Upsilon(1S)$ meson, i.e. from quark and gluon fragmentation. The most pronounced difference is the rate of baryon production. The Lund Monte Carlo program gives a reasonable qualitative description, although it cannot reproduce our data in detail.

245. ARGUS Collaboration (H. Albrecht, et al.)
RESULTS ON BARYON-ANTIBARYON CORRELATIONS IN e^+e^- ANNIHILATION.

The correlations between $p\bar{p}$, $\Lambda\overline{\Lambda}$, $\Xi^-\overline{\Lambda}$ and $\Lambda(1520)\overline{\Lambda}$ baryons have been measured in e^+e^- continuum events and in direct Υ decays. The observed correlations exclude the production of point-like diquark-antidiquark pairs as the dominant source of baryons. Information concerning angular momentum compensation follows from the observed $\Lambda(1520)\overline{\Lambda}$ correlation.

246. ARGUS Collaboration (H. Albrecht, et al.)
TWO PHOTON PRODUCTION OF FINAL STATES WITH A $p\bar{p}$ PAIR.

Two-photon production of the exclusive final states $p\bar{p} + n\pi$ (n = 0, 1, 2, and 3) has been investigated using the ARGUS detector at the e^+e^- storage ring DORIS II at DESY. The reactions $\gamma\gamma \to p\bar{p}\pi^0$ and $\gamma\gamma \to p\bar{p}\pi^+\pi^-\pi^0$ have been observed for the first time, as have the Δ^{++} and $\overline{\Delta}^{++}$ baryons in the final state $p\bar{p}\pi^+\pi^-$. No evidence was found for $\Delta^{++}\overline{\Delta}^{++}$ production. Topological cross sections for two-photon production of $p\bar{p}$, $p\bar{p}\pi^0$, $p\bar{p}\pi^+\pi^-$ and $p\bar{p}\pi^+\pi^-\pi^0$, as well as the cross section for $\gamma\gamma \to \Delta^{++}\bar{p}\pi^- + c.c$, have been measured. Upper limits are given for the cross sections for $\gamma\gamma \to \Delta^0\overline{\Delta}^0$, $\gamma\gamma \to \Delta^{++}\overline{\Delta}^{++}$ and $\gamma\gamma \to \Lambda\overline{\Lambda}$.

247. ARGUS Collaboration (H. Albrecht, et al.)
OBSERVATION OF THE D^{*0} (2459) IN e^+e^- ANNIHILATION.

Using the ARGUS detector at the DORIS II storage ring at DESY, we have observed a charmed meson of mass $(2455 \pm 3 \pm 5)$ MeV/c^2, decaying to $D^+\pi^-$. The natural width of this state is determined to be $(15\ ^{+13+5}_{-10-10})$ MeV/c^2. The fragmentation function is hard, as expected for a leading charmed particle from nonresonant e^+e^- annihilation. Analysis of the decay angular distribution supports the hypothesis that the observed state is an $L = 1$ excited charmed meson with spin-parity 2^+.

248. ARGUS Collaboration (H. Albrecht et al.)
MEASUREMENT OF D^{*+} POLARIZATION IN THE DECAY $\overline{B}^0 \to D^{*+}\ell^-\bar{\nu}$.

Using the ARGUS detector at the e^+e^- storage ring DORIS II, we have measured the average polarization of D^{*+} mesons originating from exclusive decays $\overline{B}^0 \to D^{*+}\ell^-\bar{\nu}$. By comapring the D^{*-} decay angular distribution to a functional form $1+\alpha\cos^2\theta$, we determine α to be $\alpha = 0.7 \pm 0.9$. This value of α leads to a ratio of of longitudinal to transverse decay width $\Gamma_L/\Gamma_T = 0.85 \pm 0.45$, which allows the determination, $|V_{cb}| = 0.052 \pm 0.011$. The importance of the transverse helicity component is manifested in the q^2 and lepton spectra, which are also presented.

249. ARGUS Collaboration (H. Albrecht, et al.)
MEASUREMENT OF D^0 DECAYS INTO $\overline{K}^0\omega$, $\overline{K}^0\eta$ AND $\overline{K}^{*0}\eta$.

Using the ARGUS detector at the electron-positron storage ring DORIS II at DESY, we have measured the branching ratios of the D^0 meson decays into $\overline{K}^0\omega$, $\overline{K}^0\eta$ and $\overline{K}^{*0}\eta$ relative to the $K^-\pi^+$ mode. Using the known branching ratio $Br(D^0 \to K^-\pi^+) = (4.2 \pm 0.4 \pm 0.4)\%$ we find $Br(D^0 \to \overline{K}^0\omega) = (4.2 \pm 1.6 \pm 0.9)\%$, $Br(D^0 \to \overline{K}^0\eta) < 2.7\%$ (90% CL) and $Br(D^0 \to \overline{K}^{*0}\eta) < 2.9\%$ (90% CL).

250. ARGUS Collaboration (H. Albrecht, et al.)
A MEASUREMENT OF $\gamma\gamma \to \rho^+\rho^-$.

The reaction $\gamma\gamma \to \rho^+\rho^- \to \pi^+\pi^-\pi^0\pi^0$ has been studied with the ARGUS detector at the e^+e^- storage ring DORIS II at DESY. Near threshold, the cross section for this reaction is about four times smaller than for the reaction $\gamma\gamma \to \rho^0\rho^0$.

251. ARGUS Collaboration (H. Albrecht, et al.)
MEASUREMENT OF INCLUSIVE B MESON DECAYS INTO BARYONS.

The decay of B mesons into the baryons p, Λ and Ξ^- has been studied. The measured inclusive branching ratios for these decays are $Br(B \to pX) = (8.2 \pm 0.5\ ^{+1.3}_{-1.0})\%$, $Br(B \to \Lambda X) = (4.2 \pm 0.5 \pm 0.6)\%$ and $Br(B \to \Xi^- X) < 0.51\%$ at the 90% confidence level. In addition investigations on $p\bar{p}$, $\Lambda\bar{p}$ and $\Lambda\overline{\Lambda}$ correlations were performed, yielding an approximately equal rate of protons and neutrinos. From this one can derive a total baryonic branching ratio $Br(B \to \text{baryons})$ of $(7.6 \pm 1.4)\%$.

252. ARGUS Collaboration (H. Albrecht, et al.)
B MESON DECAYS TO $D\pi$ AND $D\rho$.

The decays $\overline{B} \to D\pi^-$ and $\overline{B} \to D\rho^-$ are observed in data taken by the ARGUS detector at DORIS II. The measured branching ratios of the decays $B^- \to D^0\rho^-$ and $\overline{B}^0 \to D^+\rho^-$ are $(2.1 \pm 0.8 \pm 0.9)\%$ and $(2.2 \pm 1.2 \pm 0.9)\%$ respectively, while those of the decays $B^- \to D^0\pi^-$ and $\overline{B}^0 \to D^+\pi^-$ are $(0.19 \pm 0.10 \pm 0.06)\%$ and $(0.31 \pm 0.13 \pm 0.10)\%$ respectively.

253. ARGUS Collaboration (H. Albrecht, et al.)
SEARCH FOR EXCLUSIVE RADIATIVE DECAYS OF $\Upsilon(1S)$ AND $\Upsilon(2S)$ MESONS.

Using the ARGUS detector at the DORIS II storage ring we have searched for radiative decays $\Upsilon(1S) \to \gamma X$, where X decays into two oppositely charged pions, kaons, or protons. Upper limits on branching fractions are presented. In the $\pi^+\pi^-$ invariant mass region 290–570 GeV/c^2 the upper limit obtained varies from 3 to 4.5×10^{-5}. These values are lower than the theoretical prediction of about 5×10^{-5} for Higgs production, which includes perturbative QCD radiative corrections to the Wilczek formula and the estimate $Br(H^0 \to \pi^+\pi^-) \approx 45\%$, accounting for Higgs coupling to gluons via heavy quark loops. Previous experimental results do not exclude the existence of a standard minimal Higgs in this mass region.

254. ARGUS Collaboration (H. Albrecht, et al.)
UPPER LIMITS FOR THE DECAY OF τ LEPTONS INTO η-MESONS.

The production of η-mesons in τ-decays has been studied with the ARGUS detector at the DORIS II storage ring. $\tau^+\tau^-$ events were selected by their characteristic 1-versus-3 charged particle topology. The $\pi^+\pi^-$ invariant mass distribution was used to search for evidence of η production in τ-decays. No signal was observed, leading to upper limits at 95% CL of $BR(\tau^- \to \eta\pi^-\pi^0\nu_\tau) < 1.1\%$, $BR(\tau^- \to \eta\eta\pi^-\nu_\tau) < 0.8\%$, $Br(\tau^- \to \eta\eta\pi^-\pi^0\nu_\tau) < 0.9\%$ and $Br(\tau^- \to \eta\pi^-\pi^0\pi^0\nu_\tau) < 1.2\%$. A search for the decay $\tau^- \to \eta\pi^-\nu_\tau$, where the π^0 from the decay $\eta \to \pi^+\pi^-\pi^0$ is also reconstructed, yielded an upper limit at 95% CL of $Br(\tau^- \to \eta\pi^-\nu_\tau) < 0.9\%$. In addition, the inclusive value $(1.65 \pm 0.3 \pm 0.2)\%$ for $BR(\tau^- \to \omega X^-\nu_\tau)$ follows from this analysis.

255. WA59 + E180 Neutrino Collaboration (A. Asratyan, et al.)
EVIDENCE FOR ANTINEUTRINO PRODUCTION OF D_s^* MESONS.

Production of vector charmed strange D_s^{*-} mesons in $\bar{\nu}_\mu N$ collisions is studied using the combined data of two bubble chamber experiments.

256. ARGUS Collaboration (H. Albrecht, et al.)
MEASUREMENT OF THE DECAYS $\tau^- \to K^{*-}\nu_\tau$ AND $\tau^- \to \rho^-\nu_\tau$.

Using the ARGUS detector at DORIS II, we have determined the branching ratios for the decays $\tau^- \to K^{*-}\nu_\tau$ and $\tau^- \to \rho^-\nu_\tau$ to be $(1.23 \pm 0.21\ {}^{+0.11}_{-0.21})\%$ and $(21.5 \pm 0.4 \pm 1.9)\%$, respectively. These results are in agreement with theoretical expectations based on the Standard Model.

257. ARGUS Collaboration (H. Albrecht, et al.)
FIRST OBSERVATION OF $\gamma\gamma \to K^{*+}K^{*-}$.

The final states $K_S^0 K_S^- \pi^+\pi^-$ and $K_S^0 K^\mp \pi^0\pi^\pm$, produced in two-photon reactions, have been studied using the ARGUS detector at the e^+e^- storage ring DORIS II at DESY. The reaction $\gamma\gamma \to K^{*+}K^{*-}$ has been observed for the first time. Its cross section is about eight times larger than that for $\gamma\gamma \to K^{*0}\overline{K}^{*0}$, but it has a similar $W_{\gamma\gamma}$ dependence.

258. A. Cherubini, R. Odorico (Bologna U. & INFN, Bologna)
EXPLOITING NONLEPTONIC DECAYS IN TOP SEARCHES.

We show that a massive use of statistical event correlations, learned by means of Monte Carlo event generators, makes it feasible to employ a pure jet signature in top quark searches. A relevant ingredient for that is also the less diffuse angular distribution of energy deposition and of particles in top events, which are dominated by quark jets. On the other hand, we show that the utilization of mass cuts on jet combinations to identify top non-leptonic decays meets severe limitations already at the parton level, largely because of QCD radiation.

259. TOPAZ Collaboration (L. Adachi, et al.)
A SEARCH FOR A FOURTH-GENERATION CHARGE $-1/3(b')$ QUARK USING INCLUSIVE MUONS IN e^+e^- ANNIHILATIONS AT $\sqrt{s} = 56.5$ to 60.8 GeV.

A search for fourth-generation charge $-1/3(b')$ quarks using inclusive muon events in e^+e^- annihilations at $\sqrt{s} = 56.5$ to 60.8 GeV is presented, considering its decays through loop-induced flavor-changing neutral-current as well as its charged current decays. Production of b' (and *top*) was excluded as the cause of slightly high R-values observed at $\sqrt{s} \geq 56.0$ GeV, if they decay dominantly through charged-current ($> 68\%$). One large aplanarity event ($A = 0.17$) with an isolated muon, observed at $\sqrt{s} = 60.8$ GeV, is also presented.

260. B. Margolis, P. Valin (McGill U.), R.R. Mendel (Western Ontario U.)
QUARK MASSES AND MIXINGS FROM RADIATIVE CORRECTIONS.

The striking quark mass hierarchy between families and the structure of the Kobayashi-Maskawa mixing matrix are known to have an origin outside the Standard Model. We propose a general dynamical scheme whereby the quarks acquire their masses by coupling to heavy fermions through emission and reabsorption of bosonic quanta, which are common to all families. By allowing the form factors at the vertices of these generic radiative correction processes to be slightly dependent on flavor and family indices, we can reproduce in a very natural and stable fashion the observed mass spectrum and the mixing scheme as well as CP violation. The model form factors used here require large top masses to obtain small ratios $|V_{ub}/V_{cb}|$.

261. AMY Collaboration (T. Kumita, et al.)
MEASUREMENTS OF R FOR e^+e^- ANNIHILATION AT TRISTAN.

The R ratio for e^+e^- annihilations into hadronic final states is measured to be approximately ten percent higher than Standard Model predictions for CM energies between 56 and 61.4 GeV.

262. TPC/Two-Gamma Collaboration (H. Aihara, et al.)
A MEASUREMENT OF THE TOTAL HADRONIC CROSS SECTION IN TAGGED $\gamma\gamma$ REACTIONS.

We present a measurement of the total cross section for $\gamma\gamma \rightarrow$ hadrons, with one photon quasi-real and the other a spacelike photon of mass-squared $-Q^2$. Results are presented as a function of Q^2 and the $\gamma\gamma$ center-of-mass energy W, with the Q^2 range extending from 0.2 GeV^2 to 60 GeV^2, and W in the range from 2 to 10 GeV. The data were taken with the TPC/Two-Gamma facility at the SLAC e^+e^- storage ring PEP, which was operated at a beam energy of 14.5 GeV. The cross section exhibits a gentle fall-off with increasing W. Its Q^2-dependence is shown to be well-described by an incoherent sum of vector-meson and point-like scattering over most of the observed W range. Agreement at high Q^2 is improved if a minimum p_T cutoff (motivated by QCD) is imposed on the point-like contribution.

263. AMY Collaboration (S.S. Myung, et al.)
A STUDY OF MULTIHADRON EVENTS WITH ISOLATED LEPTONS PRODUCED IN e^+e^- ANNIHILATIONS AT $\sqrt{s} = 50$ TO 61.4 GeV.

The production rate of multihadron events with isolated leptons in e^+e^- annihilations at $\sqrt{s} = 50$ to 61.4 GeV is measured using the AMY detector at TRISTAN. Our observations are consistent with expectations based on the production of five quark flavors. The 95% confidence-level mass limit for the t quark is 30.4 GeV/c^2: that for a fourth generation charge $-1/3$ b' quark that decays via conventional charged current interactions is 28.9 GeV/c^2.

264. AMY Collaboration (H. J. Kim, et al.)
MEASUREMENTS OF THE CROSS SECTION FOR $e^+e^- \rightarrow \gamma\gamma$ AT TRISTAN.

We report measurements of the cross section for $e^+e^- \rightarrow \gamma\gamma$ at center-of-mass energies from 50 to 61.4 GeV using the AMY detector at TRISTAN. The results are compared to predictions from QED.

265. ARGUS Collaboration (H. Albrecht, et al.)
A STUDY OF CABIBBO-SUPPRESSED D^0 DECAYS.

Using the ARGUS detector at the e^+e^- storage ring DORIS II, we have measured branching ratios for the Cabibbo-suppressed decays $D^0 \to \pi^-\pi^+$, $D^0 \to K^-K^+$, and $D^0 \to K_s^0 K^-\pi^+$, observed the Cabibbo-allowed decay $D^0 \to K_s^0 K_s^0 K_s^0$ and obtained an upper limit on the branching ratio for $D^0 \to K^0\overline{K}^0$.

266. ARGUS Collaboration (H. Albrecht, et al.)
OBSERVATION OF THE CHARGED ISOSPIN PARTNER OF THE $D^{*0}(2459)$.

Using the ARGUS detector at the DORIS II e^+e^- storage ring at DESY, we have observed a new charmed meson of mass $(2469 \pm 4 \pm 6)$ MeV$/c^2$, decaying to $D^0\pi^+$. This state is a strong candidate for the charged isospin partner of the $D^{*0}(2459)$. The isospin mass splitting is measured to be $(14 \pm 5 \pm 8)$ MeV$/c^2$.

267. ARGUS Collaboration (H. Albrecht, et al.)
MEASUREMENT OF THE DECAY $B^0 \to D^-l^+\nu$.

Using the ARGUS detector at the e^+e^- storage ring DORIS II, we have investigated the decay $B^0 \to D^-l^+\nu$, where l^+ is e^+ or μ^+. The B^0 mesons were produced in 150000 $\Upsilon(4S)$ decays. Assuming electron-muon universality we obtain a branching ratio $BR(B^0 \to D^-e^+\nu) = BR(B^0 \to D^-\mu^+\nu) = (1.8 \pm 0.6 \pm 0.5)$ %.

268. ARGUS Collaboration (H. Albrecht, et al.)
SEARCH FOR $b \to s\gamma$ IN EXCLUSIVE DECAYS OF B MESONS.

Using the ARGUS detector at the e^+e^- storage ring DORIS II at DESY, a search for decays $B \to K^*\gamma$ has been performed where K^* represents 7 different K mesons. No evidence for such decays has been found, and upper limits are quoted. These provide valuable constraints on the top quark mass, the Higgs sector of the Standard Model, and a variety of extensions thereto.

269. Hai-Yang Cheng, Sheng-Nan Lai (Academia Sinica, Taiwan)
SPIN ASYMMETRY IN PROTON-PROTON COLLISIONS AS A PROBE OF SEA AND GLUON POLARIZATION IN A PROTON.

Quark and gluon spin densities in a proton are phenomenologically parametrized based on the EMC data and on some plausible theoretical arguments. Four different characteristic values of gluon and sea polarizations suggested by various theoretical conjectures are considered. The sea polarization in a proton is probed by measuring the spin-spin asymmetry A_{LL}^{DY} in the Drell-Yan process, while the helicity asymmetry A_{LL}^{γ} in the direct photon production at high p_t is employed to test the gluon spin content. Helicity asymmetries in both processes are quite sizable. A_{LL}^{DY} is positive and of order 10^{-1} if the sea is polarized opposite to the proton spin, as suggested by the EMC data. However, even in the absence of the sea polarization at the EMC energies, we find A_{LL}^{DY} to be large and negative. Experimental measurements of A_{LL}^{DY} and A_{LL}^{γ} together will not only provide a clean probe of sea and gluon polarizations, but also test whether the combination $\Delta s - (\alpha_s/4\pi)\Delta G$ inferred from the EMC data is valid i.e. whether gluons contribute to the spin-dependent structure function $g_1^p(x, Q^2)$ via a triangle anomaly.

270. L.W. Whitlow, et al. (Rochester U. & American U. & CIT & LLL, Livermore & Mass. U. & Tel Aviv U. & Fermilab & SLAC & Stanford U.)
A COMBINED ANALYSIS OF SLAC EXPERIMENTS ON DEEP INELASTIC $e-p$ AND $e-d$ SCATTERING.

We report recent work on the extraction of $R = \sigma_L/\sigma_T$ and the structure function F_2 over a large kinematic range, which is based on a reanalysis of deep inelastic $e-p$ and $e-d$ scattering cross sections measured at SLAC between 1970 and 1985. All these data were corrected for radiative effects using improved versions of external and internal radiative correction procedures. The data from seven individual experiments were normalized to those from the recent high-precision SLAC experiment E140.

We find that $R_p = R_d$, as expected in QCD. The value of R is higher than predicted by QCD even when target-mass effects are included. This difference indicates that additional dynamical higher-twist effects may be present.

The structure functions F_2^p and F_2^d were also extracted from the full data sets of normalized cross sections using an empirical fit to R. These structure functions were then compared with data from the CERN muon scattering experiments BCDMS and EMC. We find that our data are consistent with the EMC data, if the latter are multiplied by a normalization factor of 1.07. No single, uniform normalization factor can be applied to the BCDMS data that will bring them into agreement with the SLAC data.

271. Mark II Collaboration (G.S. Abrams, et al.)
FIRST MEASUREMENTS OF HADRONIC DECAYS OF THE Z BOSON.

We have observed hadronic final states produced in the decays of Z bosons. In order to study the parton structure of these events, we compare the distributions in sphericity, thrust, aplanarity and number of jets to the predictions of several QCD-based models. The data and models agree within the present statistical precision.

272. CLEO Collaboration (M.S. Alam, et al.)
SEARCH FOR A NEUTRAL HIGGS BOSON IN B MESON DECAY.

Using the CLEO detector at the Cornell Electron Storage Ring we have searched for neutral Higgs boson production in B decay, both through the exclusive modes $B \to H^0K$ and $B \to H^0K^*$ using the decay of the H^0 into a pair of muons, pions, or kaons, and through the inclusive decay $B \to H^0X$ using only the muon decay of the H^0. We find no evidence for a Higgs boson with mass between $2m_\mu$ and $2m_\tau$.

273. Mark II Collaboration (G. S. Abrams, et al.)
INITIAL MEASUREMENTS OF Z BOSON RESONANCE PARAMETERS IN e^+e^- ANNIHILATION.

We have measured the mass of the Z boson to be 91.11 ± 0.23 GeV/c^2, and its width to be $1.61^{+0.60}_{-0.43}$ GeV. If we constrain the visible width to its Standard Model value, we find the partial width to invisible decay modes to be 0.62 ± 0.23 GeV, corresponding to 3.8 ± 1.4 neutrino species.

274. R.R. Volkas, A.J. Davies, G.C. Joshi (Melbourne U.)
NATURALNESS OF THE INVISIBLE AXION MODEL.

We discuss the choice of parameters that is necessary to establish the enormous hierarchy between the doublet and singlet Higgs vacuum expectation values in the invisible axion model. We show that this choice is assoicated with an approximate symmetry of the theory, thus demonstrating that the hierarchy in this model is natural in the sense of 't Hooft. It is then shown that this choice also leads to the existence of Higgs particles with masses $\sim M_W$, which is particularily relevant for invisible axion models featuring spontaneous CP-violation. We comment on how our conclusions are affected by the introduction of right-handed neutrinos with large Majorana masses within the context of the see-saw mechanism.

275. CLEO Collaboration (W.Y. Chen, et al.)
MEASUREMENT OF THE MUONIC BRANCHING FRACTIONS OF THE $\Upsilon(1S)$ AND $\Upsilon(3S)$.

Using the CLEO detector at the Cornell Electron Storage Ring, we have measured the muonic branching fractions, $B_{\mu\mu}$, of the $\Upsilon(1S)$ and $\Upsilon(3S)$ to be $(2.52 \pm 0.07 \pm 0.07)\%$ and $(2.02 \pm 0.19 \pm 0.33)\%$, respectively.

276. CLEO Collaboration (D. Bortoletto, et al.)
A SEARCH FOR $b \to u$ TRANSITIONS IN EXCLUSIVE HADRONIC B MESON DECAYS.

From a data sample of 242,000 $B\overline{B}$ pairs collected with the CLEO detector at the $\Upsilon(4S)$ resonance, we set upper limits for the branching ratios for exclusive hadronic decays of B mesons arising from $b \to u$ transitions. Model dependent limits for the ratio of the Kobayashi-Maskawa matrix elements $|V_{ub}/V_{cb}|$ are presented.

277. CLEO Collaboration (P. Avery, et al.)
A SEARCH FOR EXCLUSIVE PENGUIN DECAYS OF B MESONS.

We have measured upper limits on branching fractions for rare exclusive decays of B mesons arising from one-loop diagrams in the Standard Model of electroweak interactions. We also obtain an upper limit for the lepton-number-violating decay $B^0 \to \mu^{\pm}e^{\mp}$.

278. CLEO Collaboration (D. Bortoletto, et al.)
THE DECAY $D^0 \to K^0\overline{K}^0$.

Using the CLEO detector we have observed a signal for the decay mode $D^0 \to K^0\overline{K}^0$ via the decay chain $D^0 \to K_S^0K_S^0$, $K_S^0 \to \pi^+\pi^-$. We determine $B(D^0 \to K^0\overline{K}^0) = 0.11{}^{+0.06}_{-0.04}{}^{+0.02}_{-0.02}\%$.

279. CLEO Collaboration (S.E. Csorna, et al.)
SEARCH FOR NEUTRINOLESS DECAYS OF THE τ LEPTON.

We have searched for neutrinoless τ decays into three charged particles. Evidence of such decays would demonstrate non-conservation of lepton flavor and, in some cases, lepton number. We see no signal for any such neutrinoless τ decays and set upper limits on their branching fractions.

280. CLEO Collaboration (M.S. Alam, et al.)
A MEASUREMENT OF THE B^0 SEMILEPTONIC BRANCHING RATIO.

We have identified a sample of $\Upsilon(4S) \to B^0\overline{B}^0$ decays by partially reconstructing events from the decay $\overline{B}^0 \to D^{*+}\pi^-$. The $\Upsilon(4S)$ events were produced at the Cornell Electron Storage Ring (CESR) and recorded with the CLEO detector. We use these events to make the first measurement of the $\overline{B}^0$ semileptonic branching ratio: $B(\overline{B}^0 \to l\nu X) = 10.6 \pm 2.8 \pm 1.1\%$.

281. CLEO Collaboration (M. Artuso, et al.)
$B^0\overline{B}^0$ MIXING AT THE $\Upsilon(4S)$.

We have measured $B^0\overline{B}^0$ mixing by observing like-sign dilepton events in $\Upsilon(4S)$ decay. Assuming that the semileptonic branching fractions of the charged and neutral B mesons are equal and that the $\Upsilon(4S)$ decays to B^+B^- 55% of the time and to $B^0\overline{B}^0$ 45% of the time, we measure the mixing parameter, r, to be $0.19 \pm 0.06 \pm 0.06$, where the first error is statistical and the second is systematic.

282. CLEO Collaboration (T. Bowcock, et al.)
SEARCH FOR THE PRODUCTION OF FRACTIONALLY CHARGED PARTICLES IN e^+e^- ANNIHILATIONS AT $\sqrt{s} = 10.5$ GeV.

We report the results of a search for particles with electric charge $Q = \frac{2}{3}e$ produced in e^+e^- annihilations at $W = 10.5$ GeV. The ratio of the cross section for the inclusive production of fractionally charged particles (q) to the dimuon point cross section, $R_q = \sigma(e^+e^- \to qX)/\sigma(e^+e^- \to \mu^+\mu^-)$, is found to have an upper limit at 90% confidence level of 1.0×10^{-4} for masses less than 4 GeV/c^2.

283. CLEO Collaboration (W.Y. Chen, et al.)
B DECAYS TO CHARM: EXCLUSIVE DECAYS AND INCLUSIVE RATES.

We have measured the exclusive branching ratios of several decay modes of the B meson. $\overline{B}^0 \to D^{*+}D_s^-$ and $D^+D_s^-$, and $B^- \to D^0D_s^-$ as well as $\overline{B}^0 \to D^{*+}\rho^-$ have been observed for the first time. The mass difference between $\overline{B}^0$ and B^+ has been measured to be 0.4 ± 0.6 MeV.

284. CLEO Collaboration (W.Y. Chen, et al.)
MEASUREMENT OF D_s DECAY MODES.

Using the CLEO detector, we have measured the branching ratios for the exclusive decays of the D_s^+ meson into $\overline{K}^{*0}K^+$, $\overline{K}^0K^+$, and $K^{*+}K^+$ relative to the decay $D_s^+ \to \phi\pi^+$ to be close to unity. We have also measured the mass difference between the D_s^+ and D^+ meson to be 98.5 ± 1.5 MeV/c^2.

285. CLEO Collaboration (M.S. Alam, et al.)
MEASUREMENT OF THE ISOSPIN MASS SPLITTING $\Xi_c^+ - \Xi_c^0$.

Using the CLEO detector, we present a measurement of the mass of the charged charmed strange baryon Ξ_c^+ and its neutral counterpart the Ξ_c^0. The Ξ_c^+ is observed through its decay to $\Xi^-\pi^+\pi^+$, and its mass is measured to be $2467 \pm 3 \pm 4$ MeV. The isospin mass splitting of the Ξ_c system is found to be $M(\Xi_c^+) - M(\Xi_c^0) = -5 \pm 4 \pm 1$ MeV.

286. CLEO Collaboration (R. Fulton, et al.)
RADIATIVE Υ(1S) DECAYS.

We report on a study of exclusive decays of the Υ(1S) resonance. We have considered decays into final states of the form $\Upsilon \to \gamma n(\mathrm{h^+h^-})$, with $(\mathrm{h^+h^-})$ denoting a charged hadron pair (either $\pi^+\pi^-$, K^+K^-, or $\mathrm{p\bar{p}}$), and n denoting the number of such pairs in the event. We measure branching ratios for $n = 2$, 3 and 4, and search for structure in the recoiling hadronic system. The structure which we observe in the $n = 1$ case is consistent with our expectations for the continuum process $e^+e^- \to \gamma X$. No evidence is observed for two-body radiative decays of the Υ(1S) meson in any of the above channels.

287. CLEO Collaboration (D. Bortoletto, et al.)
STUDY OF THE DECAY $\overline{B}^0 \to D^{*+}l^-\bar{\nu}$.

Using a sample of 484,000 B mesons collected with the CLEO detector at CESR, we have measured $B(\overline{B}^0 \to D^{*+}l^-\bar{\nu})$ and found that the polarization of the D^{*+} is small. These results are compared with models of semileptonic B decay.

288. B.A. Arbuzov (High Energy Physics, Protvino), E.E. Boos, V.I. Savrin, S.A. Shichanin (Moscow State U.)
THE CONVENTIONAL QED INTERPRETATION OF THE GSI e^+e^- NARROW RESONANCES.

The explanation of the GSI e^+e^- narrow resonances as being due to non-perturbative effects of the conventional QED is proposed. An application of the quasi-potential approach discloses a set of new resonances in a system of two charged particles. Numerical calculations agree with data on the e^+e^- and pp narrow resonances. Additional new resonances are predicted for e^+e^-, pp, e^-e^- systems.

289. TPC/Two-Gamma Collaboration (H. Aihara, et al.)
EXCLUSIVE PRODUCTION OF $\mathrm{p\bar{p}}\pi^+\pi^-$ IN PHOTON-PHOTON COLLISIONS.

We report a measurement of the $e^+e^- \to e^+e^-\mathrm{p\bar{p}}\pi^+\pi^-$ process with the TPC/Two-Gamma facility at the PEP e^+e^- storage ring at SLAC. Forty-five $\mathrm{p\bar{p}}\pi^+\pi^-$ events were identified in data corresponding to an integrated e^+e^- luminosity of 142 pb^{-1}. The cross section for $\gamma\gamma \to \mathrm{p\bar{p}}\pi^+\pi^-$ is given both as a function of $W_{\gamma\gamma}$, the $\gamma\gamma$ center-of-mass energy, with $W_{\gamma\gamma}$ between 2.5 and 5.5 GeV, and as a function of the invariant mass squared, q^2, of one of the photons, with $-q^2 < 7$ GeV^2. The cross section falls much less rapidly with $W_{\gamma\gamma}$ than does the cross section for a similar process, $\gamma\gamma \to \mathrm{p\bar{p}}$. No $\Delta^0\bar{\Delta}^0$ production is observed, and only a small fraction of the events at low $W_{\gamma\gamma}$ is consistent with $\gamma\gamma \to \Delta^{++}\bar{\Delta}^{--}$, $\Delta^{++}\bar{\mathrm{p}}\pi^-$ or $\bar{\Delta}^{--}\mathrm{p}\pi^+$. In an expanded search through the same data, four events compatible with either $\Lambda\bar{\Lambda}(\Lambda \to \mathrm{p}\pi^-)$ $\Sigma^0\bar{\Lambda}(\Sigma^0 \to \Lambda\gamma)$ production were found.

290. TPC/Two-Gamma Collaboration (H. Aihara, et al.)
INVESTIGATION OF THE ELECTROMAGNETIC STRUCTURE OF η and η' MESONS BY TWO-PHOTON INTERACTIONS.

The TPC/Two-Gamma facility at PEP was used to study the reactions $\gamma\gamma^* \to \eta$ and $\gamma\gamma^* \to \eta'$. The $\eta\gamma^*\gamma$ and $\eta'\gamma^*\gamma$ transition form factors were measured as functions of Q^2, the negative of the invariant mass squared of the tagged photon, in the range $0.1 < Q^2 < 7\ (GeV)^2$. These determinations of the electromagnetic structure of the η and η' mesons are consistent with both vector meson dominance and QCD. Interpreted within QCD, they can be used to provide information on the type and distribution of quarks within the pseudoscalar mesons.

291. G.S. Vidyakin, et al. Kozlov, V.P. Martemyanov, S.V. Sukhotin, V.G. Tarasenkov, S.Kh. (Kurchatov Inst., Moscow)
OBSERVATION OF THE WEAK CHARGED CURRENT IN INTERACTION OF REACTOR ANTINEUTRINOS WITH DEUTERONS.

First results are obtained for the interaction of reactor antineutrinos with deuterons via the weak charged current channel $\sigma_{\text{exp}} = (0.95 \pm 0.22) \cdot 10^{44}$ cm^2/fission event.

292. G.S. Vidyakin, et al. (Kurchatov Inst., Moscow)
STUDY OF REACTOR ANTINEUTRINO SCATTERING BY ELECTRONS, USING AN ORGANOFLUORIC SCINTILLATOR-BASED DETECTOR.

Preliminary results of experimental research into the process of reactor antineutrino elastic scattering by electrons, using an organofluoric scintillator-based detector, are presented in this work. In the recoil electron energy range of (3150-5175) keV the $\sigma_{\text{exp}} = (7.5 \pm 3.5) \cdot 10^{-46}$ cm^2/fission event cross section is obtained at the $\sin^2\theta_W = (0.29 \pm 0.07)$ value of the free parameter of the electroweak theory.

293. S.N. Gninenko, et al. (Moscow, INR)
A SEARCH FOR A KėV PSEUDOSCALAR IN TWO-BODY DECAYS OF ORTHOPOSITRONIUM.

We have searched for the exotic annihilation of orthopositronium: $e^+e^-(^3S_1) \to \gamma + a^0$, where a^0 denotes a light neutral pseudoscalar. The upper limit on the ratio of this decay rate to the 3γ decay rate obtained is $\Gamma(o - Ps \to \gamma + a^0)/\Gamma(o - Ps \to 3\gamma) < 8.2 \times 10^{-4}$ (90% C.L.), if the mass $m_a < 30$ KeV. This preliminary result excludes the hypothesis that this decay resolves the discrepancy between the theoretical and experimental values of the orthopositronium lifetime in vacuum.

294. V.V. Ammosov, et al. (Serpukhov, IFVE, GDR Academy of Science, Berlin)
STUDY OF IDENTICAL PARTICLE CORRELATIONS IN NEUTRINO-NUCLEUS INTERACTIONS.

Experimental results on the study of the interference correlation effect for identical particles produced in neutrino-nucleus interactions are presented. The experiment has been performed with the SKAT bubble chamber. The radius of the π^--meson emission region is found to be $r = 1.4 \pm 0.4\ fm$. Certain indications of the dependence of the value of the radius r on the target mass have been obtained. Two-proton correlations with the shape dependent on the proton pair momentum have been observed.

295. Yu. F. Pirogov (Serpukhov, IFVE)
ON SYMPLECTIC SPACE-TIME EXTENSION.

A multidimensional ($d = (2\ell)^2$, $\ell \geq 1$) space-time extension in the spinor framework based on the symplectic symmetry structure group $Sp(2\ell, C)$ has been proposed as an alternative for pseudo-orthogonal extension.

296. V.N. Goryachev (Serpukhov, IFVE)
FIRST RESULTS OBTAINED IN THE BEAM-DUMP EXPERIMENT WITH THE IHEP-JINR NEUTRINO DETECTOR.

Data processing of events selected in the 70 GeV/c beam dump experiment has been performed and the first results on a smaller part of statistics has been obtained. The experiment has been carried out with the Neutrino detector of Berlin-Budapest-Dubna-Serpukhov collaboration. Results have been compared with those of the first beam-dump experiment performed at Serpukhov accelerator in 1978 and with the results obtained with the BIS-2 set-up.

297. Stanislaw Jadach (Jagellonian U.), B.F.L. Ward (Tennessee U.)
YFS2 - THE SECOND ORDER MONTE CARLO FOR FERMION PAIR PRODUCTION AT LEP/SLC WITH THE INITIAL STATE RADIATION OF TWO HARD AND MULTIPLE SOFT PHOTONS.

A Monte Carlo program simulating fermion pair production is presented. It features a multiphoton bremsstrahlung out of the initial state beams. The contributions form soft photons are summed rigorously up to an infinite order using the Yennie-Frautschi-Suura method while the contributions from up to two hard photons are also properly treated. Four momenta of all soft and hard photons are explicitly generated and the total energy momentum conservation is exactly obeyed. The program is primarily aimed for LEP/SLC type experiments and will be helpful in the precise measurements of the Z^0 mass and width, the basic parameters in precision tests of standard electroweak theory. It can also be used far away from the Z^0 resonance as well. From the point of view of the QED it provides the total cross section with precision 0.1% near Z^0 and 0.5% away from Z^0 resonance. With some restriction it can provide predictions for various asymmetries.

298. TOPAZ Collaboration (I. Adachi, et al.)
A SEARCH FOR b' QUARK IN FCNC DECAY MODE AT TRISTAN.

A search for b' quark via FCNC mode decay is described. The search was carried out by looking for isolated high energy photon(s) or back to back jet pairs in 4 jet topology using the data taken by the TOPAZ detector at TRISTAN in the energy range: $\sqrt{s} = 50.0 \sim 60.8$ GeV. As the branching fraction of b' decay is not known, the experimental limit was given in a plane of mass and branching ratio. Assuming the branching fraction of FCNC to be 100%, the mass limit for b' was set as ≥ 28.3 GeV/c^2 independent of $Br(b' \to b\gamma)$.

299. TOPAZ Collaboration (I. Adachi, et al.)
A SEARCH FOR HEAVY STABLE SCALAR QUARKS AT TRISTAN.

Heavy stable scalar quarks have been searched for in e^+e^- annihilation, using energy loss measurements of particles by the TOPAZ TPC. Preliminary results of the analysis with 18 pb^{-1} data taken in the energy range $\sqrt{s} = 52.0$ to 60.8 GeV are reported.

300. Bernd A. Kniehl, Johann H. Kuhn (Munich, Max Planck Inst.)
QCD CORRECTIONS TO THE Z DECAY RATE.

Corrections to the Z decay rate are calculated in perturbative QCD. The axial vector current induced rate differs in order α_s^2 from the rates induced by the vector current and the electromagnetic current respectively. The additional term is calculated analytically. It is shown that the dominant $\mathcal{O}(\alpha_s^3)$ correction is identical with the correction for $R = \sigma_{\rm had}/\sigma_{\rm point}$ at low energies and the (small) differences are estimated.

301. CELLO Collaboration (H.J. Behrend, et al.)
HEAVY QUARK CHARGE ASYMMETRIES WITH THE CELLO DETECTOR.

The production of b and c quarks in e^+e^- annihilation has been studied with the CELLO detector in the range from 35 GeV up to the highest PETRA energies. The heavy quarks have been tagged by their semileptonic decays. The charge asymmetries for b quarks at 35 and 43 GeV have been found to be $A^b = -(22.2 \pm 8.1)\%$ and $A^b = -(49.1 \pm 16.5)\%$, respectively, using a method incorporating correlated jet variables for separation of the heavy quarks from the background of the lighter quarks. For c quarks we obtain $A^c = -(12.9 \pm 8.8)\%$ and $A^c = +(7.7 \pm 14.0)\%$, respectively. The axial vector coupling constants of the heavy quarks c and b are found to be $a_c = +(0.26 \pm 0.51)$ and $a_b = -(1.21 \pm 0.44)$ taking $B^0 - \overline{B^0}$ mixing into account. The results are in agreement with the expectations from the standard model.

302. Yu. I. Titov (Kharkov, FTI)
ON A POSSIBILITY OF USING AN UNBOUND NEUTRON TARGET IN ELECTRON STORAGE RINGS.

No abstract included.

303. Mark III Collaboration (J. Adler, et al.)
RECENT RESULTS ON HADRONIC D_s AND D MESON DECAYS FROM THE MARK III.

Recent results on hadronic D_s and D decays from the Mark III collaboration are presented. The absolute branching ratio $B(D_s^+ \to \phi\pi^+)$ is studied by searching for fully reconstructed $e^+e^- \to D_s^{*\pm}D_s^{\mp}$ events using seven hadronic decay modes of the D_s^+. A limit of $B(D_s^+ \to \phi\pi^+) < 4.1\%$ at 90% C.L. is obtained. Evidence is presented for the decay $D_s^+ \to f_0(975)\pi^+$ which agrees with a recent experimental observation. Upper limits are set for the relative branching ratios $B(D_s^+ \to \eta\pi^+)/B(D_s^+ \to \phi\pi^+) < 2.5$ and $B(D_s^+ \to \eta'\pi^+)/B(D_s^+ \to \phi\pi^+) < 1.9$, where the η is studied in both the $\gamma\gamma$ and the $\pi^+\pi^-\pi^0$ decay modes and the η' in the $\eta\pi^+\pi^-$, $\eta \to \gamma\gamma$ decay chain. The resonant substructure of $D^0 \to K^-\pi^+\pi^-\pi^+$ and $D^+ \to \overline{K}^0\pi^+\pi^-\pi^+$ is studied. The decay modes $D^0 \to a_1^+K^-$ and $D^+ \to a_1^+\overline{K}^0$ are found to be large and account for 50% of these final states. The decay mode $D^0 \to \overline{K}^{*0}\rho^0$ is found to be smaller than the theoretical prediction.

304. B. Barish, et al. (Cal Tech & UC, Santa Cruz & Cincinnati U. & CERN & Illinois U., Urbana & Oregon U. & SLAC & Washington U., Seattle)
THE TAU CHARM FACTORY: PHYSICS AND DESIGN STATUS.

The results of the Tau Charm Workshop are summarized. The authors are designing and proposing the construction of an e^+e^- collider and detector in the energy range 3–4.2 GeV with a luminosity of $10^{33}\text{cm}^{-2}\,\text{sec}^{-1}$ which is a factor of 100–1000 higher than present machines. The very high luminosities will allow precise tests of the Standard Model for tau and charm physics, and broad exploration of new physics associated with the second and third generations of particles. Particle yields for τ leptons and charmed mesons will be several times 10^7 events per year and for the J/ψ and ψ' resonances roughly 10^9 events per month. Studies indicate the feasibility of both a high luminosity machine with $L = 10^{33}\text{cm}^{-2}\,\text{sec}^{-1}$ and the detector. The currently preferred site in the United States is SLAC and some site details are given.

305. G.L. Fogli, E. Lisi (Bari U.)
BOUNDS ON A FOURTH GENERATION OF QUARKS AND LEPTONS FROM ELECTROWEAK RADIATIVE CORRECTION EFFECTS.

One-loop radiative corrections in the minimal standard model are sensitive to the parameters (masses and mixings) of a fourth generation of quarks and leptons. The contributions to the radiative corrections induced by the fermions of the fourth generation are analyzed and their effects compared with the present estimates of the low-energy neutral current couplings and vector boson masses. This allows to derive specific bounds on the parameters of the fourth generation.

306. CLEO Collaboration (P. Avery, et al.)
SEARCH FOR THE PROCESS $\Upsilon \to D^{*+}X$.

We have searched for the process $\Upsilon \to D^{*+}X$. We measure a 90% C.L. upper limit for $B(\Upsilon \to D^{*+}X)$ of 5×10^{-3} for $x_F = p/p_{\text{max}} > 0.2$.

307. CLEO Collaboration (P. Avery, et al.)
STUDY OF RARE D^0 DECAYS.

We have made measurements of rare decay modes of the D^0 meson using data collected with the CLEO detector at the Cornell Electron Storage Ring (CESR). We report the first observations of the $D^0 \to K^-\pi^+\pi^-\pi^+\pi^0$, $D^0 \to \phi\pi^+\pi^-$, and $D^0 \to \overline{K}^{*0}\eta$ decay modes of the neutral D meson, and present new measurements of the branching ratios for the $D^0 \to K^-\pi^+\pi^0$, $D^0 \to \overline{K}^0\pi^0$, and $D^0 \to \overline{K}^0\omega$. Our value for the ratio, $BR(D^0 \to K^+K^-)/BR(D^0 \to \pi^+\pi^-) = 2.22 \pm 0.36 \pm 0.30$ is a substantial improvement over previous results.

308. CLEO Collaboration (J. Alexander, et al.)
EXCLUSIVE SEMILEPTONIC DECAYS OF B MESONS INTO D MESONS.

We report new measurements of the branching fractions $B(B^- \to D^0\ell^-\overline{\nu})$ and $B(B^- \to D^{*0}\ell^-\overline{\nu})$. From these results, we find that the ratio of semileptonic widths for vector versus pseudoscalar final states is $1.6\ ^{+1.2+0.7}_{-0.6-0.5}$ and the ratio of charged to neutral B meson lifetimes is $0.85 \pm 0.20\ ^{+0.22}_{-0.16}$.

309. CLEO Collaboration (J. Alexander, et al.)
ON SEMILEPTONIC BRANCHING RATIOS OF B^0 AND B^- MESONS AND $B\overline{B}$ MIXING.

New measurements of the semileptonic branching ratio of neutral B mesons and the ratio of the semileptonic branching ratios for charged and neutral B mesons are combined with other measurements to yield best-fit values of these quantities. This results in a substantial reduction of the systematic error in the measurement of the $B^0\overline{B}^0$ mixing parameter, r. The value of r obtained by including both CLEO and ARGUS data in the fit is, $r = 0.17 \pm 0.05$.

310. CLEO Collaboration (T. Bowcock, et al.)
P-WAVE CHARMED MESONS IN e^+e^- ANNIHILATION.

Using the CLEO detector at the CESR storage ring, we have observed evidence for $L = 1$ charmed mesons in e^+e^- annihilations. We measure the $D^{**0}(2450)$ state at mass $(2461 \pm 3 \pm 1)\,MeV/c^2$ and width $(20\ ^{+\ 9+\ 9}_{-12-10})\,MeV/c^2$ decaying to $D^+\pi^-$. We also see an enhancement in the $m(D^{*+}\pi^-) - m(D^{*+})$ mass difference spectrum and by examining different D^{*+} polarization angles, we see structure consistent with two states decaying to $D^{*+}\pi^-$. We measure the lower mass state, at mass $(2428 \pm 2 \pm 1)\,MeV/c^2$ and width $(23\ ^{+8+4}_{-6-3})\,MeV/c^2$, by fixing the higher mass component from the decay $D^{**0}(2459) \to D^{*+}\pi^-$. We present arguments for the spin-parity assignments of these states. We also report observation of a candidate $L = 1$ $c\bar{s}$ state decaying to $D^{*+}K^0_s$.

311. JADE Collaboration
A STUDY OF PHOTON PRODUCTION IN HADRONIC EVENTS FROM e^+e^- ANNIHILATION.

Results are presented on an investigation of photons produced in multihadronic final states from e^+e^- annihilation at 35 GeV and 44 GeV center of mass energy. Scaling violation between 14 and 44 GeV is observed in inclusive photon spectra. Comparing inclusive π^0 spectra with charged pion spectra it is found that the average π^0 multiplicity exceeds the charged pion multiplicity scaled by factor of 0.5 by $(16 \pm 5)\%$ and $(21 \pm 7)\%$ at 35 and 44 GeV, respectively. The excess can be attributed to isospin violating decays of hadrons. The η multiplicity at 35 GeV is found to be $\langle n_\eta \rangle = 0.64 \pm 0.09 \pm 0.06$ at 35 GeV. With a significance of three standard deviations a signal from quark bremsstrahlung is observed. The measured charge asymmetry in hadronic final states due to the interference between initial and final state radiation of $A = -0.141 \pm 0.41$ in accord with QED expectations.

312. Jurgen G. Korner (Johannes-Guttenberg U.), Gerhard A. Schuler (DESY)
EXCLUSIVE SEMILEPTONIC BOTTOM MESON DECAYS INCLUDING LEPTON MASS EFFECTS.

We discuss the exclusive semi-leptonic (s.l.) bottom meson decays $B \to D(D^*) + 1 + \nu$ where we include non-zero lepton mass effects in the kinematics and dynamics. We develop the general formalism for the non-zero lepton mass case. We then look at how rates, spectra and angular correlations are affected by non-zero lepton masses in the context of a specific spectator quark model. Numerical results are presented for s.l. decays involving the $e-$, $\mu-$ and $\tau-$leptons. We also discuss the s.l. decays $B \to \pi(\rho)$, $D \to K(K^*)$ and the free quark decay model.

313. Jurgen G. Korner (Mainz U., Inst. Phys.), Gerhard A. Schuler (Hamburg U.)
LEPTON MASS EFFECTS IN SEMILEPTONIC B MESON DECAYS.

We calculate decay rates, decay spectra and polarization parameters for exclusive semi-leptonic (s.l.) $b \to c$ and $b \to u$ bottom meson and free bottom quark decays involving the τ–lepton. These are compared to the corresponding observables calculated for the light lepton modes. A surprising feature is a significant scalar current contribution in s.l. $B \to D$ and $B \to \pi$ decays which leads to an average positive longitudinal τ–polarization in the $B \to D$ mode.

314. M. Alston-Garnjost, et al. (LBL, Berkeley, Mount Holyoke Coll., New Mexico U., Idaho National Lab.)
A SEARCH FOR NEUTRINOLESS DOUBLE BETA DECAY OF ^{100}Mo.

We report the results of an experimental search for neutrinoless double beta decay using isotopically enriched foils of ^{100}Mo interleaved in a stack of 40 silicon detectors. With an exposure of 69.9 mole-days, we place lower half-life limits on the $J^P = 0^+ \to 0^+$ transition of 4×10^{21} years (68% C.L.) and on the $0^+ \to 2^+$ transition of 4×10^{20} years (68% C.L.).

315. ARGUS Collaboration (H. Albrecht, et al.)
OBSERVATION OF A NEW CHARMED STRANGE MESON

Using the ARGUS detector at the DORIS II e^+e^- storage ring at DESY, we have obtained evidence for a new charmed-strange meson which decays into $D^{*+}K^0$. Its mass is observed to be $2535.9 \pm 0.6 \pm 2.0$ MeV/c^2 and its width to be less than 4.6 MeV/c^2 at the 90% confidence level. No structure is observed at this mass in the D^+K^0 invariant mass spectrum, which suggests that an unnatural spin-parity is preferred.

316. OMEGA Photon Collaboration
CHARGE ASYMMETRY IN LOW-p_t PHOTOPRODUCTION OF HADRONIC FINAL STATES IN FORWARD DIRECTION.

The charged particle asymmetry, $(n^+ - n^-)/(n^+ + n^-)$, in inclusive photoproduction of charged particles on protons has been studied as a function of x_F and p_t in the ranges $x_F > 0$ and $0 < p_t < 2$ GeV and for photon energies in the range $65 < E_\gamma < 175$ GeV. The asymmetry is found to be positive, increasing with x_F and p_t and decreasing with E_γ. Asymmetries of π's and K's in the final state have been obtained separately. These data have been compared with a parton recombination model which provides a qualitative explanation for the asymmetry and the x_F and E_γ dependence of the effects observed.

317. G. Gerbier (Saclay)
CONVENTIONAL TECHNIQUES FOR DARK MATTER DETECTION.

Direct dark matter searches with conventional techniques have yielded their first results. Experiments using Germanium detectors have already eliminated Dirac neutrinos of masses greater than 11 GeV/c^2. Here, we describe a new experiment using Silicon, that is sensitive to lower masses particles like cosmions. Emphasis is placed on the energy calibration, an essential step to measure the rates as a function of energy. The results of this calibration, obtained for kinetic energies between 5 and 14 keV, agree well with the LSS theory. We also describe the Saclay project to construct a low pressure Hydrogen TPC (Time Projection Chamber). This detector would be sensitive to cosmions with spin dependent interactions.

318. CDHSW Collaboration (A. Blondel, et al.)
ELECTROWEAK PARAMETERS FROM A HIGH STATISTICS NEUTRINO NUCLEON SCATTERING EXPERIMENT.

The final results from the WA1/2 neutrino experiment in the 1984 CERN 160 GeV narrow band beam are presented. The ratios R_ν and $R_{\bar\nu}$ of neutral to charged current interaction rates of neutrinos and antineutrinos in iron are measured to be $R_\nu = 0.3072 \pm 0.0033$ and $R_{\bar\nu} = 0.382 \pm 0.016$. A value of the electroweak parameter $\sin^2\theta_W = 1 - m_W^2/m_Z^2$ is extracted from R_ν. The result is $\sin^2\theta_W = 0.228 + 0.013\,(m_c - 1.5) \pm 0.005\,(\text{exp.}) \pm 0.003$ (theor.) where m_c is the mass of the charmed quark in GeV for $m_t = 60$ GeV, $m_H = 100$ GeV, $\rho = 1$. Combining R_ν and $R_{\bar\nu}$, one obtains a value for $\rho = 0.991 + 0.023\,(m_c - 1.5) \pm 0.002\,(\text{exp.})$. Alternatively, R_ν and $R_{\bar\nu}$ yield a precise value of the ratio of intermediate vector boson masses $m_W/m_Z = 0.880 - 0.007\,(m_c - 1.5) \pm 0.002$ (exp.) $\pm\ 0.002$ (theor.). Comparison of these results with those from direct measurements of the vector boson masses are presented. In a model-independent analysis the left-and right-handed neutral current coupling constants, g_L^2 and g_R^2, are determined.

319. B.K. Kerimov (Moscow State U.), A.M. Mourao (Lisbon, IFM)
PARITY EFFECTS AND THE MASS AND MAGNETIC MOMENT OF THE NEUTRINO IN RADIATIVE NEUTRINO-ELECTRON SCATTERING.

We study inelastic neutrino-electron scattering and look for deviations from the Standard Model due to the mass and the magnetic moment of the neutrino. We perform a numerical analysis of the angular and energetic distributions of the bremsstrahlung photons for different values of the energy and magnetic moment of the neutrinos and of the photon energy. We analyze the parity non-conservation effects and show that the degree of the circular polarization of the photons can be used to obtain new limits on the mass and/or the magnetic moment of the neutrino. In the energy range where the neutrino magnetic moment may give an important contribution to νe scattering, it appears to be easier to detect radiative rather than elastic νe scattering.

List of Participants and Author Index

LIST OF PARTICIPANTS

The following is a list of the registered participants in the 14th International Symposium on Lepton and Photon Interactions.

Abad, Julio A.
Univ. de Zaragoza

Abe, Kazuo
KEK, Tsukuba

Abe, Koya
Tohoku Univ.

Abolins, Maris
Michigan State Univ.

Adolphsen, Chris
SLAC

Alam, Mohammad S.
State Univ. of New York

Alexander, James P.
Cornell Univ., LNS

Ali, Ahmed
DESY

Allaby, James V.
CERN

Altarelli, Guido
CERN

Ammar, Raymond G.
Univ. of Kansas

Aoki, Ken-Ichi
Kyoto Univ., RIFP

Arisaka, Katsushi
Univ. of California, Los Angeles

Arteaga-Romero, Napoleon
College de France

Atwood, William B.
SLAC

Bagger, Jonathan
Harvard Univ.

Baldo-Ceolin, Milla
Univ. di Padova

Ballam, Joseph
SLAC

Baltay, Charles
Yale Univ.

Banner, Marcel
CEN Saclay

Baranko, Gregory
Univ. of Colorado

Barbiellini, Guido
CERN

Barger, Vernon D.
Univ. of Wisconsin

Barnett, R. Michael
Lawrence Berkeley Lab.

Bartelt, John
SLAC

Bartl, Alfred
Univ. Wien

Bassetto, Antonio
Univ. di Padova

Baur, Ulrich
CERN

Bento, Luis F.L.
SLAC

Benvenuti, Alberto C.
Univ. di Bologna

Berger, Christoph
Techische Hochschule

Bernardi, Gregorio
Univ. Paris VII, LPNHE

Bertrand, Ghislaine M.
Univ. de Brussels

Bettoni, Diego
Univ. di Ferrara

Bianchi, Fabrizio
Univ. di Torino

Bisello, Dario
Univ. di Padova

Bitar, Khalil M.
Florida State Univ.

Blocker, Craig A.
Brandeis Univ.

Bloom, Elliott
SLAC

Bluemer, Hans
Univ. of Mainz

Boca, Gianluigi
Univ. di Pavia

Bock, Peter
Univ. of Heidelberg

Bogdanov, Peter V.
Atomic Energy Comm., Moscow

Boos, Edward E.
Moscow State Univ.

Booth, Christopher N.
Univ. of Sheffield

Booth, Paul S.L.
Univ. of Liverpool

Bottino, Alessandro
Univ. di Torino

Brau, James E.
Univ. of Oregon

Breakstone, Alan
Univ. of Hawaii

Breidenbach, Martin
SLAC

Brodsky, Stanley
SLAC

Brooks, Roger
SLAC

Brown, Stanley G.
Physical Review

Bruni, Graziano
Univ. di Bologna

Bulos, Fatin
SLAC

Buras, Andrzej
TU Munich

Burke, David
SLAC

Buschhorn, Gerd W.
Max Planck Inst., Munich

Busza, Wit
MIT, LNS

Butler, Joel N.
Fermilab

Cabibbo, Nicola
INFN, Rome

Caffo, Michele
Univ. di Bologna

Cahn, Robert N.
Lawrence Berkeley Lab.

Caldwell, David O.
Univ. of California, Santa Barbara

Campbell, Myron K
Univ. of Chicago

Carnegie, Robert K.
Carleton Univ.

Caso, Carlo
Univ. di Genova

Cassel, David G.
Cornell Univ., LNS

Castro, Andrea
Univ. di Padova

Cence, Robert J.
Univ. of Hawaii

Ceradini, Filippo
Univ. di Roma

Chadwick, George B.
SLAC

Chang, Ngee-Pong
City College, N. Y.

Cheng, Hai-Yang
Academia Sinica, Taipei

Chernyak, Victor L.
IYF, Novosibirsk

Chetyrkin, Konstantin
INR, Moscow

Conforto, Gianni
Univ. Urbino/INFN Firenze

Cords, Dieter
SLAC

Costa, Giovanni
Univ. di Padova

Cottingham, W.N.
Univ. of Bristol

Coupal, David
SLAC

Cousins, Robert D.
Univ. of California, Los Angeles

Cronin, James
Univ. of Chicago

Cundy, Donald C.
CERN

Dahmen, Hans Dieter
Univ. of Siegen

Dalpiaz, Paola Ferretti
Univ. di Ferrara

Danilov, Michael V.
ITEP, Moscow

Dasu, Sridhara
SLAC

Datta, Amitava
Jadavpur Univ.

Dau, Wolf Dieter
CERN

Davier, Michel
LAL, Orsay

Dawson, Sally
Brookhaven Nat. Lab.

De Alfaro, Vittorio
Univ. di Torino

Denisov, Sergei P.
IHEP, Serpukhov

DeSangro, Riccardo
Lab. Naz. di Frascati

Deshpande, Nilendra G.
Univ. of Oregon

DeSimone, Patrizia
Lab. Naz. di Frascati

DeStaebler, Herbert
SLAC

Devenish, Robin C.E.
Univ. of Oxford

Di Giacomo, A.
INFN, Pisa

Di Lella, Luigi
CERN

Diambrini-Palazzi, Giordano
Univ. di Roma

Dick, Louis
CERN

Dijkstra, Hans
CERN

Divakaran, P. P.
Tata Inst., Bombay

Dixon, Lance
SLAC

Dombey, Norman
Univ. of Sussex

Donnachie, Alexander
CERN

Dore, Ubaldo
Univ. di Roma

Dorfan, Jonathan
SLAC

Dornan, Peter J.
Imperial College, London

Dosselli, Umberto
Univ. di Padova

Dowell, John D.
Univ. of Birmingham

Drago, Elena E.
Univ. di Napoli, IFT

Drees, Manuel
CERN

Drell, Persis
Cornell Univ., LNS

Drell, Sidney
SLAC

Dumarchez, Jacques
Univ. Paris VI & VII

Dunietz, Isard
SLAC

Dydak, Friedrich
CERN

Edwards, Kenneth W.
Carleton Univ.

Eggert, Karsten
CERN

Einhorn, Martin B.
Univ. of Michigan

Einsweiler, Kevin
CERN

Eisele, Franz
DESY

Eisenberg, Yehuda
Weizmann Inst., Rehovoth

Eisenstein, Bob I.
Univ. of Illinois, Urbana

Ellis, Stephen D.
Univ. of Washington

Engler, Arnold
Carnegie Mellon Univ.

Escobar, Carlos O.
Univ. de Sao Paulo

Fadin, V.S.
IYF, Novosibirsk

Fanchiotti, Huner H.
Univ. Nac. de La Plata

Farrar, Glennys R.
Rutgers Univ.

Favart, Denis J.
Univ. Cath. de Louvain

Feldman, Gary J.
SLAC

Feltesse, Joel
CEN Saclay

Fischer, Herbert M.
Univ. of Bonn

Ford, William
Univ. of Colorado

Foster, George William
Fermilab

Fournier, Daniel A.
LAL, Orsay

Fowler, Earle C.
DOE, Washington D.C.

Franzini, Paolo
Columbia Univ.

Fraser, Gordon
CERN

Freudenreich, Klaus
ETH Zurich

Frey, Raymond E.
Univ. of Michigan

Froggatt, Colin D.
Univ. of Glasgow

Fryberger, David
SLAC

Fujii, Tadao
Tokyo Univ.

Gaemers, Karel J.
NIKHEF, Amsterdam

Gaillard, Mary Kay
Lawrence Berkeley Lab.

Galtieri, Lina
Lawrence Berkeley Lab.

Gamba, Diego
Univ. di Torino

Garavaglia, Theodore
Dublin Inst. Adv. Studies

Garbincius, Peter H.
Fermilab

Garcia-Canal, Carlos A.
Univ. Nac. de La Plata

Gerasimov, Sergo B.
JINR, Dubna

Gilchriese, Murdock
SSC Central Design Group

Gilman, Frederic
SLAC

Gittelman, Bernard
Cornell Univ., LNS

Gladding, Gary E.
Univ. of Illinois, Urbana

Gninenko, S.N.
INR, Moscow

Goldhaber, Gerson
Lawrence Berkeley Lab.

Goldman, Terrance
Los Alamos Nat. Lab.

Gonzalez-Romero, Enrique M.
CIEMT, Madrid

Goryachev, Vladimir N.
IHEP, Serpukhov

Gotsman, Errol A.
Univ. of Tel Aviv

Gourlay, Stephen
Fermilab

Graessler, Herbert
Technische Hochschule, Aachen

Green, Michael
Queen Mary College

Gribov, V.W.
Landau Inst., Moscow

Grindhammer, Guenter
DESY

Groom, Donald E.
SSC Central Design Group

Gunion, John F.
Univ. of California, Davis

Haber, Howard E.
Univ. of California, Santa Cruz

Haidt, Dieter
DESY

Haissinski, Jacques
LPNHE, Orsay

Halling, Alfred M.
Cornell Univ., LNS

Hand, Louis N.
Cornell Univ.

Hansl-Kozanecka, Traudl
MIT, LNS

Hanson, Gail G.
SLAC

Harari, Haim
Weizmann Inst., Rehovoth

Hari Dass, N.D.
Inst. Math. Sci., Madras

Harris, Frederick A.
Univ. of Hawaii

Haruyama, Masahiro
Asahikawa Medical Coll.

Harvey, J.
Princeton Univ.

Hasegawa, Katsuo
Tohoku Univ.

Hasenfratz, Peter
Univ. of Bern

Hayashi, Masaki
Saitama Medical Coll.

Hearty, Christopher
Lawrence Berkeley Laboratory

Hedberg, Vincent J.L.
CERN

Heinloth, Klaus
Univ. of Bonn

Heinzelmann, Goetz
Univ. of Hamburg

Hendry, Archibald W.
Indiana Univ.

Henri, Victor P.
NSF, Washington D.C.

Hikasa, Ken-ichi
KEK, Tsukuba

Himel, Thomas
SLAC

Hitlin, David G.
California Inst. of Technology

Hobson, Peter R.
Brunel Univ.

Hofmann, Werner
Max Planck Inst., Heidelberg

Hofstadter, Robert
Stanford Univ.

Holmgren, Sven-Olof
Univ. of Stockholm

Horisberger, Roland P.
Paul Schewer Inst.

Horn, David
Univ. of Tel Aviv

Hughes, Vernon
Yale Univ.

Huston, Joey W.
Michigan State Univ.

Huth, John E.
Fermilab

Innes, Walter
SLAC

Isakov, V.V.
INR, Moscow

Iso, Chikashi
Tokyo Inst. Tech.

Itoh, Ryosuke
KEK, Tsukuba

Izen, Joseph M.
Univ. of Illinois, Urbana

Jackson, J. David
Lawrence Berkeley Lab.

Jamin, Matthias R.
Univ. of Heidelberg, ITP

Janot, Patrick
LAL, Orsay

Jaros, John A.
SLAC

Jawahery, Abolhassan
Univ. of Maryland

Jentschke, Willibald K.
DESY

Jobes, Melvyn
Univ. of Birmingham

Jones, Keith
Nuclear Physics, Copenhagen

Jonsson, Leif S.
Univ. of Lund

Joshi, Girish C.
Univ. of Melbourne

Joshipura, Anjan S.
Phys. Res. Lab, Ahmedabad

Jung, Chang Kee
SLAC

Kadyk, John A.
Lawrence Berkeley Lab.

Kajikawa, Ryoichi
Nagoya Univ.

Kalyniak, Patricia A.
Carleton Univ.

Karchin, Paul
Yale Univ.

Karyotakis, Yannis
LAPP, Annecy

Kasemann, Matthias
CERN

Kastor, David
SLAC

Kayser, Boris
NSF, Washington D.C.

Keeler, Richard K.
Univ. of Victoria

Kendall, Henry W.
MIT, LNS

Kennedy, Bruce W.
Univ. Coll., London

Kent, Joel C.
Univ. of Calfornia, Santa Cruz

Khoze, Valerie A.
Leningrad INP

Kienzle, Maria Novella
Univ. of Geneva

Kim, Choong S.
Univ. of Durham

Kim, Peter
SLAC

Kim, Shinhong
Univ. of Tsukuba

Kirk, William
SLAC

Kirkby, Jasper
CERN

Kirpichnikov, Igor. V.
ITEP, Moscow

Klein, Max
DAW, Berlin

Kleinknecht, Konrad
Univ. of Mainz

Klem, Daniel E.
Carleton Univ.

Kluge, Eike-Erik
Univ. of Heidelberg, IHEP

Koide, Yoshio
Univ. of Shizuoka, Yada

Kolanoski, Hermann F.
Univ. of Dortmund

Komamiya, Sachio
SLAC

Korner, Jurgen G.
Univ. of Mainz

Koshiba, Masa-Toshi
Tokai Univ.

Koska, Wayne A.
Univ. of Michigan

Kotthaus, Rainer
Max Planck Inst., Munich

Kozanecki, Witold
SLAC

Kreinick, David L.
Cornell Univ., LNS

Krogh, Jurgen R. von
Univ. of Heidelberg

Kroll, Norman
Univ. of California, San Diego

Kuhn, Johann H.
Max Planck Inst., Munich

Kunori, Shuichi
Univ. of Maryland

Kuo, T.K.
Purdue Univ.

Kurimoto, Takeshi
Osaka Univ., Inst. Phys.

Kwak, Nowhan
Univ. of Kansas

Labarga, Luis
Univ. of California, Santa Cruz

Lam, C.S.
McGill Univ., Montreal

Langacker, Paul G.
Univ. of Pennsylvania

Lankford, Andrew
SLAC

Le Diberder, Francois
SLAC

LeBellac, Michel
Univ. de Nice

Lee-Franzini, Juliet
State Univ. of New York

Lee, Shih-Chang
Academia Sinica, Taipei

Lee, Wonyong
Columbia Univ.

Leith, David
SLAC

Levi, Michael E.
Lawrence Berkeley Lab.

Levy, Aharon
Univ. of Tel Aviv

Lewellen, David
SLAC

Limentani, Silvia
Univ. di Padova

Linglin, Denis
LAPP, Annecy

Lipton, Ronald J.
Carnegie Mellon Univ.

Lissauer, David A.
Brookhaven Nat. Lab.

Littenberg, Laurence S.
Brookhaven Nat. Lab.

Liu, Kehfei
Univ. of Kentucky

Loebinger, Frederick K.
Univ. of Manchester

Longo, Michael J.
Univ. of Michigan

Lord, Jere J.
Univ. of Washington

Louis, Jan
SLAC

Lu, Kan U.
Calif. State Univ., Long Beach

Lubrano, Pasquale
CERN

Ludwig, Jens
Univ. of Freiburg

Luth, Vera
SLAC

Lutz, Pierre R.
College de France

Lynch, Harvey L.
SLAC

Maalampi, Jukka
ITP Helsinki

MacFarlane, David B.
McGill Univ., Montreal

Magnussen, Norbert
Univ. of Wuppertal

Majerotto, Walter
OAW, Vienna

Maki, Akihiro
KEK, Tsukuba

Mallik, Usha
Univ. of Iowa

Mandelli, Luciano
Univ. di Milano

Mangotra, Lalit
Jammu Univ.

Mani, Sudhindra
Fermilab

Mankel, Rainer
Univ. of Dortmund

Mannel, Thomas T.
Technische Hochschule, Aachen

Mansouri, Freydoon
Univ. of Cincinnati

Mantovani, Giancarlo
INFN, Perugia

Mapelli, Livio P.
CERN

Margolis, Bernard
McGill Univ., Montreal

Martemianov, Vladimir
Kurchatov Inst., Moscow

Martin, Jean Paul
IPN Lyon

Martinec, Emil John
Univ. of Chicago

Martinelli, Guido
Univ. di Roma

Masuda, Naohiko
Yamagata Univ.

Mateev, Matey D.
Univ. of Sofiya

Matthews, John A.J.
Johns Hopkins Univ.

Matumoto, Ken-iti
Toyama Univ.

Mazzoni, M. Alessandra
CERN

McDaniel, Boyce D.
Cornell Univ., LNS

Meshkov, Sydney
NBS, Washington, D.C.

Metcalf, William J.
Louisiana State Univ.

Migneron, Roger
Univ. Western Ontario

Minakata, Hisakazu
Tokyo Metropolitan Univ.

Minkowski, Peter C.
Univ. of Bern

Minten, Adolf
CERN

Mir, Ronen
Univ. of Washington

Miyake, Kozo
Kyoto Univ.

Moffeit, Kenneth
SLAC

Mohapatra, Rabindra N.
Univ. of Maryland

Moneti, Giancarlo
Syracuse Univ.

Moniez, Marc
LAL, Orsay

Montgomery, Hugh E.
Fermilab

Moorhouse, Robert G.
Univ. of Glasgow

Morii, Toshiyuki
Kobe Univ.

Morrison, Rollin J.
Univ. of California, Santa Barbara

Mourao, Ana Maria
Lisbon, CFN

Mours, Benoit
SLAC

Muller-Kirsten, Harald J.W.
Univ. of Kaiserslautern

Muller, David
SLAC

Munger, Charles
SLAC

Murakami, Akira
Saga Univ.

Nagy, Miroslav
Acad. Science, Bratislava

Nash, Thomas E.
Fermilab

Nauenberg, Uriel
Univ. of Colorado

Navarria, Francesco L.
Univ. di Bologna

Ng, C. Ray
Purdue Univ.

Niczyporuk, Bogdan B.
CEBAF

Nieh, H.T.
State Univ. of New York

Nowak, Grazyna
INP Cracow

Noyes, H. Pierre
SLAC

Nozaki, Tadao
KEK, Tsukuba

Nuzzo, Salvatore
Univ. di Bari

O Fallon, J.R.
DOE, Washington D.C.

Oddone, Pier
Lawrence Berkeley Lab.

Odian, Allen C.
SLAC

Ogren, Harold
Indiana Univ.

Ohnemus, James E.
Florida State Univ.

Okabayashi, Masao
Shiga Univ.

Okusawa, Toru
Osaka City Univ.

Olsen, Stephen L.
Univ. of Rochester

Olsson, Martin G.
Univ. of Wisconsin

Osland, Per
Univ. of Bergen

Otter, Gerd
Technische Hochschule, Aachen

Palka, Henryk
INP Cracow

Pallin, Dominique
U. de Clermont-Ferrand

Palmer, William F.
Ohio State Univ.

Panman, Jaap
CERN

Panofsky, W.K.H.
SLAC

Partridge, Richard A.
Brown Univ.

Paul, Wolfgang
Univ. of Bonn

Paver, Nello
Univ. of Trieste

Peccei, Roberto
Univ. of California, Los Angeles

Perez-Y-Jorba, Jean P.
LAL, Orsay

Perl, Martin L.
SLAC

Peruzzi, Ida
Lab. Naz. di Frascati

Pescara, Luisa
Univ. di Padova

Peskin, Michael
SLAC

Petrossian, Zhudex V.
Yerevan Physical Inst.

Peyranere, Michel C.
Languedoc Univ.

Pfeil, Walter
Univ. of Bonn

Phua, Kok Khoo
Nat. Univ. of Singapore

Piccolo, Marcello
Lab. Naz. di Frascati

Piemontese, Livio
Univ. di Ferrara

Pierazzini, Giuseppe M.
INFN, Pisa

Pietschmann, Herbert
Univ. Wien

Pirogov, Youri F.
IHEP, Serpukhov

Pitman, Dale
SLAC

Pocsik, Gyorgy
Lorand Eotvos Univ.

Pradhan, Trilochan
Inst. Phys., Bhubaneswar

Prentice, James D.
Univ. of Toronto

Prepost, Richard
Univ. of Wisconsin

Prescott, Charles Y.
SLAC

Pullia, Antonino
Univ. di Milano

Pumplin, Jonathan
Michigan State Univ.

Quigg, Chris
SSC Central Design Group

Quinn, Helen R.
SLAC

Rabin, Monroe S.Z.
Univ. of Massachusetts

Raghavan, Rangachari
Tata Inst., Bombay

Rapuano, Frederico
Rome Univ.

Reithler, Hans H.
Technische Hochschule, Darmstadt

Reya, Ewald P.
Univ. of Dortmund

Riazuddin
King Fahd Univ., Dhahran

Richards, David G.
Univ. of Edinburgh

Richman, J.
Univ. of California, Santa Barbara

Richter, Burton
SLAC

Riordan, Michael
SLAC

Riska, Dan O.
Univ. of Helsinki

Ritchie, Jack L.
Univ. of Texas

Ritson, David
SLAC

Roberts, Richard G.
Rutherford Appleton Lab.

Robinett, Richard W.
Pennsylvania State Univ.

Rochester, Leon S.
SLAC

Rock, Stephen E.
American Univ.

Roe, Natalie
Lawrence Berkeley Lab.

Ronan, Michael T.
Lawrence Berkeley Lab.

Roosen, Robert
Vrije Univ. Brussel

Rosen, Jerome L.
Northwestern Univ.

Rosenberg, Leslie J.
Univ. of Chicago

Ross, Douglas A.
Univ. of Southampton

Ross, Graham G.
Univ. of Oxford

Rowson, Peter
Columbia Univ.

Rozanov, Alexander
CERN

Ruderman, Malvin
Columbia Univ.

Sacquin, Yves
CEN Saclay

Salecker, Helmut
Univ. of Munich

Salvini, Giorgio
Univ. di Roma

Samuel, Mark A.
Oklahoma State Univ.

Santroni, Alberto
Univ. di Genova

Sarantsev, V.V.
Leningrad INP

Sarma, Kuruganti V.L.
Tata Inst., Bombay

Schildknecht, Dieter
Univ. of Bielefeld

Schindler, Rafe
SLAC

Schmidt-Parzefall, Walter
DESY

Schmidt, Ivan
Univ. Cat., Valparaiso

Schneekloth, Uwe
MIT

Schnetzer, Stephen R.
Rutgers Univ.

Schramm, David
Univ. of Chicago

Schroder, Henning
DESY

Schuler, Gerhard
Univ. of Hamburg

Schwartz, Melvin
Digital Pathways, Palo Alto

Schwitters, Roy F.
SSC Laboratory

Sciulli, Frank
Columbia Univ.

Seiden, Abraham
Univ. of California, Santa Cruz

Selove, Walter
Univ. of Pennsylvania

Sens, Hans C.
NIKHEF, Amsterdam

Serednyakov, Sergey I.
IYF, Novosibirsk

Sessoms, Allen L.
US Embassy, Paris

Settles, Ronald D.
Max Planck Inst., Munich

Shaevitz, Michael H.
Columbia Univ.

Shen, Ben
Univ. of California, Riverside

Shephard, William D.
Univ. of Notre Dame

Shibata, Edward I.
Purdue Univ.

Shinkawa, Takao T.
KEK, Tsukuba

Shumeiko, Nikolaj M.
Byelorussian State Univ.

Shupe, Michael A.
Univ. of Arizona

Siegrist, James L.
Lawrence Berkeley Lab.

Sinervo, Pekka
Univ. of Pennsylvania

Singer, Paul
Technion, Haifa

Singh, Virendra
Tata Inst., Bombay

Sivers, Dennis
Argonne Nat. Lab.

Skillicorn, Ian O.
Univ. of Glasgow

Slater, W. James
Malaspina Coll., Nanaimo

Smith, A.J. Stewart
Princeton Univ.

Smith, Wesley H.
Univ. of Wisconsin

Soding, Paul H.
DESY

Soergel, Volker
DESY

Song, Hi Sung
Seoul Nat. Univ.

Sonnenschein, Jacob
SLAC

Soper, Davison E.
Univ. of Oregon

Stairs, Douglas G.
McGill Univ., Montreal

Stamenov, Dimitar
INR, Sofiya

Stanfield, Kenneth C.
Fermilab

Stefanski, Ray
Fermilab

Steger, Herbert
DESY

Steiner, Herbert
Lawrence Berkeley Lab.

Strauch, Karl
Harvard Univ.

Strong, John A.
Royal Holloway College

Stroot, Jean Pierre
Univ. de Brussels

Stroynowski, Richard A.
California Inst. of Technology

Sugano, Katsuhito
Argonne Nat. Lab.

Swartz, Morris
SLAC

Takeuchi, Tatsu
Yale Univ.

Tapper, Robert J.
Univ. of Bristol

Taylor, Frank E.
MIT, LNS

Taylor, Geoffrey N.
Univ. of Melbourne

Taylor, Richard E.
SLAC

Thompson, John C.
Rutherford Appleton Lab.

Thorndike, Edward H.
Univ. of Rochester

Titov, Yuri
FTI, Kharkov

Toge, Nobukazu
SLAC

Toki, Walter
SLAC

Tornqvist, Nils A.
Univ. of Helsinki

Trentadue, L.
Univ. di Milano

Trilling, George H.
Lawrence Berekely Lab.

Tsai, Yung-Su
SLAC

Tsarev, V.A.
Lebedev Inst., Moscow

Tscheslog, Evelin
DESY

Tung, Wu-Ki
Illinois Inst. of Technology

Valle, Jose W. F.
Univ. de Valencia

van de Walle, Remy T.
Univ. of Nijmegen

Vilain, Pierre
Univ. de Brussels

Vinogradov, Vladimir
JINR, Dubna

Visser, Jan
North-Holland Publishing

Volodko, Anton G.
JINR, Dubna

Voss, Gustav-Adolf R.
DESY

Wachsmuth, Horst W.
CERN

Wagner, Albrecht
Univ. of Heidelberg

Wagner, Friedrich
Univ. of Kiel

Wagner, Stephen R.
Univ. of Colorado

Wagoner, Robert V.
Stanford Univ.

Wahl, Heinrich
CERN

Walsh, Thomas F.
Univ. of Minnesota

Walter, Michael
DAW, Berlin

Ward, Benjamin F.L.
Univ. of Tennessee

Webber, Bryan R.
Univ. of Cambridge

Weber, Gustav
DESY

Weidberg, Anthony R.
CERN

Weill, Raymond
Univ. de Lausanne

Weinstein, Alan J.
California Inst. of Technology

Weinstein, Marvin
SLAC

Weseler, Siegfried
KFZ, Karlsruhe

Whisnant, Kerry L.
Iowa State Univ.

White, James T.
Texas A & M Univ.

Wick, Klaus K.
Univ. of Hamburg

Wicklund, Arthur Barry
Argonne Nat. Lab.

Wicklund, Eric
California Inst. of Technology

Wiik, Bjorn
DESY

Willenbrock, Scott
Brookhaven Nat. Lab.

Williams, David
Univ. of California, Santa Cruz

Williams, P.K.
DOE, Washington, D.C.

Wilson, R.
Harvard Univ.

Wilson, Robert J.
Boston Univ.

Wimpenny, Stephen J.
Univ. of California, Riverside

Winstein, Bruce
Univ. of Chicago

Winter, Klaus H.
CERN

Winter, Marc
CRN Strasbourg

Witherell, Michael S.
Univ. of California, Santa Barbara

Wittek, Wolfgang
Max Planck Inst., Munich

Wojcicki, Stanley G.
Stanford Univ.

Wu, Ji Min
IHEP Beijing

Wu, Sau Lan
Univ. of Wisconsin

Wuensch, Bernhard
Univ. of Bonn

Yager, Philip M.
Univ. of California, Davis

Yamada, Sakue
Univ. of Tokyo

Yamaguchi, Yoshio
Tokai Univ.

Yamamoto, Hitoshi
Univ. of Chicago

Yamamoto, Richard K.
MIT, LNS

Yan, Tung-Mow
Cornell Univ., LNS

Yao, York-Peng E.
Univ. of Michigan

Ye, Minghan
IHEP, Beijing

Yekutieli, Gideon
Weizmann Inst., Rehovoth

Yodh, Gaurang B.
Univ. of California, Irvine

Young, Bing-Lin
Iowa State Univ.

Young, Charles
SLAC

Yuan, Tzu-Chiang
Vanderbilt Univ.

Yuta, Haruo
Tohoku Univ.

Zacek, Gabriele
CERN

Zallo, Adriano
Lab. Naz. di Frascati

Zhu, Ren-Yuan
California Inst. of Technology

e ν Fit:
$N_\nu = 3.0 \pm 0.9$
$N_\nu < 4.4$ at 95% c.l.
nconstrained Fit:
$N_\nu = 3.5 \pm 0.8$
The uncertainty in the overall lumi
contributes 0.5 statistical and 0.25 sys
atic
LEPTON PHOTON

AUTHOR INDEX

Altarelli, G.	286
Cahn, R.N.	60
Campbell, M.	274
Cronin, J.W.	403
Danilov, M.V.	139
DiLella, L.	318
Dydak, F.	249
Eggert, K.	305
Feldman, G.J.	225
Feltesse, J.	13
Fournier, D.	168
Harvey, J.A.	433
Huston, J.	348
Huth, J.	368
Karchin, P.E.	105
Khoze, V.A.	387
Kirpichnikov, I.V.	83
Kreinick, D.L.	129
Littenberg, L.S.	184
Maki, A.	203
Panman, J.	258
Panofsky, W.K.H.	1
Perl, M.L.	89
Ross, G.G.	41
Ruderman, M.	420
Sciulli, F.	452
Sinervo, P.K.	328
Voss, G.A.	442
Weidberg, A.R.	266
Weinstein, A.J.	236
Winstein, B.	155
Ye, M.	122
Zheng, Z.	122